THE
ILLUSTRATED DICTIONARY OF ELECTRONICS

3RD EDITION

RUFUS P. TURNER
& STAN GIBILISCO

TAB Professional and Reference Books

Division of TAB BOOKS Inc.
P.O. Box 40, Blue Ridge Summit, PA 17214

THIRD EDITION

SECOND PRINTING

Printed in the United States of America

Reproduction or publication of the content in any manner, without express
permission of the publisher, is prohibited. No liability is assumed with respect to
the use of the information herein.

Copyright © 1985 by TAB BOOKS Inc.

Library of Congress Cataloging in Publication Data

Turner, Rufus P.
The illustrated dictionary of electronics.

Includes index.
1. Electronics—Dictionaries. I. Gibilisco, Stan.
II. Title.
TK7804.T87 1985   621.381′03′21   85-4623
ISBN 0-8306-0866-4
ISBN 0-8306-1866-X (pbk.)

Contents

How to Use This Book

The *Illustrated Dictionary of Electronics—3rd Edition* is similar to any other dictionary that you might use. The terms are alphabetically arranged from A to Z. Illustrations are used to clarify and enhance the terms, and each illustration has a caption that identifies its corresponding definition. Many of the definitions are cross-referenced for additional information or a contrasted definition. Cross-referenced terms appear in italics.

The Symbols, Charts, and Data section at the back of this book contains a great deal of information that will aid your understanding of the definitions. The Schematics Symbols Chart will help you interpret the schematics that appear throughout the *Dictionary*. There are also charts and tables to help you interpret the equations that appear throughout the book.

A 1. Symbol for *gain*. 2. Symbol for *area*. 3. Symbol for *ampere* (SI unit for current).

Å Symbol for *angstrom*.

A— Symbol for negative terminal of filament-voltage source in a vacuum-tube circuit.

A + Symbol for positive terminal of filament-voltage source in a vacuum-tube circuit.

a 1. Abbreviation of *atto-* (prefix). 2. Abbreviation of *are*. 3. Abbreviation of *acceleration*. 4. Abbreviation of *acre*. 5. Abbreviation of *anode*. 6. Now-obsolete abbreviation of cgs prefix *ab*.

A0 Federal Communications Commission designation for radio emission consisting solely of an unmodulated carrier.

A1 Federal Communications Commission designation for radio emission consisting of a continuous-wave carrier keyed by telegraphy.

A2 The FCC designation for radio emission consisting of a tone-modulated continuous wave.

A3 The FCC designation for radio emission consisting of amplitude-modulated radiotelephony.

A4 The FCC designation for radio emission consisting of amplitude-modulated facsimile signals.

A5 The FCC designation for radio emission consisting of amplitude-modulated television video signals.

aA Abbreviation of *attoampere*.

AAAS Abbreviation of American Association for the Advancement of Science.

AAC Abbreviation of *automatic aperture control* (NASA).

AAM Abbreviation of *air-to-air missile*.

AAS Abbreviation of *advanced antenna system* (NASA).

AASR Abbreviation of *airport and airways surveillance radar*.

AB Abbreviation of *acquisition beacon* (NASA).

A-B In sound and acoustics, the direct comparison of two sources of sound by turning on one and the other alternately.

ab 1. Prefix which transforms the name of a practical electrical unit to that of the equivalent electromagnetic cgs unit (e.g., *abampere, abohm, abvolt*). See individual entries of such cgs units. Also see *cgs*. 2. Abbreviation for *absolute*.

abac A graphic device for the solution of electronic problems. Also see *alignment chart*.

abacus An ancient sliding-bead-type calculating device, and by extension its electronic equivalent.

abampere The unit of current in the cgs electromagnetic system. 1 abampere approximately equals 10 amperes and corresponds to 1 abcoulomb per second.

ABA number The code number assigned to members of the American Banking Association for the purpose of expediting the clearing of checks.

A-battery In battery-operated tube circuits, the A battery supplies power to tube filaments or heaters.

abbreviated dialing In telephone systems, special circuits requiring fewer-than-normal dialing operations to connect subscribers.

abc 1. Abbreviation of *automatic bass compensation*, a system for boosting the volume of bass notes at low amplifier gain. 2. Abbreviation of *automatic bias control*. 3. Abbreviation of *automatic brightness control*. 4. Abbreviation of *automatic brightness compensation*.

abcoulomb The unit of electrical quantity in the cgs electromagnetic system. 1 abcoulomb equals 10 coulombs and is the quantity of electricity which flows past any point in a circuit in 1 second when the current is 1 abampere.

aberration A fault in solid optical lenses and reflectors and in similar electronic devices which causes malfunction, such as poor focus, halation, and color distortion.

ABETS Acronym for *airborne beacon electronic test set* (NASA).

abfarad The unit of capacitance in the cgs electromagnetic system. 1 abfarad equals 10^9 farads and is the capacitance across which a charge of 1 abcoulomb produces a potential of 1 abvolt.

abhenry The unit of inductance in the cgs electromagnetic system. 1 abhenry equals 10^{-9} henry and is the inductance across which a current which changes at the rate of 1 abampere per second induces a potential of 1 abvolt.

ABL Abbreviation of *Automated Biology Laboratory* (NASA).

ABMEWS Abbreviation of *antiballistic-missile early warning system.*

abmho The unit of conductance and of conductivity in the cgs electromagnetic system. 1 abmho equals 10^9 mhos and is the conductance through which a potential of 1 abvolt forces a current of 1 abampere.

abnormal dissipation Power dissipation higher or lower than the customary level, usually an overload.

abnormal glow The unusual glow completely surrounding the cathode of a gas tube, such as a glow tube, when the tube current is excessive.

abnormal oscillation 1. Oscillation where none is desired or expected, as in an amplifier. 2. Oscillation at two or more frequencies simultaneously when single-frequency operation is expected. 3. Oscillation at an incorrect frequency. 4. Parasitic oscillation.

abnormal propagation 1. The chance shifting of the normal path of a radio wave, as by displacements in the ionosphere, so that reception is degraded. 2. Unintentional radiation of energy from some point other than the transmitting antenna.

abnormal reflections Sharp, intense reflections at frequencies higher than the critical frequency of the ionosphere's ionized layer.

abnormal termination The shutdown of a computer program run or other process caused by the detection of an error by the associated hardware and indicating that some ongoing series of actions cannot be executed correctly.

abnormal triggering The false triggering or switching of a circuit or device, such as a flip-flop, by some undesirable source instead of the true trigger signal. Electrical noise pulses often cause abnormal triggering.

abohm The unit of resistance and of resistivity in the cgs electromagnetic system. 1 abohm equals 10^{-9} ohm and is the resistance across which a steady current of 1 abampere produces a potential difference of 1 abvolt.

abort To curtail volitionally, for any reason whatever, an operation, experiment, process, or project before it has run its normal course.

AB power pack 1. A portable dry-cell or wet-cell array containing both A and B batteries in one package. 2. An ac-operated unit in one package for supplying A and B voltages to battery-operated equipment, in lieu of the batteries.

abrasion machine An instrument for determining the abrasive resistance of a wire or cable.

abrasion resistance A measure of a wire's or wire covering's resistance to mechanical damage.

ABS A computer keyboard abbreviation for absolute value (of a number, variable, or expression).

abscissa A point located on the horizontal axis of a graph or screen.

absence-of-ground searching selector A rotary switch that searches for an ungrounded contact in a dial telephone system.

absolute 1. A temperature scale based on zero being the complete absence of heat. Units of measure are same as units on Celsius and Fahrenheit scales. (See *absolute scale.*) 2. Independent of any arbitrarily assigned units of measure or value.

absolute accuracy The full-scale accuracy of a meter with respect to the absolute standard.

absolute address In a digital computer program, the location of a word in memory storage, as opposed to location of the word in the program.

absolute altimeter An altimeter which employs transmitted and reflected radio waves for its operation and thus does not depend upon barometric pressure for its indication of altitude. Also see *absolute altitude.*

absolute altitude Distance above the surface of the terrain, rather than height above sea level.

absolute code A computer code in which the exact address is given for storing or locating the reference operand.

absolute coding In computer practice, coding that uses absolute addresses.

absolute constant A mathematical constant that has the same value wherever it is used.

absolute delay The time elapsing between the transmission of two synchronized signals from the same station or from different stations, as in radio, radar, or loran. By extension, the time interval between two such signals from any source, as from a generator.

absolute digital position transducer A digital position transducer whose output signal indicates absolute position. (See *encoder.*)

absolute efficiency The ratio Xx/Xs, where Xx is the output of a given device, and Xs the output of an ideal device of the same kind under the same operating conditions.

absolute encoder system A system which permits the encoding of any function (linear, nonlinear, continuous, step, and so on) and supplies a nonambiguous output.

absolute error The difference indicated by the approximate value of a quantity minus the actual value. This difference is positive when the approximate value is higher than the exact value, and is negative when the approximate value is lower than the exact value. Compare *relative error.*

absolute gain Antenna gain for a given orientation when the reference antenna is isolated in space and has no main axis of propagation.

absolute humidity The mass of water vapor per unit volume of air. Compare *relative humidity.*

absolute instruction A computer instruction that states explicitly, and causes the execution of, a specific operation.

absolute magnitude The absolute magnitude of a complex quantity is the vector sum of the real and imaginary components of the quantity, i.e., it is the square root of the sum of the squares of those components: $|x| 1 = \sqrt{a^2 + b^2}$, where $|x|$ is the absolute magnitude, a is the real component, and b is the imaginary component. Also see *absolute value* and *impedance.*

absolute maximum rating The highest value which a quantity may have before malfunction or damage occurs.

absolute maximum supply voltage The highest supply voltage that can be applied to a circuit without permanently altering its characteristics.

absolute measurement of current Measurement of a current directly in terms of defining quantities. 1. *Tangent galvanometer method.* Current is proportional to the tangent of the angle of deflection of the needle of this instrument. Deflection depends upon torque resulting from the magnetic field produced by current in the galvonometer coil acting against the horizontal component of the earth's magnetic field. 2. *Electrodynamometer method.* With this 2-coil instrument, current is determined from the observed deflection, the torque of the suspension fiber of the movable coil, and the coil dimensions.

absolute measurement of voltage Measurement of a voltage directly in terms of defining quantities. 1. *Calorimetric method.* A current-carrying coil immersed in water raises the temperature of the water. The difference of potential which forces the current through the coil then is determined in terms of the equivalent heat energy. 2. *Disk-electrometer method.* In this setup, a metal disc attached to one end of a balance beam is attracted by a stationary disk mounted below it, the voltage being applied to the two disks. The other end of the beam carries a pan into which accurate weights are placed. At balance, the voltage is determined in terms of the weight required to restore balance, the upper-disk area, and the disk separation.

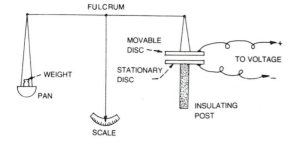

Disk-electrometer method of voltage measurement.

absolute minimum resistance The resistance between the wiper and resistance terminals of a potentiometer adjusted for minimum resistance.

absolute Peltier coefficient The product of the absolute Seebeck coefficient and absolute temperature of a material.

absolute pitch A tone in a standard scale determined by its rate of vibration and independent of other tones in the entire range of pitch.

absolute power Power expressed in absolute units.

absolute pressure Pressure (force per unit area) of a gas or liquid determined with respect to that of a vacuum (taken as zero).

absolute-pressure transducer A transducer actuated by pressure from the outputs of two different pressure sources, and whose own output is proportional to the difference between the two applied pressures.

absolute scale A temperature scale with its zero point at -273.1 degrees C, so called because its zero point is at absolute zero. Also called Kelvin scale (Lord *Kelvin*, 1824–1907). (The unit of thermodynamic temperature is the *kelvin*; thus, any temperature in the absolute scale may be written as *kelvins* rather than *degrees*.)

absolute Seebeck coefficient The quotient, as an integral from absolute zero to the given temperature, of the Thompson coefficient of a material divided by its absolute temperature.

absolute spectral response The frequency output or response of a device in absolute power.

absolute systex of units A system of units in which the fundamental (*absolute*) units are those expressing length (l), mass (m), charge (q), and time (t). All other physical units, including practical ones, are then derived from these absolute units.

absolute temperature Temperature measured on the absolute scale.

absolute tolerance The value of a component as it deviates from the specified or nominal value. It is usually expressed as a percentage of the specified value.

absolute units Fundamental physical units (see *absolute system of units*) from which all others are derived. See, for example, *abampere*.

absolute value The magnitude of a quantity without regard to sign or direction. The absolute value of a is written $|a|$. The absolute value of a positive number is the number itself; thus $|10|$ equals 10. But the absolute value of a negative number is the number with its sign changed: $|-|10|$ equals 10.

absolute-value computer A computer in which data is processed in its absolute form; i.e., every variable maintains its full value. (Compare to *incremental computer*.)

absolute-value device In computer practice, a device that delivers a constant-polarity output signal equal in amplitude to that of the input signal. The output signal thus always has the same sign.

absolute zero The temperature -273.16 degrees C (-459.72 degrees F); also, zero kelvins. The coldest possible temperature.

absorbed wave A radio wave which becomes lost in the ionosphere as a consequence of molecular agitation and the accompanying energy loss it causes there. Absorption is most pronounced at low frequencies.

absorber 1. Any material, body, or device which absorbs radiated energy and dissipates it. 2. A substance used in a nuclear reactor to absorb neutrons without propagating them.

absorptance The amount of radiant energy absorbed in a material; equal to 1 minus the transmittance.

absorptiometer Device to measure the reduction of pressure in a gas as it is absorbed by a liquid to determine the rate of absorption.

absorption The taking up of one material or medium by another into itself, as by sucking or soaking up. Also the retention of one medium (or a part of it) by another medium through which the first one attempts to pass. See, for example, *absorbed wave, absorption coefficient, dielectric absorption*. Compare *adsorption*.

absorption circuit A circuit which absorbs energy from another circuit or from a signal source, especially a resonant circuit such as a wavemeter or wavetrap.

absorption coefficient 1. *Linear absorption coefficient.* Symbol, μ. The fractional decrease in intensity of an X-ray beam per centimeter of absorbing material through which it passes. 2. *Mass absorption coefficient.* Symbol, μ_m. The fraction of energy absorbed from an X-ray beam of unit cross section by 1 gram of the material through which it passes.

absorption control In a nuclear reactor, control by means of absorber material (see *absorber, 2*) usually contained in rods moved in or out of the region near the core.

absorption current In a capacitor, the current resulting from dielectric absorption.

absorption dynamometer A power-measuring instrument in which a brake absorbs energy from a revolving shaft or wheel.

absorption edge Definite long-wavelength absorption band boundary in an X-ray spectrum.

absorption fading Fading of a radio wave, resulting from (usually) slow changes in the absorption of the wave in the line of propagation.

absorption frequency meter See *wavemeter*.

absorption loss 1. Transmission loss due to dissipation of electrical energy, or conversion of it into heat or other forms of energy. 2. Loss of all or part of a skywave because of absorption by the ionosphere.

absorption marker A small pip introduced onto an oscilloscope trace to indicate a frequency point. It is so called because it is produced by

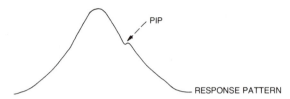

Absorption marker.

the action of a frequency-calibrated tuned trap similar to an absorption wavemeter.

absorption modulation Amplitude modulation of a transmitter or oscillator by means of an audio-frequency-actuated absorber circuit. In its simplest form, the modulator consists of a few turns of wire coupled to the transmitter tank coil and connected to a carbon microphone. The arrangement absorbs energy from the transmitter at a varying rate as the microphone changes its resistance in accordance with the sound waves it receives.

absorption spectrum For electromagnetic waves, a plot of absorption coefficient (of the medium of propagation) versus frequency. Also called *emission spectrum*.

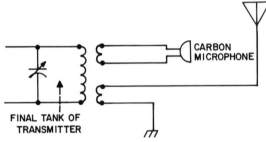

Absorption modulation.

absorption trap See *wavetrap*.

absorption wavemeter. A resonant-frequency indicating instrument that couples inductively to the device under test.

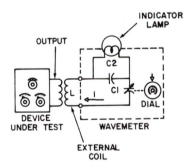

Absorption wavemeter.

absorptivity The capacity to absorb. In sound practice and microwave

practice, a measure of the energy absorbed by a given volume of absorber material.

A-B test Comparison of two sounds by reproducing first one, and then the other, immediately afterward. The two sounds may be alternated several times.

abvolt The unit of potential difference in the cgs electromagnetic system. 1 abvolt equals 10^{-8}V and is the difference of potential between any two points when 1 erg of work is required to move 1 abcoulomb of electricity between them.

abwatt The unit of power in the cgs electromagnetic system. 1 abwatt equals 10^{-7}W and is the power corresponding to 1 erg of work per second.

ac 1. Abbreviation of *attitude control* 2. Abbreviation of *aerodynamic center*. 3. Abbreviation of *alternating current*. 4. A suffix meaning *automatic calculator* or *automatic computer*. It appears as the end syllable in many coined names applied to electronic computers, examples being Eniac, Oarac, Seac, and Univac.

a/c 1. Abbreviation of *aircraft*. 2. Abbreviation of *air conditioning*.

ACA Abbreviation of *automatic circuit analyzer*.

ac base current Symbol, $I_{b(ac)}$. The ac component of base current in a bipolar transistor.

ac base resistance Symbol, $R_{b(ac)}$. The dynamic base resistance in a bipolar transistor.

ac base voltage Symbol, $V_{b(ac)}$. The ac component of base voltage in a bipolar transistor. It is the ac input signal voltage in a common-emitter amplifier or emitter-follower amplifier.

ac bias In a tape recorder, the high-frequency current which is passed through the recording head to linearize operation.

acc 1. Abbreviation of *automatic chroma control*. 2. Abbreviation of *automatic color compensation*. 3. Abbreviation of *acceleration*.

ac cathode current Symbol, $I_{k(ac)}$. The ac component of cathode current in an electron tube.

ac cathode resistance Symbol, $R_{k(ac)}$. The dynamic cathode resistance in an electron tube. $R_{k(ac)}$ equals dE_k/dI_k for a constant value of E_g.

ac cathode voltage Symbol, $E_{k(ac)}$. The ac component of cathode voltage in an electron tube. It is the ac output signal voltage in cathode-follower and grounded-grid amplifiers.

accelerated life test A test program that simulates the detrimental effects of time to some device, apparatus, or equipment in an abbreviated period.

accelerated service test A service or bench test in which equipment or a circuit is subjected to an extreme condition to simulate the effects of average use over a long time.

accelerating conductor or relay A conductor or relay that prompts the operation of a succeeding device in a starting mode according to established conditions.

accelerating electrode In a cathode-ray tube or klystron, the electrode to which the accelerating voltage is applied.

accelerating time The elapsed time beginning when voltage is applied to a motor, and ending when the motor shaft reaches its full or maximum speed.

accelerating voltage A positive high voltage applied to the accelerating electrode of a cathode-ray tube to increase the velocity of electrons in the beam.

acceleration The rate of increase of velocity. Often expressed in centimeters, inches, or *feet per second per second*, it may also be expressed in other units of distance and time as well. The SI unit of acceleration is m/s/s.

acceleration at stall The angular acceleration of a servomotor at stall, determined from stall torque and moment of inertia of the rotor of the motor.

acceleration derivative Acceleration (a) expressed as the second derivative of distance (s) with respect to time (t): a equals d^2s/dt^2.

acceleration potential See *accelerating voltage*.

acceleration switch A switch that operates automatically when the acceleration of a body to which it is attached exceeds a predetermined rate in a given direction.

acceleration time The time required by a computer to take in or deliver information after interpreting instructions. Compare *access time*.

acceleration torque During the accelerating period of a motor, the difference between the torque demanded and the torgue actually produced by the motor.

acceleration voltage The potential between accelerating elements in a vacuum tube, the value of which determines average electron velocity.

accelerator 1. A type of atom smasher: a machine or device in which charged particles, such as neutrons, are given high velocity for use in atomic disintegration and kindred processes. 2. A substance which speeds up a chemical reaction, e.g., the catalyst used with certain resins for encapsulating electronic equipment.

accelerograph An instrument that records the acceleration in velocity of earthquakes.

accelerometer A transducer whose output voltage is proportional to the acceleration of the moving body to which it is attached.

accentuation Emphasis of a desired band of frequencies, usually in the audio-frequency spectrum.

accentuator A circuit or device, such as a filter, tone control, or equalizer, used to emphasize a band of frequencies, usually in the audio-frequency spectrum. Also see *accentuation*.

acceptable-environmental-range test A test to disclose the environmental conditions equipment can endure while maintaining at least the minimum desired reliability.

acceptable quality level A percentage representing an acceptable average of defective components allowable for a process or the lowest quality a supplier is permitted to regularly present for acceptance. (Abbreviation, *aql*.)

acceptance sampling plan A probabilistic method of sampling a quantity of objects from a lot and determining from the sample whether to accept or reject the lot or to take a second sample.

acceptance test A test performed on incoming equipment or on submitted samples to determine if they meet tester's or supplier's specifications.

acceptor 1. Any device or circuit, such as a series-resonant circuit, which provides relatively easy transmission of a signal, in effect accepting the signal. 2. A hole-rich impurity added to a semiconductor to make the latter p-type. It is so called because its holes can accept electrons. Compare *donor*.

acceptor circuit See *acceptor, 1*.

acceptor impurity See *acceptor, 2*.

access 1. As a verb, to gain entrance to. 2. Port or opening into an equipment, used for ease of servicing. 3. In a computer, the action of going to a specific memory location for the purpose of data retrieval.

access arm A mechanical device that positions the read/write mechanism in a computer storage unit.

access control register A register that is part of a computer protection system that prevents interference between different software modules.

access, immediate The direct and quick access of data in storage without serial delay caused by other data units.

access method A method of transferring information or data from main storage to an input/output unit.

access mode A COBOL technique used to retrieve or input a logic record that is part of a mass storage device's assigned file.

access right The access status given to computer system users that indicates the method of access permitted; e.g., read a file only or write to a file.

access time The time required by a computer to begin delivering information after the memory or storage has been interrogated.

accidental error A chance error made by an inexperienced operator making measurements and recording data. In a large quantity of measurements, such errors tend to neutralize each other, as some are positive and others negative.

accidental triggering The undesired chance operation of a flip-flop or other switching circuit, as by a noise pulse or other extraneous signal.

ac collector current Symbol, $I_{c(ac)}$. The ac component of collector current in a bipolar transistor.

ac collector resistance Symbol, $R_{c(ac)}$. The dynamic collector resistance of a bipolar transistor. $R_{c(ac)}$ equals dV_c/dI_c for a constant value of I_b (common-emitter circuit) or I_e (common-base circuit).

ac collector voltage Symbol, $V_{c(ac)}$. The ac component of collector voltage in a bipolar transistor. It is the ac output signal voltage in a common-emitter or common-base amplifier.

accompanying audio channel The rf signal that supplies television sound. (Also called *cochannel sound frequency*.)

ac component In a complex wave (i.e., one containing both ac and dc), the alternating, fluctuating, or pulsating member only of the combination. Compare *dc component*.

accordion A printed circuit connector contact with a Z-shaped spring that allows high deflection with low fatigue.

ac-coupled flip-flop A flip-flop that is operated by the rise or fall of a clock pulse.

ac coupling Inductive coupling or capacitance coupling. Coupling of these types, unlike direct coupling, transmits ac, but not dc, since capacitors and transformers both block dc. Compare *direct coupling*.

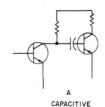

A
CAPACITIVE

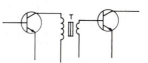

B
INDUCTIVE

AC coupling methods.

accumulator 1. In a digital computer, a circuit or register device which receives numbers, totals them, and stores them. 2. Storage battery.

accuracy 1. Precision in the measurement of quantities and in the statement of physical characteristics. 2. Degree of precision. Usually expressed, in terms of error, as a percentage of the specified value (e.g., 10V plus or minus 1 percent), as a percentage of a range (e.g., 2 % of full scale), or as parts (e.g., 100 parts per million).

accuracy rating The maximum amount of errors in an instrument, given as a percentage of the full-scale value.

accw Abbreviation of *alternating-current continuous wave*.

ac/dc Abbreviation of *alternating current/direct current*. Pertains to equipment which will operate from either ac or dc power lines.

ac directional overcurrent relay A relay that works on a specific value of alternating overcurrent rectified for a desired polarity.

ac drain current Symbol, $I_{D(ac)}$. The ac component of drain current in a field-effect transistor.

ac drain resistance Symbol, $R_{D(ac)}$. The dynamic drain resistance in a field-effect transistor; $R_{D(ac)}$ equals dV_d/dI_D for a constant value of V_G.

ac drain voltage Symbol, $V_{D(ac)}$. The ac component of drain voltage in a field-effect transistor. It is the ac output signal voltage in a common-source FET amplifier.

ac dump The removal of all ac power from a system or component.

ac emitter current Symbol, $I_{e(ac)}$. The ac component of emitter current in a bipolar transistor.

ac emitter resistance Symbol, $R_{e(ac)}$. The dynamic emitter resistance of a bipolar transistor; $R_{e(ac)}$ equals dV_e/dI_e for a constant value of I_b (emitter-follower circuit) or V_{CC} (common-base circuit).

ac emitter voltage Symbol, $V_{e(ac)}$. The ac component of emitter voltage in a bipolar transistor. It is the ac input signal voltage in a common-base amplifier, and the ac output signal voltage in an emitter-follower amplifier.

ac equipment Apparatus designed for operation from an ac power source only. Compare *dc equipment* and *ac/dc*.

ac erasing In tape recording, the technique of using an alternating magnetic field to erase material already recorded on the tape.

ac erasing head Also called *ac erase head*. In tape and wire recording, a head carrying alternating current to erase material already recorded on the tape or wire. Also see *ac erasing*.

acetate Cellulose acetate, a tough thermoplastic material which is an acetic acid ester of cellulose. It is used as a dielectric and in the manufacture of phonograph records and photographic films.

acetate base 1. The cellulose acetate film which served as the base for the magnetic oxide coating in early recording tape. (Most such tapes today are of polyester base.) 2. The cellulose acetate substrate onto which certain photosensitive materials are deposited for lithographic reproduction. Also see *acetate* and *anchorage*.

acetate disk A phonograph record disk made from cellulose nitrate. *acetate film* See *acetate base, 2*.

acetate tape Recording tape consisting of a magnetic oxide coating on a cellulose acetate film. Also see *acetate base*.

ac gate voltage Symbol, $V_{G(ac)}$. The ac component of gate voltage in a field-effect transistor. It is the ac input signal voltage.

ac generator 1. A rotating electromagnetic machine which provides alternating current, e.g., an ac dynamo, or alternator. 2. An oscillator or combination of oscillator and output amplifier.

ac grid voltage Symbol, $E_{g(ac)}$. The ac component of control grid voltage in an electron tube. It is the ac input signal voltage in a common-cathode amplifier or cathode follower.

A channel The left channel of a two-channel stereo system.

achieved reliability A statement of reliability based on the performance of mass-produced parts or systems under like environmental conditions. Also called *operational reliability*.

achromatic 1. Without color. In a TV image, the tone shades from black through gray to white. 2. Occasionally, pertaining to black-and-white television, although *monochromatic* has tended to replace the term in this sense.

achromatic lens A lens that has been corrected for chromatic aberration. It therefore has the same focus for light rays of all colors.

Achromatic lens cross section.

achromatic locus (region) An area on a chromaticity diagram that contains all points representing acceptable reference white standards.

achromatic scale A musical scale without accidentals.

ACIA Abbreviation of *asynchronous communications interface adapter*.

acicular A term describing the shape of a magnetic particle on a recording tape. Such particles are shaped like thin rods.

acid A substance which dissociates in water solution and forms hydrogen (H) ions. Example: sulfuric acid. Compare *base, 2*.

acid depolarizer Sometimes called *acidic depolarizer*. An acid, in addition to the electrolyte, used in some primary cells to forestall polarization.

ac line A power line delivering alternating current only.

ac line filter A filter designed to remove extraneous signals or electrical noise from an ac power line while causing virtually no reduction of the power-line voltage or power.

ac line voltage The voltage commonly delivered by the commercial power line to a consumer. In the U.S. this usually is 115, 117, or 120V. In Europe, 220V is common.

aclinic line An imaginary line, drawn on a map of the world, or of an area, connecting points of equal inclination (dip) of the magnetic needle.

ACM Association for Computing Machinery.

ac magnetic bias See *ac bias*.

ac meter A meter that is intended to work only on alternating current or voltage. Such meters include the *iron-vane* and *rectifier* types.

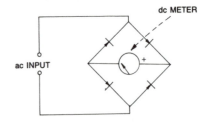

ac meter of the rectifier type.

ac noise Electrical noise of a rapidly alternating or pulsating nature.

ac noise immunity In computer practice, the ability of a logic circuit to maintain its state in spite of excitation by ac noise.

acorn tube A tiny electron tube made of glass and having terminal pins extending from its ends and sides for the most direct access to the internal elements. It takes its name from its characteristic shape.

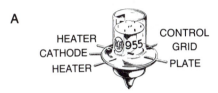

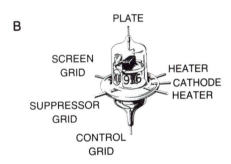

Acorn tubes of triode (A) and pentode (B) construction.

acous Abbreviation for *acoustic; acoustical.*

acoustic Pertaining to heard sound (versus audio-frequency currents or voltages) or hearing.

acoustic absorption The assimilation of energy from sound waves passing through a given medium or reflected by the latter.

acoustic absorption loss That portion of sound energy lost (as by dissipation in the form of heat) because of acoustic absorption.

acoustic absorptivity The ratio of sound energy absorbed by a material to sound energy impinging upon the surface of the material.

acoustic attenuation constant The real component of the complex acoustical propagation constant, expressed in nepers per unit distance.

acoustic burglar alarm An alarm that picks up the noise made by an intruder. The alarm device responds to the impulses from concealed microphones.

acoustic capacitance The acoustic equivalent of electrical capacitance. Expressed in cm^5/dyne.

acoustic clarifier In a loudspeaker system, a set of cones attached to the baffle which vibrate to absorb and suppress sound energy during loud bursts.

acoustic compliance Compliance in acoustic transducers, especially loud-speakers. It is equivalent to electrical capacitive reactance.

acoustic coupler Any device, such as a modem (see *modulator demodulator*), for achieving *acoustic coupling.*

acoustic coupling Transmission of information via a sound link, generally between a telephone and a pickup/reproducer in remote computer-terminal, facsimile, and control operations.

acoustic cutter See *acoustic scriber.*

acoustic damping The deadening or reduction of the vibration of a body so as to eliminate sound waves arising from it, or to cause those sound waves which do arise to die out quickly.

acoustic delay line Any equivalent of a special transmission line, which introduces a useful time delay between input and output signals. In one form, it consists of a crystal block or bar with an input transducer at one end and an output transducer at the other. An electrical input signal in the first transducer sets up sound waves which travel through the interior of the crystal; the piezoelectric reaction of the crystal to sound vibrations sets up an output voltage in the second transducer. The delay is due to the time required for the acoustic energy to travel the length of the crystal bar.

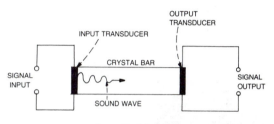

Acoustic delay line.

acoustic depth finder A direct-reading device for determining the depth of a body of water or for locating underwater objects, by means of sonic or ultrasonic waves transmitted downward and reflected back to the instrument.

acoustic dispersion Variation of the velocity of sound waves depending on their frequency.

acoustic elasticity In a loudspeaker enclosure, the compressibility of air behind the vibrating cone of the speaker. In general, the compressibility of any medium through which sound passes.

acoustic electric transducer A transducer, such as a microphone or hydrophone, which converts sound energy into electrical energy. Compare *electrical/acoustic transducer.* Also see *acoustic transducer.*

acoustic feedback Feedback resulting from sound waves from a loudspeaker or other reproducer reaching a microphone (or other input transducer) in the same system. Also, feedback resulting from such sound waves setting some part of an amplifier circuit into vibration and thus modulating the currents flowing in the circuit. Acoustic feedback usually causes howling or whistling.

acoustic filter Any sound-absorbing or transmitting arrangement, or combination of the two, which transmits sound waves of desired frequency while attenuating or eliminating others.

acoustic frequency response The sound-frequency range as a function of sound intensity: a means of describing the performance of an acoustic device.

acoustic generator A device that produces sound waves, usually of a desired frequency or intensity. Included are electrical devices (headphones or loudspeaker operated from a suitable oscillator, buzzers, bells, flames) and mechanical devices (tuning forks, bells, strings, whistles).

acoustic homing system 1. A system employing a sound signal for guidance purposes. 2. A guidance method in which a missile homes in on noise generated by a target.

acoustic horn So called to distinguish it from a microwave horn, a tapered tube (round or rectangular, but generally funnel-shaped) which directs sound and, to some extent, amplifies it.

acoustic howl See *acoustic feedback*.

acoustician 1. (A person skilled in acoustics (an acoustics technician). An audiologist.

acoustic impedance Unit, *acoustic ohm*. The acoustic equivalent of electrical impedance. Like the latter, acoustic impedance is the *total* opposition encountered by acoustic force. Also like electrical impedance, acoustic impedance has resistive and reactive components: acoustic resistance and acoustic reactance.

acoustic inductance Also called *inertance*. The acoustic equivalent of electrical inductance expressed in grams/cm^2/cm^2.

acoustic inertance Acoustic inductance.

acoustic inhibition See *auditory inhibition*.

acoustic intensity See *sound intensity*.

acoustic interferometer An instrument which evaluates the frequency and velocity of sound waves in a fluid (liquid or gas) in terms of a standing wave set up in the fluid by a transducer and reflector as the frequency or transducer-to-reflector distance is varied.

acoustic labyrinth A loudspeaker enclosure whose internal partitions form a usually rectangular maze-like tube lined with sound-absorbing material. The tube effectively connects at its top to the back of the speaker, and at its bottom it terminates in a mouth which opens to the front of the enclosure. The labyrinth provides an extremely efficient reproduction system because of its excellent acoustic-impedance matching capability.

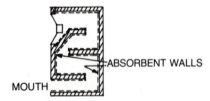

ABSORBENT WALLS

MOUTH

Acoustic labyrinth.

acoustic lens A system of barriers for refracting sound waves, acting on them in the way an optical lens does on light waves.

acoustic line Baffles or other such structures within a speaker that act as the mechanical equivalent of an electrical transmission line to enhance the reproduction of very low bass frequencies.

acoustic load A device which serves simultaneously as the output load of an amplifier and as a transducer of electrical energy into acoustic energy, e.g., headphones, loudspeaker.

acoustic memory In a computer, a volatile memory element employing an acoustic delay line, often incorporating quartz or mercury as the transmission and delay element.

acoustic mine An explosive naval mine detonated by vibrations from a ship's engines or propellers.

acoustic mirage A type of sound distortion in which the hearer experiences the illusion of two sound sources when there is only one. The phenomenon is due to the effect of a large temperature gradient in the air or water through which the sound passes.

acoustic mode Crystal-lattice vibration without producing an oscillating dipole.

acoustic noise Interferential (usually disagreeable) sounds carried by the air (or other propagation medium) to the ear or to an acoustic transducer. This is in contrast to electrical noise, which consists of extraneous current or voltage impulses and is inaudible until converted into sound.

acoustic ohm The unit of acoustic resistance, reactance, or impedance. 1 acoustic ohm equals volume velocity of 1 cm/s produced by a sound pressure of 1 microbar (0.1 Pa). Also called acoustical ohm.

acoustic phase constant The imaginary component of the complex acoustic propagation constant expressed in radians per second or radians per unit distance.

acoustic phase inverter A bass reflex loudspeaker enclosure.

acoustic phonograph A record player (now obsolete) in which the needle sets a thin diaphragm into vibration, the diaphragm in turn causing the air in a horn to vibrate reproducing the recorded sound.

acoustic pickup In acoustic phonographs, a pickup consisting of a needle, needle holder, and vibrating diaphragm.

acoustic pressure 1. The acoustic equivalent of electromotive force—expressed in dynes per square centimeter; also called *acoustical pressure*. 2. Sound pressure level.

acoustic radiator Literally, a radiator of sound; the cone of a loudspeaker, diaphragm of a headphone, vibrating reed of a buzzer, the flame of a thermal reproducer.

acoustic radiometer An instrument for measuring the intensity of a sound wave (see *sound intensity*) in terms of the unidirectional steady-state pressure exerted at a boundary as a result of absorption or reflection of the wave.

acoustic reactance Unit, *acoustic ohm*. The imaginary component of acoustic impedance. It is due either to acoustic capacitance, acoustic inductance, or to a combination of both.

acoustic recorder A recorder in which a horn collects sound waves that set into vibration a thin diaphragm at its apex, to which a needle holder is attached. The vibrating needle cuts a groove in the record disk rotating beneath it. The action of the *acoustic phonograph* is reversed in the acoustic recorder.

acoustic reflectivity The ratio Fr/Fi, where Fr is the rate of flow of sound energy reflected from a surface and Fi is the rate of flow of sound energy . . . incident to the surface.

acoustic refraction The deflection of sound waves being transferred obliquely between media which transmit sound at different speeds.

acoustic regeneration See *acoustic feedback*.

acoustic resistance Unit, *acoustic ohm*. The real component of acoustic impedance. As is true of its counterpart—electrical resistance—acoustic resistance is that opposing force which causes energy to be dissipated in the form of heat. It is attributed to molecular friction in the medium through which sound passes. See *acoustic ohm*.

acoustic resonance Intensification of sound, due to peak response of a reproducer or acoustic filter at a particular frequency of narrow band of frequencies, or to the coincidence of two or more sound waves in phase. Also see *acoustic feedback*.

acoustic resonator 1. A chamber—such as a box, cylinder, or pipe—in which an air column resonates at a particular frequency. 2. A piezoelectric, magnetostrictive, or electrostrictive body which vibrates at a resonant audio frequency that is governed by the mechanical dimensions of the body when an audio voltage at that frequency is applied to the body.

acoustics 1. The physics of sound; the study and applications of acoustic phenomena. 2. The qualities of an enclosure or sound chamber (room, auditorium, or box) which describe how sound waves behave in it.

acoustic scattering The dispersion of a sound wave in many directions as a result of diffraction, reflection, or refraction.

acoustic scriber In mechanical (i.e., nonelectric disk-cutting recorders, an assembly comprising a vibrating diaphragm, needle holder, and needle, which cuts the record disk. Also see *acoustic recorder.*

acoustic shock A morbid physiological condition resulting from exposure to a sudden, loud sound. Its manifestations range from simple alarm through dizziness and nausea to physical pain, and its effects may be transient or protracted.

acoustic suspension A loudspeaker design that allows exceptional low-frequency reproduction for a fairly small physical size. An airtight enclosure is used to increase the tension on the speaker cone.

acoustic system 1. A coordinated array of acoustic components (e.g., acoustic filters, resonators, and so on) which responds to sound energy in a predetermined manner. 2. An audio-frequency system in which sound energy is converted into electrical energy, processed, and then reconverted into sound energy for a clearly defined purpose.

acoustic telegraph A telegraph giving audible signals, as opposed to one giving visual signals (blinder) or printed messages.

acoustic transducer 1. Any device, such as headphones or loudspeaker, for converting audio-frequency electrical signals into sound waves. 2. Any device, such as a microphone, for converting sound waves into alternating, pulsating, or fluctuating currents.

acoustic transmission The direct transmission of sound energy without the intermediary of electric currents.

acoustic transmission system A set of components designed for generation of acoustic waves.

acoustic transmissivity The ratio e_t/e_i, where e_t is the sound energy transmitted by a medium, and e_i, the incident sound energy reaching the surface of the medium. Acoustic transmissivity is proportional to the angle of incidence. (Also called *acoustic transmittivity.*)

acoustic treatment Application of sound-absorbing materials to the interior of an enclosure or room to control reverberation.

acoustic wave The traveling vibration, consisting of particle compression and rarefaction, whereby sound is transmitted through air and other media.

acoustic wave filter See *acoustic filter.*

acoustoelectric effect The generation of a voltage across the faces of a crystal by sound waves traveling longitudinally through the crystal.

acoustoelectronics A branch of electronics concerned with the interaction of sound energy and electrical energy in devices such as surface-wave filters and amplifiers. In such devices, electrically induced acoustic waves travel along the surface of a piezoelectric chip and generate electrical energy. Also called *praetersonics* and *microwave acoustics.*

ac plate current Symbol, $I_{p(ac)}$. The ac component of plate current in an electron tube.

ac plate resistance Symbol, $R_{p(ac)}$. The dynamic plate resistance of an electron tube. $R_{p(ac)}$ equals dE/dI_p for a constant value for E_g.

ac plate voltage Symbol, $E_{p(ac)}$. The ac component of plate voltage in an electron tube. It is the ac output-signal voltage in a common-cathode amplifier.

ac power Symbol, P_{ac}. Unit, watt. The power acting in an ac circuit, P_{ac} equals $EI \cos \theta$, where E is in volts, I in amperes, and θ is the phase angle. Compare *dc power.* Also see *power.*

ac power supply A power unit which supplies ac only, e.g., ac generator, vibrator-transformer, oscillator, inverter. Compare *dc power supply.*

acquisition 1. The gathering of data from transducers or a computer. 2. Locating the path of an orbiting body for purposes of collecting telemetered data. 3. Orienting an antenna for optimum pickup of telemetered data.

acquisition and tracking radar An airborne or ground radar which locks in' on a strong signal and tracks the body reflecting or transmitting the signal.

acquisition radar A radar that spots an oncoming target and supplies position data regarding the target to a fire-control or missile-guidance radar, which then tracks the target.

acr Abbreviation of *audio cassette recorder; audio cassette recording system.*

ac reclosing relay The controlling component in an alternating-current circuit breaker. It causes the breaker to reset after a specified period of time.

ac relay A relay designed to operate on alternating current without chattering or vibrating.

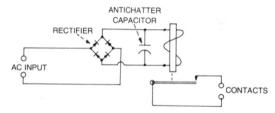

AC relay (rectifier-type).

ac resistance Pure resistance in an ac circuit. Unlike reactance and impedance, which are also forms of opposition to the flow of current, ac resistance introduces no phase shift.

acronym A word formed from letters or syllables taken from other applicable words of a multiword term. Acronyms are convenient for the naming of new devices and processes in electronics. Usually, a term is considered an acronym only when it is spelled in all-capital letters; once the term is accepted and popularized, it is written as a conventional word and is no longer thought of as an acronym. Example: *Laser* was once an acronym for *l*ight *a*mplification by the *s*timulated *e*mission of *r*adiation. By the popularization process, the acronym became a conventional word from which other terms (such as the verb "lase") were derived by back-formation.

acrylic in A synthetic resin, used as a dielectric and in electronic encapsulations, which is made from acrylic acid or one of its derivatives.

ACS Abbreviation of *automatic control system.*

ac source current Symbol, I_s (ac). The ac component of source current in a field-effect transistor.

ac source resistance Symbol, $R_{S(ac)}$. The dynamic source resistance in a field-effect transistor; R_S (ac) $= dV_S/dI_S$ for a constant value of V_G.

ac source voltage Symbol, V_S (ac). The ac component of source voltage in a field-effect transistor. It is the ac output-signal voltage in a source-follower (grounded-drain) FET amplifier.

acss Abbreviation of *analog computer subsystem.*

ac time overcurrent relay A device with a certain time characteristic, which breaks a circuit when the current exceeds a certain level.

actinic Exhibiting *actinism:* the property whereby radiant energy (such as visible and ultraviolet light, X-rays, etc.) causes chemical reactions.

actinic rays Short-wavelength light rays in the violet and ultraviolet portion of the spectrum that give conspicuous photochemical action.

actinium Symbol, Ac. A radioactive metallic element. Atomic number, 89. Atomic weight, 227.

actinodielectric Exhibiting a temporary rise in electrical conductivity during exposure to light.

actinoelectric effect The property whereby certain materials (such as selenium, cadmium sulfide, germanium, silicon) change their electrical resistance or generate a voltage on exposure to light. Also see *actinodielectric*.

actinometer An instrument for measuring the direct heating power of the sun's rays or the actinic power of a light source.

action 1. A chemical process (e.g., *local action* on the plate of a battery cell, electrolytic action in marine corrosion, galvanic action in cells). 2. A unique electronic phenomenon; e.g., photoelectric action, transistor action, tunneling action, antenna action, switching action. 3. A performance; e.g., abortive action, initiating action.

action current A small transient current flowing in a nerve during excitation.

activate To start an operation, usually by application of an appropriate enabling signal.

activated cathode The electron-emitting element of an electron tube, which has been coated with a material, such as thorium oxide, to increase the electron-emission efficiency.

activation 1. Treatment of the cathode of an electron tube for increased electron emission. 2. Creating artifical radioactivity in a substance. 3. Supplying electrolyte to a battery cell to prepare the latter for operation. 4. Causing the acceleration of a chemical reaction.

activation time In the activation of a battery cell (see *activation, 3*), the interval between addition of the electrolyte and attainment of full cell voltage.

active A circuit that requires a power supply for its operation. This differs from the passive circuit, which operates with no external source of power.

activator A substance added to an accelerator (see *accelerator, 3*) to speed the action of the latter.

active area The forward-current-carrying portion of the rectifying junction of a metallic rectifier.

active arm See *active leg*.

active balance The sum of return currents at a terminal network balanced against the local circuit or drop resistance in telephone repeater operation.

active communications satellite A communications satellite carrying receivers which pick up beamed signals from a ground point and amplify them, and transmitters which send the signals back to another ground point. Also called *active comsat*. Compare *passive communications satellite*.

active component 1. A device capable of some dynamic function (such as amplification, oscillation, signal control) and which usually requires a power supply for its operation. Examples: tube (except diode), transistor, tunnel diode, silicon controlled rectifier, magnetic amplifier. Compare *passive component*. 2. An in-phase quantity in an ac circuit, i.e., the quantity which contains no reactance.

active component of current See *active current*.

active computer The computer in an installation that is processing data.

active comsat See *active communications satellite*.

active control system A device or circuit that compensates for irregularities in the operating environment.

active current In an ac circuit, the current component that is in phase with voltage, in contrast to reactive current, which is out of phase,

powerless and, therefore, "inactive" with respect to power in the circuit. The active current is equal to the average power divided by the effective voltage.

active decoder An automatic ground station device that gives the number or letter designation of a received radio beacon reply code.

active device 1. An electronic component such as a tube or transistor that is capable of amplifying. 2. Broadly, any device (including electromechanical relays) that can switch (or amplify) by application of low-level signals.

active electric network A network containing one or more active elements, usually amplifiers or generators, in addition to passive elements.

active element The driven or rf-excited element in a multielement antenna or antenna array.

active file A computer file in use, i.e., one which is being updated or referred to.

active filter A filter, consisting of an amplifier and suitable tuning elements, usually inserted into a feedback path.

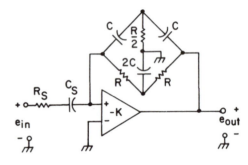

Active filter.

active guidance See *active homing*.

active homing The system whereby a missile homes in on a target by means of a radar aboard the missile. Also called *active guidance*.

active infrared detection Detection of infrared rays reflected from a target to which they were beamed.

active ingredient In a mixture, a component which is responsible for—or contributes to—desired behavior of the mixture; distinguished from inert ingredients, which serve only as supporting vehicles to hold the mixture together.

active jamming Transmission or retransmission of signals for the express purpose of disrupting communications.

active junction A pn or np junction in a semiconductor device, as a result of diffusion.

active leg An element within a transducer, which changes one or more of its electrical characteristics in response to the input signal of the transducer. Also called *active arm*.

active lines In a U.S. television picture, the lines (approximately 488) which make up the picture. The remaining 37 of the 525 available lines are blanked and are called *inactive*.

active material 1. In a storage cell, the chemical material in the plates which provides the electrical action of the cell, as distinguished from the supporting material of the plates themselves. 2. A radioactive substance. 3. The phosphor coating of a cathode-ray tube screen. 4. The material used to coat an electron-tube cathode.

active mixer A signal mixer employing one or more active components, such as tubes or transistors. An active circuit provides amplification, input-output isolation, and high input impedance, in addition to the mixing action. Compare *passive mixer.*

active modulator A modulator employing one or more active components, such as tubes or transistors. An active circuit provides gain, input-output isolation, and high input impedance, in addition to modulation. Compare *passive modulator.*

active network See *active electric network.*

active pressure The electromotive pressure that produces a current in an ac circuit, as distinguished from voltage impressed on it.

active pullup An arrangement using a transistor as a pullup resistor replacement in an IC, providing low output impedance and low power consumption.

active RC network 1. A resistance-capacitance circuit which contains active components (transistors, tubes), as well as passive components (capacitors, resistors). 2. An *RC* network in which some or all of the resistors and capacitors are simulated by the action of active devices (tubes or transistors).

active repair time The time during which maintenance is done on a system, and the system is out of operation.

active satellite See *active communications satellite.*

active substrate In an integrated circuit, a substrate consisting of single-crystal semiconductor material into which the various IC components are formed; it acts as some or all of the components. This is in contrast to a substrate consisting of a dielectric, on the surface of which the various components are deposited.

active systems Radio and radar systems which require transmitting equipment to be carried in a vehicle.

active tracking system A system in which a transponder or responder on board a vehicle retransmits information to tracking equipment; e.g., azusa, secor.

active transducer 1. A transducer which contains an active device, such as a tube, transistor, or magnetic amplifier, for immediate amplification of the sensed quantity. 2. A transducer which is itself an active device, e.g., *transducer-triode*: a vacuum tube in which movement of a pickup stylus extending from the tip of the tube causes the plate inside the tube to move, altering tube characteristics.

active wire In the armature of a generator, a wire experiencing induction and which, therefore, is delivering voltage.

activity 1. Intensity of, as well as readiness for, oscillation in a piezoelectric crystal. 2. Radioactive intensity. 3. Intensity of thermal agitation. 4. Thermionic emission of electrons.

activity curve Also called *activity plot.* A plot of radioactive activity vs time.

activity ratio The ratio of active to inactive records in a computer file.

ac transducer A transducer which (1) requires an ac supply voltage, or (2) delivers an ac output signal even when operated from a dc supply.

ac transmission The use of an alternating voltage to transfer power from one point to another, usually from generators to a distribution center, generally over a distance.

actual cathode In an electron-coupled oscillator, the true cathode of the tube, as distinguished from the *virtual* cathode, that results from use of the screen grid as the plate of the oscillator.

actual ground Ground as "seen" by an antenna. It is not always true ground (i.e., the earth itself); it may be an artificial ground, such as that provided in some antenna structures or resulting from nearby rooftops, buildings, etc.

actual height The highest altitude where radio wave refraction actually occurs.

actual power Also called *active* or *average power.* Symbol, P_{avg}. In a resistive circuit under sine-wave conditions, average power is the product of the rms voltage and the rms current. It is also equal to half the product of the maximum current and maximum voltage.

actuating device A device or component that operates electrical contacts to effect signal transmission.

actuating system 1. An automatic or manually operated system that starts (sometimes, stops or modifies) an operation. 2. A system which supplies energy for actuation.

actuating time Also called *actuation time.* The time interval between generation of a control signal, or mechanical operation of a control device, and the resulting actuation.

actuation 1. The starting or step modification (and sometimes the stopping) of an operation or process. 2. Activation of a mechanical or electromechanical switching device.

actuator An electromechanical device that uses electromagnetism to give a longitudinal or rotary thrust for mechanical work. It is often the end (load) device of a servosystem.

ACU Abbreviation of automatic calling unit.

ac voltage A voltage, the average value of which is zero, which periodically changes its polarity. In one cycle, an ac voltage starts at zero, rises to a maximum positive value, returns to zero, rises to a maximum negative value, and finally returns to zero. The number of such cycles repeated per second is termed the ac frequency.

ac voltmeter See *ac meter.*

ac vtvm Abbreviation of *alternating-current vacuum-tube voltmeter.* A voltmeter using internal vacuum-tube amplifiers intended expressly for the measurement of ac voltage.

acyclic machine Also called *acyclic generator.* A dc generator in which voltage induced in the active wires of the armature is always of the same polarity.

a/d Abbreviation for *analog-to-digital* (often capitalized). See *analog-to-digital conversion.*

Adam Communications code word for phonetic verbalizing of the letter *a.*

adapter 1. A fitting used to change either the terminal scheme or the size of a jack, plug, or socket to that of another. 2. A fitting used to provide a transition from one type or style of conductor to another (e.g., waveguide to coaxial line). 3. An auxiliary system or unit used to extend the operation of another system (e.g., citizens-band adapter for broadcast receiver).

UG-212C/U

ADAPTER

Adapter.

adaptive communication A method of communication that adjusts itself according to the particular requirements of a given time.

adaptivity The ability of a system to respond to its environment by changing its performance characteristics.

adc Abbreviation of *analog-to-digital converter*.

Adcock antenna A directional antenna system consisting of two spaced vertical antennas, which behaves like a loop antenna. Its members are connected and positioned so that it discriminates against (cancels) horizontally polarized waves, but delivers output which is proportional to the vector difference of signal voltages induced in the two vertical arms.

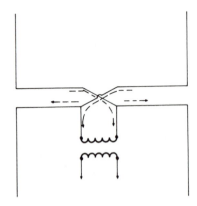

Adcock antenna.

Adcock direction finder A radio direction-finding system based on the directivity of the Adcock antenna.

Adcock radio range A radio range system based upon four Adcock antenna, each situated at the corner of a square; a fifth antenna is situated at the center of the square.

add-and-subtract relay A stepping relay which may be switched either uprange (add) or downrange (subtract).

addend In a calculation, any number to be added to another. Compare *augend*.

addend register In a digital computer, the register which stores the addend.

adder 1. In a digital computer, the device or circuit which performs binary addition. A *half adder* is a two-input circuit that can produce a sum output and a carry output, but cannot accommodate a carry signal from another adder. A *full adder* can accommodate a carry input as well as two binary signals to be added. Also see *analog adder*. 2. A circuit in a color TV receiver that amplifies the receiver primary matrix signal.

additive 1. The character or characters added to a code to encipher it. 2. In a calculation, material which is to be added. 3. An ingredient, usually in a small quantity, which added to another material to improve the latter in quality or performance.

additive color A color formed not by filtering but by combining the rays from two or three primary-colored lights onto a single neutral surface. For example, by projecting a red and a green beam onto a neutral screen, a yellow additive color results.

additive primaries Primary colors which form other colors through mixing of light (see *additive color*), but are not themselves formed by mixing other additive primaries. For example, red, green, and blue are the additive primaries used in color television. Through appropriate mixing, these colors will form an almost unlimited variety of other col-

ors. Compare *subtractive primaries*, which form the color spectrum by mixing of pigments rather than lights. In additive systems, each superimposed primary color increases the total light output from the reflecting (viewing) surface; in subtractive systems, each superimposed primary decreases the total reflectivity. Thus, equal combination of additive primaries produces white, and equal combination of subtractive primaries produces black.

addition record An extra data store created in a computer in processing.

additron A beam-switching tube (focused electrostatically) once used as a binary adder in high-speed digital computers.

add mode In an electronic calculator, a mode in which two decimal places are inserted without the need for the opposite operator to put in a decimal point. Useful for financial computations.

address 1. In computer practice, a (usually) numerical expression designating the location of material within the memory, or the destination of such material. 2. The accurately stated location of information within a computer; a data point within a grid, matrix, or table; or a station within a network. 3. In computer practice, to select the location of stored information.

address comparator A device that assures that an address being read is the right one.

address computation In digital computer practice, the technique of producing or modifying only the address part of an instruction.

address, direct See *absolute address*.

address, dummy A fictitious address used for illustration.

address field In a computer, the part of the instruction that gives the address of a bit of data (or a word) in the memory.

address generation The programmed generation of numbers or symbols used to retrieve records from a randomly stored direct access file.

address indirect An address that specifies a storage location that contains another address; succeeding addresses so located can be indirect.

address, memory The memory sections in a digital computer that contain each individual register.

address modification In computer practice, altering only the address portion of an instruction; if the command or instruction routine is then repeated, the computer will go to the new address.

address part In a digital computer instruction, that part of an expression specifying location. Also called *address field*.

address register In a computer, a register in which an address is stored.

add/subtract time In a digital computer, the time required to perform addition or subtraction, excluding the time required to get the quantities from storage and to enter the sum or difference into storage.

add time In digital computer operation, the time required to perform addition, excluding the time required to get the quantities from storage and to enter the sum into storage.

a/d encoder Analog-to-digital encoder: a device that changes an analog quantity into a digital corollary.

adf Abbreviation of *automatic direction finder*.

adiabatic damping In an accelerator (see *accelerator. 1*), reduction of beam size as beam energy is increased.

adiabatic demagnetization A technique using a magnetic field to keep a substance at a low temperature, sometimes within a fraction of a degree of absolute zero.

adiactinic Not transmitting actinic rays.

A-display A radar scope display in which the target appears displaced perpendicular to the (usually) horizontal time base.

adjacency A character recognition condition in which the spacing reference lines of two characters printed consecutively in line are closer than specified.

adjacent- and alternate-channel selectivity The selectivity of a receiver or rf amplifier, with respect to adjacent-channel and alternate-channel signals. That is, the extent to which a desired signal is passed and signals lying one and two channels widths away are rejected.

adjacent audio channel See *adjacent sound channel.*

adjacent channel The channel (frequency band) immediately above or below the channel of interest.

adjacent-channel attenuation The reciprocal of the selectivity ratio of a receiver. The selectivity ratio is the ratio of the sensitivity of a receiver tuned to a given channel to its sensitivity in an adjacent channel or on a specified number of channels removed.

adjacent-channel interference In television reception, the interference from stations on adjacent channels. A common form arises from the picture signal in the next higher channel and the sound signal in the next lower channel.

adjacent-channel selectivity The extent to which a receiver or tuned circuit can receive on one channel and reject signals from the nearest outlying channels.

adjacent sound channel In television, the rf channel containing the sound modulation of the next lower channel.

adjacent video carrier In television, the rf carrier containing the picture modulation of the next higher channel.

adjustable component Any circuit component the main electrical value of which may be varied at will, e.g., adjustable capacitor, inductor, resistor, load.

adjustable instrument 1. An instrument whose sensitivity, range, or response may be varied at will, e.g., multirange meter, wideband generator. 2. An instrument which requires adjustment or manipulation to measure a quantity, e.g., bridge, potentiometer, attenuator.

adjustable motor tuning An arrangement that allows the motor tuning of a receiver to be confined to a portion of the frequency range.

adjustable resistor A resistor in which the resistance wire is partially exposed to allow varying the component's value.

adjustable voltage divider A wirewound resistor with terminals that slide on exposed resistance wire to produce various voltage values.

adjusted circuit A circuit in which leads normally connected to a circuit breaker are shunted so that current can be measured under short-circuit conditions without breaker tripping.

adjusted decibels Noise level (in decibels) above a reference noise level (designated arbitrarily as zero decibels), measured at any point in a system with a noise meter which has previously been adjusted for zero (at reference) according to specifications.

admittance Symbol, *Y*. Unit, siemens (mho). The property denoting the comparative ease with which an alternating current flows through a circuit or device. Admittance (*Y*) is the reciprocal of impedance (*Z*): $Y = 1/Z$.

adp 1. Abbreviation of *ammonium dihydrogen phosphate*, a piezoelectric compound used for sonar crystals. 2. Abbreviation of *automatic data processing.*

absorption Adhesion of a thin layer of molecules of one substance to the surface of another without absorption. An example is absorption of water to the surface of a dielectric. Compare *absorption.*

adu Abbreviation of automatic dialing unit.

advance ball In a mechanical recorder (see *acoustic recorder*), a ball attached to the cutter to maintain mean depth of cut.

advanced license An amateur-radio license, conveying all

radiotelephone privileges except for a few small bands allocated to extra-class licensees. The second highest class of amateur license.

advance feed tape Paper tape in which the feed (sprocket) and character holes are tangent to a line drawn across their leading edges and perpendicular to the tape's length; thus, it is possible to tell the front end from the tail end of such a tape. Compare *center-feed tape.*

advance wire A resistance wire used in space-heater elements in addition to resistors. It is an alloy of copper and nickel, which has high resistivity and a negligible temperature coefficient of resistance.

aeolight A glow lamp using a cold cathode and a mixture of inert gases; because its illumination can be regulated with an applied signal voltage, it is often used as a modulation indicator for motion picture sound recording.

aerial See *antenna.*

aerial cable A wire or cable run through the air, using support structures such as towers or poles.

aerodiscone antenna A miniature discone antenna designed for use on aircraft.

aerodrome Variation (Brit.) of airport.

aerodrome control radio station See *airport control station.*

aerodynamics The science dealing with forces exerted by air and other gases in motion, especially upon bodies (such as aircraft) moving through these gases.

aerogram See *radiogram.*

aeromagnetic Pertaining to terrestrial magnetism as surveyed from a flying aircraft.

aerometeorograph A meteorograph designed for operation aboard an aircraft.

aerometer An instrument for measuring the density and weight of air and other gases.

aerometry The science of air measurements, which is often aided by electronics.

aeronautical advisory station A civil defense and advisory communications station in service for the use of private aircraft stations.

aeronautical broadcasting service The special service that broadcasts information regarding air navigation and meteorological data pertinent to aircraft operation.

aeronautical broadcast station A station of the aeronautical broadcasting service.

aeronautical fixed service A fixed radio service that transmits information regarding air navigation and flight safety.

aeronautical fixed service station A station which operates in the aeronautical fixed service.

aeronautical ground station A land station providing communication between aircraft and ground stations.

aeronautical marker-beacon signal A distinctive signal designating a small area above a beacon transmitting station for aircraft navigation.

aeronautical marker-beacon station A land station which transmits an aeronautical marker-beacon signal.

aeronautical mobile service A radio service consisting of communications between aircraft, and between aircraft and ground stations.

aeronautical radio-beacon station An aeronautical radionavigation land station transmitting signals used by aircraft and other vehicles to determine their position.

aeronautical radionavigation services Services provided by stations transmitting signals used in the navigation of aircraft.

aeronautical radio service Service embracing aircraft-to-aircraft, aircraft-to-ground, and ground-to-aircraft communications important to the operation of aircraft.

aeronautical station A station on land, and occasionally aboard ship, operating in the aeronautical mobile service.

aeronautical telecommunication agency The agency which administers the operation of stations in the aeronautical radio service.

aeronautical telecommunications Collectively, all of the electronic and nonelectronic communications used in the aeronautical service.

aeronautical utility land station A ground station in an airport control tower, providing communications having to do with the control of aircraft and other vehicles on the ground.

aeronautical utility mobile station At an airport, a mobile station that communicates with aeronautical utility land stations, and with aircraft and other vehicles on the ground.

aeronautics The art and science of designing, building, ground-testing and flying aircraft.

aerophare See *radio beacon.*

aerophysics The physical science of the air.

aerospace 1. The region encompassing the earth's atmosphere and extraterrestrial space. 2. Pertaining to research into and utilization of aerospace.

AES Abbreviation of *Audio Engineering Society.*

AEW Abbreviation of *airborne* (or *aircraft*) *early warning.*

aF Abbreviation of *attofarad.*

af Abbreviation of *audio frequency.*

afc Abbreviation of *automatic frequency control.*

AFIPS American Federation of Information Processing Societies.

afpc Abbreviation of *automatic frequency/phase control.*

afsk Abbreviation of *audio-frequency shift keying.*

afterglow The tendency of the phosphor of a cathode-ray-tube screen to glow for a certain time after the cathode-ray beam has passed. Also see *persistence.*

afterheat In a nuclear reactor, the heat generated by the decay of radioactive atoms; it continues after fission has stopped.

afterpulse An extraneous pulse in a multiplier phototube (photomultiplier), induced by a preceding pulse.

a/g Abbreviaiton of *air-to-ground.*

agc Abbreviation of *automatic gain control.*

AGE Abbreviation of *aerospace ground equipment.*

agenda A schedule of the major procedural operations for a solution or computer run; a group of programs used to manipulate a problem matrix in linear programming.

agendum call card In linear programming, a card punched to represent an item of an agenda.

agent An active force, condition, mechanism, or substance which produces or sustains an effect. Thus, a sudden voltage rise is a *triggering agent* in certain bistable circuits; arsenic, a *doping agent* in semiconductor processing; slow cooling of a heated metal, an *annealing agent.*

aging 1. An initial run of a component or circuit over a certain period of time shortly after manufacture to stabilize its characteristics and performance. 2. The changing of electrical characteristics or of chemical properties over a protracted period of time.

agonic line An imaginary line connecting points on the earth's surface, at which a magnetic needle shows zero declination (i.e., makes no angle with the meridian) and points to true north.

AGREE Avisory Group on Reliability of Electronics Equipment.

Ah Abbreviation of *ampere-hour.* Also see *amp-hr.* Depending on standard, abbreviation may be a-h, a-hr, A h.

aH Abbreviation of *attohenry.*

aided tracking In radar and fire control, a system in which manual correction of target tracking error automatically corrects the rate of movement of the tracking mechanism.

AIEE American Institute of Electrical Engineers, now consolidated with the IRE, forming the IEEE.

AIP American Institute of Physics.

air The mixture of gases which constitutes the earth's atmosphere and figures prominently in the manufacture and operation of numerous electronic devices. By volume, air contains about 21 percent oxygen, 78 percent nitrogen, and lesser amounts of argon, carbon dioxide, helium, hydrogen, krypton, neon, and xenon. It also contains varying amounts of water vapor, and in smoggy areas, carbon monoxide and the oxides of sulfur and nitrogen.

airborne intercept radar A type of short-range radar used aboard fighter and interceptor aircraft for tracking their targets.

airborne long-range input Equipment aboard aircraft, for the purpose of facilitating the use of long-range missiles.

airborne noise See *acoustic noise.*

airborne radar platform Surveillance and altitude-finding airborne radar.

air capacitor A capacitor in which air is the dielectric between the *conductive* plates.

aircarrier aircraft station On an aircraft, a radio station that is involved in the carrying of people for hire, or in the transportation of cargo.

air cell A primary cell (see *cell*) in which the positive electrode is depolarized by reduction of oxygen in the air.

air cleaner See dust *precipitator*

air column The airspace inside an acoustic chamber, pipe, or horn.

air-cooled component A component, such as a tube or power transistor, that is cooled by air (still or forced), compared with one cooled by a circulating liquid, such as water or oil.

air-cooled transistor A transistor (particularly a power transistor) from which the heat of operation is conducted away through radiation directly into the surrounding air, which may be still or moving; the transistor is usually mounted on a heatsink or provided with radiating fins.

air-cooled tube An electron tube from which heat produced by its operation is dissipated directly into the surrounding air, which may be still or moving. The tube may or may not be provided with a fin-type radiator.

air-core coil An inductor having no magnetic core material, so called because the only material at the center of the coil is air.

Air-core coils.

air-core transformer A transformer without the usual magnetic core material, so called because air is the only material at the center of and surrounding the transformer coils.

aircraft bonding The practice of solidly connecting, for electrical purposes, the metal parts of an aircraft, including the engine.

aircraft flutter A (usually fast) repetitive fading and intensifying of an rf signal resulting from reflections of the signal by passing aircraft.

aircraft station A nonautomatic radio station installed on board an aircraft.

air-dielectric coax Coaxial cable in which the space between inner and outer conductors is essentially empty, i.e., air-filled. The inner conductor is held away from the inner wall of the outer conductor by spaced beads or washers, or by a spiral-wound filament of a high-grade, low-loss dielectric.

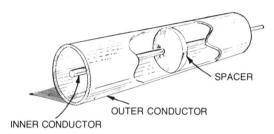

SPACER

OUTER CONDUCTOR

INNER CONDUCTOR

Air-dielectric coaxial cable.

air environment Any airborne communications equipment.

airflow The path or movement of air in, through, or around electronic equipment, especially in forced-air cooling.

air gap A narrow space between two parts of a magnetic circuit, e.g., the gap in the core of a filter choke. Often, this gap is filled with a nonmagnetic material, such as wood, for mechanical support. 2. The space between two or more magnetically coupled or electrostatically coupled components. 3. A device that gets its name from the narrow gap between two small metal balls, needle points, or blunt rod tips therein. When an applied voltage is sufficiently high, a spark is discharged across the gap.

air/ground control radio station A station for aeronautical telecommunications related to the operation and control of local aircraft.

air-insulated line 1. An open-wire feeder or transmission line. Typically, the line consists of two parallel wires held apart by separators (bars or rods of high-grade dielectric material) situated at wide intervals. 2. Air-dielectric coax.

air lock A small chamber whose opposing doors can only be opened one at a time, thus maintaining the condition of the air within, and isolating neighboring areas beyond the doors.

air-moving device A device, such as a specially designed fan or blower, for the forced-air cooling of electronic components.

airport beacon A radio or light beacon that marks the location of an airport.

airport control station A station that provides communications between an airport control tower and aircraft in the vicinity.

airport surveillance radar An air-traffic-control radar which scans the airspace some 30 to 60 miles about an airport and displays in the control tower the location of all aircraft below a certain altitude and all obstructions in the vicinity.

air-position indicator An airborne computer system that, using airspeed, aircraft heading, and elapsed time, furnishes a continuous indication of position of the aircraft.

air-space coax Coaxial cable in which air serves as the basic dielectric.

air-to-air communication Radio transmission from one aircraft to another in flight. Compare *air-to-ground communication* and *ground-to-air communication*.

air-to-air missile A guided missile fired from one aircraft at another target aircraft.

air-to-ground communication Radio transmission from an aircraft in flight to ground. Compare *air-to-air communication* and *ground-to-air communication*.

air-to-ground radio frequency The carrier frequency, or band of such frequencies, allocated for transmissions from an aircraft to a ground station.

air-to-surface missile A guided missile fired from an aircraft in flight at a target on the surface of the earth.

airwaves 1. Radio waves. The term is slang, but has wide currency and probably came from the public's mistaken notion that radio signals are propagated by the air. 2. Skywaves.

Al Symbol for *aluminum*.

alabamine See *astatine*.

alacratized switch A mercury switch in which the tendency of the mercury to stick to the parts has been reduced.

alarm A circuit that alerts personnel to a system malfunction, a detected condition, or an intruder.

alarm hold A device that keeps an alarm sounding once it has been actuated.

alarm relay A relay that is actuated by an alarm device.

albedo For an unpolished surface, the ratio of reflected light to incident light.

albedograph An instrument for measuring the albedo of planets.

alc Abbreviation of *automatic level control*.

alerting device An audible alarm which includes a self-contained solid-state audio oscillator, and which when powered from the ac line or a battery delivers a usually raucous signal.

Alexanderson alternator (E.F.W. *Alexanderson*). An early, dynamo-type machine for generating radio-frequency power. It consists of a rapidly spinning steel disk having many teeth or slots around its rim, mounted so that teeth pass through the gap between the poles of a field-coil assembly. Interruption of the magnetic field by the teeth produces frequencies on the order of 200 kHz at a power level of 200 kW.

Alexanderson antenna A vlf antenna consisting of a horizontal wire grounded at equally spaced points by vertical wires having base-loading coils.

Alford antenna A loop antenna, in a square configuration, with the corners bent toward the center to lower the impedance at the current nodes.

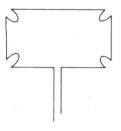

Alford antenna.

algebraic adder In computer practice, an adder that provides the algebraic rather than arithmetic sum of the entered quantities.

alegebraic operation Calculator operation which permits keystrokes for quantities and operators to follow the order in which an equation is normally written. Compare *reverse Polish notation*.

algebraic sum The sum of two or more quantities with consideration of their signs. Compare *arithmetic sum*.

ALGOL Acronym for *algo*rithmic *l*anguage; a problem-oriented, high-level computer language using algebraic notation to express information and following the rules of Boolean algebra.

algorithm A step-by-step procedure for solving a problem, e.g., the procedure for finding the square root of a number.

algorithmic language A computer language, used for describing a numeral or algebraic process.

alias A label that is an alternate (synonymous) term for items of the same type; a label and the several aliases can identify the same data element in a computer program.

align 1. To adjust (i.e., to preset) the circuits of an electronic system, such as a receiver, transmitter, or test instrument, for predetermined response. 2. To arrange elements in line with each other and to space them, as in an electron tube. 3. To orient antennas so that they are in line of sight with respect to each other.

aligned-grid tube An electron tube, such as a beam-power tube, in which at least two grids are supported, one concentric to the other, so that the wires of one grid shade the wires of the other.

alignment The process of aligning equipment or electrodes. Also see *align* and *aligned-grid tube*.

alignment chart A line chart for the simple solution of electronic problems. It is so called because its use involves aligning numerical values on various scales, the lines intersecting at the solution on another scale. Also called nomograph.

alignment pin A pin or protruding key, usually in the base of a removable or plug-in component, to insure that the latter will be inserted correctly into a circuit. Often, the pin mates with a keyway, notch, or slot.

alignment protractor A device that is used for calibration of the pickup arm on a phonograph. The arm movement should be orthogonal to the grooves on the disk at all points.

alignment tool A specialized screwdriver or wrench (usually nonmagnetic) used for adjusting padder or trimmer capacitors or inductor cores.

alive See *live*.

alkali See *base, 2*.

alkali metals Metals whose hydroxides are bases (alkalis). The group includes cesium, francium, lithium, potassium, rubidium, and sodium; some are used in phototubes.

alkalin battery 1. A battery composed of alkaline cells and characterized by a relatively flat discharge curve under load.

alkaline cell 1. A *primary cell* in which the negative electrode is granular zinc mixed with a potassium hydroxide (alkaline) electrolyte; the positive electrode is a polarizer in electrical contact with the outer metal can of the cell. A porous separator divides the electrodes. This type of cell delivers a terminal potential of 1.5V, but has much higher capacity than does a 1.5V carbon-zinc cell. 2. See *Edison battery*.

alkaline-earth metals The elemental metals barium, calcium, strontium, and sometimes beryllium, magnesium, and radium, some of which are used in vacuum tubes.

alkaline earths Substances which are oxides of the alkaline-earth metals. Some of these materials are used in vacuum tubes.

all-diffused Microcircuit type in which both active and passive elements have been fabricated by diffusion and related processes.

Allen screw A screw fitted with a six-sided (hexagonal) hole, into which fits an Allen wrench.

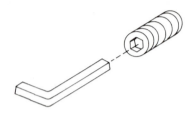

Allen screw and wrench.

Allen wrench A device used for tightening or loosening an Allen screw. It is a hexagonal rod, and may have various sizes.

alligator clip A clip lead with jagged teeth, designed to be used with temporary electrical connections.

allocate 1. To assign (especially through legislation) operating frequencies or other facilities or conditions needed for scientific or technical activity; see, for example, *allocation of frequencies*. 2. In computer practice, to assign locations in the memory or registers for routines and subroutines.

allocated channel A channel assigned to an individual or group.

allocated-use circuit 1. A circuit in which one or more channels have been authorized for the exclusive use of one or more services. 2. A communications link assigned to users needing it.

allocation of frequencies See *radio spectrum*.

allocator A telephone system distributor associated with the finder control group relay assembly that reserves an inactive line-finder for another call.

alloter relay A telephone system line-finder relay that reserves an inactive line-finder for the next incoming call from the line.

allotropic Pertaining to a substance existing in two forms.

alloy A metal which is a mixture of several other metals (e.g., brass from copper and zinc) or of a metal and a nonmetal.

alloy deposition In semiconductor manufacture, despositing an alloy on a substrate.

alloy-diffused transistor A transistor in which the base is diffused and the emitter is alloyed. The collector is provided by the semiconductor substrate into which alloying and diffusion are effected. Compare *alloy transistor* and *diffused transistor*.

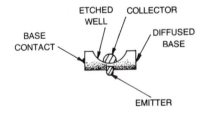

Alloy-diffused transistor (microalloy type).

alloy diode A junction-type semiconductor diode in which a suitable substance (such as p-type) is alloyed into a chip of the opposite type (such as n-type) to form the junction. Also called *alloy-junction diode*.

alloy junction In a semiconductor device, a positive/negative (pn) junction formed by alloying a suitable material (such as indium) with the semiconductor (silicon or germanium).

alloy transistor A transistor whose junctions are created by alloying. Also see *alloy junction*.

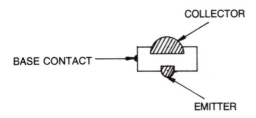

COLLECTOR

BASE CONTACT

EMITTER

Alloy transistor.

all-relay central office In telephone service, an automatic central-office switchboard that uses relay circuits to make line interconnections.

all-pass filter Also called *all-pass network*. A filter which (ideally) introduces a desired phase shift or time delay, but has zero attenuation at all frequencies.

all-wave Pertaining to a wide operating-frequency range. Few systems are literally *all* wave. For example, a so-called all-wave radio receiver might cover the various services between 500 kHz and 40 MHz only.

all-wave antenna An antenna which may be operated over a wide frequency range with reasonable efficiency and (preferably) without needing readjustment; a log periodic antenna array.

all-wave generator A signal generator which will supply output over a wide range of frequencies.

all-wave receiver A radio receiver which may be tuned over a very wide range of frequencies, such as 10 kHz to 70 MHz.

allyl plastics Plastics, sometimes used as dielectrics or for other purposes in electronics, based on resins made by polymerization of monomers—such as diallyl phthalate—containing allyl groups.

alcino Coined from the words *al*uminum, *nic*kel, and *co*balt. An alloy used in strong permanent magnets, it contains the constituents noted plus (sometimes) copper or titanium.

alpha Symbol, α. The current gain of a common-base-connected bipolar transistor. It is the ratio of the difference of collector current to the differential of emitter current; $\alpha = dIc/dIe$. For a junction transistor, alpha is always less than unity, but very close to it.

alphabet The set of all characters in a language.

alphabetic coding In computer practice, an abbreviation system for coding information to be fed into the computer. The coding contains letters, words, and numbers.

alphabetic-numeric Also called alphabetical-numerical and alphanumeric. In computer practice, pertaining collectively to letters of the alphabet and special characters, and to numerical digits. See also *alphanumeric*.

alphabetic string (word) A character string containing members of a character set, i.e., characters other than numerals.

alpha cutoff frequency The high frequency at which the alpha of a common-base-connected bipolar transistor falls to 70.7 percent of its low-frequency value.

alphanumeric See *alphanumeric*.

alphanumeric Pertaining to combinations of letters and numerals. Also called *alphameric*.

alphanumeric code In computer practice or in communications, a code in which letters of the alphabet are represented by numbers.

alphanumeric readout A type of digital readout which displays both letters and numerals.

alpha particle A nuclear particle bearing a positive charge. Consisting of two protons and two neutrons, it is given off by certain radioactive substances. Compare *beta rays* and *gamma rays*.

alpha radiator A radioactive substance that emits alpha rays.

alpha ray See *alpha particle*.

alpha system An alphabetic code signaling system.

alteration An inclusive-OR operation.

alternate channel In communications, a channel situated two channels higher or lower than a given channel. Compare *adjacent channel*.

alternate-channel interference Interference caused by a transmitter operating in the channel beyond an adjacent channel. Compare *adjacent-channel interference*.

alternate frequency A frequency allocated as an alternate to a main assigned frequency and used under certain specified conditions.

alternate mode The technique of displaying several signals on an oscilloscope screen by rapidly switching the signals in sequence at the end of each sweep.

alternate routing (routine) A secondary (backup) communications path used when primary (normal) routing is impossible.

alternating-charge characteristic In a nonlinear capacitor, the relationship between the instantaneous charge and the instantaneous value of an alternating voltage.

alternating current Abbreviation, ac. A current which periodically alternates its direction of flow. In one cycle, an alternation starts at zero, rises to a maximum positive level, returns to zero, rises to a maximum negative level, and again returns to zero. The number of such cycles completed per second is termed the ac frequency. Also see *current*.

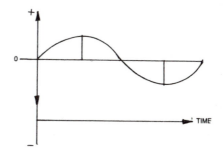

Alternating current (sine wave).

alternating-current continuous wave An amplitude-modulated signal resulting from the operation of an oscillator or rf amplifier with raw ac voltage.

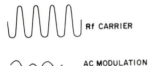

Rf CARRIER

AC MODULATION

WAVE PATTERN FOR MORSE CODE
REPRODUCTION OF LETTER A (· –)

Alternating-current continuous wave.

alternating current/direct current See *ac/dc.*

alternating-current erasing head See *ac erasing head.*

alternating-current pulse A short-duration ac wave.

alternating-current transmission 1. The transmission of alternating currents along a length of conductor, especially for power-transfer purposes. 2. A means of picture transmission in which a given signal strength produces a constant value of brightness for a very short time.

alternating voltage Also called *alternating-current voltage*. See *ac voltage.*

alternation In ac practice, a half-cycle. In a complete cycle, there are two alternations, one in the positive direction and one in the negative direction.

alternative denial A NOT-AND operation.

alternator 1. Any mechanically driven machine for generating ac power. Sometimes specifically one having a permanent-magnet rotor (as in a magneto). 2. An early dynamo-type machine for generating radio-frequency power. See *Alexanderson alternator, Bethenod alternator, Goldschmidt alternator, Telefunken alternator.*

alternator transmitter A radio transmitter that generates power with an rf alternator.

altimeter An instrument for measuring elevation (height above sea level). Altimeters of the aneroid and thermistor types depend on barometric pressure for their operation. Compare *absolute altimeter.*

altimeter station An airborne transmitter whose signals are used to determine the altitude of aircraft.

altitude The vertical distance above sea level, or above the actual terrain.

altitude delay In a plan-position-indicating type of radar, the sync delay introduced between transmission of the pulse and start of the trace on the indicator screen, to eliminate the altitude circle in the display.

altotroposphere An atmospheric zone about 40 to 60 miles in elevation.

ALU Abbreviation of *arithmetic and logic unit.*

alumel An alloy used in thermocouples, it is composed of nickel (3 parts) and aluminum (1 part).

alumina An aluminum-oxide ceramic used in electron tube insulators and as a substrate in the fabrication of thin-film circuits.

aluminium See *aluminum.*

aluminum Symbol, Al. An elemental metal. Atomic number, 13. Atomic weight, 26.98. Aluminum is widely used in electronics, familiar instances being chassis, wire, shields, semiconductor doping, and electrolytic-capacitor plates.

aluminum antimonide Formula, AlSb. A crystalline compound useful as a semicondctuor dopant.

aluminized screen A TV picture-tube screen having a thin layer of aluminum deposited on its back to brighten the image and reduce ion-spot formation.

Am Symbol for *americium.*

A/m *Ampere per meter*: the SI unit of magnetic field strength.

AM Abbreviation of *amplitude modulator, amplitude modulation.*

amalgam An alloy of a metal and mercury. Loosely, any combination of metals.

amateur 1. A nonprofessional, usually noncommercial devotee of any technology (as a hobby). 2. A licensed amateur radio operator legally authorized to operate a radio station in the amateur service.

amateur band Any one of several bands of radio frequencies assigned for noncommercial use by licensed radio amateurs (see *amateur, 2*). In the United States, there are 9 such bands between 1.80 MHz and 450 MHz. Also see *amateur service* and *amateur station.*

amateur call letters Call letters assigned by the licensing authority, especially to amateur stations. Call-letter combinations consist of a letter prefix denoting the country in which the station is situated, plus a number designating the location within the country, and two or more letters identifying the particular station. Example: W6ABC (here K or W = United States, 6 = California, and ABC = identification of individual licensee (issued alphabetically except under special circumstances). Also see *amateur, 2, amateur band, amateur service;* and *amateur station.*

amateur callsign See *amateur call letters.*

amateur extra-class license The highest class of amateur-radio operator license in the United States. It conveys all operating privileges.

amateur radio Broad term used to mean the practice of hobbying in the amateur service. Also, the equipment used for this purpose.

amateur radio operator Individual licensed to transmit radio signals in the amateur service.

amateur service Two-way radio communication engaged in purely for hobby purposes, i.e., without pecuniary interest. Also see *amateur, 2, amateur band, amateur call letters,* and *amateur station.*

amateur station A radio station licensed in the amateur service. Also see *amateur, 2, amateur band; amateur call letters,* and *amateur service.*

amauroscope An electronic aid to the blind in which photocells in a pair of goggles pick up light images. Electric pulses proportional to the light are impressed upon the visual receptors of the brain through electrodes in contact with nerves above each eye.

amber A yellow or brown fossil resin which is historically important to electronics for two reasons: (1) It is the first material reported to be capable of electrification by rubbing (Thales, 600 BC); (2) the words *electricity, electron,* and *electronics* are derived from the Greek name for amber—*elektron.*

ambience The acoustic characteristic of a room, in terms of the total amount of sound reaching a listener from all directions.

ambient An adjective meaning *surrounding*. Often used as a noun in place of the adjective-noun combination (thus, "10 degrees above *ambient*," instead of "10 degrees above *ambient temperature*").

ambient humidity The amount of moisture in the air at the time of measurement or operations in which dampness must be accounted for.

ambient level The amplitude of all interference (acoustic noise, electrical noise, illumination, etc.) coming from sources other than that of a signal of interest.

ambient light Also called *ambient illumination*. Room light or outdoor light incident to a location at the time of measurement or operations.

ambient-light filter In a TV receiver, a filter mounted in front of a picture-tube screen to minimize the amount of ambient light reaching the screen.

ambient noise 1. In electrical measurements and operation, background electrical noise. 2. In acoustical measurements and operations, audible background noise.

ambient pressure Surrounding atmospheric pressure.

ambient temperature The temperature surrounding apparatus and equipment (room temperature).

ambient-temperature range 1. The range over which ambient temperature varies at given location. 2. The range of ambient temperature which will cause no malfunction of, or damage to, a circuit or device.

ambiguity 1. Any unclear, illogical, or incorrect indication or result. 2. The seeking of a false null by a servo. 3. In digital computer practice, error resulting from improper design of logic.

ambiguous case In trigonometric calculations involving the law of sines, a condition in which two sides of a triangle and an angle opposite one of them are given, and in which there is the possibility of more than one solution for the remaining sides and angles.

ambiguous count In digital counters, a clearly incorrect count. See *accidental triggering*.

ambisonic reproduction A close approximation of the actual directional characteristics of a sound in a given environment. The reproduced sound almost exactly duplicates the sound in the actual environment in which it was recorded.

American Morse code (Samuel F. B. *Morse*, 1791-1872). A telegraph code, at one time used principally on wire telegraph lines in the United States. It differs from the International Morse code used in radiotelegraphy, which often employs a dot group, where the International code uses a dot-and-dash group.

American Radio Relay League A worldwide organization of amateur radio operators, headquartered in Newington, Connecticut. The official publication is the monthly magazine, *QST*.

American Standards Association Abbreviation, ASA. At one time, the name of the national association in the U.S., devoted to the formation and dissemination of voluntary standards of dimensions, performance, terminology, etc. See *ANSI*.

American wire gauge Abbreviation, AWG. Also called *Brown and Sharpe* or *B & S* gauge. The standard American method of designating the various wire sizes. Wire is listed according to gauge number from 0000 (460 mils diameter) to 40 (3.145 mils diameter).

americium Symbol, Am. A radioactive elemental metal first produced artificially in the mid-forties. Atomic number, 95. Atomic weight, 243.

AM/FM receiver A radio set that can receive either amplitude-modulated or frequency-modulated signals. Usually, a band switch incorporates the demodulation-selection circuitry so that as the frequency range is changed, the appropriate detector is accessed.

AM/FM transmitter A radio transmitter whose output signal may be frequency- or amplitude-modulated by a panel selector switch.

AM/FM tuner A compact radio receiver unit which can handle either amplitude- or frequency-modulated signals and which delivers its low-amplitude output to a high-fidelity audio power amplifier. Compare *AM tuner* and *FM tuner*.

A-minus Also, A−. The negative terminal of an A-battery, or pertaining to the part of a circuit connected to that terminal.

AMM Abbreviation of *antimissile missile*.

ammeter Instrument used for measuring the amount of current (in amperes) flowing in a circuit.

ammeter shunt A resistor connected in parallel with an ammeter to increase its current range. Also see *Ayrton-Mather galvanometer shunt*.

ammeter-voltmeter method The determination of resistance or power values from the measurement of voltage (E) and current (I). For resistance, $R = E/I$; for power, $P = EI$.

ammonium chloride Formula, NH4Cl. The electrolyte in the carbon-zinc type of primary cell. Also called *sal ammoniac*.

AMNL Abbreviation of *amplitude-modulation noise level*.

amortisseur winding A winding that acts against pulsation of the magnetic field in an electric motor; a winding that acts to prevent oscillation in a synchronous motor.

amorphous substance A noncrystalline substance.

amp One style (often verbal) of abbreviating *ampere*. Also see *A, 1*.

ampacity Current-carrying capacity expressed in amperes.

amperage The strength of an electric current, i.e., the number of amperes.

ampere (Andre Marie *Ampere*, 1775-1836). Abbreviations, A (preferred) a, amp. The SI base unit of current intensity (I). The ampere is the constant current which, if maintained in two straight parallel conductors of infinite length and of negligible circular cross section, and placed 1 meter apart in a vacuum, would produce between the conductors a force of 2×10^{-7} newton per meter. One ampere flows through a 1-ohm resistance when a potential of 1 volt is applied; thus $I = E/R$. Also see *microampere, milliampere, nanoampere, picoampere*.

ampere-hour The quantity of electricity that passes through a circuit in one hour when the rate of flow is one ampere. Also see *battery capacity*.

ampere-hour meter An instrument for measuring ampere-hours. It contains a small motor driven by the current being measured and which moves a point on an ampere-hour scale. The motor speed is proportional to the current level; the position taken by the pointer is proportional to current and elapsed time.

Ampere's law Current flowing in a wire generates a magnetic flux which encircles the wire in the clockwise direction when the current is moving away from the observer.

ampere-turn Symbol, NI. A unit of magnetomotive force equal to 1 ampere flowing in a single-turn coil. The ampere-turns value for any coil is obtained by multiplying the current (in amperes) by the number of turns in the coil.

Amperian whirl The stream of electrons in a single-turn, current-conducting wire loop acting as an elementary electromagnet.

Amperite tube A simple voltage or current regulator consisting of an iron wire filament in a glass bulb filled with a nonoxidizing gas. The filament resistance increases with decreased filament voltage.

amp-hr One style of abbreviating *ampere-hour*. Also see *Ah*.

amplidyne A dynamo-like rotating dc machine which can act as a power amplifier, since the response of the output voltage to changes in field excitation is quite rapid. Used in servosystems.

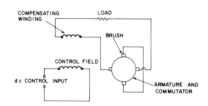

Amplidyne.

amplification 1. The process of apparently increasing the amplitude of a signal. This entails an input signal controlling a local power supply to give a larger output signal. Depending upon the kind of input and output signals, amplification may be categorized as *current, voltage, power,* or some combination of these. 2. The signal increase resulting from amplification.

amplification factor 1. Symbol, μ. In the operation of an electron tube, the ratio of the differential of plate voltage to the differential of grid voltage for a zero change in plate current; $\mu = dEp/dEg$. 2. The alpha or beta of a bipolar transistor.

amplified agc An automatic-gain-control system embodying amplification of the fed-back control signal.

amplified back bias Declining voltage developed across a fast-time-constant circuit in an amplifier stage and fed back into a preceding stage.

amplifier Any device which acts to increase the magnitude of an applied signal. It receives an input signal and delivers a larger output signal which, in addition to its increased amplitude, is a replica of the input signal. Also see *current amplifier, power amplifier, voltage amplifier.*

amplifier diode Any semiconductor which will provide amplification in a suitable circuit or microwaveplumbing setus. See *diode amplifier.*

amplifier distortion Distortion (amplitude, harmonic, nonlinear, phase) of a signal, arising within an amplifier operated in compliance with specified conditions.

amplifier input 1. The terminals and section of an amplifier to which is presented the signal to be amplified. 2. The signal which is to be amplified.

amplifier noise Collectively, all extraneous signals present in the output of an amplifier when no working signal (see *amplifier input, 2*) is applied to the amplifier input terminals.

amplifier nonlinearity The condition in which the amplifier output signal exhibits a nonlinear relationship to the corresponding input signal. Also see *amplifier distortion.*

amplifier output 1. The terminals and section of an amplifier which deliver the amplified signal for external use. 2. The amplified signal.

amplifier power The power level of the output signal delivered by an amplifier (also called *output power*), or the extent to which the amplifier increases the power of the input signal (also called *power amplification*).

amplifier response The performance of an amplifier throughout a specified frequency band. Factors usually included are gain, distortion, amplitude vs frequency, and power output.

amplify To perform the functions of amplification (see *amplification, 1*).

amplifying delay line A delay line that causes amplification of signals in a circuit intended for pulse compression.

amplistat A self-saturating magnetic amplifier.

amplitron Backward-wave amplifier used in microwave circuits.

amplitude The extent to which an alternating or pulsating current or voltage swings from zero or from a mean value.

amplitude-controlled rectifier A thyratron- or thyristor-based rectifier circuit.

amplitude density distribution The distribution function of a voltage, giving the probability that, at a certain instant, the voltage has a certain value.

amplitude distortion In an amplifier or network, the condition in which the output-signal amplitude exhibits a nonlinear relationship to the input-signal amplitude.

amplitude factor For an ac wave, the ratio of the peak value to the rms value. The amplitude factor of a sine wave is equal to the square root of 2, or 1.414 213 5.

amplitude fading Fading in which the amplitudes of all components in an amplitude-modulated wave (i.e., carrier and sidebands) wax or wane uniformly. Compare *selective fading.*

amplitude/frequency response Performance of an amplifier throughout a specified range, as exhibited by a plot of output-signal amplitude vs frequency for a constant-amplitude input signal.

amplitude gate Transducer which transmits only those portions of an input wave that lie within two close-spaced amplitude boundaries; also called *slicer.*

amplitude limiter A circuit, usually affording automatic gain control, which holds the amplitude of an output signal constant at a predetermined level in spite of large excursions of input-signal amplitude. A dc-biased diode also performs limiting action, but by clipping action.

amplitude-modulated generator A signal generator whose output is amplitude modulated. Usually, this instrument is an rf generator modulated at an audio frequency.

amplitude-modulated transmitter A radio-frequency transmitter whose carrier is varied in amplitude according to the rate of change of a suitable information wave (such as audio frequencies for speech, music, or tone signals; or picture signals and control pulses for television and facsimile).

amplitude modulation A method of modulation in which the amplitude of the carrier voltage is varied in proportion to the changing frequency value of an applied (usually audio) voltage, the carrier frequency remaining unaltered by the process. Compare *frequency modulation.*

amplitude-modulation noise level Spurious amplitude modulation of a carrier wave by noise (i.e., by extraneous signals, not by the intended modulating signal).

amplitude noise The signal amplitude fluctuations returned by the target that affect radar accuracy.

amplitude of noise The level of random noise signals in a system. The amplitude of noise is measured in the same way that signal amplitude is measured.

amplitude range The maximum-to-minimum amplitude variation of a program. May be expressed as a direct numerical ratio, or in decibels.

amplitude response The maximum output obtainable at various frequencies over the range of an instrument operating under rated conditions.

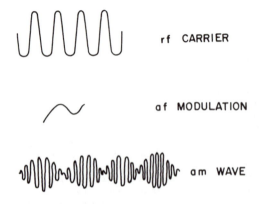

rf CARRIER

af MODULATION

am WAVE

Amplitude modulation.

amplitude selection Selection of a signal according to its correspondence to a predetermined amplitude above or below a certain threshold, or between two limits.

amplitude separator In a television receiver, a circuit which separates the control pulses from the composite video signal.

amplitude suppression ratio The ratio of undesired output of an FM receiver to desired output when the test signal is amplitude modulated and frequency modulated simultaneously.

amplitude-vs-frequency distortion Distortion resulting from varying gain or attenuation of an amplifier or network with respect to signal frequency.

AM tuner A compact radio receiver unit which handles amplitude-modulated signals and delivers its low-amplitude audio output to a high-fidelity amplifier. Compare *AM/FM tuner* and *FM tuner*.

amu Abbreviation of *atomic mass unit*.

AN Prefix designator used by American military services to indicate commonality.

anacoustic Pertaining to the lack of sound or absence of reverberation or transmission of sonic waves.

anacoustic zone In space, a silent region due to the great distances between air molecules.

analog Anything which corresponds, point for point or value for value, to an otherwise unrelated quantity. Thus, voltage is the analog of water pressure, and current is the analog of water flow. From this similarity, certain water-system problems may be solved with an *analog computer* if the water factors are represented by currents and voltages. Also see *analog computer*.

analog adder An analog circuit or device which receives two or more inputs and delivers an output equal to their sum.

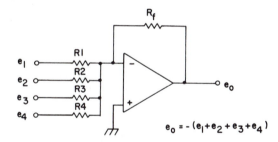

$$e_0 = -(e_1 + e_2 + e_3 + e_4)$$

Analog adder of the inverting type.

analog adder/subtracter An analog circuit or device which receives two or more inputs and delivers an output equal to their sum or difference (in any combination), as desired.

analog channel In an analog computer, an information channel in which the extreme limits of data magnitude are fixed and the data may have any value between the limits.

analog communications A communications system employing an initially continuous signal (such as a carrier wave) which is varied proportionally in amplitude, frequency, phase, or in other ways by an information signal. The information may be supplied directly by a transducer, as when a thermistor or microphone varies a direct current, or through a modulator, as in an AM or FM radio transmitter. The system takes its name from the fact that the affected carrier is analogous to the information signal.

analog computer A computer in which input and output quantities are represented as points on continuous (or small-increment) scales. To represent these quantities, the computer employs voltages or resistances which are proportional to the numbers to be worked on. When the quantities are nonelectrical (such as pressure or velocity), they are made analogous by proportional voltages or resistances.

analog data 1. Data represented in a quantitatively analogous way. Examples are the deflection of a movable-coil meter, the positioning of a slider on a slide rule, the indication of a clock, the setting of a variable resistor to represent the value of a nonelectrical quantity, and so on. Also see *analog*. 2. Data displayed along a smooth scale of continuous values (as by a movable-coil meter) rather than in discrete steps (as by a digital meter).

analog differentiator An analog circuit or device whose output is the differential of the input-signal voltage with respect to time.

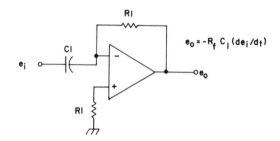

$$e_0 = -R_f C_1 (de_i / dt)$$

Analog differentiator.

analog divider An analog circuit or device which receives two inputs and delivers an output equal to their quotient.

analog electronics Electronic techniques and equipment based upon uniformly changing signals, such as sine waves, and often employing continuous-scale indicators, such as D'Arsonval meters. Compare *digital electronics*.

analog information Approximate numerical information as opposed to digital information, which is assumed to be exact.

analog integrator An analog circuit or device whose output is the integral of the input signal voltage with respect to time.

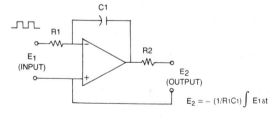

$$E_2 = -(1/R_1 C_1) \int E_1 \delta t$$

Analog intergrator.

analog inverting adder An analog adder that delivers a sum having the opposite sign to that of the input quantities.

analog meter An indicating instrument that employs a movable-coil arrangement or the equivalent, to display values along a graduated scale. Compare *digital meter*.

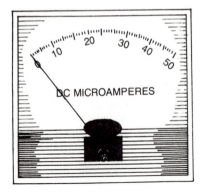

Analog meter.

analog multiplexer 1. A multiplexer used with analog signals (see *multiplexer*). 2. An analog time-sharing circuit.

analog multiplier An analog circuit or device which receives two or more inputs and delivers an output equal to their product.

analog network A circuit analog of physical variables that permits mathematical relationships to be shown directly by electric or electronic means.

analog output An output quantity which varies smoothly over a continuous range of values rather than in discrete steps.

analog record Also called *analog recording*. A record or recording method in which the recorded material varies smoothly, i.e., proportional to time or to some other factor.

analog recorder Any recorder, such as a recording oscillograph, potentiometric recorder, electroencephalograph, electrocardiograph, or lie detector, which produces an analog record. The counterpart is a digital recorder, which gives a readout in discrete numbers (printed or visually displayed).

analog representation Representation of information within a smooth, continuous range rather than as separate (discrete) steps or points.

analog signal A signal that attains an infinite number of different amplitude levels, as opposed to one that can attain only a finite number of levels as a function of time.

analog subtracter An analog circuit or device which receives two inputs and delivers an output equal to their difference.

analog summer See *analog adder*.

analog switch A switching device that will only pass signals that are faithful analogs of transducer parameters.

analog-to-digital conversion The conversion of an analog quantity (e.g., position of a rotating shaft) into digital units, such as binary bits, for digital computer processing.

analog-to-digital converter Any circuit or device which performs analog-to-digital conversion.

analysis The rigorous determination of the constants and mode(s) of operation of an electronic circuit or device. Compare *synthesis*.

analytical engine An early calculator invented in 1833 by Charles Babbage (1792-1871).

analyzer 1. Any instrument that permits analysis through close measurements and tests, e.g., *distortion analyzer, wave analyzer, gas analyzer*. 2. One of a pair of polarizing plates (*polarizer* and *analyzer*) through which light is passed. The polarizer transmits only the part

of the light that vibrates in a certain plane. The analyzer, then, transmits or shuts out the "selected" light when it (the analyzer) is rotated correctly with respect to the polarizer; it will pass it for other rotations. 3. A computer program used for debugging purposes, it analyses other programs and summarizes references to storage locations. 4. A logic analyzer or analysis interface to an oscilloscope.

anastigmatic yoke Also called *full-focus yoke*. In a TV receiver, a deflection yoke having a cosine winding for better focus at the edges of the picture.

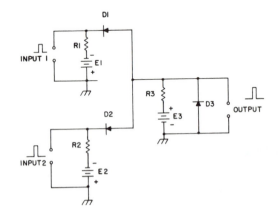

AND circuit.

anastigmat lens A compound lens system (consisting of one converging and one diverging lens) for correcting astigmatism, a fault in some single lenses.

anchorage In plastic recording tape, the adhesion of the magnetic oxide coating to the surface of the tape.

ancillary equipment Equipment that does not enter into the operation of a central system, but is auxiliary. Examples are input/output components of a computer, and test instruments and indicators attached to a system.

AND circuit In digital systems and other switching circuits, a type of gate which delivers an output signal only when *all* of several input signals occur simultaneously. Compare *OR circuit*.

Anderson bridge An ac bridge circuit having six impedances, permitting the value of an unknown inductance to be determined in terms of a standard capacitance.

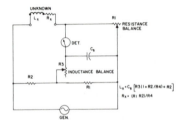

Anderson bridge.

AND gate 1. AND circuit. 2. In a TV receiver, an AND circuit which holds the keyed-agc signal off until a positive horizontal flyback pulse and a horizontal sync pulse appear simultaneously at the input.

anechoic Pertaining to the absence of echos. Examples: anechoic chamber, anechoic enclosure, anechoic room.

anechoic chamber An enclosure that does not reflect sound waves that approach its walls. Such a chamber is used for testing certain audio devices.

anemograph A recording anemometer.

anemometer An instrument for measuring or indicating wind velocity.

angel 1. An extraneous image, usually of short duration, on a cathode-ray screen. The term applies particularly to a radar image due to low-atmospheric reflection or to a bird or other mobile object. 2. Air-deployed chaff suspended from a balloon or parachute, designed to create radar reflection images as a decoy tactic.

angle 1. The geometric figure formed by the meeting of two lines. 2. The circular distance through which a vector moves. 3. The elapsed interval measured in degrees or radians along the horizontal axis of a wave pattern. 4. The arc between vectors.

angle jamming A radar jamming technique, in which the return echo is jammed with a signal containing improper azimuth or elevation angle components.

angle modulation Variation (modulation) of the angle of a sine-wave carrier in response to the modulating source, as in *frequency modulation* and *phase modulation*.

angle noise In radar reception, the interference resulting from variations in the angle at which an echo arrives from the target.

angle of arrival The angle which the line of propagation of a radio wave makes with the surface of the earth. Compare *angle of departure*.

angle of azimuth The horizontal angle between viewer and object or target, usually measured clockwise from north.

angle of beam The angle enclosing most of the transmitted energy in the radiation from a directional antenna.

angle of conduction 1. Also called *angle of flow*. The number of degrees of an excitation-signal cycle during which output (plate or collector) current flows in an amplifier circuit. 2. The number of degrees of any sine wave at which conduction of a device (e.g., a triac) begins.

angle of convergence 1. In any graphical representation, the angle formed by any two lines or plots which come together at a point. 2. The angle formed by the light paths of two photocells focused upon the same object.

angle of declination The angle between a horizontal line and a descending line. Compare *angle of elevation*.

angle of deflection In a cathode-ray tube, the angle between the electron beam at rest and a new position resulting from deflection.

angle of departure The angle made by the line of propagation of a transmitted radio wave with the (horizontal) surface of the earth at the transmitting station. Compare *angle of arrival*.

angle of depression See *angle of declination*.

angle of divergence In a cathode-ray tube, the angle formed by the spreading of an undeflected electron beam as it extends from the gun to the screen.

angle of elevation The angle made with the horizontal axis by an ascending line. Compare *angle of declination*.

angle of flow See *angle of conduction*.

angle of incidence The angle made by a ray or the line of propagation of a wave with a perpendicular to the surface struck by the ray. Compare *angle of reflection*.

angle of lag The phase difference (in degrees or radians) whereby one component follows another in time, both components being of the same frequency. Compare *angle of lead*. Also see *phase angle*.

angle of lead The phase difference (in degrees or radians) whereby one component precedes another in time, both components being of the same frequency. Compare *angle of lag*. Also see *phase angle*.

angle of radiation 1. The angle, with respect to the surface of the earth, at which the principal lobe of a wave leaves a transmitting antenna. 2. The angle of a receiving or transmitting antenna's optimum sensitivity.

angle of reflection The angle made by a ray or the line of propagation of a wave and a perpendicular to the surface the ray leaves. Compare *angle of incidence*.

angle of refraction The angle made by a ray or the line of propagation of a wave and a perpendicular to the surface of a medium of propagation, as the ray passes out of the medium.

angle tracking noise Noise in a servo system that results in a tracking error.

angstrom (Anders J. *Angstrom*, 1814-1874). Symbol, A unit of length used to describe certain extremely short waves and microscopic dimensions; 1 angstrom equals 10^{-4} micron, or 10^{-10} meter.

angular acceleration Unit, rad/s². The rate of change of angular velocity, i.e., the second derivative of angular velocity.

angular accelerometer An accelerometer capable of measuring angular acceleration.

angular aperture The aperture of an objective lens (optical or electronic), expressed by the angle between lines from the principal focus to two points on the rim at the ends of a diameter. Thus, "a lens of 30 degrees *aperture*."

angular deviation loss The ratio, expressed in decibels of microphone or loudspeaker response on the principal axis of response to the response at a designated angle from that axis.

angular difference See *phase angle*.

angular displacement In an ac circuit, the separation, in degrees, between two waves. See *phase angle*.

angular distance In a vectorial representation of quantities, the angle A0B, which depicts the angular separation of points A and B relative to point 0.

angular frequency Ac frequency expressed in radians per second, and equal to $\theta 1f$, where f is the frequency in Hz and θ equals $2\pi f$.

angular length Length, as along the horizontal axis of an ac wave or along the standing-wave pattern on an antenna, expressed as the product of radians and wavelength.

angular-mode keys On a calculator or computer, the DEG, RAD, and GRAD keys, for expressing or converting angles in *degrees, radians,* and *grads*, respectively.

angular momentum The product of angular velocity and mass of an object.

angular phase difference For two sinusoidal waves, the phase difference expressed in degrees or radians.

angular rate In navigation, the rate of bearing change, expressed in degrees or radians.

angular resolution The ability of a radar to distinguish between two targets by angular measurement.

angular velocity The number in radians per second of a rotating vector or radius. For a periodic quantity, such as a sine-wave ac, angular velocity equals ωf, where f is the frequency in Hz, and ω equals $2\omega f$.

angus pen recorder An instrument that makes a permanent record of the time a channel is used.

anharmonic oscillator An oscillating device in which the force toward the balance point is not linear with respect to displacement.

anhysteresis The magnetization of a material by a unidirectional field containing an alternating field component of gradually decreasing amplitude.

anion A negative ion. Also see *ion*.

anisotropic Pertaining to the tendency of some materials to display different magnetic and other physical properties along different axes.

anl Abbreviation of *automatic noise limiter*.

anneal To heat a metal to a predetermined temperature and then let it cool slowly. The operation prevents brittleness and often stabilizes electrical characteristics.

annealed laminations Core laminations for transformers or choke coils, which have been annealed.

annealed shield A magnetic shield for cathode-ray tubes, which has been processed by annealing.

annealed wire Soft-drawn wire that has been subjected to annealing.

annealing A method of increasing the ductility of a substance by heating and cooling.

annotations 1. Marking on copies of original engineering-installation documents to show changes made during the installation. 2. Any set of comments or notes accompanying a program, an equipment or system, or a process.

annular Ring-shaped.

annular conductor A number of wires stranded in three concentric layers of alternating twists around a hemp core.

annular transistor A mesa transistor in which the base and collector take the form of concentric rings around a central emitter.

annulling network A subcircuit shunting a filter to cancel reactive impedance at the extreme ends of the pass band of the filter.

annunciation relay A relay that indicates whether or not a circuit is carrying current.

annunciator A device, usually aural but occasionally visual or both aural and visual, for sounding an alarm or attracting attention.

anode The positive electrode of a device, i.e., the electrode toward which electrons move during current flow.

anode balancing coil Mutually coupled windings used to maintain equal currents in parallel anodes operating from a common transformer terminal.

anode breakdown voltage In a gas tube, the voltage at which the tube suddenly conducts, i.e., when the starter gap is nonconducting and all other electrodes are at cathode potential.

anode-bypass capacitor See *plate bypass capacitor*.

anode characteristic Also called *plate characteristic*. For an electron tube, the plot of anode current vs anode voltage (in triodes and multigrid tubes the control-grid voltage is held constant for each anode curve).

anode current Current flowing in the anode circuit of a device, especially a thyratron or glow tube. Plate current in a vacuum tube.

anode dark space In gaseous and glow discharge tubes, a dark space on the anode.

anode dissipation Also called *plate dissipation*. The power (the product of anode voltage and anode current) dissipated by the anode of a tube.

anode efficiency Also called *plate efficiency*. In an electron tube, the ratio P_o/P_i, where P_o is output power and P_i is the dc anode-power input.

anode load impedance Symbol, $Z_{L(anode)}$. Also called *plate load impedance*. In a tube circuit, the output impedance that is ac-connected between the anode and ground, or dc-connected between the anode and dc anode power supply.

anode modulation See *plate modulation*.

anode neutralization In a tube circuit, a method of neutralization in which a negative-feedback capacitor is connected between the lower (far) side of the plate tank circuit and the grid.

anode potential See *anode voltage*.

anode power input Symbol, $P_{A(input)}$. The product of anode current and anode voltage.

anode power supply The ac or positive dc power supply unit that delivers current and voltage to the anode of an electron tube.

anode rays A discharge of positively charged particles sometimes observed in a two-element vacuum tube operated at high voltage. It has been attributed to, among other culprits, certain impurities in the anode metal.

anode saturation The point beyond which further increase in anode voltage does not produce an increase in anode current.

anode sheath A coat of electrons surrounding the anode in some tubes.

anode strap In a magnetron, a metal strap connecting the anodes of a multicavity device.

anode supply See *anode power supply*.

anode terminal 1. In a diode (semiconductor or tube), the terminal to which a positive dc voltage must be applied to forward-bias the diode. Compare *cathode terminal*. 2. In a diode (semiconductor or tube), the terminal at which a negative dc voltage appears when the diode is employed as an ac rectifier. Compare *cathode terminal*. 3. The terminal which is connected internally to the anodic element of any device.

anode voltage Symbol, E_A or V_A. The difference in potential between the anode and cathode of a device.

anode voltage drop In a cold-cathode glow tube, the anode voltage when the tube is in operation, as opposed to the anode voltage when the tube is not conducting.

anodic Pertaining to the anode or to anodelike effects.

anodizing An electrolytic process in which a protective oxide film is deposited on the surface of a metallic body acting temporarily as the anode of the electrolytic cell.

anomalous dispersion Dispersion of electromagnetic radiation characterized by a decrease in refractive index with increase in frequency.

anomalous propagation 1. The low-attenuation propagation of uhf signals through atmospheric layers. 2. Rapid fluctuation of a sonar echo because of variations in propagation.

anoxemia toximeter An electronic instrument for measuring or alerting against the onset of anoxemia (deficiency of oxygen in the blood), especially in airplane pilots.

AN radio range A navigational facility entailing four zones of equal signal strength. When the aircraft deviates from course, an aural Morse-code signal, A (*dit dah*) or N (*dah dit*) is heard; but when the aircraft is on course, a continuous tone is heard.

ANSI Abbreviation of *American National Standards Institute*.

AN signal The signal provided by an AN radio range to apprise aircraft pilots of course deviation to the right or left.

answerback The automatic response of a terminal station to a remote-control signal.

answer cord In a telephone system, the cord used for answering subscriber's calls and incoming trunk calls.

answering machine A device for automatically answering a telephone and recording a spoken message from the caller.

answer jack Jack on which a station calls in and is answered by an operator.

answer lamp Telephone switchboard lamp that lights when an answer cord is plugged into a line jack; it goes out when the telephone answers, and lights when the call is completed.

ant Abbreviation of *antenna*.

antenna In a communication system, the contrivance which picks up a received signal or radiates a transmitted signal. An antenna can be a simple wire or rod, or a complicated structure.

antenna ammeter An rf ammeter, usually of the thermocouple type, employed to measure current flowing to a transmitting antenna.

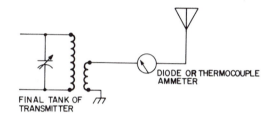

Antenna ammeter.

antenna amplifier 1. A radio-frequency amplifier, often installed at the antenna, employed to boost signals before they reach a receiver. 2. Occasionally, the first rf amplifier stage of a receiver.

antenna array See *array*.

antenna bandwidth The frequency range throughout which an antenna will operate at a specified efficiency without needing alteration or adjustment.

antenna beam width The effective width of a directional antenna's beam. It is expressed as the angle (in degrees) between opposite half-power points in the beam.

antenna coil The primary coil of the input rf transformer of a receiver, or the secondary coil of the output rf transformer of a transmitter.

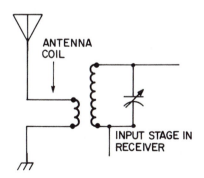

Antenna coil.

antenna coincidence The condition of two directional antennas pointed toward each other.

antenna-conducted interference Extraneous signals generated in a transmitter or receiver and presented to the antenna, from which they are radiated.

antenna core A ferrite rod or slab around which a coil of wire is wound to act as a self-contained antenna, usually in a miniature receiver.

antenna coupler A device consisting of an inductor, rf transformer, or a combination of inductor(s) and capacitor(s), employed to match the impedance of an antenna to that of a transmitter or receiver.

antenna coupling The inductive or capacitive coupling (or a combination of the two) employed to transfer energy from an antenna to a receiver or from a transmitter to an antenna.

antenna crosstalk Crosstalk between antennas, usually expressed in decibels.

antenna current 1. Radio-frequency current flowing from a transmitter into an antenna. 2. Of less practical concern, the current flowing from a receiving antenna into a receiver.

antenna detector A circuit that warns aircraft personnel that they are being observed by radar. It picks up the radar pulses and actuates a warning light or other device.

antenna diplexer A coupling device which permits several transmitters to share one antenna without troublesome interaction. Compare *antenna duplexer*.

antenna directivity The directional characteristic of an antenna, usually expressed in degrees of beam width between half-power or similiar reference points on either side of the major lobe.

antenna director In a directional antenna, an element situated in front of the radiator and separated from it by an appropriate fraction of a wavelength. Its function is to intensify radiation in the direction of transmission. Compare *antenna radiator* and *antenna reflector*.

antenna duplexer A circuit or device permitting one antenna to be shared by two transmitters without undesirable interaction.

antenna effect The tendency of wires or metallic bodies to act as antennas, i.e., to radiate or pick up radio waves.

antenna effective area In directional transmission, the area for power reception, expressed by the quotient $\lambda^2/4\pi A\mathrm{p}$, where λ is the wavelength and $A\mathrm{p}$ is the power gain in the specified direction.

antenna efficiency The ratio, usually expressed as a percentage between the radio-frequency power applied to an antenna and the power actually radiated into space.

antenna factor A factor (in decibels) added to an rf voltmeter reading to find the true open-circuit voltage induced in an antenna.

antenna field The electrostatic or electromagnetic field immediately surrounding an antenna.

antennafier Low-profile antenna/amplifier used with portable communications systems.

antenna front-to-back ratio For a directional antenna, the ratio of field strength in front of the antenna (i.e., directly forward in the line of maximum directivity) to field strength in back of the antenna (i.e., 180 degrees from the front), as measured at a fixed distance from the radiator.

antenna gain For a given antenna, the ratio of signal strength received or transmitted) to that obtained with a comparison antenna, such as a simple dipole.

antenna ground system The ground system associated with an antenna, i.e., a metallic ground plane, counterpoise, or the earth itself.

antenna/ground system An arrangement embodying both an antenna and a low-resistance connection to the earth, as opposed to a straight antenna system which involves no connection to earth.

antenna height 1. Height of an antenna above actual ground. 2. The height of an antenna above the terrain, determined as the mean of a number of points on the terrain with respect to the antenna.

antenna impedance The impedance an antenna presents to a transmitter or receiver. It is considered to be the impedance at the attachment point of the transmission line or feeder and varies from about 50 to 600 ohms, depending on antenna type and installation.

antenna-induced potential Also called *antenna-induced microvolts*. The voltage across the open-circuited terminals of an antenna.

antenna lens Also called *lens antenna*. A radiator designed to focus microwave energy in much the same manner that an optical lens focuses light rays. Lens antennas are made from dielectric materials as well as from metals.

antenna lobe A well-defined region in the radiation pattern of an antenna in which radiation is most intense, or in which reception is strongest. Also see *antenna pattern*.

antenna matching The technique of establishing a satisfactory relationship between antenna impedance and either transmission-line or transmitter-output imepdance, for maximum transfer of power into the antenna. Also, the matching of antenna impedance to receiver-input impedance for delivery of maximum energy to the receiver.

antennamitter Antenna/oscillator combination that serves as a low-power transmitter.

antenna pattern A polar plot of antenna performance that shows field strength vs angle of azimuth, with the antenna at the center.

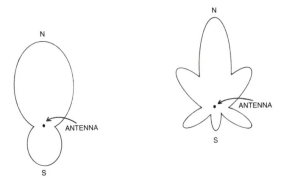

Antenna patterns.

antenna polarization The position of an antenna, with respect to the surface of the earth, that determines the wave polarization for which the antenna is most efficient. A vertical antenna radiates vertically polarized waves; a horizontal antenna radiates horizontally polarized waves broadside to itself and vertically polarized waves at high angles off its ends.

antenna power Symbol, P_{ant}. Rf power developed in an antenna by a transmitter; P_{ant} equals I^2R, where I is the antenna current and R is the antenna resistance at the point I is measured.

antenna power gain The ratio of the power required to produce a selected field strength with an antenna of interest. to the power re-

quired for the same field strength (at the same test location) with a reference antenna, such as a standard dipole.

antenna preamplifier A highly sensitive amplifier used to enhance the gain of a receiver. Usually employed at the very-high frequencies and above.

antenna radiation The propagation of radio waves by a transmitting antenna.

antenna radiator The element of an antenna that receives rf energy from the transmitter and radiates waves into space. Compare *antenna director* and *antenna reflector*.

antenna range 1. The frequency band, communication distance characteristically covered, or other continuum of values that specify the operating limits of an antenna. 2. The region immediately surrounding an antenna in which tests and measurements usually are made. Sometimes called *antenna field*.

antenna reflector In a directional antenna, an element situated behind the radiator and separated from the latter by an appropriate fraction of a wavelength. Its function is to intensity radiation in the direction of transmission. Compare *antenna director* and *antenna radiator*.

antenna relay In a radio station, a low-loss, heavy-duty relay by means of which the antenna may be switched between transmitter and receiver.

antenna resistance See *antenna impedance*.

antenna resonant frequency The frequency, or narrow band of frequencies, at which antenna impedance appears resistive.

antenna stage 1. The first rf amplifier stage of a receiver. 2. Occasionally, the final rf amplifier of a transmitter.

antenna switch In a radio station, a low-loss, heavy-duty switch by means of which the antenna may be connected to transmitter, receiver, or safety ground.

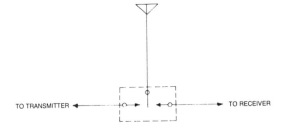

Antenna switch of single-pole, single-throw construction.

antenna system Collectively, an antenna and all of the auxiliary electrical and mechanical devices needed for its efficient operation, including couplers, tuners, transmissions lines, supports, insulators, and rotator.

antenna terminals 1. The points at which a transmission line is attached to an antenna. 2. The signal input terminals of a receiver. 3. The signal output terminals of a transmitter.

antennaverter An antenna and converter combined into a single circuit, intended for connection to the antenna terminals of a receiver to allow operation on frequencies outside the band for which the receiver has been designed.

antenna wire 1. The radiator element of a wire-type antenna. 2. A strong solid or stranded wire (usually copper or phosphor-bronze) used for antennas.

antiaircraft missile A surface-launched guided missile whose target is an aircraft in flight.

anticapacitance switch A switch whose members are thin blades and stiff wires widely separated to minimize capacitance between them.

anticathode The target electrode of an X-ray tube.

anticlutter circuit A supplementary circuit in a radar receiver, which minimizes the effect of extraneous reflections that would obscure the image of the target.

anticlutter gain control In a radar receiver, a circuit that automatically raises the gain of the receiver slowly to maximum after each transmitter pulse, to reduce the effect of clutter-producing echoes.

anticoincidence Noncoincidental occurrence of two or more signals. Compare *coincidence*.

anticoincidence circuit In computers and control systems, a circuit which delivers an output signal only when two or more input signals are not received simultaneously. Compare *coincidence circuit*. Also see *NAND circuit*.

anticoincidence operation An exclusive-OR operation.

anticollision radar Any vehicular radar system employed to prevent collisions.

antiferroelectric 1. Pertaining to the property whereby the polarization curve of certain crystalline materials shows two regions of symmetry. 2. An antiferroelectric material.

antiferromagnetic Pertaining to the behavior of materials in which, at low temperatures, the magnetic moments of adjacent atoms point in opposite directions.

antihunt The condition in which hunting is counteracted, usually by removing overcorrection in automatic control or compensation systems.

antihunt circuit 1. A circuit which minimizes or eliminates hunting. Also see *antihunt*. 2. In a TV receiver, a subcircuit (a capacitor and resistor) which stabilizes an automatic frequency control circuit by preventing hunting (the rapidly recurrent shifting above and below the desired frequency, followed by final settling at that frequency).

antihunt device A device that performs the functions of an antihunt circuit.

antijamming The counteraction of jamming.

antilogarithm Abbreviations, antilog; $\log^{-1}$. The number corresponding to a given logarithm. Thus, log 10^4 equals 4 and antilog 4 equals 10^4.

antimagnetic Pertaining to materials having low retentivity.

antimatter Matter that is the counterpart of conventional matter, i.e., having positrons instead of electrons, antineutrons instead of neutrons, antiprotons instead of protons. Also see *antiparticle*.

antimicrophonic See *nonmicrophonic*.

antimissile missile A missile whose target is another missile in flight.

antimony Symbol, Sb. A metalloidal element. Atomic number, 51. Atomic weight, 121.76. Often used as n-type dopant in semiconductor manufacture.

antineutrino The antiparticle of the neutrino, emitted as a result of radioactive decay. Compare *neutrino*.

antineutron An uncharged particle having a mass equal to that of the neutron, but with a magnetic moment in the direction opposite that of the neutron.

antinode A point of maximum amplitude in a standing wave.

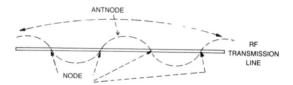

Antinode.

antinoise carrier-operated circuit A circuit that cuts off the audio output of a receiver while the station transmitter is in use . This may be accomplished in the automatic volume-control circuit of the receiver, or in the speaker or audio line.

antinoise microphone A lip or throat microphone or any microphone that discriminates against acoustic noise.

antinucleon A particle having the mass of a nucleon, but with the opposite electrical charge and direction of magnetic moment. Compare *nucleon*.

antioxidant A material, such as a lacquer coat or an inactive oxide layer, which prevents or slows oxidation of a material exposed to air.

antiparticle A subatomic particle opposite in character to conventional particles (electrons, protons). Antiparticles constitute antimatter. Also see *antineutrino, antineutron, antinucleon, antiproton, positron*.

antiphase The property of being in opposite phase.

antipincushioning magnets In some TV receivers, a pair of corrective magnets in the deflection assembly on the picture tube, which eliminate pincushion distortion (disfigurement of the raster so that it resembles a pincushion, i.e., a rectangle with its sides bowed in).

antiproton A subatomic particle having a mass equal to that of the proton, but with opposite electrical charge.

antiquark An *antiparticle* corresponding to a *quark*.

antirad substance A material that protects against damage caused by atomic radiation.

antiresonance 1. Parallel resonance. 2. Occasionally, being detuned from a resonant frequency.

antiresonant circuit See *parallel-resonant circuit*.

antiresonant frequency 1. The resonant frequency of a parallel-resonant circuit. 2. In a piezoelectric crystal, the frequency at which impedance is maximum (as in a parallel-resonant circuit).

antisidetone Pertaining to the elimination in telephone circuits of interference between the microphone and earphone of the same telephone.

antisidetone equipment Components or circuits for eliminating sidetone interference (see *antisidetone*).

antiskating device A device that creates an outward force on the pickup arm of a phonograph. The pickup arm tends to move toward the center of a disk because of friction with the disk; the antiskating device balances this inward force and reduces the condition of friction.

antistatic agents Methods or chemicals used to lower static electricity in vinyl records.

antistickoff voltage The low voltage applied to the coarse synchro control transformer rotor winding in a dual-speed servo system to eliminate ambiguous behavior in the system.

antitransmit/receive switch Abbreviation, *ATR switch*. In a radar installation, an automatic device to prevent interaction between transmitter and receiver, and vice versa.

antivoice-operated transmission Radio communications using a voice-activated circuit as a transmitter interlock during reception on the companion receiver.

apc Abbreviation of *automatic picture control; automatic phase control.*

aperiodic Not characterized by predictable periods or steps; thus, of an antenna, nonresonant at multiples of any frequency.

aperiodic current The unidirectional current that follows an electromagnetic disturbance in an LCR circuit in which *R* is equal to or higher than the critical circuit resistance.

aperiodic damping Damping of such a high degree that the damped system, after disturbance, comes to rest without oscillation or hunting.

aperiodic discharge A discharge in which current flowing in an LCR circuit is unidirectional, rather than oscillatory. For this condition, $1/LC$ is less than or equal to $R^2/4L^2$.

aperiodic function A nonrepetitive function, e.g., hyperbolic functions (sinh, cosh, tanh, etc.).

aperture 1. The larger, normally open end of a horn antenna or horn loudspeaker. 2. An opening in an opaque disc or mask that passes a predetermined amount of light or other radiant energy. 3. The portion of a directional antenna through which most of the radiated energy passes.

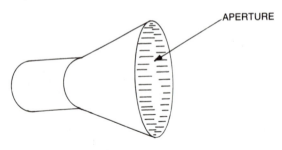

APERTURE

Aperture.

aperture antenna Antenna whose beam width depends on the size of a horn, reflector, or lens.

aperture compensation In a TV camera, the minimizing of aperture distortion by widening the video-amplifier passband.

aperture distortion In a TV camera, distortion due to the scanning beam covering several mosaic globules simultaneously in the camera tube. This situation, due to beam thickness, causes loss of resolution in the picture.

aperture illumination Amplitude/phase field distribution over the aperture.

aperture mask In a three-gun color-TV picture tube, a thin, perforated sheet mounted behind the viewing screen to insure that a particular color phosphor will be excited only by the beam for that color. Also called *shadow mask*.

aperture plate In a digital computer, a memory plane consisting of a ferrite plate containing parallel rows of holes interconnected by metal lines plated on the surface.

APL A programming language designed for ease of implementation by programmers and characterized by the requirement for a special character set. As an abbreviation, it stands for a programming language.''

apl 1. Abbreviation of *average picture level.* 2. Abbreviation of *automatic phase lock.*

A-plus Also, A +. The positive terminal of an A-battery. Also, pertaining to the part of a circuit connected to that terminal.

A-power supply A term sometimes used to denote the unit that supplies energy to a tube filament or heater. Compare *B-power supply.*

apparent bearing In radio direction finding, the uncorrected direction from which a signal arrives.

apparent error See *absolute error.*

apparent power In an ac circuit, the power value obtained by simple multiplication of current by voltage (*P* equals *IE*), with no consideration of the effects of phase angle. Compare *true power.*

apparent power loss The loss in an ammeter or voltmeter, caused by the imperfection of the instrument. At full scale, the ammeter has a certain voltage across its terminals; the apparent power loss is the current multiplied by this voltage. A voltmeter carries a small current; the apparent power loss is the product of the current and the indicated voltage.

applause meter An instrument consisting essentially of a microphone, audio amplifier, and indicating meter reading directly in sound level. It is so called because of its familiar use in measuring audience response, as indicated by loudness of applause.

Applegate diagram For a velocity-modulated tube a plot of the positions of electron bunches in the drift-space vs time.

Appleton layer Collectively, the *F*1 and *F*2 layers of the ionosphere.

apple tube A color picture tube, used in television, with the red, blue, and green phosphor in vertical strips.

appliance Electrical equipment in general. This may include any home-operated device.

application A specific program problem to which a computer is applied.

application factor A factor involved in determining the failure rate of a circuit or system, affected by unusual operating conditions.

application schematic diagram A diagram of pictorial symbols and lines illustrating the interrelationship of functional circuit blocks in a specific program mode.

applicators 1. In dielectric heating, the electrodes between which the dielectric body is placed and the electrostatic field developed. 2. In medical electronics, the electrodes applied to a patient undergoing diathermy or ultrasonic therapy.

applied voltage The voltage presented to a circuit point or system input, as opposed to the voltage drop resulting from current flow through an element.

applique circuit A circuit for adapting equipment to a specialized job.

approach-control radar A radar installation serving a ground-controlled approach (GCA) system.

approximate data 1. Data obtained through measurements. Such data can never be completely accurate, since all measurements are subject to some error. 2. Loosely, estimated data or imprecise calculations.

aql Abbreviation of *acceptable quality level.*

Aquadag Tradename of a material consisting of a slurry of fine particles of graphite. Aquadag forms a conductive coating on the inside and outside walls of some cathode-ray tubes.

aqua pura Formula, H2O. Pure water; in most instances, distilled water. Pure water is a nonconductor with a dielectric constant of 81.

Ar Symbol for *argon.*

arbitrary function fitter A circuit or device, such as a potentiometer, curve changer, or analog computer element, providing an output current or voltage which is some preselected function of the input current or voltage.

arc 1. A luminous sustained discharge between two electrodes. Because it is sustained rather than intermittent, an arc is distinguished from a spark discharge, the latter being a series of discharges (sparks), even when it appears continuous. 2. In graphical presentations, a section of curved line, as of a circle.

arc-back In a gas-filled tube, inverse current flow due to excessions in the anode-cathode space during the nonconducting half-cycle of applied ac voltage.

arc converter A radio-frequency oscillator consisting of a dc electric arc (operated in a hydrocarbon vapor) and a tuned LC circuit. The arc oscillates by virtue of its negative resistance. Before there were high-powered tube-type transmitters, the arc converter was the most widely used device in early long-distance, high-powered radiotelegraph and radiotelephone transmitters, often delivering up to 100 kilowatts at frequencies of 50 to 1000 kHz.

arc cosine Abbreviations, arc cos; $\cos^{-1}$. The angle corresponding to a given cosine. Thus, cos 60 equals 0.5 and $\cos^{-1}$ 0.5 equals 60.

arc discharge The sustained, luminous thermionic discharge between anode and cathode in a gas-filled tube. Also see *arc*.

arc-discharge tube A gas-filled or mercury-vapor tube which utilizes ionic phenomena for switching, voltage regulation, or rectification. Also see *arc, arc discharge, arc tube, thyratron*.

arc drop The anode-to-cathode voltage drop in any arc-discharge tube during conduction. Also called *arc-drop voltage*.

arc-drop loss The average product of the instantaneous values of voltage and current in a gas tube. Theoretically, this must be determined by mathematically integrating the voltage and current for at least one complete wave cycle.

arc-drop voltage The potential difference, at a given instant, between the anode and cathode of a gas tube.

arc extinction The extinguishing of the arc discharge in an arc tube, such as a thyratron. When the tube is operated from an ac power supply, the arc is automatically extinguished each time the ac cycle passes through zero. With a dc power supply, however, a special arc-examination circuit must be employed.

arc failure 1. Insulation failure due to arcover. 2. Failure of make-and-break contacts through damage caused by arcing.

arc function An inverse trigonometric function. See, specifically, *arc cosine; arc sine; arc tangent*.

arc furnace A high-temperature electric furnace in which heat is produced by one or more arcs.

architecture The functional design elements of a computer, especially its processor, and the manner in which they interact in practice.

archived file A file stored on some backing medium such as magnetic tape rather than being held permanently and under the control of a central processor; the file will be apart from the operating system's catalog of current files but can be reconstituted as needed.

arcing contacts Make-and-break contacts between which an arc occurs when they are separated.

arcing time The elapsed time between the breaking of contacts and the end of the arc between the contacts.

arc lamp An electric lamp in which a brilliant arc jumps between the tips of two rods (originally carbon).

arc oscillation 1. Oscillations produced by an *arc converter*. 2. Oscillations produced when opening dc relay contacts (not spark).

arc oscillator See *arc converter*.

arcover The (usually abrupt) creation of an arc between electrodes, contacts, or capacitive plates.

arcover voltage 1. The voltage at which arc-back occurs in a tube. 2. The voltage at which disruptive discharge occurs, typically accompanied by an arc.

arc resistance The ability of a material, usually a dielectric, to resist damage from arcing. This property is commonly expressed as the length of time between the start of the arc and the establishment of a conductive path through the material.

arc sensor A device for detecting visible arcs and excessive reflected power in microwave systems.

arc sine Abbreviations, arc sin; $\sin^{-1}$. The angle corresponding to a given sine. Thus, sin 45 equals 0.707 and $\sin^{-1}$ 0.707 equals 45.

arc suppression Extinguishing an arc discharge. Disruptive arcs in electronic circuits are suppressed by means of auxiliary diodes or resistor capacitor networks.

arc suppressor A device, or combination of devices, which provides arc suppression.

arc-suppressor diode A semiconductor diode used to prevent arcing between make-and-break contacts.

arc tangent Abbreviations, arc tan; $\tan^{-1}$. The angle corresponding to a given tangent. Thus, tan 30 equals 0.577 and $\tan^{-1}$ 0.577 equals 30.

arc-through 1. The puncturing of a material by an arc. 2. In an arc tube, the undesired flow of current during the interval when the tube should be nonconducting.

arc transmitter See *arc converter*.

arc tube See *arc discharge tube*.

area code In the United States, a three-digit number that indicates the location, according to specified assigned districts, of a telephone subscriber. When making a long-distance call, the area code of the desired station must be given in addition to the seven-digit telephone number.

area redistribution A way of measuring the duration of an irregularly shaped pulse by taking the width of a rectangle having the same peak amplitude and area as the pulse's duration.

area search The scanning of a large group of computer records for those of a major category or class for further processing.

argon Symbol, Ar. An inert gaseous element. Atomic number, 18. Atomic weight, 39.94. Argon, present in small amounts in the earth's atmosphere, is used in some voltage-regulator tubes and indicator lamps.

argon glow lamp A gas diode containing argon. Also see *gas diode*.

argon laser A laser employing argon gas.

argument 1. The direction angle of a polar vector. 2. An independent variable whose value determines the value of a function.

arithmetic address An address obtained by performing an arithmetic operation on another address.

arithmetic and logic unit Abbreviation, ALU. The part of a digital computer containing the arithmetic circuits that perform the arithmetic and logic operations; distinguished from storage, input/output, and peripheral units.

arithmetic circuit In a digital computer, a circuit that performs a logic operation. Included in this category are adders, storage registers, accumulators, subtracters, and multipliers. See separate listings under those headings.

arithmetic element See *arithmetic circuit*.

arithmetic mean The average of a group of quantities, obtained by dividing their sum by the number of quantities.

arithmetic operation In digital computer practice, the numerical processes performed: addition, subtraction, multiplication, division, comparison.

arithmetic progression A mathematical series in which each term following the first is obtained by adding a constant quantity to the preceding one. For example, $S = 1, 2, 3, 4, \ldots n$. Compare *geometric progression*.

arithmetic shift In a digital computer multiplication or division of a quantity by a power of the base employed in the notation.

arithmetic sum The sum of two or more quantities disregarding their signs. Compare *algebraic sum*.

arithmetic symmetry A filter response that is exactly symmetrical about the center frequency when the frequency scale is linear.

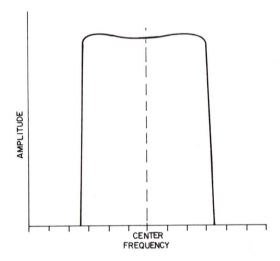

Arithmetic symmetry.

arithmetic unit The part of a digital computer in which logic and arithmetic operations are carried out.

arithmometer Obsolete term for a calculator that performs multiplication by means of a series of additions.

arm 1. Any of the distinct branches of a circuit or network. Also called leg. 2. A usually straight movable element in a device, usually containing a contact for switching.

armature 1. The rotating member of a motor. 2. The rotating member of some types of electromechanical generator. 3. The movable member of a relay, bell, buzzer, or gong. 4. The movable member of an actuator. 5. The soft-iron keeper placed across the poles of a permanent magnet to conserve power.

armature coil A coil of insulated wire wound on a ferromagnetic core to provide the electromagnetic properties of an armature. In a motor or generator, the armature coil is distinguished from the field coil.

armature core The ferromagnetic core upon which the armature coil of a motor or generator is wound.

armature gap 1. In a motor or generator, the space between an armature core and the pole of a field magnet. 2. In a relay, the space between the armature and the relay-coil core.

armature hesitation A momentary delay in the movement of a relay.

armature-hesitation contact chatter Undesired (usually rapid, repetitive) making and breaking of relay contacts that are opening or closing, due to armature hesitation.

armature-impact contact chatter Undesired (usually rapid, repetitive) making and breaking of relay contacts, due to bounce when the armature strikes the relay core (closure) or backstop (opening).

armature relay A relay that uses an electromagnet to pull a lever, made of magnetic material, toward or away from a set of fixed contacts.

armature travel The distance traveled by an armature during relay operation.

armor Protective metal cable covering.

Armstrong FM system (Edwin H. *Armstrong*, 1890-1954). A phase-shift method of frequency modulation. See *phase modulation*.

armature voltage control Controlling motor speed by changing applied armature winding volage.

armchair copy Amateur radio term for reception of exceptionally clear signals.

arming the oscilloscope sweep Enabling an oscilloscope to trigger on the next pulse by closing a switch.

Armstrong oscillator (Edwin H. *Armstrong*, 1890-1954). An oscillator circuit employing inductive feedback between the output and input of a tube or transistor. The untuned plate (collector) coil is called a *tickler*; the grid (base) coil is tuned to set the oscillator frequency. The amount of positive feedback is controlled by varying the coupling between the coils.

Armstrong superheterodyne circuit See *superheterodyne circuit*.

Armstrong superregenerative circuit See *superregenerative circuit*.

ARPA Advanced Research Projects Agency (a subsidiary of the Department of Defense).

array 1. A directive antenna consisting of an assembly of properly dimensioned and spaced elements, such as radiators, directors, and reflectors. 2. A coordinated group or matrix of components, such as diodes, resistors, memory cells, and so on, often enclosed in one capsule. 3. Subscripted variables representing data arranged so that a program can examine the array and extract data relevant to a particular subscript. The positioning of the array elements (in columns, rows, or both) aids matrix algebra computation and allows the display of data in a desired format through programming. A single row or column of singly subscripted variables (ones and zeros) is called a *one-dimensional* array; the rank and file arrangement of doubly subscripted variables is called a *two-dimensional* array.

array antenna An antenna which has more than one element, one or more of which may be driven, for the purpose of obtaining a certain directional pattern, along with power gain.

array device A group of similar or identical components that are connected together in a certain fashion, to perform a specific task.

arrester 1. A device used to protect an installation from lightning. It consists of a varistor or an air gap (see *air gap, 3*) connected between an outside antenna—or overhead power line—and ground. The device passes little or no current under ordinary conditions, but passes heavy current to ground during a lightning stroke, thereby protecting the installation. Also called *lightning arrester*. 2. A self-restoring protective device used to reduce voltage surges on power lines.

ARRL American Radio Relay League.

arrowhead A wideband, logperiodic antenna having linear polarization.

ARS Abbreviation of *amateur radio service*.

arsenic Symbol, As. A metalloidal element. Atomic number, 33. Atomic weight, 74.91. Arsenic is familiar as an n-type dopant in semiconductor processing.

ARSR Abbreviation of *air route surveillance radar.*

articulation A measure of effectiveness of voice communications, expressed as the percentage of speech units understood by the listener when the effect of context is negligible.

artificial antenna See *dummy antenna.*

artificial ear A microphone-type sensor, equivalent to the human ear, employed to measure sound pressures.

artificial echo 1. In radar practice, reflections of a transmitted pulse returned by an artificial target. 2. A signal from a pulsed rf generator, delayed to simulate an echo.

artificial ground The operating ground provided by the radials or disk of a ground-plane antenna, as opposed to actual ground which is the earth itself. Compare *true ground.*

artificial horizon In aircraft instrumentation, a device which displays lines showing the position of the aircraft in flight, with reference to the horizon.

artificial intelligence The ability of a computer to not only process data, but to learn, refine its own processes, and even to reason on an elementary level.

artificial ionization An artificial reflecting layer created in the atmosphere to provide a skip condition.

artificial language A language that is not commonly used, but which has been devised for efficiency in a particular situation, especially in a computer system.

artificial line See *artificial transmission line.*

artificial load See *dummy load.*

artificial radioactivity Radioactivity induced in a substance which ordinarily is not radioactive, as by bombardment with high-velocity particles, e.g., neutrons.

artificial transmission line. A network of capacitors and inductors, having characteristics similar to those of the more bulky transmission line it replaces in tests and measurements. It also serves as a time-delay or phase-shift device and as a pulse-forming network.

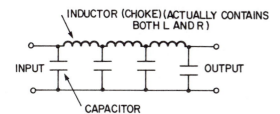

Artificial transmission line.

artificial voice A device used to test and calibrate noise-canceling microphones, consisting essentially of a small loudspeaker having a baffle whose acoustical properties simulate those of the human head.

artos stripper A machine that cuts and strips wire for the fabrication of multi-conductor cables.

artwork 1. In etched-circuit production, the scaled drawings from which the mask or etch pattern is obtained photographically. 2. Collectively, the illustrations depicting an electronic circuit, device, or system.

As Symbol for *arsenic.*

ASA Abbreviation of *American Standards Association.* See *ANSI.*

asbestos A nonflammable fibrous material, consisting of calcium and magnesium silicates, used for high-temperature insulation.

A-scan A radar-screen presentation in which the horizontal time axis displays distance or range, and the vertical axis displays the amplitude of signal pulse and echo pulses.

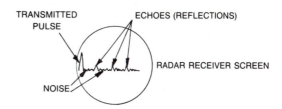

A-scan.

ASCII Abbreviation of *American Standard Code for Information Interchange.*

ASI Abbreviation of *American Standards Institute.*

A-scope A radar scope giving an A-scan.

Askarel A synthetic, nonflammable liquid dielectric.

aspect ratio 1. The width-to-height ratio of a TV picture: 3 units high by 4 units wide. 2. In an airfoil, the ratio of span to mean chord.

asperities On the surface of an electrode, tiny points at which the electric field is intensified and from which discharge is highly probable.

aspheric Also aspherical. Pertaining to a surface (as that of a lens, reflector, or cavity) which is not completely spherical.

ASR 1. Abbreviation of *airborne (or airport) surveillance radar.* 2. Abbreviation of *automatic send/receive.*

ASRA Abbreviation of *automatic stereophonic recording amplifier.*

assemble 1. To gather subprograms into a complete digital computer program. 2. To translate a symbolic program language into a machine (binary) language program by substituting operation codes and addresses.

assembly 1. A finished unit which may be either a practical working model or a dummy, a prototype or a final model; an assembly is an integrated aggregation of subunits. 2. A low-level computer source-code language employing crude mnemonics that are more easily remembered than the object-code equivalents, which are "words" consisting solely of ones and zeros.

assembly language A source code employing mnemonic instructions. (See *assembly*, 2.)

assembly program The program which operates on a symbolic language program to produce a machine language program in the process of assembly. (Also known as *assembler, assembly routine.*)

assign To reserve part of a computing system for a specific purpose, normally for the duration of a program run.

assigned frequency The radio carrier frequency or band of frequencies assigned to a transmitting station by a licensing authority. Also see *ratio spectrum.*

associative storage (memory) Computer memory in which locations are identified by content rather than by specific address.

assumed decimal point A decimal point that does not occupy an actual computer storage space but is used by a computer to align values for calculation; the decimal point is assumed to be at the right, unless otherwise specified.

astable Having two temporary states; bistable.

astable circuit A circuit (such as that of a blocking oscillator) which has two unstable states and whose operation is characterized by alternation between those states at a frequency determined by the circuit constants.

astable multivibrator A free-running multivibrator. The common circuit employs two tubes or transistors, their inputs and outputs being cross coupled. Conduction switches alternately between the two.

astatic 1. Without fixed position or direction. 2. In a state of neutral equilibrium.

astatic galvanometer The Kelvin-reading version of this instrument is the galvanometer. Its movable element consists of two identical magnetized needles mounted nonparallel on the same suspension. Each needle is surrounded by a coil, both of which are oppositely wound and connected in series to the current source. A large permanent magnet provides the field against which the needle assembly rotates. The instrument is unaffected by the earth's magnetism, since the resultant magnetic moment of the movable element is zero.

astatine Symbol, At. A radioactive elemental halogen produced artificially resulting from radioactive decay. Atomic number, 85. Atomic weight, 211 (?). Formerly called *alabamine*.

A station One of the two stations in the transmitting system of loran (long-range radionavigation).

astigmatism A focusing fault in an electron-beam tube in which electrons in different axial planes focus at different points.

ASTM *American Society for Testing and Materials.*

astrionics The design, production, and application of electronic devices and systems for use in space vehicles and space navigation. (Also, sometimes, *astronics*.)

astronautics The science and art of space travel.

astronomical unit Abbreviation, AU. A unit of distance equal to 1.496 3×10^5 km (approximately, the mean distance between the earth and sun).

astrotracker A star-tracking device.

A-supply See *A-power supply*.

asymmetrical cell A photocell exhibiting asymmetrical conductivity.

asymmetrical conductivity Conduction in one direction only, as in a diode.

asymmetrical distortion In a binary system, lengthening or shortening of one of the states, by comparison to the theoretical or ideal duration.

asymmetrical FET A field-effect transistor in which the source and drain electrodes cannot be interchanged without degrading performance.

asymmetrical multivibrator An unbalanced multivibrator, i.e., one in which the circuit halves are not identical. Thus, if the time constants of the halves are different, short, widely separated output pulses will be generated.

asymmetrical sideband See *vestigial sideband*.

asymmetrical sideband transmission See *vestigial sideband transmission*.

asymmetrical wave A wave whose upper (positive half-cycle) and lower (negative half-cycle) portions have different amplitudes or shapes. Also called *asymmetric wave*.

asymmetry control An adjustment in a device intended for measur-

ing the pH, or acidity/alkalinity. This corrects the inaccuracies resulting from differences between the electrodes.

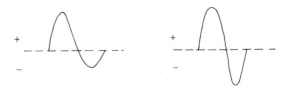

Asymmetrical waves.

asymptote In graphical presentations, a straight line which always approaches, but never meets, a given curve.

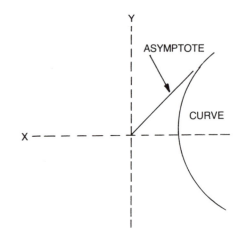

Asymptote.

asymptotic breakdown voltage A voltage which will cause dielectric breakdown if applied continuously for a sufficiently long time.

asymptotic expression An expression having a very small (by percentage) error.

asynchronous 1. Not synchronous, i.e., nonrecurrent (as in out-of-phase waves). 2. A mode of computer operation in which the completion of an operation starts another.

asynchronous computer See *asynchronous, 2.*

asynchronous device A device not regulated by the system in which it is used, as far as its operating frequency or rate is concerned.

asynchronous input In digital circuitry, any flip-flop input at which a pulse can affect the flip-flop output independently of the clock.

asychronous motor An ac motor whose speed is not proportional to the supply frequency.

asynchronous transmission Data transmission in which each character or symbol begins with a start signal and ends with a stop signal. This eliminates the need for the data to be sent at a uniform speed.

asynchronous vibrator In a vibrator-type portable power supply, a vibrator that only makes and breaks the primary circuit of the stepup transformer. This is in contrast to the synchronous vibrator, which also makes and breaks the secondary circuit in synchronism with the primary. Also called *nonsynchronous vibrator*.

asyndetic With operation symbols omitted between operands, e.g., A × B. B written as AB.

AT A quartz crystal cut wherein the angle between the *x*-axis and the crystal face is 35 degrees.

At Symbol for *astantine*.

AT-cut crystal A piezoelectric crystal cut at a 35-degree angle with respect to the optical axis of the quartz. The frequency of such a crystal does not appreciably change with variations in temperature.

atmosphere 1. The gas surrounding a planet, particularly the air sheathing the earth. 2. A unit of pressure (abbreviation atm); 1 atm equals 101 325 Pa.

astmospheric absorption See *absorption loss, 2.*

atmospheric absorption noise Noise, principally above 1 GHz, due to atmospheric absorption (see *absoprtion loss, 2*).

atmospheric bending The bending of advancing radio waves by reflection, refraction, or both, by the ionosphere. (See *atmospheric reflection.*)

atmospheric duct A tropospheric stratum through which radiation in excess of 3 GHz is propagated with marked efficiency.

atmospheric electricity Static electricity present in the atmosphere, which evidences itself in disturbance of radio communications and in displays of lightning.

atmospheric noise Receiver nosie due to atmospheric electricity. Also called *static.*

atmospheric pressure Abbreviation, atm press. Unit, atmosphere (atm). Pressure of the earth's atmosphere, as indicated by a barometer; 1 atm equals 101 325 Pa.

atmospheric radio wave See *skywave.*

atmospheric radio window The band of frequencies (approximately 10 MHz to 10 GHz) including radio waves that will penetrate the earth's atmosphere.

atmospheric reflection Return of a radio wave to earth, resulting from reflection by an ionized portion of the atmosphere.

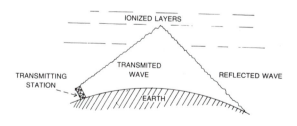

Atmospheric reflection.

atmospheric refraction Downward bending of radio waves as a result of atmospheric dielectric constant variation.

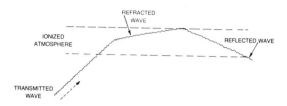

Atmospheric refraction.

atmospherics See *atmospheric noise.*

atom 1. The smallest material particle which still displays the unique characteristics and properties of an element. Atoms consist of a central nucleus around which electrons can be revolve in definite orbits, the difference between elements being due to the number of orbital electrons and nuclear particles. Also see *Bohr atom*, and *Rutherford atom*. 2. In a computer compiling operation, an operator or operand.

atomechanics The physics of electron movement.

atomic battery A battery in which atomic energy is converted into electrical energy.

atomic charge The electrification (i.e., the electron charge) exhibited by an ion.

atomic clock A highly accurate electric clock driven ultimately from a cesium-beam-controlled oscillator.

atomic cocktail Slang term for a preparation containing a radioactive substance, swallowed by a patient for diagnostic or therapeutic purposes (e.g., as it applies to cancer).

atomic disintegration The decay of the atomic nucleus of a radioactive substance, accompanied by the emission of radiant energy. Also see *alpha particle, beta rays, gamma rays.*

atomic energy Energy released by the disruption of atomic nuclei. Also see *atomic power.*

atomic fission See *fission.*

atomic frequency The natural vibration frequency of an atom.

atomic fuel Any substance, such as uranium, whose atoms may be split to provide fission in a nuclear reaction.

atomic fusion See *fusion.*

atomicity The condition or state of being composed of atoms. See *valence.*

atomic pile See *reactor, 2.*

atomic mass unit Abbreviation, amu. A unit which expresses the relative mass of isotopes. The mass of the most abundant natural isotope of oxygen (O16) is 16 amu.

atomic migration Transfer of a valence electron progressively between atoms in the same molecule.

atomic number Abbreviation, *at.No.* The number of protons in the nucleus of an atom (also the number of electrons if the atom is neutral). Thus, the atomic number for copper is 29, indicating 29 protons in the nucleus and 29 orbital electrons outside the nucleus. An isotope of an element has the same atomic number as the element, but a different atomic weight.

atomic power The energy released in a nuclear reaction, through fission or fusion.

atomic radiation The emission of radiant energy by radioactive substances. There are three principal kinds of radiation: alpha, beta, and gamma.

atomic ratio The ratio of the quality of a substance to its number of atoms.

atomic reactor See *reactor, 2.*

atomics See *atomistics.*

atomic structure See *Bohr atom* and *Rutherford atom.*

atomic theory All matter is composed ultimately of atoms, which are the smallest particles retaining the identity of an element. Atoms combine to form molecules, the smallest particles that retain the identity of a compound. Atoms themselves contain minute particles and charges (see *particle, Bohr atom,* and *Rutherford atom*).

atomic time A means of time determination that makes use of the resonant vibrations of certain substances, such as cesium. The atomic time is synchronized with astronomical time.

atomic weight Abbreviation, *at wt.* The weight of one atom of a given element compared with an arbitrarily selected number representing the weight of a reference element (such as 16 for oxygen, or 12 for carbon).

atomistics The science of the atom and atomic energy. Also called *atomics.*

ATR tube Abbreviation of *antitransmit-receive tube.* In radar systems, a gas tube which automatically switches the transmitter off during receiving intervals.

ATS IBM's remote *administrative terminal system,* which permits a terminal typewriter operator to communicate directly with a computer in creating, manipulating, or adding large volumes of textual data to an information base.

attack 1. Rise of a pulse from zero to maximum amplitude. 2. The time required for a pulse to rise from zero to maximum amplitude. 3. The initialization of a circuit voltage or current for a certain purpose, such as an automatic gain control. 4. The rise of a musical note from zero to full volume.

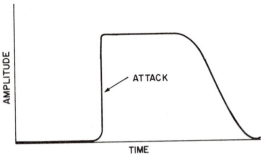

Attack.

attack time The time required for an applied signal that suddenly increases in amplitude to reach $1 - 1/e$, that is, 63.2 percent of its final, stable value.

attemperator An automatic temperature-controlling device.

attention display A computer-generated chart or graph displayed on the crt of a control facility as an alert about a particular situation.

attenuate To reduce in amplitude.

attenuation The reduction of signal amplitude.

attenuation characteristic 1. In an amplifier, network, or component, the decrease of signal amplitude vs frequency. It is usually expressed in decibels per octave. 2. In a transmission line, the decrease of signal amplitude per unit length. Usually expressed in decibels per 100 ft or decibels per mile.

attenuation constant See *attenuation characteristic.*

attenuation distortion A type of distortion characterized by variation of attenuation with frequency within a given frequency range.

attenuation equalizer An equalizer that stabilizes the transfer impedance between two ports at all frequencies within a specified frequency band.

attenuation-frequency distortion Distortion characterized by attenuation of the frequency components in a complex waveform. Frequency-sensitive RC networks (such as a Wien bridge) exhibit this type of distortion when they attenuate a fundamental and each harmonic unequally.

attenuation network A combination of components (R, C, or L singly or in any necessary combination) providing constant signal attenuation with negligible phase shift throughout a frequency band.

attenuation ratio The ratio indicating relative energy decrease propagation ratio magnitude.

attenuator A device for reducing signal amplitude in precise, predetermined steps, or smoothly over a continuous range. Attenuators are networks of resistors, capacitors, or both. The simplest attenuator is a resistive or capacitive potentiometer.

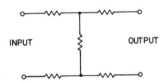

Attenuators.

attenuator tube An independently controlled gas tube used to control rf power.

attitude The position of an aircraft or space vehicle relative to a (usually terrestrial) reference point, often determined with electronic instruments.

atto Abbreviation, a. A prefix meaning 10^{-18}, e.g., attofarad.

attofarad Abbreviation, aF. A unit of low capacitance; 1 aF equals 10^{-18}F.

attraction The drawing together or pulling toward, as in the attraction between electric charges or magnetic poles. Unlike charges and poles attract each other (+ to −, N to S). Compare *repulsion.*

ATV Abbreviation of *amateur television* (amateur radio service).

AU Abbreviation of *astronomical unit.*

Au Symbol for *gold.*

audibility The quality of being able to be heard. In a healthy listener, the threshold of audibility is extremely low; at the threshold, the pressure of a sound wave varies from normal by approximately 10^{-4} pascal. The frequency range of audibility extends roughly from 16 Hz to 16 kHz.

audibility curves Graphs such as the Fletcher-Munson curve which show the range of variation in human hearing in terms of frequency vs sound intensity.

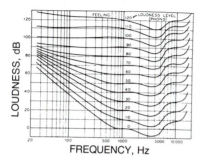

Audibility curves.

audible Heard or capable of being heard.

audible frequency See *audio frequency*.

audible tone A vibration of air molecules that can be detected by the human ear, and with periodic properties such as a sine-wave vibration.

audio 1. Pertaining to the audio-frequency spectrum or to equipment or performance associated with that spectrum. 2. Sound or any other audio-frequency factor, quality, component, or channel.

audio amplifier See *audio-frequency amplifier*.

audio band The range (band) of audio frequencies as it pertains to equipment operating on it.

audio channel 1. The portion of a complex signal or waveform used for conveying audio information exclusively. 2. The audio section of a transmitter or receiver (as opposed to rf). 3. A radio channel of fixed frequency, reserved for voice communications.

audio component The audio-frequency portion of any wave or signal.

audio converter A frequency converter stage in which a received rf signal is heterodyned by a local rf oscillator signal to give an af beat-note output. The beat note is then amplified by an af amplifier. The method is sometimes used for cw radiotelegraphy.

audio frequency A frequency lying within the audible spectrum. See *audio-frequency spectrum*.

audio-frequency amplifier An amplifier which operates on part or all of the frequency range 20 Hz to 20 kHz, and intended for such use. (A amplifier would not be classed as one, even though it is used in the same frequency range.)

audio-frequency choke An inductor (usually having a ferromagnetic core) which blocks audio-frequency current, but passes direct current.

audio-frequency feedback 1. Feedback (positive, negative, or both) affecting audio-frequency circuits. 2. Acoustical feedback.

audio-frequency filter A filter of any type that operates on any part of the frequency range 20 Hz to 20 kHz.

audio-frequency meter An instrument for measuring frequencies in the audio-frequency spectrum (approximately 20 Hz to 20 kHz). Three types are in common use: 1. *Analog type*. Gives direct indications of frequency on the scale of a D'Arsonval meter; usual range, 20 Hz to 100 kHz. 2. *Digital type*. Gives direct indications of frequency by means of readout lamps; usual range, 1 Hz to 15 MHz. This instrument is useful also as a radio-frequency meter. 3. *Bridge type*. Consists of a frequency-sensitive bridge, such as a Wien bridge, with a null-indicating meter. The operator balances the bridge and reads the unknown frequency from the dial of the balance control.

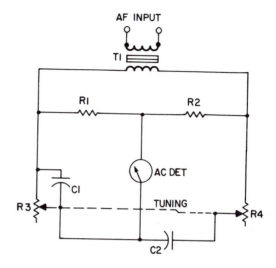

CI = C2,   R3 = R4

AT NULL, $f = 1/(2\pi R3\,C1)$

Audio-frequency meter.

audio-frequency noise Any electrical noise signal causing interference within the audio-frequency spectrum.

audio-frequency oscillator See *audio oscillator*.

audio-frequency peak limiter Any circuit or device, such as an agc amplifier or biased diode, which performs the function of audio limiting.

audio frequency-shift keying A type of A2 emission in which the rf carrier is continuously modulated by tone (e.g., 850 Hz); keying is accomplished by changing the tone to another frequency. This type of keying is used in facsimile and radioteletype communications.

audio-frequency-shift modulator A modulator for audio-frequency-shift keying of a signal.

audio-frequency spectrum The band of frequencies extending from roughly 20 Hz to 20 kHz. Hi-fi component specifications often extend this range somewhat in both directions.

audio-frequency transformer A transformer designed for operation in the audio-frequency spectrum. A high-quality af transformer is characterized by high-efficiency core material, low-capacitance windings, low leakage reactance, and excellent magnetic shielding.

audio-frequency transistor A lowfrequency transistor usually employed at audio frequencies.

audiogram A graph used to rate hearing, employed by audiologists and audiometrists.

audio image In superheterodyne cw reception, an undesired interferential signal at a frequency f to one side of the desired signal, where f is the audio frequency created by the cw oscillator's output beating with the desired signal. Audio images are reduced or eliminated in single-signal superhet circuits.

audio-level meter An ac meter for monitoring signal amplitude in an audio-frequency system. It may indicate in volts, decibels volume units

(VU), or arbitrary units, and is often permanently connected in the circuit.

audio limiter A limiter or clipper operated in the af channel of a receiver or transmitter to hold the output-signal amplitude constant, or to minimize the effect of noise peaks.

audiologist A person skilled in testing hearing, i.e., in using audiometers and other electronic instruments and evaluating their indications for the fitting of hearing aids.

audiometer An instrument used for hearing tests, consisting of a specialized af amplifier with calibrated attenuators, output meter, and signal source.

audiometrist A person skilled in the use of audiometers and other electronic instruments that measure sound and human hearing, and deals with attendant health and behavior problems. Compare *acoustician* and *audiologist*.

audio mixer An amplifier circuit for blending two or more af signals, such as those delivered by microphones or receivers.

audion The name for the first triode, originally a trade name.

audio oscillator 1. An oscillator which delivers an output signal in the frequency range 20 Hz to 20 kHz. 2. An af signal generator. Some instruments of this type operate above and below the limits of the common audio-frequency spectrum, e.g., 1 Hz to 1 MHz.

audio output The output of an audio-frequency oscillator or amplifier. It may be measured in terms of peak or rms volts, amperes, or watts.

audiophile A sound-reproduction hobbyist.

audio power Alternating-current power at frequencies roughly between 20 Hz and 20 kHz. When used in connection with transmitters and other modulated rf equipment, the term refers to modulator power output.

audio response unit Recorded digitized responses held in computer storage; they can be linked by an audio response unit to a telephone network to answer inquiries audibly.

audio signal generator See *audio oscillator, 2*.

audio spectrum The range of sine-wave frequencies detectable by the human ear when they occur as acoustic vibrations. This range is about 20 Hz to 20 kHz.

audio squelch A squelch circuit which operates only on the audio channel of a receiver.

audio system 1. That portion of any electronic assembly that is used to process sound. 2. Special computer equipment capable of storing and processing vocalized data.

audiotape Magnetic tape for the recording and reproduction of sound.

audio taper A semilogarithmic variation in resistance vs angular rotation in volume and tone controls used in audio circuits. At midposition (50% point in potentiometer rotation) the counterclockwise portion has one-tenth the resistance of the clockwise portion. The ear will

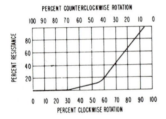

Audio taper.

hear at approximately half-volume owing to the ear's subjective audibility curve.

audio visual Pertaining to a combination of sound and sight (e.g., television and sound motion pictures), as in audio visual aids to education and commerce.

auditory backward inhibition A type of auditory inhibition in which a sound is erased from the memory of the hearer by a second sound that arrives about 60 milliseconds after the first.

auditory inhibition The involuntary ability of a hearer to cancel or partially reject sound waves, depending on their intensity, direction of impact, or the interval between their emission and arrival.

auditory mirage See *acoustic mirage*.

audit trail A history of the processes relating to a record, transaction, or file in a computer system. Created during the routine processing of data, the trail is stored as a file. The audit trail allows auditing of the system or the subsequent recreation of files.

augend In a calculation, the number to which another is to be added. Compare *addend*.

augend register In a digital computer, the register which stores the augend. Compare *addend register*.

augmented operation code In digital computer practice, an operation code augmented by information elsewhere in the introduction.

augmenter A quantity added to another that must be brought to a required value.

AUM Abbreviation of *air-to-underwater missile*.

aural Pertaining to hearing or sound. The term is often used to distinguish between sound which is actually heard and sound which is represented by audio-frequency currents.

aural signal 1. Also called *aural component*. The sound portion of a television signal. 2. The audio-frequency component of a modulated rf wave.

aural transmitter The sound transmitter in a television broadcasting system.

aurora A phenomenon sometimes called the *northern lights* seen in the night sky. In the Northern Hemisphere, it is *aurora borealis*; in the Southern Hemisphere, *aurora australis*. The aurora, which appears to be colored lights, results from atmospheric electricity.

auroral absorption Radio wave absorption by an aurora.

auroral reflection Aurora borealis or aurora australis (see *aurora*).

authorized channel The carrier frequency or band assigned to a transmitting station by a licensing authority. Also see *radio spectrum*.

autoalarm Any of several circuits or devices operated from a receiver to alert a radio operator when there is a message for him.

autocondensation The application of radio-frequency energy into the body for medical purposes. The living organs serve as the dielectric for a capacitor, to which rf is applied.

autoconduction The application of radio-frequency currents into the body, by placing the living organ inside a coil and supplying the coil with rf. Used for medical purposes.

autocorrelation function A measure of the similarity between delayed and undelayed versions of a signal, expressed as a delay function.

autodyne detector A tube or transistor detector providing autodyne reception.

autodyne reception Radio reception of cw signals by means of an oscillating detector. This is in contrast to *heterodyne reception*, in which a local rf oscillator generates an audio beat note with the cw signal in a separate detector.

autoelectronic effect In a vacuum tube or gaseous tube, the emission of electrons by a cold cathode when the anode-to-cathode voltage is sufficiently high.

automated communications The transfer of data without the use of operating personnel; it is done by computer or electronic machines.

automatic Self-regulating, independent of human intervention. Some periodic adjustment may or may not be needed.

automatic abstract Key words taken from a document and arranged in meaningful order. The automatic selection and arranging follows programmed criteria.

automatic bass compensation Use of a resistor-capacitor network in conjunction with the gain control in an audio system to boost the volume of bass notes to compensate for the ear's inefficiency at low frequencies.

automatic bias Dc operation bias (grid bias, base bias, or gate bias) obtained from the voltage drop produced by dc output-electrode current flowing through a resistor which is common to the input and output circuits of a stage (e.g., plate current through the cathode resistor of a tube; collector current through the emitter resistor of a bipolar transistor; drain current through the source resistor of a field-effect transistor).

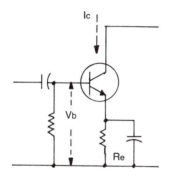

Automatic bias.

automatic brightness control A circuit that uses the same principles employed in automatic gain control to hold steady the average brightness of a TV picture.

automatic calculator See *analog computer* and *digital computer*.

automatic carriage Typewriters, automatic key punches, and other control devices that can control automatically the spacing and feeding of paper, cards, forms, and so forth.

automatic check 1. In a digital computer, the automatic inspection of operation and performance by a self-contained subsystem. 2. The circuit or device for performing this inspection.

automatic chrominance control In a color-TV receiver, a subcircuit that controls the gain of the chrominance bandpass amplifier by automatically adjusting its bias.

automatic circuit breaker Any device that opens a circuit automatically when the flow of current becomes excessive. The breaker generally resets automatically after a specified length of time, or after power has been temporarily removed from the circuit.

automatic coding Using a computer to describe the steps for solving a problem before the program for the problem is written.

automatic connection A connection between users made automatically by electronic switching equipment.

automatic contrast control A circuit that automatically adjusts the gain of the video if and rf stages of a TV receiver to preserve good picture contrast.

automatic controller In servo systems, any of several circuits or devices that samples a variable signal, compares it with a standard (reference) signal, and delivers a control or correction signal to an actuator.

automatic crossover 1. Current limiting in a power supply. 2. A device that switches a circuit from one operating mode to another, automatically, when conditions change in a predetermined manner.

automatic current limiter A circuit or device for holding the output current of a power supply to a safe value during overload.

automatic current regulator A circuit or device which holds the output current of a generator or power supply to a predetermined value in spite of wide variations in load resistance.

automatic cutout A device that shuts down a circuit or system when the safe limits of operation are exceeded. A circuit breaker is an example of such a device; so is a thermostat in a power amplifier.

automatic data processing Abbreviation, *ADP*. The use of digital computers and accessories for calculations and tabulations using information sometimes gathered automatically by the system.

automatic degausser A system for automatically demagnetizing the picture tube in a color-TV receiver.

automatic dialing unit Abbreviation, *adu*. A device that automatically generates dialing digits.

automatic dictionary A computer system component that substitutes codes for words and phrases in information retrieval systems. In language translating systems, it provides word-for-word substitutions.

automatic direction finder Abbreviation, *adf*. A specialized receiver/antenna combination for automatically showing the direction from which a signal arrives.

automatic error correction A technique of correcting transmission errors automatically using error detecting and correcting codes and, usually, automatic retransmission.

automatic exchange A transmission exchange in which interterminal communications are accomplished without operators.

automatic focusing A method of focusing a picture tube automatically, in which a resistor connects the focusing anode to the cathode; thus, no external focusing voltage is necessary.

automatic frequency control A system which keeps a circuit automatically tuned to a desired signal frequency. A detector (such as a discriminator) operated from the tuned circuit delivers dc output voltage only when the circuit is operating above or below the signal frequency; otherwise, it has zero dc output. The dc output, when present, alters the value of a voltage-sensitive capacitor or a reactance tube connected across the tuned circuit to retune the stage to the desired frequency.

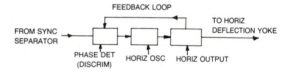

Automatic frequency control used in typical TV receiver.

automatic gain control Abbreviation, *agc*. A system which holds the gain and accordingly, the output of a receiver or amplifier substantially constant, in spite of input-signal amplitude fluctuations. A rectifier samples the ac signal output and delivers a dc voltage proportional to that output. The dc voltage is then applied in correct polarity as bias to the early stage(s) in the system to reduce the gain when the output swings beyond a predetermined level, and vice versa.

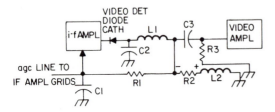

Automatic gain control used in typical TV receiver.

automatic grid bias A method of obtaining grid bias in a tube, wherein a resistor develops a voltage drop because of current flowing through it. The resistor is usually placed in the cathode circuit of the tube, raising the cathode above ground potential, while the grid remains at direct-current ground.

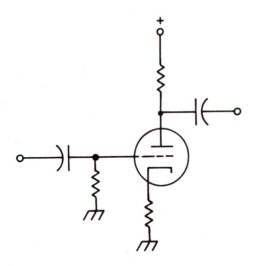

Automatic grid bias.

automatic hardware dump The assignment of data from memory to a peripheral, caused by a system error. It is used to provide diagnostic information, e.g., concerning the condition of an instruction code process; the information dumped goes to a medium that allows convenient analysis.

automatic height control In a TV receiver, a system that automatical-

ly maintains the height of the picture, in spite of signal-amplitude fluctuations, power-line voltage changes, and gain variations. The action is obtained with a voltage-dependent resistor in the grid circuit of the vertical output tube, where it automatically regulates the gain of the tube.

automatic intercept A telephone answering machine. Allows messages to be recorded when the subscriber is not able to answer the telephone.

automatic interrupt A program interruption caused by hardware or software acting on an event independent of the interrupted program.

automatic level compensation See *automatic gain control.*

automatic level control A circuit that adjusts the input gain of a magnetic-tape recording device compensatorily. (Abbreviation, alc.)

automatic modulation control Abbreviation, *amc.* In a radio transmitter, a circuit for automatically varying the gain of the af speech/modulator channel to compensate for fluctuating input audio. The action prevents overmodulation.

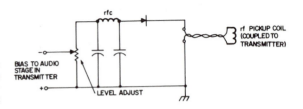

Automatic modulation control.

automatic noise limiter Abbreviation, *anl.* Any of several circuits for clipping noise peaks exceeding a predetermined maximum received-signal amplitude.

automatic phase control In a color-TV receiver, a circuit which synchronizes the burst signal with the 3.58 MHz color oscillator.

automatic pilot An electronic device which automatically steers a ship, airplane, or space vehicle.

automatic polarity In an electronic meter, such as an analog or digital dc *tvm* or *vtvm*, a means of automatically switching the input polarity of the instrument when the input signal polarity is shifted. Also called *bipolar operation.*

automatic programming See *automatic coding.*

automatic protective device A circuit or device (such as a fuse, circuit breaker, varistor, or limiter) which protects another circuit or device by automatically removing, reducing, or increasing current or voltage during overload or underload.

automatic radio compass See *automatic direction finder.*

automatic ranging Automatic range switching, especially in electronic meters.

automatic regulation 1. Voltage regulation. In a power supply, the automatic holding of the output current to a constant value, in spite of variations in the input voltage or load resistance. 2. Current regulation. In a power supply, the automatic holding of the output current to a constant value, in spite of variations in the input voltage or load resistance.

automatic relay The relaying of messages automatically from one station to another via intermediate points, without the need for human operators.

automatic repeater station A station which receives signals and simultaneously retransmits them, usually on a different frequency.

automatic reset 1. The self-actuated restoration of a circuit or device to a given state, e.g., the state of rest. 2. A circuit or device which accomplishes automatic reset.

automatic-reset relay A relay used for automatic reset.

automatic scanning 1. The automatic (usually repetitive) tuning or adjustment of a circuit or system throughout a given frequency range. 2. The repetitive sweep of a cathode-ray beam.

automatic scanning receiver Also called *panoramic receiver*. A receiver which is automatically tuned (usually repetitively) over a frequency band. Such a receiver either homes-in on a signal when one is found, or displays on a cathode-ray screen the distribution of signals in the band.

automatic secure voice communications A wideband and narrowband voice-digitizing application to a security network providing encoded voice communication.

automatic send/receive set A teletypewriter capable of receiving and transmitting.

automatic sensitivity control 1. A self-actuating circuit employing principles similar to those used in automatic gain control, which varies the sensitivity of the rf and i-f sections of a receiver in inverse proportion to the strength of a received signal. 2. In a bridge null detector, a circuit similar to the one described above, which operates ahead of the detector, varying the sensitivity of the latter automatically.

automatic sequencing The ability of a digital computer to perform successive operations without additional instructions from the operator.

automatic short-circuiter Device that automatically short-circuits the commutator bais in some single-phase commutator motors.

automatic short-circuit protection A circuit that allows the output of a power supply to be short-circuited without damage to the components in the supply. Usually consists of a current-limiting device.

automatic shutoff A switching arrangement that automatically shuts off a device or circuit under certain specified conditions.

automatic switch center A switching network, in a telephone system, that routes calls to their destinations without the need for a human operator.

automatic target control For a vidicon TV-camera tube, a circuit which automatically adjusts the target voltage in proportion to brightness of the scene.

automatic telegraph reception Telegraph reception providing a direct printout of the received information, without direct intervention by an operator.

automatic telegraph transmission Telegraph transmission originating from tapes, disks, or other records, rather than from a hand-operated key.

automatic telegraphy Communication utilizing automatic telegraph transmpssion and automatic telegraph reception.

automatic time switch A time-dependent circuit or device that opens or closes another circuit at the end of a predetermined time interval.

automatic tracking A method of keeping a radar beam automatically fixed on a target.

automatic trip A circuit breaker that automatically opens a circuit.

automatic tuning A process whereby a circuit tunes itself to a predetermined frequency upon receiving a command signal.

automatic voltage regulator A circuit that keeps the output of a power supply, or the input voltage to a system, constant, no matter what the load resistance or input voltage to the supply.

automatic volume control Abbreviation, *avc*. See *automatic gain control*.

automation zero In an electronic meter, such as a *tvm* or *vtvm*, a means of automatically setting the indicator to zero in the absence of an input signal.

automation The automatic control of machines or processes by self-correcting electronic systems. See *robot*.

automaton A device that operates automatically, and may have lifelike qualities, such as a robot.

automonitor In digital computer practice, to require the machine to supply a record of its information-handling operations. Also, the program for such instructions.

autopatch A remotely controllable device that "patches" a radio communications system into a telephone land-line network.

autopilot A self-correcting control and guidance device for automatic management of an aircraft or missile.

autoradiograph A photograph of a radioactive source, obtained by direct exposure of the film to the source.

autoranging See *automatic ranging*.

autosyn A device or system operating on the principle of the synchronous ac motor, in which the position of the rotor in one motor (the *transmitter*) is assumed by the rotor in a distant motor (the *receiver*) to which the first is connected.

auto tracking A method of controlling the output voltages of many different power supplies simultaneously.

autotransducter A type of magnetic amplifier whose power windings serve also as control windings.

autotransformer A single-winding transformer in which the primary coil is a fraction of the entire winding for voltage stepup, or the secondary coil is a fraction of the entire winding for voltage stepdown.

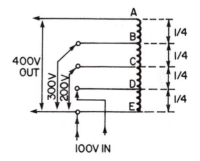

Autotransformer.

auxiliary circuit A circuit which is supplementary to the main circuit.

auxiliary contacts In switches and relays, contacts which are supplementary to the main contacts and are usually actuated with them.

auxiliary equipment 1. Apparatus not directly governed by the central processing unit of a digital computer, such as memory storage. 2. Peripheral equipment in any system. 3. Backup equipment.

auxiliary memory In a digital computer, a unit supplementary to the main memory, which it augments.

auxiliary receiver A standby receiver.

auxiliary relay 1. A standby relay. 2. A relay whose operation supports that of another relay. 3. A relay which is actuated in turn by the operation of another relay.

auxiliary switch 1. A standby switch. 2. A switch wired in series or parallel with another switch. 3. A switch which is operated by another switch.

auxiliary transmitter A standby transmitter.

a/v Abbreviation of *audio-visual*.

aV Abbreviation of *attovolt*.

availability The proportion of time during which an apparatus is operating correctly. It is usually given as a percentage.

available conversion gain The ratio of the input power to the output power of a transducer or converter. Generally given in decibels.

available gain The ratio Po/Pi, where Pi is the available power at the input of a circuit and Po, the available power at the output.

available line The percentage of the length of a facsimile scanning line which is usable for picture signals.

available power The mean square of the open-circuit terminal voltage of a linear source divided by four, times the resistive component of the source impedance. The available power is the maximum power delivered to a load impedance equal to the conjugate of the internal impedance of the power source.

available power gain In a power transistor, the ratio of available transistor output power to the power available from the generator; PGa equals Po/Pg. It is dependent upon the generator resistance (Rg), but not upon the transistor load resistance (RL).

available signal-to-noise ratio The ratio Ps/Pn, where Ps is the available signal power at a given point in a system and Pn, the available random-noise power at that point.

available time 1. The time during which a computer is available and ready for immediate use. 2. The amount of time a computer is available to an individual.

avalanche The phenomenon in semiconductors operated at high inverse bias voltage, whereby carriers acquire sufficient energy to produce new electron—hole pairs as they collide with atoms. The action causes the inverse current to increase sharply.

avalanche breakdown In a reverse-biased semiconductor, the sudden, marked increase of reverse current at the bias voltage at which avalanche begins. The action resembles a breakdown, but it is nondestructive when the current is limited by external means.

avalanche conduction Enhanced conduction due to avalanche.

avalanche current The high current that flows through a semiconductor junction in response to an avalanche voltage.

avalanche diode A silicon diode utilizing the mechanism of avalanche breakdown. When the diode is operated in its avalanche region, the voltage drop across it is substantially constant, being independent of the avalanche current. Because of this, the diode is useful for voltage regulation and amplitude limiting. Also called *zener diode*.

avalanche impedance The reduced impedance of a diode during avalanche.

avalanche noise Electrical noise generated in a junction diode operated at the point at which avalanche just begins.

avalanche transistor A transistor that operates at a high value of reverse-bias voltage, causing the pn junction between the emitter and base to conduct because of avalanche breakdown.

avalanche voltage See *breakdown voltage, 3*.

avc Abbreviation of *automatic volume control*.

average absolute pulse amplitude The average (disregarding algebraic sign) of the absolute amplitudes of a pulse, taken over the duration of the pulse.

average brightness The average brilliance of a TV picture or oscilloscope image.

average calculating operation The operating time considered typical for a computer calculation, i.e., one which is longer than an addition and shorter than a multiplication; it is frequently taken as the average of nine additions and one multiplication.

average current The average value of alternating current flowing in a circuit. It is equal to 0.637 times the maximum value of current; $Iavg$ equals 0.637 Im (for sine waves).

average life See *mean life*.

average noise figure The ratio of the total noise output from a circuit to the thermal noise output at 290 degrees Kelvin. Usually expressed in decibels, with the noise taken at all frequencies.

average power The average value of power in an ac circuit. In a resistive circuit, it is the square of the effective (rms) current times the resistance; $Pavg$ equals $(Irms)^2 R$ (for sine waves).

average pulse amplitude The integral, or average, of the pulse amplitudes at all instants during a pulse.

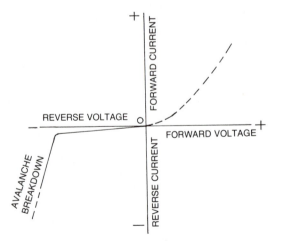

Avalanche breakdown.

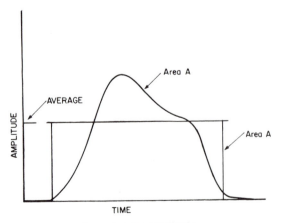

Average pulse amplitude.

average rectified current The average value of rectifier output current before filtering. For a half-waverectifier and resistive load, the average rectified current is the maximum current divided by pi; I_{avg} equals I_m/π (for sine-wave ac input).

average rectified voltage The average value of rectifier output voltage before filtering. For a half-waverectifier and resistive load, the average rectified voltage is the maximum voltage divided by pi; E_{avg} equals E_m/π (for sine-wave ac input).

average value 1. The arithmetic mean of two or more quantities. 2. The geometric mean of two or more quantities. 3. The harmonic mean of two or more quantities. 4. In ac practice, the average current, voltage, or power.

average voltage The average value of ac voltage. It is equal to 0.637 times the maximum voltage; $E_{avg} = 10.637\ E_m$ (for sine waves).

avg Abbreviation of *average*.

aviation channels Frequency channels assigned to the aviation services.

aviation services The radiocommunication services of the aeronautical-mobile and radionavigation personnel.

avigation Acronym for *avi*ation and nav*igation*. Aircraft navigation by means of electronic equipment.

avionics Acronym for *avi*ation and electr*onics*. The design, production, and application of electronic devices and systems for use in aviation, avigation, and astronautics.

Avogadro's constant (Amedo *Avogadro*, 1776-1856.) Symbol, *NA*. The number of molecules in a kilogram-molecular-weight of any substance; *NA* equals 6.0223×10^{23} kmol.

aW Abbreviation of *attowatt*.

AWG Abbreviation of *American wire gauge*.

axial leads The centrally located leads coming out of the ends of cylindrical components such as electrolytics.

axial ratio The ratio of the minor to major axes of a waveguide's polarization ellipse.

axis 1. A coordinate in a graphical presentation or display (e.g., horizontal and vertical axes in a rectangular coordinate system). 2. The real or imaginary straight line around which a body rotates, or the line that passes through the center of a symmetrical arrangement (line of symmetry).

axis of abscissas The horizontal axis (*x*-axis) of a rectangular-coordinate graph or screen. Compare *axis of ordinates*.

axis of imaginaries The vertical axis of the complex plane in which rectangular vectors lie. Compare *axis of reals*.

axis of ordinates The vertical (*y*-axis) of a rectangular-coordinate graph or screen. Compare *axis of abscissas*.

axis of reals The horizontal axis of the complex plane in which rectangular vectors lie. Compare *axis of imaginaries*.

Ayrton-Mather galvanometer shunt A step-adjustable universal shunt resistor for varying the sensitivity of a galvanometer. It has the virtue of keeping the galvanometer critically damped. The shunt is also useful in multirange milliammeters, microammeters, and ammeters: the sensitive meter movement is never without a shunting resistor during range switching.

Ayrton-Perry winding A noninductive winding comprising two inductors conducting current in opposite directions, the opposing flow canceling the magnetic field.

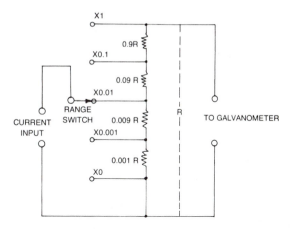

Ayrton-mather galvanometer shunt.

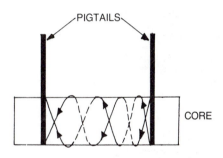

Ayrton-Perry winding.

azel display A plan-position display incorporating two different radar traces on a single crt, one giving bearing, the other elevation.

azimuth Angular measurement in the horizontal plane, clockwise from north; it is important in navigation, direction finding, land surveying, and radar.

azimuth alignment The alignment of record and playback tape head gaps so that their centerlines are parallel.

azimuth blanking Blanking of a radar receiver's crt screen as the antenna sweeps an azimuth region.

azimuth rate The change rate of true bearing.

azimuth resolution The angle or distance of separation of two equirange targets, required for differentiation by radar.

azimuth stabilization The technique of presenting any specific reference line of direction (compass point) at the top of a radar display screen.

azon bomb Acronym for *az*imuth *on*ly. A bomb which may be guided to the right or left by means of radio control.

azusa An electronic tracking system operating on the C-band. One station provides slant range and two direction cosines.

B 1. Symbol for *susceptance*. 2. Symbol for *magnetic flux density*. 3. Abbreviation of *battery*. 4. Symbol for *boron*. 5. Symbol for base of transistor (see *base, 1*). 6. Abbreviation of *bass*. 7. Abbreviation of *bel*. 8. Anode voltage or main operating voltage in any circuit (when used with sign). (Also see *B-voltage*.)

b 1. Symbol for *susceptance*. 2. Symbol for base of transistor (see *base, 1*). 3. Abbreviation of *bass*. 4. Symbol for *barns*.

B&S See *American Wire Gauge*.

B5-cut crystal A piezoelectric plate cut from a quartz crystal in such a way that the face of the plate is at an angle with respect to the Z axis of the crystal. This cut has a low temperature coefficient of frequency.

BA Abbreviation of *battery*. See also *B* and *BAT*.

Ba Symbol for *barium*.

babbit A relatively soft, tin-base alloy of various compositions. One composition contains 7.4% antimony, 3.7% copper, and 88.9% tin.

babble Interference caused by crosstalk from a number of channels.

babble signal A jamming signal containing babble components. See *babble* and *jamming*.

BABS Abbreviation of *blind approach beacon system*.

back bias 1. A feedback signal (negative or positive). 2. Reverse bias (also see *bias*). 3. A negative bias voltage obtained from a voltage divider connected in series with B-minus and ground.

backbone A form of transmission line with capacitive connections between the generator and the load.

back conduction Conduction of current in reverse-polarity direction, as across a semiconductor junction that is reverse biased.

back contact A contact that closes a circuit when a relay, switch, or jack is in its normal rest position.

back current Symbol: Ib. The normally small current flowing through a reverse-biased pn semiconductor junction. Also called *reverse current* and *inverse current*. Compare *forward current*.

back diode A semiconductor diode that is normally operated in reverse conduction.

back echo An echo resulting from the rear lobe of an antenna radiation pattern.

back emf See *back voltage*.

back emission Emission, as in a glow-discharge tube or rectifier tube, when a device is operating with reverse voltage (e.g., when a rectifier plate is negative).

backfire See *arc-back*.

Back-Goudsmit effect See *Zeeman effect*.

background 1. Context or supporting area of a picture; e.g., the background of a TV picture. 2. Background noise.

background control In a color TV receiver, a potentiometer used to set the dc level of the color signal at one input of the three-gun picture tube.

background count Residual response of a radioactivity counter in an environment as free as practicable of radioactivity. This background is caused largely by cosmic rays and inherent radioactivity of the surrounding building and other bodies.

background job A low-priority, relatively long-running computer program that can be interrupted so that a higher-priority program can be run.

background noise Electrical noise inherent to a particular circuit, system, or device, and which remains when no other signal is present.

background processing In a computer, the running of programs having low priority.

background radiation Radiation from materials in the test environment. Also see *background count*.

background response The response of a radiation detector to background radiation.

backlash 1. Slack or lag in action of moving parts. Example: delay between initial application of a force (such as that required to turn a knob) and movement of a part or device (e.g., a potentiometer or variable capacitor). 2. On a tuning dial, arc within which slack or lag is discernible.

backloaded horn A loudspeaker enclosure in which the front of the speaker cone feeds sound directly into the listening area while the rear of the cone feeds sound into same area through a folded horn.

back lobe In the pattern of a directional antenna, the lobe directly behind the major lobe, which represents the signal emanating from the antenna in a direction opposite the direction of the major lobe.

backplate Flat electrode in a TV camera tube that receives the stored-charge image via capacitance coupling.

back porch In a TV horizontal sync pulse, the time interval between the end of the rise of the blanking pedestal and the beginning of the rise of the sync pulse. It is that portion of the flat top of the blanking pedestal behind the sync pulse. Compare *front porch*.

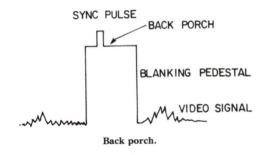

Back porch.

back-porch effect In transistor operation, continuation of collector-current flow for a short time after the input signal has fallen to zero.

back-porch tilt Departure of the top edge of a back porch from true horizontal.

back resistance Symbol, r_b. Resistance of a reverse-biased pn semiconductor junction. Also called *reverse resistance*.

back scatter Scattering of a wave back toward the transmitter from point beyond the skip zone. This phenomenon is caused by reflections back from those points. Compare *forward scatter*.

back-shunt circuit A circuit and system of keying originally employed with an arc converter, which allows the converter to remain in a constant-load operating condition. When the keying relay is closed for signal, the arc supplies the antenna, but when the relay is open, the converter is connected to an equivalent *LCR* load.

backstop A contact or barrier (such as a screw or post) that serves to limit backswing of the armature of a relay.

backswing The tendency of a pulse to overshoot, or reverse direction after completion. Backswing is measured in terms of the overshoot amplitude as a percentage of the maximum amplitude of the pulse.

back-to-back circuit An ac control circuit employing thyristors or thyratrons. To pass both half-cycles of ac, the tubes are connected back to back (i.e., the anode of one to the cathode of the other).

back-to-back connection Connection of diodes or rectifiers back to back (i.e., the anode of one to the cathode of the other) to pass both half-cycles of ac in certain control circuits.

back-to-back sawtooth A symmetrical sawtooth wave in which the rise slope is equal to the fall slope. Also called *triangular wave* and *pyramidal wave*.

backup An element, such as a circuit component, which is used to replace a similar component in case of failure of the latter.

backup facility In an electrical or communications system, a facility intended for use when the primary, or main, facility is not operational.

back voltage 1. Voltage induced in an inductor by the flow of ac through the inductor, so called because its polarity is opposite to that of the applied voltage that it bucks. Also called *counter emf*. 2. A voltage employed to obtain bucking action (e.g., the voltage used to zero the meter in an electronic voltmeter circuit). 3. Reverse voltage applied to a semiconductor junction.

backwall In a pot core, the plate or disk that connects the sleeve and center post to close the magnetic circuit.

backward diode A semiconductor diode so processed that its high current flow takes place when the junction is reverse biased. Such a diode is also a negative-resistance device.

backward-wave oscillator Abbreviation, *bwo*. A microwave oscillator tube similar to the traveling-wave tube. Like the traveling-wave tube, the bwo contains a helical transmission line. In the electron beam, electron bunching results from interaction between the beam and the rf field, and reflection takes place at the collector. Accordingly, the wave moves *backward* from collector to cathode, and oscillation is sustained because the backward wave is in phase with the input. Output is taken from the cathode end of the helix. Tuning, by varying the helix dc volt, affords a typical 2:1 bandwidth in the 1 to 40 GHz range.

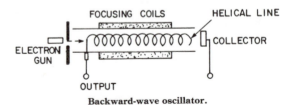

Backward-wave oscillator.

backward-wave tube See *backward-wave oscillator*.

back wave In some continuous-wave transmitters, the undesirable signal radiated when the key is open.

baffle 1. A board on which a loudspeaker is mounted to separate acoustic radiation emanating from the back of the cone from radiation emanating from the front. The baffle improves bass response by increasing the wavelength (lowering the frequency) at which phase cancellation occurs.

baffle plate 1. See *baffle*, 2. A metal plate mounted in a waveguide to reduce the latter's cross-sectional area.

baffle shield See *baffle*, 2.

bail A wire loop or chain holding one member of a two-member assembly to prevent loss, e.g., the short chain holding the dust cap of a jack.

Bakelite A plastic dielectric material. Chemical composition, *phenol-formaldehyde resin*.

Baker A phonetic alphabet code word for letter *B* (now generally obsolete).

baking-out In the process of evacuating a system, the procedure of heating the system to a high temperature to drive out gases occluded in the glass and metal parts.

balance 1. See *bridge*. 2. To null a bridge or similar circuit. 3. To equalize loads, voltages, or signals between two circuits or components.

balance coil 1. A type of autotransformer that enables a three-wire ac circuit to be supplied from a two-wire line. A series of taps around the center of the winding enables the circuit to be compensated for unequal loads. 2. See *balancing coil*.

balance control A variable component, such as a potentiometer or variable capacitor, used to balance bridges, null circuits, or loudspeakers.

balanced Having the same impedance with respect to ground.

balanced amplifier Any amplifier with two branches having the same impedance with respect to ground. Usually, this means that the two branches are in phase opposition.

balanced bridge Any four-leg bridge circuit in which all legs are identical in all electrical respects.

balanced circuit 1. A circuit which has its electrical midpoint grounded, as opposed to the single-ended circuit, which has one side grounded. 2. A bridge circuit in the condition of null.

balanced converter See *balun*.

balanced currents Currents having the same value. In the two conductors of a balanced line, these currents are at every point equal in amplitude and opposite in phase.

balanced delta A connection of coils or generators in a three-phase system in such a way that the three combine additively with a 120° time angle between the emf's of any two coils.

balanced detector A symmetrical demodulator, such as a full-wave diode detector or an FM discriminator.

balanced electronic voltmeter An electronic voltmeter circuit in which two matched tubes or transistors are connected in a four-arm bridge arrangement. Drift in one half of the circuit opposes that in the other half, with the result that drift of the zero point is virtually eliminated.

balanced filter A filter consisting of identical sections, one of these sections being inserted in each of the halves of a balanced system, such as a balanced line.

balanced input An input circuit whose electrical midpoint is grounded. Compare *single-ended input*.

balanced input transformer An input transformer in which the center-tap of the primary winding is grounded.

balanced line A pair of parallel wires that possess a uniform characteristic impedance because the two leads are of the same material and diameter and the distance between them is constant. In a balanced two-wire line, the currents in the two conductors are of equal amplitude and opposite phase.

balanced loop antenna A loop antenna having a grounded electrical midpoint determined by the junction of two identical series-connected capacitors shunting the loop.

balanced low-pass filter A low-pass filter used in a balanced system or balanced transmission line.

balanced method A system of instrumentation in which a zero, or balanced, reading is employed, and the reading may be either side of the zero reading.

balanced modulator A symmetrical circuit (using tubes, transistors, an integrated circuit, or diodes as principal components) which delivers an output signal containing the frequency sum and frequency difference of two input signals. One input signal is applied to the components in a push-pull arrangement, and the other input signal is applied to the components in parallel.

balanced multivibrator A switching oscillator circuit in which the two halves are identical in configuration and as nearly identical as practicable in performance.

balanced network Any network intended to be used with a balanced system or balanced transmission line. It is characterized by a pair of terminals, each of which shows the same impedance with respect to ground.

balanced oscillator A push-pull oscillator.

balanced output Output which is balanced against ground, e.g., where the electrical midpoint of the output circuit is grounded.

balanced output transformer 1. A push-pull output transformer having a center tapped primary winding. 2. An output transformer with a grounded centertap on its secondary winding.

balanced probe A probe, such as one for an electronic voltmeter or oscilloscope, which has a balanced input and (usually) a single-ended output.

balanced-tee trap A wavetrap constructed in a T-configuration with a resonant section in each on a balanced transmission line.

balanced telephone line A telephone transmission line which has two sides, similar to a balanced radio-frequency transmission line. Either side has the same impedance with respect to ground.

balanced termination A load device (or the practice of using such a device) in which its sections provide identical termination for each of the sections or conductors of a balanced system, such as a balanced line.

balanced-to-unbalanced transformer See *balun*.

balanced transmission line A transmission line whose two sides are symmetrical with respect to ground; e.g., a horizontal two-wire line.

balanced varactor tuning A two-varactor back-to-back circuit for adjusting the value of an inductor using an applied dc voltage. This arrangement has the advantage over a single-varactor (unbalanced) circuit in that high tuned-circuit Q is maintained and harmonic generation is reduced.

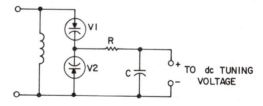

Balanced varactor tuning.

balanced voltages In any symmetrical system, such as a balanced line or push-pull circuit, two or more input or output voltages adjusted to have the same amplitude and (usually) opposite phase.

balanced-wire circuit A circuit or conductor system having identical halves which are symmetrical with respect to ground and to other conductors.

balancing circuit See *bucking circuit*.

balancing coil In a receiver, a centertapped antenna coil which is balanced to ground to eliminate the Marconi effect.

ballast 1. A component that is used to stabilize the current flow through or operation of a circuit, stage, or device. 2. An iron-core choke used in series with one of the electrodes in a fluorescent or other gas-discharge lamp.

ballast lamp See *ballast tube.*

ballast resistor 1. A nonlinear inductive power resistor whose *EI* characteristic is such that current through the resistor is independent of voltage over useful range. This feature enables the ballast resistor to act as an automatic voltage regulator when it is simply connected in series with a power supply and load. 2. A small (usually high-resistance) resistor operated in series with a glow lamp, such as a neon lamp, to prevent overload.

ballast transformer A misnomer often used in place of *ballast, 2.*

ballast tube A tube, usually containing only an iron filament, which has properties similar to those of the ballast resistor.

ballistic galvanometer An undamped galvanometer used particularly to observe electric charges by noting the single throw resulting from the momentary flow of current through the galvanometer coil.

ballistic missile A missile that is guided electronically during the first part of its flight, but becomes free falling as it approaches its target.

ballistic missile early-warning system Abbreviation, BMEWS. An electronic system for detection of enemy ballistic missiles and giving early warning of their approach.

ballistics The electronics-supported science concerned with the motion of projectiles and similar bodies in air or space.

ballistic trajectory The path of a missile after thrust has been removed.

balloon antenna A vertical antenna consisting of a wire or wires held aloft by a captive balloon.

balop Balopticon.

balopticon An opaque-picture projecting system in which the picture is viewed by a TV camera and displayed by a picture tube. Also called *balop.*

balun A type of impedance-matching rf transformer. It is a wideband device, usually providing a 4:1 impedance ratio and available in several different forms.

banana jack The female half of a two-part quick-connector combination. Splicing of a circuit is completed by inserting a *banana plug* into this jack.

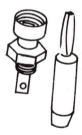

Banana jack and plug.

banana plug The male half of a two-part quick-connector combination, having sides usually composed of flat springs which expand to fill the orifice of the mating female *banana jack* into which it is inserted.

band 1. A continuous range of frequencies, usually designated by the lower and upper frequency (or wavelength) limits of the range. 2. A continuous microgroove track on a recording disk when there are several such tracks thereon.

band center In a given frequency spectrum, the geometric mean between the band limits.

band-elimination filter See *band-rejection filter.*

band gap In any atom, the difference in electron energy between the conduction and valence bands.

bandpass 1. The frequency limits between which a device or circuit transmits ac energy with negligible loss. 2. Possessing the ability to allow passage of signals at a given frequency or band of frequencies while disallowing passage of other signals. Compare *bandstop.*

bandpass amplifier An amplifier that is tuned to pass only those frequencies between preset limits.

bandpass coupling A coupling circuit having a flat-topped frequency response so that a band of frequencies rather than a single frequency is coupled into a succeeding circuit. Also see *bandpass, 1.*

bandpass filter A filter designed to transmit a band of frequencies with negligible loss while rejecting all other frequencies. Compare *band-rejection filter.*

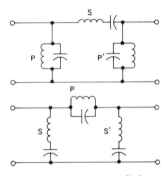

Bandpass filter. Resonant frequency finds parallel-resonant circuits (P and P') to be high-impedance path and series-resonant circuit (S) a low-impedance path. All other frequencies find P circuits low in impedance and S high.

bandpass flatness The degree to which the response of a band-pass amplifier or filter is flat within the pass band.

bandpass response The amplitude-versus-frequency response of a bandpass filter. The signal is transmitted between two definite frequencies, and rejected at other frequencies.

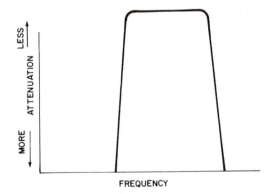

Bandpass response.

band-rejection filter A circuit or device (usually consisting of coils and capacitors) designed to shunt signals at resonance to ground and to pass all other signals with about equal efficiency.

band selector Any switch or relay which accomplishes bandswitching.

bandset capacitor In a bandspread receiver (see *bandspreading*), a variable capacitor employed to preset the tuning range in each band to correspond to graduations on the tuning dial. This capacitor is a trimmer *or* padder operated in conjunction with the main tuning capacitor.

bandspreading The process of widening the tuning range within a given frequency band over the entire dial of a radio receiver. Otherwise, the entire band would occupy only a portion of the dial, the tuning would be difficult. Also see *bandset capacitor.*

bandspread tuning control The adjustment in a communications receiver that allows tuning over a desired band of frequencies. This control is separate from the main tuning control.

bandstop 1. Band-rejection filter. 2. Possessing the ability to suppress or block signals of a given frequency or band of frequencies while allowing signals of other frequencies to pass with little or no attenuation. Compare *bandpass.*

bandstop filter See *band-rejection filter.*

band suppression 1. The frequency limits between which a device or circuit rejects or blocks ac energy and transmits with negligible loss energy at other frequencies. 2. The act of blocking signals at a given frequency or within a frequency band.

band-suppression filter A circuit or device that is designed to eliminate or block passage of a specific frequency or band while allowing other frequencies to pass with negligible loss; thus, any band rejection or bandstop filter.

bandswitch A (usually rotary) low-reactance selector switch for accomplishing bandswitching.

bandswitching In a receiver, transmitter, or test instrument, the process of switching self-contained tuned circuits to change from one frequency spectrum to another within the range of the device's intended operation.

bandwidth See *bandpass, 1.*

bank A collection of usually similar components used in conjunction with each other or in some sequence, usually parallel. Examples: resistor bank, lamp bank, transformer bank.

banked transformers Parallel-operated transformers.

bankwound coil A coil wound in such a way that most of its turns are not side by side, thus reducing the inherent distributed capacitance.

bantam tube An octal-base miniature tube that is smaller than full-sized version of the same type. The bantam type number includes the letter *B*; thus: 6SN7GTB.

bar 1. Abbreviation, b. The cgs unit of pressure, in which $1 \text{ b} = 10^5$ pascals per square centimeter. 2. Abbreviation of *buffer address register* (capitalized). 3. A horizontal or vertical line produced on a TV screen by a bar generator and used to check linearity. 4. A thick plate of piezoelectric crystal. 5. A solid metal conductor, usually uninsulated, of any cross section. 6. A silicon ingot from which semiconductor devices may be fabricated.

bare conductor A conductor having no insulating covering, a common example being bare copper wire.

bar generator A special type of radio-frequency signal generator which produces horizontal or vertical bars on the screen of a TV receiver, for use in adjustment of horizontal and vertical linearity.

bar graph A graphical presentation of data in which numerical values

are represented by horizontal or vertical bars of size corresponding to the values.

barium Symbol, Ba. An elemental metal of the alkaline-earth group. Atomic number, 56. Atomic weight, 137.36. Barium is present in some compounds used as dielectrics (e.g., barium titanate).

barium-strontium oxides The combined oxides of barium and strontium employed as coatings of vacuum-tube cathodes to increase electron emission at relatively low temperatures.

barium strontium titanate A compound of barium, strontium, oxygen, and titanium used as a ceramic dielectric material. It exhibits ferroelectric properties and is characterized by high dielectric constant.

barium titanate Formula, $BaTiO2$. A ceramic employed as the dielectric in ceramic capacitors. It exhibits high dielectric constant and some degree of ferroelectricity.

Barkhausen criterion In an oscillator (positive-feedback amplifier), the condition in which the feedback factor $BA = 11$, and the gain of the circuit is theoretically infinite.

Barkhausen effect The occurrence of minute jumps in the magnetization of a ferromagnetic substance as the magnetic force is increased or decreased over a continuous range.

Barkhausen interference Interference resulting from oscillation because of the Barkhausen effect.

Barkhausen-Kurz oscillator A positive-grid (negative-plate) triode used as a microwave oscillator. Oscillation results from the cloud of electrons being alternately moved back and forth through the grid between the cathode and plate.

Barkhausen oscillator See *Barkhausen-Kurz oscillator.*

Barkhausen tube A positive-grid oscillator tube.

bar magnet A relatively long permanent magnet in the shape of a bar, with rectangular or square cross section.

barn Symbol, b. A non-SI unit of nuclear cross section equal to 100 square femtometers. This unit is approved as compatible with SI (International System of Units).

Barnett effect The development of a small amount of magnetization in a long iron cylinder that is rotated rapidly about its longitudinal axis.

barograph A recording barometer.

barometer An instrument for measuring atmospheric pressure.

barometer effect The condition indicating that the intensity of cosmic rays is inversely proportional to the barometric pressure. The barometer effect is said to be approximately equal to 1 or 2% per centimeter of mercury.

barometric pressure The atmospheric pressure, usually given in inches of mercury. The average barometric pressure at the surface of the earth is just under 30 inches of mercury.

bar pattern A series of spaced lines or bars (horizontal, vertical, or both) produced on a TV picture screen by means of a bar generator and useful in adjusting horizontal and vertical linearity of the picture.

barrage array An antenna array in which a string of collinear elements are vertically stacked. The end quarter-wavelength of each string is bent in to meet the end quarter-wavelength of the opposite radiator to improve balance.

barrage jamming The jamming of many frequencies at the same time.

barrell distortion TV picture distortion consisting of horizontal and vertical bulging.

barretter A low-voltage, filament-type voltage regulator similar to the ballast tube, but containing a tungsten filament having a positive temperature coefficient of resistance. It is used to regulate transistor biases and as a bolometer in microwave measurements.

barretter mount A fixture for holding a barretter inside a waveguide.

barrier 1. The carrier-free space-charge region in a semiconductor pn junction. 2. An insulating partition placed between two conductors or terminals to lengthen the dielectric path.

barrier balance The state of near equilibrium about a semiconductor pn junction (after initial junction forming) entailing a balance of major and minor conduction currents due to random migration of carriers across the junction.

barrier capacitance 1. The reactive element formed by the value of capacitance in a bipolar transistor between emitter and collector, which varies with changes in applied volt and with junction temperature. 2. The capacitance across any pn junction that is reverse biased.

barrier height The difference in voltage between opposite sides of a barrier in a semiconductor material.

barrier layer See *barrier, 1.*

barrier-layer cell A photovoltaic cell, such as the copper oxide or selenium type, in which photons striking the barrier layer produce the output volt.

barrier potential The apparent internal dc (contact) potential across the barrier (see *barrier, 1*) in a pn junction.

barrier strip A terminal strip having a barrier (see *barrier, 2*) between each pair of terminals.

Barrier strip.

barrier voltage The emf required for the initiation of current flow through a pn junction.

Barrow oscillator A tunable triode oscillator for the 70 to 700 MHz range, employing coaxial lines for the plate tank and for tuning the tube-filament lines.

baryon A *subatomic particle* made up of three *quarks.*

base 1. In a bipolar transistor, the intermediate region between the emitter and collector, which usually serves as the input or controlling element of the transistor's operation. 2. A substance which dissociates in water solution and forms hydroxyl (OH) ions. Example: sodium hydroxide. 3. The constant figure upon which logarithms are computed (10 for common logs, 2,71828 for natural logs). 4. The radix of a number system (thus, base 10 for the decimal system). 5. A fixed nonportable radio communications installation; a *base station.*

base address The number in a computer address that serves as the reference for subsequent address numbers.

baseband The frequency band of the modulating signal in a transmitter. This is generally equivalent to the range of voice frequencies, although it may be restricted depending on the bandwidth of the transmitted signal. The baseband generally ranges from about 10 Hz at the minimum to anywhere between 3 and 20 kHz at the maximum.

baseband frequency response The amplitude-versus-frequency characteristic of a transmitter within the baseband, or modulating, frequency range.

base bias The steady dc voltage applied to the base electrode of a transistor to determine the operating point along the transistor characteristic.

base-bulk resistance The resistance of the semiconductor material in the base layer of a bipolar transistor.

base-charging capacitance In the common-emitter connection of a bipolar transistor, the internal capacitance of the base-emitter junction.

base current Symbol, I_b. Current flowing through the base electrode of a bipolar transistor. Also see *ac base current* and *dc base current.*

base electrode See *base, 1.* Also called *base element.*

base element 1. Base electrode. 2. One of the basic metals, such as iron or tin, that are not generally considered precious (as opposed to *noble*).

base film The plastic substrate of a magnetic recording tape.

base frequency 1. The principal frequency in a wave; also called *basic frequency.* 2. The frequency of operation of a base-station transmitter when the receiver is tuned to a second channel.

base-input circuit A common-collector circuit, common-emitter circuit, or emitter follower.

base insulator A stout dielectric insulator used to support a heavy conducting element and keep the conductor isolated from other possible conductors or conductive paths.

base line In visual alignment procedures involving an oscilloscope and rf sweep generator, a zero-voltage reference line developed by the generator as a horizontal trace on the oscilloscope screen.

baseline stabilizer A clamping circuit which holds the reference voltage of a waveform to a predetermined value. Also called *dc restorer.*

base-loaded antenna A usually vertical antenna or radiating element, the electrical length of which is adjusted by means of a loading coil in series with and positioned at the bottom of the antenna or radiator.

base material In printed circuits, the dielectric material used as a substrate for the metal pattern. Also called *base medium.*

base notation The numbering or radix system used in any application (as octal, decimal, duodecimal, hexadecimal).

base number See *base, 4.*

base pin One of the straight prong-like terminals on an electrical or electronic component; it is used to provide support for the device and to allow a physical connection between the socket terminal into which it fits and one of the internal electrodes of the device.

base plate The chassis plate upon which components are mounted before wiring.

base potential See *base voltage.*

base region See *base, 1.*

base resistance Symbol, R_b. Resistance associated with the base electrode of a bipolar transistor. Also see *ac base resistance* and *dc base resistance.*

base resistor The external resistor connected to the base of a bipolar transistor. In the common-emitter circuit, the base resistor is analogous to the grid resistor of a tube amplifier and the gate resistor of a field-effect transistor amplifier.

base spreading resistance Symbol, r_{bb}. In a bipolar transistor, the bulk-material resistance of the base region between the collector junction and emitter junction.

base station The head station or fixed home station in a communication network.

base voltage Symbol, V_b. The voltage at the base electrode of a bipolar transistor. Also see *ac base voltage* and *dc base voltage.*

BASIC Acronym for *beginner's all-purpose symbolic instruction code,* a primitive but versatile and easy-to-learn computer language developed at Dartmouth College.

basic frequency 1. Fundamental frequency (as opposed to one of its harmonics). 2. Base frequency.

basic protection Lightning protection devices that should be included in any electronic or communications installation.

basket-weave coil A type of single-layer inductor in which adjacent turns do not parallel each other around the circumference, but zigzag oppositely as a strand does in the woven pattern of a basket. This non-parallelism of adjacent turns reduces distributed capacitance.

bass Low audio frequencies corresponding to the musical pitches so named.

bass boost 1. The special emphasis given to low audio frequencies (the bass notes) by selective circuits in audio systems. 2. The technique of boosting the bass response, as above, for more faithful reproduction of sound, usually at low amplitude levels.

bass compensation See *bass boost, 1.*

bass control 1. A manually variable potentiometer for adjusting bass boost of an amplifier or sound system. 2. The arrangement of components required to achieve amplitude variation of bass in an audio signal.

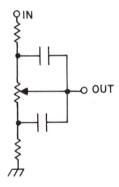

Bass control, 2.

bass reflex enclosure A loudspeaker cabinet with a critically dimensioned duct or port that allows back waves to be radiated in phase with front waves, thus averting unwanted signal cancellation.

bass reflex loudspeaker A loudspeaker mounted in a bass reflex enclosure. Also see *acoustical phase inverter.*

bass resonant frequency The low frequency at which a loudspeaker or its enclosure displays resonant vibration.

bass suppression In speech transmission, the removal of all frequencies below 600 Hz or so, on the assumption that those frequencies contribute little to intelligibility. This suppression permits the speech level to be increased without overmodulating a transmitter; it also allows smaller audio transformers to be used, since transformer core size must increase as the frequency it passes decreases.

bassy In audio and high-fidelity applications, a sound in which the low-frequency components, below about 500 Hz, are predominant.

BAT Abbreviation of battery. Also abbreviated B and BA.

batch fabrication process The manufacture of devices in a single batch from materials of uniform grade. Particularly, the manufacture of a large number of semiconductor devices from one batch of semiconductor material by means of carefully controlled, identical processes.

batch processing In digital-computer practice, the processing of quantities of similar information during a single run.

bat-handle switch A toggle switch, the lever of which is relatively long and thick and is shaped more or less like a baseball bat.

Bat-handle switch.

bathtub capacitor A (usually oil-filled) capacitor housed in a metal can resembling a bathtub.

bathyconductorgraph An instrument used to measure the conductivity of sea water.

bathythermograph An instrument which delivers a graph of temperature plotted against depth of sea water.

Batten system A system for document identification in a computer.

battery A multicell device which generates dc electricity by means of electro-chemical action. A battery consists of a group of cells connected either in series to supply a desired voltage or in parallel to supply a desired current, or both. Also see *cell, Edison battery, lead-acid battery, primary battery,* and *storage battery.*

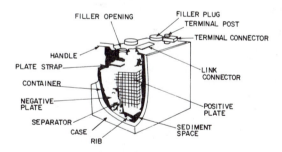

Battery consisting of three integral cells.

battery acid An acid, such as sulfuric acid, used as the electrolyte of a battery. Colloquially, any battery electrolyte, whether acid, base, or salt.

battery capacity The current-supplying capability of a battery, usually expressed in ampere-hours (A-h).

battery cell See *cell, 1.*

battery charger 1. A dc power supply usually but not necessarily embodying a stepdown transformer, rectifier, and filter; it is used to charge a storage battery from an ac power line. 2. A motor-generator combination employed for the same purpose. 3. A combination of solar cells or other voltaic transducers used to charge a storage battery with dc obtained from a nonelectrical energy source.

battery clip 1. A heavy-duty metallic clamp used for quick, temporary connection to a large cell terminal such as that of a lead-acid storage battery. 2. A small connector of the snap-fastener type, used for quick connection to a small power source such as a transistor-radio battery.

battery eliminator A dc power supply usually but not necessarily embodying a transformer, rectifier, and filter, which permits battery-powered equipment to be operated from an ac power line.

battery holder 1. A case or container of any kind for holding a cell or battery. 2. A shelf for holding a cell or battery. 3. A small, metal bracket-type device for holding a cell or battery between two contacts.

battery life 1. The ampere-hour or watt-hour capacity of a battery. 2. The number of times that a rechargeable battery can be cycled before the battery becomes unusable.

battery receiver A usually portable radio or TV receiver operated from self-contained batteries.

battery substitute See *battery eliminator*.

battery tube An electron tube designed for operation from batteries. A filament-type emitter is employed for low A-battery drain.

bat wing On a TV or FM antenna, a metallic element resembling a bat's wing.

baud A unit of communications processing speed in telegraphy and digital data communications systems, usually equal to a rate of one bit per second.

Baudot code A machine communications code employing five parallel binary digits of equal length, the interpretation of which depends on the history of the previous transmission or an additional case bit.

Baume' (Antione *Baume'*, 1728-1804). Abbreviation, Be. Unit, °Be'. Pertaining to the Baume' scaled for hydrometers. There are two such scales—one for liquids heavier than water and one for liquids lighter than water. The heavier-than-water hydrometer sinks to 0° in pure water and to 15° in a 15% salt solution; the lighter-than-water hydrometer sinks to 0° in a 10% salt solution and to 10° in pure water. From the Baume' readings, specific gravity at 60 °F = 145 (145 − n) for liquids heavier than water, or 140 (140 + n) for liquids lighter than water (in each case, n is the reading on the Baume' scale).

bay One of several sections of a directional antenna array.

bayonet base The insertable portion of a plug-in component (e.g., tube, lamp, etc.), having a projecting pin which fits into a slot or keyway in the shell of the socket into which the component is inserted.

bayonet socket A socket having a suitably slotted shell for receiving the bayonet base of a plug-in component.

bazooka A linear balun in which a quarter-wavelength of metal sleeving surrounds a coaxial feeder and is shorted to the outer conductor of the feeder to form a shorted quarter-wave section.

bb Abbreviation of *blackbody*.

B-battery In battery-operated tube circuits, the battery that supplies power to the tube plates and screens. Compare *A-battery* and *C-battery*.

BBC Abbreviation of *British Broadcasting Corporation*.

BBM Abbreviation of *break before make*.

b-box Index register of a computer.

BC Abbreviation of *broadcast*; also abbreviated BCST.

BCD Abbreviation of *binary-coded decimal*.

BCFSK Abbreviation of *binary code frequency-shift keying*.

B-channel One of the channels of a two-channel stereophonic system. Compare *A-channel*.

BCI Abbreviation of *broadcast interference*.

BCL Amateur radio abbreviation of *broadcast listener*.

BCN Abbreviation of *beacon*.

BCO Abbreviation of *binary-coded octal*.

BCST Abbreviation of *broadcast*; also abbreviation *BC*.

BDC Abbreviation of *binary decimal counter*.

B-display A radarscope display in which the target is represented by a bright spot on a rectangular-coordinate screen. Compare *A-display* and *J-display*.

Be Symbol for *beryllium*.

Be' Abbreviation of *Baume'*.

beacon 1. A beam of radio waves employed for navigation and direction finding. 2. A transmitter that radiates such a beam.

beacon direction finder A direction finder using a signal received from a beacon station.

beacon receiver A receiver that is specially adapted for the reception of beacon signals (see *beacon, 1*).

beacon station 1. A station broadcasting beacon signals (see *beacon, 1*) for direction finding or navigation. 2. Sometimes, a radar transmitting station.

beacon transmitter A transmitter specially adapted for the transmission of beacon signals (see *beacon, 1*).

bead 1. A small ferrous ferrule that is used as a passive decoupling choke by slipping it over the input power leads of a small circuit or stage. 2. A magnetic memory element in a ferrite-core matrix.

beaded coax A coaxial line in which the inner conductor is separated rigidly from the outer conductor by means of spaced dielectric beads.

beaded support A plastic or dielectric bead used for the purpose of supporting the inner conductor of an air-insulated transmission line of coaxial construction.

bead thermistor A thermistor consisting essentially of a small bead of temperature-sensitive resistance material into which two leads are inserted.

beam 1. The more or less narrow pattern of radiation from a directional antenna. 2. A directional antenna, especially a yagi. 3. The stream or cloud of electrons emitted by the cathode in an electron tube, especially a beam power tube.

beam alignment 1. The lining-up of a directional transmitting antenna with a directional receiving antenna for maximum signal transfer. 2. In a TV camera tube, the lining-up of the electron beam so that it is perpendicular to the target. 3. In a cathode-ray tube, the positioning of the electron rays so that they converge properly on the screen regardless of the deflection path.

beam angle In the radiation from an antenna, the direction of most intense radiation, the side limits of which are determined by the points at which the field strength drops to half the value in the principal direction.

beam antenna A directional antenna used for transmitting and receiving rf signals.

beam bender 1. In a TV picture tube, the ion-trap magnet. 2. Deflection-plate correction device or circuit.

beam bending Deflection of an electron beam by electric or magnetic fields.

beam blanking See *blank, 2*.

beam convergence The meeting, at a shadow-mask opening, of the three electron beams in a three-color TV picture tube. See *beam alignment, 3*.

beam crossover The half-power point of a directional antenna. The reference point is considered to be the direction of maximum radiation.

beam current The current represented by the flow of electrons in the beam of a cathode-ray tube.

beam cutoff In an oscilloscope or TV picture tube, complete interruption of the electron beam usually as a result of high negative control-grid bias.

beam-deflection tube A tube in which an electron beam may be switched between the common cathode and any one of a number of plates. Also see *beam-switching tube.*

beam deflector 1. A deflection plate in an oscilloscope tube. 2. A *beam-forming* plate.

beam efficiency In a cathode-ray tube, the ratio of the number of electrons generated by the gun to the number reaching the screen. The efficiency is high in electromagnetic-deflection tubes, lower in electrostatic-deflection tubes.

beam electrode Beam-forming plate.

beam-forming plate In a beam power tube, one of the deflector plates which concentrates the electron emission into beams as the electron stream passes the screen.

beam-forming tetrode A four-element beam power tube.

beam hole In a reactor shield (see *reactor, 2*), an aperture which allows a beam of radioactive particles to emerge for external use.

beam lead In an integrated circuit, a relatively thick and strong lead which is deposited in contact with portions of the thin-film circuit and provides stouter connections than continuations of the thin film would provide.

beam-lead isolation In an integrated circuit, reduction of distributed capacitance and other interaction through use of beam leads.

beam modulation See *intensity modulation.*

beam multiplier tube See *photomultiplier tube.*

beam parametric amplifier A parametric amplifier in which the variable-reactance component is supplied by a modulated electron beam.

beam-positioning magnet In a three-gun color TV picture tube, a permanent magnet employed to position one of the electron beams correctly with respect to the other two.

beam power tube A tetrode or pentode in which special deflector plates concentrate the electrons into beams in their passage from cathode to plate. The beam action greatly increases plate current at a given plate voltage.

beam-rider control system A missile-guidance system in which a control station sends a radio beam to a missile. The beam is moved in such a way that, as the missile stays within the beam, it hits the target.

beam-rider guidance 1. An aircraft landing guidance system, in which the aircraft follows a radio beam in its glide path. 2. The circuitry in a guided missile using a beam-rider control system.

beam-splitter A device used for dividing a light beam (as by a transparent mirror) into two components, one transmitted and the other reflected; hence, a *beam-splitting mirror.*

beam-splitting In radar, a method of calculating the mean azimuth of a target from (a) the azimuth at which the target is first revealed by one scan, and (b) the azimuth at which the target information ceases.

beam-splitting mirror In an oscilloscope-camera system, a tilted, transparent mirror which allows rays to pass horizontally from the oscilloscope screen to the camera and to be reflected vertically to the viewer's eye.

beam-switching tube A cold-cathode tube having a central cathode around which are circumferentially arranged a series of targets and switching grids. The electron beam passes from the cathode to one target, completing a circuit. A voltage applied to the next adjacent grid then will switch the beam to the associated target, and so on sequentially around the circle, as the switching signal is applied successively to the various grids.

beam tube See *beam-power tube* and *beam-switching tube.*

beamwidth of antenna The angular wideness of the main lobe of the pattern of radiation from a directional antenna.

bearing The direction of an object or point expressed in degrees within a 360° horizontal clockwise boundary, with the center of the circle serving as the observation point.

bearing resolution In radar practice, the minimum horizontal separation of two targets, in degrees, which will permit the individual targets to be displayed accurately.

beat Any one of the series of pulsations constituting a beat note, which result from heterodyning one signal against another.

beat frequency The frequency (f_c) resulting from the mixing of two signals (f_a and f_b) of different frequency. Sum and difference beat frequencies result: $f_{c1} = f_a + f_b$; $f_{c2} = |f_a - f_b|$.

beat-frequency oscillator An oscillator employed to set up an audible beat note with the signal generated by another oscillator. In sideband reception the beat is modulated at an audio rate to form speech; in continuous-wave reception the beat may be adjusted by a panel control to a suitable pitch.

beating 1. The combination of signals of different frequencies resulting in sum and difference frequencies. 2. The fluttering noise heard when two audio tones, very close in frequency and very similar in amplitude, are emitted at the same time.

beat marker In the visual (oscilloscopic) alignment of a tuned circuit, a marker pip resulting from the beat note between the sweep-generator signal and the signal from a marker oscillator.

beat note The sum or difference frequency resulting from the heterodyning of two signals or, under some conditions, of more than two signals.

beat-note reception 1. Reception in which a radio-frequency carrier is made audible by heterodyning it with a local oscillator to produce an audible beat note. 2. By extension, superheterodyne reception (see *superheterodyne circuit*).

beat tone A beat note in which the frequency is within the range of hearing.

beaver tail A flat, or elongated radar beam, wide in the azimuth plane. Primarily used for determining the altitude of a target. The beam is moved up and down to find the target elevation.

bedspring A directional antenna consisting of a broadside array having a flat reflector and resembling somewhat a bedspring.

beep A test or control signal, usually of single tone and short duration.

beeper 1. Any device for producing a *beep.* 2. A pocket- or hand-carried transceiver, especially one for maintaining two-way contact with personnel who are away from their base.

beetle A urea formaldehyde plastic used as a dielectric material and as a container material.

begohm Obsolete unit of high resistance, reactance, or impedance, denoting 1 billion ohms.

bel Abbreviation, B. The basic logarithmic unit (named for Alexander Graham Bell) for expressing gain or loss ratios. 1 bel is equivalent to a power gain of 10. Also see *decibel.*

B-eliminator An ac-operated power supply which replaces the B batteries in battery-operated tube equipment. Also see *battery eliminator.*

bell An electric alarm device consisting of a metallic gong which emits a ringing sound when struck by an electrically vibrated clapper.

Bellini-Tosi direction finder An early direction finder in which the sensing element consists of two triangular vertical antennas crossed at right angles, the antennas being open at the top and accordingly not acting an conventional coil antennas.

bell-shaped curve A statistical curve (so called from its characteristic shape) which exhibits a normal distribution of data. Typically, the curve describes the distribution of errors of measurement around the real value.

bell transformer A (usually inexpensive) stepdown transformer for operating an electric bell or similar alarm or signaling device from the ac power line.

bell wire Low-voltage cotton-covered 18-gauge (AWG) copper wire, so called because of its principal early use in the wiring of electric-bell circuits.

belt generator A very-high-voltage electrostatic generator, a principal part of which is a fast-traveling endless belt of dielectric material. At the lower end, charges of one sign are sprayed on the belt at 10 to 100 kV dc and are carried to the inside of a hollow metal sphere at the upper end, where they are removed and spread to the surface of the sphere, which they raise to a potential up to several million volts.

benchmark A test standard for measuring product performance.

benchmark routine A routine (program) designed to evaluate computer software or hardware performance (execution time, accuracy, etc.).

bench test An extensive checkout of a piece of equipment in the test laboratory, either to find an intermittent problem, or to check for reliability.

bend An angular shift in the lengthwise direction of a waveguide.

bending effect 1. The downward reflection of a radio wave by the ionosphere. 2. The low-atmosphere turning of a radio wave downward by temperature discontinuity and atmospheric inversions.

Benito A continuous-wave method of measuring the distance of an aircraft from the ground, involving the transmission of an audio-modulated signal from ground and the retransmission back to ground by the aircraft. The phase shift between the two signals is proportional to the distance to the aircraft.

bent antenna An antenna having the ends of its horizontal radiator bent downward to conserve horizontal space.

bent gun A TV picture tube neck arrangement having an electron gun that is slanted so as to direct the undesired ion beam toward a positive electrode, but which allows the electron beam to pass to the screen.

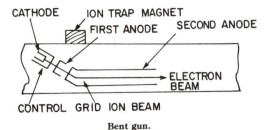

Bent gun.

BeO Formula for beryllium oxide. Also see *beryllia*.

berkelium Symbol, Bk. A radioactive elemental metal produced artificially. Atomic number, 97. Atomic weight, 247.

beryllia Formula, BeO. Beryllium oxide, used in various forms as an insulator and structural element (as in resistor cores).

beryllium Symbol, Be. An elemental metal. Atomic number, 4. Atomic weight, 9.013. Beryllium is present in various dielectrics and alloys used in electronics.

Bessel functions Functions for dealing with periodic electronic phenomena in which the waveform often displays decrement. These functions are solutions of the differential equation (Bessel's equation):

$$x^2 \, (d^2y/dx^2) + x \, (dy/dx) +$$
$$(x^2 - n^2)y = 0$$

Also called *cylindrical functions*.

beta Symbol, β. The current gain of a common-emitter bipolar transistor stage. It is the ratio of the induced change of collector current to the applied change of base current: $\beta = dI_c/dI_b$.

beta circuit The output-input feedback circuit in an amplifier.

beta cutoff frequency The frequency at which the current amplification of a bipolar transistor falls to 70.7% of its low-frequency value.

beta function The transcendental function
$$\beta(m, \, n) = \int_0^1 x^{m-1} \, (1-x)^{n-1} \, dx$$
where m and n are greater than zero.

beta particles Minute radioactive subatomic bits identical with the electron and emitted by some radioactive materials. Also see *beta rays*.

beta rays Rays emitted by the nuclei of radioactive substances, consisting of a stream of beta particles (i.e., electrons or protons) which move at velocities up to 299.8 million meters per second. Compare *alpha particle* and *gamma rays*.

beta-to-alpha conversion For a bipolar transistor, the conversion of current amplification expressed as beta β to current amplification expressed as alpha α: $\alpha = \beta/(\beta + 1)$.

betatron A type of accelerator (see *accelerator, 1*) in which injected electrons are given very high velocity by being propelled through a doughnut-shaped glass container (forming a 1-turn secondary of a 600 Hz transformer) by induction effects.

beta zinc silicate phosphor Formula, (ZnO + SiO2):Mn. A phosphorescent substance used to coat the screen of a cathode-ray rube. The fluorescence is green-yellow.

Bethenod alternator An early French dynamo-type system for generating radio-frequency power. Actually, the Bethenod system consisted of several machines similar to the Alexanderson alternator mounted on one drive shaft and interconnected electrically. This arrangement reduced the number of rotor teeth or slots needed at a particular drive speed for a given frequency. Power levels up to 500 kW were reached. Sometimes called *Latour alternator*.

BeV Abbreviation of *billion electronvolts*. Also see *electronvolt, GeV, MeV*, and *million electronvolts*. This abbreviation has been supplanted by the SI (International System of Units) abbreviation GeV, for *gigaelectronvolts*.

bevatron An accelerator (see *accelerator, 1*) similar to the synchrotron which accelerates particles to levels greater than 10 GeV.

Beverage antenna (Harold H. *Beverage*). A nonresonant, directional antenna having a diamond shape (or in the form of a long, parallel-wire line) terminated at its far end by a noninductive resistor equal to the characteristic impedance of the antenna.

beyond-the-horizon propagation See *forward scatter*.

bezel A faceplate for an electronic instrument, usually having a fitted rim and cutouts for knobs, switches, jacks, etc.

bfo Abbreviation of *beat-frequency oscillator*.

BG Abbreviation of *Birmingham wire gauge*. Also abbreviated BWG.

B-H curve A plot showing the *B* and *H* properties of a magnetic material. Magnetizing force *H* is plotted along the horizontal axis, flux density *B* along the vertical axis.

B-H loop See *box-shaped loop*.

B-H meter Any instrument for displaying or evaluating the hysteresis loop of a magnetic material.

bhp Abbreviation of *brake horsepower.*

Bi Symbol for *bismuth.*

bias Any parameter of which the value is set to a predetermined level to establish a threshold or operating point. While it is common to think of bias currents and bias voltages, other parameters (e.g., capacitance, resistance, illumination, magnetic intensity, etc.) may serve as biases.

bias battery A small battery designed to supply bias voltage to a vacuum-tube stage. The positive electrode is connected to the cathode and the negative terminal to the grid, usually through a grid resistor.

bias cell A tiny button-type dry cell designed especially to supply control-grid bias to a tube.

bias compensator A device that balances the inward force on the pickup arm of a phonograph, reducing the friction.

bias current A steady, constant current which presets the operating threshold or operating point of a circuit or device, such as a transistor, diode, or magnetic amplifier. Compare *bias voltage.*

bias distortion Distortion due to operation of a tube or transistor with incorrect bias so that the response of the device in nonlinear.

biased diode A diode having a dc voltage applied in either forward or reverse polarity. Current flows readily through the forward biased diode; the reverse biased diode appears as an open circuit. The biased diode is the basis of clippers, limiters, slicers, and similar circuits.

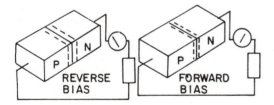

Biased diodes.

biased off In a circuit or device, the state of cutoff due to application of a control-electrode bias. Example: collector-current cutoff when the dc base bias of a bipolar transistor reaches a critical value, or drain-current cutoff when the dc gate bias reaches a critical value in a field-effect transistor.

bias modulation Generically, amplitude modulation obtained by superimposing the modulating signal on the dc bias of the control grid, suppressor, or screen of a radio-frequency power amplifier.

bias oscillator In a magnetic recorder, an oscillator operated at a frequency in the 40 to 100 kHz range to erase prerecorded material and bias the system magnetically for linear recording.

bias resistor A usually fixed resistor, such as the cathode resistor in a tube circuit or the emitter resistor in a bipolar transistor circuit, across which a desired bias voltage is developed by current flowing through the resistor.

bias set 1. A motor-generator once employed solely to supply dc control-grid bias to a tube-type transmitter. 2. A control, such as a potentiometer or variable autotransformer, used to adjust manually the dc bias of a circuit.

bias stabilization 1. The maintenance of a bias supply steady against load or line variations, as by means of automatic voltage regulation.

2. The stabilization of transistor dc bias voltage by means of resistance networks or through the use of barretters, diodes, or thermistors.

bias supply 1. Batteries providing bias voltage or current for tubes or semiconductors. 2. A motor-generator used as a bias set. 3. A line-operated unit for supplying dc bias and consisting of a transformer, rectifier, and high-grade filter.

bias voltage A steady voltage which presets the operating threshold or operating point of a circuit or device, such as a vacuum tube or transistor. Compare *bias current.*

bias windings The dc control windings of a saturable reactor or magnetic amplifier.

biconcave Pertaining to a lens that is concave on both faces. Also called concavoconcave.

biconical antenna A form of broadband antenna, consisting of two conical sections joined at the apexes. The cones are at least 1/4 wavelength in diagonal height. The vertex angles of the cones may vary.

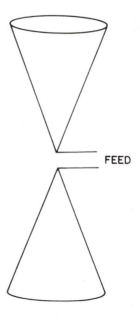

Biconical antenna.

biconical horn antenna A double-horn microwave antenna radiating relatively sharp front and back beams.

biconvex Pertaining to a lens that is convex on both faces.

bidecal base The 20-pin base of a cathode-ray tube. Also see *diheptal, duodecal,* and *magnal.*

bidirectional Radiating or receiving (usually equally) from opposite directions, e.g., front-and-back radiation from an antenna or loudspeaker or front-and-back pickup with an antenna or microphone.

bidirectional antenna An antenna having a directional pattern consisting of maximum lobes 180 degrees apart.

bidirectional bus In computers, a data path over which both input and output signals are routed.

bidirectional bus driver In a microcomputer, a signal-driving device which permits direct connection of a buffer-to-buffer arrangement on one end (the interface to I/O, memories, etc.) and data inputs and outputs on the other. This device permits bidirectional signals to pass and provides drive capability in both directions.

bidirectional counter A counter that can count consecutively up from a given number or down from that number. Also called *up-down counter*.

bidirectional current A current that flows in both directions.

bidirectional loudspeaker A loudspeaker that delivers sound waves to the front and rear.

bidirectional microphone A microphone that picks up sound waves equally well from the front and rear.

bidirectional transistor A symmetrical transistor such as some FETs i.e., one in which the two main current-carrying electrodes may be interchanged without influencing device performance.

bifilar electrometer An electrometer in which the sensitive element comprises two long platinized-quartz fibers. When an electric potential is applied, the fibers separate by a distance proportional to the voltage.

bifilar resistor A wirewound resistor with two oppositely wound filaments. The nature of the winding tends to cancel the inductance, making the device useful at a much higher frequency than an ordinary wirewound resistor.

bifilar transformer A transformer in which unity coupling is approached by interwinding the primary and secondary coils; i.e., the primary and secondary turns are wound side by side and in the same direction.

bifilar winding 1. A method of winding a coil (such as a resistor coil) in the shape of a coiled hairpin, so that the magnetic field is self-canceling and the inductance minimized. 2. A method of winding audio transformers in a bifilar manner to minimize leakage reactance.

bifurcated contact A forked contact whose parts act as two contacts in parallel for increased reliability.

bilateral Any graph that is symmetrical with respect to the origin. Such a curve shows symmetry in a mirror-like manner relative to the line $y = -x$.

bilateral amplifier A tube or transistor amplifier that transmits or receives in either direction equally well; i.e., the input and output may be exchanged at will.

bilateral antenna A bidirectional antenna such as a loop antenna.

bilateral elements A circuit element or component (as a capacitor, resistor, inductor) that transmits energy equally well in either direction. Compare *unilateral elements*.

bilateral network A network, such as the symmetrical twin-tee circuit, the input and output terminals of which may be exchanged without affecting the performance of the network.

billboard antenna A phased group of dipole antennas that lie in one plane. There may or may not be a reflector behind the entire array.

bilobe pattern A radiation pattern consisting of two lobes.

bimetal A union of two dissimilar metals, especially those having a different temperature coefficient of expansion. The two are usually welded together over their entire surface.

bimetallic element A strip or disk of bimetal. When the element is heated, it bends in the direction of the metal having the lower temperature coefficient of expansion; when cooled, it unbends. Usually, an electrical contact is made at one extreme or the other so the element may serve as a thermostat.

bimetallic switch A temperature-sensitive switch based upon a bimetallic element.

bimetallic thermometer A thermometer based upon a bimetallic element which is mechanically coupled (as through a lever and gear system) to a pointer which moves over a temperature scale.

bimetallic thermostat A thermostat in which a bimetallic element closes or opens a pair of switch contacts.

bimorphous cell A piezoelectric transducer consisting of two crystal plates, such as Rochelle's salt, bound intimately face to face. In a crystal microphone or phonograph pickup, vibration of the transducer results in a voltage output; in a crystal headphone or loudspeaker, an ac signal voltage impressed on the transducer causes vibratory mechanical motion.

BiMOS A combination of bipolar and MOSFET transistors in an integrated circuit. Thus, a typical BiMOS device may have MOSFET input for high impedance, and bipolar output for low impedance.

binant electrometer An electrometer in which a thin platinum vane ("the needle") is suspended within two halves of a metal pillbox-shaped container. The halves or binants are biased with a dc voltage of 1-12V and the unknown voltage is applied to the vane. Also called *Duant electrometer* and *Hoffman electrometer*.

binary 1. In notation, *two-numbered*. Thus, binary arithmetic employs only two digits: 0 and 1. 2. Pertaining to two-element chemical compounds.

binary arithmetic Mathematical operations performed using only the digits 0 and 1.

binary cell In a computer memory, an element which can display either one of two stable states.

binary chain A cascade of binary elements, such as flip-flops, each unit of which affects the stable state of the succeeding unit in sequence.

binary channel Any channel whose use is limited to two symbols.

binary code A system of numbers representing quantities by combinations of 1 and 0; a binary-number system.

binary-coded decimal notation In digital computer practice, a system of notation in which each digit of a decimal number is represented by its binary equivalent. Thus, the decimal number 327 in BCD notation becomes 0011 0010 0111. (By contrast, in pure binary notation, 327 is 101 000 111.)

binary-coded octal notation A method of numbering in which each base-8 digit is represented by a binary number from 000 to 111.

binary-controlled gate circuit A gate circuit controlled by a binary stage. An example is a gating transistor which receives its on/off pulses from a flip-flop.

binary counter A counter circuit consisting of a cascade of bistable stages. Each stage is a scale-of-two counter, since its output is on for every second input pulse. At any instant, the total binary count in a multistage counter thus is shown by the on and off states of the various stages in sequence.

binary decoder A device or stage which accepts binary signals on its input lines and provides a usually exclusive output (representing a decimal digit, for example).

binary number system The base-2 system of notation. This system employs only two symbols—0 and 1—and accordingly is easily applied to two-position switches, relays, and flip-flops.

binary numeral A combination of numerals 1 and 0 in succession, representing a number in binary notation.

binary preset switch In a binary counter or binary control circuit, a selector switch that allows the circuit to be preset to deliver an output pulse only after a predetermined number of input pulses.

binary relay See *bistable relay*.

binary scaler In its simplest form, a single two-stage device such as a flip-flop, which functions as a divide-by-two counter, since one output pulse results from every two input pulses. Higher-order scaling is obtained by cascading stages.

binary search A system of search entailing the successive division of a set of items into two parts and the rejection of one of the two until all items of the sought-for kind are isolated.

binary signal Any signal that can attain either of two states. Such a signal is always a digital signal.

binary-to-decimal conversion 1. The automatic conversion of a number represented by a series of binary pulses into the corresponding decimal number which then is displayed by a readout device. 2. The arithmetic operating of converting a binary number into a decimal number; this can be done by noting the powers of 2 represented by the various binary digits in a number and then adding the decimal values of these powers. For example, the binary number 101 000 111 is converted as follows, starting with the first significant digit:

$$
\begin{array}{lll}
1 & 1 \times 2^8 = & 256 \\
0 & 0 \times 2^7 = & 000 \\
1 & 1 \times 2^6 = & 64 \\
0 & 0 \times 2^5 = & 000 \\
0 & 0 \times 2^4 = & 000 \\
0 & 0 \times 2^3 = & 000 \\
1 & 1 \times 2^2 = & 4 \\
1 & 1 \times 2^1 = & 2 \\
1 & 1 \times 2^0 = & \underline{1} \\
& & 327
\end{array}
$$

Thus, 101 000 111 = 327.

binary word A binary numeral that has a particular meaning, agreed upon by convention. For example, the letters A through Z may be represented by binary numbers 00001 through 11010; a word may be represented by several blocks of five digits.

binaural Literally, *two-eared*. In sound recording and reproduction, the transcription of a broad sound source using two microphones spaced at approximately the distance between the ears on a human head, and played back using headphones to re-create the stereo effect. The technique evolved into multichannel stereophonic reproduction.

binaural sound Binaural sound is the equivalent of a listener hearing a concert through a pair of earholes; it takes earphones to reproduce the signal. If speakers are substituted for the earphones, the listener hears monophonically, as if he were standing back 10 feet or so from the two earholes in the wall.

binder A material (such as lacquer) which acts as a holder and cohesive medium for the particles of another material. Example: Binders are used in carbon resistors, ceramic dielectric bodies, powder cores, and resistive and metallic paints.

binding energy A property of the nucleus of an atom. The binding energy of a nucleus is equal to the difference between the nuclear weight and the sum of the weights of the lighter particles making up the nucleus. (Relativity theory shows mass and energy to be interchangeable.) The nucleus is stable when the binding energy is high.

binding force Any one of the electrostatic forces which bind crystals together. See separate entries under *covalent binding forces, ionic binding forces,* and *metallic binding forces.*

binding post A screw-type terminal of various styles, having a hole into which a wire or tip may be inserted and gripped.

binistor A semiconductor switching device which exhibits two stable states and also negative resistance.

binomial An algebraic expression containing two terms joined by a plus (+) or minus (-) sign; thus, any polynomial with only two terms.

binominal array A form of directional array with very small minor-lobe amplitude and a bidirectional radiation pattern.

binomial theorem The theorem (of Newton) which permits a binomial to be raised to any desired power without performing the multiplications. Thus, the expansion of $(a \pm b)^n$ is: $(a \pm b)^n = a^n \pm na^{n-1}b + (n(n-1)/2!) (a^{n-2}b^2) \pm (n(n-1) (n-2)/3!) (a^{n-3}b^3) + \ldots na^{n-1}b^{n-1} + b^n$

where there are $n + 1$ terms at the right side of the equation. In electronics, power series are convenient for expressing such binomials.

biochemical cell A fuel-cell energy source in which electricity is generated chemically through the oxidation of biological substances. Also called *biochemical fuel cell.*

bioelectricity 1. Electric currents in living tissues, generated by the organism and not applied by external means. 2. The science or study of such currents.

bioelectrogenesis The study and application of electricity generated by living animals, including man, in the powering and control of electronic devices, as in man-machine systems.

bioelectronics Electronics in relation to the life sciences, especially the electronic instrumentation of biological experiments.

bioengineering The engineering of equipment (such as electron microscope, electroencephalograph, centrifuge, irradiator, etc.) for study and experiment in the life sciences, or of equipment (such as pacemaker, hearing aid, X-ray apparatus, shock-therapy unit, etc.) for aid or support of life processes.

biological shield A protective absorbent shield which minimizes the radiation from a radioactive source to prevent injury.

bioluminescence 1. The emission of light by a living organism. 2. The light itself so produced by living organisms.

biometrics Mathematics, and in particular statistics and probability, applied to biology.

bionics The study, design, and application of microelectronic systems which simulate the functions of living organisms.

biotelemetry The use of *telemetry* to collect data from living organisms or to direct their movement.

biotelescanner An instrument that monitors body functions via radio, from a great distance.

Biot/Savart law The law which expresses the intensity of the magnetic field (H) in the vicinity of a long, straight wire carrying a steady current (I) as $H = 2I/r$, where H is in oersteds, I in amperes, and r (the distance) in centimeters. Ampere's law may be derived from the Biot-Savart law, and vice versa.

bip Abbreviation of *binary image processor.*

biphase half-wave rectifier An alternative term for the *full-wave rectifier*; also, each leg of a two-diode full-wave rectifier.

BIPM Abbreviation of *International Bureau of Weights and Measures.*

bipolar The condition of possessing two pole sets. In a conventional (non-FET) transistor, one pole set exists between base and collector and another pole set exists between base and emitter.

bipolar driving unit A magnetic headphone or loudspeaker in which both poles (north and south) of a magnet actuate a diaphragm or lever.

bipolar operation See *automatic polarity.*

bipolar transistor A two-junction transistor whose construction takes the form of a pnp or an npn "sandwich." Such devices are current-operated compared with field-effect transistors, which are voltage-operated. The bipolar transistor (of which the familiar npn and pnp types are examples) uses both electron and hole conduction.

biquadratic equation See *quartic equation*.

biquinary code A variety of binary-coded-decimal notation in which seven bits are used to represent each decimal digit. The bits are written in two groups (one two-bit followed by one five-bit), and the positional values are 5 and 0 for the two-bit group, and 4, 3, 2, 1, and 0 for the five-bit group, as shown below for the decimal digit 7:

Biquinary

Decimal	5	0	4	3	2	1	0
7	1	0	0	0	1	0	0

biquinary decade A decade counter consisting of a binary stage followed by a quinary stage.

biradial stylus A stylus used for a phonograph, having an elliptical cross section.

bird 1. Slang for orbiting *satellite*. 2. Slang for *guided missile*.

birdie 1. A spurious beat note. 2. A parasitic oscillation.

Birmingham wire gauge Abbreviation, BWG. Also called *Stubs gauge*. A method of designating the various sizes of solid wire. BWG diameters are somewhat larger than corresponding American wire gage diameters for a given wire size.

bismuth Symbol, Bi. A metallic element. Atomic number, 83. Atomic weight, 209.

bismuth flux meter A flux meter in which the sensor contains a length of bismuth wire which acts as a magnetoresistor.

bismuth thermocouple A thermocouple employing the junction between bismuth and antimony wires and used in thermocouple -type meters.

bistable Having two stable states.

bistable device Any device, such as a flip-flop, the operation of which exhibits two stable states and which may be switched at will from one state to the other.

bistable multivibrator A multivibrator the operation of which exhibits two stable states. Also called *Eccles/Jordan circuit* (when a vacuum tube is employed) and *flip-flop*. Compare *astable multivibrator* and *monostable multivibrator*.

bistable relay A relay having two stable states: open and closed. Successive actuating pulses open and close the relay, two consecutive pulses being required to return the relay to a given state. Also called *binary relay*, *relay flip-flop*, and *electromechanical flip-flop*.

bistate Having two states. Example: the performance of a *flip-flop*.

bistatic radar A radar set in which the transmitting and receiving antennas are separated.

bit 1. In the binary system, the smallest unit of information, consisting of a 0 or a 1. (Formed from *binary digit*.) 2. Abbreviation of *built-in test* (capitalized as an acronym).

bit density Bits per unit area of volume, as the number of bits per square centimeter of magnetic tape.

BITE Abbreviation of *built-in test equipment*.

bit rate The speed at which bits are transmitted or handled. In data transmission the bit rate per second is referred to as *baud*.

bit-slice processor A microprocessor whose word or byte capacity is achieved through the use of interrelated smaller capacity processors, e.g., a 16-bit unit derived from eight 2-bit *slices*.

Bjerknes' equation An expression for the total (primary plus secondary) decrement of a tuned circuit, based upon measurements of the tank current at the resonant frequency and at a frequency near resonance:

$$\delta^1 + \delta = \pi \pm \frac{(\omega^2 - \omega^2 r)}{\omega\omega r} \sqrt{\frac{I_1^2}{I_r^2 + I_1^2}}$$

BK 1. Radiotelegraph signal for *break*. 2. Abbreviation of *break-in*.

B/K oscillator See *Barkhausen/Kurz oscillator*.

Bk Symbol for *berkelium*.

black-and-white Any system of image reproduction, transmission, or reception in which the image is composed of opaque elements (black) and white or bright areas, as in noncolor television reception.

black area An area in which there is only an encrypted signal.

blackbody An ideal black surface completely absorbing energy of any wavelength impinging upon it and reflecting no energy.

blackbody radiation Radiation from a heated ideal blackbody. This radiation is conceived as starting in the infrared region and shifting to shorter wavelengths, including the visible spectrum, as the temperature is increased.

black box 1. Any "box" or "block" which may be included in an analysis or synthesis based upon the *black-box concept*. 2. Any functional unit (such as a module) whose operating characteristics are known and which can be inserted into a system in development or maintenance operations. 3. Any subcircuit or stage which may be specified *in toto* as needed, in terms of its known or prescribed performance, in a system, but whose structure (or indeed present existence) need not be known.

black-box concept A technique for development of equivalent circuits and of considering their operation. The "box" has a pair of input terminals and a pair of output terminals, and one input terminal is often common to one output terminal. The contents of the box need not be known; but from the input and output current and voltage relationships, its nature may be determined. Moreover, from the available input signal and desired output signal, the internal circuit of the box can be specified.

black compression Attenuation of the level of dark areas in a television picture.

blacker than black The video-signal amplitude region above the level that just darkens the TV screen. Signal information (such as control pulses) in this region therefore are not seen.

black light A lamp which produces a principal portion of its radiation in the ultraviolet region.

blackout 1. Complete obliteration of signals or power. 2. Complete blanking of the screen of an oscilloscope or picture tube.

blackout A sudden loss of ac line power usually as result of an outage resulting from an overload or other power failure upstream (nearer the power source than where the blackout occurs).

black reference In a TV signal, the blanking level of pulses, beyond which the sync pulse is in the blacker-than-black region.

black reference level The voltage threshold of the black reference, i.e., its distance above zero volts.

black transmission A system of picture of facsimile transmission in which the maximum copy darkness corresponds to the greatest amplitude (in an amplitude-modulated transmitter) or the lowest instantaneous frequency (in a frequency-modulated transmitter). The opposite of white transmission.

blank 1. A piezoelectric plate cut from a quarter crystal, but not yet finished to operate at a desired frequency. 2. To obscure or interrupt a signal or electron beam (usually momentarily), as in Z-axis blanking in an oscilloscope. 3. A silicon wafer cut from a large slab, but containing dopants only. 4. An unrecorded phonograph record. 5. A location (such as a character or space) that is used to verify proper data processing character grouping and values.

blanketing Radio interference consisting of a complete blocking or "washing out" of signals throughout an entire frequency band, unaffected by tuning.

blanking Obscuring or momentary elimination of a signal (see *blank, 2*).

blanking interval The short period during which the electron beam of a crt is cut off so that it can return to its start position without creating a trace on the screen.

blanking level The discrete, predetermined level (usually a threshold voltage at which blanking takes place).

blanking pedestal In the horizontal pulse of a TV signal, the lower portion between zero volts and the blanking level.

blanking pulse A pulse which produces momentary blanking (see *blank, 2*).

blanking time The time interval during which the electron beam of a cathode-ray tube is interrupted by a blanking signal.

blank record A recording disk on which no material whatever has been recorded.

blank tape Magnetic tape that has never been subjected to the recording process and that is substantially free from noise.

blank wire Magnetic recording wire that has never been subjected to the recording process and that is substantially free from noise.

blasting Severe overloading of a sound system including the loudspeaker.

bleeder A resistor used permanently to drain current from charged capacitors, as in a power supply; it establishes the predetermined initial load level for a power supply or signal source and serves as a safety device.

bleeder current The current normally flowing through a bleeder.

bleeder divider Several resistors series-strung across the output of a power supply or its regulator. As a load resistor the bleeder improves regulation and protects against no-load voltage surges. The resistor junctions allow various voltages to be drawn from the supply.

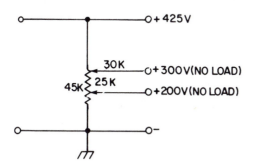

Bleeder divider.

bleeder power Power lost in a bleeder.

bleeder resistor See *bleeder*.

bleeder temperature Temperature rise in a bleeder due to normal power dissipation.

bleeding whites A flowing of the white areas of a TV picture into the black areas; an overload condition.

blemish See *burn*.

blind approach See *blind landing*.

blind flight The flying of aircraft entirely by means of instruments and electronic communications.

blind landing Landing of an aircraft entirely by means of instruments and electronic communications.

blind zone 1. In radar practice, an area which gives no echoes. 2. Skip zone (see *zone of silence*).

blip 1. The pulse-like figure on a radarscope scan, indicating the transmission or reflection (see *A-scan* and *J-scan*). Also called *pip*. 2. In visual alignment of a tuned circuit, using a sweep generator and marker generator, the pulse or dot produced on the response curve by the marker signal. 3. A short, momentary signal pulse, such as a single Morse dot. 4. Abbreviation for *background-limited infrared photoconductor* (capitalized).

blip-scan ratio The number of radar scans necessary to show a visible blip, or echo, on a radar screen.

Bloch wall The transition layer between adjacent ferromagnetic domains (see *domain*).

block 1. A group of data words or digits. 2. A group of memory storage spaces. 3. A circuit that operates as an identifiable unit. 4. The symbol for a circuit in a block diagram.

block diagram A simplified diagram of an electronic system in which stages are shown as functional two-dimensional boxes with the wiring and detail circuitry omitted.

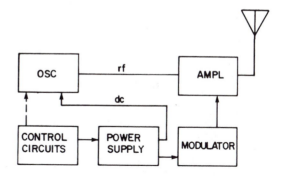

Block diagram of radio transmitter.

blocked grid The condition in which a vacuum tube goes out of operation because of electrons accumulated on the control grid. Operation is resumed when these electrons leak off.

blocked-grid keying Keying of the rf output amplifier of a radiotelegraph transmitter by means of cutoff grid bias. When the key is open, the bias is high enough to reduce the amplifier output to zero; but when the key is closed, the bias is lowered and the amplifier delivers full output.

blocked impedance The input impedance of a transducer whose output load is an infinite impedance.

blockette In a computer, the subdivision of a character block that will be handled as a unit during data transfer.

blocking action Obstruction of circuit action (usually abruptly) through internal action or by the application of an external signal. Thus, the operation of an amplifier may be blocked (output reduced to zero) by an input signal or by excessive feedback, either of which overloads the input.

blocking capacitor A capacitor inserted into a circuit to prevent the passage of direct current while easily passing alternating current.

blocking choke Any inductor, such as a choke coil, employed to prevent the flow of an alternating current while allowing direct current to pass with little resistance.

blocking interference Radio interference from signals strong enough to reduce the receiver output through blocking action.

blocking oscillator An oscillator which turns itself off after one or more cycles. It does this as a result of an accumulation of negative charge on its input electrode (grid of a tube, base of a bipolar transistor, gate of a field-effect transistor). The action is repetitive. In the *self-pulsing* type of blocking oscillator, a series of pulses consisting of trains of sine waves with intervening spaces is generated. In the *single-swing* type of blocking oscillator, the output consists of a series of single cycles with long intervals between them.

blocking oscillator synchronization 1. In the blocking oscillator used in the vertical deflection circuit of a TV receiver, synchronization of the oscillator with vertical sync pulses arriving in the video signal. 2. Synchronization of the repetition rate of any blocking oscillator with a suitable external control signal.

block length The number of characters, bits, or words comprising a defined unit word or character group.

block transfer The conveyance of a word or character grouping in a computer register to another register or a peripheral device.

blooming On a cathode-ray screen, the (usually fuzzy) enlargement of the electron-beam spot, due to defocusing.

blooper 1. A radio receiver that is in oscillation, and is transmitting a signal that causes interference. 2. A parasitic oscillation in a radio transmitter. 3. A statement in which a radio announcer makes an embarrassing error or breach of etiquette.

blow The opening of a fuse or circuit breaker because of current that is too great.

blower A fan employed for the purpose of removing heat from electronic circuits. Used especially in tube type amplifiers, where much heat is generated.

blowout 1. An alternate term for *burnout*. 2. The usually forceful opening of a circuit breaker. 3. The extinguishing of an arc.

blowout coil An electromagnet which provides a field to extinguish an arc.

blowout magnet A permanent magnet which performs the same function as a blowout coil.

blst Abbreviation of ballast.

blue-beam magnet In a color TV picture-tube assembly using three electron guns, a small permanent magnet for adjusting the static convergence of the beam for blue phosphor dots.

blue box An accessory device (sometimes unlawfully used) which generates tones that switch a telephone circuit in the placing of calls.

blue glow 1. The bluish glow between anode and cathode of a gassy vacuum tube. 2. The normal color of the gas discharge in an argon glow lamp. 3. The normal color of the discharge that fills a mercury-vapor tube or mercury-vapor rectifier. 4. In a neon lamp (which normally exhibits a red glow), a bluish light resulting from high-voltage arcing.

blue haze See *Blue glow, 1.*

blue gun The electron in a three-gun color TV picture tube, the beam from which strikes the blue phosphor dots.

blueprint 1. A type of contact-print reproduction in which usually a sheet of sensitized paper is exposed to an image on a translucent or transparent film, under strong light, and is then developed and fixed. Although this process is still used to reproduce electronic illustrations and typescripts, it has been superseded largely by other (dry) processes. 2. Loosely, any plan or design for the development of a system.

blue restorer In a three-gun color TV circuit, the dc restorer in the blue channel.

blue ribbon program A computer program that has been hand-prepared and debugged completely before its first computer run.

blue video voltage The signal voltage presented to the grid of the blue gun of a three-gun color TV picture tube.

blurring 1. Blooming. 2. A defocusing of a TV picture or oscilloscope trace. 3. An obscuring of a signal by echoes or trailing (as in the slow dying out of a dot or dash).

BM Abbreviation of *ballistic missile.*

BMEWS Abbreviation of *ballistic missile early warning system.*

B-minus The negative terminal of a B-power supply.

B-minus bias See *back bias, 3.*

BNC A type of antenna feedline connection, especially common in VHF and UHF radio communications. It is an abbreviation of *bayonet Neill-Concelman,* who invented the connector.

B-negative See *B-minus.*

BNL Abbreviation of *Brookhaven National Laboratory.*

BO Abbreviation of *beat oscillator.* Also abbreviated *bfo.*

board 1. A panel containing patch jacks. 2. A circuit board.

boat A type of crucible in which a semiconductor material is melted and sometimes processed. The material of which the boat is made (e.g., graphite) does not react with or contaminate the semiconductor.

bobbin A usually nonmetallic spool on which a coil is wound.

Bode plot A pair of curves plotted to the same frequency axis, one showing the gain of a network or amplifier and the other its phase shift. Phase and amplitude of active and passive networks can be exhibited. Also called *Bode curve* and *Bode diagram.*

body-antenna effect The tendency of the human body to act as a receiving antenna when a finger is touched to the antenna input terminal of a receiver or when a hand (or the whole body) is brought close enough to the circuit to provide capacitive coupling.

body capacitance Capacitance between the body of the operator (as one plate of an equivalent capacitor) and a piece of electronic equipment (as the other plate). This phantom capacitance is often the cause of detuning and of the injection of interfering signals and noise (the body acting as a pickup antenna).

body electrode 1. An electrode attached to the human body (or to the body of a laboratory animal) to conduct body-generated currents to an instrument, as in cardiography, electroencephalography, and myography. 2. An electrode attached to the human body (or to the body of a laboratory animal) to conduct currents into the body, as in shock therapy and skin-resistance measurement.

body leakage Leakage of current through the bulk or body of a dielectric material, as opposed to *surface leakage.*

body temperature 1. In a thermistor, a rating which is the temperature measured on the surface of the device and is any combination of ambient temperature, power dissipation, and operation of the internal heater element if the thermistor has one. 2. Normal temperature of the human body, generally considered as 98.6°F or 37°C.

bof Abbreviation of *barium oxide ferrite.*

boffle A loudspeaker enclosure made up of stretched screens which are sound absorbing and elastic.

bogey See *bogie.*

bogie 1. The exact value of a specified characteristic. Thus, if resistance is given as 1K + 1%, − 0.5%, the bogie value is 1K. 2. The average value, i.e., the arithmetic mean. 3. A false or unidentified echo on a radar screen.

Bohr atom The concept of the nature of the atom, proposed by Niels Bohr in 1913 partly to explain why the electrons in the Rutherford atom do not fly off into space or fall into the nucleus. The Bohr theory places the electrons in permissible orbits where they cannot radiate energy (see *Bohr radius*). They can radiate or absorb energy, however, if they go to a lower orbit or to a higher orbit, respectively. Compare *Rutherford atom*.

Bohr magneton Symbol, μ B. Unit of the quantized magnetic moment of a particle. 1 μB = 9.274 096 × 10^{-24} J/T.

Bohr model See *Bohr atom*.

Bohr radius Symbol, $a0$. A physical constant whose value is 5.291772 × 10^{-11} meter (to an error of about 1.5 ppm).

boiling point Abbreviation, bp. The temperature at which a liquid vaporizes. The boiling point of water is 100 °C or 212 °F.

bolometer Any device which is essentially a small, nonrectifying, temperature-sensitive resistor that can be used for heat sensing, rf power measurement, curve changing, demodulation, etc. Included in this category are *barretters, thermistors,* and wire-type *fuses.*

bolometer bridge A dc bridge in which a bolometer is one of the four arms. The bridge is balanced first with the bolometer cold. The bolometer then is excited with rf current, whereupon the resultant heating changes the bolometer resistance and the bridge next is rebalanced for this new resistance. The rf power driving the bolometer may be determined from a previous calibration of the bridge.

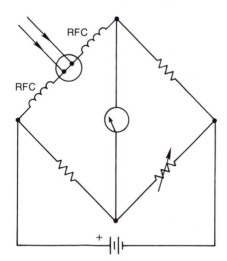

Bolometer bridge.

Boltzmann constant Symbol, k. A figure which enters into the calculation of thermionic emission and of thermal noise factor. It represents the temperature equivalent of work function, in joules per kelvin; k = 8.63 × 10^{-5} V/K = 1.38 × 10^{-23} J/K.

Boltzmann's principle A description of the statistical distribution of large numbers of tiny particles under the influence of a force, such as an electric or magnetic field. When the system is in statistical equilibrium, the number of particles in any portion of the field is

$$N\text{E} = No \text{ e } - \text{ E/kt}$$

where E is the potential energy of a particle in the observed area, No is the number of particles per unit volume in a part of the field where E is zero, k is the Boltzmann constant, T is the absolute temperature of the system of particles, and e = 2.718.

bombardment The usually forceful striking of a target with rays or a stream of particles.

bond 1. An area in which two or more items are securely and intimately joined. 2. The attractive force that holds an atomic or subatomic particle or particle group together.

bonded-barrier transistor A bipolar transistor in which the connection at the base region is alloyed.

bonded negative-resistance diode A diode that displays a negative-resistance characteristic over part of its current curve. This results from avalanche breakdown.

bonding 1. The formation of bonds between adjacent atoms in a crystalline material, such as a semiconductor. See specifically *covalent binding forces, ionic binding forces,* and *metallic binding forces.* 2. The secure fastening together of conducting surfaces, as by soldering or brazing, to give a high-conductance, leak-free continuum.

bond strength The minimum stress required to separate a material from another to which it is bonded.

bone conduction Also called conductive hearing. The process that allows hearing because of acoustic conduction through the bones of the skull.

bone-conduction transducer A device used in place of the earphone in a hearing aid to convey sound energy to the bone structure of the head.

book capacitor A variable capacitor in which the metal plates are bonded along one edge and separated from each other by means of mica sheets. The capacitance is varied by opening and closing the assembly book fashion.

Boolean algebra A system of symbolic logic in which modern digital-computer techniques have many roots.

Boolean calculus A form of Boolean algebra, in which time is a parameter.

Boolean function In mathematical logic, a function that makes use of Boolean algebra or Boolean calculus.

boom 1. A horizontal support for a microphone, enabling the latter to be suspended over a sound source but out of sight of a camera. 2. A horizontal support for a small antenna which is undergoing tests or sampling the field of another antenna. 3. The principal and usually central horizontal element of a yagi-type antenna, which establishes the center of gravity, ground reference, and directional axis of the radiation pattern; the reflector and directors become the cross members.

boost B-plus See *B-plus boost.*

boost capacitor In the damper circuit of a TV receiver, a capacitor employed to boost the B+ voltage (see *B-plus boost*). Also called *booster capacitor.*

boost charge A high-current, short-interval charge used to revitalize a storage battery quickly. Also called *booster charge.*

booster 1. Any device serving to increase the amplitude of a signal (e.g., as an amplifier or preamplifier) or of an energy source (e.g., to boost the output of a power supply). 2. An rf amplifier used ahead of a television receiver.

booster battery 1. A battery used to forward bias a diode detector into a favorable region of its conduction curve or to bias a bolometer into the square-law region of its response. 2. A battery supplying power to a *booster*.

booster gain The amplification (usually in terms of voltage gain) provided by a booster (see especially *booster, 2*).

boot A usually flexible protective nipple or jacket pulled over a cable or connector, so called from its resemblance to a foot boot.

boot loader A form of computer program that operates on the bootstrap routine.

bootstrap A technique for making a device or process achieve a condition through its own actions; see *bootstrap circuit,* for example.

bootstrap circuit A specialized form of follower circuit presenting very high input impedance. Its chief feature is the return of the control-element resistor to a tap on the cathode-element resistor. The technique may be employed with tubes or transistors, and takes its name from the idea that such a circuit lifts its input impedance by its own bootstraps.

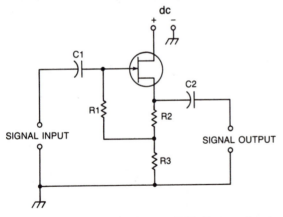

Bootstrap circuit (with junction-type field-effect transistor).

bootstrap routine A computer routine in which the first few instructions put in storage are later used to complete the routine, as supplemented by some sort of operator instruction. Also, a program part used to establish an alternate version of the program.

bootstrap voltage See *B-plus boost.*

borax-aluminum cell An electrolytic cell consisting essentially of an aluminum electrode and a lead electrode in a saturated solution of sodium tetraborate (borax). After electroforming, such a cell can be used either as a rectifier or as an electrolytic capacitor.

boresighting The alignment of a radar antenna, or other highly directional antenna system that has a pattern in both the azimuth and elevation planes. The boresight is the direction in which maximum gain takes place.

boric acid Formula, H_3BO_3. A compound used variously in electronics, especially as the electrolyte in electrolytic capacitors.

bornite Formula, Cu_5FeS_4. A natural mineral which is a sulfide of copper and iron. Its crystalline structure made it important in early

semiconductor diodes (crystal detectors).

boron Symbol, B. A metalloidal element. Atomic number, 5. Atomic weight, 10.82. Boron is familiar as a dopant in semiconductor processing.

bot 1. Abbreviation for *beginning of tape*. 2. Abbreviation of *bottom*.

bottom bend rectification In triode and pentode tubes, signal rectification (detection) obtained by operating the tube near plate-current cutoff. The rectification efficiency is low, due to the slow curvature of the plate-current characteristic in the vicinity of cutoff; but the amplification provided by the tube compensates for this deficiency.

bottoming Excessive movement of the cone of a loudspeaker or of the diaphragm of a headphone so that the magnet or supporting structure is struck by the moving-coil piston assembly.

bounce 1. The springback or vibration of the armature of a relay on closure. 2. An abnormal, abrupt change in the brightness of a TV picture.

boundary 1. In a polycrystalline substance, the area of contact between adjacent crystals. 2. The area of meeting of two regions (such as n and p) in a semiconductor.

boundary defect A condition in which a piezoelectric crystal has two regions, intersecting in a plane, with different polarizations.

bound charge The portion of a charge on a conductor which, because of induction from neighboring charge, will not escape to ground when the conductor is grounded. Compare *free charge.*

bound electron An electron which is held tightly in its orbit within an atom so that it is not ordinarily free to drift between atoms and thus contribute to electric current flow.

bow-tie antenna A center-fed antenna in which the two horizontal halves of the radiator are triangular plates resembling a bow tie. A flat reflector consisting of closely spaced horizontal wires is mounted back of the triangles.

bow-tie test An oscilloscope-display checkout of a single-sideband signal in which the appearance of the display gives an indication of signal quality. The transmitter output signal is fed to the scope's vertical deflection plates and the exciter's audio output is fed to the horizontal sweep input of the scope.

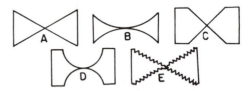

Bow-tie test patterns. A is normal indication; all lines are perfectly straight. Too much bias produces concave-sided pattern such as B. Overload produces pattern C, with chopped-off peaks rather than points. Combination of overdrive and too much bias produces Pattern D. Inadequate carrier suppression puts ripples on sides of pattern as at E.

boxcars Long pulses with short separating spaces between them.

box-shaped loop The characteristic square-loop hysteresis curve (B-H loop) resulting when a sine wave of current is used to magnetize a sample of magnetic material. In this plot, which covers all four quadrants, the horizontal axis (H) displays magnetizing force, and the vertical axis (B) displays magnetization. Also see *hysteresis.*

bp 1. Abbreviation of *boiling point.* 2. Abbreviation of *bandpass.*

bpi Abbreviation of *bits per inch*.

B-plus 1. Symbol, B+. The positive dc voltage required for certain electrodes of tubes, transistors, etc. Compare *A-plus*. 2. The positive terminal of a B-power supply.

B-plus boost The additional positive voltage which is added to the low dc B+ voltage in a TV receiver by the action of the damper tube.

B-plus-plus Symbol, B++. See *B-plus boost*.

B-positive See *B-plus*.

B-power supply A name used sometimes for the unit that supplies high-voltage dc energy to a tube plate or screen circuit. Compare *A-power supply*.

bps Abbreviation of *bits per second*.

Br Symbol for *bromine*.

bracketing A troubleshooting routine characterized by isolating progressively smaller areas in a circuit or chain of stages until the defective subcircuit or stage is located.

Bradley detector A locked-oscillator circuit formerly used as an FM detector.

Bragg angle See *Bragg's law*.

Bragg's law The law that expresses the condition under which a beam of X-rays is reflected with maximum directness by a crystal, and the angle of reflection: $\sin \theta = n\lambda/2d$, where θ is the Bragg angle (the complement of the angle of incidence or angle of reflection), d is the distance the layers or planes of atoms are spaced apart, and λ is the X-ray wavelength. It is assumed that n is a whole number.

braid A woven network of fine metal wires used for grounding purposes. It is usually made of fine copper conductors. The increased surface-area-to-volume ratio improves the conductivity, at radio frequencies, over a single conductor having the same cross-sectional area.

braided wire A length of braid. Used for grounding or shielding purposes.

brain waves Alternating or pulsating potentials due to brain activity and picked up by electrodes attached to the scalp. The waves may be amplified to be viewed on a cathode-ray screen, heard by headphones or speakers, or traced by an electroencephalograph.

branch 1. Any one of the separate paths of a circuit. With respect to the layout of its components, a branch may be series, parallel, series-parallel, parallel-series, or any combination of these. Also called *leg*. 2. A *branch circuit*.

branch circuit In electrical wiring, a group of outlets served through a single cutout from a source of power-line ac voltage. The source can be a distribution center, subdistribution center, main, or submain. Interior lighting circuits are usually branch circuits since many lights are connected to one circuit controlled by a single fuse or circuit breaker.

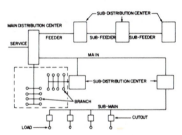

Branch circuit (enclosed in broken lines).

branch current Current flowing through a branch of a circuit and whose magnitude with respect to the total current of the circuit depends upon the nature of the branch.

branched In molecular polymers, the condition of side chains being attached to the main chain.

branched windings Forked windings of a polyphase transformer.

branch point See *junction point*.

branch voltage The voltage, or voltage drop, across a branch of a circuit.

brass 1. An alloy of copper and zinc widely used in electronics. Compared to annealed copper, this metal has four times the resistivity, half the temperature coefficient, more than twice the tensile strength, and a lower melting point (900 °C). 2. Colloquialism for *telegraph key*.

brass pounder Telegraph operator or radiotelegraph operator.

Braun electroscope An electroscope consisting essentially of a fixed metal vane to which is fastened at a pivot a movable needle. The repulsion between the two when an electric charge is applied causes the needle to move over a calibrated scale.

brazing The joining of two metal (usually iron or steel) parts together with a suitable melted copper-alloy metal. Compare *soldering*.

breadboard 1. A perforated board, a chassis, or any basic framework on which electronic components may be mounted and quickly wired for the preliminary test of a circuit, so called because the first such foundation units of this sort actually were wooden breadboards. Hence, any preproduction prototype circuit. 2. To set up a circuit on a breadboard.

breadboard model 1. The preliminary model of an electronic device, built on a breadboard (see *breadboard*, 1). 2. Loosely, any prototype.

break 1. An open circuit. 2. To open a circuit. 3. In communications, a word implying a desire to transmit on a wavelength already occupied by radio traffic. 4. *Break-in* 1.

break-before-make contacts Contacts, especially in a rotary selector switch, which open one circuit before closing the next one.

breakdown 1. Failure of a circuit or device, due principally to excessive voltage, current, or power. A sudden high current, however, does not always indicate failure (see, for example, *avalance breakdown* and *zener breakdown*). 2. The separation of an electronics problem or project into its constituent parts for more ready solution.

breakdown diode See *avalanche diode* and *zener diode*.

breakdown impedance See *avalanche impedance*.

breakdown region The region, in a pn junction, in which avalanche breakdown occurs.

breakdown strength See *dielectric strength*.

breakdown voltage 1. The voltage at which current suddenly passes in destructive amounts through a dielectric. 2. The voltage at which a gas suddenly ionizes, as in a gas tube. 3. The voltage at which the reverse current of a semiconductor junction suddenly rises to a high value (nondestructive if the current is limited). See *avalanche breakdown* and *zener breakdown*.

break-in 1. A technique of radio communication in which one station interrupts a transmission from another station, rather than waiting until the end of the latter's transmission. 2. Burn-in.

break-in keying A system of radiotelegraph keying in which the receiver is in operation whenever the key is open, permitting break-in operation (see *break-in* and *break-in operation*).

break-in operation In radiotelegraph, the practice of interrupting at any time to "talk back" to the other transmitting station. This operation is made possible by some forms of break-in keying.

break-in relay An electromechanical or solid-state relay which enables break-in operation.

breakoff voltage The voltage at which the discharge in a gas tube suddenly ceases.

breakover point In a silicon controlled rectifier, the source-voltage value at which the load current is suddenly triggered to its steep climb. Also called *triggering point*.

breakover voltage In a silicon controlled rectifier with open gate circuit, the anode voltage at which anode current is initiated.

breakpoint A point in a computer program when, for the purpose of obtaining information for the program's analysis, the sequence of operations is interrupted by an operator or a monitor program.

breakpoint frequencies The upper- and lower-frequency points at which the gain-versus-frequency response of an amplifier or network departs from flatness.

breakpoint instruction An instruction that stops a computer.

breakthrough 1. A new discovery, insight, or solution to a problem, which results in an advancement in the state of the art. 2. Punchthrough. 3. Breakdown.

break time The time taken for a relay to drop out completely, or a switch to open. Compare *make time*.

breathing Slow, rhythmic pulsations of a quantity, such as current, voltage, brightness, beat note, etc.

breezeway That part of the back porch of the sync pulse in NTSC color television between the trailing edge of the pulse and the color burst.

B-register An index register in a computer for storing words that are used to change an instruction before it is executed by the program.

Bremsstrahlung radiation The radiation emitted by a charged particle whose speed is altered by the presence of a proton.

brevity code A form of code that is not intended to conceal information, but simply to shorten the length of a message. The *Q signals* are an example of a brevity code.

Brewster angle From Brewster's law, the polarizing angle at which (when light is incident) the reflected and refracted rays are perpendicular to each other. Also see *pseudo-Brewster angle*.

Brewster's law (Sir David *Brewster*, 1781-1868). For any dielectric reflector, the relationship in which the refractive index is equal to the tangent of the polarizing angle.

bridge 1. A network of usually four "leg" components connected so that an input signal may be applied across two branches in parallel and an output signal taken between two points, one on each of the parallel branches. At some ratio of the resultant four arms of the circuit the output points are equipotential and the output voltage accordingly zero. The bridge is then said to be *balanced* or *set to null*. 2. A circuit such as that described in *1*, above, used for electrical measurements. 3. An audio or servo amplification system in which the load is driven from two oppositely polarized outputs, neither of which are at ground potential.

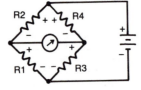

Bridge, 1. X-Y is at null (zero voltage) only when R1:R3 = R3:R4.

bridge balance control The potentiometer or variable capacitor used to adjust a bridge circuit to null.

bridge-balanced dc amplifier A circuit in which a bridge is employed to buck out the quiescent dc voltage of a tube or transistor from the amplifier output.

bridge-connected amplifier 1. A dc amplifier stage in which the transistors and resistors are connected in a four-arm bridge circuit with respect to dc. When the bridge is initially balanced, all dc is eliminated in the output load. The input signal unbalances the bridge and this results in an amplified output signal in the load. 2. An amplifier pair having opposing outputs across which a load may be bridged for double the power output of either amplifier alone.

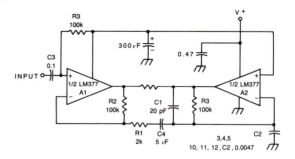

Bridge-connected amplifier.

bridged differentiator See *Hall network*.

bridge detector The output-indicating device (e.g., meter, oscilloscope, headphones) that shows whether the bridge is balanced or unbalanced. Also called *null detector* or *null indicator*.

bridged integrator A null network consisting of a two-stage resistance-capacitance integrator circuit bridged by a capacitor. This network gives a shallow null at a single frequency determined by the R and C values in the integrator. Compare *bridged differentiator*.

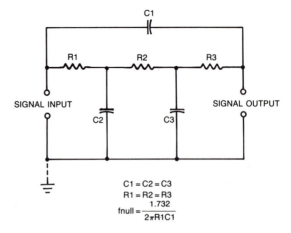

Bridged integrator.

bridged-tee attenuator An attenuator consisting of a tee section between the input and output of which is bridged a single series arm.

bridged-tee circuit Any circuit (of resistors, capacitors, inductors, or a combination of these) which consists of a tee section bridge by a single series section from input to output.

bridged-tee null network A bridged-tee circuit of resistances and reactances so proportioned that at some setting of the R and C values, the output of the circuit (like that of a balanced bridge) is zero.

bridged-tee oscillator A low-distortion oscillator circuit whose frequency is determined by a bridged-tee null network inserted into the negative-feedback path of the circuit.

bridge feedback A combination of current feedback and voltage feedback around an amplifier circuit. It is so called because in the feedback circuit the feedback resistors and the output resistance of the amplifier form a four-arm bridge.

bridge generator The power source (e.g., a battery or oscillator) that supplies the signal to a bridge used for electrical measurements. (see *bridge, 2*).

bridge indicator See *bridge detector*.

bridge oscillator See *bridge generator*.

bridge rectifier A full-wave rectifier circuit in which four rectifying diodes are connected in a bridge configuration. Each half-cycle of ac input is rectified by a pair of diodes in opposite quarters of the bridge and in series with each other.

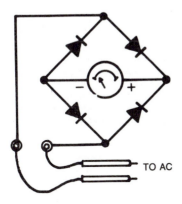

Bridge rectifier.

bridge source See *bridge generator*.

bridge-type of meter A frequency-sensitive bridge (such as the Wien bridge) which can be used to measure audio frequency. Since the bridge can be balanced at only one frequency at a time, its adjustable arm may be calibrated to read directly in frequency.

bridge-type impedance meter An impedance-measuring circuit in which the unknown impedance Z is connected in series with a calibrated variable resistor R and an ac voltage is applied to the series circuit. The separate voltage drops across the resistor and impedance are measured successively as the resistor is varied. When the two voltage drops are identical, Z equals R and may be read from the dial of the calibrated resistor.

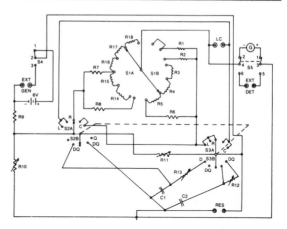

Bridge-type impedance meter.

bridge-type oscillator A resistance-capacitance tuned oscillator in which a Wien bridge is used as the frequency-determining circuit in the feedback loop.

bridge-type power meter 1. See *bolometer bridge*. 2. A four-arm bridge specially designed to operate at radio frequencies. At null, the impedance of the unknown is read directly from the balancing dial or calculated from bridge constants. This instrument is used to measure the impedance of circuit components, antennas, and transmission lines.

bridge-type swr meter A four-arm bridge specially designed to operate at radio frequencies. At null, the voltage standing-wave ratio (swr) is calculated from the bridge resistance values or read from a direct-reading scale on the null-indicating meter.

bridging amplifier An amplifier whose input impedance is so high that the amplifier may be connected across a load or line with virtually no disturbance.

bridging coupler A voltage-dependent resistor which permits an occasionally used device (such as a bell) to be connected permanently across a regularly used device (such as a telephone) without continuously short-circuiting the latter. Thus, the bridging coupler ordinarily has very high resistance; but when the line voltage is momentarily raised, the resistance lowers and the bell rings.

bridging gain The gain of a bridging amplifier expressed as the ratio (in decibels) of the power developed in the amplifier load to the power in the load to which the input terminals of the amplifier are connected.

bridging loss Loss resulting from the shunting of a speaker, microphone, earphone, or other transducer by a resistor, capacitor, or inductor. Generally, the loss is expressed as a power ratio in decibels.

Brierly ribbon pickup A phonograph pickup based upon two gold ribbons backed with a self-damping material, and similar to the ribbon microphone.

Briggsian logarithm (Henry *Briggs*, 1556-1631). A logarithm to the base 10. Also called *common logarithm*. Compare *natural logarithm*.

Briggs system of logarithms See *Briggsian logarithm*.

brightness SI unit, candelas per square meter (cd/m^2); cgs unit, lambert (Lb). The quantity of light given out perpendicular to the surface emitting it (whether by radiation or reflection) per unit area of surface.

brightness control 1. In a TV receiver or oscilloscope, the potentiometer that varies the negative bias voltage on the control grid of the cathode-ray tube, the brightness of the image being inversely proportional to the voltage. 2. The control of the brightness of an illuminated area.

brilliance See *brightness.*

brilliance control 1. The brightness control in a television receiver. 2. The brightness control in a cathode-ray oscilloscope. 3. A control for adjusting the level of the tweeter output in a speaker system.

British thermal unit Abbreviation, Btu. That amount of heat required to raise the temperature of a pound of water by a single degree (Fahrenheit) in an ambient environment of slightly greater than 39 °F.

broadband Possessing a characteristic wide bandwidth relative to other circuits or devices of a given genre.

broadband amplifier An amplifier having very wide frequency response, such as dc to 10 MHz. Examples: *instrument amplifier* and *video amplifier.*

broadband antenna An antenna that operates satisfactorily over a comparatively wide band of frequencies without requiring retuning at individual frequencies.

broadband electrical noise Electrical noise which is present over a wide frequency spectrum.

broadband i-f A comparatively broadly tuned intermediate-frequency amplifier. The wide frequency response is important when an increased bandpass is preferred to high selectivity, as in high-fidelity radio tuners.

broadband interference Interference other than noise that is present over a wide band of frequencies.

broadband klystron A klystron oscillator with a broadbanded tuned circuit.

broadband tuning Tuning characterized by a selectivity curve having a pronounced flat top or broad nose which passes a wide band of frequencies. Also called broadband response.

broadcast 1. A radio-frequency transmission of an intelligence-bearing signal that is directed not to one receiving station but to all receivers. 2. To make such a transmission.

broadcast band Any band of frequencies allocated for broadcasting (see *broadcast service, 1*), but particularly the U.S. AM and FM radio broadcast bands (535 to 1605 kHz (AM) and 88 to 108 MHz (FM).

broadcast endorsement An FCC modification to a Third Class operator permit, which allows the license holder to operate AM stations with an output power of 10 kW or less and FM stations with 25 kW or less of output power.

broadcasting The dissemination of signals for reception by the general public, and not for communications purposes.

broadcast interference Abbreviation, BCI. Interference to normal reception by broadcast receivers, usually arising from signals emitted by other stations.

broadcast receiver A receiver intended primarily for picking up standard broadcast stations. Also see *broadcast band.*

broadcast service 1. Generically, any radio transmitting service (including television) which sends out emissions for general reception rather than addressing them to specific receiving stations. 2. Particularly, the service provided by a station operating in the *broadcast band.*

broadcast station Any station in the broadcast service but especially one assigned to operate in the standard U.S. broadcast bands. Also called *broadcasting station.*

broadcast transmitter A (particularly radio) transmitter specially designed for and operated in the broadcast service.

broad response Slow deflection of an indicator, such as a meter, over a relatively wide range of values of the input quantity.

broadside In a perpendicular direction; for example, broadside radiation from an antenna.

broadside array An antenna array so designed that maximum-signal radiation is in the direction broadside to the array.

broad tuning Tuning characterized by pronounced signal width which results in adjacent-channel interference. A common cause of such impaired selectivity is low Q in the tuned circuit(s).

bromine Symbol, Br. A nonmetallic element of the halogen family. Atomic number, 35. Atomic weight, 79.92.

bronze An alloy of copper and tin which has various uses in electronics. Also see *phosphor bronze.*

Brown and Sharpe gauge See *American wire gauge.*

Brownian movement (Robert *Brown*, 1773-1858). Random movement of microscopic particles, especially in solutions. Einstein showed a connection between this movement and the Boltzmann constant.

brownout A deliberate lowering of line voltage by a power company to reduce load demands.

Bruce antenna An alternately bent antenna in the vertical plane, consisting of alternate quarter- and eighth-wave sections fed in such a way that current in adjacent vertical sections is in phase. Radiation is broadside to the antenna and is vertically polarized.

brush A usually metal or carbon strip, blade, or block providing sliding contact with another part, as a motor commutator.

brush discharge A discharge of electricity (often giving a glow of fine, hairy lines) into the air from a pointed conductor. It consists of a wind of repelled ions.

brush holder The housing for a *brush* in a motor, generator, rheostat, sliping junction in rotating data-transmission systems, and so on.

brute force 1. The transmission of a signal of excessive or unnecessary power. 2. An inefficient approach to a problem, which may solve the problem, but requires far more energy, effort, or computer memory space than the minimum needed to accomplish the same result.

brute-force filter A pi-type dc power supply filter of the low-pass type, so called because of large *L* and *C* values often selected through rule-of-thumb methods.

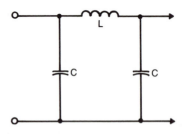

Brute force filter.

brute supply An unregulated power supply.

B-scope A radarscope that presents a B-display.

B service A teletype communication system operated by the Federal Aviation Administration (*FAA*).

bst Abbreviation of *beam-switching tube.*

B-substitute See *B-eliminator* and *battery eliminator.*

B-supply The dc power supply which provides anode operating voltages, such as tube plate and screen voltages. Compare *A-supply.*

BT-cut crystal A piezoelectric plate cut from a quartz crystal at an angle of rotation (about the X-axis) of $-49°$. It has a zero temperature coefficient of frequency at approximately 25 °C. Also see *crystal axes* and *crystal cuts.*

Btu Abbreviation of *British thermal unit.*

BuAer Abbreviation of *Bureau of Aeronautics.*

bubble memory In digital-computer practice, a special type of static magnetic memory. The magnetic material is divided into regions which are magnetized in different directions. The resultant domains can be formed into small bubbles which then constitute the memory elements.

bubble shift register A shift register employing a magnetic bubble (see *bubble memory*) which can be moved sequentially from electrode to electrode on a wafer.

bubbling See *motorboating.*

bucket A computer memory, or a designated location in such a memory.

bucking The process of counteracting one quantity, such as a current or voltage, by opposing it with a like quantity of equal magnitude but opposite polarity.

bucking circuit 1. A circuit used to obtain bucking action. The simplest form is perhaps a battery and potentiometer to supply a variable voltage of polarity opposite to that of the voltage to be bucked. Another such circuit is the bridge. 2. The zero-set circuit in an electronic voltmeter.

bucking coil A coil which is placed and positioned so that its magnetic field bucks and cancels the field of another coil. Troublesome hum fields sometimes are neutralized with such a coil.

bucking voltage See *back voltage, 2.*

buckling The warping of storage-battery plates usually from excessive charge or discharge.

buckshot In an AM radio transmission, broadband signal splatter caused by excessive modulation or detuned multiplier circuits.

buffer 1. An amplifier used principally to match two dissimilar impedance points and isolate one stage from a succeeding one in a cascaded system, and thus to prevent undesirable interaction between the two. 2. In a digital computer, a storage site used temporarily during data transfers to compensate for differences in data flow rates. 3. In digital-computer practice, a follower stage employed to drive a number of gates without overloading the preceding stage.

buffer amplifier See *buffer, 1.*

buffer capacitor A high-voltage fixed capacitor employed across a transformer secondary to suppress voltage spikes and sharp waveforms, especially when the input is a square wave.

buffer circuit 1. In a data system employing a keyboard, an electronic circuit that allows the operator to type ahead of the data output. 2. A buffer.

buffered output An output (power, signal, etc.) that is delivered from the generating device through an isolating stage, such as a *buffer amplifier.* This arrangement protects the device from variations in the external load. Compare *unbuffed output.*

buffer storage 1. A buffer employed for the purpose of interfacing between two data systems having different rates of transmission. 2. A buffer.

bug 1. Slang for *wiretap, 1.* 2. Slang for *circuit fault, 1.* 3. Contraction of *bug key.* 4. Slang for an error in a computer program.

bug key Semiautomatic telegraph key, either *mechanical* or *electronic.*

building-block technique The process of assembling an electronic equipment by quickly connecting together already completed stages (in the form of boxes or blocks) and supplying power and signals to the setup. Also called *modular technique.*

building-out circuit A short section of transmission line shunting another line; it is used for impedance matching. Also called *building-out section.*

building-out section See *building-out circuit.*

buildup 1. The process whereby the voltage of a rotating generator starts at a point determined by the residual magnetism of the machine and gradually increases to a voltage representing the point at which the resistance line crosses the magnetization curve. 2. The (usually gradual) accumulation of a quantity, e.g., the buildup of charge in a capacitor.

bulb A globe-like container having any of a number of characteristic shapes from spherical to tubular and usually evacuated, for enclosing the elements of an electron device, such as a vacuum tube, gas tube, photocell, or lamp.

bulge 1. A nonlinear attenuation-versus-frequency curve in a transmission line. 2. A localized nonlinearity in a function.

bulk The body of a semiconductor specimen (with respect to conduction through it, its resistivity, etc.), as opposed to junctions within the specimen. Thus, current flows through a junction, but it can also flow, more or less, through the *bulk* of the semiconductor wafer into which the junction has been formed.

bulk effect An effect, such as current, resistance, or resistivity, observed in the overall body of a sample of material, as opposed to a region within the material or on its surface. Thus, a silicon diode may display junction resistance (i.e., resistance offered by a junction processed in a wafer of silicon), as well as bulk resistance (i.e., the effective resistance of all paths *around* the junction, through the body of the wafer). Compare *surface effect.*

bulk-erased tape Recording tape whose signal content has been removed by means of a bulk eraser.

bulk-erase noise Noise generated by bulk-erased tape when the latter passes through deenergized record or erase heads in a tape machine.

bulk eraser A type of power-line-frequency degausser which erases an entire reel of magnetic tape without requiring that the latter be unreeled and passed continuously under an erase head. Also called *bulk degausser.*

buncher In a klystron, a cavity resonator containing two grids mounted parallel to the electron stream. The electrostatic field of the grids alternately accelerates and retards the electrons, velocity-modulating the stream into bunches.

buncher grids In a klystron, the closely spaced grids that velocity-modulate the electron beam into successive bunches.

buncher resonator In a velocity-modulated tube such as a klystron, the input cavity resonator.

buncher voltage The rf grid-to-grid voltage in the buncher resonator of a klystron.

bunching The production of electron bunches in a velocity-modulated tube, such as a klystron. Also see *buncher.*

bunch stranding A technique for combining several thin wires into a single thick wire. Often used in guy wires and electrical conductors to improve tensile strength and flexibility. At radio frequencies, bunch stranding also improves electrical conductivity by increasing the ratio of surface area to cross-sectional area. This minimizes losses caused by skin effect.

Bunet's formula A formula for calculating the inductance of a multilayer air-core coil having a diameter less than three times the length:

$$L = \frac{a^2 N^2}{9a + 10l + 8.4c + 3.2cl/a}$$

where N is the number of turns, and the other symbols represent the dimensions shown in the accompanying illustration.

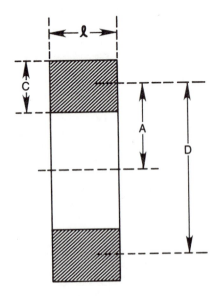

Bunet's formula.

burden See *voltage burden*.

burn A blemish on the screen of a cathode-ray tube caused by destruction of the phosphor there, which itself results from prolonged focusing of an intense electron beam. Burns on the screen of a TV picture tube usually result from ions that reach the screen when the ion trap is not working correctly.

burn-in A usually long and carefully controlled preliminary operation of a device to stabilize its electrical characteristics after manufacture, as in such "seasoning" of an electron tube before it is released for use.

burning In the reactivation of a vacuum tube having a thoriated tungsten filament which has lost its emission, the process of operating the filament briefly at a voltage somewhat higher than normal in order to form a new thorium layer. Burning follows the process of flashing.

burning voltage A voltage, somewhat higher than normal, at which the thoriated tungsten filament of a disabled vacuum tube is operated briefly after flashing to form a new thorium layer.

burnout 1. Failure of a conductor or component due to overheating from excess current or voltage. 2. The open-circuiting of a fuse. 3. Electrical failure of any type.

burst 1. The abrupt ionization of the gas in an ionization chamber, by cosmic rays. 2. An abrupt increase in the amplitude of a signal. Also, the type of signal resulting from burst action. 3. Color burst.

burst amplifier In a color-TV receiver, the amplifier that separates the burst pulse from the video signals and amplifies the former. See *color burst*.

burst gate timing In a color-TV receiver, the timing of the gating pulse with the input signal of the burst amplifier.

burst generator A signal generator delivering a burst output (see *burst*, 2) for testing a variety of equipment. Its output is intermediate between continuous waves and square waves and is convenient for rapidly appraising the performance of amplifiers, filters, electronic switches, transducers, loudspeakers, and so on.

burst transmission A short transmission at high speed. This method of transmission saves time, but increases the necessary bandwidth of a signal by the same factor as the ratio of speed increase.

bus 1. A main conductor in a circuit. A bus may be *high* in the sense that its potential is above or below ground, or it may be *low* or at ground reference. 2. In computer practice, a common group of paths over which input and output signals are routed.

bus bar See *bar, 4*.

bus driver A buffering device designed to increase the driving capability of a microprocessor (such as MOS), which itself may be capable of driving no more than a single TTL load.

business machine Any piece of electronic equipment used for business purposes.

busing Parallel interconnection of circuits.

busy test A check conducted to find out whether or not a certain telephone subscriber line is in use.

busy tone An intermittent tone that indicates that the subscriber line being called is in use.

Butler oscillator An oscillator which consists of a two-stage amplifier with a quartz crystal in the positive-feedback path from output to input.

butterfly capacitor A plate-type variable capacitor having two stator sections and what is essentially a single rotor section common to the two stators. External connections are made to the stators only, so that no wiping contact is required to the rotor and the troubles associated with such a contact are avoided. The butterfly capacitor thus is actually two variable capacitors in series. The unit is so called from the shape of its rotor.

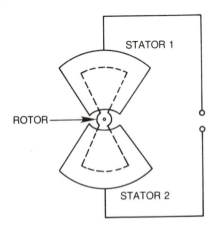

butterfly circuit A combination of butterfly capacitor and a ring of which the stator plates of the capacitor are an integral part. The resulting structure is a compact variable-frequency tuned circuit in which the ring supplies the inductance, and the butterfly the capacitance. Also called *butterfly tank* and *butterfly resonator*.

butterfly resonator See *butterfly circuit*.

Butterworth filter A high-pass, bandpass, or low-pass filter characterized by a absence of peaks in the pass band and usually designed for audio systems because of its unique response flatness.

Butterworth function A function that is employed in the design of a Butterworth bandpass filter.

button 1. A tiny lump of impurity material placed on the surface of a semiconductor wafer for alloying with the latter to form a junction. See *alloy junction*. 2. Single-button carbon microphone element. 3. A usually small switch actuated by finger pressure. Also called *pushbutton* and *pushbutton switch*. Sometimes, the term *button* is applied only to the insulated knob or pin which is pushed to operate the switch.

button capacitor A button-shaped ceramic or silvered-mica fixed capacitor which, because of its disk shape and mode of terminal connection, offers very low internal inductance.

button microphone A microphone in which a carbon button (see *button, 2*) is attached to a diaphragm, which is set into vibration by sound waves. This motion causes the button resistance to vary, modulating a direct current passing through the button. A *single-button microphone* employs only one button, whereas a *double-button microphone* has two, one mounted on each side of the center of the diaphragm.

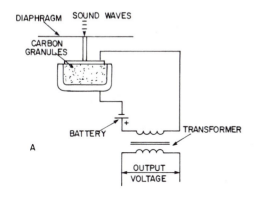

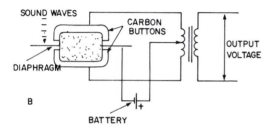

Button microphone.

buzz 1. A low-pitched rough sound with high-frequency components, usually the result of electrical interference from nonsinusoidal voltages generated by neighboring equipment or devices. 2. The waveform associated with such a sound. 3. The fastening of two conducting surfaces by *Kellie bonding*.

buzzer A nonringing device used principally to generate sound other than that achievable with sine waves. In an electromechanical vibrating-reed buzzer, the reed acts as an armature which is mounted close to the core of an electromagnet. At quiescence the reed rests against a stationary contact. When voltage is applied to the electromagnet, the reed is attracted to the core, moving away from the contact; but this breaks the circuit, the magnetism ceases, and the reed springs back to the contact. The action is repeated continuously at a frequency dependent upon the reed dimensions and its distance from the core.

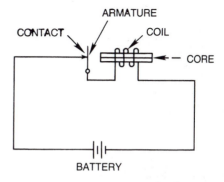

Buzzer.

BV Abbreviation of *breakdown voltage*.

B-voltage The dc voltage required by certain electrodes of tubes, transistors, etc.—especially the plate and screen voltage of a tube, as opposed to the filament voltage (*A-voltage*) and control-grid voltage (*C-voltage*).

bw 1. Abbreviation of bandwidth. 2. Abbreviation *black-and-white* (also abbreviated *B&W*).

bwa Abbreviation of *backward-wave amplifier*.

BWG Abbreviation of *Birmingham wire gauge*.

bwo Abbreviation of *backward-wave oscillator*.

BX Symbol and abbreviation for armored and insulated flexible electrical cable.

bypass A route (either intended or accidental) through which current easily flows around a component or circuit instead of through it.

bypass capacitor A capacitor employed to conduct an alternating current around a component or group of components. Often the ac is removed from an ac/dc mixture, the dc being free to pass through the bypassed component.

B-Y signal In a color TV receiver, the color-difference signal which, when combined with a luminance (*Y*) signal, forms a blue primary signal for the three-gun picture tube.

byte In digital-computer practice, a sequence of adjacent bits treated as a unit. A word consists of one or more bytes.

C 1. Abbreviation of *capacitance*. 2. Symbol for *collector of a transistor*. 3. Symbol for *carbon*. 4. Abbreviation of *Celsius*. 5. Symbol for *Coulomb*. 6. Abbreviation of *calorie*.

c. 1. Abbreviation of *centi*. 2. Abbreviation of *cents*. 3. Symbol for *capacitance*. 4. Symbol for *speed of light* in a vacuum.

Ca Symbol for *calcium*.

cabinet An enclosure for a piece of apparatus. It may or may not incorporate electromagnetic shielding.

cable 1. A usually flexible but sometimes rigid conductor, often of appreciable length. Although the term is occasionally applied to a single conductor, especially when it is a braid or weave of a number of wires, *cable* usually means a bundle of separate, insulated wires. 2. *Cablegram*.

cable address The address as a code word, of the recipient of a cablegram.

cable assembly A special-purpose cable with connectors.

cable attenuation Reduction of signal intensity along a cable, usually expressed in decibels per foot, hundred feet, mile, etc.

cable capacitance Capacitance between conductors in a cable or between conductors and the outer sheath of a cable. 2. Sometimes, capacitance between a cable and earth.

cable clamp A support device for cable runs in equipment and systems.

cable communications Telegraphy or telegraphy via a (usually undersea) cable.

cable connector A connector, such as a coaxial fitting, that joins cable circuits or connects a cable to a device.

cabled wiring Insulated leads connecting circuit points; they are tied together with lacing cord or with spaced fasteners.

cablegram A (usually printed) message transmitted or received via undersea cable. Compare *radiogram* and *telegram*.

cable loss See *cable attenuation*.

cable Morse code A code used principally in undersea cable telegraph. It is a three-element character system in which dots and dashes of equal length are represented by positive and negative pulses, respectively.

cable run The path taken by a cable.

cable splice To electronically interconnect the individual connectors in two separate cables.

cable TV See *community-antenna television*.

cache memory A short-term, high-speed, high-capacity computer memory. Similar to a scratch-pad or read-write memory.

cactus needle A phonograph stylus made from the tip of a cactus thorn.

cadmium Symbol, Cd. A metallic element. Atomic number, 48. Atomic weight, 112.41. Many electronic structures are cadmium plated for protection and ease of soldering.

cadmium borate phosphor Formula, $(CdO + B_2O_3)$: Mn. A substance used as a phosphor coating on the screen of cathode-ray tubes. The characteristic fluorescence is green-orange.

cadmium cell A cell used as a source of reference voltage. The cell voltage is approximately 1.1 volt under normal conditions.

cadmium-plated To plate with cadmium, a process applied to a number of electronic components.

cadmium selenide photocell A photoconductive cell in which cadmium selenide is the light-sensitive material.

cadmium silicate phosphor Formula, $(CdO + SiO_2)$: Mn. A substance used as a phosphor coating on the screen of cathode-ray tubes; the characteristic fluorescence is orange-yellow.

cadmium standard cell See *standard cell*.

cadmium sulfide photocell A photoconductive cell in which cadmium sulfide is the light-sensitive material.

cadmium tungstate phosphor Formula, $CdO + WO_3$. A substance used as a phosphor coating on the screen of cathode-ray tubes; the characteristic fluorescence is light blue.

cage A completely shielded enclosure, such as a screen room, which is covered with a grounded fine-mesh conductive screen on all sides.

cage antenna An antenna whose parallel-connected elements are connected to cylindrical supports, making them, as a group, take the form of a cone.

CAL An acronym for *c*onversional *al*gebraic *l*anguage, a general-purpose problem-oriented computer programming language used in time-sharing systems.

calcium Symbol, Ca. A metallic element of the alkaline-earth group. Atomic number, 20. Atomic weight, 40.08.

calcium phosphate phosphor Formula, Ca3 (PO4)2. A substance used as a phosphor coating on the screen of long-persistence cathode-ray tubes; the characteristic fluorescense is white, as is the phosphorescence.

calcium silicate phosphor Formula, (CaO + SiO2): Mn. A substance used as a phosphor coating on the screen of cathode-ray tubes; the characteristic fluorescence ranges form green to orange.

calcium tungstate phosphor Formula, CaWO4. A substance used as a phosphor coating on the screen of short-persistence cathode-ray tubes; the characteristic fluorescence is blue, as is the phosphorescence.

calculate To perform the steps of an intricate mathematical operation. Compare *computer*.

calculating punch A data processing peripheral that reads punched cards, makes calculations, and punches new data into those cards or new cards.

calculator A device or machine for performing mathematical operations, especially arithmetic. The distinction between the electronic calculator and the electronic computer is that the latter is a dedicated general-purpose device, one which can be adapted to a number of applications; this distinction even holds true for programmable calculators.

calculus See *differential calculus* and *integral calculus*.

calendar age The age of a piece of equipment, measured since the date of manufacture. Specified in years, months, and days. The actual manufacture date may alternatively be given.

calendar time The time available in a working period, i.e., a 40-hour work week represents a calendar time of 120 hours (5 × 24).

calibrate To compare and bring into agreement with a standard.

calibrated measurement 1. A measurement made with an instrument that has been calibrated with a standard reference source. 2. A measurement that is corrected for instrument error.

calibrated meter An analog or digital meter that has been adjusted to agree as closely as possible with a reference source.

calibrated scale A scale whose graduations have been carefully checked for accuracy, i.e., they correspond to the true values of the quantity they represent. The scale may be graduated to read directly in units of the quantity or it may consist only of arbitrary numbers which may be compared to a graph or chart of the quantity's values.

calibrated sweep In an oscilloscope, a sweep circuit calibrated to indicate sweep frequency or time at all control settings.

calibrated triggered sweep In an oscilloscope, a triggered sweep circuit calibrated in terms of sweep time or frequency.

calibration 1. Determining the accuracy with which an instrument indicates a quantity. 2. Determining the degree to which the response of a circuit or device corresponds to desired performance.

calibration accuracy 1. The amount of agreement between the value of a quantity, as indicated by an instrument, and the true value, expressed as a percentage of error (e.g., ±1%). 2. The precision of a direct-reading meter in terms of its full-scale deflection (e.g., ±2% of full scale).

calibration curve A plot of standard values of a quantity versus the corresponding response of an instrument or component.

calibration marker A pip or blip superimposed on a pattern displayed on a radar or oscilloscope screen to identify a point closely as to frequency, voltage, distance, or some similar term.

calibrator A device used to perform a calibration e.g., a signal generator.

calibrator crystal A highly accurate and stable quartz crystal used in an oscillator for frequency calibrations. A common example is the 100-kHz crystal used in frequency-standard oscillators.

californium Symbol, Cf. A radioactive element produced artifically. Atomic number, 98. Atomic weight, 251.

call 1. A transmission by a station for the purpose of (a) alerting a particular receiving station for which there is a message, or (b) alerting all receiving stations to prepare them for a general broadcast message. 2. In a computer program, a branch to a closed subroutine; also, to branch to such a subroutine.

call direction code Abbreviation, CDC. In telegraph networks, a special code that when transmitted to a terminal causes the teleprinter there to be automatically switched on.

calling sequence Computer program instructions needed to establish the conditions for a call *(call, 2)*. Sometimes, subroutine instructions providing a link to the main program.

call instruction A computer program instruction that makes a program controller (a unit within a CPU) branch to a subroutine; it also locates and identifies the parameters needed for the subroutine's execution. Also called *subroutine call*.

call letters Letters used to identify licensed radio stations.

calomel electrode A standard electrode for a pH meter containing mercury, calomel (mercurous chloride), and potassium chloride.

calorie Abbreviation, cal, C. The amount of heat at 1 atmosphere of pressure which will raise the temperature of 1 gram of water 1 degree Celsius.

calorimeter An instrument for measuring heat. By adaptation, a calorimeter may be used to check rf power, especially at microwave frequencies (see *calorimetric power meter*).

calorimeter system See *calorimetric power meter*.

calorimetric power meter A wattmeter in which the power to be measured heats an oil or water bath, the power value being determined from the consequent temperature rise of the liquid.

CAM Abbreviation of *content-addressed memory*.

cambric Finely woven cotton or linen used for insulation. One type of spaghetti (conductor insulation), for example, is varnished cambric tubing.

camera cable A multiwire cable that conducts the video signal from a TV camera to control equipment.

camera chain In television, the camera and the equipment immediately associated with it, excluding the transmitter and its peripherals.

camera signal The output signal delivered by a TV camera.

camera tube Any video pickup tube, such as an iconoscope or orthicon, that converts light reflected by a scene into a corresponding television signal.

camp-on In a telephone system, a method of engaging a line that is busy until it becomes available for use.

CAN Abbreviation of *cancel character*.

can A metal enclosure or container roughly resembling a tin can (though not necessarily cylindrical), used for shielding or potting components. 2. Sometimes, *headphone*.

Canada balsam A transparent cement derived from the turpentine distilled from balsam fir resin. It is useful in optical technology and in certain areas of electro-optics.

Canadian Standards Association The Canadian equivalent of the National Bureau of Standards in the United States. An agency that publishes agreed-on standards for industries.

canal rays In a vacuum tube, rays consisting of positive ions arising from the anode electrode. Canal rays travel in the direction opposite to that of cathode rays (which arise at the same time from the cathode electrode).

cancel character 1. *Ignore character.* 2. A control character indicating that the associated data is erroneous.

cancellation The elimination of one quantity by another, as when a voltage is reduced to zero by another voltage of equal magnitude and opposite sign.

candela Symbol, cd. The SI unit of luminous intensity; 1 cd is the luminous intensity of 1/600 000 square meter of a perfect radiator at the temperature of freezing platinum.

candle Abbreviation, c. Also called *international candle*. A unit of light intensity which is the value of emission by the flame of a sperm candle 7/8″ in diameter burning at the rate of 7.776 grams per hour.

candlepower Abbreviation, cp. Luminous intensity in international candles: the luminous intensity resulting from the burning of a 7/8-inch-diameter sperm whale oil candle at 7.776 grams per hour.

candoluminescence White light produced without extreme heat.

cannibalization The deliberate use of parts from operational equipment to temporarily repair or maintain other equipment. It is a last-resort, emergency measure.

cap. 1. Abbreviation of *capacitance*. 2. Abbreviation of *capacitor*.

capacimeter See *capacitance meter.*

capacitance Symbol, C. Unit, farad. The property exhibited by two conductors separated by a dielectric, whereby an electric charge becomes stored between the conductors. Capacitance is thought of as analogous to mechanical elasticity. Also see *farad*.

capacitance bridge A four-arm ac bridge for gauging capacitance against a standard capacitor. In its simplest form, it has a standard capacitor in one arm and resistors in the other three.

capacitance-bridge neutralization A circuit for neutralizing a triode amplifier, in which a four-capacitor bridge is formed by the variable neutralizing capacitor, tube gridplate capacitance, tube grid-cathode capacitance, and an external fixed capacitor.

capacitance coupling The transfer of ac energy between two circuits or devices by a capacitor or capacitance effect. Also see *coupling*.

capacitance diode *Varactor.*

capacitance divider An alternating-current voltage divider that uses capacitors rather than resistors. Often employed in oscillators such as the Colpitts type.

capacitance filter A filter consisting of only a high-capacitance capacitor. Because of the capacitor cannot discharge instantaneously, it tends to maintain its voltage and thus smooth out the ripples in the voltage applied to it.

capacitance-inductance bridge A combination ac bridge which, by means of switching or patching, can be used for either capacitance or inductance measurement. Both capacitance and inductance can be

measured in terms of a standard capacitance; however, some of these bridges employ standard inductors in the inductance measuring position.

capacitance meter A direct-reading meter for measuring capacitance. In most of the several types available, a stable ac voltage is applied to the meter circuit to which an unknown capacitor is connected in series; meter deflection is roughly proportional to the reactance of the capacitor. Also called *microfarad meter.*

capacitance ratio In a variable capacitor, the ratio of maximum to minimum-capacitance.

capacitance relay A relay circuit which operates from a small change in its own capacitance. It consists essentially of an rf oscillator whose tank capacitance is very low. When a hand or just a finger is brought near the circuit's short pickup antenna, the attendant increase in capacitance detunes the oscillator activating the relay. Also called *proximity relay* and *proximity switch*.

capacitance-resistance bridge A combination ac bridge which, by means of switching or patching, can be used for either capacitance or resistance measurement. The unknown resistance is measured against a standard resistor; the unknown capacitance against a standard capacitor.

capacitance sensor See *capacitance transducer.*

capacitive amplifier See *dielectric amplifier.*

capacitive attenuator An ac attenuator whose elements are capacitors in any desired combination of fixed and/or variable units. The desired attenuator is afforded by the capacitance ratio.

capacitive coupling A means of coupling between circuits that employs a series capacitor for direct-current blocking. The signal passes through the capacitor, but the blocking effect allows different bias voltages to be applied to the two stages.

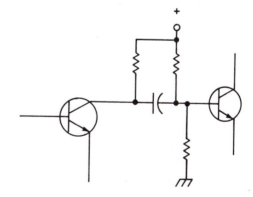

Capacitive coupling.

capacitive diaphragm A metal plate deliberately placed in a waveguide, to introduce capacitive reactance and thereby cancel an inductive reactance.

capacitive-discharge ignition An electronic ignition system for automotive engines which provides nearly constant high voltage regardless of engine speed. A dc-to-dc step-up converter charges a large capacitor (typically to 300 volts) when the distributor breaker points are closed; when they are open, the capacitor discharges through the standard 90 to 1 turns ratio ignition coil thereby theoretically generating an ignition pulse of 27,000 volts.

capacitive division Reduction of an ac voltage by a capacitive voltage divider.

capacitive feedback Feeding energy back from the output to the input of an amplifier or oscillator through a capacitor.

capacitive-input filter A smoothing filter for ac power supplies in which the element closest to the rectifier is a capacitor, regardless of the components or circuits placed subsequently.

capacitive load A load consisting of a capacitor or a predominantly capacitive circuit.

capacitive post A protrusion inside a waveguide, for the purpose of introducing capacitive reactance to cancel an inductive reactance.

capacitive potentiometer See *capacitive voltage divider*.

capacitive reactance Symbol, X_c. Unit, ohm. The reactance exhibited by an ideal capacitor; $X_c = 1/(2\pi fC)$. In a *pure* capacitance reactance, current leads voltage by 90 degrees. Also see *capacitance, capacitor,* and *reactance*.

capacitive speaker See *electrostatic speaker*.

capacitive transducer A transducer consisting essentially of a refined variable capacitor whose value is varied by a quantity under test, such as pressure, temperature, liquid level, etc.

capacitive tuning Variable-capacitor tuning of a circuit.

capacitive voltage divider A capacitive attenuator usually consisting of two series-connected capacitors whose values are such that an applied ac voltage is divided across them in the desired ratio.

capacitive welding An electronic welding system in which energy stored in a capacitor is discharged through the joint to be welded, which develops the heat necessary for the operation.

capacitive window A pair of capacitive diaphragms used in a waveguide to introduce capacitive reactance.

capacitor A passive electronic-circuit component consisting of, in basic form, two metal electrodes or plates separated by a dielectric (insulator).

capacitor amplifier See *dielectric amplifier*.

capacitor antenna See *condenser antenna*.

capacitor bank A network of capacitors connected in combination yielding a desired characteristic.

capacitor braking The connection of a capacitor to the winding of a motor after the removal of power, to speed up the process of braking.

capacitor color code See *color code*.

capacitor decade See *decade capacitor*.

capacitor-discharge ignition *Capacitive-discharge ignition*.

capacitor filter In a direct-current power supply, a filter consisting simply of a capacitor connected in parallel with the rectifier output.

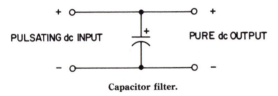

Capacitor filter.

capacitor-input filter A filter whose input component is a capacitor.

The capacitor-input power-supply filter is distinguished by its relatively high dc output voltage but somewhat poorer voltage regulation compared with the *choke-input filter*.

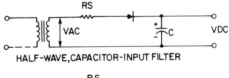

HALF-WAVE, CAPACITOR-INPUT FILTER

FULL-WAVE, CENTER-TAP, CAPACITOR-INPUT

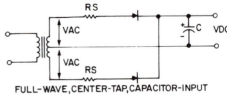

FULL-WAVE BRIDGE, CAPACITOR-INPUT

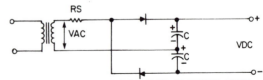

FULL-WAVE, VOLTAGE DOUBLER, CAPACITOR-INPUT

Capacitor-input filters.

capacitor leakage Direct current flowing through the dielectric of a capacitor. In a good nonelectrolytic capacitor this current may be less than 1 microampere. In an electrolytic capacitor, it may normally be several milliamperes, depending upon capacitance and applied voltage.

capacitor loudspeaker See *electrostatic speaker*.

capacitor microphone See *condenser microphone*.

capacitor motor An ac motor which employs a capacitor in series with an auxiliary field winding for starting purposes. Initially out-of-phase current in the auxiliary field (starting winding) causes a rotating field which turns the rotor. When the rotor reaches a safe speed, a centrifugal switch disconnects the capacitor and auxiliary field, and the motor continues running as an induction motor.

capacitor pickup A phonograph pickup whose vibrating stylus varies the capacitance of a small air capacitor. Its action is similar to that of the condenser microphone.

capacitor series resistance The ohmic loss in a capacitor. It results partly from conductor losses, and partly from losses in the dielectric material.

capacitor-start motor See *capacitor motor.*

capacitor substitution box An enclosed assortment of selected-value capacitors arranged to be switched one at a time to a pair of terminals. In troubleshooting and circuit development, any of several useful fixed capacitance values may be thus obtained.

capacitor transducer See *capacitive transducer.*

capacitor voltage The voltage at the terminals of a capacitor. Sometimes, the voltage rating of a capacitor.

capacitor voltmeter See *electrostatic voltmeter.*

capacity 1. A measure of a cell's or battery's ability to supply current during a given period. 2. *Capacitance.* 3. The number of bits or bytes a computer storage device can hold. 4. The limits of numbers that a register can process.

capacity lag In an automatic control system, the delay caused by the storing of energy by parts. For example, in a heating system, capacity lag results from the time taken to heat the air or fluid after the thermostat turns on the heat.

capillary electrometer A sensitive voltage indicator consisting essentially of a column of mercury in a transparent capillary tube in which is suspended a small drop of acid. When a voltage is applied to both ends of mercury column the acid drop moves toward the low-potential end of the column over a distance proportional to the voltage.

capstan The driven spindle or shaft of a magnetic tape recorder or transport.

capture area The effective ability of a radio antenna to pick up electromagnetic signals. The larger the capture area, the greater the antenna gain.

capture effect 1. In frequency-modulation receivers, the effect of domination by the stronger of two signals, or by the strongest of several signals, on the same frequency. 2. In an automatic-frequency-control system, the tendency of the receiver to move toward the strongest of several signals near a given frequency. 3. In general, the tendency of one effect to totally predominate over other effects of lesser amplitude.

capture ratio A measure of FM tuner selectivity: the ratio in decibels, of signal strengths between rejected unwanted signals and the one being tuned in.

carbon Symbol, C. A nonmetallic element. Atomic number, 6. Atomic weight, 12.011. Carbon, besides being an invaluable material for electronics, is the major constituent of compounds that organic chemistry deals with. It is also the element that is used to describe life forms.

carbon arc The arc between two electrified pencils of carbon or, as in an arc converter, between a carbon pencil and a metal electrode.

carbon brush A brush (contact) made of carbon or some mixture of carbon and another material, used in motors, generators, variable autotransformers, rheostats, and potentiometers.

carbon button See *button, 2.*

carbon-button amplifier An audio-frequency amplifier having as the active component an earphone whose diaphragm is attached to a carbon microphone button (see *button, 2).* The input signal applied to the earphone makes its diaphragm vibrate. The vibrating button then modulates a local direct current. Amplification results from the large ratio of modulated local current to input-signal current.

carbon-button oscillator See *hummer.*

carbon/disk rheostat A rheostat consisting of a stack of carbon disks arranged so that a controllable pressure can be exerted on the stack.

As a knob is turned, a screw increases or decreases the pressure, varying the total resistance of the stack.

carbon filament A usually thin conductor consisting of a carbonized thread that characterized early electric light bulbs. The carbon filament has a negative temperature coefficient of resistance. Carbon filaments have employed as varistors and bolometers.

carbon-film resistor A stable resistor whose resistance element is a film of carbon vacuum-deposited on a substrate, such as a ceramic.

carbonize The application of a coat of carbon onto an electrode, either by electroplating or by any other means.

carbonized filament A vacuum-tube filament of thoriated tungsten upon which a layer of tungsten carbide has been formed, a treatment that increases electron emission and allows higher operating temperature than can be afforded by the plain filament.

carbonized plate A vacuum-tube plate coated with carbon for increased heat dissipation.

carbon microphone A microphone employing one or two carbon buttons. Also see *button, 2* and *button microphone.*

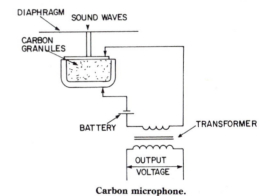

Carbon microphone.

carbon-paper recorder A recorder in which a signal-actuated stylus writes, by impression only through a sheet of carbon paper onto a plain sheet underneath. This technique obviates the need for an ink-carrying stylus.

carbon pickup A phonograph pickup in which stylus vibration varies the resistance of carbon granules in a button. The action is similar to that of a carbon microphone.

carbon-pile regulator A voltage regulator in which a stack of carbon workers is in series with the shunt field. The pile resistance (and this field current) depends on pressure applied to the pile by a wafer spring acting through a movable iron armature. Voltage drops increase the pressure and voltage rises decrease the pressure, thus regulating the generator with which it is associated.

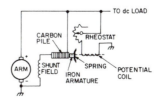

Carbon-pile regulator.

carbon-pile rheostat See *carbon-disk rheostat.*

carbon recording 1. A record made with a *carbon-paper recorder.* 2. The use of a carbon-paper recorder in data acquisition, facsimile, communications, and similar applications.

carbon resistor A resistor made from carbon, graphite, or some composition containing carbon.

carbon/silicon-carbide thermocouple A thermocouple that is a junction between carbon and silicon carbide. Typical output is 353.6 mV at 1210 °C.

carbon transfer recording A method of facsimile reception in which the image is reproduced by carbon particles sprayed on the paper, a process controlled by the received signal.

carbon-zinc cell See *zinc-carbon cell.*

Carborundum Formula, SiC. Tradename for a synthetic silicon carbide used as a semiconductor, refractory, or abrasive. Also see *silicon carbide.*

Carborundum crystal Tradename for a characteristically superhard crystal of silicon carbide.

Carborundum detector An early semiconductor-diode (crystal) detector employing a Carborundum crystal and a steel contact (*cat's whisker.*)

Carborundum varistor A voltage-dependent resistor made from Carborundum.

carcinotron A special kind of oscillator tube used at ultra-high and microwave frequencies.

card 1. A usually thin, rectangular phenolic panel on which components are assembled and wired to produce a circuit or subcircuit. 2. The usually flat, thin insulating strip on which a resistor element is wound. 3. A thin (usually paper) board perforated with holes, according to a code, to store data, i.e., *punched card.* 4. Variation (Brit.) of *cardboard.*

card back The unprinted side of a punched card. Compare *card face.*

card bed In a punched-card machine, the device that holds the cards before they are punched or read.

card code The scheme whereby the holes in a punched card represent letters or numbers.

card column Punched card lines running parallel with the short edge; the columns which are punched with holes represent characters or numbers. Compare *card row.*

Cardew voltmeter An early form of *hot-wire meter* employed to measure voltage.

card face The printed side of a punched card.

card feed In a punched-card system, the mechanism that successively advances the cards.

card field The fixed columns on a punched card, containing the same type of information.

card fluff The paper burrs around the edges of holes in a punched card, often the cause of misfeeding or misreading.

card format A description of punched card content (usually specified by a program).

card hopper See *hopper.*

card image An exact, stored representation of all the characters (*not* holes) on a punched card. Compare *binary image.*

cardiac monitor An electronic device that displays or records electrical impulses from the heart for medical observation or diagnosis.

cardiac pacemaker An electrical cardiac stimulator that causes the heart to beat at certain intervals. Used when the patient has heart disease that prevents the heart from regulating itself.

cardiac stimulator An electronic device (sometimes implanted in the subject) that supplies electric pulses to stimulate heart action. Also called *defibrillator* and *pacemaker.*

card image In memory storage, the data contained on a single card.

cardiogram *Electrocardiogram.*

cardiograph *Electrocardiograph.*

cardioid diagram A polar response curve having the shape of a cardioid pattern.

cardioid microphone A microphone having a (roughly) heart-shaped sound-field pickup pattern.

cardioid pattern A pattern having the general shape of a stylized heart.

cardiotachometer A device that indicates the pulse rate.

cardistimulator See *cardiac stimulator.*

card jamming In a card reader, the accidental piling up of cards.

card loader A load routine (see *load, 3*) for punched cards.

card machine A device which translates data into a punched-hole pattern on a card or which reads the hole pattern on a previously punched card. Also called *punched-card machine.*

card pickup A photocell or feller contact which produces signals from holes in a punched card.

card-programmed Programmed (as a computer or instrument) by means of punched cards.

card punch A machine employed to punch holes in (data) cards, according to a code, for the storage of data. See *card, 3.*

card punch buffer A temporary storage area that holds card punch control signals coming from a central processor.

card reader In a punched-card system, the device which picks up information from the holes in a punched card and translates the data represented by the pattern of perforations into electrical pulses.

card reproducer An offline machine for duplicating punched cards.

card row On a punched card, a horizontal line of punching positions. Compare *card column.*

card sensing The conversion of card information into electrical impulses for use by a machine.

card stacker A machine that stacks punched cards in a deck after they have been punched or read.

card system A computer system in which the only input peripheral is a card reader and the only output peripherals are a card punch and printer.

card-to-card A method of punched card duplication in which the duplicate is made at a terminal receiving the original card's data.

card-to-magnetic-tape converter A device that writes, as logical records, data it reads from punched cards.

card-to-tape 1. Transferring punched-card data to paper tape. 2. The programmed transferal of punched-card data to magnetic tape.

card-to-tape converter A device that performs a card-to-tape operation (see *card-to-tape, 1.*).

card track In a punched-card machine, the transport that carries cards from hopper to stacker (i.e., through reading and punching stations).

card verifier 1. A punched-card machine that checks the accuracy of data on punched cards against a document containing the data; the process may be automatic or operator-implemented. 2. The human operator who checks card accuracy.

card wreck The jamming of punched cards being transported along the track in a punched-card machine.

Carey-Foster bridge A special version of the slide-wire bridge that is useful for measuring an unknown resistance whose value is close to that of a standard resistance.

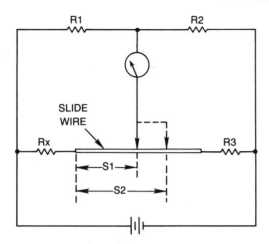

Carey-Foster bridge.

Carey-Foster mutual inductance bridge An ac bridge which permits measurement of mutual inductance in terms of a standard capacitor.

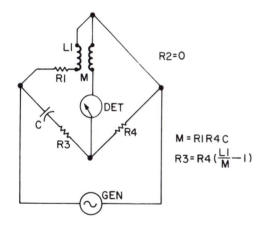

$$R2 = 0$$

$$M = R1R4C$$

$$R3 = R4\left(\frac{L1}{M} - 1\right)$$

Carey-Foster inductance bridge.

carnauba wax A wax obtained from the Brazilian wax palm that is used as an insulant and as the dielectric in some electrets.

carnotite Formula, K2O• 2UO3• V2O5• 2H2O. A radioactive mineral containing hydrous potassium uranium vanadate, which has been used as a source of radium.

Carnot theorem In thermodynamics, the proposition that in a reversible cycle, all available energy is converted into mechanical work. Also called *Carnot's principle*.

carrier 1. See *carrier wave*. 2. A particle or its equivalent, such as an electron or hole, whose movement constitutes a flow of current. 3. In the same sense as *2*, above, an *ion*.

carrier amplifier See *dielectric amplifier*.

carrier beating 1. The mixing of two radio-frequency carriers that are separated by a small amount of frequency, resulting in an audible tone in a receiver. 2. A heterodyne in a facsimile or television signal, resulting in a pattern of cross hatches in the received image.

carrier choke A usually rf choke coil inserted in a line to block a carrier component.

carrier chrominance signal For conveying color TV information, sidebands of a modulated chrominance subcarrier that are added to the monochrome signal and contain any suppressed subcarrier.

carrier color signal For conveying color information in color TV transmission, the sidebands of a modulated chrominance subcarrier (plus the unsuppressed chrominance subcarrier) added to the monochrome signal.

carrier concentration In a semiconductor material, the number of current carriers per unit volume.

carrier control 1. Modification, adjusting, or switching a carrier wave. 2. Adjustment, manipulation, or variation of a circuit or device by means of a carrier wave.

carrier current The current component of a carrier wave, or the amplitude of that current. Compare *carrier power* and *carrier voltage*.

carrier-current communication See *wired wireless*.

carrier-current control 1. Control of the current component in a carrier wave. 2. Remote control by means of wired wireless.

carrier-current receiver See *wire-radio receiver*.

carrier-current relay An rf relay circuit operated over a wire line by means of a transmitter.

carrier-current transmitter See *wired-radio transmitter*.

carrier deviation See *carrier swing*.

carrier dispersion In a semiconductor, the spreading out of electrons and holes, which leave the emitter simultaneously but arrive at the collector via many different paths.

carrier frequency The center frequency of a *carrier wave*.

carrier-frequency pulse A pulse that contains radio-frequency oscillation.

carrier-frequency range The band of carrier frequencies over which a transmitter or signal generator may operate.

carrier-frequency stereo disk A stereo phonograph disk containing two lateral-cut channels, one of which is conventional; the other is cut in a manner that allows it to frequency modulate a supersonic carrier. The pickup output contains the af signal from the first channel, and the carrier is modulated by the second channel, its information being later recovered by demodulation.

carrier injection The apparent emission *(injection)* of electrons or holes into a semiconductor when a voltage is applied to the junction.

carrier leak 1. A point at which carrier-wave energy escapes a circuit or enclosure. 2. The residual carrier voltage present in the output of a carrier-suppressing circuit.

carrier level The amplitude of an unmodulated carrier wave.

carrier lifetime In a semiconductor, the interval before an injected current carrier (see *carrier injection*) recombines with an opposite carrier and ceases to be mobile.

carrier line In carrier-current systems (see *wired wireless*), the line or cable conducting the carrier-wave energy.

carrier mobility Symbol, μ. In a semiconductor material, the average drift velocity speed of electrons and holes per unit electrostatic field. In germanium, electron mobility, μe, is $3.6 \times 10^2 cm^2/E$; hole mobility, $\mu \eta$, is $1.7 \times 10^2 cm^2/E$.

carrier noise Modulation of a carrier when there is no input from the modulator itself; unwanted modulation.

carrier noise level The noise signal amplitude that results from unintentional fluctuations of an unmodulated carrier.

carrier-on-light transmission A form of transmission in which many different signals are sent simultaneously by modulating a beam of light at multiple frequencies.

carrier-on-microwave transmission A form of transmission in which many different signals are sent simultaneously by modulating a microwave signal at multiple lower frequencies.

carrier-on-wire transmission A form of transmission in which many different signals are sent at the same time over a wire, by using radio-frequency carriers. Also called carrier-current communications or wired radio.

carrier oscillator In a single-sideband receiver, the rf oscillator that supplies the missing *carrier wave*.

carrier power The actual power represented by an rf carrier applied to an antenna so measured by either the direct or indirect method. The direct method involves determination of power according to the formula $P = I^2R$, where I is antenna current and R no antenna resistance at the point of I measurement. The indirect method involves determination of power according to the formula $P = EIF$, where E and I are antenna current and voltage and F is a factor less than 1.0 whose value depends upon modulation type.

carrier power-output rating The power delivered by an unmodulated transmitter or generator to the normal load or its equivalent.

carrier shift In an amplitude-modulated transmitter or generator, the undesired change of average carrier voltage during modulation.

carrier-shift indicator An instrument for detecting carrier shift. It usually contains only a pickup coil, semiconductor diode, and dc milliammeter in series. Meter deflection is steady until carrier shift is detected, whereupon the needle fluctuates.

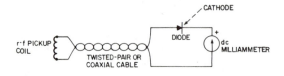

Carrier-shift indicator.

carrier signaling In wire telephony, the use of carrier-wave signals to operate such functions as dialing, ringing, busy signal, etc.

carrier storage In a semiconductor device, the tendency of mobile carriers to stay near a junction for a short time after the junction voltage has been removed or reversed in polarity.

carrier-storage amplifier See *crystal amplifier, 2.*

carrier suppression Elimination of the carrier in an amplitude-modulated signal so that only one or both sidebands is transmitted.

carrier swing In frequency-modulated or phase-modulated transmission, the total deviation (lowest to highest instantaneous frequency) of the carrier wave.

carrier system The transmission of many signals over one circuit, accomplished by modulating various different carriers at different frequencies. Different signals may employ different modulation methods.

carrier telegraphy 1. Continuous-wave telegraphy by *wired wireless.* 2. Wired-wireless telegraphy in which a radio-frequency carrier is modulated by an audio-frequency keying wave.

carrier telephony Telephone communication by *wired wireless.*

carrier terminal 1. At each end of a carrier-current line or cable, the equipment for generating, modifying, or utilizing the carrier energy. 2. In a balanced modulator, the point of carrier insertion.

carrier-to-noise ratio The ratio of carrier amplitude to noise-voltage amplitude. See *carrier noise level.*

carrier transmission Transport of information by a carrier, as by an amplitude-modulated radio wave that carries the low-frequency information as the af modulation envelope and delivers it to the demodulator at the receiving station.

carrier-type dc amplifier A high-frequency ac amplifier ahead of which is operated a generator and transducer. A dc voltage applied to the transducer modulates the carrier supplied by the generator; the amplifier boosts the modulated wave, and the resultant output is rectified at a level higher than that of the dc input signal.

carrier voltage The voltage component of a carrier wave; also, the amplitude of this component. Compare *carrier current* and *carrier power.*

carrier wave The wave which serves as the vehicle for the transmission of information applied to it (modulation).

carry 1. In adding a column of figures, the digit added to the column at the left when the sum exceeds one less than the radix value. 2. In digital computer and counter practice, a pulse corresponding to the arithmetic operation in which a figure is carried to the next column in addition.

carrying capacity The ability of a conductor, such as copper wire, to carry current safely (expressed in maximum amperes).

carry-complete signal In an arithmetic computation by a computer, and adder-produced signal indicating that the pertinent carries have been generated.

carry system A communications system in which several carries occupy one circuit.

carry time The time taken for a digital computer or counter to perform a carry operation (See *carry, 2*).

Cartesian coordinates Distances *(coordinates)* from the X and Y axes which together locate any point in a plane. Compare *polar coordinates.*

cartridge 1. The replaceable transducer assembly of a microphone or phonograph pickup (which also holds the stylus). 2. A magnetic-tape magazine. Also see *tape cartridge.* 3. An insulating tube housing a fuse, semiconductor component, resistor, capacitor, or other part.

cartridge fuse A fuse consisting of a fusible wire enclosed in a cartridge (see *cartridge, 3*) having a ferrule at each end for plug-in connection.

cascadable Capable of, or designed for, being connected in cascade with other similar or identical components.

cascade 1. Components or stages connected and operated in sequence, as in a three-stage amplifier. The components or stages are often but not necessarily identical. 2. To form a cascade.

cascade control 1. In an automatic control system, a controller whose setting is varied by the output of another controller. 2. An automatic control system in which the control units are connected in stages, so that one unit must operate before the next one can.

cascaded amplifier A multistage amplifier in which the stages are forward-coupled in succession.

cascaded carry In digital computer practice, a system of performing the carry operation (see *carry*) in which the $n + 1$ place receives a carry pulse only when the nth place has received carry information to generate the pulse.

cascade thermoelectric device A thermoelectric device consisting of several cascaded sensors (see *cascade, 1*).

cascade voltage doubler A voltage-doubler circuit (see *voltage doubler*) consisting of two diode-capacitor combinations in cascade. Unlike the conventional voltage-doubler circuit with two capacitors in the output, the cascade voltage doubler has one in the input and one in the output.

cascode A high-gain, low-noise, high-input-impedance amplifier circuit consisting of a grounded electron-emitter input stage coupled directly to a grounded-control-electrode output stage.

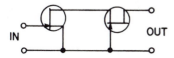

Cascode arrangement of field-effect transistors.

case temperature The temperature at a designated point on the outside surface of a component's case or housing.

Cassegrain antenna A dish antenna that uses Cassegrain feed.

Cassegrain feed A directive-antenna feed system in which rf energy is fed to a small reflector by a waveguide at the center of the main reflector, to which the energy is then reflected by the small reflector.

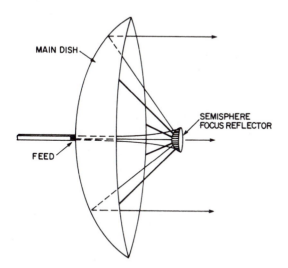

Cassegrain antenna.

cassette 1. A holder (magazine) of reels of magnetic tape that is itself a mechanical subassembly, which may be easily inserted into and removed from a tape deck. 2. A lightweight holder of photographic film or X-ray plates (before, during, and after exposure).

casting out nines Checking the validity of an operand by dividing it by nine and then using the remainder of the division (the *n* modulo) as a check digit that accompanies the result Also see *check digit; module n check.*

castor oil A viscous insulating oil extracted from castor beans. Highly refined castor oil is used as an impregnant in some oil-filled capacitors. Dielectric constant, 4.3 to 4.7. Dielectric strength, 380V/mil.

CAT Abbreviation of *computerized axial tomography.*

catalysis The process whereby an agent called a *catalyst* enhances a chemical reaction without entering into the reaction. Catalysts are used in electronics, for example, to promote the setting of resins in potting and encapsulating operations.

catalyst See *catalytic agent.*

catalytic agent A substance which accomplishes catalysis.

cataphoresis As caused by the influence of an electrostatic field, the migration toward the cathode of particles suspended in a liquid.

catastrophic failure 1. Sudden, unexpected failure of a component or circuit. 2. Failure which can result in the breakdown of an entire system. Also called *catastrophic breakdown.*

catcher In a klystron, the second reentrant cavity. (See *klystron*).

catcher diode A diode that is connected to regulate the voltage at the output of a power supply. The cathode is connected to a source of reference voltage. If the anode, connected to the source to be regulated, becomes more positive than the cathode, the diode conducts and prevents the regulated voltage from rising more than 0.3 volt above the reference voltage (for germanium diodes) or 0.6 volt above the reference voltage (for silicon diodes).

catcher grids In a klystron the grids through which the bunched electrons pass on their way from the buncher to the collector. Catcher grids absorb energy from the bunched electrons and present it to the collector circuit.

category In a computer system, a group of magnetic disk volumes containing information related by a common application.

category storage A computer file storage section containing a number of categories and used by an operating system.

catenation See *concatenation.*

cathamplifier A balanced low-distortion, two-tube power-amplifier output stage having a single-ended input and push-pull output. A distinguishing feature of this circuit is the coupling transformer between both cathodes and one grid.

cathode 1. The negative electrode of a device, i.e., the electrode from which electrons move when a current passes through the device. 2. In an electron tube, the electron-emitting electrode (filament or indirectly heated cathode sleeve).

cathode activity A measure of the effectiveness of a tube cathode as an emitter of electrons.

cathode-biased limiter A triode limiter in which positive-peak limiting is determined by cathode bias. The positive-peak cutoff is due to overdriving the grid.

cathode bombardment In a gas tube, the striking of the cathode by the many positive ions produced by electron-molecule collision. Such bombardment causes secondary electron emission by the cathode.

cathode bypass capacitor The capacitor connected in parallel with the cathode resistor of a tube. It provides an easy path to ground for the ac component of cathode current without disturbing the dc bias between cathode and ground.

cathode-coupled amplifier An amplifier or phase-inverter circuit in which two successive stages are coupled by a common cathode circuit.

cathode-coupled multivibrator A multivibrator circuit in which the two stages are feedback-coupled by a common cathode circuit.

cathode-coupled oscillator A wide-range oscillator circuit in which the two triode stages are feedback-coupled by a common cathode circuit. Only the output stage has a tuned tank.

cathode-coupled phase inverter A phase-inverter circuit in which the two tubes are coupled by a common cathode resistor.

cathode coupling Use of a device's cathode electrode as the input or output terminal. See, for example, *cathode-coupled amplifier* and *cathode-coupled multivibrator*.

cathode current Symbol *Ik*. The current flowing in the cathode circuit of a tube. Cathode current is the total of grid, plate, screen, and suppressor currents and may have an ac as well as dc component.

cathode-current density For a cathode electrode, the current per unit area expressed in amperes or milliamperes per square centimeter.

cathode dark current The electron emission from the photocathode of a camera tube when there is no illumination.

cathode dark region In a cold-cathode gas tube, the zone near the cathode at which very little illumination occurs.

cathode electrode 1. See *cathode terminal, 3.* 2. The negative terminal of a device.

cathode element In a tube, an indirectly heated emitter of electrons. Also see *cathode, 2.*

cathode emission 1. The giving up of electrons by the cathode element of a device, such as a vacuum tube. Electrons may be emitted by either hot or cold cathodes, depending on the tube. 2. Collectively, electrons released by a cathode.

cathode fall of potential In a gas tube, the major portion of the plate-to-cathode potential drop which occurs close to the cathode.

cathode follower A vacuum-tube amplifier circuit in which the output signal is taken at the cathode (and referenced to ground). The circuit is characterized by zero phase shift, a maximum voltage gain of 1, a lowimpedance output, and very low stage distortion.

cathode glow In a cold-cathode glow-discharge tube, the luminosity of the cathode surface facing the cathode dark space. (See *Crookes' dark space*).

cathode guide In a readout tube (see *counter tube*), the electrode that switches the glow from one number to the next.

cathode heating time The time required for the temperature of a tube cathode to increase from cold to maximum specified operating temperature after the cathode current has been initiated. Also called *cathode warmup time.*

cathode impedance Symbol, *Zk*. The impedance of the cathode circuit of a tube, especially a cathode follower.

cathode injection Injection of a signal (e.g., local-oscillator signal in a superheterodyne receiver) into the cathode circuit of a tube. Cathode injection has been used in signal-mixing operations, electronic switches and gates.

cathode interface A resistance-capacitance layer formed between the metal sleeve and oxide coating of an electron-tube cathode.

cathode keying In a radiotelegraph transmitter, forming code signals by operating a key connected between ground (or B-minus) and the cathode of the tube in the keyed stage.

cathode luminous sensitivity For a photomultiplier tube, the cathode's sensitivity to light. This sensitivity figure is the ratio of photocathode current to incident light flux.

cathode modulation A system of amplitude modulation in which af modulating power is applied to the cathode of the modulated rf amplifier stage. Compared with plate modulation, cathode modulation requires 30% less af power for 100% modulation of the dc power input to the rf amplifier.

cathode neutralization A scheme for increasing the gain of a two-stage amplifier having unbypassed cathode resistors to the value it would be if the resistors were bypassed with high capacitance. The gain increase results from positive feedback supplied by a resistor connected between both cathodes.

cathode particle See *electron.*

cathode protection A means of counteracting corrosion in an electrode by applying a negative voltage.

cathode pulse modulation Cathode modulation by pulses.

cathode-ray oscillograph An instrument which provides a permanent record, by photographic or other means, of the image on the screen of a cathode-ray tube.

cathode-ray oscilloscope See *oscilloscope.*

cathode rays Invisible rays emanating from the cathode element of an evacuated tube operated with a high voltage between anode and cathode. Cathode rays (electrons) cause certain substances, *phosphors*, to glow upon striking them.

cathode-ray scanning tube Any of several tubes in which an electron beam is deflected horizontally and vertically to scan an area. These include oscilloscope tubes, storage tubes, radar indicators, TV camera tubes, and so forth.

cathode-ray tube 1. An evacuated tube containing an anode and cathode and which generates cathode rays when operated at high voltage. 2. An oscilloscope tube.

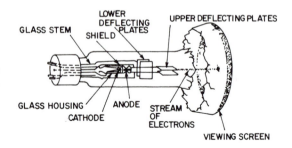

Cathode-ray tube.

cathode-ray tuning indicator See *magic-eye tube.*

cathode regeneration Regeneration obtained by applying positive feedback to the cathode circuit of a tube.

cathode resistor Symbol, *Rk*. A resistor connected between ground (or B-minus) and the cathode of a tube. Cathode current flowing through the resistor creates a voltage drop which appears at the grid of the tube as an automatic, negative bias voltage.

cathode spot The point where the starting arc is formed on the mercury pool cathode in an ignitron.

cathode sputtering See *sputtering.*

cathode terminal 1. In a diode (semiconductor or tube), the terminal to which a negative dc voltage must be applied for forward-biasing the diode. Compare *anode terminal.* 2. In a diode (semiconductor or tube), the terminal at which a positive dc voltage appears when the diode acts as an ac rectifier. Compare *anode terminal.* 3. The terminal connected internally to the cathode element of device. 4. In an electron tube, an indirectly heated electron emitter.

cathode-triggered ring counter A tube-type ring-counter circuit in which an input pulse applied to all right-hand cathodes switches the previous stage into conduction.

cathode voltage Symbol, E_k. The voltage between ground (or B-minus) and the cathode of a tube; it may have both ac and dc components.

cathodic protection A method of preventing corrosive galvanic action in underground metal pipes or the submerged hulls of ships. The part to be protected is used as the cathode of a circuit through which a direct current is passed in the direction opposite to that which caused the corrosion, thus counteracting it.

cathodofluorescence Fluorescence resulting from a material's exposure to cathode rays.

cathodograph An X-ray picture.

cathodoluminescence In a vacuum chamber in which a metal target is bombarded with high-velocity electrons (cathode rays), the emission of radiation of a wavelength characteristic of the metal.

cation A positive ion. Also see *ion*.

CAT scanner The x-ray apparatus for *computerized axial tomography*.

cat's eye 1. Pilot-light jewel. 2. *Magic-eye tube.*

cat's whisker A fine wire electrode whose pointed tip is pressed against the surface of a semiconductor wafer in a point-contact diode or point-contact transistor or against the crystal in a crystal detector.

CATV Abbreviation of *community-antenna television* (commonly, *cable TV*).

caustic soda electrolyte Symbol, NaOH. Sodium hydroxide solution, as used in some secondary cells and experimental devices.

cavitation The local formation of cavities in a fluid used in ultrasonic cleaning, due to reduction in pressure at those points.

cavitation noise In an ultrasonic cleaner, the noise resulting from the collapse of bubbles produced by cavitation.

cavity A metallic chamber (can) in which energy is allowed to reflect, sometimes resulting in resonance.

cavity filter A microwave (usually band-rejection) filter consisting of a resonant cavity and associated coupling devices.

cavity frequency meter See *cavity wavemeter*.

cavity impedance The impedance across a cavity at a particular frequency. At resonance, the cavity impedance is purely resistive.

cavity magentron A magnetron whose anode is a series of resonant cavities.

cavity oscillator An oscillator with a cavity tuned circuit.

cavity radiation Energy radiated from a tiny hole in an otherwise sealed chamber. The radiation occurs at all electromagnetic wavelengths; the greater the temperature within the chamber, the greater the frequency at which the radiation has its maximum amplitude.

cavity resonance The phenomenon whereby a hollow cavity resonates; specifically, resonance in small metal cavities at microwave frequencies.

cavity resonator See *resonant cavity*.

cavity wavemeter An absorption wavemeter whose adjustable element is a tunable resonant cavity into which rf energy is injected through a waveguide or coax. Such an instrument is useful at microwave frequencies.

CB Abbreviation of *citizens band*.

Cb Symbol for *columbium*.

Cb Symbol for *base capacitance* of a transistor.

C-band The band of radio frequencies between 3.9 and 6.2 GHz.

C-battery In battery-operated tube circuits, the battery that supplies bias voltage to the control grids.

C-bias Symbol, E_c. The negative bias voltage applied to the control grid of a tube.

C-bias supply A battery, generator, or ac-operated dc supply used as a source of grid bias.

Cc Symbol for *collector capacitance* of a transistor.

cc 1. Abbreviation of *cubic centimeter*; The International Organization for Standardization recommends cm^3. 2. Abbreviation of *cotton-covered*.

CCA Abbreviation of *current-controlled amplifier*.

CCD Abbreviation of *charge-coupled device*.

CCIF *The International Telephone Consultative Committee.*

CCIR Comite Consultatif International des Radiocommunications (*International Radio Cnsultative Committee*).

CCIT Comite Consultatif International Telegrafique (*International Telegraph Consultative Committee*).

CCITT Comite Consultatif International Telegrafique et Telephonique (*International Telegraph and Telephone Consultative Committee*).

CCS Abbreviation of *continuous commercial service*.

CCTV Abbreviation of *closed-circuit television*.

CCTV monitor A video monitor that receives a signal from a CCTV transmitter.

CCTV signal The picture signal in a CCTV system. It may be either a modulated radio-frequency signal or a composite video signal.

ccw Abbreviation of *counterclockwise*.

Cd Symbol for *cadmium*.

cd Symbol for *candela*.

CD-4 A method of obtaining quadraphonic reproduction on a phonograph disk using modulated carriers with the frequency above the human hearing range.

CDI Abbreviation of *capacitor-discharge ignition*.

C-display A radarscope display showing the target as a dot whose coordinates represent the bearing (horizontal) and angle of elevation (vertical). Compare *A-display, J-display, K-display.*

cd/m² Candelas per square meter, the SI unit of luminance.

Ce Symbol for *cerium*.

Ce 1. Symbol for *emitter capacitance* of a transistor.

ceiling 1. The maximum possible power output from a transmitter. 2. The maximum possible current or voltage that a circuit can deliver. 3. In aviation, the level of the cloud base.

ceilometer An instrument for measuring ceiling (cloud height).

celestial guidance In a guided missile, a system whose guidance section uses the positional information provided by automatic star sightings made during flight.

cell 1. A single (basic) unit for producing dc electricity by electrochemical or biochemical action, as in a battery (a group of connected cells). Also see *primary cell, standard cell, storage cell*. 2. An addressable, one-word-capacity storage element in a computer memory.

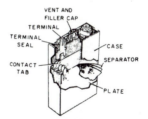

Cell from a large nickel-cadmium battery pack.

cell counter A bioelectronic instrument employed to count blood cells and other minute particles.

cell-type enclosure A screen room (a room designed to prevent the escape of rf energy) characterized by double-walled copper-mesh shielding.

cellular coil A coil having a crisscross (usually multilayer) winding. Examples: lattice-wound coil, honeycomb coil, basket-weave coil.

celluloid A thermoplastic dielectric material which is a blend of cellulose nitrate and camphor. Dielectric constant, 4 to 7. Dielectric strength, 250 to 780 V/mil.

cellulose acetate A plastic dielectric material used as a substrate for magnetic tapes, photographic film, and similar applications. Dielectric constant, 6 to 8. Dielectric strength, 300V to 1kV/mil. Also see *acetate*.

cellulose acetate base See *acetate base*.

cellulose acetate butyrate A thermoplastic dielectric material which is an acetic and butyric acid ester of cellulose.

cellulose acetate disk See *acetate disk*.

cellulose acetate tape See *acetate tape*.

cellulose nitrate The nitric acid ester of cellulose, a plastic insulating material. See *celluloid*.

cellulose nitrate disk A phonograph record consisting of a substrate coated with cellulose-nitrate. Also called *lacquer disk*.

cellulose propionate A thermoplastic molding material which is a propionic acid ester of cellulose.

Celsius scale A temperature scale in which zero is the freezing point of water and 100, the boiling point of water. Also called *centigrade scale*. Compare *absolute scale, Fahrenheit scale*.

cent An audio-frequency interval of 1/100 of a half step. A half step is the frequency difference between two immediately adjacent keys on a piano.

center-fed antenna An antenna in which the feeders are connected to the center of the radiator.

center feed 1. Attaching a feeder or transmission line to the center of the radiator of an antenna. 2. Connection of signal-input terminals to the center of a coil. 3. Descriptive of paper tape whose feed holes are aligned with character hole centers. Compare *advance feed tape*.

center frequency 1. The frequency, in a communications receiver, that is midway between the lower and upper 3-dB attenuation points. 2. The average frequency of a modulated carrier. 3. The carrier frequency of a modulated signal.

centering control In an oscilloscope circuit, a potentiometer used to position the image on the screen (particularly in the center). Separate controls are provided for horizontal and vertical centering.

center loading In a coil-loaded antenna, placement of the loading coil at the center of the radiator rather than at the (more common position) base of the radiator.

center of channel The frequency that is midway between the lowest and highest frequency components of a communications channel.

center of gravity The point on, or in, an object about which the object balances in a gravitational field.

center of mass 1. The center of gravity. 2. The center of rotation in an orbiting system of objects.

center of radiation The point from which the energy radiated by an object appears to arrive.

center tap A connection made to the centermost turn of a coil or to the center-value point of a resistor, filament, or capacitor pair.

center-tap keying In a radiotelegraph transmitter, operation of a key (or keying relay) between ground and the center tap of the secondary of the transformer supplying filament power to the keyed stage.

center-tapped coil See *center-tapped winding*.

center-tapped filament A tube or lamp filament having a tap at its center.

center-tapped inductor An inductor having a tap at half the total inductance.

center-tapped potentiometer A potentiometer having a tap at half the total resistance of the resistance element.

center-tapped resistor A fixed resistor having a tap at half the total resistance.

center-tapped transformer A transformer having one or more center-tapped windings.

center-tapped winding A winding having a tap at half the total number of turns.

Center-tapped winding.

center tracking frequency In three-frequency alignment (tracking) of a circuit, the frequency between the upper and lower frequency limits (alignment or tracking points of the circuit).

center wire The straight wire cathode in a gaseous voltage-regulator tube or Geiger-Mueller tube.

center-zero meter A meter having its zero point at the center of the scale, e.g., a dc galvanometer.

centi A prefix meaning *hundredth (s)*, i.e., 10^{-2}. Abbreviation, *c*.

centigrade scale *Celsius scale*.

centimeter Abbreviation, *cm*. A unit of length equal to 10^{-2} meter or 0.3937 inch.

centimeter-gram-second Abbreviation, cgs. The now-little-used system of units in which the centimeter is the unit of length; the gram, mass; and the mean solar second, time. Electrical units in the cgs system fall into two categories: *electrostatic* and *electromagnetic*. The names of cgs electrostatic units have the prefix *stat* (e.g., *statampere, statvolt*, etc.). Cgs electromagnetic units have the prefix *ab* (e.g., *abampere, abvolt*, etc.).

centimeter waves See *microwaves*.

centipoise A measure of the viscosity of liquids; 1 centipoise (10^{-2} poise) is equal to $10^{-3} \ N^{0} \ s/m^{2}$.

central processing unit In a digital computer, the section containing the arithmetic and logic, control, and internal memory units. Also called *central processor*.

Central Radio Propagation Laboratory A government laboratory that studies radio propagation and collects, correlates, and analyzes data for predicting propagation conditions. The organization also studies methods of measuring propagation.

centrifugal force The force which urges the mass of a rotating body away from the axis of rotation. Compare *centripetal force*.

centrifugal switch A switch actuated by centrifugal force, e.g., the automatic disconnection switch in a capacitor motor.

centripetal force The force which draws the mass of a rotating body toward the axis of rotation. Compare *centrifugal force*.

Ceracircuit A form of hybrid integrated circuit perfected by the Sprague Company.

ceramal See *cermet*.

ceramic-based microcircuit A tiny circuit printed or deposited on a ceramic substrate.

ceramic capacitor A capacitor employing a ceramic dielectric such as barium titanate or titanium dioxide. Such capacitors offer high capacitance in a small package.

ceramic dielectric 1. A ceramic used as a dielectric in capacitors. Examples: barium titanate, barium strontium titante, titanium dioxide. Ceramic dielectrics provide high dielectric constant. 2. A ceramic used as an insulator. Examples: isolantite, porcelain, steatite.

ceramic filter A resonant filter similar to a crystal filter but employing a piezoelectric ceramic material.

ceramic magnet A permanent magnet made of a magnetic ceramic material such as mixtures of barium oxide and iron oxide.

ceramic microphone A microphone employing a *ceramic piezoelement* to convert sound waves into electrical impulses.

ceramic piezoelement A component employing a piezoelectric ceramic material. Examples: ceramic filter, ceramic microphone, ceramic phono pickup, ceramic transducer, electrostrictive transducer. Also called *piezoelectric ceramic*.

ceramic resistor Carborundum resistor whose value is voltage-dependent. It usually displays a negative temperature coefficient of resistance (but a positive coefficient is available) and a negative voltage coefficient of resistance.

ceramics 1. Clay-based materials used as dielectrics and insulators in electronics. Examples: barium titanate, titanium dioxide, porcelain, isolantite, steatite. 2. The science and art of using and developing ceramics.

ceramic-to-metal seal A bond in which ceramic and metal bodies are joined, for example, the bonding of a metal lead to a ceramic disk through which it passes to provide leakproof seal. Also called *ceramet seal*.

ceramic transducer A transducer employing a *ceramic piezoelement* to translate such parameters as pressure and vibration into electrical pulses.

ceramic tube A high-temperature vacuum tube employing a ceramic material, instead of glass, as the envelope; the tube offers low losses at high frequencies.

Cerenkov radiation Light emanating from a transparent material that is traversed by charged particles whose speed is higher than the speed of light through the material.

Cerenkov rebatron device An apparatus for generating radio-frequency energy by passing an electron beam through a piece of dielectric having a small aperture.

ceresin wax A yellow or white wax obtained by refining ozocerite and used as as insulant and sealant against moisture. Dielectric constant, 2.5 to 2.6.

cerium Symbol, Ce. A metallic element of the rare-earth group. Atomic number, 58. Atomic weight, 140.13.

cerium metals A group of metals belonging to the rare-earth group: cerium, illinium, lanthanum, neodymium, praseodymium, and samarium.

cermet An alloy of a ceramic, such as titanium carbide and nickel, a metal. A thin film of cermet is used as a resistive element in some microcircuits. Cermet is an acronym for *ceramic metal*.

certified tape A magnetic recording tape that has been thoroughly checked and found to have no flaws.

cerusite Symbol, $PbCO_3$. A native lead carbonate whose crystals can be white, yellowish, or colorless. Also see *cerusite crystal*.

cerusite crystal A crystal of lead carbonate (see *cerusite*) employed in early semiconductor diodes (crystal detectors).

cesium Symbol, Cs. A metallic element of the alkali-metal group. Atomic number, 55. Atomic weight, 132.91. Cesium is used in some phototube elements (as the light-sensitive material) and as a getter (gas eliminator) in vacuum tubes.

cesium-vapor lamp A low-voltage arc lamp used as an infrared source.

Cf Symbol for *californium*.

Cgk Symbol for *grid-cathode capacitance* of a tube.

Cgp Symbol for *grid-plate capacitance* of a tube.

cgs Abbreviation of *centimeter-gram-second*.

chad The punched-out particle(s) constituting refuse from tape punching. Compare *card fluff*.

chadded tape Punched paper tape in which the chad is left partially attached to the tape's punched holes.

chadless Punched paper tape without *chad*.

chafe 1. An area that has been abraded by rubbing or scraping. 2. To produce a chafe.

chaff Strips of metal foil used to create radar interference or ambiguity in locating a target by multiple reflections of the beam. Also called *mirror*.

chain broadcasting Simultaneous transmissions from a number of broadcast transmitters connected together in a network by wire line, coaxial cable, or microwave link.

chain calculation As performed by a calculator, a calculation that can be entered as it would normally be written, i.e., without the need for regrouping operands.

chain code To prevent a word from recurring within a cyclic operation in a computer, a sequential arrangement of words that are related thusly: each is derived from one adjacent to it by displacing bits left or right by one position; the leading bit is dropped and one is inserted at the end.

chained record A computer file having records that are stored randomly; each record is linked to its successor in the file by its address, which is contained in the record's control field, the first record in the file being the *home* record.

chaining search A method of examining the items in a chained list through the transformation of the digits identifying the record into the address of a location containing information about the item of interest.

chain list A data set arranged so that each item within the set has the address of the next item. Also see *chaining search*.

chain printer In the readout channel of a digital computer, a high-speed printer carrying printer's type on a revolving chain.

chain radar system A number of radar stations along a missile-flight path that are connected in a communications or control network.

chain reaction A reaction (as in nuclear fission) which is self-sustaining or self-repeating. Unless controlled from outside, such a reaction runs to destruction.

chain switch A switch that is actuated by pulling a light metal chain. Successive pulls turn the switch alternatively on and off.

chalcopyrite Formula, $CuFeS_2$. Yellow sulfide of copper and iron. Also see *chalcopyrite detector*.

chalcopyrite detector An early semiconductor diode (crystal detector) in which a crystal of chalcopyrite was the semiconductor.

chance occurrence A random happening, i.e., an event that seems to be completely outside any conceivable cause-effect chain or unifying pattern of appearance.

change dump In computer operation (especially in debugging), the display of the names of locations that have changed following a specific event.

change file See *transaction file*.

change of control In a sequence of computer records being processed, a logical break that initiates a predetermined action, after which processing continues.

changer In a phonograph, a device that allows several disks to be played, on one side only, one after the other. Commonly found in high-fidelity record players.

change record A computer record that changes information in a related master record. Also called *transaction record*.

change tape See *transaction tape*.

channel 1. A frequency (or band of frequencies) assigned to a station. 2. A *keyway*. 3. A subcircuit in a large system, e.g., the rf channel of a receiver, vertical-amplifier channel of an oscilloscope, modulator channel of a transmitter. 4. The end-to-end electrical path through the semiconductor body in a field-effect transistor.

channel analyzer A (usually multiband) continuously tunable instrument, similar to a tuned-rf radio receiver, used in troubleshooting radio communications circuits by substituting a perfect a perfect channel for one that is out of order.

channel balance The state in which the apparent amplitude of two or more channels is identical.

channel capacity The fullest extent to which a channel can accommodate the information (frequencies, bits, words, etc.) to be passed through it.

channel designator A name, number, or abbreviation given to a channel in a communications system.

channel effect The *possible* current flow through a high impedance between the collector and emitter in a transistor.

channel frequency The center frequency of a communications channel.

channeling Multiplex transmission in which separate carriers within a sufficiently wide frequency band are used for simultaneous transmission.

channelizing The subdivision of a relatively wide frequency band into a number of separate bands.

channel reliability 1. The proportion of time, usually expressed as a percentage, that a communications channel is useful for its intended purpose. 2. The relative ease with which communications may be carried out over a particular channel.

channel-reversal In stereo reproduction, interchanging the two channels to bring to one speaker the signal previously going to the other and vice versa.

channel-reversing switch In a stereo system, a switch for channel reversal.

channel sampling rate The rate at which individual channels are sampled. For example, in the electronic switching of an oscilloscope, channel sampling rate is the number of times per second each input-signal channel is switched to the instrument.

channel selector A switch or relay used to put any of a series of channels into functional status in a system.

channel separation 1. The spacing between communications channels,

expressed in kilohertz. 2. In stereo reproduction, the degree to which the information on one channel is separate from the other expressed in decibels.

channel shift 1. The interchange of communications channels (e.g., the shift from a calling frequency to a working frequency). 2. See *channel reversal*.

channel strip An amplifier for a television receiver, which operates at a fixed channel.

channel-to-channel connection A device such as a channel adapter employed to transfer data rapidly between any two channels of two digital computers (at the baud rate of the slower channel).

channel-utilization index An indication of the extent to which channel capacity is used. For a given channel the index is the ratio of information rate to channel capacity, each expressed in units per second.

channel wave An acoustic wave that travels within a region or layer or a substance because of a physical difference between that layer and the surrounding material. An example of a channel wave is the propagation of sound over a still lake.

channel width In a frequency channel, the difference $f2$-$f1$, where $f1$ is the lower-frequency limit and $f2$ is the upper-frequency limit of the channel.

chapter A self-contained computer program section.

character 1. One of the symbols in a code. 2. In computer practice, a digit, letter, or symbol used alone or in some combination to express information, data, or instructions.

character code In a communications or computer system, the combination of elements (e.g., bits) representing characters.

character crowding A reduction of the time interval between successive characters, especially those read from tape.

character density The number of characters that can be stored in a given length of a medium (say, on magnetic tape).

character emitter A coded-pulse generator in a digital computer.

character generator A device that converts coded information into readable alphanumeric characters.

characteristic 1. A quantity which characterizes (typifies) the operation of a device or circuit. Examples: *plate current, emitter voltage, output power*. 2. In *floating point notation*, the exponent. Compare *mantissa*.

characteristic curve A curve showing the relationship between an independent variable and a dependent variable with respect to the parameter(s) for a device or circuit. Example: the collector voltage-collector current characteristic curve of a transistor.

characteristic distortion 1. In a digital signal, pulse distortion caused by the effects of the previous pulse or pulses. 2. Distortion in the characteristic curve of a transistor or tube.

characteristic frequency. The frequency peculiar to a given channel, service, or response.

characteristic impedance Symbol, $Z0$. The resistance that would be simulated by a given two-conductor line of uniform construction if that line were of infinite length. Characteristic impedance is determined by the materials used for the two conductors, the dielectric used to insulate the two conductors, the diameters of the conductors, and the spacing between them. The characteristic impedance of TV twinlead is 300 ohms; that of RG-8/U coaxial cable is 50 ohms. 2. In a transmission line, $\sqrt{Z1\,Z2}$, where $Z1$ is the open-circuit impedance and $Z2$ is the short-circuit impedance of the line. 3. The value of resistance which, if it terminates a transmission line, results in no reflections along the line.

characteristic overflow In using floating-point arithmetic, the condition that occurs when a characteristic exceeds the upper limit specified by a program or computer.

characteristic spread The range of values over which a characteristic extends, an output-power spread of 10W (i.e., from 15 to 25W) for a wide-tolerance amplifier.

characteristic underflow In using floating-point arithmetic, the condition that occurs when a characteristic exceeds the lower limit specified by a program or computer.

character modifier In address modification, a constant (compare *variable*) that refers to a specific character's location in memory.

character-oriented A computer in which character locations rather than words can be addressed.

character printer A computer output device that prints matter in the manner of a conventional typewriter.

character reader In a digital computer, an input device that can read printing and script directly.

character recognition The reading of a written or printed character by a computer, including its identification and encoding.

character sensing The detection of characters by a computer input device. This may be done galvanically, electrostatically, magnetically, or optically.

character set The set of characters in a complete language, or in a communications system.

character string A one-dimensional character array, i.e., a list of characters that, when printed or displayed, would appear in a row or column, but not both (as in a matrix).

character subset A classification of characters within a set, e.g., the subset *numerals*.

Charactron A cathode-ray readout tube that displays letters, numbers, and symbols on its screen.

charcoal tube In a system for producing a high vacuum, a trap containing activated charcoal, which is heated to dull red then cooled by liquid air to absorb gases.

charge 1. A quantity of electricity associated with a space, particle, or body. 2. To electrify a space, particle, or body, i.e., to give an electric charge. 3. To store electricity, as in a storage battery or capacitor. Compare *discharge*.

charge carrier A mobile particle whose movement consitutes an electric current, e.g., a mobile electron or hole in a semiconductor.

charge-coupled device Abbreviation, CCD. A high-speed, high-density computer storage medium in which the transfer of stored charges provides the method of operation.

charge density The degree of charge or current-carrier concentration in a region.

charged particle 1. See *charge carrier*. 2. See *ion*.

charged voltage 1. The voltage across a fully charged capacitor. 2. The terminal voltage of a fully charged storage cell.

charge holding See *charge retention*.

charge of electron The charge of the electron (negative) is 1.602×10^{-19} coulombs.

charger 1. See *battery charger*. 2. Any device or circuit which charges a capacitor.

charge retention 1. The holding of charge by a battery (i.e., without leakage). 2. The holding of a charge by a capacitor.

charge-storage tube A cathode-ray tube that holds a display of information on its screen until the operator removes it by pressing an erase button.

charge-to-mass The ratio of the electric charge to the mass of a subatomic particle.

charge-to-mass ratio of electron The ratio of the charge *(e)* of the electron to the mass *(me)* of the electron: *e/me*; charge $e = 1.602 \times 10^{-19}$ C, mass $m = 9.11 \times 10^{-31}$ kg.

charge transfer 1. Switching a charge from one capacitor or another. 2. The capture of an electron by a positive ion from a neutral atom of the same kind, resulting in the ion becoming a neutral atom, and the atom, a positive ion.

charging The process of storing electrical energy in a battery (a form of chemical energy). Also, the process of storing electrical energy in a capacitor.

charging current 1. The current flowing into a capacitor. 2. The current flowing into a previously discharged storage cell. Compare *discharge current*.

charging rate 1. The rate at which charging current flows into a battery, expressed in ampere-hours or millampere-hours. 2. The rate at which charging current flows into a capacitor or capacitance-resistance circuit, expressed in amperes, milliamperes, or microamperes.

charged voltage 1. The voltage across a fully charged capacitor. 2. The terminal voltage of a fully charged storage cell.

Charlie Phonetic alphabet code word for the letter C.

chassis A (usually metal) foundation on which components are mounted and wired.

chassis ground A ground connection made to the metal chassis on which the components of a circuit are mounted. When several ground connections are made to a single point on the chassis, a *common ground* results.

chatter 1. A rapidly repetitive signal due to interruption or variation of a current (usually interference). 2. Extraneous vibration, as of a needle on a phonograph record, an armature of a relay, etc.

chatter time The interval between the instant that contacts close (for example, in a relay) and the instant at which their chatter ends.

cheater cord An extension cord used to conduct power to a piece of equipment (especially a TV) by temporarily bypassing a safety switch (interlock).

check 1. A test generally made to verify condition, performance, state, or calculations; specifically, in computer practice, it applies to operands or results. 2. The usually abrupt halting of an action.

check bit Binary check digit. Also see *check digit*.

check character In a group of characters, one whose value is dependent on the other characters, which it checks when the group is stored or transferred.

check digit In computer practice, a number added to a group of digits (forming a code) that identifies entities in the system (including personnel) and used for verification. The check digit is the remainder when the number code, say 459, is divided by a fixed number, say 5; in this case, the check digit (the remainder of 459/5) is 4, and the amended code number is 4594. Also called *check number*.

check indicator An indication, made through a video screen display, that something has been shown to be invalid according to a check.

checking program For debugging purposes, a diagnostic program capable of detecting errors in another program. Also called *checking routine*.

checkout A test routine that ascertains whether or not a circuit or system is functioning according to specifications.

checkout routine A routine used by programmers to debug programs.

checkpoint A point in a digital-omputer program at which sufficient information has been stored to allow restarting the computation from that point.

checkpoint dump Recording details of the progress of a program run(s), which may be needed to reconstruct the process in case of failure.

checkpointing Writing a program so that during a program run information is frequently dumped (as insurance against loss due to a failure).

check problem A presolved problem used to check the operation of a digital computer or program.

check register In some digital computers, a register in which transferred information is stored for agreement with the same information received a second time.

check routine A special program designed to ascertain if a program or computer is operating correctly. Also see *check problem.*

check row On a paper tape, a row or rows acting as a check symbol within a field and used to make a summation check.

checksum Used as part of a summation check, a sum derived from the digits of a number. Also called *hash total.*

check symbol For a specific data item, a digit or digits obtained by performing an arithmetic check on the item, which it then accompanies through processing stages for the purpose of checking it.

check total See *control total.*

check word A check symbol in the form of a word added to, and containing data from, a block of records.

cheese antenna A directive antenna whose reflector is a parabolic cylinder enclosed by plates perpendicular to the elements and spaced so that only one mode propagates in the desired direction.

chelate Pertaining to cylic molecular structure in which several atoms in a ring hold a central metallic ion in a *coordination complex.*

chemical deposition The coating of a surface with a substance resulting from chemical reduction of a solution (see *chemical reduction*). In mirror making, for example, formaldehyde reduces a solution of silver nitrate and deposits metallic silver on the surface of polished glass. Also see *chemically deposited printed circuit.*

chemical detector See *electrolytic detector.*

chemical effect An effect—such as electrolysis, electroplating, and the reduction of ores—produced by electricity passing through a material.

chemical load An arrangement of a chemical material or device for the passage of electricity through it. Examples: electroplater, electrolytic cell for the production of hydroren gas, storage battery.

chemically deposited printed circuit A printed circuit in which the pattern of metal lines and areas are chemically deposited on a substrate.

chemically pure Abbreviation, CP. Free from impurities.

chemical rectifier See *electrolytic cell.*

chemical reduction Making a chemical compound (usually in solution) a metal by removing the nonmetallic component from the compound. For example, when copper oxide is heated in the presence of hydrogen (a reducting agent), the oxygen (the nonmetallic component) is driven out and copper (along with some water) remains.

chemical resistor See *electrolytic resistor.*

chemical switch See *electrochemical switch.*

CHIL *Current-hogging injection logic.* A form of bipolar digital logic circuit.

Child's law The plate current (i_p) of a diode tube varies directly as the three-halves power of the plate voltage (V_p) and inversely as the distance between the plate and cathode: $i_p = BV_p^{3/2}$, where B is a constant which includes the square root of the charge-to-mass ratio of the electron. ($BV_p^{3/2}$ could be written as the square root of the cube in BV_p.)

chip A small slab, wafer, or die of dielectric or semiconductor material on which a subminiature component or circuit is formed.

chip capacitor A subminiature capacitor formed on a chip.

chip resistor A subminiature resistor formed on a chip.

chip tray A chad receptacle located at the card or paper tape punching site.

Chireix-Mesny antenna A short-wave beam antenna in which each dipole section is one side of a square. Cophased horizontal and vertical components of current flow in each of the resulting diagonals, and radiation is broadside to the plane of the primary and parasitic radiators.

chirp A change in the frequency of a continuous-wave Morse-code signal. The chirp usually occurs rapidly, at the beginning of each dot or dash. The chirp may go up or down in frequency. Chirp occurs because of a change in the output impedance of an oscillator as it is keyed. Modern code transmitters do not exhibit significant chirp.

chirp modulation A form of modulation in which the frequency of a signal is deliberately changed in a systematic way. Used in some radar systems.

chirp radar A radar system that employs chirp modulation.

chlorinated diphenyl A synthetic organic substance employed as an impregnant in some oil-filled capacitors.

chlorinated naphthalene See *halowax.*

chlorine Symbol, Cl. A gaseous element of the halogen family. Atomic number, 17. Atomic weight, 35.46.

choke 1. To restrict or curtail passage of a particular current or frequency by means of a discrete component, such as a choke coil. 2. See *choke coil.*

choke air gap A fractional-inch opening in the iron core of a filter choke, usually filled with wood or plastic. The gap prevents saturation of the core when the choke coil carries maximum rated direct current.

choke coil An indicator providing high impedance to alternating current while offering virtually no opposition to direct current.

choke-coil modulation See *choke-coupled modulation.*

choke-coupled modulation An amplitude-modulation system in which the modulator is coupled to the rf amplifier through a shared iron-core choke coil through which the plate currents of the amplifier and modulator flow. Also called *constant-current modulation* and *Heising modulation.*

choke flange At the end of a waveguide, a flange in which a groove forms a choke joint.

choke-input filter A filter whose input component is an inductor (choke). The choke-input power-supply filter is distinguished by its superior voltage regulation compared with the *capacitor-input filter.*

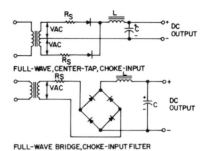

FULL-WAVE, CENTER-TAP, CHOKE-INPUT

FULL-WAVE BRIDGE, CHOKE-INPUT FILTER

Choke-input filters.

choke joint A joint connecting two waveguide sections and permitting efficient energy transfer without requiring electrical contact with the inside wall of the waveguide.

chopped dc See *interrupted dc.*

chopped mode A technique for sequentially displaying on the screen of a single-gun cathode-ray tube, several signals which are not referenced to the oscilloscope sweep.

chopped signal An ac or dc signal which is periodically interrupted, as by means of a *chopper.*

chopper A device or circuit which interrupts a current at some predetermined rate. Ideally, such a device is characterized by distinct on and off operation.

chopper amplifier An amplifier before which a chopper is operated to convert a dc signal to ac for boosting by the amplifier (see *chopper converter*). Also called *converter amplifier.*

chopper converter A chopper that interrupts a direct current, and changes it to a pulsating, rectangular-wave current or voltage which can then be handled by a stable ac amplifier and rectified to supply an amplified dc.

chopper stabilization 1. Stabilization of dc amplification by using a *chopper converter* ahead of a stable ac amplifier and rectifying the amplifier output. 2. In a regulated power supply, use of a *chopper amplifier* at the control-circuit input to improve regulation.

chopper-stabilized amplifier See *chopper amplifier* and *chopper stabilization, 1.*

chopper transistor A transistor, such as Type 2N2004, which will provide continuous interruption of a circuit in the manner of an electromechanical interrupter. See *chopper.*

chopping Clipper action (see *clipper*).

chopping frequency The frequency at which a chopper interrupts a signal.

chord 1. A harmonious mixture of tones of different frequencies. 2. A straight line that joins two points on a curve (such as an arc of a circle). 3. The width of an airfoil.

chord organ An electronic organ which will sound a chord when a key is pressed. (see *chord, 1*).

chorus At extremely low frequencies, the reception of signals, natural in origin, that appear to sweep upward in frequency. This effect results from differences in the speed of propagation of electromagnetic energy, depending on the frequency, in the range below about 10 kHz. The exact origin of the signals is unknown, but is believed to be lightning discharges.

Christiansen antenna A radio-telescope antenna for obtaining high resolution. Two straight arrays are placed at an angle, intersecting approximately at their centers. The resulting interference pattern has extremely narrow lobes.

Christmas tree A treelike pattern on the screen of a television receiver, cause by loss of horizontal sync.

chroma The quality of a color: hue and saturation.

chroma circuit In color TV, one of several circuits whose ultimate purpose is to produce a color component on the screen.

chroma-clear raster In color TV reception, the clear raster resulting from a white video signal or from operation of the chroma circuits of the receiver (as if they were receiving a white transmission). Also called *white raster.*

chroma control In a color TV receiver, a rheostat or potentiometer which permits adjustment of color saturation through variation of the chrominance-signal amplitude before demodulation.

chromatic aberration A fault resulting from dispersion by a lens, i.e., refraction, in different amounts, of light of different colors. This fault causes, among other erratic color effects, rainbows along the edges of images.

chromatic fidelity See *color fidelity.*

chromaticity 1. The state of being chromatic (see *chroma*). 2. A quantitative assessment of a color in terms of dominant or complementary wavelength and purity.

chromaticity coordinate For a color sample, the ratio of any one of the three tristimulus values (primaries) to the sum of the three.

chromaticity diagram A rectangular-coordinate graph in which one of the three *chromaticity coordinates* of a three-color system is plotted against another coordinate.

chromaticity flicker Flicker caused entirely by chromaticity fluctuation (see *chromaticity, 2*).

chromel A nickel-chromium alloy with some iron content, used in thermocouples.

chromel-alumel junction A thermocouple employing wires of the alloys chromel and alumel.

chromel-constantan thermocouple A thermocouple consisting of a junction between wires or strips of chromel and constantan. Typical output is 6.3 mV at 100 °C.

chrome plated Plated with chromium.

chrominance In color TV, the difference between a reproduced color and a standard reference color of the same luminous intensity.

chrominance amplifier In a color TV circuit, the amplifier separating the chrominance signal from the total video signal.

chrominance cancellation On a black-and-white picture tube screen, cancellation of the fluctuations in brightness caused by a chrominance signal.

chrominance-carrier reference In color TV, a continuous signal at the frequency of the chrominance subcarrier; it is in fixed phase with the color burst and provides modulation or demodualtion phase reference for carrier-chrominance signals.

chrominance channel In color TV, a circuit devoted exclusively to the color function, as opposed to audio and general control channels.

chrominance component In the NTSC color TV systems, either of the components (*I-signal* or *Q-signal*) of the complete chrominance signal.

chrominance demodulator In a color TV receiver, a demodulator which extracts video-frequency chrominance components from the chrominance signal and a sine wave from the chrominance subcarrier oscillator.

chrominance gain control A rheostat or potentiometer in the red, green, and blue matrix channels of a color TV receiver, used to adjust the primary-signal amplitudes.

chrominance modulator In a color TV transmitter, a device which generates the chrominance signal from the I and Q components and the chrominance subcarrier.

chrominance primary One of the transmission primaries (red, green, blue) upon which the chrominance of a color depends.

chrominance signal The color signal component in color TV that represents the hues and saturation levels of the colors in the picture.

chrominance subcarrier In color TV, the 3579.545 kHz signal that serves as a carrier for the I- and Q-signals.

chrominance-subcarrier oscillator A crystal-controlled oscillator in a color TV receiver, that generates the subcarrier signal (see *chrominance subcarrier*).

chrominance video signals Output signals from the red, green, and blue channels of a color TV camera or color TV receiver matrix.

chromium Symbol, Cr. A metallic element. Atomic number, 24. Atomic weight, 52.01.

chronistor An elapsed-time indicator in which current flowing during a given time interval electroplates an electrode. The duration of the interval is determined from the amount of deposit.

chronograph 1. An instrument which provides an accurate time base along the horizontal axis of its permanent record. 2. Stopwatch.

chronometer A precision clock. Electronic chronometers often employ a highly accurate and stable crystal oscillator followed by a string of multivibrators to reduce the crystal frequency to an audio frequency (such as 1 kHz) which drives the clock motor.

chronoscope An instrument for precisely measuring small time intervals.

CHU Calls letters of the Canadian time-signal station whose primary frequency is 7.335 MHz.

Ci Symbol for *input capacitance.*

CIE *International Commission on Illumination.*

cinching In a reel of magnetic tape, the slipping of tape as force is applied.

cinematograph See *kinematograph.*

cipher A code used for the purpose of preventing interception of a message by third parties.

circ 1. Abbreviation of *circuit.* 2. Abbreviation of *circular.*

circle graph A representational device consisting of a disk subdivided into various triangular areas (radiating from the center of the circle), which are proportional to represented quantities. Often called *pie chart* from its resemblance to a sliced pie.

circle of confusion A circular image of a point source of light, resulting from an aberration in an optical system.

circle of convergence of a series In the complex-number plane, a circle bounding the region of convergency of a series. Also see *convergent series.*

circle of declination The graduated circular scale of a declinometer.

circlotron amplifier A high-powered microwave amplifier of the one-port, cross-field, nonlinear type using a magnetron.

circuit 1. A closed path through which current flows from a generator, through various components, and back to the generator. (An electronic circuit is often a combination of interconnected circuit.) 2. The wiring diagram of a circuit.

circuit analysis The careful determination of the nature and behavior of a circuit and its various parts. It may be theoretical, practical, or both. Compare *circuit synthesis.*

circuit analyzer A multimeter. Also see *circuit tester.*

circuit board A panel, plate, or card on which components are mounted and interconnected to provide a functional unit.

circuit breaker A resettable fuse-like device designed to protect a circuit against overloading. In a typical circuit breaker the winding of an electromagnet is connected in series with the load circuit and with the switch contact points. Excessive current though the magnet winding causes the switch to be tripped.

circuit capacitance The total capacitance (lumped plus stray) present in a circuit.

circuit capacity 1. The ability of a circuit to handle a quantity (such as current, voltage, frequency, power, etc.) safely and efficiently, e.g., a circuit capacity of 50A. 2. The number of channels which can be accommodated simultaneously by a circuit.

circuit card See *card, 1* and *circuit board.*

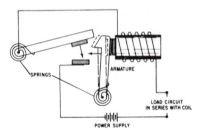

Circuit breaker.

circuit component 1. Any of the electronic devices or parts (capacitors, resistors, transistors, and so forth) which are connected through wiring to form a circuit. 2. An electrical quantity required for, or arising from, circuit operation. Examples: input voltage, feedback current, stray capacitance, circuit noise.

circuit diagram A drawing in which symbols and lines represent the components and wiring of an electronic circuit. Also called *circuit schematic, schematic diagram,* and *wiring diagram.*

circuit dropout A sometimes momentary interruption of circuit operation that may be due to a break in the circuit.

circuit efficiency The effectiveness of circuit operation, customarily expressed as the ratio of input to output signal values, or vice versa.

circuit element See *circuit component, 1.*

circuit engineer An electronic engineer who specializes in circuit analysis, circuit synthesis, or both.

circuit fault 1. Malfunction of a circuit. 2. An error in circuit wiring.

circuit hole A perforation within the conductive area of a printed-circuit board, for the insertion and connection of a pigtail, terminal, etc. or for connecting the conductors on one side of the board with those on the other.

circuit loading Intentionally or unintentionally drawing power from a circuit.

circuit noise 1. Electrical noise generated by a circuit in the absence of an applied signal. 2. In wire telephony, electrical noise as opposed to acoustic noise.

circuit noise level The ratio of circuit-noise amplitude to reference-noise amplitude, expressed in decibels above the reference amplitude.

circuit-noise meter A meter that measures the intensity of the noise generated within a circuit.

circuit parameter See *circuit component, 2.*

circuit protection Automatic safe-guarding of a circuit from damage, as by overload, excessive drive, heat, vibration, etc. Protection is afforded by various devices and subcircuits, ranging from the common fuse to sophisticated limiters and breakers.

circuit reliability A quantitative indication of the ability of a circuit to provide dependable operation as specified.

circuitry 1. Collectively, electronic and electrical circuits. 2. A detailed plan of a circuit and its subcircuits. 3. Collectively, the components of a circuit.

circuit schematic See *circuit diagram.*

circuit simplification 1. In circuit analysis the reduction of a complex circuit to its simplest representation to minimize labor and to promote clarity. Thus, through application of Kirchhoff's laws, a complicated circuit could theoretically be reduced to a single generator in series with a single load impedance. 2. In circuit synthesis the arrangement of a circuit so as to give desired performance with the fewest components and least complex wiring.

circuit synthesis The development of a circuit under the guidance of theoretical or practical knowledge of basic electronics principles and component parameters. Compare *circuit analysis*.

circuit tester An instrument—often a specialized continuity tester, but occasionally a dynamic performance tester as well—for checking electronic circuits. Also see *multimeter*.

circuit tracking Aligning or pretuning circuits for identical response. It applies especially to cascaded circuits whose variable elements (capacitors, resistors, inductors, autotransformers and the like) must follow each other in step when ganged together.

circular angle The angle described by a radius vector as it rotates counterclockwise around a circle.

circular antenna A half-wave horizontally polarized antenna whose elements are rigid conductors (such as copper tuding) bent into a circle.

circular electric wave A wave having circular electrostatic lines of force.

circular functions Trigonometric functions of the angle described by a vector rotating counterclockwise around a circle. Also see *cosine, cosecant, cotangent, secant, sine, tangent*. Compare *hyperbolic functions*.

circular magnet See *ring magnet*.

circular magnetic wave An electromagnetic wave in which the magnetic lines of force are circles.

circular mil A unit of cross-sectional area equivalent to 0.785 millionths of a square inch, or the area of a circle having a diameter of 0.001 inch. Generally, the circular mil is used to specify the cross-sectional area of a conductor such as wire.

circular-mil area For a round conductor, the diameter (in mils) squared.

circular mil foot A unit of volume in which the length is 1 foot and the cross-sectional area in 1 circular mil.

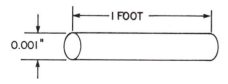

Circular mil foot.

circular polarization A form of electromagnetic-wave polarization in which the orientation of the electric lines of force rotates continuously and uniformly as the wave propagates through space. Circular polarization may occur in either a clockwise or counterclockwise sense.

circular-polarized wave A usually transverse electromagnetic wave whose electric or magnetic field vector around a point, describes a circle.

circular radian The angle enclosed by two radii of a unit circle and subtended by a unit arc. The radian is a unit of plane angular measurement. A circle contains 2π radians. 1 circular radian = 57.296°. Compare *hyperbolic radian* and *steradian*.

circular scan A radar scan in which the electron-beam spot describes a circle centered around the transmitting antenna (see *j-display*).

circular sweep In an oscilloscope, a sweep obtained when the horizontal and vertical sinusoidal deflecting voltages have the same amplitude and frequency but are out of phase by 90 degrees.

circular trace An oscilloscope pattern consisting of a circle obtained with a circular sweep of the electron beam.

circular waveguide A waveguide to circular cross section.

circulating register In a digital computer, a register in which digits are taken from locations at one end and returned to those at the other end.

circulating tank current The alternating current that oscillates between the capacitor and inductor within a tank circuit.

circulator A multiterminal coupler in which microwave energy is transmitted in a particular direction from one terminal to the next.

cis A prefix meaning *on this side of* e.g., *cislunar field*, the field on this side of the moon.

citizen band A band of radio frequencies allocated for two-way communication between private citizens (apart from amateur and commercial services).

citizens radio service Two-way radio communication in a citizen band. In the United States, the FCC licenses users of this service without requiring them to take an examination.

Ci(x) Abbreviation of *cosine integral*.

c/kg Coulombs per kilogram, the unit for electron charge-to-mass ratio.

C/kmol Coulombs per kilmol, the unit for the Faraday constant.

ckt Abbreviation of *circuit*.

Cl Abbreviation of *chlorine*.

cl Abbreviation of *centiliter*.

clamp See *clamper* and *clamp tube*.

clamper A tube or semiconductor device that restricts a wave to a predetermined dc level. Also called *dc restorer*.

clamping 1. Fixing the operation of a device at a definite dc level. Also see *clamper*. 2. In TV, establishing a fixed level for the picture signal at the start of each scanning line.

clamping circuit See *clamper*.

clamping diode A diode employed to fix the voltage level of a signal at a particular reference point.

clamp tube A tube (and circuit) employed to cut off an rf power amplifier or to limit its dissipation when its excitation is lost. Also called *clamper tube*.

clamp-tube modulation Screen-grid amplitude modulation obtained with a triode clamp operated between the screen grid and ground of the modulated rf amplifier. The audio input signal is applied to the control grid of the clamp tube.

clapper In a bell, the ball or hammer that strikes the bell; in an electric bell it is affixed to the vibrating armature.

Clapp-Gouriet oscillator A Colpitts oscillator in which a capacitor is connected in series with the inductor. The circuit offers high frequency stability in the presence of input and output capacitance variations.

Clapp oscillator A series-tuned hybrid Colpitts circuit having a tuning capacitor in series with the inductor rather than in parallel, as in conventional Colpitts circuit. The circuit allows the use of a smaller tuning capacitor, resulting in improved stability.

Clark cell See *zinc standard cell*.

class A amplifier An amplifier whose bias is set at approximately the halfway position between cutoff and saturation. Plate current flows during the complete ac driving-voltage cycle.

class AB amplifier A push-pull amplifier whose grid bias is adjusted to a level between that of a class A amplifier and class B amplifier.

class AB modulator A modulator whose output stage is a class AB amplifier.

class-AB operation The operation of a transistor, field-effect transistor, or tube, in which the collector, drain, or plate current flows for less than the whole signal cycle, but for more than half the cycle.

class-A modulator A circuit for obtaining amplitude-modulated signals; essentially a class-A amplifier with variable gain.

class-A operation The operation of a transistor, field-effect transistor, or tube, in which the collector, drain, or plate current flows during the entire signal cycle.

class A oscilloscope See *A-scope.*

class A signal area A strong TV signal area characterized by minimum field strength of 2.500 microvolts per meter.

class-A station In the Citizens' Radio Service, a station licensed in the band 460 to 470 MHz. The maximum allowed input power to the transmitter is 60 watts.

class B amplifier An amplifier whose bias is adjusted to operate at the cutoff knee in the characteristic curve. Output current flows during positive excursions of the ac input signal but not during negative excursion because a negative excursion drives the stage into immediate cutoff.

class B modulator A push-pull modulator whose output stage is a class B amplifier.

class-B operation The operation of a transistor, field-effect transistor, or tube, in which the collector, drain, or plate current flows for approximately half the signal cycle.

class B oscilloscope See *B-scope.*

class-B station In the Citizens' Radio Service, a station licensed in the band 460 to 470 MHz. The maximum allowed input power to the transmitter is 5 watts.

class C amplifier An amplifier whose input-electrode bias is adjusted to twice the plate-current cutoff level or higher. Output current, accordingly, flows during less than half the ac driving-voltage cycle.

class-C operation The operation of a transistor, field-effect transistor, or tube, in which the collector, drain, or plate current flows for significantly less than half the signal cycle. The class-C amplifier requires a large amount of drive, and cannot be used as a linear amplifier.

class-C station In the Citizens' Radio Service, a station licensed in the band 26.960 to 27.230 MHz, 27.255 MHz, and in the range 72 to 76 MHz for radio-control purposes.

class-D station In the Citizens' Radio Service, a station licensed to operate on any of 40 discrete channels in the range 26.965 to 27.405 MHz. The transmitter output power is limited to 4 watts for amplitude modulation and 12 watts peak-envelope power for single sideband.

class D telephone A telephone restricted to use by emergency services, such as fire departments, guard alarm installations, and watchmen.

class J oscilloscope See *J-scope.*

clean room A room for the assembly or testing of critical electronic equipment. The term comes from the extraordinary steps taken to remove dust and other contaminating agents. The personnel wear carefully cleaned garments (or disposable clothing), gloves, caps, and masks; in some situations, they are required to walk between ceiling and floor ducts of a vacuum system upon entry to the room.

cleanup process In the process of electron tube evacuation, a technique used to remove residual and occluded gases from the vacuum apparatus and from the device being evacuated.

clear 1. In computer practice, to restore a switching element (e.g., a flip-flop) or a memory element to its standard (e.g., zero) state. 2. In computer practice, an asynchronous input.

clearance The distance between (1) terminals, (2) two live terminals, or (3) one live terminal and ground.

clear band In optical character recognition, the part of a document that has to remain unprinted.

clear channel A channel in the standard AM broadcast band that is designated to only one station within the area covered by the signal from that station.

clear memory A function in a calculator or small computer that erases the contents of the memory.

clear raster The raster on the screen of a TV picture tube in the absence of a signal, noise, or faulty beam deflection.

cleavage Ascribed to a rock or crystal, the quality of splitting along definite planes. Also, a fragment resulting from such a cleft.

click filter See *key-click filter.*

click method 1. An emergency technique for making a current aduibly detectable by making and breaking the circuit carrying it to headphones. A single click results from each make and each break. Also see *tikker.*

click suppressor See *key-click filter.*

climate chamber A test chamber which provides accurately controlled temperatures (and sometimes humidities and pressures as well) for evaluating the performance of electronic components and circuits. Also called *environmental test chamber.*

climatometer An instrument incorporating a hygrometer and bimetallic thermometer whose pointers intersect to indicate, on a dial, "comfort zones," i.e., best temperature-to-humidity ratio.

clip A pinch-type connector whose jaws are normally held closed by a spring.

Clip.

clipped-noise modulation Modulation of a jamming signal through clipping action to increase the sideband energy and resulting interference.

clipper A circuit whose output voltage is fixed at a value for all input-voltages higher than a predetermined value. Clippers can *flat-top* the positive, negative, or both positive and negative peaks of an input voltage.

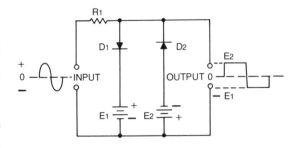

Clipper.

clipper amplifier An amplifier operated so that the positive, negative, or both positive and negative peaks are clipped in the output signal. The clipping action results from feeding a regular symmetric waveform into an amplifier so that on negative excursion extremes the stage is cut off and on positive excursion extremes the amplifier is driven into saturation.

clipper-limiter A device which delivers output signal whose amplitude range corresponds to input-signal voltages between two predetermined limits. Thus, a noise-reduction device employing an element or elements that clip all pulses whose amplitudes are greater than the signal being processed.

clipping 1. Leveling off (flat-topping) a signal peak at a predetermined level. Also see *clipper*. 2. In audio practice, the loss of syllables or words because of cutoff periods in the operation of the circuit (usually caused by overdriving a stage).

clock In a digital computer or control system, the device or circuit which supplies the timing pulses that pace the operation of the digital system.

clocked flip-flops A "master-slave" arrangement of direct-coupled flip-flops. Information entered into the master unit when the input-trigger pulse amplitude is high is transferred to the slave unit when the amplitude is low.

clock frequency In a digital computer or control, the reciprocal of the period of a single cycle, expressed in terms of the number of cycles occurring in one second of time (hertz, kilohertz, or megahertz).

clock generator A test-signal generator that supplies a chain of pulses identical to those supplied by the clock of a digital computer.

clock module A complete plug-in or wire-in digital unit whose readout indicates time of day or elapsed time. Connected to a suitable power supply, it serves as either a clock or timer.

clock pulse A time-base pulse supplied by the clock of a digital computer, expressed as a period whose reciprocal is frequency.

clock rate See *clock frequency* and *baud*.

clock track On a magnetic tape or disk for data storage, a track containing read or write control (clock) pulses.

clockwise Abbreviation, cw. Rotation in a right-hand direction around a circle, starting at the top. Compare *counterclockwise*.

clockwise-polarized wave An elliptically polarized TEM wave whose electric-intensity vector rotates clockwise as observed from the point of propagation. Compare *counterclockwise-polarized wave*.

close coupling Tight coupling, as when a primary and a secondary coil are placed as close together as possible for maximum energy transfer. Also called *tight coupling*. Compare *loose coupling*.

closed capacitance The value of a variable capacitor whose rotor plates are completely meshed with the stator plates. Compare *open capacitance*.

closed circuit A continuous unbroken circuit, i.e., one in which current can flow without interruption. Compare *open circuit*.

closed-circuit cell A primary cell, such as the early gravity cell, designed for heavy and polarization-free service.

closed-circuit communication Communication between units in a closed circuit, i.e., not extending to other units or circuits.

closed-circuit signaling Signaling accomplished by raising or lowering the level of a signaling current flowing continuously in a circuit.

closed-circuit television Abbreviation, CCTV. A usually in-plant television system in which a transmitter feeds one or more receivers through a cable.

closed core A magnetic core generally constructed in a 0 or D configuration to confine the magnetic path to the core material. Compare *open core*.

closed-core choke A choke coil wound on a closed core. Also called *closed-core inductor*; see *closed core*. Compare *open-core choke*.

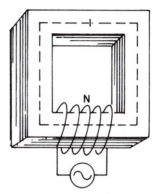

Closed-core choke.

closed-core transformer A transformer wound on a closed core; see *closed core*. Compare *open-core transformer*.

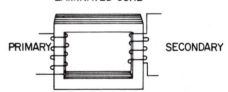

Closed-core transformer.

closed loop 1. The feedback path in a self-regulating control system. An oscillator, for example, is a closed-loop amplifier. 2. A loop within a program which would continue indefinitely except for external exit command.

closed-loop bandwidth The frequency at which the gain of a closed loop (1) drops 3 decibels from the dc or midband value.

closed-loop control system A control system in which self regulation is obtained by means of a feedback path (see *closed loop*). An example is the voltage regulator in which a rise in output voltage is fed back to the input. This changes the input voltage and reduces the output voltage to its correct value. Compare *open-loop control system*.

closed-loop input impedance The input impedance of an amplifier having feedback.

closed-loop output impedance The output impedance of an amplifier having feedback.

closed-loop voltage gain The voltage gain of an amplifier having feedback.

closed magnetic circuit A magnetic circuit in which the flux is uninterrupted, as in a ferromagnetic core, which has no air gap. Also see *closed core*.

closed subroutine In a digital computer program, a subroutine that can be accessed and left by branch instructions, such as BASIC's GOSUB and RETURN.

close-spaced array A beam antenna in which the elements (radiator, director, reflector) are spaced less than a quarter-wavelength apart.

close-talk microphone A microphone that must be placed close to the mouth. Such a microphone is less susceptible to background noises than an ordinary microphone, and is useful in environments where the ambient noise level is high.

closing rating A specification for closure conditions in a relay, including duty cycle and contact life (total guaranteed closures before contact failure).

closure 1. The act of closing or being closed, e.g., *switch closure, relay closure.* 2. Circuit completion, i.e., the elimination of all discontinuities.

cloud The mass of electrons constituting the space charge in a vacuum tube.

cloverleaf antenna An omnidirectional transmitting antenna in which numerous horizontal, four-element radiators (stacked vertically, a quarter-wavelength apart) are arranged in the shape of a four-leaf clover.

C-L/ratio See *L-C ratio.*

clutch In a punched card system, that which transfers drive to the transport.

clutch point In a punched-card system, the operational cycle point at which the clutch is engaged.

clutter Extraneous echoes that interfere with the image on a radar-scope screen.

clutter gating In radar practice, a switching process which causes the normal video to be displayed in regions free of clutter and the video indicating target movement to be displayed only in cluttered areas.

Cm Symbol for *curium.*

cm Abbreviation of *centimeter.*

c.m. Abbreviation of *circular mil.*

cm^2 Abbreviation of *square centimeter.*

cm^3 Abbreviation of *cubic centimeter.*

Cmax Abbreviation of *maximum capacitance.*

C-meter See *capacitance meter.*

Cmin Abbreviation of *minimum capacitance.*

CML Abbreviation of *current-mode logic.*

CMOS Abbreviation of *complementary metal-oxide semiconductor.*

CMR Abbreviation of *common-mode rejection.*

CMRR Abbreviation of *common-mode rejection ratio.*

C-network A circuit having three impedances connected in series, the free leads being connected to a pair of terminals and the two internal junctions, to another pair of terminals.

Co Symbol for *cobalt.*

Co Symbol for *output capacitance.*

coalesce In computer practice, to create one file from several.

coarse adjustment Adjustment of a quantity in large increments. Compare *fine adjustment.*

coarse-chrominance primary See *Q-signal.*

coastal bending A change in the horizontal direction of a line-of-sight radio wave when it crosses a coastline.

coast station In the maritime mobile service, a land station that communicates with shipboard stations.

coated cathode A vacuum-tube cathode that has been coated with a substance, such as an alkaline-earth metal to increase its electron emission.

coated filament A tube filament coated with some material, such as a mixture of barium and strontium oxides, to increase electron emission.

coating 1. The application of a substance to another substance by means of electroplating, electrophoresis, or similar process, for the purpose of protecting the material, isolating it from the environment, or improving the conductivity of an electrical connection to some other object. 2. The magnetic material on a recording tape.

coating thickness In magnetic tape, the thickness of the magnetic coating applied to the base tape.

coax Abbreviation of *coaxial cable* or *coaxial line.*

coaxial antenna A vertical antenna consisting of a quarter-wavelength metal pipe fed by a coaxial cable. The outer conductor of the cable is connected to the pipe through a shorting disk at the top, and the inner conductor is connected to a quarter-wave vertical radiator extending from the top of the pipe.

coaxial balun See *bazooka.*

coaxial cable A cable consisting of two concentric conductors: an inner wire and an outer, braided sleeve.

Coaxial cable; from left to right: insulating jacket, woven outer conductor, low-loss insulating sleeve, braided inner conductor.

coaxial cavity A cavity consisting of a cylindrical metal chamber housing a central rod; the cavity can be tuned to resonance by means of a piston.

coaxial connector A device used for splicing coaxial line, or for connecting a coaxial line to a transmitter, receiver, or other piece of apparatus.

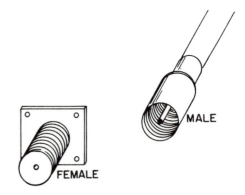

Coaxial connector.

coaxial diode A semiconductor diode housed in a cylindrical metal shell acting as one contact, and provided with a recessed, concentrically mounted end pin, which serves as the other contact.

coaxial filter 1. A filter that uses a coaxial cable as a tuned circuit. 2. A filter designed to be used in a coaxial transmission line.

coaxial jack A female connector whose concentric terminals have the same spacing as the coaxial cable connector with which it makes.

coaxial line See *coaxial cable*.

coaxial-line frequency meter A microwave absorption wavemeter (see *wavemeter*) with input and output receptacles for insertion into a coaxial line.

coaxial-line oscillator See *concentric-line oscillator*.

coaxial-line matching section See *bazooka*.

coaxial loudspeaker See *coaxial speaker*.

coaxial plug A male connector whose concentric terminals have the same spacing as the coaxial cable to the end of which the plug is attached.

coaxial receptacle A coaxial connector, such as a coaxial jack or plug. Receptacles are installed in equipment, wheras plugs are usually attached to the end of coaxial cables.

coaxial relay A relay specially designed to switch coaxial cables without disturbing their impedance.

coaxial speaker A large (low frequency) speaker and a small (high frequency) speaker mounted concentrically (the smaller within the larger). The combination, in concert with a crossover network, provides fairly good wide-range response for the space it saves.

coaxial stub A length of coaxial cable acting as a branch to another coaxial cable.

coaxial switch A switch specially designed to connect and disconnect coaxial cables without disturbing their impedance.

coaxial tank A uhf tank consisting of a rod within a cylinder. The tank is usually tuned by a small variable capacitor connected between the rod and cylinder at one end of the combination.

coaxial-tank oscillator A stable, self-excited oscillator employing a coaxial tank. Also see *concentric-line oscillator*.

coaxial transistor A transistor in which a semiconductor wafer is mounted centrally in a metal cylinder (the base connection) and is contacted on opposite faces by the emitter and collector whiskers, which are axially mounted.

coaxial transmission line A transmission line that is a coaxial cable.

coaxial wavemeter A type of absorption wavemeter in which the tunable element is a section of coaxial line (i.e., a metal cylinder surrounding a metal rod). An internal short-circuiting disk is moved along the cylinder to connect its inner wall to selected points along the rod's length, thereby varying the resonant frequency. The instrument is useful at microwave frequencies.

cobalt Symbol, Co. A metallic element. Atomic number, 27. Atomic weight, 58.94.

COBAL Acronym for *common business-oriented language*, an internationally accepted, high-level computer programming language for accurately expressing, in a standard (English) form business data-processing procedures. A COBOL compiler can interpret such statements as IF STOCK LESS THAN MINIMUM GO TO REORDER.

cochannel interference Interference between similar signals transmitted on the same channel.

CockcroftWalton accelerator An accelerator (see *accelerator, 1*) in which nuclei of an ionized gas are given high velocity, in a single acceleration through a straight tube, by a high dc voltage.

codan Any of several muting (*squelch*) systems i.e., one that suppresses noise that a sensitive receiver equipped with automatic gain control would be plagued with in the absence of a carrier. The receiver is quiet until a carrier of predetermined strength is received. The name is an acronym from Carrier-Operated Device Antinoise.

codan lamp A lamp that alerts a radio operator that a signal of satisfactory strength is being received. Also see *codan*.

code 1. A set of symbols for communications, e.g., the Morse code of radiotelegraphy and wire telegraphy in which dots and dashes correspond to letters, numbers, and marks of punctuation. 2. In a computer program, symbolically represented instructions. 3. Encode.

code character 1. The representation of character in a particular code form. 2. A sequence of dots and dashes in the Morse code.

code conversion The translation of a coded signal from one form of code to another.

code converter A circuit that performs code conversion.

coded decimal digit A decimal digit expressed in zeros and ones (computer code). Also see *binary-coded decimal notation*.

code-directing characters Characters added to a message to indicate how and where it is going.

coded passive reflector antenna A passive reflecting radar antenna whose reflection characteristics vary according to a rode.

coded program See *program*.

coded signal 1. A wire- or radiotelegraph signal in which secrecy is achieved by using letters in secret cipher groups, instead of straight language. 2. A *scrambled signal*.

coded stop See *programmed halt*.

code elements The bits comprising a character in a code.

code holes In a punched card or tape, holes representing data.

code lines A written computer program instruction.

code machine Any one of several devices for recording or reproducing code signals.

code position The part of a data medium (e.g., card row) reserved for data.

code-practice oscillator A (usually) simple audio oscillator with a key and headptones or a loudspeaker for practicing sending Morse code.

coder 1. In computer practice, one who prepares instructions from flow charts and procedures devised by a programmer. 2. A device that delivers coded signals.

code receiver A radiotelegraph receiver.

code ringing A method of ringing a telephone subscriber in a predetermined manner to convey a certain message.

code segment The instruction part of computer storage associated with a process. Compare *data segment, dump segment*.

code set The set of codes representing all of the characters in a particular code and language.

code speed See *keying speed*.

code transmitter 1. A radiotelegraph transmitter. 2. A tape-operated keyer for wire telegraphy or radiotelegraphy.

code word See *phonetic alphabet code word*.

coding 1. Performing the service of a coder (see *coder, 1*). 2. Writing instructions for a digital computer, a part of programming.

coding check A pencil-and-paper verification of a routine's validity.

coding sheet A form on which program instructions are written prior to input.

codiphase radar A radar system employing beam forming, signal processing, and a phased-array antenna.

codistor A voltage-regulating semiconductor device.

coefficient 1. A factor in an indicated product. Thus, in $4y$, 4 is the coefficient of y y, the coefficient of 4. 2. A parameter (plate current, emitter voltage, gate resistance, etc.). 3. The fixed-point (fractional) part in floating-point number representation; also called *characteristic*.

coefficient of coupling Symbol k. The ratio of mutual inductance between coupled inductors to the maximum possible (theoretical) value of mutual inductance (M_{max} = 1). The coefficient of coupling k = $M/\sqrt{L_1 L_2}$, where L_1 and L_2 are self-inductance values for the coupled inductors. See also *inductive coupling* and *mutual inductance*.

coefficient of current detection See *current-detection coefficient*.

coefficient of reflection A measure of the amount of electromagnetic field reflected in a transmission line from the load feed point. The coefficient of reflection is equal to the square root of the reflected power divided by the forward power.

coercive force The demagnetizing force required to remove the residual magnetism of a material.

coercivity See *coercive force*.

cogging Nonuniform rotation of a motor's armature; the velocity increases as an armature coil enters the magnetic field and decreases as it leaves the field.

coherence The property of electromagnetic radiation wherein all the waves are in the same phase.

coherent bundle A bundle of optical fibers, such that the individual fibers are in the same relative positions at either end of the bundle.

coherent carrier A carrier that agrees in frequency and phase with a reference signal.

coherent electroluminescent device See *diode laser*.

coherent light Light in which the phase relationship between succeseive waves is such that the beam consists of parallel rays which provide a high concentration of energy. Also see *laser*.

coherent-light radar See *colidar*.

coherent oscillator In a radar system, an oscillator that provides a *coherent reference*.

coherent-pulse operation Pulse operation characterized by a fixed phase relationship between pulses.

coherent radiation Radiation characterized by definite phase relationships between the energy components at various points across a beam.

coherent reference A stable reference frequency with which other signals are phase locked for coherence.

coherent transponder A transponder in which the frequency and phase of the input and output signals have a fixed relationship.

coherer An early radiotelegraph detector consisting of a mixture of silver and nickel filings held between the tips of metal plugs in a glass tube. When the coherer is connected to a tuned circuit and antenna, the received rf signal (a dot or dash) causes the filings to stick together (cohere) and close a local circuit through a bell or buzzer.

coil A long conductor or group of conductors wound into a tight package to take advantage of the resulting inductance. See also *inductor*.

coil antenna See *loop antenna*.

coil checker An ac meter or simple bridge for checking inductors. Such instruments usually only indicate inductance values, but some give readings of resistance or approximate inductor Q.

coil dissipation The power wasted in a coil as heat. Generally, this dissipation or loss is proportional to the resistance of the coil, and to the square of the current passing through the coil.

coil form The insulating support around which a coil is wound.

coil loading Inserting inductors into a line or antenna to alter its electrical characteristics. Also see *loading coil*.

coil magnification factor The Q of an inductor (L), i.e., the ratio X_L/R_L.

coil neutralization See *inductive neutralization*.

coil resistance The resistance of a coil (inductor), as distinct from its reactance, due almost entirely to the coil wire's resistance.

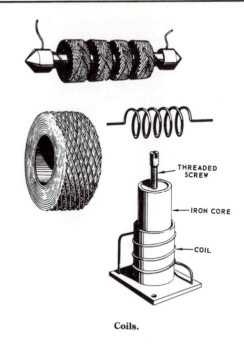

THREADED SCREW

IRON CORE

COIL

Coils.

coilshield A metal can designed expressly to provide efficient electrostatic and electromagnetic shielding of a coil.

coincidence The simultaneous occurrence of two or more signals. Compare *anticoincidence*.

coincidence amplifier An amplifier that delivers an output sugnal only when two or more input signals occur simultaneously.

coincidence circuit See *AND circuit*.

coincidence counter A circuit or device, such as a gate, which delivers an output pulse only when two or more input pulses occur simultaneously; the output pulses go to a device that counts them.

coincidence detector See *AND circuit*.

coincidence gate See AND gate.

coincident-current selection Selection of a magnetic core (in a core memory or similar device) by applying two or more currents simultaneously.

coin shooting The practice of searching for coils and similar small buried objects by means of a *metal locator*.

coke A porous material obtained from the destructive distillation of coal. It is valued for the production of carbon components for electronics, e.g., the carbon electrode of dry cells; brushes for motors, generators, and variable autotransformers; and miscellaneous electrodes.

cold 1. An electrical circuit or terminal that is at ground potential. 2. A term denoting a bad solder joint.

cold alignment The alignment of tracking a system (especially of its tuned circuits) when the system is not in operation, as when tube or transistor power is off. Also called *quiet alignment*.

cold cathode 1. In an electron tube, a cathode that emits electrons without being heated. 2. A cathode electrode operated at a temperature below ambient temperature.

cold-cathode tube A tube, such as a gas diode without a filament, in which electrons are pulled from a cold cathode by the high anode voltage.

cold chamber An enclosure in which electronic equipment can be tested at selected, precise subzero temperatures. Compare *oven*.

cold emission Electron emission by an unheated cathode, as in a cold-cathode tube.

cold flow The (usually gradual) change in the dimensions of a material, such as a plastic in a molded part.

cold junction In a thermocouple system, an auxiliary thermocouple connected in series with the hot thermocouple and either immersed in ice or operated at ambient temperature.

cold light Light produced without attendant heat, as from the ionization of a gas by a high voltage (neon bulbs, fluorescent lamps) or by electroluminescence, bioluminescence cathodoluminescence, and so forth.

cold pressure welding Welding sometimes used in the fabrication of electronic equipment, in which the metal parts to be joined are pressed together tightly to the point of deformation, whereupon they become welded.

cold resistance The resistance of an unheated electronic component. Compare *hot resistance*.

cold rolling A method of manufacturing an inductor core, such that the magnetic grains are all arranged lengthwise.

cold solder joint A solder joint in which insufficient heat has been applied, resulting in a bad connection.

cold spot 1. An area of a circuit or component whose temperature is ordinarily lower than that of the surrounding area. 2. A node of current or voltage. Compare *hot spot*.

cold weld A welded joint produced by means of *cold pressure welding*.

colidar An optical radar system using unmodulated, coherent (laser-produced) light. The term is an acronym for *coherent light detection and ranging*.

collate In data processing, to produce an ordered set from two or more similarly ordered sets (as a punched cards).

collator In a punched-card system, a device that collates (see *collate*) punched cards.

collector 1. In a bipolar transistor, the electrode toward which emitted current carriers travel. 2. In a klystron the final electrode toward which electrons migrate after passing through the buncher and catcher. 3. In an iconoscope a cylindrical electrode around the circumference of the tube, which gathers and conducts away the electrons released by the mosaic. 4. The final (target) electrode in a backward-wave or traveling-wave tube. 5. A computer program segment that collates compiled segments so they can be loaded into the computer.

collector capacitance 1. Symbol, Cc. The capacitance of the collector junction in a bipolar transistor. 2. The capacitance of the collector electrode in a klystron, iconoscope, backward-wave tube, or traveling-wave tube.

collector current 1. Symbol, Ic. The current flowing in the collector circuit of a bipolar transistor. Also see *ac collector current* and *dc collector current*. 2. Current flowing in the collector circuit of a klystron, iconoscope, backward-wave tube, or traveling-wave tube.

collector-current cutoff See *collector cutoff*.

collector cutoff In a bipolar transistor, the condition in which the transistor collector current is cut off, i.e., reduced to the residual value. Also see *cutoff current*.

collector cutoff current See *cutoff current*.

collector-diffusion isolation A method of making integrated circuits containing bipolar transistors. Provides electrical separation of the transistors in a semiconductor integrated circuit.

collector dissipation Symbol, Pc. In a bipolar transistor, the power dissipation of the collector electrode. The collector dc power dissipation is the product of collector current and collector voltage: $Pc = VcIc$.

collector efficiency Symbol, n. In a bipolar transistor circuit, the radio of ac power output to dc collector-power input: $n(\%) = 100\ Pac/(Vc\ Ic)$.

collector family For a bipolar transistor, a group of collector current vs collector voltage curves. Each is plotted for a particular value of base-bias voltage (common-emitter circuit) or emitter-bias voltage (common-base circuit).

collector junction In a bipolar transistor, the junction between collector and base layers.

collector mesh In a cathode-ray storage tube, a flat, fine wire screen which attracts and conducts away the secondary electrons knocked out of the storage mesh by the electron beam.

collector multiplication Symbol, a. In a bipolar transistor, an increase in the number of electrons at the collector electrode, due to a momentary alteration of the charge density of the collector junction by injected carriers reaching the junction.

collector resistance In a bipolar transistor, the internal resistance of the collector junction. See *ac collector resistance* and *dc collector resistance*.

collector ring 1. A rotating, brush-contacted ring electrode connected to one end of a coil in an ac generator. 2. A similar ring which, with a brush, serves as a connection to a rotating element, as in a signal-gathering system. 3. The collector electrode in an iconoscope.

collector transition capacitance The capacitance between the collector and base of a bipolar transistor under normal operating conditions. This capacitance has a limiting effect on the operating frequency of a bipolar device.

collector voltage Symbol, Vc. In a bipolar transistor, the voltage on the collector electrode. See *ac collector voltage* and *dc collector voltage*.

collimated rays Electromagnetic waves made parallel or nearly parallel. This may be done by means of a reflector, a lens, or a laser.

collimation 1. The rendering of light rays (or rays of other radiation) parallel. 2. Adjustment of the line of sight of an instrument, such as a level or transit.

collimation equipment Optical-alignment equipment.

collimator A device for producing parallel rays of light or other radiation. In one version, a small light source is placed behind a slit or pinhole at one focal point of a converging lens; parallel rays emerge from the opposite side of the lens.

collinear antenna A broadside directional antenna consisting of two or more half-wave radiators; the current is kept in phase in each section by quarter-wave stubs between each radiating section. The radiators are stacked end to end horizontally or vertically. Also called *Franklin antenna*.

Collins coupler A single-section, pi-filter circuit used to match a radio transmitter to a wide variety of antenna. Also called *pi coupler* and *Collins network*.

collodion A viscous solution of pyroxylin and a solvent (such as acetone, alcohol, or ether) sometimes used as a binding agent for coils and other components.

cologarithm Abbreviation, colog. The logarithm of the reciprocal of a number; $\operatorname{colog} N = \log 1/N = -\log N$.

color The characteristic that arises from the wavelength of light and is seen by the eye as various phenomena ranging from red at one end of the visible spectrum to violet at the other end. See *hue*.

color balance In a color TV receiver, adjustment of the beam intensities of the individual guns of a three-gun picture tube to compensate for the difference in light emissivity of the red, green, and blue phosphors on the tube screen.

color bar-dot generator A special rf signal generator which develops a bar or dot pattern on the screen of a color TV picture tube, for test and alignment purposes.

color-bar pattern A color TV test pattern of vertical bars, each a different color.

color breakup A transient separation of a color TV picture into its red, green, and blue components, as a result of a sudden disturbance of viewing conditions (blinking of eyes, moving of head, intermittent blocking of screen, etc.).

color burst As a phase reference for the 3 579.545 kHz oscillator in a color TV receiver, approximately nine cycles of the chrominance subcarrier added to the back porch of the horizontal blanking pedestal in the composite color signal.

color carrier See *chrominance subcarrier*.

colorcast A *color* tele*cast*.

color code 1. A system employing colored stripes or dots to mark the nominal values and other characteristics on capacitors, resistors, and other components. 2. A code representing the various frequencies being used by RC modelers at meets, and used on flags attached to transmitters, for example, as a safeguard against jamming.

color coder See *color encoder*.

color contamination In a color TV system, faulty color reproduction resulting from incomplete separation of the red, green, and blue channels.

color-coordinate transformation In a color TV system, the computation (performed electrically in the system) of the tristimulus (primary) values with reference to one set of primaries, from the same colors derived from another set of primaries.

color-difference signal. Designated *B-Y, G-Y,* and *R-Y*. The signal resulting from reducing the amplitude of a color signal by an amount equal to the luminance-signal amplitude. Also see *B-Y signal, G-Y signal,* and *R-Y signal*.

color dot 1. A phosphor spot on the screen of a color TV picture tube. 2. One of the colored dots indicating the capacitance, voltage, and tolerance of a capacitor (see *color code, 1*). 3. A colored dot used in the color code applied to some resistors to indicate the number of zeroes to be added to the value indicated by preceding colors.

color edging In a color TV picture, an aberration consisting of false color at the boundaries between areas of difference color.

color encoder In a color TV transmitter, the circuit or channel in which the camera signals and the chrominance subcarrier are combined into the color-picture signal.

color fidelity The faithfulness with which a color TV system, lens, or film reproduces the colors of a scene.

color filter A transparent plate or film which transmits light of a desired color and eliminates or attenuates all other.

color flicker In a color TV system, flicker resulting when both luminance and chromaticity fluctuate.

color fringing In a color TV picture, false color around objects, sometimes causing them to appear separated into different colors.

color generator A special rf signal generator for adjusting or troubleshooting a color TV receiver. The color signals it delivers are identical to those produced by a TV station.

color graphics Computer *graphics* displayed in color on a cathode-ray screen.

colorimeter A meter used to measure the color intensity of a sample relative to a standard.

colorimetric A characteristic of visible light, representing the wavelength concentration. Refers to the perceived color of a light beam.

colorimetry The science and art of color measurement.

color killer In a color TV receiver, a circuit that, in the absence of a color signal, delivers a negative bias to cut off the bandpass amplifier.

color match In photometry, the condition in which color agreement exists between the halves of an area. Also see *color matching*.

color matching The art of selecting colors—by instrument or eye—so that they are identical in hue, saturation, and brilliance.

color media Substances that transmit essentially one color of visible light, while blocking other colors.

color meter A photoelectric instrument for measuring color values and comparing and matching colors. One type tests samples for comparison through red, green, and blue light filters in succession; if the same reading is obtained for each sample, they match in hue, saturation, and brilliance.

color mixture An additive combination of two or more colors. Thus, red + yellow = orange, blue + red = violet, red + blue + green = white, etc.

color oscillator The oscillator in a color television receiver that coordinates the color response. This oscillator is operated at a precise frequency of 3.579545 MHz, to within plus or minus 10 Hz.

color phase The phase difference between an *I* or *Q* chrominance primary signal and the chrominance carrier reference.

color-phase diagram For color TV, a quadrant diagram showing (for each of the three primary and complementary colors) the difference in phase between the color-burst signal and the chrominance signal as well as the peak amplitude of the latter. Also shown are the peak amplitude and polarity of both in-phase and quadrature components required for the chrominance signals. For color TV receiver adjustment, the color-phase diagram is displayed, in effect, by a *vectorscope* when a suitable signal from a color generator is applied to the receiver.

color-picture signal 1. A signal containing the electrical components corresponding to the color hue, saturation, and brilliance of a scene. 2. The combination of chrominance and luminance signals minus blanking and sync signals.

color-picture tube A specialized cathode-ray tube used in a color television receiver. Three different images are produced, in red, blue, and green.

color primaries In color TV, the (*additive*) primaries are red, green, and blue, which, on being correctly mixed, produce all colors. Color printing uses subtractive primaries: (roughly, red, blue, and yellow).

color purity The ratio of wanted to unwanted components in a color. In a pure color there are no components other than those required to produce the color (*color*, in this context, includes white).

color-purity magnet A permanent magnet on the neck of a color TV picture tube to improve color purity by proper displacement of the electron beam.

color registration In color TV reception, the precise superimposition of the red, green, and blue so that the composite is free from *color edging*.

color rendering index A mathematical expression defining the effect of the color of a light source on an object. For example, in red light, a blue object appears nearly black.

color sampling rate The number of times per second that each primary color is sampled in a color TV receiver.

color saturation The extent to which a color is without a white component, 100% saturation indicating a complete absence of white.

color sensitivity 1. The degree of which a photosensitive device, such as a photocell or TV camera tube, responds to different colors of light. 2. The degree to which photographic film responds to different colors of light.

color signal See *color-picture signal*.

color spectrum The electromagnetic frequency band containing visible light; it extends from red on one end to violet on the other. In order of decreasing wavelength the colors are red, (0.65 μm), orange, yellow, green, blue, and violet (0.41 μm).

color subcarrier A modulated monochrome signal whose sidebands convey color information.

color-sync signal See *color burst*.

color television Television in which the picture approximates natural color. It operates on the basis of mixing three primary colors (red, blue, green) of phosphors on the picture tube screen.

color television receiver A television receiver designed to reproduce color pictures.

color television signal The signal transmitted by a color television transmitter, containing all of the information needed to reproduce a complete, full-color picture of any subject.

color temperature Unit: K. The temperature (Kelvin) to which a black body must be raised to match the color of an observed source. Color temperature is important in identifying star types.

color transmission The television transmission of a picture in color.

color triad On the screen of a color TV picture tube, one of the color cells, each of which contains one of the three phosphor dots: red, green, and blue.

color triangle A triangle which may be inscribed on a chromaticity diagram to reveal the chromaticity range resulting from adding the three primaries.

color TV signal The complete signal—video, color, and sync components—required for transmitting a picture in color.

Colpitts oscillator A self-excited oscillator circuit in which the tank is divided into input and feedback portions by a capacitive voltage divider (two capacitors connected in series across the tank coil) rather than by a center tap on the coil.

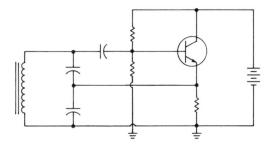

Colpitts oscillator.

columbium Symbol, Cb. The former name of the metallic element *niobium*. Atomic number, 41. Atomic weight, 92.91.

column See *card column*.

column binary Binary number representation on punched cards wherein consecutive digits correspond to consecutive column punching positions. Compare *row binary*.

column speaker An acoustic speaker with a long cabinet, such that a large column of air is used for resonating or reinforcing purposes. This type of speaker radiates over a wide azimuth angle, while providing a narrow beam in the elevation plane.

column split On a punched card machine, the device for reading, as two separate characters or codes, two parts of a single column.

COM Abbreviation of *computer output on microfilm*.

coma An aberration which causes the beam spot on the screen of a cathode-ray tube to resemble a comet.

coma lobe An aberration in the radiation or response pattern of a dish antenna that occurs when the radiating element is not exactly at the focal point of the reflector. When the directional pattern is altered by moving the driven element rather than turning the entire antenna, the coma lobes appear.

comb amplifier An arrangement of several sharply tuned bandpass amplifiers whose inputs are connected in parallel and whose outputs are separate; the amplifiers separate various frequencies from a multifrequency input signal. The name is derived from the comblike appearance of the response pattern of various output peaks displayed along a frequency-base axis.

comb filter A selective device which passes a series of frequencies (actually, several very narrow bands of frequencies) within a band while rejecting frequencies in between, so called because its frequency-response curve resembles the teeth of a comb. Also see *comb amplifier*.

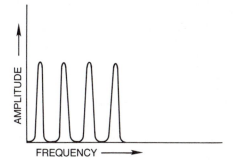

Comb filter response.

comb generator 1. A signal generator that provides outputs at evenly spaced frequencies. So called because, on a spectrum analyzer, its output looks like the teeth of a comb. 2. A transmitter with many spurious signals at its output.

combination 1. A functional (usually stationary) setup of two or more pieces of equipment. Examples: *transmitter-receiver combination, motor-generator combination, radio-phonograph combination*. 2. In mathematics, a selection of several factors from a group, without regard to order. Thus, from the group ABC, the three possible combinations are *AB, AC,* and *BC.* Compare *permutation*.

combinational circuit Two or more basic logic circuits combined so that the output of the arrangement depends wholly on the inputs.

combination bridge A bridge that affords two or more classes of measurement, usually selectable by a function switch. Examples: *capacitance-inductance bridge, capacitance-resistance bridge*.

combination cable A cable having conductors grouped in pairs, threes, quads, or similar arrangements.

combination feedback See *current-voltage feedback*.

combination microphone Two or more microphones combined into one unit.

combination speaker Two or more louspeakers combined into one, e.g., a coaxial speaker.

combination tube An electron tube containing several complete sets of elements; therefore, actually two or more separate tubes in one envelope. Examples: *diode-triode, triode-pentode, diode-triode-pentode*, etc.

combinatorial logic A form of logic in which the output states depend on the input states, but on no other factor.

combined head See *read-write head*.

combined reactance The net reactance (X) in a circuit: $X = X_L - X_C$.

combiner A circuit or device for mixing various signals to form a new signal. Also see *mixer*.

combiner circuit In a color TV camera, the circuit that combines the chroma and luminance with the sync.

comeback A spurious response in a bandpass or band-rejection filter, at a frequency well above or below the passband or stopband.

command 1. In computer practice, the group of selected pulses or other signals which cause the computer to execute a step in its program. 2. Instruction.

command chain Part of a computer operation carried out independently as a series of input/output instructions.

command control In automation, electronic control, and computer practice, the performance of functions in response to a transmitted signal.

command destruct signal A signal for instigating the destruction of a missile in flight.

command guidance system A system in which a guided missile and its target are both tracked by radar.

command language A computer language made up of command operators.

command link In a command guidance system the section that transmits missile-steering commands.

command network A radio net in which the chain of command is rigorously defined and followed.

command reference The current or voltage to which a feedback signal is referenced in a control system or servomechanism.

comment A statement written into a program for a documentation rather than implementation, e.g., to describe the purpose of a step or subroutine.

comment field A record or file in which instructions or explanations are given.

commercial data processing A commercial rather than industrial or scientific application of data processing.

commercial killer A usually remote-controlled, electronic relay for disabling a radio or TV receiver during commercials.

commercial language A computer programming language for commercial applications (payroll, for example).

common 1. Grounded. 2. Directly connected to several different points

in a circuit or system.

common area A computer storage area usable by several programs or segments within a program.

common-base circuit A bipolar transistor circuit in which the transistor base is the common (or grounded) electrode. Also called *grounded-base circuit*.

common battery 1. A battery that supplies both filament and plate voltage to an electron tube. 2. A battery shared by two or more different circuits or equipments. 3. In wire telephony, a central office battery which supplies the entire system.

common-battery office In wire telephony, a central office providing a common battery.

common business-oriented language See *COBOL*.

common-capacitor coupling The process of coupling one tuned circuit to another by means of a capacitor which is common to both circuits.

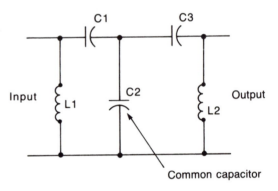

Common-Capacitor Coupling.

common-carrier fixed station A radio station providing public service and whose location is fixed.

common-cathode circuit A tube circuit in which the cathode is the common (or grounded) electrode. Also called *grounded-cathode circuit*.

common-channel interference Radio interference resulting from two stations transmitting on the same channel. It is characterized principally by (1) beat-note (heterodyne whistle) generation and (2) supression or *capture* of the weaker signal by the stronger one.

common-collector circuit A bipolartransistor circuit in which the collector is the common (or grounded) electrode. Also called *grounded-collector circuit* and *emitter follower*.

common communications carrier A communications company authorized by the licensing agency to furnish public communications.

common-component coupling See *common-capacitor coupling, common-inductor coupling*, and *common-resistor coupling*.

common-drain circuit A field-effect transistor circuit in which the drain terminal is the common (or grounded) electrode. Also called *grounded-drain circuit* and *source follower*.

common-emitter circuit A bipolar transistor circuit in which the emitter is the common (or grounded) electrode. Also called *grounded-emitter circuit*.

common-gate circuit A field-effect transistor circuit in which the gate is the common (or grounded) electrode. Also called *grounded-gate circuit*.

common-grid circuit A tube circuit in which the control grid is the common (or grounded) electrode. Also called *grounded-grid circuit*.

common ground A single ground-point connection shared by several portions of a circuit.

common impedance A single impedance shared by parts of a circuit. Because currents from the various parts flow through this impedance simultaneously, coupling (desired or undesired) can take place between them.

common-impedance coupling See *common-capacitor coupling*, *common-inductor coupling*, and *common-resistor coupling*.

common-inductor coupling The process of coupling one tuned circuit to another by means of an inductor which is common to both circuits.

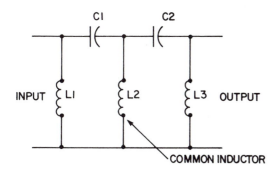

Common-inductor coupling.

common language A language recognized by all the equipment in a data processing system.

common logarithm Abbreviation, log10. A logarithm to the base 10. Also see *logarithim*.

common mode Pertaining to signals or signal components that are identical in amplitude and duration.

common-mode characteristics In an operational amplifier, characteristics denoting amplifier performance when a common signal is applied to inverting and noninverting inputs.

common-mode gain The voltage gain of a differential amplifier with a common-mode input.

common-mode impedance input The input impedance between ground and one of the inputs of a differential amplifier. Compare *common-mode input impedance*.

common-mode input capacitance In a differential amplifier, the internal capacitance of the common-mode input circuit.

common-mode input circuit In a differential amplifier, the input circuit between ground and the inputs connected together.

common-mode input impedance In a differential amplifier, the open-loop impedance between ground and the inputs connected together. Compare *common-mode impedance input*.

common-mode input signal A signal applied to the common-mode input circuit of a differential amplifier, i.e., to both inputs connected together. Compare *common-mode signal*.

common-mode input voltage In a differential amplifier, the maximum voltage which may be applied safely between ground and the inputs connected together.

common-mode interference A form of interference that occurs across

the terminals of a grounded system.

common-mode rejection The extent to which a differential will reject a signal presented simultaneously to both inputs in phase, or of two signals identical in amplitude, frequency, and phase applied separately to the two inputs. Also see *common-mode rejection ratio*.

common-mode rejection ratio In a differential amplifier, the extent to which the amplifier cancels undesirable signals. It is the ratio of the differential gain to the common-mode gain. Also see *common-mode rejection*.

common-mode signal The algebraic average of two signals applied simultaneously to the two ends of a balanced circuit, such as a differenttial amplifier. Compare *common-mode input signal*.

common-mode voltage In a balanced amplifier, the voltage common to both inputs. Also see *common-mode input voltage*.

common-mode voltage gain See *common-mode gain*.

common-mode voltage range The range limited by the maximum nonsaturating input voltage which may be applied to both inputs of an operational amplifier.

common pool An assigned memory store, utilized by two or more circuits or systems.

common-resistor coupling The process of coupling one circuit to another by means of a resistor which is common to both circuits.

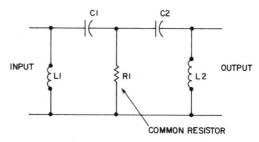

Common-resistor coupling.

common-source circuit A field-effect transistor circuit in which the source terminal is the common (or grounded) electrode. Also called *grounded-source circuit*.

common-user channels Communication channels open to all licensees.

communication band A band of frequencies whose use is authorized expressly for communications rather than for other services such as broadcasting, education, remote control, etc.

communication channel 1. In radio or wire service, a (usually auxiliary) channel for direct exchange of information between units of the service, e.g., a *talking circuit* between a broadcast studio and the transmitter house. 2. A data transmission channel between distant points, as between a remote terminal and a central computer system.

communication link 1. Collectively, the equipment providing a communication channel between two transmitters. 2. Data terminal equipment.

communications The science and art of using and developing electronic equipment and processes for the transmission and reception of intelligence.

communications common carrier An organization licensed to provide public communication services.

communications network An organization of transmitting and receiving stations for the reliable exchange of intelligence. Also called *net*.

communications receiver A (usually multiband) radio receiver designed expressly for listening to other than standard radio broadcasts, i.e., for short wave listening (to amateur, weather, or other stations).

communications satellites Satellites (in orbit around the earth), which provide propagation paths (e.g., by reflection or retransmission) for radio waves between terrestrial transmitters and receivers. Also see *active communications satellite* and *passive communications satellite*.

community-antenna television Abbreviation, CATV. A system in which an advantageously located receiving station picks up TV signals, which are amplified (if necessary), and distributed in the community served by the system. Commonly called *cable TV*.

commutating capacitor 1. In a flip-flop circuit, a capacitor connected in parallel with the cross-coupling resistor to accelerate the transition from one stable state to the other. Also called *speedup capacitor*. 2. The capacitor connected in parallel between SCR stages to momentarily reverse the current going through the SCR to cut it off.

commutation 1. In a generator, armature coil current changing direction as the coils alternately pass the north and south poles of the field. When the ends of each coil are connected to opposite bars of a commutator, properly situated brushes in contact with the commutator show no change in polarity, since a particular brush is always connected to a "north coil," and the opposite brush to a "south coil." 2. In a thyratron or SCR circuit, momentarily reversing the polarity to cut the tube or rectifier off.

commutator 1. In a motor, generator, or rotating selector, an arrangement of a number of separated, parallel metal bars or strips around a rotating drum. As the drum turns, the bars contact one or more brushes that are in sliding contact with the commutator. 2. An electronic circuit which (like the electromechanical commutator described above) switches a single input sequentially to a series of output terminals, or switches a number of inputs sequentially to a single pair of output terminals.

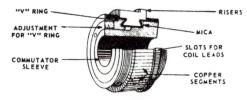

Commutator construction.

commutator ripple The pulsating voltage superimposed on the dc voltage delivered by an unfiltered generator.

Compactron A specially fabricated *combination tube*.

compander Acronym for *compress and expander*. In the transmission and reception of af intelligence, a system employing a volume compressor at the transmitter and a volume expander at the receiver. The compressor reduces dynamic range before transmission, and the expander restores it after reception.

companding The combined processes of volume compression and expansion. Also see *compander*.

companion keyboard An auxiliary keyboard connected to a regular keyboard and operated remotely.

comparator 1. An instrument for checking the condition of a component by comparing it directly with an identical component of known quality. A comparator often has a scale reading in *percent deviation* or simply go/no-go. Examples: *capacitor comparator, resistor comparator, coil comparator*. 2. An IC operational-amplifier whose halves are well balanced and without hysteresis; it is therefore suitable for circuits in which two electrical quantities are compared. 3. In a computer system, a device whose output signal depends on the result of its comparing two data items.

compare In a computer program, a relational test performed on two quantities to determine their relative magnitude, including an indication of the test's result and, sometimes, taking appropriate action (as is, say, the BASIC statement IF A > B THEN GO TO...).

comparison 1. An examination of different data bits or items, resulting in a conclusion about some aspect of their relationship. 2. An expression of the relationship between two voltages, currents, or phase angles.

comparison bridge A bridge designed especially for the quick comparison of components; e.g., the comparison of resistors with a standard resistor, inductors with a standard inductor, capacitors with a standard capacitor.

comparison measurement A measurement in which a quantity or component is compared with a known like quantity or component value rather than having the measurement displayed directly by a meter. Examples: bridge measurements, potentiometric, measurements, frequency matching.

compass 1. Any of several instruments for determining direction on the earth's surface, e.g., magnetic (mariner's) compass, gyrocompass. 2. A radio *direction finder*. 3. An instrument for drawing circles.

compatibility The condition relating two computers in which both can operate on the same program without its modification.

compatible color TV A color-television system whose transmissions can be received in black and white by any ordinary monochromatic receiver.

compatible IC A hybrid integrated circuit which has an active element inside the integraded structure and a passive element deposited on its insulated outer surface.

compensated amplifier A wideband amplifier whose frequency range is extended by special components and circuit modifications. Also see *compensating capacitor* and *compensating coil*.

compensated diode detector A diode detector in which a positive dc voltage from the agc rectifier is applied to the diode plate. The voltage is always proportional to the signal carrier. The arrangement allows the diode to handle a heavily modulated AM signal without producing excessive distortion.

compensated-impurity resistor A resistor consisting of a diffused semiconductor material to which are added controlled amounts of n- or p-type dopants (impurities).

compensated-loop direction finder A direction finder whose loop antenna is complemented by another antenna for polarization-error compensation.

compensated semiconductor A doped semiconductor material in which the acceptor impurity cancels the effects of the donor impurity.

compensated volume control A combination volume-tone control which provides bass boost at low volume levels to compensate for the ear's deficiency at low frequencies.

compensating capacitor 1. A capacitor having a temperature coefficient of capacitance numerically equal to but having the opposite sign that of another capacitor of opposite polarity in a tank or other circuit. When the capacitors are connected in parallel, a temperature-induced value change in the main capacitor is balanced by an equal and opposite change in the compensating capacitor; and net capacitance of the circuit does not change. Tuned circuits, then, can be compensated for the frequency drift in this manner. 2. In a video amplifier, a large capacitance connected between ground and tap on the plate, collector, or drain resistor to boost low-frequency response. Compare *compensation coil.* 3. A usually low-capacitance capacitor or known temperature coefficient, operated in combination with a main capacitor to reduce capictance/temperature drift of the latter to zero or to some desired positive or negative value.

compensating diode A junction diode used to temperature-stabilize a transistor current. It is usually forward biased in the base-bias network of the transistor.

compensating filter 1. A selective filter employed for the purpose of eliminating some irregularity in the frequency distribution of received energy. 2. A filter used to change the wavelength distribution of electromagnetic energy.

compensating resistor 1. A, usually low-resistance, resistor of known temperature coefficient, operated in combination with a main resistor to reduce the resistance/temperature drift of the latter to zero or to some desired positive or negative value. 2. Sometimes, a *trimmer resistor.*

compensation Adjusting a quantity, manually or automatically, to obtain precise values or to counteract undesired variations. Example: temperature compensation of electronic components. For illustration, see *compensating capacitor, 1.*

compensation coil In a video amplifier, an inductor connected in series with the tube plate resistor or transistor collector (or drain) resistor, or in the coupling path between stages, or both, to boost high-frequency response.

compensation filter See *compensating capacitor, 2.*

compensation signal A signal recorded on a tape track containing computer data, which automatically corrects playback speed error.

compensation theorem An impedance Z in a network may be replaced by a generator having zero internal impedance and whose generated voltage equals the instantaneous potential difference produced across Z by the current flowing through it. Compare *maximum power transfer theorem, Norton's theorem, reciprocity theorem, superposition theorem,* and *Thévenin's theorem.*

compensator A device or circuit which provides compensation.

compilation The process of using a *complier.*

compilation time The period during which a program is compiled, as distinct from *run time.*

compile 1. To unify computer subroutines into an all-encompassing program. 2. To gather information or data together into a single file or file set.

Compiler A compiler may be hardware or software. Its function is to translate the high-level computer language used by the human programmer into the machine language that is understood by the computer.

compiler language Any computer language that serves as an interface between the operator and the computer.

compiler program A program that converts compiler language into machine language. An assembler program.

compiling routine In digital computer operation, a routine permitting the computer itself to construct a program to solve a problem.

complement 1. Generally, the difference between any number and the radix of that number system. Thus, the complement of 9 in the decimal

system is 1, since the radix is 10. 2. A number produced from the radix of the applicable system less one, the most common types being (1) the *radix* or *true complement* (in the decimal system, *nines complement,* which is derived by subtracting each of the digits in the number being complemented from one less than the radix and adding one to the result; the (*true, radix*) complement of 365 is 635 (999 minus 365 plus one); and (2) the *diminished radix complement,* which is derived by the first method described, with the exception that one is not added to the result of the subtraction (thus, the diminished radix complement of 365 is 634). As applied to binary numbers, this is called the *ones complement,* which is commonly used in computers to represent a number's negative value; thus 101 (binary five) becomes 111 minus 101, or 10 (binary two).

complementary A Boolean operation whose result is the same as that of another operation but with the opposite sign; thus, OR and NOR operations are complementary.

complementary colors In the *additive* color system, two colors which together produce white; in the subtractive system, the result is gray. In either system, complements are opposite each other on the color wheel.

complementary constant-current logic A form of bipolar logic with high operating speed and high component density.

complementary metal-oxide semiconductor Abbreviation, CMOS. A semiconductor device consisting of two, complementary MOSFET's (i.e., one n-channel type and one p-channel type) integrated into a single chip.

complementary operator The logical negation operation.

complementary pushpull circuit A complementary-symmetry amplifier circuit.

complementary rectifier In the output circuit of a magnetic amplifier, nonsaturating half-wave rectifier elements.

complementary silicon-controlled rectifier A silicon-controlled rectifier that has polarity opposite from the usual silicon-controlled rectifier.

complementary-symmetry circuit A bipolar transistor circuit employing an npn and pnp transistor. The transistors conduct during opposite half-cycles of the input signal, the result being that push-pull output is provided with a single-ended input; no phase-splitting input circuit is required. The complementary-symmetry circuit offers very low output impedance, permitting a loudspeaker voice coil (or other low-impedance load) to be operated directly without a coupling transformer.

complementary-symmetry device A device, such as a transistor amplifier or a transistor pair, employing a complementary-symmetry circuit.

complementary tracking A control system in which several secondary (slave) devices are controlled by a primary (master) device.

complementary transistors A transistor pair of opposite polarity operated in a complementary-symmetry circuit or its equivalent.

complementary wave An electromagnetic wave in a transmission line that occurs as a result of reflection. Any impedance discontinuity will result in complementary waves.

complementer A logic circuit which provides an output pulse when there is no input pulse, and vice versa. Also called *inverter* and *NOT circuit.*

complementing Representing the negative value of a number by using its complement (see item 2 under *complement*).

complement number In a base-n number system, the number m which, when added to another number p, equals n. The whole numbers m and p are called complement numbers.

complement-number handling A computer system in which the operations are carried out via the complements of the input numbers.

complement-setting technique The technique of determining the number of pulses required to complete the switching of a counter circuit when it is started at some state higher than full zero. The number of pulses required for completion is equal to the number representing the starting state's complement. Thus, in a four-stage binary counter, the total count is 16. To give its full count after only 6 input pulses, the counter must be started in state 10, since 10 is the complement of 6 with respect to 16.

complete carry In digital computer operation, a system permitting all carries to generate carries.

complete circuit See *continuous circuit*.

complete modulation Modulation to the maximum extent. In amplitude modulation, this means 100% modulation, i.e., variation of the carrier amplitude between zero and twice its unmodulated value.

complete operation By a computer, totally obeying a program instruction.

complete routine A vendor-supplied computer program that is usable as is.

complex function 1. A function of a complex variable. 2. An IC type whose several circuits are integrated and interconnected on a single chip to accomplish an action more complicated than that afforded by one of the circuit alone.

complex integer See *complex number*.

complex notation Notation taking into consideration both the real and imaginary components of a quantity. Thus, impedance *(Z)* is a complex quantity including a resistive (real) component *(R)* and a reactive (imaginary, because it includes $j \sqrt{-1}$, an imaginary number) component *(X)*: $Z = R + jX$ (first quadrant), $-R + jX$ (second quadrant), $-R - jX$ (third quadrant), or $R - jX$ (fourth quadrant).

complex number A number expressed in complex notation, e.g., $a + jb$ (or $a + b\sqrt{-1}$).

complex operator The value $\sqrt{-1}$, which is represented in engineering by the letter j, and in mathematics by i.

complex parallel permeability An expression of complex permeability of an inductor core under actual operating conditions, assuming zero loss in the conductors of the coil winding. A parallel combination of reactance and resistance.

complex periodic wave A periodic wave composed of a sine-wave fundamental and its harmonics. A square wave, for example, is a complex periodic wave consisting of a sinusoidal fundamental (which has the same frequency as the square wave) plus at least 10 odd-numbered sinusoidal harmonics,

complex permeability An expression of inductor-core permeability, obtained from the mathematical ratio of the magnitudes of the vectors representing the induction and electromagnetic field strength within the core.

complex plane The plane in which rectangular vectors lie.

complex quantity A quantity containing both real and imaginary components. Example: Impedance *(Z)* is a complex combination of resistance R (a real component) and reactance jX (an imaginary component): $Z = R + jX$.

complex radar target A radar target that is large enough in theory to be detected by radar, but, because of its geometry, cannot be detected. This effect is the result of phase combinations of signal components reflected from various surfaces on the target.

complex series permeability An expression of complex permeability of an inductor core under actual operating conditions, assuming zero loss in the conductors of the coil winding. A series combination of reactance and resistance.

complex steady-state vibration Periodic vibration with more than one sine-wave component.

complex tone A tone made up of more than one sine-wave component.

complex variable A variable having real and imaginary parts.

complex waveform The shape of a *complex periodic wave*. It is the resultant of the individual sine-wave components, i.e., of the fundamental and all the harmonics.

complex-wave generator A signal generator whose output signal is any of several selectable waveforms and frequencies (or repetition rates). Also see *function generator*.

compliance Ease in bending, an important characteristic of phonograph styli and in rating transducers, such as loudspeakers. Expressed in cm/dyne, compliance is the reciprocal of stiffness and may be thought to be the accoustical or mechanical equivalent of capacitance.

compliance range The voltage range required to maintain a constant current throughout a load-resistance range.

compliance voltage The range over which the output voltage of a constant-current power supply must swing in order to maintain a steady current in a varying load.

compliance-voltage range The output voltage range of a constant current supply.

component 1. A device or part employed in a circuit to obtain some desired electrical action, e.g., resistor (passive component) tube (active component). Also see *active component* and *passive component*. 2. An attribute inherent in a device, circuit, or performance, e.g., the *reactive component* of an inductor. 3. A specified quantity or term, e.g., the *wattless component* of ac power. 4. A piece of equipment in an audiophile's music system.

component density The number of components (see *component, 1*) in an electronic assembly.

component failure rate 1. The percentage of components, out of a specified group, that fail within a specified length of time. 2. The frequency with which a given component, in a certain application, can be expected to fail.

component layout The mechanical arrangement of components (see *component, 1*) in an electronic assembly.

component stress The electrical or mechanical strain to which a component is subjected. In general, the greater the stress, the higher the component failure rate.

composite cable A cable containing other cables of different types.

composite circuit A circuit handling telegraphy and telephony simultaneously without causing mutual interference.

composite color signal The complete color TV signal, including all picture, color, and control components.

composite conductor A set of wires connected in parallel. May consist of different metals in each wire.

composite current A current having both ac and dc components, i.e., an alternating current superimposed on a direct current. Also called *fluctuating current*.

composite curve A curve or pair of curves showing two modes of operation, as of biased and unbiased conditions.

composite filter A filter having sections.

composite video signal The television picture signal containing picture information and sync pulses.

composite video-signal distortion Distortion of the *composite video signal* as evidenced by overshooting, ringing, and sync-pulse shortening.

composite voltage A voltage having both ac and dc conponents, i.e., an ac voltage superimposed on a dc voltage. Also called *fluctuating voltage.*

composite wave filter Two or more wave filters (not necessarily of the same type) operated in cascade.

composition resistor A resistor made from a mixture of materials, usually finely divided carbon and a binder.

compound A substance in which the atoms of two or more elements have united chemically to form a molecule of the substance. For example, an atom of cadmium (Cd) and one of sulfur (S) combine to form a molecule of cadmium sulfide (CdS).

compound connection A direct connection of two transistors, the amplified output of the first being further amplified by the second. The connection provides extremely high beta. Also called *Darlington pair.*

compound generator A generator having both series and shunt fields. Also called *compound-wound generator.*

compound horn A horn reflector used for transmission of microwave energy. The faces of the horn approach four geometric plane surfaces as the distance from the center increases.

compound modulation A system of successive modulation, the modulated wave from one step becoming the modulating wave in the next. Also called *multiple modulation.*

compound motor A motor having both series and shunt fields. Also called *compound-wound motor.*

compound transistor Two or more transistors directly coupled in the same envelope for increased amplification. Also see *compound connection.*

compound-wound generator See *compound generator.*

compound-wound motor See *compound motor.*

compress To reduce the bandwidth or dynamic range of a signal.

compressed-air capacitor A high-voltage air-dielectric capacitor enclosed in a case in which the air pressure is held at several hundred psi. The device exploits the dielectric strength of compressed air, which is higher than that of air at normal pressures.

compressed-air speaker A speaker that makes use of an airtight chamber to enhance the acoustic reproduction at certain frequencies.

compression The reduction of output-signal amplitude as input-signal amplitude rises. Compare *expansion.*

compression ratio 1. In a system containing compression, the ratio A_1/A_2, where A_1 is the gain (or transmission) at a reference-signal level and A_2 is the gain (or transmission) at a specified higher signal level. 2. In an automobile engine, the ratio of cylinder volume when the piston is at bottom dead center to that when the cylinder is at top dead center; the higher the figure, the better the performance, but the lower the gas mileage.

compression wave A wave disturbance that travels by means of longitudinal compression and expansion of the medium carrying the wave. Sound waves through air are compression waves.

compressor A circuit or device which limits the amplitude of its output signal to a predetermined value in spite of wide variations in input-signal amplitude. Unlike a *clipper* a compressor does not flat-top output signal peaks.

DIRECTION OF PROPAGATION

Compression wave.

compressor driver unit A loudspeaker transducer which works into an air space connected by a throat to a horn rather than by driving a diaphragm.

Compton diffusion An effect that occurs when a photon and electron collide. Some of the energy from the photon is transferred to the electron. On a large scale, such collisions result in diffusion of electromagnetic waves.

Compton effect The increase in wavelength (decrease in frequency) of X-rays scattered by the electrons of lighter atoms bombarded with the X-rays.

Compton shift See *Compton effect.*

compute To perform a mathematical operation by means of a relatively simple process. Thus, a digital *computer* solves intricate problems using simple arithmetic steps. Compare *calculate.*

computer A device or machine for performing mathematical operations on data and giving the results as information or control signals. Electronic computers are of two major types: *analog* and *digital.*

computer code In digital computer practice, the language the computer uses to perform its operations. The *only* language a computer "understands." Also called *machine code.*

computer-controlled catalytic converter A microprocessor-controlled system for automatically supervising gaseous emissions exhausted by a motor vehicle. An oxygen sensor monitors the exhaust stream, and the associated electronic system adjusts the air-to-fuel ratio of the carburetor to reduce smog-producing pollutants in the exhaust. Example: Chevrolet C-4 System.

computer diode A semiconductor diode having low capacitance and fast *recovery time,* thus suiting it to rapid switching in computer circuits and to very-high-frequency applications. Example: 1N914.

computer engineer An electronic engineer who is skilled in the theory and application of computers, related equipment, and associated mathematics.

computer file See *file.*

computer game See *video game.*

computer instruction See *instruction.*

computer interfacing apparatus The equipment used to connect a computer to other systems, such as monitoring devices or instruments.

computerized axial tomography Abbreviation, CAT. A multiple x-ray system which enables the observer to obtain cross-sectional images of the internal organs of the body.

computer program See *program.*

computer programmer A person who is skilled in devising (flow-charting and, usually, writing) the routines a digital computer uses to solve problems or process data.

computer storage tube A cathode-ray tube in which the electron beam scans and stores information in thousands of memory cells on a target. A cell "remembers" by acquiring and holding an electrostatic charge when it is struck by the beam from the writing gun. Information taken is *read* out of a cell by a second beam from the *reading gun.*

computer system. A central processor and its associated online and offline peripherals.

computer technician A professional who is skilled in building, repairing, and maintaining computers, and who, occasionally, designs them. A computer technician usually works under the supervision of a computer engineer.

computer terminal A teleprinter or video display unit and keyboard, used by the operator of a computer. An interface between the computer and a person.

computer/tv interface A device or circuit for delivering the output of a digital computer to a standard television receiver so that the latter can serve as a *graphic terminal.*

computer word See *word.*

computing amplifier See *operational amplifier.*

computing machine See *computer.*

concatenation 1. A method of speed control for a 3-phase motor in which two induction motors are operated with their shafts coupled together. The stator of the first motor is connected to the 3-phase supply, and the slip rings of this motor are connected to the field of the second motor. The slip rings of the second motor are connected to the three ganged sections of a Y-rheostat used for adjusting the speed. 2. Arrangement of a set into a series.

concavo-convex Pertaining to a lens having a concave face of greater curvature than its convex face.

concentrated-arc lamp A very brilliant low-voltage lamp containing nonvaporizing electrodes in an inert-gas atmosphere. An arc drawn across the electrodes creates the light source.

concentrated winding A coil winding having a large number of turns in a small space.

concentration gradient Between points in a semiconductor, the difference in electron or hole concentration.

concentric cable See *coaxial cable.*

concentric capacitor A fixed or variable capacitor whose plates are concentric cylinders. Also called *concentric-plate capacitor.*

concentric jack See *coaxial jack.*

concentric line See *coaxial line.*

concentric-line oscillator A stable, self-excited uhf oscillator whose frequency-determining grid tank consists principally of a section of concentric (coaxial) line.

concentric plug See *coaxial plug.*

concentric receptacle See *coaxial receptacle.*

concentric tank See *coaxial tank.*

concentric transmission line See *coaxial transmission line.*

concentric-wound coil A combination of two or more coils wound on top of, and insulated from, each other.

conceptual modeling A technique for solving problems by devising a mathematical model based on the results of an experiment; experiments performed on the model are used to verify its validity.

concurrent conversion In computer practice, running conversion and conventional programs together. Also see *conversion program.*

concurrent processing See *multiprogramming.*

condenser 1. An obsolete term for *capacitor.* 2. A mirror or lens for concentrating light (on an object, for example). 3. Something that condenses a gas or vapor. 4. A type of microphone.

condenser antenna A two-wire horizontal antenna system in which the radiator is a wire situated above a counterpoise.

condenser microphone A microphone in which a tightly stretched metal diaphragm forms one plate of an air capacitor and a closely situated metal plug forms the other. A dc bias voltage is applied to the arrangement. Impinging sound waves cause the diaphragm to vibrate, varying the capacitance of the capacitor described and making e output current alternate accordingly. Also called *capacitor microphone.*

condensing routine A computer program that takes an object (user written) program from an internal or external memory to punched cards in a way that maximizes the cards' storage capacity.

condensite A plastic insulating material whose base is phenol formaldehyde resin.

conditional Dependent on some external factor, and therefore subject to change.

conditional branch The point in a computer program where a relational test is performed and the statement line in which the test is made is left so that an out-of-sequence instruction can be implemented. Such a branch might be made, for example, following the BASIC statement IF Z = Y THEN GO TO (another line in the program).

conditional branch instruction The instruction in a computer program that causes a conditional branch.

conditional implication operation A Boolean operation in which the result of operand values a and b are as follows:

Operands		Result
a	b	
0	0	1
1	0	0
0	1	1
1	1	1

Also called *inclusion, if-then operation.*

conditional jump See *conditional branch.*

conditional stop instruction In a computer program, an instruction that can cause a halt in the run, as dictated by some specified condition.

conditional transfer See *conditional branch instructions.*

condition code A set of constraints for a computer program. The condition code sets the limits on what can be done with the computer under certain circumstances.

conditioning 1. The process of making equipment compatible for use with other equipment. Generally involves some design or installation changes. 2. Interfacing.

Condor A continuous-wave navigational system giving a cathode-ray-tube display for automatically determining the bearing and distance from a ground station. Compare *Benito.*

conductance Symbol, G. Unit, siemens. The ability of a circuit, conductor, or device to conduct electricity. Conductance is the reciprocal of resistance $G = 1/R = I/E$.

conducted heat Heat transferred by conduction through a material substance, as opposed to convection (movement of matter) and radiation (which occurs through empty space). A heat sink conducts dissipated energy away from a transistor, for example.

conducting layer See *Kennelly-Heaviside layer.*

conduction 1. The propagation of energy through a medium, depending on that medium for its travel. 2. The transfer of electrons through a wire. 3. The transfer of holes through a P-type semiconductor material. 4. Heat transfer through a material object (see conducted heat).

conduction angle See *angle of conduction*.

conduction band In the arrangement of energy levels within an atom, the band in which a free electron can exist; it is above the valence band in which electrons are bound to the atom. In a metallic atom, conduction and valence bands overlap; but in semiconductors and insulators, they are separated by an energy gap.

conduction current 1. The electromagnetic-field flow that occurs in the direction of propagation. A measure of the ease with which the field is propagated. 2. Current in a wire or other conductor.

conduction-current modulation In a microwave tube, cyclic variations in the conduction current; also, the method of producing such modulation.

conduction electron See *free electron*.

conduction error In a temperature-acutated transducer, error caused by conduction of heat between the sensor and the mounting.

conduction field An energy field that exists in the vicinity of an electric current.

conductive coating A conducting layer applied to the glass envelope of a cathode-ray tube, such as an oscilloscope tube or picture tube. Also see *Aquadag*.

conductive coupling See *direct coupling*.

conductive material See *conductor*.

conductive pattern The pattern of conductive lines and areas in a printed circuit.

conductivity Symbol, σ. Unit, S/m (siemens per meter). Specific conductance, i.e., conductance per unit length. Conductivity is the reciprocal of resistivity: $\sigma = 1/\rho$.

conductivity meter A device for measuring electrical conductivity. Generally, such a device is calibrated in mhos.

conductivity modulation In a demiconductor, the variation in conductivity resulting from variation of charge-carrier density.

conductivity-modulation transistor A transistor in which the bulk resistivity of the semiconductor material is modulated by minority carriers.

conductor 1. A material which conducts electricity with ease, such as metals, electrolytes, and ionized gases. Various materials vary widely in their suitability as conductors; the conductivity of commercial copper, for example, is almost twice that of aluminum. Compare *insulator*. 2. An individual conducting wire in a cable, insulated or uninsulated.

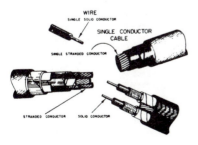

Conductors.

conduit A hollow tube, made of plastic or metal, through which wires,

cables, and other transmission media are fed.

cone The conical diaphragm of a (usually dynamic) loudspeaker.

cone antenna An antenna in which the radiator is a sheet-metal cone or a conical arrangement of rods or wires.

conelrad A protective system to prevent enemy aircraft or missels from using broadcast and television stations to locate cities. In the event of an emergency, all stations leave the air except those delegated to transmit intermittently on 640 kHz or 1.240 MHz. The system gets its name from *con*trol of *el*ectromagnetic *rad*iation.

cone marker A uhf marker beacon whose conical energy lobe radiates vertically from a radio-range beacon station. Aircraft in flight use such markers to accurately locate the beacon station.

cone of silence A small zero-signal zone directly over a low-freqeuncy radio-range beacon. The zone is the product of the combined directive properties of the beacon transmitting antenna and the antenna on an aircraft.

cone speaker A loudspeaker having a sound-producing cone (diaphragm) made of specially treated paper or other material, as opposed to a loudspeaker having a flat diaphragm.

confetti On a color TV screen, color spots caused by chrominance-amplifier noise.

confidence The probability that a predicted result will occur.

confidence factor Confidence, expressed either as a fraction (between 0 and 1) or as a percentage.

confidence interval The range over which a parameter may vary, such that a given confidence factor is maintained.

confidence level The confidence factor.

confidence limitations The maximum and minimum points of a confidence interval. Outside the confidence-limitation points, the confidence level drops below the required minimum.

configuration 1. The characteristic arrangement of components in an electronic assembly or of the equipment symbols in the corresponding circuit diagram. 2. Computer system.

configuration state In a computer system, the availability status of a device for a given application. Such status is indicated by *configured-in*: available; *configured-out*: available but restricted to certain users; and *configured-off*: unavailable.

configuration table Within a computer's operating system, a table giving the configuration state for various system units.

configured-in See *configuration state*.

configured-off See *configuration state*.

configured-out See *configuration state*.

conformance The degree to which a quantity or variable corresponds to a standard or to expectations.

conformance error The extent (usually expressed as a percent) to which conformance is lacking.

conical antenna See *cone antenna*.

conical horn A horn (antenna, loudspeaker, or sound pickup) having the general shape of a cone: the cross-sectional area varies directly as the square of the horn's axial length.

conical scanning In radar transmission, a method of scanning in which the beam describes a cone, at the apex of which is the antenna.

conic sections The figures obtained when a cone is sliced at various points by a plane; they are *circle*, *ellipse*, and *parabola*. A hyperbola is formed by slicing two cones joined at their apexes.

conjugate For a given complex number $a + bi$, the quantity $a - bi$. When complex conjugates are multiplied together, the result is $a^2 + b^2$.

conjugate branches In a network, two branches of such a nature that a signal in one has no effect on the other.

conjugate bridge A bridge in which the detector and generator occupy positions opposite to those in a conventional bridge of the same general type.

conjugate impedances Separate impedances in which the resistive components are equal and the reactive components are equal in magnitude but opposite in sign.

conjunction The logical AND operation.

connect To provide an electrical path between two points.

connection The point at which two conductors are physically joined.

connective An operation symbol written between operands.

connector 1. A device which provides electrical connection. 2. A fixture (either male or female) attached to a cable or chassis for quickly making and breaking one or a number of circuits. 3. A symbol connecting points on a flowchart.

conoscope A polariscope which employs focused polarized light to examine crystals (as in checking the optical axis of a quartz crystal).

consequent poles The poles of an equivalent single magnet that is formed when two magnets are aligned with their two identical poles together. Thus, when the two north poles are placed together, the consequent poles are a south pole at each end and a north pole at the center.

conservation of energy The preservation of the potential for work by a given quantity of energy, even when it undergoes a change in form within a system. The law of conservation of energy shows that energy can be neither created nor destroyed, but only changed in form. Consider a system in which water falling from a height drives a generator providing electricity to energize a coil whose heat converts water into steam for driving an engine that turns an electric generator. Here, energy is changed from one form to another. Losses at various points in the process can be accounted for as the unintentional generation of heat.

console 1. The main station or position for the control of equipment, an audio system, atom smasher, etc. Also the equipment present at such a station. 2. The cabinet housing console equipment and which stands on the floor rather than on a table. 3. Equipment permitting communication with a computer, e.g., switch panel, keyboard-display, etc.

consonance 1. Harmony between tones. 2. Acoustical or electrical resonance between divorced bodies or circuits.

constant 1. A quantity whose value remains fixed, such as pi. Compare *variable*. 2. The value of a component specified for use in a particular electronic circuit; sometimes, the component itself. 3. In a computer program, data items that remain unchanged for each run.

constant-amplitude recording In sound recording, the technique of holding the maximum amplitude of the signal steady while the frequency changes. In disk recording, the constant-amplitude method is most effective for low frequencies. Compare *constant-velocity recording*.

constantan An alloy of copper and nickel used in some thermocouples and standard resistors.

constantan-platinum thermoscouple A thermocouple employing the junction between constantan and platinum wires and used in thermocouple-type meters.

constant area As allocated by a computer program, an area of memory that holds constants.

constant bandwidth In a broadband tuned circuit, bandwidth which does not change with frequency. It is commonly achieved by special coupling circuits.

constant current A current which undergoes no change in value as it flows through a changing resistance. Compare *constant voltage*.

constant-current characteristic A condition in which the current through a circuit remains constant, even if the voltage across the circuit increases or decreases.

constant-current curve A plot of an independent variable against a dependent variable, in which the dependent variable remains constant after reaching a certain value. Examples: E_pI_p curve of a pentode, V dc curve of a bipolar transistor.

constant-current drive Driving power obtained from a constant-current source.

constant-current modulation See *choke-coupled modultion*.

constant-current power supply See *constant-current source*.

constant-current sink See *current sink*.

constant-current source A power supply whose current remains steady during variations in load resistance. Also called *constant-current supply* and *current-regulated supply*. Compare *constant-voltage source*.

constant-current supply See *constant-current source*.

constant-current transformer A transformer supplied from a constant-voltage source, which automatically delivers a constant current to a varying secondary load.

constant-k filter A filter section in which Z_1Z_2 equals k^2 at all frequencies. Here, Z_1 is the impedance of the series element and Z_2 is the impedance of the shunt element.

constant-power dissipation line A line connecting points on a family of current-voltage characteristic curves, the points corresponding to the maximum power which may safely be dissipated by a device (e.g., tube or transistor) to which the curves apply.

constant-resistance network A circuit of resistors which, when terminated in a resistance load, presents a constant resistance to a driving source under various conditions of operation.

constant-speed motor 1. A shunt motor (so called because the curve showing its speed vs armature current is reasonably flat). 2. A motor that runs at an unvarying speed throught the action of associated automatic electronic control circuitry.

constant-velocity recording In recording sound on disks, the technique of holding steady the maximum transverse velocity of the stylus tip at the zero axis as the frequency changes. The constant-velocity method is most effective at medium and high frequencies. Compare *constant-amplitude recording*.

constant voltage A voltage which undergoes no change in value as the resistance across which it is applied is varied. Compare *constant current*.

constant-voltage, constant-current supply A combination current-regulated and voltage-reguleted power supply providing constant current to a low load resistance and constant voltage to a high load resistance.

constant-voltage drive Driving power obtained from a constant-voltage source.

constant-voltage source A power supply whose output voltage remains steady during variations in load current. Also called *constant-voltage supply* and *voltage-regulated supply*.

constant-voltage transformer A special transformer used to smooth variations in power-line voltage. A capacitor in the device causes a winding to resonate at the line frequency (e.g., 60 Hz); heavy current circulting in the resulting tank, because it is frequency-dependent, is independent of line voltage. One type employs a resonant secondary winding; another, a resonant tertiary; and yet another, a combination of the two.

winding; another, a resonant tertiary; and yet another, a combination of the two.

construct A source (user's) computer program statement that, when implemented, produces a predetermined effect.

consumer reliability risk 1. The chance a consumer takes when he buys something that has not been tested. 2. An expression of the failure rate for a consumer item.

contact 1. A conducting body—such as a button, disk, or blade—which upon being pressed against another conductor serves to close an electric circuit. Example: switch contact, spring contact. 2. The state of being touched together, as when two conductors are brought into contact to close a circuit.

contact arc The arc that results when current-carrying contacts are separated.

contact area 1. The face of a contact. 2. The common area shared by two conductors in mutual contact.

contact bounce The springing apart or vibration of contacts upon making or breaking.

contact chatter The abnormal vibration of mating contacts, which may be caused by contact bounce or by an extraneous alternating current.

contact-closure input The input circuit of a device, such as a control-system amplifier, which is actuated by the closing of switching contacts. Compare *contact-open input.*

contact combination The set of contacts on a switch or electronic relay.

contact detector A detector (i.e., rectifier or demodulator) composed of two dissimilar materials in contact with each other. Crystal detectors (including modern semiconductor diodes) are of this general type. Some slight contact-detector action can even be obtained with two disimilar fine wires (such as copper and iron) by touching their tips lightly together.

contact emf A tiny dc voltage generated by the contact of two dissimilar materials.

contact follow The tendency of relay contacts to follow the actuating signals.

contact force 1. The force with which relay contacts close with a given amount of coil current. 2. The force with which a pair of relay contacts are held together when current flows through the coil of a relay. 3. In a mercury-wetted relay, the force exerted by the mercury on the contacts as the relay closes.

contact gap The distance between contacts when they are open.

contact load The value of electric power to which closed contacts are subjected.

contact microphone A microphone placed in contact with a vibrating surface for pickup; as applied to electric guitars, it's a *pickup.*

contact miss 1. The improper alignment of contacts in a switch or relay. 2. The condition of relay contacts not lining up properly.

contact modulator An electromechanical chopper. See *chopper* and *chopper converter.*

contact-open input The input circuit of a device, such as a control-system amplifier, which is actuated by the opening of switching contacts. Compare *contact-closure input.*

contactor noise 1. Electrical noise that is the product of make-and-break contact action or by fluctuations in conduction when the contacts are closed. 2. Disturbing sounds coming directly from contacts that are opening and closing.

contact potential The small dc voltage resulting from the bombardment of an electrode by electrons when the electrode has no external

voltage applied to it. Thus, a negative voltage appears on the plate of a tube diode as soon as the cathode begins emitting electrons.

contact-potential bias Negative dc voltage on the control rid of a vacuum tube, resulting from contact potential between rid and cathode electrodes (which form an equivalent diode).

contact pressure The pressure holding contacts together.

contact protector A component—such as a diode, capacitor, resistor, or combination of these—which serves to suppress the firing of contacts, thus protecting them from excessive wear, even destruction.

contact rating The maximum current, voltage, or power specified for a given set of contacts.

contact rectifier A rectifier consisting of two dissimilar materials in intimate mutual contact. Examples: copper-copper oxide, magnesium-copper sulfide, selenium-aluminum, germanium-indium.

contact resistance The usually very low resistance of the closed contacts of switches, relays, and other similar devices.

contact separation See *contact gap.*

contact strip See *terminal strip.*

contact switch A switch which uses a form of contact (see *contact, 1*) to make and break an electric circuit, as compared with an electronic switch (see *electronic switch, 1*), which uses the on-off action of a bistable tube or transistor.

contact travel The distance over which a relay or switch contact must move to close a circuit.

contact wetting The wetting of a contact surface with mercury.

contact wipe A sliding motion between closed contacts for good connection and for contact-surface cleaning.

container file See *controlling file.*

contaminated material 1. A semiconductor material containing some undesired substance. 2. A material unintentionally made radioactive.

contamination 1. The presence of an impurity in a substance. 2. The addition of a radioactive material to a substance.

content-addressed storage In a computer, memory locations identified by content (see *contents*) instead of by address. Also called *associative storage.*

contention The result of interference among more than one transmitting station on the same communications channel.

contents The information held in a computer system's randon-access memory. Also, the data held in a specific storage location.

Continental code A version of the Morse code used internationally in radiotelegraphy. Also called *International Morse code* and *general service code.* Compare *American Morse code.*

continuity The state of being continuous, as when an electric circuit is uninterrupted.

continuity test A test of the completeness of an electrical path. Ideally, the only concern is whether the circuit is open or closed, but sometimes circuit resistance is also of interest. Common continuity testers are ohmmeter, battery and buzzer, battery and lamp.

continuity tester A device (such as an ohmmeter, battery and buzzer, battery and lamp) with which a continuity test can be made,

continuity writer The person who prepares copy for a radio or television broadcaster.

continuous carrier. A medium (such as a radio-frequency wave) which will carry information (as when the carrier is modulated) with no disruption of the medium itself.

continuous circuit An uninterrupted circuit.

continuous commercial service Abbreviation, ccs. A category in which safe operating parameters are listed for electronic components and communications equipment operated over long, uninterrupted periods. Compare *intermittent commercial and amateur service.*

continuous duty The requirement of a device to sustain a 100-percent duty cycle for a prolonged period of time.

continuous-duty rating A maximum current, voltage, or power rating for equipment operated for extended periods.

continuous function A function which yields a smooth unbroken curve.

continuous load A load that requires continuous feed for a prolonged period of time.

continuous memory See *nonvolatile memory.*

continuous power The maximum sine-wave power that an amplifier can deliver for 30 seconds.

continuous recorder An instrument which provides an *uninterrupted recording.*

continuous recording A record made on a continuous sheet or tape instead of on separate sheets or tapes. An example is a continuous-playing tape used for repeated public announcements.

continuous spectrum A spectrum exhibiting an unbroken range of phenomena from one limit to the other.

continuous stationery The pack of paper a line printer uses; it consists of sheets connected by a perforated edge and folded accordian fashion.

continuous variable A variable which may take any value within a range of values.

continuous wave Abbreviation, cw. A periodic wave such as an rf carrier which is not interrupted at any point between its normal start and ternination and which is generally unmodulated. 2. An unmodulated rf "carrier" that is interrupted as with a telegraph key according to some code (such as Morse) to convey intelligence.

continuous-wave laser See *CW laser.*

continuous-wave radar See *CW radar.*

contour A control on an audio reproduction system which increases the base and treble amplitudes at low levels to compensate for the ear's natural losses in these ranges. The *contour* control sometimes attenuates signals in the 3 kHz region, the area of maximum sensitivity to the human ear.

contours of equal loudness See *audibility curves.*

CONTRAN A computer language that requires no compiler, or translating, interface between the operator and the machine. The programming is done in a language similar to machine language.

contrast 1. In TV or facsimile, the degree to which adjacent areas of a picture are differentiated. Insufficient contrast makes for a "flat" picture; excessive contrast, a "hard" one. 2. In optical character recognition, the degree to which a character is distinguishable from its background.

contrast control A potentiometer for adjusting the gain of the video channel of a TV receiver and, accordingly, the picture's contrast.

contrast range In an image or pattern, the value range from the lightest to the darkest parts.

contrast ratio In an entire or partial TV picture, the ratio of maximum to minimum luminance.

control 1. An adjustable component—such as a rheostat, potentiometer, variable capacitor, or variable inductor—which allows some quantity to be varied at will. 2. A test or experiment conducted simultaneously with another similar test conducted under conditions lacking the factor under consideration. Thus, if 10 resistors coated with a special var-

nish are tested at 120 °F, 10 identical unvarnished resistors could be tested (as a control) under the same conditions, in this way making the effect of the varnish ascertainable. 3. As a computer function, understanding and implementing instructions or carrying out tasks according to specific conditions.

control ampere-turns The ampereturns of the control winding in a magnetic amplifier.

control block A storage block for control information in a computer.

control bus In a computer, the path (conductors) linking the CPU's control register to memory.

control card A card that provides control information for a computer.

control character A character (bit group) used to start the control of a peripheral.

control characteristic A representation (such as a grid-voltage vs plate-current curve) depicting the extent to which the value of one quantity controls the value of another.

control circuit 1. A circuit in which one signal or process is made to control another signal or process. 2. In a digital computer, a circuit which handles and interprets instructions and commands the arithmetic and logic unit (and other operating circuits) accordingly.

control computer A computer in which a process being controlled supplies signals to the computer, which replies with its own signals to control the process.

control counter See *control register.*

control data In a computer record having a key, information used to put the records in some sequence. 2. Information affecting a routine's selection or modification.

control electrode An electrode (such as a grid, a base, or a gate) to which an input signal may be applied to control an output signal.

control field 1. In direct-current generators of the amplifying type, an auxiliary field winding used for feedback and regulation, in contrast to the self-excited field winding (which is the conventional field winding of the generator). 2. A computer record field containing control data.

control flux In an amplidyne, magnetic flux generated by current flowing through the control winding.

control grid See *grid, 1.*

control-grid bias The negative dc voltage applied between ground and the control grid of a tube to establish the operating point of the tube.

control-grid injection Injection into a stage via the control grid of the tube(s) in the stage.

control-grid modulation See *grid-bias modulation.*

control-grid-to-plate transconductance See *transconductance.*

control hole One of the holes in a punched card that specifies how other information on the card is to be handled. Also called *function hole.*

control language Within the operating system of a computer, the command set that the operator or programmer uses to control the running of a program or the operation of peripherals. Also called *job control language, system control language* (JCL and SCL).

control language interpreter See *control language* and *interpreter.*

controlled avalanche diode A diode that has a well-defined avalanche voltage. A semiconductor-diode device used primarily for voltage regulation in power supplies.

controlled-carrier modulation See *quiescent carrier operation.*

controlled-carrier transmission See *quiescent carrier operation.*

controlled rectifier A rectifier whose dc output may be varied by adjusting the voltage or phase of a signal applied to the device's control element. See *silicon controlled rectifier* and *thyratron.*

controller 1. The control signal of an electronic control (or servo) system. 2. A device, such as a specialized variable resistor, used to adjust current or voltage.

controller function The operation of a servo-system controller. The control of the movements of a servo-system device.

controlling file A computer storage area encompassing several complete magnetic disk cylinders; its size can be changed to accommodate a number of files.

control locus For a thyratron, a plot depicting critical bias.

control loop See *control tape.*

control mark See *tape mark.*

control panel The panel on which are mounted the controls, switches, monitoring devices, and other devices essential to regulating and supervising an electronic system; specifically, the switch panel a computer operator or programmer uses to communicate with a computer's central processing unit, i.e., by which he can, using machine language, address locations, for example.

control plate The metallic plate or disc that serves as the antenna of a *capacitance relay* or *touchplate relay.*

control program A program that arranges computer-operation programs in a certain order. The control program puts information in the computer memory for later use.

control punching See *control hole.*

control ratio For a thyratron the ratio dEa/dEg, where dEg is the critical grid voltage change and dEa is the resulting anode voltage change.

control rectifier A semiconductor diode device, used for the purpose of switching large currents. A small control signal can provide switching of high-power devices.

control register In a computer, the register that stores the address of the next instruction in the program being run.

control sequence The order in which instructions are executed in a digital computer.

control stack In a computer system, a unit of hardware having storage locations and used to perform arithmetic, assist in allocating memory to programs, and to control internal processes.

control statement In a programming language, an instruction that causes some action to be taken, as specified by a condition; also applicable to source program statements that affect the compiler's operation without modifying the machine code.

control tape Punched paper or plastic tape in the form of a closed loop and used to control printing divices. Also called *control loop.*

control total For a file or record group, a total derived during an operation; it is used to verify that all the records have been processed similarly.

control transfer The situation in which the control unit of a digital computer leaves the main sequence of instructions and takes its next instruction from an out-of-sequence address.

control transfer instruction See *branch instruction.*

control transformer A displacement transducer having a control transformer resembling a synchro, except that no voltage is applied to the rotor of the control transformer; the position of the rotor determines the output voltage available at the rotor terminals, which is usable for control purposes.

control unit Within a computer, hardware that takes program instructions in order, interprets them, and starts the indicated operation.

control-voltage winding In a servomotor, the winding that receives a varying voltage of a phase different from that applied to the fixed-voltage windings.

control winding In a magnetic amplifier, the winding that conducts the control-signal current.

control word A word, i.e., a bit group, stored in a computer's memory and used for a control function.

convection The flow of a gas or liquid that results in the transfer of heat from one location to another.

convection cooling Cooling a component, such as a tube or power transistor, by the upward movement of surrounding air that has been heated by the component.

convection current 1. The motion of current carriers or a charge across the surface of a conductor or dielectric. 2. Air currents rising above a heat source or heated body.

convective discharge The continuous high-voltage current discharge across a spark gap.

convenience outlet 1. A wall outlet providing 120 volts alternating current, for household purposes. 2. A outlet in a laboratory that provides power for a certain application.

convergence 1. The eventual meeting of values or bodies at some point (sometimes at infinity, as in some convergent series). 2. The intersection point of the beams from separate electron guns in a cathode-ray tube.

convergence coil One of a pair of coils used in a color TV receiver to produce dynamic beam convergence (see *convergenece, 2).*

convergence control In a color TV receiver, a potentiometer in the high-voltage circuit for convergence adjustment (see *convergence, 2).*

convergence electrode An electrode that provides an electrostatic field for converging electron beams. Compare *convergence magnet.*

convergence frequency The frequency of the last member of a spectrum series.

convergence magnet An assembly that provides a magnetic field to converge electron beams. Compare *convergence electrode.*

convergence phase control In a three-gun color TV receiver, a variable resistor or variable inductor used to adjust the phase of the dynamic convergence voltage.

convergence plane 1. In a color-television picture tube, the plane in which the red, green, and blue beams all focus. 2. In a cathode-ray tube, the plane in which the electron beam reaches its sharpest focus.

convergent series A mathematical series which approaches a finite value as the number of terms increases. Thus, the series $0.3 + 0.03 + 0.003 + \ldots$ approaches 1/3 as a limit. Compare *divergent series* and *infinite series.*

converging lens A lens having a real focus for parallel rays. Compare *diverging lens.*

conversational compiler A compiler that, using the *conversational mode* of operation, tells the programmer whether or not each statement entered into the computer is valid and whether or not he can proceed with the next instruction.

conversational mode Computer operation in which the user communicates directly with the machine, i.e., he is given responses to what he is inputting.

conversion 1. The mixing of signals to produce sum and difference frequencies. Also see *converter, 1.* 2. The process of changing dc to ac. Also see *converter, 2.* 3. The process of changing low-voltage dc to highvoltage dc. Also see *converter, 3.* 4. Converting a computer file to another format and, possibly, transferring it to a different storage medium (e.g., from tape to internal memory). 5. Converting a program devised for one computer to a form acceptable by another.

conversion efficiency In a converter (see *converter, 1*), the ratio of input-signal amplitude to output-signal amplitude. For example, in a superheterodyne converter, a large i-f output for a low rf input indicates high conversion efficiency.

conversion equipment In a computer system, an offline device for transferring data from one medium to another, e.g., card-to-tape converter. Also called *converter*.

conversion exciter An exciter for transmitters, in which an output signal of a desired frequency is obtained by beating the output of a variable-frequency self-excited oscillator with the output of a fixed-frequency oscillator (such as a crystal oscillater).

conversion gain Amplification as a byproduct of conversion. See *converter, 1, converter tube, conversion efficiency*.

conversion loss Conversion gain of less than 1.

conversion program A program for data conversion (see *conversion, 4*) or program conversion (see *conversion, 5*).

conversion rate The number of readings per second that can be produced by an *analog-to-digital converter*.

conversion time In digital computer operation, the time required for the machine to read out all the digits in a coded word.

conversion transconductance See *conversion efficiency*.

conversion transducer A transducer for frequency conversion. Also see *converter, 1* and *converter, 2*.

convert 1. To perform frequency conversion (see *conversion, 1*). 2. To perform voltage conversion (see *conversion, 2* and *conversion, 3*). 3. In computer practice, to change information from one number base to another. 4. To perform conversion as in *conversion, 4 & 5*.

converter 1. A heterodyne mixer in which two input signals of different frequency are mixed to yield a third (output) signal of yet a different frequency. Also see *converter tube*. 2. A machine for converting dc to ac, e.g., a rotary converter. 3. A transistor circuit for converting a low-voltage dc to higher-voltage dc. 4. Conversion equipment. 5. A circuit or device that changes analog data to digital data or vice versa.

converter amplifier See *chopper amplifier*.

converter stage A special-purpose transistor or electron tube used principally to mix two signals (such as a received signal and local-oscillator signal in a superheterodyne receiver), and deliver the resultant signal. See also *heptode, hexode* and *pentagrid converter tube*.

convexo-concave Pertaining to a lens having a convex face or greater curvature than its concave face.

Cook system A stereo recording system in which the two channels are cut in parallel grooves. Not used today.

coolant A liquid (often water or oil) used to remove heat from an electronic component.

Coolidge X-ray tube An X-ray tube containing a heated filament (with focusing shield) and a slanting tungsten target embedded in a heavy copper anode.

cooling Reducing the operating temperature of an electronic component to a safe level. Common devices for cooling are (1) heatsinks, (2) still air, (3) forced air, (4) circulating water, (5) circulating oil, or (6) other liquids.

coordinate bond A covalent bond comprised of a pair of electrons supplied by only one of the atoms joined by the bond.

coordinate digitizer A device or circuit that encodes a coordinate graph into digital signals for storage or transmission.

coordinate of chromaticity See *chromaticity coordinate*.

coordinates The axes which points may be located in space. See *Cartesian coordinates* and *polar coordinates*.

coordinate system A mathematical means of uniquely defining or locating a point on a line, in a plane, or in space. The most common coordinate system is the Cartesian system, consisting of number lines intersecting at right angles.

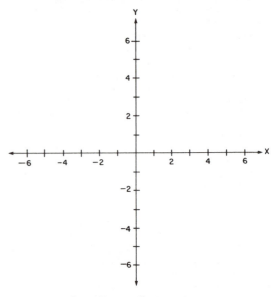

Cartesian coordinate system.

coordination complex An ion or compound having a central (usually metallic) ion combined by coordinate bonds with a definite number of surrounding groups, orns, or molecules. Also see *chelate*.

coplanar-grid tube A vacuum tube having two control grids between its cathode and plate. In a grid-leak detector using this tube, the rf signal is applied in phase opposite to the two grids from a center-tapped input transformer.

copper Symbol, Cu. A metallic element. Atomic number, 29. Atomic weight, 63.54. Copper is one of the better conductors of electricity and heat.

copper-clad wire Iron or steel wire plated with copper.

copper-constantan thermocouple A thermocouple consisting of a junction between wires or strips of copper and constantan. Typical output is 4.24 mV at 100 °C.

copper loss Power (I^2R) loss in copper-wire lines or coils.

copper-oxide diode A small diode in which the semiconductor material is copper oxide. Such diodes, widely used before the ready availability of selenium and silicon, are still occasionally found in meter-rectifier service.

copper-oxide modulator An amplitude modulator whose action is derived from the nonlinear conduction characteristic of copper-oxide diodes.

copper-oxide photocell A photoelectric cell in which the light-sensitive material is copper oxide.

copper-oxide rectifier A rectifier in which the semiconductor material is copper oxide. Rectifiers of this type are suitable for low-voltage service; they were widely used before the advent of germanium, silicon, and silicon, rectifiers.

copper pyrites See *chalcopyrite.*

copper-sulfide rectifier A rectifier in which the unilateral junction is between copper sulfide and magnesium elements. Like the copper-oxide rectifier, the copper-sulfide unit was once widely used in low-voltage applications.

copy 1. Text (printed or written matter). 2. The degree to which transmitted information is coherent, as might be expressed in a radio operator's query of a station receiving his signals (How do you *"copy?"*). 3. To duplicate information in a storage device, the original being in another device or in a different location in the same device. 4. A duplicate.

copying telegraph A descriptive term for a facsimile system.

CORAL A high-level computer programming language for real-time applications.

Combino disk A variable resistor consisting of a semiconductor disk capable of exhibiting the Corbino effect. The disk is inserted into an adjustable magnetic field, which serves as the control medium.

Corbino effect A phenomenon similar to the Hall effect, in which a current flows around a disk carrying a radial current when the disk is inserted into a magnetic field, perpendicular to it. Compare *Hall effect.*

cord 1. A length of flexible, insulated cable having usually, one to three conductors. 2. Tough, insulating string, such as dial or lacing cord.

cordless Descriptive of a plug without a flexible cord.

cordwood A type of construction in which electronic components are sandwiched perpendicularly between layers of components, so called because it looks like stacked cordwood.

cordwood module A module containing discrete components mounted perpendicularly between two parallel printed circuits.

core 1. The body upon which a coil or transformer is wound. Coil cores may be made of magnetic material or, when used only to support a winding, of dielectric material. Low-frequency transformer cores must always be of a magentic metal, such as iron, steel, or some magnetic alloy. 2. At one time, the term for a computer's main memory (from *core memory*).

core dump Dumping core memory content to an output peripheral. Also see *dump.*

cordless induction heater An induction heater in which the body to be heated receives energy directly from the field of the energizing coil (there is no intervening core). Compare *core-type induction heater.*

core loss Loss of energy in a magnetic core due to eddy currents and hysteresis in the core material.

core memory A computer memory consisting of a series of small ringshaped magnetic cores into or out of which information may be read by changing the magnetization of the cores.

core plane A usually flat assembly of special magnetic cores through which pass associated current-conducting wires to provide a core memory.

core saturation The condition in which a core of magnetic material accommodates the maximum number of magnetic lines characteristic of that material. Increasing the magnetizing force produces no additional magnetization.

core shift register A shift register employing special magnetic cores as bistable components. (See description of cores under *core memory.*)

core storage A high-speed magneticcore storage unit in a digital computer. Also see *core memory* and *core plane.*

core transformer A transformer whose coils are wound around a core.

core wrapping The placing of an insulating layer over an inductor or transformer core. This minimizes the chances of short-circuiting between the windings and the core material.

core-type induction heater An induction heater in which the body to be heated is magnetically linked to the energizing coil by a core. Compare *coreless induction heater.*

corner An abrupt turn in the axis of a waveguide.

corner effect A rounding off of the frequency response of a filter at the corners, i.e., at the upper and lower limits of the passband.

corner frequency A point (frequency) at which the slope of a response curve changes abruptly.

corner-reflection antenna A directional antenna consisting of a dipole radiator situated at the apex formed by two nonparallel flat reflecting sheets or a single folded sheet.

corona A luminous discharge in the space surrounding a high-voltage conductor caused by ionization of the air. The discharge constitutes a loss of energy.

corona effect The production of a corona, especially at the end of a pointed terminal, when the voltage gradient reaches a critical value (for air it is 75 kV/in.).

corona failure A type of high-voltage failure resulting from a body's erosion by corona.

corona loss Loss due to energy dissipation through a corona. It occurs as a result of the emission of electrons from the surface of electrical conductors at high potentials and is dependent upon the curvature of the conductor surface, with most emission occurring from sharp points and the least from surfaces with a large radius of curvature. Corona loss is usually accompanied by a blue glow and a crackling or hissing sound.

corona resistance The length of time an insulating material can withstand a specified level of field-intensified ionization before completely breaking down.

corona shield A shield surrounding a high-voltage point to prevent corona by redistributing the electric lines of force.

corona starting voltage The minimum voltage between two electrodes, or on a single electrode in free space, at which corona occurs.

corona voltmeter A voltmeter for measuring the peak value of a voltage in terms of corona discharge. In consists of a metal tube in which a central wire is mounted, the parts being connected to the voltage source. The air density in the tube is varied until corona occurs.

corpuscle In electronics, a tiny particle; it was the name given to the electron by some early experimenters and philosophers.

correction 1. The addition of a factor that provides greater accuracy in a measurement. 2. A change in the calibration of an instrument to increase the accuracy.

correction factor A percentage, or numerical factor, added to or subtracted from a reading to provide a greater degree of accuracy. Often used with instruments known to be inaccurate by a certain amount.

corrective feedback Feedback used to correct (bring to a prescribed level) a quantity constituting the input to the system being corrected.

corrective maintenance The repair of a circuit or system after it has malfunctioned or broken down.

corrective network A network which improves the performance of the circuit into which it is inserted.

corrective stub A combination tuning-matching stub (see *quarterwave transformer* permitting an antenna to be matched to a feeder line and allowing the reactance of the radiator to be tuned out.

corrector magnets See *antipincushioning magnets.*

correed relay A sealed reed relay used as a high-speed switching device in communications equipment.

correlation The degree of correspondence between data. *High correlation* means a close correspondence; *low correlation*, less close correspondence; *zero correlation*, no correspondence at all. Correlation, however, does not necessarily imply causation; events can correlate perfectly and yet have no connection with each other. Thus, if event *A* always ocurs when event *B* does, they are highly correlated. But this can mean a number of things: (1) event A causes event B; (2) event B causes event A; (3) both are caused by some undetermined external force; or (4) A and B occur coincidentally. Correlation is a rating against a scale of zero to 1 (no correlation to perfect correlation).

correlation detector A detector which compares a signal of interest with a standard signal at every point, delivering an output proportional to the correspondence between the two signals.

correlation distance The smallest distance between two antennas that results in fading of signals under conditions of tropospheric propagation. Employed at frequencies in the very high range and above, to determine the maximum range over which communications can be carried out with total reliability.

correlation tracking A method of target tracking in which phase relationships are employed to determine positions.

corrosion-resistant Materials that are treated to be immune to corrosion by the elements. Such substances are preferable for use in marine or tropical environments, where corrosion is especially severe.

corruption The altering of data or a code as a result of a program error or machine fault.

COS Abbreviation of *complementary-symmetry device.*

cosecant Abbreviation, csc. A trigonometric function; csc Θ = *c/a*, where *c* is the hypotense of a right triangle and *a* is the side opposite Θ. Consecant is the reciprocal of sine: csc *a* = 1/sin *a*.

cosecant-squared antenna A radar antenna which radiates a *cosecant-squared beam.*

cosecant-squared beam A radar beam whose intensity varies directly with the square of the cosecant of the angle of elevation.

cosech Abbreviation of *hyperbolic cosecant.* (Also abbreviated csch.)

cosh Abbreviation of *hyperbolic cosine.*

cosine Abbreviation, cos. A trigonometric funciton; cos Θ = *b/c*, where *b* is the side adjacent to Θ and *c* is the hypotenuse of the right triangle.

cosine integral Abbreviation, *Ci(x)*. The integral $-\int_x^\infty$ (cos μ/μ)dμ. Compare *sine integral.*

cosine law The brightness in any direction from a perfectly diffusing surface is proportional to the cosine of the angle between the direction vector and a vector perpendicular to the surface.

cosine wave A periodic wave which follows the cosine of the phase angle. Compare *sine wave.*

cosine yoke A magnetic-deflection yoke having nonuniform windings for improved focus at the edges of a TV picture. Also called *anastigmatic yoke* and *full-focus yoke.*

cosmic noise Radio noise produced by signals from extraterrestrial space.

cosmic rays Extremely penetrative rays consisting of streams of atomic nuclei entering the earth's atmosphere from outer space. Their energy has been estimated to be on the order of a tera-electronvolt. (10^{12}eV).

cosmology A branch of astronomy concerned with the structure, origin, and space-time relationships of the universe.

COS/MOS intergrated circuit An IC, such as an *operational amplifier,* utilizing *metal-oxide field-effect transistors* in a *complementary-symmetry circuit.*

cosmotron See *proton-synchrotron.*

cost analysis In a commercial or industrial organization, ascertaining the expense associated with a service, process, or job.

cot Abbreviation of *cotangent.*

cotagent Abbreviation, cot. A trigonometric funciton; cot Θ = *b/a*, where *a* is the side adjacent to Θ and *b* is the side opposite Θ (in a right triangle) Contangent is the reciprocal of tangent: cot *a* = 1/tan *a*.

coth Abbreviation of *hyperbolic contangent.*

Cotton-Mouton effect See *Kerr magneto-optical effect.*

Cottrell process Dust precipitated by high voltage: dust in the air is made to flow through a grounded metal chamber containing a wire maintained at high voltage; the dust particles become charged and adhere to the chamber walls from which they are later collected.

coul-cell A coulometer of the electrolytic-cell type.

coulomb (Charles Augustin *Coulomb,* 1736-1806). Abbreviation, C. The unit of electrical quantity: the number of electrons (electrical quantity) passing any point in 1 second constitutes a current of 1 ampere; 1C = 6.24 × 10^{18} electrons.

Coulomb's laws The laws which shows that the force *F* of attraction (or repulsion) between two electric charges *Q* is directly proportional to the product of the charges (or to the magnetic pole strengths) and inversely proportional to the square of distance *d* between them:

$$F = \frac{Q_1 Q_2}{d^2}$$

coulometer An instrument giving measurements in coulombs. A typical version keeps a cumulative count of coulombs (ampere-seconds) by integrating current with time. Also called *coulombmeter.*

Coulter counter See *cell counter.*

count 1. The number of pulses tallied by a counting system. 2. A single response by a radioactivity counter. 3. A record of the number of times an instruction or subroutine in a computer program is executed (by increasing the value of a variable by one, as stated in a FOR-NEXT loop, for example).

countdown A decreasing count of time units remaining before an event or operation takes place showing time elapsed and time remaining.

counter 1. A circuit, such as a cascade of flip-flops, which keeps track of the number of pulses applied to it and usually displays the total number of pulses. 2. A mechanism, such as an electromechanical indicator, which keeps track of the number of impulses applied to it and displays the total. 3. An electronic switching circuit, such as a flip-flop or steeping circuit, which responds to sequential input pulses applied to it, giving one output pulse after receiving a certain number if input pulses. 4. *opposite* or *contrary* to. Example: *counter emf.*

counterclockwise Abbreviation, ccw. Rotation to the left (from the top). Compare *clockwise.*

counterclockwise-polarized wave An elliptically polarized TEM wave whose electric-intensity vector rotates counterclockwise as observed from the point of propagation. Compare *clockwise-polarized wave.*

counter efficiency The sensitivity of a radiation or scintillation counter to incident X rays or gamma rays.

counterelectromotive cell A cell employed to counteract a direct-current voltage.

counter emf See *back voltage* and *kickback*.

counter-meter A radioactivity instrument, such as a Geiger counter, which indicates the number of radioactive particles per unit time detected by the instrument.

countermodulation A method of counteracting cross modulation in a radio receiver by bypassing the cathode resistor of the first rf amplifier tube for rf, but not for af. The af components caused by cross modulation accordingly appear across the resistor and modulate the carrier with signals of equal amplitude and opposite phase to cancel cross-modulation components.

counterpoise An artificial antenna ground composed of a long metal conductor usually stretched close to the surface of the earth and insulated from it. The length of a counterpoise should at least be equal to the height (above ground) of the antenna with which it is associated.

counterpoise ground system An antenna arrangement including a counterpoise.

counter tube 1. A tube, such as the Geiger-Meuller tube, in which a penetrating radioactive particle ionizes a gas and produces an output pulse. 2. A flip-flop tube. 3. A tube operated so that it delivers one or more output pulses after receiving a certain number of input pulses.

counter voltage See *back voltage* and *kickback*.

counting-type frequency meter A direct-reading analog or digital frequency meter which indicates the number of pulses (or cycles) per second applied to it.

count-remaining technique See *complement-setting technique*.

couple Two dissimilar metals in contact with each other or immersed in an electrolyte.

coupled circuits Circuits between which energy is transferred electrostatically, electromagnetically, by some combination of the two, or by direct connection.

coupled impedance The impedance which a circuit "sees" when it is coupled to another circuit. Thus, when the secondary of a transformer is terminated with an impedance, the primary sees a combination of that impedance and its own.

coupler A device for transferring energy between two circuits and employing capacitive coupling, direct coupling, inductive coupling, or some combination of these.

coupling The linking of two circuits or devices by electrostatic lines of force (*electrostatic*, or *capacitve*, coupling) or electromagnetic lines of force (*electromagnetic*, or *inductive*, coupling), or by direct connection (*direct* coupling) for the purpose of transferring energy from one to the other. Also see *capacitive coupling, coefficient of coupling, direct coupling, inductive coupling, mutual inductance*.

coupling aperture A hole in a waveguide that is employed for the purpose of transmitting energy to the waveguide, or receiving energy from outside the waveguide.

coupling capacitor A capacitor employed to conduct ac energy from one circuit to another. Also see *capactive coupling*.

coupling coefficient See *coefficient of coupling*.

coupling diode A semiconductor diode connected between the stages of a direct-coupled amplifier. Correctly poled it acts as a high resistance between the stages when there is no signal and, hence does not pass the high dc operating voltage from one stage to the next. When a signal is present, however, the diode resistance decreases and the signal gets through.

coupling efficiency A measure of the effctiveness of a coupling system, i.e., the degree to which it delivers an undistorted signal of correct amplitude and phase.

coupling loop 1. A usually one-turn coil constituting one wind of a coupling transformer. 2. A small loop inserted into a waveguide to induce a microwave energy into it.

coupling probe A usually short, straight wire or pin protruding into a waveguide to couple microwave energy electrostatically into the latter, somewhat in the manner of an antenna.

coupling transformer A transformer employed primarily to transfer ac energy electro—magnetically into or out of a circuit.

covalent binding forces In a crystal, the binding forces resulting from the sharing of valence electrons by neighboring atoms.

covalent bonding The binding together of the atoms of a material as a result of shared electrons or holes.

coverage 1. The area within which a broadcast or communication station can be reliably heard. 2. The shielding effectiveness of a coaxial cable.

coversed sine Abbreviation, covers. The trigonometric functional equivalent of the *versed sine* of the complement of an angle, i.e., the difference between the sine of an angle and unity (1): covers $a = 1 - \sin a$. Also see *versed sine*.

CP Abbreviation of *chemically pure*.

cp 1. Abbreviation of *candle power*. 2. Abbreviation of *central processor*.

Cpk Symbol for *plate-cathode capacitance* of a tube.

C-power supply See *C-bias supply*.

cps 1. Abbreviation of *cycles per second*. (*cycles per second*, to denote ac frequency, has been supplanted by *hertz*). 2. Abbreviation of *characters per second*.

CPU Abbreviation of *central processing unit*.

CQ A general call signal used in radio communication, especially by amateur stations, to invite a response from any station that hears it.

Cr Symbol for *chromium*.

cracked-carbon resistor A high-stability resistor in which the resistance material is particulate carbon.

cradlephone A telephone in which the microphone and earphone are mounted on opposite ends of a handle which rests on the crossmember of a stand connected to a base containing the bells and dial. Also called *French phone, French telephone,* and *handset*.

crate A foundation unit into which modules are plugged to establish a circuit.

crater lamp A glow-discharge tube whose light-emitting element is a crater instead of the usual plate.

crawl 1. See *creeping component*. 2. The credits (names of staff and their contribution to content) superimposed and moving usually vertically on a TV picture at the end of a program.

crazing The formation of tiny cracks in materials, particularly in such dielectrics as plastic and ceramic.

creep See *cold flow*.

creepage Current leakage across the surface of a dielectic.

creeping component A quantity, such as current, voltge, or frequency, which slowly changes in value with time.

crest factor See *amplitude factor*.

crest value The maximum amplitude of a composite current or voltage.

crest voltmeter A peak-reading (or sometimes peak-responsive) voltmeter.

crippled mode The mode of operation for a computer or other hardware in which some of the components are inoperable. Compare *graceful degradation*.

crisscross neutralization See *cross-connected neutralization*.

crisscross rectifier circuit A conventional bridge rectifier circuit configured in such a way that two of the diodes are connected crisscross fashion between input and output terminals.

critical angle 1. The angle of radition with respect to a tangent to the earth at the transmitting point, above which the ionisphere will not reflect the signal back to earth. 2. The minimum angle of incidence at which rays are totally reflected.

critical characteristic A parameter that has a disproportionate effect on other variables. A small change in a critical characteristic can result in a large change in the operating conditions of a circuit or system.

critical component A component or part that is especially important in the operation of a circuit or system.

critical coupling The value of coupling at which maximum power transfer occurs. Tigthtening coupling beyond the critical value decreases power transfer. At the point of critical coupling, $k \sqrt{Q1Q2} = 1$, where k is the coefficient of coupling; $Q1$, the primary-circuit Q; and $Q2$, the secondary-circuit Q.

critical coupling factor The value of the coefficient of coupling at the point of critical coupling; $k = 1/\sqrt{Q1Q2}$. Also see *critical coupling*.

critical damping The value of damping that yeilds the fastest transient response without overshoot.

critical dimension The cross-sectional size of a waveguide that determines its minimum usable frequency.

critical failure A component or circuit failure that results in shutdown of a system, or a malfunction resulting in improper operation.

critical field The smallest magnetic-field intensity in a magnetron that keeps an electron, emitted from the cathode, from reaching the anode.

critical frequency For a particular layer of the ionosphere, the high frequency at which a vertically propagated wave is no longer reflected back to the earth.

critical grid current For a thyratron, the instantaneous value of grid current at which anode current starts to flow.

critical grid voltage For a thyratron the instantaneous value of grid voltage at which anode current starts to flow.

critical inductance In a choke-input power-supply filter, the minimum inductance which will maintain a steady value of average dc load current. At this value of inductance, the current is on the verge of discontinuity.

critical potential The potential difference required for an electron to excite or ionize an atom with which it collides.

critical voltage 1. The voltage at which a gas ionizes. 2. In a thyratron, the voltage at which conduction begins.

critical wavelength The wavelength corresponding to critical frequency.

CRO Abbreviation of *cathode-ray oscilloscope*.

Crookes dark space In a glow-discharge tube, the narrow dark space next to the cathode. Also see *Crookes tube*.

Crookes tube A glow-discharge tube containing an anode, cathode, and a small amount amount of gas under low pressure.

Crosby circuit A frequency-modulation circuit in which a reactance tube shunts a radio-frequency tank; the reactance is varied at an audio-frequency rate by the modulating voltage applied in series with the control-grid bias of the reactance tube.

cross antenna An antenna in which two (usually equal-length) horizontal radiators cross each other at right angles and are conneted together and to a feeder at their point of intersection. It takes its name from its horizontal-cross shape.

cross assembler A program used with one computer to translate instructions for another computer.

crossband operation Communications in which two bands are used. Two stations X and Y engaged in crossband operation transmit and receive on alternate frequencies, separated by a wide increment.

crossbar switch A three-dimensional array of switch contacts in which a magnetic selector chooses individual contacts according to their coordinates in the matrix.

cross bearings Radio-compass bearings taken on two transmitting stations. Lines corresponding to the stations' location of the receiver (radio compass) being at the intersection of the lines.

cross beat A spurious frequency arising from cross modulation.

cross-check To compare the result of a calculation or (computer) routine with the result obtained by a different method.

cross color In the chrominance channel of a color TV receiver, crosstalk interference caused by monochrome signals.

cross-connected neutralization Neutralization of a push-pull amplifier by feedback through two capacitors, each connected from the output circuit of one tube (or transistor) to the input circuit of the other.

cross-coupled multivibrator A multivibrator circuit in which feedback is provided by a coupling capacitor between the output of the second stage and the input of the first stage; the stages are forward-coupled by a capacitor of the same value.

cross coupling 1. The state of being cross-coupled (see, for example, *cross-coupled multivibrator*). 2. Undesired coupling between two circuits.

cross current A current that flows counter to another.

crossed-field amplifier A tube which employs crossed electric and magnetic fields to convert dc power into high microwave energy.

crossed-pointer indicator A two-pointer meter used in aircraft to show the position of the aricraft relative to the glide path.

crossed-wire thermoelement Two wires or strips of dissimilar metals joined or twisted at a point that constitutes a thermeoelectric junction. In usual operation, a high-frequency current is passed through one wire, and a proportional dc voltage generated by thermoelectric action is taken from the other wire.

cross flux The magnetic flux component which is perpendicular to the flux produced by field magnets.

cross-hair pattern A TV test pattern consisting of a single vertical line and a single horizontal line forming a simple cross and resembling the crossed hairs of an optical instrument.

crosshatch generator A modulated rf signal generator which produces a crosshatch pattern on a TV picture-tube screen.

crosshatch pattern A grid of horizontal and vertical lines produced on a TV picture-tube screen by a cross-hatch generator and employed in checking horizontal and vertical linerarity.

cross modulation 1. A type of radio interference between two strong stations that are close in frequency. When the desired carrier is unmodultaed, it becomes modulated by the undesired signal. 2. The production of signals by junctions in pipes and wiring near a radio receiver. These objects pick up waves and deliver energy at a different frequency, which finds its way into the receiver. Also called *external cross modulation*. 3. The interaction between signals of different frequency when they magnetize a core of nonlinear magnetic material. Also see *crosstalk*.

cross-modulation factor An expression of the amount of cross modulation (or crosstalk) present in a particular instance. It is equal to $M1/M2$, where $M1$ is the percent modulation a modulated wave produces in a superimposed unmodulated wave, and $M2$ is the percent modulation of the modulated wave.

cross-neutralized circuit See *cross-connected neutralization.*

crossover 1. In a circuit diagram, a point at which lines representing wires intersect but are not connected. 2. In a characteristic curve, point at which the plot crosses an axis or operating point. 3. See *crossover network.*

crossover distortion Distortion of a characteristic at a crossover point (see *crossover, 2)* for example, a bend in the curve where the plot of an ac cycle passes through zero.

crossover frequency The frequency at which a *crossover network* delivers equal power to the two circuits it supplies.

crossover network Following final amplification in a sound-reproduction system, an active outboard filter circuit or a passive one within a speaker enclosure that usually blocks or passes high frequencies, the purpose being to deliver each band of signals to the correct loudspeaker.

crossover point See *crossover, 2.*

crossover S-curve The S-shaped image obtained on an oscilloscope screen during sweep-generator alignment of a FM detector. In correct alignment, the exact center of the S-curve (the crossover point) coincides with the zero point on the screen.

cross-sectional area 1. The surface area of a face of conductor after cutting through it at a right angle. Specified in square inches, square millimeters, or circular mils. 2. The total of the cross-sectional areas of all the wires in a stranded conductor.

crosstalk Undesired transfer of signals between systems or parts of a system. Crosstalk causes interference between telephone lines, antennas, parts, etc.

crosstalk coupling Undesired coupling between circuits, due to crosstalk.

crosstalk factor See *cross-modulation factor.*

crosstalk level The amplitude of crosstalk, usually expressed in deceibels above a reference level.

crosstalk loss Loss of energy due to crosstalk.

crowbar An action producing a high overload on a circuit protection device.

crowfoot 1. A pattern formed by the cranking or crazing of solid plastics of solidified encapsulating compounds, so called from its resemblance to a bird's footprint. 2. In a gravity battery cell, the zinc electrode, so called from its resemblance to a bird's foot.

crt Abbreviation of *cathode-ray tube* (often capitalized).

crud Colloquialism for the usually broadband mixture of electrical noise (from within and without a system) and undesired signals that interferes with a desired signal.

cryogenic device A device, such as a cryotron, which exhibits unique electrical characteristics such as superconductivity at extremely low temperatures.

cryogenic motor A motor designed for operation at very low temperatures.

cryoelectronics The study of the behavior of electronic devices, circuits, and systems at extremely low temperatures.

cryogenics The branch of physics dealing with the behavior of devices at extremely low temperatures (approaching absolute zero) and with producing such low temperatures.

cryosar A semiconductor switch utilizing low-temperature avalanche breakdown.

cryoscope An instrument used to determine freezing point.

cryostat A chamber for maintaining a very low temperature for cryogenic operations. Also see *cryogenics.*

cryotron A switching device consisting essentially of a straight tantalum wire around which a single-layer control coil is wound. The magnetic field generated by control current flowing through the coil causes the tantalum wire to become superconductive at a temperature of approximately 4.4K. See *superconductivity.*

cryotronics Low-temperature electronics, concerned with such phenomena as superconductivity. The term is an acronym from *cryogenics* and *electronics.* Also see *cryogenics.*

crypto A prefix added to words, that implies encoding for the purpose of changing or hiding the meaning of a message or signal.

crystal 1. A material distinguished by the arrangement of its atoms into a redundant pattern called a *lattice* and which presents a (in however small a fragment) a characteristic geometric shape (cubical, hexagonal, pyramidal, etc.). 2. A crystal fragment. 3. A plate or bar cut from a piezoelectric crystal.

crystal amplifier 1. A transistor. 2. A semiconductor diode circuit using carrier storage. Transistor action and, accordingly, pulse amplification is obtained by alternately making one electrode of the diode an emitter or collector. Also see *Holt amplifier.*

crystal audio receiver An audio radar receiver, consisting of a crystal detector and audio-amplifier stages.

crystal axes The imaginary lines traversing a piezoelectric crystal, along which (or perpendicular to which) plates are cut for oscillators, resonators, or transducers.

crystal calibrator A refined crystal oscillator used to generate harmonic checkpoints for frequency calibration. Common fundamental calibrator frequencies are 100 kHz and 1 MHz.

crystal capacitor See *varactor.*

crystal control The control of the operating frequency of a circuit by means of a piezoelectric crystal.

crystal-controlled receiver A superheterodyne radio receiver whose local oscillator is crystal controlled.

crystal-controlled transmitter A radio transmitter whose master oscillator is crystal controlled.

crystal current Current flowing through a crystal; specifically, the rf current flowing through a quartz plate in a crystal-controlled oscillator.

crystal cuts The classification of piezoelectric plates according to the angle at which they were cut from a quartz crystal. Common cut designations are AT, BT, CT, DT, X, Y, and Z. Various cuts afford such complementary factors as frequency, temperature, and thickness. Also see *crystal axes.*

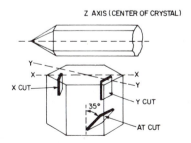

Crystal cuts and axes.

crystal detector A rudimentary form of semiconductor diode consisting of a mounted lump of mineral (the *crystal)* in contact with a springy *cat's whisker*. The point of the catwhisker is moved to various points of contact on the crystal surface until the most sensitive rectifying spot is found.

crystal diode Semiconductor diode. Also see *gallium-arsenide diode, germanium diode, junction diode, laser diose, point-contact diode, selenium diode, signal diode, silicon diode.*

crystal earphone An earphone in which the transducer is a piezoelectric crystal. Electrical impulses applied to the crystal vary its shape and cause a vibration which is transmitted to a diaphragm that produces corresponding sound waves.

crystal filter See *crystal resonator.*

crystal headphone See *crystal earphone.*

crystal holder A fixture specially designed to hold a piezoelectric crystal; it insures minimum distortion of crystal dimensions and minimum residual capacitance, inductance, and resistance.

crystal imperfection A variation in the lattice structure of an otherwise perfect crystal.

crystal lattice The orderly, redundant pattern of atoms and molecules within a crystalline mterial; it is a characteristic of a given material.

crystal-lattice filter A crystal resonator in which piezoelectric crystals are used to give a desired shape to the filter response curve.

crystalline material A material exhibiting the chracteristic properties of a crystal (see *crystal, 1*).

crystallogram An X-ray photograph or other record of crystal structure.

crystallography The science dealing with crystals and their properties (see *rystal, 1*).

crystal loudspeaker A loudspeaker whose transducer is a piezoelectric crystal. Electrical impulses applied to the crystal vary its shape and cause vibrations that are transmitted to a diaphragm or cone, which produces corresponding sound waves.

crystal meter A rectifier-type ac meter employing a semiconductor diode (as a crystal rectifier) in series with a dc milliammetter or microammeter.

crystal microphone A microphone whose transducer is a natural or synthetic piezoelectric crystal. Sound waves striking the crystal (directly or via a diaphragm) vary its shape, making it produce an af output voltage.

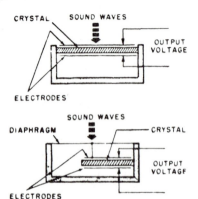

Crystal microphones.

crystal mixer A mixer (converter) circuit utilizing the nonlinearity of a semiconductor diode to mix signals.

crystal operation 1. The operating characteristics of a piezoelectric crystal in a particular circuit. 2. Crystal control.

crystal oscillator An oscillator whose operating frequency is determined by the dimensions of an oscillating piezoelectric quartz-crystal plate. Compare *self-excited oscillator.*

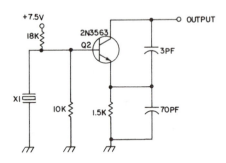

Crystal oscillator.

crystal oven A constant-temperature chamber for stabilizing the frequency of a quartz crystal by maintaining its operating temperature at a fixed point.

crystal photocell A photoelectric cell in which the light-sensitive material is a crystalline substance material is a crystalline substance, such as germanium, selenium, silicon, etc.

crystal pickup A phonograph pickup whose transducer is a natural or synthetic piezoelectric crystal. The crystal is attached (either directly or through a mechanical linkage) to a stylus, whose movement in the record groove varies the shape of the crystal. The resultant vibration generates a corresponding af output voltage across the crystal.

crystal probe An ac (rf) probe whose rectifying element is a semiconductor diode.

crystal pulling Extracting a single crystal from a molten mass of crystalline material. Single crystals are used for high-quality semiconductor devices. Also see *Czochralski method, single crystal,* and *single-crystal material.*

crystal receiver See *crystal set.*

crystal rectifier A semiconductor diode. Sometimes, by extension, a semiconductor power rectifier.

crystal resistor A temperature-sensitive resistor made from silicon and exhibiting a positive temperature coefficient of resistance.

crystal resonator A highly selective resonant circuit in which the center frequency is the resonant frequency of a piezoelectric quartz-crystal plate.

crystal sensor See *crystal transducer.*

crystal set A simple radio receiver employing a tuned circuit, semiconductor-diode detector, and earphones.

crystal slab See *quartz bar.*

crystal socket 1. A low-capacitance low-loss socket for a piezoelectric crystal. 2. A socket for a semiconductor diode.

crystal tester 1. A (usually simple) oscillator used to check quartz crystals. Most such units check only the crystal's ability to oscillate; more elaborate ones also check crystal current, frequency, temperature

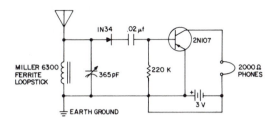

Crystal set.

coefficient, activity, filter action, and so on. 2. An instrument for checking the electrical characteristics of semiconductor diodes. 3. An instrument for checking the performance of piezoelectric ceramics.

crystal tetrode A transistor having four elements: emitter, collector, and two bases.

crystal transducer A transducer employing a piezoelectric crystal as the sensitive element. Examples: crystal earphone, loudspeaker, microphone, pickup, vibration transducer.

crystal triode See *transistor*.

Cs Symbol for *cesium*.

CS Abbreviation of *complementary symmetry*. (Also *COS*.)

Cs Symbol for *standard capacitance* and *source capacitance*.

csc Abbreviation of *cosecant*.

C-scan See *C-display*.

csch Abbreviation of *hyperbolic cosecant*.

C-scope A radarscope giving a C-display.

CT-cut crystal A piezoelectric plate cut from a quartz crystal at an angle of rotation around the X axis of $+38°$. Such a plate has a zero temperature coefficient of frequency at 25 °C. Also see *crystal axes* and *crystal cuts*.

CTL Abbreviation of *complementary-transistor logic*.

Cu Symbol for *copper*.

cube 1. A six-sided geometric solid. 2. The third power of a number (N^3).

cube tap A type of portable electric outlet in which a pair of male prongs and three pairs of female contacts are on the sides of a molded cube.

cubical antenna An antenna in which the elements form the outline of a geometric cube or rectangular prism.

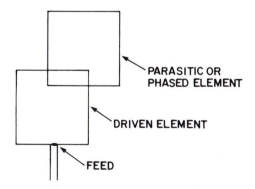

PARASITIC OR PHASED ELEMENT

DRIVEN ELEMENT

FEED

Cubical antenna.

cubical quad antenna See *quad antenna*.

cubic equation A polynomial equation of the third degree. Its general form is $a_0x^3 + a_1x^2 + a_2x + a_3 = 0$.

cue Information, a state, or a situation capable of alerting an operator or preparing a circuit to act in some appropriate manner.

cue circuit A device for transmitting cues employed in program control.

cueing receiver 1. A (usually miniature) radio receiver used to pick up cues. Example: a receiver carried by a technician, actor, or lecturer. 2. A receiver or other pickup circuit that receives a cueing pulse, which it uses to set another circuit.

cueing signal A signal that provides a cue.

cu ft Abbreviation of *cubic foot*.

cu in Abbreviation of *cubic inch (inches)*.

cumulative error In a sum or other final value, the total error which has accumulated from the individual errors in the various terms. Also called *systematic error*.

cumulative grid detector See *grid-leak detector*.

cup core A coil core which also forms a magnetic shield around the coil.

cuprous-oxide rectifier See *copper-oxide rectifier*.

cur Abbreviation of *current*.

curie Abbreviation C. A unit of radioactivity; 1 curie is the amount of radiation from (or in equilibrium with) 1 gram of radium. Also, 1 curie is the amount of radiation which will produce 3.71×10^{10} disintegrations per second (3.71×10^4 rutherfords).

Curie point 1. The temperature above which a ferromagnetic material exhibits paramagnetism. 2. The temperature at which the ferroelectric properties of a substance disappear.

Curie temperature As a magnetized substance is heated, the lowest temperature at which magnetization is lost. Generally measured in degrees Celsius or degrees Kelvin.

Curie's law For a paramagnetic substance the ratio of the magnetization to the magnetizing force is inversely proportional to the absolute temperature.

Curie-Weiss law Above the Curie point the susceptibility of a paramagnetic material varies inversely as the excess of temperature above the Curie point increases. This law is invalid for applications at or below the Curie point.

curium Symbol, Cm. A radioactive metallic element produced artificially. Atomic number, 96. Atomic weight, 242 (?).

current Symbol, I or i. The flow of electricity, i.e., the characteristic drift movement of carriers, such as ions, electrons, or holes. $I = E/R$. Also see *ampere*.

current amplification 1. Amplification of an input current (the result being a larger, output current). 2. The signal increase I_{out}/I_{in} resulting from this process. Also called *current gain*.

current amplifier An amplifier which is operated primarily to increase a signal current. Compare *power amplifier* and *voltage amplifier*.

current antinode A current loop (maximum point). Compare *current node*.

current attenuation 1. The reduction of current amplitude along a line. 2. A value less than 1 of the ratio of current output to current input of device.

current-balance switch A switch or relay, operated by the existence of a difference between two currents.

current carrier See *carrier, 2*.

current-carrying capacity The maximum current (usually expressed in amperes) that a conductor or device can safely conduct.

current coil The series coil in a nonelectronic wattmeter. Compare *potential coil*.

current-controlled amplifier Abbreviation, CCA. An amplifier in which gain is controlled by means of a current applied to a control-input terminal.

current density The current (usually expressed in amperes per square centimeter) passing through a cross-sectional area of a conductor in a given interval of time.

current-detection coefficient For a tube circuit, the second derivative of the $E_p I_p$ characteristic. When the plate circuit contains resistance (R_L) only, the current-detection coefficient $a = (\mu^2 r_p r_p')/2(r_p + R_L)^2$. Here, r_p is the variation in plate resistance r_p' due to curvature.

current drain 1. The current supplied to a load (device) by a generator or generator-equivalent. 2. The current required by a device for its operation; also, the current taken by the device during standby periods.

current echo Reflected current in a transmission line that is not terminated in an impedance exactly matching its characteristic impedance.

current-fed antenna An antenna in which the feeder or transmission line is attached to the radiator at a current loop (voltage node). Compare *voltage-fed antenna*.

current feed The delivery of current to a device or circuit, as to a point where current rather than voltage dominates. Compare *voltage feed*.

current feedback 1. A feedback signal consisting of current fed from the output of the input circuit of an amplifier. 2. A system or circuit for obtaining current feedback.

current-feedback pair A two-stage, direct-coupled transistor amplifier having dc shunt-series feedback.

current flow Current carriers (see *carrier, 2*) passing through a solid, liquid, gas, or vacuum. Also see *current* and *current density*.

current gain See *current amplification*.

current hogging 1. An undesirable condition that sometimes takes place when two or more transistors or tubes are operated in parallel. One device tends to do all the work, taking all the current. The result can be destruction of that device. 2. The tendency of one component in a group of identical series-connected components to dissipate most of the power.

current-hogging injection logic A form of bipolar digital logic, similar to current-hogging logic but having the greater density characteristic of injection logic.

current instruction register A register in which are held instructions ready for execution by a program controller.

current lag The condition in a given circuit in which current changes follow corresponding changes in voltage. Compare *current lead* and *voltage lag*.

current lead The condition in which current changes in a circuit precede corresponding changes in voltage. Compare *current lag* and *voltage lead*.

current limiting Controlling current so that it is mainained at a desired value.

current-limiting resistor A series resistor inserted into a circuit to restrict the current to a prescribed value.

current loop The point on a transmission line at which the current reaches a local maximum.

current meter A usually direct-reading instrument, such as an ammeter, milliammeter, or microammeter, employed to measure current strength. Also see *electronic current meter*.

current-meter operation The operation of a voltmeter as a current meter by connecting it to respond to the voltage drop across a shunt resistor carrying the current of interest.

current-mode logic In computer practice, transistor logic in which the transistors operate in the unsaturated mode.

current node A minimum current point. Compare *current antinode*.

current noise Electrical noise produced by current flowing through a resistor.

current probe A pickup transformer usually having a snap-around, one-turn coil that picks up energy from a conductor and couples it into an ac ammeter.

current rating 1. A specified value of operating current. 2. See *current-carrying capacity*.

current-regulated supply See *constant-current source*.

current regulation Stabilizing current at a predetermined level.

current regulator See *barretter*.

current relay A relay actuated by specific values of pickup and dropout current.

current saturation In the operation of a device (such as a tube, transistor, saturable reactor, or magnetic amplifier), the leveling off of current at a value beyond which no further increase occurs, even though an input parameter is varied further. Also, the particular current point at which saturation begins.

current sense amplifier An amplifier employed to increase the sensitivity of current sensing or to decrease the loading of a current-sensing component.

current sensing Sampling a current, e.g., when the voltage drop across a series resistor is employed as a proportional indication of the current flowing through the resistor.

current-sensing resistor A low-value resistor inserted into a circuit primarily for current sensing.

current sensitivity In a current meter or galvanometer, current (in amperes or fractions thereof) per scale division.

current-sheet inductance Symbol, L_s. Low-frequency inductance of a single-layer coil, calculated with the formula $L_s = (0.100\,28aN^2)/l$, where L_s is in μH, a is the radius in inches, and l is the length in inches.

current shunt 1. A resistor connected in parallel with a voltmeter to convert it into an ammeter. 2. A resistor connected in parallel with the input of a voltage amplifier to make the response of the amplifier proportional to input-signal current.

current sink A circuit or device through which a constant current may be maintained; a subcircuit of this kind in an IC.

current-sinking logic A form of bipolar digital logic. Current flows from one stage to the input of the stage immediately before.

current-squared meter An ammeter or milliammeter whose deflection is proportional to the square of the current.

current-stability factor In a common-base connected bipolar transistor, the radio dI_e/dI_c, where I_c is the collector current and I_e is the emitter current.

current strength The magnitude of electric current (see *current*), i.e., the number of carriers flowing past a given point per unit time, expressed in coulombs per second or in amperes.

current transformer 1. A transformer employed primarily to stepup or stepdown current. A stepup turns ratio between primary and secondary coils provides a stepdown current ratio, and vice versa. 2. A particular transformer (as above) used to change the range of an ac milliammeter or ammeter.

current vector In a vector diagram, a line with an arrowhead (vector) showing the magnitude and phase of a current. Compare *voltage vector*.

current-voltage feedback In an amplifier or oscillator, combined current and voltage feedback.

cursor 1. A marker that indicates the position of type in a video alphanumeric display. Commonly used in computers and word processors. 2. The sweeping line on a radar display. 3. The movable marker on a slide rule.

curve tracer 1. A device which supplies a special variable test voltage to a component or circuit under test, at the same time supplying a sweep voltage to an oscilloscope. The component's resultant output voltage is also presented to the oscilloscope. As a result, the response curve of the component appears on the oscilloscope screen. 2. A device which produces a permanent record (photographic or graphic) of an electrical phenomenon. Also called *oscillograph* or *recorder* (see *recorder*, 2).

curvilinear trace. A trace made on curvilinear recording paper, i.e., paper on which the vertical lines rather than being straight are curved according to the arc through which the recording pen swings.

cut-in angle In a semiconductor rectifier circuit, the angle greater than zero degrees at which current conduction begins. Compare *cut-out angle*.

Cutler antenna A parabolic-dish antenna, in which the driven element consists of a waveguide having two apertures on opposite sides of a resonant cavity.

Cutler feed An aircraft antenna feed system in which rf energy is fed to the reflector by a resonant cavity at the end of a waveguide.

Cutler tone control A dual RC filter circuit of the general bridged-tee variety. Variation of the series leg provides adjustable treble boost; variation of the shunt leg, adjustable bass boost.

cutoff The quality of reducing some operating parameter, such as collector current or plate current to zero by increasing the negative bias at the input electrode.

cutoff attenuator A variable nondissipative-attenuator consisting of a variable length of waveguide used at a frequency below cutoff.

cutoff bias In a tube or transistor circuit, the value of control-electrode bias that produces output current cutoff.

cutoff current Symbol, i_{co}. In a transistor, the small collector current that flows when the emitter current is zero (common-base circuit) or when the base current is zero (common-emitter circuit).

cutoff frequency 1. Symbol, f_{co}. The high frequency at which the current amplification factor of a transistor drops to 70.7% of its 1 kHz value. 2. In a filter, amplifier, or transmission line, the frequency point at which transmission or rejection begins. Examples: the high-frequency cutoff of an amplifier; the upper and lower cutoff points of a bandpass filter.

cutoff limiting Limiting (output-peak clipping) resulting from overdriving a tube or transistor into output current cutoff. Compare *saturation limiting*.

cutoff potential See *cutoff bias*.

cutoff voltage See *cutoff bias*.

cutoff wavelength 1. The wavelength corresponding to cutoff frequency. 2. For a waveguide, the ratio of the velocity of electromagnetic waves in free space (3×10^8m/s) to the cutoff frequency of the waveguide.

cutout 1. A device, such as a circuit breaker, that automatically disconnects a circuit, usually to prevent overload but occasionally to prevent underload. 2. Emergency switch. 3. Fuse.

cutout angle In a semiconductor rectifier circuit, the angle less than 180 degrees at which current conduction stops. Compare *cut-in angle*.

cutout base A fuse block.

cut rate 1. The speed at which a cutter moves across the surface of a blank disk during the recording process. 2. The number of cut lines per inch in a disk recording.

C-voltage Negative control-grid bias in a tube circuit.

cw 1. Abbreviation of *continuous wave*. 2. Abbreviation of *clockwise*.

cw filter In a radiotelegraph receiver, an audio-frequency filter usually consisting of a selective amplifier or bridge, which boosts a desired signal or rejects an undesired one.

cw laser A laser that emits energy in an uninterrupted stream rather than in spurts.

cw monitor See *keying monitor*.

cw oscillator 1. In a radio receiver, a variable-frequency oscillator which heterodynes a radiotelegraph signal in the i-f amplifier to make audible the continuous-wave dits and dahs. 2. Sometimes, an external variablefrequency rf oscillator whose output beats against the actual carrier of a continuous-wave radiotelegraph signal, making it audible as dits and dahs. 3. An unmodulated, unkeyed oscillator.

cw radar A radar system in which rf energy is transmitted over an uninterrupted period to the target.

cw reference signal A sinusoidal reference signal used to control the conduction time of a synchronous demodulator in color TV.

Cx Symbol for *unknown capacitance*.

cyan The dark blur of the tricolor combination in color TV.

cybernetics The study of control system theory in terms of the relationship between animal and machine behavior.

cycle 1. Abbreviation, c. One complete alternation of an ac current or voltage (as from zero, through maximum positive, back to zero, through maximum negative, and back to zero). An ac frequency of 1 cycle per second is 1 Hz (see *hertz*). 2. One complete sequence of operations.

cycle counter A device which totals the number of cycles of a phenomenon repeated during a given period. Also see *counting-rate meter*.

cycle index The number of times a particular cycle has been, or must be, iterated in a computer program.

cycle index counter That which indicates how often a cycle of computer program instructions has been executed. In a program, for example, this can be accomplished by increasing, through instruction, the value of a location's content every time a loop operation is performed.

cycle life The total number of charge-discharge cycles a rechargeable battery cell can telerate before becoming useless.

cycle reset To change the value of a *cycle count* (making it zero or some other value).

cycle shift See *cyclic shift*.

cycle time Pertaining to an operation, the duration of a complete cycle.

cycle timer A timer which switches a circuit or device on and off according to a predetermined cycle. Also called *programmed timer*.

cyclic code *Gray code*.

cyclic memory In computer practice, a memory whose locations can only be accessed points in a cycle, as of a floppy disk.

cyclic shift The moving of data out of one end of a storage register and reentering it character-by-character or bit-by-bit at the other end in a closed loop, e.g., 87654 cyclically shifted one place to the right becomes 48765.

cyclic variations (ionosphere) Periodic changes in the features of the ionosphere due to variations in its ionization. A number of factors, including sunspot activity, underlie these changes.

cycling The tendency of a parameter to oscillate back and forth between two different values.

cyclogram A method of showing the relationship between two signals on an oscilloscope. The two signals must have a fixed phase relationship.

cyclotron A type of atom smasher (see *accelerator, 1*) in which an applied rf field and the intense magnetic field of a large, external magnet cause particles to travel with increasing velocity in a spiral path be-

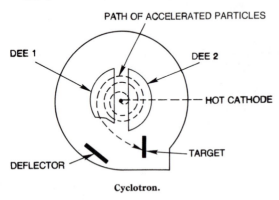

Cyclotron.

tween two semicircular metal boxes called *dees*. When the particles go fast enough in the correct path, they are expelled and strike a target in their path.

cyclotron frequency The angular frequency of a charged particle in a cyclotron. The cyclotron frequency depends on the number of times per second the magnetic field of the device is reversed.

cyclotron radiation An electromagnetic field produced by the circular movement of charged particles in a fluctuating magnetic field.

cylinder See *seek area.*

cylinder magnet A permanent magnet in the shape of a cylinder.

cylindrical capacitor See *concentric capacitor.*

cylindrical contour The most common curvature of the face of a magnetic tape recording heat; it is a section of a cylinder having a constant radius of 1/2 to 1 inch.

cylindrical functions See *Bessell functions.*

cylindrical magnet See *cylinder magnet.*

cylindrical wave An electromagnetic wave whose field surfaces are nearly perfect cylinders.

cylindrical waveguide A waveguide resembling a round pipe.

Czochralski method A technique for obtaining a relatively large single crystal from a substance, such as the semiconductors germanium and silicon. The method consists essentially of dipping a *seed crystal* into a molten mass of the same substance and then slowly withdrawing it while rotating it.

D 1. Symbol for *deuterium*. 2. Symbol for *electric displacement*. 3. Symbol for *electric flux density*. 4. Symbol for *dissipation factor*. 5. Symbol for *drain* (see *drain, 3*). 6. Abbreviation of *dissipation*. 7. Symbol for *determinant*. 8. Symbol for *diffusion constant*.

d 1. Abbreviation of *deci*. 2. Symbol for *differential*. 3. Symbol for *distance*. 4. Symbol for *density*. 5. Symbol for *drain* (see *drain, 3*). 6. Abbreviation of *dissipation*. 7. Abbreviation of *day*. 8. Abbreviation of *degree*. 9. Abbreviation of *diameter*. 10. Abbreviation of *drive*.

DA Abbreviation of *digital-to-analog*.

dA 1. Symbol for *differential of area*. 2. Symbol for *differential of amplification*. 3. Seldom-used abbreviation of *deciampere*.

da Abbreviation of *deka*.

DAC Abbreviation of *digital-to-analog converter*.

DACI Abbreviation of *direct adjacent-channel interference*.

dag Abbreviated form of *Aquadag*.

dagc Abbreviation of *delayed automatic gain control*.

daisy chain A method of transferring a signal in a computer from one stage to the next.

daisy wheel A form of printing device consisting of a disk having several dozen radial spokes, each of which has a character molded on its face. The disk rotates to the proper position in the printing process, and a hammer strikes the spoke to press the molding against the ribbon and paper.

DAM Abbreviation of *data-addressed memory*.

Damon effect The change that the susceptibility of a ferrite undergoes under the influence of high rf power.

damped galvanometer A galvanometer with a provision for overswing limiting or oscillation prevention.

damped loudspeaker A loudspeaker in which undesirable excursions are prevented by damping in the associated amplifier or speaker circuit.

damped meter 1. A meter with a provision for overswing limiting or oscillation prevention. 2. A meter which is protected during transport by a shorting bus between the two meter terminals.

damped natural frequency 1. The frequency at which a damped system having one degree of freedom will oscillate after momentary application of a transient force. 2. In the presence of damping, the rate at which a sensing element oscillates freely.

damped oscillations Oscillations in which the amplitude of each peak is lower than that of the preceding one; the oscillation eventually dies out (the frequency becomes zero). Compare *continuous wave*.

damped speaker See *damped loudspeaker*.

damped wave A wave whose successive peaks decrease in amplitude (i.e., it decays), eventually reaching a frequency of zero. Compare *continuous wave* and *undamped wave*.

damped-wave decay See *decrement, 1*.

dampen To cause the amplitude of a signal to decay.

damper See *damping diode*.

damper diode See *damping diode*.

damper tube See *damping diode*.

damper winding A special short-circuited motor winding that opposes pulsation or rotation of the magnetic field.

damping See *damping action*.

damping action 1. Quenching action. 2. The prevention of overswing, dither, or flutter in a meter or loudspeaker (see *damped galvanometer, damped loudspeaker, damped meter*). 3. The prevention of oscillation or ringing in a circuit (see *damping tube*).

damping coefficient A figure expressing the ratio of the damping in a system to critical damping.

damping constant Logarithmic decrement. Also see *decrement*.

damping diode A diode employed to prevent oscillation in an electric circuit, e.g., the diode that prevents ringing in the power supply of a TV receiver. Also called *damper* and *damper tube.*

damping factor 1. Symbol, a. For a coil of inductance L and rf resistance r in a damped-wave circuit, the value $r/2L$, where L is in henrys and r in ohms. 2. Abbreviation, Fo. For a torque motor, the ratio of stall torque to no-load speed expressed in in.-oz./rad/s.

damping magnet A permanent magnet so situated with respect to a moving conductor, disk, or plate that the resulting field opposes the movement.

damping ratio See *damping coefficient.*

damping resistance 1. The value of shunt resistance required to prevent ringing in a coil. 2. The value of resistance required for critical damping of a galvanometer.

damping resistor 1. A shunt across a coil to prevent ringing. 2. A resistor used to provide critical damping of a galvanometer.

damping tube See *damping diode.*

Daniell cell A nonpolarizing primary wet cell with zinc (negative) and copper (positive) electrodes. The zinc plate is in a porous cup containing a weak zinc-sulfate solution; the cup is in a jar filled with a saturated copper-sulfate solution in which the copper electrode is immersed. Typical voltage for the cell is 1.1V.

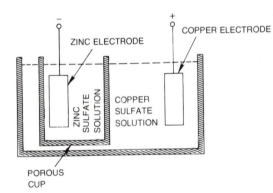

Daniell cell.

daraf The unit of elastance. It is the reciprocal of farad, its name being *farad* spelled backward.

dark conduction The flow of dark current in a photoconductive or glow-discharge device.

dark current The usually tiny current flowing through a darkened photoconductive cell, phototransistor, or glow-discharge device.

dark discharge The occurrence of a discharge in a gas, that does not produce any visible light.

dark heater An electron-tube heater which has been specially coated to provide the required heat at temperatures lower than those associated with incandescence.

dark resistance The electric resistance of a photoelectric tube when no light is present.

dark space See *Crookes' dark space.*

dark-spot signal A spurious signal generated by some TV camera tubes and arising from secondary-emission effects.

dark-trace tube An oscilloscope tube on whose white screen a long-

persistence magenta image is traced by the electron beam. Illuminating the screen with bright light intensifies the image.

Darlington amplifier A high-gain amplifier employing a Darlington pair (see *compound connection*) to provide the direct-coupled transistor stages.

Darlington pair See *compound connection.*

D'Arsonval current A large low-voltage, high-frequency current at one time thought to be therapeutic.

D'Arsonval meter An indicating meter in which a coil turns between the poles of a strong magnet against the force of spiral springs. A pointer attached to the coil, which turns on jeweled pivots, moves over a calibrated scale.

D'Arsonval movement The mechanism of a D'Arsonval meter.

DART Abbreviation of *data analysis recording tape.*

dash The longer of the two characters (*dot* and *dash*) of the telegraph code. The duration of the dash is three times longer than that of a dot.

dash—dot telegraphy See *dot-and-dash telegraphy.*

dashpot A delayed-action device in which the movement of a piston is slowed by air or a liquid in a closed cylinder.

dashpot relay A time-delay relay assembly in which the delay is obtained with a *dashpot.*

DAT Abbreviation of *diffused-alloy transistor.*

data Collectively, all factual material pertaining to an event or process. Specifically, in calculation and computation, numbers, symbols, and letters representing quantities, as opposed to symbols representing operations and memory addresses.

data acquisition The gathering of data (see *data collection* and *data system, 1*).

data-acquisition system A computer terminal intended for the purpose of gathering data from one or more external points.

data analysis display unit A video display peripheral for online data analysis.

data area A computer memory area that holds data only, i.e., one that does not contain program instructions.

data bank A data file stored in a direct access storage device, which can be drawn from by many system users through remote terminals.

data base A data file arranged so that its use does not change its content or configuration.

data block A set of data bits, comprising an identifiable item.

data bus A conductor or medium over which digital data is transmitted from one place to another.

data carrier storage A medium of data storage outside of a computer, e.g., a magnetic disk.

data code A set of abbreviations or codes for data characters or words.

data collection The pickup of signals representing test data and their transmission to a computer, data processor, or recorder. Also see *data system, 1.*

datacom Acronym for *data communications.*

data communication The transmission and reception of data signals between points in a system.

data communication terminal A computer peripheral providing an input and output link to a central computer system, and which can be used offline for other functions.

data compression 1. Minimizing the length of a data transmission by eliminating all redundancies. 2. The process of shortening the time required for a transmission. 3. The process of reducing the bandwidth of a data transmission. 4. The process of reducing the dynamic amplitude range of a data transmission.

data control Control of incoming and outgoing data in a data processing system.

data conversion Changing data from one form to another, e.g., from analog to digital.

data converter 1. A circuit or device for performing data conversion. 2. Analog-to-digital converter; digital-to-analog converter.

data description The description of a unit of data as given in a computer source program.

data display A display device, such as a special crt terminal, which presents data for visual examination. Compare *data printout*.

data element 1. A component of a data signal, e.g., a number, letter, symbol, or the equivalent electrical pulses. 2. A device or circuit for acquiring or processing data. 3. A unit of data, e.g., a field in a file.

data-flow diagram A flow diagram showing the movement of data through a data-processing system.

data format The form of data in a record or file, e.g., character format; numerical format.

data gathering See *data collection*.

data-handling capacity 1. The amount of information that can be stored in a memory circuit. 2. The amount of data that can be transmitted over a certain medium. 3. The rate at which data can be transferred under certain conditions.

data-handling system A system that gathers, routes, transmits, or receives data, but does not necessarily process it.

data item A logical element (character, byte, bit) describing a characteristic of a record used by a system for which there is a specific application.

data level Descriptive, through a programming language, of the relative weight of logical elements (data items) in a computer record. Also called *data hierarchy*.

data link The portion of a computer system which gathers data and, if necessary, converts it to a form acceptable by a computer.

data logger A recorder for analog or digital data or both.

data matrix Variables and their possible values stored as a series of columns and rows of values in a computer memory.

data name An operand specified in a computer source program.

data pickup 1. A transducer which collects data signals from a source; it converts nonelectrical data into corresponding electrical signals and delivers its output to a data processing system. 2. Data acquisition.

data playback The reproduction of data signals stored by some method of data recording.

data plotter See *X−Y plotter*.

data printout 1. A device which prints a record of data or the results of a computation. 2. A permanent printed record, usually of a calculation or computation, especially the printed output of a computer peripheral device.

data processing Work done on acquired data—as in solving problems, making comparisons, classifying material, organizing files—by (usually) automatic equipment.

data-processing equipment A digital computer and the peripheral equipment needed to collate, store, analyze, and reduce data.

data-processing machine A computer or other device for processing and storing data, as opposed to one used primarily to solve problems or perform routine tasks. Also called *data processor*.

data-processing system An electronic system for automatic data processing. It may be based on analog or digital techniques or both.

data processor See *data-processing machine*.

data receiver At a particular point in a data-processing system, a circuit or device for receiving data from a data transmitter. Also see *data reception*.

data reception Receiving data signals from some point within or outside a data-processing system.

data-reception system A data receiver and its associated equipment.

data record A computer-program processed record containing a data unit.

data recorder A machine for storing data acquired in the form of electrical signals (see *data recording*).

data recording 1. For future use, the preservation of data signals by some process, such as magnetic-tape recording, paper-tape punching, etc. 2. A record of data signals, as on magnetic or paper tape.

data reduction The summarization of a mass of electronically gathered data.

data-reduction system A system employed for the purpose of minimizing the amount of data necessary to convey given information.

data representation Values and data as described by numerals, symbols, and letters, e.g., computer program instructions.

data segment As related to a particular computer process, a subunit of allocated storage containing data only.

data set A device connecting a data processor to a telegraph or telephone line.

data signal 1. A signal (such as one of binary bit combinations) which can represent data as numbers, letters, or symbols. 2. A signal current or voltage proportional to some sampled quantity and which may be used to actuate indicating instruments during tests or measurements.

data statement A computer source program statement identifying a data item and specifying its format.

data synchronizer A device employed to synchronize data transmission within a computing or processing system.

data system 1. An arrangement for collecting, recording, and routing data in the form of electrical signals. 2. An arrangement for processing data, i.e., for correlating, computing, routing, storing, etc.

data terminal A remote input/output device connected to a central computer.

data transducer In tests and measurements, a transducer that converts a monitored phenomenon into electrical quantities which may be used for computer analysis or calculations. Also see *data signal*.

data transmission Sending data signals from a pickup point or processing stage to another point within a data-processing system; also, sending such signals to points outside the system.

data-transmission system A data transmitter and its associated equipment.

data transmission utilization measure The ratio of the useful data output of a data-transmission system to the total data input.

data transmitter A circuit or device for sending data from point to point within or outside of a dataprocessing system. Also see *data transmission*.

data unit Characters in a group that are related in a way that makes them a meaningful whole.

data value A measure of the amount of information contained in a certain number of data bits. The greater the ratio of the actual information to the number of bits, the higher the data value.

data words In digital computer practice, words (bit groups) representing data rather than program instructions.

davc Abbreviation of *delayed automatic volume control*.

David Phonetic alphabet code word for letter *D*.

daylight effect The modification of transmission paths during the day because of ionization of the upper atmosphere by solar radiation.

daylight lamp An incandescent lamp whose filament is housed in a blue glass bulb, which absorbs some red radiation and transmits most of the violet.

daylight range The distance over which signals from a given transmitter are consistently received during the day.

daylight visible range The maximum distance over which—under specified conditions—a large black body can be seen against a white sky. Also called *daylight visual range*.

DB 1. Abbreviation of *diffused base* of a transistor. 2. Abbreviation of *double break* (relay).

dB 1. Abbreviation of *decibel*. 2. Symbol for *differential of susceptance*.

dBa Abbreviation of *adjusted decibels*.

dBc Abbreviation of *decibels referred to the carrier*.

DBD Abbreviation of *double-base diode*.

dBd The power gain of an antenna in the direction of maximum radiation, compared to the radiation in the favored direction of a half-wave dipole in free space receiving the same amount of power. Expressed in decibels.

dBi The power gain of an antenna in the direction of maximum radiation, compared to the radiation from a theoretical isotropic antenna in free space receiving the same amount of power. Expressed in decibels.

dBj The level of an rf signal, in decibels, relative to 1 millivolt.

dBk Abbreviation of *decibels referred to 1 kilowatt*.

DBM Abbreviation of *data base management*.

dBm Abbreviation of *decibels referred to 1 milliwatt*, where the resistive load impedance is 600 ohms. Thus, 0 dBm is 1 mW across 600 ohms.

dB meter A usually high-impedance ac voltmeter with a scale reading directly in decibels.

dBmV Abbreviation of *decibels referenced to 1 millivolt*.

dBrap Abbreviation of *decibels above reference acoustic power* (10^{-6}W).

dBrn Abbreviation of *decibels above reference noise*.

dBV Abbreviation of *decibels referred to 1 volt*.

dBW Abbreviation of *decibels referred to 1 watt*.

dBx Abbreviation of *decibels above reference coupling*.

dC Symbol for *differential of capacitance*.

dc 1. Abbreviation of *direct current*. Abbreviation of *direct-coupled*.

dc—ac converter A circuit which converts a dc input voltage into an ac output voltage, with or without stepup or stepdown. Also called *inverter*.

dc alpha The current amplification factor (*alpha*) of a common-base transistor stage for a dc input (emitter) signal. Compare *dc beta*.

dc amplifier 1. Direct-coupled amplifier. 2. An amplifier for boosting dc signals.

dc balance 1. Adjustment of a circuit or device for dc stability or dc null. 2. Adjustment of a circuit for dc stability during gain changes. 3. A potentiometer or other variable component used to obtain dc balance.

dc bar See *dc bus*.

dc base current Symbol, $Ib(dc)$. The static direct current in the base element of a bipolar transistor.

dc base resistance Symbol, $Rb(dc)$. The static dc resistance of a bipolar transistor's base element; $Rb(dc) = Vb/Ib$.

dc base voltage Symbol, $Vb(dc)$. The static dc voltage at the base element of a bipolar transistor.

dc beta The current amplification factor (*beta*) of a common-emitter-connected transistor for a dc input (base) signal. Compare *dc alpha*.

dc block A coaxial section having a capacitance in series with the inner or outer conductor, or both, to block dc while passing rf. Compare *dc short*.

dc bus A supply conductor carrying direct current only.

dcc Abbreviation of *double cotton covered* (wire).

dc cathode current Symbol, $Ik(dc)$. The static direct current in the cathode element of an electron tube.

dc cathode resistance Symbol, $Rk(dc)$ The static dc resistance of the cathode path of an electron tube.

dc cathode voltage Symbol, $Ek(dc)$. The static dc voltage at the cathode of an electron tube.

dc circuit breaker A circuit breaker operated by direct-current overload or underload, depending on its design and application.

dc collector current Symbol, $Ic(dc)$. The static direct current in the collector element of a bipolar transistor.

dc collector resistance Symbol, $Rc(dc)$. The static dc resistance of a bipolar transistor's collector element; $Rk(dc) = Vc/Ic$.

dc collector voltage Symbol, $Vc(dc)$. The static dc voltage at the collector element of a bipolar transistor.

dc component In a complex wave (i.e., one containing both ac and dc), the current component having an unchanging polarity. The dc component constitutes the mean (average) value around which the ac component alternates, pulsates, or fluctuates.

dc converter A dynamoelectric machine for converting low-voltage dc into higher-voltage dc. It is essentially a low-voltage dc motor coupled mechanically to a higher-voltage dc generator. Compare *dc inverter*.

dc coupling See *direct coupling*.

dc drain current 1. Symbol, $ID(dc)$. The static direct current in the drain element of a field-effect transistor. 2. Symbol, IL. The current drawn by a dc load.

dc drain resistance Symbol, $RD(dc)$. The static dc resistance of an FET's drain element; $RD(dc) = VD/ID$.

dc drain voltage Symbol, $VD(dc)$. The static dc voltage at the drain element of a field-effect transistor.

dc dump In digital computer practice, removing dc power from a computer, which would eradicate material stored in a volatile memory.

dc emitter current Symbol, $Ie(dc)$. The static direct current in the emitter element of a bipolar transistor.

dc emitter resistance Symbol, $Re(dc)$. The static dc resistance of a bipolar transistor's emitter element; $Re(dc) = Ve/Ie$.

dc emitter voltage Symbol, $Ve(dc)$. The static dc voltage at the emitter element of a bipolar transistor.

dc equipment Apparatus designed expressly for operation from a dc power supply. Compare *ac equipment* and *ac – dc*.

dc erase head In a magnetic recorder, a head supplied with a dc erase current.

dc error voltage In a TV receiver, the dc output of the phase detector, which is used to control the frequency of the horizontal oscillator.

dc gate current Symbol, $IG(dc)$. The very small static direct current in the gate element of a field-effect transistor.

dc gate resistance Symbol, $RG(dc)$. The very high, static dc resistance of an FET's gate element; $RG(dc) = VG/IG$.

dc gate voltage Symbol, $VG(dc)$. The static dc voltage at the gate element of a field-effect transistor.

dc generator 1. A rotating machine (dynamo) for producing direct current. Also see *dynamoelectric machinery*. 2. Generically, a device which produces direct current: batteries, photocells, thermocouples, etc.

dc generator amplifier A special type of generator which provides power amplification. The input signal essentially energizes the field winding of a constant-speed machine; because the output voltage is proportional to field flux and armature speed, a high output voltage is obtained. Also see *amplidyne*.

dc grid bias Steady dc control-grid voltage used to set the operating point of an electron tube.

dc grid current Symbol, $I_{G(dc)}$. The static direct current in the control-grid element of an electron tube.

dc grid voltage Symbol, $E_{g(dc)}$. The static dc voltage at the control grid of an electron tube.

dc inserter In a television transmitter, a stage which adds the dc pedestal (blanking) level to the video signal.

dc inverter An electrical, electronic, or mechanical device which converts dc to ac. Also called *inverter*.

dcl Abbreviation of *dynamic load characteristic*.

dc leakage The unintended flow of direct current.

dc leakage current 1. The direct current that normally passes through a correctly polarized electrolytic capacitor operated at its rated dc working voltage. 2. The zero-signal reverse current in a *pn* junction.

DCM Abbreviation of *digital capacitance meter*.

dc motor A motor that operates from direct current only.

dc noise Noise heard during the playback of magnetic tape which was recorded while direct current was in the record head.

dc noise margin In a digital or switching circuit, the difference $E_o - E_i$, where E_o is the output-voltage level of a driver gate and E_i is the input threshold voltage of a driven gate.

dc operating point For a tube or semiconductor device, the static, zero-signal dc voltage and current levels.

dc overcurrent relay A relay or relay circuit actuated by dc coil current rising above a specified level. Compare *dc undercurrent relay*.

dc overvoltage relay A relay or relay circuit actuated by dc coil voltage rising above a specified level. Compare *dc undervoltage relay*.

dc patch bay A patch bay in which the dc circuits of a system are terminated.

dc picture transmission In television, transmission of the dc component of the video signal; this component corresponds to the average illumination of the scene.

dc plate current Symbol, $I_{p(dc)}$. The static direct current in the plate element of an electron tube.

dc plate resistance Symbol, $R_{p(dc)}$. The static dc resistance of the internal plate—cathode path of an electron tube; $R_{p(dc)} = E_p/I_p$.

dc plate voltage Symbol, $E_{p(dc)}$. The static dc voltage at the plate electrode of an electron tube.

dc positioning Positioning of the spot on the screen of an oscilloscope tube by means of dc voltages adjusted with position-control potentiometers and applied to the horizontal and vertical deflecting plates.

dc power Symbol, P_{dc}. Unit, watt. The power in a dc circuit; $P_{dc} = EI$, where E is in volts and I is in amperes. Compare *ac power*. Also see *power*.

dc power supply A power unit which supplies dc only. Examples: battery, transformer—rectifier—filter circuit, dc generator, photovoltaic cell. Compare *ac power supply*.

dc relay A relay having a simple coil and core system for closure by direct current, which may be rectified ac.

dc resistance Resistance offered to direct current, as opposed to in-phase ac resistance.

dc resistivity The resistivity of a sample of material measured using a pure dc voltage under specified conditions (physical dimensions, temperature, etc.).

dc restoration The reinsertion of the dc component into a signal from which the component has been extracted through a capacitor or transformer. Also see *dc restorer*.

dc restorer A circuit that reinserts the average dc component of a signal after the component has been lost because the signal passed through a capacitor or transformer. Also called *clamper*.

DCS Abbreviation of *dorsal column stimulator*.

dc screen current Symbol, $I_{s(dc)}$. The static direct current in the screen element of an electron tube.

dc screen resistance Symbol, $R_{s(dc)}$. The static dc resistance of the screen element of a tube; $R_{s(dc)} = E_s/I_s$.

dc screen voltage Symbol, $E_{s(dc)}$. The static dc voltage at the screen element of an electron tube.

dc shift A shift in the *dc operating point*.

dc short A coaxial fitting providing a dc path between the center and outer conductors while permitting rf current to flow through the coaxial section with ease. Compare *dc block*.

dc signaling A signaling procedure employing direct current as the medium; e.g., simple wire telegraphy or telephony.

dc source 1. dc generator. 2. A live circuit point from which one or more direct currents can be taken.

dc source current Symbol, $I_{S(dc)}$. The static direct current in the source element of a field-effect transistor.

dc source resistance Symbol, $R_{S(dc)}$. The static dc resistance of an FET's source element.

dc source voltage Symbol, $V_{S(dc)}$. The static dc voltage at the source element of a field-effect transistor.

dc suppressor current Symbol, $I_{sup(dc)}$. The static direct current in the suppressor element of an electron tube.

dc suppressor voltage Symbol, $E_{sup(dc)}$. The static dc voltage at the suppressor element of an electron tube.

dc thyratron A thyratron operated to provide large amounts of controllable ac power from a dc power source (inverter function) or to control large amounts of direct current.

DCTL Abbreviation of *direct-coupled transistor logic*.

dc-to-dc inverter See *dc inverter*.

dc transducer 1. A transducer that depends upon direct current for its operation; i.e., it has a dc power supply whose output is modulated by the sensed phenomenon. 2. A transducer that converts a direct current into some other form of energy, such as, heat, pressure, or sound.

dc transformer A dc-to-dc converter providing voltage stepup. The applied dc is usually first converted to ac, which is then stepped up by a transformer. The higher-voltage ac is then rectified to give a high dc output voltage. The name comes from the apparent stepup of dc by the circuit.

dc transmission 1. Sending dc power from a generating point to a point of use. 2. In television transmission, the retention of the dc component in the video signal (see *dc picture transmission*).

dc tuning voltage The capacitance-varying dc voltage applied to a varactor acting as the capacitor in an *LC* tuned circuit, which is smoothly tuned by the adjustable dc voltage.

dcu Abbreviation of *decimal counting unit*.

dc undercurrent relay A relay or relay circuit which is actuated by dc coil current dropping below a specified level. Compare *dc overcurrent relay*.

dc undervoltage relay A relay or relay circuit which is actuated by dc voltage dropping below a specified level. Compare *dc overvoltage relay*.

dcv Abbreviation of *dc volts* or *dc voltage*.

dc vacuum-tube voltmeter Abbreviation, *dc vtvm*. A vacuum-tube voltmeter for measuring dc voltage. It may be adapted for ac voltage measurements by connecting a diode rectifier circuit ahead of its input terminals.

dc voltage Abbreviation, dcv. A voltage which does not change in polarity, an example being the voltage delivered by a battery or dc generator. Also see *voltage*.

dc working voltage Abbreviation, dcwv. The rated dc voltage at which a component may be operated continuously with safety and reliability.

dc working volts Abbreviation, dcwv. The actual value, expressed in volts, of a *dc working voltage*.

dcwv Abbreviation of *dc working voltage*.

dD Symbol for *differential of electric displacement*.

DDA Abbreviation of *digital differential analyzer*.

DDD Abbreviation of *direct distance dialing* (telephone).

D-display See *D-scope*.

DE Abbreviation of *decision element*.

dE Symbol for *differential of voltage*. (Also de and *dV*.)

deac In FM receivers, a device employed for deemphasis. The name is the abbreviation of *deaccentuator*.

deaccentuator See *deac*.

deactuating pressure For an electrical contact, the pressure at which contact is made or broken as the pressure reaches the level of activation. The contact is made or broken in opposition to the desired function.

dead 1. Unelectrified or free of signals or fields. 2. Electrically or mechanically inoperative.

dead band 1. A radio frequency band which is not being used by stations; i.e., it is silent. 2. A range of values for which an applied control quantity (e.g., current or voltage) has no effect on the response of a circuit.

deadbeat The state of a moving body's (such as the pointer of a meter or the voice coil of a loudspeaker) coming to rest without overswing or oscillation.

deadbeat galvanometer See *deadbeat instrument*.

deadbeat instrument A meter or recorder which is highly damped to insure deadbeat.

deadbeat meter See *deadbeat instrument*.

dead break An unreliable contact of a relay, caused by insufficient pressure.

dead center 1. Nonrevolving center. 2. The position of a crank or cam when the turning moment on it is zero. 3. The limit of travel of a reciprocating device, as of a piston.

dead circuit A circuit which is electrically disabled, even temporarily.

dead end The unused end of a tapped coil; i.e., the turns between the end of the coil and the last turn used.

dead-end tower A supporting tower for an antenna or transmission line which can withstand stresses due to loading or pulling.

dead file A computer file that is not in use but is being kept in a record.

dead-front panel A metal panel that, for safety and desensitization, is completely insulated from voltagebearing components mounted on it; it is often grounded.

dead interval See *dead time*.

dead line A deenergized line or conductor.

dead period See *dead time, 1, 2*.

dead room An anechoic room in which acoustic tests and studies are made.

dead short A short circuit having extremely low (virtually no) resistance.

dead space See *dead band*.

dead spot 1. On a tube cathode (directly or indirectly heated), a spot from which no electrons are emitted. 2. An area in which radio waves from a particular station are not received.

dead stretch The tendency of insulating materials to permanently retain their approximate dimensions after having been stretched.

dead time 1. Down time. 2. An interval during which there is no response to an actuating signal. 3. In a computer system, an interval between related events that is allocated to prevent interference between the events.

dead volume In a pressure transducer, the zero-stimulus volume of the pressure port cavity.

dead weight The weight of an inert body.

dead zone See *zone of silence*.

debatable time Computer time that cannot be placed in any other category.

debounced switch A switch in sensitive computer or control systems which has circuitry for eliminating the electrical effects of bounce (see *bounce, 1*).

de Broglie's equation An equation for the wavelength of an electron. The nonrelativistic form of this equation is $\lambda = h/(m_0 v)$, where λ is the wavelength in centimeters, *h* is equal to 6.54×10^{-27} erg-seconds, *m₀ (electron rest mass)* is 9.03×10^{-28} grams, and *v* (electron velocity) is in centimeters per second. In 1924, de Broglie predicted that an electron should have a wave characteristic as well as mass. Also see *de Broglie waves*.

de Broglie waves The waves which are thought to surround a moving particle, such as an electron or atom. Also see *de Broglie's equation*.

debug Generally, to remove errors or inperfections from an equipment, design, or procedure; especially, to uncover, evaluate, and remove errors from a computer program.

debugging aid routine A computer program used to test other programs.

debugging period The interval following completion of a software design, a hardware interconnection, or the manufacture of an equipment, during which errors and imperfections are sought and corrected.

debunching In a velocity-modulated tube, such as a klystron, a beamspreading space-charge effect which destroys electron bunching.

Debye length The maximum distance between an electron and a positive ion over which the electron is influenced by the field of the ion.

Debye shielding distance See *Debye length*.

deca A prefix (combining form) meaning *ten*; e.g., *decagram*.

decade 1. A frequency band whose upper limit is 10 times the lower limit. Example: 20—200 Hz decade. 2. A set of 10 switched or selectable components in which the total value is 10 times that of individual values. Examples: *decade capacitor, decade inductor, decade resistor*. Also called *decade box*. 3. A group, sometimes a unit of access, of 10 computer storage locations.

decade amplifier An amplifier or preamplifier whose gain can be adjusted in increments of 10 (thus, × 1, × 10, × 100, etc.).

decade band 1. A frequency band in which the difference between upper and lower frequency limits is 10 (i.e., $f2 - f1 = 10$). 2. A band in which all frequencies are 10 times those in a lower-frequency reference band; thus, the 20—30 MHz band referred to the 2—3 MHz band.

decade box A group of components which provide values in 10 equal steps selected by a switch or jacks. For compactness, the components and the associated hardware are enclosed in a box or can. See, for example, *decade capacitor, decade inductor,* and *decade resistor*.

decade capacitance box See *decade capacitor*.

decade capacitor A capacitor whose value is variable in 10 equal steps. Thus, a *units decade* might have ten 1 μF steps; a *tens decade*, ten 10 μF steps; and so on. Compare *decade inductor* and *decade resistor*.

decade counter A counter (see *counter, 1, 2*) in which the numeric display is divided into sections, each having a value ten times that of the next and displaying a digit from zero to nine. Thus, the display of the number 15790 is a combination of 1 in the ten-thousands section, 5 in the thousands section, 7 in the hundreds section, 9 in the tens section, and 0 in the units section.

decade inductance box See *decade inductor*.

decade inductor An inductor whose value is variable in 10 equal steps. Compare *decade capacitor* and *decade resistor*.

decade resistance box See *decade resistor*.

decade resistor A resistor whose value is variable in 10 equal increments. Compare *decade capacitor* and *decade inductor*.

decade scaler A scale-of-10 electronic counter, i.e., a circuit delivering one output pulse for each group of 10 input pulses.

decalescent point In a metal whose temperature is being increased, the temperature at which heat is suddenly absorbed. Compare *recalescent point*.

decametric waves Waves in the 10 to 100 meter band (30 to 3 MHz).

decay 1. The decrease in the value of a quantity, e.g., current decay in an *RC* circuit. 2. The gradual, natural loss of radioactivity by a substance.

decay characteristic 1. The decay of a parameter; usually an exponential function. 2. The persistence time in a storage oscilloscope.

decay constant The probability that a radioactive atom will decay in a given amount of time.

decay rate For a given radioactive substance, the rate at which spontaneous disintegration takes place.

decay time The time required for pulse amplitude to fall from 90% to 10% of the peak value. Also called *fall time*.

Decca A 70–130 kHz cw radionavigation system (British).

decelerated electron A high-speed electron which is abruptly decelerated upon striking a target, causing X rays to be emitted.

decelerating electrode A charged electrode which slows the electrons in an electron beam.

decelerating element See *decelerating electrode*.

deceleration Acceleration that results in a decrease in speed.

deceleration time The time taken by magnetic tape in a recorder to stop moving after the last recording or playback has occurred. Also applicable to a computer storage medium that moves while it is being read from or written into.

decentralized data processing Data processing in which the computing equipment is distributed among and serves various managerial subgroups.

deception A method of producing misleading echoes in enemy radar.

deception device A radar device, or radar-associated device, for deception.

deci Abbreviation, d. A prefix meaning one tenth, i.e., 10^{-1}. Examples: *decibel, decimeter*.

decibel Abbreviation, dB. A practical unit of gain derived from the bel; 1 dB = 0.1 B. The dB value (*N* dB) is equal to 10 log10 *Po*/*Pi* or 20 log10 *Eo*/*Ei*, where *Ei* is input voltage, *Eo* is output voltage, *Pi* is input power, and *Po* is output power.

decibel meter See *dB meter*.

decibels above reference acoustic power Abbreviation, *dBrap*. The ratio of a given acoustic power level to a lower reference acoustic power level, specified in decibels.

decibels above reference noise Abbreviation, *dBrn*. The ratio of the noise level at a selected point in a circuit to a lower reference noise level, in decibels.

decibels referenced to 1 millivolt Abbreviation, *dBmV*. The relative voltage level of a usually rf signal when compared with a 1 mV signal measured at the same terminals.

decibels referred to 1 milliwatt Abbreviation, *dBm*. The ratio of an applied power level to the power level of 1 mW, when measured across a 600-ohm resistance (specified in decibels).

decibels referred to 1 volt Abbreviation, *dBV*. The ratio of a given voltage to 1V, expressed in decibels.

decibels referred to 1 watt Abbreviation, *dBW*. The ratio of a given power level to the power level of 1W, expressed in decibels.

decider See *decision element*.

decigram A unit of weight equal to 0.1 gram.

deciliter A unit of volume equal to 0.1 liter (10^{-4}m^3).

decilog A unit equal to 1/10 the common logarithm of a ratio.

decimal 1. Pertaining to the base-10 number system (see *decimal number system*). 2. A decimal fraction, i.e., a fraction represented by figures to the right of the decimal point and arranged serially according to negative powers of 10. Examples: $0.1 = 10^{-1}$, $0.001 = 10^{-3}$ etc.

decimal attenuator An attenuator circuit whose resistances are chosen for attenuation in decimal steps. Thus, one section provides attenuation in steps of tenths of applied voltage, another in hundredths, and so forth.

decimal code A method of defining numbers, in which each place has a value of ten times that immediately to the right.

decimal-coded digit 1. A numeral from 0 to 9. 2. A numeral in the decimal system. 3. A binary representation of a decimal value from 0 to 9.

decimal digit A numeral from 0 to 9.

decimal equivalent The decimal number equal to a given fraction; e.g., the decimal equivalent of 21/64 is 0.3281.

decimal fraction See *decimal, 2*.

decimal notation See *decimal number system*.

decimal number system The familiar base-10 number system in which the digits 0 through 9 can be used to represent values according to their position relative to the radix (here, decimal) point. Positions left of the point represent successive positive powers of ten beginning with 10^0, while those to the right represent successive negative powers of 10 beginning with 10^{-1}. Thus, the decimal number 365.87 equals $3 \times 10^2 + 6 \times 10^1 + 5 \times 10^0 + 8 \times 10^{-1} + 7 \times 10^{-2}$.

decimal point The radix point in a decimal number. It serves to separate the integral part from the fractional part of the number. Thus, 1 is written 1.75.

decimal-to-binary conversion 1. The automatic conversion of a decimal number to a series of pulses representing the corresponding binary number. This conversion is accomplished with electronic circuitry. 2. The arithmetic operation of converting a decimal number into the corresponding binary number. This can be done as follows: if the decimal number (*n*) is even, write it as *n* + 0; if it is odd, write it as (*n* – 1) + 1. Divide the number written by two and write the result (*P*) as *P* + 0 if even, or (*P* – 1) + 1 if odd. Continue the process until *P* or *P* – 1 reaches zero. The resulting column or row of ones and zeros (read from bottom to top or right to left) gives the binary equivalent of the original decimal number *n*. For example, the decimal number 961 is converted as follows:

$$960 + 1$$
$$480 + 0$$
$$240 + 0$$
$$120 + 0$$
$$60 + 0$$
$$30 + 0$$
$$14 + 1$$
$$6 + 1$$
$$2 + 1$$
$$0 + 1$$

Thus, decimal 961 is binary 1 111 000 001.

decimeter waves See *microwaves*.

decimetric waves Waves in the 0.1 to 1 meter frequency band (3000 to 300 MHz).

decineper A natural-logarithmic unit equal to 0.1 neper.

decinormal-calomel electrode For use with a pH meter, a calomel electrode containing a 0.1 normal solution of potassium chloride.

decipher See *decoding, 3*.

decision 1. A choice based upon the evaluation and comparison of data and the identification of a specified objective. 2. In digital computer practice, the automatic selection of the next step in a sequence, on the basis of data being compared by a relational test.

decision box A block on a computer flowchart indicating the point at which a decision (see *decision, 2*) must be made as to which of several branches the program will take.

decision elements See *logic circuits*.

decision instruction A computer program instruction to compare the values of operands and take an appropriate action, as per the BASIC instruction IF A = B THEN GO TO (line number).

decision procedure In decision theory, a series of calculations made to minimize risk, failure, cost, or the like.

decision table A table showing what action should be taken according to the relationship between variables under certain conditions. For example, if a number of people were asked to report to an office on a day corresponding to the initial of their last names, the decision table representing the variables, conditions, and actions might be:

A to J	yes	no	no	no
K to N	no	yes	no	no
P to S	no	no	yes	no
T to Z	no	no	no	no
Go on Mon	X	-	-	-
Go on Tue	-	X	-	-
Go on Wed	-	-	X	-
Go on Fri	-	-	-	X

decision theory A statistical discipline concerned with identifying and evaluating choices and alternatives, and determining the best sequence of steps to take in reaching an objective.

decision tree In decision theory, a diagram showing alternative choices, so called from its resemblance to a tree with branches.

deck 1. Tape deck. 2. A pack of punched cards in a computer file.

declarative macroinstruction As part of an assembly language, in-

structions to the compiler to do something or record a condition without affecting the object program.

declarative statement A computer source program instruction specifying the size, format, and kind of data elements and variables in a program for a compiler.

declination The angle subtended by the deviation of magnetic north from true north; it is therefore the angle subtended by a freely turning magnetic needle and the imaginary line pointing to true north. Compare *inclination*.

declinometer An instrument for measuring declination.

decode 1. To unscramble a coded message. 2. In digital computer practice, to deliver a specific output from character-coded inputs. 3. In a multiplex system, the separation of the subcarrier from the main carrier.

decoder A circuit or device for performing decoding (see *decoding, 1, 2, 3, 4*).

decoder—driver An IC containing a decoder and driver (see, for example, *driver, 3*).

decoding 1. In computer and data processing practice, digital-to-analog conversion. 2. The conversion to English of a message received in a code. 3. Translating a message from a secret code, i.e., deciphering a message. 4. The automatic conversion of a signal into the appropriate switching action (as the enabling of a transmitter or receiver by a tone in a selective calling system.)

decoding circuit A circuit intended for the purpose of translating a code into ordinary language.

decoherer A device used to break up the cohered filings in a coherer, thus preparing it for the next signal. In early coherer-type wireless receivers, the decoherer was the clapper of the electric bell driven by the coherer; it tapped the glass tube of the coherer to loosen the filings.

decollator An offline computer device for separating the parts of output continuous stationery sets. Also see *continuous stationery*.

decommutation The extraction of a signal component from the composite signal resulting from commutation.

decommutator A circuit or device for performing decommutation, including demodulators, demultiplexers, and signal separators.

decoupler A device that isolates two circuits, so that a minimal amount of coupling exists between them.

decoupling The elimination or effective minimization of coupling effects, as in decoupling amplifier stages to prevent interaction through a common power-supply lead.

decoupling capacitor 1. A capacitor which provides a low-impedance path to ground to prevent common coupling between the stages of a circuit. 2. The capacitive member of an *RC* decoupling filter.

decoupling filter A resistance—capacitance filter usually inserted into a common dc line in a multistage amplifier to prevent interstage feedback coupling through the common impedance of the line.

decoupling network One or more decoupling filters.

decoupling resistor The resistive member of an *RC* decoupling filter.

decoy In radar deception, an object which provides misleading reflections. Also see *chaff*.

decreasing function A function whose curve has a negative slope, i.e., *y* varies inversely with x.

decrement 1. The rate at which the successive cycles of a damped wave die away. The decrement value is the natural logarithm of the ratio of two successive peaks of the same polarity. Except for large values, decrement is approximately the fractional difference between successive peaks. Also called *logarithmic decrement*. 2. A quantity used to lessen the value of a variable. 3. To lower the value (of a register, for example) by a single increment.

decremeter An instrument for measuring the decrement of a radio wave. Also see *Kolster decremeter*.

decremeter capacitor A variable capacitor for use in a decremeter. The rotor plates are shaped so that equal angular rotations correspond to the same decrement at all settings; i.e., as a percent, the capacitance change for a given rotation is constant throughout the capacitor range.

dedicate To assign to a certain purpose.

dee One of the D-shaped chambers in and between which particles accelerate in a spiral path to high velocity in a cyclotron.

dee line In a cyclotron, a support for the dee, with which it forms a resonant circuit.

deemphasis In frequency modulation, introducing a falling-response characteristic (response falls as modulating frequency increases) to complement the rising response of preemphasis. Also called *postemphasis* or *postequalization*. Compare *preemphasis*.

deemphasis amplifier An amplifier employed to remove the highfrequency preemphasis applied to signals prior to broadcasting, multiplexing, tape recording, or telemetering. Also see *deemphasis* and *preemphasis*.

deemphasis circuit A low-pass *RC* filter which provides deemphasis in an FM receiver.

deemphasis network See *deemphasis circuit*.

deenergize To take a circuit or device out of operation; i.e., to remove its power or signal excitation.

deep-diffused junction A pn junction made by diffusing the impurity material deep in the semiconductor wafer. Compare *shallow-diffused junction*.

deep discharge Complete discharge of a cell or battery; usually done prior to recharging.

deep-space net A radar system intended for constant monitoring of spacecraft.

defect 1. Absence of an electron (hence, presence of a hole) in the lattice of a semiconductor crystal. 2. An abnormality of design, construction, or performance of an electronic circuit or device. 3. In a computer system, a hardware or software fault that could be the eventual cause of a failure.

defect conduction In a semiconductor material, conduction via holes.

deferred addressing Indirect addressing in which a preset counter makes several references to find a desired address.

deferred entry An entry into a computer subroutine, delayed because of a delay in the exit from a control program.

deferred exit An exit from a computer subroutine, delayed because of a particular command.

defibrillation Use of a cardiac stimulator to halt fibrillation of the heart, as caused by electric shock.

defibrillator See *cardiac stimulator*.

definite integral An integral for which the lower and upper limits of the quantity to be integrated are stated. For example, in the area integral $A = \int_a^b y\, dx$, the limits between which the area is calculated are a and b. Compare *indefinite integral*. Also see *integral, integral calculus,* and *integration*.

definite-purpose component A component designed for a specific use rather than for a wide range of possible applications, e.g., a video detector diode (as opposed to a generalpurpose diode). Compare *generalpurpose component*.

definition 1. Clarity of a TV image, i.e., one having good contrast and faithful tones. 2. Good intelligibility of reproduced sounds.

deflecting coils See *deflection coils*.

deflecting electrode See *deflection electrode*.

deflecting plate In a cathode-ray tube, a plate that attracts or repels the electron beam, causing the spot to move horizontally or vertically on the screen. Also called *deflection plate*.

deflecting torque The torque required to move the pointer of a meter or the pen or mirror of a recorder.

deflection 1. In a cathode-ray tube, movement of the electron beam by electric or magnetic fields. 2. Movement of the pointer of a meter or the pen or mirror of a recorder by an applied current or voltage.

deflection coils External coils carrying sawtooth currents, which provide electromagnetic deflection of the cathode-ray beam in TV picture tubes, camera tubes, radar display tubes, sonar display tubes, and some oscilloscopes.

deflection electrode An electrode, such as a deflection plate, employed to deflect an electron beam.

deflection factor Symbol, *G*. The reciprocal of deflection sensitivity, i.e., $G = 1/S$. Also see *deflection sensitivity*.

deflection plane For a cathode-ray tube, the plane that is perpendicular to the axis of the tube and contains the deflection center.

deflection plate See *deflecting plate*.

deflection polarity In a cathode-ray tube, the polarity of the voltage applied to a particular deflecting plate to move the electron beam in a particular direction.

deflection sensitivity Symbol, *S*. Deflection voltage per unit displacement of the electron beam at the screen of a cathode-ray tube, expressed in V/cm or V/in.

deflection voltage The potential difference between the deflection plates of a cathode-ray tube. Used to control the direction of the electron beam striking the phosphor screen.

deflection yoke An assembly of deflection coils in TV picture and camera tubes, and in some magnetically deflected oscilloscope tubes. The usual combination is two series-connected horizontal deflection coils and two series-connected vertical deflection coils.

deflector 1. A beam-forming plate in a beam-power tube. 2. A deflection plate in a cathode-ray tube. 3. A deflection coil or yoke in a TV picture-tube assembly (or similar assembly for a camera tube or magnetic-deflection oscilloscope tube). 4. A mechanical attachment for improving the angle of radiation of a loudspeaker by spreading the higher-frequency. waves.

defocusing Throwing the image on the screen of an oscilloscope tube or TV picture tube out of focus by any force that causes the electron beam to spread.

defruiting The elimination of non-synchronized echoes in a radar system.

deg Abbreviation of *degree*.

degassing During the evacuation of an electron tube or similar device, the removal of gas, including that which has bonded to the glass and metal parts.

degauss See *demagnetize*.

degausser 1. Demagnetizer. 2. A device for bulk erasing magnetic tape. 3. Degaussing circuit.

degaussing 1. Demagnetizing a body. 2. For protection against detonating marine mines, counteracting the earth's magnetic field with the opposing field of current flowing through a coil wound around a ship's hull.

degaussing circuit In a color TV receiver, a circuit including a thermistor, voltage-dependent resistor, and degaussing coil for automatically demagnetizing the picture tube when the receiver is switched on.

degaussing coil A coil carrying an alternating current; the resulting magnetic field demagnetizes objects which have become accidentally magnetized.

degeneracy In microwave practice, the appearance of a single resonant frequency for two or more modes in a resonator.

degenerate modes In microwave practice, a set of modes with the same resonant frequency or propagation constant.

degenerate parametric amplifier An inverting parametric amplifier in which the two signals are of the same frequency, which is half the pump frequency.

degeneration In an amplifier, the technique of feeding a portion of the output back to the input out of phase with the input signal to improve fidelity at the expense of gain; negative feedback. Also called *inverse feedback*. Compare *regeneration*.

degenerative feedback Negative feedback. Also called *inverse feedback*. Compare *regenerative feedback*.

degenerative resistor An unbypassed cathode resistor in a common-cathode tube circuit, an unbypassed emitter resistor in a common-emitter bipolar-transistor circuit, or an unbypassed source resistor in a common-source field-effect transistor circuit. Signal current flowing through the resistor produces negative feedback current (degeneration), which reduces the gain of the stage, but increases the linearity of the transfer characteristic.

degradation 1. Gradual deterioration in condition or performance of a circuit or device. 2. Limited computer system operation due to the failure of some equipment.

degradation failure Failure occurring at the terminal point of degradation.

degraded operation See *degradation*.

degreaser See *ultrasonic cleaning tank*.

degree 1. A unit of circular angular measurement equal to 1/360 the circumference of a circle. Also called *geometric degree*. 2. A unit of temperature measurement. See *degree absolute, degree Celsius, degree centigrade, degree Fahrenheit*, and *degree Reaumur*.

degree absolute Symbol, K. The unit of temperature on the *absolute scale*. Also see *absolute scale*.

degree Celsius Symbol, °C. The unit of temperature on the Celsius scale.

degree centigrade Symbol, °C. The unit of temperature of the centigrade scale (now called *Celsius scale*).

degree Fahrenheit Symbol, °F. The unit of temperature on the Fahrenheit scale.

degree of current rectification For a rectifier, the ratio of the average dc output current to the rms ac input current.

degree of voltage rectification For a rectifier, the ratio of the average dc output voltage to the rms ac input voltage.

degree Reaumur Symbol, °R. The unit of temperature on the Reaumur scale.

degrees of freedom The ways in which a point may move or a system may change. In three-dimensional space, a rigid body, for example, has six degrees of freedom: possible motion in three coordinate directions, and rotation around any of the three coordinate axes extending through its center of mass; a nonrigid body, on the other hand, has an infinite number of degrees of freedom. In a system of electrical quantities, there are generally as many degrees of freedom as there are variables.

degrees-to-radians conversion The conversion of angles in degrees to angles in radians. To change degrees to radians, multiply degrees by 0.017 453 3. Thus, 45° = 0.7854 radian. Compare *radians-to-degrees conversion*.

deion circuit breaker A circuit breaker in which the arc occurring when the contacts open is quickly extinguished by an external magnetic device.

deionization The conversion of an ionized substance, such as gas, to a neutral (un-ionized) state. The process changes the ions into unchanged atoms.

deionization potential The voltage at which an ionized substance becomes deionized; for example, the voltage at which a glow discharge is extinguished when the gas ions become neutral atoms at that voltage. Also called *extinction potential*.

deionization time The time (in μs or ms) required for an ionized gas to become neutral after removal of the ionizing voltage.

deionization voltage See *deionization potential*.

deionize To restore to a neutral condition (without charge); i.e., to convert ions to neutral atoms, as in the deionization of the gas when the discharge in a glow tube is extinguished.

deka A prefix meaning ten(s), i.e., 10^1, e.g., *dekameter*.

dekahexadecimal number system See *hexadecimal number system*.

Dekatron A cold-cathode counter tube.

del Symbol, ∇ (inverted delta). In differential calculus, a differential operator. Del denotes the symbol vector expression $i(\delta/\delta x) + j(\delta/\delta y) + k(\delta/\delta z)$.

delamination The splitting apart, in layers, of an insulating material, such as mica or bonded plastic film.

delay 1. The interval between the instant at which a signal or force is applied or removed and the instant at which a circuit or device subsequently responds in a specified manner. 2. The time required for a signal to traverse a given medium, such as air, mercury, or quartz.

delay action Response occurring some time after a stimulus has been applied or removed, e.g., the retarded opening of a delayed-dropout relay.

delay circuit 1. A circuit, such as a resistance—capacitance or resistance—inductance combination, which introduces a time delay. 2. Delay line circuit.

delay coincidence circuit A coincidence circuit (see *AND circuit*) triggered by two pulses, one of which lags behind the other.

delay counter In a digital computer, a device which halts a program run long enough for an operation to be completed.

delay distortion 1. Distortion resulting from variations in the phase delay of a circuit or device at different points in its frequency range. 2. In a facsimile signal, variations in the delay of different frequency components of the signal.

delayed agc See *delayed automatic gain control*.

delayed automatic gain control An automatic gain control circuit which operates only when the signal exceeds a predetermined threshold level, thus providing maximum amplification of weaker signals.

delayed automatic volume control See *delayed automatic gain control*.

delayed avc See *delayed automatic gain control*.

delayed break In relay or switch operation, contacts separating some time after the switch has been thrown or the relay deenergized. Compare *delayed make*.

delayed close See *delayed make*.

delayed closure See *delayed make*.

delayed contacts Contacts that open or close at a predetermined instant after their activating signal is applied or removed.

delayed drop-in See *delayed make.*

delayed dropout See *delayed break.*

delayed make In relay or switch operation, contacts closing some time after the switch has been thrown or the relay energized. Compare *delayed break.*

delayed open See *delayed break.*

delayed ppi Plan-position indicating radar having a delayed time base.

delayed pull-in See *delayed make.*

delayed repeater A repeater that receives and stores information which it retransmits later in response to a switching or interrogation signal.

delayed repeater satellite An active communications satellite that acts as a delayed repeater, i.e., it receives and records information at one location and retransmits it at another location.

delayed sweep 1. In an oscilloscope or radar, a sweep which starts at a selected instant after the signal under observation has started. 2. The (usually calibrated) circuit for producing such a sweep.

delayed updating Updating a computer record or record set so that the record fields are left unchanged until all other changes attendant to the pertinent event are processed.

delay equalizer A delay distortion correcting network (see *delay distortion, 1*).

delay—frequency distortion Distortion due to variation of envelope delay within a frequency band.

delay line A device (not always a line) which introduces a time lag in a signal; the lag is the time required for the signal to pass through the device. Thus, a pulse passing through the delay line will arrive at a given point later than a pulse transmitted outside the line.

delay-line memory In a digital computer, a memory employing a delay line, associated input and output coupling devices, and an external regenerative-feedback path. Information is kept stored by causing it to recirculate in the line by regeneration.

delay-line register In a digital computer, a register that operates in the manner of a *delay-line memory* and has a register length (capacity) of an integral number of words.

delay-line storage See *delay-line memory* and *delay-line register.*

delay multivibrator See *monostable multivibrator.*

delay-power product Unit, watt-second. The figure of merit for an IC gate. Increasing gate power reduces propagation delay. Also called *propagation delay-power product.*

delay relay A relay that opens or closes at the end of a predetermined time interval.

delay switch A switch having delayed make, delayed break, or both.

delay time 1. The interval between the instant a voltage or current is applied and the instant a circuit or device operates. 2. In an output pulse, the interval (td) between the instant (to) an ideal pulse is applied to the input of a system and the instant the output pulse reaches 10% of its maximum amplitude. 3. The time elapsed between the presentation of a pulse to the input of a delay line and the appearance of the pulse at the output.

delay timer A timer that starts or stops an operation after a prescribed length of time. In some senses, a delay relay or switch.

delay unit In a radar system, a circuit for delaying pulses.

delete 1. To erase or blank out a signal. 2. The elimination from a computer file of a record or record group. 3. To remove a computer program from memory.

deletion record In the master file of a digital computer, a new record that causes existing ones to be deleted.

delimiter In digital computer practice, a character limiting a sequence of characters of which it is not itself a member.

Dellinger effect The sudden disappearance of a radio signal as a result of an abrupt increase in atmospheric ionization caused by a solar eruption.

deliquescent material A material which absorbs enough moisture from the air to get wet. For example; calcium chloride, a deliquescent material, is often used to keep electronic equipment dry. Compare *hygroscopic material.*

delta 1. Symbol, Δ. An increment. Example: ΔIe signifies an increment of emitter current (i.e., $Ie2 - Ie1$). 2. Delta connection. 3. A triangular section for matching a feeder line to an antenna (see *wye-matched impedance antenna*).

delta circuit A three-phase electrical circuit with no common ground.

delta corciot See *delta connection.*

delta connection A triangular connection of coils or load devices in a three-phase system, so called from its resemblance to the Greek letter *delta.* Compare *wye-connection.*

delta-matched antenna See *wye-matched impedance antenna.*

delta-matched impedance antenna See *wye-matched impedance antenna.*

delta matching transformer In a delta-matched antenna (see *wye-matched impedance antenna*), the fanned-out (roughly delta-shaped) portion of the two-wire feeder at its point of connection to the radiator, it matches the impedance of the feeder to that of the radiator.

delta modulation The conversion of an analog signal into a digital pulse train which subsequently may be decoded to yield the original analog signal.

delta network See *delta connection.*

delta pulse-code modulation In wire or radio communication, the conversion of an af signal into equivalent digital pulse trains.

delta quantity An increment, i.e., the difference between two values of a variable. Also see *delta, 1.*

delta rays Emission of secondary electrons as a result of radioactivity.

delta tune In communications transceivers, a control that allows the receiver frequency to be offset slightly from the transmitter frequency.

delta waves Brain waves of a frequency less than 9 Hz. Also see *electroencephalegraph* and *electroencephalogram.*

Deluc's pile See *dry pile.*

dem Abbreviation of *demodulator.*

demagnetization curve The second quadrant of a hysteresis curve (see *box-shaped loop*) showing reduction of demagnetization.

demagnetization effect The phenomenon in which uncompensated magnetic poles at the surface cause a reduction of the magnetic field inside a sample of a material.

demagnetize To remove magnetism from an object either temporarily or permanently.

demagnetizer A device which will divest a body of its magnetism. In its simplest form, it is essentially a multiturn coil carrying alternating current. The body to be demagnetized is inserted into the coil, where it is exposed to the ac field for a short time. Also called *degausser.*

demagnetizing current The half-cycle of an alternating current (or polarity of a direct current) flowing through a coil wound on a permanent magnet (as in a headphone, permanent-magnet loudspeaker, or polarized relay) and which reduces the magnetic field. Compare *magnetizing current, 2.*

demagnetizing force 1. A magnetic force whose direction reduces the residual induction of a magnetized material. 2. That which tends to destroy or impair a permanent magnet, such as heat or a sudden blow.

demand factor In the use of electric power, the ratio of the consumer's maximum demand to the actual power comsumed.

demand processing Descriptive of a system that processes data almost as soon as it is available (without storing it).

demarcation strip An interface between a terminal unit and a carrier line.

Dember effect The appearance of a voltage between regions in a semiconductor when one of the regions is illuminated.

demodulation The process of retrieving information (modulation) voltage from a modulated carrier voltage. In receivers and certain test instruments, demodulation is *detection*.

demodulator A device or circuit which performs the process of demodulation. AM demodulators include diode and triode detectors; FM demodulators include discriminators, quadrature detectors, and ratio detectors.

demand read (write) Inputting or outputting data blocks to or from a central processor as needed for processing.

demodulator probe A diode probe which removes the modulation envelope from an applied amplitude-modulated rf signal and presents the envelope to a voltmeter or oscilloscope.

De Moivre's theorem A theorem that gives any power of a complex number in polar form: $[r (\cos \theta + j \sin \theta)]^n = r^n (\cos n \theta + j \sin n \theta)$, where n may be positive or negative, integral or fractional. The expression $(\cos \theta + j \sin \theta$ can function as a trigonometric operator which rotates a vector through θ degrees. When the operator is applied twice in succession, the total rotation is 2θ, when it is applied three times, 3θ and so on.

demonstrator A device for teaching and showing principles of operation, of an electronic circuit, for example. Also see *dynamic demonstrator*.

DeMorgan's theorem A rule of sentential or digital logic. It states that the negation of (A AND B), for any two statements A AND B, is equivalent to NOT A OR NOT B. Also, the negation of (A OR B) is equivalent, logically, to NOT A AND NOT B. This can be extended to any number of statements in combination, by inductive logic.

demountable tube A power tube which can be dismantled and reassembled for inspection and electrode replacement.

demultiplexer A circuit or device which separates the components of a multiplexed signal transmitted over a channel.

demultiplexing circuit See *demultiplexer*.

denary band A band in which the highest frequency is 10 times the lowest, i.e., $f2/f1 = 10$.

dendrite 1. The branching (tree like) structure formed by some materials, such as semiconductors, as they crystallize. 2. The branching portion of a nerve cell; hence, the corresponding circuit element in the electronic model of such a cell.

dendritic growth 1. Dendrite (see *dendrite, 1*). 2. Growing long, flat semiconductor crystals.

dendron See *dendrite, 2*.

denominator The term below or to the right of the division of a fraction. The denominator is the divisor in a fraction representing division. Compare *numerator*.

dens Abbreviation of *density*.

dense binary code A binary representation system, in which any possible combination of characters is assigned some correspondent.

densitometer An instrument for measuring the density of a body.

density 1. Weight per unit volume of a material. 2. Concentration of charge carriers or of lines of force. 3. The number of items per unit volume or area, as in *population density*.

density modulation Modulation of the density, with respect to time, of electrons in an electron beam.

density of electrons The concentration of electrons, i.e., the number per unit area or volume.

density packing A figure indicating the quantity of bits per inch or per centimeter, stores on a magnetic tape.

dentiphone An early hearing aid which transmitted sounds through the user's teeth. Also called *audiophone*.

dentophonics The pickup of speech impulses directly from the mouth and throat without the need for intervening air.

dependent equations Equations which are alike and have an infinite number of solutions. Compare *independent equations* and *inconsistent equations*.

dependent linearity Linearity (especially in its deviation from an ideal slope) as a dependent variable.

dependent variable A changing quantity whose value at any instant is governed by the value at that instant of another changing quantity (the independent variable). For example, the value of tube plate current depends on the value of plate voltage. Compare *independent variable*.

depletion—enhancement mode Operation which is characteristic of the *depletion—enhancement-mode MOS-FET*.

depletion—enhancement-mode MOSFET A metal-oxide semiconductor field-effect transistor designed for zero gate-bias voltage. An ac gate signal voltage drives the MOSFET alternately into the depletion mode (negative signal halfcycle) and enhancement mode (positive signal half-cycle). Compare *depletion-type MOSFET* and *enhancement-type MOSFET*.

depletion field-effect transistor A field-effect transistor whose operation is based on control of depletionlayer width. Also see *barrier, 1, depletion—enhancement mode, depletion—enhancement-mode MOSFET, and depletion-type MOSFET*.

depletion layer See *barrier, 1*.

depletion-layer capacitance See *junction capacitance*.

depletion-layer rectification Rectification provided by a semiconductor junction.

depletion-layer transistor A transistor whose action depends on modulation of current carriers in a space-charge region (depletion layer).

depletion mode Operation which is characteristic of the depletion-type MOSFET.

depletion region The region around a semiconductor junction which is devoid of current carriers.

depletion-type MOSFET A metaloxide semiconductor field-effect transistor in which the channel directly under the gate electrode is narrowed (*depleted*) by a negative gate voltage in the n-channel transistor or by a positive gate voltage in the p-channel transistor.

depolarization 1. In a primary cell, the removal of the agents which have caused polarization. 2. The addition of a polarization-inhibiting substance to the electrolyte of a primary cell.

depolarizer A substance which retards polarization in a battery cell. An example is the manganese dioxide used in dry cells.

depolarizing agent See *depolarizer*.

deposition The application of a layer of one substance (usually a metal) to the surface of another (the substrate), as in evaporation, sputtering, electroplating, silk-screening, etc.

depth finder See *acoustic depth finder*.

depth indicator 1. A sounding instrument for determining the depth of a body of water. 2. On a depth finder, the meter which indicates the depth of water.

depth of cut On a phonograph disk, the depth of the recorded groove.

depth of heating In dielectric heating, the depth heat penetration in the sample when both electrodes are applied to one of its faces.

depth of modulation The degree to which a carrier wave is modulated.

depth of penetration The extent to which a skin-effect current penetrates the surface of a conductor.

depth sounder See *acoustic depth finder*.

de-Q 1. To reduce the Q of a component or tuned circuit. 2. To apply a Q-spoiler.

derate See *derating*.

derating The practice of reducing an operating parameter (e.g., current, voltage, power) as another factor (such as temperature) increases, to insure efficient, reliable, and safe operation.

derating curve A curve that shows the extent to which a quantity (such as allowable power dissipation) must be reduced as another quantity (such as temperature) increases.

derating factor The amount by which a current, power, or voltage must be decreased to insure safe and efficient operation of a circuit or device in a given environment (temperature, altitude, humidity, etc.). Also see *derating* and *derating curve*.

derivative The ratio of two differentials (see *differential, 1*), e.g., the derivative of collector current (Ic) with respect to collector voltage (Vc) is dIc/dVc. (A derivative shows the slope of a curve at a point of interest.)

derivative action In a control system, an action producing a corrective signal proportional to the rate of change of the controlled variable (i.e., proportional to the first derivative of the variable).

derivative control A method of automatic control, actuated according to the number of errors per second.

derivative of constant Where the equation is $y = c$, the derivative $dy/dx = dc/dx = 0$.

derivative of consecant Where the equation is $y = \csc \theta$, the derivature $dy/dx = d(\csc \theta)/d\theta = -\csc \theta \cot \theta$.

derivative of cosh Where the equation is $y = \cosh \theta$, the derivative $dy/dx = d(\cosh \theta)/d\theta = \sinh \theta$.

derivative of cosine Where the equation is $y = \cos \theta$, the derivative $dy/dx = d(\cos \theta)/d\theta = \sinh \theta$.

derivative of cotangent Where the equation is $y = \cot \theta$, the derivative $dy/dx = d(\cot \theta)/dx = -\csc^2\theta$.

derivative of curve The tangent to a curve at a point of interest. The derivative is the slope of the curve at that point and may be expressed as dy/dx, where x and y are the coordinates of the point.

derivative of exponential function Where the equation is $y = e^x$, the derivative $d(e^x)/dx = e^x$.

derivative of inverse function Where x conventionally is the independent variable and y, the dependent variable, the derivative of the inverse function dx/dy is equal to the reciprocal of the derivative of the direct function, i.e., $dx/dy = 1/(dy/dx)$.

derivative of natural logarithm Where the equation is $y = \ln x$, the derivative $dy/dx = d(\ln x)/dx = 1/x$.

derivative of product Where the equation is $y = u\nu$, the derivative $dy/dx = d(u\nu)/dx = u(d\nu/dx) - \nu(du/dx)$.

derivative of quotient Where the equation is $y = u\nu$, the derivative $dy/dx = d/dx (u/\nu) = \nu (du/dx) - u(d\nu/dx)/\nu^2$.

derivative of reciprocal Where the equation is $y = 1/x$, the derivative $dy/dx = d(1/x)/dx = -x^{-2}$.

derivative of secant Where the equation is $y \sec \theta$, the derivative dy/dx

$= d(\sec \theta)/dx = \sec \theta \tan \theta$.

derivative of sine Where the equation is $y = \sin \theta$, the derivative $dy/dx = (d \sin \theta)/d \theta = \cos \theta$.

derivative of sinh Where the equation is $y \sinh \theta$, the derivative $dy/dx = (d \sinh \theta)/d\theta = \cosh \theta$.

derivative of sum Where the equation is $y = u + \nu$, the derivative $dy/dx = du/dx + d\nu/dx$.

derivative of tangent Where the equation is $y = \tan \theta$, the derivative $dy/dx = (d \tan \theta)/d\theta = \sec^2 \theta$.

derivative of tanh Where the equation is $y = \tanh \theta$, the derivative $dy/dx = (d \tanh \theta)/d\theta = \sec^2 \theta$.

derivative of variable multiplied by constant Where the equation is $y = cx$, the derivative $dy/dx = c$.

deriviative of variable with constant exponent Where the equation is $y = x^n$, the derivative $dy/dx = d(x^n)/dx = nx^{n-1}$.

derived center channel The sum or difference of the left and right channels in a stereophonic system. The derived center channel is fed to an amplifier, and then to a central speaker.

derived function See *derivative*.

Dershem electrometer A variation of the quadrant electrometer. In the Dershem instrument, the needle (to which a small mirror is attached) rotates within slots cut in the quadrant plates and, therefore, can never accidentally touch the plates.

description A data element that is part of a record and is used to identify it.

desensitization The process of making a circuit or device less responsive to small values of a quantity (see *desensitize*).

desensitize 1. To reduce the sensitivity of a receiver. 2. To reduce the gain of an amplifier. 3. To reduce the small-quantity response of an instrument.

desiccant A compound, such as cobalt chloride, which is used for the purpose of keeping enclosed items dry.

design 1. A unique, planned arrangement of electronic components in a circuit, in accordance with good engineering practice, to achieve a desired end result. 2. A unique layout of components or controls, in accordance with good engineering practice, esthetics, and (often) ergonomics. 3. Invention. 4. Plan. 5. To produce a design, as in 1, 2, 3, and 4, above.

designation 1. Within a computer record, coded information identifying the record so it can be handled accordingly. 2. In a specific punched card column, punching indicating the card's type.

designation hole See *control hole*.

design-center rating A specified parameter which, if not exceeded, should provide acceptable average performance for the greatest number of the components so rated.

design compatibility The degree to which a transmitter and receiver are designed for the rejection of unwanted electromagnetic noise.

design engineer An engineer who is skilled in the creation of new designs and in the comparative analysis of designs. Also see especially *design, 1*.

design-maximum rating See *maximum rating*.

design-proof test A performance test made on a newly completed circuit or device to determine the suitability of the design (see *design, 1, 2, 3*).

desk calculator An (often sophisticated) electronic calculator no larger than most mechanical or electrical calculators. Most give a readout-lamp display; others, a printout, too.

Desk-Fax A facsimile transceiver, which can be placed on a desktop, employed for wire or radio transmission and reception of still pictures.

Trade name of Western Union.

desk-top computer A digital computer small enough to be operable on a desk top, in the manner of an adding machine or electric calculator. Also see *desk calculator*.

desolder To unsolder joints, especially with a special tool which protects delicate parts and removes the melted solder by suction.

destaticization Chemical process employed to minimize the retention of electrostatic charges by certain substances.

destination The point in a system to which a signal of any sort is directed.

destination file A computer file that receives data output during a specific program run.

destination register In a digital computer, a register into which data is entered.

Destriau effect Light emission resulting from the action of an alternating electric field on phosphors imbedded in a dielectric.

destructive addition A computer logic operation in which the sum of two operands appears in the memory location occupied by one of the operands.

destructive breakdown A breakdown (see *breakdown, 1*) in which the deteriorating effects are irreversible, e.g., the permanent rupture of a dielectric by excessive applied voltage.

destructive interference Interference resulting from the addition of two waves whose amplitudes are of different polarities.

destructive read In a computer or calculator, the condition in which reading the answer erases the data (as from a location) used in the calculation.

destructive test A test which unavoidably destroys the test sample. Compare *nondestructive test*.

DETAB A COBOL-based computer programming language permitting the programmer to present problems as decision tables. Also see *decision table*.

detail See *definition, 1*.

detail constant Pertaining to a video signal, the ratio EH/EL, where EH is the amplitude of high-frequency components and EL is the amplitude of the low-frequency reference component of the signal.

detected error In a computer system, an error which is disclosed but uncorrected until final output is available.

detection See *demodulation*.

detectophone A "bug," or device for eavesdropping on a conversation. The device may employ a tape recorder or a tiny radio transmitter.

detector 1. Demodulator. 2. A device which senses a signal or condition and indicates its presence.

detector balanced bias In a radar system, bias obtained from a controlling circuit and used for an anticlutter operation. Also see *anticlutter circuit* and *anticlutter gain control*.

detector bias Steady dc voltage applied to a detector to set its operating point.

detector blocking In a regenerative detector, the action in which a strong signal pulls the detector oscillation frequency into step with itself, thereby forcing the detector to oscillate at the signal frequency.

detector circuit A demodulator circuit, i.e., one employed to recover the intelligence from a modulated carrier.

detector probe See *demodulator probe*.

detector pull-in See *detector blocking*.

detector stage In a receiver or instrument, the separate stage containing the detector circuit. Some systems, such as a superheterodyne receiver, have more than one detector. Also see *first detector* and *second detector*.

detent A mechanical stop used on a rotary switch to hold the switch pole securely in each selected position.

determinant Symbol, D. The difference of the products of the elements in the two diagonals of a matrix. Thus, for the matrix

$$\begin{vmatrix} a11 & a12 & k1 \\ a21 & a22 & k2 \end{vmatrix} \quad \text{the determinant D} =$$

$$\begin{vmatrix} a11 & a12 \\ a21 & a22 \end{vmatrix} = a11a22 - a12a21.$$

detune 1. To adjust a tuned circuit to some frequency other than its resonant frequency. 2. To set the frequency of a receiver or transmitter to some point other than that normally used. 3. To stagger-tune a receiver intermediate-frequency system.

detuning Tuning to a point above or below the frequency to which a device or system is normally (or initially) adjusted (usually the resonant frequency of the device).

detuning stub A device employed for the purpose of coupling a feed line to an antenna, while choking off currents induced on the feed line as a result of the near-field radiation of the antenna.

deupdating Producing an earlier form of a computer file by substituting older records for current ones.

deuterium Symbol, D. Heavy hydrogen: the hydrogen isotope having an atomic weight of approximately 2(i.e., ^{1}H2.).

deuterium oxide Symbol, D2O. Heavy water. This compound has wide use in nuclear reactors.

deuteron The nucleus of a deuterium atom.

deuton See *deuteron*.

deutron See *deuteron*.

deviation 1. In an FM signal, the amount of carrier frequency shift away from its unmodulated frequency during modulation. Deviation is usually expressed in kilohertz and is directly proportional to the amplitude of the modulating signal. 2. The amount by which a quantity drifts from its proper value, e.g., an allowable frequency deviation of $\pm$ 10 Hz in a 1 MHz carrier.

deviation absorption In radio waves slowed after reflection from the ionosphere, absorption at frequencies near the critical value.

deviation distortion In an FM receiver, distortion resulting chiefly from discriminator nonlinearity and restricted bandwidth.

deviation ratio In an FM signal, the ratio between the modulating frequency and the deviation. Here, the deviation is the peak frequency shift at full modulation, and the modulating frequency is the highest frequency to be transmitted.

deviation sensitivity For an FM receiver, the smallest frequency deviation which will produce a specified af output power. Deviation sensitivity is expressed in kilohertz or as a percentage of rated deviation of the receiver measured with the receiver set for maximum gain.

device A simple or complex discrete electronic component. Sometimes, a subsystem employed as a unit and, therefore, thought of as a single component.

device complexity The degree to which a device is complicated. Particularly, the number of circuit elements in an integrated circuit.

device control character In a paper tape, a coded character that controls equipment only; i.e., it has no other significance.

device independence A characteristic of a computer, that allows operation independent of the types of input/output devices used.

Dewar flask A double-walled glass flask whose inner walls are silvered; the space between walls is evacuated. The flask keeps liquids at a desired temperature for several hours. Also called *thermos bottle*.

dew point For a gas containing water vapor (typically air), the highest temperature at which the vapor condenses as the gas is cooled. The dew point depends on the amount of vapor in the gas.

dew-point recorder An instrument for determining and recording the temperature at which water vapor in the air condenses to a liquid.

df Abbreviation of *direction finder.*

df antenna An antenna which is mechanically rotatable or has an electrically rotatable response pattern for use with a direction finder.

df antenna system Two or more df antennas arranged for maximum directivity and maneuverability, together with associated feeders and couplers.

dF 1. Symbol for *differential of force.* 2. Symbol for *differential of field intensity.*

df Symbol for *differential of frequency.*

D flip-flop A delayed flip-flop. The state of the input determines the state of the output during the following pulse, rather than during the current pulse.

dG Symbol for *differential of conductance.*

dg Abbreviation of *decigram.*

dH Symbol for *differential of magnetic field intensity.*

dh Symbol for *differential of height.*

dI Symbol for *differential of current.* (Also, d *i.*)

dia Abbreviation of *diameter.* (Also, *diam.*)

diac A two-terminal, bilateral, threelayer semiconductor device which exhibits negative resistance. When the applied voltage exceeds a critical value, the device conducts; a gateless triac.

diagnosis Determination of the cause and location of a malfunction or the cause of error.

diagnostic routine 1. An efficient sequence of diagnostic tests for rapid, foolproof troubleshooting. 2. A sometimes vendor-supplied computer program for debugging other programs or finding the cause of hardware failure. Also called *diagnostic program.*

diagnostic test 1. A test made primarily to ascertain the cause of dysfunction in electronic equipment. Compare *performance test.* 2. To apply a diagnostic routine (*diagnostic routine, 2*) to hardware faults, or to implement one to prevent such a fault.

diagnotor In digital computer practice, a troubleshooting routine combining both diagnosis and editing.

diagram A (usually line) drawing depicting a circuit, assembly, or organization. See, for example, *block diagram* and *circuit diagram.*

dial 1. A graduated scale—arranged horizontally, vertically, in a circle, or over an arc—used to show the distance through which a variable component (such as a potentiometer, variable capacitor, or switch) has been adjusted. A pointer may move over the scale, or the scale may be moved past a stationary pointer. 2. The graduated face of a meter.

dial cable A flexible cable or belt conveying motion on the shaft of an adjustable component (such as a potentiometer or variable capacitor) to a dial.

dial-calibrated attenuator A variable attenuator with a dial reading directly in decibels.

dial-calibrated capacitor A variable capacitor with a dial reading directly in picofarads or microfarads.

dial-calibrated inductor A variable inductor with a dial reading directly in microhenrys or henrys.

dial-calibrated potentiometer A potentiometer with a dial reading directly in output volts, percentage of input voltage, number of turns (when resistance is linear function), or other quantity.

dial-calibrated resistor A variable resistor with a dial reading directly in ohms or megohms.

dial-calibrated rheostat See *dialcalibrated resistor.*

dial cord A dial cable. *Cord* usually designates a fabric string, whereas a cable is a flexible, braided wire.

dial knob The knob used to turn a dial under a pointer, or to turn a pointer over a dial scale.

dial lamp See *dial light.*

dial light The usually small lamp illuminating the dial of an electronic equipment. Sometimes the dial light also serves as a pilot light.

dial lock A small mechanism employed to lock a dial at a particular setting to prevent further turning.

dialer See *automatic dialing unit.*

dialing key In a telephone system, a dial that employs keys rather than a rotary dial.

dial jack In a telephone system, a set of jacks that facilitates interconnections between dial cords and external lines.

dial light A lamp or light-emitting diode placed in the dial mechanism of a radio receiver, transmitter, or transceiver. Allows the dial to be read in dim light or in darkness.

dialog equalizer In sound transmission and recording, an equalizer that reduces low-frequency response during dialog and extreme closeups. (Also spelled *dialogue.*)

dial pulse An interruption of the direct current in a telephone system when the dial contacts of the calling telephone open. The number of such interruptions corresponds to the digit dialed.

dial scale The graduated portion of a dial.

dial system 1. Dial telephone system. 2. The unique arrangement of dials (and sometimes also of dial-less knobs) for adjusting electronic equipment.

dial telephone A telephone in which a numbered rotatable disk is used for producing the switch interruptions that cause generation of the transmitted multidigit telephone numbers.

dial telephone system The complete automatic circuit, including central-office facilities, for dial telephone operation.

dial tone In a telephone system, a constant hum or whine heard before dialing, indicating that the system is operational.

dial-up In a telephone system, the calling of one subscriber by another, using a dial system.

diam Abbreviation of *diameter.* (Also, *dia.*)

diamagnet A diamagnetic substance (see *diamagnetism*). Compare *paramagnet.*

diamagnetic Possessing diamagnetism. Compare *paramagnetic.*

diamagnetic material A material exhibiting diamagnetism.

diamagnetism The state of having a magnetic permeability of less than one. Compare *paramagnetism.*

diamond antenna A nonresonant wideband directional antenna whose horizontal wire elements are arranged in a parallelogram. The arrangement is fed at one corner, the opposite corner being terminated with a noninductive resistor. Also called *rhombic antenna.*

diamond-grid radiator antenna See *Chireix-Mesny antenna.*

diamond lattice The orderly internal arrangement of atoms in a redundant pattern in crystalline materials, such as germanium or silicon.

diamond stylus A phonograph "needle" having as its point a small, ground diamond.

diapason 1. Either of the two principal stops (*open diapason* and *closed diapason*) of an organ that cover the entire range of the instrument. When one is used, a note played is automatically sounded in several octaves. 2. Tuning fork; pitch pipe.

diaphony See *dissonance.*

diaphragm A usually thin metal or dielectric disk employed as the vibrating member in headphones, loudspeakers, and microphones, and as the pressure-sensitive element in some pressure sensors and barometers.

diaphragm gauge A sensitive gaspressure gauge using a thin metal diaphragm stretched flat. Increments of pressure move the diaphragm in relation to a nearby electrode, accordingly varying the capacitance between the two. Electronic circuitry converts the capacitance change into direct readings of pressure as low as 10^{-1} psi (about 10^{-3} pascal).

diathermic Able to transmit heat rays effectively.

diathermotherapy The use of diathermy in the treatment of various physiological disorders.

diathermy 1. In medicine and physical therapy, producing heat in subcutaneous (below the skin) tissues by means of high-frequency radio waves. 2. The radio-frequency power oscillator and associated equipment used in the operation described in 1, above.

diathermy interference Radio interference from the operation of unshielded and unfiltered diathermy machines.

diathermy machine See *diathermy, 2.*

diatomic Having two atoms, e.g., a diatomic molecule.

diatomic molecule A molecule (such as that of oxygen) composed of two atoms. Compare *monatomic molecule.*

dibble Doubling a number and adding one. Thus, dibble $n = 2n + 1$.

dibit A combination of two binary digits. There are four possible dibits: 00, 01, 10, and 11.

dice Plural of *die,* 1, 3.

dichotomizing search In digital computer practice, locating an item in a table of items which are arranged by key values in serial order. The required key is compared with a key halfway through the table; according to this relational test, half of the table is accepted and again divided for comparison, and so on until the keys match and the item is found.

dichotomy Characterized by the usually repetitive branching into two sets, groups, or factions.

dichroic mirror A mirror which reflects light of one color and transmits that of other colors.

dichroism The property of (1) a crystal, showing different colors, depending on which axis is the line of sight, (2) a solid taking on different colors with transmitting layer thickness changes; (3) a liquid changing color according to solution concentration. Also called *dichromatism.*

dicing The cutting of a semiconductor melt, crystal wafer, or other material into dice (see *die*).

Dictaphone A sound recording and playback machine employing cylindrical records (in its early form) and used principally for dictation.

die 1. A small wafer of useful electrical material, such as a semiconductor or a precision resistor chip. 2. A casting designed to mold molten metal into a specific configuration until the metal hardens. 3. Any small object of roughly cubical proportions. 4. To lose power or energy completely, usually unintentionally. 5. Of a computer program, to produce unpredicted and useless results following an initial run.

die bonding The bonding of dice or chips to a substrate.

die casting Making a casting by forcing molten metal (such as an aluminum alloy, lead, tin, or zinc) under high pressure into a die or mold.

dielectric A material which is a nonconductor of electricity (see *insulator, 1*). Although such common nonconductors as glass, wood, and plastic might first come to mind, dry air, too, is a dielectric, as is *pure* water.

dictionary Specifications for the size and format of computer file operands, and data names for field and file types, given in a table.

dielectric absorption The phenomenon whereby a dielectric material retains a portion of an electric charge after being discharged. This apparent charge absorption requires that some capacitors be discharged a number of times before the terminal voltage finally reaches zero.

dielectric amplifier An amplifier circuit in which the active component is a capacitor having a nonlinear dielectric. A signal voltage applied to the capacitor varies the capacitance and, accordingly, capacitor current (supplied by a high-frequency local oscillator). This modulated current flowing through a load resistor develops an output signal voltage higher than the input-signal voltage; thus, amplification is the result.

dielectric antenna An antenna (usually of the directive microwave type) in which the radiating element, or most of it, is made of a dielectric material, such as polystyrene.

dielectric breakdown The usually sudden destructive conduction of a high current through a dielectric when the applied voltage exceeds a certain critical value for the material.

dielectric breakdown voltage The voltage at which dielectric breakdown occurs.

dielectric capacity See *dielectric constant.*

dielectric constant Symbol, k. For a dielectric material, the ratio of the value of a two-plate capacitor using the dielectric material, to the value of the equivalent capacitor with dry air as a dielectric. Also called *inductivity* and *specific inductive capacity.*

dielectric current 1. Current flowing over the surface of a dielectric material in response to a varying electric field. 2. Current flowing through a dielectric as a result of its finite insulation resistance.

dielectric dissipation For a dielectric material in which an electric field is set up, the ratio of the lost to the recoverable part of the introduced electrical energy.

dielectric dissipation factor The cotangent of the dielectric phase angle.

dielectric fatigue In some dielectric materials subjected to a constant voltage, the deterioration of dielectric properties with time.

dielectric guide A waveguide made from a solid dielectric, such as polystyrene.

dielectric heater A high-frequency power generator for dielectric heating.

dielectric heating The heating and forming of a dielectric material, such as a plastic, by making the material temporarily the dielectric of a twoplate capacitor connected to a highpower rf generator. Losses in the dielectric cause its heating. Compare *induction heating.*

dielectric hysteresis See *dielectric absorption.*

dielectric isolation In a monolithic IC, the isolation of circuit elements from each other by a dielectric film, as opposed to isolation by reversebiased pn junctions.

dielectric lens A lens made of a dielectric material for focusing microwaves. Its operation is analogous to that of an optical lens.

dielectric loss For a dielectric material subjected to a changing electric field, the rate of transformation of electric energy into heat.

dielectric loss angle Symbol, δ. Ninety degrees minus the dielectric phase angle.

dielectric loss factor For a die r For a dielectric material, the product of the dielectric constant and the tangent of the dielectric loss angle.

dielectric loss index See *dielectric loss factor.*

dielectric matching plate A dielectric plate used in some waveguides for impedance-matching.

dielectric mirror A reflector containing a number of layers of dielectric material. Its action depends on light being partially reflected from the interfaces between materials having unequal indexes of refraction.

dielectric phase angle For a dielectric material, the angular phase difference between a sinusoidal voltage applied to the material and the component of the resultant current having the same period as that of the voltage.

dielectric phase difference See *dielectric loss angle*.

dielectric polarization The effect characterized by the slight displacement of the positive charge in each atom of a dielectric material, with respect to the negative charge, under the influence of an electric field.

dielectric power factor The cosine of the dielectric phase angle or the sine of the dielectric loss angle.

dielectric puncture voltage See *dielectric breakdown voltage*.

dielectric rating The breakdown voltage, and sometimes the power factor, of the dielectric material used in a device, such as a relay, motor, or switch.

dielectric ratings Electrical characteristics of a dielectric material: breakdown voltage, power factor, dielectric constant, etc.

dielectric resistance See *insulation resistance*.

dielectric rigidity See *dielectric strength*.

dielectric-rod antenna A unidirectional antenna that employs a dielectric substance to obtain power gain.

dielectric soak See *dielectric absorption*.

dielectric strain The distorted internal state of a dielectric as caused by the influence of an electric field. Also called *dielectric stress*.

dielectric strength The highest voltage a dielectric can withstand before breakdown occurs (see *breakdown voltage, 1*). Dielectric strength is usually expressed in volts or kilovolts per mil of material thickness.

dielectric stress The stress existing in the dielectric of a charged capacitor, due to electrostatic energy. It is evidenced by a distortion of the electron orbits in the dielectric's atoms.

dielectric susceptibility For a polarized dielectric, the ratio of polarization to electric intensity.

dielectric tests Tests made to determine dielectric characteristics (see *dielectric ratings*), specifically, dielectric constant and breakdown voltage.

dielectric waveguide See *dielectric guide*.

dielectric wedge A wedge-shaped dielectric slug placed inside a waveguide for impedance matching.

dielectric wire A small *dielectric waveguide* that acts as a wire; i.e., to carry signals between points in a circuit.

Dietzhold network A four-terminal, shunt m-derived circuit employed in the compensation on wideband amplifiers.

Dietzhold peaking In the frequency compensation of amplifiers, compensation obtained with a shunt m-derived network (see *Dietzhold network*).

difference The result of a subtraction.

difference amplifier See *differential amplifier*.

difference channel In a stereophonic amplifier, a third audio channel that handles the difference between signals in the right channel and those in the left channel.

difference detector A detector whose output is the difference between two simultaneous input signals.

difference frequency A third frequency resulting from mixing two frequencies and equal to their difference.

difference of potential The algebraic sum of voltages at two points of different electrical potential. Thus, the potential between a $+7V$ point and a $-3V$ point is 4V. Also see *electromotive force, potential difference, volt,* and *voltage*.

difference quantity See *increment*.

difference signal 1. The resultant signal obtained by subtracting, at every instant for at least one full cycle, the amplitudes of two signals. 2. The difference of the left and right channel outputs in a stereo system.

differential 1. Symbol, d. An infinitesimal difference between two values of a quantity. Thus, the differential of plate voltage is represented by dEp. Also see *derivative, differential calculus,* and *differentiation*. 2. A device, consisting of a gear system, which adds or subtracts angular motions and delivers the result. 3. A gear system in which the motion of a shaft is transferred to two other shafts aligned with each other and perpendicular to the first shaft. 4. One of two coils arranged to produce opposite polarities at a point in a circuit.

differential amplifier An amplifier consisting of two identical sections, each having input terminals. Two separate output terminals or a single output terminal common to both amplifiers is provided. Ground is common to both sections.

differential analyzer An analog computer that solves differential equations using integrators.

differential angle For a mercury switch, the angle between operation and release positions.

differential calculus The branch of mathematics concerned with the theory and applications of differentiation. Also see *derivative* and *differential, 1*. Compare *integral calculus*.

differential capacitance The derivative of capacitance change with respect to voltage; dC/dE.

differential capacitance characteristic The function relating differential capacitance to voltage.

differential capacitor A dual variable capacitor with two identical stator sections and a single rotor section which turns into one stator section and out of the other. The capacitance of one section, therefore, decreases while that of the other increases.

differential coefficient See *derivative*.

differential coil See *differential, 4*.

differential comparator A dualamplifier, linear IC which delivers an output that is proportional to the difference between two input signals.

differential compound dc generator A compound-wound dc generator in which the magnetomotive force of the series field opposes that of the shunt (main) field; the generator has poor regulation.

differential compound dc motor A compound-wound dc motor in which the magnetomotive force of the series field coil opposes that of the shunt (main) field coil.

differential cooling Reducing temperature at different points on a surface at different rates.

differential delay The difference $dmax - dmin$ across a frequency band, where $dmax$ is the maximum frequency delay and $dmin$ is the minimum frequency delay.

differential discriminator A device that passes pulses whose amplitudes are between two predetermined values above or below zero.

differential distortion In an agc circuit, distortion from effects which cause shunting of the diode load resistor.

differential equations Broadly, equations containing differentials (see *differential, 1*). A differential equation is classified by order (first order, second order, etc.) in accordance with the highest-order derivative it contains. A differential equation of the nth order is represented by $F = (x,y,y'y''...y^n) = \phi(x)$. Also see *ordinary differential equation* and *partial differential equation*.

differential flutter Fluctuations in the speed of a magnetic tape, that are nonuniform in different parts of the tape.

differential gain For a differential amplifier, the average gain of the two sections of the amplifier; $Ad = (A1 + A2)/2$. Compare *differential unbalance*.

differential gain control A circuit or device for setting the gain of a radio receiver in terms of an anticipated change in signal strength, to reduce the receiver output signal differential.

differential galvanometer A galvanometer in which currents in two similar coils neutralize each other; thus, there is zero deflection when the currents are equal.

differential gap The smallest range of values a controlled variable must take to change a three-position controller's output from on to off, or vice versa.

differential gear See *differential, 3*.

differential generator A servo-driven selsyn differential generator. Also see *differential transmitter*.

differential heating Increase of temperature at different points on a surface at different rates.

differential impedance See *differential-input impedance*.

differential induction coil An induction coil having two differentially wound primary coils.

differential input In a differential amplifier, the circuit between input terminals 1 and 2, as opposed to the input circuit between inputs 1 or 2 and ground.

differential-input amplifier A differential amplifier whose output is proportional to the differences between two input signals, each applied between an input terminal and a common ground.

differential-input capacitance In a differential amplifier, the capacitance between the input terminals.

differential-input impedance In a differential amplifier, the input impedance between the differential input terminals.

differential-input measurement For a differential amplifier, a *floating* measurement made between the input terminals.

differential-input rating In an operational amplifier, the greatest difference signal that can be placed between the inputs while allowing proper operation.

differential-input resistance In a differential amplifier, the resistance between the input terminals.

differential-input voltage In a differential amplifier, the signal voltage presented to the floating input terminals.

differential-input voltage range In a differential amplifier, the range of signal voltages which may be applied between the differential input terminals without overdriving the amplifier.

differential input-voltage rating The maximum differential-input voltage which may be applied safely to a differential amplifier.

differential instrument A galvanometer or other meter in which deflection results from the differential effect of currents flowing in opposite directions through two identical coils. Also see *differential galvanometer*.

differential keying A system of break-in keying in which the oscillator stage of a transmitter containing a keyed amplifier is disabled when the key is open to prevent interference with the receiver at the keying station, and is enabled when the key is closed.

differential microphone See *double button microphone*.

differential-mode gain In an operational amplifier, the ratio, in decibels, between the output voltage and the differential input voltage.

differential-mode input In an operational amplifier in differential mode, the difference between the two input signal voltages.

differential-mode signal In a balanced three-terminal circuit, such as the input of a differential amplifier, a signal applied between the floating (ungrounded) input terminals.

differential modulation A method of modulating a carrier, according to rate of change in the amplitude of the signal.

differential permeability The derivative of normal induction with respect to magnetizing force.

differential phase In a television system tested with a low-level, high-frequency sine-wave signal ($f1$) superimposed on a low-frequency, sine-wave signal ($f2$), the difference in phase shift of $f1$ throughout the system for two specified levels of $f2$.

differential phase-shift keying Keying of a carrier by varying the carrier phase at one instant with respect to the phase at the previous instant.

differential pressure The difference in pressure between two points.

differential-pressure transducer A transducer that delivers an output proportional to the difference between two sensed actuating pressures.

differential protective relay A differential relay which operates to protect equipment or personnel when the difference between the two actuating quantities reaches a prescribed level.

differential receiver A synchro differential that receives the electrical output of two synchro transmitters. The receiver can subtract one input voltage from the other.

differential relay A relay actuated by the difference between two currents or voltages.

differential selsyn A selsyn (see *autosyn* and *synchro*) in which the position assumed by the rotor is proportional to the sum of rotor and stator field values.

differential sign The letter d as used in calculus to indicate the differential of the quantity which it precedes. Thus, dx is the differential of x.

differential stage An amplifier circuit that employs two inputs, with the output signal amplitude depending on the difference between the two input amplitudes.

differential synchro See *differential receiver* and *differential transmitter*.

differential transducer A dual-input, single-output sensor, such as a differential pressure transducer, which is actuated by two sensed quantities and delivers an output proportional to their difference.

differential transformer A variableinductance transformer having a (usually cylindrical) core which is moved in and out to provide adjustable linkage between the interwound primary and secondary windings by changing their inductance. This operation permits adjustment of the amplitude and phase of the transformer output voltage with respect to the input voltage.

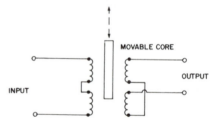

Differential transformer.

differential transmitter A synchro differential connected to a synchro transmitter. In a synchro receiver supplied by this combination, the change in rotor position is the algebraic difference between the transmitter-rotor-position and the differential-rotor position.

differential triangle In a differential graph, the triangle whose hypotenuse is tangent to the curve of the function under study, the other two sides being dx and dy, respectively.

differential unbalance For a differential amplifier, the average difference in gain between the two amplifier sections; $(A1 – A2)/2$. Compare *differential gain*.

differential voltage 1. The voltage difference between the input signals to a differential device. 2. The breakdown voltage minus the operating voltage for a lamp.

differential voltage gain 1. The ratio, in decibels, between the differential output and differential input voltages of an amplifier. 2. The instantaneous ratio, in decibels, between the rate of change of the output signal voltage and the rate of change of the input signal voltage in an amplifier.

differential winding See *differential coil*.

differential-wound field In a motor or generator, a field winding having series and shunt coils whose fields are opposing.

differentiate 1. To determine the first derivative of a mathematical function. 2. To produce an output signal, the instantaneous amplitude of which is proportional to the instantaneous rate of change of the input amplitude. 3. To determine the second, third, or higher derivative of a mathematical function.

differentiating circuit See *differentiating network*.

differentiating network A four-terminal *RC* network (series capacitor, shunt resistor) whose output voltage is the derivative of the input voltage with respect to time. Compare *integrating network*.

differentiation The process of computing a derivative. Also see *derivative; differential, 1;* and *differential calculus*.

differentiator 1. Differentiating network. 2. An operational amplifier whose output voltage $Eo = – RC(dEi/dt)$, where Ei is the input voltage; Eo, the output voltage, R, the feedback resistance; C, the input capacitance; and dt the differential of time. Compare *integrator*.

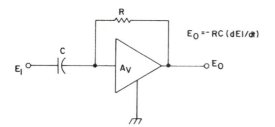

Differentiator, 2.

diffracted wave A wave which, upon striking an object, becomes bent. Also see *diffraction*.

diffraction A modification which rays undergo due to interference of one part of a beam with another when the beam is deflected, the light breaking up into dark and light bands or colored bands. The effect can be seen by looking obliquely at a phonograph record; the concentric grooves act as *diffracting grating*, the transmission type being used in simple spectroscopes. Compare *refraction*.

diffraction camera See *X-ray diffraction camera*.

diffraction grating A transparent plate containing up to 40,000 equally spaced parallel lines or grooves. Light passing through the slits between these lines produces a spectrum (similar to that set up by a prism) as a result of interference patterns.

diffraction of radio waves The bending of radio waves around an object. The curvature of a ground wave around the earth is an example, i.e., disregarding the effect of refraction also present.

diffraction pattern See *X-ray diffraction pattern*.

diffractometer An instrument for measuring the diffraction of radiation, such as light or X-rays.

diffuse 1. To produce diffusion. 2. Energy that is diffused.

diffused-alloy transistor See *drift field transistor*.

diffused-base transistor A bipolar transistor in which the base region has been diffused into the semiconductor wafer. Also see *diffused junction*.

diffused device A semiconductor device in which the junction is produced by diffusion (see *diffusion, 1*). Examples: *diffused-base transistor, diffused diode, diffused-junction rectifier,* and *diffused-mesa transistor*.

diffused diode A semiconductor diode having a diffused junction.

diffused-emitter-and-base transistor A transistor in which *n* and *p* materials both have been diffused into the semiconductor wafer to provide emitter and base junctions. Also see *diffusion, 1* and *diffused transistor*.

diffused junction In a semiconductor device, a pn junction formed by diffusing a gas into a semiconductor at high temperature that is below the melting point of the semiconductor. Typically, a gas containing an n-type impurity is diffused into p-type semiconductor material. Compare *alloy junction*.

diffused-junction rectifier A semiconductor rectifier employing a diffused junction.

diffused-junction transistor See *diffused-base transistor, diffused-mesa transistor,* and *diffused transistor*.

diffused-layer resistor In an integrated circuit, a resistor produced by diffusing a suitable material into the substrate.

diffused-mesa transistor A mesa transistor whose base is an n-type layer diffused into a p-type wafer (the remaining p-type material serving as the collector); its emitter is a small p-type area diffused into or alloyed with the n-layer. Unwanted diffused portions are etched away, leaving the transistor in the familiar mesa shape.

diffused planar transistor A diffused transistor in which emitter, base, and collector electrodes are exposed at the face of the wafer, which is passivated (has an oxide layer grown on it) to forestall leakage between surface electrodes.

diffused resistor See *diffused-layer resistor*.

diffused sound 1. Sound distributed so that its energy flux is the same at all points. 2. Sound whose source is difficult to locate or seems to shift, as that heard from out-of-phase stereo channels.

diffused transistor A transistor in which one or both electrodes are created by diffusion. See *diffused junction*.

diffused transmission The transmission of diffused energy through a substance.

diffusion 1. In the fabrication of semiconductor devices, the slow, controlled introduction of a material into the semiconductor, for example, the high-temperature diffusion of an n-type impurity (from a gas containing it) into a p-type wafer to form a diffused pn diode. 2. The random velocity and movement of current carriers in a semiconductor, resulting from a high-density gradient (possible without an electric field). 3. The characteristic spreading of light reflected from a rough surface or transmitted through a translucent material, producing a "soft" effect. 4. The migration of atoms from one substance to another, as in the diffusion of one gas throughout another.

diffusion bonding A method of joining different substances by diffusing atoms of one into the other. This technique is employed in the manufacture of certain semiconductor diodes, transistors, and other devices.

diffusion capacitance The current-dependent capacitance of a forward-biased semiconductor junction.

diffusion constant Symbol, D. Unit, cm^2/s. A figure expressing the ability of carriers to diffuse (see *diffusion, 2*); $D = (\mu kT)/e$, where k is the Boltzman constant (1.4×10^{-23}J/K); μ, the mobility constant ($cm^2/V.s$); T, absolute temperature (K); and e, electron charge (1.6×10^{-19}C).

diffusion current Current resulting from the diffusion of carriers within a substance (see *diffusion, 2*).

diffusion length In a semiconductor junction, the distance a current carrier travels to the junction during carrier life.

diffusion process 1. The technique of processing semiconductor devices by diffusion (see *diffusion, 1*). 2. Producing a high vacuum by means of diffusion (see *diffusion pump*).

diffusion pump A pump for fast, efficient creation of a high vacuum in electron tubes and similar devices. In one form, the pump, in conjunction with a force pump, uses mercury vapor as the pumped medium. Gas molecules evacuated from the device diffuse into a chamber where condensing mercury vapor traps and carries them off.

diffusion transistor A transistor whose operation is based principally on the diffusion of current carriers (see *diffusion, 2*).

dig-in angle A stylus angle of 90° to the surface of a phono disk. Compare *drag angle*.

DIGIRALT Abbreviation for *DIGItal Radar ALTmetry*. A system that utilizes digital techniques to enhance the accuracy of an altimeter employing radar.

digit A single symbol in a numbering system (e.g., 0 through 9 in the decimal system, or 0 or 1 in the binary system) whose value is dependent on its position in a group and on the radix of the particular system used.

digital Descriptive of that which uses signals representing characters or numbers, the signals being of discrete rather than continuously variable values, or those produced by pulses of one current or voltage value.

digital annunciator An annunciator which gives an alphanumeric digital display of information, as well as sounding an alarm.

digital barometer An electronic barometer providing a digital readout.

digital capacitance meter Abbreviation DCM. A microfarad meter with a digital readout for measured capacitance values.

digital circuit A circuit affording a dual-state switching operation, i.e., on or off, high or low, etc. Also called *binary circuit*.

digital comunications Radio or wire communications using a dual-state mechanism (on—off, positive—negative, modulated—unmodulated) to represent information.

digital comparator A comparator which presents two digital values, one for each of the quantities being compared.

digital computer A high-speed, electronic machine for performing mathematical operations, file management, machine control, or other "intelligent" functions, and whose basic internal operations (data storage, comparing, and computaion) are based on semiconductor devices assuming one of two states (on or off, high or low). Compare *analog computer*.

digital data Information represented and processed in the form of combinations of digits (8 and 1 in the binary system).

digital-data cable A cable specially designed to conduct high-speed digital pulses with minimal distortion and loss.

digital data-handling system A system which accepts, sorts, modifies, classifies, records, or otherwise handles *digital data*, displaying the final result of passing the data to a computer.

digital device 1. A digital integrated circuit. 2. Any circuit or system that operates by digital means.

digital differential analyzer Abbreviation, DDA[6] A digital computer that can perform integration using specialized circuitry.

digital display A presentation of information (such as the answer to a problem) in the form of actual digits, as opposed to one in the form of, say, a meter deflection. See, for example, *digital-type meter*.

digital electrometer An electrometer having a digital current or voltage indicator.

digital electronics Electronic techniques based upon pulse-type signals, as opposed to uniformly varying, continuous signals. Compare *analog electronics*.

digital frequency meter A direct-reading frequency meter employing high-speed electronic switching circuits and a digital readout. Such instruments read frequency from less than 1 Hz to many GHz.

digital divider In a computer, a device that can divide, i.e., provide a quotient and remainder using dividend and divisor signals.

digital HIC A hybrid integrated circuit designed for digital applications. Also see *digital integrated circuit*.

digital IC See *digital integrated circuit*.

digital incremental plotter A device that can draw, according to signals received from a computer, graphs depicting solutions to problems.

digital indicator See *readout lamp*.

digital information See *digital data*.

digital information display See *digital display*.

digital integrated circuit An integrated circuit for on—off operations, such as switching, gating, and so on.

digital integrator A device that can perform integration, in which increments in input variables (x and y) and an output variable (z) are represented by digital signals.

digital logic A form of Boolean algebra, consisting of negation, conjunction, and disjunction, in which the binary digit 1 has the value "true" and 0 the value "false" (in positive logic) or vice-versa (in negative logic). Digital logic is the basis by which all digital devices function.

digital-logic module 1. A circuit that performs digital operations. 2. A logic gate.

digital multimeter Abbreviation, dmm. A voltohm-milliammeter giving a digital readout of measured values.

digital multiplier In a digital computer, a device that can multiply, i.e., it produces a product signal from multiplier and multiplicand signals.

digital output An output signal of digital pulses representing a number equal or proportional to the value of a corresponding input signal.

digital panel meter A readout-display meter whose relatively small size allows mounting it on a panel.

digital phase shifter A phase shifter actuated by a digital control signal.

digital photometer An electronic photometer giving a digital readout of illumination values.

digital power meter An electronic wattmeter providing a digital readout of measured power.

digital readout An indicating device tha displays a sequence of lamps or numerals that represent a measured value. Also see *digital meter* and *readout lamp*.

digital recording A system for tape-recording data which is to be processed by a digital computer. A varying quantity is sampled and converted into bits, which are then recorded on magnetic tape.

digital representation The use of digital signals (see *digital*) to represent information as characters or numbers.

digital rotary transducer A device which delivers a digital output signal proportional to the rotation of a shaft.

digital shaft-position indicator See *digital rotary transducer*.

digital signal A signal made up of a series of digital pulses. Also see *digital*.

digital sound Sound recording and reproduction accomplished with digital pulses instead of sinusoidal signals. One of the advantages of such a system is wider frequency response than that obtained with analog methods. Another is greater dynamic range.

digital speech communications A system of voice communications, in which the analog voice signal is encoded into digital pulses at the transmitter, and decoded at the receiver.

digital subtractor In a compuer, a device tha can perform subtraction, i.e., it produces a difference signal from sudtrahend and minuend signals.

digital television 1. A television system in which the picture information is encoded into digital form at the transmitter, and decoded at the receiver. 2. A form of television picture transmission that functions according to picture motion rather than absolute brightness.

digital temperature indicator See *digital thermometer*.

digital thermometer An electronic thermometer giving a digital readout of temperature.

digital-to-analog conversion The conversion of a digital quantity into an analog presentation, such as shown by a performance curve. Compare *analog-to-digital conversion*.

digital-to-analog converter A circuit or device that performs digital-to-analog conversion. Compare *analog-to-digital converter*.

digital transmission 1. A method of signal transmission in which the modulation occurs in defined increments, rather than over a continuous range. 2. A message that is sent in digital form.

digital-type meter An indicating instrument in which a row of numeral indicators (such as readout lamps, *q.v.*) displays a value. Compare *analog-type meter*.

digital voltmeter Abbreviation, dvm. An electronic voltmeter having a digital readout rather than a meter.

digital voltohmmilliammeter Abbreviation, dvom. See *digital multimeter*.

digital wattmeter See *digital power meter*.

digital compression In digital computer operation, the process of representing data with an economy of characters to reduce file size.

digit current In digital computer practice, the current associated with writing or reading a digit into or out of a memory cell.

digit delay element A logic element (gate) whose output signal lags the input signal by one digit period.

digit filter A device for detecting designations. See *designation*.

digitize 1. To express the results of an analog measurement in digital units. 2. To convert an analog signal into corresponding digital pulses.

digitizer See *analog-to-digital converter*.

digit period As determined by the pulse repetition rate of a computer, the duration of consecutive digital signals in a series.

digit place See *digit position*.

digit plane In a matrix-type computer memory, the plane within a three-dimensional array of memory storage elements representing a digit position. Also see *digit position*.

digit position The ordinal position of a digit in a number, the first place being occupied by the least significant digit; e.g., 7 is the third digit in 756,

digit pulse A pulse that energizes magnetic core memory elements representing a digit position in several words.

digitron A display in which all of the characters lie in a single, flat plane.

digit selector In a machine that processes punched cards, a device which, upon activation by card designations (see *designation*), instigates the operations needed for the card type.

digit time The duration of a digit signal in a series of signals.

digit-transfer bus In a digital computer, a main line (of conductors) that transfers information among various registers; it also does not handle control signals.

diheptal crt base The 14-pin base of a cathode-ray tube. Also see *bidecal*, *duodecal*, and *magnal*.

DIIC Abbreviation for dielectric-isolated integrated circuit. Several separate integrated-circuit wafers are contained in a single package, and kept electrically insulated by layers of dielectric.

dialtometer An instrument used to measure expansion.

dimensional analysis Analysis in which all terms represent acual physical quantities. Also see *dimensionless quantity* and *physical quantity*.

dimensional ratio In magnetic studies, the ratio of the longest diameter of an elongated ellipsoid of revolution to the shortest.

dimensional stability Nonvariance or little variance in the shape and size of a medium (such as film) during the processing of that material.

dimensionless quantity A quantity which, rather than being physical, is merely a number. Example: logarithm, exponent, numerical ratio, etc. In contrast are the *physical quantities*: 3 volts, 5000 hertz, 10 amperes, etc.

diminished radix complement See *complement*.

dimmer An electronic device usually for dimming incandescent lamps. Employing amplified control, the device enables high-wattage lamp loads to be smoothly controlled with a small, light-duty (often, volume-control-type) rheostat, or potentiometer. A photoelectric-type dimmer automatically controls lamps in accordance with the amount of daylight.

dimmer curve The function of a light-dimmer voltage output as a function of setting on a linear scale.

DIN Abbreviation for *Deutsche Industrie Normenausschuss*. A German association that sets standards for the manufacture and performance of electrical and electronic equipment, as well as other devices.

D-indicator In radar practice, an indicator combining type B and C indicators (see *B-display* and *C-display*).

Dingley induction-type landing system An aircraft landing system providing lateral and vertical guidance; instead of radio it employs the magnetic field surrounding two horizontal cables laid on or under either side of the runway.

diode A device containing an anode and a cathode (as a tube) or a *pn* junction (as a semiconductor device) as principal elements and providing unidirectional conduction.

diode action 1. The characteristic behavior of a diode, i.e., rectification and unidirectional conduction. 2. Two-electrode rectification or unidirectional conductivity in any device other than a diode (e.g,, rectification between the grid and cathode of a triode, or asymmetrical conductivity between the collector and base of a transistor).

diode amplifier 1. A parametric amplifier employing a varactor. 2. An amplifier utilizing hole-storage effects in a semiconductor diode (see *crystal amplifier, 2*). 3. A negative-resistance amplifier employing a tunnel diode.

diode array A combination of several diodes in a single housing.

diode assembly See *diode array*.

diode bend Triode plate-current saturation occurring at the point where positive grid voltage equals plate voltage. Diode bend occurs before normal plate-current saturation when the plate voltage is lower than that required for full cathode emission.

diode bias 1. In a vacuum tube, dc bias resulting from diode rectification in the control-grid—cathode circuit. 2. A steady dc voltage applied to a diode to establish its operating point.

diode capacitance The total capacitance shunting a diode. In a semiconductor diode, the capacitance between terminals and electrodes and the internal, voltage variable capacitance of the junction. In a tube diode, the capacitance between terminals and the internal plate—cathode capacitance.

diode capacitor 1. A capacitor normally operated with a diode. 2. A voltage-variable capacitor utilizing the junction capacitance of a semiconductor diode.

diode—capacitor memory cell A high-value capacitor in series with a high-back-resistance semiconductor diode. An information pulse forward-biases the diode and charges the capacitor, which remains charged (holding an information bit) because of the long time constant of the high capacitance and the high back resistance of the diode.

diode characteristic 1. The current—voltage curve of a diode (tube or semiconductor). 2. In tube testing, the current—voltage characteristic when all electrodes except the cathode are connected to act as the plate of an equivalent diode; the technique provides a rough measure of tube condition.

diode checker An instrument for testing semiconductor diodes. There are two forms: A static checker, which measures dc forward and reverse current; and a dynamic checker (see *dynamic diode tester*), which displays the entire diode response curve on an oscilloscope screen.

diode chopper A chopper employing an alternately biased diode as the switching element.

diode clamping Clamping achieved with diodes. Also see *clamping diode*.

diode clipper A clipper employing one or more diodes. A single, dc-biased diode will limit the positive or negative peak of an applied ac voltage, depending on diode polarity and bias; two biased diodes with opposing polarity will clip both peaks. Also see *limiter*.

diode converter See *diode mixer*.

diode current 1. The forward or reverse current of a diode. 2. In a tube, current flowing in the grid—cathode circuit as a result of diode action between the elements when the grid signal voltage is positive.

diode current meter A dc milliammeter or microammeter with a semiconductor-diode rectifier for measuring ac.

diode curve changer A diode or network of diodes employed to make a linear current—voltage curve take some other shape.

diode demodulator See *demodulator probe* and *diode detector*.

diode detector A detector circuit in which a diode (semiconductor or tube) demodulates the signal. The diode, a simple device, provides linear response at high signal amplitudes, but affords no amplification.

diode equation The equation for semiconductor diode current:

$$Id = Is \left[1\mathrm{n} \left(eV/kT \right) - 1 \right]$$

where Id = total diode current (amperes); Is reverse saturation current (A); e, charge of electron (1.6×10^{-19}C); V, anode-to-cathode potential (V); k, the Boltzman constant (1.38×10^{-23} J/K); and T, absolute temperature of diode (°K).

diode feedback rectifier 1. In a rectified carrier, negative feedback system for an AM transmitter, the diode that rectifies the modulated carrier and provides the audio envelope (as diode output) for use as negative-feedback voltage, which is applied to the speech-amplifier—modulator channel to reduce distortion, noise, and hum, at the same time providing automatic modulation control. 2. The diode that rectifies a part of the signal at the output of an audio amplifier and provides a proportional dc voltage for use as agc bias.

diode field strength meter A radio field-strength meter. The circuit consists essentially of an *LC* tuned circuit, diode detector, and dc milliammeter or microammeter. The deflection of the meter is roughly proportional to the rf signal voltage.

diode gate A passive switching circuit of biased diodes. Also see *AND circuit* and *OR circuit*.

diode impedance The vector sum (resultant) of the resistive and reactive components of a diode. In a semiconductor diode, the inductive component of reactance is almost entirely the inductance of leads and electrodes, whereas the capacitive component of reactance is the shunting capacitance between leads and electrodes plus the voltage-variable capacitance of the junction. The resistive component is almost entirely the voltage-variable resistance of the junction.

diode isolation A means of insulating an integrated—circuit chip from its substrate. The chip is surrounded by a pn junction that is reverse-biased. This prevents conduction between the chip and the substrate.

diode lamp See *laser diode*.

diode laser See *laser diode*.

diode light source See *laser diode*.

diode limiter See *diode clipper*.

diode line In tube-characteristic curves, the line showing the point(s) where the positively driven grid loses control, allowing grid current to flow as a result of equivalent diode rectification.

diode load 1. The current drawn from a diode acting as a rectifier or demodulator. 2. The output (load) resistor into which a diode operates.

diode load resistance The required value for a diode load resistor.

diode load resistor A resistor usually operated at the output of a diode rectifier or diode detector. Also see *diode, 2*.

diode logic Logic circuitry, such as AND and OR circuits, employing diodes as the principal components.

diode matrix In counters and computers, an array (horizontal and vertical) of wires, the intersections of some being interconnected through diodes whose polarities determine circuit operation. A series of AND circuits is provided by this arrangement, which acts as a high-speed rotary switch when it is supplied with input pulses, the output being switched successively to the various vertical wires.

diode mixer A signal mixer (see *mixing*) which operates by virtue of the nonlinearity of a (usually forward-biased) semiconductor diode.

diode noise limiter A noise limiter circuit having one or more biased diodes.

diode oscillator 1. An oscillator based on the negative resistance or breakdown characteristics of certain diodes, such as high-reverse-biased germanium diodes, tunnel diodes, four-layer diodes, and neon bulbs. 2. An oscillator exploiting the negative plate-to-cathode resistance of a tube diode operated at ultrahigh frequencies.

diode pack A device containing more than one diode. An example is the full-wave bridge-rectifier integrated circuit.

diode peak detector A diode detector whose load resistance is high at modulation frequencies; the voltage across the resistance is proportional to the modulation envelope (peak value of the modulated carrier).

diode peak voltmeter A diode-type ac vtvm (or semiconductor version of the instrument) in which the deflection of the dc milliammeter or microammeter is proportional to the peak value of the applied ac voltage.

diode—pentode A tube combining diode and pentode sections in a single envelope; a 1S5, for example.

diode probe A test probe containing a diode employed as either a rectifier or demodulator.

diode pulse amplifier See *crystal amplifier, 2*.

diode recovery time The interval during which relatively high current continues to flow after the voltage across a semiconductor junction has been abruptly switched from forward to reverse. Recovery time is attributable to *diode storage*.

diode rectification ac to dc conversion by diode action (see *diode action, 1*).

diode rectifier 1. A small-signal diode rectifier in a light-duty power supply. 2. A small-signal diode rectifier in the automatic-gain-control circuit of a superheterodyne receiver. Also called *agc rectifier* or *avc rectifier*. 3. Generically, a semiconductor diode rectifier, as opposed to a tube-type rectifier.

diode—remote-cutoff pentode A tube combining a diode section and a remote-cutoff pentode section in a single envelope, e.g., a 6CR6.

diode resistor 1. A resistor usually operated with a diode. 2. A voltage-variable resistor utilizing the (usually forward) resistance of a semiconductor diode.

diode—sharp-cutoff pentode A tube combining a diode section and a sharp-cutoff pentode section in a single envelope. Example: 6BY8.

diode storage The carriers remaining within a pn junction for a short time after the forward bias voltage they have been injected with has been either removed or switched to reverse polarity.

diode storage time See *diode recovery time*.

diode switch See *diode gate*.

diode sync separator A diode employed in a TV receiver circuit to separate and deliver the sync pulses from the composite video signal.

diode temperature stabilization 1. Keeping the temperature of a diode at a constant level. 2. Using the temperature—resistance characteristic of a forward-biased semiconductor diode to stabilize a circuit (such as a transistor amplifier stage), i.e., to prevent variations due to temperature change.

diode tester See *diode checker*.

diode—tetrode A combination tube consisting of a diode section and a tetrode section; e.g., a 12EM6.

diode transistor 1. Unijunction transistor. 2. A semiconductor diode whose operation simulates that of a transistor (by means of pulsed operation that alternately makes the single junction an emitter or collector). See *crystal amplifier, 2*. 3. A transistor connected to operate solely as a diode.

diode—transistor logic Abbreiation, DTL. Logic circuitry in which a diode is the logic element and a transistor acts as an inverting amplifier.

diode—triode A combination tube consisting of one or more diode sections and a triode section. Example: 12FK6.

diode tube See *tube diode*.

diode-type meter A rectifier-type ac meter consisting of a semiconductor diode (usually point-contact type) or a combination of diodes and a dc milliammeter or microammeter. The diode rectifies the ac input, the resulting dc deflecting the meter.

diode vacuum tube A vacuum tube having two elements: cathode and plate (anode). Also see *diode tube*.

diode varactor A conventional semiconductor diode or rectifier employed as a makeshift varactor (voltage-variable capacitor).

diode variable resistor See *diode varistor*.

diode varistor A conventional diode employed as a makeshift varistor (voltage-variable resistor).

diode voltage reference See *zener voltage reference*.

diode voltage regulator See *zener voltage regulator*.

diode vtvm An ac vacuum-tube voltmeter consisting principally of a diode rectifier followed by a dc milliammeter or microammeter. Meter deflection is proportional to the average value of the applied ac.

Diophantine analysis In number theory, a method (using squares) of obtaining rational solutions for indeterminate algebraic equations and problems. Also see *Diophantine equations*.

Diophantine equations Equations of the form $f(x, y, z...) = 0$ in which x, y, z... are rational numbers and f is a polynomial.

diopter 1. A unit of optical refractive power; it is equal to f^{-1} in meters, where f is focal length. 2. A lens placed in front of another (as on a camera) to increase the latter's effective focal length.

dioptrics The branch of physics concerned with the refraction of light by lenses.

DIP Abbreviation of *dual-inline package*, an integrated circuit contained within a standard housing characterized by its low profile, rectangular body, and symmetrical placement of leads along two opposing sides of the device. Also see *dual-inline package*.

dip 1. Inclination. 2. A distinct decrease in the value of a varying quantity, followed by an increase, e.g., the dip in plate current when a crystal oscillator is tuned to resonance.

dIp Symbol for *differential of plate current*.

dip adapter An external accessory which allows an rf signal generator to be used as a dip meter.

dip coating 1. Applying a protective coat of insulating material to a conductor or component by dipping it into the liquid material, then draining and drying it. Compare *spray coating*. 2. The coat applied in this way.

dip encapsulation Imbedding a component or circuit in a protective block of insulating material (such as a plastic) while the material is in a liquid state, and then allowing the material to harden in ambient air or in an oven.

dip impregnation Saturating a component or material (such as absorbent film) with a substance (such as oil or wax) by dipping or vacuum forcing.

diplexer A coupler which permits several transmitters to operate simultaneously into a single antenna.

diplex operation 1. Simultaneous transmission or reception of two signals using a single antenna. 2. Simultaneous transmission or reception of two signals on a single carrier.

diplex reception Reception in diplex operation.

diplex transmission Transmission in diplex operation.

dip meter A tunable rf instrument which, by means of a sharp dip of an indicating meter, indicates resonance with an external circuit under test. Specific names are derived from the active component used: *grid-dip meter, gate-dip meter*, etc.

dip needle See *inclinometer*.

dipolar Possessing two poles. (Also *bipolar*.)

dipolarization See *depolarization*.

dipole 1. A molecule having an electronic moment, i.e., one in which the center of the negative charges is not also the center of the positive charges. 2. Dipole antenna.

dipole antenna A fundamental form of antenna from which many others are derived. It consists of a center-fed single wire a half-wavelength long (at the operating frequency) that is suitably corrected for end effects.

dipole disk feed A method of coupling radio-frequency energy to a disk-shaped antenna. The energy is applied to a dipole located adjacent to the disk.

dipole feed A method of coupling radio-frequency energy to an antenna by means of a half-wave dipole. The dipole is directly fed by the transmission line, and the dipole radiates energy to the rest of the system.

dip oscillator See *dip meter*.

dipotassium tartrate Abbreviation, DKT. An organic piezoelectric material.

dipped component A device (coil, capacitor, resistor, etc.) which has been given a protective coat by dipping it into a suitable material (such as oil, varnish, or wax) and draining off the surplus.

dipper Collective term for resonancetype instruments, such as a *dip-meter* or *dip adapter*.

dipper interrupter A cyclic switching device in which a contact pin is part of a revolving wheel partially immersed in mercury.

dipping The application of a protective coating or impregnant to a component by dipping it into a suitable material. Also see *dip coating, dip encapsulation*, and *dip impregnation*.

dipping needle See *inclinometer*.

dip soldering 1. Soldering leads or terminals by dipping them into molten solder and then removing excess solder. 2. Tinning printed-circuit patterns by dipping the pc boards into molten solder or placing them in contact with the surface of a solder bath. Also, soldering leads in printed circuits by the same methods.

DIP switch A switch (or group of miniature switches) mounted in a *dual-inline package* for easy insertion into an integrated-circuit socket or printed-circuit board.

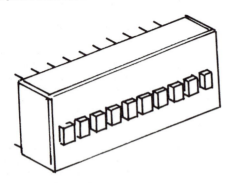

DIP Switch.

direct-access storage device A computer memory device with which data access time is unaffected by the data's location. Also called *random-access storage device*.

direct-acting recorder See *graphic recorder*.

direct-acting recording instrument See *graphic recorder*.

direct address The *actual* address of a computer storage location, i.e., the one designated by machine code. Also called *absolute* or *real* address.

direct capacitance The capacitance between two points, such as two terminals, as opposed to the subsidiary capacitance also associated with those points; i.e., each terminal also has capacitance to ground and, perhaps, to a nearby object.

direct allocation In digital computer practice, to specify the necessary memory locations and peripherals for a particular program when it is written.

direct coding Computer programming in machine language.

direct control In a computer system, control of one machine by another.

direct-conversion receiver A heterodyne receiver in which the incoming rf signal beats with the rf output of a local oscillator, producing an af beat note, which is then amplified by an af amplifier that can provide high gain and (if tuned) selectivity. Although the direct-conversion receiver somewhat resembles the superheterodyne type, it doesn't have an i-f amplifier and second detector. Also see *zero-beat reception*.

direct-coupled amplifier An amplifier in which the output circuit of one stage is wired directly to the input circuit of the following stage; i.e., there is no intervening capacitor or transformer. Such an amplifier can handle ac or dc signals and has wide frequency response.

direct-coupled transistor logic Abbreviation, DCTL. In digital computer and switching circuit practice, a logic system employing only direct-coupled transistor stages.

direct coupling Direct connection of one circuit point to another for signal transmission, i.e., without intermediate capacitors or transformers. Because coupling devices aren't used, direct coupling provides transmission of dc as well as ac.

direct current Abbreviation, dc, a current which always flows in one direction, e.g., the current delivered by a battery. Also see *current*.

direct-current amplifier An amplifier for boosting direct-current signalsn as opposed to dc voltage signals. Also see *dc amplifier, 2*.

direct-current bar See *dc bar*.

direct-current beta See *dc beta*.

direct-current block See *dc block*.

direct-current bus See *dc bus*.

direct-current circuit breaker See *dc circuit breaker*.

direct-current component See *dc component*.

direct-current converter See *dc converter*.

direct-current coupling See *dc coupling*.

direct-current dump See *dc dump*.

direct-current equipment See *dc equipment*.

direct-current erase head See *dc erase head*.

direct-current generator See *dc generator*.

direct-current inverter See *dc inverter*.

direct-current leakage See *dc leakage*.

direct-current motor See *dc motor*.

direct-current noise See *dc noise*.

direct-current power See *dc power*.

direct-current relay See *dc relay*.

direct-current resistance See *dc resistance*.

direct-current shift See *dc shift*.

direct-current short See *dc short*.

direct-current signaling See *dc signaling*.

direct-current source See *dc source*.

direct-current thyratron See *dc thyratron.*

direct-current transducer See *dc transducer.*

direct-current transformer See *dc transformer.*

direct-current transmission See *dc transmission.*

direct digital control In a digital computer, multiplexing or time sharing among a number of controlled loops.

direct display unit A crt peripheral that displays data recalled from memory.

direct-distance A form of telephone service that allows direct dialing of long-distance numbers.

direct drive Pertaining to electromechanical accessories for electronic equipment, the transmission of power directly from a source (such as a motor) to a driven device without intermediate gears, belts, or clutches.

direct-drive torque motor In a position.ng or speed-control system, a servoactuator connected directly to the driven load.

direct-drive tuning A tuning (or adjusting) mechanism in which the shaft of the variable component (such as a potentiometer or variable capacitor) is turned directly by a knob, i.e., without gearing, dial cable, or similar linkage.

directed number A number having direction as well as magnitude. A vector quantity. Examples: $+5$, -3, $0.5 + j4$.

direct electromotive force A dc voltage that doesn't fluctuate or does so negligibly.

direct emf See *direct electromotive force.*

direct grid bias See *dc grid bias.*

direct ground A ground connection made by the shortest practicable route. Compare *indirect ground.*

direct illumination See *direct light.*

direct induced current A transient current induced in the same direction as the induction current when it is interrupted.

directing antenna See *directional antenna.*

direct-input circuit A circuit, especially an amplifier, whose input terminal is wired directly to the input electrode of the input device (tube, transistor), i.e., without a coupling capacitor or transformer.

direct-insert subroutine In digital computer practice, a subroutine which is directly inserted into a larger instruction sequence, and which must be rewritten at every point it is needed.

direct instruction A computer program instruction that indicates the location of an operand in memory that is to be operated upon.

direction The position of one point with respect to another from which it is viewed.

directional 1. Depending on the direction or orientation. 2. Having a concentration in an identifiable direction. 3. A form of transducer in which radiation, or sensitivity, is concentrated in certain directions at the expense of radiation or sensitivity in other directions.

directional antenna An antenna which transmits and receives signals in a general direction. Also called *beam, beam antenna,* and *directive antenna.*

directional array 1. A directional antenna having a set of elements assembled in such a way that their combined action shapes the radiation into a unidirectional pattern. 2. A group of antennas spaced and phased to produce unidirectional radiation and reception patterns.

directional beam 1. An antenna whose radiation or reception pattern strongly favors a direction. 2. The radiation or reception pattern of such an antenna.

directional characteristic The precise directional properties of an antenna or transducer.

directional CQ In amateur radio, a special CQ call that invites reply only from stations in a certain direction or in a particular city, state, or country.

directional coupler A microwave device which couples an external system to waves traveling through the coupler in one direction.

directional diode A high-backresistance semiconductor diode inserted into a dc signal circuit or control circuit to permit unidirectional current flow (the direction in which the diode is forward biased by the current).

directional filter In carrier-current transmission, a filter which halves the frequency band, one half being for east—west transmission, and the other for west—east transmission.

directional gain Symbol, ks. The ratio (in dB) of the power which would be radiated by a loudspeaker if the free-space axial sound pressure were constant over a sphere, to the actual radiated power.

$$Ks = 10[\log 10\ 2\pi/\int_0^p \int_0^{2\pi}\ (p/pa^2\ \sin\theta d\theta]$$

where Ks is the directional gain (directivity index) in dB; pa, the axial free-space sound pressure (dynes/cm^2) at the same distance and for pa; and θ and ϕ the angular polar coordinates of the system with the speaker axis at $\theta = 0$.

directional homing The pursuit of a path in a manner which maintains the target or guiding station at a constant relative bearing.

directional horn See *directive horn.*

directional hydrophone A hydrophone whose response pattern strongly favors one direction.

directional lobe In the spatial response pattern of a device, such as an antenna or loudspeaker, a portion showing emphasized response in a given direction.

directional microphone A microphone which strongly favors sound coming from the front.

directional pattern See *directivity pattern.*

directional phase shifter A phase-shifting circuit in which the characteristics are different in one direction as compared with the other direction.

directional power relay A relay which is actuated when the monitored power reaches a prescribed level in a given direction.

directional relay See *polarized relay.*

directional response For any form of transducer, a radiation or sensitivity pattern that is concentrated in certain directions.

directional separation filter See *directional filter.*

directional variation of radio waves. Variations in the field strength of radio waves in various directions. These changes seem to have a variety of causes, including antenna directivity, ground characteristics, ionospheric factors, the presence of interfering objects, and so forth.

direction angle In radar practice, the angle between the center of the antenna baseline and a line going to the target.

direction cosine The cosine of the direction angle.

direction finder A receiver specially adapted to show the direction from which a signal is received, thus revealing the direction of the receiver with respect to the transmitting station, and vice versa. In its simplest form, a direction finder is a radio receiver with a loop antenna which is rotatable over map or compass card. For increased accuracy, checks are made with signals from two transmitting stations; the exact location of the receiver is pinpointed by triangulation.

direction finding The art and science of taking bearings by means of a direction finder.

direction of current flow Electron drift from a negative point to a positive point. Because the drift is the basis of current flow, current is said to flow from negative to positive.

direction of lay In a multistrand cable, the lateral direction of winding of the top strands—called *left-hand lay* or *right-hand lay*—away from the observer.

direction of polarization The direction of the electrostatic field in a linearly polarized wave.

direction of propagation The direction in which energy moves from a transmitter or between equivalent points in a sector of space under consideration.

direction rectifier In a control system, a rectifier whose dc output voltage has a magnitude and polarity dependent upon the magnitude and relative polarity of an ac selsyn error voltage.

directive In a computer source program, a statement directing the compiler in translating the program into machine language without being translated itself. Also called *control statement*.

directive antenna An antenna designed for best propagation or reception in the direction of one of its horizontal dimensions. Also called *beam antenna* and *directional antenna*.

directive gain For a directive antenna, a rating equal to $4\pi\,(Pr/Pt)$, where Pr is the radiated power per steradian in a given direction and Pt is the total radiated power.

directive horn A microwave antenna having the shape of a (usually rectangular) horn.

RECTANGULAR   PYRAMIDAL   CONICAL

Directive horn.

directivity 1. In an antenna, a directional response. 2. The degree to which the radiation or sensitivity of a transducer is concentrated in certain directions. 3. The angle between the half-power points of a directive antenna in the azimuth plane. 4. In an antenna system, the ratio, in decibels, between the power in the favored direction and the power in the exact opposite direction; also called front-to-back ratio. 5. The forward power gain of an antenna, with respect to a dipole in free space.

directivity diagram The illustrative response pattern of a beam antenna or other directional device. Also see *directivity pattern*.

directivity factor 1. A measure of the directivity of an antenna or transducer. 2. In acoustics, the directivity factor is the ratio, in decibels, between the gain in the maximum direction and the gain in the minimum direction, for a transducer such as a speaker or microphone.

directivity index For a sound emitting transducer, the ratio (in dB) of $E1$ to $E2$, where $E1$ is the average intensity over an entire sphere surrounding the transducer and $E2$ is the intensity on the acoustic axis. For an acoustic pickup transducer, $E1$ is the average response over the entire sphere; $E2$ the response on the acoustic axis.

directivity of antenna For a beam antenna, the ratio $Emax/Eavg$, where $Emax$ is the maximum field intensity at a selected distance from the antenna and $Eavg$ is the average field intensity at the same point.

directivity of directional coupler The ratio (in dB) of $P1$ to $P2$, where $P1$ is the power at the forward-wave-sampling terminals (measured with a forward wave in the transmission line) and $P2$ is the power at the terminals when the forward wave is reversed in direction.

directivity pattern The calculated or measured response pattern (transmission or reception) of an antenna, microphone, loudspeaker, or similar device, with particular attention to the directional features of the pattern.

Directivity pattern.

directivity signal A spurious output signal due to finite directivity in a coupler.

direct light Light rays traveling directly from a source to a receptor or target without reflection. Compare *indirect light*.

direct lighting A form of lighting in which the source of illumination shines directly on the objects to be illuminated, without intermediate reflectors.

directly grounded Connected to earth or to the lowest-potential point in a circuit by a direct ground. Compare *indirectly grounded*.

directly heated cathode A tube filament. It is so called because, when heated, it becomes the cathode of the tube, i.e., the emitter of electrons. Compare *indirectly heated cathode*.

directly heated thermistor A thermistor whose temperature changes with the surrounding temperature and the dissipation of power by the thermistor. Compare *indirectly heated thermistor*.

directly heated thermocouple A meter thermocouple heated directly by signal currents passing through it. Compare *indirectly heated thermocouple*.

direct measurement Immediate measurement of a quantity, rather than determining the value of the quantity through adjustments of a measuring device, e.g., measuring capacitance with a microfarad meter, rather than with a bridge. Compare *indirect measurement*.

direct memory access The transfer of data from a computer memory to some other location, without the intervention of the central processing unit (*CPU*). Abbreviated *DMA*.

direct numerical control In a computer or data system, the capability for distributing information among numerically-controlled machines whenever desired.

director In a multielement directive antenna, an element that is mounted in front of the radiator element and phased and spaced to direct the radiation forward. The director functions in conjunction with the reflector element, which is mounted behind the radiator.

directory See *dictionary*.

direct pickup Televising events as they happen.

direct piezoelectricity The production of a piezoelectric voltage by mechanically stressing a suitable crystal.

direct playback In sound reproduction, directly reproducing an actual recording with no additional processing of the latter. Example: playing an original recorded phonograph disk rather than a disk massproduced from the original recording.

direct-point repeater A relay-operated telegraph repeater. The received signals actuate the relay, which switches the second line.

direct-radiator loudspeaker A loudspeaker whose cone or diaphragm is directly coupled to the air.

direct ray A ray (or wave) which reaches a receiver directly without reflection, refraction, or (ideally) without encountering obstructions.

direct recording 1. A record produced by a graphic recorder. 2. The technique of producing such a record.

direct-recording instrument A device, such as a graphic recorder, which directly produces a permanent record (such as an inked trace) of the variations of a quantity.

direct resistance coupling A form of coupling in which the collector, drain, or plate of the first amplifying device is connected through a resistor directly to the base, gate, or grid of the second device. The resistance value may vary; sometimes the connection is a direct short circuit.

directrix A fixed line to which a curve is referred, e.g., the directrix of a parabola.

direct scanning In television, the sequential viewing of parts of a scene by the camera (even though the entire scene is continuously illuminated).

direct serial file organization A technique of organizing files stored in a direct access device, in which a record can be chosen by number and amended where it is without altering other members of the file.

direct sound wave A sound wave arriving directly from its source, especially a wave within an enclosure that hasn't been affected by reflection.

direct substitution 1. An exact component replacement. 2. Installing an exact component replacement.

direct synthesizer A device for producing random, rapidly changing frequencies for security purposes. A reference oscillator provides a comparison frequency; the output frequency is a rational-number multiple of this reference frequency.

direct voltage See *dc voltage.*

direct wave A wave that travels from a transmitter to a receiver without being reflected by the ionosphere or the ground. Compare *skywave.*

direct Wiedemann effect Twisting that takes place in a wire carrying current in a longitudinal magnetic field and resulting from the helical resultant of the longitudinal field and the circular field around the wire.

direct-wire circuit A communications or control line of wires connecting a transmitter (or control point) and a receiver (or controlled point) without an intermediary, such as a switchboard.

direct-writing recorder See *graphic recorder.*

direct-writing telegraph 1. Printing telegraph. 2. Telautograph.

dis A prefix meaning *deprived of.* For the formation of electronic terms, the prefix must be distinguished from *un,* meaning *not.* Thus, a *discharged* body is one which was charged, but has been emptied of its charge. An uncharged body is one that ordinarily or presently is not charged. The same distinction applies to *disconnected* and *unconnected.*

disable In digital computer practice, to defeat a software or hardware function.

disc See *disk.*

discharge The emptying or draining of electricity from a source, such as a battery or capacitor. The term also denotes a sudden, heavy flow of current, as in *disruptive discharge.* Compare *charge.*

discharge current 1. Current flowing out of a capacitor. 2. Current flowing out of a cell, especially a storage cell. Compare *charging current.*

discharge key See *discharge switch.*

discharge lamp A gas-filled tube or globe in which light is produced by ionization of the gas between electrodes. Familiar examples are the neon bulb and fluorescent tube.

discharge phenomena The effects associated with electrical discharges in gases, such as luminous glow.

discharge potential See *ionization potential.*

discharger 1. A short-circuiting tool for discharging capacitors. 2. A spark gap or other device for automatically discharging an overloaded capacitor.

discharge rate The current which can be supplied by a battery reliably over a given period. Expressed in milliamperes, amperes, milliamperehours, or ampere-hours.

discharge switch A switch for connecting a charged capacitor to a resistor or other load through which the capacitor discharges. In some circuits, when the switch is in its resting position it connects the capacitor to the charging source.

discharge tube See *gas diode.*

discharge voltage See *ionization potential.*

discharging tongs See *discharger, 1.*

discone antenna An antenna comprising a metal disk attached to the apex of an up-pointing metal cone. The disk is parallel to the base of the cone. The discone has an omnidirectional radiation pattern.

disconnect 1. To separate leads or connections, thereby interrupting a circuit. 2. A type of connector whose halves may be pulled apart to open a cable or other circuit quickly. 3. Disconnect switch.

disconnector See *disconnect, 2* and *disconnect switch.*

disconnect signal A signal sent over a telephone line, ending the connection.

disconnect switch A switch whose main function is to open a circuit quickly (either manually or automatically) in the event of an overload.

discontinuity 1. A break in a conductor. 2. Discontinuous function.

discontinuous function A function which is not continuous over the entire range of an independent variable, but which breaks at some value in the range. Thus, the function $y = \tan x$ shows discontinuity at the point $x = \pi/2$, where it is one and thereafter approaches zero.

discontinuous wave trains See *damped waves.*

discrete 1. Complete and self-contained, as opposed to a part of something else. 2. Composed of individual, separate members.

discrete capacitor Capacitance which is entirely self-contained, rather than being electrically spread out. Also called *lumped capacitor.* Compare *distributed capacitance.*

discrete circuit A circuit comprised of discrete components, such as resistors, capacitors, diodes, and transistors, not fabricated into an integrated circuit.

discrete component A self-contained device which offers one particular electrical property in lumped form, i.e., concentrated at one place in a circuit, rather than being present as a pervasive quality. A discrete component is built especially to provide that property and exists independently, not in combination with other components. Example: capacitor, inductor, resistor. Compare *distributed component.*

discrete device Any component or device that operates as a self-contained unit.

discrete element A discrete device that forms part of a larger system.

discrete inductor An inductive component which is entirely self-contained, rather than being electrically spread out. Also called *lumped inductor.* Compared *distributed inductance.*

discrete information source A source of data containing a finite number of individual elements, rather than a continuously variable parameter.

discrete part See *discrete component*.

discrete random variable A random variable having only a finite number of possible values.

discrete resistor A resistive component which is entirely self-contained, rather than being electrically spread out. Also called *lumped resistor*. Compare *distributed resistance*.

discrete sampling Sampling of individual bits or characters, one or more at a time.

discrete thin-film component A discrete component produced by the thin-film process, e.g., thin-film capacitor, thin-film potentiometer, etc.

discretionary wiring A method of interconnecting the components and circuits on a semiconductor wafer, for optimum performance. This required a separate analysis and wiring pattern for every chip.

discriminant A mathematical expression such as the numerator $b^2 - 4AC$ in the formula for solving quadratic equations, which provides a criterion for the performance of another, complicated expression.

discrimination 1. Sharp distinction between electrical quantities of different value. 2. The demodulation of an FM signal; i.e., the delivery of an af signal corresponding to the frequency variations in the FM carrier.

discriminator An FM second detector in which two diodes are operated from the center-tapped secondary of a special i-f transformer. The circuit is balanced for zero output when the received signal is at the carrier frequency, value, but delivers an output when the carrier swings above or below the value. Also see *Foster-Seeley discriminator* and *Travis discriminator*.

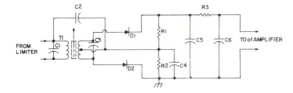

Discriminator.

discriminator transformer The special input transformer in an FM discriminator circuit.

discriminator tuner A device that tunes a discriminator to a selected subcarrier.

discriminator tuning device See *discriminator tuner*.

dish See *dish antenna*.

dish antenna A transmitting or receiving antenna whose reflector has the general shape of a shallow, circular dish of parabolic or other cross section.

dish-type construction A type of panel-and-chassis construction in which the chassis is fastened vertically to the back of the panel.

disintegration 1. The destructive breakdown of a material. 2. The stripping of a tube cathode of its emissive coating (see *disintegration voltage*). 3. The decay of a radioactive material.

disintegration constant Symbol, λ. The ratio of the number of atoms of a radioactive substance that break up each second to the total number of atoms; $\lambda = n/N$, where n is the number breaking up each second and N is the total number.

disintegration voltage The anode voltage at which the cathode of a gas tube begins to become stripped of its electron-emitting material. For safety and reasonable tube life, the anode working voltage must be between the ionization and disintegration values.

disintegrator An ultrasonic device for reducing crystals or particles to fine suspensions.

disjoint The state of having no outcome in common, as in *disjoint sets*.

disjoint sets In set theory, mutually exclusive sets, i.e. sets whose intersection is an empty set.

disjunction The logical inclusive-OR operation.

disk 1. A flat, circular plate, e.g., rectifier disk. 2. A phonograph record or the equivalent unrecorded blank. 3. Magnetic disk.

disk capacitor A fixed (usually twoplate) capacitor consisting of a disk of dielectric material on whose faces are deposited metal-film plates.

Disk capacitor.

disk coil See *disk winding*.

disk dynamo A rudimentary dc generator in which a copper disk rotates between the poles of a permanent magnet. The outer edge of the disk becomes positively charged; the center, negatively charged.

disk engraving 1. Recording sound by cutting a groove in a record disk. 2. The groove resulting from such a process.

diskette A smaller-than-standard magnetic recording disk, especially the flexible type used for microcomputer data storage.

disk files An information-storage system in which data are recorded on rotating magnetic disks (see *disk memory*).

disk generator 1. Disk dynamo. 2. A disk-type electrostatic generator.

disk memory In a digital computer systems, an online or offline device in which information is recorded on the magnetic coating (similar to that on magnetic tape) of a rotating disk. Also see *floppy disk*.

disk pack In disk files, a set of disks which may be handled as a single unit.

disk recorder A device for recording (and usually also playing back) sound or other signals on record disks. Also see *disk recording*.

disk recording 1. Recording sound or other signals on record disks. 2. A disk resulting from such a recording.

disk rectifier A semiconductor rectifier (such as copper-oxide, selenium, magnesium-coppersulfide, or germanium type) in which the active material is deposited on a metal disk.

disk resistor A resistor consisting of a resistive material deposited on a metal disk; or a disk of resistive material. In the latter, electrodes are plated on the faces of the disk, one or more of which are held between clips or screws for connections.

disk-seal tube A high-frequency vacuum tube whose cathode, grid, and plate are closely spaced parallel disks. This construction allows the tube to be inserted into a concentric line and reduces inherent inductance.

disk storage See *disk files, disk memory, and disk pack.*

disk system A sound-motion-picture system employing audio disks synchronized with the film.

disk telegraph An early visual telegraph in which letters of the alphabet and numbers were incribed around the circumference of a dial. When a pointer was moved to a particular letter on the transmitter dial, the pointer on the receiver dial moved to the same letter.

disk thermistor A thermistor having the general shape of a disk.

disk-type memory See *disk memory.*

disk varistor A varistor having the general shape of a disk.

disk winding An armature or coil winding which is flat. Also called *disk coil, pancake coil,* and *spiral coil.*

dislocation A crystal region in which the arrangement of atoms does not have the perfect lattice structure of the crystal.

disperse 1. To separate into constituents. 2. To redistribute data bits over a larger memory space.

dispersion 1. The property of a material that causes energy at certain wavelengths to pass through it at different speeds. 2. The separation of a wave into its various component frequencies (as when white light is dispersed into the color spectrum by a prism). 3. The scattering of a microwave beam when it strikes an obstruction. 4. A suspension of finely divided particles within another substance.

dispersive medium A medium which will disperse a wave passing through it.

displacement 1. Change in the position of a point, particle, figure, or body, and thus the vector representing this change. 2. Movement of a member through a specified distance.

displacement current 1. An alternating current proportional to the rate of change of an electric field and existing in addition to usual conduction current. 2. The current flowing into a capacitor immediately after application of a voltage. Displacement current continues to flow, although continually diminishing in value, until the capacitor becomes fully charged.

displacement laws Laws that can be used to determine the location of a radioactive element in the periodic table. One states that a new element resulting from alpha ray transformation has an atomic weight of 4 less than that of the parent element, and an atomic number of 2 less than that of the parent. Another states that a new element resulting from beta ray transformation has the same atomic weight as that of the parent element, and an atomic number of 1 more than that of the parent.

displacement of porches In a television signal, the amplitude difference between the front porch and back porch of a horizontal sync pulse.

displacement of vectors Rotation of vectors.

displacement transducer A transducer in which movement (displacement) of a rod, armature, core, reed, or the like converts mechanical energy into proportionate electrical energy.

display A visually observable presentation of information, such as data entered into a computer, an answer to a problem solved by a computer, the value of a measured quantity, or a plot of a variable.

display blanking See *display inhibit.*

display console In a computer system, a peripheral that is used to access and display data being processed or stored; often it is a crt—

keyboard unit with a light pen.

display control An interface device between a central processor and several visual display units (crt terminals).

display dimming See *display inhibit.*

display inhibit In a digital meter, the blanking or dimming of the readout display when the instrument is not being used.

display loss The ratio P_1/P_2, where P_1 is the minimum input-signal power that can be detected by an ideal output device at the output of a receiver, and P_2 is the minimum input-signal power value seen by an operator using an output device with the same receiver. Also called *visibility factor.*

display mode 1. A particular method of presenting a display. For example, a character display on a video unit may consist of light characters on a dark background, or dark characters on a light background. 2. An operating mode for a particular device, in which a display is employed.

display module A module containing circuitry and readout lamps for indicating a count.

display primaries In a color TV receiver, the primary colors (red, green, blue); when mixed correctly, they produce a wide range of colors over nearly the entire visible spectrum.

display-storage tube A special cathode-ray tube in which patterns and other information may be stored for later viewing. The tube has two electron guns: a *writing gun* and a *reading (viewing) gun.*

display unit A device which presents information for visual reading. Included are analog and digital meters, cathode-ray tubes, data printers, graphic recorders, and so on. Also see *display console.*

display visibility The ease with which a display can be read by an operator.

display window 1. In a panoramic display, the width of the presented frequency band in hertz. 2. The panel opening through which the indication of a display unit appears.

displayed part That portion of a number displayed in the readout of a calculator or computer. On the unseen end of the number may be several digits that are not displayed, but which the machine takes into consideration in the calculation.

disposable component A circuit component or machine part which is so inexpensive that it is better to discard instead of repair it when it fails.

disruptive discharge Sudden, heavy current flow through a dielectric material when it fails completely under electric stress.

dissector A transducer that samples an illuminated image point by point.

dissector tube A TV camera tube employing a flat photocathode upon which the image is focused by the lens system; electromagnetic deflection from external coils provides scanning: electrons pass sequentially from the image cathode to a scanning tube at the opposite end of the camera tube. Also called *Farnsworth dissector tube* and *orthoiconoscope.* See illustration on next page.

dissimilar dual triode A combination tube consisting of two triode sections with dissimilar characteristics, e.g., a 6CY7.

dissipation The consumption of power, often without contributing to a useful end, and usually accompanied by the generation of heat.

dissipation constant Symbol, δ. For a thermistor, the ratio of the change in power dissipation to a corresponding change in body temperature; $\delta = dP_d/dT_b$.

dissipation factor 1. For a dielectric material, the tangent of the dielectric loss angle. Also called *loss tangent.* 2. Symbol, D. For an impedance (such as a capacitor), the ratio of resistance to reactance; $D =$

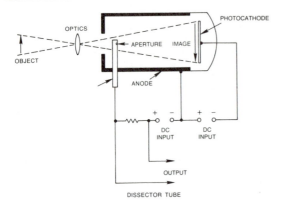

Dissector tube.

R/X. It is the reciprocal of the figure of merit (Q).

dissipation line A resistive section of transmission line, used for dissipating power at a certain impedance. Two parallel lengths of resistance wire are terminated by a large, noninductive resistor having a value equal to the characteristic impedance of the line.

dissipator 1. A device employed primarily to consume power, i.e., a power sink. 2. A device for removing heat generated by a device's operation, e.g., a heatsink attached to a power transistor.

dissociation The condition that characterizes electrolytes (certain acids, bases, or salts in water solution) in which the molecules of the material break up into positive and negative ions.

dissonance The unpleasant effect (especially in music) produced by certain nonharmonious combinations of sounds.

dissymmetrical network A network having unequal input and output image impedances.

dissymmetrical transducer A transducer having unequal input and output image impedances.

distance mark On a radar screen, a mark indicating the distance from the radar set to the target.

distance-measuring equipment In radionavigation, a system which measures the distance of the *interrogator* to a transponder beacon in terms of the transmission time to and from the beacon.

distance protection The use of a protective device within a specified electrical distance along a circuit.

distance relay In circuit protection, a relay which operates to remove power when a fault occurs within a predetermined distance along the circuit.

distance resolution In radar practice, the minimum distance over which targets can be separated.

distant control See *remote control*.

distorted-drive multiplier A frequency multiplier whose excitation signal is a peaked wave which has been predistorted to decrease the angle of flow in the device, thus increasing its efficiency.

distorted nonsinusoidal wave A nonsinusoidal wave whose ideal shape (square, rectangular, sawtooth, etc.) has been altered.

distorted sine wave A wave which is approximately of sinusoidal shape, i.e., it is not an exact plot of $y - \sin\theta$ because of the presence of harmonics.

distortion The deformation of a signal waveform, however slight. Also,

the additional deformation of a signal exhibiting less than ideal waveshape when it passes through a circuit. Some distortion originates within the signal generator itself, other forms come from circuits and devices transmitting the signal.

distortionless 1. Having no distortion. 2. Having a propagation velocity that does not depend on frequency.

distortion meter An instrument for measuring harmonic distortion. It consists of a highly selective filter which removes the fundamental frequency of the signal under test, and a sensitive voltmeter which measures the total voltage of the harmonics when they remain. The distortion percentage is determined from the ratio between filter-output and filter-input voltages, which is indicated directly by the meter if it is set to 100% deflection for the filterinput voltage.

distortion tolerance The maximum amount of distortion which may be present in a signal without making it useless. Naturally, this varies over wide limits. The maximum harmonic distortion which might be acceptable in a high-fidelity sound system could be less than 0.1% total, whereas in some applications of ac power, 10% would be okay.

distress frequency A radio frequency on which a distress signal is transmitted. Ships at sea and aircraft over the sea use 500 kHz (by international agreement). In Citizen Band communications, channel 9 has been set aside for emergency use.

distress signal A signal indicating that trouble exists at the transmitting station and imploring aid from the recipient. The international radiotelegraph distress signal is the three-letter combination *SOS*; the international radiotelephone distress signal is the word *mayday* the phonetic equivalent of the French *m'aidez* (help me).

distributed Existing over a measurable interval, area, or volume; not concentrated in a single place or places.

distributed amplifier A wideband (usually vhf or uhf), untuned amplifier whose vacuum tubes are spaced (*distributed*) along parallel, artificial delay lines consisting of coils which act in combination with the input and output capacitances of the tubes. Adding tubes to the lineup increases the gain. Distributed amplifiers are very useful as preamplifiers for TV receivers.

distributed capacitance Symbol, Cd. Capacitance that is dispersed throughout a component or system, rather than being lumped in one place. An example is the distributed capacitance of a coil.

distributed component An electrical property which is spread throughout a circuit or device (in which it is secondary), rather than being concentrated at one point (in a discrete component). Thus, a capacitance effect may spread over a two-wire transmission line; in this case it is *distributed capacitance*. Another example is the resistance of an inductor. Distributed components are often unintended, even unavoidable; however, they can be useful. Compare *discrete component* and *lumped component*. Also see *distributed capacitance, distributed inductance,* and *distributed resistance*.

distributed constant See *distributed component*.

distributed-constant delay line A delay line whose capacitance and inductance are distributed throughout the line. Compare *lumped-constant delay line*.

distributed inductance Symbol, Ld. Inductance which is dispersed throughout a system or component, rather than being lumped in one place, such as in a coil, e.g., the inductance of an antenna or capacitor.

distributed network 1. A network in which electrical properties, such as resistance, inductance, and capacitance, are distributed over a measurable interval, area, or volume. 2. A network whose characteristics do not depend on frequency within a given range.

distributed-parameter network A network composed of distributed components, rather than one of lumped components.

distributed pole In a motor or generator, a pole having a *distributed winding*.

distributed resistance Symbol, *Rd*. Resistance that is dispersed throughout a component or circuit, rather than being lumped in one place, suct as in a resistor. An example is a tank coil's or antenna's resistance to high frequencies.

distributed-shell transformer A transformer having two complete closed cores that are perpendicular to each other.

distributed winding In a motor or generator, a winding which is placed in several slots (rather than in one slot) under a pole piece.

distributing amplifier An amplifier having a single input and two or more outputs that are isolated from each other; it distributes signals to various points.

distributing cable 1. In a CATV system, the cable connecting the receiver to the transmission cable. 2. In power service, the cable running between a feeder and the consumer's home.

distribution 1. The selective delivery of a quantity; e.g., *power distribution*. 2. In statistical analysis, the number of times particular values of a variable appear. Also called *frequency distribution*.

distribution amplifier A low output impedance power amplifier which distributes a radio, TV, or audio signal to a number of receivers or speakers.

distribution cable See *distributing cable*.

distribution center 1. The central point from which a signal is routed to various points of use. 2. In electric power practice, the point at which generation, conversion, and control equipment is operated to route power to points of use.

distribution factor For a polyphase alternator, the factor *kd* by means of which the total voltage (*ET*) may be determined using the coil voltage (*Ec*) and the number (*N*) of coils; $ET = NEc$. Distribution factor $kd = [\sin(sd/2)]$, where s = number of slots per phase per pole; and d, the angle between adjacent slots.

distribution function In statistical analysis, the function F(*x*) expressing the probability that X takes on a value equal to or less than x.

distribution switchboard 1. A switchboard through which signals may be routed to or between various points. 2. A switchboard for routing electric power to points of use.

distribution transformer A stepdown transformer used to supply low-voltage ac power to one or more consumers from a high-voltage line.

distributive law The arithmetic law that states that $x(y + z)$ is the equivalent of $xy + xz$.

distributor 1. Electronic commutator (see *commutator, 2*). 2. A rotary switching device consisting of a rotating blade and a number of contacts arranged in a circle for sequentialls switching a voltage to a number of points in a circuit, e.g., the distributor in the ignition system of an automotive engine.

disturbance An undesired variation in, or interference with, an electrical or physical quantity.

disturbed-one output In digital computers, the *one* output of a magnetic cell which has received only a partial write pulse train since it was last written into. Compare *undisturbed-one output*.

disturbed-zero output In digital computers, the *zero* output of a magnetic cell which has received only a partial write pulse train since it was last read from. Compare *undisturbed-zero output*.

dither 1. Vibrate; quiver. 2. The condition of dithering, e.g., the dither

of a meter's pointer or of an oscilloscope trace.

divergence loss Loss of transmitted sound energy, resulting from spreading.

divergent series A mathematical series which approaches infinity as the number of terms increases. Thus, the series $Sn = 2 + 4 + 8 + 16...$ has no limit, because an additional term twice the value of the last, is always conceivable. Compare *convergent series* and *infinite series*.

diverging lens A lens having a virtual focus for parallel rays. Compare *converging lens*.

diversity 1. The property of being made up of two or more independent components or media. 2. A form of reception in which two or more independent receivers are connected to two or more independent antennas, and their outputs are fed together to provide a signal less likely to fade. 3. A system of transmission in which two or more independent paths are employed simultaneously to increase intelligibility.

diversity factor 1. A measure of the degree to which a system displays unity among its constituents. 2. The sum of the requirements of each constituent of a system, divided by the total requirement of the system.

diversity gain Signal gain achieved by using two or more receiving antennas.

diversity reception A method of minimizing or preventing the effects of fading by using receivers whose antennas are 5 to 10 wavelengths apart. Each receiver, tuned to the same signal, feeds a common audio amplifier. The extent of signal fade is different at the various antennas, at one of which it is highly probable that the signal will be strong.

diverter-pole generator A well regulated dc generator whose shunt winding is on the main field pole, the series winding being on a diverter pole whose flux opposes that of the main pole.

divide-by-seven circuit A three-stage binary circuit having feedback from stage three to stage one. Stage three is turned on by the fourth input pulse; at that time the feedback pulse turns on stage one, simulating one input pulse and reducing the usual counting capacity from eight to seven.

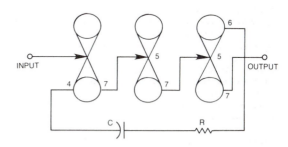

Divide-by-seven circuit.

divide-by-two circuit A circuit that delivers one output pulse for each two successive input pulses, i.e., a flip-flop.

divided-carrier modulation Modulation by adding two identical frequency carriers that are 90 degrees out of phase. The resulting signal is amplitude and phase modulated, but retains the frequency.

divided circuit A parallel circuit.

divided equipment A system of modular electronic components interconnected with cables. A simple example is a radio receiver having an external power supply and external loudspeaker.

dividend In mathematical division, the term which is to be divided. Compare *divisor* and *quotient.*

divider 1. Voltage divider. 2. Frequency divider. 3. Pulse-count divider. 4. A computing circuit or device for performing mathematical division.

divider probe A test probe which divides an applied signal voltage by some factor, such as 2, 5, or 10, to place it within the range of the instrument with which the probe is used.

dividing network See *crossover network.*

divine proportion The ratio of 1:1.618. It is derived from the division of a length into two parts so that the smaller is to the greater as the greater is to the whole length. The name comes from the ancient Egyptian and Greek belief that this represents the most beautiful proportion; it is the proportion between the various dimensions of the Parthenon, for example. It is also the ratio between line segments in a pentagram. Also called *golden section, scared quotient,* and *seqt.*

division 1. Separating a quantity into a number of equal parts as indicated by the divisor. 2. Frequency division (see *frequency divider*). 3. Voltage division (see *voltage divider*).

division of vectors 1. The quotient of two rectangular vectors determined by the principle of rationalization in algebra, i.e., by multiplying the numerator and denominator of the indicated division by the conjugate of the denominator, simplifying, and performing the division. 2. To find the quotient of two polar vectors: the quotient of their moduli and the difference of their arguments.

divisor In mathematical division, the term that indicates the number of parts into which another term (the dividend) is to be divided. Compare *dividend* and *quotient.*

dj Abbreviation of *diffused junction.*

DKT Abbreviation of *dipotassium tartrate.*

dL Symbol for *differential of inductance.*

D-layer A layer of the ionosphere which is below the E-layer; its altitude is approximately 60 kilometers.

dM Symbol for *differential of mutual inductance.*

dm Abbreviation of *decimeter.*

DMA Abbreviation of *direct memory access* and *direct memory addressing.*

DME Abbreviation of *distance-measuring equipment.*

dmm Abbreviation of *digital multimeter.*

DNS Abbreviation of *doppler navigation system.*

doctor To use unconventional (sometimes substandard) methods in fixing a circuit or device or in correcting a bad design.

document 1. In digital computer practice, especially in file maintenance, a form giving information pertinent to a transaction. Also see *transaction.* 2. To perform documentation (see *documentation, 2*).

documentation 1. Paperwork explaining the scope of programs and how they can be optimized. 2. Anotating a computer program at critical points, during its writing, e.g., so that the purpose of various segments are understood; a measure of good programming practice, documentation becomes especially valuable for program modification or debugging.

document reader An electronic device that reads printed cards, usually for data entry into a computer.

dog A malfunctioning circuit or device or the cause of the malfunction. Also see *tough dog.*

doghouse An enclosure for antenna loading inductors and other resonating components, placed at the base of a vertical broadcasting tower.

dog whistle See *ultrasonic whistle.*

Doherty amplifier A highly efficient linear rf amplifier in which a *carrier tube* and a *peak tube* operate jointly, both receiving amplitude-modulated rf excitation. During unmodulated intervals, the carrier tube supplies carrier power to the load, while the peak tube, biased to cutoff, idles. When the excitation is modulated, however, the peak tube (on positive modulation peaks) supplies output power which combines with that of the carrier tube, the increase in power corresponding to the condition of full modulation of the carrier. On negative modulation peaks the peak tube doesn't supply power, and the output of the carrier tube is reduced to zero.

dolby An electronic method for improving the audio reproduction quality of magnetic-tape systems. The gain is increased during the recording process for low-level sounds. Then, during playback, the gain of the low-level sounds is reduced back to its original level.

dolby A A dolby system with four frequency ranges, operated independently. Used mostly by recording professionals.

dolby B A modified form of dolby A, with only one band of noise-reducing circuitry. Used primarily by hobbyists.

Dolezalek electrometer See *quadrant electrometer.*

dolly 1. A low, wheeled frame or platform for transporting electronic equipment. 2. A tool with which one end of a rivet is held while the head is hammered out of the other end.

dom Abbreviation of *digital ohmmeter.*

domain 1. A region of unidirectional magnetization in a magnetic material. 2. A region of unidirectional polarization in a ferroelectric material. 3. A region in which a variable is confined.

domestic electronics The branch of electronics concerned with appliances, automatic controls, protective devices, entertainment systems, communications devices, and other equipment for the home.

domestic induction heater A household cooking utensil heated by currents induced in it. A primary coil (connected to the ac power line) is imbedded in the utensil, which acts as a short-circuited secondary coil.

dominant In statistical analysis, the nature of any quantity that imposes its effects even in the presence of other quantities.

dominant mode For a waveguide, the mode ($TE0,1$ for a rectangular guide; $TE1,1$ for a circular guide) having the lowest cutoff frequency.

dominant series A series which is absolutely convergent (see *convergent series*) and in which the absolute value of each term is greater than that of the corresponding term in a given series.

dominant term In a set or a series, the term whose absolute value is greater than that of any other term.

dominant wave In a waveguide, the wave having the lowest cutoff frequency.

dominant wavelength The preponderant wavelength in the light making up a color.

donor An electron-rich impurity added to a semiconductor to make it an *n*-type. It is so called because it donates its excess electrons. Compare *acceptor, 2.*

donor atom An atom having an excess electron. When a substance having such atoms is added to an intrinsic semiconductor, the extra electron is *donated,* making the semiconductor an *n*-type.

donor impurity A substance whose atoms have excess electrons and which donates electrons to the atomic structure of the semiconductor crystal to which it is added. Donor elements are pentavalent and make semiconductors *n*-types. Also see *donor atom.* Compare *acceptor impurity.*

do-nothing instruction A computer program instruction that causes no action to be taken; it is used to provide space for (1) future program updating or (2) filling out a block of instructions as needed by a compiler. Also called *dummy instruction*.

donut capacitor A flat, ring-shaped capacitor.

donut coil See *toroidal coil*.

donut crystal See *doughnut crystal*.

donut magnet See *ring magnet*.

donut pattern The three-dimensional rf radiation pattern around a straight wire radiator.

doodad See *doohickey*.

doohickey A usually unnamed device, especially one used to achieve some significant modification of circuit performance. Also see *gadget* and *gimmick, 1*.

doorknob capacitor A high-voltage fixed capacitor, so called from its round package which somewhat resembles a doorknob.

doorknob tube A special uhf vacuum tube, so called from its characteristic shape. The unique design provides short electron-transit time and low interelectrode capacitance—very desirable features for ultrahigh frequency applications.

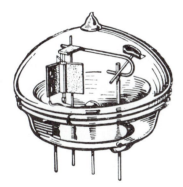

Doorknob tube.

dopant An impurity added in controlled amounts to a semiconductor to make it an *n*-type or *p*-type. Also see *acceptor, 2* and *donor*.

dope To add impurities to a semiconductor material. Doping allows the manufacture of n-type or p-type seminiconductors with varying degrees of conductivity. In general, the greater the extent of doping, the higher the conductivity.

doped junction In a semiconductor device, a junction produced by adding a dopant to the semiconductor melt.

doping Adding a dopant to a semiconductor to alter the way it conducts current.

doping agent See *dopant*.

doping compensation Opposite doping, i.e., adding a donor impurity to p-type semiconductor material or adding an acceptor impurity to n-type semiconductor material.

doping gas A gas diffused into a semiconductor material to dope it. For example, phosphorus pentoxide gas is used to create an *n* region in *p*-type silicon.

Doppler cabinet A loudspeaker enclosure with which a vibrato effect is achieved by rotating or reciprocating either the loudspeaker or a baffle board; the length of the sound path is altered cyclically.

Doppler effect The phenomenon of an apparent change in signal frequency when the source and observer are in relative motion, the change being an increase as the two approach each other, and a downward shifting as the bodies separate. This is often witnessed with sound waves, for example, as when the pitch of an automobile horn seems to rise as the car approaches and falls as the car passes. The Doppler effect is observable also in other forms of radiation, including light and radio waves.

Doppler enclosure See *Doppler cabinet*.

Doppler radar A radar that employs the change in carrier frequency of the signal returned by a moving target (approaching or receding) to measure its velocity.

Doppler ranging See *doran*.

Doppler shift See *Doppler effect*.

Doppler's principle See *Doppler effect*.

doran A continuous-wave trajectory measuring system utilizing Doppler shift (see *Doppler effect*). The name is an acronym from *Doppler ranging*.

dorsal column stimulator Abbreviation, DCS. A radio-frequency for temporarily relieving pain.

dosage meter See *dosimeter*.

dose The total quantity of radiation received upon exposure to nuclear radiation or X rays.

dosimeter An instrument for measuring the amount of exposure to nuclear radiation or X rays.

dot 1. The shorter of the two characters (dot and dash) of the telegraph code. The dot, a short sound, mark, or perforation, is a third the length (duration) of a dash. Compare *dash*. 2. One of the small spots of red, green, or blue phosphor on the screen of a color TV picture tube. 3. A small spot of material alloyed with a semiconductor to form an alloy junction. 4. The junction of two lines on a schematic diagram representing a wired connection; also called *solder dot*.

dot AND Externally connected circuits or functions whose combined outputs result in an AND function. Compare *dot OR*.

dot-and-dash telegraphy Telegraphy (wire or radio) by means of dot and dash characters. Also see *dash* and *dot, 1*.

dot cycle One period of an alternation between two signaling conditions, each of which is of unit duration, e.g., a unit mark followed by a unit space.

dot encapsulation A method of packaging cylindrical components by pressing them into the holes of perforated disks; interconnections are made, to complete a circuit, on each face of the disks.

dot generator A special rf signal generator employed to produce a pattern of red, green, and blue dots on the screen of a color TV receiver.

dot matrix A rectangular array of spaces, usually 5 wide by 7 high, some of which are filled in to form alphanumeric and punctuation characters.

dot-matrix display A display that shows characters in dot-matrix form.

dot-matrix printer A computer output peripheral that prints characters as a matrix of dots using a number of styli; a common character matrix size is 5×7, i.e., 35 dots.

dot movement pattern The movement of the red, green, and blue dots on the screen of a color TV picture tube as the red, green, and blue magnets and the lateral magnet are adjusted for convergence of the dots at the center. The blue dots move horizontally or vertically; the red and green dots, diagonally.

dot OR Externally connected circuits or functions whose combined outputs result in an OR function. Compare *dot AND*.

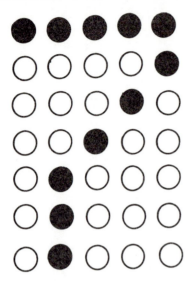

Dot Matrix.

dot pattern In color TV testing with a dot generator, dots of color (a red group, green group, and blue group) produced on the screen. With overall beam convergence, the three groups blend to produce white.

dot product The product (*ab*) of two vectors (*a* and *b*). The product is a scalar whose magnitude is the product of the magnitudes of *a* and *b* times the cosine of the angle between them.

dot-sequential system The color TV system in which the image is reproduced by means of primary-color dots (red, green, blue) sequentially activated on the screen of the picture tube. Compare *field-sequential system* and *line-sequential system*.

double-amplitude-modulation multiplier A modulating system in which a carrier is amplitude-modulated first by one signal and then by a second signal. The resulting signal is fed to a detector, the output of which contains the product of the two modulating signals.

double-anode diode A semiconductor diode having two anodes and a common cathode.

double armature An armature (such as that of a dynamotor or a two-voltage generator) which has two separate windings on a single core and has two separate commutators.

double-balanced mixer See *balanced-mixer*.

double-balanced modulator See *balanced modulator*.

double-base diode See *unijunction transistor*.

double-base junction transistor A junction transistor having the usual emitter, base, and collector electrodes, plus two base connections, one on either side of the base region. The additional base connection acts as a fourth electrode to which a control voltage is applied. Also called *tetrode transistor*.

double-beam crt See *dual-beam oscilloscope*.

double-beam oscilloscope See *dual-beam oscilloscope*.

double-bounce calibration In radar practice, a calibration technique for determining zero-beat error. Round-trip echoes are observed, the correct range being the difference between the two echoes.

double-bounce signal A signal which is received after having been reflected twice.

double-break contacts The member of a set of contacts which is normally closed on two others. Compare *double-make contacts*.

double-break switch A switch which opens a previously closed circuit at two points simultaneously on closing. Compare *double-make switch*.

double bridge See *Kelvin double bridge*.

double buffering During a computer peripheral's input/output operation, the use of two memory areas for temporary storage.

double-button microphone A carbon microphone having two buttons and mounted on each side of the center of a stretched diaphragm and connected in push-pull. Also see *button microphone*.

double-channel duplex Two-way communication over two independent channels. One station transmits on one channel, and the other station transmits on the other channel. The result is conversation-mode communications, in which one operator may interrupt the other at any time; both receivers are always operational.

double-channel simplex A system of communication in which two channels are used. One station transmits on one channel, and the other station transmits on the other channel. But interruption is not possible, since whenever either operator transmits, the station receiver is cut off.

double-checkerboard pattern In a magnetic core memory, the maximum noise which appears when half of the half-selected cores are in the one state and the others are in the zero state. Also called *worst-case noise pattern*.

double circuit tuning A circuit whose output and input are tuned separately. Such tuning provides increased selectivity when input and output are resonant at the same frequency, and decreased selectivity (flat-topping) when they are tuned to different frequencies. Also see *double-tuned amplifier*.

double clocking A phenomenon that occurs in some digital circuits when the input pulse is nonuniform, and appears as two pulses to the device. The device is thus actuated at twice the desired frequency.

double-coil direction finder A direction finder employing an antenna comprising two identical, perpendicular coils. The directivity of the antenna is the resultant of the directivity of individual coils.

double conversion Two complete frequency conversions in a superheterodyne system. For example, a 7 MHz signal might be first converted to a 1550 kHz intermediate frequency, which in turn could be converted to a 455 kHz second intermediate frequency. This action provides both good signal-to-image ratio and high selectivity; the first conversion widely separates the signal and image because of the high intermediate frequency, and the second conversion affords optimum selectivity of a lower intermediate frequency. Also called *dual conversion*.

double-conversion receiver See *double-conversion superheterodyne*.

double-conversion superheterodyne A superheterodyne receiver employing double conversion.

double-current generator 1. A dynamo-type generator supplying both ac and dc from one armature winding. 2. A rotary converter operating on direct current and delivering alternating current.

double-diamond antenna A broadband antenna consisting of two rhomboid plates attached to each side of the feeder.

double-diffused epitaxial mesa transistor A transistor in which a thin mesa crystal is overlaid on another mesa crystal. Also called *epitaxial-growth mesa transistor*.

double-diffused transistor See *diffused-emitter-and-base transistor*.

double diode See *duodiode*.

double-diode limiter A limiter in which two diodes are connected back to back in parallel, to limit both peaks of an ac signal.

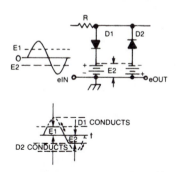

Double-diode limiter.

double-doped transistor See *grown-junction transistor.*

double emitter follower See *compound connection.*

double-ended amplifier See *push-pull amplifier* and *double-ended circuit.*

double-ended circuit A symmetrical circuit, i.e., one having identical halves, each operating on a half-cycle of the input signal. Example: a push-pull amplifier.

double-extended Zepp antenna A type of collinear center-fed antenna in which each flat-top section is 0.65λ. This antenna gives increased gain over that of the Zepp and double Zepp (see *double Zepp antenna*).

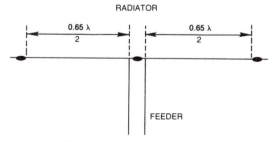

Double-extended Zepp antenna.

double-frequency-changer superheterodyne See *double-conversion superheterodyne.*

double-hump resonance curve A resonance curve which has been flattened by double tuning; it exhibits two resonance peaks. Also see *double-tuned amplifier.*

double-hump wave See *double-pulse wave.*

double image Two overlapping TV pictures, an effect caused by the signal arriving over two different paths (one possibly attributable to reflection of the wave) and, hence, at different instants. Also called *ghost.*

double insulation The use of two layers of insulation on a conductor, made of different materials.

double integral Symbol, ∫ ∫. An integral involving successive integration performed twice. Thus, the area of a sector under a curve is $A =$

∫ ∫ dy dx. Also see *integral, integral calculus,* and *integration.*

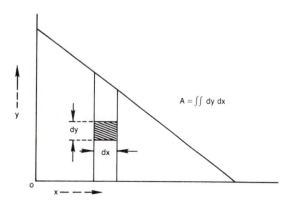

Double integral.

double integration The use of double integrals especially in determining the area of a sector under a curve. Also see *integral, integral calculus,* and *integration.*

double ionization Ionization resulting from an electron colliding with an ion. In a gas, for example, a neutral atom may collide with an electron, which could knock an electron out of the atom. The atom, then, would become a positive ion, which may in turn be bombarded by an electron, releasing still another electron.

double-junction photosensitive semiconductor See *phototransistor.*

double layer See *Helmholtz double layer.*

double-length number See *double-precision number.*

double local oscillator A mixer system in which a local oscillator generates two accurate rf signals separated by a few hundred hertz. The difference frequency is employed as a reference in some applications.

double-make contacts A set of normally open contacts of which one closes against two others simultaneously. Compare *double-break contacts.*

double-make switch A switch which closes a previously open circuit at two points simultaneously. Compare *double-break switch.*

double moding In microwave practice, the abrupt changing of frequency at irregular intervals.

double modulation Using a modulated carrier to modulate another carrier of a different frequency.

double-play tape A thin magnetic recording tape that has approximately twice the playing time of the usual tape. Although the playing time is longer, double-play tape is more subject to jamming and stretching than standard-thickness recording tape.

double-pole Having two poles, e.g. a double-pole switch.

double-pole, double-throw switch or relay A switch or relay having two contacts which may be *thrown* (closed) simultaneously in one of two directions to close or open two circuits.

double-pole, single-throw switch or relay A switch or relay having two poles which may be thrown in one direction (compare previous entry) to simultaneously close or open two circuits.

double precision The use of two independent representations for a single character.

double precision hardware Within a computer, arithmetic units permitting the use of double precision operands, sometimes also accommodating floating-point arithmetic.

double-precision number In digital computer practice, a number represented by two words for greater accuracy.

double pulse reading Pertaining to a magnetic core in a computer memory, recording bits as two states held simultaneously by one core having two areas that can be magnetized with alternate polarities, e.g., positive—negative could be, say, zero, and negative—positive, one.

double-pulse wave An ac wave having two successive positive humps followed by two successive negative humps within a cycle. The output voltage of a varistor bridge has such a waveshape for an ac input.

double-pulsing station A loran station which transmits at two pulse rates upon receiving two pairs of pulses.

double pumping A method of obtaining increased peak output power from a laser by pumping it for a comparatively long interval and then immediately pumping it for a short interval.

doubler 1. A circuit or device for multiplying a frequency by two (see *frequency doubler*). 2. A circuit or device for multiplying a voltage by two (see *voltage doubler*).

double probe A test probe which multiplies an applied signal voltage by two so it may be handled more effectively by the instrument with which the probe is used.

double punching In perforating a punched card, putting two holes in one column; it is an error if it occurs in a field of a card that is part of a record.

double-rail A form of logic system in which two lines are employed, with three possible states. The output may be high, low, or undecided.

double response 1. Two-point response, as that associated with tuning a receiver to a signal and then to its image. 2. Double-hump tuning.

double screen A cathode-ray tube having a two-layer screen on which there is an additional, long-persistence coating of a different color.

double shield Two independent electromagnetic shields for a circuit enclosure or cable. The shielding structures are concentric, and may be connected together at a single point (the common point).

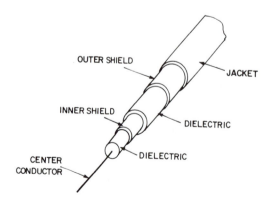

Double Shielded Cable

double sideband Abbreviation, DSB. In a modulated signal, the presence of both sidebands.

double sideband, suppressed carrier Abbreviation, DSSC. A transmission technique in which both sideband products of modulation are transmitted and the carrier is suppressed. Compare *lower sideband, suppressed carrier* and *upper sideband, suppressed carrier*.

double-sideband system A modulation or demodulation system utilizing both sidebands, with or without the carrier.

double-sideband transmitter A modulated transmitter employing a double-sideband system.

double-spot tuning In a superheterodyne receiver, tuning in the same signal at two different places on the dial, a condition caused by image response.

double squirrel-cage induction motor A polyphase induction motor having a double squirrel-cage rotor. The rotor slots contain two bars, an upper bar having low reactance (being near the air gap) and high resistance, and a lower bar having high reactance and low resistance. This motor has low starting current, high starting torque, and a full-load slip of less than 5%.

double-stream amplifier A traveling-wave tube in which microwave amplification results from the interaction of two electron beams of different average velocity.

double-stub tuner Two stubs (see *stub*) connected in parallel with a transmission line and usually spaced 3/8-wavelength apart; it is used as an impedance matcher.

double superheterodyne See *double-conversion superheterodyne*.

double superheterodyne reception See *double-conversion superheterodyne*.

double-surface transistor See *coaxial transistor*.

doublet 1. A polarized atom (*electric doublet*). For example, an atom which is under electric stress in the dielectric of a capacitor behaves as if it has a positive charge on one side and a negative charge on the other. 2. A system of two equal and opposite magnetic poles situated close together (*magnetic doublet*). 3. The waveform of the output voltage delivered by a linear-response delay line which is excited by a current step function (see *unit function*). 4. Doublet antenna.

doublet antenna A center-fed half-wave antenna.

double-throw Operating in opposite directions, as selected, e.g., a double-throw relay or switch.

double-throw circuit breaker A circuit breaker which closes in both its pull-in and dropout positions.

double-throw switch (relay) A switch or relay having two ganged poles. Also see *double-pole, double-throw switch or relay* and *double-pole, single-throw switch or relay*.

double-trace recorder See *double-track recorder, 2*.

double tracing Displaying two signals simultaneously on the screen of an oscilloscope through the use of an electronic switch.

double-track recorder 1. A tape recorder whose head is positioned so that separate recordings may be made as two tracks on the tape. 2. A graphic recorder which produces two separate parallel tracings.

double triode See *dual triode*.

doublet trigger A two-pulse, constant-spaced trigger signal used for coding.

double-tuned amplifier An amplifier whose input and output circuits are tuned.

double-tuned circuit A circuit, such as an amplifier or filter, employing separate input and output tuning. Also see *double circuit tuning* and *double-tuned amplifier*.

double-tuned detector A form of frequency-modulation discriminator with two resonant circuits. One is tuned slightly higher than the channel center frequency, and the other is tuned an equal amount below the center.

double-vee antenna A broadband, modified-dipole antenna resembling two vees in line. Also see *vee antenna*.

double-winding generator A dynamo-type generator having separate armature windings for supplying two voltages (both either ac or dc, or one ac and one dc voltage).

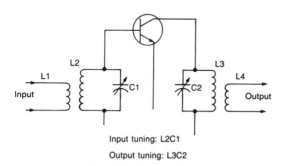

Input tuning: L2C1

Output tuning: L3C2

Double-tuned circuit.

double-wye rectifier A heavy-load circuit employing six rectifier diodes, each conducting for 120 degrees, and an interphase winding. The circuit is equivalent to two three-phase, half-wave rectifiers connected in parallel.

double-Y rectifier See *double-wye rectifier*.

double Zepp antenna A center-fed antenna whose flat-top consists of two half-waves in phase. It is so called from its resemblance to two back-to-back Zepp antennas.

doubling 1. To produce the second harmonic of a signal. 2. In a conversation, unintentional simultaneous transmission by both operators, resulting in missed information. 3. In a speaker, distortion resulting in large amounts of second-harmonic output.

doubly balanced modulator See *balanced modulator*.

doughnut capacitor See *donut capacitor*.

doughnut coil See *toroidal coil*.

doughnut crystal A relatively large, zero-temperature-coefficient piezoelectric quartz crystal cut in the form of a torus with the *Y*-axis passing through the center of rotation.

doughnut magnet See *ring magnet*.

down-convert In superheterodyne conversion, to heterodyne a signal to an intermediate frequency lower than the signal frequency. Compare *up convert*.

down lead See *lead-in*.

down time The period during which electronic equipment is completely inoperable (for any reason).

downturn A usually sudden dip in a performance curve. Compare *upturn*.

downward modulation Modulation in which the average carrier component decreases during modulation. Example: amplitude modulation of a transmitter in which the antenna current decreases during modulation. Compare *upward modulation*.

Dow oscillator See *electron-coupled oscillator*.

DP Abbreviation of *data processing*.

dP Symbol for *differential of power*. (Also *dp*.)

dp Symbol for *differential of pressure*.

dpdt Abbreviation of *double-pole, double-throw* (switch or relay).

dpm 1. Abbreviation of *digital power meter*. 2. Abbreviation of *digital panel meter*. 3. Abbreviation of *disintegrations per minute*.

dps Abbreviaton of *disintegrations per second*.

dpst Abbreviation of *double-pole, single-throw* (switch or relay).

dQ 1. Symbol for *differential of quantity or charge*. 2. Symbol for *differential of figure of merit*.

dR Symbol for *differential of resistance*. (Also, *dr*.)

dr Abbreviation of *dram*.

drag 1. A retarding force (due to friction) acting on a moving body in contact with another moving or stationary body or medium. 2. A retruding force (similar to 1, above) introduced by an applied magnetic field or similar cause.

drag angle In disk recording, an angle of less than 90 degrees between the stylus and the disk. The acute angle causes the stylus to drag instead of digging in. Compare *dig-in angle*.

drag cup A cup of nonmagnetic metal (usually copper or aluminum) that, when rotated in a magnetic field, acquires an emf proportional to the speed of rotation. The device is often used as a brake.

drag-cup motor A servomotor whose shaft has a copper or aluminum drag cup which rotates in the field of a two-phase stator. Eddy currents set up in the cup by the field winding produce torque; braking action, direction control, and speed control are obtainable by means of associated electronics.

drag magnet In a motor-type meter, a braking magnet, i.e., one employed to reduce speed through eddy-current effects. Also called *retarding magnet*.

drain 1. The current or power drawn from a signal or power source. 2. The load device which absorbs the current or power in 2, above. 3. The drain electrode of a field-effect transistor; it is equivalent to the plate of a tube or the collector of a bipolar transistor.

drainage equipment Devices and systems for protecting circuits against transients generated by circuit breakers and similar safety devices.

drain-coupled multivibrator A multivibrator employing two field-

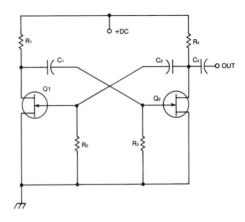

Drain-coupled multivibrator.

effect transistors in the circuit equivalent of a plate-coupled tube-type multivibrator. The drain of one stage is capacitance-coupled to the gate of the other stage.

D-region A low region of the ionosphere under the E-region and whose ionization varies during the day with the inclination of the sun.

dress The (usually experimental) arrangement of leads for optimum circuit operation (minimum capacitance, best suppression of oscillation, minimum pickup, etc.).

dressed contact A contact having a permanently attached locking spring member.

drift 1. Within a conductor or semiconductor, the controlled, directed movement of charge carriers resulting from an applied electric field. Compare *diffusion, 2.* 2. A usually gradual and undesirable change in a quantity, such as current, as a result of a disturbing factor, such as temperature or age.

drift current In a semiconductor, the current resulting from a flow of charge carriers in the presence of an electric field. The charge carriers are electrons in n-type material and holes in p-type material.

drift field The inherent internal electric field of a drift-field transistor.

drift-field transistor An alloy-junction bipolar rf transistor for which the impurity concentration is graded from *high* on the emitter side of the base wafer to *low* on the collector side. This creates an internal drift field which accelerates current carriers and raises the upper frequency limit of the transistor.

drift-matched components Active or passive components which have been closely matched in terms of the drift of one or more parameters with respect to time, temperature, or the like. Also see *drift, 2.*

drift mobility Symbols, μe, μh. For current carriers in a semiconductor, the average drift velocity per unit electric field. At 300K (about 25 °C), the drift mobility (μe) for electrons is 3.9×10^3 cm^2/V • s for germanium, and 1.3×10^3 cm^2/V • s for silicon. For holes, the drift mobility (μh) is 1.8×10^3 cm^{2}V • s for germanium, and 500 cm^2/V • s for silicon.

drift space 1. In an electron tube, a space that is free (or nearly so) of ac fields from the outside, and in which the repositioning of electrons is thus governed by the space-charge forces and the velocity distribution of the electrons. 2. In a klystron the space between buncher and catcher cavities in which there is no field.

drift speed The average velocity of electrons or ions moving through a medium.

drift transistor See *drift-field transistor.*

drift velocity The net velocity of a charged particle (electron, hole, ion) in the direction of the field applied to the conducting medium.

drift voltage The usually gradual change in voltage resulting from such causes as internal heating. Also called *voltage drift.*

drip loop In a transmission line for an antenna or power service, a loop near the point of entry to the building for the purpose of precipitating moisture.

drip-proof motor A motor with ventilating apertures arranged so that moisture and particles cannot enter the machine.

drip-tight enclosure A housing designed to prevent entry of falling moisture and dust; it also prevents accidental contact with the enclosed apparatus or machinery.

drive 1. To excite, i.e., to supply with input-signal current, power, or voltage (see *driving current, driving power,* and *driving voltage*). 2. Input-signal excitation (see *driving current, driving power,* and *driving voltage*). 3. A device that moves a recording medium, e.g., tape transport, floppy-disk drive.

drive belt A continuous belt used to transmit mechanical power from a driving pulley to a driven pulley.

drive circuit 1. A circuit employed for the purpose of providing the excitation to a motor. 2. A driver amplifier.

drive control In a TV receiver, the potentiometer used to adjust the ratio of horizontal pulse amplitude to the level of the linear portion of the sawtooth scanning-current wave.

driven element In a multielement antenna, an element to which rf energy is fed directly, as opposed to a *parasitic element*, which is excited by a nearby radiator element.

driven-element directive antenna A multielement directional antenna whose elements are driven from the feeder line; i.e., no element is parasitic. Compare *parasitic-element directive antenna.*

driven multivibrator A multivibrator whose operation or frequency is controlled by an external synchronizing or triggering voltage. Compare *free-running multivibrator.*

driven single sweep A single oscilloscope sweep which is initiated by the signal under observation.

driven sweep An oscilloscope sweep which is initiated by the signal under observation.

drive pattern A pattern of interference in a facsimile system that is caused by improper synchronization of the recording spot.

drive pin A pin used to prevent a record from slipping on the rotating turntable of a recorder or reproducer. It is similar to and located near the center pin of the turntable.

drive pulse In digital computer practice, a pulse that magnetizes a cell in a memory bank.

driver 1. A device that supplies a useful amount of signal energy to another device to insure its proper operation. Examples: a current driver for a magnetic-core memory, an oscillator driving a loudspeaker (a siren), etc. 2. A power amplifier stage (tube or transistor) that supplies signal power to a higher-powered amplifier stage. 3. In a digital computer, a stage that increases the output current or power of another stage, e.g., a clock driver.

driver element In a multielement directive antenna, the element excited directly by the feeder, the other elements (directors and reflectors) being parasitic.

driver impedance 1. The output impedance of a driver stage; e.g., Ep/Ip for a tube, Vc/Ic for a transistor. 2. The impedance seen looking from the driven stage of an amplifier, through the driver transformer, to the driver stage. It is the vector sum of driver reactance and resistance.

driver inductance In an amplifier's driver transformer, the inductance as seen looking through the transformer from the driven stage into the driver stage. Driver inductance $Ld = Ls + (Ns/Np)^2 Lp$, where Lp is the primary-winding inductance; Ls the secondary-winding inductance, Np the primary turns; and Ns the secondary turns.

driver resistance In an amplifier's driver transformer, the resistance (Rd) seen looking through the transformer from the driven stage into the driver stage.

$$Rd = s + (Ns/Np)^2 Rp1 + Rp$$

where Rp is the primary-winding resistance, Rs the secondary-winding resistance, $Rp1$ the internal plate resistance of the driver tube, Np the number of primary turns, and Ns the number of secondary turns.

driver stage An amplifier stage whose chief purpose is to supply excitation (input-signal current, power, or voltage) to the next stage. Also see *driver.*

driver transformer The transformer that couples a driver stage to a driven stage. Example: the interstage transformer inserted between the plate of a single-ended driver tube and the two grids of a push-pull power-output stage in an audio amplifier.

driver transistor A transistor operating in a driver stage (see *driver, 2*).

driver tube An electron tube in a drive stage (see *driver, 2*).

driving current Input-signal current required for a given output from a driven device. Example: In a certain transistor dc amplifier stage, 1 mA of base driving current produces a collector output current of 150 mA.

driving-point admittance The reciprocal of driving-point impedance.

driving-point impedance The input impedance of a network.

driving power The input-signal power required for a given output form a driven device. Example: In the output stage of a certain tube-type audio amplifier, 0.25W of grid driving power is required for 5W output.

drive wire The wire forming the coil around the toroidal cell in a magnetic core memory and supplying drive pulses. Also see *drive pulse*.

driving-range potential In cathodic protection, the difference of potential between the anode and (protected) cathode.

driving signal 1. Drive (see *drive, 2*). 2. In television, time-scanning signals (line-frequency pulses and field-frequency pulses) at the pickup location.

driving spring In a stepping relay, the spring that moves the wiper blades.

driving voltage Input-signal voltage required for a given output from a driven device. For example, in a certain beam-power rf amplifier stage, a grid-input (driving voltage) of only 0.25V rms produces an rf power output of 50W.

DRO Abbreviation of *digital readout*.

drone A pilotless, i.e., radio-controlled, aircraft.

drone cone A loudspeaker cone which is mouthed in a bass-reflex enclosure with other speakers but is undriven. Also called *passive radiator*.

droop 1. A dip (see *dip, 2*) in the curve of a variable. 2. In a pulse train, the decrease in mean amplitude (in percent of maximum amplitude) at a given instant after attainment of maximum amplitude.

drooping radials In a *ground-plane antenna* employing wire radials as the ground plane, radials which are bent downward to provide a transmission-line impedance match and a lower angle of radiation.

drop 1. In wire communications, the line connecting a telephone cable to a subscriber's building. 2. Voltage drop.

drop bar A device which automatically grounds or short-circuits a capacitor when the door of a protective enclosure is opened.

drop cable See *distributing cable, 1*.

drop channel In a communications system utilizing several channels, a channel that is not used.

drop-in The unintentional creation of bits when a magnetic storage device is being read from or written into. Compare *dropout, 4*.

drop indicator In a signaling system, such as an annunciator, a hinged flap that drops into view when the signaling device is actuated.

dropout 1. The opening of a relay or circuit breaker. 2. In digital-computer practice, variation in signal level of the reproduced tape-recorded data. Such variation can result in errors in data reduction. 3. In the production of monolithic circuits, a special image placed at a desired point on the photomask. 4. Digit loss during a read or write operation involving a magnetic storage device.

dropout current See *dropout value*.

dropout error The loss of a bit, or other unit of data, in a re-

cording/reproducing process. It is caused by defects in the recording system, the playback system, or the magnetic tape. An error resulting from dropout.

dropout power See *dropout value*.

dropout value The level of current, power, or voltage at which a device, such as a circuit breaker or relay, is released.

dropout voltage See *dropout value*.

dropping resistor A series resistor providing a voltage reduction equal to the voltage drop across itself. For example, aa 1K resistor in series with a 45V battery and carrying a current of 10 mA will provide a voltage reduction equal to 10V (IR = 0.01 × 1000 = 10V), thus dropping the 45V to 35V.

drop relay In a telephone system, a relay that is activated by the ringing signal. The relay is used to switch on a buzzer, light, or other device.

drop repeater A repeater intended for a termination of a communications circuit in a telephone system.

dropsonde A parachute-supported *radiosonde* dropped from a high-flying aircraft.

drop-tracks The tracks of radioactive particles made visible by moisture in an ionization chamber.

drop wire A wire that runs from a building to a pole (for line extension) or to a cable terminal (for cable extension).

drum 1. A rotating cylinder coated with a magnetic material on which digital information may be recorded in the form of tiny magnetized spots that later may be *read* as the drum rotates under pickup heads, or erased when the stored information is no longer needed. 2. In some graphic recorders, facsimile receivers, etc., a rotating cylinder carrying the recording sheet.

drum capacitor See *concentric capacitor*.

drum controller The device that regulates the recording process on a drum memory.

drum magnet See *cylinder magnet*.

drum mark On a track of a magnetic drum, a character that signifies the end of a character group.

drum memory In digital computers, a memory based upon a magnetic drum (see *drum, 1*).

drum parity The degree of accuracy in a drum recording/reproducing system.

drum programmer A device for sequencing operations. Its heart is a rotating drum around whose surface contacts or points may be placed to actuate or terminate operations at selected times.

drum receiver A facsimile receiver using recording paper or photographic film wound around a revolving drum.

drum recorder A graphic recorder in which the record sheet is wound around a rotating drum.

drum resistor A resistor consisting of a hollow cylinder of resistive material. Such a resistor can be cooled by circulating air or liquid through it.

drum speed The speed (in rpm) of the rotating drum in a graphic recorder or a facsimile transmitter or receiver.

drum storage The storage of data as magnetic impulses on a cylindrical, or drum-shaped, medium.

drum switch A sequential switch whose contacts are pins or teeth placed at points around the outside of a revolving drum.

drum transmitter A facsimile transmitter in which the sheet bearing the material to be transmitted is wound around a revolving drum.

drum-type controller A motor-driven drum switch arranged to time various operations through sequential switching.

drum-type memory See *drum memory*.

drum varistor A varistor that is a hollow cylinder of nonlinear resistance material. This varistor can be cooled by circulating air or liquid through it.

drum winding In a motor or generator, an armature whose conductors are on the outer face of the core, the two branches of a turn lying under adjacent poles of opposite polarity.

drunkometer An instrument for testing the extent of alcoholic intoxication. It electronically measures blood alcohol content through analysis of the subject's breath.

dry In an electric cell, a term used to describe an electrolyte that is semiliquid or solid.

dry battery A battery of dry cells.

dry cell 1. A Leclanché primary cell in which the positive electrode is carbon; the negative electrode, zinc, and the electrolyte, a gel of ammonium chloride and additives. Also see *cell* and *primary cell*. 2. A battery cell whose electrolyte is a gel or paste.

dry circuit A circuit in which the maximum voltage is 50 mV and the maximum current 200 mA.

dry-contact rectifier See *dry-disk rectifier*.

dry contacts Contacts that neither make nor break a circuit.

dry-disk rectifier A solid-state rectifier, such as a copper-oxide, magnesium-copper-sulfide, or selenium type, which consists of a metal disk coated with a semiconductor material. The name was originally used to distinguish this rectifier from the wet electrolytic rectifier.

dry electrolytic capacitor An electrolytic capacitor whose electrolyte is a paste or solid. Compare *wet electrolytic capacitor*.

dry flashover voltage Complete-breakdown voltage between electrodes in dry air when all insulation is clean and dry.

dry pile A voltaic pile containing numerous disks silvered or tinned on one face and covered with manganese dioxide on the other.

dry reed A metal contact, generally used as a relay, that moves toward or away from another fixed contact under the influence of a magnetic field.

dry-reed relay See *dry-reed switch*.

dry-reed switch A switch consisting of two thin, metallic strips (reeds) hermetically sealed in a glass tube. When an external magnet is brought close by, it attracts one of the reeds, which then contacts the other reed, closing the circuit. Compare *mercury-wetted reed relay*.

Dry-reed switch.

dry run 1. The preliminary operation of equipment for testing and appraisal. Such a procedure precedes putting the equipment into regular service. 2. A step-by-step paper-and-pencil "run" of a computer program before it is machine-implemented.

dry shelf life The life of a battery cell stored thout its electrolyte.

dry-transfer process A method of transferring printed-circuit patterns and panel labels from sheets by rubbing them onto the substrate or panel.

dry-type forced-air-cooled transformer A dry-type transformer which is cooled by forced air circulation.

dry-type self-cooled-forced-air-cooled transformer A dry-type transformer whose self-cooling rating is based on cooling obtained from forced air circulation.

dry-type self-cooled transformer A dry-type transformer which is cooled by natural air circulation.

dry-type transformer A transformer which, rather than being immersed in oil, is cooled entirely by the circulation of air.

DSB Abbreviaton of *double sideband*.

ds Symbol for *differential of distance*.

DSBSC Abbreviation of *double-sideband suppressed-carrier*. (Also abbreviated *DSSC*.)

dsc Abbreviation of *double silk covered* (wire).

D-scope A radarscope whose display resembles that of a C-scope, the difference being that blip height gives an approximation of the distance.

D-service Federal Aviation Agency service providing radio broadcasts of weather data, notices to airmen, and other advisory messages.

DSP Abbreviation of *double silver plated*.

DSR Abbreviation of *dynamic spatial reconstructor*.

DSS Abbreviation of *direct station selection* (telephone).

DSSC Abbreviation of *double-sideband suppressed-carrier*. (Also, *DSBSC*.)

DT Abbreviation of *data transmission*.

dT Symbol for *differential of temperature*.

dt Symbol for *differential of time*.

dta Abbreviation of *differential hermoanalysis*.

DT-cut crystal A piezoelectric plate cut from a quartz crystal at an angle of rotation about the Z-axis of $-53°$. It has a zero temperature coefficient of frequency at approximately $30°C$. Also see *crystal axes* and *crystal cuts*.

DTL Abbreviation of *diode—transistor logic*.

DTn Abbreviation of *double tinned*.

DTS 1. Abbreviation of *data-transmission system*. 2. Abbreviation of *digital telemetry system*.

DU Abbreviation of *duty cycle*.

dual 1. A combination of two components—such as diodes, transistors, etc.—in a single housing. The components are often carefully matched. Compare *quad, 1*. 2. A device or circuit which behaves in a manner analogous to that of another operating with component and parameter counterparts. Thus, a current amplifier may be the dual of a voltage amplifier; a series-resonant circuit, the dual of a parallel-resonant circuit; a field-effect transistor, the dual of a bipolar transistor, and so on.

dual-beam crt A cathode-ray tube having two separate electron guns, for use in a *dual-beam oscilloscope*.

dual-beam oscilloscope An oscilloscope having two electron guns and deflection systems; it can display two phenomena on the screen simultaneously for comparison.

dual capacitor 1. Two fixed capacitors combined in a single housing, sometimes sharing a common capacitor plate. 2. a two-section, ganged variable capacitor.

dual-channel amplifier An amplifier having two separate, often identical channels.

dual-cone speaker A speaker designed for a wide range of audio frequencies. One cone responds to the bass and midrange audio frequencies, and a smaller cone responds to the higher frequencies.

dual-control heptode A heptode having two signal grids, e.g., a 12EG6.

dual diode A combination tube consisting of two diode sections in a single envelope, e.g., a 6AL5. Also called *duodiode* and *duplex diode*.

dual diode—dual triode A combination tube consisting of two diode sections and two triode sections in a single envelope, e.g., a 6B10.

dual-diode—high-mu triode A combination tube consisting of two diode sections and a high-mu triode section in a single envelope, e.g., a 6AV6.

dual-diode—low-mu triode See *duodiode—low-mu triode*.

dual-diode—medium-mu triode See *duodiode—medium-mu triode*.

dual-diode-pentode See *duodiode—pentode*.

dual-diode—tetrode See *duodiode—tetrode*.

dual-diode—triode See *duodiode—triode*.

dual-diversity receiver A receiver or receiver system for dual-diversity reception.

dual-diversity reception Diversity reception using two receivers.

dual double-plate triode A combination tube consisting of two triode sections, each having two plates, in a single envelope, e.g., a 12FQ8.

dual-emitter transistor A low-level silicon pnp chopper transistor of the planar passivated epitaxial type; it has two emitter electrodes.

dual-frequency calibrator A secondary frequency standard providing two fundamental test frequencies, e.g., 100 kHz and 1 MHz.

dual-frequency induction heater An induction heater whose work coils carry energy of two different frequencies. The coils heat the work either simultaneously or successively.

dual gate 1. A digital IC consisting of two gate units. 2. Containing two gates, as a *dual-gate FET*.

dual-gate FET A field-effect transistor having two gate (input) electrodes.

dual-gate MOSFET A metal-oxide-semiconductor field-effect transistor having two gate (input) electrodes.

dual-groove record An early stereophonic record on which two discrete signals were cut (recorded) as concentric spirals that were decoded by separate pickup styli.

dual high-mu triode A combination tube consisting of two high-mu triode sections in a single envelope, e.g., a 12AD7.

dual-inline package Abbreviation, DIP. A flat, molded IC package having terminal lugs along both long edges.

duality 1. The condition of being a dual (see *dual, 2*). Duality is sometimes an aid in the design of certain circuits requiring complementary parameters, e.g., current-operated circuit analogs of voltage-operated circuits. 1. See *duality of nature*.

duality of nature 1. The state of phenomena exhibiting two natures— e.g., light as particles *or* waves, electrostatic effects in terms of fields *or* point charges, and so on. 2. The tendency of a set of principles to be duplicated in sense by predictable analogies, as between inductance and capacitance, electrostatics and magnetics, etc.

dual local oscillator See *double local oscillator*.

dual medium-mu triode A tube combining two medium-mu triode sections in a single envelope, e.g., a 12AU7A.

dual meter A meter having two meter movements and scales in a single case; the arrangement permits simultaneous monitoring of two quantities.

dual modulation The modulation of a single carrier or subcarrier by two different types of modulation, each carrying different kinds of information.

dual network A network which is the dual of another network having complementary parameters. Thus, a common-emitter current-sensitive transistor circuit is the dual of a common-cathode voltage-sensitive vacuum tube circuit. Also see *duality*.

dual operation In digital logic, the operation resulting from inverting all of the digits. Every 1 is replaced with a 0, and vice-versa.

dual-output power supply A power supply with two outputs; often one is positive and the other negative, or one ac and the other dc.

dual pentode A combination tube consisting of two pentode sections in a single envelope, e.g., a 6BU8. Also called *twin pentode*.

dual pickup In disk reproduction, a{pickup having two styli, one for large-groove records and one for fine-groove records.

dual potentiometer A ganged assembly of two potentiometers.

dual preset counter A preset counter that will set alternately to two different numbers.

dual rail See *double rail*.

dual resistor See *dual potentiometer* and *dual rheostat*.

dual rheostat A ganged assembly of two rheostats.

dual stereo amplifier 1. A two-channel audio amplifier for stereophonic applications. 2. A two-channel linear IC for stereophonic af applications.

dual-system loudspeakers See *two-way speaker*.

dual trace In a cathode-ray oscilloscope, the use of two separate electron beams, which can show two different signals simultaneously on a single screen.

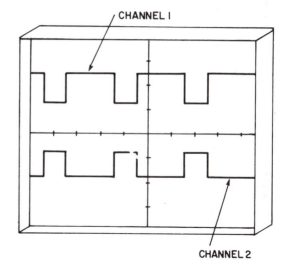

Dual trace.

dual-trace recorder See *double-track recorder, 2*.

dual-track recorder See *double-track recorder*.

dual recording In digital computer practice, updating replica master files simultaneously.

dual triode A tube combining two triode sections in a single envelope, e.g., 26BX8. Also called *twin triode*.

dual use The use of a communications system for two modes of data transfer at the same time.

Duant electrometer See *binant electrometer*.

dub 1. To insert something into a recording. 2. To make a copy of a recording. 3. A copy of a recording.

dubbing The adding of sound to a recorded magnetic tape, record disk, or film, e.g., replacing the sound track of a film in one language with that of another language.

duct 1. A narrow propagation path, sometimes traveled by microwaves, created by unusual atmospheric conditions. 2. A pipe or channel for cables and wires.

duct effect The creation of a duct (see *duct, 1*) through which radio waves seem to travel when temperature inversions are present in the atmosphere.

ductilimeter An instrument used for measuring the ductility of metals.

ducting The confinement of a radio wave to a duct (see *duct, 1*) between two layers of the atmosphere or between an atmospheric layer and the earth.

Duddell arc A carbon—copper arc circuit that produces audible continuous waves. In essence it consists of a series LC circuit shunting an arc.

Duerdoth's multiple feedback system In an amplifier, feedback through several paths to improve response over that afforded by single-path feedback. In a simple application of multiple feedback, a single external loop is augmented with unbypassed emitter resistors in the amplifier stages.

Duerdoth's stability margin A feedback-amplifier stability margin equal to 6 dB increase in gain at low and high frequencies over βA values between 0.3 and somewhat less than 2. For higher βA values, Duerdoth adopts an angular margin (say, 15°); below $\beta A = 0.3$, no danger of instability is present.

dull emitter In an electron tube, a filament or cathode that normally glows dimly.

dull-emitter filament A tube filament that emits electrons in useful quantities, even though it glows dimly.

dummy 1. A nonoperative model of a piece of equipment, usually assembled with dummy components (see *dummy component, 1*) for the purpose of developing a layout. 2. A *dummy antenna, dummy component,* or *dummy load* (all of which, *q.v.*). 3. Part of a computer program that, rather than being useful for the problem at hand, only serves to satisfy some format or logic requirement.

dummy antenna 1. A nonradiating device which serves as a load for a transmitter; i.e., it takes the place of the regular antenna during tests and adjustments of the transmitter. 2. A device containing a network of discrete L, C, and R elements inserted between an rf signal generator and receiver to simulate a standard antenna.

dummy component 1. A nonoperative component used in developing a layout or package. 2. A nonoperative component fraudulently included in a piece of equipment, e.g., an unwired transistor in a receiver circuit, a common occurrence during the early days of the transistor, when a 10-transistor radio brought more money than an 8-transistor radio without regard to the circuit itself.

dummy instruction In a computer, a command that serves no operational purpose, other than to fill a format requirement.

dummy load A (usually resistive) load device employed to terminate a power generator or power amplifier during adjustments and tests. The load resistance is equal to the output impedance of the generator or amplifier.

dummy resistor A power-type resistor employed as a dummy load.

dump 1. In digital-computer practice, to transfer, completely or partially, the contents of memory into a peripheral. 2. Sometimes, to switch off all power to a computer, deliberately or accidentally, thereby losing what is in RAM.

dump and restart During a halt in a computer program run, to backtrack to the last dump point and use the data there to resume the run. Also see *dump point.*

dump check In digital-computer practice, the checking of all digits be-

ing transferred (see *dump, 1*) to forestall error when they are retransferred.

dumping To transfer the output at various stages in a computer program run to an external storage medium, so it will be available (in case of a failure) for the program's resumption from a point other than at the beginning.

dump point In writing a computer program, a point at which instructions are given to transfer data processed thus far to a storage medium that would be unaffected by a software or hardware failure. Also see *dump and restart* and *dumping.*

dumping resistor 1. Bleeder. 2. A resistor having the minimum resistance permissible under individual circumstances and used to discharge a capacitor, i.e., it acts to provide an alternative path to a potentially destructive short circuit.

duo Any pair of matched components, usually in a single package.

duodecal crt base The 12-pin base of a cathode-ray tube. Also see *bidecal, diheptal,* and *magnal.*

duodecal socket A 12-pin tube socket. Also see *duodecal crt base.*

duodecimal 1. Having 12 possibilities, states, choices, etc. 2. Duodecimal number.

duodecimal number system A system of numbering in which the radix is 12 . The system uses the digits 0 through 9 plus two other characters, usually A and B, to represent 10 and 11; e.g., decimal 1000 is duodecimal 6B4: $(6 \times 144) + (11 \times 12) + 4 = 1000$.

duodiode See *dual diode.*

duodiode—high-mu triode A combination tube consisting of two diode sections and one high-mu triode section in a single envelope, e.g., a 6BT6.

duodiode—low-mu triode A tube combining two diode sections and one low-mu triode section in a single envelope, e.g., a 6BU6.

duodiode—medium-mu triode A combination tube consisting of two diode sections and one medium-mu triode secton in a single envelope, e.g., a 6BJ8.

duodiode—pentode A combination tube consisting of two diode sections and one pentode section in a single envelope, e.g., a 6BT8.

duodiode—tetrode A tube combining two diode sections and one tetrode section in single envelope, e.g., a 12DL8.

duodiode—triode A combination tube consisting of two diode sections and one triode section in a single envelope, e.g., a 12DV7.

duolateral coil A multilayer, lattice-wound coil (see *universal winding*) in which the turns in successive layers are staggered slightly. Also called *honeycomb coil.*

duopole A two-pole all-pass device.

duotriode See *dual triode.*

duplex 1. A mode of communication in which two channels are used, so that either operator in a conversation may interrupt the other at any time. 2. The transmission of two messages over a single circuit, at the same time.

duplex artificial line In wire telephony, a balancing network that simulates the impedance of the actual line and the remote terminal equipment; it prevents an outgoing transmission from interfering with the local receiver.

duplex cable A cable which is a twisted pair of insulated stranded-wire conductors.

duplex channel A channel used for wire or radio duplex operation.

duplex communication See *duplex operation.*

duplex diode See *dual diode.*

duplex diode—high-mu triode A combination tube consisting of two diode sections and one high-mu triode section in a single envelope, e.g., a 6BK6.

duplexer In radar practice, a device operated by the transmitted pulse to automatically switch the antenna from the receiver to the transmitter.

duplexing assembly In a radar system, a device which automatically makes the receiver unresponsive to the outgoing transmitted signal while allowing incoming signals to reach the receiver easily. Also called *transmit—receive switch*.

duplex computer system A computer installation of two computer systems, one standing by to take over in case of the other's failure.

duplex operation The simultaneous operation of a transmitter and receiver at two locations. This becomes possible (without mutual interference) through the use of two sufficiently separated carrier frequencies.

duplex system A system composed of two identical equipment sets, either one of which will perform the intended function while the other stands by.

duplex tube An electron tube that is two separate tubes in a single envelope.

duplication check In digital-computer practice, the checking of an operation by doing it twice by alternate methods to insure the accuracy of results.

duplicate To transfer data from one storage location to another. Compare *dump, 1*.

dural See *duralumin*.

duralumin An alloy of aluminum, copper, magnesium, manganese, and silicon. It offers strength with minimal weight.

duration control A potentiometer or variable capacitor for adjusting the duration of a pulse.

duration time The period during which a pulse is sustained, i.e., the interval between turn-on and turn-off time.

during cycle The interval during which a timer is in operation.

durometer An instrument for measuring the hardness of a material.

dust collector See *dust precipitator*.

dust core A magnetic core for radio-frequency coils consisting of very minute particles of iron or an alloy, such as Permalloy.

dust cover A (usually metallic) removable, boxlike enclosure for electronic equipment, or a similar transparent version in the form of a lid on an audio component.

dust-ignition-proof motor A motor whose housing completely prevents the entry of dust, virtually eliminating the danger of fine dust sparking inside the machine.

dust precipitator An electrostatic device for removing dust, lint, and other particles from the air. It consists essentially of a pair of screens or wires through which the air passes; a potential of several thousand volts is maintained between them. The particles acquire a charge and then stick to the oppositely charged wire.

Dutch metal A copper—zinc alloy.

duty cycle In the operation of a device, the ratio of on time to idle time. Duty cycle is expressed as a decimal or percentage.

duty cyclometer A direct-reading instrument for measuring duty cycle.

duty factor 1. The ratio P_{avg}/P_m, where P_{avg} is average power in a system and P_m is peak (maximum) power. 2. The product of duration and repetition rate of regularly recurring pulses comprising a carrier.

duty ratio See *duty factor, 1*.

dV Symbol for *differential of voltage*. (Also *dE* and *de*).

dv Symbol for *differential of velocity*.

dvm Abbreviation of *digital voltmeter*. (Often cap.)

dvom Abbreviation of *digital voltohm-milliammeter*. (Often cap.)

dwell meter An instrument which shows the period (or angle) during which contacts remain closed.

dwell switching Switching action in which the contacts are held closed (or a circuit kept on) for specified periods, as opposed to *momentary switching*.

dwell tachometer A combination dwell meter—tachometer for automobile engine testing and adjustment. The dwell meter allows observation and adjustment of the ignition point cam angle and the tachometer shows motor speed in rpm.

DX 1. Radiotelegraph abbreviation of (usually appreciable) *distance*. 2. Abbreviation of *duplex*.

dX Symbol for *differential of reactance*.

DXer An amateur radio operator who prefers to talk with faraway stations (see *DX*).

Dy Symbol for *dysprosium*.

dY Symbol for *differential of admittance*.

dyadic operation A binary operation, i.e., one using two operands.

dyn Abbreviation of *dyne*.

dyn See *dyna*.

dyna A prefix (combined form) meaning *power*, e.g., *dynamometer*, *dynatron*.

dynameter An instrument used to determine the magnifying power of a telescope.

dynamic 1. Dependent on variable conditions. 2. Movable. 3. A form of transducer that operates by means of the effects of moving conductors in magnetic fields.

dynamic acceleration Acceleration whose magnitude and direction are constantly changing.

dynamic allocation In multiprogramming, a system in which a monitor program assigns peripherals and areas of memory to a program.

dynamic analogy A mathematical similarity between various phenomena involving the motion of particles.

dynamic base current See *ac base current*.

dynamic base resistance See *ac base resistance*.

dynamic base voltage See *ac base voltage*.

dynamic behavior The behavior of a device or system involving the motion of particles, over a period of time.

dynamic braking A technique for stopping a motor quickly using a resistor (the dynamic braking resistance) connected across spinning armature. The resistor dissipates the energy generated by the motor, producing a damping action that results in braking.

dynamic cathode current See *ac cathode current*.

dynamic cathode resistance See *ac cathode resistance*.

dynamic cathode voltage See *ac cathode voltage*.

dynamic characteristic The performance characteristic of a device or circuit under actual ac-signal operating conditions, as opposed to the static characteristic. Example: dynamic emitter current vs static emitter current.

dynamic check 1. A test made under actual operating conditions of a device or circuit. 2. A test made with ac signal current or voltage, rather than with dc quantities.

dynamic collector current See *ac collector current*.

dynamic collector resistance See *ac collector resistance*.

dynamic collector voltage See *ac collector voltage*.

dynamic contact resistance In relay or switch contacts, variation in the electrical resistance of the closed contacts because of variations in contact pressure.

dynamic convergence In a color TV picture tube, the meeting of the three beams at the aperture mask during scanning.

dynamic curve For a tube or transistor, a characteristic curve which takes into account the presence of resistance in series with the device to which the curve applies.

dynamic debugging Any debugging operation performed on a computer system during a normal-speed program run.

dynamic decay Decay resulting from such factors as ion charging in a storage tube.

dynamic demonstrator A teaching aid consisting of a board displaying an electronic circuit, behind which is mounted the actual circuit. Various circuit components (especially adjustable ones) are mounted on the front of the board, in clear view at places where their circuit symbols appear. Pin jacks at important test points in the circuit allow connection of a meter, signal generator, and oscilloscope leads for testing or demonstrating the circuit.

dynamic deviation The difference between ideal output and actual output of a circuit or device operating with a reference input that changes at a constant rate and which is free of transients.

dynamic diode tester An instrument which displays the response curve (or family of curves) of a diode on a calibrated oscilloscope screen. The horizontal axis of the screen indicates voltage; and the vertical axis, current; zeros for both quantities are at center screen. Also see *dynamic rectifier tester*.

dynamic drain current See *ac drain current*.

dynamic drain resistance See *ac drain resistance*.

dynamic drain voltage See *ac drain voltage*.

dynamic dump A dump that occurs during a program run. See *dumping*.

dynamic electric field An electric field whose intensity is constantly changing, either periodically or randomly.

dynamic emitter current See *ac emitter current*.

dynamic emitter resistance See *ac emitter resistance*.

dynamic emitter voltage See *ac emitter voltage*.

dynamic equilibrium 1: The state of balance between constantly varying quantities. 2. The tendency of two current-carrying circuits to maintain at a maximum the magnetic flux linking them.

dynamic error In a periodic signal delivered by a transducer, an error due to the restricted dynamic response of the device.

dynamic flip-flop A flip-flop (bistable multivibrator) which is kept on by recirculating an ac signal. The device may be switched on or off, however, by a single pulse. Compare *static flip-flop*.

dynamic focus Compensation for defocusing caused by the electron beam sweeping in an arc across a flat color TV picture tube screen; the method employs ac focusing-electrode voltage.

dynamic gate voltage See *ac gate voltage*.

dynamic grid voltage See *ac grid voltage*.

dynamic impedance The dynamic resistance (dE/dI) of a device, as opposed to its static resistance (E/I). Also called *dynamic resistance*.

dynamic limiter A limiter, such as is employed in FM circuits, that maintains the output-signal level in spite of appreciable excursions of input-signal amplitude.

dynamic loudspeaker See *dynamic speaker*.

dynamic magnetic field A magnetic field whose intensity is constantly changing, either periodically or randomly.

dynamic memory A usually random-access data storage method in which the memory cells must be electrically refreshed periodically to avoid loss of held data.

dynamic microphone A microphone in which a small coil attached to a vibrating diaphragm or cone moves in a uniform magnetic field to generate the ac output voltage.

dynamic mutual conductance See *dynamic transconductance*.

dynamic-mutual-conductance tube tester See *dynamic tube tester*.

dynamic noise suppressor A noise limiter consisting of an af filter whose bandwidth is directly proportional to signal strength; i.e., it is varied automatically by signal amplitude.

dynamic operating line A curve displaying the control function of a device. For example, the *Eg-Ip* characteristic curve of an electron tube is drawn between the limits of saturation and cutoff.

dynamic output impedance The output impedance of a power supply as seen by the load. It is dEo/dIL, where dIL is the differential of load current and dEo is the resulting differential of output voltage.

dynamic pickup A phonograph pickup whose stylus causes a small coil to vibrate in the field of a permanent magnet.

dynamic plate current See *ac plate current*.

dynamic plate impedance See *ac plate resistance*.

dynamic plate resistance See *ac plate resistance*.

dynamic plate voltage See *ac plate voltage*.

dyanmic printout A printout that takes place as a single function, actuated by one command, and completing itself in one operation.

dynamic problem checking A method of checking the solution obtained by an analog computer, to see that it makes sense (is not absurd).

dynamic programming A method of problem solving, in which continual checks are made to ensure accuracy or conformance to a certain set of rules.

dynamic range In sound reproduction, the ratio (in dB) of the faintest to the loudest sounds reproduced without significant distortion or noise.

dynamic rectifier tester An instrument which displays the response curve of a rectifier on a calibrated oscilloscope screen. During the test, the rectifier receives an ac voltage which has a low positive peak and high negative peak, both corresponding to the rated forward and reverse voltages (respectively) of the rectifier. The horizontal axis of the screen indicates voltage; the vertical axis, current; zeros for both quantities are at center screen.

dynamic regulation In an automatically regulated system, such as a voltage-regulated power supply, the transient response of the system. Dynamic regulation is determined from maximum overshoot and recovery time when the load or line value is suddenly changed.

dynamic regulator A circuit or device providing dynamic regulation.

dynamic relocation In a computer, the capability of rearranging the memory in any desired manner.

dynamic reproducer 1. Dynamic microphone. 2. Dynamic pickup. 3. Dynamic speaker.

dynamic resistance See *dynamic impedance*.

dynamic run See *dynamic check*, 1. see also *dynamic debugging*

dynamics The study of bodies, charges, fields, forces, or pulses in motion. Compare *statics*.

dynamic sequential control In digital computer operation, the computer's changing the sequence of instructions during a run.

dynamic shift register A form of shift register utilizing temporary storage. All of the information can be moved, as a unit, among various blocks of memory.

dynamic source current See *ac source current*.

dynamic source resistance See *ac source resistance*.

dynamic source voltage See *ac source voltage*.

dynamic spatial reconstructor Abbreviation, DSR. An advanced x-ray machine, developed at the Mayo Clinic, that displays organs in three-dimensional views in motion, and allows them to be electronically dissected without actually operating on the patient.

dynamic speaker A loudspeaker in which a small coil (*voice coil*), attached to a diaphragm or cone and carrying an audio-frequency signal current, moves back and forth in a permanent magnetic field and, accordingly, causes the diaphragm or cone to vibrate (emit sound). Compare *magnetic speaker*.

dynamic stop As caused by a computer program instruction, a loop indicating the presence of an error.

dynamic storage See *dynamic memory*.

dynamic subroutine A form of computer subroutine that allows the derivation of other subroutines in various forms.

dynamic test See *dynamic check*.

dynamic transconductance Transconductance determined from ac parameters (Gm = dIp/dEg), rather than from static parameters (Gm = Ip/Ep). Compare *static transconductance*.

dynamic-transconductance tube tester See *dynamic tube tester*.

dynamic transfer characteristic An input-output characteristic determined with respect to the load of a transfer device. Also see *dynamic characteristic*.

dynamic transistor tester 1. An instrument for checking the ac beta of a transistor, rather than its (simple) dc beta. 2. An instrument for determining the condition of a transistor from its performance in a simple oscillator circuit. 3. An instrument which displays the response curve (e.g., Ic vs Vc for constant Ib) or a family of such curves on a calibrated oscilloscope screen. Also see *dynamic diode tester* and *dynamic rectifier tester*.

dynamic tube tester An instrument that checks the dynamic characteristics (such as ac transconductance) of electron tubes, rather than just verifying static characteristics or emission.

dynamo A mechanical generator of electricity—typically, a rotating machine.

dynamoelectric machinery Rotating electric machinery; specifically, generators. Examples: amplidynes, generators, dynamotors, rotary converters.

dynamometer 1. Electrodynamometer. 2. A device for mechanically measuring the output power of a motor.

dynamometer ammeter See *electrodynamometer*.

dynamometer voltmeter See *electrodynamometer*.

dynamophone A dynamometer (see *dynamometer, 2*) which employs two telephone circuits to measure the twist of a shaft.

dynamostatic machine A machine driven by ac or dc power for the generation of static electricity.

dynamotor A (usually small) self-contained motor-generator. The motor and generator portions are enclosed in a common housing, giving the machine the appearance of a simple motor.

dynaquad A pnpn four-layer semiconductor device with three terminals, similar to the silicon-controlled rectifier or thyristor.

dynatron A form of vacuum tube that displays a negative-resistance characteristic, resulting in oscillation at ultra-high and microwave frequencies.

dynatron effect A negative-resistance region evidenced in the *Ep-Ip* characteristic of a tetrode when dc screen voltage exceeds dc plate voltage. The negative slope of the plate-current curve is due to (1) the screen's attracting secondary electrons from the plate when screen voltage exceeds plate voltage (2) the attendant reduction of plate current during the interval.

dynatron frequency meter A heterodyne-type frequency meter employing a dynatron oscillator.

dynatron kink The dip in tetrode plate current caused by the dynatron effect. Tetrode amplifiers must be operated beyond this kink to forestall negative-resistance oscillation.

dynatron oscillator A tetrode oscillator using the negative resistance resulting from the dynatron effect.

dyne Abbreviation, d. A unit of work. One dyne (10^{-5} newton) is the force which will give a mass of 1 gram an acceleration of 1 centimeter per second per second. Compare *newton*.

dyne-centimeter See *erg*.

dyne-five In the Giorgi mks system, a unit of force equal to 1 newton.

dyne per square centimeter Abbreviation, d/cm^2. A unit of pressure equal to 0.1 pascal (0.9869 × 10^{-6} atmosphere).

dyne-seven A unit of force equal to 10^7 dynes.

dynistor A semiconductor diode that continues to conduct after the forward voltage is reduced below the normal threshold point. To stop the conduction, a reverse voltage must be applied, or voltage must be entirely removed from the device. Used in switching applications.

dynode In a photomultiplier tube, a slanting electrode which receives a beam of electrons generated by the light-sensitive cathode and reflects it (with added secondary electrons) to another dynode, and so on from dynode to dynode to the high-voltage plate (collector). Several dynodes are spaced around the inner circumference of the tube, each receiving the reflected beam from the previous dynode and adding secondary electrons as it reflects the beam to the next dynode. In this way, the emission from the light-sensitive cathode becomes greatly amplified by the time it reaches the plate electrode.

dysprosium Symbol, Dy. An element of the rare-earth group. Atomic number, 66. Atomic weight, 162.51. Dysprosium is a highly magnetic substance.

dZ Symbol for *differential of impedance*.

E 1. Symbol for *voltage*. 2. Symbol for *electric field strength*. 3. Abbreviation of *emitter*. 4. Symbol for prefix *exa*. 5. Symbol for *energy*.

e 1. Symbol for *voltage*. 2. Abbreviation of *emitter*. 3. Symbol for *electron charge*. 4. Symbol for the *natural number 2.71828...* (base of natural log). 5. Symbol for *eccentricity*. 6. Abbreviation of *erg*.

EAM Abbreviation of *electronic accounting machine*.

E and M terminals The output and input leads in some signaling systems. Also called *E and M leads*.

early-failure period The period immediately after manufacture of a device, during which the failure rate due to defects in equipment or workmanship is high.

early-warning radar Abbreviation, EWR. A radar system which gives immediate warning when enemy aircraft enter the monitored area.

earphone 1. Headphone (usually a single unit). 2. Telephone receiver. 3. A plug-type earphone that is small enough to be inserted into the ear.

earpiece See *earphone, 3*.

earth 1. The ground. 2. An electrical connection to the earth (see *ground connection, 2*).

earth connection See *ground connection, 2*.

earth currents 1. Electric currents induced in the earth by current flowing through underground or underwater cables. 2. Electric currents flowing through the earth between ground connections of electrical equipment.

earth ground 1. A common connection to an electrode buried in the earth, such that good conductivity is maintained between the common circuit point and the earth itself. 2. A rod driven into the surface of the earth for use as a common circuit connection.

earth inductor A type of magnetometer consisting essentially of a coil which is spun in a magnetic field, such as that of earth, delivering a voltage proportional to field strength. Also called *generating magnetometer*.

earth's magnetic field The natural magnetic field whose lines of force extend from north to south, but do not coincide with earth's north and south geographical poles. The field somewhat resembles that of a bar magnet.

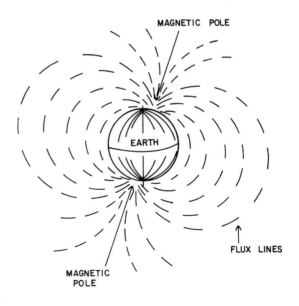

Earth's magnetic field.

Eastern Standard Time Abbreviation, EST. Local mean time at the 75th meridian west of Greenwich.

east-west effect The phenomenon in which the number of cosmic rays approaching earth near the equator from the west is greater than that from the east by 10 percent.

Eavg Symbol for *average voltage.*

Eb Symbol for *battery voltage.*

EBB Symbol for *plate-supply voltage.*

E bend In a waveguide, a smooth change in the direction of the axis, which remains parallel to the direction of polarization.

EBI Abbreviation of *equivalent background input.*

ebiconductivity Conductivity resulting from electron bombardment.

ebonite Hard rubber used as an insulant. Dielectric constant: 2.8. Dielectric strength: 30-110 kV/mm.

EBR Abbreviation of *electron beam recording.*

EBS Abbreviation of *electron-bombarded semiconductor.*

EBS amplifier An amplifying device employing an electron-bombarded semiconductor. The electron beam is modulated by the input signal, and the modulated resistance of the semiconductor target modulates a relatively heavy current to provide an amplified output. Current gains on the order of 2000 are possible with this amplifier.

ec Abbreviation of *enamel-covered* (in reference to wire).

ECC Symbol for *grid-supply voltage.*

eccentric circle See *eccentric groove.*

eccentric groove On a phonograph record, an off-center groove in which the stylus rides at the end of the recording, where it causes the tone arm to trip the record-changing mechanism.

eccentricity 1. The condition of being off center, intentionally or not. It is often a consideration in the behavior of dials, potentiometers, and servomechanisms. 2. On a phonograph record, the condition in which the spiral recording groove and the center hole of the disk are not concentric.

Eccles-Jordan circuit A bistable multivibrator containing two cross-coupled tubes. Also called *flip-flop.*

ECCM Abbreviation of *electronic counter-countermeasures.*

ECDC Abbreviation of *electrochemical diffused-collector*, and description of a type of transistor.

ECG Abbreviation of *electrocardiogram.* (also, *EKG.*)

ECG telemetry Use of a radio telemetering system for continuously monitoring a heart patient who is free to move about unencumbered by trailing wires (he bears a subminiature radio transmitter operating on signals from sensors actuated by his heart action).

echelon 1. A level of calibration accuracy, the highest echelon being the national standard for the particular measurement involved. 2. A level of maintenance in which lower ordinal numbers refer to less critical tasks, and higher ordinal numbers refer to tasks requiring progressively higher skills and technological expertise.

echo 1. A signal which is reflected back to the point of origin. 2. A reflected or delayed signal component which arrives at a given point behind the main component. 3. Loosely, a reflected signal.

echo area The area of a target which will return a radar signal as an echo.

echo attenuation In a bidirectional wire-communication circuit equipped with repeaters or multiplexers, the attenuation of echo currents set up by conventional operation.

echo box A resonant-cavity device used to test a radar set. Part of the transmitted energy enters the box, which retransmits it to the receiver. The signal reaching the receiver is a slowly decaying transient whose intensity eventually falls below the level that can be displayed on the

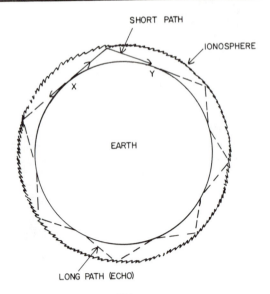

Echo.

screen; the time required to reach this level indicates radar performance.

echo chamber A reverberation chamber, electronic recording device, or room for acoustic tests or for simulating sonic delays.

echo check In telegraphic communication (wire and radio) and numerical data transfer, checking the accuracy of received material by sending it back to the transmitting station for comparison with the original material.

echo depth sounder See *acoustic depth sounder.*

echo eliminator 1. A device which quiets a navigational instrument after the latter has received a pulse, so as to prevent reception of a subsequent, delayed pulse. 2. In a two-way telephone circuit, a voice-operated device which suppresses echo currents caused by conversation currents going in the opposite direction.

echoencephalograph An ultrasonic medical instrument that allows viewing of internal organs. Employed for diagnostic purposes in certain situations, instead of the X-ray machine.

echo intensifier A device employed at a radar target to boost the intensity of reflected energy.

echo interference Radio interference resulting from a reflected signal arriving slightly later than the direct signal.

echo matching In an echo-splitting radar system, the trial-and-error orientation of the antenna to find the direction from which the pulse indications are identical.

echo ranging An ultrasonic method of determining the bearing and distance of an underwater object.

echo sounder A fathometer (see *acoustic depth sounder*).

echo splitting Separating a radar echo into two parts so that a double indication appears on the radar screen. Before taking a reading, the operator brings the two images to equal height by making specific antenna adjustments.

echo suppression In a telephone circuit, a device that chokes off reflected waves, thereby minimizing audible echo.

echo suppressor See *echo eliminator*.

echo talk Echo in a telephone system that results in distracting interference.

echo wave A reflected wave, such as a radio wave reflected alternately between earth's surface and the ionosphere.

ECL Abbreviation of *emitter-coupled logic*.

eclipse effect A decrease in the critical frequency of the E and $F1$ layers of the ionosphere during a solar eclipse.

ECM Abbreviation of *electronic countermeasures*.

eco Abbreviation of *electron-coupled oscillator*.

econometer An instrument for continuously monitoring the amount of carbon dioxide in (factory) flue gases.

E-core A transformer or transducer core having the shape of an E. Coils may be wound on one or all of the E's crosspieces.

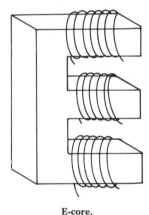

E-core.

ECPD Engineer's Council for Professional Development.

ECTL Abbreviation of *emitter-coupled transistor logic*.

EDD Abbreviation of *envelope-delay distortion*.

eddy current A circulating current induced in a conducting material by a varying magnetic field, often parasitic in nature. Such a current may, for example, flow in the iron core of a transformer.

eddy-current device A brake, coupling, clutch, drag cup, drive unit, or similar device whose operation is based upon the generation of torque, pull, or opposition by the action of eddy currents. See, for illustration, *drag cup*, *drag-cup motor*, and *drag magnet*.

eddy-current heating Heating due to eddy-current losses in a material.

eddy-current loss Power loss resulting from eddy currents induced in nearby structures by an ac field. Thus, eddy currents in the core of a transformer give rise to such loss.

edge connector A terminal block with a number of contacts, attached to the edge of a printed-circuit board for easy plugging into a foundation circuit.

edge control In the manufacture of paper, an electronic system for constantly maintaining the width of a sheet by sensing the edges and correcting the machine accordingly. Transducers that sense the passing edges deliver output signals proportional to variations from standard width.

edge effect The extension of electric lines of force between the outer edges of capacitor plates: this portion of the interplate field contributes a small amount of capacitance. Because the lines of force are not confined to the space between plates, they can cause capacitive coupling with external bodies.

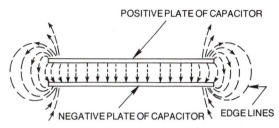

Edge effect.

edge-punched card In digital computer practice, a punched card whose edge is perforated in a rather narrow column, the center being used for written annotation.

edgewise meter A meter having a curved horizontal scale; this arrangement allows mounting the instrument edgewise in a panel.

edgewise-wound coil A coil made of a flat metal strip cut in the shape of a coil spring. The design allows the use of clips to vary the inductance, but this advantage is often offset by the coil's high distributed capacitance.

edging In a color TV picture, extraneous color of a different hue than the objects around whose edges it appears.

Edison base A threaded base on light bulbs, cone-type heaters, and some pilot lamps.

Edison battery A group of *Edison cells* connected in series, parallel, or both and contained in a single package with two electrodes.

Edison cell A secondary (storage) cell in which the active positive plate material consists of nickel hydroxide held in steel tubes assembled into a steel grid; the active negative plate material is powdered iron oxide mixed with cadmium, and the electrolyte, potassium hydroxide. The open-circuit voltage of the cell is typically 1.2V at full charge.

Edison distribution system The traditional three-wire, 110-250V dc distribution system for use in buildings not supplied with alternating current.

Edison effect The (thermionic) emission of electrons (negative particles) from a hot filament sealed in an evacuated bulb; they are attracted by a cold, positively charged metal plate in the bulb. Edison discovered this action in 1888, when he was looking for the cause of his newly invented incandescent lamp's becoming black inside. This discovery formed the basis of the vacuum tube upon which electronics rested until the coming of modern semiconductor devices in the 1950s.

E-display A radar display in which a target is represented by a blip whose horizontal coordinate indicates distance, and vertical coordinate, elevation.

edit 1. In tape recording, the modifying of the recorded material by deleting (cutting out or erasing), adding (splicing or overrecording), or changing the sequence of the material by physically or magnetically altering the tape. 2. In digital computer practice, to do all that is necessary to make data ready for processing.

editing 1. Alteration of data stored in memory, either by adding information, removing information, changing information, or (usually) a combination of these operations. 2. Alteration of a magnetic-tape recording by means of splicing.

EDM Abbreviation of *electrical-discharge machining*.

EDP Abbreviation of *electronic data processing*.

EDPC Abbreviation of *electronic data processing center*.

EDPM Abbreviation of *electronic data processing machine*.

EDPS Abbreviation of *electronic data processign system*.

EDT Abbreviation of *ethylene diamine tartrate* (a synthetic piezoelectric material).

EDU Abbreviation of *electronic display unit*.

EDVAC Acronym for *e*lectronic *d*iscrete *v*ariable *a*utomatic *c*omputer, a development of the University of Pennsylvania.

Edward Phonetic alphabet code word for the letter *E*.

EE Abbreviation of *electrical engineer* or *electronic engineer*.

Eeff Symbol for *effective voltage*.

EEG Abbreviation of *electroencephalogram*.

EEPROM Abbreviation of *electrically erasable programmable read-only memory*. See *PROM* and *ROM*.

Ef Symbol for *filament voltage*.

effective acoustic center The apparent point of propagation of spherically divergent sound waves radiated by an acoustic generator.

effective actuation time The total actuation time of a relay, i.e., the sum of the initial actuation time and subsequent intervals of contact chatter.

effective address The address a computer uses in implementing an instruction, i.e., one not necessarily coinciding with the address given in the instruction.

effective ampere An effective current of 1 ampere. Also see *effective current*.

effective antenna length See *electrical length*.

effective antenna resistance The rf resistance of an antenna as measured at the input point.

effective area A measure of the directional properties of an antenna system. Given a specified direction, the effective area is equal to $\lambda^2 g/4\pi$, where λ is the wavelength in meters, g is the antenna power gain expressed as a ratio, and the resultant effective area is given in square meters.

effective bandwidth The bandwidth of an ideal bandpass filter which, at a reference frequency, has the same transfer ratio as an actual bandpass filter under consideration; it also has the same current and voltage characteristics.

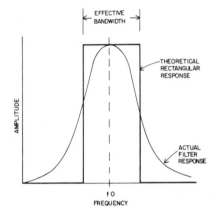

Effective bandwidth.

effective capacitance The actual capacitance between two points in a circuit due to stray, inherent, and lumped capacitance.

effective conductivity Conductivity measured between the parallel faces of a unit cube of a material.

effective confusion area In a radar system, an area in which interference makes it impossible to tell whether or not a target is present.

effective current Symbol, *Ieff*. The root-mean-square value of alternating current (see *effective value*). For a sinusoidal current, *Ieff* = 0.707*Imax*, where *Imax* is the maximum value of the current. Also called *rms current*.

effective cutoff See *effective cutoff frequency*.

effective cutoff frequency For a filter or similar device operated between specified impedances, the frequency at which insertion loss is higher than the loss at a specified reference frequency in the pass band.

effective field intensity The rms value of the inverse-distance field-strength voltage in all horizontal directions 1 mile from a transmitting antenna.

effective height The above-ground height of an antenna in terms of its performance as a transmitter or receiver of electromagnetic energy. For a quarter-wave vertical wire, the effective height *he* is *ha*/π, where *ha* is the actual height above ground.

effective input capacitance 1. The actual operative capacitance present at the input terminals of a circuit or device due to the shunt capacitance of the terminals themselves and the net capacitance of the circuit connected to the terminals. 2. In a vacuum tube, the capacitance between grid and cathode terminals. It is a combination of the internal grid—cathode capacitance and a network of other internal capacitances in the tube.

effective isolation The condition of components or circuits being so well isolated or shielded that no significant direct coupling, capacitive coupling, or inductive coupling exists between them.

effective instruction The machine language version of an instruction given in a computer program, as produced by resident software.

effectively bonded The condition afforded by an extremely low-resistance union between two conducting surfaces which are solidly fastened together.

effectively grounded The condition of being connected to earth or to the low-potential end of a circuit by means of an extremely low-resistance connection.

effective parallel capacitance Inherent capacitance that manifests itself in parallel with two circuit points in combination with any lumped capacitance.

effective parallel resistance 1. The leakage resistance which manifests itself in parallel with a dielectric, e.g., the leakage resistance of a capacitor. 2. Parallel-resistance effects due to stray shunt-resistance components.

effective percentage of modulation For a complex waveform, an expression of the equivalent percentage of modulation by a pure sine wave. Given a certain proportion of power in the sidebands with modulation by a complex signal, the effective percentage of modulation is that percentage which, when the modulating signal is sinusoidal, results in the same proportion of power in the sidebands.

effective phase angle In ac circuits, phase angle with respect to waveforms for current and voltage. When both waveforms are sinusoidal, the effective phase angle is the actual phase angle (i.e., power factor = cos ω). But when harmonics are present in current or voltage, the angles differ, the difference being greater in capacitive circuits than in inductive circuits.

effective radiated power Symbol, *Per*, abbreviation, ERP. The rf power delivered by an antenna in terms of antenna gain; *Per = PiAp*, where *Per* is the effective radiated power (watts); *Pi*, the antenna input power (watts); and *Ap*, the power gain of the antenna.

effective resistance 1. In a coupled circuit, the sum of the actual resistance of the circuit and the reflected resistance of the load. 2. Effective antenna resistance.

effective series inductance Inherent (distributed) inductance acting in series with other components in a circuit. The inherent inductance of the wire in a wirewound resistor, for example, manifests itself in series with the resistance of the device.

effective series resistance Inherent (distributed) resistance acting in series with other components in a circuit. Thus, the inherent resistance of the wire in a coil appears in series with the inductance of the coil. Likewise, a capacitor has an effective series resistance due to the resistance of leads, plates, and connections.

effective shunt capacitance See *effective parallel capacitance*.

effective shunt resistance See *effective parallel resistance*.

effective sound pressure The rms value of instantaneous sound pressure at one point in a sound cycle.

effective speed of transmission In telegraphy (wire or radio) and in electronic data transmission, the transmission speed (characters per minute, bits per second, etc.) that can be reliably maintained for a given period.

effective thermal resistance The effective temperature rise (in degrees per watt of dissipation) of a semiconductor junction above an external reference temperature that is at equilibrium.

effective time For a computer, the time during which useful work is done.

effective transmission speed See *effective speed of transmission*.

effective value The root-mean-square value of an ac quantity. The effective value an alternating current produces in a pure resistance has the same heating effect the equivalent direct current does. Also called *rms value*.

effective volt An effective potential of 1 volt (rms). Also see *effective volt*.

effective voltage Symbol, *Eeff*. The root-mean-square value of ac voltage (see *effective value*). For a sinusoidal voltage, *Eeff = 0.707Emax*, where *Emax* is the maximum value of the voltage. Also called *rms voltage*.

effective wavelength Wavelength in terms of measured frequency and effective propagation velocity.

efficiency See *electrical efficiency*.

efficiency modulation A system of amplitude modulation in which the efficiency of the rf amplifier is varied at an audio-frequency rate.

efficiency of rectification Symbol, η. For a rectifier, the ratio of dc output voltage (*Edc*) to the peak value of ac input voltage (*Eac*); $\eta = Edc/Eac$. (For percent efficiency, multiply the result by 100.)

efflorescence The giving up of water by a substance upon exposure to air. Some materials employed in electronics exhibit this property. Common efflorescent compounds are hydrated ferrous carbonate, ferrous sulfate, and sodium carbonate.

efflorescent material A material exhibiting efflorescence. Compare *deliquescent material*.

E field 1. An electric field. 2. The electric-field component of an electromagnetic wave.

EFL Abbreviation of *emitter-follower logic*.

Eg 1. Symbol for *grid voltage*. 2. Symbol for *generator voltage*.

Eh Symbol for *heater voltage*.

EHD Abbreviation of *electrohydrodynamic(s)*.

ehf Abbreviation of *extremely high frequency*.

EHP Abbreviation of *effective horsepower*.

E-H tee A waveguide junction in which E- and H-plane tee junctions intersect the main waveguide at the same point. Also see *waveguide tee*.

E-H tuner An impedance-transforming E-H tee with two arms that are terminated in tunable plungers for critical adjustments. See *waveguide plunger*.

ehV Abbreviation of *extra-high voltage*.

Ei Symbol for *input voltage*.

EIA Electronic Industries Association.

eight-level code A code, such as the American Standard Code for Information Interchange (ASCII), in which each character is represented by eight bits.

E indicator A radar elevation display, in which the horizontal scale shows range and the vertical scale shows elevation. The display screen itself, therefore, lies in a vertical plane containing the target and the radar antenna.

Einstein equation The mass-energy equation depicting the interconversion of mass and energy; $E = m^2$, where *E* is energy (ergs), *m* is mass (grams), and *c* is the velocity of light in a vacuum (centimeters per second).

einsteinium Symbol, Es. A radioactive element produced artificially. Atomic number, 99. Atomic weight, 254.

Einstein shift The decrease in frequency and loss of energy experienced by quanta acted upon by gravitational attraction.

Einstein's theory The theory of relativity put forth by Albert Einstein. It—both the general and special versions—expresses a number of ideas, one of which is easily applied to the electron: the mass (*m*) of the electron (see *electron mass*) varies with the velocity at which it moves:

$$m = m/(1 - \sqrt{v^2/c^2})$$

where *m* is the mass (grams) of the electron; *mo*, the rest mass (grams) of the electron; *v*, the initial velocity (cm/s) of the electron; and *c*, the velocity of light (3×10^8 m/s).

Einthoven string galvanometer A simple galvanometer in which a silvered glass filament carrying current is mounted in a magnetic field set up by either a permanent magnet or an electromagnet. The current causes the filament to be deflected through a distance proportional to current strength, the deflection being observed through a microscope.

EIT Abbreviation of *engineer-in-training*.

either-or operation The logical inclusive-OR operation.

EJC Engineer's Joint Council.

Ek Symbol for *cathode voltage*.

eka-aluminum See *gallium*.

eka-silicon See *germanium*.

EKG Abbreviation of *electrocardiogram*. (Also, *ECG*.)

EKG telemetry See *ECG telemetry*.

EL Abbreviation of *electroluminescent*.

elapsed time 1. In data processing and computer practice, what *seems* to be the duration of a process, compared with actual processing time as measured by internal clocks, for example. 2. The accumulated time, usually expressed in hours, minutes, and seconds, that an operation takes or a machine runs.

elapsed-time meter An instrument which indicates the time an electronic device or system has been in operation. Most such meters are based on electric clockwork that runs only while the system is in operation, holding the count during shutdown periods. Also see *electrolytic elapsed-time meter.*

elastance Unit, daraf. The opposition of a capacitor to being charged. It is the reciprocal of capacitance. One daraf is the elastance of a capacitor which requires 1V/C displacement. (Daraf is *farad* spelled backward.)

elastic collision Collision between two charged particles in which neither loses energy even though they are deflected from their normal paths.

elasticity 1. The ability of a body to return to its original shape after being deformed. See *Young's modulus.* 2. Electric elasticity (see *elastivity, 1, 2*).

elastic limit The maximum stress that can be tolerated by a material without being permanently deformed.

elastic wave A wave in an elastic medium such as air or water; thus, a wave that is mechanically produced.

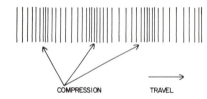

Elastic waves.

elastivity 1. Specific elastance, i.e., the elastance in darafs per cubic unit of a dielectric. 2. The ratio of electric stress to displacement.

E-layer A radio-wave-reflecting layer in the ionosphere; it is 50 to 90 miles above the surface of the earth.

elbow bend A 90-degree bend in a waveguide.

ELD Abbreviation of *edge-lighted display.*

electra A radionavigational system in which equal-intensity signal zones (usually 24) are provided.

electralloy A nonmagnetic alloy used in the manufacture of radio hardware, such as chassis. A nonmagnetic alloy useful in electronic foundation structures such as chassis.

electre See *electrum.*

electrepeter A device used to change the direction of an electric current.

electret A device whose heart is a dielectric disk or slab which is permanently polarized electricalaclly and so possesses a permanent electric field. The electret is the electric equivalent of the permanent magnet. Certain waxes, ceramics, and plastics acquire permanent

polarization after they have been heated and then cooled slowly in an intense electric field.

electret microphone A microphone in which sound waves cause an electret to vibrate and, accordingly, generated an af output voltage.

electric 1. Pertaining to electricity and its various manifestations. 2. Electrostatic.

electric absorption See *dielectric absorption.*

electric accounting machine A data processing machine that is neither a computer nor a computer peripheral.

electric chair The electrode-bearing chair used for execution by means of high-voltage electricity. See *electrocution.*

electrical-acoustical transducer A transducer, such as a headphone, sonic applicator, or buzzer, which converts electrical energy into sound energy. Compare *acoustical-electrical transducer.*

electrical angle The angle assumed at any instant by the rotating vector representing an alternating current or voltage (360° in an ac cycle). Thus, for an ac sine wave, the angle is 90° ($\pi/2$ rad) for positive maximum, 270° ($3\pi/2$ rad) for negative maximum, and 0°, 180° (π rad) and 360° for zero. Also called *phase angle.*

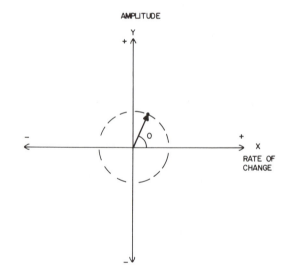

Electrical angle.

electrical attraction The attraction between two oppositely charged bodies or particles. Compare *electrical repulsion.*

electrical axis In a quartz crystal, the axis through opposite corners of the hexagonal cross section. The various electrical axes are *x, x',* and *x"* (or *x-x, x'-x',* and *x"-x"*). The electrical axis is perpendicular to the mechanical axis, which runs through the crystal's length. Also see *crystal axes* and *x-axis, 2.*

electrical bail An action in which a special switch changes contact position and locks itself in that position after a station has been actuated, at the same time releasing a previously actuated station.

electrical bandspread In a tuned circuit, bandspread obtained by changing values of inductance or capacitance rather than by mechanical gearing.

electrical bias A current maintained in a relay coil (sometimes an auxiliiary coil) to keep the relay partially closed, thus sensitizing it. Compare *mechanical bias.*

electrical boresight In radar practice, the tracking axis as determined by an electrical test, such as one involving a sharp null or sharp peak response.

electrical center The point at which an adjustable component (variable resistor, variable inductor, etc.) has exactly half its total value. This point does not always coincide with the physical center.

electrical conductance See *conductance.*

electrical conduction The flow of current carriers through a material. The degree of conduction is indicated by the material's value of conductance.

electrical conductivity See *conductivity.*

electrical coupling The coupling of two or more circuits or elements by means of electric-field effects.

electrical degree 1. In a periodic waveform, the length of time corresponding to 1/360 of the time for completion of one cycle. 2. In space, that distance representing 1/360 of the wavelength in the medium through which electromagnetic energy travels.

electrical discharge The flow of current out of a voltage reservoir such as a battery or capacitor.

electrical discharge in gases The phenomenon of electric conduction (current) by a gas, caused by sudden breakdown due to gas ionization. The discharge is often accompanied by light, as in the red glow of a neon bulb.

electrical-discharge machining A method of machining metals in which the metal is vaporized by an arc formed between an electrode and the metal workpiece (anode). In this way, metal is removed in tiny bits from the surface of the workpiece.

electrical distance Distance in terms of the time required for an electromagnetic wave to travel between two points in free space.

electrical drainage Diverting electric currents away from underground pipes to prevent corrosion by electrolysis.

electrical efficiency Symbol, η. The ratio of the output of an electrical device to the input. Efficiency may be expressed as a decimal or percentage. For example, for a vacuum tube, power efficiency η is $100(Po/Pi)$, where η is efficiency (%); Pi, input power (W); and Po, output power (W).

electrical elasticity See *capacitance.*

electrical element See *element, 2.*

electrical energy Energy in the form of electricity (see *electricity, 1*). The term is often used in place of *electricity.*

electrical engineer Abbreviation, EE. A trained professional skilled in applying physics and mathematics to electricity and in the theory and application of basic engineering and related subjects. Of particular interest to the EE are the generation and distribution of electrical energy and the design and application of electromechanical devices. Compare *electronic engineer.*

electrical erosion In electrical contacts, loss of metal as a result of the evaporation or transfer of metal during switching.

electrical filter A bandpass, band-rejection, highpass, or lowpass filter that operates by electrical means.

electrical forming See *electroform, 1.*

electrical gearing In an electromechanical system, such as a servo, the condition in which an output shaft is electrically rotated at a speed different from that of an input shaft.

electrical glass High-temperature insulating materials made from glass fibers.

electrical inertia See *inductance.*

electrical initiation 1. Starting an action (electrical or nonelectrical) by means of an electrical signal. 2. Using an enabling pulse.

electrical instrument A device such as a frequency, voltage current, or power meter for measuring an electrical quantity. An electrical instrument *per se* doesn't employ tubes or transistors; an electronic instrument does.

electrical interlock A door- or lid-operated switch connected in series with the on-off power switch of a piece of equipment so that power is removed (for shock prevention) whenever the door is opened, the lid lifted, or the case removed.

electrical length The wavelength of an antenna or transmission line in terms of actual performance. The electrical length is less than the actual length as a result of ground-capacitance effects, end effects, and the velocity of waves in wire.

electrical load A device connected to a source of electricity (generator, amplifier, network, etc.) for a useful purpose (heat, work, etc.).

electrically connected Connected via direct path, such as through a wire, resistance, inductance, or capacitance.

electrically erasable PROM A PROM that can be erased by an electrical signal rather than by exposure to ultraviolet light. Also see *PROM* and *ROM.*

electrically variable capacitor A varactor (see *voltage-variable capacitor*).

electrically variable inductor An inductor whose value varies inversely with the amount of direct current which is caused to flow through it or through an auxiliary winding on the same core.

electrically variable resistor A varistor (see *voltage-dependent resistor*).

electrical nature of matter The ultimate reality of matter as a complex interplay of waves and particles. Also see *electron theory of matter, wave mechanics,* and *wave theory of matter.*

electrical network A circuit containing two or more components (including generators and loads), usually arranged in some pattern.

electrical noise Extraneous undersirable currents or voltages which interfere with desirable electrical quantities. Compare *acoustic noise.*

electrical polarity The distinct difference observable in electrification, and designated *positive* (+) and *negative* (–). Negative electrification is characterized by an excess of electrons, positive electrification by a deficiency of them.

electrical repulsion The mutual repulsion of bodies or particles having like electric charges; i.e., two positive entities will repel each other, as will two negative entities. Compare *electric attraction.* Also see *electrical polarity.*

electrical reset An electromechanical device for resetting a relay that ordinarily remains in the position resulting from actuation.

electrical resistance The in-phase current-retarding effect all conductors exhibit to some extent. Also see *resistance.*

electrical resistivity See *resistivity.*

electrical resolver A synchro whose rotor has two perpendicular windings in addition to another winding.

electrical scan A method of changing the orientation of the major lobe of an antenna. The antenna is kept physically stationary, but the phase/amplitude relationships of the signals applied to different driven elements are varied.

electrical sheet Sheet iron or steel used for motor laminations.

electrical system 1. The overall configuration of electrical elements for a set of apparatus. 2. The wiring system that supplies power to a set of devices.

electrical taste See *galvanic taste*.

electrical technology The theory and practical application of electricity. Electrical technology is taught as a subengineering major, usually in two-year colleges that award the degree of associate in arts (AA) or associate in science (AS)

electrical time constant For a torque motor, the ratio of armature inductance to effective armature resistance. Compare *mechanical time constant*.

electrical transcription 1. A phonograph record made electrically, as opposed to one made mechanically. 2. A radio program in which such a record is played. 3. Any direct mechanical or electrical recording of an audio signal.

electrical transducer 1. A transducer that converts a nonelectrical phenomenon into a proportional current, voltage, or frequency. 2. A transducer that converts electricity in one form to electricity in another, e.g., a transducer actuated by dc voltage and delivering an ac voltage whose frequency is proportional to the dc voltage.

electrical twinning A defect in which two quartz crystals intergrow in such a way that the electrical sense of their axes becomes reversed. Compare *optical twinning*.

electrical unit A standard for measuring an electrical quantity. Examples of practical units are ampere, ohm, volt, watt, siemens, etc.

electrical wavelength The distance over which an electromagnetic wave propagates in the time required for completion of one cycle. In free space, the electrical wavelength in meters is equal to 300 divided by the frequency in megahertz. In other media, the wavelength may be shorter, depending on the velocity factor.

electrical zero 1. A zero-output or minimum-output point resulting from the adjustment of a bridge or other zero-set circuit. 2. In a meter whose pointer is mechanically set to some point above or below the zero on the scale, the zero setting obtained when the meter is deflected to scale zero by a current or voltage. 3. For a synchro, the position at which the amplitudes and time phase of the outputs are defined.

electric and magnetic double refraction See *Kerr electro-optical effect* and *Kerr magneto-optical effect*.

electric arc A sustained luminous discharge in the space between two electrodes. Also see *arc, 1*. Compare *electric spark*.

electric aura See *electric wind*.

electric balance See *bridge, 1*.

electric bell See *bell*.

electric brazing A method of brazing in which electric current generates the required heat.

electric breakdown 1. The usually sudden ionization of a gas by an electric field and the accompanying heavy current flow through the gas. 2. The (destructive) puncture of a dielectric by the strain produced by high voltage. Also see *dielectric strength*. 3. The usually nondestructive, abrupt increase in semiconductor junction current at a high reverse voltage. See, for example, *avalanche breakdown*.

electric breakdown voltage 1. The voltage at which avalanching occurs. 2. Dielectric strength.

electric breeze See *electric wind*.

electric buzzer See *buzzer*. Compare *electronic buzzer*.

electric calculator An electrically driven machine for performing mathematical operations. Its electromechanical nature distinguishes it from the electronic calculator, which employs no moving parts other than keys. Also see *calculator*.

electric catfish A fish native to tropical and northern Africa, which can deliver a strong electric shock.

electric cell See *cell, 1*.

electric charge Potential energy as the electrification of a body or component. For a capacitance of C farads charged to a potential of E volts, the charge Q, in coulombs, is CE. Also see *energy stored in capacitor*.

electric chronograph An instrument for accurately recording time intervals.

electric chronometer A precision electric or electronic timepiece. Also see *electric clock* and *electronic clock*.

electric circuit A network of interconnected components and devices, often including a source of electric power. Current flowing through a circuit is acted upon by the components, which produce a desired end.

electric clock A clock driven by electric current. Electric clocks fall into two categories: those driven by synchronous ac motors and those driven by (usually dc) stepping mechanisms.

electric column See *voltaic pile*.

electric conduction The flow of current carriers through a conductor.

electric constant Symbol, ϵ_0. The fixed electrical permittivity of free space, the value of which is $8.854\,188 \times 10^{-12}$ farad per meter.

electric contact See *contact, 1, 2*.

electric controller An adjustable device for modifying the operating voltage or power of a component or system. Compare *electronic controller*.

electric cooling 1. Cooling using the Peltier effect. 2. Electrostatic cooling. 3. Forced-air cooling (of equipment) by electric blowers or fans.

electric current The phenomenon wherein electrons move in a directed manner through a material or vacuum. In a semiconductor material, electric current may be due to the movement of holes as well as electrons, depending on the nature of the semiconductor. In a gas or electrolyte, current is the flow of ions (see *ionic current*).

electric current density See *current density*.

electric delay line See *delay line*.

electric density See *electric space density* and *electric surface density*.

electric dipole A pair of equal charges having opposite polarity and separated by a fixed distance.

electric discharge See *electrical discharge*.

electric-discharge lamp See *discharge lamp*.

electric disintegration See *electric dispersion*.

electric dispersion In a colloidal suspension, dispersion accomplished by passing an electric current through the material.

electric displacement The movement of a body or particle in response to an electric current or field.

electric double refraction See *Kerr electro-optical effect*.

electric doublet See *doublet, 1*.

electric dust precipitator See *dust precipitator*.

electric eel An eel (fish) capable of delivering a disabling shock on contact.

electric elasticity See *elastivity, 1, 2*.

electric endosmosis See *electro-osmosis*.

electric eye See *photoelectric cell* or *phototube*.

electric fidelity The frequency response of a circuit or device.

electric field The space surrounding an electric charge or charged body, in which electric energy acts (electric lines of force fill the space). Also see *flux* and *line of force, 1*.

electric field intensity See *electrostatic field intensity*.

electric field strength 1. Symbol, E. In an electromagnetic wave, the amplitude of the electric component of the field, expressed in volts per meter. 2. Dielectric strength.

electric-field vector See *electric-field strength, 1*.

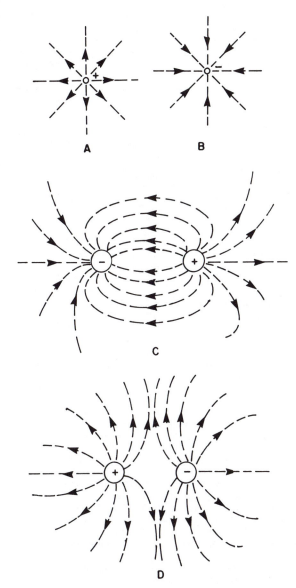

ELECTRIC FIELD AROUND A POSITIVE
CHARGE (A), A NEGATIVE CHARGE (B),
UNLIKE CHARGES (C) AND LIKE CHARGES
(D).

Electric field.

electric filter See *electric wave filter*.

electric fish Fish capable of generating intense electric shocks, e.g., electric catfish and electric eel.

electric forces The forces exerted by electric charges or electric fields. Also see *unit electrostatic charge*.

electric flux See *electrostatic flux*.

electric flux density Symbol, *D*. In an electric field, the number of lineenes of force per unit area, usually expressed in coulombs per square meter.

electric focusing See *electrostatic focusing*.

electric force The force exerted by an electrically charged particle or an electric field.

electric furnace An electrically heated chamber (sometimes heated by an electric arc) employed in ore reduction, carbide manufacture, and other high-temperature processes.

electric generator A device for producing electricity. Thus, many different devices, such as batteries, dynamos, oscillators, solar cells, and thermocouples, are classed as generators.

electric glow The light (usually pale blue) that occasionally accompanies an electric discharge in air.

electric guitar A guitar whose sound (vibrations) is converted by a pickup (transducer) to an electrical analog for amplification.

electric hygrometer An instrument for measuring humidity in terms of the moisture-sensitive resistance of a sensor. Only a relatively simple dc circuit is used. Compare *electronic hygrometer*.

electric hysteresis See *electrostatic hysteresis*.

electrician A tradesman who installs and services electrical equipment and wiring.

electric image For solving certain problems involving electricity, an array of electrical points forming an image of certain other electrical points.

electricity 1. The phenomenon due to stationary or moving electric charges. 2. The branch of physics concerned with the nature of electricity and its production. 3. Electrical energy or power.

electric lamp An electric-powered device used primarily as a source of light. Common types are arc, incandescent, and fluorescent.

electric light 1. Light produced by means of electricity. 2. Electric lamp.

electric lines of force Lines of force associated with an electric charge and constituting the charge's electric field. Used interchangeably with *electrostatic lines of force*. Also see *electric field* and *line of force, 1*.

electric machine A mechanical device for generating static electricity. See, for illustration, *electronic generator*, *Van de Graaf generator*, and *Wimshurst machine*.

electric meter 1. An instrument—e.g., ammeter, voltmeter, wattmeter—for indicating an electrical quantity (usually directly). 2. Service meter (see *kilowatt-hour meter*).

electric mirror See *electron mirror*.

electric moment In an electric field of unit intensity, the maximum torque exerted on an electric dipole.

electric motor A machine that converts electrical energy into mechanical work. The familiar form is the machine in which an armature rotates between the poles of a field magnet, mechanical energy being produced at the armature's revolving shaft.

electric needle A needle electrode carrying high-frequency current; it is used in surgery to cut tissue and sear it immediately afterwards to prevent bleeding.

electric network See *electrical network*.

electric organ See *electronic organ*.

electric oscillations The alternate flow of electric charges in opposite directions.

electric osmosis See *electro-osmosis*.

electric piano See *electronic piano*.

electric potential See *electrostatic potential*.

electric power Symbol, *P*. Unit, watt. The rate at which electrical energy is used. Power is energy per unit time; in the context of electricity, it is expressed as the product of current and voltage. In terms of heat losses, it is often expressed as I^2R.

electric precipitator See *dust precipitator*.

electric probe A pin or rod inserted into an electrostatic field to sample it, or into an electromagnetic field to sample its electric component. See, for example, *waveguide probe*. Compare *magnetic probe*.

electric radiation 1. The radiation of energy by means of electric waves. 2. The energy so radiated.

electric recording Inkless recording on paper by direct use of an electric current. There are two principal types: (1) A current-carrying stylus burns away (in a fine line) the metallic coating of the recording paper, exposing the dark underlying layer. (2) A stylus delivers current which produces a line by means of electrolysis in a special paper (see *electrolytic recorder*). Compare *electrostatic printer*.

electric reset See *electrical reset*.

electric residue A residual electric charge.

electric rings Colored rings formed on a plate by the electrolytic deposition of substances, such as copper and some peroxides.

electric screen See *electrostatic screen*.

electric shield See *electrostatic screen*.

electric shock The (potentially lethal) physiological reaction caused by electricity. When slight, it is characterized by involuntary contractions of the muscles and tingling sensations; a severe shock causes paralysis, cardiac arrest, and burns.

electric spark A momentary, luminous discharge of electricity in the space between two electrodes. Compare *arc, 1*.

electric steel Steel which has been processed in an electric furnace.

electric strain See *dielectric strain*.

electric strain gauge A device for detecting the strain that a certain stress produces in a body. Typically, such a gauge consists of one or more fine insulated wires cemented to the surface under test. As the surface becomes strained, the wire stretches, undergoing a change in electrical resistance that is proportional to the change in strain.

electric strength See *dielectric strength*.

electric stress See *dielectric stress*.

electric stroboscope See *electronic stroboscope*.

electric surface density The ratio of the electric charge on a surface to the area of the surface. Compare *electric volume density*.

electric tachometer See *electronic tachometer*.

electric telemeter A device that sends metered data over a wire transmission line to a remote point for monitoring.

electric thermometer See *electronic thermometer*.

electric transcription See *electrical transcription*.

electric transducer A transducer which responds to nonelectric energy and then delivers a proportional electric current or voltage to another circuit; an electrical transducer.

electric tuning A means of adjusting the frequency of a receiver, transmitter, transceiver, or oscillator, without the use of mechanical devices. An example is digital tuning, in which a frequency is chosen by selecting the digits independently.

electric vane A small demonstration device consisting of a rotor having several spokes terminating in points. The rotor is mounted on a pivot bearing and, when it is connected to a source of high voltage,

spins from the force of electricity escaping from the points into the surrounding air.

electric vector In an electromagnetic field, the vector representing the electric component. It is perpendicular to the magnetic vector.

electric volume density The ratio of the electric charge in a space to the volume of the space. Compare *electric surface density*.

electric watch A watch driven by a tiny, self-contained cell which drives an electrical escapement or other stepping mechanism. Compare *electronic watch*.

electric wave See *electromagnetic waves*.

electric-wave filter A circuit or device for separating signals of one frequency from those of other frequencies. Also see *filter, 1*.

electric whirl See *electric vane*.

electric wind 1. Air currents set up by electrons escaping from the sharp point of a high-voltage electrode. 2. The outward-rushing plasma (solar wind) ejected by the sun and traveling through space.

electrification 1. Generating an electric charge in a body, as in charging a glass rod by rubbing it with a silk cloth. 2. Providing electric service, e.g., *rural electrification*. 3. The conversion of a system from purely mechanical to electrical or electromechanical.

electroacoustic Combining electricity and sound. Thus, loudspeakers and microphones are electroacoustic devices.

electroacoustic amplifier See *carbon-button amplifier* and *surface-wave amplifier*.

electroacoustic device A device that transfers energy by converting it from electrical to acoustic form.

electroacoustic transducer A transducer that converts sound vibrations into electrical pulsations or, conversely, one that converts electricity into sound.

electroanalysis Chemical analysis performed by electrolytic methods.

electroanesthesia Anesthesia produced by an electric current going through some part of the body.

electroballistics The art and science of electrically or electronically measuring the velocity of bullets or other projectiles.

electrobath Electroplating solution.

electrobiology Biology concerned with electrical phenomena in living organisms.

electrobioscopy The examination of a body for viability by inducing muscular contractions with an electrical impulse.

electrocapillarity The production of capillary effects by means of electricity. For an application of the phenomenon, see *capillary electrometer*.

electrocardiogram Abbreviation, ECG or EKG. A record made by an *electrocardiograph* of the changes in potential caused by the heartbeat and used as a diagnostic aid.

electrocardiograph An instrument which records changes in electrical potential caused by the heartbeat.

electrocardiophonograph A medical instrument that detects and records the impulses of the heart.

electrocatalysis Catalytic action produced by electricity (see *catalysis*).

electrocautery 1. In medicine, a cauterizing instrument consisting essentially of a platinum wire (at the tip of an insulated probe) which is heated by an electric current. Also see *electric needle*. 2. Cauterizing with an electrocautery.

electrochemical deterioration An electrochemical reaction that results in the permanent or temporary failure of a device.

electrochemical diffused-collector transistor Abbreviation, ECDC. A high-current pnp transistor in which metal has replaced the etched-away mass of *p* material, providing a built-in heatsink.

electrochemical equivalent In electrolysis or electroplating, a constant (Z) for the metal in plates. For a given metal, Z is the mass (in grams) of the metal deposited by 1 coulomb of electricity.

electrochemical junction transistor See *surface-barrier transistor*.

electrochemical measurements 1. Measurements made on chemical substances with electrical instruments to determine such factors as conductivity, pH, dielectric strength, dielectric constant, and so forth. 2. Measurements of electrical or electronic phenomena in terms of electrochemical response, e.g., current drain in terms of weight of plated metal, or voltage in terms of gas breakdown.

electrochemical polarization The disabling of a primary cell caused by gas products deposited around or on one of the electrodes.

electrochemical recording See *electric recording*.

electrochemical reduction Extracting a metal from a compound by electrolysis.

electrochemical series See *electromotive series*.

electrochemical switch A static, ionic ac switch consisting of an anode, cathode, and a control electrode, all immersed in an electrolyte. A positive control-electrode voltage switches the device on, initiating ion current from anode to cathode through the liquid.

electrochemical transducer A transducer that converts chemical changes into electrical quantities, or vice versa. Examples: soil-acidity probe, electrolytic elapsed-time meter.

electrochemistry The branch of chemistry concerned with chemical action arising from the effect of electricity on substances, and electrical effects produced by chemical action.

electrochromic display A display that operates by means of electric fields, which control the light-transmission and light-reflection characteristics in different regions of the material.

electrochronometer A precision electric or electronic clock. Also see *electric clock* and *electronic clock*.

electrocoagulation Use of a high-frequency current to solidify tissue, as in arresting bleeding.

electrocorticogram See *electroencephalogram*.

electroculture Acceleration or modification of plant growth through the application of electricity to plants, seeds, or soil.

electrocution Death or execution from electric shock.

electrode A body, point, or terminal in a device or circuit, which delivers electricity or to which electricity is applied. A positive electrode is an *anode*; a negative electrode, a *cathode*.

electrode admittance The admittance encountered by current flowing through an electrode; the property is entirely that of the electrode and is the reciprocal of electrode impedance.

electrode capacitance The capacitance between an electrode and a reference body, such as ground or another electrode.

electrode characteristic The mathematical function of electrode current versus electrode voltage.

electrode conductance The conductance encountered by current flowing through an electrode; the property is entirely that of the electrode and is the reciprocal of electrode resistance.

electrode current Current entering or leaving an electrode.

electrode dark current See *dark current*.

electrode dissipation The power lost in the form of heat in an electrode.

electrode drop Voltage drop resulting from electrode resistance.

electrode impedance The impedance encountered by alternating current flowing through an electrode; the property is entirely that of the electrode and is the reciprocal of electrode admittance.

electrode inverse current In a vacuum tube, a flow of current in the reverse direction through a given electrode (either the cathode or the plate).

electrodeless discharge Discharge in a gas tube that is not directly connected to a power source. A familiar example is the glow of a neon lamp held in a strong rf field.

electrodeposit 1. To deposit a substance by electrical action. Also see *electrophoresis* and *electroplating*. 2. A deposit that is formed on an electrode by electrophoresis or electroplating.

electrodeposition The electrical application of a layer of one material (such as a metal) on the surface of another (the substrate), e.g., electroplating, evaporation, sputtering.

electrode potential See *electrode voltage*.

electrode reactance The imaginary component of electrode impedance.

electrode resistance The resistance encountered by current flowing through an electrode; the property is entirely that of the electrode and is the reciprocal of electrode conductance.

electrodermography A method of monitoring the functions of the human body by measuring the resistance between two electrodes placed on the surface of the skin.

electrode voltage The voltage between an electrode and a reference point, such as ground or another electrode.

electrodiagnosis 1. The diagnosis of a disease or disorder through the use of electromedical instruments. 2. Troubleshooting the electrical portion of electromechanical equipment.

electrodialysis See *electro-osmosis*.

electrodissolution Dissolving a constituent substance of an immersed electrode by electrolysis.

electrodynamic Pertaining to electricity in motion, i.e., current flow and its accompanying electric and magnetic fields.

electrodynamic braking Stopping a tape-deck motor quickly by applying a braking voltage. In this method, dc braking current flows through the shaded-pole ac reel motor.

electrodynamic instrument See *electrodynamometer*.

electrodynamic loudspeaker See *dynamic speaker*.

electrodynamics The branch of electricity concerned with moving electric charges.

electrodynamic speaker See *dynamic speaker*.

electrodynamism See *electrodynamics*.

electrodynamometer An indicating meter whose movable coil rotates between two stationary coils which produce the effect of the permanent magnet found in the D'Arsonval movement. All three coils are connected in series, and the magnetic fields of the two stationary coils are additive. This type of meter operates on either ac or dc. Electrodynamometers are available as ammeters, voltmeters, and wattmeters.

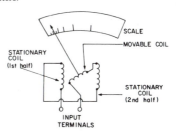

Electrodynamometer.

electroencephalogram Abbreviation, EEG. A record made by an encephalograph showing changes in electric potential resulting from bioelectric action in the brain. The record is used as a diagnostic aid. Also see *brain waves*.

electroencephalograph An instrument which produces a record of voltage changes resulting from the brain's bioelectricity.

electroencephalography The monitoring of brain impulses for medical diagnostic purposes.

electroencephaloscope A type of oscilloscope employed to pick up, amplify, and display changes in potential due to the brain's bioelectric action.

electroextraction Extracting a substance from a mixture (e.g., a metal from an ore) by an electrical process such as electrolysis.

electroform 1. To precondition a material or device (e.g., a semiconductor junction) by passing a current through it for a specified period. 2. To form articles by electrodepositing material on a mold or core.

electrogalvanized Electroplated with zinc.

electrogastrogram A recording of the electrical impulses and other functions of the stomach, for medical diagnostic purposes.

electrograph 1. A picture transmitting or receiving device (see *facsimile receiver* and *facsimile transmitter*). 2. A device used for the electrolytic etching or transfer of designs.

electrographic recording A method of producing a visible pattern or record, using electrodes to create discharge through an insulating material.

electrographite Synthetic graphite prepared by heating carbon in an electric furnace.

electrojet A region of high current concentration in the sky near bright auroral displays or along the magnetic equator.

electrokinetic energy The energy of electricity in motion, a form of kinetic energy. Thus, an electric current, a flow of electrons, manifests electrokinetic energy. Compare *electrostatic energy*.

electrokinetics A branch of electricity concerned with (1) the behavior of moving charged particles (such as ions and molecules) and bodies in motion, and (2) the generation of static charges by moving liquids or solids in contact with each other. Some of the topics in this category are electrolysis, electrophoresis, and electro-osmosis.

electroless process Plating a metal from a solution of one of its salts without using electricity.

electroluminescence The ability of certain phosphors to emit light continuously when an ac voltage is applied to them.

electroluminescent cell A device for generating light by electroluminescence. It consists of a sandwich of a luminescent-phosphor layer and two transparent metal films. An ac voltage applied between the films causes the phosphor to glow through the transparent metal.

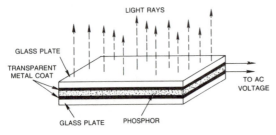

Electroluminescent cell.

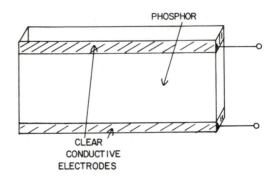

Electroluminescent cell.

electroluminescent lamp A lamp employing one or more electroluminescent panels.

electroluminescent panel A panel (available in a wide variety of sizes and shapes) which is a complete electroluminescent cell. It delivers low-intensity light when an ac voltage is applied to it.

electrolysis 1. The action whereby a current passing through a conductive solution (electrolyte) produces a chemical change in the solution and the electrodes. 2. An electrical method of destroying hair roots.

electrolyte A substance which ionizes in solution. Electrolytes conduct electricity, and in batteries they are instrumental in producing electricity by chemical action.

electrolytic 1. Containing an electrolyte substance. 2. An electrolytic capacitor.

electrolytic capacitor A capacitor (commonly called an electrolytic) in which one "plate" is an electrolyte (a liquid, such as boric acid, or a paste); the other plate is an aluminum can containing liquid electrolyte, or an aluminum foil against which paste electrolyte presses. A thin oxide film electroformed on the foil or the can serves as the dielectric, which, because it is very thin, gives a small electrolytic a high value of capacitance. Also see *dry electrolytic capacitor* and *wet electrolytic capacitor*.

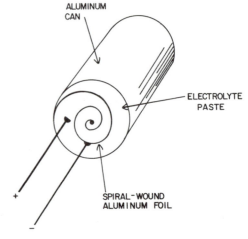

Electrolytic capacitor.

electrolytic cell In general, a cell containing an electrolyte and at least two electrodes. Included in this category are voltaic cells, electrolytic capacitors, and electrolytic resistors.

electrolytic conduction Electric current flowing through an electrolyte, an action characterized by (1) positive electrolyte ions migrating to the negative electrode, where they acquire electrons; (2) negative ions migrating to the positive electrode, where they lose electrons; and (3) current flow in the external circuit, which therefore consists of conventional electron flow (current in the electrolyte is a movement of ions).

electrolytic conductivity Conductance of an electrolyte. It is the conductance of a 1 cm cube of the electrolyte at a specified temperature.

electrolytic corrosion Corrosion caused by an applied voltage or accelerated by the voltage. Compare *galvanic corrosion*.

electrolytic current meter See *voltameter*.

electrolytic deposition Electrodeposition by electrolysis, as in electroplating.

electrolytic detector An early radio detector in which signal rectification occurred between the point of a fine wire (see *Wollaston wire*) and the surface of dilute acid held in a small metal cup.

electrolytic dissociation See *dissociation*.

electrolytic elapsed-time meter An instrument which indicates the time equipment has been in operation in terms of the amount of metal electroplated on the cathode of an electrolytic cell by energy consumed during the period.

electrolytic gas A gas produced by electricity. An example is hydrogen and oxygen generated in a ratio of two to one according to the formula H_2O, by the electrolysis of water.

electrolytic interrupter An interrupter once used to make and break the primary circuit of transformers operated from a dc source. It is of occasional interest today, especially in emergency applications. The device is essentially a small container of dilute sulfuric acid in which two platinum wires are immersed. Direct current passing through the interrupter, which is connected in series with a battery and a load (such as the primary winding of a transformer), is rapidly chopped at a frequency dependent upon the cavitation rate of the electrolyte.

electrolytic iron Very pure iron obtained by electrolytic refining.

electrolytic potential The difference of potential which appears between a metal electrode in an electrolyte and the electrolyte immediately surrounding it. Also see *electromotive series*.

electrolytic recorder A data recorder which employs a paper impregnated with a chemical that turns dark when an electric current passes through the paper from the point of a stylus.

electrolytic rectifier A rectifier consisting of an aluminum electrode and a lead or carbon electrode in a solution of borax or sodium bicarbonate, or in a solution of ammonium citrate, ammonium phosphate, and potassium citrate. Also called *chemical rectifier*.

electrolytic refining Extracting or purifying metals by electrolysis.

electrolytic resistor An emergency resistor made by immersing two wire leads in an electrolyte; the weaker the solution, the higher the resistance (pure water is a nonconductor).

electrolytic switch See *electrochemical switch*.

electrolyze To subject something to electrolytic action.

electrolyzer A cell used in the production of various materials by electrolysis. See, for example, *electrochemical reduction* and *electrolytic refining*.

electromagnet Any device which exhibits magnetism only while an electric current flows through it.

electromagnetic Exhibiting both electric and magnetic properties, e.g., an electromagnetic wave.

electromagnetic attraction 1. The attraction of iron or steel to an electromagnet. 2. The attraction of an electromagnetic pole to the opposite pole of another electromagnet (north pole pulling south, south pulling north). A unit pole attracts another unit pole 1 cm away with a force of 10^{-5}N (1 dyne). Compare *electromagnetic repulsion*.

electromagnetic communications 1. Broadly, communication employing a combination of electric and magnetic phenomena. This would include wire telegraphy (excluding the optical type, which employs electricity only), wire telephony, radiotelegraphy, radiotelephony, facsimile (radio and wire), and television (radio and wire). 2. Specifically, electronic communications based upon electromagnetic waves, i.e., radiocommunications.

electromagnetic compatibility In radio communications practice, the relative immunity of a device or devices to the effects of electromagnetic fields.

electromagnetic complex A system that produces electromagnetic radiation.

electromagnetic component 1. The magnetic component of an electromagnetic wave, which is perpendicular to the *electrostatic component*, and can be thought of as the wave's current component. 2. A device operated by electromagnetism, such as a coil-type relay or a current-operated field magnet.

electromagnetic constant Symbol, *c*. The propagation velocity of electromagnetic waves, i.e., the speed of light, through a vacuum; c = 299 792 458 meters per second.

electromagnetic coupling See *inductive coupling*.

electromagnetic crack detector An instrument which uses electromagnetic fields to find cracks in iron or steel.

electromagnetic crt A cathode-ray tube employing electromagnetic deflection.

electromagnetic cylinder Solenoid coil.

electromagnetic deflection In a TV picture tube and some oscilloscopes, deflection of the electron beam by the magnetic fields of external horizontal- and vertical-deflection coils. Compare *electrostatic deflection*.

electromagnetic deflection coil See *deflection coils*.

electromagnetic delay line See *decay line*.

electromagnetic energy Energy in the form of electric and magnetic fields. A radio wave traveling through space has electric and magnetic components, between which energy oscillates.

electromagnetic energy conversion The conversion of electrical energy into mechanical work and vice versa, through the intermediary of an electromagnetic field.

electromagnetic environment A region in which electric and magnetic fields are present.

electromagnetic field A combination of alternating electric and magnetic fields. The electric lines of force are perpendicular to the magnetic lines of force at every point in space. The field propagates in a direction perpendicular to both the electric and magnetic lines of force. The frequency of oscillation may be as low as a fraction of 1 Hz; there is no defined upper limit. See figure on next page.

electromagnetic field intensity See *magnetic intensity*.

electromagnetic flux The magnetic lines of force surrounding a coil or conductor carrying an electric current.

electromagnetic focusing In a TV picture tube, electron-beam focus-

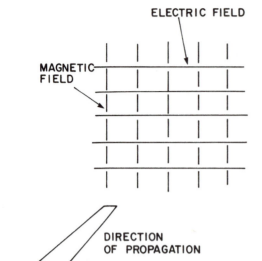

Electromagnetic field.

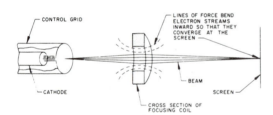

Electromagnetic focusing.

ing obtained by varying the direct current flowing through an external focusing coil.

electromagnetic force The force which causes a conductor to be displaced from its position in a magnetic field when it conducts current.

electromagnetic frequency spectrum The frequency range of electromagnetic radiation—including radio waves, light, and X rays. At the low-frequency end are subaudible frequencies (below 10 Hz); at the other, extremely high frequencies (associated with cosmic rays).

electromagnetic horn radiator A horn used to radiate microwave energy. Also see *horn antenna*.

electromagnetic induction Inducing a voltage in a circuit or conductor by causing alternating current to flow in another nearby circuit or conductor. Compare *electrostatic induction*.

electromagnetic inertia The tendency for the current in a circuit to lag the voltage at high frequencies.

electromagnetic interference Disturbances of equipment operation caused by electromagnetic fields from outside sources.

electromagnetic lens A coil or coil system whose magnetic field causes

an electron beam passing through it to converge or diverge as a light beam does in passing through an optical lens. Compare *electrostatic lens*. Also see *electromagnetic focusing*.

electromagnetic loudspeaker See *magnetic speaker, 1, 2*.

electromagnetic mass The mass a moving electric charge is thought to possess.

electromagnetic microphone A microphone in which sound energy is converted into proportionate electrical energy by electromagnetism. Common examples are the dynamic microphone and velocity microphone.

electromagnetic mirror A reflector of electromagnetic waves, e.g., antenna elements, ionospheric layers, buildings, hills, and so on.

electromagnetic momentum The momentum of a moving electric charge, comparable to that of matter in motion. Electromagnetic momentum is thus the product of electromagnetic mass and charge velocity.

electromagnetic oscillograph 1. An electromechanical data recorder (see *recorder, 2*) for tracing the waveform or variations of a signal. 2. Electromechanical oscilloscope.

electromagnetic oscilloscope 1. An oscilloscope using electromagnetic deflection. 2. Electromechanical oscilloscope.

electromagnetic pulse In electromagnetic induction, the displacement of an electron in a conductor by the magnetic field.

electromagnetic radiation The propagation of electromagnetic energy into space. Electromagnetic radiation (waves) moves at the speed of light.

electromagnetic reaction The reaction between magnetic fields. Also see *electromagnetic attraction* and *electromagnetic repulsion*.

electromagnetic reconnaissance The use of electromagnetic apparatus to detect potential enemy activity in a certain geographic region.

electromagnetic relay See *electromechanical relay*.

electromagnetic repulsion The repulsion of a pole of an electromagnet by the pole of another electromagnet (north pole opposing north pole, south opposing south). A unit pole repels another unit pole 1 cm away with a force of 10^{-5}N (1 dyne). Compare *electromagnetic attraction*.

electromagnetics The theory and application of electromagnetism.

electromagnetic screen See *electromagnetic shield*.

electromagnetic shield A partition, can, or box made of magnetic material (iron, steel, or special alloy) enclosing a magnetic component. The magnetic flux generated by the component is confined by the shield, thus preventing interference with external components. Likewise, external magnetic fields are prevented from reaching the component. Compare *electrostatic screen*.

electromagnetic speaker See *magnetic speaker, 1, 2*.

electromagnetic spectrum See *electromagnetic frequency spectrum*.

electromagnetic switch 1. A switch actuated by magnetism produced by control current flowing through a coil wound on an iron core. 2. Electromechanical relay.

electromagnetic theory of light The theory that light consists of electromagnetic waves that are similar to radio waves but of shorter wavelength.

electromagnetic tube A cathode-ray tube using electromagnetic deflection, e.g., the usual TV picture tube.

electromagnetic unit Abbreviation, emu. A unit of measure in the electromagnetic system of cgs units. Also see *centimeter-gram-second*.

electromagnetic vibrator See *interrupter*.

electromagnetic waves Waves produced in a conductor or in space

WAVELENGTH,
METERS

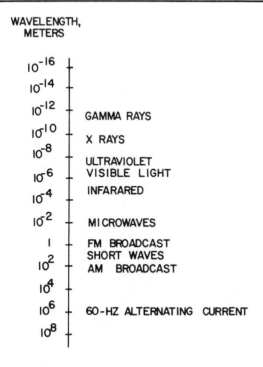

Electromagnetic frequency spectrum.

by an electric charge's oscillation. Such a wave has an electric and a magnetic component acting at right angles to each other.

electromagnetism 1. Magnetism resulting from the flow of an electric current, e.g., current flowing through a coil of wire wound on an iron core. 2. Electromagnetics.

electromagnetizer A magnetizer employing continuous dc as the magnetic-field source.

electromechanical A term used to describe any device that converts energy from electrical to mechanical form, or vice-versa. Examples are the motor and the generator.

electromechanical amplifier An amplifier that converts an electrical input signal into mechanical motion (vibratory or rotary), which it then converts back into an electrical output signal of higher current, voltage, or power. Examples are the *amplidyne, carbon-button amplifier,* and *electroacoustic amplifier.*

electromechanical bell See *electric bell.*

electromechanical chopper A vibrator-type interrupter used primarily to chop direct current, converting it into a square-wave signal whose amplitude is proportional to current strength. Also see *chopper* and *chopper converter.*

electromechanical counter A device which indicates the number of pulses that have been applied to it. Typically, it has a series of dials, each capable of displaying the numerals 0 to 9 in sequence, one for each decade in the count. The dials are geared together, the train being operated by the stepping action of an electromagnetic escapement. Compare *electronic counter.*

electromechanical energy The energy stored by an inductor or capacitor in an electromechanical device.

electromechanical filter See *ultrasonic filter, 1.*

electromechanical flip-flop See *bistable relay.*

electromechanical frequency meter A usually direct-reading instrument for measuring frequency in the lower and middle portions of the af spectrum and using mechanical motion resulting from the applied signal. There are two varieties: the *movable-iron type* and the *reed-type.* Also see *power-frequency meter, 1,2.*

electromechanical modulator 1. Chopper. 2. Interrupter. 3. A meter-type modulator resembling a D'Arsonval movement. An alternating supply current in coils wound on the poles of the movement's permanent magnet induces an ac output signal in the movable coil. This latter signal is coupled to the output through a transformer and a dc-blocking capacitor. A dc input signal is also applied to the movable coil, which is deflected to a position proportional to the current. The ac output voltage is proportional to the position of the coil in the ac field and to the dc input voltage.

electromechanical oscillator An oscillator consisting of an electromechanical amplifier provided with regenerative feedback. An example is the *hummer.*

electromechanical oscilloscope A galvanometer-type instrument for displaying a varying or alternating current or voltage. The signal is applied to a meter movement having a movable coil, which swings or vibrates in response to the signal. A tiny mirror cemented to the coil reflects a beam of light to a rotating mirror which sweeps the beam across a translucent screen on which the image is produced.

electromechanical recorder An instrument in which a pen or stylus is moved on a sheet of paper by the varying signal current or voltage being recorded. Also see *oscillograph, 1* and *recorder, 2.*

electromechanical rectifier A rectifier in which a moving part, such as a vibrating reed or rotating commutator-slip-ring unit, is driven by alternating current to close the circuit during positive or negative ac half-cycles, thus rectifying the ac. Compare *electrolytic rectifier* and *electronic rectifier.*

electromechanical relay An electromagnetic switch consisting, in one form, of a multiturn coil wound on an iron core near an armature with a movable end-contact. When control current flows through the coil, it becomes magnetized and attracts the armature, closing the movable contact against a stationary one.

electromechanical timer A device for automatically timing a process or an observed event. Most such timers are based upon an accurate clock (electric or spring driven) which opens or closes contacts at predetermined instants. Compare *electronic timer.*

electromechanical transducer 1. A transducer that translates mechanical signals into electrical ones or vice versa without the intermediary of electronic devices (tubes, transistors, etc.). 2. A special triode having a stylus attached to the plate and extending beyond the envelope. Pressure on the tip of the stylus moves the plate, changing the gap between plate, grid, and cathode and altering the electrical characteristics of the tube as an amplifying device. The tube's output for a standard input signal, then, is proportional to the applied pressure. The device has many uses, among them the checking of surface roughness.

electromechanical valve A usually poppet-type valve for gases or liquids. The valve is operated by electromagnetic action and is often aided by an electronic (servo) circuit.

electromechanics The theory and application of electromechanical devices.

electromedical engineering The branch of electronic engineering concerned with the theory, design, and application of electronic equipment to medical diagnosis or treatment.

electromedical equipment Electrical or electronic equipment used in medical diagnosis or treatment.

electromerism Ionization in gases.

electrometallurgy The branch of metallurgy concerned with the use of electricity (especially in electrolysis) to separate or purify metals or to furnish heat for metallurgical processes.

electrometer A specially designed, highly sensitive electronic voltmeter used to measure extremely low voltages and, indirectly, extremely low currents. It is sometimes used as a galvanometer.

electrometer amplifier A stable low-noise amplifier for increasing the sensitivity of an electrometer.

electrometer tube For use in an electrometer, a specially selected vacuum tube having high grid-cathode resistance, low input-circuit leakage, and negligible photoelectric emission. Tube characteristics are chosen for current gain rather than voltage gain.

electrometry The science of electrical measurements.

electromigration The movement of atoms in a substance from one place to another, because of interaction between electrons and ions in the presence of electric currents. This effect can cause the eventual deterioration of certain semiconductor devices.

electromotion Motion produced by electric charges or electrons.

electromotive force Abbreviation, emf. Electrical pressure, the potential that causes electrons to flow through a circuit. See *voltage*.

electromotive series A list of metals arranged according to the potential between the surface of the metals and an electrolyte into which they are immersed. Some metals acquire a positive potential (with respect to hydrogen, for which the potential is zero) and others, a negative potential.

electromotor A generator or motor, depending upon the context in which the term is used.

electromyogram The record produced by an electromyograph.

electromyograph An instrument for indicating the weak voltages generated by muscular activity.

electromyography The monitoring and analysis of the electrical activity of human muscles.

electron The subatomic particle that carries the unit negative charge of electricity. The electron has a mass of $9.109\ 534 \times 10^{-31}$ kilogram and carries a charge of $1.602\ 189 \times 10^{-19}$ coulomb.

electron acceleration See *electron motion, 2*.

electronarcosis Loss of consciousness caused by passing a weak current through the brain. Useful in treating certain mental disorders, electronarcosis is somewhat similar to electroshock (see *electroshock, 1*).

electron attachment The bonding of an electron to a neutral atom to form a negative ion. Also see *anion* and *ion*.

electron avalanche See *avalanche* and *electron multiplication*.

electron band 1. An emission line in the spectrum of an element or compound, caused by the movement of electrons from higher to lower energy levels within the atoms. 2. An absorption line in the spectrum of an element or compound, caused by the movement of electrons from lower to higher energy levels within the atoms.

electron beam See *electron stream, 1*.

electron-beam bender Any element that causes intentional deflection of the electron stream in a crt.

electron-beam focusing Reducing the size of the spot produced by the electron beam in a cathode-ray tube or TV picture tube. This is accomplished by adjusting the dc bias voltage on a focusing electrode.

electron-beam generator 1. Electron gun. 2. A tube such as a klystron in which velocity modulation of the electron beam generates extremely high radio frequencies.

electron-beam instrument An instrument, such as an oscilloscope, based on a cathode-ray tube.

electron-beam machining Welding or shaping materials by controlled electron beams.

electron-beam magnetometer A magnetometer in which the magnetic field under measurement deflects the electron beam in a cathode-ray tube over a distance proportional to field intensity.

electron-beam recording In digital computer practice, a technique whereby the output of a computer is recorded on microfilm by an electron beam.

electron-beam scanning tube A tube in which an electron beam strikes a sensitized screen to produce a spot of light, which is deflected electrically or magnetically across a screen. Examples are oscilloscope tubes, storage tubes, TV picture tubes, and radarscope tubes.

electron-beam tube Electron tubes—such as beam-power tubes, klystrons, oscilloscope tubes, and TV picture tubes—in which an electron beam is generated and controlled.

electron-beam welding A method of welding in which an electron beam is focused on the workpiece to heat it.

electron-bombarded semiconductor A semiconductor wafer, plate, or junction which is acted on by an electron beam; it alters the resistance of the semiconductor to control the current in an external circuit.

electron-bombarded semiconductor amplifier See *EBS amplifier*.

electron bunching See *bunching*.

electron charge See *elementary charge*.

electron cloud A mass of free electrons in a vacuum or solid. Example: the space charge in a vacuum tube.

electron-coupled multivibrator A pentode-based multivibrator that is coupled to the load through the electron path in the tubes.

electron-coupled oscillator A stable self-excited circuit in which the screen of a vacuum tube is used as the anode of a Hartley oscillator. The electron stream passes through the screen to the plate, which serves as the output electrode. The electron-stream coupling between screen and plate provides buffering, reducing the loading effect.

electron coupling Coupling by means of the electron stream in a vacuum tube. See, for illustration, *electron-coupled multivibrator* and *electon-coupled oscillator*.

electron density See *density of electrons*.

electron drift 1. The movement of an electron from atom to atom in a conductor, as caused by the influence of an applied voltage. 2. In a semiconductor, directed electron movement. Also see *drift, 1*.

electronegative Having negative electrification or polarity (see *electrical polarity*). Compare *electropositive*.

electron emission The giving off of electrons into surrounding space by a material. Depending on the material, electron emission may be initiated by application of heat, light, torsion, electron impact, a high-voltage field, and other actions.

electron flow See *electron drift*.

electron gas See *electron cloud*.

electron g-factor Symbol, $ge/2$. A physical constant that expresses the ratio of electron magnetic moment to the Bohr magneton ($\mu e/\mu B$): 1.001 159 657. Also called *free electron g-factor*.

electron-gas binding forces. See *metallic binding forces.*

electron-gas bonding See *bondong,* 1 and *metallic binding forces.*

electron gun A composite electrode for generating an electron beam (see *electron stream, 1*) in a vacuum. In a cathode-ray tube, the gun comprises a heated cathode, control electrode, accelerating electrodes, and a focusing electrode.

electron-hole pair In a semiconductor, an electron and a related hole. Each electron in the conduction band has a counterpart in the valence band, a vacancy (hole) left by the electron's moving to the conduction band.

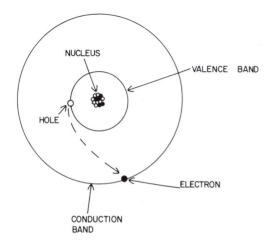

Electron-hole pair.

electronic Descriptive of any network of a system whose operation involves the use of nonmechanical amplification or switching devices such as tubes, transistors, thyristors, and integrated circuits.

electronic adder A tube or transistor circuit (such as an operational amplifier) for performing arithmetic addition. In such a circuit, output-signat amplitude is the sum of input-signal amplitudes. Also see *adder* and *analog adder*.

electronic aid An electronic device or circuit which contributes to the operation of a nonelectronic device or system; a pH meter, for example, is an electronic aid to chemistry.

electronic aide See *electronic technician.*

electronic attenuator An attenuator in which the variable resistor is the output section of a tube or transistor. Varying the dc bias of the input section varies the resistance of the tube or transistor output section.

electronic autopilot A servo-type device for detecting and automatically correcting an aircraft's flight path.

electronic balance An electronic scale, which employs a sensitive current-measuring device in conjunction with a movable tension device.

electronic brain Electronic computer, especially a sophisticated one.

electronic breadboard 1. A thin, usually nonmetallic board or card having prepunched holes for the quick assembly of electronic circuits for test and evaluation. 2. Any circuit prototype that is manually wired during the experimental phase of product development.

electronic bug 1. A telegraph key employing an electronic circuit to generate dits and dahs. Also see *electronic key, 1.* 2. An undetermined source of problems or improper operation in an electronic circuit.

electronic buzzer 1. A mechanical buzzer driven by a simple transistorized dc amplifier. 2. An oscillator circuit that produces a sound similar to a mechanical buzzer.

electronic calculating punch A machine that punches on a card the result of calculations it has performed on data it has read from another punched card.

electronic calculator A fully electronic (tube or transistor) machine for performing mathematical operations. It is considerably faster and smaller than the electric calculator.

electronic camouflage The use of electronics by a target craft to minimize or prevent the reflection of radar echoes.

electronic carillon An electronic system which produces sounds (through one or more loudspeakers) resembling those of a bell carillon.

electronic chime See *electronic carillon.*

electronic circuit An electric circuit containing active electronic components (semiconductor devices, tubes, and so on), as opposed to a circuit containing only passive electrical components (such as resistors, switches, heating elements, etc.).

electronic clock 1. An electric clock whose motor is driven by a constant-frequency oscillator (crystal or tuning fork type) followed by multivibrators and amplifiers. 2. Any electronic timing circuit that produces pulses at predetermined intervals for the purpose of regulating the operation of other circuits, subsystems, or assemblies.

electronic commutator See *commutator, 2.*

electronic conduction The conduction of electricity (i.e., current flow) viewed as the movement of free electrons between the atoms of the conductor.

electronic control 1. The science and art of automatically controlling machines and devices by means of electronic circuits. 2. A circuit or device which provides automatic electronic control.

electronic controller A controller (see *controller, 2*) having no moving parts. For automatic operation, such a controller often contains a circuit that senses control signals, compares them with a signal standard, and automatically adjusts output control power accordingly.

electronic counter A fully electronic circuit that indicates the number of pulses that have been applied to it (see *counter, 1*). Unlike the electromechanical counter, the electronic counter has no moving parts and is therefore capable of extremely high-speed, noiseless operation.

electronic counter-counter measures Abbreviation, ECCM. Procedures for interfering with a foe's electronic countermeasures.

electronic countermeasures Abbreviation, ECM. Interference with enemy radio and radar emissions usually but not necessarily by electronic means. Also see *jamming.*

electronic coupling Coupling via an electronic beam. Also see *electron coupling.*

electronic crowbar A switch that prevents destructive currents from flowing through the components of a circuit.

electronic current meter A current meter (ac or dc) which employs a tube or transistor amplifier ahead of the analog or digital indicator to provide increased sensitivity. Also see *FET current meter, transistor current meter,* and *vacuum-tube current meter.*

electronic data processing Abbreviation, EDP. See *data processing* and *electronic information processing.*

electronic data-processing center Abbreviation, EDPC. An installation of electronic equipment and accessories for processing and storing data in the form of (usually digital) signals, (see *data processing* and *electronic information processing*).

electronic data-processing machine Abbreviation, EDPM. A device, such as an electronic computer, used in the automatic processing of (usually digitized) data.

electronic data-processing system 1. A unique arrangement of machines for processing data. 2. The sequence of steps in, and the underlying rationale for, the processing of data by automated equipment.

electronic deception See *deception* and *deception device*.

electronic device Any device that operates by means of the conduction of charge carriers.

electronic differential analyzer See *analog differentiator, digital differential analyzer,* and *differential analyzer*.

electronic differentiator A tube- or transistor-type device for performing mathematical differentiation. Also see *differentiator, 2*.

electronic digital computer A fully electronic digital computer as contrasted with an electromechanical one (e.g., one using relays instead of flip-flops).

electronic divider 1. An electronic device for performing arithmetic division. In a digital computer such a divider may be a sequence of flip-flops, each of which produces a single output for every two input pulses. In an analog computer, the output signal amplitude is equal to the quotient of two input-signal amplitudes. 2. A frequency divider. 3. A voltage divider using active components rather than resistors.

electronic dust precipitator See *dust precipitator*.

electronic efficiency The efficiency η of an electron beam as a medium of power transmission. Electronic efficiency, $\eta\%$, is $(100Pi)/Po$, where Pi is the power supplied to the beam and Po is the power delivered by the beam.

electronic engineer A trained professional skilled in the physics and mathematics of electronics and in the theory and application of basic engineering and related subjects. Also called *electronics engineer*. Compare *electrical engineer*.

electronic equivalent of gravity In equations for the acceleration, velocity, and distance (traveled) of an electron, the factor equal to $(eF)/m$, where e = electron charge, F = potential gradient of field; and m = electron mass. Also see *electron motion, 2*.

electronic flash 1. A device containing a circuit employing an electronic flash tube as a light source for photography or other purposes. Also called *photoflash*. 2. A bright momentary light burst produced by the equipment described in 1, above.

electronic flash tube A tube used to produce brilliant bursts of light in photoflash units, stroboscopes, and laser exciters. A flash tube usually contains xenon (a gas), which is fired by a high-voltage pulse.

electronic frequency meter 1. An instrument which gives direct readings of frequency in hertz, kilohertz, or megahertz on an analog scale or as a digital readout. 2. Any device that indicates the operating frequency of another device, directly or indirectly, when used for such purpose.

electronic frequency synthesizer An instrument which supplies a number of selectable frequencies derived from one or more internally generated fixed frequencies.

electronic gas A collection of free electrons whose behavior resembles that of a gas.

electronic gate A logic gate that operates by electronic means.

electronic guitar See *electric guitar*.

electronic heating Producing heat in a workpiece by means of high-frequency energy. The two principal methods are *dielectric heating* and *induction heating*.

electronic hygrometer An electric hygrometer whose sensitivity and stability have been increased by the addition of tube or transistor circuitry.

electronician See *electronic technician*.

electronic induction See *electrostatic induction*.

Electronic Industries Association Abbreviation, EIA. An American association of electronic manufacturers and others which sets standards, disseminates information, provides industry-government liaison, and maintains public relations for the industry.

electronic information processing The use of electronic equipment (especially digital computers and attendant devices) to perform mathematical operations on data entered into the system in the form of electrical signals. Also see *data processing*.

electronic instrument An instrument whose circuit employs active devices (tubes, transistors, ICs) for increased sensitivity over that of the electrical counterpart, and for minimum loading of a device under test. Compare *electrical instrument*.

electronic integrator A tube or transistor device (such as an operational amplifier) for performing mathematical integration. Also see *integrator, 2*.

electronic intelligence 1. Information exchanged by electronic means. Examples: radio messages, radar information, computer data. 2. The faculties of reasoning and decision, as apparently simulated by an electronic computer.

electron interference The malfunctioning of a device because of nearby currents, voltages, or electromagnetic fields.

electronic inverter An electronic device for converting dc to ac. Typically, an inverter is a tube or transistor square-wave oscillator inductively coupled to ac output terminals. The dc to be inverted energizes the tubes or transistors, which themselves perform the switching function at the rate determined by, for example, the RC components of the circuit. Also see *inverter,1*.

electronic jamming The deliberate transmission of electromagnetic energy for the purpose of interfering with the operation of a device or devices.

electronic key 1. For telegraphy (radio or wire), an electronic circuit that generates a continuous string of accurately spaced and timed dots or dashes, depending upon its lever's position (right or left). 2. An electronic device for opening an electronic lock.

electronic keyer See *electronic key, 1*.

electronic lock A lock which will open only after application of a special coded sequence of signals (of the correct amplitude, frequency, width or modulation, for example).

electronic microammeter See *FET current meter, transistor current meter,* and *vacuum-tube current meter*.

electronic microphone A special vacuum tube in which one electrode vibrates in correspondence with an audio signal. The vibration changes interelectrode spacing and, accordingly, the electrical characteristics of the tube (operated at dc), whose output reproduces the sound wave.

electronic microvoltmeter See *microvoltmeter*.

electronic milliammeter See *FET current meter, transistor current meter,* and *vacuum-tube current meter*.

electronic millivoltmeter A millivoltmeter (ac or dc) which employs an amplifier ahead of an analog or digital indicator to provide high input impedance and increased sensitivity.

electronic multimeter A tube or transistor voltohmmilliammeter. Also see *electronic instrument.*

electronic multiplier 1. A device, such as a Hall generator, whose output is equal (or proportional) to the product of two inputs; i.e., it can perform arithmetic multiplication. 2. Electron multiplier.

electronic music 1. Music produced by a combination of electronic oscillator, amplifier, and loudspeaker. A number of successful instruments have been developed. See, for example, *electronic carillon, electronic organ, electronic piano,* and *theremin.* 2. The electronically amplified sounds of conventional musical instruments.

electronic organ A musical instrument with a keyboard similar to that of a conventional organ, in which tones produced by oscillators or electrically driven reeds are processed and amplified for delivery to a system of loudspeakers.

electronic packaging See *encapsulation.*

electronic part A lowest replaceable unit, or component, in an electronic circuit.

electronic phase meter A tube or transistor instrument for measuring phase difference. Direct readings, in degrees of lead or lag, are given on an analog scale or as a digital readout.

electronic photoflash A transistorized light-intensity meter. Also see *electronic instrument.*

electronic piano A musical instrument having the keyboard of a conventional piano and provided with electronic amplification.

electronic picoammeter See *FET current meter, transistor current meter,* and *vacuum-tube current meter.*

electronic picovoltmeter See *picovoltmeter.*

electronic power supply A power supply (ac, dc, or ac-dc) employing tubes or transistors for stabilization and output control.

electronic precipitator See *dust precipitator.*

electronic product Any commercially manufactured electronic device, intended for purchase by the public, by industry, or by the government of any country.

electronic profilometer A tube or transistor instrument for measuring surface roughness.

electronic ratchet A stair-step circuit or other arrangement functioning in the manner of an equivalent electromechanical stepping switch. See also *commutator, 2.*

electronic reconnaissance The use of electronic means to locate enemy installations, such as radio stations, guided-missile sites, radar bases, and so on.

electronic rectifier A rectifier employing tubes or semiconductor devices to change ac to dc, as opposed to the electromagnetic rectifier and electrolytic rectifier.

electronic regulator A tube or transistor current or voltage regulator, as opposed to a reactor-type or electromechanical device. See, for example, *voltage regulator.*

electronic relay 1. A switching circuit employing one or more tubes or transistors, which performs the relay function without moving parts. 2. An electronic component designed to switch on application of appropriate gating signals (the triac, diac, silicon controlled rectifier are examples).

electronic resistor The internal plate-cathode circuit of a vacuum tube, the internal collector-emitter circuit of a common-emitter bipolar stage, or the internal drain-source circuit of a field-effect transistor (FET); each of these acts as a resistance whose value can be varied by adjusting dc input electrode bias.

electronics The branch of electricity concerned with the behavior and

application of tubes and semiconductor devices, and with the circuits in which they are used.

electronic serviceman An electronic technician skilled in repairing and maintaining electronic equipment. Also called *electronic service technician.*

electronic shutter See *Kerr cell.*

electronics technology The theory and practical application of electronics. Electronics technology is taught as a subengineering major, usually in two-year junior colleges or technical institutes awarding the degree of associate in arts (AA) or associate in science (AS).

electronic stethoscope A stethoscope employing a miniature microphone, amplifier, and earphones. The amplifier gain is continuously controllable, and its bandwidth often selectable for emphasizing particular heart sounds and other body noises.

electronic stimulator A device for applying controlled electrical pulses to the body to stimulate muscles or nerves during diagnosis or therapy.

electronic stroboscope A stroboscope which employs a rate-calibrated oscillator rather than a mechanical contactor to generate pulses that strobe the lamp.

electronic subtracter An electronic circuit for performing arithmetic subtraction.

electronic surge A sudden, large increase in the voltage in a conductor, usually caused by a strong electromagnetic pulse. Sometimes called a transient or transient spike.

electronic switch 1. A nonmechanical device, such as a flip-flop or gate, whose characteristic on-off operation can be used to make and break an electric circuit. Compare *contact switch.* 2. A device employing tube or transistor gating and sequencing circuits to present several signals alternately to the single input of an oscilloscope, allowing simultaneous viewing of the signals.

electronic tachometer An instrument for measuring angular velocity, usually in revolutions per minute. Ideally, the response is independent of sensor voltage amplitude (showing only the number of pulses per unit time reaching the meter circuit).

electronic technician A professional skilled in building, testing, repairing, or maintaining electronic equipment, and sometimes in its design.

electronic thermal conductivity The thermal-conductivity component resulting from the transfer of heat by electrons and holes.

electronic thermometer Any one of several types of instruments for measuring temperature as a result of variations in a temperature-sensitive component, such as a resistor, thermocouple, thermistor, or varistor.

electronic timer An electronic circuit or device for automatically timing a process or observed event. Most are based on the time constant of a stable *RC* circuit. Compare *electromechanical timer.*

electronic tube See *electron tube.*

electronic tuning Variation of the resonant frequency of a device or circuit by changing the bias voltage or current of a controlling electronic component.

electronic voltmeter A voltmeter (ac or dc) that employs electronic amplification ahead of the indicating meter to provide high input impedance and increased sensitivity. Also see *FET voltmeter, transistor voltmeter,* and *vacuum-tube voltmeter.*

electronic voltohmmeter A tube or transistor voltohmmeter.

electronic warfare The use of electronic means—such as radio, radar, fire control, missile guidance, etc.—for military purposes, including interfering with an enemy's use of similar means.

electronic watch 1. A watch whose movement is a tiny high-frequency ac motor driven by a stable oscillator. 2. Any miniature timepiece incorporating solid-state circuitry, but especially one employing a digital form of readout.

electronic wattmeter A wattmeter in which an amplifier is employed for increased sensitivity. Also see *wattmeter*.

electronic waveform synthesizer A signal generator which delivers an alternating or pulsating signal whose waveform may be tailored by means of adjustable circuit components.

electron image tube See *dissector tube*.

electron lens A device which focuses an electron beam in a manner similar to the focusing of light rays by a glass lens. Also see *electrostatic lens, electromagnetic lens,* and *waveguide lens*.

electron magnetic moment Symbol, μ_e. The energy per unit flux density available in an electron: $9.284\ 832 \times 10^{-24}$ joule per tesla.

electron mass Symbol, m_e. The quantity of matter thought to be present in an electron at rest: $m_e = 9.109\ 534 \times 10^{-31}$ kg.

electron microscope A microscope in which light as a source of illumination is replaced by an electron beam focused by electromagnetic lenses. The electron microscope affords magnifications far in excess of those obtained with the most powerful optical microscopes.

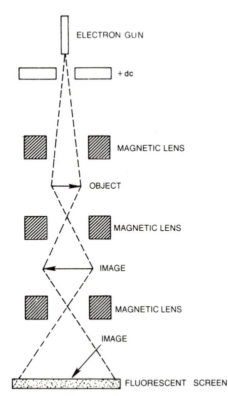

Electron microscope.

electron mirror A reflector of electrons, especially a dynode element in a photomultiplier tube or electron-multiplier tube.

electron motion 1. The movement of electrons in a conductor, semicon-

ductor, or space as the result of electric or magnetic attraction or repulsion. 2. The movement of an electron as a charged mass. In an electric field, this movement simulates that of a free-falling body in a gravitational field.

electron multiplication 1. In a gas discharge, the production of additional electrons as a result of collisions between electrons, atoms, and molecules. 2. The increased production of electrons in a semiconductor when avalanche occurs.

electron-multiplier tube 1. A tube utilizing a sequence of secondary emissions for increased current amplification. Electrons from the cathode strike a positively biased dynode with a force that dislodges secondary electrons which, upon joining those first emitted, are reflected to a second positive dynode which contributes more secondary electrons, reflecting the total to a third positive dynode, and so on. The last dynode in the chain reflects the enhanced beam to an anode collector which passes the high current to an external circuit. 2. Photomultiplier tube.

electronography Printing by means of the electrostatic transfer of ink from a printing plate across a gap to an impression cylinder.

electron optics See *electro-optics*.

electron-orbit oscillator See *Barkhausen-Kurtz oscillator*.

electron orbits In an atom, the various paths around the nucleus in which an electron may revolve. An electron moves from an inner to an outer orbit upon absorbing energy, and releases energy when it falls from an outer to an inner orbit. Also see *electron shells*.

electron oscillator 1. A positive-grid tube oscillator (see *Barkhausen-Kurtz* oscillator). 2. A device in which oscillation is obtained by causing electrons to move in an oscillatory path, to move in an oscillatory path, to travel in bunches, etc. Examples: klystron, magnetron, traveling-wave tube.

electron pair Two electrons, from adjacent atoms, which sometimes share the same orbits but always produce a bond between two adjacent atoms.

electron-pair bond The bond between an electron pair.

electron physicist A scientist specializing in electron physics.

electron physics The physics of electronics, usually from a highly theoretical viewpoint.

electron-proton magnetic moment ratio Symbol, μ_e/μ_p. A physical constant whose value is $658.210\ 688$, derived from the division of the magnetic moment of the electron by that of the proton.

electron-ray indicator See *magic-eye tube*.

electron recoil The recoil of an electron from a photon it has collided with.

electron rest mass Symbol, m_e. The mass of an electron at rest: $m_e = 9.109\ 534 \times 10^{-31}$ kg.

electron scanning Deflection of an electron beam. See, for example, *electrostatic deflection* and *electromagnetic deflection*.

electron shells The imaginary spherical shells, concentric with the nucleus of an atom, in which the electrons are thought to have their orbits—or which, if you can be so persuaded, represent the levels (distance from the nucleus) at which electrons migrate randomly, *periodically* appearing at points the locus of which is an "orbit."

electron spin The rotation of an electron (i.e., around its axis). This motion is independent of the electron's revolution (orbital movement around the nucleus of an atom).

electron stream 1. The beam of electrons generated by the electron gun in a cathode-ray tube. 2. The electrons moving between cathode and plate in an electron tube.

electron-stream instrument See *electron-beam instrument.*

electron-stream meter An oscilloscope (or cathode-ray tube alone) employed as a "meter."

electron-stream transmission efficiency The ratio of the current through a positive electrode to the current impinging on it. In a tube, for example, some electrons are absorbed by the plate, while others are reflected.

electron telescope A telescope employing a combination of a glass lens, photocathode, and electrostatic focusing. Light from the object is focused on the photocathode by the lens, the electrons emitted being focused electrostatically upon a phosphorescent viewing screen.

electron theory of matter The general belief that all matter is composed of atoms having electrons (negatively charged particles) that travel in discrete orbits around a central (net positively charged) nucleus. The nature of matter is determined by the number of electrons in the component atoms. Particles in the atom are cohesive because the net positive charge of the nucleus is equal to the total negative charge of the electrons. However, electrons in outer orbits are more loosely bound and, as free electrons, can escape the influence of the nucleus, and leaving the atom under some circumstances. In conductors, free electrons are plentiful; in insulators, virtually none are available.

electron transit time The time required for an electron to travel a given distance. For a vacuum tube, the upper frequency limit of operation is governed by the time it takes an electron to reach the plate after leaving the cathode. Transit time is usually stated in fractions of a microsecond.

electron tube A usually easily replaceable evacuated or gas-filled chamber in which electrons are emitted (usually by a hot cathode) and controlled (usually by a voltage applied to a grid electrode).

electron-tube meter See *vacuum-tube voltmeter.*

electron-tube static characteristic An expression of the relation among parameters of a vacuum tube, at a constant level of voltage and current.

electron unit See *elementary charge.*

electron velocity The velocity (v) acquired by an electron that moves between two points of potential difference V. Velocity $v = \sqrt{(2Ve)/me}$, where me is the mass of the electron. Also see *electron motion, 2.*

electronvolt Abbreviation, eV. A basic unit of electrical energy. One electronvolt is the energy acquired by a unit charge moving through a potential difference of 1V, or $1.602\ 189 \times 10^{-19}$ joule. Also see *BeV, GeV,* and *MeV.*

electron-wave tube A tube, such as a klystron or traveling-wave tube, in which electrons traveling at different velocities interact with each other, modulating the electron stream.

electron weight See *electron mass.*

electro-oculogram A recording of the voltage that is found between the anterior and posterior parts of the eyeball.

electro-optical transistor A phototransistor or pair of phototransistors in a single package and used in electronic circuits to sense changes in light levels.

electro-optical valve See *Kerr cell.*

electro-optic radar A form of radar that makes use of visual apparatus for locating a target.

electro-optics The branch of electronics dealing with related electrical and optical phenomena: photoelectricity, light generation, laser technology, light amplification, etc. It is concerned, too, with those electronic phenomena that are analogous to optical phenomena, such

as electronic focusing, reflection, refraction, diffraction, and so forth. Also called *electron optics* and *optoelectronics.*

electro-osmosis Causing liquids to flow by applying an electric field across the walls of a porous plug. The force exerted by the field on ions in the liquid causes it to flow.

electropad The skin-contacting electrode of an electrocardiograph.

electropathy See *electrotherapy.*

electrophilic Pertaining to the tendency to seek electrons.

electrophobia The irrational fear of electricity, a psychological condition sometimes exhibited by victims of serious electric shock, but it may be attributable to other causes as well.

electrophonic effect Sound heard by someone when an alternating current is passed through some part of his body.

electrophoresis The movement of dielectric particles through a liquid in which they are suspended, produced by the electric field between electrodes immersed in the suspension.

electrophoresis equipment 1. Any device intended for the purpose of depositing a dielectric material onto a metal, by means of electrophoresis. 2. Any equipment in which electrophoresis takes place.

electrophoresis scanner A device that senses the movement of charged particles caused by electrophoresis effects.

electrophoretic deposition A type of deposition in which a low-voltage direct current passing through a colloidal suspension of dielectric polymer particles deposits them as a coating on a metallic body (the anode in the process). Electrophoretic deposition can provide a better coating than one obtained with spray painting or dipping.

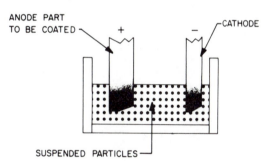

Electrophoretic deposition.

electrophorus A simple device used to demonstrate electrostatic generation and induction. It consists essentially of a smooth metal plate at the end of an insulating handle and an accompanying cake of resin or hard rubber. The cake is rubbed with cat's fur, making it negatively charged. The metal plate is touched to the charged cake; by induction, it acquires a bound positive charge on the face that touched the cake and a free negative charge on the opposite face. When the plate is lifted and its top face touched momentarily with the finger or grounded, the negative charge leaks off, often with a sharp spark, but the positive charge remains.

electrophotographic process See *xerography.*

electrophotography The production of photographs by means of electricity. See *xerography.*

electrophotometer A light-intensity meter employing a photoelectric sensor and a meter, but usually not incorporating an amplifier. Compare *electronic photometer.*

electrophrenic respiration A system of inducing respiration in which one or both of the phrenic nerves (i.e., of the diaphragm) are stimulated electrically to produce contractions of the diaphragm muscles.

electrophysiology 1. The study of electrical processes in the human body. 2. The study of how electrical impulses affect, and are produced by, body organs.

electroplaques In electric fish, small voltage-generating cells which are connected in series-parallel networks.

electroplate 1. To cause one metallic substance to adhere to the surface of another through the effects of electrolysis. 2. A metal plating deposited via electrolysis.

electroplating Depositing one metal on the surface of another by electrolytic action. Also see *electrophoresis.*

eletropolar Having electrical polarity.

electropolishing An electrolytic method of smoothing a rough metal surface. The workpiece to be polished becomes the anode of an electrolytic cell in which electrolytic action dissolves tiny surface irregularities.

electropositive Having positive electrification or polarity (see *electrical polarity*). Compare *electronegative.*

electropotential series See *electromotive series.*

electropsychrometer An electronic instrument for humidity measurements.

electroreduction In electrolysis, reduction of the cathode electrode.

electrorefining The refining of metals by means of electrolysis. Also see *electrochemical reduction, electrodeposition, electroextraction, electrolytic iron,* and *electrolytic refining.*

electroretinograph An instrument used to measure the electrical response of the human retina to light.

electroretinography The process of detecting and measuring electrical impulses from the retina.

electroscope An instrument for detecting electric charges and fields. The common type uses a pair of gold-leaf strips hung from the end of a metal rod in a glass tube or jar. When the exposed end of the rod is brought near a charged object, the leaves repel each other and spread apart.

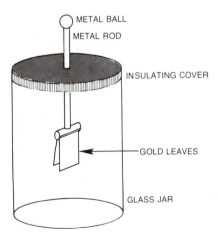

Electroscope.

electrosection The use of an arc-generating device for making surgical incisions.

electrosensitive recording See *electric recording.*

electroshock 1. The practice of creating a controlled electric shock in the brain as a treatment for certain mental disorders. 2. The electric shock used in the therapy described in 1, above.

electrospinograph An instrument which senses and records electrical impulses in the spinal cord.

electrostatic Pertaining to stationary electric charges and fields and their application.

electrostatic actuator A device for measuring the sensitivity of a microphone. Electrostatic charges produce forces on the diaphragm of the microphone, and the resulting output is recorded.

electrostatic amplifier See *dielectric amplifier.*

electrostatic capacitor A capacitor.

electrostatic charge See *electric charge.*

electrostatic component The electric component of an electromagnetic wave. It is perpendicular to the electromagnetic component and can be thought of as the wave's voltage component.

electrostatic constant See *electric constant.*

electrostatic convergence See *electrostatic focusing.*

electrostatic cooling Accelerated cooling of a body through the application of an intense electrostatic field. For this process to succeed, the body must be in a free convection state, a corona discharge must be present, and the field must not be uniform.

electrostatic copier A document-copying apparatus, which makes use of electrostatic effects to reproduce printed material.

electrostatic coupling See *capacitance coupling.*

electrostatic deflection In oscilloscope tubes and some early TV picture tubes, deflection of the electron beam by the electrostatic fields between pairs of internal horizontal and vertical deflecting plates. Compare *electromagnetic deflection.*

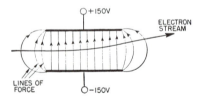

Electrostatic deflection.

electrostatic electrometer See *Kelvin absolute electrometer.*

electrostatic electrophotography See *xerography.*

electrostatic energy The potential for electrical energy. Thus, electrostatic energy is present in the stationary charge of a capacitor. Compare *electrokinetic energy.*

electrostatic field A stationary electric field.

electrostatic field intensity The voltage between two given points in an electric field.

electrostatic flux The flux existing around an electric charge or a charged body.

electrostatic focusing In a TV picture tube, electron-beam focusing achieved by varying the dc bias voltage on a focusing electrode. Compare *electromagnetic focusing.*

electrostatic galvanometer A galvanometer operating on the principle of the electrostatic voltmeter.

electrostatic generator A device for producing high-voltage electric charges—a belt generator, for example.

electrostatic headphone A device similar to an electrostatic speaker, but held against the head for private listening. Incoming audio signals cause attraction and repulsion among charged plates, resulting in acoustic vibration.

electrostatic hysteresis The tendency of some dielectrics (especially ferroelectric materials) to saturate and retain a portion of their polarization when an alternating electric field to which they are exposed reverses polarity. This causes the charge to lag behind the charging force.

electrostatic induction The charge acquired by a body inserted into an electric field. Compare *electromagnetic induction.*

electrostatic instrument An indicating meter whose movement consists of a stationary metal plate near a parallel, rotating metal plate. A voltage applied to the plates charges them (their proximity gives the arrangement capacitance), and the attraction between them causes the movable member to rotate against the torque of a returning spring over an arc proportional to the voltage. Example: electrostatic voltmeter.

electrostatic lens An assembly of deflecting plates or cylinders whose electric field causes an electron beam passing through it to converge or diverge in much the way a light beam passing through an optical lens does. Compare *electromagnetic lens.*

electrostatic line of force See *line of force, 1.*

electrostatic loudspeaker See *electrostatic speaker.*

electrostatic memory A memory unit in which an information bit is stored as a charge, as in an electrostatic memory tube.

electrostatic memory tube A cathode-ray tube in which information bits are stored in capacitive cells swept by the scanning electron beam.

electrostatic microphone See *capacitor microphone.*

electrostatic phase shifter See *phase-shifting capacitor.*

electrostatic picture tube A TV picture tube in which the electron beam is deflected by charged plates rather than by induction coils.

electrostatic potential In an electric field, the potential energy represented by the voltage between the two elements creating the field or between any two points within the field.

electrostatic precipitator See *dust precipitator.*

electrostatic printer A computer output peripheral in which the printing medium, a fine dust, is fused by heat onto paper which has been charged according to the data being represented.

electrostatic process. 1. Any process that makes use of electrostatic action. 2. A method of photography in which visual images are converted to electrostatic images.

electrostatic recording A method of recording that employs a signal-controlled electric field.

electrostatic relay A high-input-impedance relay consisting of two polarity-controlled contacts; opposite charges impressed on the contacts close the relay, and like charges open it.

electrostatics The branch of electricity concerned with electrical charges at rest and their effects. Compare *electrodynamics* and *electrokinetics.*

electrostatic screen A shield against electric flux consisting of a number of straight, narrowly separated rods or wires joined at only one end. The shield has little effect on magnetic flux. Also called *Faraday shield.*

electrostatic separator A device for separating finely divided particles from a mixture exposed to an intense electrostatic field.

electrostatic series A list of materials arranged in this sequence: any one of them becomes positively electrified when rubbed with another lower in the list, or negatively electrified when rubbed with another higher in the list. Compare *electromotive series.*

electrostatic shield Any metallic enclosure designed to confine an electric field.

electrostatic speaker A loudspeaker whose vibrating diaphragm is one of two plates in an air capacitor, the other being a closely situated metal plate (or plug). An af voltage applied to the plates causes them to vibrate. Also called *capacitive loudspeaker* and *capacitor loudspeaker.*

electrostatic sprayer An equipment for spray painting, in which fine droplets of paint are attracted by an electrostatic field to the surface to be coated.

electrostatic storage See *electrostatic memory.*

electrostatic stress 1. Stress in the vicinity of a charged body or particle. 2. Dielectric stress.

electrostatic transducer See *capacitance transducer.*

electrostatic tube of force The space between electric lines of force going through adjacent points on the boundary of a given area in an electric field.

electrostatic tweeter A small electrostatic speaker for reproducing high-frequency sounds. Compare *woofer.*

electrostatic unit Abbreviation A *esu,* unit of measure in the electrostatic system of cgs units. Also see *centimeter-gram-second.* Compare *electromagnetic unit.*

electrostatic vector See *electric vector.*

electrostatic voltmeter A voltmeter utilizing the attraction between two closely spaced metal plates (one stationary, the other rotating) to which the unknown voltage is applied. A pointer attached to the rotating plate moves over a voltage scale. A spiral spring returns the rotating plate to rest (and the pointer to zero) when the voltage is removed.

electrostatography See *xerography.*

electrosteel See *electric steel.*

electrostenolysis The deposition of certain metals from a solution in capillary tubes when an electric current passes through the solution.

electrostimulation Electrical excitation of the spinal cord for the relief of pain. A tiny radio receiver is surgically inserted into the body near the spine, and its output electrodes are implanted in the dorsal area of the spinal cord. The patient wears a miniature transmitter in his belt; and when he feels pain, he holds the transmitter antenna over the skin area where the receiver is implanted, and switches the transmitter on.

electrostriction The contraction of a ceramic plate (such as barium titanate) when a voltage is applied across its parallel faces. Compare *magnetostriction.*

electrostrictive ceramic A ceramic exhibiting electrostriction.

electrostrictive relay A relay in which the movable contact is carried by a bar of electrostrictive material, such as an electrostrictive ceramic. A control voltage deforms the material, causing the contacts to close.

electrosurgery Surgery, sometimes bloodless, achieved with diathermy-like equipment. See *diathermy, 2.*

electrosynthesis Chemical synthesis produced by means of electric currents or fields.

electrotape Also called electronic tape measure. Any device that measures distance by electronic means, such as radar or sonar.

electrotechnology See *electrical technology* and *electronics technology.*

electrotellurograph An instrument for measuring ground currents.

electrotherapeutics See *electrotherapy.*

electrotherapy The treatment of disorders or diseases by electrically induced heat, especially by diathermy.

electrothermal 1. Pertaining to electrically generated heat. 2. Pertaining to a combination of electricity and heat.

electrothermal expansion element A thermostatic element, such as a bimetallic strip, whose expansion is used in heat-sensitive switches.

electrothermal recorder See *electric recording, 1* and *thermal recorder.*

electrothermic See *electrothermal.*

electrothermic device A device whose operation depends on the heat of an electric current, e.g., bolometer, hot-wire ammeter, thermocouple, varistor,

electrothermic instrument A hot-wire or thermocouple-type meter.

electrothermics The study and application of the heating effects of electricity in conductors and junctions.

electrotitration In chemistry, the completion of titration as indicated by an electrical measurement, such as of the resistance of the solution being titrated.

electrotonic Pertaining to electrotonus.

electrotonus Modification of a nerve's sensitivity by passing a constant current through it.

electro-ultrafiltration In physical chemistry, filtering a colloidal suspension by electro-osmosis.

electrovalence 1. The number of charges acquired by an atom gaining electrons or, conversely, the number of charges forfeited by an atom losing electrons. 2. Valence resulting from electron transfer between atoms and the resulting creation of ions.

electrovalency See *electrovalence.*

electrovalent bond See *ionic bond.*

electrowin To recover (win) a metal from a solution of its salts by means of electrolysis.

electrum A natural alloy of gold and silver.

element 1. Electrode. 2. Circuit component. 3. A fundamental, unique substance whose atoms are of only one kind (examples: aluminum, carbon, silicon, sulfur). There are more than 100 elements, some man-made. Elements combine to form compounds. 4. A circuit, such as an AND gate, that can be taken as a unit because it performs a special function. 5. In digital computer practice, a subunit of a category and one which cannot be further categorized, e.g., a word element is a bit; a file element is a record.

elemental area In a facsimile or television picture, a scanning line segment as long as the line's width.

elemental charge The negative charge associated with an electron, or the positive charge associated with a proton. See, for example, *elementary charge.*

elemental semiconductor A semiconductor containing one undoped element (see *element, 3*).

elementary charge Symbol, *e.* The electric charge of the electron. This charge is equal to $1.602\ 189 \times 10^{-19}$ coulomb.

elementary particle A minute charged or uncharged particle within the atom, i.e., electron, proton, neutron, meson, etc.

element error rate In communications or data transfer, the ratio Nr/Nt, where Nr is the number of elements received incorrectly and Nt is the number of elements transmitted.

element spacing 1. The spacing between radiator, director, and reflector elements in a directional antenna. 2. The spacing between the in-

ternal electrodes of an electron tube.

elevation Angular position (in degrees) of a point above the horizontal.

elevation-position indicator A type of radar display simultaneously indicating the elevation of and line-of-sight distance to the target.

elevator control 1. An electronic system for automatically stopping an elevator even with a floor and opening the doors. Various safety functions may also be included, an example being the reopening of a closing door when a passenger steps into the car. 2. In an aircraft, the mechanical, electronic, or electromechanical devices or circuits involved in actuation of the elevators.

eliminator 1. Generically, a device or circuit acting as a surrogate for an inconvenient or undesirable component, e.g., battery eliminator; tube eliminator. 2. Generically, a device for removing or minimizing an undesirable signal or quantity, e.g., harmonic eliminator; interference eliminator.

ell A coaxial fitting which is a right-angle line section with a coaxial connector at each end. It takes its name from its shape (L).

ellipse A geometric figure having the Cartesian-plane formula $(x - x_0)^2/a^2 + (y - y_0)^2/Ib^2 = 1$, where a and b are constants, and x_0 and y_0 represent the center point.

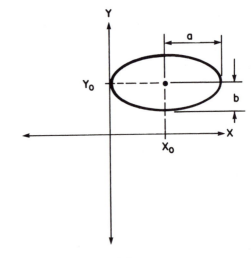

Ellipse.

elliptical function A function, similar to the Chebyshev and Butterworth functions, employed in the design of bandpass filters. The elliptical function results in a somewhat less desirable filter response than the Chebyshev or Butterworth functions.

elliptical load line For any amplifier with an output transformer, a load line in the shape of an ellipse, obtained when the load connected to the output element is reactive rather than purely resistive.

elliptically polarized wave An electromagnetic wave in which the rotation of the electric-intensity vector at one point describes an ellipse.

elliptical polarization Polarization characterized by elliptical rotation of the wave vector at a given point.

elongation A form of modulation distortion resulting from multipath propagation. Some of the paths result in greater propagation delay than other paths; this causes the modulation envelope to spread out. The higher the modulating frequency, the greater the effect.

ELSE A word used in a BASIC computer program that gives an instruction based on a relational test and, in this respect, is related to IF-THEN, ON-GOTO, etc. It specifies the operation to be done if the conditions given in the same program line don't occur.

ELSIE Abbreviation of *electronic letter-sorting and indicator equipment.*

EM 1. Abbreviation of *efficiency modulation.* 2. Abbreviation of *electromagnetic(s).* 3. Abbreviation of *electromagnetic iron.* 4. Abbreviation of *electromagnetizer.* 5. Abbreviation of *electron microscope.* 6. Abbreviation of *exposure meter.* 7. Abbreviation of *electromotive.*

Em 1. Symbol for *maximum voltage.* 2. Symbol for *maximum junction field.*

e/me The ratio of the elementary electron charge to its mass: 1.758 805 $\times 10^{11}$ coulombs per kilogram. Also see *charge-mass ratio.*

emanation Emission of electrons, radioactive particles (X rays, light, or radio waves).

Emax Symbol for *maximum voltage.*

embedding See *encapsulation.*

embossed-foil printed circuit A printed circuit made by pressing the pattern from metal foil into the insulating substrate and then removing the surplus foil.

embossed-groove recording 1. A phonograph record into which grooves are embossed rather than scribed. 2. Recording sound by embossing grooves on record disks.

embossing stylus The rounded-tip stylus used to make an embossed-groove recording.

EMC Abbreviation of *electromagnetic compatibility.*

emergency channel A communication channel allocated for emergency service.

emergency communication Radio or other electronic transmission and reception of urgent messages, e.g., distress signals, storm warnings, and the like.

emergency equipment 1. Apparatus kept in standby status for immediate operation when regularly used equipment fails. 2. Equipment, especially vehicular, for use in emergency situations. Examples are ambulance, fire-fighting trucks and equipment, and the like.

emergency power supply An ac or dc power unit kept in standby status for immediate use when the regular power supply fails.

emergency service A radio service devoted exclusively to emergency communication.

emergent ray The ray leaving a lens, prism, or similar processing device.

emf Abbreviation of *electromotive force.*

emf standard See *standard cell.*

EMG 1. Abbreviation of *electromyogram.* 2. Abbreviation of *electromyograph.*

EMI Abbreviation of *electromagnetic interference.*

E microscope See *electron microscope.*

emission 1. The ejection of particles, especially electrons, from a material. 2. Waves radiated from any source (as from a transmitting antenna or from an amplifier stage). 3. The emanation of radiant, electromagnetic, electrical, or magnetic energy.

emission characteristic The curve depicting the emission current of a tube as a function of its filament or cathode voltage.

emission code A system of abbreviating the various types of radio emission. See *emission mode.*

emission current In an electron tube, the current resulting solely from the emission of electrons by the cathode, i.e., without the aid of an external plate battery.

emission efficiency The rate at which electrons are emitted by a hot cathode, usually expressed in milliamperes per watt of cathode power.

emission mode Any of various official classifications of emission types. For AM emissions there are six emission modes, numbered A0 through A5; for FM there are six, numbered F0 through F5. Pulse emissions are designated with a P prefix.

emission power 1. The rate at which energy is radiated from an object. 2. Transmitter output power.

emission saturation In an electronic tube, the point at which all the electrons emitted by the cathode are attracted to the plate. Beyond this point, increasing plate voltage causes no increase in plate current.

emission spectrum The radiation spectrum of a substance that emits energy, e.g., the light spectrum of an incandescent metal.

emission tester A tester used to rate electron tubes in terms of electron emission. In using the instrument, the tube's electrodes, except the cathode, are joined to act as the plate of a simulated diode. Applying a test voltage results in "diode" current that indicates tube condition.

emission types See *emission mode.*

emission-type tube tester See *emission tester.*

emission velocity The initial velocity of an electron as it leaves an emitting surface.

emissive power The rate at which a surface emits energy of all wavelengths in all directions, per unit area of radiating surface, regardless of temperature.

emissivity For a radiating source, the ratio W_1/W_2, where W_1 is the energy emitted by the source at a particular temperature, and W_2 is the energy emitted by a blackbody (i.e., a theoretically perfect radiator) at the same temperature.

emitron A form of cathode-ray tube.

emitron camera tube A form of television camera tube.

emittance For an energy-radiating source, the radiated power per unit area of radiating surface.

emitted electron An electron that has left an atom of a material and has escaped into surrounding space or entered a neighboring material. Electrons are emitted by hot bodies, light-energized photoconductors, and other agents. Also see *work function.*

emitter 1. A body which discharges particles or waves (see *emission*). 2. In a semiconductor device, the area, region, or element from which current carriers are injected into the device. In a transistor symbol, the emitter is that electrode shown with an arrowhead. 3. In a punched-card machine, a device that produces signals simulating holes, i.e., nonexistent perforations.

emitter—base junction In a bipolar transistor, the boundary between base and emitter regions.

emitter bias Emitter current or voltage maintained to set the operating point of the transistor.

emitter bulk resistance The portion of the resistance of the semiconductor material in a transistor that affects emitter resistance.

emitter-coupled logic A bipolar form of digital logic, abbreviated ECL.

emitter-coupled multivibrator A two-transistor multivibrator circuit in which the emitters share a common resistor. The circuit is analogous to the cathode-coupled tube multivibrator.

emitter-coupled phase inverter A transistor phase inverter in which the out-of-phase component is taken from the collector and the in-phase component from the emitter resistor (of the same transistor). Another transistor is often used to amplify the in-phase component so that both outputs are equal in magnitude.

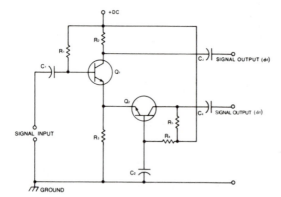

Emitter-coupled phase inverter.

emitter current Symbol, *Ie*. The current in the emitter electrode of a bipolar transistor.

emitter degeneration In a transistor amplifier, current degeneration obtained by use of an unbypassed emitter resistor. The arrangement is comparable to cathode degeneration in a tube circuit, and results in virtually distortion-free amplification at a sacrifice in voltage gain.

emitter follower A transistor amplifier in which the input signal is applied to the base, and the output signal is taken from the emitter resistor. The circuit is analogous to a tube-type cathode follower. See also *emitter degeneration*.

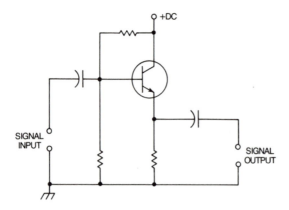

Emitter follower.

emitter-input circuit See *common-base circuit*.
emitter junction See *emitter—base junction*.
emitter resistance Symbol, *Re*. 1. The resistance of the emitter electrode in a bipolar transistor. 2. External resistance connected to a transistor's emitter terminal.

emitter stabilization In a common-emitter transistor stage, an emitter resistor that stabilizes the circuit against temperature variations. A temperature-induced rise in collector current causes an increase in emitter current, resulting in a corresponding increase in base bias voltage developed by collector-current flow through the resistor, which in turn reduces collector current to its original level.

emitter-to-base junction See emitter—base junction.

emitter voltage Symbol, *Ve*. The voltage at the emitter electrode of a bipolar transistor.

emp Abbreviation of *electromagnetic power*.

emphasizer An audio-frequency device with a specially tailored response, intended to maximize intelligibility of a voice.

empire cloth Varnished cambric employed as an insulating sheet or tape.

empirical Observable; derived from experimentation.

empirical curve A curve plotted from data acquired from observations, tests, and calculations rather than from mathematical laws or other theory.

empirical design The design of electronic circuits by cut-and-try methods and, to some extent, through intuition arising from experience, i.e., practical as opposed to theoretical design.

empirical probability Probability estimated from experience and observations. This method is often used in quality-control and reliability procedures.

empty medium A computer storage medium, such as magnetic or paper tape, that is ready to accept data, i.e., rather than being completely blank, it contains the signals necessary for processing the to-be-added data.

empty set In set theory, a set which contains no events, members, or terms. It is important to note that a set does not disappear when its members are removed, but becomes an empty set, a useful concept in probability and statistics.

emu Abbreviation of *electromagnetic unit(s)*.

emulator In computer engineering, a sophisticated device which substitutes for a similar device or stage in the computer and thereby provides a basis for experimenting and troubleshooting without disturbing the equivalent part of the computer.

En Symbol for *voltage remaining at null*.

enable To initiate the operation of a circuit or device by applying a pulse or trigger signal.

enable pulse 1. A pulse that initiates the operation of a circuit or device. 2. A binary pulse that augments a write pulse to make a magnetic core change state.

enabling gate A digital device that regulates the length of a pulse for specialized use.

enameled wire Wire that is insulated by a coat of baked enamel.

encapsulant A material, such as potting resin, used to encapsulate a circuit or device.

encapsulated circuit A circuit encapsulated in plastic or wax (see *encapsulation*).

encapsulated component An electronic part which is encapsulated in plastic or wax (see *encapsulation*).

encapsulating material See *encapsulant*.

encapsulation The embedding of a circuit or component in a solid mass of plastic or wax.

encephalogram See *electroencephalogram*.

encephalograph See *electroencephalograph*.

enciphered facsimile Facsimile communications which has been rearranged, or scrambled, at the transmitting location, so that it cannot be intercepted by a third party. Written material or pictures may be sent in this manner. A deciphering device is needed at the receiver end of the circuit.

enclosure 1. A cabinet, case, or other housing for electronic equipment, such as a receiver, transmitter, or test instrument. 2. A specially designed housing for a loudspeaker.

encode 1. To convert signals or data into a desired (usually digital) form. Also called *code*. (Compare *decode, 1, 2*.) 2. To equip a transmitter with a tone-producing device (encoder). 3. To develop and apply an encoding system to a group of transceivers or transmitters of a communications network.

encoder 1. An analog-to-digital or digital-to-analog converter. 2. An electromechanical device for translating the angular position of a rotating shaft into a corresponding series of digital pulses. Also see *shaft-angle encoder*. 3. A device for encoding data (see *encode*). 4. A machine with a keyboard for printing characters that can be read by OCR equipment. 5. A tone generator used as a receiver enabler in the transmitters of a communications network.

end-around carry In a computer, a carry produced in the most significant position, causing a carry into the least significant position.

end-around shift In digital-computer practice, the transfer of characters from one end to the other of a register. Also called *logical shift*.

end bell 1. The part of a motor housing that supports the bearing and protects internal rotating parts. 2. A clamping part fastened to the back of a plug or receptacle. 3. Either of the two frames of a transformer that contains the mounting lugs.

end bracket See *end bell, 2*.

end cell A cell intended for series operation in conjunction with a storage battery. As the voltage of the battery drops, the end cell can be added into the circuit.

end effect 1. In a tapped coil, losses due to induced currents flowing in the inductance and distributed capacitance of the unused end of the coil. 2. Edge effect in a capacitor.

end-fed antenna An antenna whose lead-in or feeders are attached to an end of the radiator.

end-fire antenna An antenna that exhibits maximum radiation or best reception at its ends. Also see *end-fire array*.

end-fire array A beam antenna of multiple elements providing end-fire directivity. Compare *broadside array*.

end-fire directivity In a directive antenna, beaming a signal along the plane of the antenna, i.e., off its ends.

end instrument A device capable of converting intelligence into electrical signals, and vice versa, and which needs to be connected to only one terminal of a loop.

end item A final, completed product or component.

end mark In digital-computer practice, a signal or code indicating the close of an information unit.

endodyne reception A British term for homodyne reception (see *zero-beat reception*).

end-of-data mark A code or character signaling that all the data in a computer storage medium has been read or used.

end-of-field mark In computer practice, a "flag" code which signals that the end of a field has been reached.

end-of-file mark In computer practice, a code instruction which signals that the last record in a file has been read.

end-of-job card At the end of a pack of punched cards ready for input to a computer, a card that (1) alerts the program that the job is done and (2) starts another action.

end-of-message character A character or code signaling the end of a message.

end-of-run The end of a computer program or program run as indicated by the program.

end-of-tape mark A physical marker at the end of a magnetic tape, e.g., something that can be sensed by methods other than that used to read the tape.

end-of-tape routine A computer program that handles the processing needed after the last record on a reel of magnetic tape has been reached.

end-on armature A relay armature that moves in the direction of the core's axis.

end-on directional antenna See *end-fire antenna* and *end-fire array*.

endoradiograph An X-ray picture, derived or enhanced by the introduction of substances into the body.

endoradiosonde A tiny pill-enclosed transducer and radio transmitter for sensing physiological conditions in the stomach and intestines and transmitting corresponding signals to instruments outside.

endothermic reaction A chemical reaction producing cold, i.e., one in which kinetic energy is lost. Compare *exothermic reaction*.

end-plate magnetron A magnetron whose oscillation intensity is increased by a positive and a negative end plate, the electric field between them causing the electrons to move axially while spinning.

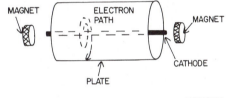

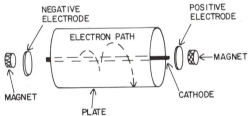

End-plate magnetron.

end point 1. For a precision potentiometer, the shaft position between the last and first positions of measurement. 2. The point at which the useful life of a device may be considered spent. 3. The point at which a time interval or operational sequence ends. 4. The end-point voltage of a primary or Edison storage cell. 5. For a lead-acid storage cell, the specific-gravity value of the electrolyte at which the cell is considered in need of recharging (1.150 to 1.175).

end-point control A form of quality control in which the end item is checked for defects.

end-point sensitivity A means of expressing the sensitivity of a meter or other indicating device. End-point sensitivity is the ratio, in decibels, between the input signal required to produce a full-scale or maximum reading and the smallest detectable input signal.

end-point voltage The voltage of a battery or cell terminal when the device is no longer useful.

end resistance In a rheostat or potentiometer, the resistance between the wiper and the end terminal when the wiper is set to the end point of the device.

end-resistance offset In a potentiometer, the resistance between the wiper and an end terminal when they are in contact.

end-scale deflection See *end-scale value*.

end-scale value For an indicating meter, the electrical quantity indicated at the last graduation on the scale.

end section Either the input or output section of a *multisection filter*.

end setting 1. The fully clockwise or fully counterclockwise setting of a rotatable control. 2. The minimum or maximum setting of a control.

end shield In a magnetron, a shield confining the space charge to the interaction space.

end spaces The cavities at either end of the anode block in a multicavity magnetron tube; they terminate all the anode-block cavity resonators.

end use The intended (final) application of a circuit or device.

energize To apply operating power and input signals to a circuit or device.

energized The condition of a circuit or device that is powered or excited.

energy Symbol, *W*. Unit, *joule*. 1. The capacity for doing work. Some common forms of energy are electrical, mechanical, and chemical. Also see *conservation of energy, kinetic energy,* and *potential energy*. 2. The work done by electric power. The unit used by utility companies is the kilowatt-hour (kWh). Here, energy kWh = Pt, where P is power in kilowatts and t is the period (hours) during which the power is used. One kWh could represent the use of 100W for 10 hours, 10W for 100 hours, and so on.

energy-band diagram A diagram depicting the various energy levels within the atom of a conductor, semiconductor, or insulator.

energy barrier The natural potential gradient across a semiconductor junction. In the absence of an applied voltage, the gradient, not measurable from the outside, prevents total interaction between the n- and p-type materials. Also see *barrier, 1*.

energy cell 1. A usually small primary or secondary cell, especially the kind used in hearing aids and electronic watches. 2. A capacitive-type dc source (see *energy-storage device, 2*).

energy consumption The use of energy by a device or system. The energy, rather than being depleted, is transformed (see *conservation of energy*). Thus, when an electronic device *uses* energy, it converts it into work and heat.

energy conversion The transformation of energy from one form to another. See, for illustration, *conservation of energy* and *energy transformation*.

energy-conversion device A device which accomplishes energy conversion. For illustration, see *conservation of energy*.

energy density For an energy-producing cell, the ratio of available energy to the cell's mass.

energy factor See *Q*.

energy gap In the energy-level diagram for a semiconductor or insulator, the region between valence and conduction bands representing the minimum energy required to make the electron pass from the valence to the conduction band, i.e., to become a current carrier. Also called *forbidden band*.

energy level A constant-energy state such as one of the energy levels an electron is thought to occupy in an atom.

energy-level diagram 1. A diagram showing the energy level (in electronvolts) of electrons in the various shells of an atom. 2. A diagram showing variations in power that correspond to variations in current in a channel.

energy loss In any system, the energy which is unavoidably lost; i.e., it is not converted into useful work. Also see *entropy* and *power loss*.

energy of a charge The energy level of an electrostatic charge. It is $QV/2$ ergs, where Q is the quantity of electricity and V is the potential.

energy product An expression of the effectiveness of a permanent magnet. The magnetic flux density is multiplied by the magnetic field strength to obtain the energy product, specified in gauss-oersteds.

energy-product curve See *box-shaped loop*.

energy redistribution A mathematical process for determining the effective duration of a pulse. The instantaneous power output of the pulse is integrated from the start to the end of the pulse. Then, a rectangular pulse is constructed having the same peak power and the same total energy content. The length of this rectangular pulse is considered to be the effective duration of the actual pulse.

energy state The condition of an electron, as expressed by its position and velocity with respect to the position and velocity of other electrons.

energy-storage capacitor A usually high-value capacitor employed primarily to store the charge used to fire a lamp (as in a photoflash unit), create a spark discharge (as in electronic ignition), or perform some similar function.

energy-storage device 1. Capacitor. 2. A small, electrochemical component offering very high capacitance (e.g., several farads) and low leakage current (less than 1 pA). It has a number of applications, including long-interval timing, power-supply filtering, and energy-cell service. Its active ingredients are compressed powders.

energy stored in capacitor The electrical energy in the field between the plates of a charged capacitor. In this instance, energy $W = CE^2/2$, where W is energy in joules; C, capacitance in farads; and E, voltage in volts.

energy stored in inductor The magnetic energy in the field surrounding an inductor carrying current. There, energy $W = (LI^2)/2$, where W is energy in joules; L, inductance in henrys; and I, current in amperes.

energy transformation The conversion of one form of energy into another, as with a transducer.

engine analyzer An instrument for checking the performance of an automobile engine. In addition to measuring voltage and resistance throughout a car's electrical system, the instrument measures engine speed, cam dwell angle, and other factors.

engineer 1. A person who designs machines, circuits, and other devices. 2. A person who develops methods of utilizing machines, circuits, or other devices more efficiently, or for new applications. 3. To design or implement an apparatus.

engineering the science and art of applying scientific laws to technical problems and designing practical devices. Also see *electrical engineer* and *electronic engineer*.

enhanced-carrier demodulation A method of reducing distortion during AM demodulation in which a properly phased and synchronized local carrier is added to the signal in the demodulator.

enhancement mode Operation characteristic of enhancement-type MOSFETs.

enhancement-type MOSFET A metal-oxide semiconductor field-effect transistor in which the channel directly under the gate electrode is widened (enhanced) by a negative gate voltage in the n-channel unit or by a positive gate voltage in the p-channel unit. Compare *depletion-enhancement-mode MOSFET* and *depletion-type MOSFET*.

ENIAC An electronic computer developed at the University of Pennsylvania. The name is an acronym for *electronic numerical integrator and computer*.

ENIC Abbreviation of *voltage negative-impedance converter*.

ensemble 1. A collection of devices that functions together as a complete unit. 2. A set of random mathematical functions, all starting at the same point.

ensi Abbreviation of *equivalent-noise-sideband input*.

entladungsstrahlen Intense ultraviolet radiation by electric sparks. At atmospheric pressure, the frequency of radiation in wavelengths is between 4×10^{-8} and 9×10^{-8} meter (400 and 900 angstroms), depending upon spark length. The rays are rapidly absorbed by air, but can be reflected by glass. The term is derived from the German word for *discharge rays*.

entropy 1. Symbol, ϕ. Unit, J/K. In all closed physical systems, the measure of energy wasted. According to the second law of thermodynamics, for example, supplied heat can never be converted entirely into work. 2. In communications, the amount of information in a message, based on $\log n$, where n is the number of possible equivalent messages. 3. The state of matter and energy being inert and uniform.

entry 1. A unit of computer input or output information. 2. A data item in a table or list. 3. A computer source program statement. 4. In a computer program, the address of the first instruction.

entry condition A condition that has to be specified before a computer program is run, e.g., establishing operand values.

entry point In a computer program, the first instruction to be implemented, or a point during the run when data can be entered.

envelope 1. On a graph, the imaginary line joining successive ac signal peaks. In the graph for an amplitude-modulated signal, the line reproduces the ac modulating wave. 2. The shell of an electron tube or the enclosure of a transistor.

envelope delay In a tuned amplifier, time delay introduced in the envelope of a modulated signal by varying the phase of the envelope with the modulating frequency. This delay varies directly with the amount by which the sidebands shift with respect to the carrier frequency.

envelope-delay distortion Distortion occasioned by envelope delay.

enveloped file A computer file with labels permitting it to be handled by a computer of a type different from that used to make the file.

environmental conditions See *environmental factors*.

environmental factors Aspects of the space immediately surrounding and sometimes influencing electronic equipment. Examples: altitude, dust, light, moisture, noise, pressure, shock, temperature, vibration.

environmentally sealed Sealed against the effects of adverse environmental factors.

environmental test chamber See *climate chamber*.

Eo 1. Symbol for *output voltage*. 2. Symbol for *zero reference voltage*.

EOF Abbreviation of *end of file*.

EOL Abbreviation of *end of life*.

EOLM Abbreviation of *electro-optical light modulator*.

EOR Abbreviation of *end of* (program) *run*.

EOS Abbreviation of *electro-optical system(s)*.

EOT Abbreviation of *end of tape*.

EOTS Abbreviation of *electro-optical tracking system*.

Ep 1. Symbol for *plate voltage*. 2. Symbol for *peak voltage*.

ep See *extended play*.

ephemeris time Time measured with respect to the orbit of the earth around the sun. Initiated in the year 1900 AD.

episcotister A mechanical light beam modulator. The device consists essentially of a series of rotating disks having transparent and opaque sections which alternately interrupt and pass the light beam at an audio-frequency rate.

epitaxial Exhibiting epitaxy. Also see *epitaxy*.

epitaxial deposition The tendency of certain materials to grow on a semiconductor substrate under certain conditions.

epitaxial device A semiconductor device built by means of epitaxial growth.

epitaxial film A film of single-crystal semiconductor material deposited onto a single-crystal semiconductor substrate.

epitaxial growth Growing monocrystalline silicon on a silicon wafer by precipitating silicon from a gas in which the wafer is placed. Epitaxy is secured between the precipitate and the wafer.

epitaxial-growth mesa transistor See *double-diffused epitaxial mesa transistor*.

epitaxial growth process In the manufacture of semiconductor devices, growing a single-crystal semiconductor material on a *seed crystal* or single-crystal wafer so that epitaxy results.

epitaxial layer A semiconductor layer exhibiting epitaxy. Also see *epitaxial growth*.

epitaxial mesa transistor An epitaxial transistor in which a thin mesa layer is overlaid on another mesa layer.

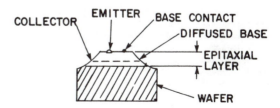

Epitaxial mesa transistor.

epitaxial planar transistor A planar transistor having an epitaxially grown collector on a low-resistivity substrate, and a diffused base and emitter.

epitaxial process See *epitaxial growth process*.

epitaxial transistor A transistor in which an epitaxial layer (into which a base region later is diffused and an emitter region alloyed) is grown on the face of a semiconductor wafer which serves as the collector. Also see *double-diffused epitaxial mesa transistor*.

epitaxy The condition in which atoms in a thin film of single-crystal semiconductor material grown on the surface of the same kind of wafer continue their characteristic alignment. Also see *epitaxial growth*.

E plane The plane of an antenna containing the electric field.

E-plane bend See *E bend*.

E-plane tee junction A waveguide junction (see *waveguide tee*) whose structure changes in the plane of the electric field.

epoxy resin A synthetic resin used to encapsulate and pot electronic equipment, or as a cement. Epoxy resins are based on ethylene oxide or its derivatives.

Epp Symbol for *peak-to-peak voltage.*

EPROM Abbreviation of *electrically programmable read-only memory.*

epsilon The fifth letter (ϵ) of the Greek alphabet. It is sometimes used as the symbol for the natural number 2.718 828 (often written *e*); also, it symbolizes permittivity, especially that of a free-space vacuum: 8.854 188 × 10^{-12} farad per meter.

EPU 1. Abbreviation of *electronic power unit.* 2. Abbreviation of *emergency power unit.*

Eq. Abbreviation of *equation.*

equal alternations Positive and negative half-cycles of a wave which have identical shape and amplitude.

equal-energy source A light source that has a constant emission rate (energy per unit wavelength).

equal-energy white The color of light emitted by a source radiating equally the wavelengths of the visible-light spectrum.

equal heterodyne In a beat-frequency system, the condition in which the outputs of the two heterodyning oscillators are identical.

equality circuit A logic circuit that, when two numbers are put into it, generates a one if the numbers are equal, or a zero if they aren't.

equalization The use of a circuit or device to make the frequency response of a line, amplifier, or other device uniform over a given frequency range, or to modify that response as desired.

equalizer A circuit or device, such as a compensated attenuator, for achieving equalization.

equalizer circuit breaker A form of circuit breaker that trips in the event of unbalance in an electrical system.

equalizing current An output-equalizing current that flows in the circuit of two compound generators connected in parallel.

equalizing network A circuit employed to equalize a line.

equalizing pulses In a TV signal waveform, several pulses (preceding and following the vertical sync pulse and having a repetition rate of twice the power-line frequency) which start the vertical retrace at the correct instant for good interlace.

equal-loudness curves See *audibility curves.*

equal vectors Vectors which have the same magnitude and direction but do not necessarily originate at the same point. Compare *identical vectors.*

equation solver A (usually analog) computer for solving linear simultaneous equations or for determining the roots of polynomials.

equation of first degree See *first-degree equation.*

equation of second degree See *quadratic equation.*

equation of third degree See *cubic equation.*

equations of higher degree Equations in which variables have exponents greater than one. See, for example, *cubic equation* and *quadratic equation.*

equimolecular Having the same number of molecules.

equiphase surface Any surface in a wave, over which the field vectors at a particular instant have either 0° or 180° phase difference.

equiphase zone The space region in which two radionavigation signals show no phase difference.

equipment 1. Collectively, apparatus or components designated for a specific purpose, such as electronic equipment, emergency vehicular equipment, radio equipment. 2. A functional electronic unit, such as a test instrument, receiver, memory unit, and the like. Expressed in the singular, *an* equipment.

equipment chain A system consisting of series-connected circuits. All circuits must be functional for the chain to operate.

equipment ground An electrical ground connection intended to reduce the chances of electric shock. An equipment ground does not necessarily constitute a good radio-frequency ground; it serves only to eliminate potential differences among the individual units in a system.

equipment life The period during which electronic equipment functions according to specifications; it is terminated at an end point (see *end point, 2*).

equipment test A usually preliminary, qualifying test of electronic equipment.

equipotential Having the same potential.

equipotential cathode See *indirectly heated cathode.*

equipotential line Between two charged plates, the locus (an imaginary line) of points having the same potential with respect to the plates.

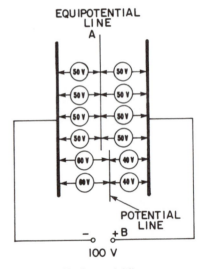

Equipotential line.

equipotential surface A surface on which all points have the same electrical potential.

equisignal Pertaining to signals having equal intensity.

equisignal localizer See *tone localizer.*

equisignal radio-range beacon For aircraft guidance, a radio-range beacon that transmits two distinct signals which are received by aircraft with equal intensity only in certain directions.

equisignal surface The "surface" around a transmitting antenna formed by points of equal field intensity.

equisignal zone The region in which two radionavigation signals have identical amplitude.

equivalence The condition existing when one network can be substituted for another without disturbing currents, impedances, and voltages at the terminals.

equivalent absorbing power See *equivalent stopping power.*

equivalent absorption Unit, sabin. The rate at which a surface absorbs sound energy.

equivalent binary digits For a given decimal number or specific character, the corresponding binary digits (bits), e.g., 5 are required to represent the alphabet; decimal 26 = binary 11 010, a number having 5 digits.

equivalent capacitance The value of a single lumped capacitance which would cause the same action as the capacitance that is distributed throughout a circuit.

equivalent circuit A circuit which has the same overall current, impedance, phase, and voltage relationships as a more complicated counterpart which it usually replaces for analysis. Thus, a circuit of four resistances series-parallel may be replaced with an equivalent circuit containing one resistance (equal to the series-parallel resistance of the original).

equivalent component density For a circuit in which discrete components are not used or are not evident, the volume of that circuit divided by the number of discrete components which would be required if the circuit employed them.

equivalent conductance For a mounted ATR tube, the resonant-frequency conductance of the unit.

equivalent conductivity The conductivity of a solution that contains 1 gram equivalent of the solute in the space between electrodes 1 centimeter apart.

equivalent dark-input current For a photoelectric device, the illumination required for an output current equal to the *dark current* (q.v.) of the device.

equivalent decrement The value of decrement in a damped wave which would result in the same amount of interference at a receiver as the interference caused by the sidebands of an amplitude-modulated signal.

equivalent delay line A comparatively simple network (such as an *RC* circuit) which will provide the attenuation and phase characteristics of an ideal delay line.

equivalent delta In a three-phase system, a delta-connected circuit that is equivalent to a given wye-connected circuit, from the standpoint of impedance and phase. Also see *delta connection* and *wye connection*. Compare *equivalent wye* and *wye-equivalent circuit*.

equivalent diode 1. A hypothetical diode representing a triode in terms of the combined effects of grid and plate charges on the space charge. In some types of circuit analysis, this imaginary diode replaces the triode. 2. A simulated diode. An example appears when the grid of a vacuum tube is biased positive with respect to the cathode. The grid then acts as a diode plate against the cathode, and will provide rectification.

equivalent differential input capacitance For a differential amplifier, the equivalent input capacitance (see *equivalent capacitance*) at one input (inverting or noninverting) when the opposite input is grounded.

equivalent differential input impedance For a differential amplifier, the equivalent input impedance at one input (inverting or noninverting) when the other input is grounded.

equivalent differential input resistance For a differential amplifier, the equivalent input resistance at one input (inverting or noninverting) when the other input is grounded.

equivalent equations Two equations for an unknown that have the same root.

equivalent four-wire system A two-wire line over which full-duplex operation is obtained by use of frequency division.

equivalent grid voltage Symbol, $Eg\,eq$. For an electron tube, the ratio $(Eg + Ep)/\mu$, where Eg is grid voltage (volts); Ep plate voltage (volts); and μ, amplification factor.

equivalent height The virtual height of an ionospheric layer above the earth.

equivalent impedance 1. The value of a single lumped impedance which would cause the same action as the impedance distributed throughout a circuit. 2. An impedance which draws current of the same strength and phase as that drawn by an impedance it replaces. Thus, a series impedance may be equivalent to a given parallel impedance.

equivalent inductance The value of a single lumped inductance which would cause the same action as the inductance distributed throughout a circuit.

equivalent input offset current For a differential amplifier, the difference between currents flowing into the inverting and noninverting inputs when the output voltage is zero.

equivalent input offset voltage For a differential amplifier, the input voltage required to reduce the output voltage to zero.

equivalent input wideband noise voltage For a differential amplifier, the ratio Vo/Av, where Vo is the output-noise voltage (rms volts) and Av is the dc voltage gain.

equivalent loudness The actual intensity, in decibels, of a given sound whose apparent loudness changes with frequency. The threshold of hearing, for example, is 3 phons regardless of frequency. The *equivalent loudness* of 3 phons, however, is about 70 dB at 20 Hz, 10 dB at 225 Hz, – 3 dB at 3.5 kHz, and 15 dB at 10 kHz.

equivalent network A network that can replace a more complex network for analysis purposes.

equivalent noise input The value of modulated luminous flux which, when applied to a photoelectric device, produces an rms output current equal to the device's rms noise current.

equivalent noise resistance Symbol, Req. The resistance imagined to be connected between grid and cathode of a noiseless vacuum tube; it expresses the noise of random electron movement in the tube. For a triode, $Req = 2.5/gm$, gm is the transconductance (siemens) of the tube at the operating point.

equivalent-noise-sideband input Abbreviation, *ensi*. A specification for receiver noise characteristics. Numerically, ensi $= 0.3Es\sqrt{Pn/Ps}$, where Es is the voltage of an unmodulated rf carrier applied to the receiver; Pn, the resulting noise-output power of the receiver (measured with an rms meter); and Ps, the noise-output power measured with the rf signal 30% amplitude modulated at 400 Hz with a 400 Hz bandpass filter inserted between the receiver output terminals and the meter.

equivalent noise temperature For a component having resistance, the temperature (degrees absolute) at which a theoretically perfect resistor having the resistance of the component would generate the same noise the component generates at room temperature.

equivalent optics The analogy between certain optical lenses and prisms and the electrostatic deflection of an electron beam. Thus, when the upper deflecting plate in an electrostatic deflection system is made negative and the lower plate is positive, the beam is deflected downward, like horizontal light rays bent by a prism. When both plates are made equally negative, the beam converges to a point, as light rays do when they pass through a double convex lens. When both plates are made equally positive, the beam spreads out, as do light rays passing through a double concave lens.

equivalent permeability The permeability of a component made of certain materials, compared with that of a component having the same reluctance, shape, and size but made of different materials.

equivalent plate-circuit theorem The theorem holding that the currents and voltages in the plate (load) circuit of a vacuum tube may be determined from an equivalent plate circuit consisting of (1) a constant-voltage generator (μEg) in parallel with the plate resistance (rp) of the tube, or (2) a constant-current generator ($gmEg$) in parallel with the plate resistance (rp) of the tube.

equivalent plate voltage For an electron tube, the product $Ep(\mu Eg)$, where Eg is grid voltage (volts); Ep plate voltage (volts); and μ, amplification factor.

equivalent reactance The value of a single lumped reactance which would cause the same action as the reactance distributed throughout a circuit.

equivalent resistance The value of a single lumped resistance which would cause the same action as the resistance distributed throughout a circuit.

equivalent series and parallel circuits Series and parallel circuits in which current, voltage, phase, and frequency relationships are identical. Any series circuit can be transformed into an equivalent parallel circuit.

equivalent series resistance The equivalent resistance acting in series with circuit components.

equivalent sine wave A sine wave of the same frequency and effective voltage as a given wave.

equivalent stopping power For a material in the path of radioactive particles, the thickness of the material that produces the same energy loss as that produced by 1 centimeter of air.

equivalent time 1. Uncalibrated time. 2. The effective duration of some phenomenon, such as a pulse.

equivalent volt See *electronvolt*.

equivalent wye In a three-phase system, a wye-connected circuit which is equivalent to a given delta-connected circuit from a standpoint of impedance and phase. Also see *delta connection, wye connection,* and *wye-equivalent circuit.* Compare *equivalent delta*.

equivalent Y See *equivalent wye*.

equivocation A condition in which the meaning of data is dependent on certain parameters.

ER Abbreviation of *echo ranging*.

Er Symbol for *erbium*.

Er Symbol for *voltage drop across a resistance*.

erasable storage In computer practice, any storage medium holding information that can be erased.

erasable PROM A PROM that can be erased, as by exposure to ultraviolet light. Also see *PROM*.

erase To obliterate or remove a signal, especially a recorded one, as in the erasure of recorded material from a magnetic tape or the image from a storage tube.

erase button A pushbutton that actuates the circuit supplying a signal which erases stored material (as the display on a storage oscilloscope).

erase current In an electromagnetic erase head the current flowing through the coil of the head. In most instances, it is a high-frequency current (usually the regular bias current), but it can be as low as 60 Hz, as long as the speaker does not respond to what remains of it on the tape after erasure.

erase head In a tape recorder, a head employed to erase recorded material from tape. It may contain a permanent magnet (see *erase magnet*) or an electromagnet whose coil carries erase current.

erase magnet In a tape recorder, a magnet employed to erase recorded material from tape. Because the strength of the magnet is greater than that of the magnetized areas on the tape, erasure is complete (the tape left demagnetized).

erase oscillator In a tape recorder, a high-frequency (typically 30-80 kHz) oscillator which supplies erase current.

eraser A device for bulk erasing recorded tape. The reel of tape is placed on the device's spindle and a penetrating ac magnetic field does the erasing. Often called *bulk tape eraser*.

erase signal A signal that causes recorded material to be erased (see *erase* and *erase current*).

erasing speed The rate at which successive storage elements are erased as in a charge-storage tube.

erasure 1. In tape-recording and digital-computer practice, the process of erasing a recorded signal (see *erase*). 2. An erasure accomplished by the process described in 1, above.

erbium Symbol, Er. A metallic element of the rare-earth group. Atomic number, 68. Atomic weight, 167.27.

E region See *E layer*.

e-register In a computer, a register used in double-precision calculations.

Er-Ey signal In color television, the resultant signal that is the difference between the original full-red and Ey signals.

ERG Abbreviation of *electroretinogram*.

erg Abbreviation, *e*. A unit of work. It is the work done by a force of 1 dyne (10^{-5}N) acting through a distance of 1 centimeter.

ergograph An instrument employed to measure and record work done by muscles.

ergometer An instrument for measuring energy consumed or work accomplished.

ergon See *erg*.

Erms Symbol for *root-mean-square voltage*.

ERP Abbreviation of *effective radiated power*.

error 1. In calculations and measurements, the difference between a true value and an observed or calculated value. 2. In electronic circuits, especially those of automatic control systems, the difference between a required (or reference signal) level and the actual signal level.

error amplifier An amplifier for boosting error current or voltage.

error-checking code An error-correcting or error-detecting code.

error-correcting code An error-detecting code that, in addition to the function indicated by its name, indicates the correct code.

error-correcting telegraph A digital communications system in which an improbable or incorrect character is not accepted. In the event such a character is received, the receiver instructs the transmitter to send that character again.

error-correction routine In computer practice, a series of programmed instructions to detect and correct errors in files.

error current An error signal that is a feedback current for automatically correcting a system.

error curve A bell-shaped curve which describes the distribution of errors in measurement around a true value.

error-detecting code In computer operation, a character-coding system which insures that an impossible combination (*forbidden characters*) will be generated by an error (for the error's detection).

error-detecting routine A computer program that detects errors by checking the validity of data.

error detection and feedback In computer practice, a system in which an error (sensed by an error-detecting code) automatically generates a request to repeat the suspect signal.

error detector A sensor which responds to an error signal by delivering a signal proportional to the error.

error diagnostics As performed by a compiler, detecting and indicating the presence of errors in source language statements.

error interrupt A computer program halt caused by a software or hardware error and accompanied by a display of what has happened.

error list As produced by a compiler, a list of source language statement faults.

error message During a computer program run, a statement (displayed on a peripheral) of what is in error.

error of measurement The positive or negative difference between the value of an actual measurement and the true (or most probable) value.

error range For a data item, the range of values over which it will cause an error.

error rate In data transmission, the ratio of *errors* transmitted to the *data* transmitted.

error-rate damping Damping that involves adding to an error signal another signal that is proportional in rate of change.

error routine A computer program segment that is input when an error is detected, so that an appropriate action is taken (correct the error, repeat the process, etc.).

error-sensing circuit A circuit that samples the output current or voltage of a power supply, amplifier, or control system; compares this output with a standard value; and delivers a feedback (correction) signal whose amplitude is proportional to the difference (error).

error signal In a servo system, an output signal whose value is proportional to the difference between the actual operating quantity of the system and a standard reference quantity. The signal is fed back to the input of the system for automatic correction.

error tape In data processing, a record tape designed and employed for storing errors for subsequent study.

error voltage An error signal that is a feedback voltage for automatically correcting a system.

Es Symbol for *einsteinium*.

Es Symbol for *screen voltage*.

Esaki diode See *tunnel diode*.

E scope See *E display*.

escape character In computer practice, a character indicating that the next character belongs in a new group.

escapement A (usually oscillating) mechanical or electromechanical device which stores energy (often in a spiral spring) on one swing, and returns that energy on the next swing. Such a mechanism advances a shaft progressively in a clock or watch and in some control equipment.

escape velocity The minimum velocity required for a body such as a rocket to escape the gravitational pull of earth or of another heavenly body.

escutcheon A usually decorative plate which frames an opening or covers a panel in an equipment, e.g., the escutcheon of a radio tuning dial.

ESD Abbreviation of *energy storage device*.

ESG Abbreviation of *electronic sweep generator*.

Esnault-Pelterie formula A formula for calculating the inductance of a single-layer coil; $L = 0.0018(a^2N^2)/(l + 0.92a)$, where L is inductance in microhenrys; a, radius of coil (inches); I, length of coil (inches), and N, number of turns.

ESS Abbreviation of *electronic switching system*.

EST Abbreviation of *eastern standard time*.

esu Abbreviation of *electrostatic unit(s)*.

Esup Symbol for *suppressor voltage*.

ET Abbreviation of *ephemeris time*.

eta The seventh letter (η) of the Greek alphabet. Symbol for *efficiency* and

electric susceptibility.

ETC Abbreviation of *electronic temperature control*.

etchant Any substance such as cupric chloride, ferrous chloride, or hydrochloric acid, used in etching.

etched circuit A circuit produced by etching the metallic coating of a substrate to provide the required pattern of conductors and terminals to which discrete components are soldered.

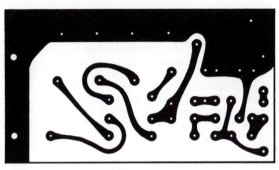

Etched circuit.

etch factor The ratio of the depth to the width of an etched track in a printed-circuit board.

etching 1. Chemically eating away a metal to form a desired pattern such as an etched circuit. 2. Thinning a quartz-crystal plate by slowly eroding one or both of its faces with hydrofluoric acid, to fine-tune the plate's frequency.

ET-cut crystal A piezoelectric plate cut from a quartz crystal at an angle of $+66°$ with respect to the Z-axis. Also see *crystal axes* and *crystal cuts*.

ether 1. A nonviscous fluid once thought to fill space, convey waves (radio, light, etc.), and sustain fields. 2. A volatile liquid occasionally used in electronics as a solvent, e.g., ethyl oxide $(C_2H_5)_2OH$. It is well known as a surgical anesthetic.

ether drift The postulated motion between a material body and the ether (see *ether, 1*). The concept was checked by Michelson and Morley, who failed to find that earth moves through the ether, much to the discredit of the whole idea of an ether.

E-transformer A differential transformer whose primary is wound on the center leg of an E-core, the secondaries being wound on the outer legs.

Ettinghausen effect A phenomenon somewhat like the Hall effect. It occurs when a metal strip carrying current longitudinally is placed into a magnetic field perpendicular to the plane of the strip: corresponding points on opposite edges of the strip exhibit different temperatures.

Eu Symbol for *europium*.

eudiometer 1. An instrument for measuring the amount of oxygen in the air. 2. An instrument for analyzing gases.

Euler constant Symbol, γ. The constant 0.577 215 665, which is the limit approached by $(1 + 1/2 + ... 1/m - \log m)$ as m approaches infinity.

eureka 1. Constantan. 2. The ground transponder beacon in the British rebecca-eureka radar navigational system (see *rebecca-eureka system*).

europium Symbol, Eu. An element of the rare-earth group. Atomic number. 63. Atomic weight, 152.

eutectic 1. A form of reaction in which mixed liquids solidify when cooled. 2. The solid substance resulting from such a reaction.

eutectic alloy A metallic alloy with a specific melting point, made via eutectic process.

eutectic bond A connection between two dissimilar metals, facilitated by a third metal alloyed, via eutectic process, to the adjoining faces.

eV Symbol for *electronvolt.*

evacuation The removal of air or other gases from a tube or chamber, specifically, the envelope of a vacuum tube that houses the internal elements.

evaporation 1. A technique for electrically depositing a film of a selected metal on a metallic or nonmetallic surface. A filament of the metal to be deposited is heated by an electric current in a vacuum chamber, which makes filament particles travel to the (nearby) object to be coated, where they condense as a film. In an alternate method, a piece of the metal to be deposited is laid on or wrapped around a filament of some other metal. 2. Electron emission by a hot cathode.

evaporation theory The theory that electrons will acquire sufficient escape velocity to leave a material when the energy acquired by (or imparted to) the electron exceeds the work function of the material. Also see *work function.*

E vector The vector that represents the electric component of an electromagnetic wave.

even harmonic In a complex waveform, an even-numbered multiple of the fundamental frequency. Compare *odd harmonic.*

even line In a TV picture, an even-numbered member of the 262.5 horizontal lines scanned by the spot in developing the even-line field. Compare *odd line.*

even-line field On a TV screen, the complete field obtained when the spot has traced all the even-numbered lines. Compare *odd-line field.*

Even-line field.

even parity check A check to verify presence of an even number of ones or zeros in a group of bits.

event A happening affecting the state of a computer file.

event counter Any device that measures the number of specified events taking place within a certain interval of time.

evolution Extracting a root of a number, e.g., square root, cube root, etc. Compare *involution.*

E wave In microwave practice, the transverse magnetic (TM) wave. Also see *waveguide modes.*

EWR Abbreviation of *early-warning radar.*

EWS Abbreviation of *early-warning system.*

Ex 1. Symbol for *voltage drop across a reactance.* 2. Symbol for *excitation energy.*

exa Symbol, E. A prefix meaning 10^{18} (International System of Units).

exact differential equations Ordinary differential equations of the first degree and first order. Also see *differential equations* and *ordinary differential equations.*

exalted-carrier reception In radio reception, overcoming the effects of selective fading by maintaining the carrier at a high amplitude. This is accomplished before demodulation by removing the carrier from an amplitude-modulated or phase-modulated signal, amplifying it, and reinserting it with the sidebands at a higher amplitude.

exc 1. Abbreviation of *exciter.* 2. Abbreviation of *excitation.*

except gate A gate which delivers an output pulse when an input pulse is present at one or more of a set of input terminals, and absent from one or more of another set of input terminals. Also called *exclusive-OR element.*

excess charge The amount of overcharge for a storage battery.

excess conduction In a semiconductor, current conduction by excess electrons.

excess electron 1. An electron which, when introduced into an atom, results in a negative ion. 2. An electron resulting from the addition of a donor impurity to a semiconductor substance.

excess fifty In computer practice, the designation of a number N by the notation $N + 50$. (Also *excess 50.*)

excess meter A meter that integrates the amount of power in excess of some predetermined level.

excess minority carriers The number of minority carriers in excess of the normal equilibrium number in a semiconductor material.

excess modified index of refraction Symbol, M. For waves transmitted through a refracting medium, a modified index of refraction of greater than one.

excess noise Electrical noise caused by current in a semiconductor material.

excess sound pressure Unit, dyne/cm^2. In a medium conducting sound waves, the quantity $Pi - Ps$, where Pi is total instantaneous pressure at a given point in the medium, and Ps is static pressure in the absence of the sound waves.

excess-three code A computer code derived from binary notation by adding binary three (i.e., 0011) to each four-bit group. Thus, decimal seven is 1010 in the code (it is 0111 in binary). Unlike the binary representation for zero, the excess-three representation (0011) contains two ones, a feature which distinguishes actual zero from a machine fault.

exchange 1. To reverse the contents of two memory banks. For example, if the memory banks are called A and B, an exchange is the placing of the contents of memory A into memory B, and the placing of the contents of memory B into memory A. The original contents are removed. 2. A two-way sequence of data transmissions. 3. A designated location in a telephone circuit.

exchange line A telephone line.

exciplex In a laser, a method of adjusting the color by means of chemical reactions in organic dyes.

excitant The electrolyte in a voltaic cell.

excitation 1. Supplying input-signal driving current, driving power, or driving voltage. 2. Input-signal driving current, driving power, or driving voltage.

excitation anode In a mercury-pool tube, an auxiliary anode whose operation maintains the cathode spot when no output current is being drawn from the tube.

excitation current 1. Grid current in an excited tube stage, or input-electrode current in an excited transistor stage. 2. Current flowing in the circuit of the excitation anode of a mercury-pool tube. 3. Current flowing in the exciter circuit of an alternator. 4. Shunt-field current in a motor.

excitation energy 1. Symbol, E_x. In artificial transmutation, the energy of a nucleus when protons of less than maximum energy have been emitted from the atom. 2. Electrical energy required by a transducer.

excitation purity In color TV, complete saturation of a hue; i.e., there is no contamination by other colors, and the saturated hue is distributed uniformly.

excitation voltage 1. The signal voltage which achieves or is required for excitation (see *excitation, 1*). 2. The value of driving voltage.

excitator An electrical discharger.

excite See *drive, 1*.

excited atom An atom in which one or more electrons have been pushed out of their normal orbits into higher ones by energy applied from the outside.

excited-field speaker A dynamic speaker in which the magnetic field is provided not by a permanent magnet but by direct current flowing through a large coil of wire wound around the speaker core. The coil usually acts as a filter choke in the dc power supply of the attendant amplifier or receiver.

excited state In artificial transmutation, the state of the nucleus when protons of less than maximum energy have been emitted from the atom. The energy of the protons, in this instance, is greater than the ground state.

exciter 1. An amplifier or oscillator (or a system of such units) which supplies the input (driving) signal to the output amplifier in a radio transmitter or similar device. 2. A small dc generator which supplies direct current to the field winding of an ac generator. 3. Induction coil.

exciter lamp 1. A concentrated-filament, high-intensity incandescent lamp employed in sound-on-film recording and reproduction and in some types of electromechanical television. 2. In a facsimile transmitter, the lamp illuminating what is being scanned.

exciter relay In an electromechanical generator, the relay that activates the dc field excitation during machine startup.

exciter response 1. A change in the exciter voltage of a motor when the field-circuit resistance changes. 2. A change in the operating conditions of a radio frequency exciter, as a result of a change in the impedance at the input of the final amplifier.

exciter unit See *exciter*.

exciting current 1. The output current produced by the exciter of a generator (see *exciter, 2*). 2. The field current of a dynamo-type generator. 3. Primary current in an unloaded transformer.

exciting power 1. The output power produced by an exciter. 2. The input-signal power required for full output from a power amplifier. Also called *driving power*.

exciting voltage 1. Input-signal voltage. 2. The input-signal-voltage amplitude required for full rated output from a power amplifier. Also called *driving voltage*. 3. The output voltage produced by an exciter.

exciton In a semiconductor or dielectric, a bound electron—hole pair.

excitron A mercury-pool rectifier whose arc is initiated mechanically, e.g., by means of a magnetic plunger in the tube. Earlier mercury-pool tubes had to be tipped momentarily to splash the mercury between the electrodes to start the arc.

exclusion principle The rule that only one particle of a particular kind can occupy a given quantum state at one time.

exclusive-OR A logic function in which the output is 1 whenever either of two inputs is 1, but is 0 when both inputs are 1 or when both inputs are 0.

excursion 1. A frequently appreciable change in the value of a quantity in a given direction. 2. In an oscillatory system, a body's moving away from the point of equilibrium or mean position.

execution A computer's performance of the operations required by an instruction.

execution time The length of time required for a computer to complete a designated operation.

executive routine In computer practice, a program which controls and processes other routines. Also called *monitor program*. Compare *monitor system*.

exhaust analyzer An instrument for examining the exhaust fumes of an internal combustion engine, to measure the presence of noxious materials and to evaluate air-to-fuel ratio and combustion efficiency.

exhaust-emission control See *computer-controlled catalytic converter* and *exhaust analyzer*.

exhaustion See *evacuation*.

exit 1. In computer operation, the last instruction in a program or program segment often taking a subroutine back to the main program. 2. To leave a routine or subroutine.

exosphere The extreme outer layer of earth's atmosphere.

exothermic Pertaining to a chemical or electrochemical reaction in which heat is given off. Compare *endothermic*.

exothermic reaction In an artificial transmutation (chemical reaction), the production of positive reaction energy; i.e., kinetic energy is gained. Compare *endothermic reaction*.

exp 1. Symbol for *exponential*. 2. Abbreviation of *experiment(al)*.

expand 1. To increase signal-amplitude range. 2. To widen the scale of a meter. 3. To widen (or magnify a portion of) the trace of an oscilloscope beam.

expandable Capable of being built up into larger and larger circuits, without limit (in theory).

expandable gate In digital logic, a gate that can be provided with an unlimited number of input lines by electrical interconnection with other gates.

expanded-scale meter A meter having a scale devoted principally to displaying a narrow range of values. Thus, such a meter employed for closely monitoring the 115V power line might have a scale reading 105-120V instead of the conventional 0-115V scale.

expanded sweep 1. In an oscilloscope, speeding up the deflection of the beam during a selected portion of the trace. 2. The circuit for the action described above.

expander A circuit for automatically raising the amplitude of a quantity. A typical example is the volume expander, a device that greatly increases the amplitude of strong signals while having a negligible effect on weak ones.

expansion The automatic increase in the gain of strong signals, and decrease in gain (or no decrease) of weak signals. See, for example, *expander*.

expansion chamber A cloud chamber (see *Wilson cloud chamber*) for viewing the paths of radioactive particles. It consists essentially of a closed glass cylinder containing highly humid air and a piston. An electrostatic field is applied through the cylinder, the piston is pulled quickly, and the volume of the chamber expands. The temperature inside falls below the dew point, a cloud is formed, and droplets of water condense on ions, making their paths visible for observation or photography through the cylinder walls.

expansion ratio For an expansion chamber, the ratio V_2/V_1, where V_1 is the volume of the chamber before expansion and V_2 is the volume after expansion.

expansion time For an expansion chamber, the interval during which expansion occurs. The interval is kept short to insure a drop in temperature low enough for vapor condensation and to obviate the possibility of continuing gas motion distorting the track of a particle.

expectation In probability theory, the middle value (average or mean) of a random variable.

expendable A component or system that for economy is best discarded instead of repaired when it fails. Also called *disposable component*.

experiment One or a series of carefully planned tests carried out under controlled conditions to obtain data or to check performance.

experimental chassis See *electronic chassis*.

experimental model A prototype of an electronic circuit or device produced solely for operational tests or as a model against which theory and design can be checked.

experimental service A special, nonamateur radio service in which operation is aimed at on-the-air testing of new methods and equipment.

experimental station A station specially licensed to operate on specific frequencies in the experimental service.

exploring coil A pickup coil for sensing a signal or magnetic field. Sometimes called a *sniffer*.

exploring electrode 1. A sampling electrode sealed in a discharge tube for measuring ionization at the point of insertion. 2. Broadly, a test probe.

explosion-proof device A device that is housed and operated so that its sparking, heating, or production of radiant energy will not cause materials in the environment to explode.

exponent A number written as a superscript indicating the power to which another number (called the *base*) is to be raised. Thus, 2^2 is the square (second power) of 2; x^3, the cube (third power) of x; and so on.

exponential 1. A base (such as the natural number e) modified by an exponent. 2. Related to a change in value as determined by an exponent. Thus, using increments for x in the equation $x = y^2$ produces an exponential curve.

exponential curve A curve based upon powers of a number (such as for $y = e^x$). Also see *exponential*, *exponential decrease*, and *exponential increase*.

exponential damping Damping action described by an exponential curve.

exponential decay See *exponential decrease*.

exponential decrease The continuous reduction in the value of a quantity according to the equation $y = e^{-x}$, which depicts the natural decay curve.

exponential function A function, such as e^x, which varies exponentially. See, for example, *exponential decay*, *exponential growth*, and *exponential series*.

exponential horn A horn-type loudspeaker or horn assembly driven by an audio driver of either circular or rectangular cross section, whose cross-sectional area(s) at any point x feet along its axis is $S_o e^{mx}$, where S_o is the cross-sectional area at the throat; e, 2.718 28; and m, the flaring constant.

exponential increase The continuous increase of a quantity according to the equation $y = e^x$ which depicts the natural growth curve.

exponential line A two-wire transmission line whose characteristic impedance varies exponentially with its electrical length.

exponential quantity A quantity involving an exponential, e.g., ae^x.

exponential series A mathematical series based on exponential expressions. Example: $e^x = 1 + x + (x^2/2!) + (x^3/3!) + (x^4/4!) + \ldots$

exponential sweep In cathode-ray devices, such as oscilloscopes, a beam sweep that starts fast and slows exponentially.

exponential transmission line Any transmission line in which the characteristic impedance varies with the line length in an exponential manner.

exponential waveform Any waveform in which the rate of change in the amplitude is directly or inversely proportional to the instantaneous amplitude. The absolute value of the derivative of such a waveform is equal to the absolute value of the instantaneous amplitude, multiplied by a constant that depends on the amplitude units.

exposure The total amount of radiation received in a given area over a specified length of time.

exposure meter 1. A usually simple instrument for measuring light intensity, especially for photographic purposes. A common form consists of a self-generating photocell connected to a dc microammeter. 2. Dosimeter.

exposure time The interval during which film in a camera or paper under an enlarger must be exposed to the image for a picture to be recorded. Exposure time is inversely proportional to lens aperture and film speed, i.e., if 1/8 second of $f/2.8$ will produce a proper exposure, so will 1/4 second at $f/4$.

expression A mathematical or logical statement represented symbolically, e.g., $A + B$.

expression control A volume control in an electronic organ.

extended class-A amplifier A push-pull af amplifier in which a triode and pentode are connected in parallel on each side of the circuit. At low signal levels, the circuit operates entirely with the push-pull triodes, the pentodes being cut off. At high signal levels, however, the af output is almost entirely that of the push-pull pentodes.

extended-cutoff tube A vacuum tube in which the amplification does not vary directly with the control-grid bias, and which cannot be completely cut off by this bias. Structurally, this tube is characterized by closer spacing of wires at the center of the grid than at the ends. Also called *remote cutoff tube* and *variable-mu tube*.

extended double Zepp antenna See *double-extended Zepp antenna*.

extended octaves Audio-frequency tones above or below the normal range of an electronic musical instrument. Special circuits must be added to make the extended octaves available.

extended-play Pertaining to a recorded phonograph disk that provides a longer playing time than conventional disks of the same size and recording speed.

extender A substance added to an encapsulant to make it go further.

extensimeter See *extensometer*.

extension cable A flexible, low-capacitance (usually concentric) cable for connecting part of one circuit to part of another. Extension cables are available with a variety of end connectors.

extension cord A flexible power cord having a male plug on one end and female receptacle on the other.

extension loudspeaker An auxiliary loudspeaker serving areas in which the main speakers can't be adequately heard.

extensometer An instrument used to measure small amounts of expansion, contraction, or deformation.

exterior label On a reel of magnetic tape (computer storage), a written identification of the tape, as opposed to the *tape label*, which is recorded on the tape.

external armature In a dynamo-type machine, an armature that rotates around the outside of the field magnets, as opposed to the usual (inside) arrangement.

external capacitor A high-value capacitor connected externally to an oscillator or sweep generator to lower its frequency.

external circuit A circuit or subcircuit connected and external to a main equipment.

external controls 1. Control devices which are connected to, but operated away from, a main circuit. 2. Manual or screwdriver-adjusted controls which are mounted on the panel of an equipment, as opposed to those mounted in the case or behind the panel.

external critical damping resistance The value of external resistance which must be connected to a galvanometer or other meter to produce critical damping.

external cross-modulation See *cross-modulation, 2.*

external damping device A resistor or short-circuiting bar connected temporarily between the terminals of a meter to keep its movement immobile during transportation.

external feedback Negative or positive feedback through a separate path outside of and around the main circuit. Example: negative feedback through an *RC* path between the output terminals and the input terminals of an amplifier.

external impedance Load impedance, i.e., an impedance connected to the output terminals of a generator or amplifier.

external load See *external impedance.*

external loudspeaker See *external speaker.*

externally caused chatter In a relay, contact chatter caused by mechanical vibration outside of the relay.

externally caused failure Failure of a circuit or component resulting from unfavorable environmental factors.

external memory In computer practice, a memory unit outside of the computer mainframe.

external power supply A power supply unit situated apart from the powered equipment. Such separation is helpful in eliminating the disturbing effects of heat, hum, and vibration associated with integral power units.

external Q For a microwave tube, the quantity $1/1(1/Q_1 + 1/Q_2)$, where Q_1 is the loaded Q and Q_2 is the unloaded Q.

external S-meter A signal-strength meter connected to a receiver but not installed in its panel.

external speaker A loudspeaker that doesn't share an enclosure with an amplifier, receiver, or other device which drives it. Such isolation is helpful in eliminating the vibration and feedback.

external storage In computer operation, storage media (disks, tapes, cards, etc.) that are outside of the computer.

extinction potential See *deionization potential.*

extinction voltage See *deionization potential.*

Extra-class license An amateur-radio license that conveys all available amateur operating privileges in the United States. The highest class of amateur license.

extract 1. To remove a signal or quantity from some product containing it, or from its source. Examples: extracting a fifth harmonic from a complex signal, extracting the dc component from a signal containing ac and dc. 2. To derive a factor, e.g., to extract a root. 3. To separate certain classes of information from an aggregate of information.

extract instruction In computer practice, the instruction to generate a new word by the serial arrangement of designated segments of specified words.

extractor 1. A circuit or device for removing a signal (or a signal component) from another circuit or device. A demodulator probe, for ex-

ample, extracts the modulation from a modulated signal. 2. A device for removing used active devices from a circuit board. Such extractors may also employ heat to desolder as well as remove the devices.

extraneous component A usually undesired inherent effect which results from the physical nature of a component or device and which it is the concern of engineering to minimize. Examples: distributed capacitance of a coil, internal inductance of a capacitor.

extraneous emission Undesired emission form a transmitter.

extraneous response The unintended response of a circuit or device.

extraneous root In the solution of an equation derived from another equation, one or more roots which satisfy the derived equation but not the original one.

extraneous signal A superfluous and usually interferential signal.

extranuclear Outside the nucleus of an atom.

extraordinary ray Of the two rays resulting from the double refraction of electromagnetic waves, the one that does not follow the usual laws of refraction. Also see *X wave.* Compare *ordinary ray.*

extraordinary wave See *X wave.*

extrapolar 1. Outside of the poles. 2. Not between the poles.

extrapolation The estimation of values beyond the range of data on hand, assuming that conditions will hold for extrapolated values, as when a curve is extended beyond its final plotted point to determine a value for a variable.

extrared See *infrared.*

extraviolet See *ultraviolet.*

extreme A remote limit, especially the lowest or highest value of a quantity, or the lowest or highest point in a range.

extremely high frequency Abbreviation, ehf. A frequency near the upper limit of the radio-frequency spectrum, especially one in the 30-300 GHz band.

extremely low frequency Abbreviation, elf. A frequency within the audio range but not used for audio.

extrinsic base-resistance—collector-capacitance product Unit: seconds, milliseconds, microseconds. For a bipolar transistor, the product $RbCc$, where Rb is base resistance and Cc is collector capacitance. This product is a time constant that determines the high-frequency operating limit of the transistor.

extrinsic conductance For a material, the conductance resulting from impurities or such external factors as environmental conditions.

extrinsic properties For a semiconductor material, properties resulting from doping (altered resistivity, majority and minority carrier differentiation, and so on). Also see *extrinsic semiconductor.*

extrinsic semiconductor A semiconductor material, such as germanium or silicon, to which a controlled amount of a suitable impurity material has been added to give the semiconductor a desired resistivity and polarity. Compare *intrinsic semiconductor.*

extrinsic transconductance For a bipolar transistor, the first derivative of collector current with respect to base—emitter voltage. It is the ratio of a small change in collector current to the small change in base—emitter voltage that produced it, collector voltage being constant; $gm = dIc/dVbe.$

extrusion The process of forming a material such as metal or plastic, by forcing it through dies. Many pieces of electronic hardware are mass produced in this manner. Examples are insulating rods and tubes, metal cans, and metal tubing.

eyelet connection A connection made by fastening conductors together with an eyelet or by soldering leads or pigtails to an eyelet.

eyepiece A small lens system for viewing an oscilloscope screen through a camera setup.

eye tube See *magic-eye tube*.

Ez Symbol for *voltage drop across an impedance*.

E-zone A portion of the earth including most of the eastern hemisphere. When propagation forecasts are made, the E zone is one of three longitude zones specified.

F 1. Symbol for *force*. 2. Symbol for *fluorine*. 3. Abbreviation of *Fahrenheit*. 4. Abbreviation of *farad*. 5. Abbreviation of *fermi*. 6. Symbol for *focal length*. 7. Symbol for (on drawings) for *filament; fuse*. 8. Symbol for *Faraday constant*.

f 1. Abbreviation of *femto*. 2. Symbol for *frequency*. 3. Symbol for *function*.

Fo Symbol for *damping factor*.

F1 The designation for radio emission consisting of frequency-shift-keyed waves.

F1 layer The lower part of the ionosphere's *F-region*. Also called *F1 region*.

F2 layer The upper part of the ionosphere's *F-region*. Also called *F2 region*.

fA Abbreviation of *femtoampere*.

fabrication tolerance The amount of variation that can be tolerated in the manufacture of components.

Fabry-Perot interferometer A resonant cavity, often used with lasers, that has mirrors at each end; the interferometer produces the optical equivalent of standing waves.

face 1. A flat crystal surface whose orientation can be expressed as its position relative to other faces. 2. The viewer's side of a screen. 3. The scale part of a meter. 4. The side of a punched card that has been imprinted.

face-down feed The face-down (see *face, 4*) placement of a punched card in a hopper.

face material In a tape recorder, the plastic used to coat the face of a head.

face-parallel cut See *Y-cut crystal*.

face-perpendicular cut See *X-cut crystal*.

face side The side of pressure-sensitive insulating tape that is coated with adhesive.

facom A radionavigation system that operates by means of phase comparison at low frequencies. Effective over long distances and under poor conditions.

facsimile The transmission and reception, through the medium of radio or by wire, of permanent pictures, writing, and other graphic material.

facsimile receiver The complete device or system which selects, amplifies, and demodulates a picture signal picked up from the air, wires, or cable, and employs the elements of this signal to reproduce the picture. Also see *facsimile*.

facsimile recorder The machine that puts a transmitted facsimile image on paper.

facsimile transmitter The complete device or system that generates signals depicting graphic material (pictures, writing, printing, etc.) and sends them to a distant point—via cable, wire lines, or radio—for subsequent reproduction. Also see *facsimile*.

factor 1. A data element that is an operand in an arithmetic operation. 2. To find the two or more numbers whose product is the number being factored. 3. One of two or more numbers whose product is the number being factored.

factorial Symbol !. The factorial of a number *n*, written *n*!, is the product of all positive integers up to and including *n*. Thus, 5! equals 1 ×2 × 3 × 4 × 5 or 120. The term *n*! is read either *n factorial* or *factorial n*. The factorial can be used to determine the number of permutations that can be developed from the members of a series; e.g., the letters in a 5-letter word can be arranged 5!, or 120, different ways.

factor of merit See *figure of merit*.

factor of safety See *safety factor*.

fade in To gradually increase an audio or video signal, especially for recording.

fadeout The complete, gradual disappearance of a signal. See *fading.*

fader In sound amplification systems, an attenuator circuit that enables the operator to fade out one signal and fade in another. Ordinarily, a fader does not provide mixing action.

fading A waxing and waning of a signal at the reception point. Fading of a radio wave results from various parts of the wave taking different paths from transmitter to receiver; arriving at the receiver in various phase relationships, some parts are mutually intensifying while others cancel each other.

Fahnestock clip A flat, sheet metal spring clip for holding a wire (usually in a temporary breadboard setup).

Fahrenheit scale A temperature scale on which the freezing point of water is 32°, and the boiling point of water is 212°. Compare *absolute scale* and *centigrade scale.*

failsafe Pertaining to devices or circuits which, upon failure, cause no damage or serious malfunction.

failsoft In a computer, a system in which operation is maintained even in the event of partial failure. Efficiency is reduced but the computer does not completely shut down.

failure The condition wherein a circuit, system, or device is not operating correctly.

failure analysis 1. The process of determining the failure rate for a component, system, or device. 2. The process of determining the cause of a failure.

failure mode The particular way in which a failure of equipment or a method occurs.

failure unit A unit of machine or device failure: one failure per 10^9 hours of operation.

fall-in The time when synchronous speed is attained in a synchronous motor.

falling characteristic A negative resistance characteristic. Also see *negative resistance.*

fall time 1. Decay time. 2. The time required for the amplitude of a pulse to decrease from maximum to zero. Compare *rise time.*

false add A logic add, i.e., addition without carries.

false alarm 1. Improper operation of an electronic alarm system, resulting in actuation of the device. 2. In radar, the presence of a false echo which causes the attendant circuits or personnel to act as though an enemy target is present.

false error A condition in which a computer system erroneously signals the existence of an error.

false precision See *midleading precision.*

false retrieval The incorrect specification of criteria for information to be selected for retrieval so that an unwanted item of data is selected. Also called *false drop.*

family Any group of components, circuits, ratings, characteristics, and so on classed together because of some common or analogous feature, application, or the like. Examples: *IC family, family of curves,* family of equations, etc.

family of curves A group of curves plotted on the same axes, which depict the performance of a circuit or device at several levels of a third parameter, e.g., curves showing transistor collector current vs collector voltage for several levels of base voltage.

fan antenna See *double-V antenna.*

fan-in 1. A number of inputs entering a common input terminal. 2. In digital computer practice, the number of inputs that can be accommodated by a logic circuit. Compare *fan-out.*

fan-in circuit A circuit having a number of input lines entering a common input point. Also see *fan-in, 1.*

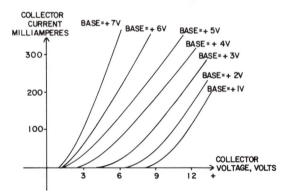

Family of curves.

fan-out 1. One common output terminal feeding a number of output lines. 2. In digital computer practice, the number of outputs that can be fed by a logic circuit. Compare *fan-in.*

fan-out circuit A circuit in which a number of output lines leave a common output terminal. Also see *fan-out, 1.*

farad (Michael *Faraday,* 1791-1867). Abbreviation, *F.* The basic unit of capacitance. A capacitor has a capacitance of 1F when a change of 1 volt per second across the capacitor produces a current of 1 ampere through it.

faraday An electrical quantity approximately equal to 9.65×10^4 coulombs; it is the quantity of electricity required in electrolysis to free 1 gram atomic weight of a univalent element. The equivalent, and preferred, SI unit is the coulomb.

Faraday cage See *electrostatic screen.*

Faraday dark space In a gas-discharge tube, the dark portion in the glow, following the luminous positive column at the anode. Compare *Crookes' dark space, negative glow,* and *positive column.*

Faraday effect See *magneto-optical rotation.*

Faraday rotation A change in the polarization of an electromagnetic wave as it travels through certain substances.

Faraday's disk dynamo See *disk dynamo.*

Faraday shield See *electrostatic screen.*

Faraday's law The voltage induced in a conductor moving in a magnetic field is proportional to the rate at which the conductor cuts magnetic lines of force.

Faraday's laws of electrolysis 1. In electrolysis, the mass of a substance liberated from solution is proportional to the strength and duration of the current. 2. For different substances liberated by the same current in a certain time, the masses are proportional to the electrochemical equivalents of the substances. Also see *electrochemical equivalent; electrolysis, 1;* and *electrolyte.*

faradic current The lopsided alternating current produced by an induction coil.

faradmeter An alternate term for *microfarad meter.*

far field 1. The region beyond the near field of an antenna (see *near field, 1*). 2. The region beyond the near field of a loudspeaker (see *near field, 2*).

far IR The lower-frequency portion of the infrared spectrum.

Farnsworth dissector tube See *dissector tube.*

far zone See *far field, 1, 2.*

fast access storage In a computer memory, the section from which information may be most quickly accessed, depending on the relative speed of other system devices.

fast-break, fast-make relay A relay that opens and closes rapidly.

fast-break, slow-make relay A relay that opens rapidly and closes slowly.

fast diode See *computer diode.*

fast drift The rapid change of a quantity or setting (usually in one direction). Compare *slow drift.*

fast-forward (control) Abbreviation, ff; ffwd. Symbol, > > . In a tape recorder, a mechanism for running the tape through the machine rapidly.

fast groove The informationless groove between tracks on a disk recording.

fast-make/fast-break relay A relay that closes and opens rapidly.

fast-make, slow-break relay A relay that closes rapidly and opens slowly.

fast-operate, fast-release relay See *fast-make, fast-break relay.*

fast-operate, slow-release relay See *fast-make, slow-break relay.*

fast-release, fast-operate relay See *fast-break, fast-make relay.*

fast time constant 1. The property of responding quickly to changes in input parameters. 2. In radar, a method of defeating attempts at jamming by modification of the receiving circuitry.

fast-release, slow-operate relay See *fast-break, slow-make relay.*

fathometer See *acoustic depth finder.*

fathom 1. To measure the depth of a body of water, as in the use of sonar for this purpose. 2. A unit of length (distance) equal to 0.549 meter.

fatigue 1. The degradation of the performance of circuits or materials with time. 2. The tendency of bodies and materials to weaken, deform, or fracture under repeated strain.

fault 1. A defective point or region in a circuit or device. 2. A failure in a circuit or device.

fault current 1. A momentary current surge. 2. A leakage current.

fault finder A troubleshooting instrument or device, e.g., a multimeter.

Faure plate A storage battery plate consisting of a lead grid containing a chemical electrolytic paste.

fax Abbreviation of *facsimile.*

fc Abbreviation of *foot-candle.*

fc Abbreviation of *carrier frequency.*

FCC See *Federal Communications Commission.*

fco Abbreviation of *cutoff frequency.*

F-display See *F-scan.*

FDM Abbreviation of *frequency division multiplex.*

FDS Abbreviation of *Faraday dark space.*

FE Abbreviation of *ferroelectric.* See *ferroelectricity.*

Fe Symbol for *iron.*

feasibility study The procedures for evaluating the potential gains in applying a computer system to a job or to an organization's process, or in modifying or replacing an existing system.

FEB Abbreviation of *functional electronic block.*

Federal Communications Commission Abbreviation, *FCC.* Established in 1934, the U.S. Government agency that regulates electronic communications. The FCC succeeded the Federal Radio Commission (FRC), which was established in 1927; the FRC suceeded the Radio Division of the Bureau of Navigation in the Department of Commerce, whose jurisdiction over radio began in 1912.

feed 1. To supply power or a signal to a circuit or device. 2. The method of supplying such a signal or power. See, for example, *parallel feed* and *series feed.* 3. To cause data to be entered into a computer for processing.

feedback 1. The transmission of current or voltage from the output of a circuit or device back to the input, where it interacts with the input signal to modify operation of the device. Feedback is positive when it is in phase with the input, negative when it is out of phase. 2. To input the result at one point in a series of operations to another point; the method allows a system to monitor its actions and make necessary corrections.

feedback admittance The admittance between the output and input electrodes in a tube.

feedback amplifier 1. An amplifier whose performance (especially frequency response) is modified by means of positive, negative or both positive and negative feedback. 2. An amplifier placed in the feedback path of another circuit to increase the amplitude of feedback.

feedback attenuation 1. In an operational-amplifier circuit, the attenuation in the voltage from output to input. 2. In an audio-frequency or radio-frequency amplifier circuit, the reduction of feedback by electronic means.

feedback bridge A bridge circuit in the feedback channel of an amplifier or oscillator. See, for example, *Meacham oscillator* and *Wien-bridge oscillator.*

feedback capacitance 1. A capacitance through which feedback current is coupled from the output to the input of a system. 2. The interelectrode capacitance of an electron tube. See *plate—grid capacitance.*

feedback control 1. The variable component (potentiometer, variable capacitor) used to adjust the level of feedback current or voltage. 2. The control of circuit performance by feedback.

feedback cutter A device employed for the purpose of cutting grooves in phonograph disks. Feedback is employed to provide a flat frequency response.

feedback factor For a feedback amplifier, the quantity $1 - \beta A$, where A is the open-loop gain of the amplifier and β is the *feedback ratio.*

feedback loop The part of a circuit that provides controlled feedback in an operational-amplifier circuit.

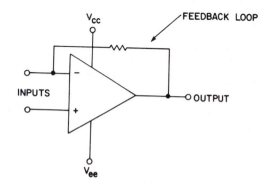

feedback oscillator A circuit in which oscillation is obtained by feeding a portion of the output of a tube or transistor back to the input circuit by inductive coupling. Also called tickler oscillator. Included in the category of feedback oscillator (an arbitrary term) is the tuned-plate, tuned-grid oscillator. See *TPTG oscillator.*

feedback path A path over which feedback, either positive or negative, may occur in a circuit. The feedback may be intentionally produced, or it may be undesirable.

feedback percentage Symbol, n. In a feedback circuit, the percentage of output voltage that is fed back; n equals $100(ef/eo)$, where ef is the feedback voltage and eo is the open-loop output voltage. Compare *feedback ratio.*

feedback ratio Symbol β. For a feedback system, the ratio ef/eo, where ef is the voltage that is fed back and eo is the open-loop output voltage of the system.

feedback rectifier See *diode feedback rectifier.*

feedback regulator 1. In a controlled-feedback circuit, the device that determines the amount of feedback.

feedback resistance 1. The internal base resistance of a point-contact transistor. 2. The resistance in a feedback loop.

feedback transfer function The transfer function of a feedback loop exclusively.

feedback winding A special winding on a magnetic amplifier or saturable reactor, for the introduction of feedback currents.

feeder 1. A conductor or conductor pair that carries electric power from one point to another. 2. The transmission line connecting a transmitter to an antenna.

feeder cable 1. A communication cable running in a primary route from a central station—or in a secondary route from a main feeder cable—as a means of making connections to distribution cables. 2. In a CATV system, the cable carrying transmission from head end to trunk amplifier. Also called *trunk cable.*

feeder loss Loss of energy resulting from resistance in, or radiation from, feeder lines.

feed holed In paper tape or punched cards, the holes engaged by sprockets; they convey no information but are sometimes used for indexing when paper tape, for example, is driven by something other than a sprocket wheel. Also called *sprocket holes.*

feeding In character recognition, a system in which documents go into the transport of a character reader at a steady, specified rate.

feeding, multiread See *multiread feeding.*

feed pitch The distance between feed holes.

feed reel The tape supply reel of a tape recorder.

feedthrough 1. The usually undesirable transmission of a signal through a circuit without being processed by the circuit, because of unavoidable capacitive coupling, for example. 2. Contraction of *feedthrough component.*

feedthrough capacitor A capacitor whose design is like that of a feedthrough terminal; it is mounted in a hole in a chassis. The center screw or wire is the "high" terminal of the capacitor, to which connections may be made above or below the chassis. The body of the device is the "low" terminal of the capacitor; it is soldered to the chassis or secured with a nut.

feedthrough component A passive device permanently installed in a panel or plate, e.g., a feedthrough capacitor or feedthrough insulator.

feedthrough insulator An insulator mounted tightly in a hole in a wall or chassis and provided with a center hole for a lead.

feedthrough terminal A terminal mounted tightly in a hole in a chassis or wall; it consists of a screw going through a feedthrough insulator. Connections can be made to either end of the screw.

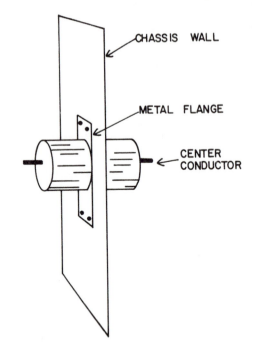

Feedthrough capacitor.

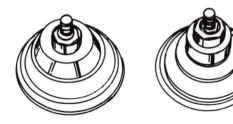

Feedthrough insulator.

FE-EL Abbreviation of *ferroelectric-electroluminescent.*

feeler 1. A wire or blade contact, e.g., a finger that senses holes in a punched card. 2. A cat's whisker.

Felici mutual-inductance balance An inductive null circuit for determining mutual inductance (Mx) in terms of a standard mutual inductance (Ms). The secondary coils of two mutual-inductance circuits are connected in phase-bucking. The standard mutual inductor, which is variable, is adjusted for null. At null, Mx equals Ms. Compare *Carey-Foster mutual-inductance bridge.*

female plug A plug whose contacts are separated by a recess into which the prongs of a mating *male plug* are inserted. Compare *hermaphrodite plug* and *male plug.*

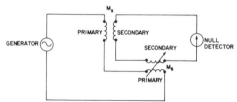

Felici mutual-inductance balance.

femto Abbreviation, f. A prefix meaning *quadrillionthF20, i.e., 10^{-15}*.

femtoampere Abbreviation, fA. A unit of low current; 1 fA equals 10^{-15} A.

femtofarad Abbreviation, fF. A unit of low capacitance; 1 fF equals 10^{-15}F.

femtovolt Abbreviation, fV. A unit of low voltage; 1 fV equals 10^{-15}V.

fence A system or string of early warning radar stations.

fermi Abbreviation, F. A small unit of length and wavelength; 1 F equals 10^{-15} meter.

fermium Symbol, Fm. A radioactive metallic element that is artificially produced. Atomic number, 100. Atomic weight, 253.

Fernico In the manufacture of metal electron tubes, an iron—cobalt—nickel alloy used as an intermediate metal between the iron shell of the tube and the glass beads through which the leads pass.

ferpic Acronym for *ferr*oelectric ceramic *pic*ture device. An image-storing device containing a photoconductive film, transparent electrodes, and a ferroelectric ceramic, in layers.

ferreed A form of magnetic switching device, similar to a reed relay, that maintains its position indefinitely without the need for a continuous current.

ferret A vehicle or craft equipped for determining the locations of enemy radar transmitters.

ferri A prefix used to denote magnetic properties.

ferric oxide Formula, Fe2O3. A red oxide of iron used to coat magnetic recording tape.

ferristor A high-frequency magnetic amplifier employing a ferroresonant circuit.

ferrite A high-resistance magnetic material consisting principally of ferric oxide and one or more other metals. After being powdered and sintered, ferrites exhibit low eddy-current loss at high frequencies and make ideal core material for inductors and switching elements. It is also used in TV deflection yokes in miniature antennas. Also see *ferrospinels*.

ferrite antenna See *ferrite-rod antenna*.

ferrite bead 1. A magnetic storage device in the form of a bead of ferrite powder fused onto the signal conductors of a memory matrix. 2. A tiny bead that may be slipped over certain current carrying leads to choke out rf.

ferrite core A coil or switching-element core made from a ferrite; specifically, in a core memory, a small magnetic toroid that can retain its polarity when charged by a pulse.

ferrite core memory A magnetic memory in which ferrite cores are interconnected by a network of input and output wires.

ferrite isolator A microwave device that permits energy to pass with negligible loss in one direction through a waveguide or coaxial line, while absorbing energy passing in the opposite direction.

ferrite limiter A device employed in the antenna circuit or front end of a receiver to prevent overload while maintaining a linear response.

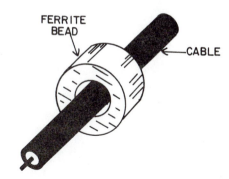

Ferrite bead.

Used mostly at ultra-high and microwave frequencies.

ferrite loop See *ferrite-rod antenna*.

ferrite memory A static memory employing ferrite cores. See *core memory*.

ferrite-rod antenna A small receiving antenna consisting of a coil of wire wound on a ferrite rod a few inches long. Also see *ferrite*.

ferrite switch A device that regulates the flow of power through a waveguide. The electric-field vector is rotated, resulting in a high degree of attenuation when actuated, but little or no attenuation when not activated.

ferroelectric 1. Producing ferroelectricity. 2. A ferroelectric material.

ferroelectric amplifier See *dielectric amplifier*.

ferroelectric capacitor A capacitor in which a ferroelectric material is the dielectric.

ferroelectric cell See *ferroelectric capacitor*.

ferroelectric crystal A crystal of ferroelectric material.

ferroelectric flip-flop A flip-flop based on the hysteresis of a ferroelectric capacitor. See *bistable multivibrator*. Compare *ferroresonant flip-flop*.

ferroelectricity Electric polarization in certain crystalline materials. The effect is analogous to the magnetization of a ferromagnetic material by a magnetic field.

ferroelectric-luminescent Pertaining to a ferroelectric cell that emits light.

ferroelectric material A nonlinear dielectric material capable of producing ferroelectricity. Examples: barium titanate, barium strontium titanate, potassium dihydrogen phosphate, guanadine aluminum sulfate hexahydrate (GASH), Rochelle salt, triglycene sulfate.

ferromagnetic 1. A type of material that conducts a magnetic field with relative ease. 2. A type of material in which a magnetic-field change causes a voltage, which in turn results in a measurable current flow.

ferromagnetic amplifier An amplifier that operates on the principle of ferromagnetism.

ferromagnetic material Highly magnetic material, e.g., iron.

ferromagnetic resonance The point at which the permeability of a magnetic material peaks at a microwave frequency.

ferromagnetic spinels Highly permeable and resistive ceramic-like materials. The low eddy-current losses and high permeability of these materials suit them to use as cores in rf transformers and inductors. Also see *ferrite*.

ferromagnetic tape Magnetic tape used for winding closed transformer cores.

ferrometer An instrument for testing hysteresis and permeability in steel and iron.

Ferron detector See *iron-pyrites detector*.

ferroresonant circuit An LC circuit in which the coil is a saturable reactor. Because of the coil's nonlinearity the circuit is resonant at only one value of ac voltage and exhibits both negative resistance and bistable operation.

ferroresonant counter A digital counter employing ferroresonant flip-flops instead of tubes or transistors.

ferroresonant flip-flop A flip-flop employing one or two ferroresonant circuits instead of tubes or transistors. See *bistable multivibrator*. Compare *ferroelectric flip-flop*.

ferroresonant shift register A shift register employing ferroresonant circuits instead of tubes or transistors.

ferrosoferric oxide See *magnetite*.

ferrospinels See *ferromagnetic spinels*.

ferrous Containing iron and being magnetizable.

Ferroxcube A nonmetallic ferromagnetic material having high permeability and resistivity, and a Curie point near room temperature. These characteristics make the material suitable for the cores of i-f and rf inductors and transformers, and for high-frequency magnetic shields.

FET Abbreviation of field-effect transistor.

fetch An operation in a computer run in which the location of the next instruction is taken from memory and changed if necessary; it then goes to the control register.

FET current meter An ammeter, milliammeter, or microammeter having a self-contained amplifier that uses field-effect transistors. Also see *electronic current meter*.

FET op-amp 1. An operational amplifier composed of field-effect transistors and associated components. 2. An operational amplifier having a field-effect transistor in its input stage.

Fetron Teledyne's field-effect transistor that has a base-pin arrangement identical to that of a vacuum tube (which the device replaces).

FET voltmeter A voltmeter employing a field-effect transistor amplifier for high-impedance input. Also see *electronic voltmeter*.

FET VOM A vom employing a field-effect transistor amplifier for increased sensitivity and input impedance. The FET vom is identical in performance—and equivalent in its basic arrangement—to a vacuum-tube voltmeter but often offers current and resistance readings.

FF Abbreviation of *flip-flop*.

fF Abbreviation of *femtofarad*.

FFI Abbreviation of *fuel-flow indicator*.

fhp Abbreviation of *fractional horsepower*.

fiber 1. A tough, vulcanized insulating material. Dielectric constant, 2.5 to 5. Dielectric strength, 2 kV/mm. 2. A thin thread of a material. 3. A light-conductive transparent filament; see *fiber optics, 1*.

fiber electrometer An instrument for measuring small quantities of electricity. It consists of a thin thread, such as one of plasticized quartz, hanging freely between two knife-edged metal pieces that are charged by the electricity being measured. The charge draws the fiber away from its position of rest, the movement being observed with a microscope. A special form of this instrument, using two fibers, is the bifilar electrometer.

fiber needle A soft phonograph needle made from a fiber, it produces less record wear than other styli, but it is short-lived.

fiber metallurgy A process in which metallic fibers or filaments are grown.

fiber-optic bundle A cable of optical fibers. See *fiber optics, 1*.

fiber-optic coupling Optical signal coupling by means of a light-conducting fiber situated between a light (signal) source and a photoreceptor. Also see *fiber optics, 1, 2*.

fiber optics 1. Extruded materials, such as certain plastic filaments, which provide paths for light. 2. The art and science of developing and using such devices.

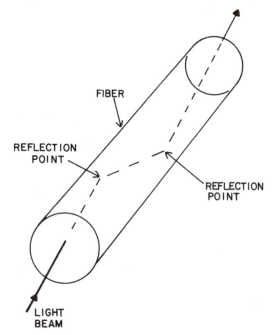

Fiber optics.

fiber optic scrambler A fiberscope in which a center section of fibers in the core is deliberately disoriented before the bundle is encapsulated; when cut, one half can decode the image encoded by the other half. See *fiberscope*.

fiberscope A flexible bundle of optical fibers having a lens at each end; it is used to view areas otherwise inaccessible to view.

fiber stylus See *fiber needle*.

Fibonacci series (Leonardo *Fibonacci 1170-1248*). A mathematical series in which each term after the second is the sum of the two preceding terms. The ratio of any term to the next preceding term

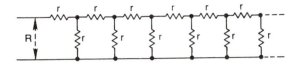

Fibonacci series. R approaches $r(1 + \sqrt{5/2})$ or $1.618r$.

approaches 1.618 as a limit. For example, in a ladder network of equal resistors, connected alternately in series and parallel, the total resistance approaches 1.618 times the value of individual resistors.

fibre Alternate (Brit.) spelling of *fiber*; also see *fiber, 1*.

fibrillation Dangerous, irregular beating of the heart that often follows electric shock. Also see *cardiac stimulator* and *defibrillation*.

fidelity Faithfulness: the degree to which a circuit or device transmits a signal without distorting it.

field 1. A volume of space in which a force is operative. See, for example, *electric field* and *magnetic field*. 2. Half of a television image (262.2 lines). 3. A computer record subdivision containing an information unit; e.g., a bank account record might have deposits as a field.

field circuit breaker A circuit breaker designed to control the field excitation of a motor or other device.

field coil 1. The winding on the field pole of a motor or generator. 2. The winding on the pole of an electrodynamic speaker. 3. The main coil of a relay. 4. The fixed coil in an electrodynamometer.

field direction The direction in which an electric field or magnetic field exerts its force.

field effect The phenomenon in which the flow of current carriers in a solid (and, accordingly, the resistance of the material) is controlled by an external electric field. A useful application is the field-effect transistor.

field-effect tetrode A field-effect transistor in which the gate electrode is split into two parts, each connected to a separate external lead. The reverse bias between the channel and either gate lead affects the conductivity through the device.

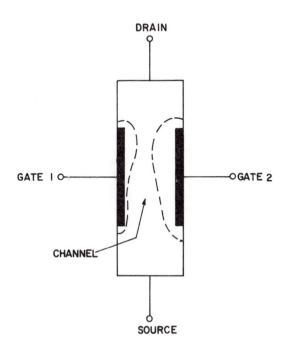

Field effect tetrode.

field-effect transistor Abbreviation, FET. A monolithic semiconductor amplifying device in which a high-impedance *gate* electrode controls the flow of current carriers through a thin bar of silicon (rarely, germanium) called the *channel*. Ohmic connections made to the ends of the channel constitute *source* and *drain* electrodes. Also see *junction field-effect transistor, metal-oxide semiconductor field-effect transistor, n-channel junction field-effect transistor*, and *p-channel junction field-effect transistor*.

field-effect tube A form of vacuum tube, with the control grid replaced by a gate, which draws current from the cathode. The gate has a positive charge.

field-effect varistor A nonlinear dual-terminal semiconductor device capable of maintaining a value of current for a range of voltages.

field emission In a tube diode, an increase in electron emission beyond the saturation value, due to an intense electric field between the plate and cathode at high plate voltage.

field-emission microscope An instrument for examining the atomic structure of high-melting-point metals; it magnifies more than 2 million times. The metal to be examined is made into a needle that is subjected to 5 to 30 kV; electrons emitted by the tip of the needle form an image on a fluorescent screen.

field forcing A method of controlling a motor by changing the magnetic field in the windings.

field-free emission current In an electron tube, the cathode current resulting solely from the emission of electrons by the hot cathode, without voltage being applied to plate or grid(s).

field frequency In television, the product of frame frequency and fields per frame (in the United States, 60/s).

field intensity 1. The strength of an electric or magnetic field. 2. The strength of a radio wave, usually expressed in microvolts, millivolts, microvolts per meter, or millivolts per meter, but sometimes in volts or volts per meter.

field-intensity meter See *field-strength meter*.

field-ion microscope A high-resolution field-emission microscope that uses helium ions instead of electrons. The ions are repelled by the tip of the metal needle under observation, forming an image on a fluorescent screen. Also see *field-emission microscope*.

fieldistor An early version of the field-effect transistor in which limited control of current carriers near the surface of a semiconductor bar or film was obtained by an external electric field applied transversely.

field length Record field size in applicable units—on a magnetic tape record, characters or words; on a punched card record, card columns.

field magnet 1. The permanent magnet in a dynamic speaker. 2. A similar magnet in an earphone, generator, microphone, motor, phono pickup, transducer, etc.

field, operation See *operation field*.

field pickup 1. A probe or sensor for insertion into an electric or magnetic field. 2. An on-location radio or television program, i.e., one coming from outside the studio. Also called *remote* or *nemo*.

field resistor A resistive component consisting of an insulated form with a thin layer of conductive material.

field rheostat The rheostat whose setting determines the amount of current flow through the field coil of a motor or generator.

field scan A form of television scanning in which the lines are scanning alternately.

field-sequential system A color-television system in which the image is reproduced by means of primary color fields (red, green, blue) flashed sequentially on the screen of the picture tube. Compare *dot-*

sequential system and *line-sequential system.*

field strength See *field intensity, 1, 2.*

field-strength meter An instrument for measuring the rf voltage of a signal reaching a chosen location. The instrument consists essentially of a radio detector equipped with a portable antenna and an output meter reading directly in rf microvolts and millivolts.

field telephone A rugged, portable telephone system for outdoor use.

field test A test of equipment under actual operating conditions, i.e., outside the laboratory or factory.

field winding See *field coil, 1, 2.*

fig. Abbreviation of *figure.*

figure-8 pattern 1. An antenna directivity pattern resembling an eight. 2. A Lissajous figure resembling an eight.

figure of merit 1. For a capacitor, inductor, or tuned circuit, the ratio (Q) of reactance to resistance (X/R). 2. For a magnetic amplifier, the ratio of power amplification to control time constant (A^p/t). 3. For a transistor, the gain-bandwidth factor $fT = hfeo\ (hfe)$, where fT is the frequency at which hfe becomes zero. 4. For a vacuum tube, $fH = (2\pi Req\ CT = Cgk + Cpk$, and $Req = \Delta\ /\ gm.$

fil Abbreviation of *filament.*

filament In a vacuum tube or incandescent lamp, the thin wire heated by electric current; it emits electrons, light, and heat. The filament is the cathode in a filament-type tube, but (as a *heater*) serves only to heat indirectly the cathode sleeve in an indirectly heated tube. Also see *heater, 1.*

filament battery See *A-battery.*

filament choke A radio-frequency choke operated in the filament lead of an electron tube. Such chokes are necessary in filament-type tubes in rf amplifiers and oscillators.

filament circuit The circuit carrying filament current.

filament coil See *filament winding.*

filament current Symbol, *If.* The current flowing through the filament of an electron tube.

filament emission Electrons emitted directly by the filament in an electron tube (thermionic emission) or the amount of such emission.

filament hum A hum signal due to voltage induced in a circuit by the ac-operated filaments (heaters) of tubes or by the filament wiring.

filament lag The time delay in the heating and cooling of an ac tube or lamp filament as filament-current changes polarity.

filament power supply A source of power, usually alternating-current, for heating the filament of a vacuum tube or tubes.

filament resistance 1. The resistance of the filament in an electron tube or incandescent lamp. 2. The resistance of an external dropping resistor in the filament circuit of a tube or lamp.

filament rheostat A variable resistor for adjusting filament current.

filament saturation For a given plate voltage, the point at which an increase in filament voltage does not cause an increase in plate current.

filament transformer A stepdown transformer that supplies power exclusively to the filament (heater) of an electron tube.

filament-type bolometer A *bolometer* in which the sensitive element is a wire filament. Examples: *barretter, incandescent lamp,* and wire fuse.

filament voltage Symbol, *Ef.* The voltage across the filament of an electron tube.

filament winding On a power transformer, the coil that supplies heating power to the filament of a tube. Also called *filament coil.*

file An organized collection (in sequence or not) of records related by a common format, data source, or application.

file conversion Converting data files from one form to another, often for the purpose of making them compatible with other computers.

file gap An area of a data medium (cards, for example) that signifies the end of a file; it may also mark the start of another file.

file identification A code that identifies a file.

file label File identification in which the first record in the file is a set of characters unique to the file; it conveys such information about, say, a tape file as a description of content, generation number, reel number, date of writing, etc. Also called *header label.*

file layout How the contents of a file are organized (usually defined by the system or specified by a program).

file maintenance To delete, add, or correct records in a file; unlike updating, which is done to reflect changes in events recorded in the file, maintenance insures that the contents of the file are accurate records of the necessary data.

file management A method of storing and recalling data from computer files such as magnetic tapes or disks.

file name In a file label, the alphanumeric character set that identifies and describes the file.

file organization The way words, bits, or records are physically arranged in the storage medium for a file, possibly including the method of access (serial, random, etc.).

file-oriented programming Computer programming that uses a general file and record control program to simplify I/O coding.

file-oriented system A (generally) commercial system having file storage as its basis.

file print A printout of the contents of a file in a storage device.

file processing The operations associated with making and using files.

file protection Preventing the possibility of writing over data files before they are made available for use—through software (having a program check file labels) or by hardware. Also see *file protection ring.*

file protection ring A ring that can be fitted to the hub of a magnetic tape reel. The two types are *write enable* and *write inhibit.* Depending on the type installed, data can or cannot be written to (recorded on) a tape file; the ring protects a data file by preventing accidental erasures.

file reconstitution Restoring a partially or completely damaged file by updating a previous generation of the file using a file of interim transactions.

file recovery Following the interruption of file processing due to system failure, the procedure for reestablishing the file's condition as necessary for the resumption of processing without losing accuracy.

file section Part of a file in certain consecutive locations on a storage medium.

file security Protective and security measures—the issuance of clearances, status markers, and the like—as they relate to computer files.

file set A collection of interrelated files stored consecutively in a magnetic disk volume (package).

fill The percentage of lines in a cable that are actually in use at a given time.

filler A nonessential data part used, for example, to bring a record to a standard size.

film See *thin film memory.*

film badge A simple dosimeter which is essentially a jacketed piece of photographic film worn by a person who risks exposure to hazardous radiation. The developed film indicates the amount of the wearer's exposure.

film capacitor A capacitor in which the electrodes are plated or deposited on the faces of a thin film of plastic or other dielectric material.

film chain A system designed for the transmission of movies over a television system. This requires synchronization of the movie frame rate with the television scanning rate.

film integrated circuit A monolithic circuit whose elements are films formed on an insulating substrate.

film frame A single picture on a strip of motion picture film.

film-frame blanking interval The interval during which a film frame is blanked out as motion-picture film moves through a camera, projector, or pickup. The blanking action allows a frame to move into position without that movement being observed.

filmorex system An electronic selection system for microfilm cards that contain a field of 20 five-digit library code numbers.

film pickup A photocell, photodiode, phototransistor, or phototube circuit employed to pick up recordings from the sound track of motion-picture film.

film reader A device for converting data on film into digital form for a computer.

film recorder An apparatus that records data as a sound pattern on film. Compare *film reproducer*.

film reproducer An apparatus that plays back data recorded on photographic film. Compare *film recorder*.

film resistor A resistor whose resistance material is a thin film (of tin oxide, e.g.) deposited on a substrate.

film scanning The conversion of a movie into a form suitable for transmission by television.

film speed 1. The speed at which motion-picture film moves intermittently through a camera, projector, or pickup, measured in feet or frames per second. 2. A measure of film's light sensitivity, given as an ASA /American Standards Association) or DIN (European) number; in either system, the higher the number, the greater the light sensitivity.

filter 1. A circuit or device which passes one frequency or frequency band while blocking others, or vice versa. Also see *save filter*. 2. An *LC* or *RC* circuit for removing the ripple from the output of a power-supply rectifier. 3. A transparent disk with special optical properties placed in front of a camera lens for a special photographic effect. 4. A character pattern used to control the elimination or selection of characters in another pattern. 5. A device or program that separates information according to specifications. 6. A machine word that specifies the elements to be treated in another machine word; also called *mask*.

filter attenuation In a selective filter, the power, current, or voltage loss, in decibels, that takes place within the passband.

filter attenuation band The frequency band rejected by a band-suppression filter.

filter capacitor A capacitor that provides capacitive reactance in a wave filter or power-supply filter while also blocking direct current.

filter center A place where information is modified for transmission to aircraft pilots. Such information may include weather data, course changes, or other instructions.

filter choke An inductor that provides inductive reactance in a wave filter or power-supply filter while affording relatively easy conduction of direct current.

filter crystal A piezoelectric crystal employed in a crystal resonator. Also see *crystal resonator*.

filter cutoff The frequency at which the transmission figure of a filter

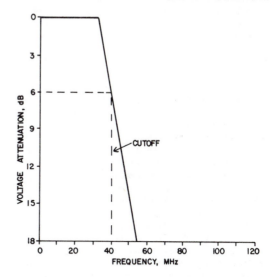

is below its maximum value by a prescribed amount.

filter discrimination The amount of fluctuation in the insertion loss of a bandpass, band-rejection, highpass, or lowpass selective filter. The fluctuation is measured at various points in the filter passband.

filter factor Symbol, a. For a power-supply filter, the quantity ($w^2 LC - 1)^n$, where $w = 2 \pi f$ (here, f is the ripple frequency in hertz), L is the choke inductance in henrys, C is the filter capacitance in farads, and n is the number of identical filter sections.

filter inductor See *filter choke*.

filter pass band The frequency band defined by the upper and lower cutoff frequencies of a bandpass filter.

filter reactor See *filter choke*.

filter slot In a waveguide, a slot that acts as a choke to suppress undesirable modes.

filter stop band The attenuation band of a band-suppression filter (see *filter attenuation band*).

filter transmission band See *filter pass band*.

filter tube A vacuum tube replacement for a choke in a power-supply

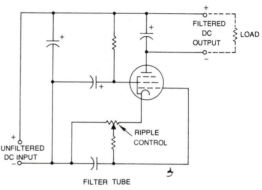

Filter tube.

filter. A small amount of the ripple in the unfiltered dc is applied to the grid of the tube, which amplifies it. The amplified ripple then appears across the load in the phase that will cancel the ripple in the original dc.

fin 1. A metal disk or plate attached to a component for the purpose of radiating heat. 2. A projection in an irregular heat sink.

final amplifier The last amplifier in a cascade of amplifier stages, as in a transmitter or instrument circuit. Also called *output amplifier*.

final result A result displayed at the end of a data processing operation. Compare *intermediate result*.

finder The switch or group of relays that selects the path for a call going through a telephone switching system. Also called *line finder*.

fine adjustment Adjustment of a quantity in small increments or as a smooth, continuous variation. Compare *coarse adjustment*.

fine-chrominance primary See *I-signal*.

fine frequency control A variable component such as a potentiometer or variable capacitor that permits a signal or response frequency to be varied over a small increment; it is often used in conjunction with a coarse frequency control. Also alled *vernier frequency control*.

fine-groove record See *microgroove record*.

fine index In computer practice, a secondary, supplemental index used with a main, or *gross*, index when the latter does not adequately detail the differences between the items being indexed.

fine-wire contact See *cat's whisker*.

finger See *feeler, 1*.

finger plethysmograph A device that senses and records the resistance through the human finger during various parts of the heart cycle.

finger rules See *Fleming's left-hand rule, Fleming's right-hand rule,* and *right-hand rule for wire*.

finish lead The lead attached to the last turn of a coil. Also called outside lead. Compare *start lead*.

finished blank The end product in the crystal manufacturing process, often including electrodes.

finishing The careful handwork and testing involved in bringing a crystal blank to a condition that is acceptable as finished according to specifications.

finishing rate The rate of charging a battery as the battery approaches a full charge. Generally, the finishing rate is less than the normal charging rate.

finite Pertaining to that which has defined limits. Compare *infinite*.

finite increment Symbol, Δ. An easily recognizable difference between successive values of a variable. Thus, an increment of voltage ΔD equals $E_2 - E_1$. Compare *differential, 1*.

finite sample space In statistics, a sample space having definite limits. Compare *infinite sample space*.

finite series A *mathematical series* having a limited number of terms, e.g., $(a + b)^2 = a^2 + 2ab + b^2$, a finite series containing only three terms. Compare *infinite series*.

finned surface The irregular surface of a heat sink. The ratio of surface area to volume is greater than with a flat surface; this increases the rate of heat loss.

fins Metal vanes radiating from usually high-current parts for heat dissipation.

FIR Abbreviation of *far infrared*.

fire A transition from non-conduction to conduction in an ionizing switching device.

fire control Aiming and firing guns automatically through the use of radar and associated electronic systems.

Fire Underwriter's regulations See *National Electric Code*.

firing 1. The ionization of a switching tube, resulting in conduction. 2. The pulse that initiates conduction in an ionization switching device.

firing angle 1. Symbol, Φ. For a magnetic amplifier, the angular distance through which the input-voltage vector rotates before the core is driven into saturation. 2. Symbol, α. For a thyratron with ac anode voltage, the point, as an angle (in degrees or radians), along the anode voltage half-cycle at which the tube fires. 3. Symbol, α. For a thyratron with dc anode voltage and ac control voltage, the point, as an angle (in degrees or radians), along the control voltage cycle at which the control voltage cycle at which the tube fires. 4. For a silicon controlled rectifier, the point, as an angle (in degrees or radians), along the control voltage half-cycle at which the SCR fires.

firing circuit Any circuit, such as a phase shifter, which permits adjustment of the firing angle of a thyratron magnetic amplifier, silicon controlled rectifier, or similar device, or which delivers the required pulse or other signal to initiate firing.

firing curve For a thyratron a curve showing ionization potential vs grid bias.

firing point In an electron tube, the ionization point of the gas (the point at which current starts to flow).

firing potential For a discharge tube, the voltage at which current through the tube becomes self-sustaining.

firing voltage See *breakdown voltage, 2* and *firing potential*.

firmware 1. Programs (software) in a nonvolatile internal computer storage, e.g., in a read-only memory (ROM), which can only be changed by replacement with an alternate unit. 2. Inalterable internal interconnections that determine what a computing device or system can do. Also called *microprogram*.

first-degree equation An equation of the first degree, i.e., one in which no term has an exponent greater than one.

first detector In a superheterodyne circuit, the signal frequency detector. Compare *second detector*.

first filter capacitor The input capacitor in a capacitor-input power-supply filter.

first Fresnel region A portion of a directional transmitted electromagnetic ray, shaped generally like a paraboloid with the apex at the transmitter and the axis in the direction of transmission. Any point in the first Fresnel zone is in such a position that the sum of the lengths of the paths from the point to the receiver, and the point to the transmitter, is no greater than 0.5 wavelength more than the distance from the transmitter to the receiver.

first generation computer A tube-type rather than solid-state computer.

first harmonic The fundamental frequency in a complex waveform from which multiples are generated.

first law of thermodynamics Quantities of heat may be converted into mechanical work, and vice versa ($W = JQ$). Also see *mechanical equivalent of heat*.

first level address See *absolute address*.

first selector The selector that responds to the first-digit dial pulses when a telephone number is called.

fishbone antenna An untuned, wideband directional antenna of the general end-fire type, which consists of a number of collector antennas, each loosely capacitively coupled to the resistor-terminated transmission line in collinear pairs. It is so called from its resemblance to the skeleton of a fish.

fishpaper A chemically treated, vulcanized-fiber paper used for electrical insulation.

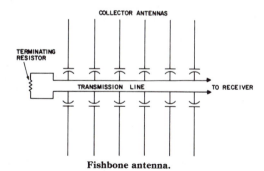

Fishbone antenna.

fission In a nuclear reaction, the splitting of an atomic nucleus into smaller bits. Also see *nuclear fission*. Compare *fusion*.

fist In radiotelegraphy and wire telegraphy, an operator's manual sending style.

fitting A device intended to mechanically fasten a wire or cable in place.

five-element code A five-impulse telegraph code that describes a character, e.g., Baudot, which also includes start and stop elements.

five-element tube See *pentode*.

five-layer device A semiconductor device containing four PN junctions. Examples: *diac, triac*.

five-level code A teletype code that utilizes five binary elements to define a character.

fix 1. In direction finding, the point at which two lines of direction intersect. 2. In electronics maintenance, to repair successfully. 3. To subject an in-process circuit board to a solution or other medium to stop a photographic action permanently. Also, the solution in which such photosensitive materials may be immersed to halt development.

fixed bias Bias voltage or current supplied from a fixed external source (such as a battery or ac power supply) and which is, therefore, independent of the operation of the biased device. Compare *automatic bias*.

fixed block length Blocks of data having a fixed number of words or characters, as required because of hardware limitations or a program instruction. Compare *variable block length*.

fixed capacitor A nonadjustable capacitor, i.e., one having a single unalterable value.

fixed component Any single-value component, e.g., a fixed capacitor, inductor, or resistor.

fixed contact The stationary contact in a relay or switch. Compare *movable contact*.

fixed-crystal detector A simple crystal detector in which the point of the cat's whisker is permanently placed in contact with a sensitive spot on the surface of the crystal.

fixed field Fields in records organized so that those containing similar information in each record are the same length and in the same relative position in the record. Compare *variable field*.

fixed form coding Coding source languages so that each part of the instruction on a card or paper tape record is in a fixed field.

fixed-frequency amplifier An amplifier that is pretuned to operate on one frequency.

fixed-frequency oscillator An oscillator that is preset to operate on one frequency. Such an oscillator may either be self-excited or controlled (crystal, fork, magnetostriction, etc.).

fixed-frequency receiver A receiver that is pretuned to receive signals of one frequency.

fixed-frequency transmitter A transmitter that is pretuned to radiate signals of one frequency. Such a transmitter may contain either a self-excited or crystal-controlled oscillator.

fixed inductor A nonadjustable inductor, i.e., one having an unalterable value of inductance.

fixed-length record A record in which word or character size is constant. Compare *variable length record*.

fixed logic Applicable to computers or peripherals whose logic can only be altered internally by changing connections.

fixed memory A nonvolatile readout computer memory that can only be altered mechanically.

fixed placement file A file which has been allocated a fixed location in storage.

fixed-point system A notation system in which a single set of digits (unlike floating point notation) represents a number, and the radix point (in the decimal system, the decimal point) can only be placed in one position for the value being expressed. Also see *floating-point calculation*.

fixed resistor A nonadjustable resistor, i.e., one having an unalterable value of resistance.

fixed station A radio station operating from a stationary point; one that isn't mobile.

fixed-step potentiometer A potentiometer whose output is varied in one or more discrete steps by fixed-resistor sections. Also see *potentiometer, 1*.

fixed word length Applicable to the organization of information in storage in which each computer word stored has a fixed number of characters or bits.

fixture A piece of hardware used in equipment setups; e.g., microwave couplers, joints, sections, etc.

fL Abbreviation of *foot-lambert*.

flag 1. A piece of information added to a data item that gives information about the data item. 2. A bit added to a character or word to delineate a field boundary. 3. An indication that an operation is complete and needn't be done by the program. 4. An indicator identifying the members of mixed sets. 5. A character that signals the presence of some condition; e.g., an error flag indicates that a data item caused an error.

flag event A program condition that causes a flag to be set.

flag line An input pulse to a microprocessor that depends on specific external instructions. Indicates a certain condition or change of state.

flagpole antenna Any of several vertical uhf or vhf antennas consisting of a radiator mounted atop a coaxial pipe or cable (see, for example, *coaxial antenna*). It takes its name from its resemblance to a flagpole.

flag terminal A form of terminal that does not require soldering for electrical contact. A protruding "flag" is crimped around the conductor.

flame alarm A (usually photoelectric or thermoelectric) device or circuit for detecting a flame and actuating an alarm.

flame control A (usually photoelectric or thermoelectric) device or circuit for sensing and automatically controlling the height of a flame, such as a gas pilot.

flame-failure control A flame control that automatically cuts off the fuel if the flame goes out.

flame microphone A microphone in which two electrodes in a flame undergo a change in electrical resistance when the flame is influenced by sound waves, thus modulating current passing between the electrodes.

flange 1. A flat, protruding edge used for fastening a connector or plug

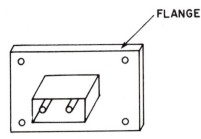

FLANGE

Flange.

to the chassis of a piece of equipment. 2. In a waveguide, a coupling employed for connection to another section of waveguide, or to a horn or other external device.

flange focus The focal length of a lens, based on the distance from the mounting flange to the focal plane.

flanking Modification of the response of a selective filter, resulting from the parallel connection of two or more similar filters.

flap See *drop indicator*.

flap attenuator A waveguide attenuator consisting of a sheet of resistance material, it is inserted transversely into the waveguide through a slot.

flare 1. The hyperbolic (cross sectional) portion of a horn (antenna or loudspeaker). 2. A transient or stationary bright area with (usually) a central pip on the screen of an otherwise blank oscilloscope or TV picture tube.

flare angle The gradual change in a waveguide's diameter over its length.

flare factor The angle at which the faces of a horn speaker are curved or turned outward.

flaring constant Symbol, *m*. A number expressing the degree of flare (see *flare, 1*) in a horn (antenna or loudspeaker). The constant *m* (in reciprocal feet) is equal to $0.6931/d$, where *d* is the distance (in feet) along the axis required for the cross-sectional area of the horn to double.

flash 1. A photoflash (see *electronic flash, 1, 2*). 2. To vaporize a metal, (such as magnesium), in an electron tube being evacuated, to absorb gases. 3. Flashover.

flash arc In a vacuum tube, a sudden high-current arc between cathode and plate at high plate voltages; it can short-circuit the plate power supply.

flashback voltage The maximum inverse voltage that causes the gas in a tube to ionize.

flash delay A device that automatically postpones the operation of a flashtube until a predetermined instant, such as the moment when a moving object arrives at a particular point before a camera.

flasher An electrical or electronic device or circuit that flashes a light or a series of lights sequentially.

flasher LED A *light-emitting diode* which, when connected to a low-voltage dc source, emits light that flashes at a basic rate of a few pulses per second (e.g., 3 pps).

flashing In the reactivation of a vacuum tube having a thoriated-tunsten filament that has lost its emission ability, the process of operating the filament briefly at a voltage much higher than normal. Flashing precedes the step of burning.

flashing voltage The voltage at which flashing is carried out.

flashlamp 1. See *flashtube*. 2. A small hand-carryable light operated from self-contained cells; a flashlight.

flashlight See *flashlamp 2*.

flashover The sudden discharge of electrical energy between electrodes or conductors, often accompanied by light; it is usually the result of excessive voltage.

flashover voltage 1. The peak voltage at which flashover occurs. 2. The voltage at which disruptive discharge occurs between electrodes and across the surface of an insulating material.

flash plating Electroplating in which a thin layer is deposited quickly.

flash test Insulation testing by applying a higher-than-normal voltage for a short time.

flashtube A straight or coiled glass tube filled with gas and provided with electrodes. When a high voltage is applied to the electrodes, the tube emits a brilliant flash of light.

flat cable A cable whose flexible conductors are molded side by side in a flexible, flat ribbon of plastic such as polyethylene.

flat-compounded generator A compound-wound generator whose windings are proportioned so that the full- and no-load voltages are identical.

flat fading Fading of a radio signal that occurs independently of frequency; all frequency components of the signal fade to the same extent at the same time.

flat file A computer file containing unfolded documents.

flat frequency response Relatively equal response to all fixed-point frequencies within a given spectrum, exhibited by an amplifier or other circuit which must transmit a band of frequencies.

flat line 1. A transmission line in which there are no standing waves, or for which the standing-wave ratio is very low. 2. Flat-ribbon line.

flat pack An integrated circuit package consisting of a square or rectangular flat housing with leads (usually straight outward) from its sides.

flat response A response characteristic in which the dependent variable is substantially constant over a specified range of values of the independent variable. For example, in amplifier operation, an output signal whose component fundamental frequencies and their harmonics are in the same proportion as those of the input signal being amplified.

flat-ribbon line A transmission line (feeder) consisting of two flexible conductors molded in a flexible, flat ribbon of plastic, such as polyethylene. Also called *twinlead*.

flattening The flattening out or blunting of a normally peaked or curved response, often caused by signal saturation within a circuit. Sine-wave clipping is an example.

flat top 1. The horizontal radiating portion of an antenna. 2. Of an amplifier, to distort by clipping of the positive half-cycles.

flat-top antenna An antenna having wires in its flat top.

flat-top beam A bidirectional, end-fire antenna consisting of two close-spaced dipoles center-fed out of phase. Also see *Kraus antenna*.

flat-top response The ability to uniformly transmit frequencies in a given band.

flat transmission line 1. A transmission line that is free of standing waves. Also see *matched transmission line*. 2. A flat-ribbon line.

flaw An irregularity in a substance that might result in poor operating characteristics.

flaw detector An instrument employing ultrasonic waves to detect internal flaws in metal. The waves are reflected by flaws.

F-layers See *F-region*.

fLb Abbreviation of *foot-lambert*.

Fleming's generator rule See *Fleming's right-hand rule.*

Fleming's left-hand rule A simple way of indicating certain relationships in the behavior of electric generators and motors: If the thumb, index finger, and middle finger of the left hand are separated as much as possible, the thumb will point in the direction of force or motion when the index finger is pointed in the direction of flux; the middle finger points in the direction of current flow. Compare Fleming's *right-hand rule.*

Fleming's motor rule See *Fleming's left-hand rule.*

Fleming's right-hand rule A simple way of indicating certain relationships in the behavior of electric generators and motors: If the thumb, index finger, and middle finger of the right hand are separated as much as possible, the middle finger points in the direction of an induced voltage, the thumb in the direction of the motion of a conductor, and the index finger in the direction of the magnetic field. Compare *Fleming's left-hand rule.*

Fleming valve Early name for the tube diode. The tube made direct use of the Edison effect for detecting radio waves. The name "valve" is suggestive of the rectifying property of the diode and has survived as the British term for electron tube.

Fletcher-Munson curves A set of curves depicting the uneven frequency response of human hearing. Also called *audibility curves.*

Flewelling circuit An early one-tube superregenerative detector circuit employing three capacitors in a feedback-voltage divider.

flexible collodion A viscous solution of pyroxylin (cellulose nitrates) used sometimes as a binder for coils (see *collodion*).

flexible contact A contact made from flat, metal spring stock; it is usually bent or curved. Also called *spring contact.*

flexible coupling A device for joining two shafts and conveying rotary motion from one to the other; it is springy so that the shafts need not be exactly aligned with each other.

flexible flat cable See *flat cable.*

flexible resistor An insulated, wirewound resistor that may be bent, coiled, or knotted.

flexible shaft A control shaft that can be bent somewhat while still allowing easy adjustment.

flex life A measure of how much bending a conductor or other flexible object can take without breaking.

flexode A diode that is flexible in that its junction can be changed, i.e., reversed without reversing its leads, its resistance being variable from the forward- to backward-resistance value.

flicker 1. To appear, disappear, and reappear, or increase and decrease in intensity frequently. 2. The effect created by such action (as in a flickering light).

flicker effect In an electron tube, noise caused by large, erratic variations in emission current due to irregular coating of the cathode.

flicker frequency The number of times the screen illumination flashes on and off in the projection of a motion picture. It is 48 per second (twice the frame rate) in conventional systems, since for each frame the screen is blanked once when the frame is pulled into position and once again during projection of the frame.

flight control Electronic monitoring and control of an aircraft in flight.

flight path The course planned for an aircraft's flight.

flight-path computer A computer that controls the course of an aircraft in flight, from takeoff to landing.

flight-path deviation The departure of an aircraft in flight from the course in the flight plan. Also see *flight path.*

flight-path-deviation meter An instrument that gives a visual indication of flight-path deviation.

flight test 1. To test airborne electronic equipment in actual flight. 2. A test made as in 1, above.

Flinders bar In a magnetic compass, a metal bar that corrects for the vertical component of the geometric field. The Flinders bar must be designed differently in different geographic locations.

flint glass A hard, bright, lead glass. Dielectric constant, 7 to 9.9. Dielectric strength, 30 to 150 kV/mm. Also see *glass.*

flip chip A monolithic semiconductor device—such as a diode, transistor, or integrated circuit—in which beadlike terminals are provided on one face of the chip for flip-flop bonding.

flip-chip bonding A system for making connections between a semiconductor chip and a header, in which leads are not run between chip and header. Instead, beadlike projections are electrodeposited as terminals around one face of the chip, which is then registered with the header terminals and bonded to them.

flip-flop 1. A bistable multivibrator. 2. A two-position relay that locks in alternate positions upon receiving successive actuating pulses.

flip-flop key In a video display, a key that, when pressed, allows viewing of one half of the screen and then the other. Often used in calculators and word processors.

flip-flop memory A bistable computer memory that stores bits of data as flip-flop states.

flip-flop relay See *bistable relay.*

flipover cartridge A phono cartridge virtually of only historical interest, which can play a 78 rpm record in one position or be flipped over to play a microgroove record; each side of the cartridge has its own needle.

floated battery A storage battery connected in parallel with a generator, which supplies the load; the battery, always completely charged, helps during high-current demands.

floating 1. To float a storage battery; see *float.* 2. An ungrounded device or circuit that is not connected to a potential. 3. Not loaded or driven. 4. Not fixed in position. 5. A dedicated ground connection that remains isolated from the common circuit ground.

floating address See *relative address.*

floating charge See *trickle charge.*

floating control 1. A potentiometer, such as a gain control, installed with its shaft insulated from ground and, accordingly, subject to body-capacitance effects. 2. A type of automatic control in which the rate of final control element movement depends on the amount the controlled variable deviates from a prescribed value.

floating grid A vacuum-tube control grid that is not connected to a signal point in the circuit.

floating ground See *floating, 5.*

floating input An ungrounded input circuit.

floating-input measurement See *differential-input measurement.*

floating instrument An instrument whose signal terminals are above ground.

floating I/O port An input/output terminal that is not loaded or being driven.

floating junction A junction (*pn*, for example), that has no net current flowing through it.

floating neutral A circuit with a variable common voltage reference.

floating paraphase inverter A dual-tube or dual-transistor adaptation of the paraphase inverter. The second tube receives its grid-input signal from a tap on the load resistor of the first tube and provides the additional phase shift that is required. See next page.

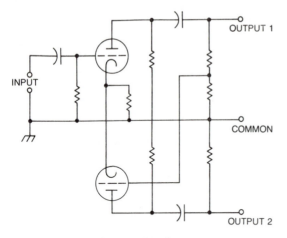

Floating paraphase inverter.

floating-point calculation In digital computer practice, a calculation using a floating point number, i.e., a number whose value is represented by two sets of digits, a fixed-point part (see *fixed-point system*) and a radix (base number) with an exponent. To add two floating-point numbers, their exponents must be equalized by shifting the mantissa of the smaller expression to the right the number of places corresponding to the difference of the exponents. For example, before adding 0.5×10^3 to 0.4×10^7, 0.5×10^3 is written 0.0005×10^7; then the numbers are added, giving 0.4005×10^7.

floating-point number A number expressed in the floating-point system.

floating-point package Computer-vendor software that enables that computer to perform floating-point calculations.

floating-point system A system of notation in which a number n is represented by two sets of numbers: a fixed-point part (see *fixed-point system*) a, the radix (base number) r, and an exponent b as follows: $n = a \times r^b$. For example, in the floating-point system, 623 can be written 6.23×10^2. Floating-point numbers can be stored economically (in terms of memory) and in magnitudes that might otherwise (i.e., if they were fixed-point numbers) be beyond the capacity of the computer to operate upon with relatively consistent accuracy. Compare *fixed-point system*.

floating probe A test electrode (wire or plate) inserted in a discharge tube at a desired point to sample the potential gradient, but which acquires a misleading negative charge with respect to the gas cloud, because electrons (traveling faster than the positive ions) tend to accumulate on the probe.

floating zero A control system in which the reference point is easily moved.

floating zone In a semiconductor ingot undergoing purification, a molten zone in which impurities float. The material in the zone is melted by the rf field of an external heating coil, which is passed along the ingot to move the molten zone to one end, picking up impurities along the way and concentrating them at the end that is later sawed off.

float switch A switch operated by a float, such as in a sump pump.

flocking 1. Particulate felt used on phonograph turntables, for example, as scratch protection. 2. To coat with flocking.

flood gun In a storage (image-holding) oscilloscope, the electron gun that sprays the storage target with low-velocity electrons and makes the image visible on the viewing screen. The gun is mounted next to one pair of deflection plates. Compare *writing gun*.

floppy disc A flexible magnetic disc employed in recording, as in computer and data system storage.

flow 1. The movement of current carriers under the influence of an electric field. 2. Angle of conduction. 3. A series of interrelated events in a time sequence.

flow angle See *angle of conduction*.

flowchart 1. A diagram depicting the logic steps in a digital-computer program. 2. A diagram showing the flow of material through a sequence of processes.

flow direction The method of delineating antecedent and successor events on a flowchart; it could be arrows or flowlines connecting the events in the way a page is read (top to bottom, left to right).

flowed-wax disk A form of recording disk, in which wax is melted onto a plastic or metal base. The grooves are cut in the wax layer.

flowline A line showing flow direction on a flowchart.

flowmeter An instrument for measuring liquid flow rate.

flow relay A relay that is actuated by a predetermined rate of fluid flow.

fluctuating current See *composite current*.

fluctuating voltage See *composite voltage*.

fluid absorption See *liquid absorption*.

fluid analogy The comparison of electric current flow to the movement of a simple fluid. Also see *water analogy*.

fluid capacitor See *water capacitor*.

fluid computer A digital computer that uses fluid logic elements, i.e., one that contains no electronic circuits or moving parts.

fluid damping Use of a viscous fluid to damp a mechanical member's movement.

fluid-flow alarm An electronic circuit that actuates an alarm when fluid flowing through pipes or other channels changes from a predetermined rate.

fluid-flow control A servo system that automatically maintains or adjusts liquid flow through pipes or other channels.

fluid-flow gauge See *fluid-flow meter*.

fluid-flow indicator See *fluid-flow meter*.

fluid-flow meter An instrument that indicates fluid flow rate through pipes or other channels.

fluid-flow switch In a fluid-cooled system, a switch that actuates an alarm when the fluid slows or stops.

fluidics 1. A form of digital logic in which circuits operate by means of fluid flow. 2. Fluid dynamics.

fluid-level control A servo system that automatically maintains the level of a fluid a tank.

fluid-level gauge An electronic system that gives direct readings of the level of a fluid in a tank.

fluid-level indicator See *fluid-level gauge*.

fluid logic Logic operations carried out by varying the flow and pressure of a gas or liquid in a circuit of channels. Also see *fluid computer*.

fluid ounce (U.S.) A unit of volume equal to 2.957×10^{-5} cubic meters.

fluid-pressure alarm An electronic circuit that actuates an alarm when fluid pressure rises or falls beyond set limits.

fluid-pressure control A servo system that automatically maintains or adjusts fluid pressure in pipes or other channels.

fluid-pressure gauge See *fluid-pressure meter*.

fluid-pressure indicator See *fluid-pressure meter.*

fluid-pressure meter An instrument that indicates the pressure of a fluid in a pipe or other channel.

fluid valve See *electromechanical valve.*

fluorescence The property of some materials to glow when excited by a stimulus, such as light or an electron beam. Compare *phosphorescence.*

fluorescent lamp See *fluorescent tube.*

fluorescent materials Materials that glow when irradiated, but cease to glow when the source of excitation is removed—the phosphor coating on crt screens, for example.

fluorescent screen A transparent or translucent plate (such as the end of a cathode-ray tube or fluoroscope) coated with phosphors that glow when struck by an electron beam or by X-rays.

fluorescent tube A mercury-vapor glow lamp distinguished by having a glass tube whose inner wall is coated with a phosphor that emits light when excited by the ultraviolet glow discharge in the vapor.

fluorescent X-rays X-rays reradiated by the atoms of a material that has absorbed X-radiation. During initial exposure, energy absorbed from the radiation raises the energy level of electrons in the atoms; when the electrons return to their normal energy levels, they reradiate some of the absorbed energy.

fluorine Symbol, F. A gaseous element of the halogen family. Atomic number, 9. Atomic weight, 19.

fluoroscope A device used for the observation of certain X-ray effects. In a simple form, an opaque (to light) screen coated with barium platinocyanide, a material which fluoresces when exposed to X-rays, is mounted in one end of a light-tight viewing hood. When an object, such as part of the human body, is placed between the screen and an X-ray tube, internal structures (such as bones) are made visible on the screen.

fluoroscopy The art of using a fluoroscope in the inspection of materials and parts or in medical examinations.

flush A form of mounting in which there is little or no protrusion from the panel surface.

flutter 1. In a high-frequency superheterodyne receiver, a rapid fluctuation in signal strength caused by tuning and detuning of the oscillator stage resulting from poor dc power-supply regulation. 2. Repetitive fluctuations in the output of a sound reproducer. Also see *wow.*

flutter bridge A bridge-type instrument for measuring flutter in constant-speed machines, such as sound recording and reproducing devices.

flutter rate The frequency of flutter, in cycles per second.

flux 1. The lines of force which are believed to extend in all directions from an electric charge (electric flux) or from a magnetic pole (magnetic flux). 2. A material that makes metals more amenable to being joined by soldering. 3. The number of photons that pass through a surface for a given time.

flux density Symbol, B. Unit, tesla. The degree of concentration of magnetic lines of force; a tesla is the equivalent of $V \cdot s/m^2$.

flux gate A device that controls the azimuth bearing of a directional system by means of interaction with the geomagnetic field.

flux graph A device that graphically records the intensity of a magnetic field around a permanent magnet or electromagnet, or around an inductor carrying a current.

flux leakage See *magnetic leakage.*

flux lines The lines of force in an electric or magnetic field.

flux linkage The passage of lines of force set up by one component through another component, so as to enclose most of the penetrated component's volume.

fluxmeter An instrument for measuring magnetic flux density. Also called *gaussmeter.*

flyback 1. The abrupt fall or reversal of a current or voltage that was previously increasing, e.g., the rapid fall of a sawtooth wave. Also see *kickback.* 2. The duration of the flyback of a sawtooth or similar wave. 3. In an oscilloscope or TV picture tube, the rapid return of the beam to its starting position.

flyback checker An apparatus that senses the presence of short or open circuits in motors, transformers, and generators, by measuring the amount of flyback (kickback).

flyback power supply See *kickback power supply.*

flyback time The time taken for the electron beam in an oscilloscope tube, TV picture tube, or camera tube to return to its starting point after it has reached the point of maximum deflection.

flyback transformer In a TV receiver circuit, the horizontal output transformer. The unit supplies horizontal scanning voltage and kickback voltage, which is rectified to produce the high-voltage dc anode potential. Also see *flyback* and *kickback power supply.*

flying-spot scanner Abbreviation, *fss.* See *flying-spot tube.*

flying-spot tube A tube, such as a TV camera tube, in which a rapidly deflected spot of light scans an image on a transparent screen; the spot is projected through the picture to a photomultiplier.

fly's-eye lens A lens comprising hundreds of small lenses; it is used in microelectronic circuit fabrication to produce many images of the same circuit.

flywheel effect In an *LC* tank circuit, the action in which energy continues to oscillate between the capacitor and inductor after an input signal has been applied; the oscillation stops when the tank-circuit finally loses the energy absorbed. The lower the inherent resistance of the circuit, the longer the oscillation will continue before dying out.

flywheel synchronization A form of television scanning synchronization employed when the received signal is very weak. The synchronization signals from the transmitter are sensed by the receiver, which then produces its own local pulses based on the rate of received pulses.

flywheel tuning A tuning dial mechanism in which the control shaft has a flywheel for the smoother tuning action afforded by the added momentum.

Fm Symbol for *fermium.*

FM Abbreviation of *frequency modulation.*

fm Abbreviation of *modulation frequency.*

FM-AM Pertaining to equipment that will operate with either amplitude-modulated or frequency-modulated signals.

FM-AM multiplier A method of frequency multiplication using both amplitude and frequency modulation of a carrier wave.

FM broadcast band The 88 to 108 MHz frequency band, within which channels spaced 200 kHz apart occupy positions from 88.1 to 107.9 MHz.

FM detector See *discriminator, ratio detector,* and *slope detector.*

FM-FM Frequency modulation by one or more FM subcarriers.

FM limiter In a frequency-modulation circuit, a stage which holds the amplitude of the FM signal to a constant value. The limiter may be active (e.g., an amplifier-limiter tube or transistor) or passive (e.g., a diode clipper).

FM multiplex See *multiplex adapter.*

FM noise Unintentional modulation of a frequency-modulated transmitter, resulting from noise in the audio-input stages.

FM-PM A system of modulation in which a carrier is phase modulated by frequency-modulated subcarriers.

FM radar A radar system in which the signal is frequency modulated and the distance to the target is measured in terms of the beat note between transmitted and reflected waves.

FM repeater A two-way radio system composed of a simultaneously operating receiver and transmitter, the latter of which retransmits (usually on a different frequency) all signals picked up by the receiver. The system is usually tower- or hilltop-mounted and used to extend the range of other two-way units in a communications network.

FM stereo The use of multiplex methods to transmit and receive stereophonic programs in an FM channel. Also see *multiplex adapter*.

FM tuner A compact radio receiver unit which handles FM signals and delivers its low-amplitude audio output to a high-fidelity system. Compare *AM tuner* and *AM-FM tuner*.

f-number The focal ratio (focal length to diameter) of a lens. The expression of the *f*-number is the symbol *f* followed by the appropriate number (e.g., *f*/3.5). For a camera, the larger the *f*-number, the dimmer the image and therefore the longer the required exposure. Some lens makers prefer a colon instead of a solidus in the *f*-number (*f*:3.5).

focal length Symbol, *F*. The distance from the center of a lens to the principal focus. Also see *principal focus*.

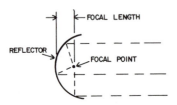

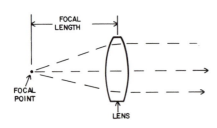

Focal length

focal ratio For a lens, the ratio *F/d*, where *F* is the focal length and *d*, the diameter of the lens. Also called *f-number*.

focometer An instrument for measuring the focal length of a lens or system of lenses.

focus 1. The point at which rays converge. Also see *principal focus*. 2. To bring rays to a point of convergence.

focus coil See *focusing coil*.

focus control In an oscilloscope or TV circuit, the potentiometer that controls the voltage on the focusing electrode of the cathode-ray tube and, accordingly, the sharpness of the image.

focus grid 1. The focusing electrode in an electrostatic TV picture tube. 2. The focusing electrode in an oscilloscope tube.

focusing anode See *focusing electrode*.

focusing coil An external coil used to focus an electron beam in a cathode-ray tube. Also see *electromagnetic focusing*.

focusing electrode The internal electrode (grid or ring) used to focus the electron beam in a cathode-ray tube. Also called *focus electrode*. Also see *electrostatic focusing* and *focus grid, 1, 2*.

focusing magnet A permanent magnet assembly for focusing the electron beam in a crt.

foil capacitor A capacitor whose plates are sheets or strips of metal foil separated by a dielectric film.

foil coil See *foil-wound coil*.

foil conductor A conductor which is a strip of metal foil rather than wire. Also see *foil pattern*.

foil electroscope See *leaf electroscope*.

foil pattern The pattern of thin metal circuit paths that constitute the "wiring" of a printed circuit. Also see *etched circuit* and *printed circuit*.

foil-wound coil A coil wound with metal foil (usually aluminum or copper) instead of wire. Such coils substantially reduce the weight of large transformers and filter chokes.

foldback characteristic See *current limiting*.

foldback current limiting In a power supply, a method of automatically reducing the output current to a safe level when the load current exceeds the maximum recommended value. This action protects both the power supply and the powered equipment.

folded dipole A doublet antenna having two wires (or tubes) which are closely spaced dipole radiators joined at their ends, one being center fed.

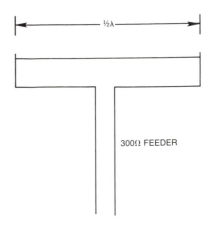

Folded dipole.

folded filament a strand of bent resistance wire, inserted into the cathode of a vacuum tube.

folded horn A loudspeaker having a horn whose flare is divided into several zigzagging chambers; that is, the horn is in effect folded to squeeze a required length into a small cabinet.

folded-horn enclosure See *labyrinth speaker*.

folded pattern An oscilloscope image having an elongated time axis obtained by successive horizontal sweeps, each placed slightly lower on the screen than the preceding one. The folded-pattern technique pro-

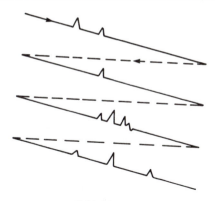

Folded pattern.

vides a time axis several times longer than the screen width.

folding frequency In a system where sampling is made at uniform frequency increments, the frequency corresponding to half the sampling rate in hertz.

foldover Distortion characterized by the horizontal or vertical overlapping of a TV picture.

follower A single-stage amplifier characterized by zero phase reversal, unity (or lower) voltage gain, and impedance stepdown, e.g., cathode follower, emitter follower, source follower.

follower drive In a servo system, the drive that mechanically follows the master drive.

following blacks In a TV picture, the effect in which a moving white object has a black border following it.

following whites In a TV picture, the effect in which a moving dark object has a white border following it.

follow-up motor See *servomotor.*

font The physical shape and size of the letters and numbers in an alphanumeric system.

font reticle In optical character recognition, an overlay reference pattern of lines used to check the size and configuration of an input character, the size of punctuation marks, and spacing between lines and characters.

foot Abbreviation, ft. A unit of linear measure in the English system; 1 ft equals 0.3048 meter.

foot-candle Abbreviation, fc. A unit of illuminance; 1 fc is the amount of direct light emitted by 1 candela (see *candlepower*) that falls on 1 square foot of a surface on which every point is 1 foot away from the source. In the International System of Units, the unit is lux (lumens per square meter). Compare *meter-candle.*

foot-candle meter A light meter whose scale reads directly in foot-candles.

foot-lambert Abbreviation, fLb or fL. A unit of luminance; the average brightness of a surface that emits or reflects 1 lumen per square foot. The S.I. (preferred) unit is candelas per square meter; 1 flb = 3.426 cd/m^2.

foot-pound Abbreviation, ft-lb. In the English system, a unit of energy equal to 1 pound displaced through a distance of 1 foot in the direction of the exerting force; the S.I. (preferred) unit is the joule; 1 ft-lb is approximately equal to 1.356 joules.

foot-pound-second system See *fps system of units*

foot switch A switch operated by the foot, generally used for the purpose of turning a playback system on and off. Often used for taking dictation.

forbidden band See *energy gap.*

forbidden character code An error-finding code using *forbidden characters:* combinations of prohibited bits. Also called *forbidden combination.*

forbidden energy band See *energy gap.*

force 1. Symbol, *F*. Units, newton, dyne, poundal. The agency or influence that accomplishes work. 2. An operator interjection made during a program run that causes the computer to execute a branch instruction; forcing is usually necessary when a condition responsible for halting a program must be bypassed.

forced coding Programming that minimizes the time required to retrieve information from storage. Also called minimum latency programming; minimum access programming.

forced oscillations Oscillations in a circuit, such as in an *LC* tank, that result from continuously applied ac excitation. Compare *free oscillations.*

foreground job A relatively high-priority short-running program that is carried out by interrupting a low priority, long-running (background) job. Compare *background job.*

force pump In a multistage vacuum system, the first pump that reduces the pressure considerably below atmospheric pressure. Also see *diffusion pump* and *vacuum pump.*

force summing device The transducer element that is moved by the force being transduced.

foreshortened addressing In control computers, the mixing of available storage by using simplified addressing instructions.

fork oscillator An audio-frequency oscillator controlled by a tuning fork. The dimensions of the fork determine its vibration frequency and, accordingly, the frequency of the oscillator.

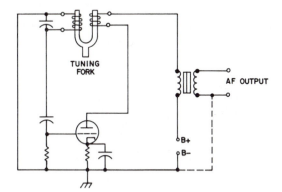

Fork oscillator.

form 1. The core or frame upon which an inductor is wound. 2. A vessel, such as a mold, used in the shaping stage of a manufacturing process.

formaldehyde Formula, HCHO. A colorless, pungent gas that is a constituent of many well known plastic insulating materials (see *phenol-formaldehyde plastics*).

formant 1. The frequency range in which the sound of a syllable is concentrated. 2. Any general group of audio frequencies.

formant filter In an electronic organ, an audio filter that changes the waveshape of a tone so that the tone will have the desired characteristics.

format The form in which data is presented e.g., the arrangement of characters, fields, words, totals, etc.

formatting Programming for a desired format.

form factor 1. Shape factor for a filter or tuned circuit. See *shape factor, 1, 2*. 2. For a half-cycle of an ac quantity, the ratio of the rms value to the average value.

form feed 1. A mechanical system that positions paper being supplied to a line printer. 2. The FF character that initiates advancement of printout paper in a printer. 3. The advancement of printout paper in a printer.

form feed character In a control loop, a character (symbol, FF) used on printing devices for controlling form feed.

forming See *electroform, 1*.

form stop An automatic device that stops a printer when the paper runs out.

FORTRAN Acronym for *formula translation*. A high-level procedure-oriented computer language developed by IBM. The source program is written as a combination of statements in English and algebraic formulas.

fortuitous conductor A medium that creates an unwanted electrical path.

fortuitous distortion Waveform distortion that results from causes other than characteristic effects or bias effects.

forty-five—forty-five recording See *Westrex system*.

forward agc Automatic gain control provided by special transistors whose transconductance decreases with increasing emitter current, and vice versa. Compare *reverse agc*.

forward-backward counter A counter that runs forward to perform addition and backward to perform subtraction.

forward bias Forward voltage or current in a transistor or semiconductor diode.

forward-blocking state For a silicon controlled rectifier, the off state during which the forward bias is so much less than the forward breakover voltage that only small off-state current flows.

forward breakover voltage For a silicon controlled rectifier, the forward voltage value at which the device abruptly switches on.

forward characteristic The current—voltage response of a semiconductor junction that is biased in the forward (high-conduction) direction. Compare *reverse characteristic*.

forward compatibility standards Standards developed to make programs for one system usable for additional or replacement equipment.

forward conduction The increased current conduction through a pn junction that is forward biased. Compare *reverse conduction*.

forward current Symbol, If. The increase in current flow through a pn junction that is forward biased. Compare *reverse current*.

forward current-transfer ratio The current gain of a bipolar transistor (*alpha* for the common-base connection, *beta* for the common-emitter connection).

forward power 1. In a transmission line, the power leaving the generating source, as measured by a directional wattmeter at that location. 2. The power arriving at the load at the terminating end of a transmission line.

forward propagation by tropospheric scatter Abbreviation, FPTS. A method of transmitting part of a radio signal beyond the horizon using the scattering effect of the troposphere. Also see *forward scat-*

ter and *troposphere*.

forward resistance Symbol, Rf. The lower resistance of a forward-biased pn junction. Also see *forward bias*. Compare *reverse resistance*.

forward-reverse ratio See *front-to-back ratio*.

forward scatter The scattering of a radio wave in the normal direction of propagation to points beyond the skip zone. The phenomenon results from reflections from nonuniform regions in the ionosphere and from points beyond the skip zone. Compare *back scatter*.

forward transconductance Symbol, gfs. For a common-source-connected FET, the ratio of a drain-current differential to the differential of gate-to-source voltage that produces it; $gfs = 10^3 \, (dID/dVGS)$, where gfs is in microsiemens, ID is drain current in milliamperes, and VGS is gate-to-source voltage in volts.

forward voltage Symbol, Ef or Vf. Voltage whose polarity causes maximum current to flow through a pn junction. Compare *reverse voltage*.

forward voltage drop The voltage across a semiconductor junction that is biased in the forward (high-conduction) direction. Compare *reverse voltage drop*.

FOSDIC Abbreviation of *film optical scanning devices for input to computer*.

Foster theorem A theorem concerning impedances, stating that the driving-point impedance is composed of certain negative and positive components.

Foster-Seeley discriminator A discriminator circuit in which the diodes are operated from a single-tuned, center-tapped secondary of the input transformer. The center tap is also capacitively coupled to the top of the transformer's primary coil. Compare *Travis discriminator*.

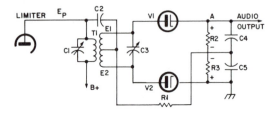

Foster-Seeley discriminator.

Foucault currents See *eddy currents*.

four-address instruction A computer instruction in which the address comprises four addresses: two for operands, one for the result of the operation, and one for the upcoming instruction.

four-element tube See *tetrode*.

Fourier analysis Use of the Fourier series to evaluate the dc, fundamental, and harmonic components of a complex wave.

Fourier series A mathematical series which shows any periodic function to be a combination of sine terms and cosine terms; any complex wave (e.g., a square wave) consists of fundamental and harmonic sine-wave components. The series is as follows: $y = f(x) = Ao/2 + A1 \cos x + A2 \cos 2x + A3 \cos 3x + ... + B1 \sin x + B2 \sin 2x + B3 \sin 3x + ...$

four-layer diode A dual-terminal npnp device which is usable as a bistable switch, sawtooth or pulse generator, memory device, etc.

four-layer transistor A transistor in which the wafer or block has four processed regions; however, the device may have only three terminals. Some examples are silicon controlled rectifier, silicon controlled switch, thyristor.

four-level laser A laser identical to the three-level laser, except for the addition of one excited state.

four-phase system A two-phase system in which the center taps of the coils are interconnected. Also called *quarter-phase system.*

four-terminal network A network having two input terminals and two output terminals. One input terminal may be internally connected to one output terminal (as when a common ground is present), but this is not mandatory.

fourth dimension Time used as a dimension with the three spatial dimensions (height, width, length) to locate point. For example, it's not enough to say that an airplane flew at a certain altitude over a given city; it must be noted *when* it was there.

four-track recording A tape recording in which four channels are recorded in two adjacent tracks on the tape, tracks one and three usually in the forward direction, two and four in the reverse direction.

four-track tape A magnetic tape with four parallel sound paths.

four-wire wye system A three-phase system in which three wires supply the respective phases, a fourth being the neutral conductor.

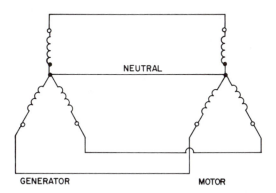

NEUTRAL

GENERATOR   MOTOR

Four-wire wye system.

Fp Symbol for *power-loss factor.*

fp Abbreviation of *freezing point.*

FPC Federal Power Commission.

FPI Abbreviation of *fuel-pressure indicator.*

FPIS Abbreviation of *forward propagation by ionospheric scatter* (see *forward propagation by tropospheric scatter*).

fpm Abbreviation of *feet per minute.*

fps 1. Abbreviation of *feet per second.* 2. Abbreviation of *frames per second.* 3. Abbreviation of *foot-pound-second,* a chiefly British system of units.

fps system of units The British system of units of measurement that uses the foot for length, pound for mass, and second for time. Compare *centimeter-gram-second, International System of Units.*

FPTS Abbreviation of *forward propagation by tropospheric scatter.*

Fr Symbol for *francium.*

fr Abbreviation of *franline.*

fractional exponent An exponent (number or letter) indicating that a number is to be raised to a fractional power, e.g., $10^{1/3}$. For any fractional exponent of a number n, $n^{a/b}$ equals $\sqrt[b]{n^a}$. The numerator of the exponent indicates the power to which n must be raised and the denominator indicates the root that must be taken of that power of n.

fractional gain Amplification less than 1. A notable example is the transfer function of a cathode follower or emitter follower.

fractional horsepower Any power rating lower than 1 hp. Also see *horsepower.*

fractional uncertainty See *relative uncertainty.*

frame 1. A single, complete television picture, scanned in 1/30 second in conventional receivers. 2. The row of bits across the width of magnetic or paper tape. 3. One of a recurring cycle of pulses. 4. In PCM, a cyclic word group including a sync signal. 5. In PSM and PDM, a complete commutator cycle.

frame frequency The number of frames of a motion-picture film that come into position per unit of time in a camera, projector, or pickup.

frame grid In an electron tube, a rectangular grid with close wires. The construction permits closer spacing of elements than is possible with a concentric grid, and results in shorter transit time and improved plate-current control.

frame of reference Geometric relationships used to describe the location of a body in space.

frame rate See *frame frequency.*

frame-repetition rate See *frame frequency.*

frame roll Momentary vertical roll in a TV picture.

frame synchronizing signal A coded PAM pulse indicating initiation of a commutation frame, or a PCM signal used to identify an information frame.

framing 1. Synchronization of the vertical component of a video signal, such that the top and bottom of the transmitted and received pictures line up. 2. The process of lining up the top and bottom of a movie picture. 3. Alignment of the characters in a digital alphanumeric transmission.

francium Symbol, Fr. A radioactive metal element of the alkali-metal group it is produced artificially through radioactive disintegration. Atomic number, 87. Atomic weight, 223.

Frank Phonetic alphabet code word for the letter *F.*

Franklin antenna A vertical colinear array that produces omnidirectional gain because of phasing among the individual components.

frankline (Benjamin *Franklin,* 1706-1790). Abbreviation, fr. A name that has been suggested for the unit of electric charge; 1 fr is the charge that exerts a force of 1 dyne on an equal charge at a distance of 1 centimeter in a vacuum.

Franklin oscillator A dual-terminal af-rf oscillator circuit consisting of a two-stage, *RC* coupled vacuum-tube amplifier with a tuned *LC* tank in the first (input) grid circuit, and with capacitive feedback from the second plate to the tank.

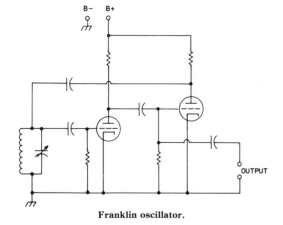

Franklin oscillator.

Fraunhofer region The area surrounding a radiating antenna, throughout which the energy appears to come from a single point located near the actual antenna.

free carrier A free electron or, in a semiconductor material, the equivalent hole. Also see *carrier, 2*; *electron*; and *hole*.

free charge The portion of a charge on a conductor which, being unaffected by a neighboring charge, will escape to ground when the conductor is grounded. Compare *bound charge*.

free electron An electron situated in one of the outer orbits of an atom, which is held only loosely within the atom. Because free electrons can escape the attraction of the atomic nucleus, they can drift between atoms, in that way forming the basis of electric current. Also see *electron* and *bound electron*.

free field Data organized in a storage medium in such a way that a data item or field can be anywhere in the medium. Compare *fixed field*.

free-floating grid See *floating grid*.

free grid See *floating grid*.

free impedance For a transducer, the input impedance produced by a perfectly short-circuited load.

free magnetic pole A magnetic pole that is so well isolated from its opposing pole that it experiences little or no influence from the latter.

free net A communications network in which stations are free to communicate with other stations in the net, i.e., without direction from the control station.

free oscillations Oscillations in a circuit, such as an *LC* tank, that continue after ac excitation has been removed. Also see *flywheel effect*. Compare *forced oscillations*.

free path In a gas tube, the path taken by an electron as it collides with atoms. Also see *mean free path*.

free-power supply 1. A simple tuned-rf detector diode used to rectify a radio signal and supply small amounts of direct current for the operation of low-powered transistor circuits. 2. Solar battery.

free reel The supply reel of a magnetic-tape recorder.

free-running frequency The frequency at which a synchronized generator, such as a multivibrator or self-excited oscillator, will operate when the synchronizing voltage is removed.

free-running multivibrator See *astable multivibrator* and *uncontrolled multivibrator*.

free space Empty space; a theoretical ideal.

free-space loss Radio transmission loss disregarding variable factors (a theoretical condition).

free-space pattern The ideal directivity pattern of an antenna that is situated a number of wavelengths above ground, which in practice is modified by reflections from ground.

free speed The angular velocity of an unloaded motor.

free-standing display In a computer system, a remote display unit for prompting peripheral operators.

freezing point Abbreviation, fp. The temperature at which a liquid starts becoming a solid at normal pressure. Compare *melting point*.

F-region The second lowest of the principal regions of the ionosphere, with an altitude at night of approximately 175 miles. In daytime, the region is divided into the lower F1 region and higher F2 region. Also called *F-layer*.

Fremodyne detector An FM detector which is essentially a conventional AM circuit detuned to one side of resonance (slope-tuned) to demodulate a frequency-modulated signal. Also see *slope detector*.

French phone See *cradlephone*.

freqmeter Contraction of *frequency meter*.

frequency Symbol, *f*. The rate at which a phenomenon is repeated. The basic unit of frequency is the hertz (Hz), which is 1 cycle per second.

frequency-agile radar A radar system in which the transmitter frequency is shifted in a predetermined pattern for the purpose of avoiding detection. A frequency-agile radar system, with a complex frequency control program, is very difficult to jam.

frequency allocation 1. The assignment of frequencies to radio and allied services by the licensing authority (in the United States, the Federal Communications Commission). 2. A frequency assignment. Also see *radio spectrum*.

frequency band A given range of frequencies, usually specified for some application, e.g., the band allocated for standard radio broadcast service. Also see *band*.

frequency bias An intentional change in the frequency of a transmitted signal.

frequency bridge 1. Any ac bridge, such as the Wien bridge or resonance bridge, which can be nulled at only one frequency for a given set of bridge-arm values. 2. A bridge (such as in 1, above) used to measure unknown frequencies.

frequency calibrator A device, such as a crystal oscillator, which provides a signal of precise frequency with which other signals may be compared. Also see *secondary frequency standard*.

frequency changer 1. A superheterodyne converter (see *converter, 1*). 2. A motor-generator in which the output voltage has the same value as the input voltage, but is of a different frequency. 3. Frequency-multiplying transformer. 4. Frequency multiplier.

frequency-change signaling See *frequency-shift keying*.

frequency channel A relatively narrow segment of a frequency band allocated to a station in a particular service.

frequency comparator A device, such as an oscilloscope or zero-beat indicator, used to check one frequency against another. Also see *frequency comparison*.

frequency comparison The observation of a current or voltage of one frequency for similarities in that of another frequency. Comparisons (as in frequency matching) may be made visually or aurally. Common instruments used are oscilloscopes, beat-note detectors, and beat-note meters.

frequency-compensated attenuator An attenuator, such as one in an electronic voltmeter or wideband oscilloscope, which has been modified by the addition of capacitors or inductors to achieve reasonably flat response over a wide range of frequencies.

frequency compensation The modification of a circuit, such as an amplifier or attenuator, by the addition of capacitors or inductors to tailor its response at specified frequencies.

frequency control 1. An adjustable component (potentiometer, variable capacitor, variable inductor) with which the frequency or frequency response of a circuit is controlled. 2. A device, such as a quartz crystal or tuning fork, that automatically sets the frequency of an oscillator.

frequency converter 1. An active or passive device for changing the frequency of a signal. 2. The mixer in a superheterodyne circuit.

frequency correction Manual or automatic resetting of a deviated frequency to its original value.

frequency counter An instrument that counts signal cycles or pulses over a standard time base (a frequency measurement).

frequency cutoff See *cutoff frequency, 1, 2*.

frequency detector See *Fremodyne detector*.

frequency deviation 1. The degree to which a frequency changes from a prescribed value. Thus, if the frequency of a 1 kHz oscillator drifts between 990 and 1010 Hz, the deviation is ± 10 Hz. 2. In an FM signal, the amount of frequency shift above and below the unmodulated carrier frequency.

frequency-deviation meter A meter that gives a direct reading of frequency deviation resulting from modulation amplitude. It employs either a tuned circuit or a frequency comparator.

frequency difference 1. In a superheterodyne circuit, the difference between the signal frequency and the oscillator frequency. 2. In any beat-frequency operation, the quantity $f2 - f1$, where $f2$ is the higher frequency and $f1$, the lower. Compare *frequency sum*.

frequency discriminator See *discriminator*.

frequency distortion A form of distortion in which the amplification of some frequencies is different from that of others.

frequency distribution See *distribution, 2*.

frequency diversity The transmission and reception of signals at two or more frequencies for the purpose of reducing the effects of fading. Generally employed in long-distance, high-frequency circuits.

frequency divider A circuit or device whose output frequency is a fraction of the input frequency. Compare *frequency multiplier*.

frequency-dividing network See *crossover network*.

frequency-division multiplex A form of multiple-signal parallel transmission in which a single carrier is modulated by two or more signals simultaneously.

frequency doubler A circuit that multiplies an input frequency by two. A doubler's input circuit is usually tuned to frequency f and its output circuit to $2f$. Frequency doubling is performed by various nonlinear devices, including tubes, transistors, varactors, and biased semiconductor diodes.

frequency drift Undesired, usually gradual, of a signal from its intended frequency or channel, expressed in hertz.

frequency indicator A device that indicates when a phase or frequency is common to two alternating currents.

frequency keying See *frequency-shift keying*.

frequency meter An instrument for measuring ac frequency. Also see *audio frequency meter, power-frequency meter,* and *wavemeter*.

frequency-modulated radar See *FM radar*.

frequency modulation Abbreviation, *FM*. A method of modulation in which the frequency of the carrier voltage is varied with the frequency of the modulating voltage, the amount of variation determined by the amplitude of the modulating signal. Compare *amplitude modulation*.

frequency modulation deviation 1. The largest value for a carrier frequency minus instantaneous modulation frequency. 2. The maximum bandwidth of an FM signal at its audio modulation amplitude peak.

frequency modulator 1. A circuit or device, such as a reactance tube or the equivalent transistor, which will modulate the frequency of an oscillator. Also called *wobulator*. 2. The modulator section of an FM transmitter.

frequency monitor A device employed (often continuously) to check the frequency of a signal; e.g., a frequency-deviation meter used in radio broadcast stations, or a frequency meter used in electric generating stations.

frequency multiplier A circuit or device whose output frequency is a multiple of the input frequency. See, for example, *frequency doubler; quadrupler, 2; quintupler, 2;* and *tripler, 2*.

frequency-multiplying amplifier See *multiplier amplifier*.

frequency-multiplying transformer A magnetic amplifier which, due to the nonlinearity of its core material, generates harmonics of the supply frequency.

frequency offset The difference between an actual frequency and the desired frequency.

frequency overlap 1. A common band of frequencies between two adjacent channels in a communications system. 2. A common frequency region between two assigned bands. 3. A condition in which parts of the sidebands of two signals occupy the same range of frequencies.

frequency pulling A change in the frequency of a circuit (especially of a self-excited oscillator) because of the detuning effects of an external circuit, device, or condition (such as temperature).

frequency pushing An effect in which a current change in a source oscillator causes a shift in source frequency.

frequency quadrupler See *quadrupler, 2*.

frequency quintupler See *quintupler, 2*.

frequency range 1. A system's frequency transmission limits beyond which power is attenuated below a specified tolerance. 2. The frequency band or bands within which a receiver is designed to operate.

frequency ratio counter See *frequency ratio meter*.

frequency ratio meter A meter that indicates the ratio between two frequencies and is particularly useful in the quick identification of harmonics.

frequency record A phonograph test disk containing recordings of various frequencies at specified amplitudes.

frequency rejection The elimination of a single frequency (or narrow band of frequencies) from a mixture of frequencies transmitted by a filter or other circuit. Compare *frequency transmission*.

frequency relay A frequency-sensitive relay (see *selective relay, 1*).

frequency response A performance characteristic that describes the operation of a device or circuit over a specified range of signal frequencies, e.g., the gain—frequency characteristic of an amplifier.

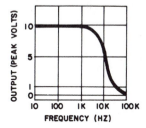

Frequency response.

frequency-response recorder A graphic recorder that automatically plots a frequency-response curve for a device under test.

frequency run A test, or test sequence, that determines the loss characteristics of a circuit as a function of the operating frequency.

frequency scanning 1. A controlled fluctuation of the transmitter frequency in a frequency-agile radar or communications system. 2. In a programmable, digital communications receiver or transceiver, a form of simultaneous digital monitoring of two or more channels. 3. The frequency-response change in a spectrum analyzer.

frequency scaler See *scaler*.

frequency-selection sensor A sensor that passes or rejects phenomena at certain frequencies while ignoring those at others.

frequency-selective relay See *selective relay, 1*.

frequency-sensitive bridge A bridge, such as the Wien bridge or resonance bridge, which can be balanced at only one frequency for a given set of bridge-arm values.

frequency separator In a TV set, the circuit that separates horizontal- and vertical-scanning sync pulses.

frequency-shift keying Abbreviation, FSK. Keying a transmitter for telegraph or Teletype communications (radio or wire) by shifting the carrier frequency over a range of a few hundred hertz.

frequency-shift radar See *Doppler radar*.

frequency-shift ranging See *doran*.

frequency span The difference $f2 - f1$, where $f1$ is the lowest frequency in a given range of frequencies and $f2$ is the highest frequency. Compare *frequency spread*.

frequency spectrum All electromagnetic radiation, from longest to shortest wavelengths, within a set of specified limits.

frequency spotting The setting up of spot frequencies (usually harmonics of a standard-frequency oscillator) and their use in identifying unknown frequencies. Also see *frequency calibrator*.

frequency spread The ratio $f2/f1$ where $f1$ is the lowest frequency in a given range of frequencies and $f2$ is the highest frequency. Compare *frequency span*.

frequency stability The degree to which a frequency remains constant during variations in temperature, current, voltage, and similar factors. It is specified in frequency units hertz (and multiples thereof) or in parts per million per unit of the variable parameter.

frequency standard A signal source of a precise frequency, against which other signal sources may be calibrated. See specifically *primary frequency standard* and *secondary frequency standard*.

frequency sum In a beat-frequency system, the quantity $f1 + f2$, where $f1$ is the lower frequency and $f2$ is the higher frequency. Compare *frequency difference, 2*.

frequency swing See *frequency deviation, 1, 2*.

frequency synthesizer A generator of highly accurate signals, often indiscrete frequency steps, for test purposes. The signals are often derived from a single-frequency source, such as a crystal oscillator. Also see *signal synthesizer*.

frequency tolerance The acceptable amount by which a frequency may vary from its intended value. The tolerance may be specified as $\pm$ a percent of the stated frequency, $\pm$ so many parts per million; or $\pm$ a number of frequency units (hertz or fractions thereof), for example, 1 MHz $\pm$ 10 Hz.

frequency-to-voltage converter A device or circuit that delivers an output voltage (usually dc) which is proportional to input frequency.

frequency translation 1. The conversion of a given frequency band from one part of the electromagnetic spectrum to another, without changing the actual separation of channels or the overall width of the band. 2. Frequency conversion.

frequency transmission The passage of a frequency or band of frequencies from a mixture of frequencies transmitted by a filter or other circuit. Compare *frequency rejection*.

frequency tripler See *tripler, 2*.

frequency-variation method A method of determining the Q of a tuned circuit by varying the frequency of the applied test voltage from resonance (fr) to a high point ($f2$) and a low point ($f1$), at which the circuit voltage is 0.707 the voltage at resonance. The figure of merit then is calculated; $Q = fr/(f2 - f1)$.

frequency-voltage converter See *frequency-to-voltage converter*.

frequency-wavelength conversion See *wavelength-period-frequency relationships*.

fresnel (A.J. *Fresnel.*, 1788-1827). A unit of frequency equal to 10^{12} Hz.

fresnel lens A usually square, rather flat plastic lens with progressively thicker concentric areas; its effect is similar to that of an automotive headlight lens.

Fresnel number A measure of the relative effects of diffraction in an optical lens. The Fresnel number is equal to the radius of the lens divided by the product of the light wavelength and the lens focal length, all measured in the same units.

Fresnel region For a radio-frequency transmitting antenna, the zone between the antenna and the Fraunhofer region (see *Fraunhofer region*). The size of the Fresnel region depends on the wavelength of the radiated energy.

friction The resistance to mechanical motion when one material is rubbed against another. Friction was one of the earliest sources of man-made electricity (see *frictional electricity* and *electric machine*). Electrical resistance, opposing the flow of current, is analogous to friction.

frictional electricity Static electricity generated by rubbing one material with another.

frictional electric machine See *electric machine*.

frictional error The change in parameters of a phonograph pickup, resulting from friction with the disk surface.

frictional loss A decrease in energy's efficiency in doing work, caused by friction between moving parts.

fringe area The region in which a signal falls to the minimum field strength necessary for satisfactory communication.

fringe howl In a regenerative detector, a howl that appears when the tube or transistor just begins to oscillate, obscuring the signal. The term is used because the circuit is operated at the fringe of oscillation.

fringing See *edge effect*.

Fritch Trade name (American Telephone & Telegraph Co.) for *frequency-selective switch*.

fritting A condition in which electrical contact corrosion creates a small hole through which molten contact material passes to form a conductive bridge.

front contact The movable contact of a relay.

front end The converter portion of a superheterodyne receiver, i.e., the rf amplifier, first detector, and local oscillator. Compare *rear end*.

front porch In a TV horizontal sync pulse, the interval between the end of the sync pulse and the fall of the blanking pedestal. Compare *back porch*.

front-surface mirror A mirror that has its reflective material on the front instead of on the back.

front-to-back ratio 1. For a semiconductor junction, the ratio of forward (positive) current to reverse (negative) current for the same value of voltage. Also called *forward-reverse ratio*. 2. In a directional antenna, the ratio of forward signal strength to back signal strength.

frost alarm A device or circuit that responds to the presence of frost and actuates an alarm. Such alarms are sensitive to temperature, moisture, or both.

FRUGAL Acronym for *fortran rules used as a general applications language*.

FRUSA Abbreviation of *flexible rolled-up solar array* (a "window shade" solar cell array for spacecraft).

F-scan In radar practice, a display in which a central blip represents the target at which the antenna is pointed; horizontal and vertical displacement of the blip indicate corresponding horizontal and vertical aiming errors.

FSK Abbreviation of *frequency-shift keying*.

FSM Abbreviation of *field-strength meter.*

FS meter See *field-strength meter.*

FSR Abbreviation of *feedback shift register.*

ft Abbreviation of *foot, feet.*

FT-cut crystal A piezoelectric plate cut from a quartz crystal at an angle +57° with respect to the Z-axis. Also see *crystal axes* and *crystal cuts.*

ft-Lb Abbreviation of *foot-lambert.*

ft-lb Abbreviation of *foot-pound.*

Fuchs antenna A simple antenna consisting of a single-wire radiator without feeder or transmission line, connected directly to the transmitter. Its disadvantage is that part of its radiated field is often inside the transmitter building.

fuel alarm A sensing circuit which actuates an alarm when the fuel in a tank or reservoir falls to a prescribed level.

fuel cell A generator that produces electricity directly from the reaction between fuel substances, such as hydrogen and oxygen.

fuel-flow alarm An electronic circuit which actuates an alarm when fuel flow changes from a prescribed value.

fuel-flow control A servo system that automatically maintains or corrects the flow rate of a fuel.

fuel-flow gauge See *fuel-flow meter.*

fuel-flow indicator See *fuel-flow meter.*

fuel-flow meter An instrument for measuring fuel flow rate.

fuel-flow switch A switch that is actuated by fuel flowing in pipes or other channels.

fuel gauge An instrument consisting of a transducer that senses the level of liquid fuel in a tank and delivers a proportional output current or voltage, and an electric meter whose needle is deflected in proportion to the current or voltage and, therefore, to the fuel level.

fuel meter See *fuel gauge.*

fuel-pressure indicator An instrument for measuring fuel pressure in pipes or other channels.

fuel-pressure meter See *fuel-pressure indicator.*

full adder In a digital computer, an adder circuit that can handle the carry signal as well as the binary elements that are to be added. Also see *adder* and *carry.* Compare *half adder.*

full bridge A bridge-rectifier circuit in which each of the four arms contains a diode. By comparison, the three-quarter bridge contains a resistor in one arm; the half bridge, resistors in two arms; and the quarter bridge, resistors in three arms.

full-duplex system In data communications, a system which transmits data in both directions simultaneously. Compare *half-duplex system.*

full-focus yoke See *cosine yoke.*

fullhouse A multichannel radio-control model plane system that allows the use of a realistic complement of working control surfaces.

full-load current The output current from a course when the load is maximum.

full-load power The power drawn from a source when the load is maximum.

full-load voltage The output voltage of a course when full power is drawn, i.e., when the loading is maximum.

full-load wattage See *full-load power.*

full-range speaker See *monorange speaker.*

full scale 1. The operating range of an instrument. 2. Transducer output as a function of highest allowable input stimulus.

full-scale error For an electrical indicating instrument, the rated full-scale input signal minus the actual input signal that causes a full-scale deflection. Thus, the predictable error in an instrument expressed as a percentage of the full-scale reading.

full-scale sensitivity The current, voltage, or power required to deflect a meter mechanism to full scale.

full track A recording track covering the full width of a magnetic tape.

full-track head A tape-recorder head having a gap that covers the full width of the tape.

full-track recording Usually applicable to quarter-inch or narrower magnetic recording tape, a one-track recording made by a head that magnetizes essentially the width of the tape.

full-wave bridge rectifier See *bridge rectifier.*

full-wave, center-tap rectifier A circuit in which the center-tapped secondary winding of a transformer operates two rectifier diodes, each on an alternate half-cycle of secondary voltage. The ripple in the dc output voltage is equal to twice the supply frequency. Compare *bridge rectifier.*

full-wave detector A detector circuit employing two diodes in a full-wave, center-tap rectifier circuit.

full-wave doubler See *full-wave voltage doubler.*

full-wave rectifier A rectifier that delivers a half-cycle of output voltage for each half-cycle of applied ac voltage. The successive output half-cycles have the same polarity. See, specifically, *bridge rectifier* and *full-wave, center-tap rectifier.* Compare *half-wave rectifier.*

full-wave vibrator In a vibrator-type power supply an interrupter that closes contacts on both ends of its swing, thus directing dc through the transformer in alternate directions. Also, a vibrator-type rectifier that closes in both directions.

full-wave voltage doubler A voltage-doubler circuit whose dc output has a ripple of twice the ac supply frequency. Compare *half-wave voltage doubler.*

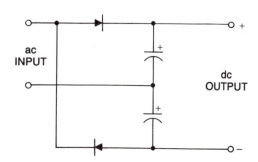

Full-wave voltage doubler.

function 1. A variable (*x*) that is so related to a second variable (*y*) that the value of the second may always be determined in terms of the first. Here, *x* is said to be a function of *y*; this relationship is written *x* = *f(x).* 2. The mathematical expression, using symbols, relating variables; e.g., the expression *x* − *y* = *z* is a function of variables *x, y,* and *z.* 3. The behavior and application for which a device or system is designed. 4. Part of a computer instruction specifying the operation to be done.

functional blocks Combinations of substances or components, which perform specific functions in an electronic circuit. An example of a functional block is a tuned circuit, containing inductive reactance, capacitive reactance, and resistance.

functional character See *control character.*

functional design Design specifications encompassing a description of how system elements will interrelate and what their logic design will be.

functional diagram Functional design represented diagrammatically. Also see *functional design*.

functional electronic block Abbreviation, FEB. A complete integrated circuit.

functional test A performance test of a device or circuit, to see that it behaves as intended in the environment in which it is to be employed.

function generator 1. A signal generator whose output is any of several selectable waveforms (e.g., sine, square, triangular, step-pulse) and frequencies (or repetition rates). 2. An analog computer circuit that produces a variable based on a mathematical function and one or more input variables.

function hole See *control hole*.

function key A keyboard key used to control the form in which a message will be received.

function polling A polling technique in which a disabled device signals its condition and specifies the remedy.

function switch In a multifunction instrument, such as a voltohm-milliammeter, the switch that permits selection of the various functions.

function table 1. A table of mathematical function values. 2. Hardware or software that translates one representation of information into another. 3. A routine that allows a computer to use the values of independent variables to determine the value of a dependent variable.

fundamental Contraction of *fundamental frequency*.

fundamental component The fundamental frequency of a complex wave.

fundamental group A set of trunk lines, in a telephone system, through which zone centers are interconnected.

fundamental frequency The lowest frequency in a complex wave.

fundamental mode See *dominant mode*.

fundamental suppression Removal of the fundamental frequency from a complex wave, leaving only the harmonics, as in the operation of a null network adjusted to the fundamental frequency.

fundamental units Base units of an absolute system of units.

fundamental wavelength The wavelength that is equal to the fundamental frequency.

fuse A safety device consisting of a wire of low-melting-point metal. When current passing through the wire exceeds a prescribed (safe) level, the resulting heat melts the wire and opens the circuit, protecting equipment from damage. See *proximity fuse*.

fuse box A set of electrical fuses, usually enclosed in a metal box.

fused junction In a semiconductor, a junction produced by alloying metals to the semiconductor material.

fused junction See *alloy junction*.

fuse resistor See *fusible resistor*.

fuse wire The low-melting-point wire used in fuses. See *fuse*.

fusible resistor A low-value resistor which also serves as a fuse in certain appliances, such as TV receivers.

fusing current The specified current level at which a wire of given diameter and metal will melt.

fusion In a nuclear reaction, the uniting of two atomic nuclei, accompanied by tremendous heat. Also see *nuclear fusion*. Compare *fission*.

future labels Program instruction labels that refer to locations which have not been allocated absolute addresses by a compiler or assembler (both, computer programs).

fuzz A form of deliberate distortion of an electric guitar's tone by use of an electronic device called a fuzzbox.

fuzz buster Slang for a radio-type device employed by drivers of vehicles to signal the presence of a *radar speed trap*.

fV Abbreviation of *femtovolt*.

G 1. Symbol for *conductance*. 2. Abbreviation of *giga*. 3. Symbol for *deflection factor*. 4. Symbol for *perveance*. 5. Symbol for *gravitational constant*. 6. Symbol for *generator* (see *generator, 1*). 7. Symbol for *gate* (see *gate, 2*).

g 1. Symbol for *gravity* (see *gravity, 2*). 2. Symbol for *conductance*. 3. Abbreviation of *gram*. 4. Subscript for *gate* (see *gate, 2*). 5. Subscript for *generator* (see *generator, 1*). 6. Unit of acceleration force equal to earth's gravitational force on a mass at rest.

GA Radiotelegraph abbreviation of *go ahead*.

G/A Abbreviation of *ground-to-air*.

Ga Symbol for *gallium*.

GaAs 1. Formula for *gallium arsenide*. 2. Integrated circuit devices based on GaAs rather than silicon.

GA coil A special form of coil, wound with extra space among the turns, to reduce the distributed capacitance.

gadget Colloquialism for any device or component. Also used pejoratively to denote a superfluous or makeshift device.

gadolinium Symbol, Gd. A metallic element of the rare-earth group. Atomic number, 64. Atomic weight, 157.26.

gage See gauge.

gain See *amplification, 2; current amplification, 2; voltage amplification, 2;* and *power amplification, 2*.

gain—bandwidth product Symbol, fT. The product of an amplifier's gain and bandwidth.

gain control 1. To adjust the gain of an amplifier. 2. A potentiometer used to adjust amplifier gain.

gain function A function between two currents or voltages in a circuit with gain.

gain reduction The drop in gain of an amplifier at high- and low-frequency extremes.

gain sensitivity control See *differential gain control*.

gain stability The degree to which the gain of a system remains constant during changes in related factors, such as temperature, supply power, loading, and the like.

galactic noise Radio noise propagated from the center of our galaxy.

galena Formula, PbS. Natural lead sulfide, which in nature takes the form of bluish-gray, cubical crystals.

galena detector A widely used, sensitive, early crystal-diode (galena) detector employing a fine wire (cat's whisker) as a tuning element.

gallium Symbol, Ga. Atomic number, 31. Atomic weight, 69.72. Gallium is one of the constituents of the semiconductor compound gallium arsenide.

gallium arsenide Formula, GaAs. A compound of gallium and arsenic used as a semiconductor material.

gallium-arsenide diode A diode in which the semiconductor material is processed *gallium arsenide*.

gallium-arsenide varactor A low-noise, microwave varactor in which the semiconductor material is gallium arsenide.

gallium-phosphide diode A light-emitting diode in which the semiconductor material is processed gallium phosphide.

galloping ghost A form of radio-control system in which the elevation and rudder can be moved to the desired extent.

galvanic cell Generic term for any electrochemical primary voltaic cell.

galvanic corrosion Corrosion of one of two dissimilar metals immersed in an electrolyte (e.g., the sea), caused by a battery action between them. Compare *electrolytic corrosion*.

galvanic couple See *voltaic couple*.

galvanic current A very small direct current such as that produced by dissimilar metals in acid or by nervous reaction in living tissue.

galvanic pile See *voltaic pile*.

galvanic series A list of metals and alloys arranged in order of the most to least likely to oxidize in a given environment.

galvanic skin response Abbreviation, GSR. The variations in electrical resistance of the (usually human) skin. This phenomenon is a useful indicator in physiology, psychology, and criminology.

galvanic taste A sharp, metallic taste experienced when a small electric current is passed through the tip of the tongue.

galvanism (After Luigi *Galvani*, 1737-1798). The production of an electric current by chemical action, as in a battery.

galvanize To coat steel with zinc to forestall corrosion.

galvanometer A sensitive current meter (but sometimes, voltmeter) used in various electrical tests and specifically as a null indicator in bridge operation. Also see *microammeter*.

galvanometer constant The number by which a galvanometer reading must be multiplied in order to obtain the current in microamperes, milliamperes, or amperes.

galvanometer modulator See *electromechanical modulator, 3*.

galvanometer recorder A graphic recorder in which a mirror in a movable-coil galvanometer reflects a beam of light to a passing strip of photographic film.

galvanometer sensitivity See *sensitivity, 3*.

galvanometer shunt A resistor placed in parallel with the input to a galvanometer to increase deflection. Also see *shunt resistor*.

galvanometry The use of galvanometers to determine the intensity and direction of electric currents.

galvanoplastics The art of electroplating.

galvanoscope An instrument for detecting and showing the direction of very weak electric currents.

galvanotherapy The use of electric currents to produce heat in the body of a human or animal.

game theory A mathematical process of picking the best strategy when confronted by an opponent who has his own strategy.

gamma The third letter of the Greek alphabet. 1. A unit of magnetic flux density (the SI unit, tesla, is preferred). 2. Capital gamma (Γ) is the symbol for *complex Hertzian vector, gamma function,* and *reciprocal inductance*. 3. Lower-case gamma (γ) is the symbol for *angle, electrical conductivity, Euler constant, gamma ray, pressure coefficient, propagation constant, proton gyromagnetic ratio,* and *surface tension*. 4. A number expressing the degree of contrast in a TV picture. 5. A number expressing the degree of contrast in a photographic print. 6. A photomicrograph (photograph taken through a microscope). 7. Resembling an upper-case gamma in physical appearance, as in *gamma section*.

gamma ferric oxide A form of coating used in formulation of magnetic recording tape.

gamma function Symbol, Γ. For any positive integer n, the factorial of n-1. Also see *factorial*.

gamma match A linear transformer for matching a coaxial feeder to a half-wave antenna. The outer conductor of the cable is connected to the center of the radiator, and an extension of the center conductor runs for a short distance parallel to the radiator, making a right-angle bend before connecting to the radiator. In this way, a short section of impedance-matching transmission line is formed by the extension of the center conductor and the portion of the radiator that is parallel to it. This arrangement takes its name from its resemblance to a capital gamma (Greek letter) lying on its side.

gamma radiation See *gamma rays*.

gamma rays Rays emitted by radioactive substances; they are similar to X-rays, but are of a shorter wavelength. Compare *alpha particle* and *beta rays*.

gamma section That portion of a gamma matching device that resembles the Greek letter *gamma; gamma match*.

gang To mechanically couple components (pots, switches, etc.) for operation by a single knob.

gang capacitor A variable capacitor consisting of sections mounted on the same shaft for simultaneous variation. Usually specified by the number of sections (four-gang capacitor, e.g.). Compare *ganged capacitors*.

ganged capacitors Separate variable capacitors mechanically connected together (e.g., by belt or gear drive) for simultaneous variation. Compare *gang capacitor*.

ganged potentiometers Separate potentiometers mechanically connected together (e.g., by belt or gear drive) for simultaneous variation. Compare *gang potentiometer*.

ganged rheostats See *ganged potentiometers*.

ganged switches Separate switches mechanically connected together for simultaneous operation. Compare *multiswitch*.

ganged tuning Simultaneous tuning of separate circuits by means of ganged capacitors or ganged potentiometers.

gang potentiometer A potentiometer consisting of sections mounted on the same shaft for simultaneous variation. Usually specified as dual potentiometer, two-section potentiometer, and so on.

gang printer In digital computer and data processing practice, an electromechanical printer capable of printing an entire line at one time.

gang punch 1. To punch identical or nonvarying information into the cards of a group. 2. A machine for this operation.

gang rheostat See *gang potentiometer*.

gang switch See *multiswitch*.

Gantt chart A chart of activity vs time used in industry as an aid in making decisions regarding the allocation of resources for specific activities, e.g., as applied to PERT (project evaluation and review techniques).

gap 1. A space between electrodes. 2. A device consisting essentially of separated electrodes providing a gap, e.g., *spark gap*. 3. A relatively narrow space cut in iron cores to provide a break in a magnetic circuit. Also see *slot, 1*. 4. The opening between the opposite poles of a tape recorder or playback head.

gap arrester A lightning arrester consisting of a number of metal cylinders separated by air gaps.

gap coding A system in which silent periods are inserted, according to a specific timing code, into a transmission.

gap converter See *spark-gap oscillator*.

gap depth In a magnetic recording head, the depth of the gap (taken perpendicular to the face). Compare *gap width*.

gap digit A digit that contributes no intelligence to the word in which it appears, e.g., a parity bit.

gap energy The energy represented by the forbidden gap between the *M*-valence band and the *N*-conduction band in a material, for example.

gap filling Modification of an antenna, for the purpose of eliminating nulls in the directional pattern.

gap insulation See *slot insulation, 1, 2*.

gap loss In a reproducing head, the loss that takes place because of the gap length.

gap oscillator See *spark-gap oscillator*.

gap-type protector A spark gap employed to protect equipment from high-voltage transients.

gap voltmeter See *needle gap* and *sphere gap*.

gap width In a magnetic recording head, the width of the gap (taken parallel with the face). Compare *gap depth.*

garbage 1. In digital computer practice, a colloquialism for useless or incorrect data. 2. Colloquialism for an unsound theory.

garble 1. Garbled matter. 2. To render communications or data unintelligible.

garbled matter Confused communications or data, usually resulting from distortion in a circuit or system. Also called *garble.*

garbler See *scrambler circuit.*

garnet maser A maser employing natural or synthetic garnet as the stimulated material. Also see *yttrium-iron-garnet* and *YIG.*

gas One of the states of matter, a gas is characterized by its widely separated molecules, which are in continual motion and can, because it is a fluid, conform to a container of any shape. Gases may readily be compressed and liquefied. Compare *liquid, plasma,* and *solid.*

gas amplification In a radiation-counting device, the ratio, in decibels, of the charge collected to the charge produced in the gas.

gas cell A cell whose operation is dependent on gas absorption by the electrodes.

gas cleanup Loss of pressure in a gas-filled tube, which will eventually lead to failure; it is due to gas ions forming compounds with metal parts or with the glass envelope.

gas current In a vacuum tube, grid current due to the presence of gas.

gas detector A device for sensing presence of various gases in the air.

gas diode A tube diode which is partially filled with a gas that ionizes during tube operation and enhances anode (plate) current. Also see *diode tube.*

gas-discharge tube See *gas tube.*

gaseous conduction The conduction of an electric current through an ionized gas, as in a gas tube.

gaseous phototube A phototube containing a small amount of a gas suitable for ionic conduction. Also see *phototube.*

gaseous regulator See *gaseous voltage regulator.*

gaseous tube See *gas tube.*

gaseous voltage regulator A gas-filled diode across which the voltage drop is substantially constant during the gas discharge and which accordingly delivers a constant output voltage. Also called *VR tube.*

gas-filled cable A cable, such as a coaxial line, that is filled with a gas, such as nitrogen, which serves as a dielectric and moisture barrier.

gas-filled counter tube A radiation counter tube containing a gas that ionizes when irradiated. See *Geiger-Mueller tube.*

gas-filled lamp 1. An incandescent lamp filled with a gas, such as nitrogen, for improved performance. 2. Discharge lamp.

gas-filled rectifier A tube diode containing gas at low pressure for increased anode current by ionic conduction.

gas-flow alarm An electronic circuit which actuates an alarm when the flow of gas through a pipe changes from a predetermined rate.

gas-flow control A servo system for automatically maintaining or adjusting the flow of gas through pipes.

gas-flow gauge See *gas-flow meter.*

gas-flow indicator See *gas-flow meter.*

gas-flow meter An instrument which indicates the rate of gas flow through a pipe.

gas-flow switch In a gas-circulating system, a switch that actuates an alarm when the gas flow rate changes.

gas focusing In a cathode-ray tube, a technique by which a gas is used for the purpose of focusing an electron beam. The ionization of the gas causes the electron beam to be made more narrow.

GASH An organic crystalline material used as the dielectric in certain ferroelectric capacitors and ferroelectric memory elements. GASH is an acronym of the chemical name of the substance: *g*uanidine *a*luminum *s*ulfate *h*exahydrate. Also see *ferroelectricity.*

gas laser A laser that employs a gas or mixture of gases (instead of a solid rod) as the stimulated medium. Some of the gases used are argon, carbon dioxide, helium, krypton, and neon.

gas maser A maser in which the stimulated material is a gas, such as ammonia.

gas noise Electrical noise resulting from the undirected motion of gas molecules in a gas tube or defective vacuum tube.

gas-pressure alarm An electronic circuit which actuates an alarm when gas pressure rises or falls.

gas-pressure control A servo system for automatically maintaining or adjusting gas pressure in pipes or other channels.

gas-pressure gauge See *gas-pressure meter.*

gas-pressure indicator See *gas-pressure meter.*

gas-pressure meter An instrument that indicates gas pressure in a pipe or other channel but provides no means for automatically correcting the pressure.

gas ratio For a gaseous tube, the ratio i_i/i_e, where i_i is ion current and i_e is the ionization potential.

gas sensor Any element, such as the filament in a *hot-filament gas detector,* which responds to the presence of a gas in the environment and activates the detector or alarm circuit.

gassiness The presence of undesirable gas in a vacuum tube. Also see *gassy tube.*

gassing 1. The generation of gas by a storage battery, especially while being charged. 2. The generation of gas during electrolysis.

gas sniffer See *gas detector.*

gassy tube Also called *soft tube.* 1. A vacuum tube that has been incompletely evacuated and, therefore, contains some gas (which degrades performance). 2. A vacuum tube in which gas has appeared after evacuation.

gas tetrode See *thyratron.*

gaston A device intended for the purpose of modulating an aircraft signal, making the signal difficult to jam. The signal is randomly modulated by noise from the device.

gas triode 1. Thyratron. 2. A cold-cathode, gas-filled, three-element tube. The two types are *grid* (such as an 884) and *starter-anode* (such as a 0A4G).

gas tube An electron tube that contains a small amount of a gas at low pressure, which ionizes during tube operation. Also see *gas diode, gas triode,* and *thyratron.*

gas-tube lightning arrester A lightning arrester consisting of a special gas diode. The tube has virtually infinite resistance at low voltages but provides a low resistance path to ground when the high voltage of a lightning stroke ionizes the gas.

gas-tube oscillator A relaxation oscillator employing a two-element gas tube as the breakdown device. See *neon-bulb oscillator.*

gas-tube regulator See *gaseous voltage regulator.*

gas valve See *electromechanical valve.*

gas X-ray tube An X-ray tube in which the positive ions of a gas bombard the cathode, which emits electrons.

gate 1. A device or circuit which has no output until it is triggered into operation by one or more enabling signals, or until an input signal exceeds a predetermined threshold amplitude. 2. The input (control) electrode of a field-effect transistor or thyristor device (e.g., SCR). 3. A

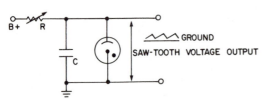

Gas-tube oscillator.

signal that triggers the passage of other signals through a circuit.

gate array Basic gates arranged in a pattern on a chip; the gates may be interconnected during manufacture to form a unit that performs whatever function is needed.

gate circuit 1. An electronic switching circuit (see *gate, 1*). 2. The circuit associated with the gate electrode of a field-effect transistor.

gate-controlled switch A device similar to a silicon-controlled rectifier or thyristor. A negative current, applied to the gate, switches the device off.

gate current Symbol, I_G. Current flowing in the gate (control) circuit of a semiconductor device. The current is finite in SCRs and other thyristors, but is almost zero in insulated-gate field-effect transistors.

gated amplifier An amplifier whose input is effectively switched on and off by gating signals.

gated-beam detector See *quadrature detector*.

gated-beam tube A special three-grid tube of the type employed in a quadrature detector (e.g., 6BN6). Because of the position of the two signal grids with respect to other tube elements, and because of the internal space-charge coupling, a 90-degree phase shift is built into the tube.

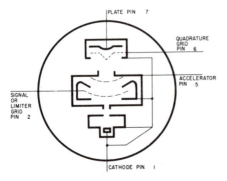

Gated-beam tube.

gated buffer A low-voltage, high-current driver, used for differentiation in a multivibrator circuit.

gated flip-flop A flip-flop in which it is impossible for both outputs to be low simultaneously.

gate-dip meter A dip meter employing a field-effect transistor with the indicating dc microammeter in the gate circuit.

gate-dip oscillator See *gate-dip meter*.

gated multivibrator A rectangular-wave generator that when triggered produces a gate voltage.

gate-drain voltage Symbol, V_{GD}. In a field-effect transistor, the max-

imum voltage permitted between the gate and drain electrodes.

gated sweep 1. In radar, a sweep whose initiation and duration are closely controlled as a measure to eliminate echoes in the image. 2. A circuit providing the action described in 1, above.

gate electrode See *gate, 2*.

gate impedance 1. The impedance of the gate electrode of a field-effect transistor with respect to the other electrode, which serves as the return. 2. The impedance of the gate winding of a magnetic amplifier.

gate leakage current See *gate reverse current*.

gate nontrigger voltage Symbol, V_{gnt}. For a thyristor, the dc voltage applied between gate and cathode, above which the device fails to maintain rated blocking voltage.

gate power dissipation Symbol, P_G. In a silicon controlled rectifier, the power consumed by the gate-cathode path.

gate-protected MOSFET A *metal-oxide semiconductor field-effect transistor* in which the gate electrode is protected from accidental burnout by means of built-in zener diodes connected back to back.

gate pulse 1. A pulse applied to the gate electrode to actuate a gate-controlled semiconductor device. 2. An actuating pulse in a gate circuit.

gate recovery time Symbol, t_{gr}. For a silicon controlled rectifier, an extension of the reverse recovery time—the interval following application of the reverse voltage required before the forward blocking voltage can be reapplied and then blocked by the device.

gate reverse current Symbol, I_{GSS}. In a field-effect transistor, reverse current in the gate-source circuit. Also called *gate leakage current*.

gate signal 1. The input or control signal applied to the gate electrode of a semiconductor device. 2. An actuating signal in a gate circuit.

gate-source breakdown voltage Symbol, BV_{GSS}. The voltage at which the gate junction of a junction field-effect transistor (JFET) enters avalanche.

gate-source pinchoff voltage Symbol, V_P. In a field-effect transistor, the gate-source voltage at which the conduction channel just closes.

gate-source voltage Symbol, V_{GS}. In a field-effect transistor, the dc voltage between the gate and source electrodes.

gate terminal 1. The terminal connected to the gate semiconductor in a field-effect transistor. 2. The terminal, or terminals, connected to the input or inputs of a digital-logic network.

gate trigger current In a gate-controlled semiconductor switch, the current flowing in the gate circuit when the device is being switched on by a gate trigger voltage.

gate trigger voltage In a gate-controlled semiconductor switch, the trigger voltage required to actuate the device.

gate tube Any electron tube designed or operated so that it conducts only if two independent control signals are applied to it simultaneously.

gate turn-off current In a gate-controlled semiconductor switch, the low value of gate current that flows when the device is being switched off. Turn-off current varies with collector (anode) current. Also see *gate turn-off voltage*.

gate turn-off voltage In a gate-controlled semiconductor switch, the low value of gate voltage which causes the device to switch off. Also see *gate turn-off current*.

gate voltage 1. The voltage applied to the gate electrode of a field-effect transistor. See *gate—source voltage*. 2. The instantaneous gate-cathode voltage in a silicon controlled rectifier. 3. The voltage across the gate winding of a magnetic amplifier.

gate winding In a magnetic amplifier, a winding that produces gating action.

gating 1. The process of using one signal to switch another (or part of another) on or off for a desired interval. 2. Selecting a part of a wave for observation or for control purposes.

gating window See *window, 2.*

gauge 1. A meter. 2. Wire data and measurements (see *wire gauge 1, 2, 3*). 3. Sheet metal thickness (e.g., 10 gauge).

gauss (Karl F. *Gauss*, 1777-1855). Unit of magnetic flux density; 1 gauss equals 1 line per square centimeter. The SI (preferred) unit of magnetic flux density is the tesla (webers per square meter); 1 gauss equals 10^{-4} teslas (symbol, T). Also see *flux density.*

Gaussian curve See *bell-shaped curve.*

Gaussian distribution In statistics, the symmetrical distribution described by a bell-shaped curve. Also called *normal distribution.*

Gaussian function A function used in the design of bandpass filters. Similar to a Bessel function.

Gaussian noise Electrical noise whose frequency distribution is described by the Gaussian curve. See *bell-shaped curve.*

Gaussian waveform A waveform that results in minimal side lobes in a pulse-compression system.

gaussmeter See *fluxmeter.*

Gauss' theorem Across any closed surface within an electric field, the total flux of force is 4π times the enclosed quantity of electricity.

gauze resistor See *woven resistor.*

GAVRS Abbreviation of *gyrocompass attitude vertical reference system.*

GCA Abbreviation of *ground-controlled approach.*

GCI Abbreviation of *ground-controlled interception.*

GCM Abbreviation of *gyrocompass module.* See *gyrocompass.*

GCT Abbreviation of *Greenwich civil time.*

G-curves Triode response curves of constant *g*m (transconductance) and *g*p (plate conductance), useful in calculating the voltage gain of resistance-coupled stages:

$$Av = -gmRL(1 + gp\,RL).$$

Gd Symbol for *gadolinium.*

G-display See *G-scan.*

gdo 1. Abbreviation of *grid-dip oscillator.* 2. Abbreviation of *gate-dip oscillator.*

Ge Symbol for *germanium.*

gear 1. Collectively, electronic equipment. 2. A toothed wheel.

gearmotor An electric motor with a gear train for speed-changing.

gear-wheel pattern A frequency-identifying *wheel pattern* produced on an oscilloscope screen by intensity-modulating a circular trace. The circular trace is produced by applying a frequency-standard signal to

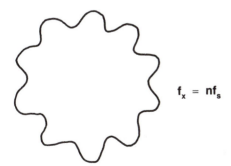

$$f_x = nf_s$$

Gear-wheel pattern.

the horizontal and vertical input terminals 90 degrees out of phase. A sine-wave signal of unknown frequency is then applied to the intensity-modulation (*Z*-axis) input terminals. The sine wave molds the circle into a number of teeth or corrugations. The unknown frequency *fx* equals *nfs*, where *n* is the number of teeth and *fs* is the standard frequency. Compare *spot-wheel pattern.*

Geiger counter A radioactivity rate-counting instrument based on the Geiger-Mueller tube. Pulses from the GM tube drive a tube or transistor, which in turn drives a meter or digital counter to indicate the count.

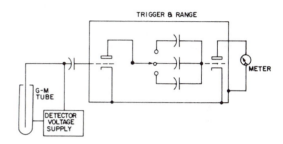

Geiger counter.

Geiger-Mueller counter See *Geiger counter.*

Geiger-Mueller region For a Geiger-Mueller tube the range of voltages within which the output pulse amplitude is constant regardless of ionizing radiation intensity.

Geiger-Mueller threshold The lowest voltage in the Geiger-Mueller region.

Geiger-Mueller tube A gas-filled radiation detector/counter tube consisting of a straight wire as an anode, surrounded by a cylindrical cathode. The tube is biased by high dc voltage. Radiation penetrating the tube ionizes the gas, each ionizing event causing a corresponding output pulse.

Geiger point counter See *point counter.*

Geiger region See *Geiger-Mueller region.*

Geiger threshold See *Geiger-Mueller threshold.*

Geissler tube A simple gas-filled glow-discharge tube with metal electrodes sealed in each end. When a sufficiently high voltage is applied between the electrodes, the highly rarefied gas ionizes and glows with the color associated with the particular gas used.

gel A substance equivalent to colloidal solution in the solid phase, e.g., silica gel.

gen Abbreviation of *generator.*

genemotor Acronym for *gene*rator *motor,* a (usually battery-driven) dynamotor having separate motor and generator windings on the same armature core.

general class license An amateur-radio license that conveys some privileges in the high-frequency bands, and all operating privileges in the very-high-frequency region and above. An examination of moderate difficulty is required.

general-purpose bridge See *universal bridge.*

general-purpose component A component design or used for a wide range of applications. For example, a general-purpose germanium diode is useful as a detector, limiter, clipper, meter rectifier, agc rectifier, and curve changer.

general-purpose computer A computer that can be used in a number of applications for which it was not specifically designed.

general-purpose diode A small-signal semiconductor diode that is useful for a variety of applications, such as detection, light-duty rectification, limiting, logic switching, and so forth. Example: 1N34A.

general-purpose function generator A nonspecialized function generator that is capable of generating a variety of different waveforms.

general-purpose program A program for the solution of a class of problems or for a specific problem according to certain parametric values. Also called *general routine*.

general-purpose relay Any relay that can be used in various situations, such as for switching alternating or direct currents.

general-purpose tester An instrument, such as a voltohm-milliammeter, which offers several test capabilities.

general-purpose transistor A transistor that may be used in several applications, such as audio amplifications, detection, oscillation.

general service code See *Continental code*.

generate 1. To develop various subroutines from parameters applied to skeletal coding. 2. To use a program generator to produce a specialized version of a general-purpose program.

generated address An address developed by program instructions for later use by that program.

generated noise 1. Electrical noise due to battery action (i.e., between dissimilar metals) in a component, such as in a potentiometer. 2. Electrical noise due to small output variations of generating devices (rotating machines, vibrators, etc.). Also called *generator noise*.

generating magnetometer See *earth inductor*.

generating station An electric power station.

generating voltmeter An instrument based on a rapidly spinning variable capacitor. Dc voltage applied to the capacitor is converted into an alternating current by the varying capacitance; the ac is proportional to the voltage.

generation 1. The production of a signal by a generator. 2. The number of recording steps between a master recording and a copy.

generation number A number that identifies the age of a file and is included in the file label on the reel of magnetic tape containing the file.

generator 1. Symbol, G. Any signal source. 2. A rotating machine for producing ac or dc electricity. 3. An electronic device for converting dc voltage into ac of a given frequency and waveshape. 4. In computer operation, a routine (akin to a compiler) that will produce a program to perform a specific version of some general operation by implementing skeletal coding according to specific parameters, e.g., *sort* generator.

generator efficiency The ratio of consumed power to delivered power in a generator. Usually expressed as a percentage.

generator noise Electrical noise caused by a rotating generator. Also see *generated noise, 2*.

generator-type microphone A microphone that produces an output voltage without the need for a supply voltage. Examples: ceramic, crystal, dynamic, electret, and velocity types.

generator-type transducer A transducer that converts mechanical motion into an electrical signal of a proportional voltage. In such a transducer, an armature or conductor moves in a magnetic field.

generic A form of software collection. Several specialized software packages may be derived from the generic collection, for use in different systems.

geodesic On a surface, the shortest line between two points.

geodesy The branch of applied mathematics concerned with the earth's measurements.

geodetic system The application of a computer to seismographic studies for the purpose of reducing drilling and mining costs.

geomagnetism Earth's magnetism. Also see *earth's magnetic field*.

geometric capacitance The ratio of the free charge of a capacitor to the voltage across its terminals.

geometric degree See *degree, 1*.

geometric mean The nth root of the product of n quantities.

geometric progression A mathematical series in which each term after the first is obtained by multiplying the preceding one by a constant quantity; e.g., $S = 1, 2, 4, 8, 16... 2n$. Also called geometric sequence.

geometric symmetry In a bandpass or band-rejection filter, a condition in which the response is identical on either side of the center frequency. Also called *mirror-image symmetry*.

George Phonetic alphabet code word for the letter *G*.

george box In an intermediate-frequency amplifier, a device used for the purpose of rejecting jamming signals. Any jamming signal with an amplitude lower than a certain minimum is rejected.

germanium Symbol, Ge. A metalloidal element. Atomic number, 32. Atomic weight, 72.6. Germanium is used in semiconductor diodes, photocells, rectifiers, and transistors.

germanium diode A diode in which the semiconductor material is specially processed germanium. Also see *germanium junction diode* and *germanium point-contact diode*.

germanium dioxide Formula, GeO2. A gray or white powder obtainable from various sources; it is reduced in an atmosphere of hydrogen or helium to yield germanium, a semiconductor material.

germanium junction diode A germanium diode employing a pn junction. Compare *germanium point-contact diode*.

germanium photocell A photoconductive cell consisting of a reverse-biased germanium point-contact diode or germanium junction diode.

germanium point contact The contact between a pointed, metal cat's whisker and a germanium wafer, as in a point-contact diode or point-contact transistor.

germanium rectifier A power rectifier employing a germanium pn junction.

germanium transistor A transistor in which germanium is the semiconductor material. Such a transistor has lower internal resistance and greater temperature drift than the silicon transistor.

German silver A copper—nickel—zinc alloy used in some resistance wires. Also called *nickel silver*.

getter A small piece of metal (such as magnesium) that is flashed (by an external rf field) in a vacuum tube to absorb gases during the process of evacuation. Also see *flash, 2*.

GEV Abbreviation of *ground effect vehicle*.

GeV Abbreviation of *gigaelectronvolt*.

gfi Abbreviation of *ground-fault interrupter*.

g-force See *gravity, 2*.

gfs Symbol for *forward transconductance*.

G/G Abbreviation of *ground-to-ground*.

ghost In television reception, a slightly displaced image appearing on the screen simultaneously with its twin (the false member of a double image).

ghost image See *ghost*.

ghost signal Any signal (such as an undesired reflection) that produces a ghost.

GHz Abbreviation of *gigahertz*.

Gi Symbol for *input conductance.*

Gibson girl A portable radio transmitter—powered by an integral crank-operated generator—introduced during World War II for pilots forced down at sea. The name was suggested by the hour-glass shape of the device, which was also the idealized shape of the girl drawn by Charles Dana Gibson in the 1890s.

giga Abbreviation, G. A prefix meaning billion, i.e., 10^9.

gigacycle Abbreviation, Gc. A billion complete cycles. An ac frequency of 1 Gc per second is 1 GHz (see *gigahertz*).

gigaelectronvolt Abbreviation, GeV. A large unit of voltage; 1 GeV equals 10^9 eV. Also see *BeV, electronvolt, MeV,* and *million electron volts.*

gigahertz Abbreviation, GHz. A unit of high frequency; 1 GHz equals 10^9 Hz.

gigaohm Symbol, GΩ. A unit of high resistance, reactance, or impedance; $1 \, G\Omega = 10^9 \Omega$.

gigo Abbreviation of *garbage in, garbage out,* a term signifying that incorrect input to a computer can only result in worthless output.

gilbert (William *Gilbert*, 1540-1603). Symbol, Σ. A unit of magnetomotive force; 1 Σ equals $0.4\pi NI$, whre NI is ampere-turns. The SI (preferred) unit of magnetomotive force is the ampere (symbol, A); 1 gilbert equals 0.795 8A.

gilbert per centimeter See *oersted.*

Gill-Morrell oscillator See *Barkhausen-Kurtz oscillator.*

gill selector A telegraph sending device used for specialized applications.

gimbal A suspension device whose orientation can be changed without affecting the attitude of the body being suspended.

gimmick 1. Colloquialism for any unnamed device. Also see *gadget.* 2. Colloquialism for any tricky manipulation or design. 3. A low-value capacitor made by twisting two short pieces of insulated wire together.

gimp Colloquialism for the tinsel and cloth conductor used in earphone cords.

Giorgi system The absolute mks system of units. Also called *mksa system.*

GJD Abbreviation of *germanium junction diode.*

glass A hard, brittle, amorphous, and usually transparent substance that is largely silicon dioxide. Glass has a multitude of uses in electronics, and there are several kinds, each having different electrical properties: its dielectric constant, for example, can be from 4 to 10; dielectric strength, from 20 to 300 kV/mm.

glass arm A stiffness of the wrist or forearm, somewhat resembling writer's cramp, experienced by some radiotelegraph operators or wire telegraph operators after prolonged use of a hand key.

glass bulb The glass enclosure of electron tubes and incandescent lamps.

glass capacitor A capacitor employing thin glass plates as the dielectric films, and usually having plates consisting of metal electroplated or electrodeposited on opposite faces of each film. Also see *molded glass capacitor.*

glass diode A semiconductor diode molded in glass.

glass electrode A probe used with a pH meter; it consists of a thin-walled glass tube containing potassium chloride and mercurous chloride. Also see *calomel electrode.*

glass envelope See *glass bulb.*

glassivation A procedure for encapsulating semiconductor devices in glass or other dielectric material.

glass-metal seal See *glass-to-metal seal.*

glass plate capacitor A capacitor in which the dielectric is a plate of glass. Also see *glass capacitor.*

glass shell See *glass bulb.*

glass-to-metal seal A bond between glass and metal in electronic devices, such as vacuum tubes, feedthrough terminals, glass plate capacitors, and the like.

glass tube An electron tube whose elements are housed in an evacuated glass envelope, e.g., 5R4GY. Compare *metal tube.*

glide path The guidance beam used by aircraft making instrument landings.

glide-path transmitter A radio-frequency transmitter that produces a guidance beam for aircraft landing purposes. The aircraft follows the beam toward the runway.

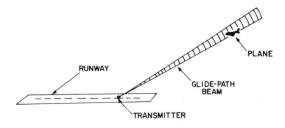

Glide-path transmitter

glide slope See *glide path.*

G-line A microwave conductor consisting of a round wire coated with a dielectric.

glitch 1. In a TV picture, a narrow, horizontal interferential bar that moves vertically. 2. A very short and unwanted high-amplitude transient that recurs irregularly in an electronic system.

glitter 1. In radar, an echo or set of echoes that fluctuates rapidly in intensity because of motion in the target. 2. A system in which moving devices are used to confuse enemy radar systems.

gloss factor For a reflecting surface, the ratio of reflected light in a selected direction to reflected light in all directions.

glossmeter An instrument for determining gloss factor.

glow discharge The luminous electrical discharge resulting from the passage of current through ionized gas in a partially evacuated tube. The color of the glow is characteristic of the particular gas used.

glow-discharge microphone A device that produces audio-frequency currents from the action of sound waves in a glow-discharge tube.

glow-discharge tube See *discharge lamp, fluorescent tube,* and *neon bulb.*

glow-discharge voltage regulator See *gaseous voltage regulator.*

glow lamp See *discharge lamp.*

glow modulator tube A gas tube whose luminous output can be modulated by an af input signal, e.g., IB59.

glow potential The voltage at which glow discharge just begins in a gas-filled tube.

glow switch In fluorescent light circuits, an electron tube containing two bimetal strips that make mutual contact when heated by the glow discharge.

glow-transfer counter tube A neon-filled counter tube employing a central anode around which are spaced 10 pin cathodes. Successive input pulses transfer the glow from cathode to cathode around the circle. An external circular scale numbers the cathodes zero to nine.

glow tube See *discharge lamp, fluorescent tube, glow modulator tube, neon bulb,* and *strobotron.*

glow-tube regulator See *gaseous voltage regulator*.

glow voltage See *breakdown voltage, 2*.

glucinium See *beryllium*.

gluon A *subatomic particle* that binds *quarks* together (coined by Prof. Murray Gell-Mann of California Institute of Technology).

GM 1. Abbreviation of *Geiger-Mueller* (see, for example, *Geiger-Mueller tube*). 2. Abbreviation of *Gill-Morrell* (see, for example, *G-M oscillator*).

gm Abbreviation of *gram*. Also abbreviated *g* (preferred).

gm Symbol for *transconductance*.

gm-cal Abbreviation of *gram-calorie*.

gm-cm Abbreviation of *gram-centimeter*.

G-M counter See *Geiger counter*.

gm-m Abbreviation of *gram-meter*.

G-M oscillator A positive-grid triode oscillator for ultrahigh frequencies. (Gill-Morrell oscillator). Also see *Barkhausen-Kurtz oscillator*.

GMT Abbreviation of *Greenwich mean time*.

G/M tube See *Geiger-Mueller tube*.

gnd Abbreviation of *ground*.

go Symbol for *output conductance*.

gold Symbol, Au. A precious metallic element. Atomic number, 79. Atomic weight, 197. Electrical contacts that must have low rf resistance are often plated with gold.

gold-bonded diode A germanium point-contact diode having a gold cat's whisker whose point is bonded to the germanium wafer. Its principal features are high forward current and almost constant, low reverse current.

gold doping The diffusion of gold into the base and collector regions of a diffused-mesa transistor; it shortens carrier storage time.

golden rectangle A rectangle having a base-to-altitude ratio of 1:1.618. The golden rectangle provided the proportions for classical Greek architecture; the ratio of its sides is also the ratio between progressive line segments in a pentagram, no doubt a factor contributing to the figure's mystery. See *divine proportion*.

golden section See *divine proportion*.

gold-leaf electroscope See *electroscope*.

Goldschmidt alternator An early dynamo for generating radio-frequency power. The machine differed from the Alexanderson alternator and Bethenod alternator in that the high frequency was not generated directly by the machine but by resonant circuits and frequency-multiplying interaction between components.

Golf Phonetic alphabet code word for the letter G

goniometer 1. Generically, any radio direction finder. 2. An inductive coupler having a secondary coil rotated by a dial calibrated to read azimuth. The coupler with a suitable antenna system comprises a direction finder. 3. A device for electrically varying the directional pattern of an antenna.

go-no test A test that indicates only acceptance or rejection of a device. No diagnosis is made.

googol One followed by a hundred zeros, i.e., 10^{100}.

googolplex The number 1 followed by googol zeros. See *googol*.

GOTO Abbreviation, GTO. In computers and programmable calculators, an instruction which, followed by a suitable label, directs the program to that label.

goto circuit In a digital-logic circuit, a device that senses the direction of electric current.

goto pair A pair of diodes connected in reverse series used in digital-logic circuits.

governor 1. A device that prevents a motor or engine from running faster

than a certain speed. 2. Any device that limits a circuit parameter.

gp Symbol for *plate conductance*.

g-parameters Conductance parameters obtained for the equivalent-pi model of a transistor: *gbe, ggc, gce,* and *gm*.

gpc Abbreviation of *germanium point-contact*.

GPI Abbreviation of *ground-position indicator*.

gr Abbreviation of grains(s).

graceful degradation A computer programming technique used to prevent debilitating breakdown by operating the system even though several subsystems have malfunctioned; also known as operating in crippled mode.

grad A unit of angular measurement equal to 0.9 degree.

graded-base transistor See *diffused-base transistor*.

graded filter A power-supply filter that supplies dc output at various points in the filter sequence. Thus, the points in the powered equipment that can tolerate the least ripple are connected to the filter output, while those which can tolerate appreciable ripple are connected to the filter input; fairly critical points are connected to an intermediate position in the filter, such as at the junction of two chokes.

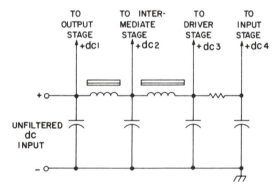

Graded filter.

graded-junction transistor A grown-junction transistor in which the temperature of the melt and the rate at which the crystal is pulled from it are closely controlled as the n- and p-layers are formed.

gradient The rate at which a variable quantity increases or decreases. See, for example, *voltage gradient*.

gradient microphone A microphone whose output varies with sound pressure. Also see *pressure microphone*.

Graetz bridge A full bridge rectifier, i.e., one having a diode in each arm.

grain boundary In a polycrystalline solid, a boundary between single crystalline regions.

gram 1. Abbreviation, g. A unit of mass and weight, $(10^{-3}$ kg); 1 gram equals 0.0353 ounce. 2. A suffix meaning something drawn (written), or recorded, e.g., radiogram, electrocardiogram.

gram atom See *gram atomic weight*.

gram atomic weight That quantity of an element equal to the atomic weight of the element; for copper (at. wt. 63.54), 1 gram atomic weight equals 63.54 grams.

gram-calorie Abbreviation, gm-cal. The amount of heat required to raise the temperature of 1 gram of water 1 degree Celsius.

gram-centimeter Abbreviation, gm-cm. A cgs unit, 1 gm-cm is the work done by a force of 1 gram exerted over a distance of 1 centimeter. Also see *joule.*

gram-equivalent See *gram atomic weight.*

grammar The sequence of words in a communication or part of a communication.

gramme armature See *Gramme ring.*

Gramme ring A type of armature for a motor or generator, consisting of an iron ring onto which is wound a coil of wire, each turn being connected to a commutator bar.

gram-meter Abbreviation, *g-m.* A unit of work equal to a force of 1 gram exerted over a distance of 1 meter. Compare joule.

gram-molecular weight See *mol.*

gram molecule A gram-molecular weight. (See *mol.*)

gramophone A phonograph.

grandfather cycle A form of backup scheme in a magnetic reproduction system. The original recordings are retained for a period of time, so that new copies of high precision can be made in case of loss.

grandfather tape An original copy of a file on tape that is retained as a source for its reconstruction as needed. Usually, three generations of a tape file—grandfather, father, and son—are kept, each identified by a generation number. See *generation number.*

granular carbon Carbon in the form of fine granules, used in the button of a carbon microphone. Also see *button, 2.*

granularity 1. In a digital device, the smallest increment that can be differentiated. 2. The limit of detail in a reproduction system.

graph A presentation of data—particularly a depiction of the manner in which one variable or set of variables changes with respect to another—in the form of curves, bars, pie charts, and the like.

graphechon A form of electron tube designed for storing information, and employed in certain computer devices.

graphical analysis The solution of problems through the use of graphic devices, such as vector diagrams, load lines, Nyquist plots, topological flow diagrams, and so forth.

graphical harmonic analysis See *schedule method.*

graphic documentation Records of data on graphs, charts, tables, diagrams, and the like.

graphic instrument See *graphic recorder.*

graphic panel In process control, a panel of illuminated lights or dials that display the status of a process.

graphic recorder An instrument in which a signal-driven pen or stylus makes a permanent record of a quantity on graph paper passing at a controlled rate under the pen. Also see *oscillograph, 1.*

graphics 1. Diagrams, charts, photos, tables, or similar, often symbolic, artwork used to convey operating data. 2. The display by a computer of graphic material on a cathode-ray screen, which may be worked on with a *light pen.*

graphic solution A graphic or diagrammatic solution to a problem as compared to a tabulation.

graphic terminal A crt display or plotter that provides visual output of a computer run.

graphite A soft form of carbon used widely in electronics, in resistors, attenuators, contacts, brushes, vacuum-tube plates, cathode-ray tube coatings, etc.

graphite-line resistor An emergency, makeshift resistor consisting of a pencil line drawn on a piece of paper. The heavier the line for a given width and length, the lower its resistance.

graphophone A phonograph.

grass Colloquialism for interferential noise patterns on a radar screen.

grasshopper fuse A spring-operated fuse which, when blown, actuates an auxiliary circuit to alert personnel of the malfunction.

graticule Calibrated gridwork, as on the face of an oscilloscope screen. See *mask, 2.*

grating A transparent plate containing many finely ruled, closely spaced, equidistant horizontal lines. Diffraction caused by this arrangement produces a spectrum from light transmitted through the grating. Also called *diffraction grating.*

grating reflector An openwork, metal antenna reflector resembling an iron grating.

Gratz rectifier A form of full-wave rectifier circuit in a three-phase, alternating-current system.

gravitational constant Symbol, *g.* The acceleration due to attraction of a unit mass at unit distance; $g = 6.673 \times 10^{-11}$ N•m^2/kg^2.

gravity 1. The universal force of attraction between material bodies, especially that force evidenced by the earth's drawing of bodies toward its center, causing them to have weight. 2. Abbreviation, *g.* The acceleration of gravity; 1 *g* equals 9.754 m/s/s (32 ft/s/s).

gravity cell A type of Daniell cell in which the positive electrode is a sheetmetal copper star (see *star, 1*) and the negative electrode is a zinc crowfoot (see *crowfoot, 2*). The star is placed at the bottom of a jar and the crowfoot at the top. The jar is half filled with a copper sulfate solution and then filled with a zinc sulfate solution. The solutions remain separate because the copper sulfate has the higher specific gravity. Like the Daniell cell, the gravity cell does not become polarized.

gray body A radiating body exhibiting constant spectral emissivity at all wavelengths. Compare *blackbody* and *blackbody radiation.*

Gray code A computer code in which—for minimizing errors—the expressions representing sequential numbers differ in only one bit.

gray scale A reference scale for use in black-and-white television, consisting of several defined levels of brightness with neutral color.

gray tin A form of tin which is allotropic at temperatures below 18 °C, when it exhibits some properties of a semiconductor. See *allotropic.*

Greek alphabet The 24-letter alphabet of the Greek language. Virtually all of the letters are used as symbols in electronics and related sciences. See individual listings in this dictionary for the individual letters.

green gun The electron gun in a color TV picture tube whose correctly adjusted beam strikes only the green phosphors on the screen.

green video voltage In a three-gun color-TV circuit, the green-signal voltage, which actuates the green gun.

Greenwich civil time Abbreviation, GCT. Mean time counted from mean midnight at Greenwich, England, the location of zero meridian.

Greenwich mean time Abbreviation, GMT. Mean solar time of the zero meridian (Greenwich, England, the location of the Royal Observatory). The term has been supplanted by UTC, or *coordinated univeral time,* as the basis of standard time throughout the world. Also called *Greenwich time.* Compare *time zone.*

grid 1. The prime control electrode in an electron tube. The grid is ordinarily a coil or mesh but takes other forms in some special tubes. Also called *control grid.* 2. Any electrode in a tube placed between cathode and anode: screen grid, suppressor grid, etc. 3. A network of sets of equally spaced parallel lines, one set perpendicular to the other. 4. In optical recognition, a scale (*grid, 3*) for measuring characters.

grid battery The battery that supplies dc grid bias to a tube. Also called *C-battery.*

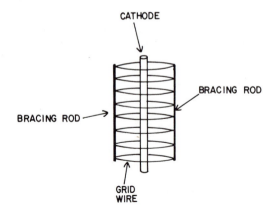

Control grid.

grid bias See *dc grid bias.*

grid-bias cell See *bias cell.*

grid-bias modulation Amplitude modulation by varying, at an audio-frequency rate, the dc control-grid bias of an rf amplifier. Also called *grid modulation.*

grid blocking Blocking action in a vacuum-tube circuit due to the large-signal charging of the grid coupling capacitor and the subsequent slow leaking off of accumulated electrons.

grid-block keying A form of amplifier keying in a continuous-wave transmitter. During key-up periods, a high negative grid bias (blocking voltage) cuts off the tube. During key-down periods, the voltage is removed, and the amplifier functions normally.

grid cap A small metal cap on the tops of some electron tubes, used for connection to the control grid. Bringing the grid lead through the top reduces internal capacitance between it and the tube and the other leads, which enter from the base.

grid capacitance See *grid-cathode capacitance.*

grid capacitor 1. A capacitor in series with the grid of a tube, employed for blocking purposes. 2. A bypass capacitor in a grounded-grid tube type amplifier. 3. The capacitor in the grid tank circuit of a tube type oscillator or amplifier.

grid-cathode capacitance Symbol, C_{gk}. The internal capacitance between the control grid and cathode of an electron tube. Also called *input capacitance.*

grid characteristic The grid-current-vs-grid-voltage performance curve for an electron tube.

grid circuit The external circuit associated with the control grid of an electron tube.

grid-circuit impedance See *grid-input impedance.*

grid-circuit tester A device that measures the rid impedance of a vacuum-tube circuit.

grid clamping Clamping action obtained from the signal rectified by the grid of a vacuum tube.

grid clip A spring clip for attaching a lead to a grid cap.

grid conductance The reciprocal of grid resistance in a vacuum-tube circuit.

grid control In an electron tube (vacuum or gaseous), the control of plate (anode) current by means of grid voltage.

grid-controlled rectifier See *thyratron.*

grid control ratio Symbol, ϱ. For a thyratron, the ratio of anode voltage to grid striking voltage.

grid current Symbol, I_g. Current flowing between the control grid and cathode in an electron tube.

grid cylinder The metal cylinder that acts as a control grid in an oscilloscope crt.

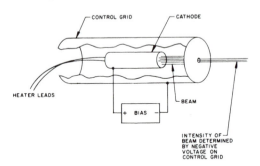

Grid cylinder.

grid detection In a vacuum tube, detection resulting from diode action between the control grid and cathode. The rest of the tube amplifies the detected signal. Also see *grid-leak detector.*

grid-dip adapter An external coupling device combined with a semiconductor-diode rf meter, for converting a signal generator into a dip meter.

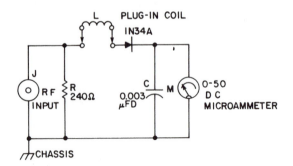

Grid dip adapter.

grid-dip meter 1. A dip meter employing a vacuum-tube oscillator; the indicating dc microammeter is in the grid circuit. 2. Loosely, any frequency-sensitive wavemeter that indicates resonance by a marked dip in input (base, grid, gate) current.

grid-dip oscillator See *grid-dip meter.*

grid dissipation 1. The amount of power given up as heat in the grid circuit of a vacuum-tube amplifier. 2. The maximum amount of power that a tube can safely dissipate as heat in the grid.

grid drive See *grid excitation.*

grid-driving power The signal power required by the control grid of a power tube.

grid emission Electron or ion emission by the control grid of an electron tube.

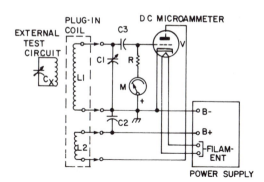

Grid-dip meter.

grid excitation Signal voltage or power applied to the control grid of a vacuum tube in an amplifier, detector, or control circuit.

grid-glow tube See *thyratron*.

grid impedance Symbol, Z_g. The internal impedance of the grid—cathode path in an electron tube.

grid input impedance The impedance of the grid input section of a vacuum-tube circuit. It is a complex combination of grid impedance and the impedance of input-circuit components.

grid injection Application of a signal to the control grid of a vacuum tube in a mixer (converter) circuit, (rather than to one of the other grids).

gridistor A special form of field-effect transistor with several channels.

grid leak A bypassed resistor in series with the control grid of a vacuum tube. The applied signal is rectified by diode action between grid and cathode; the resulting dc develops a dc grid-bias voltage across the resistor.

grid-leak detector A vacuum-tube detector circuit in which a high resistance is connected in series with the control grid. Diode action between grid and cathode rectifies the input signal and creates negative bias as a dc voltage drop across the grid-leak resistor. See *grid leak*.

grid-limited squarer A triode whose grid is driven positive to produce plate-current saturation in one direction, and negative to produce plate-current cutoff in the other direction. The result is a square-wave output signal.

grid-limiter resistor A resistor connected in series with the grid of a tube to limit grid current during the positive half-cycle of grid-signal voltage.

grid limiting The cutting off of plate current in an electron tube (and consequent limiting action) by means of a high, negative grid voltage developed by overdriving the grid.

grid loading effect The tendency of the internal grid—cathode path of a vacuum tube to load a tuned circuit, especially when the grid draws current.

grid locking A tube fault in which the grid potential has become permanently positive due to excessive grid electron emission.

grid locus plot For an ac thyratron, a curve depicting the relationship between critical grid voltage and the quantity ωt.

grid mesh The mechanical structure of a grid, e.g., gauze or a metal screen.

grid modulation See *grid-bias modulation*.

grid neutralization See *grid-neutralized amplifier*.

grid-neutralized amplifier A neutralized amplifier in which the neutralizing capacitor is connected from the plate of the tube to the free end of a center-tapped grid-tank coil. Compare *plate-neutralized amplifier*.

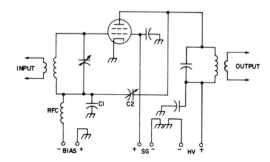

Grid-neutralized amplifier.

grid north In the grid system of navigation, the direction most nearly corresponding to geographic north.

grid-plate capacitance See *plate-grid capacitance*.

grid-plate crystal oscillator An oscillator in which a crystal is connected in the control-grid circuit of a vacuum tube; a tuned tank, connected in the plate circuit, is resonated at the crystal frequency. The arrangement is equivalent to the tuned-plate—tuned-grid self-excited oscillator circuit. Also see *TPTG oscillator*.

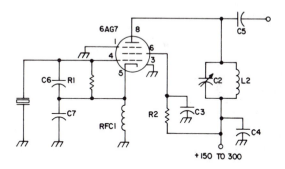

Grid-plate crystal oscillator.

grid-plate transconductance See *transconductance, 1*.

grid pool tube A gas-discharge tube in which the cathode is a pool of mercury.

grid power loss Driving-power loss in the grid-input circuit of a power amplifier.

grid-pulse modulation Modulation by pulsing a grid circuit in an amplifier or oscillator.

grid-regulation resistor See *swamping resistor, 1*.

grid resistor 1. A high-value resistor (0.5M or more) connected between control grid and ground in a vacuum-tube circuit. 2. (Sometimes) a *grid leak*.

grid return The circuit path through which the control grid on an electron tube is returned to round or to B-minus.

grid-screen capacitance Symbol, C_{gs}. The internal capacitance between control and screen in an electron tube.

grid-separation circuit A grounded-grid amplifier. See *common-grid circuit*.

grid stopper See *stopper resistor*.

grid suppressor See *stopper resistor*.

grid swing The peak-to-peak variation of a grid excitation signal.

grid tank A resonant LC circuit operating into the control grid of a vacuum tube. Compare *plate tank*.

grid tank capacitance The capacitance required to tune a grid tank. Compare *plate tank capacitance*.

grid tank inductance The inductance of the coil in a grid tank. Compare *plate tank inductance*.

grid tank voltage The ac voltage (af or rf) developed across the grid tank of an electron tube circuit. Compare *plate tank voltage*.

grid-to-cathode capacitance See *grid-cathode capacitance*.

grid-to-plate capacitance See *grid-plate capacitance*.

grid-to-plate transconductance See *transconductance, 1*.

grid-to-screen capacitance See *grid-screen capacitance*.

grid tuning Tuning of a vacuum-tube circuit by varying the capacitance, inductance, or both in the grid tank. Compare *plate tuning*.

grid tuning capacitance See *grid tank capacitance*.

grid tuning inductance See *grid tank inductance*.

grid voltage 1. Symbol, Eg. The dc bias voltage applied to the control grid of an electron tube. 2. Symbol, $Eg(ac)$. Grid excitation.

grille A covering for an acoustic speaker, used primarily for the purpose of protecting the speaker cone, but also for esthetic appeal.

grommet An elastic washer inserted through a hole in a chassis to prevent accidental grounding of a conductor or to reduce wear on a cord or cable exiting the chassis.

groove 1. See *keyway*. 2. The fine, spiral line cut into disk recording.

groove angle On a phonograph disk, the angle between the walls of the unmodulated groove.

groove depth See *depth of cut*.

groove speed In a phonograph recording or reproducing system, the speed of the cutter or needle with respect to the disk. The groove speed is greatest near the outer edge of the disk, and least near the center.

gross content The overall amount of information contained in a message. May be expressed in bits, bytes, words, or other units.

gross index The first of a pair of indexes, the gross index is used to give a reference in the second (fine) index, a supplement; both indexes are used to locate computer records in storage.

ground 1. The earth in relation to electricity and magnetism. 2. An electrical connection to the ground. Also see *ground connection, 1*. 3. The (usually negative) return point in a circuit. Also see *ground connection, 2*. 4. A short-circuit to the earth or to a circuit return point. 5. A short-circuit to the metal chassis, case, or panel of an equipment.

ground absorption The absorption (and consequential loss), of radiant energy by the earth.

ground bus The common, grounded conductor to which various parts of a system are returned for a complete circuit. Also see *bus*.

ground clamp A device that provides a mechanical and electrical bond between a conductor and a ground rod or pipe. Generally capable of passing a large amount of current.

ground clutter In radar practice, an interference pattern on the screen, caused by accidental grounding in the system (see *ground, 5*).

ground conduit A pipe housing one or more ground leads.

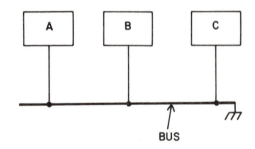

Ground bus.

ground connection 1. A low-resistance connection to the earth. 2. The point corresponding to B-minus, to which the low point of each part of a circuit is connected. Also see *common ground*.

ground constants The conductivity and dielectric constant of the earth for a particular kind of terrain at a given location. Dielectric constant has been measured between 3 and 81, conductivity between 4.64×10^{-11} and 3×10^{-15} EMU (electromagnetic units).

ground-controlled approach Abbreviation, GCA. In air navigation, a ground radar system that provides information for radio-directed aircraft approaches.

ground-controlled interception Abbreviation, GCI. A ground radar system by which an aircraft may be directed to intercept enemy aircraft.

ground current 1. An electric current flowing into the earth from an equipment. 2. An electric current flowing through the earth between two points. 3. A current flowing in the normal ground (low-potential) line of a circuit.

ground detector A device that indicates whether or not a given circuit point is at direct-current ground.

grounded antenna See *Marconi antenna*.

grounded-base circuit See *common-base circuit*.

grounded-cathode circuit See *common-cathode circuit*.

grounded-collector circuit See *common-collector circuit*.

grounded-drain circuit See *common-drain circuit*.

grounded-emitter circuit See *common-emitter circuit*.

grounded-gate circuit See *common-gate circuit*.

grounded-grid circuit See *common-grid circuit*.

grounded outlet An outlet with a receptacle having a ground contact which can be connected to equipment-grounding conductors.

grounded-plate circuit See *cathode follower*.

grounded-source circuit See *common-source circuit*.

grounded system A set of electrical conductors, or a transmission line, in which one conductor is deliberately grounded.

ground effect Distortion of the ideal free-space directivity pattern of an antenna by reflections from, and absorption by, the earth.

ground environment 1. The characteristics of the earth, in terms of electrical conductivity and loss, in a particular location. 2. The ground characteristics in the vicinity of an unbalanced antenna working against ground. 3. In aviation, the set of ground-based installations.

ground fault 1. Loss of a ground connection. 2. A short-circuit to ground.

ground-fault interrupter Abbreviation, gfi. A fast-acting electronic circuit breaker that disconnects equipment from the power line to prevent electric shock or other damage when the safety ground connection is broken.

grounding electrode A device, such as a ground plate or ground rod, which facilitates making low-resistance connections to the earth.

grounding plate A metal plate connected to the earth, on which a person stands to discharge static electricity from his body.

grounding rod See *ground rod*.

ground insulation Insulation employed between adjacent energized and grounded parts, such as transformer windings and metal cores.

ground level See *ground state*.

ground loop An undesirable current path through a grounded body, such as a metal chassis. A ground loop can result when various parts of a circuit are returned to separate ground points on a chassis. A ground loop sometimes provides a common impedance.

ground mat A grid or network of conductors, connected to earth ground, for the purpose of improving the earth conductivity.

ground noise 1. Electrical noise due to a faulty ground connection. 2. Background noise. 3. In wire circuits, such as a telephone system, electrical noise due to fluctuations in ground current. See *ground current, 2*.

ground plane 1. A metal plate or a system of horizontal rods or wires mounted high on a mast, at the base of a vertical antenna, to provide ground potential at a point several wavelengths above the surface of the earth. Also see *ground-plane antenna*. 2. In noise and interference tests, a sheet metal structure used to simulate the skin of an aircraft or missile. 3. On a circuit board, a thin metallic sheet, usually bound to the underside, which serves as a common ground and rf shield.

ground-plane antenna An antenna consisting of an elevated quarter-wave vertical radiator at whose base is mounted a ground plane comprising a flat, horizontal, metal plate or a system of radial wires or rods. Also see *ground plane, 1*.

ground plate A metal plate buried in the earth to provide a low-resistance ground connection.

ground-position indicator Abbreviation, GPI. A computer system that gives a continuous indication of an aircraft's position in terms of heading, elapsed time, and speed.

ground potential See *zero potential, 3*.

ground protection A circuit breaker that opens when the protected circuit is unintentionally grounded at some point.

ground-reflected wave A radio wave component resulting from ground reflection.

ground reflection The reflection of a radio wave by the earth.

ground resistance The dc resistance of a connection to the earth, or the resistance between two points through the earth. The magnitude of the resistance depends on several factors: composition of the soil, amount of moisture, soil electrolytic action, and the area of contact with the earth.

ground return 1. The point or path used to return a circuit to ground or to B-minus for completion. 2. Regarding radar, echoes returned from the earth's surface (including reflections from objects on it).

ground-return circuit A circuit, such as a single-wire telephone line, in which earth ground forms one leg of the circuit. Compare *metallic circuit*.

ground rod A strong metal rod driven deep into the earth as a point of ground connection.

ground speed The speed of an aircraft or missile relative to the ground.

ground state The least-energy level of all possible states in a system.

ground support equipment The ground equipment that supports a weapons system, i.e., upon which the system's function is dependent.

ground switch A switch for grounding an outside antenna during idle periods. Also called *lightning switch*.

ground-to-air communication Radio or radar transmission from a land station to an aircraft in flight. Compare *air-to-air communicatoin* and *ground-to-ground communication*.

ground-to-ground 1. Pertaining to communications between land stations. 2. Pertaining to ballistic actions between points on the surface of the earth.

ground-to-ground communication Communication between land stations.

ground wave 1. A radio wave that travels along the surface of the earth. Compare *skywave*. 2. A shock wave transmitted by the earth.

ground wire A conductor between an equipment and a ground connection, either for circuit completion or for safety.

group 1. A series of computer storage locations containing a specific record or records. 2. The data in these locations. 3. A record set having a common key value in a sorted file.

group busy In a telephone system, an audio signal indicating that all of the lines in a group are in use.

group code An error-detecting code used to verify a character group transferred between terminals.

group delay In a modulated signal, a delay in the transmission of data.

grouped-frequency operation In a two-wire communications system, the grouping of directional signals into certain frequency bands.

grouped records A group of records in which the key of one identifies the group.

grouping 1. The arrangement of data into blocks or sets. 2. On a phonograph disk, the insertion of gaps in the arrangement of grooves. 3. Any periodic irregularity in the spacing of a data transmission. 4. The bunching of grooves on a disk recording. 5. In a facsimile system, occasional spacing errors between recorded lines. 6. A mass of data arranged into groups according to common characteristics.

group mark 1. In telegraphy, an indicator that signals the end of a data unit. 2. A character indicating the end of a character group, which is usually a logical record that will be addressed and processed as a unit.

group printing The printing of a summary of the data contained in a group of punched cards.

group theory A mathematical study of combining groups, elements, and sets.

group velocity The velocity at which a group of waves or a pulse is propagated.

Grove cell A closed-circuit primary cell in which the positive electrode,

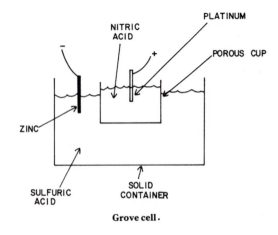

Grove cell.

platinum, is immersed in nitric acid; the negative electrode, zinc, is immersed in sulfuric acid. The nitric acid is held in a porous cup surrounded by of a larger jar of sulfuric acid.

growler 1. An electromechanical troubleshooting device that indicates the location of short circuits and grounds (especially in electric motors) by growling. 2. Any tester that gives an audible signal indicating electrical continuity.

grown-diffused transistor A transistor that is made by first growing the emitter and collector regions as a crystal, into which the base region is later diffused while the crystal is being pulled.

grown diode A type of semiconductor diode created by growing a P layer into N material (or an N layer in to P material) as the single-crystal material is being pulled from the melt. See *crystal pulling*.

grown junction A pn junction produced by adding impurities in various amounts to a crystal while it is being pulled from molten semiconductor material.

grown-junction diode See *grown diode*.

grown-junction photocell A grown-junction diode employed as a photoconductive cell. Also see *grown junction*.

grown-junction transistor A transistor made by adding n- and p-type impurities successively to a crystal in its molten state, and then slicing the resulting npn formations from the finished crystal.

G-scan A rectangular radar display consisting of a laterally centered blip that "grows wings" as a target approaches. Horizontal and vertical displacement of the blip indicate horizontal and vertical aiming errors.

G-scope See *G-scan*.

GSR Abbreviation of galvanic skin resistance.

G-string antenna In microwave practice, a transmission line consisting of a dielectric-coated wire which behaves like a coaxial line with its outer conductor removed to infinity. A horn at each end matches the line to the transmitter and receiver. Also called *surface-wave transmission line (SWTL)*. (The name comes from the first initial of Dr. George Groubau, inventor of the device, and the stringy appearance of the wire.)

G-string antenna.

GTO Abbreviation of *GOTO*.

guard band A narrow unoccupied band of frequencies at the upper and lower limits of an assigned channel; its purpose is to prevent adjacent-channel interference by preserving separation between channels.

guard circle On a phonograph disk, an inner groove that prevents collision of the pickup with the spindle at the center of the disk.

guard circuit An auxiliary circuit added to an ac bridge to compensate for the effects of stray capacitance in the bridge arms. One of its several forms is the Wagner ground.

guarded input An input-terminal arrangement in which a third terminal, maintained at the proper potential, shields the entire input-terminal combination.

guarding A method of short-circuiting a leakage current to ground. On a printed-circuit board, guarding is usually accomplished by the use of a large conducting foil surface near critical components.

guard relay A relay that insures that only one linefinder will be connected to a line circuit when other line relays are in operation.

guard ring A metal ring (or other configuration) surrounding—but separate from—a charged body or terminal, for the even distribution of the electric charge over the latter's surface.

guard shield A shield that encloses the input circuit of an amplifier or instrument.

guard terminal The third terminal in a guarded input.

guard wire A grounded wire intended to catch and ground a broken high-voltage line.

Gudden-Pohl effect The tendency of an ultraviolet irradiated phosphor to glow momentarily when subjected to an electric field.

guidance Electronic control of a missile or other vehicle.

guidance system The completely coordinated and integrated electronic and electromechanical system for the remote control of a missile or other vehicle.

guidance tape In a guided missile, a magnetic tape containing computer instructions for steering the missile in a designated course.

guide See *waveguide*.

guide connector See *waveguide connector*.

guide edge The edge of paper tape that guides it into a paper tape reader.

guided missile A missile whose progress to a target is controlled electronically by signals from a control station or by sensing equipment aboard the missile. Also see *ballistic missile*.

guided propagation Also called ducting. In the atmosphere, air masses of different temperatures or humidity level cause refraction of electromagnetic waves, guiding the signal over long distances with very little attenuation.

guide elbow See *waveguide elbow*.

guide flange See *waveguide flange*.

guide gasket See *waveguide gasket*.

guide junction See *waveguide junction*.

guide load See *waveguide load*.

guide slot See *keyway*.

guide wavelength See *waveguide wavelength*.

Guillemin line In radar practice, a special pulse-forming network for controlling modulation pulse duration.

guillotine capacitor A variable capacitor in which a sliding (instead of rotary) plate moves between two stator plates. It is so called from its resemblance to the infamous beheading machine.

gulp Several bytes of digital information.

gun See *electron gun*.

Gunn diode A semiconductor diode that exhibits the Gunn effect.

Gunn effect The appearance of rf current oscillations in a dc-biased slab of n-type gallium arsenide in a 3 kV electric field.

Gunn-effect circuit Any circuit exploiting the Gunn effect, especially a Gunn oscillator.

Gunn oscillator A discrete semiconductor rf oscillator using the Gunn effect.

gutta percha A hard, rubberlike, organic insulating material. Dielectric constant, 3.3 to 4.9. Dielectric strength, 203 to 508 kV/in.

guy wire A bracing wire for antenna masts or towers.

gyrator An active (usually cascaded-transistor) device exhibiting nonreciprocal phase shift. It provides, among other functions, the simulation of inductors from capacitors.

gyro 1. Contraction of *gyroscope*. 2. A prefix meaning pertaining to gyroscopes, containing a gyroscope, or behaving like a gyroscope.

gyrocompass A type of compass in which a spinning gyroscope, acted upon by the earth's rotation, causes the device to point to true north. Compare *magnetic compass*.

gyrofrequency The natural frequency of rotation of charged particles around the earth's magnetic lines of force.

gyromagnetic Pertaining to the magnetic properties of rotating electric charges, e.g., the gyromagnetic effect of electrons spinning inside an atom.

gyropilot See *autopilot*.

gyroscope A device which consists of a spinning wheel mounted in a gimbal. The shaft of the wheel will point in one direction, despite the movement of the earth beneath it.

gyrostabilized platform See *stable platform*.

gyrostat See *gyroscope*.

G-Y signal In a color-TV circuit, the signal representing primary green (G) minus luminance (Y). A primary green signal is obtained when the G-Y signal is combined with the luminance (Y) signal. Compare *B-Y signal* and *R-Y signal*.

H 1. Symbol for *magnetic field strength*. 2. Symbol for *magnetizing force*. 3. Symbol for *hydrogen*. 4. Symbol for *unit function*. 5. Abbreviation of *horizontal*. (Also, *hor* and *horiz.*) 6. Abbreviation of *heater*. 7. Symbol for *henry*. 8. Symbol for *harmonic*. 9. Symbol for *hydrogen*.

h 1. Abbreviation of prefix *hecto*. 2. Symbol for the *Planck constant*. 3. Abbreviation of *hour*.

hadron A *subatomic particle* made up of *quarks*.

hafnium Symbol, Hf. A metallic element. Atomic number, 72. Atomic weight, 178.5. Hafnium readily emits electrons.

hailer 1. A marine microphone-amplifier-speaker system for calling to other boats or persons ashore. 2. A comparable system for land vehicles, such as police cars. Also see *megaphone, 1*.

hair See *hairline*.

hair hygrometer A hygrometer in which a stretched hair is the moisture-sensitive element.

hairline A fine line employed as an index or a graticule marker in a precision instrument.

hairpin coil A quarter-turn coil, so called from its resemblance to a hairpin.

hairpin coupling coil A hairpin coil used as a low-impedance primary or secondary coil for input or output coupling.

hairpin match A form of impedance-matching network used at the feed point of a half-wave dipole antenna. A short length of open-wire transmission line, short-circuited at the far end, is connected in parallel with the antenna at the feed point.

hairpin pickup A short, doubled length of wire, which acts as a pickup coil at very-high and ultra-high frequencies.

hairs The crosshairs in any optical device.

hairspring A fine, usually spiral spring, especially the one in a movable-

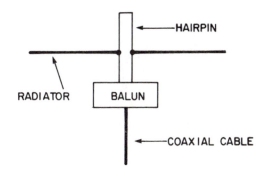

Hairpin match.

coil meter or the one connected to the balance wheel of a watch or clock.

hair-trigger Pertaining to extreme sensitivity of response, such as the tendency of a switching device to change state when excited by a weak pulse.

hair wire 1. An extremely thin wire filament in a lamp or bolometer. 2. Very small gauge wire, e.g., No. 44.

hal Abbreviation of *halogen*.

halation Blurring of an image on a photosensitive surface by unwanted light diffusion (as by poor contact between the negative and the print surface).

half-add The sum of two binary digits, in which the carry operation is omitted. Thus $0 + 0 = 0, 0 + 1 = 1, 1 + 0 = 1$, and $1 + 1 = 0$.

half-adder In digital systems, an adder circuit that can handle the two binary bits that are to be added but which cannot accommodate a carry signal. Compare *full adder*. Also see *adder* and *carry*.

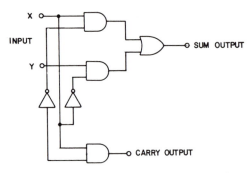

Half-adder.

half-bridge A bridge rectifier having diodes in two arms and resistors in the other two.

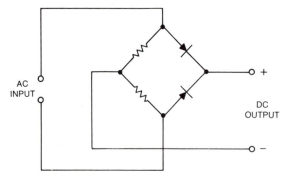

Half bridge.

half-cell A voltaic cell consisting of a single electrode immersed in an electrolyte and having a definite difference of potential; it is, in effect, half a primary cell. Also see *Helmholtz double layer*.

half-cycle Half of a complete ac alternation, i.e., 180 degrees.

half-cycle magnetizer A magnetizer employing half-cycles of rectified ac as the magnetic-field source.

half-duplex channel A communications channel in a half-duplex system.

half-duplex system In data communications, a system which transmits data in both directions, but not simultaneously. Compare *full-duplex system*.

half-lattice crystal filter A band-pass crystal-filter circuit employing two piezoelectric crystals in a four-arm bridge. Also see *crystal resonator*.

half-life See *natural disintegration*.

half-nut In a facsimile receiver, a device that guides the lead screw.

half-period Natural disintegration.

half-power point On a response characteristic curve, such as for selectivity, the points on each side of maximum at which the power is 3 dB below the peak value.

half-power width In a directional antenna system, an expression of the beamwidth of a radiated lobe. Usually expressed in the horizontal plane, but sometimes also in the vertical plane. The desired plane is first found, passing through the line corresponding to the direction of maximum radiation. In the plane, the lines are located corresponding to half the effective radiated power in the favored direction of the antenna. The half-power width is the angle, in degrees, between these two lines.

half-step In audio engineering and music, the frequency interval between two sounds whose ratio is 1.06:1.

half tap A bridging circuit or device that can shunt another circuit with the least electrical disturbance.

half-tone An expression of shading in a facsimile transmission.

half-track recorder A magnetic tape recorder that applies signals to both halves of a tape with a head that covers only half the tape's width in each of two directions. Also called *dual-track recorder*.

half-track tape Magnetic tape recorded by a half-track recorder.

half-wave Half of a complete wave, i.e., a complete rise and fall in one direction. Its graphic representation is similar in appearance to that for a half-cycle.

half-wave antenna An antenna whose radiator is effectively one-half wavelength from end to end. Such a radiator is actually about 5% less than a half-wavelength long because of ground-capacitance effects.

half-wave chopper A chopper that closes a circuit during only half the switching signal cycle.

half-wave dipole A Hertz antenna whose radiator is cut to half a wavelength. Also see *dipole antenna*.

half-wave doubler See *half-wave voltage doubler*.

half-wave feeder See *half-wave transmission line*.

half-wave radiator See *half-wave antenna*.

half-wave rectification The conversion of ac to dc during half of each ac cycle. Also see *half-wave rectifier*.

half-wave rectifier A rectifier which delivers a half-cycle of dc output for every other half-cycle of applied ac voltage. Because the successive dc half-cycles are 180 degrees apart, they have the same polarity. Compare *full-wave rectifier*.

half-wave transmission line A transmission line whose electrical length is half a wavelength at the transmission frequency.

half-wave vibrator A vibrator (see *interrupter*) whose reed operates against only one stationary contact. Compare *full-wave vibrator*.

half-wave voltage doubler A voltage-doubler circuit whose dc output has a ripple frequency equal to that of the ac supply. Although harder than the full-wave doubler to filter, this circuit has the advantage of a common ground. Compare *full-wave voltage doubler*.

halide A compound of a halogen. Examples: *sodium iodide*, used as a scintillating crystal; *ammonium chloride*, used as the electrolyte in a dry cell.

halide crystal A halogen-compound crystal such as mercuric iodide and sodium iodide, useful in detecting radioactivity.

Hall coefficient For a current-carrying conductor, the constant relationship between the Hall (transverse electric) field and the magnetic flux density.

Hall constant For a current-carrying conductor, the constant of proportionality in the equation $k = e/(im)$, where e is the transverse electric field (Hall field); i, current density; and m, magnetic field strength.

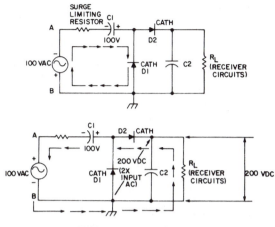

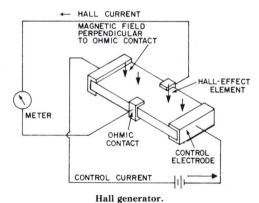

Half-wave voltage doubler.

Hall effect A phenomenon observed in thin strips of metal and in some semiconductors: When a strip carrying current longitudinally is placed in a magnetic field that is perpendicular to the strip's plane, a voltage appears between opposite edges of the strip that, although it is feeble, will force a current through an external circuit. The voltage is positive in some metals (such as zinc) and negative in others (such as gold). Also see *Ettinghausen effect, Nernst effect,* and *RighLeduc effect.*

Hall-effect modulator A device that makes use of the Hall effect for the purpose of modulating a signal, or for mixing two signals.

Hall-effect multiplier A device based upon the Hall generator and used in analog operations such as multiplication, extracting roots, etc.

Hall field The transverse electric field of a conductor carrying current in a magnetic field.

Hall generator A semiconductor device exhibiting the *Hall effect*. It is a thin wafer or film of indium antimonide or indium arsenide with leads on opposite edges.

Hall generator.

Hall mobility For a conductor or semiconductor, the product of conductivity and the Hall constant.

Hall network A resistance-capacitance null circuit whose general configuration is two cascaded high-pass tee-sections bridged by a high resistance. The circuit may be tuned with one potentiometer.

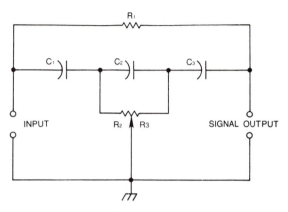

Hall network.

Hallwacks effect The phenomenon (observed by Hallwacks in 1888) in which ultraviolet light falling on a polished zinc plate causes the negatively charged electroscope to which it is connected to discharge.

halo See *afterglow* and *persistence.*

halo antenna A horizontally polarized vhf antenna having a circular dipole whose ends are capacitance loaded.

halogen Abbreviation, *hal.* A group of five very active nonmetallic elements whose members show similar chemical properties put them in group VIIA of the periodic table; they are astatine, bromine, chlorine, fluorine, and iodine.

halogen quenching The quenching of the gas discharge in a counter tube by introducing a halogen during manufacture.

Halowax A chlorinated naphthalene wax used as an impregnant for paper capacitors. Dielectric constant, 3.4 to 5.5. Resistivity, 10^{13} to 10^{14} ohm-cm.

halt A stop during the execution of a computer program run, as caused by a halt instruction, for example.

halt instruction An instruction in a computer program that causes a break in the program's execution, as by BASIC's STOP, for example.

ham Radio amateur. Also see *amateur, 2.*

Hamilton's principle The *principle of least action,* which shows, in connection with the law of conservation of energy, that motion will occur so that the integral of the product of kinetic energy (T) and elapsed time (*t*) is a minimum;

$$\int_{t1}^{t2} T\, dt = \text{a minimum.}$$

Also see *maxima and minima.*

hammer 1. The striking member in a wheel printer. 2. A large clapper in an electric bell or gong.

hammer-and-wheel See *wheel printer.*

Hamming code An error-correction code used in digital communications circuits.

hand capacitance Body capacitance, as evidenced between an operator's hand and sensitive circuit, for example.

hand generator An ac or dc generator operated by turning a hand crank.

Handie-Talkie Tradename for a portable transceiver small enough to be held in the hand during operation.

hand key A hand-operated telegraph key, as opposed to an automatic telegraph transmitter, such as a tape machine. The semiautomatic key belongs in this category.

hand-operated device A device manipulated by the operator's hand(s). Also called *manual device*.

hand punch A device for perforating data cards through the direct (electrically unassisted) pressure on a keyboard.

hand receiver 1. A single earphone that must be held against the ear. 2. A telephone receiver.

hand rules See *Fleming's left-hand rule, Fleming's right-hand rule,* and *right-hand rule for wire.*

handset See *cradlephone.*

handshaking 1. A controlled, periodic exchange of synchronizing pulses between a digital transmitter and receiver. 2. In a digital communications system, a method of error correction. The receiver detects nonstandard or improbable communications of data, and instructs the transmitter to repeat them for double-checking.

hand-type pointer In an electric meter, a spearlike pointer (like the hand of a clock), as opposed to a knife-edged pointer.

hand-wire Pertaining to electronic equipment wired by hand rather than being assembled from printed circuits or integrated circuits.

hang agc An automatic-gain-control circuit whose action is sustained for a brief interval after an actuating signal has passed, an advantage in some applications.

hangover In sound practice, the blurring or smearing of bass notes by a poorly damped or poorly mounted loudspeaker.

hangup 1. In phonograph operation, the state in which the same material is played repetitiously (i.e., the stylus does not move toward the spindle). 2. In digital-computer practice, an unexpected break during a program run due to software or hardware failure. Sometimes called *unexpected halt.*

H-antenna See *lazy-H antenna.*

hard copy 1. In digital computer practice, a readable document (printout) of material being translated to a form understood by a computer. 2. Generally, written or typed documents as opposed to data on other media such as punched cards, tape, etc.

hard-drawn wire High-tensile-strength unannealed wire.

hard dump See *hardware dump.*

hard magnetic material High-retentivity magnetic material. Also see *retentivity.*

hardness 1. The property that causes a material to resist penetration, deformation, scratches, etc.; antonym, *softness*. 2. The penetrative ability of X-rays.

hardness tester A device for measuring the hardness of a solid in terms of the force required to penetrate its surface. Also see *hardness, 1.*

hard rubber A rigid form of vulcanized rubber once used extensively as an insulator and molding material for electronic hardware (dials, knobs, sockets, terminal strips, etc.); it has been supplanted by plastics. The dielectric constant and dielectric strength of hard rubber are similar to those of Bakelite.

hard solder Solder that melts at a high temperature (usually red hot). Compare *soft solder.*

hard tube An electron tube that has been highly evacuated. Compare *soft tube.*

hard-tube sweep A sweep circuit based upon a vacuum tube, rather than on a gaseous tube. Also see *vacuum-tube sleep.*

hardware 1. Collectively, electronic circuit components and associated fittings and attachments. 2. In a computer system, the machinery associated with computation, especially a computer. Compare *software.*

hardware availability ratio A figure showing the availability of a computer system to do productive work; it is $(ta - td)/ta$, where ta is accountable time and td is downtime.

hardware check A check on data being transferred within a computer, as done by hardware, e.g., a parity check.

hardware cloth A finely woven wire screen used sometimes in place of a metal plate in antennas, reflectors, shields, and the like, and especially where air penetration is required.

hardware dump During a computer program run, data dumped to a storage device for later evaluation and occurring at the time of a failure. Also called *automatic hardware dump.*

hardware engineer A person who designs and perfects the actual electronic circuitry in a system. The hardware engineer is not involved with the programming of the system.

hardware recovery A computer system's ability (through software or hardware) to recover from a failure, i.e., to proceed from the point of failure.

hardware serviceability ratio See *hardware availability ratio.*

hardwire 1. To construct a circuit for direct-current conductivity. 2. A circuit exhibiting direct-current conductivity over a complete, closed path.

hard-wire telemetry See *wire-link telemetry.*

hardwood Tough, dense wood from an angiospermous tree, such as oak or maple. Dry hardwood is used occasionally as an insulator. Its dielectric constant lies between 2 and 7. Its dielectric strength is difficult to state, however, since it varies widely with the nature of the wood and the substance with which it is waterproofed.

hard X-rays High-frequency (shortwave) X-rays. Such radiation has high penetrating power. Compare *soft X-rays.*

Harkness circuit An early radio receiver circuit which was one of the first to employ the reflex principle: use of a single tube as an rf amplifier, detector, and af amplifier.

harmonic Symbol, H. In a complex wave, a signal component whose frequency is a whole multiple of the fundamental frequency.

harmonic accentuation Increasing the amplitude of harmonic components in a complex wave, as through the use of filters, amplifiers, or special modes of operation.

harmonic accentuator A circuit or device, such as a harmonic amplifier or bandpass filter, for emphasizing signal harmonics.

harmonic amplifier An amplifier, such as one used with a frequency standard, employed to increase the amplitude of weak harmonics. Also see *harmonic accentuation.*

harmonic analysis 1. Evaluating the harmonic content of a complex wave. See, for example, *harmonic wave analyzer; schedule method, 2; spectrum analyzer;* and *wave analyzer.* 2. Fourier analysis.

harmonic analyzer See *harmonic wave analyzer, spectrum analyzer,* and *wave analyzer.*

harmonic antenna An antenna that is operated at a whole multiple of its electrical length (wavelength).

harmonic attenuation Reduction of the amplitude of harmonic components in a complex wave, as through the use of filters, tuned amplifiers, or special modes of operation.

harmonic attenuator A circuit, device, or method of operation—such as a filter, tuned amplifier, special biasing, or special bypassing—for reducing the amplitude of harmonics.

harmonic component See *harmonic*.

harmonic composition See *harmonic distribution*.

harmonic content The amount of harmonic energy present in a complex wave. Also see *harmonic-distortion percentage* and *harmonic ratio*.

harmonic-cut crystal A quartzcrystal plate which, when operated in the proper circuit, oscillates at a harmonic of the (fundamental) frequency dictated by its thickness. In such operation, high frequencies can be obtained without grinding the plate dangerously thin.

harmonic detector A detector tuned to respond to a harmonic of a signal.

harmonic distortion 1. The generation of harmonics by the circuit or device by which the signal is processed. 2. The deformation of the original signal that results from the action described in 1, above. 3. The disproportionate reproduction of a signal's harmonic components.

harmonic distortion meter See *distortion meter*.

harmonic-distortion percentage In a signal exhibiting harmonic distortion, harmonic energy as a percentage of the total signal energy (fundamental plus all harmonics).

harmonic distribution The occurrence of harmonics by frequency, throughout a given spectrum.

harmonic elimination The complete removal of one or more harmonics from a complex wave, as through use of a filter or special modes of operation.

harmonic eliminator A circuit or device, such as a band-suppression filter, for removing harmonics.

harmonic filter 1. A bandpass filter for transmitting one or more harmonics of a complex input wave. 2. A bandsuppression filter for removing one or more harmonics of a complex input wave.

harmonic frequency 1. In a complex wave, a frequency which is a whole multiple of the fundamental frequency. 2. A frequency which is a whole multiple of another frequency to which it is referred. Compare *nonharmonic frequency*.

harmonic generator 1. An oscillator operated so that it generates strong harmonics of the fundamental frequency. 2. Frequency multiplier. 3. Sometimes, harmonic amplifier.

harmonic intensification See *harmonic accentuation*.

harmonic intensifier See *harmonic accentuator*.

harmonic interference Interference resulting from the harmonics of radio or test signals.

harmonic mean An average value (equal to the reciprocal of the arithmetic mean) of the reciprocals of several quantities. Thus, the harmonic mean of $x1$, $x2$, $x3...xn$ is $1/[1/x1 + (1/x2 + 1/x3 + ...1/xn/n)]$.

harmonic motion Periodic motion typified by a swinging pendulum and illustrated by the plot of a sine wave.

harmonic oscillator A crystal oscillator whose output frequency is a harmonic of the crystal frequency.

harmonic percentage See *harmonic-distortion percentage*.

harmonic producer 1. An oscillator that uses a tuning fork to establish the fundamental frequency. The output may be an odd or even harmonic of this frequency. 2. A frequency multiplier. 3. A nonlinear circuit employed in a calibrator to generate markers at integral multiples of the fundamental frequency.

harmonic progression A progression of the reciprocals of numbers forming an arithmetic progression. Thus, the series 1, 1/3, 1/5, 1/7, 1/9 is a harmonic progression, because 1, 3, 5, 7, 9 is an arithmetic progression. Compare *arithmetic progression* and *geometric progression*.

harmonic ratio 1. In a complex wave, the ratio of harmonic energy to total signal energy (fundamental plus all harmonics). 2. In a complex wave, the ratio of harmonic energy to fundamental frequency energy.

harmonic reducer See *harmonic attenuator*.

harmonic reduction See *harmonic attenuation*.

harmonic resonance Resonance of an antenna or a circuit at a whole multiple of the applied signal frequency.

harmonic ringing In wire telephony, the use of ac signal harmonics for selective ringing.

harmonic series The infinite divergent series $\Sigma\ 1/n$. Also see *harmonic progression*. Compare *hyperharmonic series*.

harmonic series of tones A set of audio-frequency tones in which the frequencies can be specified by f, 2f, 3f, 4f, and so on.

harmonic suppression See *harmonic elimination*

harmonic suppressor See *harmonic eliminator*.

harmonic tolerance The harmonic content permissible in a given system.

harmonic totalizer An instrument for measuring total harmonic distortion. See, for example, *distortion meter*.

harness A tied bundle of wires or cables for wiring electronic equipment.

harp antenna A vertical antenna consisting of a number of wires that fan out from point to points along a horizontal supporting wire.

Harry Phonetic alphabet code word for the letter *H*.

hartley A unit of digital information equivalent to 3.32 bits. Used in certain computer applications.

Hartley oscillator A self-excited oscillator circuit in which the two sections of a tapped tank coil supply the main coil and feedback coil required for a feedback system. The cathode of the tube (or equivalent transistor electrode) is connected to the tap.

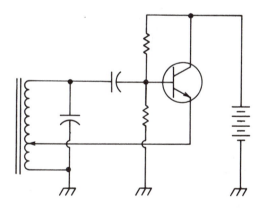

Hartley oscillator.

hash 1. Electrical noise, especially the sizzle from the discharge in gas tubes and mercury-vapor tubes. 2. Undesirable or purposefully meaningless information, as used in a hash total (checksum).

hash filter A small rf filter for eliminating the electrical noise from gas tubes and mercury-vapor tubes. Also see *hash*.

hat 1. A capacitance hat. 2. A procedure for randomizing data.

hash total See *checksum*.

hatchdot pattern A TV test pattern consisting of a *crosshatch pattern* with dots around its outer edges and one dot at its center.

hatted code A form of code in which randomization is used to maximize the difficulty of breaking the code.

hav Abbreviation of *haversine*.

haversine Abbreviation, hav. A trigonometric function equal to half the *versed sine*.

Hay bridge An ac bridge for measuring the inductance and Q of an inductor in terms of resistance, frequency, and a standard capacitance.

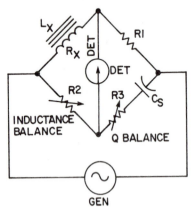

$$R_X = \frac{W^2 C^2 RI\ R2\ R3}{1 + (W^2 C^2 R_3^2)}$$

$$L_X = \frac{C_S\ R_I\ R_2}{1 + (R_3^2 W^2 C_S^2)}$$

$$Q_X = (WL_X / R_X)$$

Hay bridge.

haywire Loose, disorderly, or apparently careless wiring.

haz Abbreviation of *hazard*.

hazard Abbreviation, haz. A dangerous or potentially dangerous circuit, device, material, method, situation, or system, e.g., electric-shock hazard.

Hazeltine circuit See *neutrodyne, 1, 2*.

Hazeltine neutralization Plate neutralization (see *plate-neutralized amplifier*).

H beacon A form of homing beacon with an omnidirectional radiation pattern and a radio-frequency output of between 50 W and 2 kW.

H-bend See *H-plane bend*.

HCD Abbreviation of *hard-copy device*.

HCM Abbreviation of *half-cycle magnetizer*.

HCT Abbreviation of *heater center tap*.

HDF Abbreviation of *high-frequency direction finder*.

H-display See *H-scan*.

He Symbol for *helium*.

head 1. The top or operating portion of a device, e.g., microphone head, dynamic-speaker head. 2. In magnetic recording and reproduction, the magnetic device (transducer) that delivers or picks up recorded impulses.

head alignment 1. Positioning the cone of a dynamic speaker so that the voice coil moves freely, i.e., without rubbing against the core. 2. Positioning a magnetic-recorder head so that a proper relationship to the moving tape is maintained.

head amplifier A self-contained amplifier or preamplifier in the head of a microphone or sound-on-film pickup.

head degausser A device employed for the purpose of demagnetizing the head of a tape recorder. Unwanted magnetization can build up because of direct-current components in the driving signal.

head demagnetizer A head degausser.

header 1. A (usually glass) disk or wafer through which one or more leads pass and to which they are fully sealed. A header can be used as the bottom of the bulb in a miniature tube, or as the terminal base of an enclosed plug-in unit, such as a miniature coil, filter, or similar components. Also see *glass-to-metal seal*. 2. A data set placed before other sets as a means of identifying them and, possibly, including control data pertinent to the sets so identified.

header card In a batch of data cards, the first card which contains data about the following set of cards.

header capacitance Capacitance between the leads in a header.

header label A header recorded on a magnetic tape file.

head gap 1. In data processing equipment for reading from or writing to magnetic tape or disks, the distance between the head and the medium. 2. In audio practice, the spacing between tape-unit head electrodes; gap width.

heading The direction taken by a vehicle with reference to some point such as a radio beacon, true north, or magnetic north.

headlight In radar practice, a small rotating antenna.

headphone A transducer (miniature speaker) worn against the ear for (1) listening to music without disturbing others or (2) for monitoring live or recorded material without being disturbed by noise in the environment. Also see *receiver, 2*.

headphone amplifier An audio-frequency amplifier designed and operated primarily to supply a signal to headphones.

head room In tape recording, the region between the maximum recording level specified by the manufacturer of the equipment, and the amplitude at which tape overload occurs. It is specified in dB.

headset An assembly of usually two earphones, a headband, and a flexible cord. Also see *headphone* and *receiver, 2*.

head stack In magnetic recording, an assembly of two or more heads for multitrack service. Also see *head, 2*.

head station See *base station*.

head-to-tape contact In magnetic recording or playback, physical contact between the tape or disk and the head.

hearing aid A miniature audiofrequency device for amplifying sounds for the hard of hearing. It comprises a microphone, high-gain amplifier, and reproducer (earphone or bone-conduction transducer).

hearing loss The degree of hearing impairment expressed as the ratio (in dB) of the individual's threshold of hearing to normal threshold of hearing. Also see *audiologist*, *audiometer*, and *audiometrist*.

heart pattern See *cardioid pattern*.

heart telemetry See *ECG telemetry*.

heat A form of energy transferred—by conduction, convection, or radiation—between two bodies having different temperatures. The amount of heat is expressed in degrees, British thermal units, calories, joules, or kelvins.

heat aging 1. The degeneration of a substance, aggravated by high temperatures. 2. A test that indicates the immunity of a substance to degeneration because of high temperatures.

heat coil A device that disconnects a circuit when one temperature reaches a certain minimum level.

heat detector A sensor of heat. See, for example, *bolometer, infrared detector, microradiometer, radiometer, thermistor, thermocouple,* and *thermopile.*

heated-pen recorder See *thermal recorder.*

heated-stylus recorder See *thermal recorder.*

heated-wire flowmeter See *hot-wire flowmeter.*

heated-wire sensor A hot wire used to discriminate substances according to how they affect its heating. See, for example, *gas detector* (also usable as a vacuum gauge), *hot-wire anemometer,* and *hot-wire microphone.*

heat engine A machine that converts heat energy into mechanical energy.

heater 1. The filament of an indirectly heated tube. 2. The filament in an indirectly heated thermistor.

heater bias A dc voltage applied to the ac-operated heater of an electron tube to prevent diode rectification between the heater and another element, such as the control grid or cathode sleeve.

heater center tap 1. The center tap on the tube-heater winding of a power transformer. 2. The center tap of a resistor connected in parallel with the heater of an electron tube.

heater current Symbol, Ih. Current flowing through the heater element of an electron tube or thermistor.

heater hum Hum voltage induced in circuitry by ac fields from the heater element of a nearby electron tube or thermistor, or by heater-circuit wiring.

heater power Symbol, Ph. Power consumed by the heater element of an electron tube or thermistor.

heater power supply The ac or dc power source for the heaters of electron tubes or thermistors.

heater string A group of electron-tube heaters operated in series from a voltage that is the sum of heater voltages. Also see *series-filament operation.*

heater temperature Symbol, Th. The temperature of the heater element in an electron tube or thermistor.

heater voltage Symbol, Eh. The voltage across the heater element of an electron tube or thermistor.

heater-voltage coefficient The amount of frequency change per volt of fluctuation in the filament voltage of a klystron.

heater wattage See *heater power.*

heat exchanger A device or system that removes heat from a hot body and transfers it to another body or to the surrounding air.

heat-eye tube An infrared-sensitive device used for the purpose of locating objects in visible darkness. The tube consists of a cathode-ray device sensitive to infrared radiation.

heat gradient The temperature difference between two points on a body.

heating depth See *depth of heating.*

heating effect The production of heat (power loss) by electric current flowing in a conductor.

heating element 1. The heater in an electron tube or thermistor. 2. The resistance element (such as a strip or coil) that generates heat in an

electric-heating device.

heat loss 1. Loss of heat by its conduction, convection, or radiation from a hot body. 2. Power loss due to the heating effect of an electric current.

heat of fusion The amount of heat required to melt a unit mass of a solid that has reached its melting point.

heat of radioactivity Heat from radioactive disintegration (see *radioactivity*).

heat of reaction In a chemical or electrochemical reaction, the heat (in calories) absorbed or released.

heat of vaporization The amount of heat required to convert 1 gram of a liquid to a vapor without raising its temperature.

heat radiator See *heatsink.*

heat rays See *infrared rays.*

heat remover 1. Heatsink. 2. A forced-air or forced-liquid cooling system.

heat-resistant glass See *Pyrex.*

heat-sensitive resistor See *thermistor.*

heat-sensitive switch A make-and-break device, such as a thermostat, which is actuated by a change in temperature.

heat-shrink tubing An insulated flexible sleeving made from a plastic that shrinks permanently for a tight fit when heated; it is commonly used at the joint between a cable and connector.

heatsink A heat exchanger in the form of a heavy, metallic mounting base or a set of radiating fins; it removes heat from such devices as tubes, power transistors, or heavy-duty resistors, and radiates the heat into the surrounding air.

heatsink resistance The opposition offered by a heatsink to the flow of heat.

heat therapy 1. Using rf heating for therapeutic purposes. Also see *diathermy, 1.* 2. Using infrared rays for therapeutic purposes.

heat transfer The movement of heat from one point to another by absorption, conduction, convection, or radiation.

heatronic Pertaining to dielectric heating.

heat unit 1. British thermal unit. 2. Calorie. 3. Kelvin.

heat waves See *infrared rays.*

heat writer See *thermal recorder.*

Heaviside-Campbell bridge A form of mutual-inductance bridge. Mutual inductance is determined without regard to the operating frequency.

Heavyside layer See *Kennelly-Heavyside layer.*

heavy hydrogen An isotope of hydrogen. The term is applied to deuterium and tritium.

heavy metal A metal having a specific gravity of 5.0 or higher. Examples: iron (7.85-7.88), lead (11.3), nickel (8.6-8.9), mercury (13.6), platinum (21.4).

heavy water Formula, D20. Water in which deuterium rather than hydrogen has combined with oxygen.

hecto Abbreviation, h. A prefix meaning hundred(s), i.e., 10^2.

hectometric wave A long electromagnetic wave (100 to 1000 meters or 300 kHz to 3 MHz).

hectowatt Abbreviation, hW. A unit of power equal to 100 watts; 1 hW = 10^2W.

heelpiece A part of an electronic relay, which provides mechanical support for the armature.

Hefner candle A unit of luminous intensity equal to 0.9 candela; the standard (German) is the Hefner lamp.

Hefner lamp A standard light source whose luminous intensity is 0.9

candela. It burns amyl acetate (banana oil) and its flame has been the standard of the Hefner candle, a unit of luminous intensity devised in Germany. Also see *candle power* and *luminous intensity*.

height control In a TV receiver circuit, the potentiometer or rheostat that controls the vertical dimension of the picture by varying the amplitude of vertical scanning pulses.

height finder An altitude-measuring radar.

height-position indicator Abbreviation, HPI. A radar displaying the height of a target and its angularelevation slant range.

Heil oscillator An oscillator based on a special tube consisting of a heated cathode, first anode, metal cylinder, and second anode. Electrons emitted by the cathode pass through a hole in the first anode and become a beam which passes through the cylinder and strikes the second anode (collector). Electron bunching in the cylinder causes energy to be transferred to a tank circuit between the cylinder and anodes.

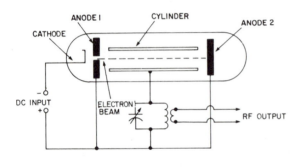

Heil oscillator.

Heisenberg uncertainty principle See *uncertainty principle*.

Heising modulation See *choke-coupled modulation*.

hekto See *hecto*.

heliacal cycle See *sunspot cycle*.

helical antenna A coil antenna mounted perpendicular to a flat metalplate reflector, an arrangement that produces circularly polarized waves in a narrow beam.

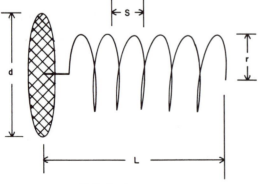

Helical antenna.

helical-beam antenna See *helical antenna*.

helical line The helix in a backwardwave oscillator or traveling-wave tube.

helical potentiometer A potentiometer whose resistance element is a wire or a coil wound into a coil of several turns. The slider moves over the wire (or the larger coil) from one end to the other as the slider or coil is turned through several complete revolutions. Also called *multiturn potentiometer*.

helical scanning Radar scanning by an antenna that moves vertically as it moves horizontally, giving a spiral motion to the radiated beam.

helical sweep See *spiral sweep, 1, 2*.

helical transmission line See *helical line*.

helicone An antenna used at ultra-high and microwave frequencies, consisting of a helical radiator within a cone-shaped reflector.

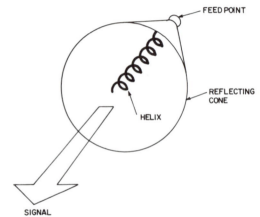

Helicone.

helionics The science of converting solar heat into electrical energy. The term is an acryonym from *helio* (the sun) and elect*ronics*.

heliostat 1. A servo-controlled motor driven device which drives a mirror to keep sunlight trained upon a specific target. 2. By extension, any similar device to keep a solar cell pointed to the sun.

helitron A form of oscillator used at ultra-high and microwave frequencies. The output frequency is variable over a wide range.

helium Symbol, He. A gaseous element. Atomic number, 2. Atomic weight, 4.003.

helium group The six inert gases in group 0 of the periodic table; they are argon, helium, krypton, neon, xenon, and radon.

helium-neon laser A laser in which the lasing substance is a mixture of helium and neon. Also see *helium* and *neon*.

helix 1. Single-layer coil. 2. That which is coil-shaped, i.e., spiral in configuration. 3. The spiral transmission line in a backward-wave oscillator or traveling-wave tube. (Also, *helical line*.)

helix line See *helical line*.

helix recorder An information recorder employing a spiral method of scanning. The recording medium is usually drum-shaped.

Helmholtz coil A device consisting of two crossed-field primary windings in which an inductively coupled secondary winding rotates.

The primary windings carry currents that differ in phase by 90 degrees. Rotating the secondary coil provides 360 degrees of continuously variable phase shift.

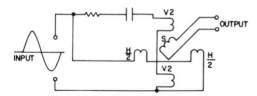

Helmholtz coil.

Helmholtz double layer An intermolecular layer between a metal and an electrolyte in which it is immersed. It is formed when the adhesive force between the metal and electrolyte decreases the surface tension of the metal, causing positive ions to migrate from the metal into the liquid. The metal, charged negatively, and the electrolyte, charged positively, form a capacitor whose dielectric is the Helmholtz layer.

Helmholtz resonator An acoustic resonance chamber whose geometry, in combination with the size of a small opening results in resonance at a specific frequency.

HEM Abbreviation of *hybrid electromagnetic* (see, for example, *hybrid electromagnetic wave.*).

hemimorphic An object with ends that have unlike faces.

He-Ne laser See *helium-neon laser.*

henry Symbol, H. The unit of inductance. It is the inductance of a closed circuit in which 1V is produced by a current changing uniformly at 1A-s. Separate entries are given for multiples of the unit.

heptode A tube having seven elements: anode, cathode, primary control electrode, and four auxiliary elements (usually grids).

hermaphroditic plug A plug that has the prongs of a male plug and the recessed contacts of a female plug. Compare *female plug* and *male plug.*

hermetic Sealed by a permanent, air-tight means.

hermetically sealed Constructed in manufacture so as to be permanently closed against the entry of air or other gases, dust, and moisture.

herringbone pattern A pattern of interference in a TV picture, so called because of its resemblance to the skeleton of a fish.

hertz Abbreviation, Hz. The unit of frequency (of periodic phenomena, such as alternating or pulsating currents); 1 Hz = 1 cycle per second.

Hertz antenna An ungrounded halfwave antenna fed by a transmission line attached to one end or to the center of the radiator. See, for example, *center-fed antenna* and *end-fed antenna.* Compare *Marconi antenna.*

Hertz effect Ionization of a gas, produced by intense ultraviolet radiation.

Hertzian antenna See *Hertz antenna.*

Hertzian oscillator See *Hertz oscillator.*

Hertzian radiation Radiation of electromagnetic (radio) waves.

Hertzian waves Radio waves.

Hertz oscillator A damped-wave generated of oscillations, employed by Hertz in his demonstration of radio waves in 1888 (verifying the earlier prediction by James Clerk Maxwell). The oscillator contains a spark gap supplied by an induction coil, attendant coils, capacitors (in the prototype, Leyden jars), and two large metal plates. Also see *sparkgap oscillator.*

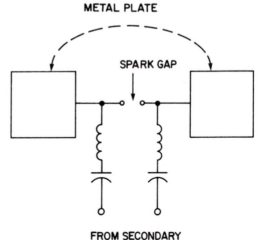

Hertz oscillator.

Hertz vector A single vector that specifies the electromagnetic field (electric and magnetic components) of a radio wave.

hesitation As distinct from a *halt*, a brief break during a computer program run during which internal operations are taking place, such as data transfer to a peripheral.

heterochromatic Made up of different frequencies, wavelengths, or colors. Compare *monochromatic, 1, 2.*

heterodyne 1. To beat one ac signal against another to produce one or more beat-frequency signals. Also see *beat frequency* and *beat note.* 2. Beat-note whistle. 3. To combine radio signals in a mixer, the purpose of which is to produce a sum or difference signal for further processing.

heterodyne detection 1. Signal detection by beating the incoming signal against one produced by a local oscillator. In this way, an unmodulated rf signal is made audible (the beat note is an audio frequency). 2. Signal detection by a superheterodyne circuit.

heterodyne detector 1. A detector that makes a radio-frequency signal audible by beating it against the rf signal of a local oscillator, the product being an audio-frequency beat note. 2. Superheterodyne detector (see *first detector, second detector,* and *superheteordyne circuit*). 3. A combination linear detector and local rf oscillator used to detect and measure the frequency of test signals. Also see *heterodyne frequency meter.*

heterodyne eliminator See *whistle filter.*

heterodyne filter See *whistle filter.*

heterodyne frequency The frequency of the beat note obtained by beating one signal against another.

heterodyne frequency meter A tunable rf meter which zero-beats an unknown frequency or one of its harmonics against a standard frequency (internally generated) or one of its harmonics.

heterodyne method See *heterodyne, 1.*

heterodyne oscillator A signal generator whose output is the beat product of outputs from two internal oscillators. The output frequency may be the sum or difference of the oscillator frequencies, as selected by output filtering or tuning. See *beat-frequency oscillator.*

heterodyne reception Radio reception (especially in telegraphy) by means of the beat-note process. Also see *heterodyne detection, 1* and *heterodyne detector, 1.*

heterodyne repeater A repeater in which the received signals are converted to another frequency before transmission. Most repeaters are of this type.

heterodyne-type frequency meter See *heterodyne frequency meter.*

heterodyne wave analyzer A wave analyzer similar to a superheterodyne receiver. The input (usually af) signal is heterodyned in a balanced modulator with that generated by an internal tunable oscillator. One of the resulting sidebands is then passed through a very sharp bandpass filter circuit (equivalent to a crystal i-f filter in a receiver) whose output actuates an ac voltmeter. A wave is analyzed with this instrument by tuning it successively to the fundamental and harmonics of the wave and noting the amplitudes indicated.

heterodyne wavemeter See *heterodyne frequency meter.*

heterodyne whistle See *heterodyne, 2.*

Heterofil A type of Wien-bridge heterodyne eliminator.

heterogeneous A group of objects or devices that have differing characteristics.

heterolysis Hydrolysis of a compound into two oppositely charged ions.

heuristic program In digital-computer practice, a program with which the computer solves a problem by trial and error, often learning in the process.

Heusler's alloys Ferromagnetic alloys of metals—such as aluminum, copper, and manganese—which are not themselves conspicuously magnetic.

hexadecimal number system An alphanumeric, base-16 system of number notation commonly used in machine language computer programming. The system employs the usual digits plus the letters *A* through *F* to represent the numbers 10 through 15 (since each place can only hold one symbol), e.g., decimal 44 154 is hexadecimal ACE $(10 \times 4096 + 12 \times 256 + 14)$.

hex inverter A collection of six digital inverters, or NOT gates, contained within one package, usually an integrated circuit.

hexode A six-element tube containing an anode, cathode, primary control electrode, and three auxiliary electrodes, e.g., a 6FG5.

Hf Symbol for *hafnium.*

hf Abbreviation of *high-frequency.*

Hg Symbol for *mercury.*

HH beacon In radionavigation, a 2 kW (minimum) nondirectional homing beacon.

hi 1. Contraction of *high.* 2. Radiotelegraph symbol for a laugh, often verbalized by radio amateurs.

HIC Abbreviation of *hybrid integrated circuit.*

hi-fi Abbreviation of *high-fidelity.*

high 1. Being of some potential above ground. 2. The on, true, or one state of a logic device. 3. Pertaining to the upper portion of a range, as in *high band* or *high frequency.* 4. Characterized by greater than normal response or performance, as in *high Q* or *high speed.*

high band 1. The higher or highest frequency band employed in communications, testing, or processing, when several such bands are available. 2. Vhf television channels 7 through 13. 3. The communications frequency range from about 144 MHz to about 170 MHz, as op-

posed to the *low-band* range of 30 to 70 MHz or so.

high boost In sound recording and reproduction, the emphasis of high frequencies in an operating spectrum. Also called *high-frequency compensation.*

high-C circuit A tuned circuit having high capacitance and low inductance at a given frequency. Such a circuit is characterized by high selectivity and low voltage. Compare *high-L circuit.* Also see *L-C ratio.*

high contrast In a picture, a limited range of gray values between black and white, or a similar condition in a color picture (overbright whites, little shadow detail). Also see *constrast.*

high definition In facsimile or television, a condition of minute detail, such that the original scene is faithfully reproduced.

high-efficiency grid modulation A system of grid-bias modulation that affords full class-C efficiency on modulation peaks. The circuit's configuration is similar to that of the Doherty amplifier, but in the grid-modulated amplifier, the grids receive a constant-amplitude rf excitation voltage with the af modulating voltage.

high-efficiency linear amplifier A linear amplifier (see *linear amplifier, 2*) affording higher plate efficiency than is obtainable with conventional class-B linear amplifiers. Efficiencies on the order of 60% at 100% modulation are possible. See, for example, *Doherty amplifier.*

high-energy materials See *hard magnetic materials.*

high-energy particle A *subatomic particle* which has been given high velocity by an accelerator (see *accelerator, 1*).

high-energy physics The discipline dealing with the characteristics, properties, and applications of *high-energy particles.*

higher-degree equations See *equations of higher degree.*

higher derivatives Derivatives of a function beyond the first (i.e., a second derivative is one derived from the first derivative, etc.). Also see *derivative* and *successive derivatives.*

higher-level language A computer language in which the operator is easily able to communicate with a computer. Generally, a higher-level language serves as an interface between a human programmer and the machine language. Examples of higher-order languages are BASIC, COBOL, and FORTRAN.

higher-level language A higher-level language.

high-fidelity Abbreviation, hi-fi. Pertaining to an audio-frequency system that is *very faithful* to the signal it is processing, i.e., one characterized by extremely low distortion and wide frequency response.

high-frequency Abbreviation, hf. Pertaining to frequencies in the 3-30 MHz band (wavelengths from 10 to 100 meters). Also see *radio spectrum.*

high-frequency alternator A dynamo for generating radio-frequency energy. Also see *alternator, 2, Alexanderson alternator, Bethenod alternator, Goldschmidt alternator,* and *Telefunken alternator.*

high-frequency bias In a tape recorder, a high-frequency sinusoidal signal superimposed upon the signal being recorded, for improving linearity and dynamic range.

high-frequency compensation See *high boost.*

high-frequency converter See *shortwave converter.*

high-frequency crystal See *harmonic crystal.*

high-frequency direction finder Abbreviation, HDF. A direction finder operated at high radio frequencies, i.e., higher than those of the standard broadcast band.

high-frequency heating Electronic heating of materials by high-frequency energy. See, for example, *dielectric heating* and *induction heating.*

high-frequency resistance See *radio-frequency resistance.*

high-frequency speaker See *tweeter.*

high-frequency trimmer 1. A low-value variable capacitor operated in parallel with a usually front-panel tuning capacitor to set the high-frequency end of the tuning range. See, for example, *oscillator trimmer.* 2. A small variable capacitor used in conjunction with a larger tuning capacitor, the function of which is to permit precision tuning of the larger device.

high-impedance voltmeter A voltmeter having a high input impedance (1M or higher).

high-L circuit A tuned circuit having high inductance and low capacitance at a given frequency. Such a circuit is characterized by low selectivity and high voltage. Compare *high-C circuit.* Also see *L-C ratio.*

high-level language A computer programming language allowing the programmer to write instructions in a more or less conventional form (i.e., math expressions with FOR-TRAN, English with COBOL or BASIC), and in which source statements correspond to several machine language instructions.

high-level modulation In an AM transmitter, introduction of the audio at the final stage of rf amplification so as to permit 100% modulation of the full-power signal.

high-level recovery Hardware recovery using data not involved in the failure, such as that on a magnetic storage medium. Also see *hardware recovery.*

highlight A bright area in a TV picture.

high-mu tube A triode having an amplification factor of 30 or higher, e.g., a 6DS4. Compare *low-mu tube* and *medium-mu tube.*

high-noise-immunity logic A form of bipolar digital logic designed for minimal sensitivity to noise. Abbreviated *HNIL.* Also known as high-threshold logic (*HTL*).

high order Descriptive of the relationship between bits or digits in a word or number: of two digits, the one holding the higher place value is the high order digit; e.g., 2 is the high order digit in 25.

high-pass filter A wave filter which transmits (with little attenuation) frequencies above a critical (cutoff) frequency and blocks frequencies below the cutoff value. Such filters may be entirely *LC, RC,* or a combination of these.

high-pass-filter method A method of measuring total harmonic distortion (percentage) using a high-pass filter to separate the harmonics from the fundamental, measuring the output voltage (E_o) of the filter, and comparing it with input voltage (E_i); distortion = 100 (E_o/E_i).

high-performance Pertaining to apparatus designed for continuous operation with maximum reliability.

high-performance navigation system Acronym HIPERNAS. A guidance system of the purely inertial and self-compensating type.

high-potential test A high-voltage test of insulation in which the applied voltage is continuously increased until the breakdown point of the dielectric is reached.

high-power rectifier A rectifier designed for high-voltage, high-current operation.

high Q For a component or circuit, a high value for the ratio X/R (reactance to resistance). This is a relative term, since a particular numerical value of Q considered high in one situation might be regarded as low under other circumstances. Also see *figure of merit, 1.*

high-resistance joint In the wiring of electronic equipment, a joint between conductors which is so poorly made it introduces a high resistance between the parts.

high-resistance voltmeter A voltmeter having a high input resistance (1M or higher).

high-speed carry In computer operation, a carry into a column causing a carry out, circumventing the usual intermediate adding circuit.

high-speed diode See *computer diode.*

high-speed flip-flop A flip-flop having short switch-on and switch-off time.

high-speed oscillscope An oscilloscope having excellent high-frequency and unit-function response and which, therefore, can reproduce high-speed pulses faithfully.

high-speed relay A relay having a short make or short break interval.

high-speed transistor See *switching transistor.*

high tension High voltage.

high-tension line A power-transmission line carrying a high voltage. Generally used for the transfer of electric power over long distances.

high-threshold logic See *high-noise-immunity logic.*

high-vacuum rectifier A tube-type diode (e.g., 5U4G) containing a filament and/or cathode and a highvoltage anode, as distinguished from a mercury vapor rectifier; now virtually obsolete, it was once used for power rectification.

high-vacuum tube An electron tube whose envelope has been highly evacuated; i.e., as much of the air as is practicable has been removed, and occluded gases have been driven from the internal electrodes and absorbed by getter material.

high voltage 1. A voltage considerably higher than those ordinarily encountered in a particular application. The term is comparative, since a few hundred volts might be high in one situation, whereas several thousand volts would be high in another. 2. In a cathode-ray tube, the voltage that accelerates the beam electrons. 3. In a TV circuit, picture-tube anode voltage.

high-voltage probe A special very high-resistance probe for measuring kilovolts with a low-range voltmeter.

hilac Abbreviation of *heavy-ion linear accelerator.*

hill-and-dale recording See *vertical recording.*

hinged-iron instrument An ac meter whose input transformer core is hinged in two parts. By means of a thumb trigger, the core may be opened and then closed around the current-carrying conductor that induces magnetism in the core; a secondary coil delivers current to the meter. Also called *clamp ammeter* or *clamp voltmeter.*

HIPERNAS Acronym for *high-performance navigation system.*

hipernick A high-permeability alloy of iron and nickel.

hipot A contraction of *high potential.* See *high-potential test.*

hiss 1. A high-pitched sound rich in sibilants (*s, sh,* and *z* sounds) produced by random high-frequency fluctuations in current. 2. The characteristic, high-pitched background noise (as in 1, above) accompanying superregeneration.

hiss filter See *hash filter.*

hit 1. Lightning stroke. (Also *direct hit.*) 2. The coincidence of two pulses.

H-lines Magnetic lines of force.

HLL Abbreviation of *high-level language.*

H-network A network of five impedances; two connected in series between upper input and output terminals, two between lower input and output terminals, and one shunted between the junctions of the seriesconnected impedances. Also called *H-pad, balanced tee-network,* and *balanced tee-pad.*

Ho Symbol for *holmium.*

hockey-stick lead On a capacitor, resistor, or other component, a pigtail

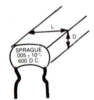

Hockey-stick lead.

lead which is given a single crimp for easy insertion into a printed-circuit board.

hodoscope   An instrument consisting essentially of closely spaced ion counters, for studying the path of an ionizing particle.

Hoffmann electrometer See *binant electrometer*.

hog horn A form of horn antenna used in microwave applications. Generally employed in the feed system of a dish antenna. The horn opening points in the direction of the feed waveguide.

hold 1. To retain data in a storage device after it has been duplicated in another location or device.2. A momentary halt of an operation or process. 3. In a TV receiver, a control that stabilizes the vertical or horizontal synchronization.

hold circuit 1. Holding circuit. 2. In a TV receiver, the circuit associated with the hold control(s). Also see *horizontal-hold control* and *vertical-hold control*.

hold control See *horizontal-hold control* and *vertical-hold control*.

hold current Symbol, Ih. The minimum current that will keep a normally open relay closed or a normally closed relay open.

hold electrode In a mercury switch, the electrode that is in permanent contact with the mercury.

holding anode The auxiliary (*keep-alive*) electrode in a mercury-arc rectifier.

holding beam In an electrostatic storage tube, the electron beam that regenerates replacement charges for those which were stored on the dielectric surface and lost. Also see *holding gun*.

holding circuit In an electromechanical relay, a separate circuit which, when energized, keeps the relay actuated.

holding coil In an electromechanical relay, the extra coil associated with the holding circuit.

holding current 1. Current in the holding coil of a relay. 2. In a gas tube, the minimum current required to maintain ionization.

holding gun In an electrostatic storage tube, the electron gun that generates the holding beam.

hold mode A condition in which the output state of a digital-logic circuit remains unchanged while the input signals are removed.

hold-off voltage The highest voltage that can be applied to a flashtube without causing it to fire.

holdover The flow of current through the ionized path created by an electric arc.

hold time The time permitted for a weld to harden in resistance welding.

hole 1. In a semiconductor atom, the vacancy resulting from the loss of an electron. When an electron is lost, so is a negative charge, leaving a hole that exhibits an equivalent positive charge which, like the electron, can apparently migrate as a current carrier. 2.The punched-out position of an information bit on a computer tape or card. 3. The punched-out portion of a chassis or panel, through which wires can be passed or components mounted.

hole conduction In a semiconductor material, electrical conduction, i.e., *hole current*.

hole current Symbol, Ih. In a semiconductor material, the current of apparently flowing holes.

hole density The degree of concentration of holes in a semiconductor. Also see *hole*.

hole-electron pair In a semiconductor, a hole and a related electron. Each electron in the conduction band has a counterpart in the valence band, a vacancy (hole) left by the movement of the electron to the conduction band.

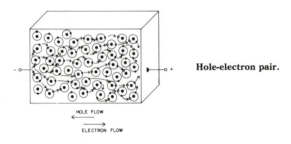

Hole-electron pair.

hole injection The creation of mobile holes in a semiconductor by applying an electric charge. Also see *hole*.

hole injector The emitter electrode of a pnp transistor; the metal whisker of a point-contact diode having an n-type wafer; or the p-layer of a forward-biased junction diode.

hole mobility Symbol, μh. The ease with which a hole moves within a semiconductor. Also see *carrier mobility*.

hole site On a computer tape or card, a place where a hole may be punched, the number of sites being the product of the number of rows and columns.

hole storage See *carrier storage*.

hole storage factor In a bipolar transistor biased to saturation, the amount of storage charge caused by excess base current.

hole trap In a semiconductor, an impurity that can cancel holes by releasing electrons to fill them.

Hollerith cards In data storage and processing, cards used for recording data in the form of punched holes which later are *read* by brushes or feelers which close a circuit through the holes. Also see *hole site* and *punched card*.

Hollerith code A system governing the positions (sites) at which holes are punched in *Hollerith cards* for recording data.

hollow-cathode tube A form of gas tube with a cylindrical cathode, having one closed end.

hollow coil A coreless inductor.

hollow conductor Tubing employed as a low-loss conductor at radio frequencies.

hollow core A core which is not solid throughout, especially one having a central mounting hole.

holmium Symbol, Ho. A metallic element of the rare-earth group. Atomic number, 67. Atomic weight, 164.94. Holmium forms highly magnetic compounds.

holocamera A camera for making holograms. Also see *hologram*.

hologram A wavefront recording made by the process of holography. By changing the frequency of the light transmitted by a hologram, various magnifications of the image can be obtained.

holography A method of producing a wavefront recording of an object

Character	12	11	0	1	2	3	4	5	6	7	8	9
A	x			x								
B	x				x							
C	x					x						
D	x						x					
E	x							x				
F	x								x			
G	x									x		
H	x										x	
I	x											x
J		x		x								
K		x			x							
L		x				x						
M		x					x					
N		x						x				
O		x							x			
P		x								x		
Q		x									x	
R		x										x
S			x		x							
T			x			x						
U			x				x					
V			x					x				
W			x						x			
X			x							x		
Y			x								x	
Z			x									x
0			x									
1				x								
2					x							
3						x						
4							x					
5								x				
6									x			
7										x		
8											x	
9												x
•		x				x					x	
'			x			x					x	

Hollerith code.

illuminated by laser light. The result, an interference pattern, is as meaningful to the viewer as the grooves on a phono disk; but when it is made to transmit (most often laser) light, an image results that is convincingly three dimensional.

Holt amplifier See crystal *amplifier, 2.*

homeostasis The condition of being in static equilibrium.

home station See *base station.*

homing 1. Guidance by means of an electronic beacon. The vehicle maintains a course toward the beacon. 2. Guidance by means of some form of emission from a target object. The emission may be sound, radio waves, infrared, visible light, or other energy.

homing antenna A direction-finding antenna, especially one on a mobile vehicle.

homing beacon A station radiating a beam for use in direction finding by mobile vehicles.

homing device A receiving device mounted on a mobile vehicle and which continuously indicates the direction of a selected transmitting station that is the vehicle's destination.

homing relay A stepping relay that returns to its starting position after each switching sequence. Also see *stepping switch.*

homing station See *homing beacon.*

homodyne reception See *zero-beat reception.*

homogeneous 1. Uniform in structure; similar at all points or locations. 2. Consisting of many identical elements.

homogous field A field whose lines of force in one plane pass through a single point.

homolysis The decomposition of a compound into a pair of neutral atoms or radicals.

homomorphism A one-to-one correspondence between the elements of two sets.

homopolar Pertaining to the union of atoms of the same polarity. *Nonionic.*

homopolar generator A dc generator whose poles have the same polarity with respect to the armature and, thus, rewuires no commutator.

homopolar magnet A magnet whose pole pieces are concentric.

honeycomb coil A multilayer coil having a universal winding.

honeycomb winding See *universal winding.*

honker A loudspeaker that favors middle frequencies. Compare *tweeter* and *woofer.*

hood A light shield for a cathode-ray tube; it allows the screen to be viewed with a minimum of interference from room light.

Hooke's law Strain is proportional to the stress that produces it, within the elastic limit. (Up to the elastic limit, Young's modulus is constant.)

hook switch A switch that closes a circuit when a headset or handset is lifted from the resting position. The common telephone receiver employs such a switch.

hook transistor A type of *pnpn* device in which the outer *p* and *n* layers serve as emitter and collector, the inner *n* layer being the base. This places a *p* layer between the base and collector, resulting in a transistor that provides high alpha as a result of carrier multiplication by the additional junction in the collector layer.

hookup See *schematic diagram.*

hookup wire Flexible wire (usually insulated) in electronic circuits.

hoop antenna See *cage antenna.*

hoot stop During a computer program run, a loop made evident by a sound signal.

hop In voice or telegraphy communications, the transmission of a wave and its subsequent reflection from the ionosphere.

hopper A device that holds punched cards ready for feeding to a reader or punch.

hor Abbreviation of *horizontal.* (Also, *H* and *horiz.*)

horiz Occasional abbreviation of *horizontal.* The usual form is *hor;* another alternate is *H.*

horizon The point where the sky and the earth seem to touch, i.e., the last visible part of the earth's surface from a given observation point. The horizon is the limit of some radio signals. Also see *artificial horizon.*

horizontal 1. Pertaining to that which is parallel to an assumed flat surface. 2. Pertaining to *width* deflection on a cathode-ray tube.

horizontal afc In a TV receiver circuit, automatic control of the horizontal sweep frequency to keep the set's horizontal scanning in step with that of the camera.

horizontal amplification Gain provided by the horizontal channel of a device, such as an oscilloscope, cathode-ray electrocardiograph, or TV receiver. Compare *vertical amplification.*

horizontal amplifier A circuit or device that provides horizontal amplification. Compare *vertical amplifier.*

horizontal angle of radiation For an antenna, the direction of maximum radiation given as an azimuth.

horizontal angle of deviation In a communications circuit, the angular difference, in degrees, between the compass direction from which a received signal arrives, and the great-circle path connecting the receiving station with the transmitting station.

horizontal axis The axis which is parallel to an assumed horizontal surface (of the earth, for example) or the one so represented in a diagram. Also see *x-axis, 1, 2*.

horizontal beamwidth In a directional antenna system, the angle, measured in the horizontal plane, between the half-power points in the major lobe.

BEAMWIDTH

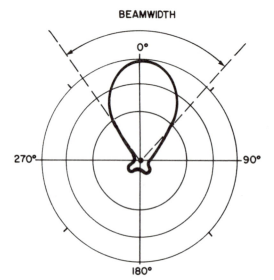

Horizontal beamwidth.

horizontal blanking See *horizontal retrace blanking*.

horizontal-blanking pulse In a television signal, the rectangular pedestal-shaped pulse which occurs between the active horizontal lines to achieve horizontal retrace blanking. Compare *vertical-blanking pulse*.

horizontal centering control See *centering control*.

horizontal channel The system of amplifiers, controls, and terminations which constitute the path of the horizontal signal in an equipment, such as an oscilloscope or graphic recorder. Compare *vertical channel*.

horizontal-convergence control In a color TV receiver, the variable component for adjusting the horizontal dynamic convergence voltage.

horizontal coordinates See *Cartesian coordinates*.

horizontal deflection In an oscilloscope or TV circuit, the lateral movement of the electron beam on the screen. Compare *vertical deflection*.

horizontal deflection coils The pair of coils in a deflection yoke that provide the electromagnetic field that horizontally deflects the electron beam in a magnetically deflected oscilloscope tube or TV picture tube. Also see *deflection coil*.

horizontal-deflection electrodes See *horizontal deflection coils* and *horizontal deflection plates*.

horizontal deflection plates In an oscilloscope tube (and some early TV picture tubes), a pair of plates which provide horizontal deflection. Compare *vertical deflection plates*.

horizontal directivity The radiation or reception pattern of a directional antenna in the horizontal plane.

horizontal-discharge tube In the horizontal-deflection circuit of a TV receiver, the tube that discharges a capacitor to generate the sawtooth scanning wave.

horizontal-drive control See *drive control*.

horizontal dynamic convergence During the scanning of a horizontal line in a color TV picture tube, convergence of the electron beams at the aperture mask. Compare *vertical dynamic convergence*.

horizontal field strength The field strength of signals passing through an antenna in a horizontal plane. Compare *vertical field strength*.

horizontal-field-strength diagram A plot of horizontal field strength, usually in polar form. Compare *vertical-field-strength diagram*.

horizontal flowcharting Flowcharting the movement of documents (rather than data) through a system.

horizontal frequency In television circuits, the horizontal scanning frequency, i.e., the frequency at which the horizontal lines are traced (15.750 kHz).

horizontal frequency response The gain-vs-frequency characteristic of the horizontal channel of an oscilloscope or graphic recorder. Compare *vertical frequency response*.

horizontal gain At a specified frequency, the overall amplification of the horizontal channel of an oscilloscope or graphic recorder. Compare *vertical gain*.

horizontal-gain control A control, such as a potentiometer, for adjusting horizontal gain. Compare *vertical-gain control*.

horizontal-hold control In a TV receiver, the control for adjusting the horizontal oscillator's frequency to prevent horizontal tearing of the picture. Compare *vertical-hold control*.

horizontal hum bars Dark, horizontal interferential bars in a TV picture, due to hum interference (see *hum, 1, 2*).

horizontal linearity Linearity of response (gain and deflection) of the horizontal channel of an oscilloscope, graphic recorder, or TV receiver. A linear picture is neither expanded nor contracted in any part. Compare *vertical linearity*.

horizontal-linearity control In an oscilloscope or TV receiver, the control with which horizontal linearity is adjusted. Compare *vertical-linearity control*.

horizontal line frequency See *horizontal frequency*.

horizontal lock See *horizontal-hold control*.

horizontally polarized wave A wave exhibiting horizontal polarization. Compare *vertically polarized wave*.

horizontal multivibrator In a TV circuit, a 15.750 kHz multivibrator which originates the horizontal sweep signal.

horizontal oscillator In a TV receiver, the oscillator (usually a multivibrator) that generates the horizontal sweep signal. Compare *vertical oscillator*.

horizontal output stage In a TV receiver, an output amplifier following the horizontal oscillator. Compare *vertical output stage*.

horizontal output transformer In a TV circuit, the output transformer in the horizontal-oscillator-outputamplifier section. Also called *flyback transformer*.

horizontal output tube In a TV circuit, the tube that boosts the horizontal-oscillator signal and drives the flyback transformer.

horizontal polarization The orientation of an electromagnetic wave whose electrostatic component is horizontal. Compare *vertical polarization*.

horizontal positioning control See *centering control*.

horizontal quantity The quantity measured along the *x*-axis of a graph represented by the horizontal deflection of an oscilloscope beam. Compare *vertical quantity*.

horizontal recording See *lateral recording*.

horizontal repetition rate See *horizontal frequency*.

horizontal resolution In a TV picture, the number of picture elements that can be discerned in a horizontal scanning line. Compare *vertical resolution*.

horizontal retrace In a cathode-ray device, such as an oscilloscope or TV receiver, the beam snapping back to its starting point after completing its horizontal sweep of the screen. Compare *vertical retrace*.

horizontal retrace blanking In oscilloscopes and TV receivers, the automatic cutoff of the electron beam during a horizontal retrace period, preventing an extraneous line on the screen during the period. Also see *blackout, 2; blank, 2; blanking interval, 1, 2; horizontal retrace;* and *horizontal retrace time.* Compare *vertical retrace blanking*.

horizontal scanning 1. The lateral sweeping of the electron beam in a cathode-ray tube. 2. Sampling *x*-axis values in a repetitive or nonrepetitive sweep of that axis.

horizontal scanning frequency See *horizontal frequency*.

horizontal sensitivity The signal voltage required at the input of a horizontal channel for full horizontal deflection. Also see *horizontal gain.* Compare *vertical sensitivity*.

horizontal signal A signal serving as a *horizontal quantity.* Compare *vertical signal*.

horizontal sweep 1. In a cathode-ray tube, the sweeping of the spot back and forth on the screen. 2. The circuit that produces horizontal sweep.

horizontal sweep frequency The frequency at which horizontal sweep takes place; in a TV receiver, it is 15.750 kHz. Also called *horizontal sweep rate.* Compare *vertical sweep frequency*.

horizontal sweep rate See *horizontal sweep frequency*.

horizontal-sync discriminator In a TV system, a circuit that compares horizontal sync-pulse phase with the phase of the signal from the horizontal sweep oscillator.

horizontal synchronization In a TV receiver, synchronization of the horizontal component of scanning with that of the transmitting camera. Also see *horizontal sync pulse.* Compare *vertical synchronization*.

horizontal sync pulse In a video signal, the pulse that synchronizes the horizontal scanning component in a TV receiver with that of the camera; it also triggers horizontal retrace and blanking. Also see *back porch.* Compare *vertical sync pulse*.

horizontal wave See *horizontally polarized wave*.

horizontal-width control See *width control, 1, 2*.

horn A radiating device that is essentially a cylindrical or rectangular pipe whose surface flares from a narrow entry to a wide exit. See, for example, *horn antenna; horn speaker;* and *megaphone, 2*.

horn antenna An antenna that resembles a round, square, or elliptical horn.

horn cutoff frequency The lowest frequency at which an exponential horn will function properly.

horn loading In a sound-transmission system, a form of propagation that makes use of a horn-shaped speaker.

horn mouth The wider (radiating) end of a horn antenna or speaker. Compare *horn throat*.

horn radiator See *horn antenna*.

horn speaker A loudspeaker employing a horn in conjunction with an acoustic transducer.

horn throat The narrower (input) end of a horn antenna or speaker. Compare *horn mouth*.

horsepower Abbreviation, hp. A unit of power equal to 746 watts.

horsepower-hour Abbreviation, hp-hr. A unit of energy or work equal to 3600 joules (746 watts/hour).

horseshoe coil See *hairpin coil* and *hairpin coupling coil*.

horseshoe magnet A (usually permanent) magnet having the shape of a horseshoe or a *U* and a rectangular cross section.

host A programmable computer that gathers and stores the information from all of the data-entry terminals in a system.

HOT Abbreviation of *horizontal output transformer.* 2. Abbreviation of *horizontal output tube*.

hot 1. Heated. 2. Electrically live. 3. Ungrounded; above ground.

hot carrier In a semiconductor, a carrier (electron or hole) whose energy is higher than that of majority carriers normally encountered in the same material.

hot-carrier diode A semiconductor diode having a metal base which receives hot carriers from a semiconductor layer. The unit has a fast switching speed, since there are virtually no minority carriers, either injected or stored.

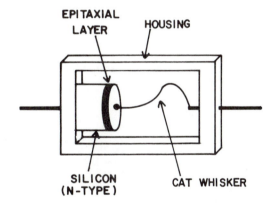

Hot-carrier diode.

hot cathode In an electron tube, a cathode that is directly or indirectly heated. Compare *cold cathode*.

hot-cathode tube A thermionic tube, i.e., one in which the source of electrons is a hot cathode. Compare *cold-cathode tube*.

Hotel Phonetic alphabet code word for the letter *H*.

hot-filament gas detector A *gas detector* in which the sensor is a heated filament acting as one arm of a Wheatstone bridge circuit. The bridge, previously balanced, becomes unbalanced when gas impinges upon the filament and changes its resistance.

hot junction The heated junction in a two-junction thermocouple circuit. Compare *cold junction*.

hot line 1. An energized wire, transmission line, or busbar. 2. A private communications channel (wire or radio) kept in constant readiness for instant use between persons of high authority.

hot-pen recorder See *thermal recorder*.

hot plate A metal device, usually heated by means of electricity, used for the purpose of conducting certain experiments.

hot resistance The resistance of a component during its operation, i.e., after is has been heated by ambient temperature or internal power dissipation. Compare *cold resistance*.

hot spark A brilliant flash seen when a capacitor discharges through a spark gap in a vacuum. This spark is a source of extremely high-frequency radiation (between 10^{-7} and 2×10^{-7} meter).

hot spot 1. In a circuit or component, an area whose temlerature is ordinarily higher than that of the surrounding area, e.g., a glowing spot on the plate of a vacuum tube. 2. Current loop. 3. In communications practice, a geographic location in which reception is markedly better than in other nearby places or from which the transmitted signal appears to be stronger.

hot-strip ammeter A current meter similar to the hot-wire meter, except that it has a heated metallic strip instead of a heated wire.

hot-stylus recorder See *thermal recorder*.

hot-tip writing The use of a heated-tip stylus in graphic recording. Also see *thermal recorder*.

hot-wire ammeter See *hot-wire meter*.

hot-wire anemometer An electrical anemometer whose indication is based on the cooling effect of the wind on a heated filament.

hot-wire flowmeter An instrument for determining the rate of flow of a gas in a pipe or other channel. The circuit is similar to that of the gas detector and hot-wire anemometer.

hot-wire gasmeter See *gas detector*.

hot-wire instrument See *hot-wire meter*.

hot-wire meter A meter in which current heats a wire, stretching it so it moves a pointer across a scale over a distance proportional to current strength.

hot-wire microphone A microphone in which sound waves vary the temperature of a heated wire and, accordingly, its electrical resistance.

hot-wire relay A time-delay relay in which actuating current heats a wire, causing it to expand, eventually opening or closing the contacts. Also see *delay relay*.

hot-wire sensor See *heated-wire sensor*.

hot-wire transducer See *hot-wire microphone*.

hour Abbreviation, h; (sometimes, hr). A unit of time measure equal to 60 minutes or 3600 seconds. Compare *minute, 1* and *second, 1*. Also see *time*.

housekeeping In digital-computer practice, the part of a program that attends to chores (setting variables to zero, e.g.,) rather than being involved in making computations for a solution.

howl A discordant sound produced in headphones or loudspeaker, as by acoustic or electrical feedback.

howler 1. An audio-frequency alarm device. 2. A sound-emitting test device (see *growler, 1, 2*).

howl repeater A form of electric feedback in which a hum or howl occurs because of oscillation. The term is used to describe an oscillating condition in a wire-communications-system repeater.

hp Abbreviation of *horsepower*.

h-p Abbreviation of *high-pressure*.

H-pad See *H-network*.

h-parameters Parameters of the fourterminal network equivalent of a transistor. They are called hybrid parameters (thus, *h*) because of their appearance in mesh and nodal equations. The basic *h*-parameters are $h11$, input resistance with output short-circuited; $h12$, reverse voltage ratio with input open-circuited; $h21$, forward current gain with output short-circuited; and $h22$, output conductance with input open-circuited.

h-particle A positive hydrogen ion or proton obtained by bombarding a hydrogen atom with alpha particles or high-velocity positive ions.

HPF Abbreviation of *highest probable frequency*.

hp-hr Abbreviation of *horsepower-hour*.

HPI Abbreviation of *height-position indicator*.

H-plane The plane of the magnetic field of an antenna. Compare *E-plane*.

H-plane bend In a waveguide, a smooth change in the direction of the axis perpendicular to the direction of polarization.

H-plane tee-junction A waveguide tee-junction whose structure changes in the magnetic-field plane. Also see *waveguide junction* and *waveguide tee*.

hr Abbreviation of *hour*. (Also *h*.)

H-scan A radar display which is in effect a modified B-display. On the screen, the target is represented by two close blips approximating a line whose shape is proportional to the sine of the target's angle of elevation.

H-scope See *H-scan*.

HSM Abbreviation of *high-speed memory*.

HTL Abbreviation of *high-threshold logic*.

hub The hole in the center of a magnetic tape reel.

hue The quality of having a particular color, i.e., *hue* is generally what is meant when the word *color* is used. Hue is the attribute of an object that is dependent upon the frequency of the light involved.

hue control The tint control in a color television receiver. The hue control allows adjustment of the color wavelength, but does not affect the saturation (intensity).

hum 1. A low-pitched sound. 2. The unheard effects of hum in terms of interference, such as moving horizontal bars on a TV screen.

human engineering The branch of engineering devoted to interfacing human beings with the machines and instruments they operate. Both a science and an art, the discipline is concerned with the safest and most efficient design, arrangement, and operation of equipment.

human interface The *interface* between a computer (or other sophisticated device) and a human operator.

hum-balance potentiometer A potentiometer connected across an ac supply, such as the filament line of a vacuum tube, with its slider grounded; at a given setting, hum interference is nulled.

hum bars See *horizontal hum bars*.

hum bucking Reducing hum interference by introducing an ac voltage of the same frequency and amplitude as the hum, but opposite in phase.

hum-bucking coil An auxiliary coil used in conjunction with the field and voice coils of an electrodynamic speaker to reduce hum interference by *hum bucking*.

hum field The magnetic field surrounding a conductor carrying hum-frequency alternating current. Also see *hum, 1, 2*.

humidity The amount of moisture in the air. Also see *absolute humidity* and *relative humidity*.

humidity meter See *electric hygrometer* and *electronic hygrometer*.

humidity sensor A pickup whose resistance or capacitance varies proportionally with ambient humidity.

hum interference Electrical interference from a hum field.

hum loop A ground loop that results in undesired hum in the output of an amplifier.

hummer A nonelectronic audio oscillator similar to the fork oscillator, but employing a thick, metal reed instead of a tuning fork. A carbon microphone button attached to the reed provides the feedback path necessary for sustained oscillation.

humming In a transformer, a noise produced by vibrations in the laminated core when a large amount of current is drawn.

hum modulation Modulating a signal by hum interference.

hump 1. A sine-wave half-cycle. 2. A curve similar to 1, above. See, for example, *double-hump resonance curve* and *double-hump wave*.

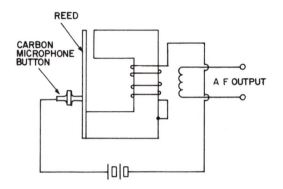

Hummer.

hunting The condition in which an electrical or mechanical system oscillates about a mean mode of operation ("hunts" for the mode), sometimes eventually settling down at the mode. It is often due to overcompensation in automatic systems.

Huygens' principle The observation that an advancing wave is the resultant of secondary waves that arise from points in the medium that have already been passed.

hv Abbreviation of *high voltage*.

H-vector A vector representing the magnetic field of an electromagnetic wave. Compare *E-vector*.

HVR Abbreviation of *high-vacuum rectifier*.

hW Abbreviation of *hectowatt*.

H-wave mode In a waveguide, a mode of transmission in which the electric lines of force are at right angles to the direction of the waveguide. Also sometimes called transverse-electric (*TE*) mode.

hy Occasional abbreviation of *henry*. The SI abbreviation and symbol, *H*, is preferred.

hybrid Descriptive of a device which is an offspring of other devices or a product of dissimilar technologies (but employing elements of each). See, for example, *hybrid junction* and *hybrid coil*.

hybrid active circuit An active circuit—such as an amplifier, oscillator, or switch—employing a combination of two dissimilar active devices, e.g., *transistor and tube*.

hybrid coil A special type of bridging transformer employed in wire telephony to prevent self-oscillation in a repeater amplifier that operates in both directions.

hybrid computer A computer system incorporating analog and digital computers.

hybrid electromagnetic wave Abbreviation, HEM wave. An electromagnetic wave whose electricfield and magnetic-field vectors are both in the direction of propagation.

hybrid IC See *hybrid integrated circuit*.

hybrid integrated circuit Abbreviation, HIC. An integrated circuit embodying both integrated and microminiature discrete components, i.e., one combining both monolithic and thin-film construction.

hybrid junction 1. Magic-tee. 2. A four-terminal device, such as a resistor circuit, special transformer, or waveguide assembly, in which a signal applied to one pair of terminals divides and appears at only the two adjacent terminals.

hybrid microcircuit A microcircuit containing diffused or thin-film

elements interconnected with separate chip elements.

hybrid parameters See *h-parameters*.

hybrid ring A hybrid waveguide junction (see *hybrid junction, 2*) consisting essentially of a reentrant line with four side arms and used as an equal power divider.

hybrid-tee See *hybrid junction, 1*.

hybrid thin-film circuit A microcircuit in which semiconductor devices and discrete components are attached to passive components and conductors that have been electrodeposited on a substrate.

hybrid transformer See *hybrid coil*.

hydroacoustic Pertaining to the sound of fluids flowing under pressure.

hydroacoustic transducer A transducer that converts energy from the high-pressure flow of a fluid into acoustic energy.

hydrodynamic pressure The pressure of a fluid in motion. Compare *hydrostatic pressure*.

hydroelectric Pertaining to the production of electricity by water power, as by a generator turned by water turbines.

hydroelectric machine A device for generating electricity from highpressure steam escaping from a series of jets.

hydroelectric power See *water power*.

hydrogen Symbol, *H*. A gaseous element. Atomic number, 1. Atomic weight, 1.008. Hydrogen is extensively used in making semiconductor materials and is found in the envelopes of some thyratrons. Compare *deuterium* and *tritium*.

hydrogen atmosphere The nonoxidizing atmosphere in which semiconductor materials are melted and processed and in which semiconductor crystals are grown. Occasionally, helium is used instead of hydrogen.

hydrogen atom A single atom of the element hydrogen. It is an extremely simple atom consisting of 1 electron and 1 proton.

hydrogen-ion concentration See *pH*.

hydrogen lamp A glow-discharge lamp that produces light by means of the ionization of rarefied hydrogen gas. Visible light is emitted at discrete wavelengths.

hydrogen thyratron A thyratron containing hydrogen under low pressure.

hydrokinetic Pertaining to fluids in motion or the forces behind such motion.

hydrolysis The process whereby chemical substances become ionized in water solution, producing electrolytes.

hydromagnetics See *magnetohydrodynamics*.

hydromagnetic wave In a fluid, a wave in which the energy is propagated via magnetic and dynamic modes.

hydrometer An instrument for measuring the specific gravity of liquids.

hydrophone An underwater sound-toelectricity transducer (microphone).

hydropower See *water power*.

hydrostatic pressure The pressure of a fluid at rest. Compare *hydrodynamic pressure*.

hygrograph A graphic recorder of humidity.

hygrometer An instrument for measuring humidity. Also see *electric hygrometer*, *electronic hygrometer*, and *hair hygrometer*.

hygroscopic material A material that absorbs moisture from the air, but not enough to get wet, e.g., lime, silk. Compare *deliquescent material*.

hygrostat A humidity-sensitive relay or switching circuit.

hygrothermograph A graphic recorder indicating humidity and temperature on the same chart.

hyperacoustical zone In the upper atmosphere, a region higher than 60 miles in which the distance between air molecules is approximately equal to the wavelengths of sound, and beyond which sound cannot be propagated.

hyperbola A conic-section curve having an eccentricity greater than 1 and satisfying the equation $x^2 - y^2 = 1$.

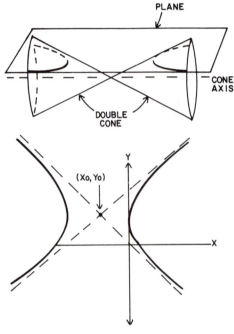

Hyperbola.

hyperbolic angle An angle subtended by a sector of a hyperbola in a manner analogous to that in which a circular angle is subtended by an arc of a circle.

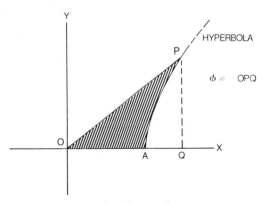

Hyperbolic angle.

hyperbolic contangent Abbreviation, coth. The cotangent of a hyperbolic angle. For a hyperbolic angle ϕ, coth ϕ = cosh ϕ/sinh ϕ. In exponential terms, coth ϕ = $(e^{\phi} + e^{-\phi})/(e^{\phi} - e^{-\phi})$. Also see *hyperbolic angle* and *hyperbolic functions.*

hyperbolic cosecant Abbreviation, cosech or csch. The cosecant of a hyperbolic angle. For a hyperbolic angle ϕ, cosech ϕ = 1/sinh ϕ. In exponential terms, cosech ϕ = $2/(e^{\phi} - e^{-\phi})$. Also see *hyperbolic angle* and *hyperbolic functions.*

hyperbolic cosine Abbreviation, cosh. The cosine of a hyperbolic angle. For a hyperbolic angle ϕ, cosh ϕ = OQ/OA, where OQ is the x-axis length of the radius projected to that axis, and OA is the x-axis distance from the origin to the hyperbola. In exponential terms, cosh ϕ = $(e^{\phi} + e^{-\phi})$. Also see *hyperbolic angle* and *hyperbolic functions.*

hyperbolic-cosine horn See *catenoidal horn.*

hyperbolic error 1. In an interferometer, a miscalculation in the direction of arrival of a signal. The signal from one antenna in the system may be assumed to be in phase with the signal from another antenna, when actually the two components differ by an integral number of whole wavelengths. 2. The angular error, in degrees, minutes, or seconds of arc, resulting from a miscalculation of phase in an interferometer.

hyperbolic exponential horn A horn whose shape follows the equation $S = S_0[(\cosh mx) + (T \sinh mx]^2$, where S is the cross-sectional area at distance x along the axis (sq ft); S_0, the cross-sectional area at the throat (sq ft); m, flaring constant; x, distance along the axis (ft); and T, shape parameter (usually 0.5 to 0.7).

hyperbolic face contour See *hyperbolic grind.*

hyperbolic functions The nonperiodic functions of a hyperbolic angle. Hyperbolic functions are related to the hyperbola as common trigonometric functions (see *circular functions*) are related to the circle. Hyperbolic functions are very useful in calculations involving traveling waves on transmission lines, attenuator design, and wave-filter design. Also see *hyperbolic cosine, hyperbolic cosecant, hyperbolic contangent, hyperbolic secant, hyperbolic sine,* and *hyperbolic tangent.*

hyperbolic grind The shape (approximate hyperbola) to which the face of a magnetic recording head is ground to provide good contact with the tape and to insure good high-frequency response.

hyperbolic horn A horn in which the increase in cross-sectional radius follows a hyperbolic law. See, for example, *catenoidal horn* and *hyperbolic exponential horn.*

hyperbolic logarithm A natural logarithm (see *Napierian logarithm*).

hyperbolic navigation A radionavigation system in which the operator of an aircraft or boat determines his position by comparison of two received signals. The two transmitters radiate signals from known positions and with known timing characteristics. The time delay from each transmitter is determined, resulting in two hyperbolic curves on a map. The point of intersection of the curves is the location of the aircraft or ship.

hyperbolic radian A unit of measure derived from a *hyperbolic angle.* A hyperbolic radian is the hyperbolic angle that encloses an area of 0.5 when the distance along the x-axis to the hyperbola is unity. Also see *hyperbolic angle.* Compare *circular radian.*

hyperbolic secant Abbreviation, sech. The secant of a hyperbolic angle. For a hyperbolic angle ϕ, sech ϕ = 1/cosh ϕ. In exponential terms, sech ϕ = $2/(e^{\phi})$. Also see *hyperbolic angle* and *hyperbolic functions.*

hyperbolic sine Abbreviation, sinh. The sine of a hyperbolic angle. For a hyperbolic angle ϕ, sinh ϕ = PQ/OA, where PQ is the y-axis distance to the intersection of the radius and the hyperbola, and OA is the x-

axis distance from the origin to the hyperbola. In exponential terms, sinh ϕ = 1/2 ($e^\phi - e^{-\phi}$). Also see *hyperbolic angle* and *hyperbolic functions*.

hyperbolic tangent Abbreviation, tanh. The tangent of a hyperbolic angle. For a hyperbolic angle ϕ, tanh ϕ = ($e^\phi - e^\phi$)/($e^\phi + e^{-\phi}$). Also see *hyperbolic angle* and *hyperbolic functions*.

hyperbolic trigonometry The branch of mathematics dealing with the theory and application of hyperbolic angles and their functions.

hyperfocal distance The shortest distance to which a lens may be focused without degrading definition at infinity.

hyperfrequency waves See *microwaves*.

hyperharmonic series The infinite series Σ $1/n^D$ The series is convergent when p is greater than one, and divergent when p is less than or equal to one. Compare *harmonic series*. Also see *mathematical series*.

hypernik See *hipernick*.

hyperon Any one of various particles having a mass greater than that of a neutron or proton.

hyperpolarization The production of an increased emf across a biological membrane.

hypersonic Pertaining to speeds appreciably higher than that of sound—approximately 344 meters (1129 feet) per second in air.

hypersyn motor A high-efficiency, high-power-factor synchronous motor combining the advantages of the dc-excited synchronous motor (stiffness), hysteresis motor (synchronizing torque), and induction motor (high starting torque).

hypervelocity Velocity in excess of 3 kilometers per second.

hypex horn See *hyperbolic exponential horn*.

hypotenuse The side of a triangle opposite the right angle.

hypothesis Something—such as an idea, impression, or system—that seems to be true or workable, but must be subjected to logical analysis and/or practical testing to prove its verity. Compare *law* and *theory*.

hypsometer An altimeter in which a thermistor (connected to a battery and current meter) is immersed in a boiling liquid. Because the liquid's boiling point is proportional to altitude, it affects the resistance of the thermistor and, hence, the deflection of the meter.

hysteresigram The hysteresis-curve record produced by a hysteresigraph.

hysteresigraph A graphic recorder that displays or records the hysteresis curve for a material. Also see *hysteresiscope*.

hysteresimeter See *hysteresis meter*.

hysteresis 1. The tendency of a magnetic material to saturate and retain some of its magnetism after the alternating magnetic field to which it is subjected reverses polarity, thus causing magnetization to lag behind the magnetizing force. 2. A similar electrostatic action in a ferroelectric dielectric material. 3. By extension, a condition in which a variable quantity decreases at a rate different from that at which it increases, the plot for this being a double-line curve. Thus, defective semiconductor diodes sometimes exhibit such a double-loop (*hysteresis*) curve.

hysteresis brake A brake whose retarding action comes from hysteresis in a permanent-magnet motor.

hysteresis clutch A magnetic clutch whose output torque (for synchronous drive or continuous slip) comes from hysteresis in a permanent-magnet motor.

hysteresis coefficient In a 1 mcc iron sample, the energy (in ergs)

dissipated during one cycle of magnetization. Also called *coefficient of hysteresis*.

hysteresiscope An oscilloscope that is specially designed to display the hysteresis curve of a material. Compare *hysteresigraph*.

hysteresis curve A response curve depicting hysteresis. Also see *box-shaped loop*.

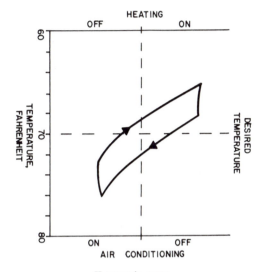

Hysteresis curve.

hysteresis cycle A complete hysteresis curve.

hysteresis distortion Signal distortion in iron-core components, such as coupling transformers, resulting from hysteresis in the iron.

hysteresis error In a meter, a difference in indications for increasing and decreasing current, an effect caused by hysteresis in iron meter parts.

hysteresis heater An induction heater in which heating results from hysteresis loss in the work (load).

hysteresis loop See *hysteresis curve*.

hysteresis loss Power loss caused by hysteresis in a magnetic material (such as iron) exposed to an alternating magnetic field. It is characterized by the generation of heat.

hysteresis meter An instrument that determines the hysteresis loss in a ferromagnetic material in terms of the torque produced when the material is rotated in a magnetic field, or vice versa.

hysteresis motor A synchronous motor that doesn't use dc excitation, nor has it salient poles; it is started by means of hysteresis losses that the rotating magnetic field causes in the secondary.

hysteretic constant For a ferromagnetic material, hysteresis loss in ergs per cubic centimeter of material per cycle of magnetization.

hysteretic loss See *hysteresis loss*.

hystoroscope A device that is used for the purpose of determining the magnetic characteristics of a material.

Hz Abbreviation of *hertz*.

I 1. Symbol for *current*. 2. Symbol for *iodine*. 3. Symbol for *intrinsic semiconductor*. 4. Symbol for *luminous intensity*.

i 1. Symbol for $\sqrt{-1}$. (Also, *j*). 2. Subscript for *instantaneous value*. 3. Symbol for *intrinsic semiconductor*. 4. Symbol for *angle of incidence*. 5. Symbol for *instantaneous current*. (Also, *Ii*). 6. Symbol for a unit vector parallel to the *x*-axis. 7. Symbol for *incident ray*.

Ia Symbol for *anode current*.

Iac Symbol for the ac component of a *composite current*. Compare *Idc*.

Iaf Symbol for *audio-frequency current*.

iagc Abbreviation of *instantaneous automatic gain control*.

IAL Abbreviation of *international algebraic language,* an early name for ALGOL 58.

iavc Abbreviation of *instantaneous automatic volume control*.

IB Symbol for *plate power-supply current*.

Ib Occasional symbol for *plate current* (usually, *Ip*).

IC 1. Abbreviation of *integrated circuit*. 2. Abbreviation of *internal connection*.

Ic 1. Symbol for transistor *collector current*. 2. Occasional symbol for *grid current* (usually, *Ig*).

ICAD Abbreviation of *integrated control and display system*.

ICAS Abbreviation of *intermittent commercial and amateur service*.

ICBM Abbreviation of *intercontinental ballistic missile*.

ICBO Symbol for the static reverse collector (leakage) current in a common-base connected transistor with an open-circuited emitter. Compare *ICEO*.

ICBS Abbreviation of *interconnected business system*.

ICEO Symbol for the static reverse collector (leakage) current in a common-emitter-connected transistor with an open-circuited base. Compare *ICBO*.

ice loading 1. In an antenna, power-line system, or other structure, the additional stress caused by accumulation of ice. 2. The weight or thickness of ice a structure can safely withstand.

ice-removal circuit A high-voltage low-frequency power supply employed to heat certain antennas to melt ice that accumulates on them.

ICET Institute for the Certification of Engineering Technicians (National Society of Professional Engineers).

ICME International Conference on Medical Electronics.

Ico Symbol for the collector cutoff current (static leakage current) of a transistor (see *cutoff current*). See, for example, *ICBO* and *ICEO*.

iconoscope A TV camera tube in which an electron beam scans a photomosaic on which the image is focused. The light-sensitive droplets of the mosaic form tiny capacitors with the insulated, metallic backplate of the mosaic, each capacitor becoming charged by the light of the picture. As the electron beam scans the mosaic, each capacitor discharges as the beam strikes it, delivering an output pulse proportional to the light intensity at that spot in the picture. (Next page.)

IC tester An instrument for checking the operation of an integrated circuit.

ICW Abbreviation of *interrupted continuous wave*.

ID Abbreviation of *inside diameter*.

Ida 1. Phonetic alphabet code word for the letter *I*. 2. Abbreviation of *integrodifferential analyzer*.

Idc Symbol for the dc component of a *composite current*. Compare *Zac*.

ideal Pertaining to a circuit, device, material, or manner of operation that is assumed to conform to the theoretical best-case example; it is not usually realized in practice. Thus, an ideal reactance has no inherent resistance.

ideal capacitor A capacitor having zero dielectric loss and a constant value of capacitance at all alternating-current frequencies.

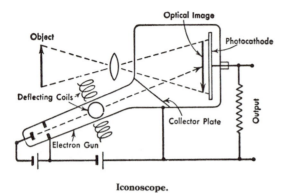

Iconoscope.

ideal component A theoretical component assumed to have no extraneous properties. Thus, a pure inductance, capacitance, or resistance. Compare *practical component*.

ideal crystal A piezoelectric crystal that acts as a theoretically perfect tuned circuit, that is, an ideal capacitor and inductor.

ideal inductor An inductor having zero loss and a constant value of inductance at all alternating-current frequencies.

I-demodulator In a color-TV receiver circuit, a demodulator which receives the chrominance and 3.58 MHz oscillator signals and delivers a video output corresponding to color in the picture.

identical vectors Equal vectors which have the same initial point or point of application.

identification 1. In radar practice, the (often automatic) determination of the target's identity. See, for example, *identification, friend or foe*. 2. In digital computer practice, a symbol or set of symbols within a label identifying a unit of data or its location.

identification beacon 1. A beacon employed for the determination of a particular geographic location. 2. An automatically transmitted station-identification signal or code, usually superimposed on the regular transmission in the form of a subcarrier or subaudible signal.

identification division The division (one of four) in a COBOL program that describes and identifies the program being compiled.

identification, friend or foe Abbreviation, IFF. A technique in which a radar station transmits an interrogating signal and the station questioned replies automatically with a suitable pulse or other signal if it is aboard a friendly aircraft or vessel. If it is aboard an enemy vehicle, the station gives no reply or gives an unsatisfactory one.

identifier A data file identification label in an input/output device, or a label that identifies a specific storage location.

identity element A logic element which, upon receipt of two input signals, provides one output signal that will be 1 only if the input signals are of the same binary level or state (both 1 or zero).

idiochromatic Possessing the photoelectric properties of a true crystal.

I-display See *I-scan*.

idle character A digital character that conveys no information, but helps maintain synchronization between the transmitter and receiver. Sometimes called a blank.

idler wheel In a tape recorder, an auxiliary, rubber-tired wheel that transfers rotary motion from the motor pulley to the rim of the capstan flywheel.

idle time The period during which data processing equipment, although operable, is out of use.

idling Standby equipment operation, as when tube filaments in a transmitter or receiver are kept hot during idle periods.

idling current The current flowing in a device during a standby period, as opposed to *operating current*. Also called *standby current*.

idling frequency In a parametric amplifier, the difference between the signal frequency and pump frequency.

idling power See *standby power*.

idling voltage The voltage required by or measured in a device that is idling, as opposed to *operating voltage*.

ID(OFF) Symbol for *drain cutoff current* in a field-effect transistor.

IDOT Abbreviation of *instrumentation online transcriber*.

IDP 1. Abbreviation of *industrial data processing*. 2. Abbreviation of *integrated data processing*. 3. Abbreviation of *intermodulation-distortion percentage*.

IDSS Symbol for *drain current at zero gate voltage* in a field-effect transistor.

Ie Symbol for *emitter current*.

IEC Abbreviation of *integrated electronic component*.

IEE Institution of Electrical Engineers (British).

IEEE Institute of Electrical and Electronics Engineers (formerly IRE).

IES Illuminating Engineering Society.

i-f Abbreviation of *intermediate frequency*.

If Symbol for *filament current*.

i-f amplifier See *intermediate-frequency amplifier*.

i-f channel See *intermediate-frequency channel*.

i-f converter The converter (first detector-oscillator) section of a *superheterodyne circuit*.

IFF Abbreviation of *identification, friend or foe*.

i-f gain The amplification of an intermediate-frequency amplifier.

i-f interference See *intermediate-frequency interference*.

IFIPS International Federation of Information Processing Societies.

i-f selectivity See *intermediate-frequency selectivity*.

i-f strip A (sometimes removable) circuit section containing a complete intermediate-frequency channel.

i-f transformer See *intermediate-frequency transformer*.

IG Symbol for *gate current*.

Ig Symbol for *grid current*.

IGFET Abbreviation of *insulated-gate field-effect transistor*.

ignition coil A small open-core transformer having a high stepup turns ratio for converting 6- or 12-volt battery potential to the high voltage needed in an automotive ignition system.

ignition interference Electrical noise generated by the ignition system of an internal combustion engine.

ignition potential See *breakdown voltage, 2*.

ignition reserve The extra voltage provided by the starter, as compared with the voltage actually needed for ignition of an internal-combustion engine.

ignition system An electrical or electronic system that supplies the high voltage in an automotive engine. See, for example, *capacitor-discharge ignition system*.

ignition voltage See *breakdown voltage, 2*.

ignitor The special starter electrode in an ignitron (see *ignitron*).

ignitor discharge A dc glow discharge—that aids rf ionization—between the ignitor and another electrode in a switching tube.

ignitor electrode In an ignitron, an electrode in contact with the mercury-pool cathode; it initiates conduction at the proper times in each cycle.

ignitor firing time In a high-voltage ignitron, the time delay following activation of the ignitor electrode before conduction begins.

ignitor oscillation In an attenuator, TR, or pre-TR tube, a kind of relaxation oscillation in the tube circuit.

ignitron A mercury-pool rectifier tube in which an auxiliary electrode (*ignitor*), is in contact with the mercury-pool cathode. Ionization in the tube is started by passing a small current pulse between ignitor and pool. This causes a small arc which quickly spreads between the anode and cathode. Also see *pool-cathode tube*.

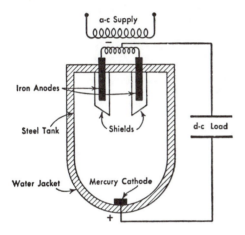

Ignitron.

ignore See *ignore character*.

ignore character 1. A character used as a signal to inhibit an action. 2. A character that is ignored.

IGSS Symbol for *gate reverse current* in a field-effect transistor.

IGY *International Geophysical Year*.

Ih 1. Symbol for *heater current*. 2. Symbol for *hold current*. 3. Symbol for *holding current*.

IHF Abbreviation of *inhibit flip-flop*.

IHFM Institute of High Fidelity Manufacturers.

ihp Abbreviation of *indicated horsepower*.

ihp-hr Abbreviation of *indicated horsepower-hours*.

Ii 1. Symbol for *input current*. 2. Symbol for *instantaneous current* (also, *i*).

Ik Symbol for *cathode current*.

IL Symbol for *current in an inductor*.

I²L Abbreviation of *integrated injection logic*. (Also *IIL*.)

illegal character 1. A character or bit group that is, according to some standard, invalid. 2. A bit group that represents a symbol in a character set.

illuminance The amount of luminous flux received per unit surface area, measured in lux (lumens per square meter).

illuminant-C In color television, the reference white that closely resembles average daylight.

illuminated pushbutton See *lighted pushbutton*.

illuminated switch See *lighted switch*.

illumination 1. Light. 2. The condition of being lighted.

illumination control A photoelectric circuit that automatically regulates electric lights according to the amount of daylight.

illuminometer A device for measuring illuminance.

ILS Abbreviation of *instrument landing system*.

IM Abbreviation of *intermodulation*.

Im 1. Abbreviation of *maximum current*. 2. Abbreviation of *meter current*.

image 1. In a superheterodyne circuit, a spurious signal related to the desired signal by twice the intermediate frequency. 2. A picture on the screen of a TV picture tube. 3. A pattern on the screen of an oscilloscope tube. 4. A picture on the mosaic of a TV camera tube. 5. A duplicate of a computer storage area that is in another part of storage or on another medium.

image admittance The reciprocal of image impedance.

image antenna An imaginary "mirror" antenna below the surface of the earth at a depth equal to the height of the true radiating antenna (above the surface), to account for the point of radiation of ground-reflected rays.

image attenuation constant Symbol, a. The real part of the *image transfer constant*; a = a1 + a2 + a3 +...an. Also see *image phase constant*.

image converter 1. A device that changes an invisible image into a visible image. Examples include the snooperscope, an infrared-to-visible converter, and photographic apparatus for infrared, ultraviolet, and X-ray wavelengths. 2. A tube that operates as an image converter.

image dissector See *dissector tube*.

image effect The effect of reflection of electromagnetic waves from the ground. An image antenna appears to radiate from a point beneath the effective ground plane. The depth of the image antenna below the effective ground plane is equal to the height of the actual antenna above the effective ground plane.

image frequency See *image, 1*.

image impedance The property of a network in which the load impedance is seen looking into the output terminals with the generator connected to the opposite end, and the generator impedance is seen looking into the input terminals with a load connected to the opposite end.

image intensification Increase of the brightness of a TV or oscilloscope image (see *image, 2, 3, 4*).

image interference A type of interference occurring in superheterodyne circuits due to the image being on the frequency of a desired signal. Also see *image, 1*.

image orthicon See *orthicon*.

image phase constant Symbol, β. The imaginary part of the *image transfer constant*; β = β1 + β2 + β3 + β4 +...βn. Also see *image attenuation constant*.

image ratio See *signal-to-image ratio*.

image rejection Elimination of image interference by a selective circuit, such as a radio-frequency preamplifier. Also see *image, 1; image interference;* and *signal-to-image ratio*.

image response In a heterodyne receiver, the response to a signal that is removed by twice the frequency of the first i-f stage from the frequency selected.

image transfer constant A number depicting the transfer of power by an impedance network. It has the same value regardless of the direction of transmission through the network; $\tanh \Omega = \sqrt{Zsc/Zoc}$, where *Zoc* is the impedance at the input terminals with the output terminals open-circuited, and *Zsc* is the impedance at the input terminals with the output terminals short-circuited ($Z = \alpha + j\beta$). Also see *image attenuation constant* and *image phase constant*.

image tube See *image converter, 2*.

imaginary axis In a vector diagram, the axis of the imaginary component (jX) of a complex quantity.

imaginary component The imaginary part of a complex number (see *complex notation.*)

imaginary number The square root of a negative quantity, or any expression containing this square root, e.g., $\sqrt{-1}$, $5\sqrt{-1}$.

IM distortion meter *intermodulation meter.*

IM distortion percentage See *intermodulation-distortion percentage.*

imitation The transmission of false signals for purposes of deception. For example, the signals from an enemy station may be recorded and retransmitted.

immediate access 1. The ability of a computer to immediately store data in memory or retrieve it. 2. Computer storage that can be accessed immediately.

immediate address An instruction address that is used as data by that instruction.

IM meter See *intermodulation meter.*

immitance Impedance *or* admittance; an acronym from *IM*pedance and ad*MITTANCE*. Example: a negative-immitance circuit.

impact excitation See *shock excitation.*

impact strength The ability of a component or material to withstand shock loading; also, the work required to fracture the material under shock loading.

IMPATT diode A microwave semiconductor (silicon or gallium arsenide) diode exhibiting negative resistance resulting from the combined effects of charge-carrier transit time and impact avalanche breakdown. It is employed as a gigahertz oscillator or amplifier, and its name is an acronym from *IMP*act *A*valanche *T*ransit *T*ime diode.

IMPATT oscillator A microwave oscillator that employs an IMPATT (impact-avalanche-transit-time) diode.

impedance Symbol, Z. Unit, ohm. The total opposition offered by a circuit or device to the flow of alternating current. It is the vector sum of resistance and reactance; $Z = \sqrt{R^2 + X^2}$. True impedance introduces a phase shift.

impedance angle The angle between the resistance and impedance vectors in an *impedance triangle.*

impedance arm The network branch that contains one or more impedances, as opposed to an arm that contains only resistance or (predominantly, reactance). Also called *impedance leg.*

impedance branch See *impedance arm.*

impedance bridge 1. An ac bridge (commonly operated at 1 kHz) used to measure resistance, inductance, capacitance, and resistive components associated with inductors and capacitors, from which impedance may be calculated. 2. Sometimes, an ac half-bridge circuit in which an unknown impedance is compared with a known resistance. Also see *Z-meter, 3.* 3. A radio-frequency bridge circuit whose balancing element reads impedance directly in ohms.

impedance bump A discontinuity in the characteristic impedance of a radio-frequency transmission line. Often caused by the use of improper splicing techniques.

impedance coil See *choke coil.*

impedance converter See *impedance transformer.*

impedance-coupled amplifier An amplifier employing capacitor—coil combinations for interstage and output load coupling.

impedance drop In an alternating-current circuit, the complex sum of the resistance drop and reactance drop.

impedance ground A ground connection in which the impedance at the operating frequency is determined by a network of resistors, capacitors, and/or inductors.

impedance leg See *impedance arm.*

impedance match The condition (for maximum power transfer) when the transmitting impedance equals the receiving impedance, or when a suitable transformer is inserted between different impedances for matching purposes. Also see *impedance matching.*

impedance matching The adjustment of impedances so they equal each other, or the insertion of a suitable transformer between different impedances to accomplish the same purpose.

impedance-matching network A network of discrete components used to match one impedance to another. An example is the *LC* coupler used to match a transmitter to an antenna.

impedance-matching transformer See *impedance transformer.*

impedance meter See *Z-meter, 1, 2, 4,*

impedance plethysmograph An electronic device used to measure changes in the chemical content of body cells.

impedance poles See *poles of impedance.*

impedance ratio The quotient of two impedances (e.g., Z_1/Z_2) which are related in some situation, such as impedance match or impedance mismatch. The impedance ratio of a transformer is equal to the square of the turns ratio.

impedance transformer 1. A transformer for converting an impedance to a different value. The turns ratio is equal to the square root of the impedance ratio. 2. A cathode follower, emitter follower, or source follower used primarily to match a high impedance to a lower impedance. 3. A short-circuited transmission line section used to match or convert impedances at radio frequencies.

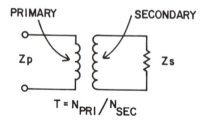

Impedance transformer.

impedance triangle A triangular vector diagram in which the impedance (z) vector is the hypotenuse and the reactance (XL or XC) and resistance (R) vectors are the perpendicular sides.

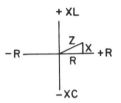

Impedance triangle.

impedance vector In a vector diagram, the resultant (vector) representing the combined reactance and resistance vectors. Also see *impedance triangle.*

impedance zeros See *zeros of impedance.*

imperative macroinstruction A macroinstruction used to create object (machine language) program instructions. See *macroinstruction.* Compare *declarative macroinstruction.*

imperative statement A source language program instruction that is converted into object program (machine language) instructions.

imperfection A fault in the lattice of a crystal. Also see *crystal lattice.*

implementation 1. Putting a (computer) system into operation and evaluating its performance. 2. Doing all that is necessary to establish a computer installation.

implosion The inward collapse of an evacuated chamber, such as the envelope of a cathode-ray tube. It is the opposite of *explosion,* an outward bursting.

impregnant A substance for the impregnation of electronic components. Examples: waxes, oils, liquid plastics, varnish.

impregnation The process of filling the spaces within a material or structure with an insulating compound. Electronic components—capacitors, inductors, transformers, transistors, diodes—are impregnated for protection and stability.

impressed voltage The voltage applied or presented to a circuit or device.

impulse A unidirectional surge in voltage (usually) or current.

impulse excitation Driving a tank circuit with a single pulse and then allowing it to oscillate at its own frequency until another driving pulse arrives.

impulse frequency In a digital telephone, the number of pulse periods per second, corresponding to a dialed digit, generated by the pulse springs.

impulse generator See *Marx generator.*

impulse noise Electrical noise from short-duration pulses such as those of an AM nature produced by automotive ignitions.

impulse ratio The ratio Ep/Es, where jep is the breakdown (or flashover or sparkover) voltage due to an impulse and Es is the corresponding voltage at the crest of the power-frequency cycle.

impulse relay A relay that is able to close or open completely when driven by a short pulse.

impulse speed The switching rate of a telephone dialing device as it transmits pulses.

impulse timer A synchronous-motor-driven timer whose cams can control many circuits; it can advance by a number of specified increments, as controlled by an integral stepping mechanism.

impulse transmission A method of transmission in which defined impulses are used to denote changes in signal content or format.

impurity A substance added to an intrinsic semiconductor to alter its electrical properties. Also see *acceptor, 2; donor impurity; dopant;* and *impurity atom.*

impurity atom In a processed semiconductor material, an atom of an impurity material which gives either n- or p-type properties to the intrinsic semiconductor.

impurity density In the manufacturing process of a semiconductor material, the amount of impurity added to the original semiconductor.

impurity ion In a crystal, an ion in a space between atoms, or one taking the place of an atom.

impurity level 1. The energy existing in a semiconductor material as a result of doping (the addition of an impurity). 2. Impurity density.

impurity material A substance introduced into a semiconductor material to alter its intrinsic electrical properties. See *acceptor, 2* and *donor impurity.*

In Symbol for *indium.*

In Symbol for the *nth value of current* in a series of values.

in. 1. Abbreviation of *input.* 2. Abbreviation of *inch.*

inaccuracy 1. The state or condition of instrument error. 2. The difference between the actual value of a parameter and the indicated value. 3. The percentage of instrument error.

inactive leg Within a transducer, an electrical component whose characteristic remains unchanged when the stimulus (quantity being transduced) is applied; specifically, a Wheatstone bridge element in a transducer.

inactive lines In a television picture, the 37 blanked lines (approximately half at the top and half at the bottom of the screen).

inactive time The period during which a radioactivity counter is insensitive to ionizing agents.

incandescence The state of glowing fron intense heat, as when a metal becomes white hot from an electric current flowing through it.

incandescent lamp A filament-type lamp.

inch Abbreviation, in. A unit of linear measure in the English system; 1 in. = 0.2540 meter.

inching See *jogging.*

inch-pound Abbreviation, in-lb. A unit of work equal to a force of 1 pound exerted over a distance of 1 inch. Compare *foot-pound.*

incident A computer system failure requiring the intervention of an operator in removing or revising the job involved.

incidental AM Undesired amplitude modulation in a frequency-modulated signal. Compare *incidental FM.*

incident field intensity The field strength of an electromagnetic field as it arrives at a receiving antenna.

incidental FM Undesired frequency modulation in an amplitude-modulated signal. Compare *incidental AM.*

incidental time Computer time devoted to other than program runs or program development.

incident light The light that strikes or enters an altering decice or medium. See *incident ray.*

incident power In a transmission line, the power that reaches the end of the line. Compare *reflected power.*

incident ray Symbol, *i.* The ray that strikes the surface of a reflecting, refracting, or absorbing body. Compare *reflected ray* and *refracted ray.*

incident wave 1. A wave propagated to the ionosphere. Compare *reflected wave* and *refracted wave.* Also see *ionosphere* and *ionospheric propagation.* 2. A wave that encounters a change (in density, for example) in a propagation medium or the transition point between media.

in-circuit tester An instrument that permits the checking of components (especially transistors) without removing them from the circuit in which they are wired.

inclination The angle between the horizon and a magnetic needle that is free to turn vertically in the plane of the magnetic meridian. Compare *declination.*

inclinometer An instrument for measuring inclination. One form consists of a magnetic needle mounted so it can swing inside a vertically mounted circular scale.

inclusive-OR operation A logical operation between two operands, the result of which depends on rules for combining bits in each position within the operands: an output of 1 results if one or both of the bits have a value of 1; zero only if both are zero.

incoherent light Light composed of nonparallel rays. Compare *coherent radiation.*

incoming inspection The inspection of equipment and materials as they enter a factory or laboratory.

incoming line A line that enters a device, facility, or stage. Compare *outgoing line.*

incomplete program A computer program of generalized steps that must be augmented with specific requirements to be implemented for a given operation. Also called *incomplete routine.*

inconsistency Contradictory computer statements, as detected by the program.

inconsistent equations Equations that cannot all be true and, therefore, have no common solution. Compare *dependent equations* and *independent equations.*

Increductor A radio-frequency magnetic amplifier or saturable reactor.

increment 1. Symbol, Δ. The difference between two successive values of a variable. Thus, an increment of current, ΔI, equals $I2$ minus $I1$. Compare *differential.* 2. A small value change. 3. A quantity to be added to another quantity.

incremental computer A computer that operates on *changes* in variables. Example: differential analyzer. Compare *absolute value computer.*

incremental digital recorder A magnetic tape recorder that moves the tape across the record head in increments.

incremental display A device that converts digital data into a form for display (characters; graphs).

incremental inductance The inductance exhibited by an inductor, such as an iron-core choke, carrying a direct current.

incremental permeability The permeability exhibited by a material when the ac magnetizing force is superimposed upon a direct current.

incremental plotter A unit which, by direction of a computer program, gives the results of a program run in the form of curves or points on a curve, along with annotational characters.

incremental representation For incremental computers, a method of representing variables in terms of *changes* in the variables.

incremental sensitivity The smallest change in a quantity under measurement that can be detected by the instrument used.

incremental sign Capital delta (Δ), a Greek letter. Also see *increment.*

ind 1. Abbreviation of *indicator.* 2. Abbreviation of *inductance.* 3. Abbreviation of *inductor.*

indefinite integral An integral in which no constant of integration is given. Example: $\int 2x\,dx = x^2$ is indefinite, since x^2 can have any number of values, depending on the missing constant of integration C. (The definite integral would be $\int 2x\,dx = x^2 + C$.) Also see *definite integral, integral, integral calculus,* and *integration.*

independent equations Different equations which have one common solution. Compare *dependent equations* and *inconsistent equations.*

independent events In probability and statistics, the case where the occurrence of one event has no effect on the occurrence of another.

independent failure A component of circuit failure not related to malfunctions elsewhere in the system.

independent mode In *tracking supplies,* an optional method of operation in which the separate units are adjustable independently of each other. Compare *tracking mode.*

independent variable A changing quantity whose value at any instant is not governed by the value of any other quantity. Compare *dependent variable.*

index 1. A reference line, hair, or point; e.g., the index of a file. 2. In mathematics, an exponent. 3. A ratio of one quantity to another, as *index of refraction.* 4. In a computer memory, a table of references in a key sequence; it can be addressed to find the addresses of other data items. 5. A number that is used to select a specific item within an array of items in memory.

index counter In a tape recorder, a (usually electromechanical) counting device that the operator can refer to find material on the tape. Also called *tape counter.*

indexed address During or preceding the execution of a computer program instruction, an address that is modified by the content of an index register.

indexing 1. An information retrieval technique used with files on a direct access storage medium or on tables in memory. 2. To modify an instruction using an index word.

index of modulation In frequency modulation, the ratio of carrier frequency deviation to modulating frequency.

index of refraction Symbol, n. The ratio $v1/v2$, where $v1$ is the velocity of light (or the form of radiation used) in the first medium through which it passes, and $v2$ is the velocity in the second medium.

index point A point of reference within the operation cycle of a punched card machine that moves cards by the use of revolving shafts.

index position See *punching position.*

index register Abbreviation, XR. In digital computer practice, a register holding a modifier that allows data to be directly addressed (each program refers to an index register when addressing storage locations). Also called *modifier register.*

index word A word (bit group) containing a modifier that will be added to a basic instruction when it is executed during a program run.

India mica High-grade mica mined in India. Its excellent dielectric properties make it useful for capacitor stacks, high-Q rf circuits, and other critical applications.

indicated horsepower Abbreviation, *ihp.* Horsepower calculated from data or ratings, as opposed to measured horsepower.

indicated horsepower-hours Abbreviation, *ihp-hr.* Horsepower-hours based on calculaion of indicated horsepower.

indicating fuse A fuse that gives some signal (such as a protruding pin) to show that it has blown.

indicating instrument An instrument, such as a meter, which gives direct readings of a measured quantity, as opposed to an instrument, such as a bridge, which must be manipulated and whose operation must often be followed by calculations.

indicating lamp A lamp that is marked or coded so that when it is on or off it conveys information.

indicator 1. Meter (see *meter, 1*). 2. Monitor. 3. Annunciator. 4. In a computer, a device that can be set by a specific condition, e.g., by a negative result or error indicator.

indicator probe A test probe having a built-in meter.

indicator tube See *magic-eye tube.*

indicial response Symbol, $I(t)$. The sum of the transient and steady-state responses to a *unit function.*

indirect addressing In computer programming, a technique in which the address in an instruction refers to a different location containing another address, which may specify yet another address or an operand. Also called *multilevel addressing.*

indirect coupling Collectively, capacitive and inductive coupling as opposed to direct coupling.

indirect ground An unintentional ground connection (e.g., accidental grounding of part of a circuit) or one obtained through a roundabout path. Compare *direct ground.*

indirect illumination See *indirect light.*

indirect light Reflected light. Compare *direct light.*

indirectly controlled Influenced by a directly controlled parameter, but not itself directly controlled.

indirectly grounded Connected to earth or to the lowest-potential point in a system inadvertently or through a roundabout path, e.g., by means of an indirect ground. Compare *directly grounded,*

indirectly heated cathode An electron-tube cathode consisting of a cylindrical or rectangular sleeve coated with a substance which is a rich emitter of electrons; it is heated by a filament (the *heater*) within the sleeve and insulated electrically from it. Also called *equipotential cathode* and *unipotential cathode.*

indirectly heated thermistor A thermistor whose temperature is changed by a built-in heater (filament) operated by the control current. Compare *directly heated thermistor.*

indirectly heated thermocouple A meter thermocouple heated by a small heater (filament) through which the signal current passes. Compare *directly heated thermocouple.*

indirectly heated tube An electron tube having an indirectly heated cathode.

indirect material A semiconductor substance in which electrons move from the conduction band to the valence band in discrete jumps or steps.

indirect measurement The measurement of a quantity by comparing it with a similar quantity, using an instrument that requires adjustment or manipulation (rather than a simple meter). For example, resistance can be measured with a bridge instead of an ohmmeter. Compare *direct measurement.*

indirect piezoelectricity In a piezoelectric crystal, the application of a voltage for the purpose of producing a strain on the crystal. A piezoelectric buzzer operates on this principle.

indirect scanning A method of TV or film scanning in which a *flying spot* of light scans the film or an object and is passed through the film (or reflected by the object) to a photocell.

indirect wave See *skywave.*

indium symbol In, A metallic element. Atomic number, 49. Atomic weight, 114.82 Indium is used as a dopant in semiconductor processing.

indoor antenna An antenna erected and operated in a building but kept away from other objects as much as possible. Compare *outdoor antenna.*

indoor radiation Electromagneic radiation from the part of an antenna feeder or lead-in that is inside the transmitter building.

indoor transformer A power service transformer which, for protection, is installed inside the building it serves.

induced Brought about by the influence of a magnetic or electric field.

induced charge An electric charge produced in a body by the electric field surrounding another charge.

induced current An alternating current established in one circuit by the alternating magnetic field of another circuit. Also see *induction.*

induced emf See *induced voltage.*

induced failure A form of component failure that occurs because of operation beyond the normal specifications.

induced radioactivity Artificial radioactivity, such as that produced by a particle accelerator (see *accelerator, 1*).

induced voltage An ac voltage set up across one circuit (especially a coil) by the alternating magnetic field of another circuit. Also see *induction.*

inductance Symbol, *L*. Unit, henry. In a conductor, device, or circuit, the inertial property (caused by an induced reverse voltage) that opposes the flow of current when a voltage is applied; it opposes a change in current that has been established. Also see *henry, induction,* and *mutual inductance.*

inductance bridge An ac bridge for measuring inductance in terms of

(1) a standard inductance or (2) a standard capacitance. See, for example, *Hay bridge, Maxwell bridge,* and *Owen bridge.*

inductance—capacitance Abbreviation, *LC,* 1. Pertaining to a circuit, a combination of inductance and capacitance, e.g., *LC* filter, parallel-resonant circuit, series-resonant circuit. 2. Pertaining to a device for measuring inductance and capacitance, e.g., *LC* bridge, *LC* meter.

inductance—capacitance bridge An ac bridge for measuring inductance and capacitance only.

inductance—capacitance filter A filter composed of inductors and capacitors. Also called *LC filter.*

inductance—capacitance meter A direct-reading meter for measuring inductance and capacitance.

inductance-capacitance-resistance Abbreviation, *LCR.* 1. Pertaining to a circuit, a combination of inductance, capacitance, and resistance, for example, a basic tuned circuit containing inductance, capacitance, and inherent resistance (losses). 2. Pertaining to a device for measuring inductance, capacitance, and resistance, e.g., *LCR* bridge, *LCR* meter.

inductance-capacitance-resistance bridge See *impedance bridge, 1.*

inductance coil See *inductor.*

inductance filter A filter employing only a choke coil.

inductance—resistance time constant The time constant (see *electrical time constant*) of a circuit containing, ideally, only inductance and resistance; $t = L/R$, where t is in seconds, L in henrys, and R in ohms. Also called *LR time constant.*

inductance standard A highly accurate, stable inductor employed in precision measuremens. Also see *primary standard* and *secondary standard.*

inductance-tube modulator See *reactance-tube modulator.*

induction 1. The ability of an alternating, pulsating, or otherwise changing current flowing in one circuit to set up a current in a nearby circuit; circuits needn't be physically connected, only linked by magnetic lines of force. Also see *self-induction.* 2. The phenomenon whereby a body becomes electrically charged by the field surrounding a nearby charged body. Also see *electric charge.*

induction coil A special high-voltage stepup transformer having an open core and a vibrator-interrupter in series with the primary winding, which carries direct current from a battery. The current is broken up into short pulses by the interrupter, and a high ac voltage is generated in the secondary winding.

induction compass A compass whose indications depend on current induced in a coil revolving in the earth's magnetic field. Compare *gyrocompass* and *magnetic compass.*

induction factor The ratio of total current to nonproductive current in an ac circuit.

induction field The portion of the electromagnetic field that returns to a radiator (as a coil carrying ac), as opposed to the *radiation field.*

induction frequency converter A mechanical device used for conversion of one fixed frequency to another fixed frequency.

induction furnace A furnace in which high-frequency magnetic fields induce currents in meal ores, melting them.

induction heater A low- or high-frequency power generator designed especially for induction heating.

induction heating Heating metallic work samples by placing them in (but insulated from) a *work coil* carrying current from a high-powered rf generator. Currents induced in the workpiece heat it. Compare *dielectric heating.*

induction loss Loss of energy from a current-carrying conductor because of inductive coupling to a nearby conductor.

induction modulator See *electromechanical modulator, 3.*

induction motor An electric motor in which the stator's rotating magnetic field makes the rotor revolve.

induction speaker An acoustic loudspeaker in which an audio-frequency current is passed through a diaphragm or coil located in a constant magnetic field. This results in movement of the diaphragm or coil.

induction transducer See *inductive transducer.*

induction-type landing system See *Dingley induction-type landing system.*

induction welding Welding in which the heating current flowing in the workpieces is induced by an electromagnetic field.

inductive capacitor A wound capacitor in which the inductance of the roll is controlled and specified. Such a capacitor is useful in compact filters and in single-frequency bypassing, where the L and C components are supplied by the capacitor. Compare *noninductive capacitor.*

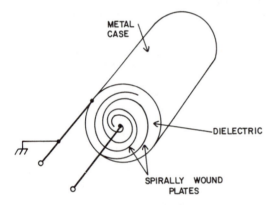

Inductive capacitor.

inductive circuit 1. A circuit in which inductance predominates. 2. A (theoretical) circuit containing inductance only.

inductive coupling The transfer of energy between two inductors (or inductive devices) by a linking electromagnetic field. Also see *coefficient of coupling, coupling, induction,* and *mutual inductance.*

inductive feedback See *magnetic feedback.*

inductive heater A radio-frequency power generator for *inductive heating.*

inductive heating The heating of a material by placing it within the magnetic field of a coil carrying rf current. Compare *dielectric heating.*

inductive kick See *back voltage* and *kickback.*

inductive load A load device which approaches a pure inductance in nature, e.g., loudspeaker, electric motor.

inductive microphone A microphone in which sound waves vibrate a conductor or coil in a strong magnetic field, producing a corresponding ac output by the resulting induction. Example: dynamic microphone.

induction neutralization Neutralization of a triode amplifier circuit by an out-of-phase voltage fed back from the output of the circuit through coupling coils.

inductive reactance Symbol, X_L. Unit, ohm. The reactance exhibited by an ideal inductor; $X_L = 2\pi f L$, where X_L is in ohms, f in hertz,

and L in henrys. In a pure inductive reactance, current lags 90 degrees behind voltage. Also see *inductance, induction, inductor,* and *reactance.*

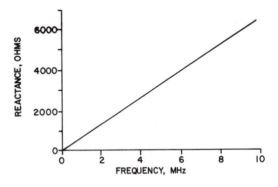

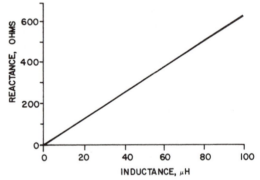

Inductive reactance.

inductive switching Switching operations in a circuit containing an inductor. Switching time is influenced by the LR time constant of the inductor; overall operation is affected by the back emf generated by the inductor. Also see *inductance—resistance time constant.*

inductive transducer A transducer in which the sensed phenomenon causes a change in inductance (or reluctance), which in turn causes a proportional change in output current, voltage, frequency, or bridge balance. Compare *capacitive transducer, crystal transducer, magnetic transducer,* and *resistive transducer.*

inductive trimmer See *trimmer inductor.*

inductive tuning Tuning a radio by changing the inductance of a coil having a movable core.

inductivity See *dielectric constant.*

inductometer An instrument for measuring inductance in terms of the resonant frequency of an LC circuit in which L is the unknown inductance and C is calibration capacitance.

inductor A coil of wire wound according to various designs, with or without a core of magnetic metal, to concentrate the magnetic field and, thus, give a higher self-inductance than a straight wire would. Also see *inductance; induction, 1;* and *self-inductance, 1.*

inductor alternator See *alternator, 2.*

inductor amplifier See *magnetic amplifier.*

inductor decade See *decade inductor.*

inductor microphone See *inductive microphone*.

inductors in parallel See *parallel inductors*.

inductors in parallel—series See *parallel—series inductors*.

inductors in series See *series inductors*.

inductors in series—parallel See *series-parallel inductors*.

inductor substitution box An enclosed assortment of common-value inductors that can be switched, one at a time, to a pair of terminals. In troubleshooting and circuit development, any of several useful fixed inductances may be thus obtained.

industrial data processing Abbreviation, IDP. The application of digital computers and associated equipment to industrial problems, through the classification, sorting, storing, and manipulation of information.

industrial electronics The branch of electronics concerned with manufacturing processes and their control, and with the operation and safeguarding of factories.

industrial instrumentation 1. Supplementing an industrial process with electrical and electronic measuring instruments. 2. The instruments so employed.

industrial television Abbreviation, ITV. Television (especially CCTV) as an adjunct to a manufacturing process or as a means of communication or surveillance within an industrial plant.

industrial tube An (often heavy-duty) highly reliable electron tube designed expressly for industrial service.

ineffective time The period during which an otherwise operational computer is not being used effectively because of delays or *idle time*.

inelastic collision A collision between charged particles in which one gains energy at the other's expense.

inert gas A gas that does not readily react with other elements. Inert gases include argon, helium, krypton, neon, and xenon. Such gases are often used in hermetically sealed devices to retard corrosion.

inertance See *acoustic inductance*.

inertia The tendency of a body at rest to remain at rest unless acted upon by a force. Compare *momentum*.

inertia in electric circuit The condition in a circuit containing inductance, in which a current change lags behind a voltage change (analogous to mechanical inertia; see *inertia*).

inertial guidance A system that automatically guides missiles and satellites in a desired trajectory without the need for continuous control by signals from a station.

inertia relay A time-delay relay whose operation is slowed by the addition of weights or other attachments.

inertia switch A switch that can sense a disturbance of its inertia.

infinite Pertaining to that which has no definite limits. Compare *finite*.

infinite baffle A loudspeaker baffle having no openings for the passage of sound from the front to the back of the speaker cone. Also see *baffle, 1*.

infinite-impedance detector A triode detector that offers the very high input impedance of a grid-cathode circuit and the large-signal capabilities of a diode detector. Audio-frequency output is taken across the rf-bypassed cathode resistor (there is no plate resistor). Plate current increases with the input signal from a very low value at zero signal level.

infinite line See *infinite transmission line*.

infinite sample space In statistics, a sample space having no definite limits. Compare *finite sample space*.

infinitesimal 1. A quantity that is too small to be measured. 2. A parameter that, because of its small magnitude, can be ignored for practical purposes.

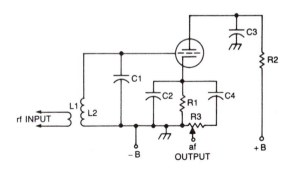

Infinite-impedance detector.

infinite series A mathematical series in which the number of terms is limitless. For example, $1/6 = 0.1 + 0.06 + 0.006 + 0.0006 + ...$

infinitesimal A quantity, such as a differential, that approaches zero as the limit; also, a negligible amount.

infinite transmission line A theoretical transmission line with normal characteristics but having infinite length.

infinity Symbol, ∞ An infinite quantity, i.e., one unlimited in duration or dimension. Thus, a quantity which may be increased or subdivided without limit is said to approach infinity.

infix notation A system of logical operation notation wherein operands are separated by operators, thus, A & B, where the ampersand means AND. Compare *prefix notation*.

inflection point On the curve of a function, a point at which a tangent crosses the curve. For a function y equal to $f(x)$, a point is an inflection point if, when the first derivative dy/dx is set to zero, the second derivative d^2y/dx^2 is zero. Compare *maxima* and *minima*.

infobond On a printed circuit board, a form of wiring on the side opposite the components. The wiring is used in place of the foil normally found on such a circuit board.

information 1. Collectively, data or communications excluding the symbols or signals used to describe, present, or store them. 2. The result of data processing; i.e., that which is derived from the compilation, analysis, and distillation of data.

information bits In an encoded signal, data characters or digits that can be treated to give information (excluding control characters).

information center A storage bank designed for use by many different subscribers, via computer.

information channel A channel through which data and associated signals are transmitted and received, as opposed to a *control channel*.

information feedback system In message transmission, a control system in which intelligence received at a terminal is returned to the sending unit for automatic verification.

information gate A gating device or circuit that opens and closes an information channel.

information retrieval In digital computer and data processing practice, the categorizing and storage of information and the automatic recall of specific file items. Also see *access time*.

information separator An indicator that separates items of information or fields in a (usually variable-length) record.

information storage In digital computer and data processing practice, holding information in memory pending retrieval.

information word A character group representing stored information and managed, as a unit, by hardware or software.

infra A prefix meaning *below* or *lower than*, e.g., *infrared.*

infrablack region In a composite video signal, the *blacker-than-black* region (see *blacker than black*).

infradyne receiver A super heterodyne receiver in which the intermediate frequency is the sum of the signal and hf oscillator frequencies, rather than their (usual) difference.

infrared Pertaining to the invisible rays whose frequencies are just lower than those of visible red light. The infrared spectrum extends from approximately 1 to 4.3×10^2 THz.

infrared communication Communication by keying or modulating infrared rays.

infrared counter-countermeasure A military tactic in which action is taken against an enemy *infrared countermeasure.*

infrared countermeasure A military tactic using counter methods to cripple enemy infrared equipment.

infrared detector A detector of infrared rays. Some infrared detectors are bolometers, radiometers, radiomicrometers, and photocells.

infrared diode A semiconductor diode, such as the gallium-arsenide type, that emits infrared rays.

infrared guidance A navigation and reconnaissance system employing infrared rays.

infrared homing The method whereby a guided missile uses infrared rays to guide it to its target.

infrared light See *infrared rays.*

infrared photography Photography in which the scene is illuminated with infrared light or gives off infrared rays and the flim is infrared sensitive.

infrared radiation See *infrared rays.*

infrared rays Radiation at frequencies in the infrared region—between the highest radio frequencies and the lowest visible light frequencies. Also called *heat rays.*

infrared spectrum The region of the electromagnetic spectrum in which infrared radiation is found.

infrared therapy The use of infrared rays by physicians and other practitioners to treat certain disorders.

infrared waves See *infrared rays,*

infrared window Any portion of the infrared spectrum in which energy is easily transmitted through the lower atmosphere of the earth.

infrasonic Pertaining to frequencies below the range of human hearing.

infrasonics The branch of physics dealing with *infrasonic* phenomena.

inharmonic distortion Distortion in which the frequencies of extraneous components are not harmonically related to the fundamental frequency. It is sometimes experienced when a tone-burst signal is applied to a loudspeaker.

inherent component A (usually extraneous) property possessed by a device because of its internal peculiarities. Thus, an inductor exhibits inherent capacitance; a capacitor, inherent resistance.

inherited error In an extended calculation, an error carried through from one of the earlier steps.

inhibit 1. In digital computer and logic practice, to prevent an action or block the input of data by means of a special *inhibit pulse*. 2. To delay an action or process.

inhibit gate A pulse-actuated gate circuit that acts as an inhibitor (see *inhibitor, 1*).

inhibitor 1. A device or circuit that produces an inhibit pulse or signal (see *inhibit, 1*). 2. An additive, such as an organic liquid, that delays the hardening of a mixture, such as an encapsulating compound.

inhibit pulse In a computer, a drive pulse that keeps other pulses from

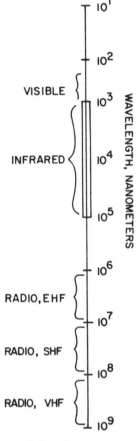

Infrared spectrum.

changing the direction of magnetization in the cells of a magnetic core memory.

inhibit signal In digital computer and logic practice, the signal that causes an inhibit action. Also see *inhibit, 1.*

initial drain The current supplied by a battery or cell at its rated voltage.

initial failure The first failure occurring in the operation of a circuit or device.

initial instructions A resident computer routine used to aid program loading. Also called *initial order.*

initial inverse voltage The peak inverse anode voltage following conduction in a rectifier tube.

initial ionizing event In the operation of a radioactivity counter, the first event that starts the chain of similar events constituting the count.

initialization A computer program instruction that sets the value of a variable to zero.

initial permeability Permeability in the low magnetization region of a material.

initial-velocity current In an electron tube, the current that flows as the result of electrons passing from the hot cathode to another element, such as a grid or plate that has no external voltage applied to it.

initiate See *trigger*.

injection 1. Introducing a signal into a circuit or device. 2. Introducing current carriers (electrons or holes) into a semiconductor.

injector 1. An element or electrode for injection (see *injection, 1, 2*). 2. A signal injector.

injector electrode See *injector, 1*.

injector grid In a multigrid converter tube, the grid to which the local-oscillator signal is applied.

ink bleed In the printing of matter for optical character recognition, ink flow around the characters, often making them unrecognizable to the reader.

ink-jet galvanometer A galvanometer whose movement controls the pressure of a jet of ink for making a recording on a paper chart. Also see *liquid-jet oscillograph*.

ink-mist recorder A graphic recorder in which the line is traced by a mist of ink.

ink recorder A graphic recorder employing a pen-and-ink stylus.

ink squeeze-out In the printing of matter for optical character recognition, the squeezing of ink from a character's center.

ink-vapor recorder See *ink-mist recorder*.

in-lb Abbreviation of *inch-pound*.

inlead The part of an electrode going through that which encapsulates a device (tube envelope; transistor base, etc.).

inline procedure The main portion of a COBOL computer program, responsible for the primary operations.

inline processing The action peculiar to a system that processes data almost immediately upon receipt, i.e., one that needn't be capable of storing a lot of unprocessed data.

inline readout In digital computer practice, a readout device that displays digits side-by-side horizontally.

inline subroutine A subroutine that must be written each time it is needed, as compared with one that can be accessed by a program branch.

inline tuning Tuning of all the stages of a channel, such as an intermediate-frequency amplifier, to the same frequency.

inner conductor The inner wire or rod of a coaxial cable or coaxial tank. Compare *outer conductor*.

inorganic Consisting of materials other than carbon compounds, and therefore not related to living things.

inorganic electrolyte Any electrolyte that is completely *inorganic*: containing no compounds of carbon.

in phase The condition in which alternating or pulsating waves or wave phenomena are in step with each other at all points. Compare *out-of-phase*.

in-phase carrier See *I-phase carrier*.

in-phase current Resistive current in an ac circuit, i.e., current in phase with voltage. Compare *quadrature current*.

in-phase feedback Feedback in phase with a main signal. Also called *positive feedback* and *regeneration*.

in-phase voltage A voltage which is in phase with another (reference) voltage.

inplant system An automatic data communications sytem within a specific building or complex.

input 1. Energy or information delivered or transferred to a circuit or device. Compare *output, 1*. 2. The terminals of a device or circuit to which energy or information is applied. Compare *output, 2*. 3. To deliver or transfer energy or information to a circuit or device (as to input data from a computer peripheral to memory).

input admittance Symbol, Y_i. The internal admittance of a circuit or device, as seen from the input terminals; the reciprocal of input impedance. Compare *output admittance*.

input area In a computer memory, an area set aside for data input from a source other than a program.

input bias current The input bias required by an operational amplifier.

input capacitance Symbol C_i. 1. The internal capacitance of a circuit or device, as seen from the input terminals. Compare *output capacitance*. 2. The grid-cathode capacitance of an electron tube.

input capacitor 1. In a capacitance-coupled circuit, the input coupling capacitor. Compare *output capacitor*. 2. The first capacitor in a capacitor-input filter, i.e., that capacitor electrically nearest the rectifier output electrode.

input choke The first choke in a choke-input filter i.e., that choke electrically nearest the rectifier output electrode, when no "preceding" capacitor is employed.

input circuit The circuit or subcircuit constituting the input section of a network or device. Also see *input, 1, 2* and *input terminals*. Compare *output circuit*.

input conductance Symbol, G_i. The internal conductance of a circuit or device, as seen from the input terminals; it is the reciprocal of input resistance. Compare *output conductance*.

input coupling capacitor See *input capacitor*.

input coupling transformer See *input transformer*.

input current 1. Symbol, I_i. The current delivered to a circuit or device. Compare *output current, 1*. 2. Symbol, I_i. Current flowing in the input leg or electrode of a circuit or device. Compare *output current, 2*.

input device 1. A device, such as an input transformer, which couples energy or information to a circuit or device. Compare *output device*. 2. A device through which another device receives data.

input equipment Collectively, input devices employed with a computer.

input error voltage In an operational amplifier, the error voltage at the input terminals when a feedback loop operates around the amplifier.

input extender A diode network that provides increased fan-in for a logic circuit. Also see *fan-in, 1*.

input gap In a velocity-modulated tube, the gap in which the electron streams is initially modulated.

input impedance Symbol, Z_i. The internal impedance of a circuit or device, as seen from the input terminals. Compare *output impedance*.

input limited The processing time limitation imposed by an input unit on the speed of a program run.

input offset current In an operational amplifier, the difference between the currents going to the input terminals when the output is zero.

input offset voltage In an operational amplifier, the potential that has to be applied between the input terminals for a zero output voltage.

input/output Abbreviated *I/O*. 1. Data transmitted to, or received from, a computer. 2. A terminal through which data is transmitted to, or received from, a device.

input/output bound A condition affecting a system in which the time consumed by input and output operations is greater than that required for other processes.

input/output buffer A computer memory area specifically reserved for the receipt of data coming from or going to a peripheral.

input/output control The part of a computer system that coordinates activity between a central processor and peripherals.

input/output equipment In digital computer practice, devices for entering information into the computer or for reading information out of it. Examples: typewriters, converters, electronic counters.

input/output isolation Arrangement or operation of a circuit or device so there is no direct path between input and output terminals around the circuit or device. Also see *isolation*.

input/output routine A routine for simplifying the programming of standard input/output equipment operations.

input/output switching The allocation of more than one channel to peripherals for communications with a central processor.

input power 1. Symbol, *Pi*. The power presented to the input terminals of a circuit or device. Also called *power input*. Compare *output power*. 2. The operating power of a circuit or device, i.e., the power-supply requirement.

input record 1. A computer record of immediate interest that is ready for processing, 2. During a computer program run, a record read into memory from an input device.

input recorder A recorder of input signals.

input register In a computer, a register that receives data from a peripheral relatively slowly and then passes it on to a central processor at a faster speed as a sequence of informational units. Also see *register*.

input resistance Symbol, *Ri*. The internal resistance of a circuit or device, as seen from the input terminals. Compare *output resistance*.

input resonator In a velocity/modulated tube, the resonator in which electron bunching takes place.

input routine A computer program section that manages data transferral between an external storage medium and a memory input area.

input section 1.Input routine. 2. Input area.

input sensitivity 1. The level of input-signal amplitude that will result in a certain signal-to-noise ratio at the output of a device. The specified signal-to-noise ratio is usually 10 or 20 dB. 2. The level of input signal in a frequency-modulated device, required to produce a specified amount of noise quieting. The specified level of noise quieting is usually 20 dB. Alternatively, 12 dB SINAD (ratio of signal to the level of noise and distortion) may be specified. 3. The minimum level of input voltage required to actuate a logic gate.

input signal The signal (current, voltage, power) presented to the input terminals of a circuit or device for processing.

input tank In a double-*LC*-tuned stage of a transmitter or power generator, the grid (in a tube) or base input (in a transistor) tank in which the input signal is resonated. Compare *output tank*.

input terminals Terminals (usually a pair) associated with the input section of a circuit or device (see *input, 1, 2*). Compare *ouput terminals*.

input transformer The input coupling transformer which delivers signal voltage or power to the input circuit of a network or device. Compare *output transformer*.

input uncertainty The combination of all parameters that result in adverse behavior in an operational amplifier.

input unit In a digital computer, the device or circuit that receives information from peripherals.

input voltage 1. Symbol, *Ei* or *Vi*. The voltage presented to a circuit or device. Compare *output voltage, 1*. 2. Symbol, *Ei* or *Vi*. The voltage across the input leg or elctrode of a circuit or device. Compare *output voltage, 2*.

input-voltage drift For an integrated circuit, the time- and temperature-dependent change in output voltage divided by the IC's open-loop voltage gain.

input-voltage offset For a differential amplifier, the input signal voltage at the differential input that results in zero output voltage.

input winding The signal winding of a magnetic amplifier.

inquiry A programmed request for information from storage in a computer.

inquiry display terminal A video display/keyboard terminal used to make an inquiry and display the response thereto.

inquiry station A terminal from which an inquiry can be sent to a central computer.

inrush The flow of current into an inductor.

inscribe To convert data to a form on a document that is readable by a character recognition device, as through the use of magnetic ink, for example.

insert A (usually metallic) bushing which may be molded into a plastic part (or pressed into it after molding is completed) to provide a bearing sleeve or threaded hole.

insert core A core whose position is adjustable to vary the inductance of the coil surrounding it.

insertion gain In a circuit or system, the gain resulting from the amplifier inserted into the system, it is usually expressed in decibels. Compare *insertion loss*.

insertion loss Loss of energy or gain by placing certain devices or subcircuits—filters, impedance matchers, or other decices in a circuit. It is usually expressed in decibels. Also see *insertion resistance*.

insertion phase shift The difference in phase produced by a circuit installed in an electrical transmission line.

insertion resistance The resistance of a component or instrument that is introduced into a circuit. Thus, the internal resistance of a microammeter becomes an insertion resistance in the circuit in which the meter is connected for current measurement.

inside antenna See *indoor antenna*.

inside diameter Abbreviation, ID. The innermost diameter of a body or figure having two concentric diameters. Compare *outside diameter*.

inside lead See *start lead*.

inside radiation See *indoor radiation*.

inside spider A voice-coil centering device within a loudspeaker.

inst 1. Abbreviation of *instrument* or *instrumentation*. 2. Abbreviation of *instant*.

instability Inconsistency in the operation of a circuit or device, in the parameters of a device, or in an electrical quantity. Instability may be attributed to a number of causes: temperature, loading, age, humidity, negative resistance, radioactivity.

installation tape number An identification number given to a reel of magnetic tape by the processing facility.

instant Abbreviation, inst. The point at which an event occurs or a quantity reaches a particular value.

instantaneous Occurring at a specified moment, or instant, of time.

instantaneous amplitude The amplitude, specified in amperes, volts, or watts, of a signal, specified at a particular moment of time.

instantaneous automatic gain control Abbreviation, iagc. An automatic gain control whose operation almost immediately follows a change in signal amplitude.

instantaneous automatic volume control Abbreviation, iavc. An *instantaneous automatic gain control* system for the immediate control of volume in receivers and af amplifiers.

instantaneous companding A form of companding that operates according to the instantaneous amplitude of the input signal.

instantaneous contacts Timer contacts that open or close almost immediately upon application of the control signal.

instantaneous current Symbol, *i* or *Ii*. The value of an alternating current at a particular instant in the cycle. For a sine wave, *Ii* = *Im*

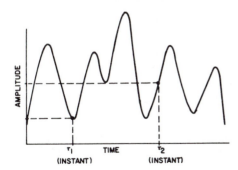

Instantaneous amplitude.

sin Ωt, where Im is the maximum current (in the units for Ii and t is time (seconds).

instantaneous disk A phonograph disk that can be played back immediately after being recorded.

instantaneous frequency The frequency of a signal at a particular moment in time. The instantaneous frequency changes in frequency-modulated or phase-modulated signals.

instantaneous power In an amplitude-modulated wave, the power at any instant in the modulation cycle. In a fully modulated (100%) AM wave the instantaneous power varies between zero and four times the unmodulated carrier power.

instantaneous power output The rate of power delivery to a load at a given instant.

instantaneous relay A relay, such as a fully electronic type (having no moving parts), which shows virtually no delay in its operation.

instantaneous sample A measurement obtained by instantaneous sampling.

instantaneous sampling Measuring the amplitude of a wave at a given instant. See, for example, *instantaneous current* and *instantaneous voltage.*

instantaneous speech power In the output of an audio amplifier, the instantaneous value of power in a speech wave, as opposed to that in a sine wave. Also see *instantaneous value* and *speech power, 1.*

instantaneous value The magnitude of a value that varies at a selected instant. See, for example, *insantaneous current, instananeous power, instantaneous speech power*, and *instantaneous voltage*. Compare *average value*, and *effective value.*

instantaneous voltage Symbol, Ei. The value of an ac voltage at a particular instant in the cycle. For a sine wave, $Ei = Em$ sin Ωt; where Em is the maximum voltage (in the units for Ei) and t is time (seconds).

instruction In digital computer practice, a set of bits defining an operation and consisting of (1) an operation code specifying the operation to be performed; (2) one or more operands or their addresses; and (3) one or more modifiers, or their addresses, to modify the operand or its address.

instruction address In a computer memory, the address of a location containing an instruction.

instruction address register As a part of a program controller, a register that holds instruction addresses that the retrieval of the instructions from memory can be controlled during a program run; program counter.

instruction code The symbols and characters that comprise the syntax of a computer programming language. Also called *instruction set,*

order code, machine code, operation code, function code.

instruction format In a computer's basic machine code, the part that specifies how characters or digits are used to represent the codes within the machine's instruction set.

instruction modification In a computer instruction, a change in the instruction code that makes the computer do a different operation when the routine containing the code is encountered again.

instruction register A register in a computer containing the address of the current instruction. Also called *control register* (abbreviation, CR).

instruction set The range of commands that form a programming language.

instruction storage A memory circuit that stores computer instructions or programs.

instruction time The time it takes a control unit to analyze and implement a computer program instruction.

instruction word In digital computer programming, a word containing (1) the instruction code (type of operation to be performed) and (2) the address part (location of the associated data in storage).

instrument A device for measuring electrical quantities or the performance of electronic equipment. Some, meters, give direct indications; others, such as bridges, must be adjusted, the measured quantities being determined from one or more adjustments (sometimes augmented with calculations).

instrumental error See *instrument error.*

instrument amplifier A high-gain wideband amplifier that increases the sensitivity of an instrument, such as an oscilloscope, meter, or graphic recorder.

instrument-approach system See *instrument landing system.*

instrumentation Planning and providing instruments and instrument systems for the collection and, sometimes, storage and analysis of data.

instrument chopper A refined chopper for converting a dc signal to ac for an ac instrument, such as a voltmeter or recorder.

instrument error Discrepancy in measured quantities due to inaccuracy of the instrument used, insertion resistance, and so on.

instrument flight Blind aircraft flight, i.e., flight guided by navigational magnetic tape by the processing facility.

instrument fuse A fast-acting low-current fuse used to protect a sensitive instrument, such as a galvanometer, milliammeter and or microammeter.

instrument lamp A light or lamp that illuminates the face of an instrument to facilitate viewing in the dark.

instrument landing A blind aircraft landing, i.e., one guided entirely by instruments.

instrument landing station The radio or radar station in a blind-landing system (see *instrument landing system*).

instrument landing system Abbreviation, ILS. The complete instrument and signal system (on the ground or in aircraft) required for an instrument landing.

instrument multiplier See *multiplier probe, 1.*

instrument preamplifier An external amplifier for an instrument that has an internal input amplifier. Also see *instrument amplifier.*

instrument relay See *meter relay.*

instrument resistance See *meter resistance.*

instrument shunt Shunt resistance used to increase an instrument's current range.

instrument transformer A transformer used to change the range of an ac meter (*current* transformer for ammeters or *potential* transformer for voltmeters).

insulant An insulating material (see *insulator, 1*), e.g., dry air, cloth, paper, wood, and so forth.

insulated Isolated from conducters by an insulant.

insulated-gate field-effect transistor Abbreviation, IGFET. See *metal-oxide silicon FET*.

insulated resistor A resistor around which is molded a nonconducting material, such as vitreous enamel or a plastic.

insulating tape Electrical insulation in the form of a thin, usually adhesive, strip of fabric, paper, or plastic.

insulation 1. A coating of dielectric material that precludes a short circuit between a conductor and the surrounding environment. 2. The application of a dielectric coating to an electrical conductor. 3. Electrical separation between or among different components, circuits, or systems.

insulation breakdown Current leakage through, and rupture of, an insulating material because of high voltage stress.

insulation ratings Collectively, the dielectric constant, dielectric strength, power factor, and resistivity of an insulating material. Sometimes included are such physical properties as rupture strength, melting point, and so on.

insulation resistance The very high resistance exhibited by a good insulating material. It is expressed in megohms (or higher units of resistance) for a sample of material of stated volume or area.

insulation system Collectively, the materials needed to insulate a given electronic device.

insulator 1. A material that, ideally, conducts no electricity; it can therefore be used for isolation and protection of energized circuits and components (also see *dielectric*). Actually, no insulator is perfectly nonconductive (see, for example, *insulation resistance*). 2. Any body made from an insulating material.

insulator arcover A sudden arc, or flow of current, over the surface of an insulator, because of excessive voltage.

integer A whole number, as opposed to a fraction or mixed number.

integral Symbol, $\int$1. The sum of an infinite series of values (increments) making up a quantity. Thus, $\int dx = x$. Compare *differential*. Also see *definite integral, indefinite integral, integral calculus*, and *integration*. 2. The part of a number to the left of the radix point.

$f(x)$	$\int f(x)\, dx$
$y = k$	$\int y\, dx = kx + c$
$y = kx$	$\int y\, dx = (k/2)x^2 + c$
$y = kx^2$	$\int y\, dx = (k/3)x^3 + c$
$y = kx^n$	$\int y\, dx = ((k/(n+1))x^{n+1} + c$
$y = \log_e (x)$	$\int y\, dx = x \log_e (x) - x + c$
$y = e^x$	$\int y\, dx = e^x + c$
$y = \sin (x)$	$\int y\, dx = -\cos (x) + c$
$y = \cos (x)$	$\int y\, dx = \sin (x) + c$
$y = \tan (x)$	$\int y\, dx = -\log_e (\cos (x)) + c$

Integrals of common functions.

integral action In automatic control practice, a control action delivering a corrective signal proportional to the time the controlled quantity has differed from a desired value.

integral calculus The branch of mathematics concerned with the theory and applications of integration. Also see *definite integral, indefinite integral, integral*, and *integration*. Like differential calculus, integral calculus is a powerful tool in electronics design.

integral contact In a relay or switch, a contact that carries current to be switched.

integral-horsepower motor A motor rated at 1 horsepower.

integral multiple A whole multiple of a number. Thus, a harmonic is an integral multiple of a fundamental frequency f, $2f$, $5f$, $10f$, and so on.

integral number See *integer*.

integrand A function or equation which is to be integrated. Thus, in the integral expression $\int_a^b y\, dx$, the integrand is $y\, dx$. Also see *integral, integral calculus*, and *integration*.

integrate 1. To perform the function of mathematical or electrical integration. 2. To construct a circuit on a piece of semiconductor material.

integrated Constructed on a single piece of material, such as a semiconductor wafer.

integrated amplifier An af amplifier having a preamplifier, intermediate amplifier, and output amplifier on a single chassis.

integrated capacitor In an integrated circuit, a fixed capacitor in which one "plate" is a layer of material diffused into the substrate; the dielectric, a thin oxide film grown on top of the first layer; and the other plate, a metal layer deposited on top of the oxide film.

integrated circuit Abbreviation, IC. A circuit whose components and connecting "wires" are made by processing distinct areas of a *chip* of semiconductor material, such as silicon. Integrated circuits are classified according to construction, a few being *monolithic, thin-film*, and *hybrid*.

integrated data processing Abbreviation, IDP. The detailed electronic classification, sorting, storage, and mathematical processing of data within a coordinated system of equipment, usually at one location.

integrated electronics That branch of electronics that is concerned with the design and fabrication of integrated circuits.

integrated resistor See *diffused-layer resistor*.

integrating circuit See *integrating network*.

integrating meter An instrument whose indication is a summation (usually) of an electrical quantity that is time-dependent, e.g., ampere-hour meter, watt-hour meter.

integrating motor An electric motor that follows the integral of the input signal. The angle of rotation of the motor shaft is equal to the integral of an input waveform.

integrating network A four-terminal *RC* nework (series resistor, shunt

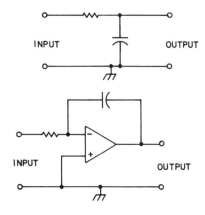

Integrating network.

capacitor) whose output voltage is (or is proportional to) the time integral of the input voltage. Compare *differentiating network*.

integrating photometer An indicating photometer whose reading is the average candlepower at all angles in one plane.

integration Finding a function when its derivative is given. Integration is the inverse of differentiation. Also see *definite integral, indefinite integral, integral,* and *integral calculus*.

integrator 1. Integrating network. 2. An operational amplifier whose output voltage, Eo, is $(1/RC)\int_0^t Ei$ dt, where Ei is the input voltage; Eo, the output voltage; R, the input resistance; C, the feedback capacitance; and d, the differential of time. Compare *differentiator, 2*. 3. A device having an output variable whose value is (a) proportional to the integral of one variable with respect to another, or (b) proportional to the integral of an input variable with respect to elapsed time.

intelligence The quality of a system or device that allows it to "learn," i.e., to better its capability by repeatedly operating on a given problem.

intelligence bandwidth 1. The bandwidth necessary to convey a specified amount of data within a certain period of time. 2. The total bandwidth of one complete signal channel in a communications or broadcast system.

intelligence signal 1. A signal that conveys data or information. 2. The modulating waveform in a communications or broadcast transmission.

intelligent terminal A computer terminal (e.g., an input/output video display/keyboard unit) that through its circuitry—i.e., by use of a microprocessor—has some data processing ability.

intelligibility tests Test that measure the coherence of electronically reproduced speech.

intensification of image See *image intensification*.

intensifier electrode See *post-accelerating electrode*.

intensifying ring In some electrostatic cathode-ray tubes, an internal metal ring serving as an extra anode to accelerate the beam and, thus, brighten the image.

intensitometer See *dosimeter*.

intensity The degree or extent of amplitude, brightness, loudness, power, force, etc.

intensity control In an oscilloscope circuit, the potentiometer that adjusts the dc voltage on the control electrode of the cathode-ray tube and, accordingly, the brightness of the image. Also called *brightness control* and *brilliance control*.

intensity level 1. A measure of sound magnitude, expressed in decibels with respect to a value of 10^{-6} W/cm^2 at sea level in the atmosphere. 2. The setting of the brightness control in a cathode-ray-tube device.

intensity modulation 1. Modulation of electron-beam intensity in a cathode-ray tube. Also called *Z-axis modulation*. 2. Sometimes, the video-signal modulation at a TV picture tube.

intensity-modulation amplifier The Z-axis amplifier in an oscilloscope. Also see *intensity modulation*.

interaction The (sometimes mutual) influence of one circuit or device on the behavior of another, as in induction.

interaction space The region in an electron tube in which electrons and an alternating magnetic field interact.

interactive display A computer display device with which its operator can supply data to the computer in response to what is displayed.

interactive graphics A computer graphics system using a cathode-ray tube for the purpose of drawing or modifying three-dimensional representations.

interactive mode See *conversational mode*.

interbase resistance The internal resistance between the bases of a unijunction transistor.

interblock A part of a computer program—or a hardware device—that will prevent interference between parts of a computer system.

interblock space On a magnetic tape, the space between recordings, caused by starting and stopping the tape; specifically, on a magnetic tape used as a computer storage medium, the interval between recorded blocks.

intercarrier receiver A TV receiver circuit in which video, sound, and sync components of the composite TV signal are amplified together in the rf, i-f, and video i-f stages; then they are separated in the video detector and video amplifier stages. Compare *split-sound receiver*.

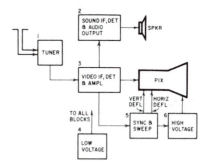

Intercarrier receiver.

intercept receiver In military service, a search receiver tuned over a wide band of frequencies to find and measure enemy signals.

interchangeability The ability of one component to substitute directly for another component of the same kind. Example: capacitor interchangeability, transistor interchangeability. Also see *replacement*.

interchangeable type bar In a punched card tabulator, a type bar that can be exchanged for another to provide a different set of symbols or characters.

intercharacter space The three-unit interval between letter symbols in telegraphy. Compare *interword space*.

intercom A comparatively simple two-way telephone or low-power radio system for use on the premises of a business.

inercommunicator See *intercom*.

interconnection 1. A mutual connection of separate circuits. 2. The interconnection of two or more separate power-generating systems.

intercontinental ballistic missile Abbreviation, ICBM. A ballistic missile capable of traveling between distant continents. Compare *intermediate-range ballistic missile*.

interdigital contacts A pair of digitated (*with fingers*) contacts that are plated, printed, or deposited on the surface of a resistor material or semiconductor subsrate. The fingers of each contact are interconnected at one end, the fingers of one contact being interleaved with those of the other.

interdigital tube A magnetron having a cathode surrounded by anode segments which are alternately interconnected at opposite ends in the manner of *interdigital contacts*.

interelectrode capacitance Capacitance between electrodes, especially between the plate and control grid of a vacuum tube.

interelement capacitance Internal pn-junction capacitance in a semiconductor device, such as a diode or transistor.

interface The meeting of surfaces or regions in a material, the surfaces of mating bodies, or of entities in a system—the circuitry that interconnects and provides compatibility between a central processor and peripherals in a computer system, for example.

interface resistance See *cathode interface.*

interface routine A computer program routine that links one system to another.

interfacial connection A connection that runs through a printed-circuit board and joins circuit joints on opposite faces of the board.

interference The disturbing effect of any undesired signal.

interference attenuator A device or mode of operation that reduces the amplitude of interference.

interference eliminator A filter, wavetrap, or similar device which removes interferential signals. Also see *interference.*

interference filter See *interference eliminator.*

interference removal See *interference eliminator.*

interference stub A length of twin-lead feeder cut to appropriate length, connected to the antenna-input terminals of a TV receiver, and short-circuited at the opposite end. A stub of correct length resonates at the frequency of an interferential signal and, acting as a wavetrap, keeps it out of the receiver. Also see *stub.*

interference trap A wavetrap that suppresses interferential signals at the rejection frequency of the trap.

interferometer 1. A radio telescope having two antennas spaced at a distance of many wavelengths, providing much greater resolution than a single antenna. Pioneered by M. Ryle of England and J. L. Pawsey of Australia. 2. Any device that displays an interference pattern for testing or experimental purposes.

interfix A method used in information retrieval systems that eliminates ambiguity in the responses to inquiries by describing the relationship between key words in a record.

interharmonic beats Beat notes between harmonics of a fundamental frequency.

interim storage See *temporary storage.*

interior label On a magnetic tape used as a computer storage medium, a label recorded at the beginning of the tape. Compare *exterior label.*

interlace A form of memory storage in which portions of the data are stored in alternate locations in the tape, disk, or drum.

interlaced field A TV picture field produced by *interlaced scanning.*

interlaced scanning In the display of a TV picture, the alternate presentation of the even- and odd-line fields.

interlace factor A number expressing the extent to which two fields are interlaced. Also see *interlaced scanning.*

interleaving In multiprogramming, the inclusion in a program of segments of another program so that both can be effectively executed simultaneously.

interlock switch See *electrical interlock.*

intermediate amplifier See *buffer.*

intermediate frequency Abbreviation, i-f. In a superheterodyne circuit, the frequency of the signal that results from beating the incoming signal with the signal produced by the local oscillator.

intermediate-frequency amplifier In a superheterodyne circuit, the fixed-frequency amplifier that boosts the intermediate-frequency signal. Also see *intermediate frequency.*

intermediate-frequency channel Usually, the intermediate-frequency amplifier in a superheterodyne circuit, but sometimes including the second detector-agc and oscillator stares.

intermediate-frequency converter See *i-f converter.*

intermediate-frequency interference Interference from signals of the intermediate frequency of a receiver or instrument.

intermediate-frequency selectivity The selectivity of an intermediate-frequency channel alone.

intermediate-frequency transformer A coupling transformer designed for use in an intermediate-frequency amplifier.

intermediate-puck drive In a tape recorder, a speed-reducing drive system in which an intermediate wheel conveys motion from the motor shaft to the rim of the flywheel.

intermediate puck wheel See *idler wheel.*

intermediate-range ballistic missile Abbreviation, IRBM. A missile or rocket having a 200 to 1500 mile range. Compare *intercontinental ballistic missile.*

intermediate repeater In wire telephony, a repeater inserted into a line or trunk at some point other than the end.

intermediate result Obtained during a program run or the execution of a subroutine, a result that is used again as an operand in deriving the final result. Compare *final result.*

intermediate section Any one of the internal sections of a *multisection filter.* Thus, the *midsection* of an odd-numbered filter.

intermediate storage In a computer system, a storage medium for temporarily holding totals or working figures. Also called *work area.*

intermediate subcarrier A modulated or unmodulated subcarrier that modulates either a carrier or another intermediate subcarrier.

intermittent An intermittent condition or signal.

intermittent commercial and amateur service Abbreviation, ICAS. Operation of equipment, such as electron tubes, for short, irregular periods, as in amateur (hobbyist) activity or infrequent commercial service. ICAS ratings are higher than continuous commercial service (CCS) ratings. Compare *continuous commercial service.*

intermittent condition A defect in a circuit or device that causes intermittent operation.

intermittent dc See *intermittent direct current.*

intermittent direct current A regularly pulsed unidirectional current. Also called *pulsating current.*

intermittent duty A duty cycle of less than 100 percent but greater than zero. Generally, an operating duty cycle of 50 percent.

intermittent-duty rating The dissipation or power rating of a component, circuit, or system, under conditions of intermittent duty (usually a 50-percent duty cycle).

intermittent operation Operation characterized by often long nonoperating intervals. Intermittent operation is often random, whereas *on-off operation* tends to be regular.

intermittent signal An interrupted signal resulting from the intermittent operation of a circuit or device.

intermodulation Abbreviation, IM. The (usually undesired) modulation of one signal by another, caused by the system's nonlinear processing of the signals.

intermodulation distortion See *intermodulation.*

intermodulation-disortion meter See *intermodulation meter.*

intermodulation-distortion percentage Abbreviation, IDP. The degree to which a low-frequency test signal modulates a higher-frequency test signal when both are applied simultaneously (in a prescribed amplitude ratio) to a device under test; $IDP = 100(b - a)/a$, where a is the peak-to-peak amplitude of the unmodulated high-frequency wave and b is the peak-to-peak amplitude of the modulated high-frequency wave.

intermodulation meter An instrument for measuring percentage of intermodulation. The instrument combines a dual-frequency signal generator, filter circuits, and percent-of-modulation meter. Also see *intermodulation-distortion percentage*.

intermodulation noise Electrical noise produced in one channel by signals in another; it is due to intermodulation.

internal amplification In a radioactivity counter tube, current enhancement resulting from cumulative ionization initiated by an ionizing particle.

internal arithmetic In digital computer operation, arithmetic operations performed in the compuer, as opposed to those performed by peripherals.

internal connection Abbreviation, IC. On a vacuum tube, the designation for a base pin connected to an electrode in the tube; it is so labeled to warn against connecting it to an external circuit or to show that no external connection is necessary.

internal impedance The impedance in a device, as opposed to that added from the outside. Compare *internal resistance*.

internal input impedance The impedance in a circuit or device, as seen from the input terminals. Compare *internal output impedance*.

internal input resistance The resistance in a circuit or device, as seen from the input terminals. Compare *internal output resistance*.

internal noise Electrical noise generated within a circuit, as opposed to that picked up from outside. Such noise comes from tubes, resistors, and other components.

internal output impedance The impedance in a circuit or device, as seen from the output terminals. Compare *internal input impedance*.

internal output resistance The resistance in a circuit or device, as seen from the output terminals. Compare *internal input resistance*.

internal resistance The resistance of a device, as opposed to added resistance. See, for example, *meter resistance*.

international broadcast station A short-wave broadcast station transmitting programs for international reception between 6 and 26.6 MHz.

international call sign The *call letters* of a station, assigned within a country according to the method of arrangement (identifying letters, or letters and numerals) prescribed by the International Telecommunication Union.

international candle See *candle*.

international coulomb A unit of electrical quantity, equal to 0.99985 absolute coulomb.

international farad A unit of capacitance, equal to 0.99952 absolute farad.

international henry A unit of inductance, equal to 1.00018 absolute henry.

international joule A unit of energy, equal to 1.00018 absolute joule.

International Morse Code See *Continental code*.

international ohm A unit of electrical resistance, equal to 1.00048 absolute ohm. The other international units are derived from this value.

International Radio Consultative Committee Abbreviation, CCIR (from Comite Consultatif Inernational Radiodiffusion). An international organization reporting to the International Telecommunications Union and studying technical operations and tariffs of radio and television.

International Steam Table calorie A unit of heat equal to 4.1868 joules.

International System of Units Abbreviation, SI (for *Syseme International d'Unites*). The international system of units of measurement established in 1960 under the Treaty of the Meter. In this system the *base* units are as follows: *meter* (m), length: 1 650 763.73 wavelengths in vacuum corresponding to the transistion 2p10 – 5d5 of krypton 86; *kilogram* (kg), mass: the mass of the protype kilogram kept at Sevres, France; *second* (s), time: the duration of 9 192 631 770 periods of the radiation that corresponds to the transition between the two hyperfine levels of the ground state of cesium 133; *kelvin* (K), thermodynamic temperature: 1/273.16 the thermodynamic temperature of the triple point of water; *ampere* (A), electric current; the current that, flowing through two infinitely long parallel wires in a vacuum and separated by 1 meter, produces a force of 2×10^{-7} newton per meter of length between the wires; *Candela* (cd), luminous intensity: the luminous intensity of 1/600 000 square meter of perfectly radiating surface at the temperature of freezing platinum. Base units combined with each other and a couple of additional, *supplementary units* (radian, steradian) form the units for expressing any known physical quantity or phenomenon; some of these *derived units* have their own special names: newton (kg • m/s^2), weber (V • s), tesla (webers per square meter), to name a few.

International Telecommunications Union Abbreviation, ITU. An international, nongovernmental organization devoted to standardizing worldwide communications practices and procedures.

International Telegraph and Telephone Consultative Committee See *CCITT*.

Internaional Telegraph Consultative Committee See *CCIT*.

international units A system of electrical units, based on the resistance through a specified quantity and configuration of the element mercury. The international ohm forms the basis for the international system of units.

international volt A unit of electrical potential, equal to 1.00033 absolute volt.

international watt A unit of power, equal to 1.00018 absolute watt.

interphone An intercom aboard a mobile vehicle.

interpolation Finding a value that falls between two values listed in a table, indicated by a dial, derived by estimate, or given by intermediate calculation. T us, if a linear variable capacitor has a value of 100 pF when its dial is set to 10, and 110 pF when the dial is set to 20, then the capacitance when the dial reads 15 may be determined by means of interpolation: $C = 100 + (15 - 10)(110 - 100)/(20 - 10) = 105$ pF.

interpolation meter See *interpolation-type instrument*

interpolation oscillator A frequency-measuring oscillator with a built-in crystal oscillator for standardizing points on the oscillator circuit dial. The dial provides a continuous range of frequencies between crystal-harmonic points. Also see *interpolation-type instrument*.

interpolation-type instrument An instrument, such as a meter or signal generator, that is used to transfer an accurate quantity point from a standard to another instrument and to provide a range of values between such points. A *secondary standard* is sometimes used as an interpolation-type instrument (see, for example, *interpolation oscillator*).

interpole motor A motor with interpoles. Also see *interpoles*.

interpoles Small auxiliary poles between the main field poles in a dc generator or motor that reduce sparking at the commutator. Also called *commutating poles*.

interpreter A computer program that can convert instructions given in a high-level language (BASIC, for example) into the machine language that a computer uses; if it is not resident in the computer's nonvolatile memory, it must be loaded each time the machine is activated.

interrecord gap See *interblock space.*

interrupt A break in a computer program, as when a background job is interrupted so that a foreground job can be run. Also see *background job; foreground job.*

interrupted commercial and amateur service See *intermittent commercial and amateur service.*

interrupted continuous wave Abbreviation, ICW. A continuous wave that is interrupted at regular intervals, as in the chopping of a wave at a regular rate. Compare *continuous wave* (CW) and *modulated continuous wave* (MCW).

interrupted dc A direct current or voltage which is periodically started and stopped by switching or chopping.

interrupter contacts Auxiliary contacts operaed directly by the armature of a stepping switch.

interruption frequency See *quenching frequency.*

interruption-frequency oscillator See *quench oscillator.*

interrupt signal The signal that causes an *interrupt.*

intersection The logical AND operation.

interstage capacitor A coupling capacitor used between two circuit stages.

interstage coupling Coupling between two circuit stares, such as those of an amplifier. Common forms of interstage coupling are direct, capacitor, choke coil, transformer, neon bulb, and zener diode.

interstage diode A semiconductor coupling diode used between two circuit stages. Also see *interstage coupling.*

interstage punching The punching of a computer data card between its usual punching positions.

interstage transformer A coupling transformer used between two circuit stages. Also see *interstage coupling.*

interstice A tiny space between regularly spaced bodies (as in a network or lattice), e.g., the interstices of a wire screen.

intersystem A power-generating network of interconnected separate systems (see *interconnection, 2*).

intersystem communications Communications between computer systems through (1) direct linking of central processors or (2) by their mutual use of peripherals and input/output channels.

intertie See *interconnection, 2.*

interval The amount of separation between successive points, events, or quantities.

intervalometer A timing device for operating equipment over a precisely defined time interval.

interval timer A device that provides power to an equipment for a precise interval upon application of a simple initiating signal or action. See also *intervalometer.*

interword space The seven-unit interval between words or code groups in telegraphy. Compare *intercharacter space.*

intoxication tester See *drunkometer.*

Intrafax Western Union's private facsimile system.

intrinsic-barrier diode See *pin diode.*

intrinsic-barrier transistor A pnip transistor or the equivalent npin transistor.

intrinsic concentration The number of minority carriers exceeding the normal equilibrium number in a semiconductor.

intrinsic conduction The flow of electron—hole pairs in an intrinsic semiconductor subjected to an electric field.

intrinsic flux A quantity equal to the product of the intrinsic flux density and the cross-sectional area in a magnet.

intrinsic flux density The increased flux density of a magnet in its

actual environment, as compared with the flux density resulting from the same magnetizing force in a perfect vacuum.

intrinsic mobility Electron mobility in an intrinsic semiconductor. Also see *carrier mobility* and *mobility.*

intrinsic Q The value of Q for an unloaded circuit. Also see *figure of merit, 1.*

intrinsic semiconductor A semiconductor whose characteristics are identical to those of a pure crystal of the material. In this condition, the semiconductor is nearly an insulator. Example: highly purified germanium or silicon before n- or p-type impurities have been added. Compare *extrinsic semiconductor.*

intrusion alarm A proximity detector used as an intrusion (burglar) signaling device.

intrusion sensor A sensitive pickup—such as a photocell, ultrasonic detector, or capacitive transducer—that responds to a nearby body by delivering an actuating signal to an intrusion alarm.

INV Abbreviation of *inverter.*

inv Abbreviation of *inverse.*

invar (Sometimes capitalized.) A nickel-steel alloy (36% nickel) having a low temperature coefficient of linear expansion (1 ppm/°C). Invar is employed in electronic equipment where mechanical distortion resulting from temperature changes must be negligible, and in magnetostrictive circuits (see *magnetostriction*).

inverse 1. Opposite in nature; e.g., an *inverse characteristic.* 2. Of opposite sign, e.g., a negative current or voltage. 3. An operation of opposite kind; thus, division is the inverse of muliplication, integration, the inverse of differentiation.

inverse beta The beta of a transistor operated with the emitter and collector interchanged.

inverse bias See *reverse bias.*

inverse characteristics The characteristics of a bipolar transistor when operated with the emitter and collector reversed.

inverse conduction See *reverse conduction.*

inverse current See *reverse current.*

inverse-distance law The inverse-square law applied to the propagation of radio waves, assuming that the waves don't encounter obstacles.

inverse feedback See *degeneration.*

inverse impedances Impedances ($Z1$ and $Z2$) that are the reciprocal of another impedance ($Z3$), satisfying the relationship $Z1 \, Z2 = (Z3)^2$. Also called *reciprocal impedances.*

inverse leakage The flow of a small static reverse current in semiconductor devices.

inverse-parallel circuit See *back-to-back circuit* and *back-to-back connection.*

inverse peak voltage See *peak inverse voltage.*

inverse piezoelectric effect Mechanical movement in a piezoelectric material, caused by application of voltage.

inverse resistance See *reverse resistance.*

inverse resonance See *parallel resonance.*

inverse-square law The intensity of a phenomenon at a point that is distant from the source is inversely proportional to the distance. This is often applied to quantitative reasoning about light, radio waves, and other forms of radiation.

inverse trigonometric function An angle expressed in terms of a given trigonometric function followed by the exponent -1 or preceded by *arc* ($\sin^{-1}$ is the equivalent of *arcsin*). Generally, $\sin^{-1} \Omega$ ($\sin \Omega = \Omega$ thus $\sin^{-1} = 30°$ (because $\sin 30° = 0.5$).

inverse voltage 1. The negative voltage at the anode of a rectifier dur-

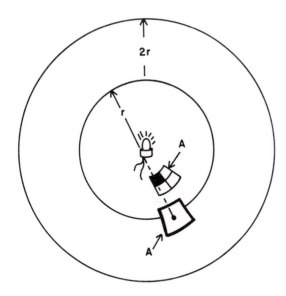

Inverse square law.

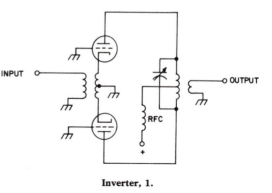

Inverter, 1.

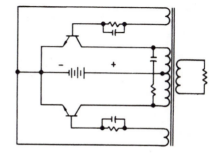

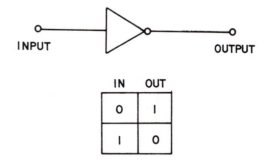

IN	OUT
O	I
I	O

Inverter 2.

ing the negative half-cycle of an input ac. 2. The voltage across a power-supply filter capacitor during the negative half-cycle of an input ac. 3. Semiconductor-junction *reverse voltage*.

inverse Wiedeman effect See *direct Wiedeman effect*.

inversion 1. A turnaround of the vertical temperature gradient of the air, causing changes in the ionosphere and, accordingly, in radio propagation. 2. Speech scrambling (see *scrambler circuit*). 3. Phase inversion (see *phase inverter*). 4. Changing dc into ac (see *incerter, 1*).

inverted amplifier A push-pull, grounded-grid, vacuum-tube amplifier.

inverted-L antenna An antenna having a horizontal radiator and a vertical feeder or lead-in attached to one end of the radiator. The entire arrangement resembles an upside-down *L*.

inverted speech See *scrambled speech*.

inverted tube A vacuum tube connected so that the plate (biased negatively) receives the input signal, and the control grid (biased positively) delivers the output signal.

inverter 1. A device that converts direct current into alternating current. 2. A logic circuit that provides an output pulse which is a negation of the input pulse. Also called a *complementer* or a *NOT circuit*. 3. Phase inverter.

inverting adder An analog adder circuit which is provided with an amplifier for a 180-degree phase shift.

inverting amplifier An amplifier providing a 180-degree phase shift between input and output.

inverting connection Connection to the inverting input terminals of a differential amplifier or operational amplifier. Also see *inverting input*. Compare *noninverting input*.

inverting input In a differential amplifier, the input circuit which produces a phase reversal between the input and output. Compare *noninverting input*.

invisible failure In a computer system, a hardware or software failure whose effect on the system is unnoticeable.

invister A unipolar semiconductor material, capable of operation at very high frequencies.

involuion Raising a number to a power: squaring, cubing, etc. Compare *evolution*.

inward-outward dialing In a telephone system, a method of dialing in which calls may be made to and from branch exchanges, without operator assistance. Sometimes called direct dialing.

I/O Abbreviation of *input/output* (see *input/output equipment*).

Io Symbol for *output current.*

iodine Symbol, I. A nonmetallic element of the halogen family. Atomic number, 53. Atomic weight, 126.91. Also see *halogen.*

ion A charged atom, i.e., one that has either gained one or more electrons (a negative ion, or *anion*) or lost one or more electrons (a positive ion, or *cation*).

ion burn A spot burned on the screen of a TV picture tube or oscilloscope by negative ions from the cathode striking a single point on the faceplate with high intensity for long periods.

ion exchange resins Granular resins containing acid or base groups, which trade ions with salts in solutions. The resins play a part in the purification of water for various industrial processes.

ionic binding forces In a crystal, the binding forces that occur when valence electrons of one atom are joined to those of a neighboring atom whose outer shell they fill.

ionic bond In a solid, an interatomic bond formed as a result of the attraction between oppositely charged atoms (ions).

ionic conduction Conduction, as in a gas or electrolyte, by ion migration (positive to the cathode, negative to the anode).

ionic crystal A crystal whose lattice is held together by electric forces between ions. Also see *ionic binding forces* and *ionic bond.*

ionic current Current caused by ion movement in a gas or liquid. Also see *ion, ionization,* and *ionic conduction.*

ionic switch See *electrochemical switch.*

ionization The loss or gain of one or more electrons by an atom. Also see *anion, cation,* and *ion.*

ionization arc The electrical discharge resulting from ionization of a material because of high voltage.

ionization chamber An enclosure containing a gas and a pair of electrodes between which a high voltage is appilied. Radiation, such as X-rays or radioactive particles, passing through the walls of the chamber ionize the gas, creating an ionization current which is proportional to the intensity of the radiation.

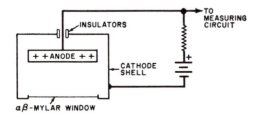

Ionization chamber.

ionization current 1. Negative grid current resulting from gassiness in a vacuum tube. 2. In a gas tube, current flowing after the ignition potential has been reached. 3. Current in a gas or electrolyte. 4. Current in an ionization chamber, Geiger-Mueller tube, or similar gaseous device.

ionization gauge A vacuum gauge that consists of a positive-grid triode whose own vacuum is part of the monitored vacuum. The plate current of the tube is inversely proportional to the hardness of the vacuum.

ionization noise In a faulty vacuum tube, electrical noise caused by gassiness.

ionization potential The voltage at which a substance (especially a gas) ionizes. Also called (for a gas) *ignition potential* (see *breakdown voltage,).*

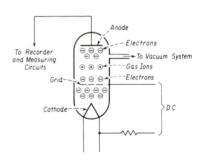

Ionization gauge.

ionization pressure In an ionized gas, the pressure increase resulting from the ionization, as compared with the same volume and mass of gas when not ionized.

ionization resistance See *corona resisance.*

ionization time The interval (in ms or µs) between the instant that an ionizing potential is applied to a gas and the instant at which the gas begins to ionize. Compare *deionization time.*

ionize To cause the electrons in a substance, particularly a gas, to move freely from atom to atom.

ionized gas A gas whose atoms, under the influence of a strong electric field or radiation, have become positive or negative ions.

ionized layer See *Kennelly-Heaviside layer.*

ionized liquid See *electrolyte.*

ionizing radiation 1. Any high-energy electromagnetic radiation that causes ionization in a gas through which the field passes. 2. High-speed atomic nuclei. 3. X-rays or gamma rays.

ion migration The movement of ions through a solid, liquid, or gas because of the influence of an electric field.

ionosphere The multilayered shell of ionized particles surrounding the earth. It is responsible for the absorption, reflection, and refraction of radio waves. Also see *Kennelly-Heaviside layer.*

ionospheric disturbance See *ionospheric storm.*

ionospheric forecasting The practice of predicting ionospheric conditions. Radio propagation data result from such predictions.

ionospheric propagation Propagation of radio waves by means of reflection or refraction by the ionosphere. Also see *hop; incident wave; ionosphere; multihop propagation; reflected wave, 1; refracted wave;* and *skywave.*

ionospheric scatter See *forward scatter.*

ionospheric storm Turbulence in the ionosphere, usually accompanied by a magnetic storm.

ion sensor A device whose operation is based on the detection of ions and the delivery of a proportionate voltage. Examples are the *Geiger counter, halogen ras leak detector, mass spectrometer,* and *vacuum gauge.*

ion sheath In a thyratron, a collection of positive ions of gas that forms around the wires of the control grid.

ion spot See *ion burn.*

ion trap See *bent-gun crt.*

ion-trap magnet An external (usually double) magnet used with a TV picture tube to deflect the ion beam away from the screen to prevent ion burn.

I/O port That part of a computer providing, by a connector, a point of input/output access to it.

Ip 1. Abbreviation of *plate current*. 2. Abbreviation of *peak current*.

I-phase carrier In color television, a carrier separated by 57 degrees from the color subcarrier.

ipm Abbreviation of *inches per minute*.

ips Abbreviation of *inches per second*.

IR 1. The product of current and resistance (see, for example, *IR drop*), 2. Abbreviation of *insulation resistance*. 3. Abbreviation of *infrared*.

Ir Symbol for *iridium*.

Ir Symbol for *current* in a resistor.

IRAC Interdepartment Radio Advisory Committee (a federal government group in the United States).

IRBM Abbreviation of *intermediate-range ballistic missile*.

IR diode See *infrared diode*.

IR drop The voltage drop (E) across a resistance (R) due to the flow of current (I) through the resistor; $E = IR$.

IRE Institute of Radio Engineers, predecessor of the IEEE.

Irf Symbol for *radio-frequency current*.

irridescence A sparkling, colorful appearance in a material, resulting from refraction, internal reflection, and interference in light waves passing through the substance. Especially noticeable in quartz and certain gems.

iridium Symbol, Ir. A metallic element of the platinum group. Atomic number, 77. Atomic weight, 192.2

iron Symbol, Fe. A magnetic, metallic element. Atomic number, 26. Atomic weight, 55,85. Iron (and its special form, steel) is perhaps the most widely used material in magnetic circuits.

iron/constantan thermocouple A thermocouple consisting of a junction between wires or strips of iron and constant. Also see *constantan*.

iron core A transformer or choke core made from iron or steel.

iron-core coil An inductor having an iron or steel core.

iron-core i-f transformer An intermediate-frequency transformer having a core of powdered iron, a form of iron having the advantage of high permeability without the fault of eddy current production.

iron-core transformer A transformer whose coils are wound on a core of iron or steel.

iron loss Power lost in the iron cores of transformers, inductors, and electrical machinery as a result of eddy currents and hysteresis.

iron oxide A compound of iron and oxygen (rust). There are several forms, depending on the number of iron and oxygen atoms in the iron oxide molecule. See, for example, *magnetite* and *red oxide of iron*.

iron pyrites Formula, FeS_2. Natural iron sulfide which in its natural state occurs as bright yellow crystals.

iron-pyrites detector An early diode (crystal) detector employing iron pyrites as the semiconductor.

iron-vane meter An ac meter whose movable element, a soft iron vane, carries the pointer and pivots near a similar, stationary vane. The vanes are mounted in a multiturn coil of wire. The current to be measured flows through the coil, the resulting magnetic field magnetizing the vanes. Because the magnetic poles of the vanes are identical, they repel each other; the movable vane is deflected (against the torque of returning springs) over an arc proportional to the current, carrying the pointer over the scale. Although, the iron-vane meter is a current meter, it may be converted into a voltmeter by connecting a suitable multiplier resistor in series with the coil.

irradiance The amount of radiant flux impinging on a unit surface area; it is expressed in watts per square centimeter or similar units (e.g., W/m^2, $\mu W/cm^2$, etc.).

irradiation 1. Exposure of a device to radioactivity or X-rays. 2. The

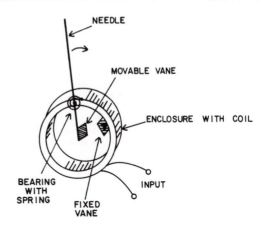

Iron vane meter.

total radiant power density that is incident upon a receiving surface.

irrational number A number that cannot be expressed as an integer or as the quotient of integers, e.g., $\sqrt{2}$.

irregularity 1. The condition of being nonuniform, or rapidly fluctuating, rather than constant. 2. A departure from normal operating conditions. 3. Nonuniformity in a surface. 4. Nonuniform distribution of matter. 5. Nonuniform distribution of data.

IR viewer An infrared viewer. See *sniperscope* and *snooperscope*.

IS 1. Symbol for *source current* in a field-effect transistor. 2. Symbol (with lower-case *s*) for *screen current* .

ISCAN Abbreviation of *inertialess steerable communications antenna*.

I-scan A radar display in which the target is shown as a complete circle whose radius is proportional to the distance to the target.

I-signal With the Z-signal, one of the two signals that modulates the chrominance subcarrier in color TV. The I-signal results from mixing a B-Y signal (with − 0.27 polarity) and an R-Y signal (with + 0.74 polarity).

isinglass Thinly laminated mica.

isobar 1. An atom whose nucleus has the same weight as that of another atom but differs in atomic number. 2. On a weather map, a line connecting points of equal pressure. Also see *bar, 1*.

isochromal phenomena Phenomena occurring at regular intervals, or of equal duration. Also called *isochronous*.

isochromatic (Also, *orthochromatic*.) 1. The quality of having or producing natural tonal values (light). 2. Color sensitivity excluding a response to red.

isochronal See *isochrone*.

isochrone On a map, a line connecting points of constant time difference in radio-signal reception.

isochronous Having identical resonant frequencies or wavelengths.

isoclinic line See *aclinic line*.

isodose Pertaining to points receiving identical dosage of radiation.

isodynamic line On a magnetic map (i.e., of the earth's magnetism), a line connecting points of equal strength.

isoelectric Having a potential difference of zero.

isoelectronic Having the same number of electrons.

isogonic line An imaginary line connecting points of equal magnetic declination on a map.

isolantite An insulating ceramic. Dielectric constant, 6.1.

isolated 1. Electrically insulated. 2. Separated in such a way that interaction does not take place.

isolated input 1. Floating input. 2. An input circuit with a blocking capacitor to exclude dc.

isolated location In a computer, a storage location which is hardware-protected from being addressed by a user's program..

isolating amplifier See *buffer, 1.*

isolating capacitor A series capacitor inserted in a circuit to pass an ac signal while blocking dc. Also called *blocking capacitor.*

isolating diode A diode used (because of its undirectional conduction) to pass signals in one direction, but block them in the other direction.

isolating resistor A high-value resistor (commonly, 1M) connected in series with the input circuit of voltmeter or oscilloscope to protect the instrument from stray pickup. In most voltmeters, the isolating resistor is built into the probe.

isolating transformer A power transformer (usually having a 1:1 turns ratio) for isolating equipment from direct connection to the power line.

isolation The arrangement or operation of a circuit so that signals in one portion are not transferred to (nor affect) another portion.

isolation amplifier See *isolating amplifier.*

isolation capacitor See *isolating capacitor.*

isolation diode 1. In an integrated circuit, a reverse-biased diode that is formed in the substrate to prevent cross-coupling and grounds. 2. Isolating diode.

isolation resistor See *isolating resistor.*

isolation transformer See *isolating transformer.*

isolator See *optoelectronic coupler.*

isolith A form of monolithic integrated circuit, in which the semiconductor is removed in certain places for the purpose of isolating different parts of the circuit.

isomagnetic Having equal magnetic intensity.

isomer A material that has the same atomic number of chemical formula as some other substance, but, because of a difference in the atomic structure, is an entirely different substance. The most common examples are graphite and diamond; both are carbon.

isophote On a graph of light intensity, a curve joing points of equal intensity.

isoplanar An integrated-circuit configuration in which insulating barriers, or metal oxides, are fabricated among the bipolar elements.

isothermal process A physical or chemical process in which there is no temperature change as other factors vary. Compare *endothermic reaction* and *exothermic reaction.*

isotope An atom having the same number of protons as the atom of a similar element, but a different number of neutrons. The atom of the element and its isotopes exhibit identical chemical properties. Thus, deuterium is an isotope of hydrogen. Some isotopes are radioactive, e.g., radioactive cobalt.

isotropic antenna See *unipole, 2.*

isotropic radiator See *unipole, 2.*

Isup Symbol for *suppressor current.*

I(t) Symbol for *indicial response.*

item 1. Component. 2. Any one of a number of similar or identical components, circuits, or systems.

iterated integral See *double integral* and *triple integral.*

iteration Repeating a series of arithmetic operations to arrive at a solution to a problem. For example, the square root of a number N can be found through iteration as follows:

$$\sqrt{N} \approx [(N/E1) + E161]/2 = E2$$

where $E1$ is an initial reasonable estimate of the square root of N. For example, because 3^2 is 9 and 4^2 rule is 16, a reasonable estimate of $\sqrt{10}$ would be 3.2. $E2$, a second, more accurate estimate derived by computation, is used to continue the iteration:

$$\sqrt{N} \approx [(N/E2) + E2]/2 = E3$$

And so on until $En = En - 1$, the square root of N. (This is usually a theoretical condition if you are using a calculator—or even a computer—because using En could give $En - 1$; in that case, using $En - 1$ will give/En again. A program for such a procedure must include a statement of acceptable accuracy, one which the iterative routine can refer to, so it knows" when to leave the iteration loop; otherwise, the machine will get hung up on an endless loop.)

iterative impedance In a network consisting of identical, cascaded sections, the input impedance of a section to which the output impedance of the preceding section is made equal.

iterative routine A program or subroutine that provides a solution to a problem by iteration.

iterative transfer constant Symbol, P. A property of iterative impedance networks defined by $I2/I1 = e^{-P}$. From this, $P = \log_e(I2/I1)$, where $I1$ is the input current and $I2$, the output current of the network. Also see *iterative impedance.*

ITU International Telecommunications Union.

ITV Abbreviation of *industrial televison.*

I-type semiconductor See *intrinsic semiconductor.*

Ix Symbol for *current in a reactance.*

Iy Symbol for *current in an admittance.*

Iz Symbol for *current in an impedance.*

ize A suffix used, with some liberty, to form verbs from nouns. Some common examples are *anodize, electroicize, plasticize,* and *transistorize.*

J

J 1. Symbol for *joule*. 2. Symbol for *jack* or *connector*. 3. Symbol for *emissive power*.

j (j operator) The square root of minus one ($\sqrt{-1}$), an imaginary number abbreviated i in other than electronics disciplines; the j operator is used in vector algebra computations for phase angle.

jack A receptacle for a plug. A plug (a male connector) is inserted into a jack (a female connector) to complete a circuit or removed from it to break a circuit.

Jack.

jack box A (usually metallic) box or can used to hold, shield, or protect a jack or group of jacks.

jacket 1. A term sometimes meaning a shield can or shield box. 2. An insulating outer case or wrapper on a component (such as a capacitor). 3. The heat-radiating or waterconducting enclosure used in cooling a power tube.

jack panel A (usually metallic) panel in which a number of jacks are mounted, usually in some order or sequence as denoted by labels.

Jack Panel.

jackscrew In a two-piece connector, a screw for mating or separating the halves of the connector.

Jacob's law The law which states an electric motor develops maximum power when $Ei = 2Ebk$, where Ei is the applied voltage and Ebk is the back-emf.

jaff Colloquial term for radar jamming that combines electronic and chaff techniques.

jag Distortion caused by temporary loss of synchronization between the scanner and recorder in a facsimile system.

jam Any snag in the flow of punched cards through a machine.

jam input 1. A means of setting a logic line to the desired condition by simply applying the desired high or low voltage. 2. The voltage so applied.

jammer A transmitter used for jamming radio communications. Also, one who jams.

jamming The deliberate use of countermeasures, such as malicious transmission of interfering signals, to obstruct communications.

jamming effectiveness The extent to which jamming is able to disrupt a service. It may be expressed quantitatively as the ratio of jamming signal voltage to jammed signal voltage.

JAN Abbreviation of *Joint Army—Navy*.

janet A vhf system for point-to-point communication via meteor-trail forward scatter.

J antenna An end-fed half-wave antenna having a quarter-wave, parallel-wire matching section resembling the letter *J*.

Janus antenna array *(Janus,* an ancient Roman God). A Doppler-navigation antenna array radiating forward and backward beams.

jar 1. (From Leyden *jar*). An obsolete unit of capacitance equal to 1/900 μF. 2. The container for the elements of a storage cell.

J-carrier system In carrier-current (wired—wireless) telephony, a broadband system which provides 12 telephone channels at frequencies up to 140 kHz.

JCET *Joint Council on Educational Television.*

JCL Abbreviation of job control language.

J-display A radar display having a circular time base. The transmitted pulse and reflected (target) pulse are spaced around the circumference; distances may be measured circumferentially between them.

JEDEC Joint Electron Device Engineering Council.

jerk The rate of change of acceleration—the third derivative of displacement.

JETEC *Joint Electron Tube Engineering Council.*

jewel bearing A low-friction bearing used in electric meters and other sensitive devices. It takes its name from a jewel pivot (such as a sapphire) in the groove of which rides the pointed end of a rotating shaft. Also called *jeweled bearing* or, simply, *jewel.*

jezebel A passive sonobuoy which detects (enemy) submarine noises and transmits them by radio to a monitoring station.

JFET Abbreviation of junction fieldeffect transistor.

JHG Abbreviation for *joule heat gradient.*

jig 1. A device constructed especially for the purpose of holding an equipment or circuit board during its repair. 2. Phonetic alphabet code word for letter J (usually capitalized).

jitter A (usually small and rapid) fluctuation in a phenomenon, such as a quantity or wave, due to noise, mechanical vibration, interfering signals, or similar internal or external disturbances, but usually specifically applied to a crt picture.

j-j coupling Interaction between spinorbit-coupled particles.

J/K Symbol for *joule(s) per kelvin,* the SI unit of entropy; also the unit for the Boltzmann constant.

J-K flip-flop A transistor—resistor flip-flop stage producing an output signal even when both inputs are in the logical one state, so called because its input terminals are labeled *J* and *K*.

J/(kg·K) Symbol for *joule(s) per kilogram kelvin,* the SI unit of specific heat capacity.

job A unit of computer work, usually consisting of several program runs.

job control language An operating system language used to describe the control requirements for jobs within the system.

job control program A program that uses control language statements and implements them as instructions controlling a job in an operating system.

job control, stacked See *sequential-stacked job control.*

job flow control To control the order of jobs being processed by a computer to make the most efficient use of peripherals and central processor time, either manually or by an operating system.

job library A series of related sets of data that will be loaded for a given job.

job-oriented terminal A data terminal that produces data in computer-ready form, e.g., a cash register that outputs paper tape which can be used by a computer.

job statement A control statement identifying the beginning of a series of job control statements for a job.

job step The execution of a computer program according to a job control statement; several job steps may be specified by a job.

job stream In a processing system, a group of consecutively run jobs.

jogging Rapid, repetitive switching of power to a motor to advance its shaft by small amounts. Also called inching.

John Phonetic alphabet code word for the letter *J*.

Johnson counter See *ring counter.*

Johnson curve A spectral curve (important in appraising solar cell performance) for air mass zero, i.e., for conditions beyond the earth's atmosphere.

Johnson-Lark-Horowitz effect The resistivity gained by a metal or degenerate semiconductor (one in which conduction is nearly equal to that of a simple metal) due to electron scattering by impurity atoms.

Johnson noise See *thermal noise.*

Johnson-Q feed system See *Q-antenna.*

join The logical inclusive-OR operation.

joined actuator A form of multiple circuit breaker in which the opening of one circuit results in the opening of all circuits.

joint See *junction, 1.*

joint circuit A communications circuit shared by two or more services.

joint communications Communication facilities being used by more than one service of the same country.

joint denial The logical NOR operation.

jolt 1. Colloquialism for *1 kilovolt.* 2. Colloquialism for *electric shock.* 3. Colloquialism for a *sudden transient.* 4. Colloquialism for a *lightning discharge.*

Joly transformer A frequency-tripling transformer whose frequency-multiplying action is dependent on the nonlinearity of the magnetic induction curve of the core material.

Jones plug A special form of polarized receptacle having numerous contacts.

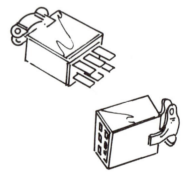

Jones Plug.

j operator See *j.*

Josephson effect The phenomenon, predicted by Brian Josephson, wherein a current flows across the gap between the tips of two superconductors brought close together and a high-frequency wave is generated.

Joshi effect The phenomenon whereby current in a gas changes as the result of irradiation by light.

joule Symbol, J. The SI unit of work. The joule is the work done when the point of application of 1 newton is moved 1 meter in the direction of force. Also see *newton*.

Joule constant See *mechanical equivalent of heat*.

Joule effect 1. The heat resulting from current flowing through a resistance. 2. Magnetostriction.

joule heat See *joule effect, 1*.

joule heat gradient The rate of change, or derivative, of the heat produced by current flowing through a resistance.

joule meter An integrating wattmeter giving readings in joules.

Joule's law The rate at which heat is produced by current flowing in a constant-resistance circuit is proportional to the square of the current.

journal A file of messages within an operating sytem providing information for restarts and historical analysis of the system's functioning.

joystick A two-dimensional potentiometer with a movable lever, allowing control of a parameter according to the position of the lever in the up/down and left/right positions. Often used in computer games for the purpose of manipulating images on a screen. The device may also be rotated clockwise or counterclockwise to obtain additional functions.

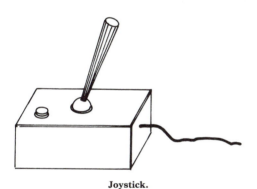

Joystick.

JPL Abbreviation of *Jet Propulsion Laboratory*.

J-rule The rule which states that during transitions of orbital electrons from higher to lower energy states (accompanied by the emission of photons), changes in the inner quantum number may only be by a factor of zero or ± 1.

J/s Symbol for *joules per second*, the unit for the Planck constant.

J-scan See *J-display*.

J-scope A radarscope giving a J-display.

JSR Abbreviation of *jump to subroutine*.

JTAC Joint Technical Advisory Committee.

judder In facsimile transmission, distortion due to movements of the transmission or reception equipment.

juice Colloquialism for *electric current*.

jukebox An automatic phonograph (usually found in public places) containing a large assortment of records.

jump 1. To provide a (temporary) circuit around a component or other circuit. 2. In digital computer practice, a programming instruction specifying the memory location of the next instruction and directing the computer to it; branch.

jump, conditional See *conditional branch*.

jump, unconditional See *unconditional branch instruction*.

jumper A short piece of wire (usually flexible, insulated, and equipped with clips) for jumping a component or circuit. See *jump, 1*.

Jumper.

jump instruction See *branch instruction*.

junction 1. A joint (connection) between two conductors. 2. The region of contact between semiconductor materials of opposite type; e.g., pn junction. 3. A waveguide fitting used to attach a branch waveguide to a main waveguide at an angle.

junction barrier See *depletion region*.

junction battery A nuclear battery in which a silicon pn junction is irradiated by strontium 90.

junction box A (usually metal) protective box or can into which several conductors are brought together and connected.

junction capacitance In a semiconductor junction, the internal capacitance across the junction; it is of special interest when the junction is reverse-biased. Also called barrier capacitance.

junction capacitor See *voltagevariable capacitor*.

junction diode A semiconductor diode created by joining an n-type region and p-type region as a wafer of semiconductor material, such as germanium or silicon.

junction field-effect transistor Abbreviation, JFET. A field-effect transistor in which the gate electrode consists of a pn junction.

junction filter A combination of separate low- and high-pass filters having a common input but separate outputs. The filter is used to separate two frequency ands and transmit them to different circuits.

junction laser See *laser diode*.

junction light source See *light-emitting diode*.

junction loss 1. The loss that occurs in a telephone circuit at connecting points. 2. Loss in a semiconductor pn junction. 3. Loss occurring at an electrical connection because of poor bonding.

junction photocell A photoconductive or photovoltaic cell that is essentially a light-sensitive junction diode.

junction point 1. A point at which two or more conductors, components, or circuits join in an electrical connection. 2. The point in a computer routine at which one of several choices is made.

junction rectifier A semiconductor rectifier which is in effect a heavy-duty junction diode or the equivalent of such diodes in combination.

junction station A microwave relay station joining one or more microwave radio legs to the main route or through route.

junction transistor A transistor in which the emitter and collector con-

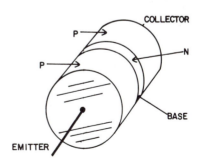

Junction transistor.

sists of junctions (see *junction, 2*) between *p* and *n* semiconductor regions. Compare *point-contact transistor.*

junctor In a crossbar system, a circuit that bridges frames of a switching unit and terminates in a switching device on each frame.

justify 1. To adjust the printing of words for aligned left- or right-hand margins. 2. In computer operation, to shift an item in a register so that the most or least significant digit is at the corresponding end of the register.

just-operate value The current or voltage level at which a relay or similar device just closes. Also called *just-close value.*

just-release value The current or voltage level at which a relay or similar device just opens. Also called *just-open value.*

just scale A musical scale of three consecutive triads, the highest note of each being the lowest note of the next. Each triad has the ratio 4:5:6 or 10:12:15.

jute Tar-saturated fiber, such as hemp, used as a protective cover for cable.

jute-protected cable A cable whose outer covering is a wrapping of tarred jute or other fiber.

juxtaposed elements Components placed or mounted side by side.

juxtaposed images Images (e.g., those on the screen of a dual-beam cathode-ray tube) that are close to each other for simultaneous viewing but do not overlap at any point.

K

K 1. General symbol for *constant*. 2. Symbol for *dielectric constant*. 3. Symbol for *potassium*. 4. Symbol for *kelvin*. 5. Radiotelegraph symbol for *go ahead* ("over"). 6. Symbol for *cathode*. 7. Symbol for computer memory storage capacity of 1024 bits. 8. Abbreviation of *kilohm*.

k 1. Abbreviation of *kilo*. 2. Symbol for the *Boltzmann constant*. 3. General symbol for constant. 4. Symbol for dielectric constant.

kA Abbreviation of *kiloampere*.

Karnaugh map A logic chart showing switching-function relationships and used in computer logic analysis to determine quickly the simplest form of logic circuit to use for a given function. The Karnaugh map is sometimes regarded as a tabular form of the Venn diagram.

Kansas City standard A frequency-shift modulation standard for computer—cassette—recorder interface. Also called *Byte standard*.

Karolus lens system A set of lenses and an electrically operated iris used in early TV receiver projection.

KB Abbreviation of *keyboard; kilobit*.

K-band The 11 to 36 GHz band of frequencies.

kc Abbreviation of *kilocycle* (obsolete).

kcal Abbreviation of *kilocalorie*.

K-carrier system A 4-wire carrier-current telephone system employing frequencies up to 60 kHz and providing 12 channels.

kCi Abbreviation of *kilocurie*.

kcs Abbreviation of a *thousand characters per second*.

K-display See *K-scan*.

KDP Abbreviation of *potassium dihydrogen phosphate*, a ferroelectric material.

keep-alive anode In a mercury-arc rectifier, the auxiliary anode in the keep-alive circuit (see *keep-alive circuit, 1*).

keep-alive circuit 1. A small filtered transformer—rectifier dc power supply used to keep the mercury ionized in a mercury-arc rectifier that supplies high-voltage dc to keyed circuits, such as a radiotelegraph transmitter. 2. A circuit in a TR or ATR tube setup, which instigates residual ionization to speed full ionization triggered by the transmitter.

keep-alive current In an indirectly heated thermistor, a small bias current passed continuously through the heater element and upon which the heater is subsequently superimposed.

keep-alive voltage 1. The dc voltage supplied to a mercury-arc rectifier by the keep-alive circuit (see *keep-alive circuit, 1*). 2. The dc voltage supplied to the keep-alive circuit of a TR or ATR tube (see *keep-alive circuit, 2*). 3. In vacuum-tube circuits, a small residual voltage applied to the filaments during "off" periods for the purpose of providing almost instant circuit operation when full power is applied; i.e., no tube warmup time is required.

keeper A small iron bar placed across the poles of a permanent magnet to forestall demagnetization.

K-electron In certain atoms, one of the electrons whose orbit is nearest the nucleus.

Kel-f *Polymonochlorotrifluorethylene,* a high-temperature insulation.

Kellie bond 1. The junction of two mated electrically conductive surfaces that are held together by an adhesive that exhibits negligible resistance thermally and electrically when set. 2. To make a buzzed joint, as from heatsink to chassis. See also *buzz*.

kelvin (Lord *Kelvin*, 1824-1907). Symbol, K. The SI unit of thermodynamic temperature; 1 K = 1/273.16 the thermodynamic temperature of the triple point of water.

Kelvin absolute electrometer An electrostatic meter consisting of a movable metal plate (or plates) and a stationary metal plate (or plates) between which voltage is applied. The movable plate(s) is displaced over a distance proportional to the potential, against the torque of a return spring. See next page for illustration.

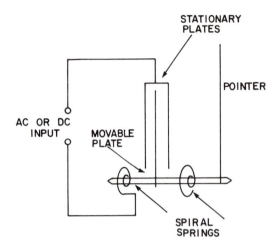

Kelvin absolute electrometer.

Kelvin balance An apparatus for measuring current in terms of magnetic pull. A coil is attached to each end of the beam of a balance, and coils ride directly above two stationary coils. Current flows through all coils, making one pair attractive and the other repellant, thus unbalancing the beam. Balance is restored by sliding a weight whose position along a graduated scale indicates the current strength.

Kelvin double bridge A special bridge for measuring very low resistance (0.1 Ω or less). The arrangement of the bridge reduces the effects of contact resistance which causes significant error when such low resistances are connected to conventional resistance bridges.

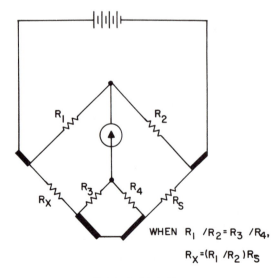

WHEN $R_1/R_2 = R_3/R_4$,

$R_X = (R_1/R_2)R_S$

Kelvin double bridge.

Kelvin replenisher A static generator consisting of two pairs of concentric, rotating semicircular conductors connected to brushes. The machine may be regarded as a rotating electric doubler.

Kelvin scale See *absolute scale*.

Kelvin temperature scale See *absolute scale*.

Kelvin voltmeter An electrostatic voltmeter in which an assembly of figure-8-shaped metal plates rotates between the plates of a stationary assembly when a voltage is applied between the assemblies. The length of the arc of rotation is proportional to the electrostatic attraction and, thus, to the applied voltage.

Kendall effect Distortion in a facsimile record caused by unwanted modulation produced by a carrier signal.

Kennelly—Heaviside layer (A.E. *Kennelly,* 1861-1939; Oliver *Heaviside,* 1850-1925). A very high ionized region surrounding the earth that reflects and refracts radio waves. Also see *ionosphere*.

kenopliotron A diode—triode vacuum tube in which the diode anode also serves as the triode cathode.

kenotron Generically, any high-vacuum, tube diode rectifier.

keraunograph A meteorological instrument for detecting distant electrical storms. In its simplest form, it consists of a galvanometer connected in series with an antenna and ground.

keraunophone A radio-receiving keraunograph.

kernel Inside an electrical conductor, a line along which there is no magnetic field. Generally, this line is near the center of the conductor.

Kerr cell A nitrobenzene-filled cell that uses the Kerr electro-optical effect and, accordingly, is useful as an electric light shutter or control.

Kerr electro-optical effect The tendency of certain dielectric materials to become double-refracting in an electric field.

Kerr magneto-optical effect The tendency of glass and some other solids and liquids to become double-refracting in a magnetic field.

Kerst induction accelerator See *betatron*.

keV Abbreviation of *kiloelectronvolt*.

key 1. A specialized hand-operated switch used to make and break a circuit repetitively to form the dot and dash signals of telegraphy. 2. A projection or pin that guides the insertion of a tube or other plug-in component into a holder or socket. 3. A digit or digits used to locate or identify a computer record (but not necessarily part of the record).

keyboard An array of lettered or numbered, low-torque pushbuttons, usually similar to the keyboard of a typewriter, used to enter information into a computer, telegraph, teletypewriter, or automatic control system.

keyboard computer A digital computer in which the input device is an electrical keyboard of the typewriter or calculator type.

keyboard entry The operation of a keyboard to enter information into a computer for processing.

keyboard lockout A keyboard interlock in a data transmission circuit that prevents data from being transmitted while the transmitter of another station on the same circuit is operating.

keyboard perforator A keyboard-operated machine that perforates paper or plastic tape for the automatic keying of a telegraph circuit, Teletype, radio transmitter, or automation system.

keyboard punch See *key punch*.

keyboard send—receive unit A teletypewriter lacking an automatic input device.

key cabinet In a telephone system, a facility that tells a subscriber which lines are busy and which lines are open.

key chirp A chirping sound in a received signal, resulting from the slight frequency shift when a radiotelegraph transmitter is keyed without some precaution preventing the shift.

key-click filter A (usually inductance—capacitance) filter for smoothing a keying wave to eliminate interference from key clicks.

key clicks 1. Clicking sounds in a received radiotelegraph signal when the transmitter being keyed produces rectangular-wave modulation (i.e., steep rises and falls); filters can eliminate the fault. 2. Clicking sounds produced by the sparking in the contacts of a radiotelegraph key or relay (interference in receivers near the offending transmitter).

keyed agc A controlled automatic gain control system in TV receiver circuits. The agc acts when the horizontal sync pulse appears; it is inactive between pulses. This action prevents "control" of the agc by noise transients and picture-signal elements.

keyed clamp A clamping circuit that uses a control signal to determine the clamping time.

keyed interval In a transmission system that is keyed periodically, an interval beginning with a change in state and having a duration of the shortest time between changes in state.

keyed rainbow generator For color-TV testing, a signal generator that produces a rainbow color pattern on the screen (i.e., a set of 10 vertical color bars representing the spectrum, with blank bars in between). The pattern results from gating the 3.56 MHz oscillator in the receiver at a frequency of 189 kHz.

keyer An automatic device for keying a radiotelegraph transmitter or wire telegraph circuit. The keyer may operate from perforated tape, an embossed disk, magnetic tape, or other similar recording.

keyer adaptor A modulated-signal detector that produces a dc signal of an amplitude sympathetic with the modulation; it provides the keying signal for a frequency-shift exciter in radio facsimile transmission.

keying The modulation of a signal—breaking it up into intervals of varying duration—by intermittently varying the frequency of the signal, or by intermittently modulating the signal's amplitude.

keying chirp A rapid change in the frequency of a continuous-wave signal, occurring at the beginning of each code element. In the receiver, the resulting sound is a chirp.

keying error rate In data transmission, the ratio of incorrectly keyed signals to the number of signals keyed.

keying filter See *key-click filter*.

keying frequency 1. In a modulated CW radiotelegraph transmitter, the audio frequency (tone) of the dot and dash signals (as opposed to the carrier frequency). 2. In CW radiotelegraphy, transmission speed (see *keying speed*). 3. The number of times per second that a black-line signal occurs while an object is scanned in a facsimile system.

keying monitor A (usually simple) detector used by an operator to listen to his keying of a radiotelegraph transmitter.

keying speed The speed (in words per minute) of a telegraph or radiotelegraph transmission.

keying transients 1. Transients arising from the keying of a radiotelegraph transmitter or wire telegraph circuit. 2. Similar transients which arise from the repetitive making and breaking of any circuit.

keying tube A vacuum tube whose plate—cathode circuit is connected in series with the cathode and ground in the rf amplifier of a radiotelegraph transmitter. The low-level keying voltage is applied to the grid of the keying tube to switch the transmitter on and off (for dots and dashes). The keying tube thereby functions as a fully electronic, high-capacity relay.

keying wave The telegraphic emission that takes place while the information part of the code characters is transmitted. It is also called a *marked move*.

keyless ringing In a telephone system, ringing that begins as soon as the calling plug is put in the appropriate jack on the jack panel.

key pulse In telephone practice, a signaling system in which the desired numbers are entered by depressing corresponding pushbuttons or keys.

key punch A keyboard-operated machine for recording information by perforating a tape or cards.

keyshelf A shelf supporting manual telephone switchboard keys.

key station The master (control) station in a communications or control network.

keystoning A type of TV picture distortion in which the top of the picture is wider than the bottom, or vice versa.

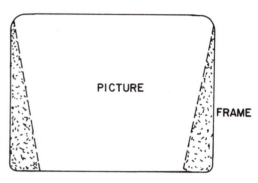

Keystoning.

key switch 1. A lockable switch that is operated by inserting and turning a key in it. 2. A switch having a long handle that transmits motion to the mechanism through a cam. 3. A telegraph key. 4. The separate short-circuiting switch sometimes mounted on the base of a telegraph key. 5. A switch that actuates the keys of an electronic organ.

key-to-disk unit A keyboard-to-magnetic-disk data processing unit that eliminates an intermediate stage of producing punched cards or paper tape.

key tube See *key tube*.

key verify To use a punched card verifier.

keyway A groove or slot into which a mating key slides to position a plug-in component, (see *key, 2*).

keyword In information retrieval systems, the significant word in the title describing a document; e.g., in the title *A Primer on French Cuisine*, the word *cuisine* would be the keyword, the others having no singular significance.

kg Abbreviation of *kilogram*.

kgC Abbreviation of *kilogram-calorie*.

kgm Abbreviation of *kilogram-meter*.

kg/m^3 Symbol for *Kilograms per cubic meter*, the SI unit of density.

K—H layer See *Kennelly—Heaviside layer*.

kHz Abbreviation of *kilohertz*.

kick 1. To place into sudden operation, as by the quick, forcible closure of a switch or the rapid application of an enabling pulse. 2. Trigger. 3. Colloquialism for abrupt, momentary electric shock.

kickback 1. In a spark-gap rf generator, the tendency of the high-frequency capacitor to discharge back through the secondary winding of the power transformer, which it might damage. Kickback can be prevented by the use of rf chokes. 2. The counter emf that appears across an inductor when current is interrupted. 3. Flyback.

kickback power supply A high-voltage power supply employing the flyback principle. See *flyback*.

kickback preventer An rf filter used to prevent kickback. See *kickback, 1*.

kidney joint A waveguide coupling used in radar. The joint is flexible, or may consist of an air gap, to allow rotation of the antenna.

Kikuchi lines A spectral pattern produced by the electrons scattered when an electron beam strikes a crystal.

killer A pulse or other signal employed to disable a circuit temporarily (e.g., a blanking pulse).

killer circuit 1. A circuit that disables some function of a system, such as the audio in a television receiver. 2. The blanking circuit in a radar receiver. 3. A circuit that prevents responses to side-lobe signals in a repeater or transponder.

kilo Abbreviation, *k*. A prefix meaning *thousand(s)*. 2. Phonetic alphabet code word for the letter K (usually capitalized).

kilobit A large unit of digital information; 1 kilobit equals 10^3 bits. See *bit, baud*.

kilobyte A unit of data, equivalent to 1,024 bytes.

kilocalorie Abbreviation, *kcal*. A large unit of heat; 1 kcal equals 10^3 calories. (See *calorie*.)

kilocurie Abbreviation, *kCi*. A large unit of radioactivity equal to 3.71 × 10^{13} disintegrations per second; 1 kCi equals 10^3 curies. Also see *curie, megacurie, microcurie, millicurie*.

kilocycle Abbreviation, *kc*. An obsolete term meaning one thousand complete cycles. (See *kilohertz*).

kiloelectronvolt Abbreviation, *keV*. A large unit of electrical energy. 1 keV equals 10^3 electronvolts. See *electronvolt*.

kilogauss A large unit of magnetic flux density; 1 kilogauss equals 0.1 tesla.

kilogram Abbreviation, *kg*. The SI base unit of mass or weight; it is equal to 10^3 grams.

kilogram-calorie Abbreviation, *kgCi*. The heat required to raise 1 kilogram of water 1 °C.

kilogram-meter Abbreviation, *kgm*. A large unit of work; 1 kgm is the work required to raise 1 kilogram 1 meter (equal to 7.2334 foot-pounds). Also see *joule*.

kilohertz Abbreviation, *kHz*. A unit of high frequency; 1 kHz equals 10^3 Hz.

kilohm Abbreviation, *kΩ*. A unit of high resistance, reactance, or impedance; 1 kΩ equals 10^3 ohms.

kilojoule Abbreviation, *kJ*. A large unit of energy or work; 1 kJ equals 10^3 joules. See *joule*.

kilolumen Abbreviation, *klm*. 1000 lumens.

kilomega Abbreviation, *kM*. A deprecated prefix meaning 1 billion, it has been replaced by *giga*.

kilomegacycle Abbreviation, *kMc*. An obsolete unit of high frequency; 1 kMc equals 10^3 megahertz. The use of gigahertz (GHz) is preferred.

kilomeghahertz An obsolete unit of high frequency; equal to 10^3 megahertz. The use of gigahertz is preferred.

kilometer Abbreviation, *km*. A large metric unit of linear measure; 1 km equals 10^3 meters (3 280.8 feet).

kilo-oersted Abbreviation, *kOe*. A unit of magnetic field strength; 1 kOe equals 10^3 oersteds. See *oersted, A/m*.

kiloroentgen Abbreviation, *kr*. A large unit of radioactive radiation; 1 kr equals 10^3 roentgens. See *roentgen, C/kg*.

kilorutherford Abbreviation, *krd*. A large unit of radioactivity equal to 10^9 disintegrations per second; 1 krd equals 10^3 rutherfords. Also see *disintegrations/second*.

kilovar A compound term coined from *kilo* and *VAR* (the abbreviation of *volt-amperes reactive*). It is equal to a reactive power of 10^3 watts.

kilovar-hour A large unit of reactive electrical energy; equal to 10^3 reactive watts per hour.

kilovolt Abbreviation, *kV*. A unit of high voltage; 1 kV equals 10^3V.

kilovolt-ampere Abbreviation, *kVa*. A unit of high power which gives the *true power* in a dc circuit and the *apparent power* in an ac circuit; 1 kVA equals 10^3W. Also see *dc power*.

kilovolt-ampere reactive See *kilovar*.

kilovoltmeter A voltmeter designed to measure thousands of volts (kilovolts).

kilowatt Abbreviation, *kW*. A unit of high power; 1 kW equals 10^3 watts. Also see *watt*.

kilowatt-hour Abbreviation, *kWh*. A large unit of electrical energy or work; 1 kWh equals 10^3 watt hours. Also see *energy, kilowatt-hour, power, watt-hour, watt-second*.

kilowatt-hour meter A motorized meter for recording (electrical) power consumption in kilowatt-hours. Also see *kilowatt-hour*.

kinematograph A motion picture camera. Also called *cinematograph* and *kinetograph*.

kine 1. Kinescope. 2. Kinescope recording.

kinescope 1. TV picture tube. 2. A motion-picture or video tape made from the screen (or taken from the circuit) of a TV picture tube. Also called *kinescope recording*.

kinescope recorder A film or tape apparatus for recording TV pictures.

kinescope recording See *kinescope, 2*.

kinetic energy The energy associated with bodies or charges in motion.

kinetograph See *kinematograph*.

kinetoscope A motion-picture projector.

King Phonetic alphabet code word for the letter *K*.

King oscillator An early microwave oscillator that employed two special tubes whose plate and grid leads extended through opposite ends of the tube envelope; included in the design were linear tanks consisting of parallel conductors bridged by capacitors.

kinkless tetrode A tetrode having an E_p—I_p characteristic curve which does not exhibit the negative-resistance kink common to tetrode operation, or one which is biased to operate far beyond the region of negative resistance.

Kirchhoff's first law The sum of the currents flowing out of a point in a circuit equals the sum of the currents flowing into that point.

Kirchhoff's laws (Gustav Robert *Kirchhoff*, 1824-1887). Two laws of electric circuits that account for the behavior of certain networks. See *Kirchhoff's first law* and *Kirchhoff's second law*.

Kirchhoff's second law The algebraic sum of all the voltage drops around a circuit (including generator voltages) equals zero.

kit A selection of components, associated equipment, supplies (such as wire and hardware), and instructions for constructing a piece of electronic equipment.

kJ Abbreviation of *kilojoule*.

Klipsch horn A loudspeaker that includes a folded low-frequency horn housed in a corner enclosure.

klm Abbreviation of *kilolumen*.

klydonograph A device that photographically records the voltage gra-

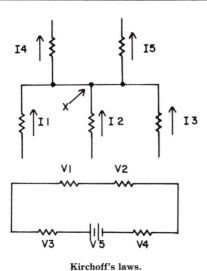

Kirchoff's laws.

dient in the presence of an electric field.

klystron A microwave tube whose operation is based on the velocity modulation of an electron beam by buncher and cavity reentrant cavities.

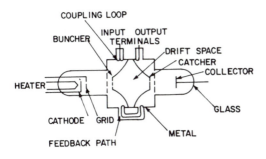

Klystron.

klystron amplifier A microwave amplifier employing a klystron.

klystron harmonic generator A frequency-multiplying rf power amplifier employing a klystron.

klystron oscillator A klystron operated as a self-excited microwave oscillator.

klystron repeater A microwave amplifier in which a klystron inserted in a waveguide boosts the amplitude of an incoming signal.

kM Abbreviation of the obsolete prefix *kilomega*. See *giga*.

km Abbreviation of *kilometer*.

kMc Abbreviation of *kilomegacycle* (obsolete); *gigahertz* is preferred.

kMHz Abbreviation of *kilomegahertz* (obsolete); *gigahertz* is preferred.

knee The bend in a response curve, usually indicating the onset of saturation, cutoff, or leveling off.

knee noise Electrical noise generated by rapidly repeating current fluctuations at the *zener knee*.

knife-edge diffraction The lessening of atmospheric signal attenua-

tion when the signal passes over a sharp obstacle and is diffracted.

knife switch A switch composed of one or more flat blades roughly resembling knife blades, which are slid firmly between the jaws of pinching contacts to close a circuit.

knob 1. A (usually round and insulated) finger dial for adjusting a variable electronic component, such as a potentiometer, variable capacitor, or rotary switch. 2. A solid round insulator usually having a low diameter to height ratio. 3. A small ball- or rod-shaped electrode or protuberance.

knocker A fire-control radar subassembly of synchronizing and triggering circuits.

knockout An area in a metal box or chassis which is relatively easily removed by tapping or knocking to provide an aperture.

knot A unit of speed, corresponding to 1 nautical mile per hour. A speed of 1 knot is about 1.15 statute miles per hour; a speed of 1 statute mile per hour is about 0.868 knots. Used by mariners for specifying speeds at sea.

k0e Abbreviation of *kilo-oersted*.

Kolster decremeter An absorption wavemeter with a movable scale; it permits measurement of the decrement of a radio wave.

Kooman antenna A unidirectional antenna consisting of stacked, full-wave, center-fed driven elements, and a reflecting screen.

Kovar An alloy of cobalt, iron, and nickel. Used mostly in glass-to-metal seals because it has characteristics of both kinds of material.

Kozanowski oscillator A positive-grid vacuum-tube uhf oscillator circuit employing two tubes having cylindrical elements, and a pair of parallel-wire tanks.

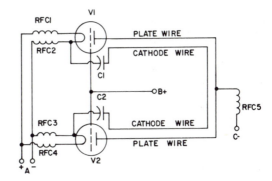

Kozanowski oscillator.

Kr Symbol for *krypton*.

kr Abbreviation of *kiloroentgen*.

K-radiation X rays emitted from an atom when an electron becomes a K-electron.

kraft paper Strong brown paper used for insulation and as the dielectric of paper capacitors.

Kramer system A system of three-phase motor control providing constant horsepower and having a dc motor coupled to the shaft of a wound-rotor three-phase induction motor. The dc supply for the motor also supplies a rotary converter. The speed-control rheostat is connected in series with the field of the motor and the dc power supply.

Kraus antenna A bidirectional, flat-top beam antenna consisting of a pair of closely spaced dipoles. Several such sections may be connected in series by crisscrossing the wires at voltage loops.

krd Abbreviation of *kilorutherford.*

kryptol A mixture of clay, graphite, and silicon carbide, which is used in heater elements because of its low resistance.

krypton Symbol, Kr. An inert, gaseous element. Atomic number, 36. Atomic weight, 83.80. Krypton is present in trace amounts in the earth's atmosphere.

K-scan In radar practice, a modified A-scan used in aiming antennas in which two pips are displayed; their relative amplitudes indicate the antenna-aiming error.

K-series A series of spectral lines for the shortest wavelengths of radiation from the innermost electron shell of a radiating atom.

KSR Abbreviation of keyboard send-receive unit.

kth term The general or representative term in a series. Thus, the sum of a simple arithmetic progression of n terms is repeated as

$$\sum_{k=1}^{n}$$

kurtosis The amount of peak curvature in a probability curve.

kV Abbreviation of *kilovolt.*

kVA Abbreviation of *kilovolt-ampere.*

kVAR Abbreviation of *reactive kilovolt-ampere.*

kVARh Abbreviation of *reactive kilovolt-ampere-hour.* (See *kilovar-hour.*)

kW Abbreviation of *kilowatt.*

kWh Abbreviation of *kilowatt-hour.*

kymograph A medical instrument for recording physiological cycles.

L 1. Symbol for *inductance*. 2. Symbol for *mean life*. 3. Abbreviation of *low*. 4. Resembling the capital letter L in physical shape. 5. Symbol for *Laplace transform*.

l 1. Symbol for *length*. 2. Abbreviation of *liter*. 3. Subscript for *low*. 4. Abbreviation of *lumen*. Also abbreviated *lm* (preferred) and *lum*.

La Symbol for *lanthanum*.

label 1. A symbolic group of characters that identifies an area of memory, an item of data, a file, or a record. 2. A label assigned to a source program instruction step to identify the step as a coding entry point or to make the step usable as a reference point for entry to the routine or subroutine in which it appears.

label group A collection of labels, usually of the same type, held in an operating system.

label identifier Within a label, a character set used to name the kind of item labeled.

label record A record identifying a file recorded on a magnetic storage medium (e.g., magnetic tape).

label set A collection of labels having a common label identifier.

labile oscillator A frequency-controlled local oscillator.

laboratory conditions The environmental, mechanical, and electrical parameters characteristic of controlled conditions. Actual operating conditions may be much different.

laboratory-grade instrument An instrument having the high accuracy and stability that suit it to precision measurements in a laboratory. Also called *precision instrument*. Compare *service-type instrument*.

laboratory power supply A regulated dc source whose adjustable output is less than 10 kV at no more than 500W.

laboratory standard See *primary standard* and *secondary standard*.

labyrinth speaker A loudspeaker whose enclosure (a wooden cubicle) includes a folded pipe or acoustic transmission line (behind the speaker); the inner walls are lined with a sound-absorbent material. When the pipe, which is open-ended, is half as long as the wavelength of the frequency being reproduced, the sound emerging from the open end is in phase with that radiated by the front of the speaker and, therefore, reinforces it. Because there is no sudden change in pressure as the sound leaves the pipe, the pipe produces no antiresonance.

laced card A punched card in which most or all card columns have been punched, so that several holes appear in each. A laced card is generally used for tests or to mark the end of a card file.

laced wiring Circuit wiring in which wires or cables run parallel in bundles that are tied together with lacing cord.

lacing cord Strong, sometimes waxed cord used to tie together wires running parallel in a bundle. Also see *laced wiring*.

lacquer disk See *cellulose-nitrate disk*.

lacquer-film capacitor A fixed capacitor with a plastic film dielectric; the film is applied as liquid lacquer to the metal foil.

lacquer master A master recording made on a lacquer disk. See *cellulose-nitrate disk*.

lacquer original See *lacquer master*.

LADAR Abbreviation of *laser doppler radar*. Also abbreviated *lopplar*.

ladder attenuator See *ladder-type attenuator*.

ladder network A cascade of **L-sections.**

ladder-type attenuator An attenuator consisting of a ladder network equipped with a switching circuit for selecting the output at various sections. See illustration on next page.

LAFOT Coded weather broadcasts aired every 6 hours by the U.S. Weather Bureau through marine radiotelephone broadcasting stations for the Great Lakes region.

lag In computations relating the phase of ac signals, the extent to which one quantity follows another in time; e.g., voltage lags behind current

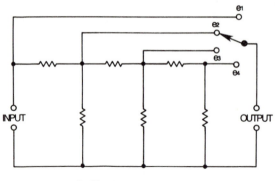

Ladder-type attenuator.

by 90° across a pure capacitance. Compare *lead.*

lagged-demand meter A meter with a built-in time delay.

lagging current Current that follows voltage (in time). Also see *lag.*

lagging load A load in which current lags behind voltage, i.e., an inductive load. Compare *leading load.*

lag network An *RC* phase-shifting circuit containing series resistance and shunt capacitance arms, which produces a lagging phase shift. Compare *lead network.*

lambda The 11th letter of the Greek alphabet. Capital lambda (Λ) is the symbol for equivalent conductivity and permeance. Lower-case lambda (λ) is the symbol for attenuation constant; charge density; linear; disintegration constant; free path; and wavelength.

lambda wave An electromagnetic disturbance that travels along the surface of an object. An example is the surface wave characteristic of low-frequency propagation.

lambert Symbol, Lb. The cgs unit of luminance, lambert is equal to the brightness of an ideal diffusing surface which radiates or reflects light at the rate of 1 lumen per square centimeter. The SI (preferred) unit of luminance is candela per square meter (cd/m^2); 1 lambert equals $10^4 cd/\pi m^2$. Also see *candela.*

Lambert's law of illumination The illumination of a surface (usually by a point light source) is inversely proportional to the square of the distance between the surface and the source; if the surface is not perpendicular to the rays, the illumination is proportional to the cosine of the angle involved.

laminated armature An armature for a motor or generator made of stacked laminations.

laminated contact A switch contact consisting of a number of laminations, each contacting a conducting counterpart.

laminated core A core for a transformer, choke, relay, or similar device, made of stacked laminations.

laminated disk A layered recording disk.

laminated pole A pole within a motor, generator, relay, electromagnet, or similar device—made of stacked laminations.

lamination 1. A relatively thin sheet of metal cut to a required shape to be stacked with other sheets of the same kind to make a laminated core or pole. 2. A relatively thin sheet of plastic that is bonded together and heat-formed with other similar sheets to produce a sheet or piece of desired thickness and strength.

lamp A device for converting electrical energy into visible light. The term includes a number of devices, e.g., arc lamp, fluorescent tube, incandescent lamp, mercury-vapor lamp, and neon bulb.

lamp-bank resistor A makeshift heavy-duty resistor consisting of several incandescent lamps arranged so they can be switched in series parallel or series-parallel to vary the resistance provided by the filaments.

lampblack Carbon obtained from soot deposited by a smoky flame. The substance is used as the basic material for some resistors.

lamp cord A two-wire insulated cord, used with low-wattage appliances at 120 volts. The wire is usually stranded copper of AWG No. 16.

lamp dimmer See *dimmer.*

lamp driver A usually single-stage circuit for amplifying a small pulse to drive an indicator lamp.

lamp extractor A special tool used to insert or extract miniature lamps for electronic equipment.

lamp jack A receptacle with a spring release which holds a small incandescent bulb. The bulb is removed and replaced by pushing and twisting.

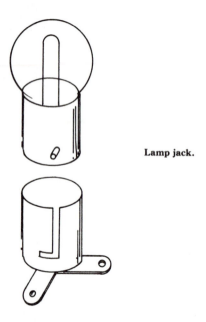

Lamp jack.

Lampkin oscillator A type of Hartley oscillator providing an approximate impedance match between grid—cathode and tank circuits.

lamp-stabilized oscillator See *Meacham oscillator.*

lamp-type expander A volume expander in which the tungsten filament in an incandescent lamp serves as the nonlinear resistor.

lamp-type readout For counters, calculators, and digital meters, a readout device in which each digit is indicated by a lamp.

land The thin vinyl wall between grooves on a phonograph record.

Land camera See *Polaroid camera.*

landing beacon The aircraft landing beam transmitter. Also see *landing beam.*

landing beam A highly directional airport radio signal beamed upward to guide aircraft landing during conditions of poor visibility.

landline A telephone or telegraph circuit completed with wires.

landmark beacon Any beacon that isn't an airway or airport beacon.

land mobile service Two-way radio service between a base station and mobile land vehicles or between the vehicles.

land mobile station A radio station aboard a ground vehicle.

land return Ground reflection of radar signals back to the transmitter.

lands Microelectronic circuit bonding points.

land station A fixed ground station.

Langevin ion An electrically charged particle, such as a grain of dust or droplet of water, resulting from the accumulation of ions.

Langmuir dark space In a luminous gas discharge, the dark region around a negatively charged probe inserted into the positive column. Compare *Crookes' dark space*.

Langmuir's equation For a diode vacuum tube with cylindrical elements, the equation

$$Ib = \frac{L\,(14 \times 10^{-6})\,Eb^{3/2}}{r\beta^2}$$

where *Ib is plate current (amp)*, L = length of cylinders (cm), Eb = plate voltage (V), r = plate radius (cm), and $\beta^2 = 1$.

Langmuir's law See *Child's law*.

language In digital-computer practice, any one of the detailed systems for representing data, instructions, and procedures through the use of symbols and symbol sequences. See *machine language, assembly language, COBOL, FORTRAN, BASIC*.

language laboratory An electronic contribution to the teaching and learning of languages, it consists of recordings in a language being studied and all the equipment associated with recording, playback, and monitoring. Students listen to the speech of experts in the language record, listen to, and later erase their own utterances in the language.

language translation The conversion of statements in one computer language to equivalent statements in another.

language translator An assembly program, compiler, or other routine used for language translation.

L-antenna See inverted-L antenna.

lanthanum Symbol, La. An elemental metal of the rare-earth group. Atomic number, 57. Atomic weight, 138.92.

lanyard A wire or cable used to quickly pull apart the halves of a quick-disconnect connector.

lap A device used for grinding piezoelectric crystals for resonance at a desired frequency.

lap dissolve The simultaneous fading out of one televised scene while another is fading up, so that one is apparently dissolving into the next. Also applicable to motion pictures.

lapel microphone A small microphone that is clipped to a lapel of the user's coat.

lap joint An overlapping splice of two conductors.

Laplace transform Symbol, L. An operator which reduces the work of solving certain differential equations by permitting them to be handled by simpler algebraic methods.

lapping Fine-tuning quartz crystal plates by moving them over a flat plate coated with a liquid abrasive.

lap winding In a motor or generator armature, a winding in which the opposite ends of each coil are connected to the adjoining segments of the commutator.

lap wrap 1. A form of asbestos cloth wire insulation. 2. A method of wrapping with electrical tape, in which there is considerable overlap among the turns of the tape.

large calorie See *kilogram-calorie*.

large-scale integration Abbreviation, LSI. The inclusion of more than 100 transistors, performing various individual but interrelated circuit functions, on a single integrated-circuit chip.

large signal A relatively high-amplitude signal that traverses so large a part of the operating characteristic of a device that nonlinear portions of the characteristic are usually encountered. Compare *small signal*.

large-signal analysis The rigorous study of circuits and devices that process large signals.

large-signal component 1. A coefficient or parameter—such as amplification, transconductance, dynamic resistance—measured under conditions of large-signal operation. Also see *large signal* and *large-signal equivalent circuit*. 2. A device designed for operation at high signal levels.

large-signal equivalent circuit For a given transistor circuit, the equivalent circuit at high signal levels (i.e., at amplitudes approaching saturation and cutoff levels). Also see *equivalent circuit*.

large-signal operation A circuit or device whose operation involves large signals. Compare *small-signal operation*.

large-signal transistor See *power transistor*.

Larmor orbit The path followed by a charged particle in a constant magnetic field. Because of interaction between the external field and the field generated by the particle, the charged particle travels in a circular path.

laryngaphone See *throat microphone*.

LASCR Abbreviation of *light-activated silicon controlled rectifier*.

LASCS Abbreviation of *light-activated silicon controlled switch*.

lase To emit coherent light. See *laser*.

laser A device for generating intense coherent light. In one form, the laser consists of a polished ruby rod having a solid mirror at one end, a transparent mirror at the other end, and high-voltage flash tube wound around the rod. The tube emits brilliant flashes that stimulate electrons in the ruby atoms, making them emit light which is then reflected between the mirrors in the rod, further stimulating the ruby electrons to generate still more light. The result is that the light, which emerges in parallel rays through the transparent mirror, is many times more intense than that of the flash tube which initiated the action.

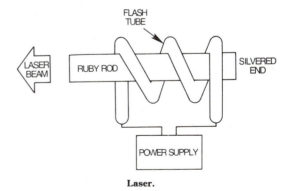

Laser.

laser-beam communication A type of light-beam communication in which a laser beam is the link between transmitting and receiving stations. Also see *laser, laser diode, light-beam communication*.

laser capacitor An energy-storage capacitor used to discharge-fire the exciter lamp of a laser. Also see *laser*.

laser cavity An optical resonant cavity that results in the emission of coherent light.

laser diode A semiconductor diode, usually of the gallium-arsenide type, which emits coherent light when a voltage is applied to its terminals. Also see *laser*.

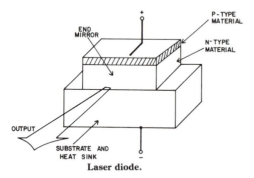

Laser diode.

laser doppler radar Acronyms, *ladar* or *lopplar*. A form of doppler radar employing the light beam of a laser instead of radio waves.

laser optical videodisc system A system in which a low-powered laser reads audio and video information from a *videodisc* and delivers it to a tv receiver.

laser ranger A radarlike device using intense light (instead of microwaves).

laser welding Welding (especially of tiny pieces) with the heat produced by laser beam.

lasing Laser action.

lat Abbreviation of *latitude*.

latch 1. A feedback loop in symmetrical digital circuits (e.g., a flip-flop) used to maintain a state. 2. A simple logic-circuit storage element comprising two gates as a unit. 3. To maintain a closed (energized) state in a pain of relay contacts after initial energization from a single electrical pulse. (See *latching relay*.)

latching current In a thyristor, the minimum value of anode current (slightly higher than the holding current) that will sustain conduction immediately after switch-on.

latching relay An electromechanical or fully electronic relay that locks into whichever mode it is energized for (on or off).

latch-on relay See *locking relay*.

latchup In a transistor switching circuit, the abnormal condition in which the collector voltage remains at its on level after the transistor is switched to cutoff from saturation.

latch voltage The input voltage at which a flip-flop changes states.

late contacts Relay contacts that are operated following the movement of other contacts during the relay's operation.

latency 1. The time taken by a digital computer to deliver information from its memory. 2. In a serial storage system, the access time less the word time.

latent image 1. In a storage tube, a stored image which is not yet visible. 2. An image stored in the mosaic of an iconoscope. 3. The image that will appear when photographic film or paper is developed.

lateral chromatic aberration An aberration affecting the sharpness of off-axis television images.

lateral compliance In phonograph reproduction, the ease with which the stylus can move laterally as it follows the groove. Also see *compliance* and *lateral recording*.

lateral-correction magnet In a color TV picture tube, a magnet operated with a set of pole pieces attached to the focus element of the blue gun; it controls horizontal positioning of the blue beam for convergence.

lateral magnet See *lateral-correction magnet*.

lateral recording A disk recording in which the groove undulates from side to side. Compare *vertical recording*.

latitude Abbreviation, lat. Angular distance measured around the earth's circumference, i.e., to the north and south from the equator. Compare *longitude*.

latitude effect The tendency of the earth's magnetic field to decrease the number of charged particles (other than photons) that reach the surface of the earth near the equator, as compared with the number reaching the surface at other latitudes.

Latour alternator See *Bethenod alternator*.

lattice 1. The orderly internal pattern (matrix) of atoms in a crystal. Also see *crystal lattice*. 2. An often symmetrical arrangement of components in a network, e.g., attenuator circuit, bridge circuit.

lattice filter A lattice network having reactance in its arms that makes it a selective circuit (see *lattice, 2*).

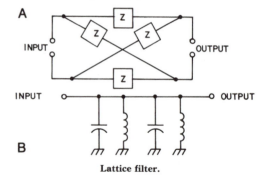

Lattice filter.

At A, one possible configuration for a lattice filter, showing the arrangement of impedances. At B, one possible circuit configuration for the bandpass impedances in the ladder filter. (Other arrangements are feasible in both cases.)

lattice network See *lattice, 2*.

lattice section See *lattice, 2*.

lattice structure See *lattice, 1*.

lattice winding See *universal winding*.

launch complex The launch, support, and control system needed for rocket launchings.

launching The energy transference from a cable into a waveguide.

lavalier microphone A small microphone that can be hung from the user's neck on a cord or chain.

law 1. A scientific law is a general, verifiable statement that describes the behavior of entities or the relationships between phenomena or concepts. Such a law is the product of inductive reasoning that follows many observations and controlled experiments; e.g., first law of thermodynamics, inverse-square law, Kirchhoff's laws, Ohm's law. 2. The nature of the change of a dependent variable, particularly as depicted by a response curve, e.g., square law.

LAWEB Civilian weather bulletins issued every 6 hours from ship and shore positions along the Great Lakes during the sailing season.

lawn mower 1. A facsimile term for a helix recording mechanism. 2. A radar receiver rf preamplifier.

law of a curve See *law, 2.*

law of averages In probability and statistics, the law stating that for a large sampling of events, the numerical probability value will be more closely approached than when the sampling is small. Compare *law of large numbers.*

law of charges Unlike charges attract and like charges repel each other.

law of cosines The square of any side of a triangle is equal to the sum of the squares of the other two sides minus twice their product times the cosine of the included angle. Compare *law of sines* and *law of tangents.*

law of divine proportion See *divine proportion.*

law of electric charges See *law of charges.*

law of electromagnetic induction See *Lenz' law.*

law of electrostatic attraction See *Coulomb's laws.*

law of electrostatic repulsion See *Coulomb's laws.*

law of induction See *Faraday's law.*

law of inverse squares See *inverse-square law.*

law of large numbers In probability and statistics, the law stating that with a large sample, the sample average is extremely likely to approximate the population average. Often erroneously called law of averages.

law of magnetism Unlike magnetic poles attract and like magnetic poles repel each other.

law of natural decay See *exponential decrease.*

law of natural growth See *exponential increase.*

law of normal distribution Gauss' law of the frequency distribution of a repetitive function describing the probability of deviations from the mean.

law of octals Of great interest in the study of semiconductors, the law that says chemical activity takes place between two atoms lacking eight valence electrons and continues until the requirement of eight electrons is satisfied for all but the first orbit, where only two electrons are required.

law of radiation See *quantum theory.*

law of reflection The angle of reflection is equal to the angle of incidence.

law of sines The length of the sides of a triangle are to each other as are the sines of the corresponding opposite angles: $a/\sin A = b/\sin B = c/\sin C$. Compare *law of cosines* and *law of tangents.*

law of tangents The difference between any two sides of a triangle is to their sum as the tangent of half the difference of the opposite angles is to the tangent of half the sum. Thus:

$$\frac{a - b}{a + b} = \frac{\tan (A - B)/2}{\tan (A + B)/2}$$

Compare *law of cosines* and *law of sines.*

Lawrence accelerator See *cyclotron.*

lawrencium Symbol, Lw. A short-lived radioactive element produced artificially from californium. Atomic number, 103. (Atomic weight unknown.)

lay See *direction of lay.*

layer 1. A complete coil winding in one plane, i.e., turns are laid side by side (not on top of each other). 2. In a semiconductor device, a region

having unique electrical properties, e.g., n-layer.

layer-to-layer transfer In a roll of magnetic tape, unwanted transfer of data between adjacent turns on the reel. If severe, this transfer can cause drop-in or drop-out in a computer. In audio applications it can sometimes be heard as a delayed echo or a faint sound occurring just prior to the actual recorded sound.

layer winding A coil winding in which the turns are arranged in two or more concentric layers.

layerwound coil An inductor wound in layers, one on top of the other. Also see *layer, 1.* Compare *single-layer coil.*

layout The arrangement of components on a chassis, printed circuit board, or panel.

lazy-H antenna An antenna consisting of two vertically stacked collinear elements, giving both horizontal and vertical directivity.

Lb Abbreviation of *lambert.*

lb Abbreviation of *pound.*

L-band A radio-frequency band extending from 390 MHz to 1.55 GHz. For subdivisions of this band, see *LC band, LF band, LK band, LL band, LP band, LS band, LT band, LX band, LY band,* and *LZ band.*

LC 1. Abbreviation of *liquid crystal*; also abbreviated *lix.* 2. Abbreviation of *inductance—capacitance.* 3. Symbol for *LC constant.*

L-carrier In a telephone system, a carrier having a frequency between approximately 68 kHz and 10 MHz. May be used in either wire-transmission or radio links.

LC band A section of the *L-band* extending from 465 MHz to 510 MHz.

LC bridge See *inductance—capacitance bridge.*

LC constant Abbreviation, *LC.* The product of the inductance (L) and capacitance (C) required for resonance at a given frequency (f).

LCD Abbreviation of *liquid-crystal display.*

LC filter See *inductance—capacitance filter.*

L-circuit See *L-network.*

LC meter See *inductance—capacitance meter.*

LCR See *inductance—capacitance—resistance.*

LC ratio In a tuned circuit, the ratio of inductance to capacitance.

LCR bridge See *impedance bridge, 1.*

Ld Symbol for *distributed inductance.*

LDF See *low-frequency direction finder.*

L-display (L-scan) A radar display in which the target appears as two horizontal traces, one extending from a vertical timebase to the right, the other to the left.

lead A conductor (usually a wire) leading to or emerging from a terminal or electrode.

lead In computations relating phase, the extent to which one quantity precedes another; e.g., current leads voltage by 90 degrees in a pure capacitance. Compare *lag.*

lead Symbol, Pb. A metallic element. Atomic number, 82. Atomic weight, 207.21. Lead can be used as a shield against atomic radiation and has many applications in electronics. See *lead-acid battery, fuse,* and *solder.*

lead-acid battery A storage battery in which the cells contain lead plates immersed in an electrolyte (sulfuric acid). The positive plate contains lead peroxide, and the negative plate, spongy lead.

lead-acid cell One cell of a lead—acid battery. See next page.

lead cell 1. A lead-acid cell. 2. A lead sulfide photocell; see *lead sulfide, 1.*

lead dress See *dress.*

leader 1. The blank section at the beginning of a magnetic or paper tape. In a magnetic tape, the leader is usually made of plastic. A paper-tape leader generally contains only the feed holes. 2. A record, preceding

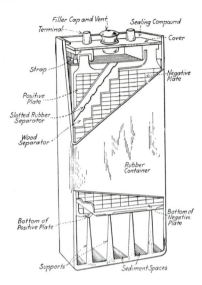

Lead-acid cell.

a group of records, that identifies the group and provides other data pertinent to the group. 3. In a lightning stroke, the initial movement of electrons or positive ions, creating the ionized path that allows discharge.

lead frame The sacrificial metal frame holding the leads of a circuit package (DIP) in place before encapsulation.

lead-in The wire connecting an antenna to a receiver or transmitter.

leading current Current that precedes voltage in time. Also see *lead*.

leading edge 1. The rising edge of a pulse; compare *trailing edge*. 2. The edge of a punched card that enters the card track of a punched card machine first.

leading end The end of a paper tape that enters a paper tape reader first.

leading ghost A twin television image to the left of the original on the screen.

leading load A load in which current leads voltage, i.e., a capacitive load.

lead-in groove Around the outer edge of a phonograph record, a blank spiral groove that leads the stylus into the first groove of the recording. Compare *lead-out groove*.

leading whites In a television picture, an abnormal condition in which the leading (left) edge of a black object has a white border.

leading-zero suppression In a digital meter, the blanking out of all zeros to the left of the decimal point.

lead-in spiral See *lead-in groove*.

lead-in tube A tube of insulating material, such as plastic or ceramic, employed to conduct an antenna lead-in through a wall.

lead-in wire 1. A single wire, used as a feed line for a shortwave receiving antenna. 2. The feed line for a television receiving antenna. 3. A feed line for a transmitting antenna.

lead network An *RC* phase-shift circuit containing series capacitance and shunt resistance; it produces a leading phase shift. Compare *lag network*.

lead-out groove Around the inner edge of a phonograph record, a blank spiral groove leading into the eccentric or locked groove.

lead-over groove On a phonograph record containing several recorded tracks, a blank groove that conducts the stylus from the end of one recording to the beginning of the next.

lead screw 1. A threaded rod that guides the cutter across the surface of a disk during its recording. 2. In facsimile transmission, a threaded shaft that moves the drum or scanning mechanism lengthwise.

lead sulfate Formula, PbS04. An insulating compound formed in a lead-acid cell (see *lead-acid battery*) by the chemical action between the lead in the plates and the sulfuric-acid electrolyte. If the sulfate is not broken down during charging of the cell, it will eventually ruin the cell.

lead sulfide 1. Formula, PbS. A compound of lead and sulfur used as the light-sensitive material in some photoconductive cells. 2. Galena.

lead zirconate titanate A synthetic piezoelectric material.

leaf electroscope An electroscope employing a pair of gold leaves or a single gold leaf and a solid strip of metal.

leak 1. A loss of energy through a stray path not intended for conduction. 2. A point from which such a loss originates. 3. Grid leak.

leakage 1. The small current that flows through an electrical insulator. 2. The electromagnetic field that is radiated or received by a feed line that should theoretically have 100 percent shielding.

leakage current The zero-signal current flowing across a reverse-biased semiconductor junction. See *ICBO* and *ICEO*.

leakage flux Collectively, magnetic lines of force around a transformer that do not link the primary and secondary coils.

leakage inductance Self-inductance due to linkage flux. Leakage inductance is effectively in series with the primary or secondary winding of a transformer.

leakage power The rf power transmitted through a TR tube or pre-TR tube.

leakage radiation Radiation from parts of a system as compared with that from the true radiator.

leakage reactance Inductive reactance due to leakage inductance in the primary or secondary circuit of a transformer.

leakage resistance 1. In an imperfect insulator, the ohmic resistance, calculated by dividing the voltage across the insulator by the current flowing through the insulator. 2. The quotient of voltage and current in a reverse-biased semiconductor junction.

leakance The reciprocal of *Rins*, insulation in ohms.

leaky 1. A term that describes a capacitor in which the dielectric material is not a perfect insulator. 2. A term that describes the condition of imperfect shielding in a coaxial transmission line. 3. A term that describes a waveguide with imperfect shielding.

leaky dielectric See *leaky insulator*.

leaky-grid detector See *grid-leak detector*.

leaky insulator An insulator that conducts significant current at a specified (test) voltage.

leaky waveguide A waveguide that has imperfect shielding, allowing some electromagnetic field to escape.

leapfrogging In radar, a phasing process that eliminates false echoes resulting from the signals of other radar sets.

leapfrog test A test performed on different locations by a computer program in memory; it moves to another memory area to continue tests on other locations.

leased line A communications circuit reserved exclusively for a specific user.

least significant bit Abbreviation, LSB. The digit with the lowest place value in a binary number.

least significant character Abbreviation, LSC. In positional notation, the extreme right-hand character in a group of significant characters. (See *positional notation.*)

least significant digit Abbreviation, LSD. The digit in a number that is at the extreme right, i.e., the one having the lowest place value.

Lecher frame A sturdy assemblage of Lecher wires.

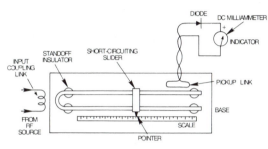

Lecher frame.

Lecher lines See *Lecher wires.*

Lecher oscillator A self-excited rf oscillator employing Lecher wires in place of an *LC* tank. Also see *line-type oscillator.*

Lecher wires A circuit segment consisting of two parallel wires or rods joined by a coupling loop on one end, the other end being open. A short-circuiting bar is moved along the wires to vary the effective length of the circuit. Rf energy is inductively coupled into the system through the loop, and the bar is slid along to various response points, as shown by meter or lamp coupled to the wires. The frequency may be determined by measuring the distance between adjacent response points. Also called *Lecher frame.*

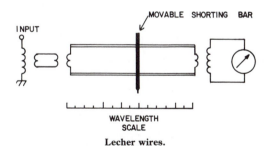

Lecher wires.

Leclanche cell See *dry cell, 1.*

LED See *light-emitting diode.*

LED—phototransistor isolator An optoelectronic isolator in which the light source is a light-emitting diode and the light-sensitive component is a phototransistor.

LEF See *light-emitting film.*

left-hand lay See *direction of lay.*

left-hand motor rule See *Fleming's left-hand rule.*

left-hand polarized wave See *counterclockwise-polarized wave.*

left-hand taper Potentiometer or rheostat taper in which most of the resistance is in the counterclockwise half of rotation as viewed from the front. Compare *right-hand taper.*

left justified An item of data that occupies consecutive locations in storage, starting at the left-hand end of its area; empty locations may appear consecutively at the right-hand end if the item needs fewer positions than have been provided.

left shift A shift operation in which the digits of a word are displaced to the left; the effect is multiplication in an arithmetic shift.

leg 1. Any one of the distinct branches of a circuit or network; also called *arm* or *branch.* 2. In a computer program, a path in a routine or subroutine.

L-electron In certain atoms, an electron whose orbit is outside of, and nearest to, those of the *K-electrons.*

Lenard rays See *cathode rays.*

length 1. The number of bits or characters in a record, word, or other data unit. 2. The end-to-end dimension of a device, circuit, line, etc.

length to fault In cable or line measurements from the home station, the distance (i.e., the cable or line length) to the point at which a fault, such as a short circuit or ground, is located.

lens 1. A usually circular piece of glass with one or both surfaces curved in cross section, used (through its refractive properties) to focus or spread rays that pass through it. Lenses for light must be transparent, but those for other radiation, such as radio waves, need not transmit light. Also see *antenna lens.* 2. In a cathode-ray tube, one or more high-voltage electrodes for focusing an electron beam to a fine point on the screen.

lens antenna See *antenna lens.*

lens disk A Nipkow disk having a lens in each hole.

lens speed The light-transmitting ability of a lens, given as an *f*-stop number: the lens' focal length divided by its diameter.

lens turret A multiple-lens mount on a camera that can be rotated for quick lens interchange.

Lenz's law (Heinrich F. E. *Lenz,* 1804-1865). A magnetic field created by an induced current is always in the direction that opposes any change in the existing field.

Lepel discharger See *quenched spark gap.*

letter-identification word See *phonetic alphabet code word.*

letters patent See *patent.*

letters shift In a teletype system, a control character that causes all of the following characters to occur in the lower case. This may be an automatic or a manual control character.

let-through current The current conducted by a circuit breaker during a short-circuit.

level 1. Operating amplitude. 2. Threshold amplitude. 3. Functional plateau or echelon, as in *first level of defense.*

level clipper See *clipper.*

level compensator 1. An automatic gain control that effectively reduces amplitude variations in a received signal. 2. An automatic gain control in telegraph receiving equipment.

level control 1. The adjustment of amplitude or threshold. 2. A potentiometer or other variable component for adjusting the amplitude or threshold of a quantity. See *attenuator; gain control, 2; volume control.*

level indicator See *volume indicator.*

level translator Any circuit or device that alters the voltage levels of input signals. An example is a converter that changes positive-logic signals to negative-logic signals.

level-triggered flip-flop A flip-flop that responds to voltage level rather than to the frequency of an input signal.

lever switch 1. A switch designed for rapid making and breaking of a circuit. 2. A radiotelegraph key.

Lewis antenna A form of antenna used at ultra-high and microwave frequencies. It resembles a horn antenna.

Leyden bottle See *Leyden jar*.

Leyden jar (*Leyden,* Holland—also Leiden—site of the invention in 1745 by Peiter van Musschenbroek, 1692-1761). The first practical capacitor. In modern form, it is a glass jar covered inside and out with metal foil and has a rod topped by a metal ball that touches the inner foil. It is still used occasionally in classrooms for demonstrating static electricity. The Leyden jar was coinvented by van Musschenbroek and invented independently by E. G. von Kleist of Pomerania, among others.

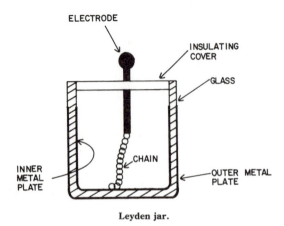

Leyden jar.

Leyden phial See *Leyden jar*.

Leyden vial See *Leyden jar*.

lf Abbreviation of *low frequency*.

LF band A section of the *L-band* extending from 1.350 to 1.450 GHz.

L-fitting See *ell*.

Li Symbol for *lithium*.

librarian program A computer program controlling a library. Also see *library*.

library In digital-computer and data-processing practice, the permanent storage of data or instructions. Also called *permanent mass storage*.

licrystal An acronym from *liquid* and *crystal*. See *liquid crystal*.

lidar See *light detection and ranging*.

lie detector See *polygraph*.

life The duration of useful service (or of operation before failure) of electronic equipment.

life test An assessment of the life of electronic equipment, either by means of full-time test runs or accelerated time tests.

lifetime See *carrier lifetime*.

lifter A device, in a magnetic tape recorder, that removes the tape from the recording and playback heads under fast-forward and rewind conditions.

light Visible electromagnetic radiation occurring in the wavelength band 0.65 μm (red light) to 0.41 μm (violet light). Included sometimes in the category of light are infrared and ultraviolet rays.

light-activated silicon controlled rectifier Abbreviation, *LASCR*.

A silicon-controlled rectifier that functions both as a photosensor and a heavy-duty bistable electronic switch allowing heavy currents to be switched by means of a light beam.

light-activated silicon controlled switch Abbreviation, *LASCS*. A pnpn device which acts simultaneously as photocell and electronic switch.

light adaption The process (also the length of the time interval) whereby the eye adjusts itself to an increase in illumination. Also sometimes, similar action in photoelectric devices.

light amplifier A solid-state amplifier employing an input electroluminescent cell and an output photocell, or some similar pair of components. The device is essentially an optoelectronic coupler with gain.

light-beam communication A system of communication in which a beam of light between transmitting and receiving stations is modulated or interrupted to convey intelligence.

light-beam meter An electric meter employing a light-beam point.

light-beam pointer A slender beam of light that replaces the pointer in a moving-coil meter. The light comes from a small incandescent lamp and is reflected by a mirror attached to the coil; when the coil moves, a spot of light moves over the scale of the meter.

light-beam receiver The receiver in a *light-beam communication system*.

light-beam recorder A graphic recorder employing a light-beam pointer. In this device, a small spot of light traces a pattern on moving photographic film, which is subsequently developed to produce a permanent record.

light-beam transmitter The transmitter in a *light-beam communication* system.

light cable A cable, consisting of numerous thin glass fibers (light wires) light-insulated from each other, through which light may be transmitted for communication or control purposes. See, for example, *light-wave telephony*.

light chopper A device that modulates a light beam by interrupting it repetitively.

light detection and ranging Acronym, *lidar*. A navigation and surveillance system in which laser light scans in a manner similar to that of radar.

light dimmer See *dimmer*.

lighted pushbutton See *lighted switch*.

lighted switch A pushbutton switch containing a pilot light that glows to show when the switch is on. Also called *illuminated switch*.

light-emitting diode A semiconductor device that emits visible light when forward biased. Also see *laser diode*.

light-emitting film A thin phosphor film which becomes luminescent when a high-frequency voltage is applied across its surface. Also see *electroluminescence* and *electroluminescent cell*.

light flasher An electron circuit or simple automatic flasher switch for flashing a lamp at regular intervals.

light flicker See *load flicker*.

light flux See *luminous flux*.

light hood See *hood*.

lighthouse tube A disk-seal tube, so called because of its resemblance to a lighthouse.

light-induced electricity See *photoelectricity*.

light load A load that is a fraction of the usual value for a given application.

light meter An instrument for measuring the intensity of light. Most consist of a photocell and dc microammeter connected in series;

however, a few employ a transistorized dc amplifier to increase the sensitivity of the microammeter.

light microsecond The unit of electrical distance; the distance light travels in free space in 1 μs (about 300 meters).

light modulation Variation of the intensity of a light beam in sympathy with a modulating signal. Also see *light modulator*.

light modulator A device with which a beam of light may be modulated by an electrical signal. See *Kerr cell*.

light negative Pertaining to negative photoconductivity, the decrease in conductivity of a photosensitive material under illumination. Compare *light positive*.

lightning The discharge that occurs between positive and negative poles in a thunderstorm. Generally, the negative pole is in the cloud and the positive pole is at the surface of the earth, resulting in a flow of electrons from cloud to ground. Some lightning occurs as a flow of electrons from ground to cloud, or between two clouds. The discharge may attain current levels of more than 100,000 amperes.

lightning arrester A device that automatically bypasses to earth a heavy nearby lightning charge, thus protecting electronic equipment connected to an outdoor antenna or line.

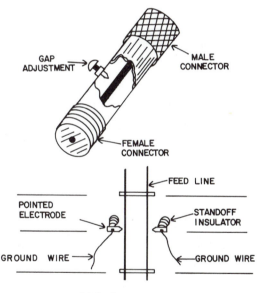

Lightning arrestor.

lightning detector See *keraunograph* and *keraunophone*.

lightning rod A protective device mounted on the outside of structures, consisting of a pointed, grounded metal rod that will conduct a lightning discharge to earth.

lightning switch See *ground switch*.

light-operated relay See *photoelectric relay*.

light-operated switch A photoelectric relay or a switch operated by such a relay.

light pen A probe containing a tiny photosensor in its tip. The tip of the light pen is touched to the screen of a cathode-ray tube to sense the beam when it passes the spot of contact.

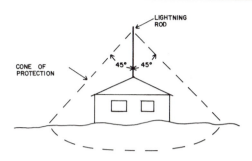

Lightning rod.

light pipe See *fiber optics, 1*.

light positive Pertaining to positive photoconductivity, when the conductivity of a photosensitive material increases under illumination. Compare *light negative*.

light quantum See *photon*.

light ray A thin beam of light. Theoretically, a ray emerges from a point source, i.e., it has no width.

light receiver See *light-beam receiver*.

light relay A photoelectric device that operates a relay according to fluctuations in the intensity of a light beam.

light-sensitive cathode A cathode that emits electrons when exposed to light.

light-sensitive diode A semiconductor diode usable as a photoconductive cell. Such diodes are available as both junction and point-contact types.

light-sensitive material A photoconductive or photoemissive substance.

light-sensitive resistor See *photoconductive cell*.

light-sensitive tube 1. A phototube. 2. A faulty electron tube whose operation has been affected by exposure to light.

light sensor 1. A light-sensitive device, such as a photocell, photodiode, phototransistor, or phototube. 2. A light-sensitive substance, such as cesium, seleniumn silicon, cadmium selenide, or lead sulfide.

light source Any generator of light. Under some conditions,, the source is regarded as a point.

light spectrum See *electromagnetic theory of light*.

light spot scanner (flying spot scanner) A television camera using as a source of illumination of spot of light that scans what is to be televised.

light transmitter See *light-beam transmitter*.

light valve 1. An electromechanical device for varying the intensity of light passing through its adjustable aperture. 2. A Kerr cell.

light-wave telephony Telephone communication by means of modulated-light transmission through thin glass fibers that may be bundled into a cable.

light-year Abbreviation, lt-yr. Pertaining to astronomy, a unit of distance equal to the distance traveled by light in 1 year in a vacuum: 9.460 55 × 10^{15} meters (5.878 × 10^{12} miles).

likelihood In probability and statistics, the chance that an event will occur or that an outcome will be realized. Also see *probability, 1, 2*.

Lilienfeld amplifier An early semiconductor-junction device described in a patent issued to Dr. Julius Lilienfeld in the early 1930s—over a decade before the discovery of the transistor.

lim Abbreviation of *limit*.

Lima Phonetic alphabet word for the letter *L*.

limen A unit that has been proposed as the minimum audible change in frequency that can be detected by at least half of a group of listeners.

limit 1. The lowest or highest frequency in a band. 2. In mathematics, a fixed value that a variable approaches. 3. The upper and lower extremes in any performance range or value range.

limit bridge A bridge used to check a component (e.g., resistance, capacitance, or inductance) in terms of the tolerance limits, rather than the nominal (named) value, of that component. Also see *bridge, 2*.

limited integrator An integrator that integrates two input signals until the corresponding output signal exceeds a certain limit.

limited stability A characteristic of a circuit or system, allowing proper operation only if the input signal and applied voltages are within certain maximum and minimum limits.

limiter A device or circuit whose output-signal amplitude remains at some predetermined level in spite of wide variations in input-signal amplitude.

limiting The restriction of the maximum peak amplitude of a signal to a designated level.

limiting amplifier An amplifier that automatically holds the output-signal level to a prescribed value.

limiting error The anticipated maximum value that the absolute error will be in a computation.

limiting resistor See *current-limiting resistor*.

limiting resolution As a measure of TV picture resolution, the most lines for picture height that can be discriminated on a test chart.

limit switch A switch that is actuated when a monitored quantity (current, voltage, illumination) reaches the limit of its range.

line 1. A wire or wires over which electrical energy travels. 2. One lengthwise path in which a force, such as electricity or magnetism, is evidenced. Such a line of force is thought to have no width, e.g., magnetic line of force.

line advance 1. Line feed in a teletype system. 2. The physical separation between the centers of adjacent scanning lines in a television system.

line amplifier An amplifier in a telephone line or similar channel, or one feeding such a line from the input end.

linear 1. In a straight line. 2. In the manner of a straight line. Thus, linear response is indicated when one quantity varies directly with another; the graph of this response is a straight line, i.e., one of constant slope. 3. The characteristic of a signal that is a replica of another (e.g., an amplifier output signal of the same waveform as that of the input signal).

linear absorption coefficient Symbol, μ. A number expressing the extent to which the intensity of an X-ray beam is reduced per centimeter of the material through which it passes.

linear accelerator An accelerator (see *accelerator, 1*) in which particles are urged in a straight line through what is essentially a long tube. This action is in contrast with that occurring in circular accelerators, such as the cyclotron.

linear amplifier 1. An amplifier for which a linear relationship exists between input and output parameters, e.g., a high-fidelity audio amplifier. 2. A class AB radio-frequency power amplifier which amplifies an AM signal.

linear array A directional antenna having equally spaced, in line elements.

linear circuit 1. A circuit whose output is a faithful reproduction of the input. See *linear amplifier, 1* and *linear detector*. 2. A circuit whose performance is linear. See *linear response, 1*.

linear decrement In a damped wave, a linear decrease in amplitude with time, as opposed to logarithmic decrement. Also see *decrement*.

linear detector A detector whose output/input relationship is linear. Also see *linear, 2, 3; linear circuit, 1; linear response, 1, 2, 3*.

linear differential transformer A device that converts the physical position of an object into an output voltage or current. The voltage or current is directly proportional to the displacement.

linear distortion Amplitude distortion in which the output and input signal envelopes are disproportionate (in the absence of spurious frequencies).

linear equation See *first-degree equation*.

linear IC See *linear integrated circuit*.

linear integrated circuit An integrated circuit designed for such operations as amplification, oscillation, nondigital regulation, analog instrumentation, and similar applications. Compare *digital integrated circuit*.

linearity 1. The degree to which performance or response approaches the condition of being linear, expressed in percent or parts per million. Also see *linear amplifier, 1; linear circuit, 1, 2; linear oscillator, 1; linear response, 1, 2, 3; linear taper*. 2. In a TV or oscilloscope image, absence of compression or stretching of any part.

linearity control In an oscilloscope or TV circuit, the potentiometer used to correct image linearity. See *linearity, 3*.

linearity error 1. The difference between a theoretically linear function and the actual function, as observed under experimental conditions. 2. The degree of nonlinearity in an amplifier that is supposed to be linear.

linear modulation 1. Modulation in which the instantaneous amplitude of the input signal is directly proportional to the instantaneous amplitude of the output signal. 2. Modulation in which the instantaneous amplitude of the input signal is inversely proportional to the instantaneous amplitude of the output signal. 3. Modulation in which the instantaneous amplitude of the input signal is directly proportional to the frequency or phase deviation of the output signal.

linear oscillator 1. An oscillator whose ac output amplitude varies linearly with its dc input. 2. A line-type oscillator.

linear reflex detector See *infinite-impedance detector*.

linear response 1. The type of response in which the value of a dependent variable is at every point equal or directly proportional to that of the independent variable. Compare *logarithmic response, 1* and *square-law response*. 2. A type of response in which a quantity (such as current) varies directly with another quantity (such as voltage). Compare *logarithmic response, 2*. 3. Low-distortion response. Also see *high-fidelity*.

linear sweep In a TV or oscilloscope circuit, the sweeping of the electron beam across the screen at a constant speed. Also see *linear, 1; linear response, 1; linearity, 2*.

linear taper In a potentiometer or rheostat, resistance variation that is directly proportional to shaft rotation. Thus, half the total resistance corresponds to movement of the shaft over half the arc of full rotation. Compare *log taper*. Also see *taper*.

linear time base For an oscilloscope, the base provided by sweeping the electron beam horizontally at a uniform rate. Also see *linear sweep*.

linear transformer A radio-frequency transformer consisting of a quarterwave section of transmission line.

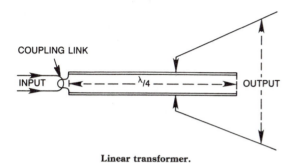

Linear transformer.

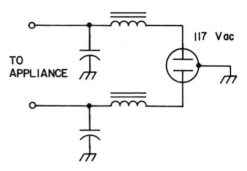

Line filter.

linear variable differential transformer Abbreviation, LVDT. A differential transformer exhibiting linear response. Also see *linear response, 1, 2.*

line balance The degree of electrical similarity between transmission line conductors or between a conductor and ground.

line-balance converter A device used to isolate the outer conductor at the end of a coaxial line from ground.

line characteristic distortion Fluctuations in the duration of received signal impulses in Teletype communications caused by changing current transitions in the wire circuit.

line circuit The telephone system relay equipment associated with stations connected to a switchboard.

line conditioning In data communications, the modification of private or leased lines by adding compensating reactances to reduce amplitude variations or phase delays over a band of frequencies.

line coordinate A symbol identifying a specific row of cells in a matrix; a specific cell can be located with an additional column coordinate.

line cord A flexible two- or three-wire insulated conductor connecting equipment to the power line by means of a plug which mates to a standard electrical outlet.

line-cord resistor A flexible resistor incorporated into a line cord to serve specifically as a filament dropping resistor. Also see *filament resistance, 2; flexible resistor; heater string; line cord.*

line current 1. Current flowing from a power line into equipment. 2. Current flowing in a transmission line. 3. Current flowing into a parallel-resonant circuit.

line diffuser A circuit that creates minor vertical oscillations of the spot on a television screen, making the individual scanning lines less noticeable.

line driver An IC capable of transmitting logic signals through long lines.

line drop The voltage drop along a line supplying power to a device.

line equalizer See *equalizer.*

line fault A discontinuity in a transmission line, resulting in signal loss at the receiving end of a circuit.

line feed In a teletype system, the movement of the paper or platen to allow for printing of an additional line of text.

line filter A circuit of one or more inductors and capacitors inserted between an ac-powered device and the power line to attenuate noise signals.

linefinder A switching device that finds one of a group of calling telephone lines and connects it to a trunk, connector, or selector.

line frequency 1. The frequency of power-line voltage—in the United States, 60 Hz. 2. The rate at which the horizontal lines are traced in a TV picture. Also see *horizontal frequency.*

line group 1. A group of signals sent by wire transmission. 2. The frequency spectrum occupied by a group of signals sent by wireless transmission.

line leakage Resistance between insulators of two wires in a telephone line loop.

line loss The sum of energy losses in a transmission line.

lineman A technician who works mainly with telephone or telegraph lines.

line noise 1. Electrical noise (as received by a radio) arising from fluctuations of current or voltage in a power line. 2. Noise in a data transmission line.

line of force 1. A line depicting the points of equal-intensity field strength about an electric charge or about a charged body. Also see *flux.* 2. A line depicting the points of equal-intensity field strength about a magnetic pole or about a magnetized body. Also see *flux.*

line-of-sight communication Communication between points within view of each other. Also see *line-of-sight distance.*

line-of-sight distance Symbol, d. The maximum distance over which an ultrahigh-frequency signal may be directly transmitted along the surface of the earth. It is slightly less than true line of sight;

$$d \text{ equals } 1.41 \sqrt{h},$$

where h is antenna height.

line oscillator See *line-type oscillator.*

line plug The plug terminating a line cord. Also see *male plug.*

line printer A machine that prints the results of a computer run line by line. Also called *printer.*

line radio See *wired radio.*

line regulation Automatic stabilization of power-line voltage.

line-sequential system The color TV system in which the image is reproduced by means of primary color lines (red, green, blue) sequentially beamed across the screen of the picture tube. Compare *dot-sequential system* and *field-sequential system.*

lines of cleavage See *cleavage.*

lines oscillator See *line-type oscillator.*

line switch 1. The main power-line switch. 2. Within a piece of electronic equipment, the switch that opens and closes the circuit to the incoming power line.

line-type amplifier A radio-frequency amplifier in which the grid or plate tanks (or equivalent transistor circuit tanks) are transmission lines comprising parallel wires, rods, or tubing.

line-type oscillator A radio-frequency oscillator in which the grid or plate tanks (or equivalent transistor circuit tanks) are transmission lines comprising parallel wires, rods, or tubing.

line unit In a wire teletype system, the terminal unit, or device that converts the teletype signals into electrical impulses and vice-versa.

line voltage 1. The voltage of a power line. 2. The voltage between the conductors of a transmission line.

line-voltage drop See *line drop*.

line-voltage monitor See *power-line monitor*.

linguistics The study of languages, including structure, symbology, and phonetics.

link 1. The small coupling coil employed in link coupling. 2. A communications path between two radio facilities for the purpose of extending the range of one, as between a remote pickup point and a broadcast transmitter. 3. In a digital computer, a branch instruction, or an address in such an instruction, used to leave a subroutine to return to some point in the main program.

linkage 1. Coupling between separated conductors or devices through the medium of electric or magnetic lines of force. 2. A connection between mechanical parts of a mechanical analog computer that perform an arithmetic function.

link circuit A closed-loop coupling circuit having two coils of a few turns of wire; each coil is placed near one to be coupled.

link coupling Low-impedance coupling via a small (usually one-turn) input or output coil fed by a twisted pair or a coaxial line.

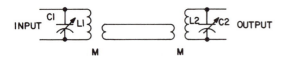

Link coupling.

linked subroutine A subroutine, entered by a branch instruction from a main routine, which executes a branch instruction returning control to the main routine.

link fuse A fuse consisting of an exposed length of fuse wire.

link neutralization Neutralization achieved by out-of-phase current fed back via link coupling from the output to the input of an amplifier. Also called *inductive neutralization*.

lin-log receiver A radar receiver whose amplitude response is linear for small signals but logarithmic for large ones.

lip microphone A small microphone operated close to or in contact with the lips.

liquid A state of matter characterized by a level of molecular motion intermediate between that of gases and solids; liquids have the ability (like gases) to take the shape of a container and are only slightly compressible. Compare *gas, plasma,* and *solid*. Also see *state of matter*.

liquid absorption For a solid material, such as dielectric, the ratio of the weight of liquid absorbed by the material to the weight of the material.

liquid capacitor See *water capacitor*.

liquid cell See *electrolytic cell*.

liquid conductor See *electrolyte*.

liquid cooling Use of circulating water, oil, or other fluid to remove heat from equipment, such as power tubes.

liquid crystal A liquid exhibiting some of the characteristics of a crystal. Also see *nematic crystal* and *smectic crystal*.

liquid-crystal display Abbreviation, LCD. For counters, calculators, digital meters, and digital clocks, a readout device in which each digit is formed by strips of liquid-crystal material. Also see *nematic crystal* and *smectic crystal*.

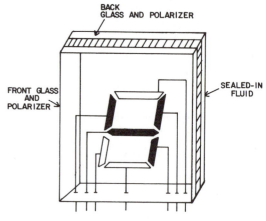

Liquid crystal display.

liquid detector 1. A device for detecting the presence of moisture. 2. Electrolytic detector.

liquid-filled transformer A transformer filled with a protective liquid insulator, such as oil.

liquid-flow alarm An electronic circuit which actuates an alarm when the flow of a liquid through pipes or other channels changes from a desired rate.

liquid-flow control A servo system that automatically maintains or corrects the rate of liquid flow through pipes or other channels.

liquid-flow gauge See *liquid-flow meter*.

liquid-flow indicator See *liquid-flow meter*.

liquid-flow meter An instrument that indicates the rate at which a liquid flows through pipes or other channels.

liquid-flow switch In a liquid-cooled system, a switch that actuates an alarm when the liquid slows or stops.

liquid interrupter See *electrolytic interrupter*.

liquid-jet oscillograph A graphic recorder (see *recorder, 2*) employing an ink-jet galvanometer to trace the pattern on a paper chart.

liquid laser A laser in which the active material is a liquid.

liquid-level alarm An electronic device which actuates visual or audio signal devices when the surface of a liquid inside a tank rises or falls to a predetermined level.

liquid-level control A servo system that automatically maintains the liquid in a tank at a predetermined level.

liquid-level gauge An electronic system that gives direct readings of the level of a liquid in a tank.

liquid-level indicator See *liquid-level gauge*.

liquid-level meter See *liquid-level gauge*.

liquid load See *water load*.

liquid-pressure alarm An electronic circuit that actuates an alarm when the pressure of a liquid changes.

liquid-pressure control A servo system that automatically maintains or corrects liquid pressure in pipes or other channels.

liquid-pressure gauge See *liquid-pressure meter.*

liquid-pressure indicator See *liquid-pressure meter.*

liquid-pressure meter An instrument that gives direct readings of liquid pressure in a pipe or other channel.

liquid-pressure switch A switch that actuates an alarm when the pressure of a liquid changes.

liquid rectifier See *electrolytic rectifier.*

liquid resistor See *electrolytic resistor* and *water resistor.*

liquid rheostat See *water rheostat.*

liquid valve See *electromechanical valve.*

liser An oscillator that produces an extremely pure microwave carrier signal.

LISP A digital-computer language used in processing lists. Also see *language.*

Lissajous figure Any one of several curves resulting from the combination of two harmonic motions. These figures are familiar in electronics, since they are obtained when signals are applied simultaneously to both axes of an oscilloscope. Also called *Lissajous pattern.*

PATTERN	RATIO
◯	1:1
⋈	2:1
⋈⋈	3:1
⋈⋈	3:2
⋈⋈	4:3

Lissajous figure.

list 1. To print serially the records in a file or in memory. 2. To print (instruct a computer to display) every item of input data in a program. 3. A one-dimensional array of numbers.

listener fatigue Tiring by prolonged listening to the (usually unfaithful) reproduction of sound.

listening angle The angle whose sides are the distances between each of a pair of speakers and the listener, the apex.

listening test The evaluation of audio equipment by listeners.

liter Abbreviation, *l.* A metric unit of volume equal to 1.0567 U.S. liquid quarts or 0.908 U.S. dry quart. A liter is the volume of 1 kilogram of water at 4 °C and under a pressure of 1 pascal.

literal operands Usually applicable to source language instructions, operands that specify the value of a constant rather than an address of a location in which the constant is stored.

lithium Symbol, Li. An element of the alkali-metal group. Atomic number, 3. Atomic weight, 6.940. Lithium is the lightest metal known.

lithium battery See *lithium cell.*

lithium cell A vented, steel-jacketed *primary cell* in which the anode element is *lithium.* The electrolyte is an organic substance without water. Nominal voltage is 2.8 V. The lithium cell has long shelf life.

Litzendraht wire See *Litz wire.*

Litz wire A woven wire having a number of copper strands, each separately enameled to insulate it from the others. The wire is woven so that inner strands come to the surface at regular intervals. Litz wire is noted for its low losses at radio frequencies.

live 1. Electrically activated, i.e., sustaining voltage or current. 2. Being broadcast as it occurs. 3. Acoustically reflective, as in a *live* room (contrasted with one that is acoustically absorbent).

live end 1. In a recording or broadcasting studio, the part of the room in which the acoustic concentration is greatest. 2. In a utility circuit, the wire or terminal that carries 120 volts alternating current (the ungrounded end).

live room A room with pronounced echoes and reverberation owing to absence of paucity of sound-insulating materials. Compare *dead room.*

LIX Abbreviation of liquid crystal.

LK band A section of the *L-band* extending from 1.150 to 1.350 GHz.

LL band A section of the L-band extending from 510 to 725 MHz.

LLL Abbreviation of *low-level logic.*

lm Preferred abbreviation of *lumen.*

lm/ft^2 Abbreviation of *lumens per square foot.* Also see *lumen.*

lm-hr Abbreviation of *lumen-hour.*

lm/m^2 Abbreviation of *lumens per square meter.* Also see *lumen* and *lux.*

lm/W Abbreviation of *lumens per watt,* a unit of luminosity. Also see *lumen.*

ln Abbreviation of *natural logarithm* (see *Napierian logarithm*). Also written *loge.*

L-network An impedance-matching circuit, filter, or attenuator which resembles an inverted letter *L.*

LO Abbreviation of *local oscillator.*

lo Abbreviation of *low,* usually as a prefix or subscript. Also abbreviated *L.*

load 1. A device or circuit that is operated by the energy output of another device or circuit. 2. The power output capability of a machine. 3. To fill an internal computer storage with information from an external storage, e.g., from a magnetic tape to a computer's random access memory.

load-and-go Automatic coding in which a user's (source) program is translated automatically into machine language and stored.

load capacitance 1. The capacitance of a load. 2. A capacitance employed as a load.

load circuit 1. The circuit that forms the load, or power-consuming portion, of a system. 2. A circuit that facilitates transfer of power to a load.

load coil See *work coil.*

load current The current flowing in a load. See *load.*

load division A method of connecting two or more power sources to a single load, for optimum power transfer.

loaded antenna An antenna having a loading coil connected to it to increase its electrical (effective) length. See *loading disk* and *loading coil.*

loaded line A transmission line in which inductors or capacitors are inserted at appropriate points to alter the characteristics of the line.

load end The end of a transmission line to which a radiator or receiver is connected.

load flicker Fluctuations in the brightness of lamps, caused by intermittent loading of the power line by other devices.

load impedance Symbol, Z_L. The impedance presented by a load connected to a generator or other source.

loading 1. The matching of source impedance to load impedance, usually by means of the introduction of an inductance or capacitance into the load itself. 2. Any form of impedance matching. 3. Modification of the acoustic impedance of a loudspeaker.

loading coil An inductor which is inserted in a circuit to increase its total inductance or to provide some special effect, such as canceling capacitive reactance. See *loaded antenna* and *Pupin coil*.

loading disk A metal disk mounted atop a vertical antenna to increase its effective length.

loading factor The ratio of source impedance to load impedance before the introduction of loading circuits.

loading inductance 1. The inductance of a load. 2. An impedance employed as a load.

loading routine (loading program) A routine permanently in memory; it allows a program to be loaded into memory from an external storage medium.

load life The longevity of a device in terms of the number of hours it can withstand its full power rating.

load line In a group of *EI* curves, a line connecting points of equal resistance (*E/I*) that are equal to a particular value of load resistance (impedance).

load power The power dissipated in a load.

load regulation Automatic stabilization of load resistance (impedance) at a constant value.

load resistance 1. The resistance of a load. 2. A resistance employed as load.

load stabilizer A device for holding load current or load voltage to a constant value.

loadstone Alternate spelling of lodestone.

load termination The load connected to the output of a circuit or device as the terminal element in a circuit or system.

load voltage The voltage developed across a load.

load-voltage stabilization Automatic regulation of load voltage.

load wattage See *load power*.

lobe In an antenna directivity pattern, a figure—such as a circle or ellipse—enclosing an area of intensified response.

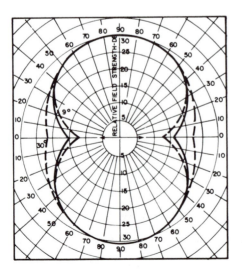

Lobe pattern of horizontal half-wave antenna.

lobing 1. In a transmitting or receiving antenna, the effect of ground reflection, resulting in phase reinforcement (lobes) at some elevation angles and phase opposition (nulls) at other angles. 2. The pattern of secondary maxima in the radiation response of a directional antenna.

local action Electrolysis between separate areas of a single electrode immersed in an electrolyte. The action is due to impurities at different spots in the electrode metal, causing one spot to act as an anode and the other as a cathode, thereby creating a small battery cell.

local battery In wire telephony, a battery installed on the subscriber's premises.

local broadcast station A standard broadcast station licensed in the local service. See *local channel* and *local station*.

local channel A radio broadcast channel intended to serve an area near the transmitting station and shared by several relatively low-powered stations (250W maximum).

local control Control of a radio transmitter from the site (in contrast to remote control).

local feedback Feedback within a circuit stage.

localizer A radionavigation transmitter whose signal guides aircraft to the centerline of a runway.

local oscillator Abbreviation, LO. An oscillator, such as a beat-frequency type, included in a piece of equipment, as contrasted to the oscillator in a distant transmitter.

local program A program that originates at the same single broadcast station from which it is transmitted.

local reception The reception of signals from local stations. Compare *long-distance communications*.

local side The group of circuits and components associated with a communications terminal at a given location.

local station A station situated within the same general area as the receiver, as opposed to a distant station.

local system library A computer program library containing standard software associated with a specific system.

local transmission The sending of communications to receivers in the same general locality as the transmitter, as opposed to long-distance transmission.

local trunk In a telephone system, the interconnecting line between local and long-distance lines.

location In digital computer operation, a memory position (often a register) specified by an address and usually described in terms of the basic storage unit a particular system uses, e.g., a character is a location in a character-oriented machine.

location counter A register in the control section of a computer containing the address of the instruction being executed.

locked groove A continuous blank groove around the inside of a phonograph record, which keeps the stylus from running into the label.

locked oscillator 1. A fixed-frequency oscillator, such as a crystal-controlled oscillator. 2. Bradley detector.

lock-in A state of synchronism, as when a self-excited oscillator is synchronized (locked-in) with a standard-frequency generator.

lock-in amplifier A detector that makes use of a balanced amplifier. The output is the difference between the plate, collector, or drain currents of the two devices.

lock-in base An electron-tube base with a central locking plug that mates with a keyed central hole in the socket to hold the tube firmly in place.

locking circuit See *holding circuit*.

locking relay See *latching relay*.

lock-in relay See *latching relay*.

lock-in socket The socket for a *Loktal tube*.

lock-in tube See *Loktal tube*.

lock-out 1. To prevent a hardware unit or routine from being activated, e.g., when there would be a conflict between operations using the same areas of memory. 2. A safeguard against an attempt to refer to a routine in use.

lock-up relay An electromagnetic relay that can be locked in the actuated state nonmechanically, i.e., by means of an electromagnet or permanent magnet.

locus The set of all points located by stated conditions; e.g., the locus of secondary points that are all equidistant from a primary point is a sphere.

lodestone A natural magnet; a form of the mineral magnetite. Also spelled *loadstone*.

Loftin—White circuit An early two-stage direct-coupled audio amplifier circuit employing a voltage-amplifying triode or pentode in the input stage and a heavy-duty triode, pentode, or beam-power tube in the output stage.

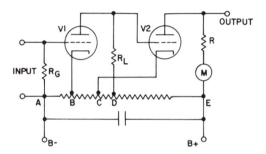

Loftin-White circuit.

log 1. Abbreviation of *logarithm*. 2. A continuous record of communications kept by a station, or a record of the operation of an equipment.

log10 Abbreviation of *logarithm to the base 10* i.e., a common logarithm. Also called *Briggsian logarithm*.

logarithm Abbreviation, log. The power to which a number, called the *base*, must be raised to equal a given number. Thus, if the given number is designated N and the base a, then $\log_a N$ equals x, because a^x equals N (example, $\log_{10} 100$ equals 2, since the base 10 must be raised to the second power to equal 100). The two bases used most often are 10 (common logarithms) and 3 (2.718 28; the base in natural, or Napierian, logarithms). Also see *antilogarithm, cologarithm, common logarithm, natural logarithm, Napierian logarithm*.

logarithmic amplifier An amplifier whose output-signal amplitude is proportional to the logarithm of the input-signal amplitude.

logarithmic curve A graphical representation of a logarithmic function, having the form $y = a \log x$. The logarithmic base may be any positive real number.

logarithmic decrement See *decrement*.

logarithmic graph A graph in which the x and y axes are incremented logarithmically. Compare *semilogarithmic graph*.

logarithmic horn A horn whose diameter varies directly as the logarithm of the length. See *horn*.

logarithmic mean See *geometric mean*.

logarithmic meter A current meter or voltmeter whose deflection is proportional to the logarithm of the quantity under measurement. The increments on the scale of such an instrument are closer together in the upper portion.

logarithmic rate of decay See *exponential decrease*.

logarithmic rate of growth See *exponential increase*.

logarithmic response 1. Response in which the value of a dependent variable is at every point equal to the logarithm of the independent variable: y equals $\log x$. 2. A type of response in which a quantity (such as current) varies directly with the logarithm of another quantity (such as voltage).

logarithmic scale A graduated scale in which the coordinates are positioned according to the logarithm of the actual distance from the origin.

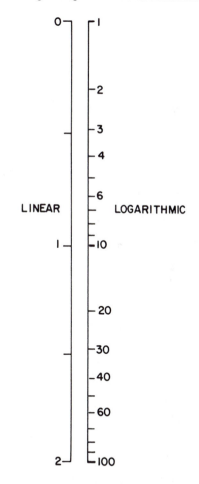

Logarithmic scale.

logarithmic series A mathematical series which is the expansion of the expression $\log_e (1 + x)$ in ascending powers of x. Thus, $\log_e (1 + x) = x - x^2/2 + x^3/3 - x^4/4 + ...x^N/N$.

logarithmic voltmeter See *logarithmic meter*.

logarithmic vtvm A vacuum-tube voltmeter having a logarithmic scale. Also see *logarithmic meter.*

loge Abbreviation of *logarithm to the base e,* i.e., a natural logarithm. Also written *ln*. Also see *Napierian logarithm.*

logic 1. In digital-computer practice, the mathematics dealing with the truth or falsity of indicated relationships and their combinations. Also see *symbolic logic.* 2. Collectively, the switching circuits and associated hardware for implementing logic functions (see 1, above), such as AND, NAND, NOR, OR, etc.

logical decision During a computer program run, a choice between alternatives based on specified conditions. For example, one alternative path in a routine might be selected because an intermediate result was negative.

logical diagram A schematic diagram showing the interconnection between gates of a logic circuit.

logical file A data set comprising one or several logical records.

logical operation 1. An operation using logical operators: AND, NOR, OR, NAND. 2. A processing operation in which arithmetic is not involved (e.g., a shift).

logical operator A word or symbol representing a logic function operating on one or more operands.

logical shift A shift operation in which digits in a word are moved left or right in circular fashion; digits displaced at one end of the word are returned at the other. Also called *circular shift, end-around shift, nonarithmetic shift, cyclic shift,* and *ring shift.*

logic array In logic circuits, a redundant arrangement of identical components in a single package.

logic circuits Gating or switching circuits which, in computer and control systems, perform such logical operations as AND, NAND, NOR, and OR. Logic circuits may employ either diodes, transistors, charge-coupled devices, tunnel diodes, thyristors, ferroelectric elements, magnetic-core elements, or a combination of these.

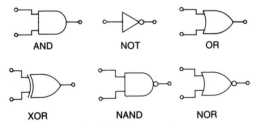

AND   NOT   OR

XOR   NAND   NOR

Logic circuits or gates.

logic comparison An operation in which two operands are compared for equal value.

logic connectives Words connecting operands in a logic statement; the truth or falsity of the statements can be determined from their content and the connectives' meanings.

logic diagram 1. A graphic representation of a logic function: AND, NAND, NOR, OR. 2. The design of a device or system represented by graphic symbols for logic elements and their relationships.

logic diode See *computer diode.*

logic flowchart The logical steps in a program or subroutine represented by a set of symbols.

logic function An expression for an operation involving one or a combination of logic operators.

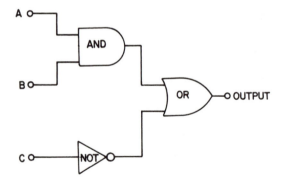

A	B	C	AB + C'
0	0	0	1
0	0	1	0
0	1	0	1
0	1	1	0
1	0	0	1
1	0	1	0
1	1	0	1
1	1	1	1

Logic diagram and truth table.

logic instruction A command to execute a logical function.

logic level 1. One of the two logic states, zero or one (on or off, high or low). 2. Of the two logic states, that which represents the "true" condition. 3. The voltage amplitude of digital signals in a logic system.

logic probe A test probe with a built-in amplifier and (usually) indicating LEDs; its tip is touched to various test nodes in a digital logic circuit to trace logic levels and pulses.

logic relay See *bistable relay.*

logic swing In a logic circuit, the difference between the voltage corresponding to the *one* state and the voltage corresponding to the *zero* state.

logic symbol 1. A symbol used to represent a logic circuit element graphically. 2. A symbol used to represent a logic connective.

log-log graph A chart on which *x* and *y* functions are logarithmically represented, as opposed to a semilog graph (in which one axis is a linear display). See also *logarithmic graph.*

log-periodic antenna A broad-spectrum, multiband directional antenna in which the lengths and spacing of the radiator and elements increase logarithmically from one end of the antenna to the other. Also called *log-periodic array.*

log taper In a potentiometer or rheostat, resistance variations that correspond to the logarithm of shaft rotation, or vice versa. Compare *linear taper.* Also see *taper.*

Loktal Trade name for a vacuum tube having an octal base with a lock-in feature. See *lock-in base.*

Loktal socket See *lock-in socket.*

Loktal tube An electron tube having a lock-in base for holding it firmly in the socket.

long Abbreviation of *longitude*.

long-distance communications Communications between transmitters and receivers in localities separated by considerable distances.

long-distance loop A direct telephone line connecting a subscriber's station to a long distance switchboard.

long-distance reception See *long-distance communications*.

long-distance transmission See *long-distance communications*.

longitude Abbreviation, *long*. Angular distance measured in degrees around the earth's circumference, to the east and west of the prime meridian at Greenwich, England. Compare *latitude*. Also see *meridian*.

longitude effect The variation (caused by the earth's magnetic field) of the strength of cosmic rays arriving at different longitudes on the surface of the earth.

longitudinal current Current flowing in the same direction in the parallel wires of a pair (the return circuit is via ground).

longitudinal parity Parity associated with bits recorded on one track of a magnetic storage medium to indicate whether the number of bits is even or odd.

longitudinal redundance A computer condition, generally affecting magnetic tape records, in which the bits in each track of a record do not meet the required parity, as determined by a *longitudinal redundancy check* (q.v.).

longitudinal redundancy check A parity check performed on a block of characters or bits (say, on a track of a magnetic disk). A parity character is generated and transmitted as the last character of the block; thus, each longitudinal block has either even or odd parity.

longitudinal wave A wave in which the oscillation is parallel with the direction of propagation.

long line 1. A single-wire antenna whose length is greater than the length of the wave fed to it for propagation. See also *long-wire antenna*. 2. In wire telegraphy, an electrical line having great physical length. 3. In electronics theory, a line of indeterminate length but whose characteristics remain stable and predictable to infinity.

long-persistence screen A cathode-ray screen on which the image remains for a time after the electron beam has passed.

long-play Abbreviation, *LP*. Descriptive of phonograph disks of the 33 1/3 rpm variety. Also see *microgroove record*.

long-play record See *microgroove record*.

long-range navigation Abbreviated *loran*. A radionavigation system that operates in any of several different frequency ranges.

long-range radar Radar that can detect targets at distances of 200 miles or more.

long skip Ionospheric propagation in which the incident ray and reflected ray are both very long. Compare *short skip*.

long-tailed pair A phase-inverter of the cathode-coupled type driven by a direct-coupled voltage amplifier.

long-term drift Gradual change in the value of a quantity, such as voltage or frequency, observed over a long period, in contrast to that noted for a brief interval (*short-term drift*).

long-term stability Stability reckoned over a long period, as contrasted to that noted for a brief interval (short-term stability).

long throw A speaker design term describing a woofer moving through long excursions; the objective is to provide good low-frequency response with low distortion.

long waves Low-frequency radio waves, particularly those in the 30 to 300 kHz region (10^3 to 10^4 meters).

long-wire antenna An antenna whose radiator is usually a number of wavelengths long.

lookup A computer programming technique in which a data item identified by a key is selected from an array.

loop 1. In a standing-wave system, a maximum-response point; e.g., current loop, voltage loop. Compare *node, 2.* 2. A loop antenna. 3. A signal path; e.g., a *feedback loop.* 4. A one- or two-turn coil for low-impedance coupling. Also see *link, 1.* 5. In a computer program run, the repetitious execution of a series of instructions that terminates when some specified condition is satisfied by a relational test, at *which point* the next instruction in the main program is obeyed.

looped amplification See *feedback factor*.

loop antenna A (usually small) portable antenna in the form of a coil or wire.

loop checking A method of checking the accuracy of data transmitted over a data link by returning signals received at one terminal to the transmitting terminal for comparison with the original data.

loop gain In a feedback system, the feedback factor.

looping plug A double phone-plug unit (for simultaneously plugging into two phone jacks which completes (loops) the circuit between the two jacks.

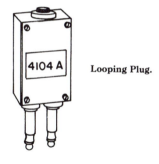

Looping Plug.

loop-input signal A signal introduced into a feedback control loop.

loop modulation See *absorption modulation*.

loop pulsing The regular, intermittent breaking of the dc path at the transmitting end of a transmission line; also called *dial pulsing*.

loopstick antenna See *ferrite-rod antenna*.

loop test A means of locating a discontinuity in a circuit by creating a closed loop including the suspected fault point.

loose coupler An early adjustable, receiver rf transformer in which a secondary coil slides within a primary coil to vary the coupling between the two. The coils are mounted horizontally, the secondary coil assembly sliding on a pair of rods.

loose coupling Coupling that is too slight to transfer any but small amounts of energy, as when a primary and secondary coil are spaced so far apart that the coefficient of coupling is small. Compare *close coupling*.

loosely coupled twin In computer practice, a system in which two processors, each having its own operating system, are used with switches so they can use common peripherals. Also see *switch*.

lopplar Acronym for *laser Doppler radar*. Also abbreviated ladar.

LORAC A radionavigation system that operates by means of phase comparison. Similar to long-range navigation (*loran*). Trade name of Seismograph Service Corporation.

loran A long-range radionavigation system in which two pairs of ground stations transmit pulsed signals, which are used by aircraft and ships to determine their position. The name is an acronym for *long range navigation.*

loran C A radionavigation system that operates at a frequency of 100 kHz. It operates on the hyperbolic principle.

loran D A radionavigation system similar to loran C. Used by aircraft, it operates independent of ground stations, and prevents unwanted enemy detection of aircraft position.

loss Energy (electrical, heat, etc.) that is lost, i.e., it is dissipated without doing useful work in a circuit or system. See *power loss.*

loss angle For an insulating material, the complement of phase angle.

loss index The product of power factor *pf* and dielectric constant *k*: $pf \times k$ = loss index.

lossless line An ideal transmission line, i.e., one having neither resistance losses nor radiation losses.

loss modulation See *absorption modulation.*

loss tangent See *dielectric dissipation* and *dissipation factor, 1.*

lossy line A line or cable having high attenuation per unit length.

loudness The amplitude of sound. Also called *volume.*

loudness control See *compensated volume control.*

loudness curves See *audibility curves.*

loudspeaker A transducer that converts electrical impulses into sound waves of sufficient volume to be heard easily by a number of listeners situated at some distance from the device. Also called *speaker.*

loudspeaker damping See *damped loudspeaker.*

loudspeaker dividing network See *crossover network.*

low band 1. The low or lowest frequency band employed in communications, testing, or processing in a given situation. 2. Television channels 2 to 6 (54 to 88 MHz). 3. In two-way radio practice, those radio channels in the vhf range from 30 to about 70 MHz.

low-capacitance probe A test probe in which capacitance has been minimized to reduce loading and detuning of the circuit under test.

low-energy criterion See *von Hippel breakdown theory.*

lower curtate A group of punching positions at the bottom of a punched card column that usually have no zone significance.

lower sideband In an amplitude-modulated wave, the component whose frequency is equal to the difference between the carrier frequency *fc* and the modulating frequency, *fm*, i.e., *f*1 equals *fc* minus *fm*. Compare *upper sideband.*

lower sideband suppressed carrier Abbreviation, LSSC. A single-sideband transmission technique in which the lower sideband is transmitted but the upper sideband and carrier are suppressed. Compare *double sideband suppressed carrier* and *upper sideband suppressed carrier.*

lowest usable frequency Abbreviation, LUF. The lowest frequency that may be used successfully at a given time for communication via the ionosphere. Compare *maximum usable frequency.*

low filter A highpass filter that removes low-frequency audio noise from the modulating waveform of a broadcast station. The result is a lower level of transmitted hum and rumble.

low-frequency Abbreviation, *lf.* 1. Pertaining to radio frequencies in the 30 to 300 kHz band (wavelengths from 10^3 to 10^4 meters). Also see *radio spectrum.* 2. Pertaining to audio frequencies below 500 Hz.

low-frequency compensation 1. In video-amplifier design, special measures, such as use of high coupling and bypass capacitances, to boost low-frequency gain. 2. Use of special circuits to increase the low-frequency response of an audio amplifier. Also see *bass boost, 1, 2.*

low-frequency direction finder Abbreviation, *LDF.* A direction finder operated in or below the standard broadcast band.

low-frequency padder See *oscillator padder.*

low-frequency parasitics Parasitic oscillations of a frequency lower than that being processed by the amplifier or generated by the oscillator in which they occur.

low-level 1. A logic term for the more negative of the two (binary) logic levels. 2. Having an amplitude that is below that normally available in comparable circuits or systems. 3. Primitive, as in *low-level language* (q.v.).

low-level contact A switch or relay contact intended for use with low values of current and voltage.

low-level language A computer programming language in which each instruction has only one equivalent machine code. *Compare high-level language.*

low-level logic Abbreviation, *LLL.* In digital-computer practice, any logic system that operates at low voltage or current levels.

low-level modulation Modulation at low power. See *cathode modulation* and *grid-bias modulation.*

low-level signal 1. A signal with small amplitudes. 2. A signal with peak-to-peak voltage so low that it does not drive an amplifier circuit out of the linear range of operation.

low-loss material A material, particularly a dielectric, having low electrical losses at a given frequency. Also see *loss.*

low-mu tube A triode having an amplification factor of eight or less, e.g., a 12B4A. Compare *high-mu tube* and *medium-mu tube.*

low order The lesser-value place of characters or digits in the hierarchy of a group (number or word); 5 and 6 are the low-order digits in the number 456.

low-order position The extreme right-hand (least significant) position in a number or word.

low-pass filter A wave filter that transmits (with little attenuation) all frequencies below a critical (*cutoff*) frequency but blocks all frequencies above the cutoff value. Such filters may be entirely *LC*, entirely *RC*, or a combination of the two.

low power Abbreviation, *LP.* Power considerably lower than that ordinarily encountered in a particular application. The term is arbitrary, since several hundred watts might be regarded as low power in one situation, whereas a fraction of a watt would be implied in another.

low-print A term for magnetic recording tape that is less susceptible to print-through than conventional tape.

low Q For a component or circuit, a low quotient for the ratio of reactance to resistance (*X/R*). This is a relative term, since a particular *Q* value considered low in one situation might be high in other circumstances. Also see *figure of merit, 1.*

low tension See *low voltage.*

low voltage 1. A voltage considerably lower than that ordinarily encountered in a particular application. The term is arbitrary, since several hundred volts might be regarded as low in one situation, while a fraction of a volt would be implied in another. 2. In a TV circuit, the supply voltage applied to all points other than the high-voltage circuit or the picture tube.

low-voltage rectifier In a TV circuit, the rectifier that supplies power for the low-voltage stages. See *low voltage, 2.*

LP 1. Abbreviation of *low power.* 2. Abbreviation of *long-playing.* 3. Abbreviation of *low pressure.*

L-pad An attenuator consisting of one series (input) arm and one shunt (output) arm, arranged in the form of an inverted letter *L*.

LPB Abbreviation of *lighted pushbutton*.

LP band A section of the L-band extending from 390 to 465 MHz.

lpm Abbreviation of *lines per minute*: the output speed of a line printer.

lpW Abbreviation of *lumens per watt*; lm/W is preferred.

L + R, L − R The sum and difference signals of stereo channels. The L + R signal is the in-phase combination of signals from both channels; the L − R signal is the out-of-phase sum. Further, L + R can be added to L − R, giving 2L, the left signal; similarly, L − R minus L + R gives − 2R, the other channel's signal.

L-regulator See *L-type voltage regulator*.

LRR Abbreviation of *long-range radar*.

LR time constant See *inductance – resistance time constant*.

LSA diode Abbreviation for limited-space-charge-accumulation diode. A solid-state diode that acts as a microwave oscillator.

LSB 1. Abbreviation of *lower sideband*. 2. Abbreviation of *least significant bit*.

LS band A section of the L-band extending from 900 to 950 MHz.

LSC Abbreviation of *least significant character*.

LSD Abbreviation of *least significant digit*.

L-section 1. A filter section generally having the shape of an inverted letter *L*. 2. An attenuator circuit having the shape of an inverted letter *L*. 3. A right-angle bend in coaxial cable (see *ell*). 4. A network section consisting of a series (input) impedance arm and a shunt (output) impedance arm. See *L-pad*.

LSI See *large-scale integration*.

LSSC Abbreviation of *lower sideband suppressed carrier*.

LT band A section of the *L-band* extending from 780 to 900 MHz.

LTROM Abbreviation of *linear-transformer read-only memory*.

L-type antenna See *inverted-L antenna*.

L-type voltage regulator A simple voltage regulator containing a series current-limiting resistor and shunt regulator (zener diode, VR tube, voltage-dependent resistor, or the like). The circuit resembles an inverted letter *L*.

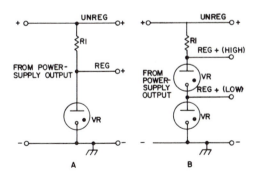

L-type voltage regulator.

lt-yr Abbreviation of *light-year*.

Lu Symbol for *lutetium*.

Lucalox General Electric's translucent ceramic; its chief constituent is polycrystalline alumina. The material has many applications in electrooptics.

Lucite See *methyl methacrylate resin*.

LUF Abbreviation of *lowest usable frequency*.

lug 1. A contact attached to the end of a wire lead to facilitate connection to a binding post. 2. A contact attached to a terminal strip, to which wire leads are soldered.

lum Abbreviation of *lumen*. The preferred (SI) form is *lm*.

lumen Abbreviation, *lm*, and sometimes *l* or *lum*. The SI unit of luminous flux; it is equal to the light that is emitted in 1 steradian (unit solid angle) by a uniform point source of a candela. Also see *candle power, illuminance, luminous intensity, solid angle, steradian*.

lumen-hour Abbreviation, *lm-hr*. The amount of light that 1 lumen delivers in one hour. Also see *lumen*.

luminaire A complete and self-contained lighting system, for television-studio use or photographic use. The kit includes all of the needed parts and accessories.

luminance The amount of light either emitted or scattered by a surface. This property is expressed in candelas per square meter (cd/m^2).

luminance channel In a color TV circuit, the channel that processes the luminance signal. See *Y-signal*.

luminance signal See *Y-signal*.

luminescence The production of cold light by a material stimulated by radiation or electron bombardment. See *electroluminescent cell* and *luminescent screen*.

luminescent cell See *electroluminescent cell*.

luminescent screen A screen—such as that of an oscilloscope tube, fluoroscope, or TV picture tube—coated with a material that glows under the influence of radiation, X rays, or electron beams.

luminiferous ether See *ether, 1*.

luminosity The luminous efficiency of radiant energy as given by the ratio of luminous flux to radiant flux (lm/W) for a specific wavelength.

luminosity factor Abbreviated *K*, and expressed in lumens per watt. The luminous intensity divided by the actual radiant intensity at a given wavelength of visible light.

luminous energy The energy in visible electromagnetic radiation.

luminous flux The rate of transfer or flow of luminous energy.

luminous intensity Luminous flux through a unit solid angle; expressed in candelas. Also see *candela*.

lumistor An amplifier or coupling device in which the input signal varies the brilliance of a lamp, electroluminescent cell, or light-emitting diode, and a photocell (or other light-sensitive device) picks up the fluctuating light and uses it to modulate an output current. In a *compact lumistor*, the light-emitting and light-sensing components are separate layers in a wafer or block of material. Compare *light amplifier*.

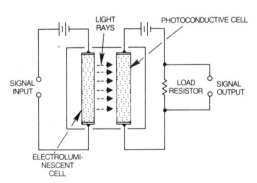

Lumistor.

lumped Pertaining to a property that is concentrated at or around a single point, rather than being distributed through a circuit; e.g., *lumped capacitance, lumped inductance.*

lumped capacitor See *discrete capacitor.*

lumped component A discrete component, i.e., one which is self-contained. Compare *distributed component.*

lumped constant The total value of any single electrical property distributed in a circuit or coil.

lumped-constant delay line A delay line having discrete capacitance and inductance components. Compare *distributed-constant delay line.*

lumped impedance A reactance and/or resistance manifested in a definite location. Examples are ordinary components, such as capacitors, inductors, and resistors.

lumped inductor See *discrete inductor.*

lumped resistor See *discrete resistor.*

lutetium Symbol, Lu. A metallic element of the rare-earth group. Atomic number, 71. Atomic weight, 174.99.

lux The unit of illuminance; lumens per square meter.

Luxemberg effect The generation of interferential signals produced by cross-modulation of two or more signals crossing the same region of the ionosphere.

luxmeter A device for measuring illumination (a photometer).

LV Abbreviation of *low voltage.*

LVDT Abbreviation of *linear variable differential transformer.*

Lw Symbol for *lawrencium.*

lx Abbreviation of *lux.*

LX band A section of the L-band extending from 950 MHz to 1.15 GHz.

LY band A section of the L-band extending from 725 to 780 MHz.

LZ band A section of the L-band extending from 1.450 to 1.550 GHz.

LZT Abbreviation of *lead zirconate-titanate,* a ceramic of use in electonics.

M

M 1. Abbreviation of prefix *mega*. 2. Symbol for *mutual inductance*. 3. Symbol for *refractive modulus*. 4. Roman numeral for 10^3.

m 1. Abbreviation of prefix *milli*. 2. Symbol for *mass*. 3. Abbreviation of *meter*. 4. Abbreviation of *mile*. (Also, *mi*.) Symbol for *modulation coefficient*.

m² Square meter. The SI unit of area.

m³ Cubic meter. The SI unit of volume.

MA 1. Abbreviation of *magnetic amplifier*. (Also, *magamp*.) 2. Abbreviation of *megampere*.

mA Abbreviation of *milliampere(s)*.

Mache unit A unit of radioactivity equivalent to 13.47 disintegrations per second (3.64×10^{-10} curie) per liter and representing the concentration of radon gas per liter (when all radiation is absorbed) that will result in a saturation current of 10^{-3} esu. (Not to be confused with *Mach number*.)

machine address See *absolute address*.

machine code See *machine language*.

machine cycle In a machine whose operation is periodic; especially one in a data processing system, a complete sequence constituting a period of operation.

machine error In a computer or data processing system, an error attributable to a hardware failure rather than to a software fault.

machine instruction A computer program instruction written in machine language.

machine language Computer program instructions and data represented in binary form. In the hierarchy of programming languages it is the lowest and the only one a computer understands. All higher level languages are translated to machine language either by an assembler, compiler, intepreter, or monitor system.

machine learning Pertains to a computer's ability to learn through repeated calculations for particular problems.

machine logic 1. The way a computer's functional parts are interrelated. 2. The facility whereby a computer solves problems.

machine operation The performance by a computer of a built-in function, e.g., subtraction.

machine operator A person participating in implementing and overseeing the processing of computer programs.

machine word In computer practice, the address of a memory location composed of the full number of bits normally handled by each register of the machine.

Mach number For a medium such as air, the ratio of the speed of a body in motion to the speed of sound in the medium. (Not to be confused with *Mache unit*.)

Maclaurin series A converging mathematical series of the Taylor type, which is useful in deriving some approximate formulas in electronics:

$$f(x) = f(o) + 1/f'(o)x + 2!/f''(o) \ x^2 + \ldots + n^1/f^n \ (o) \ x^n + \ldots$$

macro 1. Prefix denoting *extremely large*. Compare *micro*. 2. Abbreviation of *macroinstruction*.

macro assembly program An assembly program whose source statements are translated to several machine language instructions.

macrocyte An unusually large red blood cell.

macroinstruction A source program instruction which becomes several machine language instructions when operated on by a compiler.

macroprogram A computer program comprising macroinstructions.

macroscopic 1. Exceedingly large. 2. Large enough to be observed with the unaided eye.

macrosonics The theroy and applications of high-amplitude sound waves.

M

madistor A component that produces changes in current by means of magnetic-field effects. Used as an oscillator or amplifier.

MADT Abbreviation of *microalloy diffused transistor.*

MAG Abbreviation of *maximum available gain.*

magamp Acronym for *magnetic amplifier.*

magazine 1. Tape or film cartridge. 2. A hopper for holding coded magnetic or punched cards ready for input to some device.

magenta A purplish red.

magic-eye tube A tube in which electrons cause a small fluorescent screen to glow. A voltage applied to the control electrode varies the width of the glowing area in proportion to the voltage. This action permits use of the tube as a tuning indicator in radio receivers, null indicator in bridges, or as a makeshift vacuum-tube voltmeter.

magic-tee A four-branch waveguide assembly which, when its branches are correctly terminated, will effect the transfer of energy from one branch to two other branches (usually equally).

magma Molten rock within the earth.

magnal crt base The 11-pin base of a cathode-ray tube. Compare *bidecal, diheptal,* and *duodecal.*

magnesium Symbol, Mg. A metallic element. Atomic number, 12. Atomic weight, 24.32.

magnesium-copper-sulfide rectifier See *copper-sulfide rectifier.*

magnesium fluoride phosphor Formula, MgF_2. A substance used as a phosphor coating on the screen of a very-long-persistence cathode-ray tube. The fluorescence and phosphorescence are orange.

magnesium silicate phosphor Formula, $(MgO + SiO_2)$: Mn. A substance used as a phosphor coating on the screen of a cathode-ray tube. The fluorescence is orange-red.

magnesium tungstate phosphor Formula, $MgO + WO_3$. A substance used as a phosphor coating on the screen of a cathode-ray tube. The fluorescence is very light blue.

magnet A device or body of material which has the ability to attract to itself pieces of iron and other magnetic metals, and the ability to attract or repel other magnets. Also see *electromagnet, permanent magnet,* and *temporary magnet.*

magnet armature See *keeper.*

magnet battery A group of several magnets placed together in parallel (i.e., with like poles touching or resting nearby) to act as a single magnet.

magnet charger A device that produces an intense magnetic field for restoring weakened magnets or for making new magnets.

magnetic 1. Pertaining to magnetism; possessing magnetism; capable of being magnetized. 2. Magnetic material.

magnetic air-gap An empty *magnetic gap.*

magnetic amplifier An iron-core device which uses the principle of the *saturable reactor* to obtain amplification. In its simplest form, it consists of input and output coils wound on a core of square-loop magnetic metal (see *box-shaped curve*). The input coil consists of two identical windings connected in series-opposition so that currents in the output winding cannot induce voltage in the input winding. The output coil is connected in series with a load and an ac supply. A small ac signal applied to the input winding causes a large change in the impedance of the output winding and, therefore, a large change in the voltage across the load.

magnetic analysis See *mass spectrometer.*

magnetic attraction The force that causes a magnetic pole to draw to itself an unlike magnetic pole. Thus, a north pole attracts a south pole, and a south pole, a north pole. Compare *magnetic repulsion.*

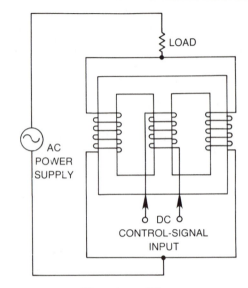

Magnetic amplifier.

magnetic axis A straight line joining the poles of a magnet.

magnetic bearing The azimuth, or compass direction, measured with respect to the north magnetic pole. Usually, the magnetic bearing is expressed in degrees, and can be read directly from a compass.

magnetic azimuth An azimuth based upon earth's magnetic north pole.

magnetic bias A steady magnetic force applied to another magnetic field to set the latter's quiescent point, e.g., sensitizing a relay by using a permanent magnet to lower the relay draw-in point.

magnetic blowout 1. The extinction of an electric arc by a strong magnetic field. 2. The apparatus for accomplishing the action described in 1, above.

magnetic bottle A container which is envisioned for atomic fusion reactions and which would consist of a magnetic field. Conventional containers cannot withstand the extremely high temperatures involved in atomic fusion.

magnetic braking See *electromagnetic braking.*

magnetic bridge An instrument comparable to the Wheatstone bridge used to measure magnetic permeability.

magnetic capacity The maximum magnetization a given material can receive.

magnetic card A computer storage medium that is a card that can be selectively magnetized or imprinted with magnetic ink to represent data.

magnetic cartridge A variable-reluctance phonograph pickup.

magnetic cell A unit consisting of one or more magnetic cores in a magnetic core memory.

magnetic centering Centering the beam in a TV picture tube by means of an electromagnetic focusing coil, a permanent magnet, or both.

magnetic character A letter, numeral, or other symbol written or printed in (visible) magnetic ink for its automatic sensing or reading in computing and signaling operations.

magnetic clutch A clutch in which the magnetism of one rotating member causes a second member to lock-in and rotate; there may be no direct physical contact between the two.

magnetic coil The winding in an electromagnet or similar device.

magnetic compass A mariner's compass employing a horizontally suspended magnetic needle as the indicator. Compare *gyrocompass*.

magnetic component See *electromagnetic component*.

magnetic conductivity See *permeability*.

magnetic controller A controller that employs electromagnets for some of its functions.

magnetic core 1. The iron core of an electromagnet, choke, transformer, relay, or similar device. 2. A small ring of square-loop magnetic material employed as a memory element in digital computers and calculators. Compare *magnetic cell*.

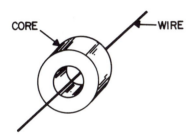

CORE   WIRE

Magnetic core.

magnetic-core storage In computer systems, a memory (storage) unit employing magnetic cores as the storage medium.

magnetic coupling See *inductive coupling*.

magnetic course In navigation, a course referenced to magnetic north rather than geographic north.

magnetic crack detector See *electromagnetic crack detector*.

magnetic creeping A gradual increase in the magnetization of a material under the influence of a steady magnetizing force.

magnetic cycle 1. For a material in an alternating magnetic field, the change in magnetic flux as a function of time. 2. The change in the magnetic-field polarity of the earth. This polarity reverses periodically.

magnetic damping The production of a damping effect or drag in a machine or meter by means of magnetic action on a moving member in accordance with Lenz' law.

magnetic declination See *declination*.

magnetic deflection See *electromagnetic deflection*.

magnetic density The concentration of magnetic flux in a region, expressed as the number of lines per unit area of cross section.

magnetic dip At a particular location on the earth's surface, the angle between the terrestrial magnetic field and a horizontal line.

magnetic dipole 1. A molecule or particle with a north and south magnetic pole. 2. Any pair of adjacent north and south magnetic poles.

magnetic direction finder Abbreviation, *MDF*. A type of compass operated by an electric signal delivered by a gyrostabilized magnetic-compass movement.

magnetic disk A rotating disk coated with a layer of magnetic material for the recording of data bits in a digital-computer system.

magnetic-disk memory In computer practice, a memory unit employing one or more magnetic disks. Also see *disk memory*.

magnetic doublet See *doublet, 2*.

magnetic drive A device in which mechanical movement is conveyed from one moving part to another by means of a magnetic clutch.

magnetic drum See *drum*.

magnetic-drum memory See *drum memory*.

magnetic effect of electric current The presence of a magnetic field around a conductor carrying electric current.

magnetic equator Passing around the earth near the conventional equator, an imaginary line along which a magnetic needle shows no dip.

magnetic feedback Feedback by means of inductive coupling between the output and input circuits of a system.

magnetic field The space around a magnetic pole or magnetized body in which magnetic energy acts; it is thought to contain flux (see *flux* and *line of force, 2*).

magnetic field intensity See *magnetic intensity*.

magnetic field strength See *magnetic intensity*.

magnetic-field viewer A device for visually examining a magnetic field. It consists of a clear plastic watchcase" filled with iron-oxide particles in liquid suspension. When it is placed within a magnetic field, the particles align themselves in the direction of the lines of force.

magnetic-film memory A magnetic memory in which memory cells consist of a thin film (thick film in some instances) of a magnetic material deposited on a substrate. Information is written into and read out of the cell through coils. Also called *thin-film memory*.

magnetic flip-flop A bistable multivibrator employing magnetic amplifiers or square-loop cores in place of tubes or transistors.

magnetic flux The lines of force (see *flux*) which exist about a magnetized body and collectively constitute a magnetic field.

magnetic flux density See *flux density*.

magnetic flux linkage The passage of magnetic lines of force through separate materials or circuits, thereby coupling them magnetically.

magnetic focusing See *electromagnetic focusing*.

magnetic force The force exerted by a magnet on a body of magnetic material (or on another magnet) within its field.

magnetic friction 1. Magnetic hysteresis (see *hysteresis, 1*). 2. The resistance experienced by a magnetic material moving in a magnetic field.

magnetic gap A space separating the materials in a magnetic circuit. This break is either an air space or one filled with a comparatively thin piece of nonmagnetic material, e.g., the gap in a choke-coil core.

magnetic head See *magnetic pickup head* and *magnetic recording head*.

magnetic heading The direction of travel according to a magnetic compass.

magnetic hysteresis See *hysteresis, 1*.

magnetic inclination See *magnetic dip*.

magnetic induction 1. The magnetization of a magnetic material such as iron or steel, when it is placed in a magnetic field. 2. The induction of an alternating voltage in a conductor by a nearby alternating magnetic field. Also see *electromagnetic induction*.

magnetic ink Writing or printing ink that is a suspension of finely divided particles of magnetic material. For applications, see *magnetic character*.

magnetic instability 1. The tendency of a magnetic recording tape to deteriorate with time. 2. Any fluctuation in the intensity of a magnetic field.

magnetic intensity The free-space strength of a magnetic field at a point of interest. Specifically, the force (in dynes) that the magnetic field would exert on a unit magnetic pole placed at that point.

magnetic iron oxide Magnetite. (Also, *ferrosoferric oxide*.)

magnetic leakage The undesired passage of magnetic flux out of a magnetic body such as filter choke, into surrounding space.

magnetic lens See *electromagnetic lens.*

magnetic line of force See *line of force, 2.*

magnetic load An electromagnetic device operating on the output of an electrical source. Such devices include actuators; alarms; electromagnets; core, disk, drum, and tape memories; relays; loudspeakers.

magnetic loudspeaker See *magnetic speaker, 1, 2.*

magnetic material 1. A material such as magnetite that exhibits natural magnetism. 2. A material such as iron or steel that is capable of being magnetized.

magnetic memory 1. In computer and data-processing systems, a device which stores data in the form of discrete magnetizations. Example: core, disk, drum, or tape memory. 2. Magnetic retentivity. Also see *retentivity.*

magnetic memory plane A two-dimensional array of cores and windings used in the memory section of a digital computer or a data-storage unit.

magnetic meridian The circle of the celestial sphere that passes through the zenith and earth's magnetic poles.

magnetic mine A naval mine detonated by a magnetic switch that is closed by the proximity of the steel hull of a ship.

magnetic modulator A core-type device somewhat similar to a magnetic amplifier employed for amplitude modulation. Modulating current passes through the control winding, and the carrier current, through the output winding.

magnetic moment Symbol, μ. Unit, joules per tesla. For a magnet, the product of pole strength and the distance between poles.

magnetic needle 1. The pivoted magnetic pointer in a magnetic compass. 2. A slender rod of magnetic material employed as a magnetic-memory element (see *magnetic memory*).

magnetic north See *north magnetic pole.*

magnetic oxide Iron oxide employed as the sensitive coating of magnetic recording tape.

magnetic pickup 1. A phonograph pickup of the variable-reluctance type (see *variable-reluctance pickup*). 2. Magnetic transducer, such as a phono cartridge, tape recording head, or similar input element.

magnetic-pickup head In a tape recorder, the head (transducer) that picks up magnetic impulses from the passing tape and converts them into alternating currents which, after amplification, reproduce the original sound. Synonymous with *magnetic recording head.*

magnetic-plate wire Wire in which a magnetic metal has been plated on top of a nonmagnetic metal.

magnetic poles 1. The extremities of a magnet at which the magnetism is concentrated. See, for example, *north magnetic pole, north-seeking pole, south magnetic pole,* and *south-seeking pole.* 2. Earth's magnetic poles. See, for example, *north magnetic pole* and *south magnetic pole.*

magnetic pressure See *magnetomotive force.*

magnetic printing 1. In a recording material such as magnetic tape, the transfer of information from one part of the material to another part (or from one medium to another) by the magnetic field of the recorded material. This phenomenon, which is also called *print-through,* sometimes takes place in recording tape on a reel. 2. Conventional lithography, letterpress, or other reproduction process in which magnetic ink is employed.

magnetic print-through See *magnetic printing.*

magnetic probe A loop or coil inserted in an electromagnetic field to sample the magnetic component. See, for example, *waveguide probe.* Compare *electric probe.*

magnetic recording Recording sounds or data by means of corresponding changes in the magnetization of a medium such as a magnetic drum, disk, tape, or wire.

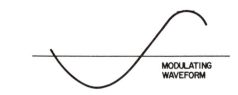

MODULATING WAVEFORM

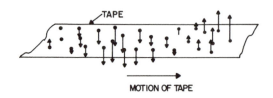

TAPE

MOTION OF TAPE

Magnetic recording.

magnetic recording head In a tape recorder, the head (transducer) that receives current impulses (analogs of the original sound vibrations) from an amplifier and converts them into magnetic impulses which magnetize spaces on the passing tape. Synonymous with *magnetic pickup head.* Also called *record head.*

magnetic recording medium 1. A magnetic cylinder, disk, drum, tape, or wire employed in the recording of sound or data. 2. The sensitive material with which any of these is coated.

magnetic relay A relay having a permanent magnet in whose field a coil, bar, or reed moves to open or close a pair of contacts.

magnetic remanence See *residual magnetism.*

magnetic repulsion The force that causes a magnetic pole to push away a like magnetic pole, although they are not in mutual contact. Thus, two north poles repel each other, and so do two south poles. Compare *magnetic attraction.*

magnetic-resonance accelerator See *cyclotron.*

magnetic rod See *magnetic needle, 2.*

magnetics 1. Collectively, magnetic components and equipment. 2. Collectively, magnetic materials. 3. The science of magnetism (see *magnetism, 2*).

magnetic saturation The condition in which a magnetic material passes all of the magnetic lines of force its permeability will permit it to accommodate; i.e., increasing the intensity of the magnetizing force will produce little or no increase in magnetization.

magnetic scan See *electromagnetic deflection.*

magnetic screen See *electromagnetic shield.*

magnetic shield See *electromagnetic shield* and *magnetic shielding.*

magnetic shielding 1. Enclosing a magnetic field to confine its flux, thus preventing interaction with outside bodies. 2. Devices such as boxes, cans, or shells of iron, steel, or a magnetic alloy for the purpose described in 1, above. Also see *electromagnetic shield.*

magnetic shift register A shift register employing magnetic flip-flops.

magnetic south See *magnetic south pole*.

magnetic speaker 1. A loudspeaker that is essentially an enlarged earphone with a horn that conveys and intensifies the sound from the vibrating diaphragm. 2. A loudspeaker in which the vibration of a diaphragm or reed in the field of a permanent magnet is conveyed by a pin to a paper or composition cone. Compare *dynamic speaker*.

magnetic storage 1. A data bank or memory that stores information in the form of magnetic fields. 2. The data on a magnetic tape or disk.

magnetic storm A disturbance in the earth's magnetism, usually correlated with sunspots and causing interference to communications. Also, the mechanism of sunspots themselves.

magnetic susceptibility See *susceptibility*.

magnetic switch A reed switch (see *reed switch, 2*) operated by bringing a magnet close by.

magnetic tape Plastic tape coated with a film of magnetic material; it can be magnetized along its length to record sounds, video signals, and computer information.

magnetic-tape core A strip of magnetic metal wound in a "donut" of the number of layers required for a desired thickness. Such construction is seen in core-memory elements and sometimes in the toroidal core of a choke or transformer. See also *toroid*.

magnetic tape deck See *tape deck*.

magnetic tape drive See *tape transport*.

magnetic tape group In a computer installation, several rack-mounted tape decks working independently or sharing interface channels.

magnetic tape head See *magnetic pickup head* and *magnetic recording head*.

magnetic tape library In a computer installation, (1) the place where magnetic tape files are kept, or (2) magnetic tape files and the records needed to utilize them.

magnetic tape parity As a safeguard against losing information bits during the transfer of information between magnetic tape and a memory device, a technique in which an extra bit is generated and added to characters under certain conditions, to make the output uniform temporarily. Lack of uniformity in output then serves as an error indicator. The original quantity is recovered by dropping the extra bit following a parity check.

magnetic tape reader A tape deck for playing back data on magnetic tape.

magnetic tape recorder A recorder-reproducer employing magnetic tape.

magnetic test coil See *search coil*.

magnetic thick film A thick film (thicker than a micrometer) of magnetic material deposited on a substrate. See, for example, *magnetic-film memory*. Compare *magnetic thin film*.

magnetic thin film A thin film (under a micrometer thick) of magnetic material deposited on a substrate. See, for example, *magnetic-film memory*. Compare *magnetic thick film*.

magnetic transducer A transducer that employs a coil, magnet, or both, to convert displacement into electrical or magnetic charges. Common varieties are the inductance-type, transformer-type, and generator-type. Compare *capacitive transducer, crystal transducer,* and *inductive transducer*.

magnetic-vane meter See *iron-vane meter*.

magnetic vector In an electromagnetic field, the vector representing the magnetic component. It is perpendicular to the electric vector.

magnetic viscosity A property of certain materials described by the

time required to magnetize them to a specified level.

magnetic whirl One of the circular magnetic lines of force around a straight conductor carrying electric current.

magnetic wire The thin wire used in wire recording and playback. See *wire recorder*.

magnetism The property whereby a device or body of material attracts to itself bodies of iron or other magnetic materials or magnets.

magnetite Formula, Fe3O4. A natural magnetic oxide of iron. Also called *lodestone*.

magnetization curve A curve depicting the magnetization of a material vs the applied magnetizing force. See, for example, *hysteresis curve*.

magnetizer A device for magnetizing magnetic materials, as in the making of permanent magnets. Also see *magnet charger*. Compare *demagnetizer*.

magnetizing current 1. A current that sets up a magnetic field of useful intensity. 2. The half-cycle of an alternating current or the polarity of a direct current flowing through a coil wound on a permanent magnet (as in a headphone, permanent-magnet loudspeaker, or polarized relay) that increases magnetic field strength. Compare *demagnetizing current*. 3. The field current of a dynamo.

magnetizing force The magnetomotive force, in gilberts, divided by the spatial distance, in meters. The intensity of a magnetic field that causes a material to become magnetized.

magnet keeper See *keeper*.

magnet meter See *magnet tester*.

magnet motor A motor in which the field is provided by a permanent magnet. Also called *permanent-magnet motor*.

magneto See *permanent-magnet generator*.

magnetocardiogram Abbreviation, *MCG*. A record made by a magnetocardiograph of the pulsating magnetic field of the heart and used as a diagnostic aid.

magnetocardiograph An instrument which produces a record of the pulsating magnetic field generated around the torso by natural ion currents in the heart.

magnetoelectric generator See *magnetogenerator*.

magnetofluid mechanics See *magnetohydrodynamics*.

magnetofluidynamics See *magnetohydrodynamics*.

magnetogasdynamics See *magnetohydrodynamics*.

magnetoionic duct A propagation path for radio waves between two points having the same geomagnetic longitude on the surface of the earth. The radio waves tend to travel with the geomagnetic lines of flux at some frequencies under certain conditions.

magnetoionics The study of the effects of the geomagnetic field on the propagation of radio waves.

magnetogenerator See *permanent-magnet generator*.

magnetograph An instrument for automatically recording a magnetic field.

magnetohydrodynamic generator A device employing magnetohydrodynamic principles to generate electric power directly from gases. In the generator, a hot gas is passed through an intense magnetic field; a pair of collector plates picks up electrons from the ionized gas.

magnetohydrodynamic gyroscope A gyroscope whose spin is obtained by a rotating magnetic field circulating a conducting fluid such as mercury, around a closed loop. Also see *magnetohydrodynamics*.

magnetohydrodynamic power generator See *magnetohydrodynamic generator*.

magnetohydrodynamics Abbreviation, *MHD*. The theory and application of phenomena produced by electrically conductive fluids and gases in electric and magnetic fields.

magnetometer An instrument for measuring the strength and direction of magnetic fields.

magnetomotive force Abbreviation *mmf*. Unit, ampere. The phenomenon that is sometimes descriptively called *magnetic pressure*. It is analogous to electromotive force (and to water pressure) and is the agent that produces a magnetic field.

magneton See *Bohr magneton*.

magneto-optical rotation The tendency of a magnetic field to rotate the plane of polarization of light passing through a substance. Also see *Kerr magneto-optical effect*.

magneto-optical valve See *Kerr magneto-optical effect*.

magnetopause The limit of the magnetosphere.

magnetoplasmadynamics See *magnetohydrodynamics*.

magnetoresistance The phenomenon whereby the resistance of a material, such as a semiconductor, changes when it is exposed to a magnetic field. Also see *magnetoresistor*.

magnetoresistor A material such as bismuth wire, indium antimonide, or indium arsenide, whose resistance becomes proportional to the strength of a magnetic field in which it is placed.

magnetosphere In the upper atmosphere, a region extending thousands of kilometers from the earth, in which charged particles are trapped by virtue of the influence of earth's magnetic field.

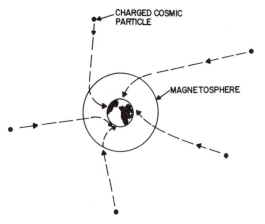

Magnetosphere.

magnetostatic field A stationary magnetic field such as that produced by a permanent magnet.

magnetostriction The expansion or contraction of a bar or rod of magnetic material (such as Invar, Monel metal, Nichrome, nickel, or Stoic metal) in proportion to the strength of an applied magnetic field. Magnetostrictive vibration in such a rod is comparable to piezoelectric vibration in a quartz crystal.

magnetostriction filter See *ultrasonic filter, 1*.

magnetostriction oscillator An oscillator whose frequency is controlled by a magnetostrictive rod (see *magnetostriction*). The dimensions of the rod and the type of metal it contains determine its vibration frequency and, accordingly, the operating frequency of the oscillator.

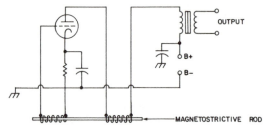

Magnetostriction oscillator.

magnetostrictive delay line A delay line in which the signal is propagated through a magnetostrictive rod. Also see *magnetostriction*.

magnetostrictive microphone A microphone in which sound vibrations produce changes in a magnetostrictive element, which in turn are converted into output-voltage changes. Also see *magnetostriction*.

magnetostrictive transducer A transducer in which some phenomenon such as vibration or pressure produces changes in a magnetostriction element, which in turn are converted into output-voltage changes. Also see *magnetostriction*.

magneto telephone A telephone with a crack type magneto, used mostly in the early days of wire communications.

magnet protector See *keeper*.

magnetrode Trademark of a radiofrequency device for externally producing hyperthermia (elevated temperature) inside the body, as in heating a cancerous tumor for therapeutic purposes.

magnetron A microwave tube consisting of a diode (with a cylindrical anode) through which the field of a powerful external permanent magnet passes. The magnetic field makes electrons leaving the cathode travel in a spiral path between the electrodes. This action gives the tube a negative-resistance characteristic, resulting in oscillation when the tube is connected in an appropriate circuit. Some magnetrons have a built-in resonant cavity.

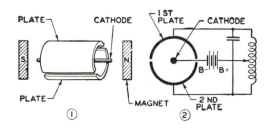

Magnetron.

magnetron beam-switching tube See *beam-switching tube*.

magnet steel A high-retentivity alloy of chromium, cobalt, manganese, steel, and tungsten, employed in the manufacture of permanent magnets.

magnet tester An instrument employed to measure the flux of a magnet. Also see *fluxmeter*.

magnet wire Insulated wire (usually solid copper) of 14 to 40 gauge, called *magnet wire* because of its original major use in winding the coils of electromagnets.

magnitude 1. Degree; extent. 2. Absolute value.

mailbox In a computer, a group of storage locations, used for data having specific addresses.

mailer 1. A container used for sending magnetic tapes or disks via the postal service. 2. A small magnetic cassette that can be easily mailed.

main anode In a vacuum tube, the plate, or electrode to which the most electrons are attracted.

main bang 1. In a radar display, the pip or pulse resulting from the actual transmitted signal. This pulse is blanked out. 2. In a spectrum analyzer, the pip corresponding to a frequency of zero, and caused by the local oscillator.

main British variation of *domestic ac supply.*

mainframe 1. The computer itself, i.e., the chassis containing the central processor and arithmetic and logic circuits. 2. Central processing unit.

main path In a computer program, the sequence of instruction execution disregarding the execution of subroutines.

main program The part of a computer program that is other than a subroutine.

main routine See *main program.*

mains In a power-distribution center, the lines that supply the entire system. An example is the set of mains leading into a house.

main storage The principal (immediate process) storage or memory unit in a digital computer or dataprocessing system. Also called *main memory.*

maintenance The process of keeping a system, circuit, or component in operating condition, with minimal down time.

maintenance routine A computer program used by computer service personnel for diagnosis during a regular service interval.

major beats The principal beats produced in a beat-note system; they are usually the sum and/or difference of two fundamental frequencies. Compare *minor beats.*

major face In a hexagonal quartz crystal, one of the three larger faces. Compare *minor face.*

majority carrier The predominant current carrier in processed semiconductor material. Electrons are the majority carriers in *n*-type material, *holes* in *p*-type material. Compare *minority carrier.*

majority logic A logic gate in which the output is high whenever the majority of its inputs is high, regardless of which of the inputs are high. Thus, in a 5-input gate of this type, the output is high when any three or more of the inputs are high.

major lobe The principal lobe in an antenna directivity pattern. Also see *lobe.* Compare *minor lobe.*

major loop The principal path for the circulation of information or control signals in an electronic system, e.g., major feedback loop. Compare *minor loop.*

make 1. The closing of contacts. 2. To close contacts.

make-before-break Abbreviation, MBB. Pertaining to a switch or relay in which the movable arm closes with the next contact before breaking with the previous one. Compare *break-before-make contacts.*

make time The time required for a relay to latch completely or a switch to close completely. Compare *break time.*

male plug A plug having one or more protruding contacts in the form of pins, blades, or prongs. Compare *female plug* and *hermaphrodite plug.*

manganese Symbol, Mn. A metallic element. Atomic number, 25. Atomic weight, 54.95.

manganese-dioxide depolarizer In a dry cell, manganese dioxide

Male plug.

mixed with powdered carbon, the mixture being a depolarizing agent. Also see *depolarizer.*

manganin A low-temperature-coefficient alloy used in making wire for precision resistors. Typical composition is copper, 84%; manganese, 12%; and nickel.

man-made interference Electrical interference generated by circuits, devices, and machines, as opposed to *natural interference.*

man-made lightning Extremely high-voltage electrical discharges in air, simulating lightning flashes.

man-made static See *man-made interference.*

manometer An instrument for measuring gas or vapor pressure, especially at low levels.

manpack A portable radio transceiver that can be used while walking.

mantissa 1. The decimal portion of a logarithm. Thus, in 3.952 502 (log 8964), the mantissa is 0.952 502. 2. The fixed point part of a number in scientific notation; thus, in 4×10^3, 4 is the mantissa.

manual 1. Actuated or operated directly by mechanical means rather than automatically. 2. The book or booklet that describes the operation and maintenance of an electronic device.

manual input Use of a keyboard peripheral to enter data into a computer program or system.

manual operation In data processing, an operation in which automatic machines are not involved.

manual telegraphy Telegraphy comprising signals transmitted by a hand-operated key and recorded by hand (pen, pencil, or typewriter).

manual telephone A dialless telephone.

manual tuning Tuning performed entirely by adjusting variable circuit components by hand.

manual word generator A device by which an operator can originate information words for input into computer memory.

MAR Abbreviation of *memory-address register.*

Marconi antenna A quarter-wave antenna grounded at the input end.

Marconi effect The undesired tendency of an entire receiving antenna system, including lead-in or feeders, to act as a Marconi antenna with respect to ground.

Marconi system A name formerly applied to wireless telegraphy and, by extension, to radio communication in general.

margin 1. A gap or space between two objects, such as adjacent plates of a capacitor. 2. Clearance. 3. The maximum amount of error that can be tolerated without risk of improper or abnormal operation. 4. In a teletypewriter, the range of adjustments in which the error frequency is acceptable.

marginal relay A relay having a small difference between its on and off currents or voltages.

marginal test As performed on equipment in a computer installation, a test (1) to determine the cause of an intermittent malfunction or (2) to verify an equipment's operating tolerances.

margin-punched card A punched card with perforations in a relatively narrow column; the unused center portion is reserved for written information.

marine broadcast station A coastal station which broadcasts information of interest to shipping: time, weather, ocean currents, etc.

marine radio Radio communications between seagoing vessels or between vessels and shore stations.

marine radiobeacon station A landbased radionavigation station whose transmitted signals are used for taking bearings.

mariner's compass See *magnetic compass.*

mark 1. In telegraphy, the dot or dash portion of a character as opposed to the dead space between such portions. 2. The intelligence part of a similar signal such as sound, light, and the like. 3. The *on* or *one* state represented by a binary bit as opposed to the *off* or *zero* state. 4. A character identifying the end of a data set. Also called *marker.*

marker A pip indicating a particular frequency on a response curve displayed on an oscilloscope screen.

marker beacons Individual coded-signal transmitters placed along a radio range and indicating features of the course marked by them.

marker frequency 1. A known frequency which may be used to identify a spot-frequency harmonic of a frequency-standard signal. 2. A known accurate signal employed to identify the limit of a radio band. 3. The frequency at some point on a response curve as identified by a marker pip (see *marker*).

marker generator An oscillator that supplies a marker pip (see *marker*).

mark hold In telegraphy, an unmodulated signal meaning information is not being sent.

mark reading The reading by an optical scanning device of marks made in specific areas of a document; the process also includes the marks' conversion to digital signals for input to a computer.

mark scanning See *mark reading.*

mark sensing A process similar to *mark reading* except that the marks, say, penciled in blocks on a form, are sensed electrically.

marker trap A wavetrap that supplies a dip-type marker pip when used in conjunction with an rf test oscillator (see *marker*).

market scanner A device which scans a black-bar type of binary label printed on a carton or other package (or magazine) and indicates the corresponding price of the merchandise on the readout of the store's checkout register.

mark-to-space ratio In radiotelegraph code transmission, the ratio of the duration (mark) of a dot to the interval (space) between successive dots.

Marx generator An impulse-type high-voltage dc generator circuit in which several capacitors are charged in parallel through a high-resistance network and, when capacitor voltage reaches a critical high value, discharge in series through spark gaps, producing a high-voltage pulse for each discharge.

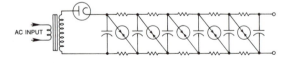

Marx generator.

Mary Phonetic alphabet code word for letter *M.*

maser A low-noise microwave amplifying device in which a microwave input signal causes high-energy-state molecules of ammonia or ruby to fall to the low-energy state and, as a result, to emit large amounts of energy as an output signal. The name is an acronym for *microwave amplification by stimulated emission of radiation.*

mask 1. A kind of stencil through which plating, electrodepositing, or diffusion can be carried on. 2. The viewing screen, or *graticule,* of an oscilloscope. 3. To provide with a mask. 4. To obliterate a signal with a stronger one. 5. A bit or character pattern used to change or extract bit positions in another pattern.

masking To use a mask (*mask, 5*).

Masonite Masonite Corporation's tough fiberboard used for panels and bases of some electronic equipment.

mass The quantity of matter in a body. Like weight, mass is expressed in kilograms in the metric (SI) system and in pounds in the English system. For a given piece of material, mass may be determined by dividing the weight by the acceleration due to gravity.

mass data Data in excess of the maximum amount that can be stored in the main (internal) storage unit of a digital computer, i.e., that which can only be accommodated by external media such as magnetic tape or disks.

mass-energy equation Energy, E, is the product of a given mass, m, and the square of the speed of light c^2, or $E = mc^2$. See also *Einstein equation.*

mass number Symbol, A. A number representing the total of neutrons and protons in the nucleus of an atom. The approximate mass of an atom is equal to $A \times mp$ where mp is the total proton (rest) mass. 2. The number indicating the sum of nuclear protons and neutrons in an atom. It is usually written as a superscript to the right of the symbol for the atom: thus, U^{238} is uranium having 238 nucleons. An *isotope* of an element will have a different mass number than that of the normal atom.

mass of electron at rest Symbol, me. The amount of matter in an electron; $me = 9.109\,558 \times 10^{31}$ kg.

mass of proton at rest Symbol, mp. The mass of the proton in the nucleus of an atom; $mp = 1.672\,614 \times 10^{-27}$ kg.

mass resistivity For wire, the resistance of a length 1 meter long having a mass of 1 gram, or of a mile-long piece having a mass of a pound.

mass spectrograph An instrument employed to analyze chemical compounds and mixtures in terms of their distinctive mass spectra, which is exhibited by ionized samples of the materials in a magnetic field.

mass spectrum An electron spectrum that can be used to identify a chemical element. Different elements have nuclei with different charge-to-mass ratios. This results in each element having a unique mass spectrum.

mass storage In a computer system, an online, usually magnetic, storage medium capable of holding large amounts of data for computer access.

mass unit See *atomic mass unit.*

master 1. Key station. 2. Master clock. 3. Master relay. 4. Master switch. 5. Master tape. 6. A mold from which phonograph records are made.

master card Usually the first or last card in a group of punched cards (or a card file) that contains information about the group.

master clock 1. In a digital computer, the primary generator of timing pulses. 2. A standard time clock that drives other (slave) clocks, or to which clocks of lesser accuracy may be referred.

master console In a computer system, an equipment with panel instruments and controls, which permits operations to be governed, monitored, and controlled by a human operator.

master control 1. The main control circuit in a system. 2. A point from which signals or programs are distributed in a communications or broadcast system.

master data In a computer record, data elements that remain unaltered for a long time.

master file A computer file of data used routinely and remaining unchanged for a long time.

master gain control The principal gain control in an amplifier system, i.e., the one used to control the gain of the entire system.

master instruction tape Magnetic tape on which is recorded related computer programs.

master library tape See *master program file*.

master oscillator Abbreviation, *MO*. The main oscillator in an electronic system, e.g., the oscillator stage in an oscillator—amplifier type of radio transmitter. This oscillator may be either self-excited or crystal-controlled.

master oscillator—power amplifier Abbreviation, *MOPA*. A type of transmitter or signal generator in which a frequency-determining oscillator drives a power amplifier which in turn delivers an output signal. Because the oscillator is isolated from the output load, this arrangement insures greater stability than one in which the oscillator alone supplies power to the load.

master pattern The etching pattern used for manufacture of a printed-circuit board.

master program file A reel of magnetic tape on which is recorded the programs regularly used in a data processing installation. Also called *master library tape*.

master record In a data processing system, the current record on magnetic tape that will be used for the next computer run.

master relay A relay that operates other (slave) relays. Compare *slave relay*.

master station See *key station*.

master switch A switch which can actuate or deactuate an entire installation or system.

master tape 1. In sound recording and reproduction, a magnetic tape which contains material from which other tapes and disks may be made. 2. In automation, a magnetic tape on which is recorded the basic signal sequence for controlling a process and other recorders. 3. In data processing, a magnetic tape which must not be erased.

masurium See *technetium*.

MAT Abbreviation of *microalloy transistor*.

match 1. To mate devices, signals, impedances, etc. for optimum compatibility in terms of signal transfer, equipment interfacing, and other optimizing qualities. 2. The condition of being compatibly mated, physically or electrically.

matched components Circuit components (capacitors, coils, diodes, resistors, transistors, tubes, etc.) that are carefully selected for similar or particularly compatible operating characteristics.

matched filter 1. A filter with input and output impedances matched to the input line and output load, respectively. 2. A filter designed for separating a signal with a particular waveform from other signals and noise.

matched impedance An impedance that has the same magnitude as that of another impedance with which it is operated. Maximum power is transferred between impedances that are matched.

matched load A purely resistive load, the impedance of which is the same as the characteristic impedance of the feed line. This results in optimum power transfer from the line to the load.

matched pair A pair of *matched components* offered in a single package. See also *duo*.

matched transmission line A transmission line which, because it is

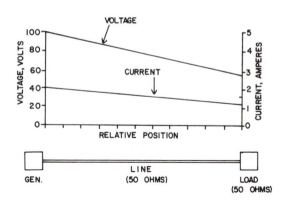

Matched transmission line.

correctly terminated, passes virtually all of its incident waves to an antenna or circuit without reflection, thus maintaining a standing-wave ratio of close to unity.

matching pad An *LC* network for matching the impedance of a load to the output impedance of a signal generator.

matching stub See *stub*.

matching transformer A transformer (af or rf) employed to match one impedance to another.

matchtone A transistorized single-frequency af oscillator used to make radiotelegraph signals audible. The keyed carrier wave from the transmitter is rectified by a small semiconductor diode whose dc output powers the oscillator during the mark portion of a dot or dash (see *mark, 1*).

mathematical analysis Through the use of algebraic and mathematical techniques, the study of operations performed on numbers and how numbers are related.

mathematical check A test of the validity of the result of an arithmetic process (by using alternate methods, for example).

mathematical logic 1. A branch of mathematics that involves the theoretical behavior of various systems of reasoning. 2. Boolean algebra. 3. Digital logic.

mathematical model See *model, 2*.

mathematical series A sequence of terms, each related to the preceding one in some way; thus $S = 1/9 = 0.1 + 0.01 + 0.001 + 0.0001 + \ldots$. Also see *convergent series, divergent series, exponential series, infinite series,* and *power series*.

mathematical subroutine Within a computer program, a subroutine serving as an arithmetic function., i.e., one for performing an operation that isn't integral to the monitor program.

matrix 1. A high-speed switching array employed in counters and computers (see *diode matrix*). 2. Generally, *two-dimensional array*. 3. Specifically, a device for solving linear simultaneous equations (by means of determinants), consisting of a rectangular array of coefficients. For *n* equations in *n* variables the matrix has *n* rows and *n* + *1* columns. For example, the two simultaneous equations $a11x + a12y = k1$ and $a21x + a22y = k2$ may be represented by the matrix:

| a11 | a12 | k1 |
| a21 | a22 | k2 |

matrix printer See *wire printer*.

matter The building material of the universe that occupies space and is perceivable by the senses. See, for illustration, *atomic theory* and *states of matter*.

matter waves See *de Broglie waves*.

Mateucci effect When the magnetization of a helically wound, ferromagnetic wire fluctuates, the tendency for a potential difference to occur.

max Abbreviation of *maximum*.

maxima Points along a curve at which a function reaches a maximum value, and on each side of which all other points are lower in value. For a function $y = f(x)$, a point is maximum if, when the first derivative dy/dx is set to zero, the second derivative d^2y/dx^2 is negative. Compare *minimal*.

maxima and minima 1. The study and solution of maximum, minimum, and inflection points on the curve of a function. See specifically *inflection point, maxima,* and *minima*. 2. In radar reflections, regions of maximum and minimum intensity.

maximal flatness For an amplifier or network, the condition in which peaks are not present in the normal pass-band response.

maximum Abbreviation, *max*. The highest member of a series of values. Also called *peak* (see *peak, 1*). Compare *minimum*.

maximum available gain Abbreviation, *MAG*. The amplification provided by a circuit or device whose input and output impedances are correctly matched to source and load.

maximum current 1. Abbreviation, *Im*. The highest value reached by an alternating-current half-cycle or by a pulse current. Also called *peak current*. 2. The highest value of current in a series of current values.

maximum power 1. Symbol, *Pm*. The highest value of power that an equipment may be called upon to supply. 2. Symbol, *Pm*. The highest value of power in a series of measurements or calculations. 3. Symbol, *Pp*. Peak power.

maximum power output See *maximum power, 1*.

maximum power transfer The condition in which the largest amount of power is delivered by a source to a load.

maximum power transfer theorem The proposition that maximum power is transferred from a generator to a load when the impedance of the load equals the internal impedance of the generator. Compare *compensation theorem, Norton's theorem, reciprocity theorem, superposition theorem,* and *Thevenin's theorem*.

maximum rating 1. The highest value of a quantity (e.g., current, voltage, or power) which may safely be used with a given device. 2. The highest value of a quantity afforded by a given device, e.g., maximum capacitance of a variable capacitor.

maximum record level 1. In a magnetic tape, magnetic disk, or phonograph disk, the highest amplitude of input signal that can be recorded with an acceptable amount of distortion. 2. The recording-head current or power that results in third-harmonic distortion of 3 percent.

maximum signal level 1. In an amplitude-modulated signal, the highest instantaneous amplitude. 2. In an amplitude-modulated facsimile or television system, the amplitude that results in a black or white picture (depending on whether the highest amplitude produces black or white).

maximum undistorted power output Abbreviation, *MUPO*. The highest power that a tube or transistor will deliver before significant distortion occurs.

maximum usable frequency Abbreviation, *MUF*. The highest frequency that may be used successfully at a given time for communication via the ionosphere.

maximum voltage 1. Abbreviation, *Em*. The highest value reached by an ac voltage half-cycle or by a voltage pulse. 2. The highest value of voltage in a series of voltage measurements or calculations.

maximum wattage See *maximum power*.

maxterm form In mathematical calculations, the factored form of a function, expressed as a product of sums. For example, the maxterm form of $f(x) = x^2 + 5x + 6$ is $f(x) = (x + 2)(x + 3)$.

maxwell Symbol, *Mx*. The cgs unit of magnetic flux; 1 maxwell = 1 line of force = 10^{-8} weber.

Maxwell bridge A four-arm ac bridge for measuring inductance against a standard capacitance.

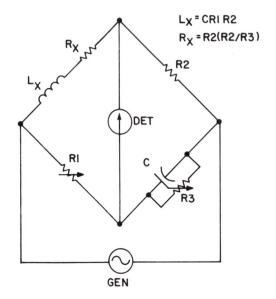

$$L_X = CRI\ R2$$
$$R_X = R2(R2/R3)$$

Maxwell bridge.

Maxwell's equation A set of four advanced equations developed by James Clerk Maxwell in 1864 and 1873, which describe vector quantities pertaining to points in space subjected to varying electric and magnetic forces. Through his classic presentation, Maxwell predicted the existence of electromagnetic waves whose later discovery made radio possible.

Maxwell's law See *Maxwell's rule*.

Maxwell's rule Every part of an electric circuit is acted upon by a force tending to move it in the direction that results in the maximum magnetic flux being enclosed.

maxwell-turn A unit of magnetic coupling (linkage) equal to 1 maxwell per turn of wire in a coil linked by magnetic flux. Also see *maxwell*.

mayday In radiotelephony, a word spoken as an international distress signal equivalent to SOS in radiotelegraphy. The word is the phonetic equivalent of the French *m'aidex* (help me).

MB Abbreviation of *midband*.

Mb Abbreviation of *megabar*.

MBB Abbreviation of *make-before-break*.

MBM Abbreviation of *magnetic bubble memory*.

MBO Abbreviation of *monostable blocking oscillator*.

MBS Abbreviation of *magnetron beam switching*.

mc 1. Obsolete abbreviation of *megacycle*. (Term has been superseded by *megahertz* [MHz]. 2. Symbol for *milicurie* (*mCi* is preferred). 3. Symbol for *meter-candle*.

Mc Symbol for *megacurie* (*MCi* is preferred).

MCG Abbreviation of *magnetocardiogram*.

McLeod gauge An instrument for measuring gas under low pressure. A measured volume of the gas under test is first compressed (to a lower known volume) to a pressure which is more easily measured by means of a mercury manometer and the application of Boyle's law.

MCM Abbreviation of *Monte Carlo method*.

McProud test A simple test for checking the tracking efficiency of a phonograph pickup and arm for microgroove disks: the pickup is required to track a 45-rpm disk on a standard turntable running at 45 rpm with 1 1/4-inch swing (maximum possible eccentricity).

MCS 1. Abbreviation of *master of computer science*. 2. Abbreviation of *misile control system*.

MCW Abbreviation of *modulated continuous wave*.

Md Symbol for *mendelevium*.

MDAS Abbreviation of *medical data acquisition system*.

m-derived filter A filter whose L and C values are derived by multiplying those of a *constant-k filter* by a factor m which lies between zero and $+1$ and is a function of the ratio foo/fc, where foo) is the frequency of infinite attenuation, and fc is the cutoff frequency. The m-derived filter exhibits sharper response than the equivalent constant-k filter.

MDI Abbreviation of *magnetic direction indicator*.

M-display See *M-scan*.

MDS Abbreviation of *minimum discernible signal*.

me Symbol for *electron rest mass*.

Meacham oscillator A highly stable rf oscillator consisting of an amplifier provided with a feedback circuit containing a four-arm bridge, one arm of which is a quartz crystal, and another, a tungsten-filament lamp acting as a nonlinear resistor. Also called *bridge-stabilized oscillator*.

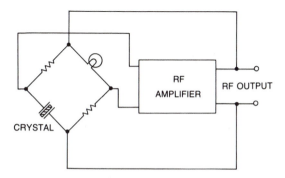

Meacham oscillator.

mean A simple average of two or more numbers (see *arithmetic mean*).

mean charge 1. In an object that is nonuniformly charged, the average charge per unit distance, area, or volume. 2. In a capacitor carrying a fluctuating current, the average amount of charge held by the plates.

mean free path In a gas tube, the average of all the free paths of electrons at a specified temperature. See *free path*.

mean life 1. Symbol, L. The average life of a radioactive substance, i.e., the time taken for $1/e$ (e = base of natural logarithms) of the substance to disintegrate. 2. The time required for excess carriers injected into a semiconductor to recombine with carriers of opposite sign. Also called *average life*.

mean proportional See *geometric mean*.

measured A quantity which is presented to an instrument for measurement.

measurand Any quantity that is measured with an instrument.

measurement 1. The process by which the magnitude of a parameter is found. 2. The value of a parameter, as obtained by such a process.

measurement error The difference between the measured value of a quantity and its true value. Also see *negative error of measurement* and *positive error of measurement*.

measurement range In a measuring device, the range within which the error is smaller than a specified value.

mechanical analogs Familiar mechanical devices, systems, or effects with which certain electrical counterparts may be compared for ease in teaching or understanding, e.g., inductance compared with mass, capacitance with elasticity, voltage with pressure, current with velocity.

mechanical axis In a quartz crystal, the axis perpendicular to the faces of the hexagon. Also see *Y-axis, 2*.

mechanical bandspread Bandspread tuning obtained by reduction-ratio gearing of the tuning mechanism. Compare *electrical bandspread*.

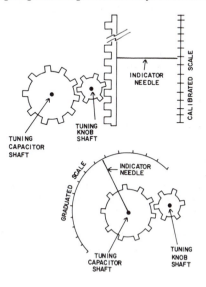

Mechanical analog of capacitor in which elastic diaphragm corresponds to the dielectric surge of water A corresponds to current flowing into the capacitor, and surge of water B corresponds to current flowing out of the capacitor.

mechanical bias 1. A steady pull applied by a spring to the armature of a relay to sensitize it by decreasing the distance the armature must move to close the contacts. 2. Sometimes, bending the relay frame to position the armature closer to the magnet for the purpose in 1, above.

mechanical damping Damping action obtained entirely by mechanical devices such as weights, dashpots, and the like.

mechanical equivalent of heat The amount of mechanical work required to produce a unit quantity of heat. For example, 4.183 joules can be converted into 1 calorie of heat.

mechanical equivalent of light The expression of luminous energy in equivalent power units. In practical measurements, this is taken as the total power output of a lamp minus the power absorbed by a transparent jacket used to remove the infrared and ultraviolet rays.

mechanical filter See *ultrasonic filter, 1.*

mechanical joint A union of electrical conductors consisting exclusively of a junction or splice made without brazing, soldering, or welding.

mechanical load An electromechanical device that uses the output of an electrical source. Such devices include actuators, brakes, clutches, meters, motors, and relays.

mechanical phonograph A nonelectronic sound recorder—reproducer used sometimes as an auxiliary device. A vibrating stylus cuts the groove directly, the sound waves being picked up (without an amplifier) by a diaphragm to which the stylus is attached.

mechanical rectifier A vibrator or commutator employed to change an alternating current into a direct current by selecting and passing only positive or negative half-cycles. Also see *electromechanical rectifier.*

mechanical register A totalizing mechanical counter. See, for example, *electromechanical counter.*

mechanical scanner A mechanical device for scanning an object or scene and breaking it up into horizontal lines that are converted to signals. Also, the similar device that scans the reproducer lamp in a mechanical TV receiver. See, for example, *Nipkow disk.*

mechanical time constant For a torque motor, the ratio of moment of inertia to damping factor. Compare *electrical time constant.*

mechanical wave filter See *ultrasonic filter, 1.*

mechanics The branch of physics concerned with forces and motion and the laws of gases and liquids. It is subdivided into kinematics and kinetics.

mechanoelectronic transducer Transducer tube (see *electromechanical transducer, 2*).

median A type of average that is the middle value in a series of terms. For example, in the series 1, 2, 3, 4, 5, 6, 7, the median is 4. Compare *arithmetic mean, geometric mean,* and *harmonic mean.*

medical electronics See *electromedical engineering.*

medium In a computer system, that storage device onto or into which data is recorded for input into memory, e.g., punched cards, magnetic tape, etc.

medium-frequency Abbreviation, *mf.* Pertaining to frequencies in the 300 kHz to 3 MHz range.

medium-mu tube A triode having an amplification factor in the 8 to 30 range (approximately). Compare *high-mu tube* and *low-mu tube.*

medium of propagation The material through which radiation is transmitted, e.g., air, rocks, water, space.

medium-scale integration A method of manufacturing integrated circuits, in which there are at least 10, but less than 100, individual gates on each chip. Abbreviated MSI.

medium tension Medium voltage.

medium-wave Abbreviation, *mw.* Pertaining to wavelengths corresponding to medium frequencies (see *medium-frequency*), i.e., those in the 100 to 1000 meter range.

meg Abbreviation of *megohm.* (Also, *M*).

mega Abbreviation, *M.* A prefix meaning *million(s)*, i.e., 10^6.

megabar Abbreviation, *Mb.* A cgs unit of high pressure. 1 Mb = 10^6 bars = 10^{11} pascals. Also see *bar, 1.*

megabit In computer and dataprocessing practice, a large unit of information. 1 megabit = 1 million bits. Also see *bit.*

megacurie Abbreviation, *MCi.* A large unit of radioactivity equal to 3.71 × 10^{16} disintegrations per second; 1 MCi = 10^6 curies. Also see *curie, kilocurie, microcurie,* and *millicurie.*

megacycle Abbreviation *Mc.* 1 million complete ac cycles. An ac frequencies of 1 Mc *per second* is 1 MHz (see Megahertz); 1 Mc = 10^6c.

megaelectronvolt Abbreviation, *MeV.* A large unit of electrical energy; 1 MeV = 10^6 eV. Also see *electronvolt.*

megahertz Abbreviation, *MHz.* A unit of high frequency; 1 MHz = 10^6 Hz.

megampere Abbreviation, *MA.* A unit of high current; 1 MA = 10^6A.

megaphone 1. A hand-held combination microphone—amplifier—loudspeaker used to amplify the voice of a person who must be heard over an appreciable area. 2. A simple horn for amplifying the voice.

megarutherford Abbreviation, *Mrd.* A large unit of radioactivity equal to 1 trillion disintegrations per second; 1 Ord = 10^6 rutherfords. Also see *kilorutherford, microrutherford, millirutherford,* and *rutherford.*

megaton 1. A unit of weight equal to 1 million tons. 2. An explosive force equal to that produced by 1 million tons of TNT.

megavolt Abbreviation, *MV.* A unit of high voltage; 1 MV = 10^6V.

megavolt-ampere Abbreviation, *MVA.* A unit of high reactive power. 1 MVA = 10^6 VA.

megawatt Abbreviation, *MW.* A unit of high power; 1 MW = 10^6 W. Also see *watt* and *kilowatt.*

megawatt-hour Abbreviation, *MWh.* A large unit of electrical energy or of work 1 MWh = 10^6 Wh = 3.6 × 10^9 joules.

megger An instrument containing an internal high-voltage dc supply, for measuring high resistance. Compare *megohmmeter.*

meg-mike Megohm-microfarads or megohm-farads.

megohm Symbol, *MΩ* or *M.* A unit of high resistance, reactance, or impedance. 1 MΩ = 10_6Ω.

megohm-farads For a large capacitor, the product of leakage resistance (megohms) and capacitance (farads). Also see *megohm-microfarads.*

megohmmeter A special ohmmeter for measuring resistances in the megohm range.

megohm-microfarads For a capacitor, the product of leakage resistance (megohms) and capacitance (microfarads). The figure is an expression for the relative insulation resistance of a capacitor.

Meissner circuit. A self-excited oscillator tuned by an inductively coupled variable *LC* feedback circuit.

Meissner effect In a superconductive material, the abrupt loss of magnetism when the temperature of the material is reduced to a value below that required for superconductivity.

Meissner oscillator A vacuum-tube oscillator, utilizing a resonant circuit between the control grid and the plate. The resonant circuit provides a 180-degree phase shift, resulting in positive feedback.

mel An expression of apparent or preceived sound pitch. A tone of 1 kHz, at a level of 40 dB with respect to the threshold of hearing, represents 1 mel. The perceived pitch depends, to some extent, on the intensity of the sound, as well as on the actual frequency.

M-electron In certain atoms, one of the electrons whose orbits are outside of and nearest to those of the L-electrons.

meltback process The technique of remelting a doped semiconductor material and allowing it to refreeze to form a grown junction.

meltback transistor A grown-junction transistor produced by the *meltback process.*

melting point Abbreviation, *mp*. The temperature at which a solid starts becoming liquid at 1 atmosphere pressure. Compare *mixture melting point* and *freezing point.*

memory The section of a digital computer that "remembers" material, i.e., the section that records and holds data until needed.

memory address register In computer storage, a register in which is stored the address of operands in other locations.

memory area A portion of computer memory reserved for a specific type of data. Also called *area.*

memory capacity As a function of the number of memory locations available, the number of bits that can be stored, usually given as a number of words comprising a number of bits, e.g., a memory capacity of 16,000 12-bit words.

memory core A small toroidal core of magnetic material having a squareloop hysteresis loop, for use inthe static magnetic memory of a computer or calculator.

memory cycle 1. The period of execution of a sequence of operations. 2. The complete operational cycle for inputting data to memory or retrieving it.

memory disk A rotating disk whose surface is coated with a magnetic oxide on which are recorded bits of computer information.

memory drum A rotating drum whose surface is coated with a magnetic oxide on which are recorded bits of information for computer storage.

memory dump In computer practice, to (1) print out what is stored in some of or all of the memory locations or (2) transfer the data from a bank of memory cells to some external storage medium.

memory guard In a computer, hardware or software which keeps certain memory locations from being addressed by a program being run.

memory location In a computer memory, a place where an information unit (word, character) can be stored; the stored information may be retrieved by appropriate addressing instructions.

memory power Computer memory efficiency in terms of data processing (cycle) speed.

memory protection A hardware device in a multiple programming computer that keeps programs from being altered by other operating programs in the installation.

memory register In a digital computer, a register employed in all instruction and data transfers between the memory and other sections of the machine. Also see *distributor, 3.*

memory tape In a data-storage system, magnetic tape on which data pulses are recorded.

memory unit See *memory.*

mendelevium Symbol, Md. A radioactive element produced artificially. Atomic number, 101. Atomic weight, 256 (approx.).

mendelian factor Gene.

menu In computer practice, a list of optional facilities for using various functions of the system.

MEP Abbreviation of *mean effective pressure.*

mercuric iodide Formula, HgI_2. A compound whose crystals are useful at room temperature as detectors in high-resolution gamma-ray spectroscopy.

mercury Symbol, Hg. A metallic element. Atomic number, 80. Atomic weight, 200.61. Mercury is the only metal that is liquid at ordinary temperatures. It is used extensively in switches, rectifier tubes, and high-vacuum pumps as well as in common thermometers.

mercury arc The arc discharge occurring in mercury vapor between solid or liquid (mercury) electrodes. The discharge is a rich source of ultraviolet radiation.

mercury-arc converter A mixer, or frequency converter, that makes use of the diode characteristics of a mercury-vapor rectifier tube. The circuit is similar to the diode mixer.

mercury-arc rectifier A heavy-duty rectifier tube utilizing ionized mercury vapor. There are two general types: *mercury-vapor rectifier* and *mercury-pool rectifier.* Also see *pool-cathode tube.*

mercury battery A series or parallel combination of mercury cells. Also see *mercury cell.*

mercury cadmium telluride Formula HgCdTe. An alloy used as a semiconductor in transistors, integrated circuits, and infrared detectors.

mercury cell A primary cell housed in a steel container and having a mercuric-oxide cathode, amalgated-zinc anode, and potassium-hydroxide-zinc-oxide electrolyte. The 1.35V or 1.4V output is reasonably constant throughout the life of the cell.

mercury delay line A delay line in which delay is obtained by propagating the signal the signal through a pipe of mercury.

mercury diffusion pump A vacuum diffusion pump employing mercury vapor. Also see *diffusion pump.*

mercury displacement relay A form of switching relay in which the electrical contact is made by moving mercury.

mercury-jet switch A multipoint switch employing a jet of mercury instead of the conventional wiper arm, for high-speed operation and reduced wear.

mercury memory A recirculating memory employing a mercury delay line. Also see *delay line* and *delayline memory.*

mercury-pool cathode In certain industrial electron tubes such as the ignitron, a cathode electrode consisting of a pool of mercury.

mercury-pool rectifier A type of mercury-arc rectifier whose cathode is a pool of mercury. In one type the arc is initiated by tilting the tube momentarily to bring the mercury into contact with a third electrode, thus causing a starting current to flow through the pool. In another type, the *ignitron,* a starter electrode is in continual contact with the mercury. Also see *excitron.*

mercury pump See *mercury diffusion pump.*

mercury rectifier See *mercury-pool rectifier* and *mercury-vapor rectifier.*

mercury relay A relay in which at least one of the contacts is mercury.

mercury storage See *mercury memory.*

mercury switch A switch consisting essentially of two or more stiff wire electrodes and a drop of mercury hermetically sealed in a glass tube. Tilting the tube causes the mercury to flow toward one end where it immerses the electrodes, providing a conductive path between them.

mercury-vapor lamp A glow lamp emitting blue-green light due to ionization of mercury vapor by an electric current. See next page.

mercury-vapor rectifier A tube-type diode rectifier containing a small amount of mercury which vaporizes and ionizes during tube operation. Also see *tube diode* and *rectifier diode.*

mercury-vapor tube 1. Mercuryvapor lamp. 2. Mercury-vapor rectifier.

mercury-wetted reed relay A reed relay in which the reeds are wetted by mercury in a pool by capillary action. The film of mercury forms a tiny bridge when the reeds open; when this bridge separates, a clean, high-speed break occurs without bounce. Compare *dry-reed switch.*

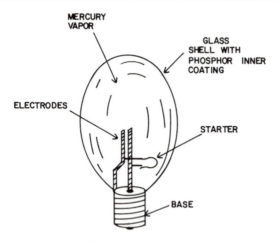

Mercury vapor lamp.

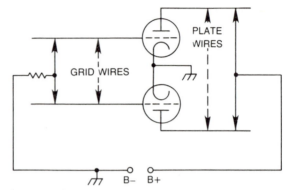

Meissner circuit.

merge 1. In computer practice, to make a single set or file from two or more record sets, as done with punched cards by a collator, for example.2. In word processing, to create a corrected master recording from two input media: the original master recording and the recording that contains the corrections.

meridian A great circle passing through earth's geographical poles and a given point on the surface of the earth, e.g., the lines of longitude on a map or globe. Also see *time zone, zero meridian,* and *zone time*.

Mershon condenser At one time, the *electrolytic capacitor,* especially the wet type.

mesa A flat-topped, protruding region in a semiconductor wafer. The mesa is produced by etching the surrounding part of the material. Some bipolar transistors are manufactured in this way.

mesa diffusion A method of manufacturing bipolar transistors. The different semiconductor materials are first diffused together. Then part of the resulting wafer is etched away, resulting in a mesa shape.

mesa transistor A diffused planar transistor in which the silicon area around the base has been etched away to reduce collector-to-base capacitance; the base—emitter region remains elevated like a high plateau, or *mesa*.

mesh 1. A combination of the elements that form a closed path in a network. 2. The closed figure (such as the *delta* or *star*) obtained by connecting polyphase windings together. 3. A grid, screen, or similar structure in an electron tube. 4. One of the flat, screenlike plates employed in a storage tube. Also see *storage mesh, storage tube,* and *viewing mesh*.

mesh current In a circuit with more than one current loop, the current at any point of a given loop.

mesh equations Equations describing fully the current and voltage relations in a network of meshes (see *mesh, 1*).

Mesny circuit A push-pull ultrahigh-frequency oscillator whose grid tank is a pair of parallel wires shortcircuited by a slider; the plate tank is a similar pair of wires. The frequency is varied by moving the sliders along the wires.

meson An unstable nuclear particle first observed in cosmic rays. A meson may be positive, negative, or neutral. Its mass lies between that of the electron and proton.

mesotron See *meson*.

message 1. A body of information communicated between transmitter and receiver. 2. Data put into a transaction processing system.

message exchange In a digital communications channel, a hardware unit that carries out certain switching functions that would otherwise have to be done by a computer.

message switching system A data communications system having a central computer that receives messages from remote terminals, stores them, and transfers them to other terminals as needed.

metadyne See *dc generator amplifier*.

metal An elemental material which exhibits several familiar properties such as luster, ductility, malleability, good electrical and heat conductivity, relatively high density, and the ability to emit electrons. Common examples are aluminum, copper, gold, lead, and silver. Compare *metalloid* and *nonmetal*.

metal-base transistor A bipolar transistor in which the base is a metal film, and the emitter and collector are films of *n*-type semiconductor material.

metal-ceramic construction The building of certain electronic components by bonding ceramic parts to metal parts. See also *cermet*.

metal-film resistor A fixed or variable resistor in which the resistance element is a thin or thick film of a metal alloy deposited on a substrate such as a plastic or ceramic.

metal finder See *metal locator*.

metal gate In a field-effect transistor, an aluminum gate electrode.

metallic binding forces In a crystal, the binding electrostatic force between cations and electrons. Also called *electron-gas binding forces*.

metallic bonding See *bonding, 1* and *metallic binding forces*.

metallic circuit A circuit, such as a two-wire telephone line, in which earth ground is not a part of the circuit. Compare *ground-return circuit*.

metallic insulator A short-circuited quarter-wave section of transmission line which acts as an insulator at the quarter-wavelength frequency.

metallicize To make a circuit fully metallic, as when two wires are employed instead of one wire and a ground connection. (Not to be confused with *metallize*.)

metallic rectifier A dry rectifier employing a metal disk or plate coated with a material such as selenium, an oxide, or a sulfide. See, for example, *copper-oxide rectifier, dry-disk rectifier, magnesium-copper-sulfide rectifier*, and *selenium rectifier*.

metallize To treat, coat, or plate with a metal. (Not to be confused with *metallicize.*)

metallized capacitor A capacitor in which each face of a dielectric film is metallized to form plates. See, for example, *metallized-paper capacitor* and *metallized-polycarbonate capacitor.*

metallized-paper capacitor A paper-dielectric capacitor whose plates are metal areas electrodeposited on each side of a paper film.

metallized-polycarbonate capacitor A fixed capacitor in which the dielectric is a polycarbonate plastic film, and the plates are metal areas electrodeposited on each face of the film.

metallized resistor See *metal-film resistor.*

metal locator An electronic device for locating underground metal deposits, pipes, or wires—or such objects hidden in walls or under floors—by means of the disturbance these objects cause to a radiofrequency or magnetic field.

metalloid An element which has *some* of the properties of a metal. Examples of metalloidal elements widely used in electronics are antimony, arsenic, germanium, silicon, and tin.

metal master See *original master.*

metal negative See *original master.*

metal-oxide resistor A resistor in which the resistance material is a film of tin oxide deposited on a substrate.

metal-oxide silicon field-effect transistor Abbreviation, *MOSFET.* A field-effect transistor in which the gate electrode is not a *pn* junction (as in the junction field-effect transistor) but a thin metal film insulated from the semiconductor channel by a thin oxide film. Gate control action, therefore, is entirely electrostatic. Drain and source electrodes are *pn* junctions. Also called *insulated-gate field-effect transistor.* Also see *depletion-type MOSFET, depletion-enhancement-type MOSFET,* and *enhancement-type MOSFET.*

metal-oxide varistor A varistor in which the resistance material is a metallic oxide such as zinc oxide.

metal-plate rectifier See *metallic rectifier.*

metal tube An electron tube that is completely housed in an evacuated metal envelope for self-shielding and mechanical ruggedness, e.g., a 6SR7. Compare *glass tube.*

metamer A visible-light beam that is identical in color (hue), but different in concentration (saturation), with respect to a reference color.

meteor-burst signals Momentary signals or increases in signal strength due to *meteor-trail reflections.*

meteorograph An instrument for the simultaneous measurement of various meteological phenomena such as temperature, humidity, etc.

meteorology The science of the atmosphere, especially the study of weather and climate. (Not to be confused with *metrology.*)

meteor-trail reflections Momentary reflection of vhf signals by the ionized trails of meteors passing through a signal path.

meter 1. An instrument for measuring and indicating the value of a particular quantity. See, for example, *current meter* and *voltmeter.* 2. A metric unit of linear measure and of electrical wavelength. A meter is equivalent to $1.650\ 763\ 73 \times 10^6$ wavelengths (in a vacuum) of the radiation corresponding to the transition between the levels 2 p10 and 5 d5 of the krypton-86 atom; 1 m = 39.37 inches. 3. To supply in specific increments or by a governed amount.

meter alignment See *visual alignment.*

meter-ampere A unit of measure of radio transmitter strength. Meter-amperes are determined by multiplying antenna current (in amperes) by the height (in meters) of the antenna above ground.

meter-candle Abbreviation, *mc.* A metric unit of illuminance. 1 mc is

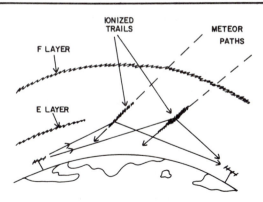

Meter-trail reflections.

the illumination on a surface 1 meter from a light source of 1 candle power. Compare *foot-candle* and *lux.*

meter equivalent The number of meters equal to a given English measure of length; e.g., the meter equivalent of 3 feet is approximately 0.9144.

meter-kilogram-second Abbreviation, *mks.* The name of the system of units in which the meter is the unit of length; the kilogram, mass; and the second, time. Compare *centimeter-gram-second* and *International System of Units.*

meter modulator See *electromechanical modulator, 3.*

meter multiplier See *multiplier resistor.*

meter protector A nonlinear resistor such as a varistor or semiconductor diode, employed to prevent overswing in an electric meter by limiting the current flowing through it.

meter rating The maximum reading on a meter, at or below which the accuracy is within certain limits, but above which the error may be greater than the specified limits.

meter rectifier A light-duty semiconductor diode (or diode bridge) employed to change ac to dc for deflection of a D'Arsonval-type dc milliammeter or microammeter. Also see *junction diode* and *semiconductor rectifier.*

meter relay A sensitive relay which is essentially a moving-coil meter whose pointer closes against a stationary contact mounted at some point along the scale.

meter resistance Symbol, *Rm.* The internal resistance of an electric meter. In a simple D'Arsonval meter, it is the resistance of the movable coil. In more complicated meter circuits, it is the resistance of the coil plus that of any internal shunt of multiplier resistors.

meter scale factor See *scale factor, 1.*

meter sensitivity See *voltmeter sensitivity.*

meter shunt A shunt resistor employed to increase the range of a current meter.

meter torque See *deflecting torque.*

meter-type modulator See *electromechanical modulator, 3.*

meter-type relay See *meter relay.*

methyl methacrylate resin A plastic insulating material. Dielectric constant, 2.8 to 3.3. Dielectric strength, 20 kV/mm. Also called *Lucite.*

metre An alternate spelling of "meter," the metric unit of length. This French spelling has been recommended as a safeguard against confusion with *meter,* the name of a measuring instrument, but its use is only sparse.

metric system The decimal system of weights and measures based upon the meter, gram, and second. Also see *centimeter—gram—second*, *meter—kilogram—second*, and *International System of Units*.

metric ton Abbreviation, *MT*. A metric unit of weight equal to 1000 kilograms or 1.1023 tons. Compare *ton*.

metric waves British designation of wavelengths between 1 and 10 meters (frequencies from 30 to 3000 MHz).

metrology The science of weights and measures, including electrical standards and electronic instruments and measurements. (Not to be confused with *meteorology*.)

metronome A mechanical or electronic device which produces audible beats (ticks) used in setting the tempo for music or for audibly timing certain processes.

MeV Abbreviation of *megaelectronvolt(s)*.

MEW Abbreviation of *microwave early warning*.

mF Abbreviation of *millifarad*.

mf 1. Abbreviation of *medium-frequency*. 2. Abbreviation of *midfrequency*.

MFD Abbreviation of *magnetofluid-dynamics* (see *magnetohydrodynamics*).

MFSK Abbreviation of *multiple frequency-shift keying*.

Mg Symbol for *magnesium*.

mg Abbreviation of *milligram*.

MGD Abbreviation of *magnetogasdynamics* (see *magnetohydrodynamics*).

MHD Abbreviation of *magnetohydrodynamics*.

MHD generator See *magnetohydrodynamic generator*.

MHD gyroscope See *magnetohydrodynamic gyroscope*.

MHD power generation See *magnetohydrodynamic generator*.

mho The now-obsolete unit for conductance, the value of which was the reciprocal of resistance in ohms. (The *mho* has been replaced by the *siemens* in the International System of Units.)

mho-centimeter Symbol, *v*-cm. Obsolete unit of conductivity denoting. the conductance of a centimeter cube of a material.

mhp Abbreviation of *millihorsepower*.

MHz Abbreviation of *megahertz*.

mi Abbreviation of *mile*. (Also, *m*.)

MIC Abbreviation of *microwave integrated circuit*.

mic Abbreviation of *microphone*.

mica A dielectric mineral of complex silicate composition, which is easily separated into numerous thin, transparent sheets. It is widely used as a capacitor dielectric and hightemperature electrical insulator. Dielectric constant, 2.5 to 7. Dielectric strength, 50 to 220 kV/mm. Also see *muscovite*.

mica capacitor A capacitor employing a thin film of mica as the dielectric. Such a capacitor exhibits high Q and good stability. Also see *mica* and *silvered-mica capacitor*.

MICR Abbreviation of *magnetic ink character recognition*.

micro 1. A prefix meaning *millionth(s)*, i.e., 10^{-6}. Symbol, μ. 2. A prefix meaning *extremely small* (as in *microstructure*). Compare *macro*.

microalloy diffused transistor Abbreviation, *MADT*. A microalloy transistor having a uniform base region that is diffused into the wafer before the emitter and collector electrodes are produced by alloying.

microalloy transistor Abbreviation, *MAT*. A transistor having tiny emitter and collector electrodes that are formed by alloying a thin film of impurity material with a collector pit and emitter pit *facing each other on opposite surfaces* of the semiconductor wafer. Also see *surface-barrier transistor*.

microammeter A usually direct reading instrument employed to measure current in the microampere range. Also see *current meter*.

microampere Abbreviation, μA. A small unit of current. $1 \mu A = 10^{-6}$ A.

microbalance A sensitive electronic weighing device. One type employs one or more servo amplifiers for the balancing operation.

microbar Abbreviation, μb. A cgs unit of low pressure. $1 \mu b = 10^{-6}$ b = 0.1 pascal. Also see *bar, 1* and *millibar*.

microbarograph A *barograph* that is sensitive to small changes in pressure.

microbeam A *beam* having extremely small cross section.

microcircuit An extremely small circuit fabricated upon and within a substrate such as a semiconductor chip. Also see *integrated circuit*.

microcode See *microinstruction*.

microcode A code for *microprogramming*.

microcomponent A tiny component in an electronic circuit. Examples are the resistors, capacitors, diodes, and transistors fabricated onto an integrated-circuit chip.

microcomputer 1. Any small computer whose central processing element is contained on a single small circuit board or within a single integrated circuit. 2. A self-contained computer designed around a microprocessor.

microcrystal A crystal that is invisible to the naked eye.

microcurie Abbreviation, μi. A small unit of radioactivity equal to 3.71 $\times 10^4$ disintegrations per second; $1 \mu ci = 10^{-6}$ curie. Also see *curie*, *kilocurie*, *megacurie*, *millicurie*, and *picocurie*.

microelectrode An electrode employed in microelectrolysis. 2. A tiny electrode, especially one of those used in integrated circuits and in certain biological applications.

microelectrolysis Electrolysis of tiny amounts of material. Also see *electrolysis, 1*.

microelectronic circuit A tiny electronic circuit other than an *integrated circuit*, i.e., one assembled in a small space with small discrete or integrated components.

microelectronic device See *microcircuit*.

microelectronics The branch of electronics dealing with extremely small components and circuits fabricated on substrates. Also see *integrated circuit*.

microelectrophoresis *Electrophoresis* of single particles.

microelectroscope A very sensitive electroscope used to detect minute quantities of electricity.

microelement A tiny component (capacitor, resistor, coil, semiconductor device, or transformer) mounted on a wafer and used in a microelectronic circuit.

microelement wafer A microwafer on which a microelement is mounted or deposited.

microfarad Abbreviation, μF. A unit of low capacitance. $1 \mu F = 10^{-6}$ F.

microfarad meter 1. A dynamometer-type meter which indicates the value of a capacitor directly in microfarads. Such instruments usually operate from a 60, 120, 400 or 500 Hz power line. 2. A direct-reading capacitance meter.

microfiche A method of storing printed information on small film cards. The pages are reduced and arranged in order from left to right and top to bottom. The card is inserted into a projecting machine to allow retrieval of the information. The photographic method is similar to that employed in microfilm.

microfilm A method of storing printed or photographic information. The pages are reduced and arranged sequentially on a strip of film, usually 35mm size. the film is inserted into a projecting device for retrieval of the information.

microgalvanometer A highly sensitive galvanometer.

microgauss One-millionth of a *gauss*.

microgram Abbreviation, μg. A metric unit of weight or mass equal to 1 millionth of a gram.

microgroove record A phonograph disk with a very fine groove (200 to 300 per inch), designed for 3331 or 45 rpm playback.

microhenry Symbol, μH. A unit of low inductance. $1 \mu H = 10^{-6}$ H.

microhm Symbol, μΩ. A unit of low resistance, reactance, or impedance. $1 \mu\Omega = 10^{-6} \Omega$.

microhm-centimeter Symbol, μΩ-cm. A unit of low resistivity. $1. \mu\Omega\text{-cm} = 10^{-6}\Omega\text{-cm}$. Also see *ohm-centimeter* and *resistivity*.

microhmmeter An instrument for measuring ultralow resistance. Such an instrument must have a special provision for canceling the effects of contact and lead resistance.

microinch Abbreviation, μin. One-millionth of an *inch*.

microinstruction A machine-code instruction that controls the operation of a computer directly, i.e., it is a "wired-in" instruction, or one set by DIP switches, independent of programs loaded into the machine.

microliter Abbreviation, μl. One-millionth of a *liter*.

microlock A special form of phase-locked-loop system, used especially with radar to improve the signal-to-noise ratio.

micromanipulator A machine which permits handling tiny parts in very small areas. An example of its use is in placing connections close together in microcircuits.

micrometer 1. An instrument for measuring very small thicknesses, diameters, etc. 2. Symbol, μ. The SI unit of length, equal to 0.000 001 meter. The micrometer has replaced the micron in the SI system.

micromho Symbol, μʊ. A now-obsolete unit of low conductance. $1 \mu\nu = 10^{-6}\nu = 10^{-6}$ siemens. Also see *conductance* and *mho*.

micromicro A prefix meaning *trillionth(s)*, i.e., 10^{-12}. Also see *pico*, which has supplanted this prefix.

micromicrofarad See *picofarad*.

micromicrohenry See *picohenry*.

micromicron Symbol, μμ. A small metric unit of linear measure equal to 1 picometer (1 millionth of a micron).

micromillimeter Symbol, μmm. A small metric unit of linear measure equal to 1 nanometer (1 millionth of a millimeter).

microminiature Pertaining to an extremely small body, component, or circuit; the last adjective in the sequence of those describing size: *standard, small, midget, miniature, subminiature, microminiature*.

micromodule A small, encapsulated circuit, comprised of smaller components. The components may be discrete, or they may consist of integrated circuits, or they may be a combination of both. The module is easily removed and replaced by means of a plug-in socket.

micron Symbol, μ. A unit of length used until recently to describe certain extremely short waves and microscopic dimensions. (The unit has been supplanted by micrometer.) $1\mu = 10^{-6}$ meter (a micrometer).

microphone A transducer which converts sound waves, especially speech and music, into electrical voltage analogs.

microphone amplifier 1. A usually high-gain preamplifier employed to boost the output of a microphone. 2. A relay-type audio amplifier employing a microphone button (see *carbon-button amplifier*).

microphone boom A device used for hanging a microphone, with the base out of the way. Often used in radio broadcasting.

microphone button See *button, 2* and *button microphone*.

microphone hummer See *hummer*.

microphone oscillator See *hummer*.

microphone relay amplifier See *carbon-button amplifier*.

microphonics Ringing (electrical noises) set up by the vibration of a com-

ponent having loose or movable elements. For example, ringing noises are generated by some vacuum tubes when they are tapped.

microphonograph A recorder of very low-intensity sound.

microphonoscope An electronic stethoscope, using amplification to enhance the response.

microphotograph An extremely small photograph, often of a pattern or mask used in producing transistors and integrated circuits. Not to be confused with *photomicrograph,* a photograph taken through a microscope.

microphotometer A sensitive instrument for measuring small-area light intensity.

microphysics The branch of physics concerned with atoms, molecules, and subatomic particles.

micropower Extremely small amounts of power, especially the very low dc supply power required by some transistors.

microprocessor A usually single-chip computer element that contains the control unit, central processing circuitry, and arithmetic and logic functions and is suitable for use as the central processing unit of a microcomputer or a dedicated automatic control system.

microprogram In computer practice, a routine of microinstructions that gives a computer a specific function independent of those established by programs being run or by the monitor program. Also see *microinstruction*. 2. In the direction of a computer, use of a routine that is stored specifically in the memory, instead of elsewhere.

microprogramming In the direction of a computer, use of a routine that is stored specifically in the memory, instead of elsewhere.

micropulsation A pulsation of extremely short duration.

microradiograph An x-ray picture showing the microscopic structure of a material.

microradiometer A sensitive detector of heat and infrared radiation, consisting essentially of a thermopile carried by the moving coil of a galvanometer.

microrutherford Symbol, μrd. A small unit of radioactivity equal to 1 disintegration per second; $1 \mu rd = 10^{-6}$ rutherford. Also see *rutherford, kilorutherford, megarutherford,* and *millirutherford*.

microscope An instrument that presents an enlarged image of an object. Also see *electron microscope*.

microscopic Exceedingly small, especially too small to be seen with the naked eye. Compare *macroscopic*.

microsecond Symbol, μs. A small unit of time measure equal to 1 millionth of a second. $1 \mu s = 10^{-6}$s.

microspectrophotometer An extremely sensitive *spectrophotometer* for examining light from tiny areas.

microstrip A microwave component which is in effect a single-wire transmission line operating above ground.

microsyn A device that translates rotational position into an electrical signal. Similar to a selsyn. Used for such purposes as rotator-direction reading.

microsystems electronics The technology of electronic systems utilizing tiny electronic components. Also see *integrated circuit, microelectronic circuit, microelement, microelement wafer,* and *microwafer*.

microvolt Symbol, μV. A unit of low voltage. $1 \mu V = 10^{-6}$V.

microvolter An accurate, external attenuator (usually for an audio signal generator) providing stepped and continuously variable output in microvolts and millivolts.

microvoltmeter A usually direct reading instrument employed to measure voltages in the microvolt range. An input amplifier boosts the test voltage sufficiently to deflect the indicating meter.

microvolts per meter Symbol, μV/m. A unit of measure of radio field strength. It refers to the rf voltage (in microvolts) developed between an antenna and ground divided by the height of the antenna (in meters) above ground. Compare *millivolts per meter*.

microvolts per meter per mile A means of expressing absolute radio-frequency field strength. Generally, the numerical value is based on the field strength, in microvolts per meter, at a distance of 1 statute mile (5,280 feet) from the source.

microwafer A wafer of insulating material such as a ceramic, on which one or more microelements are mounted and terminals deposited or plated.

microwatt Symbol, μW. A unit of low power, especially electrical power. $1 \mu W = 10^{-6} W$.

microwattage See *micropower*.

microwattmeter An instrument for measuring power in the microwatt range. Such an instrument obtains its sensitivity from a built-in input amplifier.

microwave See *microwaves*.

microwave acoustics See *acoustoelectronics* and *acoustic delay line*.

microwave dish A dish antenna for use at microwave frequencies.

microwave early warning Abbreviation, *MEW*. A high-power early warning radar system which affords large traffic-handling capacity and long range.

microwave filter A bandpass filter built into a waveguide for use of microwave frequencies.

microwave frequencies Frequencies above 1 GHz.

microwave integrated circuit Abbreviation, *MIC*. An integrated circuit designed for use at microwave frequencies.

microwave lens See *waveguide lens*.

microwave mirror A reflector of microwaves.

microwave oven An oven for rapid cooking, consisting essentially of a radio-frequency heater employing a magnetron oscillator producing microwave frequencies.

microwave plumbing Collectively, the waveguides, tees, elbows, and similar fixtures and connections used in microwave setups.

microwave radio relay The use of microwaves to relay radio, television, and control signals from point to point.

microwave refractometer An instrument employing microwaves (around 10 GHz) to measure the refractive index of the atmosphere.

microwave region The part of the electromagnetic spectrum that lies between 1 and 300 GHz (1 mm to 30 cm).

microwave relay See *microwave radio relay*.

microwave relay system A series of microwave transmitter—receiver stations for relaying communications in several line-of-sight hops.

microwaves Extremely short radio waves, especially those shorter than 0.3 m in wavelength (1 GHz or higher in frequency).

microwave spectrum See *microwave region*.

microwave transistor A usually silicon transistor whose semiconductor properties and special fabrication enable it to operate at microwave frequencies.

microwave tube A Klystron or similar tube, used for the purpose of generating or amplifying microwave radio-frequency signals.

midband Abbreviation, *MB*. The region whose limits are immediately above and below a midfrequency, the limits usually being specified for a particular case.

midband frequency See *midfrequency*.

midfrequency Abbreviation, *mf*. The center frequency in a specified band of frequencies.

midget Of reduced size (smaller than *small* and larger than *miniature*).

midrange speaker A *loudspeaker* operating most efficiently at frequencies in the middle of the audio spectrum. Such a speaker is intermediate in performance between a *woofer* and a *tweeter*.

midsection The center section of a *multisection filter* having an odd number of sections; thus, the second section of a three-section filter.

migration See *ion migration*.

mike 1. Microphone. 2. Microfarad. 3. Micrometer.

MIL Abbreviation of *military*.

mil 1. A small unit of linear measure; 1 mil = 10^{-3} inch = 0.0254 mm. 2. Thousand, as in *n* parts per *mil*. 3. A thousandth of a dollar.

mile Abbreviation, *m* or *mi*. A large unit of linear measure, 1 mi = 1.609 km = 5,280 feet.

mill A telegraph operator's typewriter.

Miller bridge A device used for measuring the amplification of a vacuum-tube circuit.

Miller effect For a vacuum tube, the variation in input capacitance as dc grid bias is varied (and the equivalent behavior of a transistor). In an *RC*-coupled amplifier circuit, input (grid—cathode) capacitance C_{input}, which actually appears larger than the true C_{gk} of the tube, is equal to $C_{gk}(A \times 1)$, where A is the voltage amplification of the circuit.

Miller oscillator A crystal oscillator circuit in which the crystal is connected between the grid of the tube and ground; the tuned tank is connected in the plate circuit; and the internal grid—plate capacitance of the tube provides feedback coupling. Also, the equivalent bipolar-transistor or FET circuit. Sometimes called *conventional crystal oscillator*.

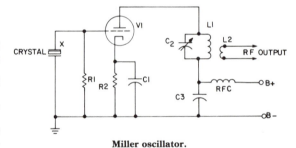

Miller oscillator.

milli Abbreviation, m. A prefix meaning *thousandth(s)*, i.e., 10^{-3}.

milliammeter A usually direct reading instrument for measuring current in the milliampere range. Also see *current meter*.

milliampere Abbreviation, *mA*. A unit of low current; 1 mA = 10^{-3} A.

milliampere-hour Abbreviation, *mAh*. A unit of low current drain or charging rate. 1 mAh = 10^{-3} Ah. Also see *ampere-hour* and *battery capacity*.

millibar Abbreviation, *mb*. A unit of low pressure. 1 mb = 10^{-3} bar = 100 pascals. Also see *bar, 1* and *microbar*.

millicurie Abbreviation, *mCi*. A small unit of radioactivity equal to 3.71 × 10^7 disintegrations per second. 1 mCi = 10^{-3} curie. Also see *curie, kilocurie, megacurie,* and *microcurie*.

millifarad Abbreviation, *mF*. A small unit of capacitance. 1 mF = 10^{-3} F. Also see *farad*.

milligram Abbreviation, *mg*. A metric unit of weight equal to 10^{-3} gram.

millihenry Abbreviation, *mH*. A small unit of inductance; 1 mH = 10^{-3} H. Also see *henry*.

millihorsepower Abbreviation, *mhp*. A small unit of power; 1 mhp = 10^{-3} hp = 746W. Also see *horsepower*.

millilambert Abbreviation, *mL*. A small unit of brightness; 1 mL = 10^{-3} L = $10^4/\pi$ cd/m^2. Also see *lambert*.

milliliter Abbreviation, *ml*. A metric unit of volume equal to 10^{-3} liter. Also see *liter*.

millimaxwell A small unit of magnetic flux equal to 10^{-3} maxwell (10^{-11} weber). Also see *maxwell*.

millimeter Abbreviaiton, *mm*. A metric unit of linear measure. 1 mm = 10^{-3} m = 0.039 37 inch. Also see *meter*.

millimeter equivalent The number of millimeters equal to a given English measure fraction; e.g., the millimeter equivalent of 5/16 inch is 7.937.

millimeter waves Wavelengths between 0.6 and 10 mm (frequencies from 30 to 500 GHz).

millimicro Abbreviation, mμ. A depricated prefix meaning *billionth(s)*, i.e., 10^{-9}. See *nano*.

millimicrofarad See *nanofarad*.

millimicrohenry See *nanohenry*.

millimicron Abbreviation, mμ. A unit of wavelength equal to 10^{-3} micron (1 nanometer). Also see *micron* and *nanometer*.

millimilliampere Abbreviation, *mmA*. A depricated term for microampere, i.e., 10^{-6} A.

millimole Abbreviation, mmol. One-thousandth of a *mole*.

milliohm Symbol, mΩ. A small unit of resistance, reactance, or impedance; 1 Ω = 10^{-3} Ω.

milliohmmeter An ohmmeter for measuring resistances in the milliohm range.

million electronvolt(s) See *megaelectronvolt*.

milliphot A unit of illumination equal to one-thousandth of a *phot*.

millipuffer See *puffer*.

milliradian Abbreviation, mrad. One-thousandth of a *radian*.

milliroentgen Abbreviation, *mr*. A small unit of radioactive dosage; 1 mr = 10^{-3} roentgen = 2.579 76 × 10^{-7} Ci/kg.

millirutherford Abbreviation, *mrd*. A small unit of radioactivity equal to 1000 disintegrations per second; 1 mrd = 10^{-3} rutherford. Also see *rutherford*, *kilorutherford*, *megarutherford*, and *microrutherford*.

millisecond Abbreviation, *ms*. A small unit of time. 1 ms = 10^{-3} s. Compare *microsecond* and *second, 1*. Also see *time*.

millitorr Abbreviation, mT. An obsolete unit of low pressure equal to 10^{-3} torr, or 0.133 322 pascal. Also see *torricelli*.

millivolt Abbreviation, mV. A unit of low voltage; 1 mV = 10^{-3} V. Also see *volt*.

millivoltmeter A usually direct reading instrument for measuring electric potential. Its sensitivity is provided by a high-gain amplifier operated ahead of the indicating meter.

millivolt potentiometer Abbreviation, *MVP*. A potentiometer-type null instrument for accurately measuring small dc voltages such as those delivered by a thermocouple. Also see *potentiometer, 2* and *potentiometric voltmeter*.

millivolts per meter Abbreviation, *mV/m*. A unit of radio field strength. It refers to the rf voltage (in mV) developed between an antenna and ground divided by the height (in meters) of the antenna above ground. Compare *microvolts per meter*.

milliwatt Abbreviation, *mW*. A small unit of electric power; 1 mW = 10^{-3}W. Also see *watt* and *microwatt*.

milliwattmeter An instrument for measuring power in milliwatts. Such instruments usually obtain their sensitivity from a built-in preamplifier.

Mills cross A radio-telescope antenna, consisting of two collinear or phased arrays with a common intersecting lobe. The result is high resolution.

min 1. Abbreviation of *minimum*. 2. Abbreviation of *minute* (see *minute, 1*).

mine An underground or undersea explosive device which is detonated by the pressure or vibration of moving troops, vehicles, or ships, or by the proximity of large bodies of magnetic material. See, for example, *acoustic mine* and *magnetic mine*.

mineral An element or compound that occurs naturally in the earth's crust. Most minerals are crystalline and many of these have found use in electronics. Some have been produced artificially.

mineral detector See *crystal detector*.

mineral oil A natural liquid insulant derived from petroleum. Dielectric constant, 2.7 to 8.0. Power factor, 0.08 to 0.2 percent at 1 kHz.

mineral-oil capacitor An oil capacitor whose paper dielectric has been impregnated with mineral oil, which is also the filler.

miniature Very small (smaller than *midget* and larger than *subminiature*).

miniature tube A small electron tube which usually has a glass envelope and button base, e.g., a 6AG5.

miniaturization The technology of minimizing the physical size of a circuit or system, while maintaining its ability to accomplish a given task.

minicalculator A pocketable electronic calculator.

minicomputer A small, comparatively inexpensive computer whose construction is made possible by the large-scale integration of circuits.

MINIDOS Acronym for *mini-disk operating system*.

minifloppy A smaller than standard flexible magnetic disk (floppy).

minigroove record A phonograph disk whose groove density (grooves per inch) is between that of a 78-rpm record and a microgroove record.

minima Points along a curve at which a function reaches a minimum value, and on each side of which all other points are higher in value. For a function $y = f(x)$, a point is a minimum if, when the first derivative dy/dx is set to zero, the second derivative d^2y/dx^2 is positive. Compare *maxima*.

minimum Abbreviation, min. The lowest member in any series of values. Compare *maximum*.

minimum detectable signal A signal whose intensity is just higher than the threshold of detection.

minimum discernible signal Abbreviation, *MDS*. The lowest input-signal amplitude that will produce a discernible output signal in a radio receiver.

miniscope A very-small-sized, lightweight oscilloscope.

minitrack A system employed to track an earth satellite, using signals transmitted to the satellite by a line of ground radio stations.

minometer A radioactivity-measuring instrument comprising an ionization chamber and a string galvanometer.

minor beats Secondary or extraneous beats produced in a beat-note system and due to various sum and difference frequency byproducts of the heterodyne process. Compare *major beats*.

minor bend A bend in a rectangular waveguide, made without twisting.

minor cycle See *word time*.

minor face In a hexagonal quartz crystal, one of the three smaller faces. Compare *major face*.

minority carrier The type of current carrier that is present in inconsequential numbers in a processed semiconductor material. Electrons are minority carriers in *p*-type material, holes in *n*-type material. Compare *majority carrier*.

minor lobe One of the lesser lobes in an antenna directivity pattern. Also see *lobe*. Compare *major lobe*.

minor loop A subordinate path for the circulation of information or control signals in an electronic system, e.g., minor feedback loop. Compare *major loop*.

minuend In subtraction, the quantity from which another (the subtrahend) is subtracted to give the difference (remainder).

minute 1. Abbreviation, min. A unit of measure of time. 1 min = 60s. Compare *second, 1* and *hour*. 2. Symbol ('). A unit of arc measure. 1 = 60" (60 seconds) = 1/60 a geometrical degree. Compare *second, 2*. 3. Very small.

MIR Abbreviation of *memory-information register*.

mirror 1. A device which consists chiefly of a highly polished or silvered surface that reflects a large part of the radiation (such as light) striking it. 2. Radar-interference material (see *chaff*). 3. To reflect, as by a mirror.

mirror galvanometer A galvanometer in which a mirror is moved by the coil. The mirror either reflects a spot of light along an external scale, or it reflects the scale, which is then read through a small telescope.

mirror-galvanometer oscillograph See *electromechanical oscilloscope*.

mirror image 1. A response curve which is identical with another. 2. For a quarter-wave Marconi antenna the extra quarter wave supplied by the earth, which acts somewhat like a mirror. 3. With the exception of a transposition of corresponding parts, a duplicate image.

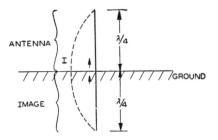

Mirror image, 2.

mirror-reflection echo A false radar echo or set of echoes, caused by reflection of the radar beam from a plane surface prior to its encountering the target or targets. The beam may also be reflected from one target to another.

mirror-type meter A meter whose movable coil carries a small mirror (rather than a pointer) which reflects a beam of light to produce a spot on a translucent scale.

MIRV Abbreviation of *multiple independently targeted reentry vehicle*, a guided missile with several warheads for separate targets.

misaligned head In a tape recorder, a record or a pickup head that is incorrectly oriented with respect to the passing tape.

misfire Failure of a gas tube or mercury-arc tube to ignite at the correct instant.

misleading precision In electronic calculations and data recording, greater precision than the instruments or conditions justify. Also see *significant figures*.

mismatch The condition resulting from joining two circuits or connecting a line to a circuit in which the impedances are substantially different.

mismatched impedances Impedances which are unequal and therefore do not satisfy the conditions for maximum power transfer.

mismatch factor For mismatched impedances, the ratio of current flowing in the mismatched load to the current that would flow in a matched-impedance load. The ratio is

$$(\sqrt{4\ Z_1\ Z_2})/(Z_1 + Z_2),$$

where 1 is the mismatched impedance and Z_2 is the matched impedance. Also called *reflectance, reflection factor, reflectivity*, and *transmission factor*.

mismatch loss For a load that is mismatched to a source, the ratio P_1/P_2, where P_1 is the power a matched load would absorb from the source and P_2 is the power actually absorbed by the mismatched load.

missile An explosive projectile. Also see *ballistic missile* and *guided missile*.

mistor A variable-resistance device, used for the purpose of detecting the presence of a magnetic field and for measuring the magnetic-field strength.

MIT Abbreviation of *master instruction tape*.

mix 1. To produce a beat signal (either the sum or the difference frequency) from two input signals. 2. The proportion of powdered iron and other inert substances in a ferromagnetic transformer core. Different mixes result in different operating characteristics.

mixed-base notation A number system in which bases are alternated between two digit positions. Also called *mixed radix notation*.

mixed calculation A mathematical calculation or expression in which more than one operation is used.

mixed modulation Modulation of several kinds coexisting in a system. Thus, a small amount of undesired frequency modulation might accompany amplitude modulation, or vice versa.

mixed number A number having integral (whole) and fractional parts, e.g., 3.141 59.

mixer A device such as a tube, transistor, or semiconductor diode, employed to mix two input signals and deliver an output equal to their difference and/or sum (see *mixing*).

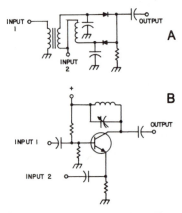

At A, a passive mixer circuit. At B, an active mixer circuit using a bipolar transistor.

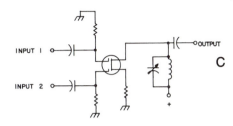

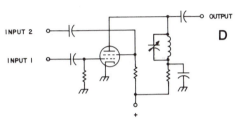

**At C, a mixer employing a
dual-gate metal-oxide-semiconductor
field-effect transistor (MOSFET).
At D, a mixer using a tetrode vacuum tube.**

mixer noise Shot-effect noise in a mixer stage (see *mixer*).

mixer tube See *converter tube*.

mixing Combining several signals so that some desired mixture of the original signals is obtained. Compare *modulation*.

mixture 1. A combination of two or more signals which retain their characteristics even when they interact to produce heat-frequency products. 2. A diffusion of one substance throughout another, without a solution or a chemical reaction resulting.

mixture melting point Abbreviation, *mmp*. The temperature at which a mixture of solid substances starts turning into a liquid at 1 atmosphere of pressure. This melting point depends upon the melting points of the substances and their relative concentration in the mixture. Also see *melting point* and *mixture*.

mks Abbreviation of *meter-kilogram-second*.

mL Abbreviation of *millilambert*.

ml Abbreviation of *milliliter*.

mm Abbreviation of *millimeter*.

mmA Abbreviation of *millimilliampere*.

mmf 1. Abbreviation of *magnetomotive force*. 2. Abbreviation of *micromicrofarad* (obsolete; see *picofarad*).

mmol Abbreviation of *millimole*.

mmp Abbreviation of *mixture melting point*.

mmv Abbreviation of *monostable multivibrator*.

mμ 1. Abbreviation of prefix *millimicro* (nano). 2. Abbreviation of *millimicron* (nanometer).

Mn Symbol for *manganese*.

mn Symbol for *neutron rest mass*.

mnemonic 1. Pertaining to memory or to memory systems. 2. A memory code or device.

mnemonic code In computer practice, a programming code such as assembly language, that, although easily remembered by the programmer, requires subsequent conversion to machine language.

mntr Abbreviation of *monitor*. (Also, *mon*.)

MO Abbreviation of *master oscillator*.

Mo Symbol for *molybdenum*.

mobile communications Communications between one mobile station and another, or between mobile and fixed stations.

mobile radio service See *mobile communications*.

mobile receiver A radio, television, or other receiver aboard a moving vehicle.

mobile-relay station A fixed station that receives a signal from a mobile station and retransmits it to one or more other mobile stations.

mobile station A station installed and operated aboard a moving or stationary vehicle. Compare *portable station*.

mobile transmitter A radio, television, or other transmitter aboard a moving vehicle.

mobility Symbol, μ. The ease with which a current carrier (electron or hole) travels. In a semiconductor material, electrons have the higher mobility. Also see *carrier mobility*.

mockup See *dummy, 1*.

mod 1. Abbreviation of *modulator*. 2. Abbreviation of *modulus*. 3. Abbreviation of *modification*.

mode 1. One of the ways a given resonant system can oscillate. 2. One of the ways electromagnetic energy may be propagated through a device or system. See *modes of propagation*.

mode coupling The exchange or interaction of energy between identical modes (see *mode, 1, 2*).

mode filter A waveguide filter that separates waves of different propagation mode but of the same frequency (see *modes of propagation*).

model 1. A sometimes small working or mockup version of a circuit, system, or device, illustrative of the final version. 2. A mathematical representation of a process, device, circuit, or system—employed in the analysis of any of these.

modem See *modulator—demodulator*.

mode purity 1. In an ATR tube, a condition in which there is no mode conversion. 2. In a modulated radio-frequency signal, the condition in which there exist no undesirable types of modulation.

moderator A substance such as graphite or heavy water, employed to slow neutrons in an atomic reactor. Also see *accelerator, 1* and *reactor, 2*.

modes of propagation The configurations in which microwave energy may be transmitted through a waveguide. For examples, see *waveguide mode*.

modes of resonance In a microwave cavity, the configurations in which resonant oscillation can exist, depending on the way the cavity is excited.

modification In computer practice, changing program addresses and instructions by performing logic and arithmetic on them as if they were data. Also see *program modification*.

modified index of refraction The index of refraction of the troposphere at any height increased by p. The factor p equals h/r, where h is elevation above sea level and r is the mean radius of the earth.

modified refractive index See *modified index of refraction*.

modifier A data item used to change a computer program instruction so that it can be used to implement different successive operations. Also see *program modification*.

modifier register See *index register*.

modify See *modification*.

moding A fault characterized by oscillation of a magnetron in undesirable modes.

modular technique See *building-block technique.*

modulated amplifier A usually highfrequency amplifier whose output is modulated. Compare *modulated oscillator.* In the amplitude modulation of an amplifier, there is little or no disturbance of the carrier frequency. Also see *modulation.*

modulated beam 1. An electron beam (as in a cathode-ray tube) which is intensity modulated by a desired signal. 2. A light beam whose intensity is modulated for communications or control purposes.

modulated carrier A carrier wave whose amplitude, frequency, or phase is modulated at a (usually) lower-frequency rate to convey intelligence.

modulated continuous wave Abbreviation, *mcw.* A high-frequency carrier wave modulated by a continuous, lower-frequency wave as in mcw telegraphy.

modulated cw See *modulated continuous wave.*

modulated electron beam See *modulated beam, 1.*

modulated light beam See *modulated beam, 2.*

modulated oscillator A usually highfrequency oscillator whose output is modulated. Compare *modulated amplifier.* Also see *modulation.*

modulated-ring pattern See *gearwheel pattern* and *spot-wheel pattern.*

modulated stage In a transmitter, an amplifier or oscillator in which the signal information is impressed on the carrier.

modulated wave See *modulated carrier.*

modulatee A term sometimes used for simplicity and expediency to designate a stage or circuit upon which modulation is impressed, e.g., the final rf amplifier stage in a radio transmitter.

modulating electrode 1. In an oscilloscope, an electrode (usually the *intensity electrode*) to which a signal may be applied to intensitymodulate the electron beam. 2. In a TV picture tube, the electrode (grid or cathode) to which the video signal is applied.

modulating signal The signal employed to modulate another signal (the carrier).

modulation Combining two signals with the result that one signal voltage (the *carrier voltage*) is varied by and in sympathy with the other (the *modulating voltage*). Usually, the carrier is a high frequency, and the modulation a low frequency. Also see *absorption modulation; amplitude modulation; frequency modulation; phase modulation; pulse modulation, 1, 2;* and *series modulation.*

modulation capability The maximum percentage of modulation a transmitter will permit before nonlinearity sets in. Also see *modulation linearity.*

modulation characteristic For an amplitude-modulated wave, the ratio of the instantaneous envelope amplitude of the modulated carrier to the instantaneous modulating voltage.

modulation code In a modulated transmitter, a system of modulation in which certain signal variations or pulses represent particular characters. Examples are the Morse code, Baudot code, and the television picture code.

modulation coefficient Symbol, m. A figure expressing depth of modulation. For an amplitude-modulated wave in which the upward modulation is equal to the downward modulation, $m = (Em - Ec)/Ec$, where Ec is the peak-to-peak voltage of the unmodulated carrier and Em is the peak-to-peak voltage of the modulated carrier. For full (100%) modulation, $m = 1$.

modulation depth See *depth of modulation.*

modulation distortion 1. In a modulated signal, sound distortion introduced by the modulation process or by the receiver circuit. 2. External cross modulation (see *cross modulation, 1*).

modulation envelope See *envelope, 1.*

modulation-envelope distortion Distortion of the envelope of an amplitude-modulated wave (see *envelope, 1*).

modulation factor See *modulation coefficient.*

modulation frequency Abbreviation, *fm.* The frequency of a modulating signal.

modulation linearity In a modulated signal, the degree to which carrier modulation follows the amplitude of the modulating signal. Ideally, this is a linear relationship.

modulation meter See *percentage-modulation meter.*

modulation monitor 1. A linear detector with a pickup coil (or antenna) and headphones for listening to a modulated signal. 2. Percentage-modulation meter (sometimes combined with a carrier-shift indicator).

modulation noise See *noise behind the signal.*

modulation percentage The extent to which a signal is amplitude modulated (see *depth of modulation*) in terms of the percentage of the unmodulated-carrier amplitude represented by the peak increase of carrier amplitude during modulation. Mod % = 100 m, where m is the modulation coefficient.

modulation ratio For a modulated signal, the quotient Mr/Mi, where Mr is the percentage of radiated-signal modulation, and Mi is the percentage of current modulation.

modulator A device or circuit for producing modulation.

modulator cell See *Kerr cell.*

modulator crystal A transparent piezoelectric crystal to which a signal voltage may be applied to modulate a beam of polarized light passing through it.

modulator-demodulator A circuit or device such as a biased diode or diode bridge, which can perform either modulation or demodulation. Also called *modem*, a blend of the two terms.

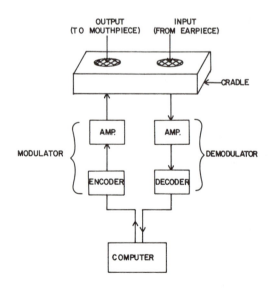

Modulator-demodulator.

modulator driver An amplifier stage that delivers excitation current, voltage, or power to a modulator stage.

modulator glow tube A cold-cathode gas tube which provides a bright point source of light that can be modulated for such purposes as facsimile transmission.

module An assembly containing a complete self-contained circuit (or subcircuit) and often miniaturized and made for plug-in operation.

modulo An arithmetic operation in which the remainder of a division is the result of interest; for example, 14 modulo 3 (the divisor) is 1 (the remainder).

modulometer Any instrument, such as a *percentage-modulation meter*, used to measure the degree of modulation of a signal and often also such modulated-signal characteristics as *carrier shift*, extraneous amplitude modulation, and extraneous frequency modulation.

modulo n check In computer practice, a technique for verifying the validity of a number used as an operand. The number being so checked is divided by another number to provide a remainder (check digit) that goes with the number. After the number is say, transmitted through some part of a computer system, it is again divided by the original divisor, and if the remainder is the check digit, the data has retained its integrity.

modulus 1. Absolute magnitude. Also see *absolute value* and *impedance*. 2. Abbreviation, *mod.* In computer practice, a whole number indicating the number of states a counter sequences through in each cycle. 3. Abbreviation, *mod.* A number (constant or coefficient) expressing the degree to which some property is possessed by a material or body; e.g., *modulus of elasticity, shear modulus, bulk modulus* 4. A constant by which a logarithm to one base must be multiplied to obtain a logarithm of the same number to another base. Thus,

$$\log_e N = 2.303 \log_{10} N,$$

where 2.303 is the modulus.

modulus of elasticity The stress-to-strain ratio in a material under elastic deformation.

moire In a television or facsimile picture, an effect produced by the convergence of straight lines. When the lines are nearly parallel to the scanning lines, the converging lines appear irregular.

moisture meter See *electric hygrometer* and *electronic hygrometer*.

mol Abbreviation of *mole*.

molar conductance See *molecular conductance*.

molar solution A solution such as an electrolyte, containing 1 mol of solute per liter of solvent. Compare *normal solution*.

mold 1. To form a mass of material into a desired shape, as by placing the material into a container having that shape and allowing it to solidify. In *hot molding*, the material is melted in the container and then cooled to hardness; in *cold molding*, the material is shaped without heat and solidifies with time. 2. The hollow contrivance employed to mold a material, as in 1, above.

molded capacitor A capacitor that is molded into a protective body of insulating material. Also see *mold, 1, 2* and *molded component*.

molded ceramic capacitor A ceramic-dielectric capacitor enclosed in a molded housing. Also see *mold, 1, 2; molded capacitor;* and *molded component*.

molded coil See *molded inductor*.

molded component A part (such as a capacitor, coil, or resistor) that is completely enclosed in a protective material (such as a plastic) which is molded around it. Also see *mold, 1, 2*.

molded electrolytic capacitor A solid-dielectric electrolytic capacitor enclosed in a molded housing. Also see *mold, 1, 2; molded capacitor; molded component;* and *solid-electrolytic capacitor*.

molded glass capacitor A glassplate-dielectric capacitor enclosed in a molded glass housing. Also see *glass capacitor; mold, 1, 2; molded capacitor;* and *molded component*.

molded inductor An inductor that is molded into a protective housing of insulating material. Also see *mold, 1, 2* and *molded component*.

molded mica capacitor A micadielectric capacitor enclosed in a molded housing. Also see *mica capacitor, mold, 1, 2; molded capacitor;* and *molded component*.

molded mud A molding compound having inferior electrical characteristics. Also see *mold, 1, 2* and *molded component*.

molded paper capacitor A paperdielectric capacitor enclosed in a molded housing. Also see *mold, 1, 2; molded capacitor; molded component; paper capacitor*.

molded-porcelain capacitor A capacitor enclosed in a body of molded porcelain. Also see *mold, 1, 2; molded capacitor; molded component;* and *porcelain*.

molded resistor A resistor that is molded in a protective housing of insulating material. Also see *mold, 1, 2* and *molded component*.

molded transistor A transistor that is encapsulated in a protective molding compound such as epoxy resin. Also see *mold, 1, 2* and *molded component*.

mole Abbreviation, *mol.* The amount of substance in a system containing as many specified entities (atoms, molecules, ions, subatomic particles, or groups of such particles) as there are atoms in 12 grams of carbon 12.

molectronics Molecular electronics.

molecular circuit See *monolithic integrated circuit*.

molecular conductance Symbol, μ. For a solution such as an electrolyte, the product of specific conductivity and the volume (in liters) of a solution that contains 1 gram molecule of the solute. Also see *solute; solution, 1;* and *solvent, 1, 2.*

molecular conductivity See *molecular conductance*.

molecular electronics The technique of processing a single block of material so that separate areas perform the functions of different electronic components, i.e., the entire block constitutes a circuit, e.g., monolithic integrated circuit.

molecular magnets According to the molecular theory of magnetism, the elemental magnets believed to be formed by individual molecules.

molecular theory of magnetism Each molecule in a piece of magnetic metal is itself a magnet (possessing a north and a south pole). These tiny magnets are thought to be normally oriented in a haphazard manner; however, when the material is magnetized by an external force, they align themselves with each other.

molecular weight Abbreviation, *mol wt.* In a molecule of a substance, the sum of the atomic weights of the constituent atoms. Thus, the molecular weight of silicon dioxide (SiO_2) is 60.09, since a molecule of this substance contains a silicon atom whose atomic weight is 28.09, and 2 oxygen atoms, each having an atomic weight of 16: 28.09 + 2(16) = 60.09.

molecule The basic particle of a *compound*; each molecule usually contains two or more atoms. For example, the formula $AgNO_3$ represents silver nitrate, each molecule of which contains 1 atom of silver (Ag), 1 atom of nitrogen (N), and 3 atoms of oxygen (0).

moletronics See *molecular electronics*.

mol wt Abbreviation of *molecular weight*.

molybdenite detector An early semiconductor diode (crystal detector) employing a crystal of molybdenite (a sulfide of molybdenum).

molybdenum Symbol, *Mo*. A metallic element. Atomic number, 42. Atomic weight, 95.95. Molybdenum is used in the grids and plates of some electron tubes.

moment The tendency to produce motion around a point, as by torque, or the product of a quantity and the distance to a point. Moment of force is *Fd*, where *F* is force and *d* is distance, e.g., the perpendicular distance between a pivot and the point at which torque is applied. In this example, the torque divided by the resultant angular velocity gives the *moment of inertia*.

momentary-contact switch A switch that maintains contact only while it is held down. Such a switch may be a pushbutton device, a toggle switch, a slide switch, or a lever switch.

momentary switching Switching of short duration, often characterized by a quick make and break immediately following activation of the switch. Compare *dwell switching*.

moment of inertia For a torque motor, the inertia of the armature around the axis of rotation. Also see *moment*.

mon 1. Abbreviation of *monitor*. 2. Abbreviation of *monaural*.

monatomic The condition of a molecule of a material that has only one atom or has only one replaceable atom or radical.

monatomic molecule A molecule having a single atom, e.g., argon, helium, neon. Compare *diatomic molecule*.

monaural Specifically, the condition of hearing with only one ear, as opposed to *binaural* or *stereo*. Compare *binaural*.

monaural recorder A single-track recorder, as opposed to a stereophonic recorder.

Monel See *Monel metal*.

Monel metal An alloy of nickel (67%), copper (28%), iron, manganese, and other metals (5%). Its resistivity is approximately 42 microhm-centimeters at 20 °C.

monimatch An amateur version of the reflected-power meter and SWR meter.

moniscope A special cathode-ray tube that produces a stationary picture for testing television equipment. Its name is a blend of *monitor* and *scope*.

monitor A device which affords the sampling of a signal or quantity. Familiar examples are *line-voltage monitor, TV monitor,* and *modulation monitor*.

monitor head A separate playback head included in some tape recorders for listening to the tape as it is being recorded.

monitoring The act, process, or technique of observing an action while it is in progress or checking a quantity while it is varying. Examples: *carrier monitoring, modulation monitoring, line-voltage monitoring*.

monitoring amplifier An auxiliary amplifier employed in monitoring an audio-frequency system.

monitoring antenna A usually small pickup antenna employed with a signal monitor or monitoring receiver.

monitoring key In a telephone system, a key used for the purpose of listening in on a two-way conversation.

monitoring receiver A radio or television receiver employed specifically to monitor a transmission directly.

monitor system A computer program usually stored in the ROM supplied by the hardware vendor. It controls (1) the implementation of programs written by the user and (2) the operation of peripherals associated with program runs and inputting or outputting data to or from memory (through the use of such program statements as CLOAD, for example). Also called *executive program*.

monkey chatter A jabbering sort of radio interference experienced when the sidebands of an undesired signal are detected, inverted in frequency, and inserted as sidebands of a desired signal.

monk's cloth A coarse drapery fabric sometimes employed to soundproof the walls and ceiling of a radio studio or recording booth.

monobrid circuit An integrated circuit in which either (1) several monolithic IC chips are interconnected to form a larger, singlepackage circuit, or (2) monolithic IC chips are interwired with thin-film components into a single-package circuit. The name is an acronym for *mono*lithic hy*brid*.

monochromatic 1. Being of one color in nature. 2. Being of a single wavelength in nature (pertaining to radiation of any kind). 3. Pertaining to black-and-white TV.

monochromaticity Consisting of one color of visible light. The brightness may vary from black to maximum.

monochromatic power density At a given temperature, the energy radiated per square centimeter of blackbody surface per second per unit wavelength range. Also see *blackbody* and *blackbody radiation*.

monochromatic sensitivity Sensitivity to light of one color only.

monochrome television Black-and-white television.

monoclinic crystal A crystal having three axes of unequal length; two of them intersect obliquely and are perpendicular to the third, e.g., the type of crystal found in one form of sulfur (*monoclinic sulfur*).

monocrystalline material See *single-crystal material*.

monode A one-element device such as a filament-type lamp, thermistor, voltage-dependent resistor, barretter, etc.

monofier A microwave tube consisting of an oscillator and an amplifier together in the same enclosure.

monograph A usually short (and sometimes learned) paper, pamphlet, or booklet written on a single topic.

monogroove stereo A method of making a stereophonic phonograph disk in which both channels are recorded as a single groove.

monolayer A thin film having a thickness of a molecule of the film material.

monolithic filter A crystal or ceramic filter constructed entirely from one piece of piezoelectric material.

monolithic integrated circuit An integrated circuit that is formed in a single block of wafer or semiconductor material. The name is derived from the Greek *monolithos (one stone)*. Compare *hybrid integrated circuit* and *thin-film integrated circuit*.

monometallic Containing or employing only one metal.

monomial A mathematical expression having only one term, e.g., πd. Compare *polynomial*.

monomolecular film See *monolayer*.

monophonic recorder See *singletrack recorder*.

monophonic system A single-channel sound system. Compare *stereo system*.

monopole antenna See *quarter-wave monopole*.

monopulse In radar and electronicnavigation practice, employing one pulse to determine azimuth and elevation simultaneously.

monorange speaker A loudspeaker that reproduces most of the full audio range. Also called *extended range speaker*. Compare *tweeter* and *woofer*.

monostable Having one stable state.

monostable blocking oscillator Abbreviation, *MBO*. A blocking oscillator which behaves somewhat like a one-shot multivibrator. That is, the oscillator delivers a single output pulse each time it receives an input (trigger) pulse.

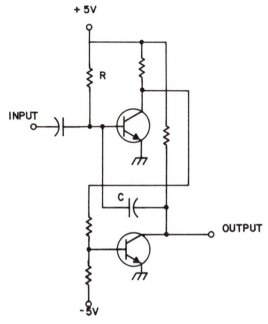

+ 5V

INPUT

R

C

OUTPUT

-5V

Monostable multivibrator.

monostable multivibrator A multivibrator that delivers one output pulse for each input (trigger) pulse. Also called *one-shot circuit* and *single-shot multivibrator*. Compare *astable multivibrator* and *bistable multivibrator*.

monostatic reflectivity The property whereby, for certain reflectors (such as a corner reflector), the angle of reflection of a ray is the same as the angle of incidence and no other.

monotone A single-pitch (singlefrequency), often sustained sound.

monotron A moniscope tube (see *moniscope)*.

monovalent See *univalent*.

Monte Carlo method 1. The use of statistical sampling in the approximate solution of an engineering problem. 2. In computer practice, the construction of mathematical models from randomly selected components taken from representative statistical populations.

Moog synthesizer An electronic device which can be made to simulate virtually any sound, including that of musical instruments and the human voice, through the use of several audio oscillators whose output can be controlled to produce tones of various harmonic content, duration, and attack and decay periods.

MOPA Abbreviation of *master oscillator-power amplifier*.

morphological electronics See *molecular electronics*.

Morse 1. *Morse code*. A general term signifying either the American Morse code or the international Morse code. 2. Telegraphy (wire or radio). 3. To signal by means of the Morse code.

Morse code A telegraph code widely used in radio and wire telegraphy. It employs dots and dashes—short sounds (dits) and long sounds (dahs)—to represent letters of the alphabet, numerals, and punctuation marks. Also see *American Morse code* and *continental code*.

MOS Abbreviation of *metal-oxidesemiconductor* or *metal-oxide-silicon* (see

metal-oxide-silicon field-effect transistor).

mosaic 1. *Photomosaic*. 2. The pattern of tiny photoelectric particles in a TV camera tube that convert the image into electric charges.

MOSFET Abbreviation of *metal-oxidesilicon field-effect transistor*.

MOSROM Abbreviation of *metaloxide-silicon read-only memory* (see *read-only memory)*.

MOST Abbreviation of *metal-oxidesilicon transistor* (see, for example, *metal-oxide-semiconductor field-effect transistor*).

most significant character Abbreviation, *MSC*. In positional number representation, the leftmost character in a significant group such as a word.

most significant digit Abbreviation, *MSD*. In a number, the leftmost digit that isn't a zero (zero being insignificant in this context).

mother 1. Generally, when used with other nouns, that which supports or appears to support other similar things, e.g., *mother board*. 2. A mold which has been electroformed from a master phonograph disk.

motherboard In a computer or data-processing device, the circuit board on which most of the main circuitry is mounted.

mother crystal A natural quartz crystal from which is produced the piezoelectric plates and other components employed in electronics.

motion The change of position of a point or body with respect to a point or reference.

motion detector A device for sensing the movement or stopping of a body such as a rotating shaft. Various sensors are used in different detectors: magnetic, photoelectric, capacitive, etc.

motion frequency The natural frequency (especially that of oscillation) of a servo.

motion-picture pickup In television practice, a camera (and the technique for using it) for picking up scenes directly from motion-picture film.

motor 1. A machine for converting electrical energy into mechanical energy. 2. The driving mechanism of a loudspeaker.

motorboard The basic mechanism of a tape recorder, embodying motor, flywheel, capstan, rollers, etc., and assembled on a board or panel.

motorboating A repetitive (usually low-frequency) popping or puffing noise in malfunctioning electronic equipment. It is so called from its resemblance to the sound of a motorboat.

motorboating filter In an audio amplifier circuit, a simple filter installed to prevent motorboating due to feedback through a common impedance such as a dc supply bus. Also see *decoupling filter*.

motor capacitor See *motor-run capacitor* and *motor-start capacitor*.

motor constant Symbol, *kM*. The ability, expressed numerically, of a torque motor to convert electric input power into torque.

motor converter An electromechanical device for converting alternating current to direct current.

motor-driven relay An electromechanical relay whose contacts are opened and closed by a rotating motor.

motor effect Magnetic force between adjacent current-carrying conductors.

motor-generator A combination of motor and generator in a single machine assembly. A common arrangement is a low-voltage motor turning a high-voltage generator. The machines' shafts may be coupled, or the motor may turn the generator through a belt, chain, or gear train.

motor meter A meter in which the movable element is essentially a continuously rotating motor. See, for example, *kilowatt-hour meter*.

motor-run capacitor A powerfactor-boosting capacitor which is present (together with its auxiliary winding) in parallel with the main winding in an induction motor. Also see *capacitor motor*.

motor-speed control A method of controlling the speed of a motor by varying the magnitude and/or phase of its current. Electronic methods involve the use of diode, thyratron, ignitron, or thyristor circuits.

motor-start capacitor A capacitor which, with an auxiliary winding, is switched into a motor circuit during the starting period and is automatically disconnected (with the winding) after the motor reaches normal running speed. Also see *capacitor motor*.

motor starter See *starting box*.

mould See *mold*.

mount 1. A mechanical device with which a component is attached to a circuit board or chassis. 2. The attachment of a circuit board to a chassis. 3. The hardware with which an antenna is attached to a mast. 4. The hardware by which a microphone is attached to a boom or other support. 5. In general, any attaching hardware.

Mountain standard time Abbreviation, *MST*. Local mean time of the 105th meridian west of Greenwich. Also see *Greenwich mean time, standard time, time zone,* and *coordinated universal time*.

mouth The radiating end of a horn (antenna, loudspeaker, etc.).

MOV Abbreviation of *metal-oxide varistor*.

movable-coil meter See *D'Arsonval type meter*.

movable-coil modulator See *electromechanical modulator, 3*.

movable contact The traveling contact in a relay or switch. Compare *fixed contact*.

movable-iron meter See *iron-vane meter*.

moving-coil galvanometer A galvanometer whose movable element is a coil of fine wire suspended or pivoted between the poles of a magnet.

moving-coil microphone See *dynamic microphone*.

moving-coil motor The driving mechanism of a moving-coil dynamic loudspeaker.

moving-coil pickup See *dynamic pickup*.

moving-coil speaker See *dynamic speaker*.

moving-conductor microphone See *velocity microphone*.

moving-diaphragm meter A headphone employed as a sensitive indicator in ac bridge and potentiometer measurements.

moving element In an electromechanical device, the portion that moves physically under variable operating conditions.

moving-film camera An oscilloscope camera in which the film is drawn past the lens continuously at a constant speed, rather than being advanced frame by frame as in a motion-picture camera.

moving-iron meter See *iron-vane meter*.

moving-target indicator Abbreviation, *MTI*. A radar device which principally displays moving-target information.

moving-vane meter See *iron-vane meter*.

mp Abbreviation of *melting point*.

mp Symbol for *proton rest mass*.

MPG Abbreviation of *microwave pulse generator*.

mpg Abbreviation of *miles per gallon*.

mph Abbreviation of *miles per hour*.

MPO Abbreviation of *maximum power output*.

mps 1. Abbreviation of *meters per second*. 2. Abbreviation of *miles per second*.

MPT Abbreviation of *maximum power transfer*.

MPX Abbreviation of *multiplex*.

MR Abbreviation of *memory register*.

mrad Abbreviation of *milliradian*.

MRBM Abbreviation of *medium-range ballistic missile*.

MRIA Magnetic Recording Industry Association.

mrm Abbreviation of *milliroentgens per minute*.

MS Abbreviation of *mass spectrometer*.

m²/s Square meters per second. The unit of kinematic viscosity.

M-scan In radar practice, a modified A-scan display in which a pedestal signal is moved along the base line to a point where it coincides with the base line of the reflected signal to determine the distance to the target.

msec Abbreviation of *millisecond*. (Also, *ms*.)

msg Abbreviation of *message*.

MSI Abbreviation of *medium scale integration*.

MST Abbreviation of *Mountain standard time*.

MT Abbreviation of *metric ton*.

MTI Abbreviation of *moving-target indicator*.

MTR Abbreviation of *magnetic tape recorder*.

mtr Abbreviation of *meter*. (Also, *m*.)

M-type backward-wave oscillator A broadband, voltage-tuned backward-wave oscillator in which the electrons interact with a backwardtraveling rf wave. Compare *O-type backward-wave oscillator*.

mu 1. Abbreviation of the prefix *micro*, i.e., a typewritten substitute for the Greek letter μ. 2. The 12th letter (M, μ) of the Greek alphabet. The symbol (μ) for *amplification factor, permeability,* the prefix *micro, micron, electric moment, inductivity, magnetic moment, molecular conductivity*. 3. Amplification factor of a tube. 4. Permeability.

μ Abbreviations and symbols beginning with μ are listed at the back of this section following the entry *myriametric waves*.

μin. Abbreviation of *microinch*.

μl Abbreviation of *microliter*.

μP Abbreviation of *microprocessor*.

MUF 1. Abbreviation of *maximum usable frequency*. 2. Occasional abbreviation of *microfarad*, now obsolete.

mu factor 1. The relative effect exerted by a given pair of electrodes on the current of an electron tube. The mu factor is $dE2/dE1)/E3$, a constant. 2. Amplification factor.

Muller tube A vacuum tube that conducts by means of ionization of an internal gas. Used for radiation detection.

multiaddress In a computer, a multiple memory location.

multianode tube An electron tube in which several main anodes operate opposite a single cathode.

multiband amplifier See *wideband amplifier*.

multiband antenna A single antenna such as one employed in amateur radio and television reception, which is operated in several bands or channels.

multiband device A device such as a tuner, receiver, transmitter, or test instrument which operates in several selectable bands.

multiband oscilloscope See *wideband oscilloscope*.

multiband receiver See *wideband receiver*.

multicasting The use of two FM stations or FM and television stations to broadcast separately the two channels of a stereo program which is reproduced by being picked up simultaneously with two receivers.

multicavity magnetron A magnetron whose anode block has two or more cavities.

multicellular horn A loudspeaker in front of which is one or more rows of rectangular cells through which the sound passes.

multichannel Pertaining to a radio-communication system that operates on more than one channel at the same time. The individual channels may contain identical information, or they may contain different signals.

multichip circuit A microcircuit composed of interconnected active and passive chip-type components.

multichip integrated circuit An integrated circuit composed of circuit elements on separate, interconnected chips.

multicontact switch A switch having more than two contacting positions.

multicoupler An impedance-matching device employed to couple several receivers to a single antenna.

multielement antenna A directive antenna having more than two elements (director and reflector).

multiemitter transistor 1. A transistor having more than one emitter. 2. A power transistor having several emitters connected in parallel in the transistor structure.

multigrid tube 1. An electron tube having several grid structures, e.g., control grid, screen, and suppressor. 2. An electron tube having several signal grids. See, for example, *pentagrid converter tube*.

multigun crt A cathode-ray tube having a number of electron guns. An example is the 64-gun tube that provides symbols, letters, and numbers for multiline information displays.

multihop propagation Propagation of a radio wave by several successive reflections between the ionosphere and the surface of the earth.

multilayer circuit A circuit consisting of several sections printed or deposited on a separate layers, which are subsequently stacked into a sandwich.

multilayer coil A coil in which the turns of wire are wound in several complete layers, one on top of the other. Compare *single-layer coil*.

multimeter A meter that allows measurement of different quantities (e.g., current, voltage, and resistance); the functions are usually made available through a selector switch.

multimode operation The operation of a device in more than one of its modes simultaneously (see *mode, 1, 2*).

multipactor A microwave switching tube capable of operating at high power levels. Characterized by high operating speed.

multipath cancellation In multipath reception, cancellation of a signal when the separate components arrive at the receiver in equal amplitude and opposite phase.

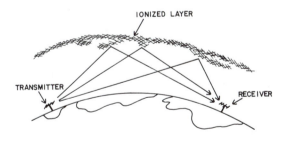

Multipath cancellation.

multipath delay In multipath reception, the lag between signal components arriving over different paths.

multipath effect At a receiver, the difference in arrival time of multipath signals (see *multipath delay*).

multipath reception Reception of a signal taking more than one path, one of them being a path directly from the transmitter, the others being the paths of reflections from objects. Such reception is the cause of ghosts in TV pictures.

multipath transmission 1. Transmission of a signal over two or more paths, one being direct and the others via reflections from some object. Also see *multipath reception*. 2. Multihop propagation.

multiphase system See *polyphase system*.

multiple access system A timesharing data processing system which can be used by a number of people at remote locations, usually through peripherals other than conventional terminals, e.g., cash registers linked to a computer system for inventory control.

multiple-address code In computer practice, an instruction code requiring the user to specify the address of more than one operand per instruction.

multiple address instruction A computer program instruction specifying the address of more than one operand.

multiple break 1. Interruption of a circuit at several points. 2. Contact bounce.

multiple-chip circuit See *hybrid integrated circuit*.

multiple connector In a flowchart, a symbol showing the mergence of several flowlines.

multiple integral See *double integral* and *triple integral*.

multiple ionization Successive ionization, as when an ion repeatedly collides with electrons. Also see *double ionization*.

multiple-length number In computer practice, a number occupying more than one register.

multiple-loop feedback system A feedback system employing more than one feedback loop. An example is a tunable audio amplifier (see *parallel-tee amplifier*) in which a negative-feedback *RC* path provides tuning, while a positive-feedback path sharpens selectivity.

multiple modulation See *compound modulation*.

multiple precision arithmetic An arithmetic process using several words (bit groups) for an operand to insure accuracy.

multiple programming 1. Computer operation in which several programs in memory share peripherals and processor time. 2. Programming a computer so that several logic or arithmetic operations can be carried out at the same time.

multiple punching Punching more than two holes in one column of a punched card.

multiple-purpose meter See *multimeter*.

multiple-purpose tester A multimeter sometimes combined with some other instrument such as a test oscillator.

multiple reel file In a data processing system, a magnetic tape data file of more than one reel.

multiple speakers A group of more than two loudspeakers, usually operated from a single amplifier system.

multiple-stacked broadside array A stacked array in which a number of collinear elements are stacked above and below each other. Also see *broadside array* and *collinear antenna*.

multiple station Descriptive of a communications network in which more than one terminal is used.

multiple-unit steerable antenna A short-wave antenna system intended to prevent or minimize the effects of fading in received signals. It consists principally of a number of rhombic antennas feeding the receiver. The system utilizes the various waves arriving at different angles. An electric steering system causes the antennas to be selected automatically in the best combination and with their outputs in the proper phase. Also see *diversity reception*.

multiple-unit tube See *multiunit tube*.

multiple winding See *drum winding*.

multiplex Pertaining to the use of a single link such as a wire line or radio carrier, for the simultaneous transmission of separate information.

multiplex adapter A special circuit (or auxiliary unit) used in FM receivers for stereophonic reception from a station transmitting a multiplex broadcast. Also see *multiplexing* and *multiplex stereo*.

multiplex code The transmission of multiple signals over a single medium, using a code such as Morse, Baudot, or ASCII.

multiplex data terminal A computer terminal acting as a modem by virtue of its accepting and transferring signals between its input/output devices and a data channel.

multiplexer A device that allows two or more signals to be transmitted simultaneously on a single carrier wave, communications channel, or data channel.

multiplexing A process in which a low-frequency carrier is modulated and then mixed with a vhf signal. Because the information is on one carrier only, a receiving operator cannot recover the information from the vhf signal unless he knows the modulation frequency. See, for example, *subcarrier* and *subsidiary communication authorization*.

multiplex stereo The use of multiplexing techniques to broadcast both channels of a stereophonic program on a single carrier wave. Also see *multiplexer* and *multiplexing*.

multiplex telegraphy 1. A system of wire telegraphy in which two or more messages are sent simultaneously in one or both directions. 2. A system of radiotelegraphy in which two or more messages may be sent simultaneously on the same carrier wave.

multiplex telephony 1. A system of wire telephony in which two or more messages may be sent simultaneously in one or both directions over the same line. 2. A system of radiotelephony in which two or more messages may be sent simultaneously on the same carrier wave.

multiplicand In arithmetic multiplication, the factor that is operated on by another factor (the *multiplier*) to yield the *product*.

multiplication 1. The arithmetic process whereby a certain factor is added to itself the number of times indicated by another factor (the *multiplier*). Compare *division, 1*. 2. A method of increasing a quantity, magnitude, or rate by some desired factor. See, for example *frequency multiplier* and *voltage multiplier*. Compare *division, 2, 3*.

multiplier 1. Frequency multiplier. 2. Voltage multiplier. 3. A circuit or device for performing arithmetic multiplication. 4. Voltmeter multiplier. 5. In arithmetic multiplication, the factor by which another factor (the *multiplicand*) is multiplied to yield the *product*.

multiplier amplifier A frequencymultiplying amplifier (such as a doubler, tripler, or quadrupler) whose output circuit is tuned to an integral multiple of the input frequency. Compare *straight-through amplifier*.

multiplier phototube See *photomultiplier tube*.

multiplier prefix A prefix such as *mega*, which, when affixed to the name of a quantity (e.g., *cycle*) indicates the amount by which that quantity is to be multiplied. Separate listings are given for other such prefixes.

multiplier probe 1. A resistor-type test probe for a voltmeter, which increases the range of the meter and, therefore, acts as a voltmeter multiplier. 2. A voltage-multiplier type of test probe for a voltmeter, which multiplies the test voltage before it is applied to the instrument, e.g., voltage-doubler probe.

multiplier register In a computer, a register that retains a multiplier (*multiplier, 5*) during a multiplication.

multiplier resistor A resistor connected in series with a current meter (usually a milliammeter or microammeter) to make it a voltmeter. Also called *voltmeter multiplier*.

multiplier tube 1. Electron-multiplier tube. 2. A tube recommended for use in a frequency multiplier (doubler, tripler, quadrupler, etc.). 3. A tube recommended for use in a voltage multiplier (doubler, tripler, quadrupler, etc.).

multipoint circuit A transmission circuit or system, in which information may be entered or retrieved at two or more locations.

multipolar Pertaining to the presence of more than two magnetic poles.

multipolar machine A motor or generator having a number of poles in its armature and field.

multiposition relay A relay having more than two positions of closure. See, for example, *selector relay*.

multiposition switch A switch having more than two contacting positions. See, for example, *selector switch* and *stepping switch*.

multiprocessing 1. In a computer system, the running of more than one program at once. 2. Parallel data processing. 3. Time sharing.

multiprogramming In a computer system, a technique that allows two or more programs to be executed at the same time. The total processing time is shorter than the time that would be needed to run each program one after the other.

multipurpose meter An electric meter that performs several functions, usually available through a function selector. Examples: *voltammeter, voltohmmeter, voltohmmilliammeter*.

multipurpose tube A general-purpose electron tube that is suitable for several different applications such as detection, amplification, and oscillation, e.g., a 6C4. (Not to be confused with *multiunit tube*.)

multirange instrument An instrument provided with several separate ranges, usually selectable by range switching, e.g., five-range voltmeter, two-scale ammeter, four-band oscillator.

multisection filter A filter having two or more selective sections connected in cascade.

multisegment magnetron See *multicavity magnetron*.

multiskip transmission See *multihop propagation*.

multistage device A device having several stages operating in cascade or otherwise coordinated with each other, e.g., a five-stage amplifier.

multistage feedback Feedback (positive or negative) between several stages in a system as opposed to feedback between the output and input of a single stage.

multistage oscillation Oscillation resulting from positive feedback between several stages of an amplifier as opposed to that occurring between the output and input of a single stage. Often identified as *two-stage oscillation, four-stage oscillation*, and so on.

multistage X-ray tube An X-ray tube providing electron acceleration by means of successive ring-shaped anodes biased to a higher voltage than the preceding one.

multiswitch A switch having a number of poles and contacts.

multitester An instrument such as a multimeter or a combined signal generator and oscilloscope, that performs a number of different test functions.

multitrack recording 1. A recording on two or more tracks, e.g., multitrack disk, multitrack tape. 2. Making a recording as in 1, above.

multiturn potentiometer A potentiometer whose shaft must be rotated through several complete revolutions to cover the full resistance range.

multiunit tube A tube envelope containing several separate sets of elements, e.g., a 12FR8 (diode-triode-pentode). (Not to be confused with *multipurpose tube.*)

multivalent Having a valence greater than 2. Compare *univalent.*

multivibrator A circuit usually containing two tubes or transistors in an *RC*-coupled amplifier, whose output is capacitance-coupled to the input. The two stages switch each other alternately in and out of conduction at a frequency determined by the *R* and *C* values. The type described is a *free-running* or *astable multivibrator.* Other types are *bistable multivibrator* (flip-flop) and *monostable multivibrator* (one-shot).

multiwire antenna An antenna having more than one wire in the radiating section. In early flat-top antennas, such wires were usually connected together at one end.

multiwire doublet antenna A doublet antenna having more than one wire in its radiator. A common form is the folded dipole, which is in effect two closely spaced dipole radiators connected together at both ends, one being center fed.

Mumetal A high-permeability alloy of iron and nickel, valued especially for use as a magnetic shield for cathode-ray tubes.

Munsell color system A system for specifying colors in terms of hue, saturation, and brilliance, according to charts. Also see *color matching.*

Muntz metal See *yellow metal.*

MUPO Abbreviation of *maximum undistorted power output.*

Murray-loop bridge A specialized form of Wheatstone bridge in which two of the resistance arms are supplied by a two-wire line (such as a telephone line). By means of resistance measurements made with the bridge, the distance from an office or station to a ground fault on the line can be determined.

MUSA Abbreviation of *multiple-unit steerable antenna.*

muscovite Formula, KH2Al2(SiO4)3. A high-grade variety of mica which is often pale-green or brown.

musical quality See *timbre.*

music chip An integrated circuit for producing various musical effects, such as tones, percussion, and the like.

music power For a power amplifier, the short-term output power obtained in the reproduction of music waveforms, in contrast to rms power output, which is the lower and more meaningful value.

music synthesizer See *Moog synthesizer.*

muting 1. Disabling a receiver or amplifier under no-signal or weaksignal conditions. 2. Softening or muffling a sound.

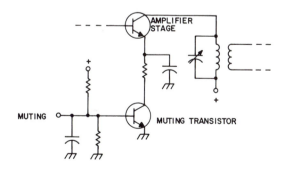

Muting.

muting circuit 1. An interferencepreventing device that automatically shuts off a radio receiver during the operation of a transmitter, and vice versa. 2. Squelch circuit. 3. In a stereo receiver, an electronic element that cuts off all audio when no signal is present or when the receiver is being tuned between carriers.

muting switch A switch or relay that cuts off a receiver during periods of transmission, or when reception is not desired. Generally, a cutoff voltage is applied to one of the intermediate-frequency stages to accomplish muting. In some cases, the audio-frequency circuit may be disabled.

mutual antenna coupling Usually undesirable coupling between antennas when they are placed too close together.

mutual capacitance Inherent capacitance between two conductors.

mutual-capacitance attenuator An attenuator which in its simplest form is essentially a shielded, two-plate variable capacitor.

mutual conductance See *transconductance.*

mutual-conductance meter See *transconductance meter.*

mutual-conductance tube tester See *transconductance meter.*

mutual impedance An impedance that is shared by two or more branches of a circuit.

mutual inductance Symbol, *M.* Unit, henry. The property shared by neighboring inductors or inductive devices, which enables induction to take place. A mutual inductance of 1 henry is present when a current change of 1 ampere per second in one inductor induces 1 volt across another second inductor. Also see *inductance.*

mutual-inductance attenuator An attenuator consisting essentially of two coupled coils (input and output) whose spacing can be gradually changed.

mutual-inductance bridge See *Carey—Foster mutual-inductance bridge.*

mutual induction The action whereby the magnetic field produced by alternating current in one conductor produces a voltage in another isolated conductor.

mutual interference 1. Adjacent-channel interference. 2. Any kind of interference between or among radio-frequency communications circuits.

mutually exclusive events Two or more events (or data points) such that the occurrence of one precludes the occurrence of the other.

mutually exclusive sets See *disjoint sets.*

MV 1. Abbreviation of *megavolt.* 2. Abbreviation of *multivibrator.*

mV Abbreviation of *millivolt.*

mv Abbreviation of *medium voltage.*

MVA Abbreviation of *megavolt-ampere.*

mV/m Abbreviation of *millivolts per meter.*

MVP Abbreviation of *millivolt potentiometer.*

MW Symbol for *megawatt.*

mW Symbol for *milliwatt.*

mw Abbreviation of *medium-wave.*

MWH Abbreviation of *megawatt-hour.*

mW RTL Abbreviation of *milliwatt resistor-transistor logic* or *low-power resistor-transistor logic.*

Mx Abbreviation of *maxwell.*

Mycalex An insulating material consisting of mica bonded with glass. Dielectric constant, 6 to 8. Resistivity, 10^{13} ohm-cm.

Mylar A tough, plastic insulating material commonly used as a magnetic tape base. Dielectric constant, 2.8 to 3.7. Dielectric strength, 7000V/mil.

Mylar capacitor A capacitor in which the dielectric film is Mylar.

Mylar tape Magnetic recording tape employing a Mylar film as the substrate.

mym Abbreviation of *myriameter*.

myoelectricity 1. Bioelectric pulses of 10 to 1000 μV amplitude produced by muscular activity and detectable by electrodes attached to the skin. Also see *electromyogram, electromyograph,* and *electromyography.* 2. The study of such phenomena.

myogram See *electromyogram*.

myograph See *electromyograph*.

myography See *electromyography*.

myriameter Abbreviation, *mym.* A metric unit of linear measure equal to 10 kilometers.

myriametric waves Radiation at very long wavelengths: 10 to 100 kilometers (3 to 30 kHz).

μ 1. Abbreviaton of the prefix *micro*. 2. Symbol for *micron*. 3. Symbol for *permeability*. 4. Symbol for *inductivity* (see *dielectric constant*). 5. Symbol for *molecular conductance*. 6. Symbol for *electric moment*. 7. Symbol for *magnetic moment*. 8. Symbol for *coefficient of friction*. 9. Symbol for *amplification factor* (see *amplification factor, 1*).

μμ 1. Symbol for prefix *micromicro* (*pico*). 2. Symbol for *micromicron* (picometer).

μ**A** Abbreviation of *microampere*.

μ**B** Symbol for *Bohr magneton*.

μ**b** Abbreviation of *microbar*.

μ**Ci** Abbreviation of *microcurie*.

μe Symbol for *electron magnetic moment*.

μ**F** Abbreviation of *microfarad*.

μ**g** Abbreviation of *microgram*.

μ**H** Abbreviation of *microhenry*.

μ**mm** Abbreviation of *micromillimeter* (nanometer).

μn Symbol for *nuclear magneton*.

μo Symbol for *free-space permeability constant*.

μξ Abbreviation of *micromho*. (Also, μ*mho*.)

μΘ Abbreviation of *microhm*. (Also, μ*ohm*.)

μΘ-**cm** Abbreviation of *microhmcentimeter*. (Also, μ*ohm-cm*.)

μp Symbol for *proton magnetic moment*.

μ**rd** Abbreviation of *microrutherford*.

μ**S** Abbreviation of *microsiemens*.

μ**s** Abbreviation of *microsecond*. (Also, μ*sec*.)

μ**V** Abbreviation of *microvolt*.

μ**V/m** Abbreviation of *microvolts per meter*.

μ**W** Abbreviation of *microwatt*.

N 1. Symbol for *number* (as in *Np*, the number of turns in a primary coil). 2. Symbol for *nitrogen*. 3. Symbol for *newton*. 4. Abbreviation of *number* (also *No.*).

n 1. Abbreviation of prefix *nano*. 2. Symbol for *number*. 3. Symbol for a *term* (e.g. n a multiplier) having an assigned value. 4. Symbol for part of an expression or operator (as in $2n$, $n-1$, $\sqrt{n}$, n^2, etc.). 5. Symbol for *index of refraction*. 6. Symbol for *amount of substance* (unit, *mol*).

NA 1. Abbreviation of *not available* (as on a specifications sheet). 2. Abbreviation of *not applicable*.

NA Symbol for the *Avogadro constant*.

Na Symbol for *sodium*.

nA Abbreviation of *nanoampere*.

NAB National Association of Broadcasters.

NAB curve In audio-frequency practice, the standard magnetic tape playback equalization curve developed by the National Association of Broadcasters (NAB).

nabla See *del*.

NAE National Academy of Engineering.

Nagaoka's constant For single-layer coils, the correction factor $2a$ which must be multiplied by the value of inductance calculated as $L = (0.039\ 48a^2N^2)/b$, where L is inductance in μH, a is coil radius in cm, b is coil length in cm, and N is the number of turns.

Nancy Phonetic alphabet code word for the letter *N*.

NAND circuit Also called *NOT-AND circuit*. In computer and control practice, a circuit which delivers a zero output signal only when two or more input signals are coincident ones. The performance of the NAND circuit is the inverse of that of the *AND circuit*.

NAND gate A gate that performs the function of a *NAND circuit*.

nano A prefix meaning *billionth(s)*, i.e., 10^{-9}. Abbreviation, n.

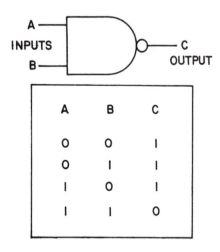

NAND circuit.

nanoampere Abbreviation, nA. A unit of current; 1 nA equals 10^{-9} ampere.

nanofarad Abbreviation, *nF*. A unit of low capacitance; 1 nF equals 10^{-9} farad.

nanohenry Abbreviaton, *nH*. A unit of low inductance; 1 nH equals 10^{-9} henry.

nanometer Abbreviation, *nm*. A unit of short wavelength or distance (length); 1 nm equals 10^{-9} meter.

nanosecond Abbreviation, *ns*. A time interval of 10^{-9} second.

nanovolt Abbreviation, *nV*. A unit of low voltage; 1 nV equals 10^{-9} volt.

nanovoltmeter A sensitive voltmeter for measuring potential in the nanovolt range.

nanowatt Abbreviation, *nW*. A unit of low power; 1 nW equals 10^{-9} watt.

nanowattmeter A meter for measuring power in the nanowatt range.

NAP Abbreviation of *nuclear auxiliary power*.

napier See *neper*.

Napierian base The number 2.718 281 828 459 as the base of natural logarithms. Symbol, *e*. Also see *Napierian logarithm* and *natural number*.

Napierian logarithm Abbreviation, *ln*. A logarithm to the base *e* (see Napierian base). Also called *natural logarithm*. Compare *common logarithm*.

NAPU Abbreviation of *nuclear auxiliary power unit*.

narrative A computer program statement that, rather then being an instruction, merely describes the purpose of what follows—usually the steps in a routine or a block of instructions—as a debugging or program modification aid. In a program written in BASIC, such a statement is preceded by the abbreviation REM, for *remark*. Also called *comment*.

narrowband 1. A frequency band in which the difference between upper and lower limits is only a few Hz or kHz, with respect to bandwidths usually encountered in the service specified. 2. Any standard bandwidth whose limits have been reduced from a prior standard. In amateur radio FM practice, for example, narrowband means 5 kHz at vhf and uhf, whereas wideband implies a 15 kHz bandwidth (a prior standard).

narrowband amplifier An amplifier whose passband is restricted to a fraction of the frequency spread common to the amplifier's application.

narrowband fm See *narrowband frequency modulation*.

narrowband frequency modulation Abbreviation, *NBFN* or *NFM*. 1. Frequency modulation in which the deviation is equal to the modulation frequency. 2. See *narrowband, 2*.

narrowband interference Signal interference whose bandwidth is narrow compared with that of the circuit affected.

narrow bandwidth For a given service, a frequency band whose upper and lower limits are only a few kHz or MHz apart, as compared with wider bandwidths ordinarily used in similar service.

narrow-sector recorder A directional radio receiver for locating sources of atmospheric noise.

NARTB National Association of Radio and Television Broadcasters.

NAS National Academy of Sciences.

National Association of Broadcasters Abbreviation, *NAB*. A countrywide organization of radio and television broadcasters.

National Association of Radio and Television Broadcasters Abbreviation, *NARTB*.

National Electrical Manufacturers Association Abbreviation, *NEMA*. A countrywide organization of manufacturers of electrical and electronic equipment and supplies.

National Electric Code Abbreviation, *NEC*. Safety regulations and procedures issued by the National Fire Protection Association for the installation of electrical wiring and equipment in the United States. Although the code is advisory from the Association's standpoint, it is enforced to various degrees by local authorities.

National Television Systems Committee Abbreviation, *NTSC*. A U.S. organization of television companies and other interested organizations which proposed recommendations for television standards that were approved by the Federal Communications Commission.

natural-decay curve See *exponential decrease*.

natural antenna frequency The fundamental resonant frequency of an electromagnetic antenna.

natural disintegration The natural decay of a radioactive substance as a result of the continuous emission of particles and rays. The time taken by one-half of a quantity of a radioactive substance to decay into a different isotope of the substance (less than 1 second to several million years) is the *half-life* of the substance.

natural electricity 1. Atmospheric electricity. 2. The electricity in living organisms, that is, *bioelectricity*.

natural frequency See *natural-resonant frequency*.

natural function generator A function generator in which the function is a natural law.

natural-growth curve See *exponential increase*.

natural interference Interference from atmospheric electricity, the sun, stars, etc., as opposed to *man-made interference*.

natural logarithm See *Napierian logarithm*.

natural magnet A natural material, such as magnetite (lodestone), which exhibits permanent magnetism.

natural magnetism Magnetism found in some natural materials (see *natural magnet*) and in the earth itself.

natural number Symbol, *e* or *ε*. The number 2.718 28.... This number is the limit of the series $e = 1 + 1 + 1/2! + 1/3! + 1/4! + ...1/n!$. The natural number *e* is the base of *Napierian logarithms*.

natural period The period corresponding to a *natural resonant frequency*.

natural radiation noise in the form of radiation emitted by natural radioactive substances, cosmic rays, etc. Also called *background radiation*.

Natural radiation Noise in the form of radiation emitted by natural radioactive substances, cosmic rays, etc. Also called *background radiation*.

natural resonance Resonance due to the unique physical constants of a body, circuit, or system. Also see *natural resonant frequency*.

natural resonant frequency 1. The frequency at which a circuit or device responds with maximum amplitude to applied signals. 2. The frequency at which a circuit or device generates maximum energy. 3. The frequency at which a body vibrates at maximum amplitude.

natural wavelength The wavelength corresponding to the natural-resonant frequency.

nautical mile A nautical unit of linear measure equal to 1.852×10^3 meters (1.1508 statute miles).

nav Abbreviation of *navigation* or *navigational*.

Navaglobe A radionavigation technique used at very-low or low frequencies over long distances.

NAVAIDS Abbreviation of *navigational aids*.

navar A radar system in which a ground radar scans the immediate vicinity of an airport, observing the flight activity, and transmits such observations to aircraft in flight. The name is an acronym from *navigation and ranging*.

navigation aid An electronic device or system—such as radar and radio direction finding—which assists in the navigation of vehicles on land or sea or in the air.

navigation beacon A beam providing aircraft and ships with navigational aid.

NAWAS Abbreviation of *national attack warning system*.

Nb Symbol for *niobium*.

nb Abbreviation of *narrowband*.

NBFM Abbreviation of *narrowband frequency modulation*. Also abbreviated *NFM*.

NBS National Bureau of Standards.

NBTDR Abbreviation of *narrowband time domain reflectometry*.

nbw Abbreviation of *noise bandwidth*.

NC 1. Abbreviation of *normally closed*. 2. On drawings, abbreviation of *no connection*. 3. Abbreviation of *numerical control*.

N/C Abbreviation of *numerical control*.

nc Abbreviation of *no connection*.

n-channel junction field-effect transistor Abbreviation, *NFET*. A junction field-effect transistor in which the gate junction is formed on a bar or die of n-type semiconductor material. Compare *p-channel junction field-effect transistor*.

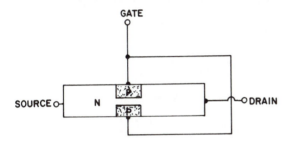

N-channel junction field-effect transistor.

n-channel MOSFET A metal-oxide silicon field-effect transistor in which the channel is n-type silicon. Also see *depletion-type MOSFET*, *depletion-enhancement-type MOSFET*, and *enhancement-type MOSFET*.

NCMT Abbreviation of *numerical-controlled machine tool*.

NCS Abbreviation of *net control station(s)*.

N-curve A negative-resistance E—I curve roughly resembling an *N*. Compare *S-curve*.

Nd Symbol for *neodymium*.

N-display A radar display in which the target is represented by a pair of vertical blips whose amplitude indicates target direction and whose position along the horizontal base line (as determined by lining up a pedestal signal with the blips) is read from the pedestal-adjustment calibration. The system is somewhat similar to a K-display (see *K-scan*.)

N-doped zinc-oxide ceramic A nonlinear-resistance material used in the manufacture of some voltage-dependent resistors.

NDRO Abbreviation of *nondestructive read*.

NDT Abbreviation of *nondestructive testing*.

Ne Symbol for *neon*.

near-end crosstalk Crosstalk originating at, or near, the telephone subscriber line in which the interference is noticed.

near field 1. The radiation field within a radius of 1 wavelength from a transmitting antenna. 2. The sound field near a loudspeaker or other reproducer.

near infrared Pertaining to radiation of 3 to 30 micrometers in wavelength.

near zone See *near field, 1*.

NEB Abbreviation of *noise equivalent bandwidth*.

NEC National Electric Code.

necessary bandwidth The minimum bandwidth needed (with a given emission) to transmit information at a required rate and of a required quality.

neck The straight portion of a crt envelope, i.e., the part between the base and the bulbous part.

NEDA National Electronic Distributors' Association.

needle 1. The stylus of a phonograph cartridge. 2. The pointer of an indicating meter. 3. One of the electrodes in a voltage-measuring spark gap (see *needle gap*). 4. The slender, pointed metal tip of a test probe. 5. A stylus used on stacks of punched cards.

needle chatter See *needle talk*.

needle drag In disk recording and reproduction, friction between the needle (stylus) and disk. Also called *stylus drag*.

needle electrode See *needle, 3, 4*.

needle gap A spark gap composed of two needles having an adjustable air gap between their points. An unknown high voltage is measured in terms of the gap width necessary for sparking.

needle memory A computer memory in which the dual-state elements are thin magnetic needles.

needle pointer The pointer of a meter or compass.

needle pressure See *vertical stylus force*.

needle probe See *needle, 4*.

needle scratch In disk recording and reproduction, noise resulting from vibration of the needle (stylus) because of an irregular groove surface. Also called *surface noise*.

needle talk Direct radiation of sound by the stylus of a phonograph pickup. Also called *needle chatter*.

needle telegraph A telegraph in which the Morse code characters are converted into magnetic needle deflections.

needle test point See *needle, 4*.

needle-tip probe A test probe which terminates in a sharp needle. Also see *needle, 4*.

needle voltmeter See *needle gap*.

neg Abbreviation of *negative*.

negate 1. To insert the NOT operation in front of a digital expression. 2. To change logic 1 to logic 0 (high to low) or vice-versa.

negation The logical NOT operation (see *NOT*) in digital systems. Also see *NAND circuit, NOR circuit, NOR gate, NOT circuit*, and *NOT-OR circuit*.

negation element In a computer system, a device that can give the reverse of a condition, event, or signal.

negative 1. Possessing *negative electrification*. 2. Opposite of positive. 3. Less than zero. 4. A photographic image whose shadings are opposite to those in the scene.

negative acceleration Retardation; deceleration.

negative acknowledgement character In a handshaking or forward-error-correction system, a response by the receiving station that indicates a missed bit or bits.

negative angle An angle in the third or fourth quadrant in a system of rectangular coordinates. Compare *positive angle*.

negative bias A steady, negative dc voltage or current applied continuously to an electrode of a device, such as a tube or transistor, to establish the operating point for example. Compare *positive bias*.

negative bus See *negative conductor*.

negative charge An electric charge consisting of a quantity of *negative electrification*. Also see *charge, 1; electric charge*; and *unit electrostatic charge*. Compare *positive charge*.

negative conductor The conductor or wire connected to the negative terminal of a current, power, or voltage source. Compare *positive conductor*.

negative crystal A crystal that exhibits negative double refraction. Compare *positive crystal*.

negative electricity See *negative charge* and *negative electrification*.

negative electrification Electrification characterized by an excess of electrons. For example, when a glass rod is rubbed with a silk cloth, the cloth becomes negatively charged because electrons are rubbed off the glass onto the cloth. Similarly, when an atom acquires an extra electron, the atom becomes negatively charged because it has an excess of electrons. Compare *positive electrification*.

negative electrode 1. An electrode connected to the negative terminal of a current, power, or voltage source. 2. The negative output terminal of a current, power, or voltage source, such as a battery or generator.

negative error of measurement An error of measurement in which the difference between a measured value and the true or most probable value is negative. Compare *positive error of measurement*.

negative exponent In mathematical notation, an exponent indicating that a number is to be raised to a negative power, e.g., 10^{-5}. Raising a number (N) to a negative power ($-n$) means taking the reciprocal of the number raised to the same positive power. That is, $N^{-n} = 1/N^n$; thus, $10^{-2} = 1/10^2 = 1/100$.

negative feedback Feedback which is out of phase with the input signal. Also called *inverse feedback, degeneration,* and *degenerative feedback.* Compare *positive feedback*.

negative-feedback amplifier An amplifier in which negative feedback is employed to improve performance or modify response.

negative function A trigonometric function having the negative sign. In a rectangular coordinate system, the sine function is negative in the third and fourth quadrants, the cosine in the second and third, and the tangent in the second and fourth. Compare *positive function*.

negative gain A misnomer for *fractional gain*, arising from the fact that fractional gain may be expressed in negative decibels.

negative ghost In a TV picture, a ghost with negative (see *negative, 4*) shading.

negative glow In a gas-discharge tube, the luminous column following the Faraday dark space and preceding the Crookes' dark space. Compare *Faraday dark space* and *positive column*.

negative-gm oscillator See *negative-transconductance oscillator*.

negative-grid generator See *negative-grid oscillator*.

negative-grid oscillator A vacuum-tube oscillator operated with negative control-grid bias and some form of positive feedback. Compare *positive grid oscillator*.

negative ground In a direct-current power system, the connection of the negative pole to common ground.

negative image 1. A picture in which the blacks, whites, and shades in between are the reverse of those in the scene (see *negative, 4*). 2. An abnormal TV picture having the reverse shading described in 1, above.

negative impedance An impedance that displays the same nonohmic behavior as that of *negative resistance*.

negative indication Specifying negative fields on a punched card (even in a column outside the field), e.g.n by overpunching the most significant field column.

negative ion An atom which has an excess of electrons and, consequently, exhibits a net negative charge. Also called *anion*. Compare *positive ion*.

negative-ion generator A device for generating negative ions and circulating them into the surrounding air for health purposes.

negative lead See *negative conductor*.

negative-lead filtering Power-supply filtering in which the choke coils and capacitors are in the negative dc lead rather than in the positive lead (the usual position). One advantage of this arrangement is the lower insulation requirement of the choke.

negative light modulation In TV transmission, the condition in which transmitted power is increased by a decrease in the initial intensity of light. Compare *positive light modulation*.

negative line See *negative conductor*.

negative logarithm The logarithm of the reciprocal of a number; $-\log N = \log 1/N$.

negative logic Binary logic in which a high negative state represents 1 and a low negative state, zero. Compare *positive logic*.

negative measurement error See *negative error of measurement*.

negative modulation In amplitude-modulated TV transmission, the decrease in transmitted power when brightness is increased. Compare *positive modulation*.

negative modulation factor For an amplitude-modulated wave having unequal positive and negative modulation peaks, a ratio expressing the maximum negative deviation from the average for the envelope. Compare *positive modulation factor*.

negative number A number less than zero, i.e., one to which the minus sign ($-$) is assigned.

negative peak 1. The negative half of an ac cycle. 2. The maximum current or voltage amplitude of a negative peak.

negative-peak voltmeter An electronic voltmeter for measuring the negative peak amplitude of an ac waveform. In its simplest form, the instrument consists of a diode and dc microammeter, with the diode oriented to pass the negative half-cycle. Compare *positive-peak voltmeter*.

negative phase-sequence relay A phase-sequence relay which responds to the negative phase sequence in a polyphase circuit. Compare *positive phase-sequence relay*.

negative picture modulation See *negative modulation*.

negative picture phase In a TV signal, the picture-signal voltage swing from zero to negative in response to an increase in brightness. Compare *positive picture phase*.

negative plate 1. The negative member of a battery cell; electron flow is from the plate through the external circuit. 2. A vacuum-tube plate which is biased negatively, as in a Barkhausen—Kurtz oscillator.

negative pole See *negative electrode, 1, 2.*

negative positive zero Abbreviation, *NPO*. Pertaining to temperature-compensating capacitors having a temperature coefficient of capacitance that changes sign within a specified temperature range.

negative potential 1. The potential measured at a negative electrode with respect to the positive electrode or to ground. 2. Potential less than that of the earth as a reference.

negative resistance The property exhibited by a few devices that is characterized by a decrease in voltage drop across the device as current through it is increased, or vice versa—or of decrease in current through the device as voltage across the latter is increased, or vice versa. This is, of course, opposite to the behavior of an ohmic (positive) resistance. Whereas a positive resistance consumes power, a negative resistance seems to supply power. Also see *N-curve, negative-resistance region, negative resistor,* and *S-curve.*

negative-resistance amplifier A simple circuit in which a negative-resistance device, such as a tunnel diode, cancels the positive resistance of the circuit, causing amplification of an applied signal.

negative-resistance device See *negative resistor*.

negative-resistance diode 1. See *tunnel diode*. 2. A reverse-biased germanium diode (and occasionally a silicon diode) which exhibits negative resistance, i.e., reverse current decreases as reverse voltage is increased. 3. A special diode tube which, when operated at ultrahigh frequencies, exhibits negative resistance due to transit-time effects.

negative-resistance magnetron A split-anode magnetron operated at a combination of anode voltage and magnetic field strength corresponding to cutoff; it exhibits negative resistance to voltage applied symmetrically between the anode halves. The frequency of oscillation is determined by an external tank circuit.

negative-resistance oscillator An oscillator which consists of a negative-resistance device connected across a tuned circuit. The arrangement oscillates because the negative resistance cancels the positive resistance (losses) of the tuned circuit. See, for example, *dynatron oscillator, negative-resistance magnetron, negative-transconductance oscillator,* and *tunnel-diode oscillator.*

negative-resistance region That portion of the $E-I$ curve of certain devices which has a negative slope, i.e., the portion in which current decreases as voltage increases, or vice versa. Also see *N-curve, negative resistance,* and *S-curve.*

negative-resistance repeater A repeater that produces gain by means of negative effective resistants.

negative resistor A device exhibiting negative resistance. See, for examples, *dynatron tube* (see *dynatron effect*), *electric arc, negative-resistance diode, negative-resistance magnetron, negative-transconductance tube,* and *tunnel diode.*

negative space charge The cloud of electrons (negative particles) in the region surrounding an emitter, such as the hot cathode of a vacuum tube.

negative temperature coefficient Abbreviation, *NTC*. A number expressing the amount by which a quantity (such as the rating of a component) decreases when the temperature is raised. The coefficient is stated as a percentage or as so many parts per million per degree temperature rise. Compare *positive temperature coefficient* and *zero temperature coefficient.*

negative terminal See *negative electrode, 2.*

negative torque In an electric motor, a torque that acts against the operating torque.

negative transconductance Symbol, $-gm$. An effect which takes place between the suppressor and screen in a pentode when the screen is more positive than the plate, and the suppressor is negative with respect to the cathode (consequently acting as a virtual cathode). The suppressor repels back to the screen electrons which have passed through the control grid and screen. An increase in the negative voltage of the suppressor causes it to repel more electrons. The action causes the screen current to increase. Because of these actions, the suppressor—screen transconductance is negative.

negative-transconductance oscillator A low-powered stable self-excited oscillator employing the negative transconductance of a pentode. In operation, the oscillator is a negative-resistance circuit similar to, but more stable than, a dynatron. Also called *transitron oscillator.*

negative-transconductance tube A pentode biased so that retarding-field negative transconductance is obtained Also see *negative transconductance.*

negative transmission In a television or facsimile system, the condition in which brighter light corresponds to lower transmitted power, and dimmer light corresponds to higher transmitted power.

negative valence The valence of a negative ion. Also see *valence.*

negator A logical NOT element, i.e., one which outputs the complement of an input bit (1 for 0 and vice versa).

negatron The term that specifically differentiates the familiar electron from a positron (positive electron).

NEI Abbreviation of *noise equivalent input.*

NEL National Electronics Laboratory.

NELA National Electric Light Association.

N-electron In certain atoms, an electron whose orbit is outside of, and nearest to, those of the M-electrons.

NEMA National Electrical Manufacturers' Association.

nematic crystal A normally transparent liquid crystal which becomes opaque when an electric field is applied to it and becomes transparent again when the field is removed. The crystal material is cut in the form of a letter or numeral and provided with a reflecting backplate for display readouts in calculators, watches, and so forth. A nematic film, placed between a light source and the observer, makes it possible to switch the display on and off by passing and interrupting light reflected from the backplate.

nematic-crystal display A display device in which an electrically controlled film of nematic-crystal material is used to transmit and interrupt light from a lamp or from a reflecting mirror, in this way displaying characters in whose shape the film has been formed.

nematic liquid In a liquid-crystal display, a normally clear liquid that becomes opaque in the presence of an electric field.

nemo A radio or television program that is picked up from a location outside the studio. Also called *field pickup* or *remote.*

neodymium Symbol, *Nd*. A metallic element of the rare-earth group. Atomic number, 60. Atomic weight, 144.27.

neon Symbol, *Ne*. An inert-gas element. Atomic number, 10. Atomic weight, 20.183. Neon, present in trace amounts in the earth's atmosphere, is used in some glow tubes, readout devices, voltage-regulator tubes, and indicator lamps.

neon bulb A (usually small) neon-filled gas diode. It has a characteristic pink glow and is ignited by a firing voltage for the particular unit. Also called *neon glow lamp* and *neon tube.*

neon-bulb flip-flop A flip-flop circuit (bistable multivibrator) employing two neon bulbs as the bistable components.

neon-bulb gate A gate circuit containing a neon bulb biased below the firing point. A trigger voltage added to the bias voltage raises the applied voltage and fires the bulb, producing an output pulse.

neon-bulb logic Logic circuits composed of neon-bulb gates.

neon-bulb memory See *neon-bulb storage.*

neon-bulb multivibrator A multivibrator employing two neon bulbs as the switching components.

neon-bulb oscillator A simple relaxation oscillator consisting essentially of a neon bulb, capacitor, resistor, and dc supply. Whereas the frequency of the sawtooth-wave output depends principally on the capacitance and resistance values, the maximum operating frequency is limited by the deionization time of the gas—to about 5 kHz, in most instances.

neon-bulb overmodulation indicator The application of a neon-bulb overvoltage indicator as a monitor for amplitude-modulated rf signals. The bulb flashes each time the modulation percentage exceeds a predetermined value. Also called *neon-bulb modulation alarm.*

neon-bulb overvoltage indicator A relatively simple circuit in which a neon bulb flashes each time a voltage monitored by the circuit exceeds a predetermined value. The flash shows that the portion of the voltage presented to the bulb has exceeded the firing potential.

neon-bulb peak indicator See *neon-bulb overvoltage indicator*.

neon-bulb ring counter A ring counter composed of neon-bulb flip-flops. The maximum counting speed is limited by the deionization time of the neon bulbs to approximately 300 events per second.

neon-bulb sawtooth generator A relatively simple relaxation oscillator employing a neon bulb, capacitor, and resistor. The output is a sawtooth wave whose frequency is determined principally by the capacitance and resistance values.

neon-bulb scale-of-two circuit A scale-of-two circuit (frequency halver) employing neon bulbs as the bistable elements.

neon-bulb storage A storage (memory) device composed of neon bulbs. A fired bulb (representing a bit of stored information) remains fired until turned off by an erase signal.

neon-bulb stroboscope A stroboscope in which a neon bulb supplies the light flashes. The circuit is essentially that of the neon-bulb oscillator, the flash rate being continuously variable by an adjustable frequency control.

neon-bulb voltage regulator A simple circuit utilizing the constant voltage drop across a fired neon bulb as a regulated voltage. The usual circuit configuration is a neon bulb and current-limiting resistor in series with a power supply.

neon-bulb volume indicator A neon-bulb overvoltage indicator used in some tape recorders to show when the volume exceeds a predetermined level, especially when the volume is high enough to cause an unacceptable amount of distortion. Some tape recorders use two bulbs, one continuously indicating normal volume level (corresponding to, say, 1% harmonic distortion), the other flashing to indicate peaks (corresponding to, say, 3% harmonic distortion).

neon glow lamp See *neon bulb*.

neon lamp See *neon bulb*.

neon-lamp readout In digital computers and counters, a readout device employing neon bulbs as indicators. An illuminated bulb signifies a 1, a turned-off bulb denotes a 0.

neon pilot lamp A neon bulb employed as a pilot lamp operated from the power-line circuit of an electronic equipment. Also called *neon pilot light*.

neon triode A three-electrode neon bulb. The third electrode acts somewhat like the grid of a thyratron; i.e., application of a control voltage to the electrode causes the gas to break down and conduct current between the two regular electrodes.

neon tube See *neon bulb*.

NEP Abbreviation of *noise equivalent power*.

NEPD Abbreviation of *noise equivalent power density*.

neper (John *Napier*). Abbreviation, *Np*. A natural-logarithmic unit expressing the ratio of two values of current, voltage, or power; nepers = $\log_e\sqrt{x1/x2}$, where $x1$ and $x2$ are the values being compared and e is 2.718 28. The neper is related to the decibel (a similar unit base on common logarithms) in the following manner:1 Np = 8.686 dB; 1 dB = 0.115 1 Np

neptunium Symbol, *Np*. A radioactive metallic element produced artificially. Atomic number, 93. Atomic weight, 239.

Nernst effect The appearance of a voltage between the opposite edges of a metal strip that is conducting heat longitudinally when the strip is placed in a magnetic field perpendicular to the plane of the strip.

Nernst-Ettinghausen effect In a piezoelectric crystal, the tendency

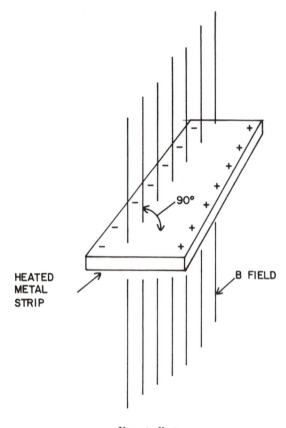

Nernst effect.

for a temperature gradient to exist as a result of applied electromagnetic fields.

Nernst lamp An incandescent lamp in which the filament material is a mixture of magnesia and certain rare oxides (such as yttria). A surrounding current-carrying coil heats the filament, lowering its resistance and causing it to glow, after which it continues to operate on a low current.

nerve cell See *neuron*.

nerve center The most essential part of a control system or communication network.

nervous breakdown The upset of a function in a digital computer (which becomes disabled as a result).

NESC National Electrical Safety Code.

nesting In digital computer and data processing practice, the including of a loop or a subroutine inside a loop or subroutine. Nested loops, for example, can be used to assign values to double-subscripted variables.

NET Abbreviation of *noise equivalent temperature*.

net See *network*.

net authentification A password or other special identification used in a radio net.

net capacitance The resultant capacitance in a circuit in which capacitances act in combination with each other.

net component The total value of two or more passive components of the same sort. See, for example, *net capacitance, net current, net impedance, net inductance, net power, net reactance, net resistance,* and *net voltage.*

net current The current flowing in a circuit in which currents aid or oppose each other.

net gain For an amplifier, the amount of gain remaining after all losses in the device have been subtracted.

net impedance The impedance of a circuit in which impedances act in combination with each other.

net inductance The inductance of a circuit in which inductances act in combination with each other.

net loss For an amplifier or other system, the algebraic sum of gains and losses between two points in the system.

net power The resultant power observed when power signals aid or oppose each other in a single circuit or system.

net reactance Symbol, Xt. The combined inductive and capacitive reactance in a circuit or device; $Xt = XL - Xc$.

net resistance The resistance of a circuit in which resistances act in combination with each other.

net voltage The resultant voltage at a point where voltages aid or oppose each other.

network 1. A circuit arrangement of electronic components, sometimes redundant in its design, e.g., *RC* network. 2. A chain of stations or other facilities, often organized for simultaneous operation.

network analog A circuit or circuits representing variables and used to express and solve a mathematical relationship between the variables.

network analysis The rigorous examination of a network to determine its properties and performance. Compare *network synthesis.*

network analyzer An analog or digital circuit for simulating and analyzing a network by means of measurements and calculations from the model so obtained.

network calculator An analog or digital device for determining the component values and performance of a given network.

network constant The value of a passive component (capacitance, inductance, resistance, etc.) used in a network.

network filter A transducer that passes or rejects waves, depending on their frequency.

network relay 1. A relay that provides protection of the circuits in a network; a circuit breaker. 2. In a communications net, the reception and retransmission of a message by a human operator.

network synthesis The design and fabrication of a network by rigorous engineering methods to achieve a prescribed performance.

network theorems See *compensation theorem, maximum power transfer theorem, Norton's theorem, reciprocity theorem, superposition theorem,* and *Thevenin's theorem.*

network topology Analyzing networks with signal-flow diagrams. Components of the diagram represent signal paths, open loops, closed loops, and node points.

network transfer function A function describing the overall processing of energy by a network; it is equal to Eo/Ei, where Ei is the input voltage and Eo the output voltage of the network.

Neumans law A property of mutual inductances. For a given orientation and environment for two inductors, the value of the mutual inductance does not change, regardless of the magnitude, frequency,

or phase of the currents in the coils. That is, mutual inductance is subject only to the physical environment surrounding the coils.

neuristor A two-terminal semiconductor device that simulates the behavior of a neuron (nerve cell) and allows machines to duplicate some of the neurological phenomena observed in the human body.

neuroelectricity Low-voltage electricity in the nervous system.

neuron A nerve cell in a living organism.

neutral 1. Having no electric charge; thus, an atom is normally neutral, since its internal positive charges neutralize its internal negative charges. 2. Devoid of voltage (e.g., a *neutral line*). 3. Pertaining to a salt (which is neither acidic nor alkaline).

neutral bus See *neutral wire.*

neutral circuit 1. A deenergized circuit, i.e., a "dead" one. 2. In a teletypewriter system, a circuit in which current flows in one direction.

neutral conductor See *neutral wire.*

neutral ground A ground connection to a neutral wire or to the neutral point of a circuit.

neutralization The process of balancing out positive feedback in an amplifier to prevent self-oscillation. Also see *neutralizing capacitor* and *neutralizing coil.*

neutralization indicator A neon bulb or rf meter used during neutralization to determine when the amplifier under adjustment is oscillating and is neutralized.

neutralize To eliminate positive feedback in a radio-frequency amplifier.

neutralized amplifier An amplifier in which neutralization has been performed to prevent self-oscillation.

neutralizing capacitor In a capacitively neutralized circuit, a small capacitor that serves as a coupler of signal energy from the output back to the input in reverse phase, to cancel self-oscillation of the circuit.

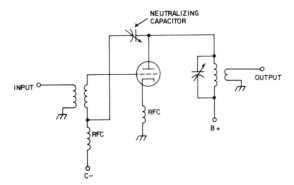

Neutralizing capacitor in simplified amplifier circuit.

neutralizing circuit Any component or set of components that is used to neutralize a radio-frequency amplifier.

neutralizing coil In an inductively neutralized circuit, a small coupling coil that picks up signal energy from the output and applies it in reverse phase to the input, to cancel circuit self-oscillation.

neutralizing tool A nonconducting screwdriver-like device for adjusting a neutralizing capacitor or coil. It is usually made of fiber or plastic.

neutralizing voltage The feedback voltage which cancels self-oscillation in the process of neutralization.

neutralizing wand See *neutralizing tool*.

neutral line See *neutral wire*.

neutral relay See *unpolarized relay*.

neutral wire In a polyphase transmission system, the line (wire) which doesn't carry current until the system is unbalanced.

neutretto An uncharged meson resulting from a positive meson that has collided with a neutron, or from a negative meson that has collided with a proton.

neutrino An uncharged particle smaller than a neutron and having zero rest mass.

neutrodon See *neutrodyne*.

neutrodyne 1. A usually triode-type radio-frequency amplifier that is neutralized by feeding energy through a small capacitor from a tap on the secondary coil of the plate output transformer back to the grid of the tube; a similar method is employed with transistors. 2. An early radio receiver employing such a neutralized amplifier, a detector, and an audio-frequency amplifier.

neutron An uncharged atomic particle having a mass approximately equal to that of the proton. The neutron is present in the nucleus of every atom except that of hydrogen.

neutron flux The magnitude of neutron radiation.

new candle See *candela*.

newton (Sir Isaac *Newton*). Symbol *N*. The SI unit of force. A force of 1 newton imparts an acceleration of 1 meter per second per second to a mass of 1 kilogram; 1N equals 10^5 dynes.

Newton's laws of motion Three natural laws discovered in 1686 by Sir Isaac Newton: *First law*: A body at rest or in motion tends to remain in that state unless it is acted upon by some force. *Second law*: A body tends to accelerate or decelerate when it is acted upon by a force, the acceleration being directly proportional to the force and inversely proportional to the mass of the body. *Third law*: For every action or acting force, there is an equal and opposite reaction or reacting force.

nexus An interconnection point in a system.

NF Abbreviation of *noise figure*.

nF Abbreviation of *nanofarad*.

NFET Abbreviation of *n-channel junction field-effect transistor*.

NFM Abbreviation of *narrowband frequency modulation*. Also abbreviated NBFM.

NFM reception Reception of narrow-band frequency modulation. Standard discriminators and ratio detectors may be used. Slope detection may be achieved with a standard AM receiver by tuning to one side of signal resonance. (detuning).

NFM transmission Transmission of a narrowband FM signal (see *narrowband frequency modulation*). A simple method consists of frequency modulating the master oscillator of the transmitter at the modulation frequency.

NFPA National Fire Protection Association (see *National Electric Code*).

NFQ Abbreviation of *night frequency*.

NG 1. Abbreviation of *negative glow* 2. Abbreviation of *no good*, used for marking inoperative or malfunctioning components.

NGT Abbreviation of *noise-generator tube*.

nH Abbreviation of *nanohenry*.

nhp Abbreviation of *nominal horsepower*.

Ni Symbol for *nickel*.

nibble In computer practice, a four-bit word.

nichrome A nickel-chromium alloy used in the form of a wire or strip for resistors and heater elements. Its resistivity is 108 microhmcentimeters at 20 °C.

nickel Symbol, *Ni*. A metallic element. Atomic number, 28. Atomic weight, 58.71. Nickel is familiar as an alloying metal in some resistance wires and as the material used in some electron-tube elements.

nickel-cadmium battery 1. A battery of nickel—cadmium cells. 2. Loosely, a nickel—cadmium cell.

nickel-cadmium cell A small 1.2V rechargeable cell in which the anode is cadmium, the cathode nickel hydroxide, and the electrolyte potassium hydroxide.

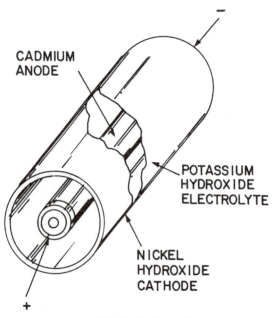

CADMIUM ANODE

POTASSIUM HYDROXIDE ELECTROLYTE

NICKEL HYDROXIDE CATHODE

Nickel cadmium cell.

nickel-iron battery See *Edison battery*.

nickel-oxide diode A diode fabricated from nickel-oxide semiconductor material.

nickel silver An alloy of copper, nickel, and zinc—used sometimes for resistance wire. Also called *German silver*.

Nicol prism An optical component for producing or analyzing plane-polarized light. It consists of two prisms of Iceland spar cemented together. Light entering the device strikes the interface, where the *ordinary ray* is totally reflected and the *extraordinary ray* passes through, both rays being plane-polarized perpendicular to each other.

NIDA Abbreviation of *numerically integrating differential analyzer*.

NIF Abbreviation of *noise improvement factor*.

night effect The phenomenon by which, at frequencies below 0.5 MHz, direction-finding signals received between sunset and the following sunrise seem to come from a transmitter that moves slowly back and forth in a line perpendicular to the line between transmitter and receiver.

night-effect errors In radio direction-finding, errors resulting from the night effect.

night range The distance over which signals from a given transmitter are consistently received after sunset.

ni junction In a semiconductor device, the junction between an n-type layer and an intrinsic layer.

nil Negligible; nothing.

nines complement Pertaining to a method for representing negative numbers (used in some computers and accounting machines). The number becomes the result of subtracting each digit from a digit one less than the radix, e.g., the decimal number − 365 would be 999 minus 365, or 634. Before output, complements are convered to the correct form.

ninety-column card A punched card having 90 columnar positions, any of which can be punched to represent a number or symbol.

niobium Symbol, Nb. A metallic element chemically resembling *tantalum*. Atomic number, 41. Atomic weight, 92.91.

Nipkow disk The perforated scanning disk of an early electromechanical television system.

NIPO Abbreviation of *negative input/positive output*.

NIR Abbreviation of *near infrared*.

nit In digital-computer practice, a choice among equally likely events; a nit = 1.44 *bit*.

nitrocellulose See *cellulose nitrate*.

nitrogen Symbol, *N*. A gaseous element. Atomic number, 7. Atomic weight, 14.008. Nitrogen is the most abundant component (78%) of the Earth's atmosphere.

Nixie tube See *readout lamp*.

Nixonite A trade name for cellulose acetate, a plastic.

Nixonoid A trade name for cellulose nitrate, a plastic.

NJCC National Joint Computer Conference.

n-layer A semiconductor layer which is doped to provide current carriers that are predominantly electrons. Compare *p-layer*.

n-level logic 1. A multilevel form of logic, with n different possible states. 2. In a computer, the connection of up to n logic gates.

NLR Abbreviation of *nonlinear resistance* or *nonlinear resistor*.

NLS Abbreviation of *no-load speed*.

N/m² Abbreviation of *newtons per square meter* (pascals).

Nm²/Kg Newton square meters per kilogram, the SI unit of the gravitational constant.

NMAA National Machine Accountants Association.

NMOS A metal-oxide semiconductor device made on a p-type substrate whose active carriers, electrons, migrate between n-type source and drain contacts.

NMR Abbreviation of *nuclear magnetic resonance*.

NMR Abbreviation of *normal-mode rejection*.

n – n junction In a semiconductor device, especially an integrated circuit, the junction between two n-type regions having somewhat different properties (sometimes designated *n*1 and *n*2).

NO Abbreviation of *normally open*.

No Symbol for *nobelium*.

No. Abbreviation of *number*.

no-address instruction In digital computer practice, an instruction requiring no reference to storage or memory for its execution.

nobelium Symbol, *No*. A radioactive element produced artificially. Atomic number, 102. Atomic weight, 255 (?).

Nobili's rings See *electric rings*.

noble Chemically inert or inactive; noble metals oxidize less rapidly than non-noble (*base*) ones.

noble gas An inert rare gas—such as argon, helium, krypton, neon, and xenon—used in electronic glow devices.

noble metal Gold, silver, platinum.

noctovision A television transmission system using infrared rays instead of visible light to scan the object.

no-charge machine fault time Unproductive computer time resulting from errors or a malfunction.

nodal point See *node*.

node 1. The terminal point at which two or more branches of a circuit meet, or a point that is common to two circuits. 2. In a standing-wave system, a zero point or minimum point, e.g., *current node*. Compare *loop, 1*. 3. A data base management system expression defining the location of information about a record, user, field, etc.

nodules 1. Oxide particles that protrude above the surface of magnetic tape. 2. In a planar pattern describing radiation or pickup characteristics (as for antennas, microphones, loudspeakers), a small peak aligned in a direction other than that of the main lobe.

noematachograph An instrument used in psychology to measure complex reaction time. Compare *noematachometer*.

noematachometer An instrument used in psychology to measure simple reaction time. Compare *noematachograph*.

no-field release In the starting box for a shunt motor, the electromagnet which normally holds the arm in *full-running* position; it is connected in series with the field winding. When the field current is lost, the arm is released, disconnecting the armature for safety. Compare *no-voltage release*.

noise 1. A (usually interferential) random-frequency current or voltage signal extending over a considerable frequency spectrum and having no useful purpose, unless it is intentionally generated for test purposes. 2. Dissonant, interferential sound; unlike *harmonious* sound, it is disagreeable. 3. Extra bits or bytes that must be removed from digital data before it can be useful.

noise abatement The elimination or reduction of noise intensity, especially a measure in a program concerned with noise pollution in the environment.

noise analysis Measurement of the amplitude and spectral distribution of noise and the determination of its character.

noise analyzer An instrument for performing noise analysis. See, for example, *noise meter*. Noise analyzers are sometimes adapted for vibration analysis.

noise-balancing system A bridge circuit inserted between a receiver and antenna for balancing out interferential signals resulting from nearby power-line leaks or similar causes.

noise bandwidth Abbreviation, *NBW*. A figure obtained by dividing the area under the power output vs frequency curve of a device by the power amplitude at the noise frequency of interest.

noise behind the signal Noise caused by, but exclusive of, a signal.

noise blanker A device that cuts off one of the intermediate-frequency stages of a radio receiver during a noise pulse. The noise blanker is effective against high-amplitude impulses of short duration.

noise canceling microphone A close-talking microphone which discriminates against distant noises.

noise clipper A biased-diode circuit used as an automatic noise limiter. The device cuts off all signals above a predetermined amplitude on the theory that noise peaks are high-level transients in an otherwise uniform signal. Noise is reduced at the sacrifice of system reproduction fidelity.

noise criteria An expression for the level of ambient acoustic noise.

noise current Noise-generated current.

noise-current generator A noise generator that supplies a useful current. Compare *noise-voltage generator*.

noise digit A digit (most often, zero) generated during normalization of a floating-point number. See *normalize*.

noise diode 1. A tube diode operated at saturation and providing a standard noise voltage as the result of random electron emission. 2. A reverse-biased silicon diode that produces a standard noise voltage.

noise elimination The total removal of noise. Compare *noise suppression*.

noise equivalent power Abbreviation, *NEP*. The power that produces an rms signal-to-noise ratio of 1 in a detector.

noise factor See *noise figure*.

noise figure Abbreviation, *NF*. Unit, dB. For any network or device, the ratio $R1/R2$, where $R1$ is the signal-to-noise power ratio of the ideal network and $R2$ is the signal-to-noise ratio of the network or device under test. Also called *noise factor*.

noise filter A filter designed to suppress noise signals which would otherwise enter an electronic circuit, e.g., a power-line noise filter.

noise generator A device for generating accurate values of noise voltage for test purposes.

noise-generator tube A special vacuum tube designed to produce a noise signal for electronic measurements. See, for example, *noise diode, 1*.

noise grade 1. The relative level of radio-frequency background noise, over all electromagnetic frequencies, in a particular geographic location. The noise grade is generally lowest near the poles and highest near the equator. 2. The mathematical function of relative electromagnetic noise intensity versus latitude and longitude. It is thus a function of two variables.

noise immunity The degree to which a circuit or device is insensitive to extraneous energy, especially noise signals.

noise-improvement factor Abbreviation, *NIF*. For a radio receiver, the ratio $SN1/SN2$, where $SN1$ is the input signal-to-noise ratio and $SN2$ is the output signal-to-noise ratio.

noise killer 1. Automatic noise limiter. 2. Noise filter.

noiseless alignment See *visual alignment*.

noise level 1. The amplitude of ambient electrical noise generated outside an electronic system of interest. 2. The amplitude of electrical noise generated in an electronic system of interest. 3. The intensity of ambient acoustic noise.

noise limiter See *automatic noise limiter*.

noise margin In a binary logic circuit, the difference between operating and threshold voltages.

noise-measuring set See *noise meter*.

noise meter An instrument for measuring acoustic noise level. It consists essentially of a sensitive, multirange voltmeter provided with a microphone, amplifier, and attenuators. The meter scale reads noise level directly in decibels.

noise power The power component of a noise signal.

noise power ratio The ratio of noise power at the output of a circuit (such as a receiver) to the noise power at the input.

noise pulse A random short-duration signal whose amplitude exceeds the average peak noise level.

noise quieting In a radio receiver, the reduction (in decibels) of background noise, with respect to a signal of interest.

noise ratio See *noise power ratio*.

noise-reducing antenna A receiving antenna having a balanced transmission line and usually some form of noise-balancing system for reducing electrical noise picked up by the antenna.

noise residue The residual output (see *null voltage, 1*) of a balanced bridge, due entirely to noise.

noise silencer A noise limiting circuit that removes noise-pulse transients with little or no effect on the signal from which the noise is removed. Compare *noise clipper*.

noise source See *noise generator*.

noise spike A sharp noise pulse.

noise suppression The reduction of noise amplitude to a level that is noncompetitive with desired signals. Compare *noise elimination*.

noise suppressor A device for eliminating electrical noise or reducing its amplitude. See, for example, *automatic noise limiter*.

noise temperature At a given frequency, the temperature of a passive system which has the same noise power per unit bandwidth as that observed at the terminals of a device under test.

noise voltage The voltage component of an electrical noise signal.

noise-voltage generator A signal generator that supplies an ac waveform containing random-frequency pulses of relatively uniform distribution over a given spectrum. Compare *noise-current generator*.

noisy mode During normalization of a floating-point number, the generation of digits, excluding zero, as part of the fixed-point part (See *normalize*).

NOL National Ordnance Laboratory. Naval Ordnance Laboratory.

no-load current 1. Output-electrode current (e.g., plate or collector current) when a device is not delivering output to an external load. 2. Current flowing in the primary winding of an unloaded transformer.

no-load losses Losses in an unloaded transformer (see *no-load current, 2*).

no-load speed The rotational speed of an unloaded motor.

no-load voltage The open-circuit output voltage of a power supply, amplifier, generator, or network.

nominal 1. Named. The nominal value of a speaker, for example, may be 8 ohms, even though the actual impedance value depends on the frequency of the applied signal. 2. Approximate, and specified as a typical example only, for the purpose of identifying the operating or value range. For example, an automotive circuit may have a nominal rating of 12V even though it may be operated within a range of 11-14.6V.

nominal band In a facsimile signal, the waveband extending between zero and the maximum frequency of modulation.

nominal bandwidth 1. For a filter, the difference $fc1 - fc2$, where $fc1$ is the nominal lower cutoff frequency and $fc2$, the nominal upper cutoff frequency. 2. For an allocated communication channel, the total bandwidth including upper and lower guard frequencies. 3. The intended and named bandwidth of a given wavelength or channel, regardless of the bandwidth of the signal at that frequency at any given time.

nominal capacitance The rated ("label") value of capacitance of a capacitor. Also see *nominal value*.

nominal current The rated ("nameplate") value of required current or of current output. Also see *nominal value*.

nominal horsepower The rated ("nameplate") horsepower of a machine, such as a motor. Also see *nominal value*.

nominal impedance The rated impedance of a circuit or device. Also see *nominal value*.

nominal inductance The rated ("label") value of a coil's inductance. Also see *nominal value*.

nominal line pitch In a TV raster, the average center-to-center separation between adjacent lines.

nominal line width 1. For a TV raster, the factor $1/N$, where N is the number of lines per unit width for the direction in which the lines progress. 2. In facsimile, the average center-to-center separation of scanning or recording lines.

nominal power factor The rated ("nameplate") value of power factor of a device. Also see *nominal value.*

nominal power rating The rated ("nameplate") value of power output, power drain, or power dissipation. Also see *nominal value.*

nominal Q The rated ("nameplate") value of *Q* of a capacitor, inductor, transformer winding, or tank circuit. Also see *nominal value.*

nominal rating See *nominal, 1, 2.*

nominal resistance The rated ("label") value of resistance of a resistor or similar device. Also see *nominal value.*

nominal speed The highest speed of a data processing unit or system, disregarding slowdowns due to other than computational operations.

nominal value The labeled value specified without reference to tolerance. The nominal value may differ significantly from the true value. Thus, the nominal capacitance of an electrolytic capacitor may be 25 μF, but the tolerance of the unit may be $-10\% + 75\%$, so the actual capacitance of a unit labeled *25 μF* may be any value between 22.5 and 43.75 μF.

nominal voltage The rated ("nameplate") value of required voltage or of voltage output. Also see *nominal value.*

nomogram See *alignment chart.*

nomograph See *alignment chart.*

nomography The geometric representation of a mathematical function or relation by means of alignment charts.

nonaccountable time The period during which a computer system is unavailable to the user (because of a power outage, for example).

nonarithmetic shift See *logical shift.*

nonblinking meter A digital meter that does not alternate, or oscillate, between two different values when the measured parameter is between two discrete values. Instead, the display is rounded off to the nearest value, and the display remains at that value continuously.

nonbridging contact In a switch or relay, a movable contact that leaves one stationary contact before contacting another.

nonchargeable battery A primary battery, i.e., one that cannot ordinarily be recharged.

noncoherent Electromagnetic radiation in which the wave disturbances are not all precisely aligned in frequency and phase.

nonconductor See *dielectric.*

noncontact temperature measurement The use of infrared or optical electronic equipment for measuring the temperature of bodies without touching them.

noncorrosive flux A solder flux that does not corrode the metals to which solder is applied.

noncrystalline Pertaining to materials which possess none of the characteristics of crystals. Compare *crystalline material.*

nondestructive read In digital computer and counter operation, the process of reading data without erasing it as a result. The name is also applied to the readout device.

nondestructive test Abbreviation, *NDT.* A test which does little or no irreversible harm to the test sample. Compare *destructive test.*

nondeviated absorption Absorption that slows waves by a negligible amount; also, normal *skywave absorption.*

nondirectional antenna An antenna that radiates (or receives) equally well in all directions.

nondirectional microphone A microphone that responds equally well

to sound from any direction; an omnidirectional microphone.

nondissipative load A purely reactive load. In such a load, the only power consumed is that which is dissipated in the inherent resistance (losses) of the load.

nondissipative stub A stub which exhibits only slight losses, since it consumes no power except that dissipated in small, inherent losses. Also see *stub.*

nonelectrical Not electrical in nature. The term is commonly used to designate the mechanical parts of electronic systems.

nonelectrolyte A substance which does not ionize in water solution. Compare *electrolyte.*

nonelectronic meter A meter which employs no tubes or transistors. Also called *conventional meter.*

nonequivalence operation See *exclusive-OR operation.*

nonerasable storage See *nonvolatile memory.* In digital-computer and data-processing practice, storage media which cannot be erased under ordinary circumstances. Examples are punched cards and perforated tape.

nonferrous metal A metal which is not iron nor related to it.

nonflammable A material that is resistant to burning.

nonharmonic frequency A frequency which has no integral numerical relationship to another frequency of interest. Compare *harmonic frequency.*

nonharmonic oscillations Parasitic oscillations that are called nonharmonic because they do not usually occur at the fundamental frequency nor at any harmonic frequency of an oscillator or amplifier in which they appear.

nonillion The number 10^{30}, so called because when written out the number contains 9 groups of three zeros each (following the first 1000).

noninductive capacitor A wound capacitor in which the edges of the spirals are short-circuited to minimize the inductance of the roll. Compare *inductive capacitor.*

noninductive resistor A wirewound resistor which is wound so that the magnetic field of the coil is self-canceling and the inductance, accordingly, is negligible.

noninverting connection Connection to the noninverting input of a differential or operational amplifier. Also see *noninverting input.* Compare *inverting connection.*

noninverting input In a differential or operational amplifier, the input circuit which assures an output signal which is in phase with the input. Compare *inverting input.*

nonionic Neutral. Possessing none of the properties of ions.

nonionizing radiation Electromagnetic radiation that does not cause ionization of gases under a given set of conditions.

nonlinear Having an output that does not coincide to the input.

nonlinear bridge See *voltage-sensitive bridge.*

nonlinear capacitor A capacitor whose value varies nonlinearly with applied voltage. Also see *voltage-variable capacitor, 1, 2.*

nonlinear coil A saturable reactor. Also see *saturated operation, 1.*

nonlinear dielectric A material (such as processed barium—strontium titanate) whose dielectric constant varies with applied voltage.

nonlinear distortion Distortion caused by nonlinear response of an amplifier or component. Nonlinearity causes different parts of the signal to be amplified or transmitted by different amounts; therefore, the amplitude variations in the output signal differ from those in the input signal.

nonlinear inductor See *nonlinear coil.*

nonlinear mixing The mixing of signals as a result of the nonlinear response of a device (such as a semiconductor diode operated in its square-law region) through which they are passed simultaneously. Also see *mixer* and *mixing*.

nonlinear network A circuit that produces distortion in an input waveform; the output and input waves are not related by a linear function.

nonlinear resistor A resistor whose value varies with applied voltage. Also see *voltage-dependent resistor*.

nonlinear response Any response for which the corresponding plot is not a straight line; doubling the independent variable, for example, does not double the dependent variable.

nonloaded Q See *unloaded Q*.

nonmagnetic 1. Possessing no magnetism. 2. Incapable of being magnetized.

nonmathematical Pertaining to materials and methods which rely upon physical description and qualitative procedures instead of mathematical development, prediction, and quantitative procedures.

nonmetal An elemental material which is devoid of the properties exhibited by metals (e.g., luster, good ductility, electrical conductivity, heat conductivity, and malleability). Examples: carbon, phosphorus, sulfur. Compare *metal* and *metalloid*.

nonmetallic conduction Collectively, ionic conduction in liquids and gases, conduction in dielectrics by small leakage currents, and thermionic conduction in a vacuum.

nonmicrophonic Without microphonic properties, e.g., a nonmicrophonic tube does not produce electrical ringing when struck.

nonnumeric character A character that is not a numeral, i.e., a symbol or letter.

nonohmic response 1. Nonlinear resistance or reactance. Compare *ohmic response*. 2. Negative resistance.

nonoscillating detector A detector which is devoid of positive feedback action and, therefore, is unable to generate a signal on its own. Compare *oscillating detector*.

nonplanar 1. Existing in three spatial dimensions. 2. A circuit that cannot be fabricated on a two-dimensional board without the use of jumper wires.

nonpolar 1. Having no pole(s). 2. Pertaining to atoms which share electrons to complete their outer shells. 3. Not polarized nor requiring polarization. A 10 μF capacitor that is nonelectrolytic, for example, is nonpolar, since it may be used in circuits without consideration of voltage polarity.

nonpolar crystal A crystal in which lattice points are identical.

nonpolarized electrolytic capacitor An electrolytic capacitor which has no definite negative and positive terminals and, consequently, can be operated on ac. See also *nonpolar, 3*.

nonpolarized reactor A saturable reactor in which the lines of force produced in the three-leg core by the coils on the two outer legs oppose each other in the center leg. When direct current is passed through the coil on the center leg to saturate the core, operation remains the same for any dc polarity. Compare *polarized reactor*.

nonpolarized relay See *unpolarized relay*.

nonprint code In telegraphy, a code used to start teleprinter functions excluding printing.

nonradioactive Pertaining to a substance that is not radioactive, or having once being radioactive has lost that property.

nonreactive circuit A circuit containing pure resistance only.

nonrechargeable battery See *nonchargeable battery*.

nonrecurrent Pertaining to phenomena which do not repeat periodically.

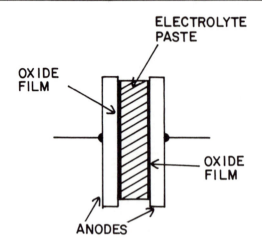

Nonpolarized electrolytic capacitor.

Thus, a single sweep in an oscilloscope is nonrecurrent.

nonrecurrent sweep See *nonrepetitive sweep*.

nonregenerative detector A detector having no regenerative feedback. Such a detector is stable but relatively insensitive. Compare *regenerative detector*.

nonprint code In telegraphy, a code used to start teleprinter functions excluding printing.

nonregenerative receiver A radio receiver in which no local signal whatever is generated. Since such a receiver is nonoscillating, it will not reproduce continuous-wave radiotelegraph signals in the conventional manner.

nonrepetitive phenomena See *nonrecurrent*.

nonrepetitive sweep In an oscilloscope, a single horizontal sweep of the screen, initiated either by the operator or by the signal under observation. Also called *single sweep*. Compare *recurrent sweep*.

nonreproducing codes In computer practice, special codes punched in paper tape to make some hardware carry out specific functions; the codes, however, do not appear on the subsequent output tape.

nonreset timer A timer that must be reset manually.

nonresident routine A computer routine which is not permanently stored in memory. Compare *resident routine*.

nonresonant A circuit in which reactance is present at the operating frequency.

nonresonant lines Transmission lines which are so dimensioned and operated that they do not resonate at the operating frequency and are untuned.

nonresonant load An ac load which is either purely resistive or is detuned from the fundamental and harmonic frequencies of the source from which it is operated.

nonreturn-to-zero In the magnetic recording of digital data, the system in which the current flowing in the write-head coil is sustained (i.e., does not return to zero) after the write pulse.

nonsalient pole A nonprojecting (often flush) pole. Compare *salient pole*.

nonsaturated color 1. Visible light that consists of energy at more than one wavelength. 2. A color that contains some white, in addition to the pure color.

nonsaturated logic A logic circuit in which transistors are prevented from saturating to attain short delay times.

non-self-maintained discharge See *Townsend discharge.*

non-self-sustaining current In a glow-discharge tube, the low current flowing during the *Townsend discharge.*

nonshorting switch A multiple-throw switch that disconnects one circuit before completing another; that is, no two poles are ever connected simultaneously.

nonsinusoidal waveform A waveform whose curve does not follow the general relationship $y = \sin x$. Examples: sawtooth wave, square wave, triangular wave, cosine wave.

nonsymmetrical wave See *asymmetric wave.*

nonsynchronous Unrelated in cyclic quality to other such qualities in the system.

nonsynchronous vibrator A power-supply vibrator which is essentially a single-pole, double-throw switch providing no mechanical rectification. A separate rectifier (either semiconductor or tube-type) must be used. Compare *vibrator-type rectifier.* Also see *vibrator-type power supply.*

nontechnical Pertaining to that which is uncomplicated and (usually) described in lay terms. An example is the simplified explanation of a complex electronic action.

nontrigger voltage For a thyristor, the maximum d gate-to-cathode voltage that can be applied without triggering the device. The amplitude of interferential signals, including noise, must be below this level to prevent accidental triggering.

nonuniform field An electric or magnetic field whose intensity is not the same at all points.

nonvolatile memory A computer storage medium whose contents remain unaltered when the power is switched off; i.e., they are available when power is switched on again.

nonvolatile storage Memory which is retained even if the main power supply is removed for an indefinite period.

no-op instruction An instruction that commands the computer to perform no operation other than to proceed to the following instruction.

NOR circuit Also called *NOT−OR circuit.* In computer and control practice, a circuit which delivers a zero output signal except when two or more input signals are zero. The NOR circuit's function is the inverse of that of the *OR circuit.*

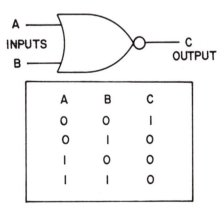

A	B	C
0	0	1
0	1	0
1	0	0
1	1	0

NOR circuit.

NOR gate A gate that performs the functions of a NOR circuit.

norm The average or ambient condition.

normal 1. Conforming to a norm, i.e., that which is expected; *usual.* 2. Perpendicular.

normal curve See *bell-shaped curve.*

normal distribution In a statistical evaluation, a probability distribution represented by the so-called bell-shaped curve. The maximum probability occurs at the 50-percent value.

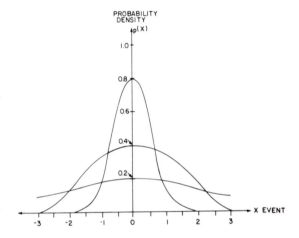

Normal distribution.

normal-distribution curve See *bell-shaped curve.*

normal electrode A standard electrode employed in electrode-potential measurements.

normal glow discharge In a glow-discharge tube, the discharge region between the *Townsend discharge* and the *abnormal glow* in which current increases sharply, but a constant voltage drop is maintained across the tube.

normal impedance A transducer's input impedance when the load impedance is zero.

normal induction curve A saturation curve for a magnetic material. Also see *box-shaped loop* and *saturable reactor.*

normalize In computer programming, to use floating-point numbers to modify the fixed-point part of a number so it is within a desired range.

normalized admittance The quantity $1/Z_n$, where Z_n is normalized impedance.

normalized frequency The unitless number represented by the ratio f_r/f, where f_r is a reference frequency and f is a frequency of interest. Response plots are sometimes conveniently drawn on the basis of normalized frequency, the reference (or resonant) frequency being indicated as 1, twice the reference frequency as 2, and so on.

normalized impedance A value of impedance divided by the characteristic impedance of a waveguide. Compare *normalized admittance.*

normally closed Abbreviation, *NC.* The condition of a switch or relay whose contacts are closed when the device is at rest. Compare *normally open.*

normally open Abbreviation, *NO*. The condition of a switch or relay whose contacts are open when the device is at rest. Compare *normally closed.*

normal mode The normal operating condition. See *normal, 1.*

normal-mode rejection Abbreviation, NMR. In a digital dc voltmeter, the level of noise on the applied voltage which will be rejected by the instrument.

normal position In a switch or relay, the state of the contacts when the device is at rest.

normal solution A solution—such as an electrolyte—in which the amount of dissolved material is chemically equivalent to 1 gram-atomic weight of hydrogen per liter of the solution. Compare *molar solution.*

normal state of atom The condition in which an atom is at its lowest energy level. For the hydrogen atom, for example, the state in which the electron is in the lowest-energy orbit (i.e., $n = 1$).

northern lights See *aurora.*

north magnetic pole The north pole of the *equivalent bar magnet* constituted by the earth's magnetic field (see *earth's magnetic field*). The north magnetic pole lies close to the geographic north pole. Compare *south magnetic pole.*

north pole 1. North magnetic pole. 2. The earth's geographic north pole. 3. North-seeking pole.

north-seeking pole Symbol, *N*. The so-called north pole of a magnet. When the magnet is suspended horizontally, this pole points in the direction of the earth's north magnetic pole. Compare *south-seeking pole.*

Norton's equivalent An equivalent circuit based on Norton's theorem replacing a Thevenin equivalent (see Thevenin's theorem) for a current-actuated device such as a bipolar transistor.

Norton's theorem With reference to a particular set of terminals, any network containing any number of generators and any number of constant impedances can be simplified to one constant-current generator and one impedance. The equivalent circuit will deliver to a given load the same current that would flow if the output terminals of the original circuit where short-circuited. Compare *compensation theorem, maximum power transfer theorem, reciprocity theorem, superposition theorem,* and *Thevenin's theorem.*

NOT In digital systems, a logic operation meaning, in effect, *not 1 but zero or not zero but 1.* Also see *NAND circuit, NOR circuit, NOR gate, NOT circuit,* and *NOT-OR circuit.*

NOT-AND circuit See *NAND circuit.*

notation The way numbers are represented, e.g., *binary notation.*

notch A dip in frequency response, typical of a band-suppression (band-elimination) filter or other frequency-rejection circuit. Compare *peak, 3.*

notch amplifier An amplifier containing a notch filter or other arrangement permitting it to reject one frequency or a given band of frequencies while passing all other higher and lower frequencies.

notch antenna An antenna with a slot in the radiating surface, for the purpose of obtaining a directional response.

notcher See *notch filter.*

notcher-peaker A circuit or device which may be set to perform either as a notch filter or peak filter.

notch filter A frequency-rejection circuit, such as a band-suppression filter, for producing a notch.

notch gate In radar, a gate that determines the minimum and maximum range.

notch sweep An oscilloscope sweep which expands only a small por-

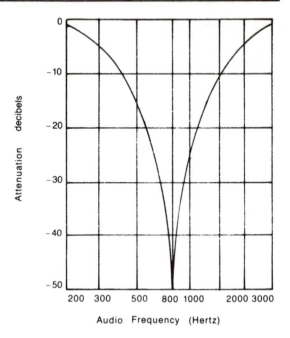

Notch.

tion (*notch*) of the pattern on the screen, leaving the portions on each side of the notch untouched. Thus, the first dozen or so cycles might appear at the normal sweep speed, the next two cycles expanded, and the remaining two or three at normal sweep speed.

NOT circuit A logic circuit which provides an output pulse when there is no input pulse, and vice versa. Also called *complementer, negator,* and *inverter.*

note See *beat note.*

NOT gate A digital circuit that inverts a logical condition, either from high (logic 1) to low (logic 0) or vice-versa. Also called an inverter.

NOT—OR circuit A logical OR circuit which, additionally, inverts the pulse.

noval base An electron-tube base having nine pins.

novar A tube with a 9-pin base.

novelty calculator See *special-purpose calculator.*

November Phonetic alphabet code word for the letter *N*.

novemdecillion The number 10^{60}, so named because there are 19 sets of three zeros following the first 1000. (*Nov* = 9, *Dec* = 10.)

novice 1. A beginner" class of amateur radio license. 2. Any beginner or inexperienced practitioner.

no-voltage release In the starting box for a shunt motor, the electromagnet which normally holds the arm in full running position, but which is connected directly across the power line to disconnect the motor in the event of power failure. When the arm is released, it falls to its off position, thereby preventing burnout, which would result if the motor were left connected to the line in the full run position when power resumed. Compare *no-field release.*

noys scale A scale of apparent acoustic noise, based on a linear function instead of the more common logarithmic function.

Np 1. Symbol for *neptunium*. 2. Abbreviation of *neper*.

Np Symbol for *number of primary turns* in a transformer.

n-phase system A polyphase system having *n* phases.

npin transistor A junction transistor having an intrinsic layer between a p-type base and an n-type collector. The emitter is a second n-type layer on the other side of the base.

N-plant See *nuclear power plant*.

n-plus-one address instruction A computer program instruction containing two addresses, one of which specifies the location of an upcoming instruction to be executed.

NPM Symbol for *counts per minute*. Compare *NPS*.

npnp device A semiconductor switching device having three junctions. Examples: four-layer diode, silicon controlled rectifier. Also called *pnpn device*.

npn transistor A bipolar transistor in which the emitter and collector layers are n-type and the base layer, p-type. Compare *pnp transistor*.

NPO Abbreviation of *negative—positive—zero*.

NPO capacitor A fixed compensating capacitor exhibiting temperature compensating ability over a wide temperature range, in which the coefficient exhibits negative, positive, and zero values.

NPS Symbol for *counts per second*. Compare *NPM*.

N-radiation X rays emitted as a result of an electron becoming an *N-electron*.

NRD Abbreviation of *negative-resistance diode*.

N-region See *n-layer*.

NRZ Abbreviaton of *nonreturn-to-zero*.

Ns Symbol for *number of secondary turns* in a transformer.

ns Abbreviation of *nanosecond*.

N-scan See *N-display*.

N-scope A radarscope for N-display.

nsec Alternate abbreviation of *nanosecond*.

Ns/m^2 *Newton-seconds per square meter*, the unit of dynamic viscosity.

NSPE National Society of Professional Engineers.

NTC Abbreviation of *negative temperature coefficient*.

nth harmonic An (often remote) any-value harmonic, i.e., a harmonic having a frequency of *n* times the fundamental frequency. Also see *harmonic* and *harmonic frequency*.

nth term A remote term in a series, visualized as having any value of numerical significance. Also see *n, 4*.

NTP Abbreviation of *normal temperature and pressure*.

NTSC National Television Systems Committee.

NTSC color signal The color-TV signal specified by the National Television Systems Committee. In the signal, the phase of a 3.58 MHz signal varies with the instantaneous hue of the transmitted color, and the amplitude varies with the instantaneous saturation of the color.

NTSC triangle On a chromaticity diagram, a triangle whose sides encompass the range of colors obtainable from the additive primaries.

NTSC-type generator A special rf signal generator for color-TV tests. It provides separate, individually selected color bars which are fully saturated. The signals are strictly in accordance with NTSC standards.

n-type conduction In a semiconductor, current flow consisting of electron movement. Compare *p-type conduction*.

n-type material Semiconductor material that has been doped with a donor-type impurity and, consequently, conducts a current via electrons. Germanium, for example, when doped with arsenic becomes n-type. Compare *p-type material*.

n-type semiconductor See *n-type material*.

nu N (capital), *v* (lower case). Thirteenth letter of the Greek alphabet.

Symbol for *reluctivity*.

nuclear battery See *atomic battery*.

nuclear bombardment In nucleonics, the bombarding of the nucleus of an atom with subatomic particles, usually neutrons.

nuclear charge The net positive charge of the nucleus of an atom.

nuclear clock A chronometer based on the rate of disintegration of a radioactive material.

nuclear energy Energy resulting from the splitting of the nucleus of an atom or from the fusion of nuclei. Also see *atomic energy, atomic power, nuclear fission, nuclear fusion, nuclear reactor,* and *nucleus*.

nuclear fission A nuclear reaction resulting from the bombardment of the nuclei in the atoms of certain radioactive materials. The bombardment with neutrons creates two new nuclei (by splitting) and several new neutrons, which split several other nuclei, producing still more nuclei and neutrons, and so on. The action is cumulative, so the result is a chain reaction which can lead to a violent explosion if not checked. Compare *nuclear fusion*.

nuclear fusion A nuclear reaction resulting from the violent collision of the nuclei of the atoms of a hydrogen isotope—such as deuterium— with each other at extremely high temperature. The process produces more energy than does nuclear fission and doesn't leave the dangerous radioactive wastes of fission.

nuclear magnetic resonance An atomic phenomenon in which a particle, such as a proton, in a steady magnetic field "flips over" when an alternating magnetic field is applied perpendicular to the steady field.

nuclear pile See *nuclear reactor*.

nuclear power plant A power-generating plant based upon an atomic pile (see *nuclear reactor*).

nuclear reaction 1. A reaction in which a heavy atomic nucleus is split into two or more lighter nuclei, with an accompanying release of radiant energy. Also called nuclear fission. 2. A reaction in which two or more light nuclei combine to form a heavier nucleus, accompanied by the release of radiant energy. Also called nuclear fission.

nuclear reactor A device in which nuclear fission can be initiated and controlled. At the center of the reactor is a core of nuclear fuel, such as a fissionable isotope of uranium (e.g., U-235 or U-238). The core is completely surrounded by a graphite moderator jacket, which is completely surrounded by a coolant jacket; the whole is surrounded by a thick concrete shield. Neutron-absorbing rods are inserted through various walls to different depths in the fuel to control the reaction. In some applications, pipes carrying water are run through the reactor, the water becoming steam (for external use in driving engines and generators) from the heat of the nuclear reaction. Also called *atomic pile*.

nuclear resonance The condition wherein a nucleus absorbs a gamma ray emitted by an identical nucleus.

nucleation A change of state in a material substance.

nucleated Having a nucleus.

nucleon An electron, proton, or neutron in the nucleus of an atom.

nucleonics The branch of physics concerned with nucleons and nuclear phenomena. The name is an acronym from *nuclear* and *electronics*.

nucleon number See *mass number*.

nucleus The center or core of an atom. The nucleus contains electrons, neutrons, protons, and other particles. The net electric charge of the nucleus is positive, and is equal to the sum of the negative charges of the orbital electrons of the atom.

null The condition of zero output current or voltage resulting from adjusting or balancing a circuit, such as a bridge.

null balance In potentiometric measuring circuits for comparing one voltage to another, the balance condition in which no current flows through the galvanometer.

null current The galvanometer current remaining at null when the null point is not fully zero.

null detection Direction finding by means of an antenna with a bidirectional or unidirectional null response.

null detector See *bridge detector.*

null frequency The frequency at which a frequency-sensitive circuit, such as a Wien bridge or twin-tee network, can be balanced.

null meter See *bridge detector.*

null method See *zero method.*

null point In a balanced circuit such as a bridge or potentiometer, the point of zero output voltage (or current) or minimum output voltage (or current).

null potentiometer 1. The variable resistor which constitutes one arm of a four-arm bridge and used to balance the bridge. 2. A potentiometric circuit employing the null method to compare one voltage with another. Also see *potentiometer, 2.*

null set See *empty set.*

null setting 1. That setting of a bridge circuit or other null device which balances the circuit. 2. The electrical zero setting of an electronic voltmeter.

null voltage 1. For a conventional bridge, the output voltage remaining when the bridge is set for its best null. 2. For a voltage-sensitive bridge, the input voltage which will produce zero output voltage.

number The mathematical representation of a quantity. In electronics, numbers are used to denote magnitude.

number cruncher A computer with great computational power, but one not necessarily able to process great amounts of data (such as payroll information).

number system A systematic sequence of numbers based a radix and a logical arrangement. See, for example, *binary number system* and *decimal number system.*

numeral A member of a digit set in a number system.

numerator The term above the bar in a fraction; the number being divided. Compare *denominator.*

numerical analysis A mathematical approach to solving problems numerically, including finding the limits of error in the results.

numerical code A code having a character set restricted to digits.

numerical control An automated system in which number sequences fed to a digital computer cause it to control machines or processes in a manufacturing operation.

numerical expectation Symbol, E_n. In probability theory (and quality-control processes based upon it), the expected number of successes in a given number of trials; $E_n = pk$, and p is the probability of success in one trial and k, the total number of trials.

Nuvistor RCA's miniature metal/ceramic vacuum tube that is especially suited to applications above 50 MHz.

nV Abbreviation of *nanovolt.*

nW Abbreviation of *nanowatt.*

nybble A piece of digital information that is larger than a bit and smaller than a byte. Compare *gulp.*

nylon A synthetic fiber-forming polyamide, useful for electrical insulation. Dielectric constant, 3.6. Resistivity, 10^{13} ohm-cm.

Nyquist criterion of stability With reference to a Nyquist diagram

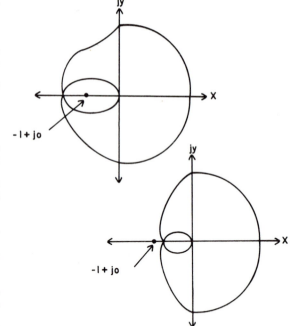

Nyquist diagram.

for a feedback amplifier, the amplifier is stable if the polar plot of loop amplification BA for all frequencies from zero to infinity is a closed curve which neither passes through nor encloses the point $1 + j0$.

Nyquist diagram A type of graph of the performance of a reactive feedback system (such as a degenerative amplifier) which depicts the variation of amplitude and phase of the feedback factor with frequency. The plot is polar and takes into account the real and imaginary components.

N-zone See *N-layer.*

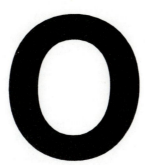

O 1. Symbol for *oxygen*. 2. Abbreviation of output.

o 1. Symbol for *output*. (Also, *out*; both are generally used as subscripts. 2. Symbol for *origin*.

O2 Symbol for *oxygen*.

O3 Symbol for *ozone*.

OAO Abbreviation of *orbiting astronomical observatory*.

OAT Abbreviation of *operating ambient temperature*.

object code 1. In a computer system, the machine-language output of the compiler, directed to the computer. 2. In a computer, the higher-order output of the compiler, directed to the operator. 3. the assembly language, directed to the compiler for translation between machine language and higher-order language and vice-versa.

objective lens In a multilens optical system, the lens nearest the object. In a number of devices it is followed by an eyepiece.

object language The computer language which a compiler derives from a higher level (source) language, such as BASIC, FORTRAN, etc.; it is usually machine language, but may be an intermediate code requiring further conversion.

object program A machine or higher order language version of a user's computer program, as produced by a compiler.

oblique-incidence transmission Transmission of radio signals via ionospheric reflection.

oboe A system of radar navigation in which a pair of ground stations measures the distance to an airborne transponder beacon and then transmits the information to the aircraft.

obsolescence-free Pertaining to a design or process hardly likely to be soon outdated. Compare *obsolescence-irone*.

obsolescence-prone Pertaining to a design or process subject to being soon outdated. Compare *obsolescence-free*.

OBWO Abbreviation of *O-type dackward-wave oscillator*.

o/c Abbreviation of *open circuit*.

occluded gas Gas that has been absorbed or absorbed by solid material, such as glass or metals, and which must be eliminated during the evacuation of an electronic device, such as a vacuum tube. Also see *outgassing*.

occupied band A waveband used by at least one transmitting station regularly.

occupied bandwidth For a given emission, the continuous band of frequencies $f2 - f1$ for which the mean (average) radiated power above $f2$ and below $f1$ is half a percent of the total mean radiated power.

occupied orbit In an atom, an orbit in which an electron is present.

OCR Abbreviation of *optical character recognition*.

oct Abbreviation of *octal*.

octal Abbreviation, *oct*. Based upon eight. See, for example, *octal number system and octal tube*.

octal base A tube base with eight pins arranged uniformly in a circle. A central plastic pin, with a notch, determines the correct tube orientation in the socket.

octal digit One of the figures in the group 0 through 7 used in the -**octal number system.**

octal notation See *octal number system*.

octal number system The base-eight system of number notation. It employs the digits 0 through 7. Compare *binary number system* and *decimal number system*. The following list compares octal with decimal numbers:

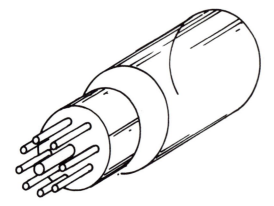

Octal base

DECIMAL	OCTAL
0	0
1	1
2	2
3	3
4	4
5	5
6	6
7	7
8	10
9	11
10	12
11	13
12	1 4
13	15
14	16
15	17
16	20

The octal system is often used as shorthand for otherwise cumbersome binary numbers: The binary number is separated into groups of three digits from right to left; each such group is then converted into its decimal equivalent. the result being the octal form of the binary number, e.g., binary 111 001 011 = octal 713 (111 = decimal 7; 001 = 1; and 011 = 3).

octal-to-decimal conversion Conversion of numbers in the octal number system to numbers in the more familiar decimal number system. This is done by expressing the octal number serially in powers of eight. Thus, octal 12 = $(1 \times 8^1) + (2 \times 8^0)$ = 8 + 2 = decimal 10.

octal tube An electron tube havinr an eight-pin base with a keyway.

octant One-eight of a circle; therefore, 40 degrees, or / quadrant.

octave The region between a given frequency f and either twice that frequency $(2f)$ or half that frequency $(f/2)$. Thus, two octaves are included in the band of frequencies $f/2 – 2f$.

octave band A band of frequencies an octave wide. Also see *octave*.

octave-band noise analyzer A noise analyzer having bandpass-filter channels whose center frequencies are an octave apart.

octave-band pressure level Sound pressure level within an *octave band*.

octave pressure level See *octave-band pressure level*.

OCTL Abbreviation of *open-circuited transmission line*.

octode A device having eight electrodes, such as an octode tube, in which the principal electrodes are anode, cathode, and primary control electrode in addition to five auxiliary electrodes.

octode tube An electron tube having eight principal electrodes. An example is the 6AZ8, a tube which contains triode and pentode sections with separate cathodes.

octonary signal An eight-level signaling code.

octonary system See *octal number system*.

OD Abbreviation of *outside diameter*.

odd-even check A method of checking the integrity of data transferred in a computer system, in which each word carries an extra digit to show whether the sum of ones in the word is odd or even.

odd harmonic In a complex waveform, a harmonic which is an odd-numbered multiple of the fundamental frequency, e.g., third harmonic, fifth harmonic, and so on. Compare *even harmonic*.

odd-harmonic intensification In a complex waveform, emphasis of the amplitude of odd harmonics with respect to that of even harmonics, a property of some multivibrators and nonsinusoidal waves.

odd line In a TV picture, one of the 262.5 odd-numbered horizontal lines scanned by the electron beam in developing the *odd-line field*. Compare *even line*.

odd-line field On a TV screen, the complete field obtained when the electron beam has traced all the odd-numbered lines. Compare *even-line field*.

odd-line interlace See *odd-line field*.

odd parity check A computer check for an odd number of ones or zeros in digital data.

odograph An automatic route and disance plotter which traces on an existing map the path taken by a vehicle carrying it.

Oe Symbol for *oersted*.

oersted Symbol, *Oe*. The cgs unit of magnetic field intensity; 1 Oe = 79.58 A/m. (Formerly he unit of reluctance.)

off-air alarm A device which gives a visible or audible indication when the carrier of a transmitter is lost. In its most rudimentary form, the device consists of a radio-frequency relay which actuates a bell, buzzer, horn, or lamp.

off-center display A ppi radar display whose center point is not the antenna position.

off-center-fed antenna An antenna in which a feeder is attached to one side of the center point of the radiator. See, for example, *Windom antenna'*.

off delay The interval during which a circuit remains on after the control signal has been switched off. Compare *on delay*.

off-ground An above- or below-ground operating voltage or ground return for such voltage that is not common with the system ground.

offhook In a telephone system, the condition in which the receiver is removed from its receptacle. For any party attempting to call the offhook subscriber, the result is a busy tone.

off-limit In a stepping relay, a condition in which the armature has gone past the limit of its travel.

offline 1. In a computer installation, equipment that is not controlled by the central processor. 2. In computer and data processing practice, operations that are not under the direct real-time control of a central processor. 3. A computer memory facility or device which is not connected to a central processor.

off-on operation See *on-off operation*.

off period 1. The interval during which an on-off circuit or device is off. 2. The time during which an equipment is shut down.

off punch To perforate a punched card so that the holes do not line up with the standard punching (hole) positions, resulting in misreading of the card during its processing.

offset Imbalance between the halves of a normally symmetrical circuit, such as that of a differential amplifier. Also see *offset current* and *offset voltage*.

offset current For an operational amplifier, the input current when the offset voltage is zero.

offset stacker A computer-controlled device that provides *offset stacking* of cards.

offset stacking In computer and data-processing practice, stacking cards so that some protrude from the pile to act as indexes.

offset voltage For an operational amplifier, the particular value of dc bias voltage required at the input to produce zero output voltage.

off state 1. The condition of an on—off circuit or device, such as a flip-flop, that is off. Conpare *on state, 1.* 2. The condition in which a circuit or device is shut down. Compare *on state, 2.*

off-state voltage The voltage drop-across a semiconductor device—such as a diode, rectifier, or thyristor—when the device is in its normal off (nonconducting) state. Compare *on-state voltage.*

off-target jamming In radio or radar jamming, to use a remote jamming transmitter so that it will not betray the location of the base station.

off time That period during which no useful work is being performed, as of an equipment when it is not functioning because of a circuit breakdown.

OGL Abbreviation of *outgoing line.*

OGO Abbreviation of *orbiting geophysical observatory.*

ohm Symbol, Ω. The basic unit of resistance, reactance, or impedance. A resistance of 1 ohm passes a current of 1 ampere in response to an applied emf of 1 volt. (Separate entries are given for multiples and fractions of the unit.)

ohmage Electrical resistance or resistivity expressed in ohms.

ohm-centimeter A unit of volume resistivity (see *resistivity*): the resistance of a centimeter cube of the material under measurement. Also see *microhm-centimeter.*

ohmic component A resistor or reactor exhibiting *ohmic response.*

ohmic contact A usually very-low-resistance connection between two materials, which provides bilateral linear conduction—that is, it exhibits none of the properties of a rectifying junction or a nonlinear resistance.

ohmic heating 1. Heating due to current passing throug a resistive material, i.e., heating due to I^2R losses in the material. 2. In an electric field, heat generated by charged particles when they collide with other particles.

ohmic loss Loss resulting from the direct-current resistance in a circuit or transmission line.

ohmic region The portion of the response curve of a negative-resistance device that exhibits positive (ohmic) resistance. The $E—I$ curve of a tunnel diode, for example, has two such positive-slope regions with a negative-slope (negative-resistance) region between them.

ohmic resistance A resistance exhibiting *ohmic response.*

ohmic response Response that follows Ohm's law: $I = E/R$. In strictly ohmic devices, neither resistance nor reactance changes with current or voltage. Compare *nonohmic response.*

ohmic value Electrical resistance expressed in ohms or multiples or frac-

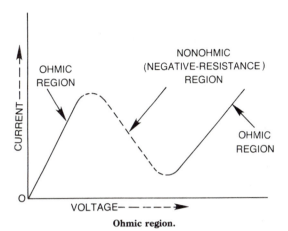

Ohmic region.

tions thereof.

ohmmeter An instrument for the direct measurement of electrical resistance.

ohmmeter zero 1. The condition of proper adjustment of an ohmmeter, indicating zero resistance for a direct short circuit. 2. The potentiometer, or other control, used for adjusting an ohmmeter to obtain a reading of zero with a short circuit.

ohm-mile A rating meaning 1 mile of wire having a resistance of 1 ohm.

ohms adjust The rheostat or potentiometer used to set the pointer of an ohmmeter before it is used to take resistance readings.

Ohm's law The statement of the relationship between current, voltage, and resistance in a dc circuit: Current varies directly with voltage and inversely with resistance, i.e., $I = E/R$, where I is current; E, voltage; and R, resistance. For ac, Ohm's law shows that $I = E/X = E/Z$, where X is reactance, and Z is impedance.

ohms per square The resistance (in ohms) between two parallel edges of a square of thin-film resistance material.

ohms-per-volt rating See *voltmeter sensitivity.*

oil-burner control An electronic system for starting and stopping the operation of an oil burner to prevent puffback and to interrupt the supply when the flame becomes erratic.

oil calorimeter A calorimeter used to measure power in terms of the rise in temperature of oil heated by the electrical energy of interest.

oil capacitor A capacitor that is impregnated or filled with oil, such as high-grade castor or mineral oil. Also see *oil dielectric.*

oil circuit breaker A circuit breaker which is filled with a high-grade insulating oil for cooling and arc elimination.

oil-cooled transformer A usually heavy-duty transformer through which oil is circulated for heat removal and arc prevention.

oil dielectric A highly refined oil employed as a dielectric, e.g., between the plates of a capacitor. Familiar examples are castor oil, mineral oil, and the synthetic oil chlorinated diphenyl.

oil diffusion pump See *oil pump.*

oiled paper Insulating paper that has been impregnated with oil for water- proofing and to increase its dielectric strength.

oil-filled cable A cable whose insulation is impregnated with oil that can be maintained at a constant pressure.

oil-filled capacitor See *oil capacitor.*

oil-filled circuit breaker See *oil circuit breaker.*

oil-filled transformer A transformer whose case is filled with an insulating oil.

oil fuse cutout A fuse cutout that is filled with an insulating oil. Compare *open-fuse cutout*.

oil-immersed transformer See *oil-filled ransformer*.

oil-impregnated capacitor See *oil capacitor*.

oil pump A vacuum diffusion pump employing oil instead of mercury. Also see *diffusion pump*.

oil switch A switch which, like the *oil circuit breaker*, is enveloped by an insulating oil.

OLRT Abbreviation of *online real time* (operation).

OM 1. Abbreviation of *optical microscope*. 2. Amateur radio radio jargon for *old man*: chief (male) operator, or husband or female operator.

omega 1. A phase-dependent radionavigation system employing single-frequency, time-shared, ICW transmissions from two or more locations. 2. The last letter of the Greek alphabet. Capital omega, Ω, is the symbol for *ohm(s)*; lower case omega, ω, is the symbol for *angular velocity* (i.e., $2\pi f$).

omnibearing A navigational bearing obtained by means of *omnirange*.

omnibearing converter An electromechanical device in which an omnirange signal (see *omnirange*) and vehicle heading information is combined, its outiut being a signal that is fed to a meter.

omnibearing indicator Abbreviation, OBI. An omnibearing converter with a dial and pointer.

omnibearing line In an omnirange system (see *omnirange*), one of the imaginary lines extending from the geographic center of the omnirange.

omnibearing selector A device which can be set manually to a selected omnibearing.

omniconstant calculator A calculator that adds or multiplies numbers in succession in such a manner as to arithmetically increase the exponent as a single button is repeatedly pressed.

omnidirectional 1. A device that responds equally well to acoustic or electromagnetic energy from any direction in three dimensions. 2. A device that radiates acoustic or electromagnetic energy equally well in any direction in three dimensions. 3. An antenna that intercepts or radiates equally well in any azimuth, or compass, direction.

omnidirectional antenna See *nondirectional antenna*.

omnidirectional hydrophone A hydrophone that picks up underwater sounds coming from practically any direction.

omnidirectional microphone A microphone which picks up sounds coming from practically any direction.

omnidirectional radio range See *omnirange*.

omnidirectional range See *omnirange*.

omnidirectional range station Abbreviation, *ORS*. A radionavigation station for omnirange service.

omnigraph A Morse-code generator that operates via marked or perforated paper tape.

omnirange A radionavigation system in which each station in a chain broadcasts a vhf beam in all directions. Pilots of aircraft home on a particular station by tuning it in and noting is bearings.

OMR Abbreviation for *optical mark recognition*.

on air See *on the air*.

on-call channel An assigned radio channel of which exclusive, fulltime use is not demanded.

on-course curvature In navigation, the rate at which the course of a vehicle deviates with reference to the distance along the true course.

on-course signal A single-tone-modulated signal indicating to the pilot following a radio beam that the flight is substantially on course.

on current See *on-state current*.

on delay An interval during which a circuit remains off after an actuating signal has been supplied. Compare *off delay*.

on-demand system A system, especially in computer and data-processing practice, that delivers information or service immediately upon request.

ondograph An electromechanical device that graphically draws alternating-current waveforms on paper.

ondoscope An rf energy detector which consists of essentially a neon bulb attached to the end of an insulating rod. When a bulb is held in an intense rf field, the field energy ionizes the gas in the bulb, causing it to glow (without direct connection to the rf circuit).

one-address code In computer programming, a code in which the address in an instruction refers to only one memory location.

one condition See *one state*.

one-digit adder See *half adder*.

one-element rotary antenna A directional antenna consisting of a radiator only (no directors or reflectors) which can be rotated.

one-for-one compiler A compiler that generates one machine language instruction fron one source language instruction.

one-input terminal In a flip-flop, the input terminal that must be energized to switch the circuit to its *one* output.

one-level address See *absolute address*.

one-level subroutine In a computer program, a subroutine in which no reference is made to other subroutines.

one-lunger Colloquialism for a radio transmitter employing only one tube or transistor.

one output See *one state*.

one-output signal The signal that results from reading a computer memory unit that is in the one state.

one-output terminal In a flip-flop, the output terminal that is energized when the circuit is in its one state.

one-plus-one address A method of computer programming in which instructions contain two addresses and an operation, the addresses referring to (1) the location of the next instruction and (2) the location of the data to be used.

ones complement Binary notation in the radix-minus-one-complement form.

one shot One-shot multivibrator (see *monostable multivibrator*).

one-shot circuit See *monostable multivibrator*.

one-shot multivibrator See *monostable multivibrator*.

one-sided wave A waveform consisting of only negative or positive half-cycles. Example: a rectified ac wave.

one state The *high*, *on*, or *true* logic state of a bistable device such as a flip-flop. Compare *zero state*. In binary notation, the one state is represented by 1.

one-third-octave band A frequency band a third-octave wide; that is, the difference between the upper-frequency limit ($f2$) and the lower-frequency limit ($f1$) is $f2 - f1 = 1/3$. Also see *octave band*.

one-to-one assembler An assembler computer program that produces a machine language instruction as a result of translating a source language statement.

one-to-one correspondence A mapping between two sets A and B, such that every element in set A has exactly one correspondent in B, and every element in B has exactly one correspondent in A.

O-network A four-impedance network containing two series (upper and lower arms and two shunt (input and output) arms.

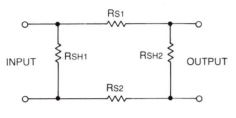

O-network.

one-way communication 1. The transmission of a message to one or more stations which receive only. Compare *two-way communication*. 2. Broadcasting.

one-way conduction See *unilateral conductivity*.

one-way radio See *one-way communication*.

one-way repeater In wire telephony, a device that amplifies and retransmits a signal in the direction the signal was traveling when it arrived at the repeater. Compare *two-way repeater*.

one-way valve A diode or rectifier (British variation).

on interval See *on time*.

online 1. Descriptive of equipment under the control of a central processor. 2. Pertaining to operations being controlled by a central processor. 3. Relating to a computer storage device independent of a cpu.

online data reduction Processing data as it comes into a computer system.

on-off keying Keying, as in radiotelegraphy and wire telegraphy, by switching a signal source on and off to form dots and dashes or a binary code, rather than alternately changing the amplitude or frequency of the signal.

on—off operaton A switching operation, especially that performed by nonmechanical (fully electronic) circuits.

on—off ratio 1. For a circuit or device, the ratio of off time to on time. 2. For a pulse, the ratio of pulse duration to the dead space between pulses.

on—off switch 1. In electronic equipment, the switch by which the equipment may be started or stopped. It can be, but is not necessarily, the power-line or B-plus switch. 2. An electronic circuit or stage that is designed to operate as a conventional switching element when triggered by an appropriate signal.

on period See *on time*.

ONR Office of Naval Research.

on resistance See *on-state resistance*.

on state 1. For a switch or switching device (such as a flip-flop), the condition of the device when it conducts current or delivers an output voltage. Compare *off state, 1.* 2. The condition of a circuit or device that is activated for operation. Compare *off state, 2.*

on-state current The current flowing through a semiconductor device, such as a diode, rectifier, or thyristor, when it is conducting. Also see *on-state voltage*.

on-state resistance The resistance of a voltage-dependent resistor that is conducting current.

on-state voltage The voltage drop-across a semiconductor device, such as a diode, rectifier, or thyristor, when the device is conducting current. Also see *on-state current*.

on the air The state of a radio station that is transmitting.

on time The length of time a switch or switching device (such as a flip-flop) remains on.

on voltage See *on-state voltage*.

op Abbreviation of *operate, operator,* or *operational*.

opacimeter 1. An instrument for measurinr he extent to which a material blocks light. 2. An instrument, such as a field-strength meter, employed to measure the effectiveness of an electrical shielding material in blocking radio waves, X-rays, or other radiation.

opacity The condition of being opaque. This applies to all forms of radiation. For example, a material may be opaque to light rays yet still be transparent to radio waves, or it may be transparent to gamma rays while being opaque to alpha particles.

op amp Abbreviation of *operational amplifier*.

op code Abbreviation of *operation code*.

open An open circuit.

open-air line See *open-wire line*.

open-back cabinet A loudspeaker enclosure in which the space behind the speaker is open to the room.

open capacitance The value of a variable capacitor whose rotor plates have been rotated completely out of mesh with the stator plates. Compare *closed capacitance*.

open-center display A ppi radar display in which a ring around the center indicates zero range.

open-chassis construction A method of assembling electronic equipment by mounting components and wiring them on a unenclosed chassis, often wihout a front panel, a kind of breadboard construction.

open circuit 1. A discontinuous circuit, i.e., one that is broken at one or more points and, consequently, cannot conduct current nor present a voltage at its extremities. Compare *closed circuit.* 2. Pertaining to no-load conditions, for example, the open-circuit voltage of a battery.

open circuit breaker A circuit breaker whose contacts are open.

open-circuit current Current flowing in the primary winding of an unloaded transformer.

open-circuited line See *open-circuited transmission line*.

open-circuited transmission line Abbreviation, *OCTL.* An unterminated transmission line.

open-circuit impedance For a transmission line or a four-terminal network, the input or driving-point impedance when the output end of the line or network is unterminated.

open-circuit jack A telephone jack that introduces a break in a circuit until a plug connected to a closed external circuit is inserted.

open-circuit For a bipolar transistor, the operating characteristics under independent input and output conditions.

open-circuit plug See *open plug*.

open-circuit resisance For a four-terminal network, the input or driving-point resistance when the output end of the network is unterminated.

open-circuit signaling A system of signaling in which current doesn't flow unil the signal circuit is in active operation. In a simple telegraph circuit, for example, current only flows when the key is depressed (to form a dot or dash).

open-circuit voltage See *no-load voltage*.

open component An open-circuit component, e.g., an open capacitor, coil, or resistor.

open core A magnetic core having a general stick shape. The disadvantage of this (simple) core is that much of the magnetic flux is outside the core, i.e., in the space between the poles. Compare *closed core*.

open-core choke A choke coil wound on an open core. Also called *open-core inductor*. Compare *closed-core choke*.

open-core transformer A transformer wound on an open core. Compare *closed-core transformer*.

open-delta connection See *vee-connection of transformers*.

open-ended That which can be built upon without modification of its existing configuration.

open-end stub A stub that is neither short-circuited nor terminated at its far end.

open-end stub tuning Adjustment of an open-end stub for optimum operation at a given frequency by pruning its length.

open-entry contact In a connector, an unprotected, opening contact of the female type.

open feeder See *open-wire line*.

open-frame machine See *open generator* and *open motor*.

open-fuse cutout A type of enclosed-fuse cutout having an exposed fuse holder and support. Compare *oil fuse cutout*.

open generator An unsealed generator, i.e., one that has openings in its housing for air circulation.

open line 1. Open-wire line, 2. Open-circuited transmission line.

open loop In a control system, a feedthrough path having no feedback around it and which, therefore, is not self-regulating. Compare *closed loop*.

open-loop bandwidth The bandwidth of an open-loop device, such as an amplifier, without feedback. Also see *open loop*.

open-loop control system A control system having only a transmission path and which, because it is without feedback, is not self-regulating. An example is the familiar fluid-level gauge that indicates the height of fluid in a tank but cannot correct the level automatically. Compare *closed-loop control system*.

open-loop differential voltare gain For a differential amplifier, the overall voltage gain—when either of the inputs is used—when the amplifier has no feedback.

open-loop gain The overall gain (ratio of output to input) of an open-loop device, such as an amplifier without feedback. Also see *open loop*.

open-loop input impedance The input impedance of an open-loop decice, such as an amplifier without feedback. Also see *open loop*. Compare *closed-loop input impedance*.

open-loop ouput impedance The output impedance of an open-loop device, such as an amplifier without feedback. Also see *open loop*. Compare *closed-loop input impedance*.

open-loop output resistance The output resistance of an open-loop device, such as an amplifier without feedback. Also see *open loop* and *open-loop impedance*.

open-loop system A circuit in which the input and output currents are independent.

open-loop voltage gain The overall voltage gain of an open-loop amplifier, i.e., one having no feedback. Also see *open loop*. Compare *closed-loop voltage gain*.

open magnetic circuit A magnetic circuit in which a complete path is not provided for magnetic flux. See, for example, *open core*. Compare *closed magneic circuit*.

open motor An unsealed motor, i.e., one that has openings in its housing for air circulation.

open-phase protection Use of an automatic device, such as an open-phase relay, to interrupt the power to a polyphase system when one or more phases are open-circulated.

open-phase relay In a polyphase system, a protective relay that opens when one or more phases are open-circuited. Also see *open-phase protection*.

open plug A phone plug to which no external connections are made, it is used to hold the blades of a jack as if they were plugged in.

open-reel A tape-recording system in which the tape, during record or playback condition, is wound onto a takeup reel that is physically separate from the tape reel. Also called a reel-to-reel arrangement.

open relay 1. A relay in its open-contact state. 2. A relay having an open-circuited coil. 3. An unenclosed relay.

open routine In computer practice, a routine that can be inserted directly into a larger routine and requires no link to the main program.

open subroutine In computer practice, a subroutine that can be inserted into a larger instructional sequence and must be recopied whenever it is required. Also called *direct-insert subroutine*.

open temperature pickup A naked temperature transducer, i,e., one that must be placed directly in contact with the monitored body.

open volume Pertaining to the maximum-gain operation of a sound-reproducing system, i.e., operation at full volume.

open wire 1. An unterminated wire. 2. A wire supported above the surface of the earth and often ungrounded.

open-wire feeder See *open-wire line*.

open-wire line A transmission line or feeder usually consisting of two straight, parallel wires held apart by bars of low-loss insulating material at regular intervals along the line.

open-wire loop A branch line connected to a main open-wire line.

open-wire transmission line See *open-wire line*.

open-wire wavemeter See *Lecher wires*.

operand In computer practice, a quantity that enters into or results from an operation.

operate 1. To manipulate according to an established procedure (e.g., to operate an instrument). 2. To perform in the sense that an electronic circuit operates.

operate current A signal current or trigger current required to actuate a device. Compare *operate voltage*.

operate delay See *operate time, 1*.

operate interval See *operate time, 2*.

operate time 1. The interval starting after the application of an operate current or voltage to a device and ending when the device operates. 2. The period during which an electronic equipment is in operation. Also see *operating time, 1*.

operate voltage The signal voltage or trigger voltage required to actuate a device. Compare *operate current*.

operating ambient temperature Abbreviation, *OAT*. The maximum or recommended temperature in the space immediately surrounding an equipment in operation.

operating angle In an amplifier circuit, the excitation-signal cycle, in degrees, during which plate or collector current flows, i.e., class A amplifiers operate during 360 degrees; class B 180 degrees; and so on.

operating bias In a tube or transistor circuit, the value(s) of dc bias required for normal operation.

operating code The code used by the operator in a computer or data-processing system.

operating conditions The environment in which a circuit or system functions in normal use.

operating current The current required by a device during its operation. Compare *idling current*.

operating cycle The sequence of events in the operation of a device. For example, the repetitive operating of a neon-bulb relaxation oscillator is a sequence of three events: (1) slow charge of capacitor, (2) firing of bulb, and (3) abrupt discharge of capacitor.

operating frequency 1. The frequency at which a circuit or device is operated. 2. The frequency of the current, voltage, or power delivered by a generator.

operating life The maximum period (from seconds to years) over which a device will operate before failure (from which it usually can't recover). Compare *shelf life.*

operating line A line drawn across a family of curves depicting the performance of a device. It intersects each curve at a single point and graphically displays the performance of the device for a given condition. Thus, an operating line on a family of output curves for a tube or transistor might depict operation with a given load resistor.

operating overload The overload to which an equipment may be exposed during customary operation.

operating point On the response curve for a device, the point indicating the quiescent level of operation (such as determined by a fixed bias coltage). An ac signal applied at this point oscillates above and below the point as a mean.

operating-point shift A movement of an operating point due to faulty operation of a circuit or device or to a value change in some critical component.

operating position 1. The control point in a system, i.e., the place where an operator (see *operator, 1*) normally functions. 2. The actual or recommended physical orientation of a device during its operation, e.g., a vertical operating position for a tube.

operating power 1. The power actually used by a device during its operation. 2. The antenna power of a radio station.

operating ratio For a given period, the ratio $t1/t2$ as a percentage, where $t1$ is the time during whic an equipment is operating correctly, and $t2$ is the duration of the period.

operating station In a computer installation, one or more consoles for the control of a data processing system by an operator.

operating system See *monitor program.*

operating temperature The actual or recommended temperature of a device during its operation.

operating-temperature range For a given device, the spread between maximum and minimum values of operating temperature.

operating time 1. The interval during which an equipment is in operation. 2. The period corresponding to *operating angle.*

operating-time characteristic For a coil-type relay, the relationship between operating time and operating power.

operating voltage The voltage required by a device, or measured at the device, during its operation. Compare *idling voltage.*

operation 1. The working of a circuit or device, i.e., its performance. 2. A process usually involving a sequence of steps, e.g., a mathematical operation.

operational amplifier Abbreviation, *op amp.* An extremely stable (usually direct-coupled) linear amplifier originally intended for performing mathematical operations; see, for example, *differentiator, 2* and *integrator, 2.* This amplifier, especially the integrated-circuit variety, has a wide range of applications outside of computation.

operational differential amplifier An operational amplifier preceded by a differential amplifier.

operational readiness In statistical analysis, the probability that a system will, at a certain time, be correctly operating or ready to operate.

operational reliability Reliability determined empirically from a study

of the actual operation of a device or system under controlled conditions. Also called *achieved reliability.*

operational transconductance amplifier Abbreviation, OTA. An integrated-circuit amplifier which differs from the conventional opamp in that its output current is proportional to its input-signal voltage.

operation code In computer practice, the part of an instruction that specifies an operation.

operation decoder In a digital computer, the circuit that reads an *operation code* and directs other circuitry in the execution of the code.

operation number In conputer programming, a number that indicates the position in the program of a particular operation or subroutine.

operation part In a computer program, the part of an instruction containing the operaton code.

operation register In a digital computer, the register storing the operation code of an instruction.

operations research A branch of computer engineering, devoted to the solution and/or optimization of functions of many variables.

operation time The interval between the instant of application of all volages to a circuit and the instant when the current reaches a specified percentage of its final, steady value.

operator 1. One who performs an operation (see *operate, 1*). 2. In mathematics, a symbol indicating an operation, e.g., j, $+$, $-$, $\times$, etc.

operator j See *j operator.*

opposition 1. The state of two quantities that are 180 degrees out of phase with each other. 2. The state of being opposed, physically, mathematically, electrically, etc.

opt Abbreviation of *optical.*

Optacon An electronic aid for the blind. It has a camera that scans printed matter and a device that forms corresponding raised letters which can be read, as would Braille, with the fingertips. The name is an acronym for *opt*ical to *tact*ile *con*verter.

optical 1. Generating or sensing visible light. 2. Visible; in the range of approximately 390 to 750 nanometers wavelength.

optical ammeter A type of optical pyrometer that measures the current flowing through the filament of an incandescent lamp.

optical axis The Z-axis of a quartz crystal (see *Z-axis, 4*).

optical character reader A device which, by OCR, can discern printed characters.

optical character recognition Abbreviaion, *OCR.* In computer and data-processing practice, the reading of alphabetical, nunerical, and other characters by photoelectric methods.

optical communications One-way or two-way communications via modulated visible light. May be conducted through transparent fibers or through the atmosphere.

optical coupler A coupling device consisting essentially of a light source (actuated by an input signal) mounted in a light-tight housing with a light-sensitive device (which delivers the output signal). In its simplest form, the arrangement consists of a lamp and photocell.

optical detector An IC-type device which provides light-to-voltage conversion. Its dc output voltage is proportional to the intensity of light impinging upon its sensor.

optical fiber A glass of plastic medium through which light is propagated for optical-communications purposes. The refractive characteristics of the fiber keep the visible light inside.

optical lever A device for amplifying the effect of a small rotation. The rotating member carries a small mirror which reflects a light beam over a curved scale, the distance through which the light spot travels on the scale being proportional to the distance between the scale and the rotating mirror. In this way, the deflection on the scale is several times the length of the arc described by the mirror, the rotation being thus amplified. A familiar application is in a light-beam meter.

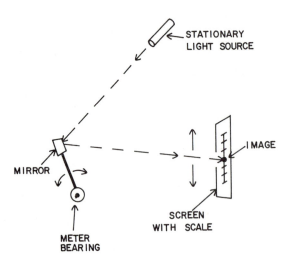

Optical lever.

optical link See *optical coupler.*

optical mark recognition A method of data transfer that involves the use of optical techniques.

optical maser See *laser.*

optical mode The vibration mode that produces an oscillating dipole in a crystal lattice.

optical pattern A Christmas-tree-like pattern produced by parallel rays of light striking a phono disk. Also see *diffraction.*

optical playback See *optical sound reproducer.*

optical pyrometer A pyrometer for measuring the temperature of a hot body in terms of the intensity and color of light it emits.

optical sound recorder A photoelectric machine for recording sound on photographic film. Also see *optical sound recording.*

optical sound recording A system for recording sound on photographic film, as in sound motion pictures. The sound is picked up by a microphone and amplified to vary the intensity of a light source. The film passing this modulated beam becomes exposed to a variable-width or variable-density track corresponding to the modulation (see *variable-density sound record* and *variable-width sound record*). When developed film is played back, its sound track modulates a light beam in the reproducer by actuating a photocell or phototube to produce the audio signal, which is then amplified.

optical sound reproducer A photoelectric machine for reproducing sound on film. A light beam in the device is modulated by the passing sound track, and in turn modulates photocell or phototube current, which is amplified to drive a loudspeaker. Also see *optical sound recording.*

optical system Collectively, the functional arrangement of lenses, mirrors, prisms, and related devices in optoelectronic apparatus.

optical tachometer An optoelectronic instrument for measuring (by means of reflected-light variations) the speed of a body, such as a rotating shaft, without electrical or mechanical attachments to the latter.

optical thermometer See *optical pyrometer.*

optical twinning A kind of defect in which two types of quartz occur in the sane crystal. Compare *electrical twinning.*

optical type font A special type (printing) style designed for use with OCR equipment.

optical wand A pencil-like optical probe, employed to read bar codes from a printed pare and translate the codes into information which is then loaded into a computer or calculator.

optic axis See *optical axis.*

optics 1. The science of light, its measurement, application, and control. 2. A system of lenses, prisms, filters, or mirrors used in electronics to direct, control, or otherwise modify light rays.

optimization The adjustment or manipulation of the elements of a process or system for the best operation or end result.

optimize 1. To manipulate a set of variables or parameters for the best possible performance of a circuit or system. 2. To maximize the value of a multivariable function.

optimum angle of radiation For a given height of the ionosphere, the angle at which a radio signal should be ransnited for the most effective reception at a location.

optinum bunching In a velocity-modulated tube, such as a klystron, the bunching associated with maximum output.

optimum collector load The ideal load impedance (see *optimum load*) for a particular transistor operated in a specified manner.

optimum coupling The degree of coupling between two circuits tuned to the same frequency that results in maximum energy transfer. Also called *critical coupling.*

optimum current The value of current which produces the most effective performance of a circuit or device.

optimum frequency See *optimum working frequency.*

optimum load The value of load impedance that produces maximum transfer of power from a generator or amplifier.

optimum plate load The ideal load impedance (see *optimum load*) for a particular electron tube operated in a specified manner.

optimum Q The most effectice figure of merit for a capacitor, inductor, or tuned circuit at a specified frequency.

optimum reliability The value of reliability that assures minimum project cost.

optimum voltage The value of voltage which results in the most effective performance of a circuit or device.

optimum working frequency In radio transmission involving reflection from the ionosphere, the freuuency of use that results in the most reliable communication between two points.

optoelectronic coupler An assembly consisting of an LED and a phototransistor. An input signal causes the diode to glow, and the light activates the transistor, which in turn delivers an output signal of higher amplitude than that of the input signal.

optoisolator A coupling device in which the coupling medium is a light beam. See, for example, *optoelectronic coupler.* See next page.

optoelectronics A branch of electronics that involves the use of visible light for communications or data-transfer purposes.

optoelectronic read head In data-processing equipment, an optoelectronic coupler used to read punched cards or paper tape.

optoelectronics See *electro-optics.*

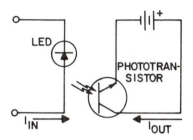

Optoisolator.

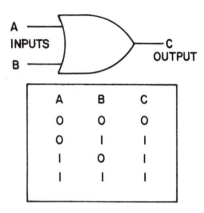

2-input OR circuit.

optoelectronic transistor A transistor having an electroluminescent emitter, transparent base, and photoelectric collector.

optophone A photoelectric device for converting light into sounds of proportionate pitch to enable blind persons to "see" by ear.

orange peel On a phonograph disk, a mottled surface that produces high background noise; so called from its resemblance to an orange peel.

orbit A closed path, especially the elliptical path of an electron around the nucleus in an atom or the path of a satellite or celestial body.

orbial-beam multiplier tube An electron-multiplying uhf oscillator or amplifier tube in which a positively charged electrode focuses electrons in a circular path.

orbital electron An electron in orbit around the nucleus of an atom.

OR circuit In digital systems and other switching circuits, a type of gate which delivers an output signal when one of several input signals is present as follows:

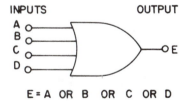

$$E = A \ OR \ B \ OR \ C \ OR \ D$$

4-input OR circuit.

order 1. An instruction to a digital computer. 2. Sequence.

OR element See *OR circuit*.

ordered pair Two terms or numbers (such as a dependent and independent variable) which are so related that for each value of one there is only one corresponding value for the other, e.g., $y = 3x$.

order of logic 1. A mathematical expression for the complexity of a system of logic. 2. The relative speed with which logical information is processed in a system.

order tone A warning signal to receiving operators in the form of a tone transmitted over a trunk preceding the transmission of an order.

ordinary differential equation A differential equation containing no partial derivatives. Compare *partial differential equation*.

ordinary ray As a result of the double refraction of electromagnetic waves, the member of a pair of rays that follows the usual laws of refraction. Also see *o-wave*. Compare *extraordinary ray*.

ordinary wave See *o-wave*.

ordinate In the rectangular coordinate system, a point located on the vertical axis.

organ 1. A computer subsystem. 2. An electronic device used for the purpose of generating music.

organic decay See *exponential decrease*.

organic electricity Electricity in the living tissues of animals and plants.

organic growth See *exponential increase*.

organic semiconductor A semiconductor material consisting of, or combined with, some compound of carbon.

OR gate See *OR circuit*.

orientation 1. The direction or position of an object in space, expressed in terms of coordinate values. 2. In a teleprinter, the calibration or alignment that determines the speed of response to a received character.

orientation of quartz plates See *crystal axes* and *crystal cuts*.

orifice An opening or window, such as that in a loudspeaker enclosure or waveguide, through which energy is transmitted.

origin 1. The starting point in a coordinate system. 2. Relative to address modification in computer practice, an address to which a modifier is added to derive a variable operand address.

original lacquer In disk-recording practice, an original recording made on a lacquer-surfaced disk that is subsequently employed to make a master.

original master In disk recording, the master disk produced from a wax or lacquer disk (see *original lacquer*) by means of electroforming.

origin distortion Change in the shape of a wave as it swings through zero (polarity).

ORS Abbreviation of *omnidirectional range station*.

orthicon A type of TV camera tube somewhat similar to the iconoscope but which provides internal amplification of light and, accordingly, can be used in dimmer places than the iconoscope can. Light amplification is provided by an arrangement similar to that of a photomultiplier tube.

orthiconoscope See *orthicon.*

orthoacoustic recording 1. A system of disk recording in which the inherent differences between high-frequency recording and low-frequency recording are compensated to give reproduction nore closely resembling the actual sound. 2. A disk made by the orthoacoustic method.

orthogonal axes Perpendicular axes.

Os Symbol for *osmium.*

osc Abbreviation of *oscillator.*

Oscar Phonetic alphabet code word for the letter O.

osciducer See *oscillating tranducer.*

oscillate 1. To fluctuate in amplitude in a uniform manner. 2. To vary above and below a specified value at a constant rate.

oscillating arc 1. Arc converter. 2. A small dc arc (especially one produced by slow-opening relay contacts) which generates high-frequency oscillations.

oscillating circuit A closed circuit containing inductance, capacitance, and inherent resistance, in which energy passes back and forth between inductor and capacitor at a frequency determined by the *L' and C* values.

oscillating crystal 1. A pieeoelectric plate which is maintained in a state of oscillation in a circuit. See, for example, *crystal oscillator* and *quartz crystal.* 2, An oscillating semiconductor diode (see *negative-resistance diode, 1, 2*).

oscillating current See *oscillatory current.*

oscillating detector A detector that is provided with positive feedback and is therefore capable of generating a signal of its own. Compare *nonoscillating detector.*

oscillating diode 1. A semiconductor diode biased into its negative-resistance region so that it oscillates in a suitable circuit. 2. An oscillating *tunnel diode,* 3. Any of several microwave diodes, such as the *IMPATT diode,* which will oscillate in a suitable system. 4. A *magnetron.* See also *diode oscillator.*

oscillating field An alternating electric or magnetic field.

oscillating rod A rod of magnetostrictive metal which is maintained in a state of oscillation in a circuit. See, for example, *magnetostriction* and *magnetostriction oscillator.*

oscillating transducer A transducer in which an input quantity varies a frequency proportionately from its center value.

oscillating wire A wire of magnetostrictive metal which is maintained in a state of oscillation in a circuit. See, for example, *magnetostriction* and *magnetostriction oscillator.*

oscillation The periodic change of a body or quantity in amplitude or position, e.g., oscillation of a pendulum, voltage, or crystal plate.

oscillation constant For an oscillating circuit, the expression $\sqrt{LC}$, where L is inductance in henrys and C is capacitance in farads. The oscillation constant is the reciprocal of the oscillation number.

oscillation control A manual or automatic device for adjusting the frequency or amplitude of the signal generated by an oscillator.

oscillation efficiency Symbol, η. The ratio, as a percentage, of the ac power output of an oscillator to the corresponding dc power input; $\eta(\%)$ = 100 ac/Pdc.

oscillation number For an oscillating circuit, the number of oscillations per 2π seconds.

oscillation test 1. A test of an oscillator to determine if a signal is being generated. 2. A test for transistors wherein the transistor is used as an oscillator to give a rough indication of its condition in terms of oscillation amplitude.

oscillation transformer A tank coil of a radio transformer, especially one that includes an output coupling coil.

oscillator A device that produces an alternating or pulsating current or voltage electronically. The term is sometimes used to describe any ac-producing device that isn't an electromechanical generator.

oscillator circuit The wiring arrangement of an oscillator. Oscillator circuits are of three general types: (1) *negative-grid oscillator* (or its transistor equivalent), (2) *negative-resistance oscillator,* and (3) *relaxation oscillator.* The first type is essentially an amplifier provided with positive feedback.

oscillator coil A tapped coil which provides the input and output windings required for an oscillator circuit. Such coils are used in signal generators, oscillators, and superheterodyne receivers.

oscillator-doubler A combination oscillator—double circuit, e.g., a crystal oscillator whose output frequency is twice the crystal frequency.

oscillator drift The usually gradual change in frequency of an oscillator, due to such factors as warmup time, voltage variations, capacitance change, inductance change, or drift in tube or transistor characteristics.

oscillator harmonic interference In a superhet receiver, interference that is the beat product of local oscillator harmonics and received signals.

oscillator interference Radio-frequency interference caused by signals from the high-frequency oscillator of a receiver.

oscillator keying Keying by making and breaking the signal output, dc power, or dc bias of the oscillator stage of a radiotelegraph transmitter.

oscillator-mixer 1. A combination stage in which a tube or transistor functions as a local oscillator and mixer in a receiver or est instrument. 2. A vacuum tube designed for oscillator-mixer service (see 1, above).

oscillator-mixer-detector 1. In a superheterodyne receiver, a stage in which the functions of high-frequency oscillator, mixer, and first detector are performed by a single tube or transistor. 2. A vacuum tube designed for oscillator-mixer-detector service (see 1, above).

oscillator-multiplier A single circuit which serves simultaneously as an oscillator and frequency multiplier. See, for example, *oscillator-doubler.*

oscillator padder In a superheterodyne receiver, a small, limited-range variable capacitor connected in series with the oscillator coil for tracking oscillator tuning at the low end of a band. Compare *oscillator trimmer.*

oscillator power supply 1. The dc or ac power supply for an oscillator. 2. Oscillator-type power supply.

oscillator radiation The emission of rf energy by the oscillator stage of a superheterodyne receiver. Also see *oscillator interference.*

oscillator-radiation voltage The rf voltage at the antenna terminals of a superheterodyne receiver that results from signal emission by the oscillator stage. This voltage is an indicator of the probable intensity of oscillator radiation. Also see *oscillator interference.*

oscillator stabilization 1. The automatic compensation of an oscillator circuit for the frequency drift resulting from changes in temperature, current, voltage, or tube or transistor parameters. 2. The automatic stabilization of the operating point of an oscillator circuit against variations resulting from changes in temperature, supply current or voltage, or tube or transistor parameters.

oscillator synchronization The locking of an oscillator in step with another signal source, such as a frequency-standard generator.

oscillator tracking In a superheterodyne receiver, the constant separation of the oscillator frequency from the signal frequency by an amount equal to the intermediate frequency at all settings of the tuning control.

oscillator transmitter A radio transmitter consisting only of a radio-frequency oscillator and its power supply. The oscillator may be keyed for radiotelegraphy or modulated for radiotelephony (AM or FM).

oscillator trimmer In a superheterodyne receiver, a small, limited-range capacitor connected in parallel with the oscillator coil for tracking oscillator tuning at the high end of a band. Compare *oscillator padder.*

oscillator tuning The separate, often ganged, tuning of the oscillator stage in a circuit.

oscillator-type power supply A type of high-voltage power supply in which an rf oscillator generates a low-voltage ac which then is stepped up by an rf transformer and finally rectified to high-voltage dc.

oscillator-type transmitter See *oscillator transmitter.*

oscillatory current A current that alternates periodically, particularly the current in an *LC* tank circuit that comes from the oscillation of energy dack and forth between the inductor and capacitor.

oscillatory discharge An electrical discharge, such as that of a capacitor, which sets up an oscillating current (see *oscillatory current*).

oscillatory surge A current or voltage surge that includes both positive and negative excursions.

oscillatory transient See *oscillatory surge.*

oscillistor A device consisting essentially of a dc-conducting bar of semiconductor material positioned in a magnetic field, and which produces oscillations.

Oscillite The picture tube in the Farnsworth TV system; synonymous with *kinescope.*

oscillogram 1. The image produced on the screen of an oscilloscope. 2. A permanent, usually photographic record made from the screen of an oscilloscope.

oscillograph 1. An instrument that makes a permanent record (photograph or pen recording) of a rapidly varying electrical quantity. Also called *recorder* (see *recorder, 2*). Compare *oscilloscope.* 2. An obsolete term for *oscilloscope.*

oscillograph recorder A direct-writing recorder (see *recorder, 2*).

oscillography The use of a graphic oscillation recorder, or oscillograph.

oscillometer A device used for determining the peak amplitude of an oscillation.

oscilloscope An instrument that presents for visual inspection the pattern representing variations in an electrical quantity. Also see *cathode-ray oscilloscope.* Compare *oscillograph.*

oscilloscope camera A special high-speed, short-focus camera with fixtures for attachment to an oscilloscope to record images from the screen. Standard and instant-film types are available.

oscilloscope differential amplifier An amplifier that processes the difference between two signals, for the purpose of displaying on an oscilloscope or oscillograph.

oscilloscope tube A cathode-ray tube for use in an oscilloscope. It contains an electron gun, accelerating electrode, horizontal and vertical deflecting plates, and a fluorescent screen.

Os-Ir Symbol for *osmiridium.*

OSL Abbreviation of *orbiting space laboratory.*

osmiridium Symbol, *Os-Ir.* A natural alloy of osmium and iridium.

osmium Symbol, *Os.* A metallic element of the platinum group. Atomic number, 76. Atomic weight, 190.2.

osmosis The diffusion of a liquid through a semipermeable membrane or porous partition. A lower-concentration solution passes through to a higher-concentration solution, the tendency of the process being to equalize the concentrations.

osmotic pressure The force that causes the positive ions to pass out of a solution toward a metal body immersed in an electrolyte. Also see *Helmholtz double layer.*

OSO Abbreviation of *orbiting solar observatory.*

osteophone A bone-conduction hearing aid.

OTA Abbreviation of *operational transconductance amplifier.*

OTL Abbreviation of *output-transformerless* (amplifier or generator that requires no output coupling transformer).

Otto Phonetic alphabet code word for the letter *o.*

O-type backward-wave oscillator Abbreviation, OBWO. A backward-wave oscillator employing harmonics having opposing phases.

ounce Abbreviation, *oz.* A unit of weight equal to 1/16 pound or 28.35 grams,

ounce-inch Abbreviation, *oz-in.* A unit of torque equal to the product of a force of 1 ounce and a moment arm of 1 inch. Compare *pound-foot.*

outage 1. Loss of power to a system. 2. Loss of a received signal.

outboard components 1. Discrete components (capacitors, coils, resistors, or transformers) which are connected externally to an integrated circuit. 2. Discrete components which are connected externally to any existing electronic device.

outcome In statistical analysis, the result of an experiment or test. An outcome can be numerical or nonnumerical.

outdoor antenna An antenna that is erected outside, usually high above the surface of the earth clear of obstacles. Compare *indoor antenna.*

outdoor booster A signal preamplifier mounted on an outdoor TV receiving antenna for improved reception.

outdoor transformer A weatherproof distribution transformer installed outside the building it services.

outer conductor The outer metal cylinder or jacket of a coaxial cable or coaxial tank. Compare *inner conductor'.*

outgassing 1. In the evacuation of electronic devices, such as vacuum tubes, the renoval of occluded gases from glass, ceramic, and metal by means of slow baking and by flashing an internal metal *getter* (such as one of magnesium). 2. The production of gases in battery cells during the final stage of charging.

outgoing line A line a which leaves a device, facility, or stage. Compare *incoming line.*

outlet A male or female receptacle which delivers a signal or operating power to equipment plugged into it.

outline flowchart In computer practice, a preliminary flowchart showing how a program will be divided into rouines and segments, input and output functions, program entry points, and so forth.

out-of-line coding Instructions for a computer program routine stored in an area of memory other than that in which the routine's program is stored.

out-of-phase Pertaining to the condition in which the alternations or pulsations of two or more separate waves or wave phenomena are out of step with each other. Compare *in-phase.*

out-of-phase current Reactive current in an ac circuit, i.e., current that is out of phase with voltage. Also see *quadrature current.*

out-of-phase voltage Voltage across a pure reactance; so called because it is out of phase with current.

outphaser A device that converts a sawtooth wave to a square wave. Used in electronic organs and synthesizers.

outphasing modulation A system of modulation in which the sideband frequencies are shifted 90 degrees from the phase position in an amplitude-modulated wave. The resulting constant-envelope wave is then amplified with high efficiency and low distortion by a class-C stage; then the signal is reconverted to an amplitude-modulated one by phase shifting the carrier with respect to the sidebands.

out-plant system A data processing system in which a central computer receives data from remote terminals.

output 1. Energy or information delivered by a circuit, device, or system. Compare *input, 1*. 2. The terminals at which energy or information is taken from a circuit, device, or system. Compare *input, 2*.

output admittance Symbol, *Yo.* The internal admittance of a circuit or device, as seen at the output terminals; the reciprocal of *output impedance.* Compare *input admittance.*

output amplifier 1. The last stage in a multistage amplifier. 2. A separate amplifier that boosts the current, voltage, or power delivered by another amplifier.

output area In a computer system, the portion of storage holding information for delivery to an output device. Also called *output block.*

output axis For a gyroscope that has received an input signal, the axis around which the spinning wheel precesses.

output block See *output area.*

output bus driver In a computer, a device that amplifies output signals sufficiently to provide signals to other devices without undue loading of the supply line (bus).

output capability The maximum power or voltage output that a circuit can deliver without distortion or other improper operating conditions.

output capacitance Symbol, *Co.* The internal capacitance of a circuit or device, as seen at the output terminals. Compare *input capacitance.*

output capacitive loading For an operational amplifier at unity gain, the maximum capacitance that can be connected to the output of the amplifier before phase shift increases to the point of oscillation.

output capacitor 1. In a capacitance-coupled circuit, the output coupling capacitor. Compare *input capacitor, 2.* The last capacitor in a filter circuit.

output capacity The maximum output capability of a device or system expressed in appropriate units, such as current, voltage, power, torque, horsepower, and the like.

output choke The last choke (inductor) in a filter circuit.

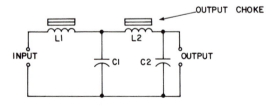

Output choke.

output circuit The circuit or subcircuit that constitutes the output portion of a network or device. Also see *output, 1, 2* and *output terminals.* Compare *input circuit.*

output-circuit distortion Distortion in the output portion of a circuit or device (such as a tube, transistor, or transformer) and usually due to an overload or nonlinear response.

output conductance Symbol, *Go.* The internal conductance of a circuit or device, as seen at the output terminals. It is the reciprocal of *output resistance.* Compare *input conductance.*

output control 1. The rain control of an amplifier. 2. The level control of a variable power supply.

output coupling capacitor See *output capacitor.*

output coupling transformer See *output transformer.*

output current 1. Symbol, *Io.* The current delived by a source, such as a battery, generator, or amplifier. Compare *input current, 1.* 2. Symbol, *Io.* Current flowing in the output leg or electrode of a circuit or device. Compare *input current, 2.*

output device 1. A load device, such as a resistor, loudspeaker, lamp, relay, motor, or the like, that utilizes the output energy delivered by a generator, amplifier, or network. 2. A device, such as an output transformer, that serves to transfer energy or information from a circuit or device. Compare *input device.* 3. In computer practice, a device which presents the results of computer operation in a comprehensible form. Examples: readouts, typewriters, tape recorders, card punchers, tape punchers, and so on.

output efficiency Symbol, *ηo.* The efficiency of a device, such as a generator or amplifier, in delivering an output-power signal. For an amplifier, $\eta o\ (\%) = 100\ Po/Pi$, where *Pi* is the dc power input and *Po* is the ac power output.

output equipment See *output device, 3.*

output filter The dc filter of an ac-operated power supply. Also see *capacitor-input filter* and *choke-input filter.*

output gap A device via which current or power is intercepted from an electron beam in a beam-power tube.

output impedance Symbol, *Zo.* The impedance looking into the output terminals of an amplifier, generator, or network. Compare *input impedance.*

output indicator A device, such as an output meter, neon bulb, magic-eye tube, cathode-ray tube, or the like, which gives a visual indication of the output-signal amplitude of an equipment.

output limiting A process for automatically maintaining the amplitude of the signal delivered by a generator or amplifier. See, for example, *automatic gain control, automatic modulation control, volume compression,* and *volume limiter.*

output load See *output device, 1.*

output load current 1. The current through the output load of an amplifier. Generally, this current is expressed in root-mean-square form. 2. The highest root-mean-square current that an amplifier can deliver to a load of a specified impedance.

output meter A meter that gives a quantitative or qualitative indication of the output of an amplifier or generator. See, for example, *output-power meter.*

output port The output terminal of a logic device.

output power Symbol, *Po.* The power deliverable by an amplifier, generator, or circuit. Also called *power output.* Compare *input power.*

output-power meter A type of direct-reading wattmeter or milliwattmeter for measuring the power output of an amplifier or generator.

output regulator A circuit or decise which automatically maintains the output of a power supply or signal source at a constant amplitude.

output resistance Symbol, *Ro.* The internal resisance of a circuit or device, as seen at the output terminals. Compare *input resistance*.

output routine In computer practice, a routine (program segment) that does the work involved in moving data to an output device, often including intermediate transferals and modifying the data as necessary.

output section See *output area*.

output stage The last stage of an amplifier.

output terminals Terminals (usually a pair) associated with output (see *output, 1, 2*). Compare *input terminals*.

output tank In an *LC*-tuned stage of a transmitter or power generator, the tank in the plate circuit of a tube, or collector circuit of a transistor, in which the output signal is developed. Compare *input tank*.

output transformer The output-coupling transformer which delivers signal voltage or power from an amplifier, generator, or network to a load or to another circuit. Compare *input transformer*.

output transistor A transistor in the final stage of an amplifier or generator; usually a power transistor.

output tube A tube in the final stage of an amplifier or generator; usually a power tube.

output unit See *output device*.

output voltage 1. Symbol, *Eo*. The voltage delivered by a source, such as a battery, generator, or amplifier. Compare *input voltage, 2*. 2. Symbol, *Eo*. The voltage across the output leg or electrode of a circuit or device. Compare *input voltage, 2*.

output winding The secondary coil of an output transformer.

outside antenna See *outdoor antenna*.

outside booster See *outdoor booster*.

outside diameter Abbreviation, *OD*. The outermost diameter of a body or figure having two concentric diameters, e.g., conduit. Compare *inside diameter*.

outside lead See *finish lead*.

outside transformer 1. Outdoor transformer. 2. A transformer mounted outside of an equipment in whose circuit it is included. External mounting can eliminate hum or prevent overheating.

oven 1. A chamber providing closely controlled operating temperature for an electronic component, such as a quartz crystal. 2. An enclosure in which electronic equipment can be tested at selected, precise high temperatures. Compare *cold chamber*.

overall feedback Positive or negative feedback around an entire system (such as an amplifier) rather than that confined to one stage or a few stages within the system,

overall gain The total gain of an entire system (such as a multistage amplifier), as opposed to that of one or several stages.

overall loudness The apparent intensity of an acoustic disturbance, generally measured with respect to the threshold of hearing,and expressed in decibels relative to the threshold level.

overbiased unit A component, such as a transistor or vacuum tube, whose bias current or voltage is higher than the correct value for a given mode of operation. Compare *underbiased unit*.

overbunching In a velociy-modulated tube such as a klystron, the condition in which the buncher voltage exceeds the value required for optimum bunching.

overcompounded generator A dynamo-type generator having a compound field winding in which the series-field winding increases the field intensity beyond the point needed to maintain the output voltage. Compare *undercompounded generator*.

overcompounding A characteristic of electromechanical motors, resulting in increased running speed with decreasing load resistance.

overcoupled transformer A transformer having greater than critical

coupling between its primary and secondary windings. In tuned circuits, such as i-f transformers, overcoupling produces double-peak response.

overcoupling Extremely close coupling (see *close coupling*).

overcurrent A current of higher than specified strength. Compare *undercurrent*.

overcurrent circuit breaker A circuit breaker that opens when current exceeds a predetermined value.

overcurrent protection Use of an overcurrent circuit breaker or overcurrent relay to protect a circuit or device.

overcurrent relay A protective relay that opens a circuit when current exceeds a predetermined value. Compare *undercurrent relay*.

overcutting In disk recording, the condition in which an excessively high amplitude signal causes the stylus to cut through the wall between adjacent grooves. Compare *undercutting*.

overdamping Damping greater than the critical value (see *action, 2*). Compare *underdamping*.

overdesign 1. To employ an unnecessarily high safety factor in the design of equipment. 2. To design equipment for performance superior to that which is required in the intended application. 3. A design that results from the practices in 1 and 2, above.

overdriven amplifier See *overdriven unit*.

overdriven unit An amplifier, oscillator, or transducer whose driving signal (current, voltage, power, or other quantity) is igher than that which the device can properly or efficiently handle for correct or intended operational performance.

overexcited Receiving higher than normal excitation, as in overexcited rf amplifiers or overexcited ac generators.

overexpose In photography and radiology, to expose a film or body for a longer time than is normal or safe. Compare *underexpose*.

overexposure The condition of being excessively exposed (see *overexpose*). This applies to radiation as well as to photographic processes.

overflow 1. In computer or calculator operation, the condition in which an arithmetic operation yields a result exceeding the capacity of the location or display for a result. 2. The carry digit that results from 1, above.

overflow indicator 1. In a digital calculator, a display that indicates that a numerical value is too large or too small to be shown with the available number of decimal places. 2. In a data-processing device, a display that indicates the presence of too many bits or characters for the available storage capacity.

overflow position In a digital computer, an auxiliary register position for developing the overflow digit (see *overflow, 1, 2*).

overflow record In data processing practice, a record that will not fit the storage area allotted for it and which must, therefore, be kept where it can be retrieved according to some reference which has been stored in its place.

overflow storage In a calculator or computer, extra storage space,allowing a small amount of overflow without loss of accuracy.

overhanging turns The turns in the unused portion(s) of a tapped coil.

overhead line A line (power, transmission) suspended above the surface of the earth between poles or towers.

over-horizon transmission See *forward scatter*.

overinsulation Use of excessive insulation for a particular application. Compare *underinsulation*.

over insulation The insulation (usually a strip of tape) laid over a wire brought up from the center of a coil. Compare *under insulation*.

overlap 1. The time during which two successive operations are performed simultaneously. 2. In a facsimile or television system, a condition in which the scanning line is wider than the center-to-center separation between adjacent scanning lines.

overlap radar A long-range radar situated in one sector and covering part of another sector.

overlay 1. A sheet of transparent or translucent material laid over a schematic diagram for the purpose of tracing connections which have been made in wiring an equipment from the diagram. 2. In computer practice, a method whereby the same internal storage locations are used for different parts of a program during a program run. It is used when the total storage requirements for instructions exceed the available main storage capacity.

overlay transistor A double-diffused epiaxial transistor having separate emitters which are connected together by means of diffusion and metallicing to increase the edge-to-area ratio of the emitters. This design raises the current-handling ability of the transistor. Also see *diffused transistor* and *epitaxial transistor*.

overload 1. Drain in excess of the rated output of a circuit or device. 2. An excessive driving signal.

overload circuit breaker See *circuit breaker*.

overloaded amplifier A power amplifier which is delivering excessive output power. Compare *underloaded amplifier, 2*.

overloaded oscillator An oscillator from which excessive power is drawn, causing instability, frequency shift, lowered output voltage, and overheating.

overload indication Any attention-catching method, such as an audible or visual alarm, for warning that a prescribed signal or power level has been exceeded.

overload level The amount of overload which may safely be applied to an equipment (see *overload, 1*).

overload protecion The use of circuit breakers, relays, automatic limiters, and similar devices to protect any equipment from overload damage by reducing current or voltage, disconnecting the power supply, or both.

overload relay A relay which is actuated when circuit current exceeds a predetermined value. Compare *underload relay*.

overload time The maximum length of time that an equipment may be subjected to an overload level of current.

overmodulation Modulation in excess of a prescribed level, especially amplitude modulation greater than 100 percent. Compare *complete modulation* and *undermodulation*.

overmodulation alarm See *overmodulation indicator*.

overmodulation indicator A device, such as a neon bulb, incandescent lamp, magic-eye tube, light-emitting diode, rf meter, rf relay, or the like, adapted to give an alarm when the modulation percentage of a signal exceeds a predetermined value.

overpotential See *overvoltage*.

overpower relay A relay that is actuated by a rise in power above a predetermined level. Compare *underpower relay*.

overpressure For a pressure transducer, pressure in excess of the maximum rating of the device.

overpunch In a computer card, a hole punched in one of the three upper rows. The overpunch determines the meaning of the pattern of lower punched holes.

override 1. To go around an automatic control system intentionally. 2. To bridge a functional stage of a system.

overscanning The deflection of the beam of a cathode-ray tube beyond the edges of the screen.

overshoot 1. The momentary increase of a quantity beyond its normal maximum value, e.g., the spike sometimes seen on a square wave, due to the overswing of a rising voltage, 2. Momentary overtravel of the pointer of a meter.

overswing See *overshoot, 2*.

overtemperature protection Use of an automatic device, such as a thermal relay or thermostat, to disconnect a device from the power supply when the device's temperature becomes excessive.

overthrow See *overshoot, 2*.

overtone See *harmonic*.

overtone crystal A piezoelectric quarts plate that oscillates at odd multiples of the frequency for which it was ground. This ability allows crystal operation at high frequencies that would otherwise be only obtainable from a fundamental-frequency plate ground so thin as to be fragile.

overtone oscillator A crystal oscillator employing an overtone crystal.

overtravel See *overshoot, 2*.

overvoltage A voltage of higher than specified value. Compare *undervoltage*.

overvoltage circuit breaker A circuit breaker that opens when voltage exceeds a predetermined value.

overvoltage protection Use of a special circuit or device to protect equipment from excessive voltage. When voltage increases beyond the overvoltage limit, the protective circuit causes shutdown.

overvoltage relay A relay that is actuated when voltage rises above a predetermined value. Compare *undervoltage relay*.

overwrite In computer practice, to write over information, i.e., to put information in a memory location that contains data already, thereby eradicating what is there; this also includes recording over data on a magnetic storage medium, such as tape.

Ovshinsky effect In thin-film solid-state devices, the tendency for switches to have the same characteristics for currents in either direction.

O-wave One (the *ordinary*) of the pair of components into which an ionospheric radio wave is divided by Earth's magnetic field, Compare *X-wave*.

Owen bridge A wide-range four-arm bridge which measures inductance in terms of a standard capacitance and bridge-arm resistances. Typical coverage is 0.1 nH to 111H.

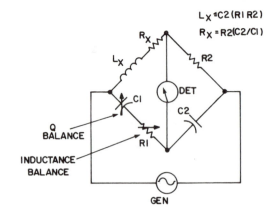

$$L_X = C2(R1\ R2)$$
$$R_X = R2(C2/C1)$$

Owen bridge.

own coding Additional program steps added to vendor-supplied software so that it can be modified to fit special needs.

ox Abbreviation of *oxygen*.

oxidation The combination of a substance with oxygen. Generally, a slow process, such as the corrosion of iron or aluminum in the atmosphere. Oxidation is aggravated by the presence of moisture and/or high temperatures.

oxidation—reduction potential The potential at which oxidation occurs at the anode of an electrolytic cell and reduction, at the cathode.

oxide-coated cathode See *oxide-coated emitter*.

oxide-coated emitter An electron-tube cathode or filament coated with a material such as thorium oxide, for increased electron emission at low emitter temperatures.

oxide-coated filament See *oxide-coated emitter*.

oxide film 1. The thin film of iron oxide which constitutes the magnetic surface of recording tape. 2. The layer of copper oxide formed on the copper plate of a copper-oxide rectifier.

oxide-film capacitor An *electrolytic capacitor*; so called because the dielectric of this capacitor is a thin oxide film.

oxide rectifier A solid-state rectifier employing a junction between copper and copper oxide. Also called *copper-oxide rectifier*.

oximeter A photoelectric instrument for neasuring the oxygen content of the blood seen when a light beam passes through the ear lobe. Also called *anoxemia toximeter*.

oxygen Symbol, *O*. Abbreviation, O2. A gaseous element. Atomic number, 8. Atomic weight, 16.000. Oxygen constitutes 21% of Earth's atmosphere.

oxygen analyzer An electronic gas analyzer designed especially to measure oxygen content. The operation of this instrument is based upon the paramagnetic properties of oxygen.

oz 1. Abbreviation of *ounce*. 2. Abbreviation of *ozone*.

oz-in Abbreviation of *ounce-inch*.

ozocerite An insulating mineral wax. Dielectric constant, 2.2. Dielectric strength, 4 to 6 kV/mm. Also spelled *ozokerite*.

ozone Symbol, *O*3. An allotropic form of oxygen. Its formula indicates that each molecule of ozone has three atoms. Ozone is produced by the action of ultraviolet rays (or electrical discharge) on oxygen; its characteristic odor (somewhat like weak chlorine) can often be detected around sparking contacts or in the air after a lightning storm.

ozone monitor An instrument for measuring the concentration of ozone in the atmosphere. One version utilizes the phenomenon of ultraviolet absorption by ozone to measure concentrations between 0.01 and 10 ppm.

P 1. Symbol for *plate* (of an electron tube). 2. Symbol for *power*. 3. Symbol for *phosphorus*. 4. Abbreviation of *pressure*. 5. Symbol for *primary*. 6. Abbreviation for prefix *peta*.

P Symbol for *permeance*.

p 1. Abbreviation of prefix *pico*. 2. Subscript for *peak*. 3. Abbreviation of *pound*. 4. Abbreviation of *point* (often capitalized). 5. Subscript for *primary*. 6. Subscript for *plate* (of an electron tube). 7. Abbreviation of *pitch*. 8. Abbreviation of *per*.

PA 1. Abbreviation of *power amplifier*. 2. Abbreviation of *pulse amplifier*. 3. Abbreviation of *particular average*. 4. Abbreviation of *pilotless aircraft*. (Also, P/A). 5. Abbreviation of *public address* (as in PA system).

Pa 1. Symbol for *protactinium* 2. Symbol for *pascal*.

pA Abbreviation of *picoampere*.

pacemaker See *cardiac stimulator*.

pacer See *cardiac stimulator*.

Pacific standard time Abbreviation, PST. Local mean time of the 120th meridian west of Greenwich. Also see *Greenwich mean time, standard time, time zone,* and *coordinated universal time*.

pack 1. A group of related punched (computer data) cards. 2. A technique for maximizing a computer memory device's storage capacity wherein more than one information item is stored in a single storage unit. Also called *crowd*.

package 1. The enclosure for an electronic device or system. This includes a wide range of housings, from the simple encapsulation of miniature transistors to forced-air-cooled enclosures for heavy power units. 2. To assemble and house an electronic equipment, or to design a housing for it in accordance with good engineering practice. 3. A computer program of general use for an application, e.g., payroll package.

package count The number of discrete packaged circuits in a system.

packaging density 1. Volumetric efficiency. 2. Computer storage capacity in terms of the number of information units that can be contained on a given segment of a magnetic medium. Also called *packing density*. 3. Within a given integrated circuit, the capacity in terms of the number of active devices that can be contained on a single silicon slab.

packet See *wave packet*.

packet switching In data communications, a standard for transmittal of information units in grouped segments.

packing In the button of a carbon microphone, bunching and cohesion between the carbon granules.

packing density The number of discrete packaged circuits within a given surface area or volume.

packing factor 1. Volumetric efficiency. 2. In computer practice, the number of bits that can be recorded in a given length of magnetic memory surface. Also called *packing density*.

pack transmitter A portable transmitter that can be strapped to the operator's back.

pack unit A portable transceiver that can be strapped to the operator's back or carried on an animal's back.

PACM Abbreviation of *pulse-amplitude code modulation*.

pad 1. An attenuator network (usually a combination of resistors) that reduces a signal by a desired amount while maintaining constant the input and output impedance. 2. In computer practice, to make a record a fixed size by adding blanks or dummy characters to it. 3. To lower the frequency of an LC circuit by adding capacitance to an already capacitively tuned network.

padder See *oscillator padder*.

padding capacitor See *oscillator padder*.

padding character In a digital communications system, a character that is inserted solely for the purpose of consuming time while no meaningful characters are sent. The insertion of such characters maintains the synchronization of the system.

paddle-handle switch A toggle switch the lever of which is a flattened rod. Compare *bat-handle switch, rocker switch,* and *slide switch.*

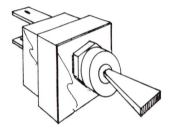

Paddle-handle switch.

PADT Abbreviation of *post-alloy-diffused transistor.*

page A display of data on a crt terminal that fills the screen.

page printer A Teletype machine or computer peripheral that prints a message in lines on a page according to an established format rather than in a single line (for the TTY machine, on a tape). Compare *tape printer.*

pager 1. A *public-address system* used for summoning purposes. 2. *Beeper, 2.*

page turning The successive display of pages (see *page*).

pair 1. Two wires, especially two insulated conductors in a cable. 2. A couple of particles.

pair annihilation The demolition of a positron and an extranuclear electron by nuclear forces. Compare *pair production.*

paired cable A cable made up of separate twisted pairs of conducting wires.

pair production The complete and simultaneous transformation of a photon (of a quantum radiant energy) into an electron and a positron when the photon is under the influence of an intense electric field. Compare *pair annihilation.*

palladium Symbol, Pd. A metallic element of the platinum group. Atomic number, 46. Atomic weight, 106.4.

Palmer scan In radar, a method of simultaneously scanning the azimuth and the elevation.

PAM Abbreviation of *pulse-amplitude modulation.*

PAN In radiotelephony, a spoken word which indicates that an urgent message is to follow. It is equivalent to the *XXX* of radiotelegraphy.

pan 1. To make a panoramic sweep, e.g., to sweep a wide area with a beam (as from an antenna), or to sweep a wide band of frequencies with a suitable tuning circuit. 2. A panoramic sweep made as in 1, above. 3. In audio engineering practice, to gradually shift from one audio channel to another or from one reproducer to another.

pan and tilt 1. An azimuth-elevation mounting for a television camera. 2. The simultaneous movement of a television camera in the vertical and horizontal directions.

pancake coil See *disk winding.*

panel A plate on which are mounted the controls and indicators of an equipment, for easy access to the operator.

panel lamp 1. Electroluminescent panel. 2. Panel light.

panel light A pilot light for illuminating the front panel of an equipment.

panel meter A usually small meter for mounting on, or through an opening in, a panel.

panoramic adapter An external device which may be connected to a receiver to sweep a frequency band and indicate carriers on the air as pips on a screen at the corresponding frequency points. Also called *pan adapter.*

panormaic display 1. A wide-angle display. 2. A spectrum-analyzer display that shows a wide range of frequencies, from zero to well above the maximum frequency in the monitored system.

panoramic radar An omnidirectional radar, i.e., one which transmits wide-beam signals in all directions without scanning.

panoramic receiver A receiver that displays pips on a screen to show carriers on the air in a given frequency band. All frequencies in the band are presented along the horizontal axis of the screen.

panpot A potentiometer by which panning can be achieved (see *pan,3.*)

pan-range A form of radar display, in which target motion can be ascertained.

pantography The transmission of radar information to a distant location for observation or recording.

Papa Phonetic alphabet code word for the letter P.

paper advance mechanism In a data processing system, the part of a printer that moves (sometimes by computer control) the paper through the printer.

paper capacitor A fixed capacitor whose dielectric is a thin film of paper. The paper is usually impregnated with wax or oil to waterproof it and to increase its dielectric constant and dielectric strength.

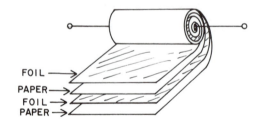

Paper capacitor.

paper tape The punched tape used to store information in a computer system or data processing system.

paper tape code On a paper tape, the pattern of holes representing information.

paper tape loop Paper tape whose ends are joined; it is used to control printing units in a computer system. Also called *control loop* and *control tape.*

paper tape punch See *tape punch.*

paper tape reader A device that senses the holes in paper tape and converts the information thereby derived into binary form.

paper tape reproducer A device for duplicating a punched paper tape.

paper tape verifier A device capable of verifying the validity of data on a paper tape by checking it against the original version.

paper throw The rapid nonincremental advance of paper tape through a printer.

PAR Abbreviation of *precision approach radar*.

par Abbreviation of *parallel*.

parabola A plane curve which is the locus of points that are equidistant from a fixed point (the *focus*) and a fixed straight line (the *directrix*). The parabola is also a conic section: it is the curve formed by the intersection of a cone by a plane parallel to the latter's axis.

parabola control See *vertical-amplitude control, 2*.

parabola generator A circuit for generating a parabolic-waveform signal.

parabolic microphone A directional microphone mounted at the principal focus of a parabolic sound reflector; the front of the microphone faces the inside of the parabola.

parabolic reflector A reflector having the shape of a parabola. It is particularly useful for focusing or directing radiation. For example, if a radiator, such as an antenna rod, is placed at the focus of the parabola, a beam of parallel rays will be emitted by the reflector.

paraboloid The surface generated by a parabola rotating about its axis of symmetry.

paraffin A relatively inexpensive, easily available, solid, white petroleum wax. At one time, it was used to impregnate capacitors and coils and to waterproof paper used for insulating purposes. Dielectric constant: 2.1. Dielectric strength: 290 kV/cm.

parallax The apparent shift in the position of a relatively nearby object when the observer changes his position or alternately blinks either eye. Thus, a pointer-type meter will seem to give different readings when viewed from different angles, the reason why some are provided with a mirrored scale.

parallel 1. Pertaining to the type of operation in a computer when all elements in an information item (bits in a word, e.g.,) are acted upon simultaneously rather than serially (one at a time). 2. The condition in which two comparably sized objects or figures are equidistant at all facing points. 3. Pertaining to the shunt connection of components or circuits.

parallel access In computer practice, inputting or outputting data to or from storage in whole elements of information items (a word rather than a bit at a time, for example.).

parallel adder In a computer or calculator, an adder in which corresponding digits in multibit numbers are added simultaneously. Also see *parallel, 1*.

parallel antenna tuning Antenna-feeder tuning in which the tuning capacitor is connected in parallel with the two feeder wires. Compare *series antenna tuning*.

parallel arithmetic unit See *parallel adder*.

parallel capacitance 1. Shunt capacitance. 2. The capacitance between the turns of a coil. Also see *distributed capacitance*.

parallel capacitors Capacitors connected in parallel. The total capacitance $C_t = C_1 + C_2 + C_3 + \ldots + C_n$. Also see *parallel circuit*.

parallel circuit A circuit in which the components are connected across each other, i.e., so that the circuit segment could be drawn showing component leads bridging common conductors as rungs would across a ladder. Compare *series circuit*.

parallel-component amplifier An amplifier stage in which tubes or transistors are connected in parallel with each other for increased power output. Also see *parallel circuit*.

parallel-component oscillator An oscillator stage in which tubes or transistors are connected in parallel with each other for increased power output. Also see *parallel circuit*.

parallel computer A computer which is equipped to handle more than one program at a time, but not through the use of multiple programming or time-sharing.

parallel-cut crystal See *Y-cut crystal*.

parallel-diode half-wave rectifier See *parallel limiter*.

parallel-fed amplifier An amplifier circuit in which dc operating voltage is applied in parallel with ac output voltage. Also see *parallel feed*.

parallel-fed oscillator An oscillator in which dc operating voltage is applied in parallel with ac output voltage. Also see *parallel feed*.

parallel feed The presentation of parallel ac and dc voltages to a device, or the presentation of a dc operating voltage in parallel with the ac output voltage of the device (as in a parallel-fed amplifier or oscillator). Also see *shunt feed*.

parallel gap welding A welding technique using two electrodes separated by a gap.

parallel gate circuit A gate circuit employing two triodes with parallel-connected plates and cathodes, and a common plate resistor. The input signal is applied to one grid, and the control signal to the other. A similar circuit may be derived using bipolar transistors or FETs.

parallel inductance Shunting inductance.

parallel inductors Inductors connected in parallel and separated or oriented to minimize the effects of mutual inductance. Also see *parallel circuit*.

parallel inverse feedback In an af amplifier circuit consisting of a pentode driver that is *RC*-coupled to a single-ended beam-power output tetrode, a simple system for obtaining negative feedback: A high resistance (e.g., half a megohm) is connected from the output-tube plate to the driver-tube plate.

parallel limiter A limiter (clipper) circuit in which the diode is in parallel with the signal. Compare *series limiter*.

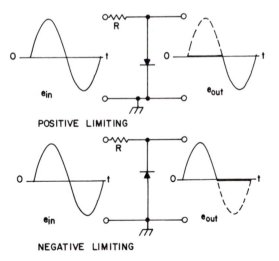

Parallel limiter.

parallel-line tuning In uhf circuits, the use of two parallel wires or rods for tuning. A straight short-circuiting bar is slid along the wires to accomplish tuning.

parallelogram A two-dimensional geometric figure having four sides. Opposite pairs of sides are parallel. Opposite interior angles have equal measure.

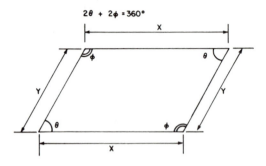

Paralleogram.

parallelogram of vectors A graphic device for finding the sum of two vectors. A parallelogram is constructed for which the two vectors are a pair of adjacent sides. The sum of the vectors is represented by the diagonal of this polygon.

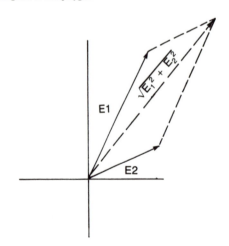

Parallelogram of vectors.

parallel operation In computer practice, the simultaneous transmission of all bits in a multibit word over individual lines, as compared with the serial transmission of a word bit by bit.

parallel output A digital output consisting of two or more lines, all of which carry data at the same time.

parallel-plane triode A triode in which the plate and cathode are two parallel planes whose areas are large compared with the distance between them.

parallel-plane tube An electron tube in which the various elements are in parallel planes.

parallel processing In computer practice, the simultaneous process-

ing of several different programs through separate channels. Compare *serial processing*.

parallel Q Symbol, Qp. The figure of merit of a parallel circuit of inductance, capacitance, and resistance; $Qp = Rp/(\theta_0 L)$.

parallel resistance 1. Shunt resistance. 2. The resistance between the plates of a capacitor. 3. The resistance between the turns of a coil.

parallel resistors Resistors connected in parallel. The total resistance $Rt = 1/(1/R_1 + 1/R_2 + ... + 1/R_n)$. Also see *parallel circuit*.

parallel resonance Resonance in a circuit consisting of a capaciator, inductor, and ac generator connected in parallel. At the resonant frequency, XL is equal to XC, capacitor current and inductor current are at maximum, line current is at minimum, and the tank-circuit impedance is at maximum. Compare *series resonance*.

parallel-resonant circuit A resonant circuit in which the capacitor, inductor, and ac generator are connected in parallel. Compare *series-resonant circuit*.

parallel-resonant trap A wavetrap consisting of a parallel-resonant LC circuit. Compare *series-resonant trap*.

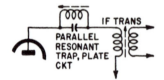

Parallel-resonant trap.

parallel-resonant wavetrap See *parallel-resonant trap*.

parallel-rod oscillator An ultrahigh-frequency oscillator tuned by means of two straight, parallel quarter- or half-wave rods—one connected to the grid of a triode and the other to the plate.

parallel-rod tuning Adjustment of the resonant frequency of a section of open-wire transmission line. A movable shorting bar allows quarter-wave resonance. The impedance at resonance is very high.

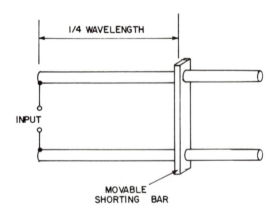

Paralleled-rod tuning.

parallel—series capacitors Capacitors connected in parallel—series. Total capacitance $C_t = 1/[1/(C_1 + C_2) + 1/(C_3 + C_4)]$. Also see *parallel-series circuit*.

parallel—series circuit A circuit consisting of parallel circuits connected in series with each other. Also see *parallel circuit* and *series circuit*.

parallel—series inductors Inductors connected in parallel—series and separated or oriented to minimize the effects of mutual inductance. Total inductance $L_t = 1/(1/L_1 + 1/L_2) + 1/(1/L_3 + 1/L_4)$. Also see *parallel—series circuit*.

parallel storage In a computer, storage in which all information items can be made available in the same amount of time.

parallel-tee amplifier A bandpass amplifier having a parallel-tee network in its negative-feedback path. The null frequency of the network determines the pass frequency of the amplifier.

parallel-tee measuring circuit A parallel-tee network employed for measuring circuit constants. Also called *twin-tee measuring circuit*.

parallel-tee network A resistance—capacitance network containing two tee sections (with R and C elements opposite in the tees) connected in parallel. The network produces a null at one frequency. Also called *twin-tee network*.

parallel-tee oscillator A resistance—capacitance tuned oscillator having a parallel-tee network in its negative-feedback path. The null frequency of the network determines oscillator frequency.

parallel transfer A form of digital information transfer, consisting of two or more lines that carry data at the same time.

parallel-wire line A transmission line or line-type tank consisting of two parallel wires supported in air.

parallel-wire tank For a uhf amplifier or oscillator, a tank consisting of two separate parallel wires connected to the tube(s) or transistor(s) at one end and short-circuited or tuned at the other end.

paramagnet A paramagnetic substance (see *paramagnetism*). Compare *diamagnet*.

paramagnetic Possessing paramagnetism. Compare *diamagnetism*.

paramagnetism The state of having a magnetic permeability of greater than 1. Compare *diamagnetism*.

parameter 1. An operating value, constant, or coefficient which may be either a dependent or independent variable (e.g., a transistor-electrode current or voltage). 2. The ratio of one coefficient to another, where both are either fixed or variable (e.g., transconductance of a tube).

parameter card A punched card having, in encoded form, the value for a parameter pertinent to a specific computer program. Also called *control card*.

parameter word In a computer memory, a place having a capacity of a word (bit group) in which is stored a parameter for a program.

parametric amplifier An rf-powered amplifier based upon the action of a voltage-variable capacitor in a tuned circuit.

parametric amplifier diode See *varactor*.

parametric converter A converter (see *converter, 1*), in which a parametric device, such as a varactor, is employed to change a signal of one frequency to a signal of another frequency. Also see *parametric down-converter* and *parametric up-converter*.

parametric diode A variable-capacitance diode (see *voltage-variable capacitor, 1*).

parametric down-converter A parametric converter in which the output signal is of a lower frequency than the input signal. Compare *parametric up-converter*.

parametric modulation Modulation in which either the inductance or capacitance of a tank circuit or coupling device is varied at the modulation frequency.

parametric oscillator An oscillator that generates visible light energy by means of a parametric amplifier and a tunable cavity.

parametric up-converter A parametric converter in which the output signal is of a higher frequency than the input signal. Compare *parametric down-converter*.

parametron See *phase-locked oscillator*.

paramistor A device consisting of several digital circuit elements that employ parametric oscillators.

paramp Abbreviation of *parametric amplifier*.

paraphase inverter A single-tube phase inverter in which the two out-of-phase output signals are obtained by taking one output from the plate and the other from the cathode; thus, the 180-degree phase difference between plate and cathode is exploited. A single transistor (bipolar or FET) can also be used.

parasitic See *parasitic oscillation*.

parasitic antenna An antenna that is not directly driven but, rather, is excited by energy radiated by another antenna.

parasitic capacitance Stray capacitance. It can be internal or external to a circuit and can introduce undesirable coupling or bypassing.

parasitic choke A small rf choke coil (with or without a shunting resistor) which acts as a parasitic suppressor.

parasitic director In a multielement beam antenna, a parasitic element acting as a director.

parasitic element In a multielement antenna, an element to which no direct connection is made; it receives rf excitation from a nearby driven element in the antenna array.

parasitic-element directive antenna A directional antenna whose directors and reflectors are parasitic elements; the radiator is the only driven element. Compare *driven-element directive antenna*.

parasitic eliminator See *parasitic suppressor*.

parasitic excitation Excitation of a beam-antenna element without direct connection to the transmitter. Thus, a director or reflector element may be excited by the field of the radiator element.

parasitic inductance Stray inductance, e.g., the internal inductance of a wirewound resistor.

parasitic oscillation Extraneous, useless oscillation present as a fault in an electronic circuit.

parasitic reflector In a multielement beam antenna, a parasitic element acting as a reflector.

parasitic resistance Stray resistance, e.g., the inherent, internal resistance of a multilayer coil.

parasitic suppressor A small resistor, coil, or combination of the two, connected in series with a tube or transistor to eliminate parasitic oscillations.

paravane A device mounted at a ship's bow and underwater to cut the lines to which mines are tethered.

PARD Abbreviation of *periodic and random deviation*.

parity 1. At par (with respect to the even-or-odd state of the characters in a group). 2. Having the quality that the number of bits (or the number of like bits) are even or odd, as intended.

parity bit 1. In computer practice, a bit added to a group of bits so that the number of ones in the group is, according to specification, even or odd. 2. In computer practice, a check bit that can be a one or zero, depending on the parity (*parity, 1*) of the total of ones in the bit group being checked.

parity check A check of the integrity of data being transferred by adding the bits in, say, a word, and then determining the parity bit needed and comparing that with the transmitted parity bit.

parity error An error disclosed by a parity check.

parity tree A digital device used for checking parity.

Parry amplifier See *cathamplifier.*

parsec Abbreviation, *pc*. The distance at which an astronomical unit subtends an angle of 1 second of arc; 1 pc = $3.085\ 7 \times 10^{16}$ meters.

part See *circuit component, 1.*

part failure The usually destructive breakdown of a circuit component.

partial carry The temporary storage of some or all of the carry information in a digital calculation.

partial derivative Symbol, δ. When two quantities both may vary, the derivative obtained when one quantity is held constant while the other varies. If, then, another partial derivative is obtained by holding the second quantity constant while the first varies, the total derivative can be found by adding the two partial derivatives. Also see *derivative; differential, 1; differential calculus;* and *differentiation.*

partial differential equation A differential equation containing partial derivatives. Compare *ordinary differential equation.*

partial differentiation The calculation of a partial derivative.

partial integration When two or more quantities may vary, the integral obtained when one quantity is held constant while the others vary for the first integration, then holding another quantity constant while the others vary for a second integration, and so on. Also see *integral, integral calculus,* and *integration.* Compare *partial differentiation.*

partial-triode operation Amplifier output pentode operation at a point intermediate between triode and pentode performance by maintaining identical B-plus voltage on screen and plate while the screen is connected to a tap on the primary winding of the output transformer.

particle 1. A tiny, discrete bit of matter. 2. One of the nonparticulate constituents of an atom. See, for example, *antiparticle, electron, meson, neutretto, neutrino, neutron, nucleon, positron, proton.*

particle accelerator See *accelerator, 1.*

particle theory of radiation A model that explains the nature of radio waves, infrared, visible light, ultraviolet, X-rays, and gamma rays in terms of discrete particles called photons.

particle velocity 1. The speed and direction of the particles from a source of atomic radiation. 2. The speed and direction of the molecules in the medium of an acoustic disturbance.

partitioning In computer practice, breaking down a large block of information into smaller blocks that can be better handled by the machine.

partition noise Noise in a pentode due to the random partition of cathode current between the screen and plate.

pascal Symbol, *Pa*. The SI (derived) unit of pressure; $1\ \text{Pa} = 1\ \text{N/m}^2 = 1.450\ 3 \times 10^{-4}\ \text{lb/sq in.}$

Pascal's triangle A triangular arrangement of numbers that are the coefficients derived from the expression $(a + b)^n$, where *n* is of the series 0, 1, 2, 3,...

Paschen—Back effect See *Zeeman effect.*

Paschen's law For a two-element, parallel-plate, gas-discharge tube, the plate-to-plate sparking potential is proportional to *Pd*, where *P* is gas pressure, and *d* is the distance between plates.

pass amplifier A tuned amplifier having the response of a bandpass filter. Like the filter, the amplifier passes one frequency (or a narrow band of frequencies) readily while rejecting or attenuating others. Compare *reject amplifier.*

pass band The continuous spectrum of frequencies transmitted by a filter, amplifier, or similar device. Compare *stopband.*

passband ripple Multiple low-amplitude resonances within the passband of a filter or tuner, resulting in a ripple pattern on the nose of the response curve.

passivation The process of growing a thin oxide film on the surface of a planar semiconductor device to protect the exposed junction(s) from contamination and shorts. See, for example, *planar epitaxial passivated transistor* and *planar transistor.*

passive circuit A circuit consisting entirely of nonamplifying components, such as capacitors, resistors, inductors, and diodes.

passive communications satellite A communications satellite which reflects a signal from one ground point to another but doesn't provide reception, amplification, or retransmission. Also called *passive comsat.* Compare *active communications satellite.*

passive component A device which is basically static in operation, i.e., it is ordinarily incapable of amplification or oscillation and usually requires no power for its characteristic operation. Examples: conventional resistor, capacitor, inductor, diode, rectifier, fuse. Compare *active component, 1.*

passive comsat See *passive communications satellite.*

passive decoder A decoder that responds to only one signal code, rejecting all others.

passive detection In reconnaissance, detecting a target without betraying the location of the detector.

passive electric network See *passive network.*

passive mixer A signal mixer employing only passive components (diodes, nonlinear resistors, nonlinear reactances), i.e., one without active components, such as tubes or transistors. Passive mixers provide no gain and often introduce a loss. Compare *active mixer.*

passive modulator A modulator employing only passive components (diodes, nonlinear resistors, nonlinear reactances), i.e., one without active components, such as tubes or transistors. Passive modulators provide no gain and often introduce a loss. Compare *active modulator.*

passive network A network composed entirely of passive components i.e., one containing no generators and providing no amplification.

passive reflector A metal surface used for the purpose of reflecting electromagnetic energy at ultra-high and microwave frequencies.

password As a security device in computer practice, a group of characters upon whose presentation to the system via a terminal the user is allowed access to memory or control of information.

paste In "dry" batteries and electrolytic capacitors, a gelatinous electrolyte.

patch 1. A temporary connection, as between a radio receiver and a telephone or, conversely, between a telephone line and a radio transmitter. 2. To make quick, usually temporary connections, as with a patch cord. 3. Instructions entered by an unconditional branch to a computer program for the purpose of correction.

patch bay A set of patch panels with terminals for all the circuits associated with a given function in a system.

patch cord A flexible line of one or more conductors with a jack or connector at each end, used to interconnect (*patch*) circuit points exposed for the purpose on a panel or breadboard.

Patchett tone control A dual tone-control circuit employing a variable series-*RC* filter for treble boost and a variable shunt-*RC* filter for bass boost. The input signal is applied in parallel to both filters. The output of each filter goes to a separate grid in a dual triode, and the signals are combined in the latter's plate circuit.

patching The interconnection of two or more signal media or lines.

patch panel A panel on which the terminals of a system are accessible for interconnection, tests, etc.

patch up 1. To replace faulty or damaged parts in an electronic system with roughly appropriate surrogates in order to restore operation quickly (usually under emergency conditions). Also see *doctor*. 2. To wire a circuit quickly using patch cords for preliminary test and evaluation.

patent 1. A document awarded by a government body, giving to an inventor the exclusive right to exploit his invention for a specified number of years. Formally called *letters patent*. 2. The monopoly granted by 1, above.

path 1. The route over which current flows. 2. In radio and navigation, the imaginary line extending directly between transmitter and receiver (or target). 3. In a computer program, the logical order of instructions.

pathometer A lie detector that indicates changes in the resistance of the body to current.

pattern 1. An established sequence of steps in a process, or arrangement of terms in a matrix. 2. The graphical representation of a varying quantity, e.g., ac wave pattern. 3. The image on the screen of an oscilloscope, or the record traced by an oscillograph. 4. The graphic polar representation of the radiation field of an antenna. 5. The arrangements of bits in a word or field.

PAV Abbreviation of *phase-angle voltmeter*.

pawl In a mechanical stepping device, as in nonelectric clock, that which is made to engage the sloping sprockets on a wheel to insure shaft rotation in one direction only.

PAX Abbreviation of *private automatic exchange*.

pay TV See *subscription TV*.

Pb Symbol for *lead*.

P-band A radio-frequency band extending from 225 to 390 MHz.

PBX Abbreviation of *private branch exchange*.

PC 1. Abbreviation of *photocell*. 2. Abbreviation of *printed circuit*. 3. Abbreviation of *positive column*. 4. Abbreviation of *point-contact*. 5. Abbreviation of *percent*. (also, *pct.*). 6. Abbreviation of *program counter*. 7. Abbreviation of *punched card*.

pc 1. Abbreviation of *picocoulomb*. 2. Abbreviation of *picocurie*. 3. Abbreviation of *parsec*.

pC Symbol for *picocoulomb*

PCB Abbreviation of *printed-circuit board*.

PC board see *printed-circuit board*.

PC diode See *point-contact diode*.

p-channel JFET See *p-channel junction field-effect transistor*.

p-channel junction field-effect transistor Abbreviation, PFET. A junction-type FET in which the gate junction has been formed on a bar or die of p-type semiconductor material. Compare *n-channel junction field-effect transistor*.

p-channel MOSFET A metal-oxide semiconductor field-effect transistor in which the channel is composed of p-type silicon. Also see *depletion-type MOSFET, depletion-enhancement-type MOSFET*, and *enhancement-type MOSFET*.

pCi Symbol for *picocurie*.

PCL Abbreviation of *printed-circuit lamp*.

PCM 1. Abbreviation of *pulse-code modulation*. 2. Abbreviation of *punched card machine*.

PCM-FM Pertaining to a carrier that is frequency modulated by information which is pulse-code modulated. Also see *frequency modulation* and *pulse-code modulation*.

PCM—FM—FM Pertaining to a carrier that is frequency modulated by one or more subcarriers which are frequency modulated by information that is pulse-code modulated. Also see *frequency modulation* and *pulse-code modulation*.

PCM level In a pulse-code-modulated signal, one of several different possible signal conditions.

PCM-PM Pulse-code modulation that is accomplished by varying the phase of the carrier wave.

PC relay See *printed-circuit relay*.

PC transistor See *point-contact transistor*.

PD 1. Abbreviation of *plate dissipation*. 2. Abbreviation of *pulse duration*. 3. Abbreviation of *proximity detector*. 4. Abbreviation of *potential difference*.

Pd Symbol for *palladium*.

PDA Abbreviation of *predicted drift angle*.

PDAS Abbreviation of *programmable data acquisition system*.

P-display See *plan position indicator*.

PDM Abbreviation of *pulse-duration modulation*.

PDM-FM Pertaining to a carrier tht is frequency modulated by one or more subcarriers which are frequency modulated by pulses which in turn are pulse-duration modulated by information. Also see *frequency modulation* and *pulse-duration modulation*.

PDM-FM-FM Pertaining to a carrier that is frequency modulated by one or more subcarriers which are frequency modulated by pulses that themselves are pulse-duration modulated by information. Also see *frequency modulation* and *pulse-duration modulation*.

PDM-PM Pertaining to a carrier that is phase modulated by pulse-duration-modulated information. Also see *phase modulation* and *pulse-duration modulation*.

PDT Abbreviation of *Pacific daylight time*.

PDVM Abbreviation of *printing digital voltmeter*.

PE 1. Abbreviation of *potential energy*. 2. Abbreviation of *professional engineer*. 3. Abbreviation of *probable error*.

peak 1. The maximum value of a quantity. 2. In an ac cycle, the maximum positive or negative current or voltage point. 3. The frequency at which the transmission by a bandpass circuit or device is maximum (attenuation is minimum), evidenced by a peak in the frequency-response curve.

peak amplitude 1. The maximum positive or negative current or voltage of a wave. 2. The maximum instantaneous power of a signal.

peak anode (plate) current The maximum instantaneous current flowing in the anode (plate) circuit of an electron tube.

peak anode (plate) voltage The maximum instantaneous voltage applied to the anode (plate) of an electron tube.

peak chopper See *peak clipper*.

peak current Abbreviation, I_p. The highest value reached by an alternating-current half-cycle or a current pulse. Also called *maximum current*.

peak detector See *peak probe*.

peak distortion 1. The maximum instantaneous distortion in a signal, generally expressed as a percentage. 2. Distortion of a modulated signal at envelope peaks.

peaked sawtooth A wave composed of a sawtooth and peaking-pulse components. The deflection voltage of a magnetic-deflection TV picture tube requires this waveform to produce a *current* sawtooth in the deflecting coils. See figure on next page.

peaked waveform An ac waveform having nearly pointed positive and negative half-cycles. Such a wave contains appreciable third-harmonic energy.

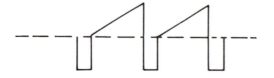

Peaked sawtooth.

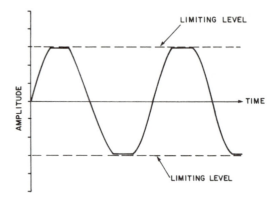

Peak limiting.

peak envelope power Abbreviation, PEP. For a linear rf amplifier handling a modulated signal, the average rf output power during a single rf cycle at the highest peak of the modulation envelope.

peaker 1. Peak filter. 2. Peaking transformer.

peaker-notcher See *notcher-peaker*.

peak factor For an ac wave, the ratio $Em/Erms$ or $Im/Irms$, where Em is maximum voltage; $Erms$, effective voltage; Im maximum current; and $Irms$, effective current.

peak filter A frequency-selective circuit, such as a bandpass filter, for producing a peak response (see *peak, 3*).

peaking The adjustment of a control or device for maximum indication on a meter or other display.

peaking coil A small inductor employed to compensate the frequency response of a circuit, such as a video amplifier or video detector. Both series and shunt peaking coils are employed.

peaking transformer A transformer whose output waveform is sharply peaked (of short duration with respect to 2 cycle). The effect is obtained by means of a special core which, because it contains little iron, saturates easily. A familiar use for such a transformer is in firing a phase-controlled thyratron.

peak inverse voltage Abbreviation, PIV. The peak value of voltage applied in the direction of normal polarity in a circuit.

peak limiting A method of limiting the maximum amplitude of a signal. When the instantaneous peak amplitude, either positive or negative, exceeds a certain value, the output is clipping at that value.

peak modulated power In an amplitude-modulated wave, the instantaneous power when the modulated carrier is at maximum upswing. In 100 percent sinusoidal modulation, the peak modulated power is four times the unmodulated carrier power.

peak point The highest current point in the current-voltage response curve of a tunnel diode. Immediately beyond this point the current decreases as the applied voltage is increased, indicating a negative-resistance region. Compare *valley point*.

peak power Symbol, Pp. Unit, watt. Ac power that is the product of peak voltage and peak current; $Pp = EpIp$, where Ep is in volts, and Ip is in amperes. Also called *maximum power* (see *maximum power, 2*).

peak probe A voltmeter test probe containing a diode circuit whose dc output voltage is close to the peak value of the applied ac test voltage.

peak recurrent forward current For a semiconductor diode, the maximum repetitive instantaneous forward current as measured under specified conditions of operation.

peak reverse voltage Abbreviation, *PRV*. In semiconductor practice, the peak value of the voltage applied in reverse polarity across the junction.

peak signal level The maximum instantaneous signal power or voltage specified for particular operating conditions.

peak to peak For an alternating-current waveform, the arithmetic difference between the peak positive and negative values of current or voltage.

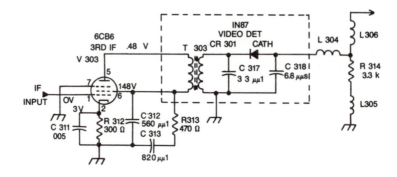

Peaking coil in a video detector.

peak-to-peak probe A voltmeter test probe containing a diode circuit whose dc output voltage is close to the peak-to-peak value of the applied ac test voltage.

peak-to-peak voltage The arithmetic sum of positive and negative peak voltages in an ac wave. Thus, a symmetrical sine-wave ac voltage of 115V rms has a peak value of 162.6V and a peak-to-peak value of 325.2V. Also see *peak voltage*.

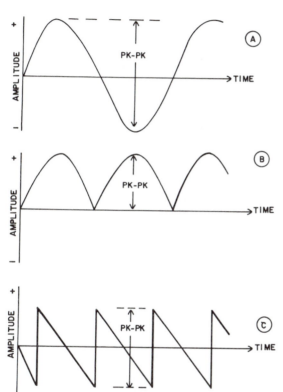

Peak-to-peak amplitudes for a sine wave (A), a rectified sine wave (B), and a sawtooth wave (C).

peak torque Symbol, Tp. For a torque motor, the maximum useful torque at maximum recommended input current, expressed in in.-oz.

peak voltage Abbreviation, Ep. The highest value reached by an ac voltage half-cycle or a pulse voltage. Also called *maximum voltage*.

peak voltmeter 1. An ac voltmeter that responds to the peak value of the applied voltage. 2. An ac voltmeter that responds to the *average* value of the applied voltage, even though its scale reads in peak volts.

pea lamp A miniature incandescent bulb.

PEC Abbreviation of *photoelectric cell*.

pecker In a paper tape reader, that which senses the holes in punched tape. See *punched type reader*.

pedestal See *blanking pedestal*.

pedestal level See *blanking level*.

Peltier effect A drop below ambient temperature at the junction between two dissimilar metals when an electric current is passed through the junction.

PEM Abbreviation of *photoelectromagnetic*.

pen-and-ink recorder A graphic recorder in which a fountain-pen-type stylus inscribes an ink line on a paper chart. Also called *pen recorder*.

pencil 1. A usually convergent or divergent beam of electrons or other particles or rays. 2. A pair of geometric entities sharing a property, e.g., lines intersecting the same point.

pencil tube A small *disk-seal tube* whose slender construction resembles a pencil.

penetrating frequency For a particular layer of the ionosphere, the high frequency at which a vertically propagated wave penetrates the layer, i.e., it is not reflected back to earth. Also called *critical frequency*.

penetrating radiation Radiation, such as X-rays, which pierces otherwise opaque materials.

penetrating rays See *cosmic rays*.

penetration depth See *depth of penetration*.

pent Abbreviation of *pentode*.

pentagrid converter tube An electron tube used mainly for signal mixing (conversion), in which the principal electrodes are cathode, control grid, screen, suppressor, plate, and two auxiliary signal-injection grids.

pentagrid gate circuit A coincidence-gate circuit employing a pentagrid converter tube. The input signal is applied to the first grid, and the control signal to the third.

pentagrid tube See *pentagrid converter tube*.

pentavalent element An element whose atoms have five valence electrons, e.g., antimony; arsenic.

pentode A five-electrode tube in which the electrodes are an anode, cathode, control grid, screen, and suppressor.

pentode field-effect transistor A field-effect transistor with three separate gates.

pentode gate circuit A gate circuit employing a pentode to which trigger signals are applied at the control-grid and suppressor electrodes. Plate current flows only when the control grid and suppressor are raised above cutoff simultaneously.

pentode transistor A bipolar transistor with three emitters.

PEP 1. Abbreviation of *planar epitaxial passivated*. 2. Abbreviation of *peak envelope power*.

PEP diode See *planar epitaxial passivated diode*.

PEP transistor See *planar expitaxial passivated transistor*.

perceived level The level of a disturbance, as sensed by a person. Generally expressed in decibels with respect to a certain threshold value. The threshold is assigned an intensity of 0 dB.

percent An expression of a fraction, in terms of hundredths. A quantity of x percent indicates a fraction of $x/100$. Percent is usually abbreviated by the symbol %.

percentage error The amount by which a measured value differs from the true value, expressed as a percentage (the number of parts per 100 that the measurement is in error).

percentage-modulation meter An instrument which gives direct readings of the modulation percentage of an amplitude-modulated signal. The meter scale or dial is graduated in increment from 0 to 110 percent.

percentage uncertainty The relative uncertainty of a measurement, expressed as a percentage; percentage uncertainty = 100 (relative uncertainty/measured value). Also see *uncertainty in measurement.*

percent distortion In the determination of harmonic distortion, the total harmonic voltage expressed as a percentage of the fundamental voltage plus total harmonic voltage; $\%D = (100Eh/Et)$, where Eh is total voltage due to harmonics, and Et is total signal voltage (i.e., the fundamental plus harmonics).

percent modulation See *modulation percentage.*

percent modulation meter See *percentaged- modulation meter.*

percent ripple The amount of ripple voltage in the dc output of a rectifier or generator expressed as a percentage of the dc output voltage.

perforated board A plastic panel provided with a number of small holes in orderly columns and rows for the insertion of the pigtails of components or of *push-in terminals* to facilitate quick assembly of prototype circuits.

perforated tape Paper or plastic tape punched with holes or slits representing encoded mathematical or other data or telegraph or Teletype signals. Also called *punched tape* or, when applicable, *paper tape.*

perforated-tape reader See *paper-tape reader.*

performance curve A curve depicting the behavior of a component or circuit under specified conditions of operation. Such a curve, for example, might display the variation of output power with input power, the variation of frequency with voltage, and so on. Compare *characteristic curve.*

perforation rate The speed at which paper tape can be punched.

perforator A manual device for punching paper tape. Also called *keyboard perforator.*

performance test A test made primarily to ascertain how a system behaves. The test is concerned with normal operation, whereas a *diagnostic test* is a troubleshooting procedure. Compare *troubleshooting test.*

perikon detector See *zincite detector.*

period Symbol, *t*. Unit, second. The duration of a complete ac cycle or of any cyclic event. Period $t = 1/f$, where f is in Hz. Also see *ac voltage, alternating current, cycle, frequency, hertz.*

periodic and random deviation Abbreviation, *PARD*. In the dc output current of a rectifier, the combined *periodic deviation*, including ripple, noise, hum, and transient spikes.

periodic curve A curve that repeats its shape in each period, e.g., sine curve.

periodic deviation Repetitive deviation of a quantity from its normal value, e.g., ripple in the dc output current of a rectifier.

periodic function A mathematical function that is represented by a periodic curve, e.g., sine function ($y = \sin v$).

periodicity In a transmission line, the tendency for power to be reflected at a point or points where the diameter of the line changes.

periodic law The observation that when the elements (see *element, 3*) are arranged in increasing order of atomic number, their physical and chemical properties recur periodically. Also see *periodic table.*

periodic table A table in which the elements (*element, 3*) are arranged according to the *periodic law.* The vertical columns in the table, labeled *groups,* contain elements possessing related properties (e.g., silicon and germanium in group IV). The rows, labeled *periods,* depict the periodic shift in the properties of the elements. By means of the periodic table, the existence of elements has been predicted, i.e., before their actual discovery, on the basis of properties they should have had. This, for example, was the case with germanium, which Dimitri Mendelejeff (originator of the table) predicted, calling it *ekasilicon.*

peripheral 1. Pertaining to equipment which is accessory to a central system, e.g., peripheral input/output devices online or offline to computers; data recorders; indicators. Also see *ancillary equipment.* 2. Peripheral equipment in a computer system.

peripheral buffer As part of a peripheral in a computer system, a storage unit in which data temporarily resides on its way to or from the cpu. Also called *input/output buffer.*

peripheral electron See *valance electron.*

peripheral equipment See *peripheral, 1, 2.*

peripheral interface adapter Abbreviation, PIA. An IC device which acts as an *I/O port* to interface a *microprocessor* with any of various peripheral devices.

peripheral transfer The transfer of a unit of data between peripherals or between a peripheral and a CPU.

permalloy A high-permeability alloy of iron and nickel.

permamagnetic speaker A permanent-magnet loudspeaker.

permanent magnet A body which is continuously magnetized, i.e., without the application of electricity and without requiring the presence of another magnet. Compare *temporary magnet.*

permanent-magnet erase Erasure of magnetic tape by the field of a permanent magnet. Typically, it is a two-step process: a magnet erases what it can of the signal, leaving any residual magnetization for a second magnet.

permanent-magnet focusing In a cathode-ray tube, the focusing of the electron beam by means of permanent magnets.

permanent-magnet generator A type of electromechanical generator in which the field (either stationary or rotating) is a multipole permanent magnet. Also called *magneto.*

permanent-magnet loudspeaker See *magnetic speaker, 1, 2.*

permanent magnet magnetizer A magnetizer employing a permanent magnet as the magnetic-field source.

permanent-magnet meter An indicating meter in which a movable coil rotates between the poles of a permanent magnet. Compare *electrodynamometer* and *iron-vane meter.*

permanent-magnet motor A motor having a permanent-magnet field.

permanent-magnet relay A polarized relay employing a permanent magnet.

permanent-magnet speaker 1. A speaker having a strong permanent magnet as the core. 2. Magnetic speaker.

permanent storage See *nonvolatile memory.*

permeability Symbol, μ. Unit, H/m. The measure of the comparative ease with which magnetic flux can be set up in a material. Permeability is the ratio of flux density in the material to flux density in air; $\mu = B/H$, where B is magnetic flux density in teslas, and H is magnetic field strength in amperes per meter.

permeability curve See *B–H curve.*

permeability-tuned oscillator A tube- or transistor-type oscillator (usually rf or if) in which the frequency is varied by permeability tuning (running a magnetic core in or out of the coil of an LC tank circuit).

permeability tuning Variation of the frequency of an LC circuit by changing the position of a magnetic core within the inductor. This type of tuning is employed in amplifiers, oscillators, filters, and wavetraps.

permeameter An instrument for measuring permeability.

permeance Symbol, P. Unit, Wb/A. The reciprocal of reluctance.

permendur A high-permeability magnetic alloy containing equal parts of iron and cobalt. At saturation, the flux density of this material can be 2 teslas (20 000 gauss).

perminvar A high-permeability magnetic alloy of cobalt, iron, and nickel. At saturation, flux density can approach 1.2 teslas (1.2×10^4 gauss).

permittivity See *dielectric constant.*

permutation A selection of several factors or objects from a group, in all possible arrangements. The number of permutations of a group of *n* different things arranged in a series is *n*! Thus, for the group *ABC*, the number of possible permutations is six: *ABC, ACB, BAC, BCA, CAB,* and *CBA.*

permutation modulation A method of modulation by means of varying the sequence of digital bits.

peroxide of lead In a lead-acid cell or battery, a compound of lead and oxygen that comprises the positive electrode or electrodes.

persistence That quality by which a quantity remains in a (decaying) state, as brought about by a stimulus after the stimulus has been removed. As it relates to electronics, for example, it is the tendency of certain phosphors to glow after the excitation has been removed. Thus, after the electron beam in a cathode-ray tube has passed over the fluorescent screen, the phosphor on the screen may continue to glow for a certain time along the path traced by the beam. Some phosphors, such as those used in high-speed oscilloscopes, have virtually no persistence, whereas others have long persistence. As the term relates to vision, it is the cause of the effect whereby one can see a brightly lit scene after closing the eyes, the same effect that translates the passage of motion picture frames into continuous motion.

persistent oscillations Successive oscillations of constant amplitude. Also called *continuous wave.*

personal equation The value of systematic error for one observing phenomena or making measurements.

perveance Symbol, *G.* For an electron tube, the ratio *ik* $(3\sqrt{ep})$, where *ik* is the space-charge-limited cathode current, and *ep* is the plate voltage.

peta Abbreviation, *P.* A prefix meaning 10^{15}.

petagram Abbreviation, Pg. A large unit of weight (force, or mass); 1 Pg = 10^{15}g.

petameter Abbreviation, *Pm.* A large unit of (astronomical) distance; 1 Pm = 10^{15} m.

Peter Phonetic alphabet code word for the letter *P.*

pF Abbreviation of *picofarad.*

pf Symbol for *power factor.*

PFET Abbreviation of *p-channel junction field-effect transistor.*

PFM Abbreviation of *pulse-frequency modulation.*

PG Abbreviation of *power gain.*

Pg Abbreviation of *petagram.*

pH 1. Symbol for *hydrogen-ion concentration.* Numerically, pH is the negative logarithm of the effective hydrogen-ion concentration in gram equivalents per liter. The scale runs from zero to 14, on which 7 denotes neutrality relative to acidity vs alkalinity; values between zero and 7 denote decreasing acidity, and values between 7 and 14 denote increasing alkalinity. See *acid* and *base, 2.* 2. Abbreviation of *picohenry.*

phanotron A hot-cathode, mercury-vapor tube diode rectifier.

phantastron A positive-grid, pentode circuit which, on receiving an input pulse, delivers an output ramp voltage of excellent linearity.

phantom Radio interference in the form of an rf beat note resulting from interference between two strong carriers, often from local radio stations. When the phantom frequency lies within the tuning range of a receiver, the phantom can be tuned in as a separate signal. But when the phantom corresponds to the intermediate frequency of the receiver, it will ride into the i-f amplifier and be present as an untunable interferential signal.

phantom channel In a properly phased stereo system, the apparent sound source centered between the left- and right-channel loudspeakers.

phantom circuit In wire telephony, a third circuit which has no wires; it results from a method (employing *repeating coils*) for making two other circuits do the work of (this third) one.

phantom signal In a radar system, a signal that does not correspond to an actual target. The origin of the phantom signal or echo cannot be readily determined.

phantom target See *echo box.*

phase angle Symbol, *v.* Unit, degree or radian. In an ac circuit, the lag or lead between the instant one alternating quantity reaches its maximum value and the instant when another alternating quantity reaches its maximum value. The instants and their difference (lag or lead) are measured in degrees or radians (a complete cycle being 360 degrees or 2π radians) along the horizontal axis of the time-vs-magnitude graph of the ac quantity. A phase angle may exist between two currents, two voltages, or between a current and a voltage.

phase-angle voltmeter An instrument that indicates both the magnitude and phase of a voltage.

phase compensation In an operational amplifier, compensation of excessive phase shift in the feedback.

phase compressor A push-pull phase-inverter circuit in which a capacitor is connected between each tube plate and the opposite output terminal to attenuate in-phase components, such as even-numbered harmonics.

phase constant A figure giving the rate (in radians per unit length) at which the phase lag of the current or voltage field component in a traveling wave increases linearly in the propagation direction.

phase-controlled thyratron A thyratron whose grid is operated from a phase shifter, such as a phase-shift bridge, to control the phase of the firing voltage and, accordingly, the tube's angle of conduction.

phased antennas Separate antennas (or antenna elements) which are excited directly or parasitically, in or out of phase with each other. The radiation pattern changes according to phasing.

phase-delay equalizer See *delay equalizer.*

phase detector See *phase-sensitive detector.*

phase difference 1. The difference (in time, angle, or fractional cycle) between the instants at which two alternating quantities reach a given value. 2. Symbol, ϑ. For a dielectric, the complement of phase angle.

phase discriminator See *discriminator, Foster—Seeley discriminator, ratio detector,* and *Travis discriminator.*

phase distortion Distortion characterized by input—output phase shift between various components of a signal passed by a circuit or device.

phase inverter An RC-coupled amplifier which has a single-ended input and push-pull output. This circuit enables a push-pull amplifier to be driven without an input transformer.

phase-locked loop A circuit containing an oscillator whose output phase and/or frequency is "steered" to keep it in sync with some reference, as a received signal. See figure on next page.

phase-locked oscillator An oscillator in which the inductance or the capacitance is varied periodically at half the driving frequency.

phase modulation Abbreviation, *PM.* A method of modulation in which the phase of the carrier current is varied with the frequency of the modulating voltage. Phase modulation is accomplished by an instantaneous carrier-frequency change (as in FM) which is proportional to the amplitude *and* frequency of the modulating voltage.

phase modulator A circuit or stage that produces phase modulation.

phase multiplier A circuit used for the purpose of phase comparison

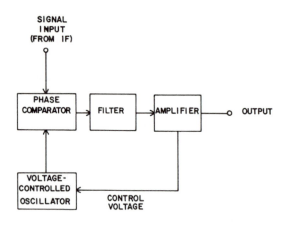

Phase locked loop.

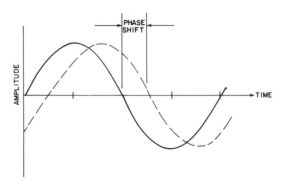

Phase shift.

between signals. The frequency of the measured signal is multiplied, resulting in multiplication of the phase difference. This improves the sensitivity of the measuring apparatus.

phase resonance See *velocity resonance.*

phase reversal Phase shift of 180 degrees, e.g., a complete reversal of a positive half-cycle of a negative half-cycle.

phase-rotation relay See *phase-sequence relay.*

phase-rotation system A system for producing single-sideband signals without using selective filters. In one such system, two balanced modulators are employed. One of these receives carrier and modulating voltages which are 90 degrees out of phase with these same voltages that are fed to the other balanced modulator.

phase-sensitive detector Abbreviation, *PSD.* A detector delivering a dc output voltage whose value is proportional to the difference in phase between a reference signal and that of an oscillator. The output voltage is employed as an error signal to keep the oscillator frequency synchronized with the reference frequency.

phase-sequence relay In a polyphase system, a relay or relay circuit that is actuated by voltages reaching maximum positive amplitude in a predetermined phase sequence. Also called *phase-rotation relay.*

phase shift 1. A change in the displacement, as a function of time, of a periodic disturbance having constant frequency. 2. The magnitude of such a change, measured in fractions of a wavelength or in electrical degrees.

phase-shift bridge A four-arm-bridge circuit for shifting the phase of an ac signal. Such a circuit is often used (with one arm variable) to shift the phase of the firing voltage for a thyratron.

phase-shift discriminator See *Foster—Seeley discriminator.*

phase shifter A circuit (such as an *LC* or *RC* network) or device (such as a Helmholtz coil or phase-shifting capacitor) which introduces a phase shift between input and output signals.

phase-shifting capacitor A special four-stator, one-rotor variable capacitor which, with a transformer-coupled *RC* circuit, provides 360 degrees of continuously variable phase shift for one rotation of the rotor. The rotor plate turns like a cam under the stators, due to the off-center insertion of the rotor shaft.

phase-shift oscillator A single-stage, *RC*-type oscillator circuit in which

the required 180-degree phase shift in the signal fed back from output to input is obtained by passing the output through a phase-shifting *RC* network.

phase-shift-type distortion meter A harmonic distortion meter of the total harmonic type, in which the distorted output signal of a device under test is compared with the distortion-free input test signal. The output signal phase is rotated to exactly 180° with respect to the input, and the two amplitudes are made equal. If there is no distortion, the signals cancel each other, and the result is zero. Any remaining signal is proportional to the *total harmonic distortion.*

phase-splitting circuit A circuit which produces from one input signal two output signals differing in phase.

phase-splitting driver A phase inverter employed as the driver of a push-pull amplifier.

phase velocity Wave velocity whose value is the product $f\lambda$, where f is the frequency, and λ is the wavelength.

phase windings In an ac generator, windings that deliver voltages differing in phase.

phasing capacitor In a crystal filter, a small variable capacitor that constitutes one arm of a four-arm bridge in which the crystal is another arm. Adjustment of this capacitor balances the bridge, thus preventing the undesirable passage of a signal through the capacitance of the crystal holder.

phasor See *complex quantity.*

phasor diagram See *vector diagram.*

PhC Abbreviation of *pharmaceutical chemist.*

phenol-formaldehyde plastics A family of plastic insulating materials made with phenolic resin. The numerous names for these materials include *Bakelite, Catalin, Durez, Durite, Formica,* and *Micarta.*

phenolic insulants See *phenol-formaldehyde plastics.*

phenolic resin A synthetic resin made by condensing phenol (carbolic acid) with formaldehyde.

phenomenon An event or circumstance that can be verified by the senses, as opposed to one subject to mere speculation, e.g., the phenomenon of magnetic attraction.

phi The 21st letter (ϕ, Φ) of the Greek alphabet. The lower case letter (ϕ) is the symbol for angle and phase; and the upper case letter (Φ), magnetic flux and scalar potential.

Phillips gate A device that allows measurement of the gas pressure in a confined chamber. A current is passed through the gas. The magnitude of the current, for a given gas, is a function of the gas pressure and temperature.

Phillips screw A screw with a pair of slots in its head. The slots are arranged like an x. Phillips screws are available in many different sizes, as are ordinary screws. The x-shaped pair of slots reduces the tendency for the screwdriver to slip out of the screw head as the screw is rotated.

Phi phenomenon The illusion of motion resulting from the rapid presentation to the eye of pictures showing objects in a succession of different positions. Television and motion pictures cause this illusion. Also see *persistence*.

pH meter An instrument used to measure the acidity or alkalinity of solutions. Also see *pH, 1*.

phon A unit of apparent change in loudness discerned by a listener, which, unlike the decibel, includes compensation for the ear's nonlinear response to attendant frequency changes. At 1 kHz a phon is the equivalent of a decibel.

phone 1. Telephone (wire or radio). 2. To establish communication via telephone. 3. A minimal, unique speech sound. Also called *sound unit*.

phone jack The female mating device for a *phone plug*.

phone monitor A simple device for listening to radiophone transmissions to test their quality. In its most rudimentary form, it consists of a pickup antenna, semiconductor-diode detector, and high-resistance headphones.

phone patch A device for establishing a connection (patch) between radio and wire-telephone facilities. Also see *patch*.

phone plug A type of plug originally designed for patching telephone circuits but now widely used in electronics and instrumentation. In its conventional form, it has a rod-shaped neck, which serves as one contact, and a ball on the tip of the neck but insulated from it, which serves as the other contact.

Phone plug.

phone test set An instrument for checking the performance of a radiotelephone transmitter. The set combines the functions of field-strength meter, modulation indicator, and aural monitor. Sometimes it includes a voltohmmilliammeter for troubleshooting the transmitter.

phonetic alphabet Words whose initial letters are used to identify the letters of the alphabet for which they stand. These words are spoken in radiotelephony to identify letters which, if spoken by themselves, might not be clearly heard.

phonetic alphabet code word In radio and wire telephony, a word chosen for its easy recognition by ear to identify the letter of the alphabet with which it begins. For example: Golf for G, Juliet for J,

X-ray for X, and so on.

phonics See *acoustics, 1*.

phonocardiogram The record made by a phonocardiograph.

phonocardiograph An instrument that makes a graphic record of heart sounds.

phono cartridge The vibration-to-electricity transducer (pickup) of a phonograph; it is actuated by the stylus (needle). Common types are ceramic, variable-inductance, and variable-reluctance.

phonocatheter A microphone that can be inserted into the body for the purpose of listening to the functions of internal organs.

phonograph A device for reproducing the sound recorded on disks or cylinders.

phonograph oscillator See *phono oscillator*.

phono jack A *phone jack* designed especially for the quick connection and disconnection of a *phonograph*.

phonon A unit of energy resulting from vibration, as of a piezoelectric crystal.

phono oscillator A small rf oscillator which is modulated by the af voltage from a phonograph. The modulated rf signal is picked up by a remote radio receiver (usually in the same room), and the sound is reproduced through a loqdspeaker connected to the receiver.

phono plug A type of *phone plug* designed especially for the quick connection and disconnection of a *phonograph*.

phonoreception The hearing of high-frequency sounds.

phonorecord A phonograph disc.

phonoselectroscope A special type of stethoscope, in which the main heartbeat is attenuated. This makes abnormal sounds more audible. The device may be adjusted in various different ways, to listen for abnormalities characteristic of different heart diseases.

phosphor A substance that glows when an electron beam impinges upon it. Such a substance is used as a coating on the screen of a cathode-ray oscilloscope tube, TV picture tube, or X-ray fluoroscope. See specific listings under *beat zinc silicate; cadmium borate, silicate, and tungstate; calcium phosphate, silicate,* and *tungstate; magnesium fluoride, silicate,* and *tungstate; zinc aluminate, ainz beryllium silicate, zinc beryllium zirconium silicate, zinc borate, zinc cadmium sulfide, zinc germanate, zinc magnesium fluoride, zinc orthosilicate, zinc oxide, zinc silicate,* and *zinc sulfide.*

phosphor bronze A kind of bronze whose elasticity, hardness, and toughness have been greatly improved by the addition of phosphorus. The metal is used for brushes, springs, switch blades, and contacts.

phosphor copper An alloy of copper and phosphorus used in the manufacture of phosphor bronze.

phosphorescence The property of some materials that ordinarily fluoresce to continue to glow after the stimulus (light or an electron beam) has been removed. Compare *fluorescence*.

phorphorescent screen A viewing screen coated with a phosphor, e.g., oscilloscope screen, fluoroscope screen.

phosphorous Exhibiting the properties of phosphor, e.g. glowing after stimulation with light, (Not to be confused with *phosphorous*.)

phosphorus Symbol, P. A nonmetallic element of the nitrogen family. Atomic number, 15. Atomic weight, 30.975. Phosphorus is used as a dopant in semiconductor processing.

phot The cgs unit of illumination: The direct illumination produced upon a 1-centimeter-distant surface by a uniform point source of 1 international foot-candle. 1 phot = 1 lumen per square centimeter.

photocathode 1. The light-sensitive cathode in a phototube. 2. The photomosaic of a TV camera tube.

photocell See *photoelectric cell.*

photocell amplifier An amplifier employed to boost the output of a photocell. With respect to the nature of the input signal, it can be an ac or dc amplifier (depending upon whether the output of the photocell is straight dc or modulated dc).

photocell card reader See *photoelectric card reader.*

photocell tape reader See *photoelectric tape reader.*

photochemical effect The phenomenon whereby certain substances undergo chemical change when exposed to light or other radiant energy. An example of such a substance is the silver bromide, silver chloride, or silver iodide on photographic film, which becomes altered when the film is exposed. The effect is also responsible for a type of smog.

photoconductive cell A photoelectric cell, such as the cadmium-sulfide type, whose resistance is proportional to the intensity of light impinging upon it. The photoconductive cell acts as a light-sensitive variable resistor in a current path. Also see *photoconductive material.*

photoconductive effect The tendency for the electrical resistance of a substance to change when infrared radiation, visible light, or ultraviolet radiation strikes it. Different substances exhibit different degrees of this effect.

photoconductive material A substance that exhibits decreased electrical resistance when exposed to light. Some photoconductive substances are cadmium selenide, cadmium sulfide, germanium, lead sulfide, selenium, silicon, and thallous sulfide. Also see *actinoelectric effect.*

photoconductivity The phenomenon whereby the electrical resistance of certain materials—such as cadmium sulfide, cadmium selenide, germanium, selenium, and silicon—is lowered upon exposure to light. Also see *photoconductive material.*

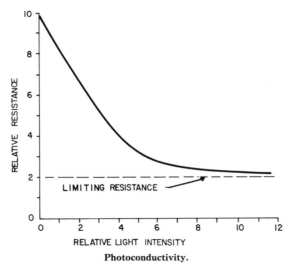

Photoconductivity.

photoconductor 1. Photoconductive material. 2. Photoconductive cell.

photocurrent See *photoelectric current.*

photo-Darlington 1. A phototransistor fabricated as a Darlington amplifier for high output current. 2. A combination of photodiode (see *light-sensitive diode*) and Darlington amplifier. (Also sometimes written *photodarlington*).

photodecomposition Chemical breakdown by the action of radiant energy. Also called *photoylsis.*

photodetector 1. A photocell-type *light meter.* 2. An *optoelectronic coupler.*

photodielectric effect The tendency for the dielectric constant of a substance to change when infrared radiation, visible light, or ultraviolet radiation strikes it. Different substances exhibit different degrees of this effect.

photodiffusion effect See *Dember effect.*

photodiode See *light-sensitive diode.*

photodisintegration In the nucleus of an atom, disintegration resulting from the absorption of radiation.

photoelasticity The tendency for the light-transmission characteristics of a substance to change with externally applied forces.

photoelectric alarm An alarm that is actuated when a light beam impinging upon a photocell is interrupted.

photoelectric amplifier 1. An amplifier for boosting the output of a photosensitive device. 2. An optoelectronic coupler possessing gain.

photoelectric card reader A punched-card reader employing a photocell, photodiode, or photo-transistor to sense light transmitted through the holes.

photoelectric cell A device which converts light energy into electrical energy by either producing a voltage (see *photovoltaic cell, selenium cell, silicon cell, solar cell, sun battery*) or by acting as a light-sensitive resistor (see *light-sensitive diode, photoconductive cell, selenium cell*).

photoelectric counter A counting device (electromechanical or fully electronic) which counts objects as they interrupt a light beam impinging upon a photocell.

photoelectric disintegration See *photodisintegration.*

photoelectric effect The phenomenon whereby temporary changes occur in the atoms of certain substances under the influence of light. Some of these materials undergo a change in their electrical resistance, whereas others generate a voltage (see, for comparison, *photoconductive material* and *photovoltaic material*).

photoelectric efficiency See *quantum yield.*

photoelectric field-effect transistor See *photo-FET.*

photoelectricity Electricity produced by the action of light on certain materials, such as cesium, selenium, and silicon. Also see *photoemission* and *photovoltaic cell.*

photoelectric material See *photoconductive material* and *photoemissive material.*

photoelectric multiplier A device that internally amplifies the current resulting from bombardment by light rays. A photomultiplier tube is an example of such a device.

photoelectric photometer An instrument that uses a photoelectric device for the purpose of measuring the intensity of infrared radiation, visible light, or ultraviolet radiation.

photoelectric pyrometer An optical pyrometer in which a photocell and appropriate filters act instead of the human eye.

photoelectric reader See *photoelectric card reader* and *photoelectric tape reader.*

photoelectric relay A relay which is actuated directly by a photocell or a photocell and amplifier. This type of relay is the basis of the photoelectric alarm.

photoelectric smoke alarm An alarm that is tripped by a photoelectric smoke detector when the density of smoke exceeds a safe level.

photoelectric smoke control A system for making automatic adjustments to a burning process when the smoke density exceeds a prescribed level. The initial element in the system is a photoelectric smoke detector.

photoelectric smoke detector A sensing device in which a photocell, photodiode, phototransistor, or phototube is excited by a light beam passing through a channel carrying smoke. The output of the cell, which is proportional to the density of the smoke, may be used to trip an alarm or deflect an indicating meter when the density of the smoke exceeds a prescribed level.

photoelectric tape reader A punched-tape reader employing a photocell, photodiode, phototransistor, or phototube to sense light passing through the holes.

photoelectric transducer A photocell, photodiode, phototransistor, or phototube employed as a sensor.

photoelectric tube See *phototube.*

photoelectric wattmeter A power-measuring instrument that is especially useful for the approximate measurement of rf power. It consists of an incandescent lamp sharing a light-tight enclosure with a photovoltaic cell. The power to be measured is applied to the lamp, which glows proportionately. The light excites the cell, causing it to deliver a dc voltage which is proportional to the power and deflects a dc milliammeter or microammeter. If known power levels are applied to the lamp, the meter may be made to read directly in watts.

photoelectromotive force The emf produced by a photovoltaic cell.

photoelectron An electron that has been displaced within an atom or ejected from the atom as the result of light energy acting on the atom.

photo-emf See *photoelectromotive force.*

photoemission The ejection of electrons from certain materials, such as cesium, when these materials are exposed to light. Also see *photoemissive material.*

photoemissive material A substance which will emit electrons when exposed to light. A typical use of such a material is in the coating of the light-sensitive cathode of a phototube. The metals cesium, potassium, rubidium, and sodium are photoemissive. Also called *photovoltaic material.*

photofabrication 1. A method of circuit-board manufacturing. The etching pattern is placed over the circuit-board material, the board is placed in a special solution, and then the assembly is exposed to visible light. The light interacts with the solution to dissolve the metal in areas exposed to the radiation, but not in areas covered by the etching pattern. 2. The above technique, applied to the manufacture of integrated circuits.

photo-FET A phototransistor of the field-effect type.

photoflash See *electronic flash, 1.*

photoglow tube See *discharge lamp.*

photogram The permanent shadow produced by an object placed between a light source and photographic paper.

photographic exposure meter See *exposure meter, 1.*

photographic recorder A graphic recorder (see *oscillograph, 1* and *recorder, 2*) which employs a light beam (deflected by galvanometer movement) that moves across photographic film or paper to produce a trace representing a varying quantity.

photographic sound recording See *optical sound recording.*

photograph reception Telephoto reception (see *facsimile*).

photograph transmission Telephoto transmission (see *facsimile*).

photoionization The ejection of electrons from atoms or molecules by the action of light.

photoisolator See *optoelectronic coupler.*

photojunction cell A photocell consisting of a semiconductor pn junction. The cell is useful mainly for its photoconductivity, though light shining on the junction produces a little photovoltaic action. Also called *junction photocell.*

photokinesis Light-induced motion, as in a radiometer.

photolithographic process A method of producing integrated circuits and printed circuits by photographing (often at considerable reduction) an enlarged pattern of the circuit on a suitable light-sensitized surface of metal or semiconductor, and chemically etching away unwanted portions of the surface.

photolysis See *photodecomposition.*

photomagnetic effect Light-sensitive magnetic susceptibility in some materials.

photomap A photo taken of terrain from a high altitude and usually overlaid with a reference grid.

photomask In photofabrication, the transparent film or template on which the etching pattern is drawn.

photometer An instrument employed to compare the luminous intensity of two light sources.

photometric measurement of power See *photoelectric wattmeter.*

photometry The science and art of visible-light measurement. The response of the human eye is used as the basis for preferred sensors (those used with photometric instruments, which have spectral sensitivity curves resembling those of the eye. Compare *radiometry.*

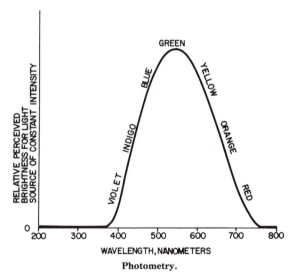

Photometry.

photomosaic In a TV-camera tube, the flat photocathode screen on which the image is projected by the lens system and scanning electron beam. The surface of the screen is covered with tiny light-sensitive silver droplets. Also see *dissector tube, iconoscope,* and *orthicon.* See figure on next page.

photomultiplier tube A special phototube which delivers high output current for a given light intensity by utilizing the secondary emission of electrons. The initial light-sensitive cathode emits electrons, which strike a specially placed metal plate with a force that dislodges more electrons, which, together with the initial emission, are reflected to

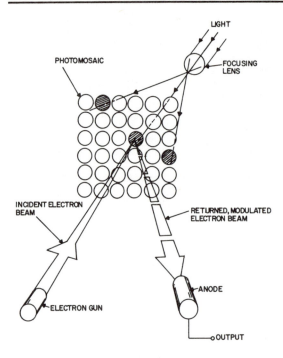

Photomosaic.

sistor, accordingly, is larger than the output of an equivalent photodiode.

phototube An electron tube which converts light energy into electrical energy by acting as a light-sensitive resistor. Characteristically, the tube contains an illuminated cathode coated with a photoemissive material, and an anode wire situated nearby. Light energy causes electrons to be emitted from the cathode in amounts proportional to light intensity; the electrons are attracted by the anode, which is connected externally to a positive dc voltage.

photovoltaic cell A photoelectric cell that generates a voltage when illuminated. The principal photovoltaic cell is the silicon cell; the selenium cell is also photovoltaic (to a somewhat lesser extent). Also see *photovoltaic material.*

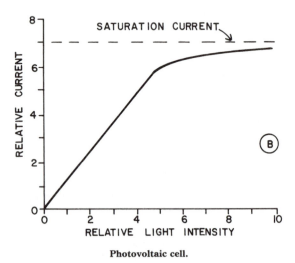

Photovoltaic cell.

a second plate where they dislodge still more electrons; and so on, from deflection plate to deflection plate, through the tube. The final plate deflects the accumulated electrons to the anode (collector electrode). (Also called *multiplier phototube*).

photon A quantum of radiant energy, especially that of light, whose energy constant W (in joules) equals hf, where h is the Planck constant, and f is frequency (hertz).

photoneutron A neutron released by photodisintegration.

photophone 1. A telephone-type communication system employing a modulated light beam transmitted between stations. 2. A process for recording sound on motion-picture film (see *optical sound recording*).

photorelay See *photoelectric relay.*

photoresistive cell See *photoconductive cell.*

photoresistive material See *photoconductive material.*

photoresistivity See *photoconductivity.*

photoresistor See *photoconductor, 1, 2.*

photosensitive device A light-sensitive electronic device. See, for example, *photoconductive cell, photodiode, photoFET, photomultiplier tube, phototransistor, phototube,* and *photovoltaic cell.*

photosphere The luminous layer at the surface of a star.

photoswitch A light-activated switch. Some photoswitches contain an electromechanical relay; others, such as the *light-activated silicon controlled switch*, have no moving parts.

phototimer An electronic timer for timing photographic processes.

phototransistor A transistor in which current carriers emitted as a result of illumination constitute an input-signal current which is amplified by the transistor. The output signal delivered by the tran-

photovoltaic material A substance which generates a voltage when exposed to light. The principal substances exhibiting this effect are silicon, selenium, and germanium. Also see *actinoelectric effect.*

photox A copper-oxide photovoltaic cell.

photran A light-sensitive, four-layer semiconductor device, used for switching purposes.

physical description An account of the operation of a circuit or device in terms of the behavior of the parts, as seen or imagined. This is in contrast to an *analytical* description, which aims to depict performance in mathematical terms.

physical properties The distinguishing characteristics of matter, apart from its chemical properties. Included are boiling point, density, ductility, elasticity, electrical conductivity, hardness, heat conductivity, index of refraction, malleability, melting point, specific heat, and state (solid, liquid, gaseous, plasma).

physical quantity A quantity expressing the actual number of physical units under consideration, as compared with a dimensionless number. Examples: 50 *volts*, 39 *kilometers*, 30 *picofarads*. Compare *dimensionless quantity*.

physics The science of energy and matter and their interactions. Physics is subdivided into the fields of mechanics, heat, acoustics, optics, and electricity and magnetism. There are, of course, a great many subdivisions within these traditional categories.

pi The 16th letter of the Greek alphabet. 1. Lower-case pi (π) is the symbol for the ratio c/d, where c is the circumference of a circle, and d its diameter. Pi is a constant that appears in a multitude of engineering formulas. Using a digital computer, French mathematicians have calculated π to 500 000 places. To the first 15 places, π is 3.141 592 653 589 793. 2. Capital pi (Π) is the symbol for the product of a number of factors. 3. The word *pi* is often used to mean *pi antenna tuner, pi-section filter, pi-section pad*, and *pi tank*, all of these devices being so called because their circuit configuration resembles the capital letter Π. 4. One of several coils spaced along a single core or form and connected in series to form a larger inductor. See, for example, *pi-wound choke*.

Pi Symbol for *input power* or *power input*.

PIA Abbreviation of *peripheral interface adapter*.

pickoff 1. To monitor a voltage, current, or other characteristic in an active circuit, without disturbing the operation of the circuit. 2. A device for electronically monitoring linear or angular displacement.

pickup 1. A device which serves as a sensor of a signal or quantity. This covers a wide variety of items, including temperature sensors, vibration detectors, microphones, phonograph pickups, and so on. 2. Collectively, energy or information that is picked up, e.g., sound pickup.

pickup arm The pivoted arm that holds the cartridge and stylus of a phonograph. Also called *tonearm*.

pickup cartridge See *phono cartridge*.

pickup current 1. The current required to close a relay. 2. Current flowing through, or generated by, a pickup.

pickup voltage 1. The voltage required to close a relay or circuit breaker. 2. The voltage delivered by a pickup.

pico 1. Abbreviation, *p*. A prefix meaning *trillionth(s)*, i.e., 10^{-12}. 2. A prefix meaning *very small*.

picoammeter A usually direct reading instrument employed to measure current in the picoampere range. Also see *current meter*.

picoampere Abbreviation, *pA*. A small unit of current; 1 pA = 10^{-12} A.

picocoulomb Abbreviation, *pC*. A small unit of electrical quantity; 1 pC = 10^{-12}C. Also see *coulomb*.

picocurie Abbreviation, *pC*. A small unit of radioactivity; 1 pCi = 10^{-12} 3.7 × 10^{10} disintegrations per second. Also see *curie, kilocurie, megacurie, millicurie, picocurie*.

picofarad Abbreviation, *pF*. A small unit of capacitance; 1 pF = 10^{-12} F.

picohenry Abbreviation, *pH*. A small unit of inductance; 1 pH = 10^{-12} H.

picosecond Abbreviation, *ps* or *psec*. A small unit of time; 1 ps = 10^{-12} s.

pi coupler See *Collins coupler*.

picovolt Abbreviation, *pV*. A small unit of voltage; 1 pV = 10^{-12} V.

picovoltmeter A usually direct reading electronic instrument employed to measure emf in the picovolt range.

picowatt Abbreviation, *pW*. A small unit of power; 1pW = 10^{-12} W.

pictorial See *pictorial wiring diagram*.

pictorial diagram See *pictorial wiring diagram*.

pictorial wiring diagram A wiring diagram in the form of a drawing or photograph of the components, as opposed to one of circuit symbols. The components are shown in their positions in the finished equipment, and the wiring as lines running between them.

picture black In facsimile or television, the signal condition resulting from the scanning of a black portion of the image.

picture detector See *video detector*.

picture diagram See *pictorial wiring diagram*.

picture element See *elemental area*.

picture information In a TV signal, the variable-amplitude component (i.e., the one carrying energy corresponding to the picture elements) that fills the space between blanking pulses.

picture reception 1. Telephoto reception (see *facsimile*). 2. Occasional term for *television reception*.

picture transmission 1. Telephoto transmission (see *facsimile*). 2. Occasional term for *television transmission*.

picture tube The special cathode-ray tube used in a television receiver to display the image. Also called *kinescope*.

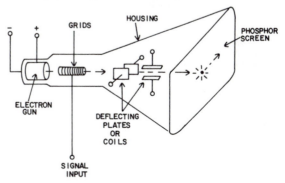

Picture tube.

pie chart See *circle graph*.

Pierce oscillator A simple crystal oscillator (tube or transistor) in which the crystal is connected directly between the input and output terminals of the tube or transistor; a tuning coil or capacitor is not required.

pie winding A method of coil winding in which two or more separate, multilayer coils are connected in series and placed along a common axis. Sometimes used in radio-frequency chokes.

piezo A prefix meaning *pressure*.

piezodielectric A substance that, when stretched or compressed, exhibits a change in dielectric constant.

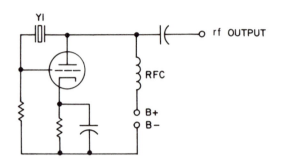

Pierce oscillator.

piezoelectric accelerometer An accelerometer employing a piezoelectric crystal whose voltage output is proportional to acceleration.

piezoelectric ceramic A ceramic which delivers a voltage when deformed; or conversely, which changes in shape when a voltage is applied to it.

piezoelectric crystal A crystal, such as quartz, Rochelle salt, tourmaline, or various synthetics, that delivers a voltage when mechanical force is applied between its faces, or which changes its shape when an emf is applied between its faces.

piezoelectric earphone See *crystal earphone.*

piezoelectric filter See *crystal filter* and *crystal resonator.*

piezoelectricity Electricity produced by deforming (squeezing, stretching, bending, or twisting) certain crystals, such as those of quartz, Rochelle salt, or tourmaline.

piezoelectric loudspeaker See *crystal loudspeaker.*

piezoelectric microphone See *ceramic microphone* and *crystal microphone.*

piezoelectric oscillator See *crystal oscillator.*

piezoelectric pickup See *crystal pickup.*

piezoelectric resonator See *crystal filter* and *crystal resonator.*

piezoelectric sensor See *crystal transducer.*

piezoelectric transducer See *crystal transducer.*

piezoid A complete piezoelectric crystal device.

piezoresistance In certain substances, the tendency of the resistance to change with stretching or compression.

piezo tweeter A *tweeter* of the piezoelectric type (see *crystal loudspeaker*).

pi filter A filter section having one series arm and two shunt arms; its configuration resembles the capital Greek letter pi (Π).

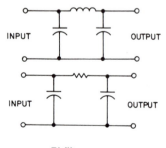

Pi filter.

piggyback component See *outboard component.*

piggyback control See *cascade control.*

piggyback tuner A separate ultrahigh-frequency TV tuner operated in conjunction with the vhf tuner of the receiver.

pigtail 1. A usually long and sometimes flexible lead, such as the *pigtail* of a fixed capacitor. 2. Descriptive of a device containing a long lead or leads, and usually mounted by such leads.

pile 1. Voltaic pile. 2. Nuclear reactor. 3. A battery of cells. 4. Any packed group of particles or granules, as in *carbon pile.*

pillow speaker A small, flat loudspeaker intended for use under a pillow.

pilot lamp See *pilot light.*

pilot light A usually small, incandescent or neon lamp which, when glowing, serves as a signal that a piece of equipment is in operation.

pilot model A preliminary model of a circuit or device constructed primarily to test the efficacy of a production process. The pilot model usually follows the prototype.

pilot production The often small-scale production of a device in a special assembly line apart from the main line in a factory.

pilot regulator A variable-gain circuit that maintains a constant output, even if the input amplitude changes.

PIM Abbreviation of *pulse-interval modulation.*

pi mode In a vane-anode magnetron, the mode of operation in which adjacent vanes have rf voltages of opposite polarity. Also written π mode.

pin 1. A semiconductor junction consisting of a layer of instrinsic semiconductor material situated between *n* and *p* layers. 2. A slender, straight, stiff prong used as a terminal or locking device (see, for example, *base pin* and *bayonet base*).

pinchoff In a junction field-effect transistor, the condition in which the gate voltage causes the two depletion regions to meet and close the channel to obstruct drain-current flow.

pinchoff voltage In a junction field-effect transistor, the lowest value of gate voltage that will produce pinchoff.

pincushion 1. A type of TV picture distortion in which each side of the raster sags toward center screen. Also see *antipincushioning magnets.* 2. The optical counterpart of the distortion described in 1, above; the opposite effect is *barrel* distortion.

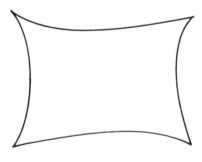

Pincushion.

pincushion-correction generator A circuit for generating a deflection signal to correct pincushion distortion (see *pincushion*). One form consists fo a *parabola generator* and opamp-type differentiator.

pin diode A silicon junction diode having a lightly doped intrinsic layer serving as a dielectric barrier between *p* and *n* layers.

pi network See *Collins coupler.*

ping A pulse of either audible or supersonic sound.

pinhole 1. A tiny hole present as a defect in a film of dielectric, semiconductor, or metal. 2. A tiny aperture which acts as a universal lens by permitting the passage of a very small bundle of light rays. The smaller the aperture, the closer to infinity the depth of field and focal point.

pinhole detector An electronic device for finding pinholes in materials. Also see *pinhole*.

pin jack A jack into which a pin plug is inserted for quick connection.

pink noise Electrical noise the amplitude of which is inversely proportional to frequency in a limited frequency spectrum. Pink noise is agreeable to the ear. Compare *white noise*.

pin plug A plug consisting of a slender metal pin which is inserted between the blades of a pin jack for a quick connection. The plug usually has a small insulated back for convenient handling.

pin straightener A device for straightening the pins of a transistor or miniature tube.

pin-usage factor For an integrated circuit, the number of gate equivalents per package pin. Also see *gate equivalent*.

PIO Abbreviation of *parallel input/output*.

pion A *subatomic particle* made up of one *quark* and one *antiquark*.

pip See *blip*.

pi pad A resistance-type attenuator having a series arm, shunt input arm, and shunt output arm; so called because of its resemblance to the Greek letter *pi*. Also see *pad*.

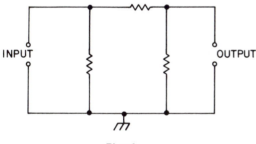

Pi pad.

pipe radiator A waveguide having an open end from which microwave energy is radiated.

Pirani gauge A type of vacuum gage in which a heated filament, which is one arm of a four-arm resistance bridge, is sealed into the vacuum system. The bridge is balanced before evacuation starts. As evacuation progresses, the heat removed from the filament becomes proportional to the pressure in the system, and the resistance of the filament changes accordingly. The bridge is then rebalanced, the difference between initial and subsequent null conditions giving an indication of the extent of the vacuum when the bridge has been appropriately calibrated.

pi section A filter or tuner section whose circuit resembles the Greek letter *pi*. See, for example, *pi, 3, 4*.

pi-section coupling Use of a pi section for coupling a radio transmitter to an antenna. Also see *Collins coupler* and *pi-section tank*.

pi-section filter A pi section employed as either a low-pass or high-pass filter, depending upon the position of the capacitors in the circuit.

pi-section tank A pi section employed simultaneously as the plate tank of an rf amplifier and as an antenna coupler.

piston 1. The movable element (cone) of a loudspeaker. 2. The movable, solid plunger of a trimmer capacitor that consists of a plug within a cylinder.

piston directivity Directivity of sound emitted by the piston of a loudspeaker (see *piston, 1*). As the frequency of the audio signal increases, radiation from a loudspeaker tends to be concentrated along the axis of the piston.

pit 1. In a printed-circuit board, a pockmark in a component or foil run. 2. A pockmark in a metallic substance, resulting from corrosion.

pitch 1. The frequency of a sound. 2. The distance between the peaks of adjacent grooves on a phonograph disk. 3. The distance between adjacent threads of a screw. 4. The distance between centers of turns in a coil (see *pitch of winding*) 5. The number of teeth or threads per unit length. 6. The distance along its axis a propeller moves in a revolution.

pitch of winding In a coil, the distance between the center of one turn to the center of the adjacent turn in a single layer of winding.

PIV Abbreviation of *peak inverse voltage*.

pivot The sometimes jewelled, stationary member of the bearing in a meter movement.

pi-wound choke A choke coil consisting of several series-connected pi's (see *pi, 4*) mounted on a single core and separated to reduce internal capacitance.

pix Abbreviatoin of *picture*.

Pixie tube A gas-filled readout tube having 10 cold cathodes under a disk-shaped anode. Around its circumference the anode has perforations shaped like the figures 0 through 9; as the glow shifts from cathode to cathode, the corresponding figure glows. Also see *readout lamp*.

pix tube See *picture tube*.

PL Abbreviation for *private line*.

place In a number, a digit position corresponding to a power of the radix.

planar diffusion In the production of a semiconductor device, the diffusion of all the elements into one face of a wafer. Consequently, connections to the elements all lie in one plane. Also see *epitaxial planar transistor* and *planar transistor*.

planar diode A semiconductor diode, having a pn junction that lies entirely within a single plane.

planar epitaxial passivated diode A junction diode which, like the planar epitaxial transistor, has been manufactured by planar diffusion and then passivated to protect the junction. Also see *epitaxial growth, epitaxy, passivation,* and *planar diffusion*.

PL/1 A computer programming language that is a hybrid of scientific and commercial types (like ALGOL and COBOL), the combined features being powerful problem-solving and mass-data-handling abilities.

planar epitaxial passivated transistor A planar epitaxial transistor which has been passivated to protect the exposed junctions. Also see *epitaxial planar transistor, planar transistor, epitaxial transistor, passivation,* and *planar transistor*.

planar transistor A transistor in which the emitter, base, and collector elements terminate on the same face (plane) of the silicon wafer. A thin film of silicon dioxide is grown on top of the wafer to insulate the exposed junctions after the leads have been attached, i.e., the transistor is *passivated*.

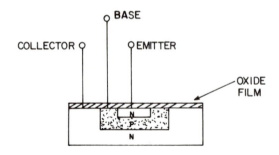

Planar transistor.

Planck constant Abbreviation, h. Unit, joule-seconds. The constant of proportionality in the fundamental law of the quantum theory, which expresses the idea that radiant energy is composed of quanta proportional to the frequency of the radiation; $h = q/v = 6.626\ 196 \times 10^{-34}$ J• s, where q is the value of the quantum and v is frequency.

plane of polarization The imaginary plane containing the direction of propagation and the electric field vector of a plane-polarized wave (see *polarization, 3* and *polarized light*).

plane-polarized light See *polarization, 3* and *polarized light*.

plane-reflector antenna A directive antenna in which the reflector is a sheet of metal or a metal screen. In a corner-reflector antenna, the reflector is a folded sheet, or two sheets joined along one edge.

planetary electron See *orbital electron*.

planimeter A mechanical instrument for measuring the area of a closed figure. The outline of the figure is traced with the pointer of the device, and the area is read from a pair of dials. In this application, the planimeter does the work of integral calculus.

planoconcave Pertaining to a lens, flat on one side and concave on the other.

planoconvex Pertaining to a lens, flat on one side and convex on the other.

plan position indicator Abbreviation, *PPI*. A radarscope on whose screen small spots of light reconstruct the scanned vicinity, revealing objects, such as buildings, boats, aircraft, and so on. The distance from the center of the screen to a spot depicts the range of an object, and the radial angle reveals its bearing.

plant In computer practice, to put the result of an operation specified by a routine in a storage location from which it will be taken for implementation of an instruction further on in the program.

plaque 1. Electroplaque. 2. Wall plaque.

plaque resistor A flat, noninductive, power resistor often employed as a dummy load during high-frequency power measurements.

plasma A usually high-temperature gas that is so highly ionized it is electrically conductive and susceptible to magnetic fields; it is now recognized as one of the states of matter. Also see *physical properties*.

plasma diode A diode in which a plasma substance produces conduction in one direction but not in the other.

plasma length See *Debye length*.

plasma oscillation In a plasma, a form of electric-field oscillation of the rapidly moving electrons.

plasma torch A torch, used for such high-heat applications as melting metal, in which a gas is heated by electricity to the high temperature at which it becomes a *plasma*.

plasmatron A form of amplifier tube sometimes used at ultra-high and microwave frequencies. Similar to a thyratron. An inert gas is excited until it becomes a plasma, producing amplification under certain operating conditions.

Plastacele See *cellulose acetate*.

plastic A synthetic material usually made from various organic compounds through polymerization (see *polymerize*). Plastics can be molded into a number of solid shapes and are available as films. Examples: Bakelite, celluloid, cellulose acetate, cellulose nitrate, polyethylene, polystyrene. Also see *thermoplastic material* and *thermosetting material*.

plasticizer A substance that is added to a plastic to make it softer or more flexible.

plate 1. The anode of an electron tube. 2. One of the electrodes of a primary or secondary battery cell. 3. One of the electrodes of a capacitor.

plateau In a response curve, a region in which an increase in the independent variable produces no further change in the dependent variable. The saturation region in the curves for pentode plate current or common-base-transistor collector current curve is such a plateau.

plate battery The battery that supplies the dc plate voltage of an electron tube. Also called *B battery*.

plate blocking capacitor A blocking capacitor connected between the plate of an electron tube and the plate tank. This capacitor allows dc voltage to be applied directly to the plate, i.e., without it passing through the tank coil, while at the same time preventing the tank coil from short-circuiting the dc plate. The capacitor freely transmits rf or af energy to the tank.

plate capacitance See *plate-cathode capacitance*.

plate-cathode capacitance Symbol, C_{pk}. Unit, pF. The internal capacitance between the plate and cathode of an electron tube. Also called *output capacitance*.

plate characteristic For an electron tube, the family of plate current-vs-plate voltage curves for various grid-bias voltages.

plate circuit The external circuit associated with the plate of an electron tube.

plate-circuit relay A milliampere-type dc relay operated in series with the plate of an electron tube.

plate conductance Symbol, g_p. Unit, mho. Conductance of the internal plate circuit of an electron tube. The value of static g_p is I_p/E_p. The value of dynamic g_p is dI_p/dE_p. Plate conductance is the reciprocal of plate resistance.

plate-coupled multivibrator A multivibrator in which the plate of the first tube is *RC*-coupled to the grid of the second tube, whose plate is *RC*-coupled back to the grid of the first tube. Compare *cathode-coupled multivibrator*.

plate current Symbol, I_p. Direct current (or the ac component of current) flowing in the plate circuit of an electron tube.

plate-current shift Change in the dc plate current of an rf amplifier during amplitude modulation. The action discloses faulty operation, since the average plate current should remain constant during modulation.

plate detector See *power detector*.

plate dissipation Abbreviation, *PD*. Unit, watt. Power expended in the plate of an electron tube. For an unloaded tube, $PD = E_pI_p$, where E_p is dc plate voltage (volts), and I_p is dc plate current (amperes). For a loaded tube, $PD = P_o - P_i$, where P_o is ac power output of an amplifier or oscillator in which the tube operates, and P_i is dc plate-power input.

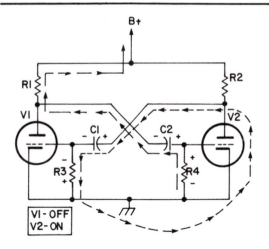

Plate-coupled multivibrator.

plated-wire memory See *wire memory*.

plate-grid capacitance Symbol, C_{pg} or C_{gp}. Unit, pF. The internal capacitance between the plate and control grid of an electron tube. Also called *interelectrode capacitance* and *feedback capacitance*.

plate keying In an oscillator or amplifier, keying that is accomplished by intermittently disconnecting the plate power supply of a vacuum tube.

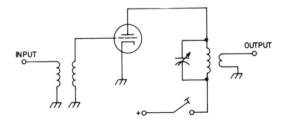

Plate keying.

plate load The power-consuming load into which the plate circuit of an electron tube operates. In an intermediate stage of a multistage amplifier, this load is the grid circuit of the following tube.

plate load impedance Symbol, Z_{Lp}. Unit, ohm. In a tube circuit, the (output) impedance that is ac-connected between the plate and ground, or dc-connected between the plate electrode and dc plate power supply. Also called *anode load impedance*.

plate meter A dc ammeter or milliammeter which indicates the plate current of an electron tube.

plate modulation A method of modulation in which a modulating voltage is superimposed upon the dc plate voltage of a higher-frequency amplifier or oscillator. Compare *absorption modulation, cathode modulation, grid-base modulation, plate-screen modulation, screen modulation,* and *suppressor modulation*.

platen The "roller" in a teletypewriter (and conventional typewriter),

i.e., that which supports the paper against impact by the type.

plate-neutralized amplifier A neutralized amplifier in which the neutralizing capacitor is connected between the free end of a center-tapped plate-tank coil and the grid of the tube. Compare *grid-neutralized amplifier*.

plate power Symbol, P_p. Unit, watt. Power in the plate circuit of an electron tube; $P_p = E_pI_p$, where E_p is plate voltage (volts, and I_p is plate current (amperes).

plate power input See *plate power*.

plate power output The output power delivered by the plate circuit of an electron tube. Compare *plate power input*.

plate power supply The usually dc power supply that furnishes energy to the plate of an electron tube.

plate pulse modulation A method of obtaining pulse modulation by injecting high-voltage pulses into the plate circuit of a vacuum-tube amplifier.

plate relay A relay operated in series with the plate of an electron tube.

plate resistance Symbol, r_p. Unit, ohm. Resistance of the internal plate circuit of an electron tube. The static value of r_p is E_p/I_p. The dynamic value of r_p is dE_p/dI_p.

plate resistor In an electron-tube circuit, the resistor connected in series with the tube plate and the plate power supply.

plate saturation In an electron tube, the point at which, while plate voltage is increasing, the plate attracts all the electrons emitted by the cathode, i.e., the point beyond which no further significant increase in plate current results from a further increase of plate voltage.

plate-screen capacitance Symbol, C_{ps}. Unit, pF. The internal capacitance between the plate and the screen of an electron tube.

plate-screen modulation A method of modulation in which modulating voltage is superimposed simultaneously upon the dc plate and screen voltages of a higher-frequency amplifier or oscillator. Compare *absorption modulation, cathode modulation, grid-bias modulation, plate modulation, screen modulation,* and *suppressor modulation*.

plate series compensation In an audio amplifier, the use of a plate decoupling circuit to obtain a fixed amount of bass boost.

plate shunt compensation The addition of a network to the plate-output circuit of a tube to boost the bass response of an amplifier.

plate spacing 1. The distance between plates in a fixed capacitor. This dimension is the same as dielectric thickness. 2. The distance between plates in a variable capacitor. Also called *capacitor air gap*.

plate supply voltage Symbol, *EBB*. The output voltage of a plate power supply.

plate tank A resonant *LC* circuit operated from the plate of an electron tube. Compare *grid tank*.

plate tank capacitance The capacitance required to tune a plate tank to resonance. Compare *grid tank capacitance*.

plate tank inductance The inductance of the coil in a plate tank. Compare *grid tank inductance*.

plate tank Q The figure of merit (see *Q*) of a plate tank. It is a function of load resistance and the *L/C* ratio of the tank.

plate tank voltage The ac voltage (af or rf) developed across the plate tank of an electron-tube circuit. Compare *grid tank voltage*.

plate tuning Tuning an electron-tube circuit by varying the capacitance, inductance, or both in the plate tank. Compare *grid tuning*.

plate tuning capacitance See *plate tank capacitance*.

plate tuning inductance See *plate tank inductance*.

plate-type capacitor A capacitor having metal plates rather than two cylinders, a cylinder and rod, or the like.

plate voltage Symbol, E_p. Dc voltage (or the ac component of voltage) applied to the plate of an electron tube.

plate winding 1. An inductor connected in series between the plate of a vacuum tube and the positive power-supply voltage. 2. The primary winding of a plate-circuit output transformer.

platiniridium Formula, *Pt-Ir.* A natural alloy of platinum and iridium.

platinotron A form of traveling-wave vacuum tube used as an amplifier at ultra-high and microwave frequencies. There are two output connections.

platinum Symbol, *Pt.* A precious metallic element. Atomic number, 78. Atomic weight, 195.09. Platinum is sometimes used for relay and switch contacts and for certain parts of electron tubes.

platinum metals The rare metals iridium, osmium, palladium, platinum, rhodium, and ruthenium. They do not react readily with other elements.

platinum-tellurium thermocouple A thermocouple employing the junction between platinum and tellurium wires and used in thermocouple-type meters.

playback The reproduction of recorded material in audio or video tape or disk systems.

playback head In a magnetic recorder—reproducer, the head that picks up the signal from the tape or wire for reproduction. Also called *read head* and *play head.*

playback loss In disk recording, the difference (at a particular point on the disk) between the recorded level and the reproduced level.

p-layer A semiconductor layer that is doped to provide current carriers which are predominantly holes. Compare *n-layer.*

player A reproducer.

playthrough The condition in which an amplifier delivers a small output signal when the gain control is set to zero.

PLC Abbreviation of *power-line communication.*

plethysmograph A medical-electronic device that allows the monitoring of the amount of blood in different parts of the body.

pliodynatron A tetrode vacuum tube in which the screen (or second) grid is supplied with a higher voltage than the plate. The result is oscillation at ultra-high frequencies.

pliotron A heavy-duty triode.

PLL Abbreviation of *phase-locked loop.*

PLM Abbreviation of *pulse-length modulation* (see *pulse-duration modulation*).

PLO Abbreviation of *phase-locked oscillator.*

plot 1. A curve depicting the variations of one quantity with respect to another. 2. To prepare such a curve.

plotter A machine that plots (*plot, 2*) automatically, often by direction of a computer.

plug A usually male type of quick-connect device which is inserted into a jack to make a circuit connection, or pulled out of the jack to break the connection. See, for example, *male plug, phone plug, power plug,* and *polarized power plug.*

plug-and-jack connection A connection made by inserting a plug into a jack.

plug fuse A fuse provided with an Edison base for screwing into a socket.

pluggable Capable of being completely removed from the rest of the system, without the need for removing any wiring. Pluggable components and circuit boards simplify the servicing of electronic equipment.

plug-in See *plug-in component* and *plug-in unit.*

plug-in capacitor A capacitor with pins or ferrules that can be quickly inserted into, or removed from, a socket.

plug-in coil A coil wound on a form having pins that can be quickly inserted into, or removed from, a socket.

plug-in coil form An insulating form with base pins that mate with socket terminals, so that a coil wound on the form may be quickly inserted into, and removed from, a circuit.

plug-in component A component or module, such as a tube, transistor, capacitor, coil, lamp, or the like, that is provided with pins, clips, or contacts for easy insertion into and removal from, a circuit. See, for example, *plug-in capacitor, plug-in coil, plug-in fuse, plug-in lamp, plug-in meter, plug-in resistor, plug-in transformer, plug-in unit.*

plug-in fuse A cartridge fuse having a metal ferrule on each end for insertion into a matching clip for easy installation and removal.

plug-in lamp A lamp with base pins for quick insertion into, or removal from, a socket.

plug-in meter A meter with pins or banana plugs for quick insertion into, or removal from, a circuit.

plug-in resistor A resistor with pins or ferrules for quick insertion into, or removal from, a socket or clips.

plug-in transformer A small transformer with pins for quick insertion into, or removal from, a socket.

plug-in unit A unit, such as a tuned circuit, amplifier, or meter, which has pins or contacts for easy insertion into, and removal from, a larger piece of equipment.

plumber's delight An antenna whose construction, including that of the mast, is entirely of metal rods or tubing, with no insulation whatever. Short circuits and grounds are prevented by making all attachments and joints at points that are *cold* with respect to the standing-wave pattern.

Plumbicon A television camera tube with a lead-oxide target. Similar to the vidicon, the Plumbicon has excellent sensitivity. The image lag is, however, shorter than in the conventional vidicon.

plumbing Collectively, the waveguides, tees, elbows, and similar pipelike devices and fixtures used in microwave setups.

plunger-type meter A meter in which an iron or steel plunger is pulled into a coil y the magnetism produced by a current flowing in the coil. The plunger is attached to a pointer which moves over the scale.

plutonium Symbol, *Pu.* A radioactive metallic element that is usually artificially produced. Atomic number, 94. Atomic weight, 239.

PM 1. Abbreviation of *permanent magnet.* 2. Abbreviation of *pulse(d) modulator.* 3. Abbreviation of *post meridian.* 4. Abbreviation of *phase modulation.*

Pm 1. Symbol for *promethium.* 2. Abbreviation of *petameter.*

Pm Symbol for *maximum power.*

PME Abbreviation of *photomagnetoelectric.*

PMG Abbreviation of *permanent-magnet generator.*

PMM Abbreviation of *permanent-magnet magnetizer.*

PMOS Abbreviation of *p-channel metal-oxide semiconductor.*

PMU Abbreviation of *portable memory unit.*

PN 1. Abbreviation of *Polish notation.* 2. Abbreviation of *positive— negative* (often not capitalized).

pn boundary See *pn junction.*

pnip transistor A junction transistor having an intrinsic layer between an *n*-type base and one of the *p*-layers.

pneumatic computer A computer that uses fluid logic, i.e., one in which information is stored and transferred by the flow of a fluid (gas or liquid) and pressure variations therein.

pn junction The boundary between *p*-type and *n*-type semiconductor materials in a single block or wafer of the materials. The junction, one of extremely intimate contact between layers, cannot be duplicated by merely touching two pieces of material (one *n*-type and one *p*-type) together, however smooth their mating faces.

pn-junction diode A diode consisting of the junction between *p*-type and *n*-type regions in the same wafer of semiconductor material.

PNM Abbreviation of *pulse-numbers modulation.*

pnpn device See *npnp device.*

pnp transistor A bipolar junction transistor in which the emitter and collector layers are *p*-type, and the base layer *n*-type. Compare *npn transistor.*

Po Symbol for *polonium.*

Po Symbol for *output power* or *power output.*

POGO Abbreviation of *polar orbiting geophysical observatory.*

point 1. A dot indicating the place of separation between the integral and fractional parts of a number. Examples: binary *point,* decimal *point.* Also called *radix point.* 2. A spot in space (e.g., where an electric charge is thought to be concentrated) which has no length, width, or thickness. 3. The place at which two or more lines or coordinates intersect.

point charge An electric charge imagined to occupy a single point in space; thus, it has neither area nor volume.

point contact The point at which the sharply pointed tip of a wire or rod conductor touches a second conductor, e.g., the contact between a catwhisker and a semiconductor wafer.

point-contact diode A semiconductor diode having a fine wire (catwhisker) whose point is in permanent contact with the surface of a wafer of semiconductor material, such as germanium or silicon.

point-contact junction The *pn* junction which is electroformed under the point at which the catwhisker touches the semiconductor wafer in a point-contact diode or transistor.

point-contact transistor A transistor composed of two fine wires (catwhiskers) which serve as the emitter and collector electrodes and whose pointed tips are nearly in contact with (a few mils apart from) the surface of a wafer of semiconductor material, such as germanium, which serves as the base electrode. The point-contact transistor was the predecessor of the junction transistor and is characterized by a current amplification factor (alpha) of greater than 1.

point counter A Geiger counter tube in which the central electrode is a pointed, fine wire. Also see *proportional counter.*

point defect 1. In a semiconductor substance or piezoelectric crystal, the absence of an atom from its place in the lattice structure. 2. The presence of an extra atom in the lattice structure.

point effect The tendency of an electrical discharge to occur more readily at a sharp point rather than along a relatively broad surface (as of an electrode).

pointer A pointed blade, stiff wire, or inscribed line on a transparent blade, which moves over a scale to indicate a setting or the value of a quantity. Also called *needle.*

pointer-type meter An indicating meter in which a pointer moves over a calibrated scale.

point impedance 1. The impedance observed at a given point in a circuit. 2. In a transmission line, the intensity of the electric field divided by the intensity of the magnetic field at a given point.

points of saturation For a magnetic core, saturation (see *saturated operation, 1*) as evidenced by a leveling off of the positive and negative halves of the magnetization curve.

point source A source from which electromagnetic radiation emanates and which is imagined to have no area or depth.

point mode Descriptive of crt terminal operation (in a computer system) in which data is displayed as plotted dots.

point-to-point communication Communication between two stations whose location can be determined.

point-to-point station A radio station providing *point-to-point communication.*

point-to-point wiring A method of wiring an electronic circuit in which wires are run directly between the terminals or components, usually by the shortest practicable route. Compare *cabled wiring.*

poise The cgs unit of absolute viscosity; 1 poise is the absolute viscosity of a fluid that requires a shearing force of 1 dyne to move a 1-sq-cm area of one of two parallel layers of the fluid (1 cm apart) with a velocity of 1 cm per second with respect to the other layer. The comparable SI unit is $N \cdot s/m^2$; 1 poise = 0.1 $N \cdot s/m^2$.

polar Pertaining to that which has reference to, or is associated with, a pole or with poles (see *pole, 1*).

polar coordinate conversion See *polar coordinate transformation.*

polar coordinates The magnitude and direction of a vector given as a radius (the magnitude) in combination with the angle (direction) between the vector and the polar axis in a plane, i.e., $r\theta$, where θ is in radians or degrees.

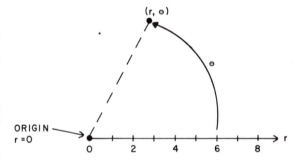

Polar coordinates.

polar coordinate transformation The transformation of polar to rectangular coordinates by the following: for rectangular coordinates x and y, $x = r \cos \theta$; $y = r \sin \theta$.

polarimeter An instrument for measuring the amount of polarized light in a ray that is only partially polarized.

polariscope An instrument used in the observation or testing of materials under *polarized light.*

polarity The condition of being positive or negative (electricity), or north or south (magnetism).

polarity blanking See *polarity inhibit.*

polarity inhibit In some instruments, especially those having *automatic polarity,* the automatic blanking of the polarity sign.

polarity-sensitive relay A dc relay that is actuated only when coil current flows in one direction. One of the simplest versions is a relay having a semiconductor diode connected in series with its coil.

polarity shifter A potentiometer connected to two dc sources so that a pair of output terminals has plus and minus polarities at one extreme of potentiometer adjustment, and minus and plus at the other extreme. At the center of the range, the output voltage is zero.

polarity switch A dpdt switch connected between a pair of dc input terminals so that the polarity of a pair of output terminals can be interchanged (switched).

polarization 1. In a radio wave, the direction of the electric lines, e.g., horizontally polarized wave, vertically polarized wave. 2. The disabling of a battery cell by the formation of insulating gas on one of the plates. 3. The condition in which transverse waves of light are confined to a specific (e.g., horizontal or vertical) plane. Also see *extraordinary ray, Nicol prism, ordinary ray, polarized light, Polaroid filter,* and *tourmaline crystal.*

polarization diversity A form of reception in which two separate receivers, tuned to the same signal, are connected to independent antennas. One antenna is vertically polarized and the other is horizontally polarized. The result is a reduction in fading caused by ionospheric effects on the polarization of the incoming signal.

polarization error In the operation of a loop antenna (e.g., that of a direction finder), null error due to waves arriving with polarization opposite that of the loop (thus, vertically polarized waves at a horizontal-plane loop, and vice versa).

polarization fading In radio reception, a form of fading that results from changes in the polarization of the arriving signal with respect to the receiving antenna. When the polarization of the arriving signal coincides with that of the receiving antenna, the received signal strength is maximum. When the received-signal polarization is at right angles to the receiving antenna, the signal strength is minimum.

polarization modulation A method of impressing information on a signal by changing the polarization of the electric lines of force.

polarization selectivity For a photoemissive surface, the condition in which the ratio of photocurrents for two different angles of plane polarization of the light incident to the surface differs from the ratio of the corresponding amounts of light absorbed by the surface. Also see *photoemission; photoemissive material;* and *polarization, 3.*

polarized capacitor A conventional electrolytic capacitor; so called because one particular terminal of this capacitor must be connected to the most positive terminal of the two connection points. Compare *nonpolarized electrolytic capacitor.*

polarized light Light whose transverse vibrations are confined to one plane by means of a filter or by certain natural conditions. A Polaroid filter is such a device, as is a Nicol prism or a tourmaline crystal.

polarized plug A plug that can be inserted into a socket or receptacle in only one way to insure safe and foolproof operation.

polarized power plug A polarized plug for connection to a power line.

polarized reactor A saturable reactor in which the lines of force produced in the three-leg core by the coils on the two outer legs are added in the center leg. Consequently, the flux may aid or oppose the controlling current which is made to flow in the coil on the center leg, depending upon the direction of the current. Compare *nonpolarized reactor.*

polarized receptacle A receptacle which is constructed so that it can receive a plug in only one way, thus preventing incorrect connections.

polarized relay 1. A relay which is actuated by one polarity (plus or minus) of current, voltage, or power; or by one particular phase. Some contain an armature-centering permanent magnet.

polarized socket See *polarized receptacle.*

polarized X-rays X-rays that are plane polarized, as when they are scattered by carbon blocks. See, for illustration, *polarization, 3* and *polarized light.*

polarizer The first of a pair of light-polarizing filters used to control the amount of light passing through the two filters. Also see *Nicol prism, Polaroid filter,* and *tourmaline crystal.* Compare *analyzer, 2.*

polar modulation Amplitude modulation in which the positive peaks of the carrier and modulated by one component and the negative peaks by another.

polarography In chemistry, a form of qualitative or quantitative analysis utilizing IE curves obtained when the voltage is gradually increased across a solution in *electrolysis.*

Polaroid See *Polaroid camera.*

Polaroid camera A camera that uses a special (Polaroid) film which, after exposure, is developed by self-contained chemicals. Such cameras are available for general-purpose photography and for recording the image on an oscilloscope screen. Also called *Polaroid Land camera* and *Land camera* (after Dr. Land, its inventor).

Polaroid filter A polarizing light filter consisting essentially of microscopic heraphathite plastic plate. The plate can be rotated to cause light passing through it to be polarized in any particular plane between horizontal and vertical, depending upon the degree of rotation. Also see *extraordinary ray, ordinary ray,* and *polarized light.*

polar planimeter See *planimeter.*

polar relay See *polarized relay.*

polar response The horizontal-plane directional response of an antenna or other transducer.

pole 1. That extremity or terminus which possesses polarity. Examples: magnetic pole, electric pole, pole of a battery, pole of a generator. 2. The movable member of a switch. 3. A frequency at which a transfer function becomes infinite (see *poles of transfer function*).

pole face The smooth end surface of a pole piece.

pole piece 1. A section of specially shaped iron or steel that is attached to a magnetic core. 2. Half of a two-piece magnetic core, which terminates in a pole.

pole shoe In an electric motor, the section of the field pole nearest the armature.

poles of impedance For a reactive network, the frequencies at which the impedance is infinite. Also called *poles.* Compare *zeros of impedance.*

poles of network function The values of p (real or complex) at which a network function is infinite. Compare *zeros of network function.*

poles of transfer function The frequencies at which a transfer function becomes infinite. Compare *zeros of transfer function.*

poling The deliberate adjustment of electromagnetic-field polarity.

Polish notation In Boolean albegra, a form of notation wherein the variables in a statement are preceded by the operators.

polling In data transmission, a technique in which channels being shared by more than one terminal are tested to find one over which data is coming in, or to ascertain which is free for transmission.

polonium Symbol, Po. A radioactive metallic element. Atomic number, 84. Atomic weight, 210 (approximately).

polycrystalline material A substance, such as a semiconductor, of which even a very small sample consists of a number of separate crystals bound tightly together. Compare *single-crystal material.*

polydirectional microphone A microphone capable of receiving sound waves from virtually all directions. Also see *omnidirectional microphone.*

polyelectrolyte An *electrolyte* having high molecular weight.

polyester A resin made by reacting a dihydroxy alcohol with a dibasic acid.

polyester backing A plastic tape (see *polyester*) on the surface of which iron oxide is deposited to yield a magnetic recording tape.

polyethylene A plastic insulating material. Dielectric constant, 2.2.

Dielectric strength, 585 V/mil.

polyethylene disk A phonograph disk made of polyethylene.

polygonal coil A coil wound on a form having a polygonal rather than circular cross section. Some polygonal forms have as many as 12 sides in cross section.

polygraph An instrument for measuring and recording electrical signals proportional to blood pressure, skin resistance, breathing rate, and other reactions that vary under emotional stress. Also called *lie detector*. Compare *pathometer*.

polymer A compound that is the product of polymerization resulting from the chemical union of monomers. Also see *polymerize*.

polymerize To unite monomers or polymers of the same kind to form a molecule having a higher molecular weight. Also see *polymerize*.

polynomial An algebraic expression containing two or more terms, e.g., $5a + 9b - 3c$. or $x^2 + 2x + 3$.

polyphase Pertaining in ac practice to the embodiment or generation of two or more phases. Compare *single phase*.

polyphase antenna A crossbar-type antenna consisting of two dipole radiators mounted perpendicular to each other at their midpoint and excited 90 degrees out of phase. The radiation pattern is approximately circular. Also called *turnstile antenna*.

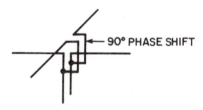

Polyphase antenna.

polyphase generator 1. A dynamo-type generator of polyphase power (two-phase, three-phase, etc.). 2. Polyphase oscillator.

polyphase oscillator An oscillator circuit that generates polyphase ac. The circuit contains separate oscillators for each phase. A three-phase circuit, for example, has three symmetrical oscillators with matched L and C values.

polyphase power Power in a polyphase circuit. In a balanced three-phase system (wye or delta), total power $P = \sqrt{3}$ (EI) cos ϕ, where E is line voltage; I, line current; and ϕ, phase angle.

polyphase rectifier A rectifier of polyphase ac power. The polyphase voltage is generally obtained from a three-phase power line through a transformer. There are several familiar circuits, each usually containing a diode for each phase. Such rectifiers offer the advantage of higher ripple frequency than is obtainable by single-phase operation. For a three-phase rectifier, for example, the ripple frequency is three times the line frequency; for a six-phase rectifier, it is six times the line frequency.

polyphase system An ac system in which voltages or currents are normally out of phase with each other by some fixed amount. Familiar types are two-phase and three-phase.

polyphase transformer An alternating-current transformer specifically designed for use in circuits in which there are two or more simultaneous current phases.

polypropylene A plastic insulating material. Dielectric constant, 2.0. Dielectric strength, 600V/mil.

polyrod antenna A tapered *dielectric antenna*, usually made of polystyrene, for directional microwave transmission.

polysilicon Polycrystalline *silicon*.

polystyrene A clear, colorless plastic of the thermosetting type which is widely used as an insulant in radio-frequency circuits and, to some extent, as a dielectric film in fixed capacitors. Dielectric constant, 2.4 to 2.9. Dielectric strength, 20 to 28 kV/mm. The name indicates that this material is a polymer of styrene (see *polymer*).

polystyrene capacitor A high-Q capacitor in which the dielectric film is polystyrene.

polyvinyl choloride Abbreviation, *PVC*. A plastic insulating material. Dielectric constant, 3.6 to 4.0. Dielectric strength, 800V/mil.

pool cathode In an industrial electron tube, a cathode consisting of a pool of mercury.

pool-cathode tube An industrial electron tube employing a pool cathode. Examples: excitron, ignitron, mercury-arc rectifier.

popcorn noise A temperature-dependent electrical noise of the random shot variety resembling the sound of popping corn, that is found in some *operational amplifiers*.

population In statistical studies, the total group of items, quantities, values, or the like under consideration. Most often, samples are taken from the population. Sometimes called *universe*.

porcelain A hard, white, usually glazed ceramic used as a dielectric and insulant. Dielectric constant, 6 to 7.5. Dielectric strength, 40 to 100V/mil. Also called *china*.

porcelain capacitor A ceramic-dielectric capacitor, in which the dielectric is comprised of porcelain or a related substance.

porcelain insulator An electric insulator fabricated from porcelain.

porch See *back porch* and *front porch*.

port 1. In a circuit, device, or system, a point at which energy or signals may be introduced or extracted in a particular manner, e.g., two-port circulator, I/O port. 2. An aperture in a loudspeaker enclosure.

portable-mobile station See *mobile station*.

portable station A station that can be carried from one location to another. A portable station differs from a mobile station in that a portable station does not usually operate in transit, whereas a mobile station does.

pos 1. Abbreviation of *positive*. 2. Abbreviation of *position*.

position 1. The location of a point or object with respect to one or more (usually fixed) references. 2. The setting of an adjustable device, such as a potentiometer, rotary switch, or variable capacitor.

positional notation A method of representing numbers in which the number is indicated by the positions and value of the component digits. The decimal number system belongs in this category, e.g., the decimal number $1284 = (1 \times 10^3) + (2 \times 10^2) + (8 \times 10^1) + (4 \times 10^0)$.

positional number system See *positional notation*.

positional representation See *positional notation*.

position-control potentiometers In an oscilloscope, potentiometers used to control the voltage applied to the horizontal and vertical deflecting plates to position the spot on the screen. Also see *centering control*.

position controls See *position-control potentiometers*.

position feedback In a servo or other control system, feedback current or voltage which is proportional to the position assumed by a member.

position fixing Determination of one's position from the intersection on a map of two lines derived from the direction-finding pickups of two transmitting stations. Also see *direction finder*.

position indicator In a tape recorder, a counter whose numbered wheels revolve when the reels do, thus aiding in locating a desired spot on the tape. Also called *tape counter.*

positioning circuit The circuit associated with a horizontal or vertical centering control (see *centering control*).

position sensor An electronic circuit that detects physical displacement, and transmits a signal proportional to the displacement.

positioning control See *centering control.*

positive 1. Possessing a positive polarity. 2. Opposite of negative. 3. Greater than zero. 4. A photographic image whose shadings are the same as those in the scene.

positive angle Referring to rectangular coordinates, an angle in the first or second quadrant. Compare *negative angle.*

positive bias A positive dc voltage or current applied continuously to an electrode of a device—as to a tube grid or transistor base—to maintain the device's operating point. Compare *negative bias.*

positive bus See *positive conductor.*

positive charge An electrical charge characterized by having relative fewer electrons than a negative charge. Also see *charge, 1; electric charge;* and *unit electrostatic charge.* Compare *negative charge.*

positive column In a glow-discharge tube, the long, luminous portion of the discharge adjacent to the anode. Compare *Crookes' dark space, Faraday dark space,* and *negative glow.*

positive conductor The conductor or line connected to the positive terminal of a current, voltage, or power source. Compare *negative conductor.*

positive crystal A crystal that exhibits positive double refraction. Compare *negative crystal.*

positive electricity See *positive charge* and *positive electrification.*

positive electrification Electrification characterized by a deficiency of electrons. For example, when a glass rod is rubbed with a silk cloth, the rod becomes positively charged, because electrons are rubbed off the glass onto the silk. Similarly, when an atom loses an electron, it becomes electrified positively, because it has a deficiency of electrons. Compare *negative electrification.*

positive electrode 1. An electrode which is connected to the positive terminal of a current, voltage, or power source. 2. The positive terminal of a current, voltage, or power source, such as a battery or generator.

positive element See *positive eletrode, 1.*

positive error of measurement An error of measurement in which the difference between a measured value and the true or most probable value is positive. Compare *negative error of measurement.*

positive exponent A positive superscript (n) indicating that a number (N) is to be raised to the positive nth power. Thus, N^n. Compare *negative exponent.*

positive feedback Feedback which is in phase with an input signal. Also called *regeneration* and *regenerative feedback.* Compare *negative feedback.*

positive function A function having the positive sign. In the rectangular coordinate system, the trigonometric sine function is positive in the first and second quadrants; the cosine in the first and fourth; and the tangent in the first and third. Compare *negative function.*

positive ghost In a TV picture, a ghost with positive shading (see *positive, 4*). Also see *ghost.*

positive-going Pertaining to a signal whose value is changing in a positive direction. This is not restricted to signals of actual positive polarity; a decreasing negative voltage, for example, is positive-going

as it falls in the direction of zero, even if it never crosses the zero line.

positive grid For an electron tube, a control grid whose bias or signal voltage is positive with respect to the cathode.

positive-grid oscillator A microwave oscillator circuit in which the control grid of a triode is operated at a positive dc potential, and the plate at a negative potential. Electrons move back and forth between cathode and plate, through the grid, and thus give rise to an oscillating current. Also see *Barkhausen—Kurz oscillator.* Compare *negative-grid oscillator.*

positive-grid oscillator tube A vacuum tube (usually a triode having concentric, cylindrical elements) especially suited for use in a positive-grid oscillator.

positive-grid-return multivibrator A monostable multivibrator circuit in which the grids are returned to a positive potential (B +) rather than to a cathode or ground. This circuit affords accurate timing.

positive ground A direct-current electrical system in which the positive power-supply terminal is connected to the common ground. Not generally used in the United States.

positive half-alternation See *positive half-cycle.*

positive half-cycle That half of an ac cycle in which current or voltage increases from zero to maximum positive and returns to zero.

positive image 1. A picture in which the blacks, whites, and grays correspond to those in the scene (see **positive, 4**). 2. A normal TV picture, i.e., one having the shading described in 1, above.

positive ion An atom that has a deficiency of electrons and, consequently, exhibits a net positive charge. Also called *cation.*

positive ion sheath In a thyratron, a layer of positive ions of gas which forms around the wires of the grid after having been attracted by the negative charge on that electrode. The sheath prevents the passage of electrons to the anode and, hence, cuts off anode current. When the grid is made less negative, the sheath thins, and at a particular point, anode current is initiated.

positive lead See *positive conductor.*

positive light modulation In TV transmission, the condition in which transmitted power increases as the light intensity increases, and vice versa. Compare *negative light modulation.*

positive line See *positive conductor.*

positive logic Binary logic in which a high positive state represents one and a low positive state represents zero. Compare *negative logic.*

positive magnetostriction A form of magnetostriction in which the physical size of a substance is directly proportional to the intensity of the surrounding magnetic field.

positive measurement error See *positive error of measurement.*

positive modulation In amplitudemodulated TV transmission, the increase in transmitted power when the brightness of the scene increases. Compare *negative modulation.*

positive modulation factor For an amplitude-modulated wave having unequal positive and negative peaks, a ratio expressing the maximum deviation from the average value of the envelope. Compare *negative modulation factor.*

positive number A number whose value is greater than zero. Compare *negative number.*

positive peak 1. The positive half of an ac cycle. 2. The maximum amplitude of a positive half-cycle or positive pulse.

positive-peak clipper A peak clipper that levels off the positive half-cycle of an ac wave to a predetermined level.

positive-peak modulation Amplitude modulation of the positive peaks of a carrier wave.

positive-peak voltmeter An electronic voltmeter for measuring the amplitude of the positive peak of an ac wave. In its simplest form, if consists essentially of a dc microammeter with a diode poled to pass the positive half-cycle. A series capacitor in the circuit is charged to approximately the peak value of the applied ac voltage. Compare *negative-peak voltmeter*.

positive phase-sequence relay A phase-sequence relay which responds to the positive phase sequence in a polyphase circuit. Compare *negative phase-sequence relay*.

positive picture modulation See *positive modulation*.

positive picture phase In a TV signal, the swinging of the picture-signal voltage from zero to positive in response to an increase in brightness in the scene. Compare *negative picture phase*.

positive plate 1. The positive member of a battery cell. Electrons flow to this plate from the negative plate, through the external circuit. 2. A vacuum-tube plate which is biased positively, as in a conventional tube circuit.

positive pole See *positive electrode, 1, 2*.

positive potential 1. The potential at a positive electrode (with respect to the negative electrode). 2. Potential greater than that at ground as a reference.

positive power See *positive exponent*.

positive rays In a tube diode, rays arising from the anode. Also called *canal rays*. Compare *cathode rays*.

positive resistance Ohmic resistance (see *ohmic response*). Compare *negative resistance*.

positive resistor A resistor whose value does not change with current or voltage changes. Compare *negative resistor*.

Positive temperature coefficient Abbreviation, *PTC*. A number expressing the amount by which a quantity (such as the value of a component) increases when temperature is increased. The coefficient is stated as a percentage of the rated value per degree, or in parts per million per degree. Compare *negative temperature coefficient* and *zero temperature coefficient*.

positive terminal See *positive electrode, 2*.

positive transmission In facsimile or television, a form of amplitude modulation in which the picture brightness is directly proportional to the signal strength at any given instant of time.

positive valence The valence of a positive ion.

positron A positively charged particle having the same mass as that of the electron and the same magnitude of electric charge but which is positive instead of negative. Sometimes called *positive electron*.

post 1. Binding post. 2. Prefix meaning (*following*) *after; subsequent (to); behind*.

post-accelerating electrode In a cathode-ray tube, the high-voltage electrode that produces postdeflection acceleration of the electron beam. Also called *intensifier electrode*.

post acceleration See *postdeflection acceleration*.

post-alloy-diffused transistor Abbreviation, PADT. A transistor in which electrodes are diffused into the semiconductor wafer after other electrodes have been alloyed.

postconversion bandwidth The bandwidth of a signal after it has been converted from one frequency to another.

postdeflection accelerating electrode See *post-accelerating electrode*.

postdeflection acceleration In a cathode-ray tube, intensification of the electron beam following beam deflection. Also see *postdeflection crt*.

postdeflection crt An oscilloscope tube provided with a high-voltage

intensifier electrode in the form of a ring encircling the inside flare of the tube, between the deflecting plates and the screen. The deflected electron beam is accelerated by this electrode. This arrangement allows the beam to be deflected at low velocity and high sensitivity and then to be accelerated for a brighter image.

postedit The editing of data in a computer output.

postemphasis See *deemphasis*.

post-equalization 1. In sound recording and reproduction, equalization during playback. Compare *pre-equalization*. 2. Deemphasis.

postmortem An investigation into the cause of failure of a circuit, device, or system.

postmortem dump At the end of a computer program run, a dump to supply information for debugging purposes.

pot 1. Potentiometer. 2. Dashpot. 3. Abbreviation of *potential*. 4. To encapsulate a circuit in a potting compound, such as epoxy resin.

potassium Symbol, *K*. A metallic element of the alkali-metal group. Atomic number, 19. Atomic weight, 39.100.

potassium chloride Formula, *KCl*. A compound used as a phosphor coating on the screen of a nearly permanent-persistance cathode-ray tube. The fluorescence is magenta or white, as is the phosphorescence.

potassium cyanide Formula, *KCN*. A highly toxic salt that is an electrolyte in some forms of electroplating.

potassium dihydrogen phosphate Abbreviation, *KDP*. An inorganic ferroelectric material.

pot core A magnetic core for a coil, often made of ferrite or of powdered iron and consisting of a central rod, a surrounding potlike enclosure, and a lid. The rod passes through the center of the coil, and the pot and lid completely enclose the coil. This arrangement provides a completely closed magnetic circuit and coil shield.

potential See *electromotive force*.

potential barrier The electric field produced on each side of a semiconductor junction by minority carriers (i.e., by holes in the n-layer and electrons in the p-layer) which face each other across the junction but cannot diffuse across the junction and recombine. Also see *barrier, 1*.

potential coil The shunt coil in a conventional wattmeter.

potential difference See *electromotive force, volt,* and *voltage*.

potential divider See *voltage divider*.

potential drop 1. A potential difference between two points in a circuit. 2. The voltage across a resistor in a direct-current circuit.

potential energy Energy resulting from the position of a body or particle (e.g., the energy stored in something lifted against gravity and held in its new position) or from the position of charges (e.g., the energy stored in a charged capacitor). Compare *kinetic energy*.

potential gradient See *voltage gradient*.

potential minimum The lowest point in the potential gradient of the negative space charge of an unbiased tube diode operating at full emission. This point, which is below ground potential, is determined by cathode-emission rate. Electrons are repelled and returned to the cathode when they are not needed to maintain this potential.

potential profile A rectangularcoordinate display of the potential gradient across a body, e.g., the cross section of a transistor.

potential transformer A small stepup transformer for increasing the range of an ac voltmeter. Compare *current transformer, 2*.

potentiometer 1. A variable resistor employed as a voltage divider. The input voltage is applied to the ends of the resistance element, and the output is taken from the slider (wiper) and one end of the element. 2. A null device whose operation is based upon the principle of the potentiometer (see 1, above) and is used for precise voltage measurements. The unknown voltage is applied to the input of a potentiometer whose resistance settings are known with great accuracy; the potentiometer is adjusted for an output voltage that exactly equals the voltage of a standard cell (as indicated by a null between the two voltages). The unknown voltage is then determined from the resistance setting of the potentiometer and the standard-cell voltage.

potentiometer noise In a current-carrying potentiometer, electrical noise generated when the wiper blade rubs against the resistance element, or by contact between the blade and element.

potentiometric recorder A type of graphic recorder (see *potentiometer, 2*). It consists essentially of a resistance-calibrated potentiometer, a standard cell, and a galvanometer. When an unknown voltage (*Ex*) is applied to the input terminals of the potentiometer and the pot is set for null, *Ex* equals *Es* (*R2/R1*), where *Es* is the voltage of the standard cell, *R1* is the input resistance of the potentiometer, and *R2* is the output resistance of the potentiometer.

Potier diagram An illustration of the phase relationship between current and voltage in an alternating-current circuit containing reactance.

potted circuit A circuit which is potted in a suitable plastic or wax (see *potting*).

potted component An electronic part which is potted in a suitable plastic or wax (see *potting*).

Potter oscillator See *cathode-coupled multivibrator*.

potting Embedding a component or circuit in a solid mass of plastic or wax held in a container. The process is similar to encapsulation, except that in potting, the container (envelope) remains part of the assembly. (In encapsulation, the mold is removed after the encapsulating material has dried and hardened.)

potting material A material, such as a resin or wax, used for potting electronic gear. Also called *potting compound*.

pound 1. Abbreviations, lb, p in commerce, where a unit of weight is implied; 16 avoirdupois ounces. 2. Abbreviation, 1 lbf. A unit of force approximately equal to 4.448 newtons. 3. Abbreviation, 1 lbm. A unit of mass approximately equal to 0.4536 kg.

poundal A unit of force. 1 poundal is the force which, when acting for 1 second, will give a 1-pound mass a velocity of 1 foot per second. 1 poundal = 13 825.5 dynes = 0.138 255 newton.

pound-foot Abbreviation, *lb-ft*. A unit of torque equal to the product of a force of 1 pound and a moment arm of 1 foot. Compare *ounce-inch*.

pounds per square inch absolute Abbreviation, *psia*. Absolute pressure, i.e., the sum of atmospheric pressure and the pressure indicated by a gage. Compare *pounds per square inch gauge*.

pounds per square inch gauge Abbreviation, *psig*. The value of pressure indicated by a gauge, without correction for atmospheric pressure. Compare *pounds per square inch absolute*.

powdered-iron core A magnetic core consisting of minute particles of iron—each coated with a film to insulate it from others—molded into a solid core. Because of its low eddy-current losses, the powdered-iron core is usable in i-f transformers and certain rf coils, where it increases the inductance of the winding. Powdered iron is also molded into shields (see, for example, *pot core*).

power 1. Symbol, *P*. Unit, watt. The rate of doing work, or producing or transmitting energy. In electronics, power is the product of current and voltage (*EI*). See, for example, *ac power, apparent power, dc power, kilovolt-ampere, power factor, reactive kilovolt-ampere, reactive volt-ampere, true power, volt-ampere, watt,* and *wattless power*. 2. The product obtained by multiplying a number *(N)* by itself a number of times specified by an exponent *(n)* which, numerically, is 1 more than the

number of times sign. Thus, $2^3 = 2 \times 2 \times 2 \times 2 = 8$; and here, 8 is the third power of 2. Compare *root, 1*. 3. The exponent in 2, above.

power amplificaton 1. Amplification of an input power to give a larger output power. 2. The signal increase *Pout/Pin* resulting from this process. Also called *power gain*.

power amplification ratio See *power amplification, 2* and *power gain*.

power amplifier An amplifier that delivers useful amounts of power to a load, such as one or more speakers. Compare *current amplifier, voltage amplifier*.

power-amplifier device A high-current tube or transistor designed especially for high power output. Such a device does not always provide significant voltage amplification but usually provides a respectable amount of power amplification. Compare *voltage-amplifier device*.

power at peak torque Symbol, *Pp*. For a torque motor, the input power in watts needed for peak torque at stall at winding temperature of 25 °C.

power attenuation 1. Reduction of power level. 2. Power loss.

power bandwidth For a high-fidelity audio amplifier, the difference between the maximum and minimum frequencies at which the amplifier can produce at least 50 percent of its maximum power output, with less than a certain amount of total harmonic distortion (usually 10 percent).

power blackout An emergency situation in which all electric power is lost in a community.

power consumption 1. For a direct-current device, the normal operating voltage multiplied by the normal drawn current. 2. For an alternating-current circuit, the root-mean-square voltage multiplied by the root-mean-square current.

power control Adjustment of the output power of a power supply, usually by means of a variable autotransformer, silicon controlled rectifier, thyratron, or similar device.

power cutoff frequency Symbol, *fco*. The frequency at which the power gain of a transistor drops 3 dB below its low-frequency value.

power derating For a higher than specific ambient temperature, the reduction of power in a component to maintain a safe dissipation level. Also see *derating, derating curve,* and *derating factor*.

power detector A vacuum-tube detector in which grid bias is set close to the plate-current cutoff value. The tube consequently operates like a class-B amplifier on the lower bend of the grid-voltage—plate-current curve, effectively demodulating the input signal. Also called *plate detector*.

power difference An expression of the power lost in a circuit when it is absorbed by a dielectric material.

power diode A heavy-duty diode usually employed in power-supply service. Also called *rectifier*.

power dissipation Abbreviation, *PD*. The power consumed by a device during normal operation and which is, therefore, not available in the electrical output of the device. An example is the dc power dissipated in the plate of an electron tube.

power divider A circuit that distributes power, in a predetermined manner, among various different loads.

power drain The amount of power drawn by a device. It may be operating power or standby power.

power dump See *dump, 2*.

power equations Variations of the basic power equation: $P = EI = E^2/R = I^2R$.

power factor Symbol, *pf*. In an ac circuit, the ratio (expressed either as a decimal or a percentage) of true power (power actually consumed) to apparent power (simple product of voltage and current). Power factor $pf = \cos \phi$, where ϕ is the phase angle. Also see *ac power*.

power-factor balance In a capacitance bridge, a separate null adjustment for the internal resistance component of a capacitor under measurement. The dial of the variable component for this adjustment reads directly in *percent power factor* in some bridges.

power-factor correction The practice of raising the power factor of an inductive circuit by inserting series capacitance. In power circuits, this affords improved economy of operation, since the current drain is brought more in line with that of a resistive circuit.

power-factor meter An instrument that gives direct readings of power factor (lead or lag). One such meter employs a dynamometer-type movement (see *electrodynamometer*) in which the rotating element consists of two coils fastened together perpendicularly.

power-factor regulator A device that regulates the power factor of an alternating-current line.

power-factor relay An ac relay that is actuated by a rise or fall in power factor with respect to a predetermined value.

power frequency 1. Power-line frequency. 2. The frequency of a ac generator or dc-to-ac inverter.

power-frequency meter An instrument for measuring power-line frequencies. In the *movable-iron type,* a soft-iron vane (to which a pointer is attached) rotates in the magnetic field of two stationary coils mounted perpendicular to each other and through which the current flows (through a resistor in series with one coil and an inductor in series with the other). The direction of the field varies with frequency because of the phase shift introduced by the resistor and inductor; this causes the deflection of the vane to be proportional to frequency. The pointer travels over a scale reading directly in hertz. In the *reed type,* the alternating current flows through the coil of a field magnet. Metal reeds cut to vibrate at a different frequency are mounted within the field of the magnet. The reed that vibrates most violently indicates the frequency of the current. A scale under the row of reeds is marked with the resonant frequency of each reed.

power gain Abbreviation, *PG*. The amount by which power is increased by the action of an amplifier. The gain is expressed either as the simple ratio of power output (*Po*) to power input (*Pi*) or in decibels: $PGdb = 10 \log_{10} (Po/Pi)$ power grid. An aggregation of power-generating stations, transmission lines, and associated equipment—usually extending over hundreds of miles and embracing several communities—so operated that individual members may deliver power to the system of draw power from it, according to their local demands.

power ground The power-supply ground for a circuit or system.

power-handling capacity 1. The amount of power that a device can dissipate, either continuously or intermittently, without suffering damage. 2. The maximum amount of input power that can be tolerated by an amplifier tube or transistor without overheating.

power hyperbola For a tube or semiconductor device, a curve plotted from the device's current and voltage values, which give the power value when multiplied, e.g., a 2-watt curve for the dc collector input of a power transistor.

power input See *input power.*

power-input control Adjustment of the output of a power supply by varying the ac input to the power transformer. A primary rheostat could be used for this purpose, but resistors consume power; so, instead, a variable autotransformer is operated ahead of the power transformer. See, for example, *variable transformer* and *Variac*.

power-level indicator 1. Decibel meter. 2. Output power meter.

power line The line through which electrical energy is received by the subscriber.

power-line communication Abbreviation, *PLC*. Carrier-current telephony or telegraphy over power lines that are common to transmitting and receiving stations. Also see *wired wireless*.

power-line filter 1. A heavy-duty rf filter inserted in the power line close to a device that creates radio interference to prevent transmission of the interference into the power line. 2. A similar rf filter inserted into the power line where it enters the power supply of a receiver or other sensitive electronic device, to prevent rf energy that might be picked up by the power line (acting as an antenna) from entering the device via the power supply.

power-line frequency The frequency of the alternating current and voltage available over commercial power lines. In the United States, it is 60 Hz.

power-line monitor An expanded-scale ac voltmeter for the continuous monitoring of power-line voltage.

power-line pickup The pickup of rf energy by usually overhead, outside power lines, which act as receiving antennas. This energy usually finds its way into a radio or TV receiver via the power supply.

power loss Power that is dissipated in a component; it generates heat while doing no useful work and therefore represents energy loss—except when the generation of heat is the end purpose.

power-loss factor Symbol, *Fp*. In interstage coupling, the ratio of available power (with the coupling network in place) to the available power when the network is disconnected.

power meter See *wattmeter.*

power modulation factor In amplitude modulation, the ratio of the peak power to the average power.

power of vectors Raising a polar vector to the *n*th power yields the *n*th power of the vector's magnitude and the product of *n* and the angle; $(r\omega)^n = r^n n\omega$.

power oscillator A heavy-duty af or rf oscillator delivering useful power output.

power output See *output power.*

power-output meter See *output power meter.*

power pack An external power-line-operated unit supplying ac and dc for the operation of an electronic equipment.

power pentode A heavy-duty pentode, i.e., one designed for relatively large amounts of output power, e.g., 4E27A.

power plug A plug for insertion into a power-line outlet.

power programmer A device that adjusts radar output power in accordance with the target distance.

power rating 1. The specified power required by an equipment for normal operation. 2. The specified power output of a generator or amplifier.

power reactive See *reactive volt-ampere.*

power rectifier A heavy-duty semiconductor diode employed to rectify ac for power supply purposes.

power relay A heavy-duty relay designed to switch significant amounts of power. The heavy contacts and armature require relatively high actuating current; this necessitates a larger coil than is found in lighter-duty relays.

power resistor A heavy-duty resistor, i.e., one designed to carry large currents without overheating.

power sensitivity For a power amplifier tube whose grid circuit consumes no power, the ratio Po/Eg, where Po is the output power, and

Eg is the corresponding ac grid-signal (excitation) voltage.

power series An infinite series (see *mathematical series*) in which the terms are arranged sequentially according to ascending powers of *x*. Thus, $Sn = a1x + a2x^2 + ... + anx^n$. If the *a* and *x* terms are independent of each other, the series may converge or diverge, depending upon their values.

powers of numbers See *power, 2.*

power stack A selenium rectifier consisting of a number of rectifier plates stacked in series for higher voltage handling.

power supply 1. A device, such as a generator or a transformer-rectifier-filter arrangement, that produces the power needed to operate an electronic equipment. 2. A reserve of available power, e.g., the power line, an installation of batteries, etc.

power-supply filter A low-pass filter employed to remove the ripple from the output of a power-supply rectifier. See, for example, *brute-force filter.*

power-supply rejection ratio The ratio of the output-voltage change for an amplifier, oscillator, or other circuit, to the change in power-supply voltage. Determined on an instantaneous basis.

power-supply sensitivity In an operational amplifier, sensitivity of the offset to power-supply variations.

power switch The switch for controlling power to an equipment. Also see *on-off switch.*

power switching Switching operating power on and off. There are two principal methods: One involves making and breaking the connections between an equipment and the power line; the other involves making and breaking the output of a line-operated or battery-type power supply.

power tetrode A heavy-duty tetrode, i.e., one designed for relatively large amounts of output power, e.g., a 7094.

power-to-decibel conversion Power level expressed in decibels referred to a reference power level; $PdB = 10 \log_{10} Px/Pref$, where *Px* is the given power level, and *Pref* is the reference power level. Various reference power levels are in use, e.g., 0.006W into 500 Ω, 0.06W into 600 Ω, and 0.01W into 600 Ω.

power transfer The passage of power from a generator to a load or from one circuit to another.

power transfer theorem See *maximum power transfer theorem.*

power transformer A transformer designed solely to supply operating power to electronic equipment, either directly or through a rectifier-filter circuit. Because a power transformer is used at low (power-line) frequencies, its core does not require the high-grade iron used in audio transformers, nor are special winding techniques needed to reduce reactances.

power transistor A heavy-duty transistor designed for power-amplifier and power-control service, e.g., a 2N255.

power triode A heavy-duty triode, i.e., one designed for relatively large amounts of output power, e.g., a 2C36.

power tube A heavy-duty electron tube designed to deliver useful amounts of power. See, for example, *power pentode, power tetrode,* and *power triode.*

power unit 1. Power supply (see *power supply, 1*). 2. A unit of power measurement. See, for example, *kilowatt, megawatt, microwatt, milliwatt, picowatt,* and *watt.*

power winding In a magnetic amplifier or saturable reactor, the output winding, i.e., the winding through which the controlled current flows.

Poynting vector In an electromagnetic wave, the vector product of instantaneous electric intensity and magnetic intensity.

PP Abbreviation of *peripheral processor.*

Pp 1. Symbol for *plate power.* 2. Symbol for *peak power.*

ppb Abbreviation of *parts per billion.*

PPI Abbreviation of *plan position indicator.*

pp junction In a semiconductor wafer, the boundary between two p-type regions having somewhat different properties.

PPM Abbreviation of *pulse-position modulation.*

ppm 1. Abbreviation of *parts per million.* 2. Abbreviation of *pulses per minute.*

pps Abbreviation of *pulses per second.*

ppt Abbreviation of *parts per thousand.*

Pr Symbol for *praseodymium.*

practical component A circuit component considered in proper combination with the stray components inherent in it. Thus, a resistor has residual inductance and capacitance, an inductor has residual capacitance and resistance. Compare *ideal component.*

practical units A set of physical/electrical units that is especially suited to a particular application.

praetersonics See *acoustoelectronics* and *acoustic delay line, surface-wave amplifier,* and *surface-wave filter.*

pragilbert The unit of magnetomotive force in the absolute mks (Giorgi) system.

pragilbert per weber The unit of reluctance in the absolute mks (Giorgi) system.

praoersted The unit of magnetizing force in the absolute mks (Giorgi) system.

praseodymium Symbol, *Pr*. A metallic element of the rare-earth group. Atomic number, 59. Atomic weight, 140.92.

preaccelerating electrode In the electron gun of a cathode-ray tube, the high-voltage electrode that provides initial acceleration to the electron beam.

preamplifier An auxiliary (usually voltage) amplifier operated ahead of a main amplifier to boost the input signal and, sometimes, provide equalization.

prebiased relay A relay through which is maintained a steady current that is just lower than that needed to close the relay. The actuating signal, then, need only be a small amount of additional current.

preburning The treating of a vacuum-tube filament by continuous operation at a given temperature, for a certain length of time, before distribution. This eliminates initial irregularities in performance.

precession Oscillation of the axis of a spinning body. This wobbling motion is exhibited by gyroscopes, spinning tops, and similar bodies. Such a change in the altitude earth's axis has resulted in *precession of the equinoxes*, causing the vernal equinox to seem to move backward against the backdrop of the constellations.

precipitation 1. Water falling from the atmosphere in some form (rain, snow, hail, sleet). See *precipitation static.* 2. The amount of precipitation. 3. A falling or floating of a solid material that separates out of a solution as a result of a chemical or physical action. Also called *precipitate.*

precipitation static Radio noise caused by atmospheric electricity arising from rain, snow, ice crystals, hail, or dust clouds through which an aircraft carrying the radio flies.

precipitator See *dust precipitator.*

precipitron See *dust precipitator.*

precision 1. Having the property of being designed with a high degree of accuracy. 2. The relative accuracy of a meter or other indicating device. 3. The accuracy of the results of an experiment, test, or measurement.

precision approach radar A radar which is aimed along the approach path to guide an aircraft during approach.

precision instrument An instrument possessing high accuracy and stability, i.e., one capable of reproducing readings or settings for various trials under set circumstances.

precision potentiometer 1. A potentiometer (see *potentiometer, 1*) possessing a highly accurate resistance calibration, linearity, and repeatability of settings. 2. A potentiometer-type voltage-measuring instrument (see *potentiometer, 2*).

preconduction current 1. Transistor cutoff current. 2. In a thyratron, the small (anode) current flowing before the tube is fired.

predetermined counter A counter that is programmed to count to a desired number and then stop.

predissociation The tendency of a molecule to break up into its constituent elements when energy is absorbed. This occurs too soon for the molecule to reradiate the energy.

predistortion See *preemphasis*.

pre-Dolby 1. To record a tape with Dolby compression. 2. A tape that has been recorded with Dolby compression.

preemphasis In frequency modulation, the introduction of a rising-response characteristic. (Response rises as modulation frequency increases.) Also called *accentuation, predistortion,* or *preequalization*. Compare *deemphasis*.

preequalization 1. In sound recording and reproduction, equalization during recording. Compare *postequalization*. 2. See *preemphasis*.

preferred tube types Electron-tube types which satisfy the largest number of application requirements and therefore reduce the number of separate types that must be stocked.

preferred values of components A number system used by the EIA for establishing the values of composition resistors and small fixed capacitors. For an entire range successive values are separated by a multiple of 10 to a power. Thus, 10%-tolerance resistors are obtainable in ratings of 0.33 Ω, 3.3 Ω, 33 Ω, 330 Ω, 3300 Ω, 33K, 330K, and 3.3M.

prefix multiplier See *multiplier prefix*.

prefix notation As used with complex expressions involving many operators and operands, a type of notation in which the expressions, rather than containing brackets, are given a value according to the relative positions of operators and operands.

preform 1. A small wafer, usually dry-pressed from powdered plastic, from which is heat molded the body of a component, such as a capacitor or resistor. Also called a *pill* or *biscuit*. 2. The preformed slab used in molding a phonograph disk. 3. To shape a moldable circuit before fixing the final configuration or package.

premedian Situated in front of the center (usually of the body).

premix A molding compound of reinforced plastic.

prerecorded disk A phonograph disk on which a recording already has been made, i.e., a *recorded disk*.

prerecorded tape Magnetic tape on which a program or data has been recorded. Also called *recorded tape*.

P-region See *P-layer*.

prescaler A device operated ahead of a counter to establish a new, usually higher-frequency, range over which frequency measurements can be made.

preselector A tuned or untuned rf amplifier operated ahead of a radio or TV receiver to boost the sensitivity of the receiver.

presence In sound reproduction the quality of being true to life. Some

listeners use the term to describe the effect of boosted upper-midrange frequencies in music.

preset counter A pulse counter that delivers one output pulse for a number of successive input pulses determined by the settings of counter-circuit controls. Thus, a preset counter might give an output pulse for each train of 125 input pulses.

preset element In automation and control, an element which can be preset to a given level or value and to which other elements may then be referred.

preset switch In the circuit of a preset counter, a multiposition rotary switch that can be set to determine the number of input pulses which must be received for the circuit to deliver one output pulse.

preset thyratron In the circuit of one type of preset counter, a thyratron operated to deliver one output pulse for a predetermined number of successive input pulses.

preshoot A downward-moving transient pip that precedes the rise of a pulse.

preshoot amplitude The peak voltage of a preshoot, measured from the zero line to the valley of the preshoot.

preshoot time The width of a preshoot, measured along the horizontal base line (time axis).

pressing 1. A process by which phonographic disks are fabricated from plastic. 2. A disk pressed from plastic.

press-to-talk switch A switch in a microphone or on the end of a control cord and used to switch-on a transmitter, telephone, or recorder when the operator wishes to speak.

pressure Abbreviation, *P* or *p*. Force per unit area. Pressure may be expressed in any appropriate units of force and area, e.g., newtons per square meter, pounds per square inch, grams per square centimeter, etc.

pressure amplitude The pressure caused by an acoustic disturbance. Usually measured in dynes per square centimeter.

pressure capacitor An enclosed fixed or variable capacitor whose breakdown voltage is increased when the air pressure inside the container is increased.

pressure contact 1. Electrical contact made by pressing two conducting surfaces together (to complete a circuit). 2. A contact (see *contact, 1*) for obtaining the condition described in 1, above.

pressure-gradient microphone See *pressure microphone*.

pressure microphone A microphone that receives sound waves at only one side of its diaphragm. This one-sided exposure results in the displacement of the diaphragm by an amount proportional to the instantaneous pressure of the sound waves.

pressure pad In a tape recorder, a small felt pad that holds the tape against one of the heads.

pressure pickup See *pressure transducer*.

pressure roller In a tape recorder, a rubber-tired roller that presses the tape against the capstan.

pressure-sensitive adhesive An adhesive, such as that found on one side of cellophane tape, press-on drafting-symbol transfer sheets, and press-on printed-circuit elements, which adheres to a surface against which it is pressed.

pressure switch A switch that is opened or closed by a change in pressure within a system.

pressure transducer A sensor for converting pressure into proportionate current or voltage. There are several types. Some employ strain gauges; others, piezoelectric crystals, potentiometers, and other variable elements.

prestore To place data in memory ahead of time, before it is intended for use.

pretuned stage A stage, such as one in an i-f amplifier or single-frequency receiver, that is preset to a frequency rather than being continuously tuned.

prf Abbreviation of *pulse repetition frequency*.

pri Abbreviation of *primary*.

primaries See *primary colors*.

primary 1. Primary coil. 2. Primary winding. 3. Primary standard.

primary battery A battery composed of primary cells.

primary capacitance 1. The distributed capacitance of the primary winding of a transformer whose secondary winding is unloaded. Compare *secondary capacitance, 1*. 2. A series or shunt capacitance used to tune the primary coil of an i-f or rf transformer. Compare *secondary capacitance, 2*.

primary cell An electrochemical cell which is self-initiating, i.e., it doesn't require charging to start its generation of electricity, e.g., a dry cell. Primary cells are not generally considered rechargable. Compare *storage cell*. Also see *cell, dry cell, standard cell*.

primary circuit 1. The circuit associated with the primary winding of a transformer. 2. Input circuit.

primary coil See *primary winding*.

primary colors See *color primary*.

primary current The current flowing in the primary winding of a transformer. Also called *transformer input current*. Compare *secondary current*.

primary electron The electron possessing the greater amount of energy after a collision between two electrons. Compare *secondary electron*.

primary emission Emission arising directly from a source, such as the cathode of an electron tube. Compare *secondary emission*.

primary frequency standard An equipment for generating precise frequencies. The heart of the device is usually a highly stable 50- or 100-kHz crystal oscilltor which can be referred to a time standard and periodically corrected. A string of multivibrators, together with harmonic amplifiers and buffers, subdivide the basic frequency to provide a variety of precise lower frequencies, some in the af spectrum. These low frequencies, in company with the harmonics of the 100 kHz oscillator, provide highly accurate frequency-spectrum check points. Compare *secondary frequency standard*. Also see *primary standard*.

primary impedance 1. The impedance of the primary winding of a transformer whose secondary winding is unloaded. Compare *secondary impedance, 1*. 2. An external impedance presented to the primary winding of a transformer. Compare *secondary impedance, 2*.

primary inductance The inductance of the primary winding of a transformer whose secondary winding is unloaded. Compare *secondary inductance*.

primary keying Keying of a radiotelegraph transmitter by making and breaking the power-line input to the power transformer. The method is practical only if the transformer does not supply the tube filaments.

primary kVA The kilovolt-amperes in the primary circuit of a transformer. Compare *secondary kVA*.

primary measuring element A detector, sensor, or transducer that performs the initial conversion in a measurement or control system. Such an element converts a phenomenon into a signal which can be transmitted to appropriate instruments for translation and evaluation.

primary power Power in the primary circuit of a transformer. Also see *primary kVA* and *primary VA*. Compare *secondary power*.

primary radiator The driven element of an antenna system that in-

corporates parasitic elements or a reflector.

primary resistance The dc resistance of the primary winding of a transformer. Compare *secondary resistance*.

primary standard A usually stationary source of a quantity (e.g., capacitance, frequency, inductance, resistance, etc.), which is so precise and is maintained with such care that corresponding quantities may be compared with it in complete confidence that any discrepancies are real. Compare *secondary standard*.

primary turns Symbol, N_p. The number of turns in the primary winding of a transformer. Compare *secondary turns*.

primary utilization factor Abbreviation, UF_p. For a transformer in a rectifier circuit, the utility factor of the primary winding (ratio of dc power output to primary volt-amperes). For half-wave rectification, $UF_p = UF_s N/P \sqrt{2}$, where UF_s is the secondary utilization factor; N, the number of rectifier diodes supplying a resistive load; and P, the number of primary phases. Numerically, the primary utilization factor is higher than the secondary utilization factor and less than one. Also see *secondary utilization factor* and *utility factor*.

primary VA The volt-amperes in the input circuit of a transformer. Compare *secondary VA*.

primary voltage The voltage across the primary winding of a transformer. Also called *transformer input voltage*. Compare *secondary voltage*.

primary winding The normal input winding of a transformer. Also called *primary coil*. Compare *secondary winding*.

prime meridian See *zero meridian*.

prime mover A machine, such as a gas engine, steam engine, or water turbine, that converts a natural force or material into mechanical power.

prime number A number that is exactly divisible only by itself and one. Examples: 2, 5, 59, 241, 487.

primitive oscillation period In a complex oscillation waveform, the shortest period for which a definite repetition occurs; the highest fundamental frequency.

principal axis The line passing through the center of the spherical part of a lens or mirror.

principal focus The focal point of rays arriving parallel to the axis of a mirror or lens.

principal mode See *dominant mode*.

principal ray The path described by an electron entering an electron lens parallel to the lens' axis, or by an electron leaving this lens parallel to the axis.

principle of uncertainty See *uncertainty principle*.

print 1. The material transferred from a typewriter onto paper. 2. The command, in a computer system, that causes data to be placed on paper or onto the output screen. 3. The alphanumeric output of a computer or data terminal.

printed capacitor A two-plate capacitor formed on a printed circuit.

printed circuit A pattern of conductors (corresponding to the wiring of an electronic circuit) formed on a board of insulating material, such as a phenolic, by photo-etching, silk-screening of metallic paint, or by the use of pressure-sensitive preforms. The pigtails of discrete components are soldered to the printed metal lines at the proper places in the circuit, or the components can be formed along with the conductors. Also see *etched circuit*.

printed-circuit board A usually copper-clad plastic board used to make a printed circuit.

printed-circuit lamp A baseless lamp having flexible leads for easy

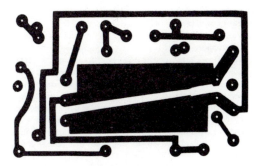

Printed circuit.

soldering or welding to a printed circuit.

printed-circuit relay A usually small relay provided with pins or lugs for easy solder-connection to a printed circuit.

printed-circuit switch A rotary switch whose contacts and contact leads are printed on a substrate.

printed coil A flat, spiral coil formed on a printed circuit.

printed component A component (see *component, 1*) that can be formed on the substrate of a printed circuit. See, for example, *printed capacitor, printed coil, printed resistor.*

printed display See *data printout, 2.*

printed element See *printed component.*

printed inductor See *printed coil.*

printed resistor A resistor printed or painted on a printed circuit.

printed wiring The printed or etched metal lines that serve as the conductors in a printed circuit.

printer In computer and calculator practice and in measurement procedures, a readout device that prints a permanent record of output data. Also see *data printout, 2* and *printing digital voltmeter.*

printing calculator An electronic calculator that supplies a printed record of the results of a calculation, or for a programmable calculator, the results, a record of program steps, and plots of curves.

printing digital voltmeter Abbreviation, *PDVM*. A digital voltmeter that delivers a printed record of a voltage reading in addition to the usual digital readout of the voltage.

printing telegraph 1. A telegraph that prints the received message on a tape or sheet. 2. Teletype.

printing wheel See *print wheel.*

print format Usually specified by a computer program, the form of information to be presented by a printer.

printout See *data printout, 1, 2.*

print-through In prerecorded magnetic tape stored on a reel, the transfer of magnetism between layers of the rolled-up tape. (Also, *print-through.*)

print wheel In a wheel printer, the rotatable wheel on whose rim the letters, numbers, and other symbols are inscribed in relief.

priority indicator A code used to specify the order of importance of a message in a group of messages to be sent (pertains to data transmission).

priority processing In multiple programming practice, a system for ascertaining the order of processing for different programs.

prism A three-flat-faced, triangular cross-section block of glass or other transparent material which refracts light passing between face through it. See *refraction.*

privacy equipment Collectively, devices, such as speech scramblers, which provide secrecy in radiotelephone communications.

privacy switch In a *telephone amplifier*, a switch (usually a pushbutton) for muting outgoing messages.

private automatic exchange Abbreviation, *PAX*. A dial telephone system for use within an organization and having no connection to the central office. Compare *private branch exchange.*

private branch exchange Abbreviation, *PBX*. A telephone system, complete with a private manually operated switchboard and individual telephone sets, installed and operated on private premises but having trunk-line connection to the central office. Compare *private automatic exchange.*

private line 1. A communications circuit in which the use is limited, by electronic means, to certain subscribers. 2. A subaudible-tone system of restricting the subscribers to a communications system. The tone frequency is predetermined. For access to the system, a transmitted signal must contain the tone of the appropriate subaudible frequency, in addition to the voice or other information.

probability 1. The branch of mathematics concerned with the likelihood of an event's occurrence. It has many applications in quality control and physics. 2. The mathematical likelihood that an event will occur. If the event can happen in p ways and can fail in f ways, and if one of the possible $(p + f)$ ways is as likely to occur as any other, the probability that it will occur is equal to $p/(p + f)$, and the probability that it will not occur is equal to $f/(p + f)$. In the simple example of tossing a perfect coin, the probability is 0.5, which means that there is a 50% chance (one in two) of heads or tails showing on each toss.

probable error Abbreviation, *PE*. The value of error above and below which all other error values are equally likely to occur.

probe 1. A usually slender pencil-like implement with a pointed metal tip and flexible, insulated lead, used to make contact with live points in a circuit under test, e.g., voltmeter probe oscilloscope probe. 2. A device, not necessarily resembling a stylus, used to pick off a voltage or current at a desired point, e.g., waveguide probe. 3. A pickup device shaped like a probe for insertion into close quarters, e.g., a probe thermistor.

probe meter See *probe-type voltmeter.*

probe thermistor A thermistor of slender construction for insertion into an area in which the temperature is to be monitored or controlled. Also called *thermistor probe.*

probe thermocouple A thermocouple in the form of a slender probe for insertion into close quarters for temperature sensing or temperature control.

probe tip See *prod.*

probe-type voltmeter A voltmeter installed in a long probe or wand. Kilovoltmeters are sometimes constructed in this fashion, with a long multiplier resistor housed in the probe.

probing A process for locating, or determining the existence of, external manmade interference in a radio communications circuit.

problem-oriented language A general term for a computer program language that allows the user to write programs as statements in terms applicable to the field of interest, e.g., COBOL's statements in English for problems relating to business.

procedure-oriented language A general term for a computer programming language with which describing algorithms for solving problems is made relatively simple.

process control The control of a process, such as one of manufacturing, by (usually analog or hybrid) computers.

processor 1. A circuit or device employed to modify a signal in response to certain requirements, e.g., clipper, waveshaper, etc. 2. A data processor (see *data-processing machine*). 3. Sometimes, *central processor*.

prod The metal tip of a probe (see *probe, 1*).

product The result obtained when two or more numbers or terms are multiplied by each other. Thus, for $e = a \times b \times c \times d$, e is the product.

product detector A detector circuit whose output is the product of two signals applied simultaneously to the circuit. In a single-sideband receiver, for example, one of the signals is the incoming signal; the other, the signal from the local beat-frequency oscillator.

production lot A manufactured set of components, circuits, or systems, intended for sale. All of the units in the production lot are identical. The finished product is suitable (presumably) for consumer use.

production unit One unit in a production lot; a finished unit, ready for use by a consumer.

product modulator A modulator whose output is equal or proportional to the product of carrier voltage and modulating voltage ($EcEm$).

product of sine waves The result of multiplying one sine wave by another with attention being paid to power factor. In the case of a resistive circuit, where the power factor is one, all values of EI are positive and equal to true power. A product wave has negative half-cycles when the circuit contains reactance.

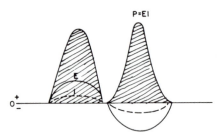

Product of sine waves.

product of vectors 1. For two vectors in polar form, their product is obtained by multiplying the vector magnitudes and adding the angles. Thus, $A\alpha \times AB(\beta + \beta)$. 2. For two vectors in rectangular form, their product is obtained algebraically: $(a + jb)(c + jd) = (ac - bd) + j(ad + bc)$.

professional engineer A person who is licensed by a state board of examiners to practice engineering independently. Also see *PE* and *registered professional engineer*.

program In computer practice, a detailed sequence of instructions representing an algorithm (the necessary steps in solving a problem) that can be implemented by a computer.

program address counter See *instruction register*.

program amplifier A broadcast preamplifier used at the studio or a remote location.

programatics The study of computer programming.

program card One of a set of cards—and usually containing one instruction—on which is punched a computer program.

program circuit In wire telephony, a wideband circuit capable of handling music and other wide-range material.

program compatibility The compatibility of a program written for one computer with another computer.

program controller In a central processor, a unit that controls the sequence and execution of program instructions.

program counter See *control register*.

program file A flexible reference system for software library maintenance.

program flowchart A representation of a computer program in the form of a flowchart. Each function and transition point is indicated by a box in the chart. A user can follow the flowchart and determine the outcome of the program for any given set of input parameters.

program library Generally, a collection of computer or programmable-calculator programs. Usually it means the collection of program used in a given computer system and is often a software package supplied by the hardware vendor. It may also be a catalog of programs with instructions for their use.

programmable calculator A calculator which, like a computer, may be programmed to perform a chain of operations in a given order repetitively. Examples: Types HP-67 and TI-59.

programmable read-only memory In a computer, a ROM that can store a program by use of a *PROM programmer*.

program maintenance The ongoing correcting, updating, and modification of computer programs belonging to a system.

programmed dump A dump that occurs during a program run per a program instruction.

programmed halt During a computer program run, a temporary cessation resulting from an interrupt or halt instruction.

programmed instruction See *macro instruction*.

programmed timer See *cycle timer*.

programmer One who writes computer programs (from specifications that are not necessarily his own).

program modification 1. In computer programming practice, a change in the effect of instructions and addresses during a program run by performing arithmetic and logical operations on them. 2. Rewiring, or adding a patch to, a computer program. Also see *patch, 3*.

program register See *control register*.

program segment A unit within a computer program that is stored with others in memory at the time of the program's execution, or sometimes, as *overlays* loaded individually when the entire program exceeds memory capacity.

program specification A description of the steps involved in the solution of a problem, from which a programmer devises a computer program.

program step An instruction in a computer program.

program tape In computer practice, a magnetic or paper tape containing programs for a system or application.

program timer 1. A programmed timer (see *cycle timer*). 2. A timing unit which controls the duration of a program.

progression An increasing or decreasing sequence of numbers, in which each term is related to the preceding term in a certain fixed way. See, for example, *arithmetic progression, geometric progression,* and *mathematical series*.

progressive scanning Television scanning in which lines are traced in succession. Modern television practice employs *interlced scanning* instead.

projected cutoff For an amplifier circuit, the operating point at which crossover distortion vanishes. The dc bias voltage (grid or gate) required for projected cutoff is somewhat lower than the value corresponding to conventional cutoff of plate or drain current.

project engineering A field of electrical or electronic engineering dealing with the coordination of a complete project.

projection television Large-screen television for viewing by a relatively large group, accomplished by means of a projection tube and an optical system.

projection tube A cathode-ray tube, especially a TV picture tube, which produces a very bright image that can be projected onto a large screen by means of a lens system.

projector 1. A device that transmits a visible image onto a surface for reproduction. 2. In general, any device that transmits a signal into space.

PROM Abbreviation of *programmable read-only memory*.

promethium Symbol, *Pm*. A metallic element of the rare-earth group, produced artificially. Atomic number, 61. Atomic weight, 147(?). Formerly called *illinium*.

promethium cell A radioactive type of battery cell employing an isotope of promethium. Radioactive particles from this substance strike a phosphor, causing it to glow. Self-generating photocells then convert this light into electricity.

prompt In computer practice, a message received by an operator from an operating system or an individual program; e.g., in a program written in BASIC, it could be the query, DO YOU WANT INSTRUCTIONS? TYPE Y FOR YES... and so forth.

prompting In computer or programmed-calculator practice, the entry of a special, required variable when the machine halts and awaits such entry.

prong See *pin*.

prony brake An arrangement for measuring the mechanical power output of a rotating machine. The prony brake is a special form of friction brake consisting of a band passed around a pulley on the rotating shaft of the machine under test and held at each end by a spring balance. The effective force of friction is equal to $F_1 - F_2$, where F_1 and F_2 are the readings given by the two balances in pounds.

PROM programmer An electronic device which, through the use of a built-in keyboard, can be made to store a computer program in a PROM.

propagation 1. Extension of energy into and through space. Thus, radiant energy is propagated from and by its source. 2. A phenomenon resulting from propagation. Thus, light rays and radio waves may be spoken of as *propagations*.

propagation constant For waves transmitted along a line, a number showing the effect the line has on the wave. This is a complex figure, i.e., one containing a real component (the *attenuation constant*) and an imaginary component (the *phase constant*).

propagation delay 1. Symbol, *tpd*. In an IC gate, the time taken for a logic signal to be propagated across the gate. 2. In digital-circuit operation, the time required for a logic-level change to be transmitted through one or more elements.

propagation delay-power product See *delay-power product*.

propagation factor The ratio Eb/Ea, where Eb is the complex electric-field strength at a point to which a wave has been propagated, and Ea is the complex electric-field strength at the point of origin. Also called *propagation ratio*.

propagation loss The path loss of an electromagnetic disturbance between the transmitting and receiving antennas.

propagation mode See *waveguide mode*.

propagation ratio See *propagation factor*.

propagation time In digital-circuit operation, the time required for a binary bit to be transferred from one point to another in the system.

propagation velocity See *velocity of propagation*.

proportion A statement of equality between two ratios. For example, four resistances can be related to each other so that $R1$ is to $R2'$ as $R3$ is to $R4$ That is, $R1/R2 = R3/R4$.

proportional action An action, such as amplificaton or conversion, that produces an output signal which is proportional to the input signal.

proportional amplifier An amplifier in which the instantaneous output amplitude is directly proportional to the instantaneous input amplitude.

proportional control A voltage-regulation system in which the feedback correction voltage is proportional to the output-voltage error.

proportional counter A Geiger tube having a pointed-wire (or ball-tipped-wire) anode. The voltage developed across the load resistor is proportional to the number of ions created by the radioactive particles entering the tube.

protactinium Symbol, *Pa*. A relatively short-lived radioactive metallic element. Atomic number, 91. Atomic weight, 231. Formerly called *proto-actinium*.

protected location In computer storage, a location whose contents are protected from mutilation or erasure by making the location only usable through some special procedure.

protection In a multiple processing computer system, preventing interference between data or programs.

protective bias In the rf amplifier of a radio transmitter, external dc grid bias applied to prevent plate-current runaway when the contact bias due to the driving signal is lost. Self-bias can also function as protective bias.

protective capacitor A power-line bypass capacitor.

protective device 1. A component that breaks a circuit in the event of excessive voltage or current from the power supply. 2. A device that prevents excessive power from being delivered to a load by a driving circuit.

protective gap 1. A spark gap connected in parallel with a component or between a line and ground as protection against overvoltage surges. 2. A spark-gap-type lightning arrester.

protective resistor 1. A bleeder resistor connected in parallel with a filter capacitor in a high-voltage dc power supply to discharge the capacitor automatically, thus preventing electric shock. 2. A series resistor that limits the current going through a device.

protector 1. A fast-acting power-disconnect device, such as a circuit breaker or fuse, that acts to protect electronic equipment. 2. A device or connection, such as a safety ground or ground-fault interrupter which acts to protect an operator from electric shock. 3. Contact protector.

protium The light isotope of hydrogen, having an atomic mass of 1.

proto-actinium See *protactinium*.

protocol The method by which a procedure is followed; a uniform set of governing regulations. The protocol ensures proper operation of a system or network.

proton A positively charged particle in the nucleus of an atom. The proton is approximately 1840 times heavier than the electron.

proton rest mass Symbol, *mp*. The mass of a proton at rest; it is a constant equal to $1.672\ 6 \times 3M1\ 10^{-27}$ kg. Compare *electron rest mass*.

proton-synchrotron A synchrotron in which frequency modulation of the rf accelerating voltage accelerates protons to energies of several billion electronvolts.

prototype The preliminary design or model of a device or system.

proustite Crystalline silver arsenide trisulfide. Artificial crystals of this compound are used in tunable infrared-ray instruments.

proximity alarm A capacitance relay employed to actuate an alerting-signal device when an area is intruded upon or a person is too close to a protected object. Also called *intrusion alarm.*

proximity detector A device that indicates the presence of a body close to it. Such a device employs some form of circuit, such as that of a capacitance relay, which changes its operating characteristics when an object enters its field.

proximity effect The influence of high-frequency current flowing in one conductor upon the distribution of current flowing in an adjacent conductor.

proximity fuse An electronic device situated in the nose of a missile or other projectile. When the missile is near the target, the fuse transmits a signal which is reflected back from the target and detonates the missile.

proximity relay See *capacitance relay.*

proximity switch See *capacitance relay.*

PRR Abbreviation of *pulse repetition rate.*

PRV Abbreviation of *peak reverse voltage.*

PS Abbreviation of *power supply.*

ps Abbreviation of *picosecond.*

PSD Abbreviation of *phase-sensitive detector.*

psec Abbreviation of *picosecond.* (Also, *ps.*)

pseudo-Brewster angle Symbol, ψB. In the progression of radio waves, the angle at which the reflection coefficient passes through a medium with vertically polarized waves. It is so called because it corresponds to the *Brewster angle* in optics.

pseudocode In a computer system, an instruction or code symbol that affects the operation of the programming in an indirect manner.

pseudo-instruction In computer programming practice, data representing an instruction and requiring translation by a compiler or assembler.

pseudo-offlining During input/output operations in a computer system, maximizing hardware by disconnecting devices that are slow from the process in question.

pseudo operation In computer practice, an operation that, rather than being performed by hardware, is carried out by special software or by macroinstruction.

pseudo-random numbers Numbers that, although produced by a computer operating on an algorithm for their generation, are useful for an application requiring random numbers; truly random numbers are seldom produced by a methodology devised for their generation.

pseudoscopy Image inversion, as in the real image of a hologram.

pseudo-stereophonic effect A somewhat heightened binaural effect obtained when two loudspeakers are situated, relative to the listener, so that a transit-time difference of 1 to 30 milliseconds results.

psf Abbreviation of *pounds per square foot.* (Also, *lb per sq ft, lb/ft^2* and $lb \cdot ft^{-2}$).

psi 1. Abbreviation of *pounds per square inch.* (Also, *lb per sq in.* and $lb/in.^2$. 2. The 23rd letter (ψ) of the Greek alphabet. It is one symbol for phase difference and is used to designate angles, electric flux of induction, and total flux of electric displacement.

psia Abbreviation of *pounds per square inch absolute.*

psig Abbreviation of *pounds per square inch gauge.*

PSK Abbreviation of *phase-shift keying.*

PSM Abbreviation of *pulse-spacing modulation (pulse-interval modulation).*

psvm Abbreviation of *phase-sensitive voltmeter.*

pswr Abbreviation of *power standing-wave ratio.*

psychoacoustics A field of acoustics that involves the effects of various sounds on listeners.

PT Abbreviation of *Pacific time.*

Pt Symbol for *platinum.*

PTC Abbreviation of *positive temperature coefficient.*

Pt-Ir Symbol for *platiniridium.*

PTM Abbreviation of *pulse-time modulation.*

PTO Abbreviation of *permeability-tuned oscillator.*

PTP Abbreviation of *paper tape punch.*

PTT Abbreviation for *press-to-talk.*

PTV Abbreviation of *public television.*

p-type conduction In a semiconductor, current flow consisting of the movement of holes. Compare *n-type conduction.*

p-type material Semiconductor material that has been doped with an acceptor-type impurity and, consequently, conducts current via hole migration. Germanium, for example, when doped with indium, becomes p-type. Compare *n-type material.*

p-type semiconductor An acceptor-type semiconductor, i.e., one containing an excess of holes in its crystal lattice.

PU Abbreviation of *pickup.*

Pu Symbol for *plutonium.*

public-address amplifier A high-gain, high-power audio amplifier designed especially for the reproduction of speech and music at large gatherings.

public-address system A system of sound reproduction especially designed for use at large gatherings indoors or outdoors. The system includes microphones, public-address amplifier, control devices, loudspeakers, and sometimes recorders and playback devices. Also called *PA systems.*

puck drive In a tape recorder, a speed-reduction system for driving the flywheel from the shaft of the (high-speed) motor. In some machines, a rubber tire mounted on the flywheel is driven, through friction, by the motor shaft. In others, an intermediate rubber-tired wheel is placed between the motor shaft and the rim of the flywheel.

puffer A meter or bridge for measuring small values of capacitance (from the spoken abbreviation *pf* for picofarad). Also called *millipuffer.*

pulldown Descriptive of a circuit, device, or individual component used to lower the value (e.g., of impedance, of a circuit to which it is connected.

pull-in current The current required to close a relay.

pulling The abnormal tendency of one circuit to cause another to slip into tune with it. This often results from coupling (intended or accidental) which is too tight. Thus, when two oscillators feed a common circuit, such as a mixer, one might *pull* the other into tune with itself.

pull-in voltage The voltage required to close a relay.

pull switch A mechanical switch actuated by a pulling action.

pullup Descriptive of a circuit or component used to raise the value (e.g., of impedance) of a circuit to which it is connected.

pulsating current See *pulsating direct current.*

pulsating dc voltage A dc voltage that periodically rises and falls between zero and a maximum value (or between two positive or negative values) without changing polarity. Thus, it is possible to have either a pulsating positive voltage or a pulsating negative voltage.

pulsating direct current A direct current that periodically rises and falls between zero and a maximum value (or between two positive or negative values) without changing polarity. Thus, it is possible to have either a pulsating positive current or a pulsating negative current. Also see *direct current.*

pulsating voltage See *pulsating dc voltage.*

pulsating wave See *pulsating direct current* and *pulsating dc voltage.*

pulse A transient signal that is usually of short duration, constant amplitude, and one polarity. A typical example is a narrow positive or negative spike. See Appendix B.

pulse amplifier An amplifier having fast rise and fall times, wide frequency response, and low distortion, for amplifying steep-sided pulses of short duration.

pulse-amplitude modulation Abbreviation, *PAM*. A type of pulse modulation in which a pulse carrier is amplitude modulated by the modulating signal.

pulse bandwidth For an amplitude pulse, the minimum bandwidth occupied. The faster the rise and/or decay times of a pulse, the greater the bandwidth. The greater the pulse frequency, the greater the bandwidth.

pulse code A code in which groups of pulses represent digits.

pulse-code modulation Abbreviation, *PCM*. A type of pulse modulation in which a signal is periodically sampled, each sample being subdivided into increments transmitted as a binary code.

pulse-count divider A circuit or device which receives an input of a certain number of pulses (or pulses per second) and delivers an output that is a function of that quantity. See, for example, *divide-by-seven circuit* and *divide-by-two circuit*.

pulse counter A circuit or device that indicates the number of pulses presented to it in a given time interval, or the total without respect to time.

pulse counting Counting pulses in a sequence. At low speed (pulse repetition rate), this may be done with an electromechanical dial-type counter. At high speed, however, a fully electronic circuit is required.

pulse delay circuit A monostable multivibrator that has been adapted to deliver its single output pulse a predetermined time after the input pulse has been applied.

pulse droop Pulse distortion represented by a sloping top on the corresponding curve.

pulsed ruby laser A laser in which flashes (pulses) of high-intensity light excite a ruby rod.

pulse duration The period during which a pulse exists, i.e., its width on a display.

pulse-duration modulation Abbreviation, *PDM*. A type of pulse-time modulation in which pulse duration is varied by the modulation. Also called *pulse-length modulation* and *pulse-width modulation*.

pulse equalizer A monostable multivibrator. It is so called because it delivers pulses of equal amplitude, shape, and width, even when it receives trigger pulses of different kinds.

pulse fall time The time required for the trailing edge of a pulse to fall from 90 to 10 percent of its peak amplitude. Compare *pulse rise time*.

pulse-frequency modulation Abbreviation, *PFM*. Modulation of the repetition rate of a train of pulses.

pulse generator A signal generator that produces pulses. A general-purpose generator of this sort will produce pulses of adjustable amplitude, duration, shape, and repetition rate.

pulse-height discriminator A circuit or device that passes only pulses whose amplitude exceeds a predetermined level.

pulse interval The interval between successive pulses.

pulse-interval modulation Abbreviation, *PIM*. Modulation of the spacing between pulses.

pulse inverter A single-stage, wideband, low-distortion, fast-rise-time amplifier of the grounded-cathode, common-emitter (or common-source) type. Such an amplifier provides 180-degree phase shift between input and output pulses (inverts the input pulses).

pulse jitter In a pulse train, a disturbance characterized by random changes in the spacing between pulses.

pulse-length modulation See *pulse-duration modulation*.

pulse load The impedance load for a pulse generator.

pulse mode Pulse modulation.

pulse modulation 1. Modulation employing pulses as the modulating signal. 2. Modulation of a "carrier" consisting of pulses. See, for example, *pulse-amplitude modulation, pulse-code modulation, pulse-duration-modulation, pulse-length modulation, pulse-numbers modulation, pulse-time modulation,* and *pulse-width modulation*.

pulse modulator 1. A modulator that delivers power or voltage pulses for modulating a carrier. 2. A device that modulates pulses (see *pulse modulation, 2*).

pulse narrower A circuit or device that reduces the duration (width) of a pulse.

pulse-numbers modulation Abbreviation, *PNM*. A type of modulation of a pulse carrier based upon variation of pulse density (i.e., pulses per unit time).

pulse operation Intermittent operation of a circuit, in the form of discrete pulses.

pulse oscillator Any oscillator with an output that consists of a series of pulses.

pulse-position modulation Abbreviation, *PPM*. Modulation of the time position of pulses.

pulse rate See *pulse repetition rate*.

pulse ratio The ratio of pulse height to pulse width.

pulse repetition frequency Abbreviation, *PRF*. See *pulse-repetition rate*.

pulse-repetition rate Abbreviation, *PRR*. The number of pulses per unit time; usually *pulses per second* (pps).

pulse rise time The time required for the leading edge of a pulse to rise from 10 to 90 percent of its maximum amplitude. Compare *pulse fall time*.

pulse scaler A circuit that is actuated by the reception of a definite, predetermined number of input pulses.

pulse-shaping circuit 1. A circuit for producing a pulse from a wave of some other shape (e.g., sine wave or square wave). 2. A circuit for tailoring a pulse to desired shape, amplitude, etc.

pulse spacing The interval between successive pulses.

pulse-spacing modulation Abbreviation, *PSM*. See *pulse-interval modulation*.

pulse-steering diode In a flip-flop circuit, a diode through which the trigger pulse must pass to switch the circuit. Because of the unidirectional conductivity of a diode, pulses of only one polarity are passed.

pulse stretcher 1. A shaping circuit that widens a pulse. 2. A circuit, such as a special monostable multivibrator, that generates a pulse that is wider than the trigger pulse.

pulse tilt The sloping of the normally flat top of a pulse either up or down. Also see *pulse droop*.

pulse time See *pulse duration*.

pulse-time modulation Abbreviation, *PTM*. Modulation of the interval between constant-amplitude, constant-duration pulses. Also called *pulse-position modulation*.

pulse train A series of successive pulses of usually one kind.

pulse transformer A transformer designed to accommodate the fast rise and fall times of pulses and similar nonsinusoidal waveforms. Such transformers often employ special core materials and are made using special winding techniques.

pulse transmitter 1. A device that transmits a series of pulses. 2. A pulse-modulated transmitter. 3. A pulse modulator.

pulse waveform The shape of a pulse. This includes various forms, ranging from positive or negative half-sinusoids, through rectangles, to thin-line spikes.

pulse width The horizontal dimension of a pulse, i.e., its duration.

pulse-width modulation Abbreviation, *PWM*. A type of pulse-time modulation in which pulse duration is varied by the modulation. Also called *pulse-duration modulation*.

pulse-width modulation—frequency modulation See *PDM—FM*.

pump 1m In a parametric amplifier, the oscillator that supplies the signal that periodically varies the reactance of the varactor. 2. The pumping signal in 1, above. 3. To perform the operation (pumping) described in 1, above. 4. To increase the energy level of an atom or molecule (by exposing it to electromagnetic radiation) to such an extent that oscillation or amplification occurrs. A ruby laser, for example, produces its intense, coherent beam as a result of pumping. 5. The radiation employed to pump an atom or molecule. 6. The device producing the radiation required to pump an atom or molecule.

pump frequency The frequency of a pump voltage.

pumping A method of laser actuation. A series of pulses, at the resonant frequency of the lasing material, is injected to cause laser output.

pump oscillator An oscillator for producing a pump voltage.

pump voltage The voltage of a pumping signal. Also see *parametric amplifier* and *pump, 1, 2.*

punch 1. A machine for punching cards or tape for data processing. 2. To perform the operation in 1, above. 3. A tool for cutting holes in metal chassis, panels, and boxes for electronic equipment. 4. High signal strength.

punch card See *punched card.*

punched card A card that is punched with holes in particular places (columns and rows) according to a code, to store information and data. Also see *card, 3.*

punched-card machine See *card machine.*

punched card verifier A machine capable of checking punched cards for the validity of data thereon.

punched tape Plastic or paper tape punched with holes according to a code to record data or Morse-code characters. Playback is effected by running the tape through a *reader* in which contact fingers sense the holes.

punched-tape reader A device for reading punched tape. In an opto-electronic type of reader, a light beam passing through the holes as they pass a lamp actuates the circuit through a photocell. In an electromechanical type of reader, a metal feeler (see *pecker*) reaches through a passing hole and touches a metal contact to close the circuit.

punched-tape recorder A device that records data in the form of holes punched into a paper tape. A reperforator.

punching positions Places for holes on a punched card, i.e., where rows and columns intersect.

punching station In a card punch (*punch, 1*), the site of the work done by the machine.

punch track In a card punch (*punch, 1*), that which moves the cards.

punchthrough In a transistor, the potentially damaging condition resulting when the reverse bias of the collector is increased to a voltage high enough to spread the depletion layer entirely through the base. This tends to connect the emitter to the collector.

punchthrough region The conduction region associated with higher-than-punchthrough voltage, in which transistor current is excessive.

Also see *punchthrough* and *punchthrough voltage*.

punchthrough voltage The voltage that causes punchthrough.

puncture voltage See *breakdown voltage, 1.*

Pupin coil One of several loading coils which are inserted at intervals in series with a telephone line to cancel line-capacitance effects and, thus, improve the clarity of speech.

pure tone A tone having no harmonics (overtones).

pure wave An undistorted wave.

purging The removal of an undesired gas or other substance from a system by introducing a material to displace it.

purifier A power-line operated ac degaussing coil which may be manually rotated in front of a color-TV picture tube to demagnetize the latter. Also called *degausser*.

purity 1. In color TV, complete saturation of a hue. 2. In a waveform, complete freedom from distortion.

purity adjustment In a color-TV picture tube, adjustment of each purity control for pure color.

purity coil A variable-current coil around the neck of a color-TV picture tube for adjusting color purity.

purity control For a purity coil, the variable resistor that controls the current for color correction.

purity magnet A ring-magnet collar around the neck of a color-TV picture tube for adjusting, by rotation, color purity.

purple plague Corrosion that occurs when aluminum and gold are placed in contact.

pushbutton See *button, 3.*

pushbutton switch See *button, 3.*

pushbutton tuner A radio or television tuner utilizing pushbutton tuning.

pushbutton tuning The tuning of a circuit to various frequencies in single steps by means of pushbutton switches.

pushdown list In data processing, a method of amending a list whereby new items entered at the top displace those occupying the list one position down.

push-in terminal A circuit contact or tie point, usually of thin, springy material, which is inserted into a hole in a *perforated board*.

push-pull A circuit in which two active devices are used. The inputs and outputs are both placed in phase opposition. In the output circuit, even harmonics are cancelled, while odd harmonics are reinforced.

push-pull amplifier An amplifier stage in which, for increased power output, two tubes or transistors are operated 180 degrees out of phase with each other in opposite halves of a symmetrical circuit. Also see *push-pull circuit.*

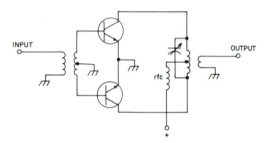

Push-pull amplifier.

push-pull circuit A symmetrical circuit in which two active elements (tubes or transistors) operate on separate halves of the input-signal cycle and deliver a combined output signal.

push-pull deflection In an oscilloscope, the application of deflection voltage to a pair of deflecting plates 180 degrees out of phase with each other. For this purpose, the output amplifier in the horizontal or vertical deflection channel is a push-pull stage.

push-pull doubler See *push-push multiplier.*

push-pull microphone A set of two microphones, in which the audio-frequency outputs are in phase opposition.

push-pull multiplier A push-pull amplifier employed as a frequency multiplier, i.e., one with an input circuit tuned to frequency f, and output to nf. This type of multiplier is unsuitable for even-harmonic operation, but has some merit as an odd-harmonic multiplier (tripler, quintupler). Also see *push-push multiplier.*

push-pull oscillator An oscillator stage in which, for increased power output, two tubes or transistors are operated 180 degrees out of phase with each other in opposite halves of a symmetrical circuit. Also see *push-pull circuit.*

push-pull—parallel amplifier An amplifier stage in which tubes or transistors are connected in push-pull—parallel for increased power output. Also see *parallel-component amplifier, push-pull amplifier,* and *push-pull—parallel circuit.*

push-pull—parallel circuit A push-pull circuit in which tubes or transistors are connected in parallel with each other on each side of the circuit. This arrangement gives increased power output over that of the conventional push-pull circuit. See, for example, *push-pull—parallel amplifier* and *push-pull—parallel oscillator.*

push-pull—parallel oscillator An oscillator stage in which tubes or transistors are connected in push-pull—parallel for increased power output. Also see *parallel-component oscillator, push-pull oscillator,* and *push-pull—parallel circuit.*

push-pull recording A type of film sound track consisting of two side-by-side images 180 degrees out of phase with each other.

push-pull transformer A transformer having a center-tapped winding for operation in a push-pull circuit.

push-push A circuit in which two active devices are used. The inputs are connected in phase opposition, but the outputs are in parallel. The result is reinforcement of the even harmonics, and cancellation of the fundamental frequency and all odd harmonics.

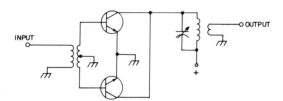

Push-push.

push-push circuit See *push-push multiplier.*

push-push multiplier A frequency-doubler circuit (two tubes or transistors) in which the inputs are connected in push-pull, and the outputs in parallel. The circuit produces twice the output afforded by one tube and doesn't produce the second harmonic cancellation peculiar to push-pull circuits.

push-to-talk switch See *press-to-talk switch.*

pushup list In data processing, a method of amending a list whereby new items are added at the end of the list; all other items retain their original positions. Compare *pushdown list.*

pV Abbreviation of *picovolt.*

PVC Abbreviation of *polyvinyl chloride.*

pW Abbreviation of *picowatt.*

PWM 1. Abbreviation of *pulse-width modulation.* 2. Abbreviation of *plated-wire memory.*

pwr Abbreviation of *power.*

pyralin See *cellulose nitrate.*

pyramidal horn antenna A rectangular horn antenna that is flared in two dimensions.

pyramidal wave See *back-to-back sawtooth.*

Pyrex A heat-resistant glass having numerous applications in electronics and chemistry.

pyroheliometer An instrument employed to measure infrared radiation.

pyroelectricity In certain crystals, electricity generated by temperature change.

pyroelectric lamp See *Nernst lamp.*

pyroelectric material A crystalline material that generates an output voltage when heated.

pyrolysis The process whereby heat changes a substance into one or several different substances by rearranging its atoms.

pyromagnetic effect In a material or circuit, the combined effect of heat and magnetism.

pyrometer An instrument, other than a thermometer, for the measurement of temperature. See, for example, *optical pyrometer.*

pyron detector An early semiconductor diode (crystal detector) utilizing a point contact between a metal (such as copper) and a crystal of iron pyrites.

Pythagorean scale A sound scale defining the relationship between various audio tones. If x and y are related by the Pythagorean scale and are adjacent in frequency, then $x = f^2$ *and* $y = f^3$, for some frequency f.

Pythagorean theorem A theorem of plane geometry. For a right triangle, with sides of lengths a, b, and c, where c is the side opposite the right angle, it is always true that $a^2 + b^2 = c^2$.

p-zone See *p-layer.*

PZT Abbreviation of *lead zirconate titanate.*

Q 1. The figure of merit of a capacitor, inductor, or *LC* circuit; = *X/R*, where *X* is reactance and *R* is resistance. 2. Symbol for *electrical charge*. 3. Occasional symbol for *selectivity*. 4. *Q-band*. 5. *Q output* (of a flip-flop).

q 1. Symbol for *quantity of electricity* (in coulombs). 2. Symbol for the *charge carried by an electron* (the charge carried by a hole is represented by −*q*). 3. Symbol for the *value of a quantum*. 4. Occasional abbreviation of *quart*.

Q adjustment The separate null adjustment for the *Q* value of a component being tested in an impedance bridge having separate resistive and reactive balances.

Q-antenna An antenna in which the transmission line (feeder) is matched in impedance to the center of the radiator by means of a *Q*-matching section.

QAVC Abbreviation of *quiet automatic volume control*.

Q-band The radio-frequency band 36 to 46 GHz. It is subdivided as follows: Qa, 36—38 GHz; Qb, 38—40 GHz; Qc, 40—42 GHz; Qd, 42—44 GHz; and Qe, 44—46 GHz.

Q bar One of the parallel metal tubes in a *Q*-matching section. Also see *Q antenna*.

Q booster See *Q multiplier*.

Q bridge An ac bridge used principally to determine the *Q* of capacitors and inductors. Bridges are usually employed for audiofrequency *Q* determinations, and resonant-type *Q* meters for radio-frequency *Q*.

QC Abbreviation of *quality control*.

QCE Abbreviation of *quality-control engineering* or *quality-control engineer*.

Q-channel In American (NTSC) color TV, the 508-kHz-wide green-magenta color information transmission band.

QCT Abbreviation of *quality-control technician*.

QCW In the local oscillator and associated circuitry of a color TV receiver, a 3.85 MHz CW signal of Q phase.

QCW signal In a color TV receiver, the component of the chrominance signal which is 90 degrees out of phase with the in-phase component.

Q demodulator In a color TV receiver, the demodulator which combines the chrominance signal and the color-burst oscillator signal to recover the Q signal (see *Q signal, 2*).

QED 1. Abbreviation of *quantum electrodynamics*. 2. Abbreviation of *quod erat demonstrandum* (which was to be demonstrated).

Q factor *See Q.*

QFM Abbreviation of *quantized frequency modulation*.

QM Abbreviation of *quadrature modulation*.

Q-matching section A linear, rf impedance-matching transformer consisting of two parallel lengths of metal tubing, used to match a feeder line to an antenna. The length and spacing of the tubes are such that the surge impedance of the matching section is the geometric mean of the feeder-line surge impedance and the radiation resistance of the radiator. Also see *Q antenna*.

Q meter A usually direct reading instrument for determining the *Q* of a capacitor, inductor, or *LC* circuit. Most *Q* meters are operated at radio frequencies, but audio-frequency instruments are available.

Q modulation Amplitude modulation obtained by varying the effective *Q* of an rf tank in step with a modulating component. See *absorption modulation*.

QMQB Abbreviation of *quick make, quick break*.

Q multiplier A positive-feedback (regenerative) amplifier which, when its input is connected across a tuned *LC* circuit, increases the effective *Q* of the circuit and greatly sharpens its tuning response. The negative resistance presented by this amplifier reduces the effective positive resistance of the tuned circuit, thereby multiplying the *Q*.

Q output The reference output of a flip-flop.

Q phase A color-TV carrier signal which is 147 degrees out of phase with the color subcarrier.

Q point The point or points at which a load line intersects a device characteristic (such as the collector curve of a transistor or plate curve of a tube) and which identifies the quiescent operating point.

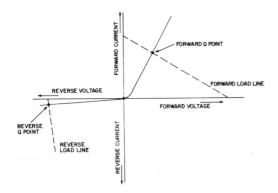

Q points on diode characteristic.

Q-section transformer See *Q-matching section.*

Q signal In color television, the quadrature component of the chrominance signal, equal to + 0.48 (R-Y) and to + 0.41 (B-Y), where *B* is the blue camera signal, *R* the red camera signal, and *Y* the luminance signal.

Q signals A series of three-letter groups, each beginning with the letter *Q*, used for simplified telegraph and radiotelegraph communication, and sometimes rapid voice communication (in radiotelephony), of commonly used phrases.

QSL card A card verifying communication with, or the reception of signals from, the station sending the card. Such verification is common in the amateur radio service and with some shortwave broadcast and CB stations.

QSO 1. Amateur radiotelegraph abbreviation for *two-way contact or communication.* 2. Abbreviation of *quasi-stellar object.*

Q spoiler A device or circuit that produces *Q* spoiling in a laser.

Q spoiling The technique of inhibiting laser action during an interval when an ion population excess is pumped up. When the laser is subsequently triggered by *Q* switching a more powerful pulse of light results than would be otherwise obtained.

Q switching A laser-switching action obtainable with Kerr cells or rotating reflecting prisms, which consists of holding the *Q* of the laser cavity to a low value during an ion-population buildup and then abruptly switching the *Q* to a higher value.

qt Abbreviation of quart. (Also, *q.*)

Q transformer See *Q-matching section.*

qty Abbreviation of *quantity.*

quad 1. A combination of four components—such as diodes, transistors, etc.—in a single housing. The components are usually carefully matched. 2. In a cable, a combination of four separately insulated conductors (sometimes, two twisted pairs) twisted together. 3. Abbreviation of *quadrant.* 4. Quad antenna. 5. Quadraphonic.

quad antenna An antenna consisting of two square loops cut to a quarter wavelength on a side and a quarter-wavelength around the periphery.

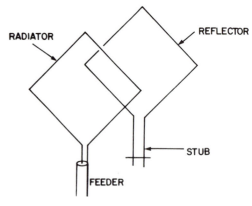

Quad antenna.

Only one loop is driven, the other acting as a parasitic reflector. Also called *cubical quad antenna.*

quadded cable See *quad, 2.*

quadding Redundancy obtained by connecting transistors or other components in series—parallel for enhanced reliability.

quad latch A set of four interconnected flip-flops, used for the storage of information.

quadrant 1. The quarter part of a circle. 2. One of the four parts formed on a plane surface by rectangular coordinates and designated I, II, III, and IV in a counterclockwise direction starting with the upper-right quadrant. 3. An altitude-measuring instrument.

quadrantal deviation That part of magnetic-compass deviation caused by the induction of transient magnetism into the horizontal soft iron of a vessel by the horizontal component of terrestrial magnetism.

quadrantal error See *quadrantal deviation.*

quadrant electrometer An electrometer whose principal parts are *quadrants* (a pillbox-shaped brass chamber split into four parts) and a *needle* (a flat, bowtie-shaped aluminum vane) suspended by a platinized quartz fiber between the quadrants.

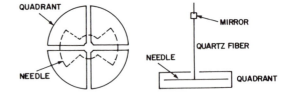

Quadrant electrometer.

quadraphonic sound Sound recording and reproduction involving four channels.

quadraphony Music recording or playback in which four distinct information channels are used. Also called *four-channel stereo.*

quadratic equation A second-degree equation, i.e., one in which the highest exponent is 2 (the square of an unknown), e.g., $ax^2 + bx + c = 0$.

quadratic formula A formula for solving quadratic equations of the form $ax^2 + bx + C = 0$, by substitution of coefficients:

$$x = \frac{-b + \sqrt{b^2 - 4ac}}{2a}$$

See *quadratic equation*.

quadrature 1. The state of (cyclic events or points) being 90 degrees out of phase. 2. In astronomy, a celestial formation of bodies 90 degrees apart.

quadrature amplifier An amplifier circuit which introduces a 90-degree phase shift. Such amplifiers are used in control devices, test instruments, transmitters, and color TV receivers.

quadrature axes The vertical axes in a complex plane, i.e., the $+j$ and $-j$ axes.

quadrature carrier See *Q phase*.

quadrature component 1. The reactive component of current or voltage. 2. A vector which is perpendicular to a reference vector. 3. The imaginary component in a complex expression.

quadrature current Reactive current ($I \sin \theta$) in an ac circuit.

quadrature detector An FM detector in which the FM signal and a fixed-frequency signal (both in quadrature, i.e., 90 degrees out of phase with each other) are applied simultaneously to separate grids of a special tube. Plate current flows in accordance with the time lag resulting from the quadrature's leading or lagging the frequency change at the control grid. Also called *gated-beam detector*.

quadrature modulation In-phase modulation of two carrier components having a 90-degree phase difference.

quadrature number An imaginary number, i.e., $\sqrt{-1}$, or the product of a real number and $\sqrt{-1}$.

quadrature-phase subcarrier signal See *QCW signal*.

quadrature portion In color TV, the portion of the chrominance signal having the same (or opposite) phase as that of the Q-signal-modulated subcarrier, and which may be 90 degrees out of phase with the in-phase portion.

quadrature sensitivity The sensitivity of a transducer to motions in a direction perpendicular to the normal axis of response.

quadrature voltage A voltage which is 90 degrees out of phase with another (reference) voltage.

quadrilateral 1. With four sides. 2. A four-sided polygon.

quadrillion One thousand trillion, i.e., 10^{15}.

quadripartite Having four parts.

quadriphonic Pertaining to a four-channel sound system. Two sets of two interrelated channels are used to obtain a panoramic sound reproduction effect.

quadripole network A four-terminal network, usually with input- and output-terminal pairs.

quadrivalent Having a valence of 4. Tin, for example, is quadrivalent. Also called *tetravalent*.

quadrupler 1. A rectifier circuit which delivers a dc output voltage approximately equal to four times the peak value of the ac input voltage. 2. An amplifier or other circuit which delivers an output signal of four times the frequency of the input signal.

quadruplex circuit A telegraph circuit in which two messages are carried in each direction simultaneously.

quadrupole 1. A combination of two dipoles, producing a force which varies inversely with the fourth power of distance. 2. A four-pole magnet used in some synchrotrons and linear accelerators to focus and bend a particle beam. 3. A system consisting of two dipoles of equal

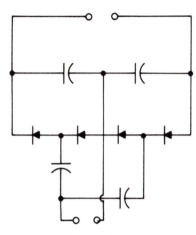

Quadrupler, 1.

and oppositely direct moment.

qual Abbreviation of *qualitative*.

qualification The quality-control or quality-assurance system in the production of components, circuits, or systems. Certain minimum requirements must be met for a device to obtain qualification.

qualitative test A test performed to determine the general mode of operation or the presence of certain factors, without regard to numerical values. Compare *quantitative test*.

quality 1. In audio-frequency practice, fidelity of transmission or reproduction. 2. The degree of conformity of a product to specifications.

quality assurance The outcome of measures taken to bring performance into conformity with specifications. See *quality, 2*.

quality control The surveillance of selection, manufacturing, and testing operations to insure conformity of a product to specifications.

quality-control engineer A professional skilled in quality-control engineering.

quality-control engineering The branch of engineering concerned principally with the technical methods of quality control and statistical methods of assessing quality (see *quality, 2*).

quality-control technician A technician whose principal duty is the performance of operations in the areas of incoming inspection, manufacturing support, and product testing, which promote quality in a product. Sometimes statistical evaluations are required. The QC tech's immediate responsibility is usually to a quality-control engineer.

quality engineering A field of electrical or electronic engineering that deals with quality assurance and control in the production of components, circuits, and systems.

quality factor See *Q*.

quality-factor bridge See *Q bridge*.

quality-factor meter See *Q meter*.

quant Abbreviation of *quantitative*.

quanta Plural of *quantum*.

quantimeter An instrument used to measure the quantity of X-rays to which a body has been exposed.

quantitative test A test performed to determine the numerical values (and their relationships) connected with observable phenomena. Compare *qualitative test*.

quantity 1. A parameter (e.g., collector current, grid voltage, etc.). 2. In calculations, a positive or negative real number.

quantization The division of the range of values of (1) a varying phenomenon, such as a wave, into a finite number of subrantes, or (2) the possible values for a variable into *quanta*, or discrete units; the verb is *quantize*.

quantization distortion Distortion introduced by the process of quantizing.

quantization error The difference between the actual values of quantities and their quantized values.

quantization noise See *quantization distortion*.

quantize To split a quantitative commodity such as energy into its smallest measurable incremental units.

quantized pulse modulation Pulse modulation involving quantization. Examples are *pulse-code modulation* and *pulse-numbers modulation*.

quantizer A circuit or device which selects the digital subdivision into which an analog quantity is placed, i.e., a sort of A/D converter.

quantometer An instrument for measuring magnetic flux.

quantum 1. Abbreviation, *q*. In quantum theory, the elemental unit of energy. The plural form is *quanta*. 2. The discrete unit derived by *quantization*.

quantum chromodynamics A term coined by Prof. Murray Gell-Mann for the theory of *quarks* and *gluons*.

quantum counter A radiation-counter tube with a window for the admission of light to the cathode.

quantum efficiency See *quantum yield*.

quantum electronics The branch of electronics concerned with energy states in matter.

quantum equivalence The principle that one electron is emitted for each photon absorbed by a material.

quantum jump The abrupt movement of a particle from one discrete energy state to another.

quantum level The orbit or ring occupied by an electron in an atom.

quantum mechanics The mechanics of quantum-theory phenomena.

quantum noise A noise signal arising from random variations in the average rate at which quanta impinge upon a detector.

quantum number A number that describes the energy level, or change in energy level, for a particle.

quantum theory The theory that the emission or absorption of energy by atoms or molecules occurs in steps rather than continuously. Each step is the emission or absorption of an energy packet $h\nu$ (called a *quantum*), where ν is the frequency of the radiation and h is the Planck constant. Thus, radiant energy is thought to be divided into quanta.

quantum transition The movement of an electron from one energy level to another within an atom.

quantum yield Symbol, *y*. The photoelectric efficiency of a light-sensitive surface in terms of the number of electrons emitted for each absorbed quantum of light.

quark A hypothetical particle having a fractional electrical charge; quarks are thought to be constituents of subatomic particles.

quarter-deflection method A method of measuring high-frequency resistance, involving the use of a sine-wave signal source, a standard noninductive variable resistor, and a square-law rf ammeter. With the standard resistance R set to zero, the current I_1 is noted. R then is adjusted to reduce the current to I_2, which is $I_1/4$. At this point, the unknown resistance R_x equals the setting of the standard resistor.

quarter-phase See *two-phase*.

quarter-phase system A two-phase system in which the generator or

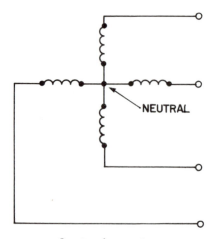

Quarter-phase system.

transformer phase windings are interconnected as shown. It is an easily unbalanced system.

quarter wave 1. The length of time corresponding to 90 electrical degrees in a wave disturbance. 2. The distance in space, or along a wire or feed line, corresponding to 90 electrical degrees in a wave disturbance.

quarter-wave antenna An antenna in which the radiator is a quarter-wavelength long at the operating frequency.

quarter-wave attenuator In a transmission line or waveguide, two energy-absorbing structures separated by an odd number of quarter wavelengths, so that the reflection from one structure is canceled by that from the other.

quarter-wave balun A balun employing quarter-wave elements. One form of this device consists of a grounded quarter-wavelength-long cylinder closed at one end and open at the other, for matching an unbalanced low-impedance line to a balanced high-impedance line.

quarter wavelength One-fourth a wavelength at a given operating frequency; $\lambda = 75/f$, where λ is a quarter-wavelength (in meters), and f is the frequency in megahertz.

quarter-wavelength line A transmission line or feeder which is a quarter-wavelength long at the operating frequency. Also called *quarter-wave line*.

quarter-wavelength matching stub An arrangement consisting of a quarter-wavelength-long parallel-wire section for matching the impedance of a nonresonant feeder line to that of an antenna. Also see *J antenna*.

quarter-wave monopole A nondirectional uhf vertical antenna requiring no ground. The radiator is a quarter-wavelength long, and so is an enlarged-diameter outer sleeve, which is connected to the outer conductor of the coaxial feeder. The two sections simulate a half-wave antenna.

quarter-wave plate A plate of double-refracting crystalline material whose thickness allows the introduction of a quarter-cycle phase difference between the ordinary and extraordinary components of light transmitted by it.

quarter-wave resonance Resonance at the operating frequency in a quarter-wave antenna.

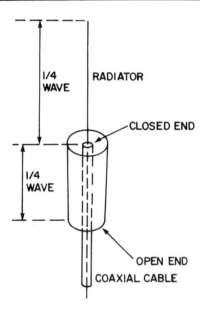

Quarter-wave monopole.

quarter-wave resonant line A section of transmission line (such as open-spaced line or coaxial cable) which is a quarter-wavelength long at the operating frequency. Such a section is useful in impedance matching and in various radio-frequency tests and measurements.

quarter-wave stub See *quarter-wave transformer.*

quarter-wave support In a coaxial line, a quarter-wave metal stub which may be used instead of an insulator to separate the inner and outer conductors.

quarter-wave termination In a waveguide, a set of two metal barriers separated by 90 electrical degrees. One barrier totally reflects the energy striking it. The other barrier allows some energy to pass through. Resonance occurs in the space between the barriers.

quarter-wave transformer A quarter-wave resonant line which is short-circuited at one end by an adjustable slider. This arrangement is useful for rf impedance matching at the frequency at which the line is a quarter-wavelength long.

quarter-wave transmission line See *quarter-wave line.*

quartic equation A fourth-degree equation of the form $a_0 x^4 + a^1 x^3 + a^2 x^2 + a^4 = 0$. Also called *biquadratic equation.*

quartz A mineral which is a variety of natural silicon dioxide, or an artificially grown material of the same sort. In the natural state, quartz occurs in hexagonal crystals having pyramidal ends. It has a number of important uses in electronics the most common of which is the manufacture of quartz crystals.

quartz bar A comparatively large, thick piezoelectric quartz plate employed in standard-frequency oscillators and in sharply tuned low-frequency filters. Common resonant frequencies are 50, 100, and 1000 kHz.

quartz crystal A natural or artificial crystal of quartz, usually self-contained in a solder-in or plug-in enclosure. The crystal serves as a highly stable LCR section and exhibits a powerful resonance at the frequency for which it is cut. It is used as the frequency determining element in precision oscillators.

quartz-crystal oscillator See *crystal oscillator.*

quartz-crystal resonator See *crystal resonator.*

quartz delay line An acoustic delay line employing quartz to transmit the sound waves.

quartz-fiber electroscope An electroscope employing a gold-plated quartz fiber instead of gold leaves.

quartz lamp A mercury-vapor lamp with a transparent quartz (instead of glass) envelope. Unlike glass, quartz readily passes the ultraviolet rays generated by the mercury discharge.

quartz oscillator See *crystal oscillator.*

quartz plate A piezoelectric plate cut from a quartz crystal. The plate is often called a *crystal.* Also see *crystal axes* and *crystal cuts.*

quartz resonator See *crystal resonator.*

quartz timepiece A watch or clock having as its control element a time-determining quartz crystal. Usually designated "quartz clock" and "quartz watch."

quasar Acronym for *quasi* stell*ar* radio source. A quasar is a starlike celestial object appearing near the limit of the observable universe, and is thought to move deeper into space at some fraction of the speed of light; it emits radio waves and intense ultraviolet and blue light.

quasi An adjective meaning *to some extent* or *like*, as in *quasi optical wave* (a radio wave that behaves like a light ray).

quasi instruction In a computer program, a data item appearing as an encoded instruction but which is not acted upon.

quasi linear feedback system A system in which the feedback elements are not entirely linear, but are substantially so.

quasi negative less positive, but not actually negative For example, a *negative-going* voltage is actually a positive voltage which has fallen to a lower value and is therefore quasi-negative, or negative with respect to a higher initial positive value.

quasi optical Behaving like light. The term is used to describe certain extremely short radio waves or other radiations which, like light rays, follow line-of-sight paths and can be directed, reflected, refracted, or diffused.

quasi optical path A line-of-sight path followed by very short radio waves, such as microwaves.

quasi positive Less negative, but not actually positive. For example, a *positive-going* voltage is actually a negative voltage that has fallen to a lower value and is therefore quasi-positive, or positive with respect to a higher initial negative value.

quasi-random A set of numbers considered to be random, but chosen according to an algorithm.

quasi scientific A term sometimes applied to the design of electronic systems or to the appraisal of circuit behavior, using an intuitive rather than analytical approach.

quasi sine wave A waveform which is not a perfect sine curve but is close enough to be considered sinusoidal for all practical purposes.

quasi single sideband A modulated waveform somewhat resembling single sideband, in which parts of both sidebands are present.

quasi rectangular wave A wave whose shape approaches that of a rectangular wave but which possesses a small amount of tilt and curvature.

quasi square wave A waveform which is not a perfect square but is close enough to be considered square for all practical purposes. Sometimes applied to a rectangular wave when a square wave is desired.

quasi-stellar radio source See *quasar.*

quasi technical A term sometimes applied to qualitative tests, as opposed to quantitative tests.

quaternary 1. The use of 4 as a base, or radix, of a number system. 2. Of an atom, joined to carbon atoms for four bonds. 3. The fourth member of a 4-unit set.

Queen Phonetic alphabet communications code word for the letter Q.

Quebec Phonetic alphabet code word for the letter Q.

quench 1. To arrest suddenly, as to quench an oscillation. 2. To cool quickly, as in the quenching of a heated metal object. 3. To extinguish the discharge in a gas tube.

quenched gap See *quenched spark gap.*

quenched-gap converter A form of rf power generator comprising a low-frequency ac source, quenched spark gap, and resonant *LC* circuit.

quenched spark gap A spark gap in which conduction is abruptly stopped by rapid deionization through quick cooling of the gap. It consists of a number of small gaps in series, each being between electrodes, such as metal disks, which are good radiators of heat.

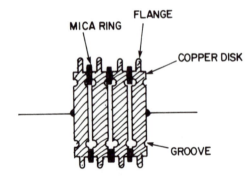

FLANGE

MICA RING

COPPER DISK

GROOVE

Quenched spark gap (remaining disk not shown).

quench frequency See *quenching frequency.*

quenching action Typical operation of a superregenerative circuit in which regeneration is increased to nearly the point of oscillation and then reduced; this action is repeated at a supersonic frequency and results in very high sensitivity. Also see *quenching frequency, quench oscillator,* and *superregenerative circuit.*

quenching circuit In a counter-tube circuit or gas-tube circuit, a subcircuit which acts quickly to inhibit multiple discharges.

quenching frequency The low frequency at which regeneration in a superregenerative circuit is increased and decreased.

quench oscillator In some superregenerative circuits, the separate supersonic rf oscillator which produces the required quenching action.

queue A list of data, steps in a process, or commands awaiting execution in a specific order.

queuing theory A branch of mathematical electronics, dealing with the optimum order in which steps should be executed to obtain a particular end result.

quibinary code In computer practice, a binary-coded decimal system in which each decimal digit is represented by seven bits occupying places whose values are 8, 6, 4, 2, 0, 1, and 0.

quibinary decade circuit A decade counter consisting of a ring-of-5 followed by a single binary stage.

quick break An operating characteristic of a switch, relay, or circuit breaker whereby the contacts open rapidly, even when the actuating current or mechanical force is slow acting.

quick-break fuse A fuse in which a spring pulls the melting wire apart rapidly. Also called *quick-blow fuse.* Compare *slow-blow fuse.*

quick-break switch A switch that opens rapidly, although its handle or lever is moved slowly by the operator. This action minimizes arcing and prevents chatter. Compare *quick-make switch.*

quick-disconnect The characteristic of a connector which enables its mating halves to be separated quickly and simply, to break the circuit in which it is situated.

quickening liquid A solution of mercuric cyanide or mercuric nitrate into which articles sometimes are dipped prior to electroplating with silver. The process insures good adhesion of the silver coat.

quick make An operating characteristic of a switch, relay, or circuit breaker whereby the contacts close rapidly, even when the actuating current or mechanical force is slow acting.

quick-make switch A switch that closes rapidly, although its handle or lever is moved slowly by the operator. Compare *quick-break switch.*

quick printer A *high-speed* printer. This device usually burns the character into the paper.

quicksilver See *mercury.*

quick-stop control A control on tape recorders and some dictating machines which allows the operator to stop the tape but keep the machine in the *play* or *record* mode. Also called *pause control.*

QUICKTRAN For multiaccess computer systems, a computer programming language based on FORTRAN and offering facilities, through the use of remote terminals, for running, testing, debugging, and compiling programs.

quiescent At rest or inactive, but still operating.

quiescent carrier operation A modulation system in which the carrier is present only during modulation, i.e., it is suppressed at all other times. Also called *controlled-carrier transmission.*

quiescent-carrier telephony A carrier-current (wired-wireless) telephone system in which the carrier is suppressed when there is no voice or alerting signal.

quiescent current Operating current (usually dc) flowing in a circuit or component during zero-signal or no-drain intervals. Also called *idling current.*

quiescent operation Zero-signal operation of a device, such as a tube, transistor, diode, magnetic amplifier, or similar component.

quiescent period The no-signal interval during which equipment is not operating, even though it is energized.

quiescent point The point on the characteristic curve of a tube, transistor, or similar device denoting the zero-signal operating conditions.

quiescent push-pull Denoting a push-pull stage, especially an audio power-output amplifier, in which the dc signal is zero or practically so.

quiescent state The inactive, or resting, state of a tube or transistor.

quiescent value The zero-signal value of current or voltage for any component supplied with operating power.

quiet agc See *delayed automatic gain control.*

quiet automatic gain control See *delayed automatic gain control.*

quiet automatic volume control See *delayed automatic gain control.*

quiet avc See *delayed automatic gain control.*

quiet battery A dc source which is specially designed and filtered to eliminate or drastically minimize all noise components in its output.

quieting Noise-voltage reduction in FM receiver output when the carrier is unmodulated.

quieting level In an FM receiver, the limiter threshold point.

quieting sensitivity In an FM receiver, the lowest input-signal amplitude at which the output signal-to-noise ratio is below the specified limit.

quiet tuning A system of tuning in which the output of a receiver is muted until a station is tuned in sharply.

quinary code A system of number representation using a pair of numbers, say, p and q, where the number is $p + q$; p can be zero or five; and q can be any of the numerals one through four. Also called *biquinary code*.

quinary counter A decade counter consisting of a five-stage ring. See *quinary code*.

quinhydrone electrode A pH meter electrode consisting of a platinum wire in a solution of quinhydrone ($C_{12}H_{10}O_4$). Also see *pH meter*.

quintupler 1. A rectifier circuit which delivers a dc output voltage approximately equal to five times the peak value of the ac input voltage. 2. An amplifier or other circuit which delivers an output signal of five times the frequency of the input signal.

R 1. Symbol for *resistance*. (Also, *r*.) 2. Radio-telegraph abbreviation for *solidly received message*. 3. Symbol for *radical* (chemistry).

R(00)—Symbol for the *Rydberg constant*.

R Symbol for *reluctance*.

r 1. Symbol for *roentgen*. 2. Symbol for *correlation coefficient*. 3. Abbreviation of *radius*. (Also, abbreviated *R*.)

re Symbol for *classical electron radius*.

RA 1. Abbreviation of *right ascension*. 2. Abbreviation of *random access*.

rabbit ears An antenna, often attached permanently to a tv receiver, consisting of two vertical *whips* (usually telescoping) the angle between which is adjustable.

RAC Abbreviation of *rectified alternating current*.

Rac Symbol for *ac resistance*. (Also, *rac*.)

race Incorrect interpretation of the clock pulses by a digital circuit. Also called racing. The circuit performs many operations during one clock pulse, rather than a single operation.

RACES Radio Amateur Civil Emergency System.

raceway See *wire duct* and *wireways*.

rack An upright frame for holding equipment of rack-and-panel construction.

rack-and-panel construction A method of building electronic equipment on a chassis attached horizontally or vertically to a vertical panel. After completion of a unit, the panel is fastened in place on a rack. Several such panels fill the rack.

rack and pinion A device used for mechanical adjustment of a control, such as the tuning control in a radio receiver. A gear engages a serrated rod. As the gear is turned, the rod moves lengthwise.

rack up In computer practice, a way of displaying data on a crt terminal in which a new line added to the already completely occupied screen bumps up what has forgone, thus eliminating the top line.

racon Acronym for *radar* bea*con*.

rad 1. A unit of ionizing radiation received by a body (dose) equal to 0.001 J/kg. 2. Abbreviation of *radiac*. 3. Abbreviation of *radian*. 4. Abbreviation of *radio*. 5. Abbreviation of *radix*. 6. Abbreviation of *radical*.

radar 1. A microwave system for detecting objects and determining their distance. Signals from the transmitter are reflected back to the transmitter site by the object, and the reflection (sometimes along with the transmission) is displayed on a cathode-ray screen. The name is an acronym for *radio detection and ranging*. Also see *plan position indicator* and *radarscope*. 2. The theory and application of radar.

radar altitude The distance of an aircraft above the surface of the earth, as determined by radar. Radar altitude varies with the terrain over which the aircraft passes.

radar antenna Any antenna employed for transmitting and/or receiving radar signals.

radar astronomy The use of radar-type equipment to observe heavenly bodies and to measure their distance from Earth.

radar beacon A radar transceiver which, upon receipt of radar signals, transmits encoded signals from which the operator can take a bearing.

radar beam The main lobe of energy emitted by a radar antenna. The radar beam is cone-shaped. The narrower the beam, the greater the resolution of the radar system.

radar clutter Visual interference on a radar screen caused by reflections from ground or sea.

radar countermeasures Abbreviations, *RCM* and *Rad CM*. Any method of interfering with enemy radar, such as *jamming* or use of *chaff*, in order to cripple the latter or to obtain information. Also see *electronic countermeasures*.

radarscope The special cathode-ray tube and associated circuit used to display a radar image (see *radar, 1*). Also see *A-display, B-display, C-*

display, D-display, E-display, F-scan, G-scan, H-scan, I-scan, J-display, K-scan, L-scan, M-scan, and *N-display.*

radar speed trap A radar-type system employed by traffic police to spot speeding vehicles.

radar telescope The transmission and reception unit employed in *radar astronomy.* Compare *radio telescope.*

RadCM Abbreviation of *radar countermeasures.* (Also, *RCM.*)

radechon A barrier-grid storage tube.

radial 1. Extending outward from a single point. 2. One of at least two straight wires extending horizontally outward from the base of a ground-mounted vertical antenna.

radial ground An earth connection composed of radials buried in the ground.

radial lead A lead (pigtail) attached perpendicular to the axis of a component, such as a resistor or capacitor.

radials Conductors arranged as spokes extending from a hub formed by the base of a vertical antenna.

radian Abbreviation, *rad.* The angle at the center of a circle subtended by an arc whose length is equal to the radius. Also called *circular radian.* 1 rad = 57.2958 degrees. There are 2 π radians in a circle.

radiance The radiant flux emitted by an object. Radiance is measured in terms of the amount of energy contained in a unit solid angle (steradian) with the source at the apex.

radian frequency See *angular velocity.*

radians-to-degrees conversion The conversion of angular measure into degrees. To change radians to degrees, multiply the number of degrees by 57.2958. Thus, 0.7854 radian = 45 degrees. Compare *degrees-to-radians conversion.*

radiant energy All forms of energy emitted by a source and moving like a wave through space. Included are radio waves, light, heat, X-rays, radioactive emissions, and so on. Sound waves are often included, but sound requires air or some other form of matter for its transmission.

radiant flux The rate at which radiant energy is emitted.

radiation 1. The emission of energy or particles, e.g., waves from an antenna, X-rays from an X-ray tube, energy from a radioactive material, heat from a body, etc. 2. Narrowly, X-rays and the emissions from radioactive substances. 3. The emission (1 or 2, above) itself.

radiation angle The horizontal or vertical angle at which electromagnetic waves are radiated from an antenna.

radiation belts See *Van Allen radiation belts.*

radiation counter An instrument used for determining the intensity of atomic-particle radiation, X-rays, or gamma rays. Operates by means of ionization of a gas in a sealed tube.

radiation-detector tube See *Geiger—Mueller tube.*

radiation-exposure meter See *exposure meter, 2.*

radiation field The portion of the electromagnetic field that is propagated by a radiator, as opposed to the induction field.

radiation intensity For a beam antenna, the radiated power per steradian in a given direction.

radiation loss Loss of energy through radiation from a conductor. Also see *loss.*

radiation pattern A graphical representation of the intensity field about a radiator, such as an antenna. Also see *lobe.*

radiation pressure Pressure exerted upon a surface by impinging electromagnetic radiation.

radiation resistance The inherent resistance at the center of a resonant antenna.

radiation sickness General physiological symptoms resulting from a short-term overdose of X-rays, gamma rays, or atomic-particle radiation.

radiator 1. The element of an antenna from which radio energy is directly radiated, as opposed to transmission line, feeder, lead-in, reflector, or director. 2. Loudspeaker.

radical 1. Sign, $\sqrt{}$. An expression indicating a root is to be extracted. The sign as shown is taken to mean a square root is involved; other roots are indicated by placing the appropriate index (3 for cube root, for example) in the vee part of the sign. Also see *radicand* and *root, 1.* 2. A group of atoms that can stay unchanged during certain chemical reactions.

radicand A number under a radical sign. Thus, in $\sqrt{2}$, $\sqrt[4]{64}$, and $\sqrt[5]{108}$, the radicands are 2, 64, and 108. Also see *radical* and *root, 1.*

radio 1. Wireless electrical communication, i.e., by means of electromagnetic waves. 2. Radio receiver. 3. Sometimes, a radio transmitter-receiver or transceiver. 4. To communicate by radio.

radio- 1. A prefix meaning "pertaining to wireless electrical communication." Example: *radiophone, radiotelegraph.* 2. A prefix meaning "employing radio waves." Example: *radiosonde, radiolocator, radiothermics.* 3. A prefix meaning "pertaining to using or possessing radioactivity," or "pertaining to X-rays." Example: *radiograph, radioisotope, radiologist.*

radioactive Having the property of emitting alpha, beta, and (sometimes) gamma rays as the result of nuclear disintegration. Also see *half-life.*

radioactive element A chemical element that is radioactive, e.g., uranium. Also called *radioelement.* Also see *radioactive.*

radioactive isotope See *radioisotope.*

radioactive transducer A pickup device for radioactivity, e.g., Geiger—Mueller tube.

radioactivity counter See *Geiger counter* and *scintillation counter.*

radio altitude See *radar altitude.*

radio amateur A hobbyist who is licensed to operate a two-way radio station in the amateur service. Also see *amateur, 2; amateur band; amateur call letters;* and *amateur service.*

Radio Amateur Civil Emergency System Abbreviation, RACES. A civil-defense organization of licensed amateur radio stations. Also see *radio amateur.*

radio astronomy The observation, study, and analysis of rf electromagnetic radiations from bodies or points in space, and the study of these bodies through their radiations.

radioautograph See *autoradiograph.*

radio beacon A radio transmitter of direction-finding or guidance signals or the signals themselves (an alternate term for the latter is *radio beam*).

radio beam 1. Antenna radiation focused in one direction. 2. Radio beacon.

radiobiology A field of biology concerned with the influence of radiant energy or radioactivity on organisms.

radiobroadcast A radio transmission directed to all receivers, especially by a station in the broadcast service. Also called *radiocast.* Also see *broadcast service, 1, 2.*

radio car An automobile equipped with a two-way radio.

radio carbon Radioactive carbon; i,e., carbon 14.

radiocast See *radiobroadcast.*

radio channel A single frequency within a band of radio frequencies in which stations are authorized to transmit signals of a specified type. Also see *channel, 1; channel separation;* and *channel width.*

radiochemistry The chemistry of radioactive substances.

radio communicaion Wireless communication carried on by means of radio-frequency electromagnetic waves.

radio compass A radio direction finder (see *direction finder*).

radioconductor A substance or body whose electrical conductivity is affected by radio waves and which can be used as a sensor of such waves.

radio control See *remote control.*

radio direction finder See *direction finder.*

radio Doppler 1. A change in the frequency of a radio signal emitted by a source having radial motion with respect to the receiver. 2. An electronic device that is used to measure radial speed by means of the Doppler effect at radio frequencies.

radio-electronics Communications-oriented electronics

radioelement See *radioactive element.*

radio engineer A trained professional skilled in the physics and mathematics of radio and in the theory and application of basic engineering and related subjects. Also see *radio engineering.*

radio engineering The branch of electronic engineering devoted to the theory and practice of *radio communication* and closely associated arts and sciences.

radio field strength The intensity of radio waves at a given point. Also see *field intensity, 2* and *radio map.*

radio frequency Abbreviation, rf. An ac frequency which is higher than the highest audio frequency, so called because of the application of such frequencies to radio communication. Also see *radio spectrum.*

radio-frequency amplifier 1. In a superheterodyne circuit, the channel in which the incoming rf signal is amplified. Compare *intermediate-frequency amplifier.* 2. Broadly, an amplifier of radio-frequency signals.

radio-frequency choke Abbreviation, RFC. A low-inductance coil employed to block rf currents. Many rf chokes have an air core; however, some have a core of ferrite or powdered iron.

radio-frequency current Symbol, Irf. 1. The intensity of a generated rf signal, usually in microamps. 2. Loosely, any measurable rf signal.

radio-frequency heating The generation of heat in an object by an intense rf field. See, for example, *diathermy, 1; dielectric heating;* and *induction heating.*

radio-frequency interference Abbreviation, RFI. 1. Annoying electrical noise in rf amplifiers, detectors, and instruments. 2. Undesired radio-frequency signals that compete with desired ones in amplifiers, receivers, and instruments.

radio-frequency meter An instrument for measuring frequencies in the rf spectrum (10 kHz and beyond). See, for example, *audio frequency meter, 2* and *wavemeter.*

radio-frequency oscillator Abbreviation, RFO. An oscillator (self-excited or crystal-controlled) for operation at radio frequencies. In such an oscillator, stray components, tube or transistor efficiencies, and general losses are of primary concern. Also see *radio frequency.*

radio-frequency power Symbol, Prf. Ac power at radio frequencies.

radio-frequency resistance The total in-phase resistance exhibited by a conductor at radio frequencies. This opposition to current includes dc resistance and the in-phase components due to skin effect, shielding, and the presence of dielectrics.

radio-frequency selectivity The selectivity of a radio-frequency channel, such as the rf amplifier and first detector of a superheterodyne circuit.

radio-frequency transformer A transformer designed for efficient operation at radio frequencies. It can be an air-core type or have a core of low-eddy-current material, such as powdered iron, and has windings designed for low distributed capacitance and low rf resistance.

radio-frequency transistor A transistor having very-high-frequency response, specifically one that is operable above 100 kHz.

radio galaxy A galaxy in which has been found a source of rf electromagnetic radiation.

radiogenic Produced by radioactivity.

radiogoniometer A radio compass (see *direction finder* and *goniometer, 1*).

radiogram A (usually printed out) message transmitted and received via radiotelegraphy or radioteletypewriter. The term is an acronym for *radio telegram.* Compare *cablegram* and *telegram.*

radiograph 1. X-ray photograph. 2. To contact by the transmission of a radiogram.

radio interference 1. Interference to radio communication, from whatever cause. 2. Abbreviation, rfi. Radio-frequency interference.

radioisotope A radioactive isotope (natural or artificial) of a normally nonradioactive chemical element, e.g., radioactive carbon. Also see *isotope.*

radio jamming See *jamming.*

radio knife A surgical instrument consisting essentially of a needle which forms a high-frequency arc that simultaneously cuts and cauterizes tissue.

radiolocation The use of radar to locate or determine the course of an object.

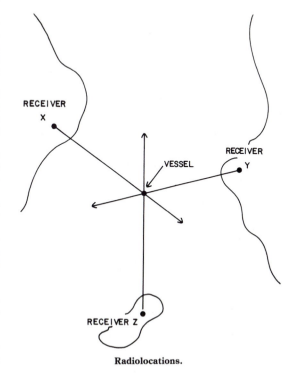

RECEIVER X

RECEIVER Y

VESSEL

RECEIVER Z

Radiolocations.

radiolocator See *radar.*

radiological system See *X-ray therapy system.*

radiologist A specialist skilled in radiology.

radiology The science embracing the theory and use of X-rays and radioactive substances in the diagnosis and treatment of diseases and ailments.

radiolucency Permeability to radiation.

radiolysis Chemical decomposition brought about by radiation.

radioman A radio technician or operator.

radio map A map of a geographic area, on which lines are drawn connecting measured points of equal field strength for signals from a radio station at the approximate center of the area.

radiometeorograph See *radiosonde*.

radiometer A device for detecting and measuring the strength of radiant energy. One form consists of a set of vanes blackened on one side and mounted on pivots in an evacuated glass bulb. Sunlight causes the vane assembly to rotate, the speed being proportional to the intensity of the light.

radiometry The science and art of measuring radiation in the infrared, visible, and ultraviolet regions of the electromagnetic frequency spectrum. Compare *photometry*.

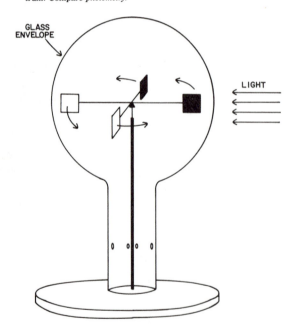

Radiometer.

radio micrometer See *microradiometer*.

radio net A group of radio stations operating together in an organization, often on or near the same frequency.

radio network See *radio net*.

radionics An acronym for *radio* and electro*nics*, suggesting the combination of the two. The term has fallen into disuse in recent years, having been supplanted by *electronics*, but it can be found in some old literature.

radio operator A technician who is licensed to operate a transmitter in the radio, television, or radar services.

radiopaque Opaque to X-rays. Compare *radioparent*.

radioparent Transparent to X-rays. Compare *radiopaque*.

radiophone See *radiotelephone*.

radiophoto A photograph transmitted and received by radio. Also see *facsimile*.

radio pill See *encoradiosonde*.

radio prospecting The use of rf devices to locate underground or underwater metals and mineral deposits. Also see *metal locator*.

radio range A radio station providing navigational aid to airplanes.

radio receiver The complete apparatus that selects, amplifies, demodulates, and reproduces a radio signal for purposes of communication, as distinct from *facsimile receiver, remote-control receiver, telemetry receiver, television receiver*, etc. A receiver having a self-contained antenna has the additional function of signal pickup.

radiosensitivity Sensitivity to X-rays.

radio serviceman An electronic technician who is skilled in the repair and maintenance of radio equipment, especially receivers. Also called *radio service technician*.

radiosonde A balloon-carried combination of radio transmitter and transducers, for sending to a ground monitoring station signals revealing such atmospheric conditions as temperature, humidity, and pressure. The name denotes a radio type of *sonde,* a device for gathering meteorological data at high altitudes.

radiosonobuoy See *sonobuoy*.

radio spectrum The coninuum of frequencies useful for radio communication and control. These frequencies have been classified in the following manner: Very low frequencies (vlf), 10 to 30 kHz; low frequencies (lf), 30 to 3000 kHz; medium frequencies (mf), 300 to 3000 kHz; high frequencies (hf), 3 to 30 MHz; very high frequencies (vhf), 30 to 3000 MHz; ultrahigh frequencies (uhf), 300 to 3000 MHz; superhigh frequencies (shf), 3 to 30 GHz.

radiostat See *crystal filter*.

radio station 1. The location at which a radio transmitter and/or receiver is installed. 2. The complete set of equipment for a radio receiving and/or transmitting installation, including the studio, linking apparatus, and antennas. 3. A standard broadcast station.

radiostrontium See *strontium 90*.

radio technician A subengineering professional who is skilled in the construction, testing, repair, and maintenance of radio equipment, and sometimes in its design, and who usually works under the supervision of a radio engineer. Also see *radio serviceman*.

radiotelegram See *radiogram*.

radiotelegraph 1. Pertaining to the theory, application, and equipment of radiotelegraphy. 2. A transmitter and receiver installation for radiotelegraphy.

radiotelegraph code See *continental code*.

radiotelegraph distress signal See *SOS*.

radiotelegraph monitor See *keying monitor*.

radiotelegraphy The transmission and reception of telegraphic communications by means of electromagnetic waves. Also see *radio, 1*.

radiotelephone 1. Pertaining to the theory, application, and equipment of radiotelephony. 2. A transmitter and receiver (or transceiver) installation for radiotelephony.

radiotelephone distress signal See *mayday*.

radio—telephone patch See *phone patch*.

radiotelephony The transmission and reception of voice, music, and other sounds (i.e., telephonic communications) over a distance by means of electromagnetic waves. Also see *radio, 1*.

radio telescope A directional antenna and associated equipment for receiving rf electromagnetic radiations from space, especially from celestial objects. See *radio astronomy*.

radioteletype See *radioteletypewriter*.

radioteletypewriter A teletypewriter adapted to radio rather than wire service.

radiotherapy The use of X-rays in the treatment of disease and disorders.

radiothermics The science of the generation of heat by means of radio-frequency current.

radiothermy Shortwave diathermy.

radio thorium Radioactive *thorium*.

radiotracer See *tracer*.

radio transmitter The complete apparatus that generates radio-frequency power, modifies it with the telegraphic or telephonic signals needed for communication, and delivers the product to an antenna for radiation into space. Here, the radio transmitter is distinguished from similar equipments: *facsimile transmitter, remote-control transmitter, telemetry transmitter, television transmitter,* etc.

radio-transparent material 1. A substance through which radio waves pass with little or no attenuation. 2. A substance through which X-rays, gamma rays, or high-speed atomic particles may pass with little or no attenuation.

radiotrician Acronym for *radio* elec*trician*. See *radio serviceman*.

radio tube 1. A vacuum tube. 2. A vacuum tube used as an amplifier, local oscillator, detector, or mixer in a radio receiver.

radiovision See *television*.

radio watch See *watch*.

radio waves Electromagnetic waves of radio frequencies (see *radio spectrum*).

radium Symbol, *Ra*. A rare radioactive metallic element. Atomic number, 88. Atomic weight, 226.05.

radius The straight-line distance from the center of a circle or sphere to its periphery.

radius vector In spherical or polar coordinates, a line segment drawn from the pole, or origin, and representing the vector magnitude.

radix The number indicating the number of symbols in a system of numerical notation, and the powers of which give the place values of the system. Thus, 10 is the radix of the decimal system, and 2 the radix of the binary system. Also called *base*.

radix complement The difference that results when each digit in a number is subtracted from one less than the radix, or base, of the pertinent system of notation, and then adding one; thus, the radix complement of decimal 365 is: 999 minus 365 plus 1, or 635. For the binary number 101, it is 11 (111 − 101 + 1 = 11).

radix-minus-one complement See *diminished radix complement*.

radix point In a number, the point (dot or period) separating the integral and fractional digits. Its significance depends upon the system of notation involved: *binary point, decimal point,* and so on.

radome A plastic shell housing a radar antenna, especially aboard an aircraft.

radon Symbol, *Rn*. A gaseous radioactive element that results from the disintegration of radium. Atomic number, 86. Atomic weight, 222.

rad/s^1 Radians per second, the SI unit of angular velocity.

rad/s^2 Radians per second squared, the SI unit of angular acceleration.

radux A continuous-wave, low-frequency radionavigation system. Position is determined by comparing the phase of two signals sent from different locations.

rainbow generator A test-signal generator that produces a full color spectrum, a pattern resembling the successive coloration of a rainbow, on the screen of a color-TV receiver. Also see *rainbow pattern*.

rainbow pattern A test pattern for servicing a color-TV receiver. It consists of a full color spectrum, thus taking its name from its resemblance to a rainbow. Also see *rainbow generator*.

RAM Abbreviation of *random-access memory*.

ramp A linearly rising sawtooth wave, so called from its resemblance to an incline.

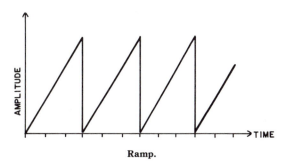

Ramp.

ramp generator A test-signal generator that produces sawtooth-wave signals. Also see *ramp*.

R and D Abbreviation of *research and development* or *research and design*. (Also, *R & D*.)

random access Abbreviation, *RA*. In computer and data-processing practice, the kind of access to storage in which data can be extracted in any order.

random-access memory In computer and data-processing systems, a memory providing access time that is independent of the address.

random deviation Irregular *ripple*.

random noise Electrical noise in which the pulses or fluctuations have no discernible pattern of occurrence, i.e., they are haphazard in frequency and amplitude.

random number A number derived by chance, as by rolling dice, or one produced by a methodology for its generation, say, through the use of a computer program; in the latter case, the number is not always random in the truest sense, but can be considered so for most applications.

random number generator Hardware or software capable of supplying random numbers.

random occurrence See *chance occurrence*.

random variable In statistics, a variable that can have a number of values, each of the same probability.

random winding A coil winding in which the turns are wound haphazardly to reduce distributed capacitance.

range 1. The limits within which a circuit or device operates, i.e., the territory defined by such limits. Examples: current range, frequency range, voltage range. 2. The difference between the upper and lower limits of deflection of a meter. 3. The distance over which a transmitter operates reliably. 4. A clear area for testing antennas. 5. The distance between a radar station and a target. 6. The possible values for a quantity or function that lie between given limits.

range capacitor See *trimmer capacitor*.

range-height indicator Abbreviation, *RHI*. A radar display in which the horizontal axis shows distance to the target, and the vertical axis shows elevation of the target.

range mark See *distance mark.*

range resistor See *trimmer resistor.*

rank 1. To arrange in a specific sequence according to significance. 2. A place in such a sequence.

Rankine scale A temperature scale on which the freezing point of water is 491.69°, and the boiling point 671.69°. For conversion to kelvins, multiply °R by 5/9.

rapid drift The fast change of a quantity or setting (usually in one direction) with time.

rapid printer See *quick printer.*

raser A radio-frequency lasing device.

raster The rectangle of light (composed of unmodulated lines) seen on the screen of a TV picture tube when no signal is present.

ratchet circuit See *electronic ratchet* and *commutator, 2.*

rate action See *derivative action.*

rate effect In a four-layer semiconductor device, the tendency for the switch to conduct undesirably as a result of a transient spike.

rate-grown transistor See *graded-junction transisor.*

rate gyro A special gyroscope for measuring angular rates.

rate of change The amount by which the value of a quantity changes with respect to another quantity (such as time). Also see *derivative* and *differential.*

rate signal A signal whose amplitude is proportional to the derivative of a variable with respect to time.

rate time In automatic-control practice, the time in minutes over which the addition of *rate action* (see *derivative action*) advances proportional action.

ratio An expression of relative magnitude given in the form N/n or $N:n;$ the ratio of 6 to 3 is 6/3 or 2 to 1 (2/1 or 2:1).

ratio-arm bridge A simple four-arm bridge in which the balancing potentiometer supplies the two ratio arms, one on each side of the slider at all settings.

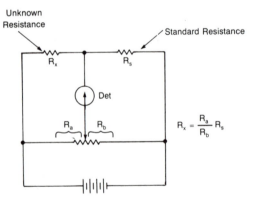

Ratio-arm bridge.

ratio arms Two impedance arms serving to establish the numerical ratio of a bridge (see *bridge, 1*).

ratio control In automatic-control practice, a system in which the controlled variable is in a prescribed ratio to another variable.

ratio detector An FM second detector resembling the *Foster—Seeley discriminator,* except that one of the two diodes is reversed and the junction point of the load resistors is grounded. In an FM circuit employing a ratio detector, no limiter is required. The ratio of the dc outputs of this detector is proportional to the ratio of the instantaneous i-f voltages applied to the two diodes. Compare *Foster—Seeley discriminator* and *Travis discriminator.*

ratio meter An instrument which compares two different signals (and indirectly their sources) and delivers a reading of their ratio.

rational number A number which is an integer or a value expressed as a bar fraction in which the divisor is not zero. Compare *irrational number.*

rational operation Multiplication, division, addition, or subtraction.

ratio of geometric progression In a geometric progression, the ratio of one turn to the next.

ratio of similitude The ratio of the lengths of corresponding sides in similar geometric figures.

rat race See *hybrid ring.*

raven red A variety of *red oxide of iron,* a commercial red paint employed as the magnetic coating of early recording tapes.

raw ac Unrectified alternating current or voltage.

raw data Data that has not been processed in any way.

rawinsonde A radiosonde tracked by a radio direction finder to determine wind velocity. The name is an acronym from *radar wind radiosonde.*

raw tape See *blank tape.*

ray 1. A line of radiant energy. Such a line (e.g., one of magnetic force) is imagined to arise from a point source and have no width. 2. One of many lines coming from a central point. 3. A line segment, including its labeled point of origin, that is not terminated by a point opposite its origin, e.g. one of the rays (or *half-lines)* forming an angle.

ray-control electrode In a magic-eye tube, a wire electrode near the cathode which deflects some of the electrons and produces the shadow image on the target.

ray-control tube See *magic-eye tube.*

Raydist A continuous-wave, medium-frequency radionavigation system. Position is determined according to the phase difference between two signals transmitted from different locations.

Rayleigh—Carson theorem An expression of the reciprocal relationship between the transmitting and receiving properties of an antenna: If a potential $E1$ applied to antenna 1 causes a current $I1$ to flow at a given point in antenna 2, then a voltage $E2$ equal to $E1$ applied at that point in antenna 2 will produce a current $I2$ equal to $I1$ (same magnitude and phase) at the point in antenna 1 where $E1$ originally was applied. Also see *reciprocity theorem.*

Rayleigh distribution A probability-density function, used to describe the behavior of sky-wave electromagnetic signals.

Raysistor A type of *optoisolator.*

Rb Symbol for *rubidium.*

Rb Symbol for *base resistance.* (Also, *rb.*)

RBC Abbreviation of *red blood count.*

RC 1. Abbreviation of *resistance—capacitance.* 2. Abbreviation of *radio-controlled.* 3. Abbreviation of *remote control.*

Rc 1. Symbol for *collector resistance.* (Also, *rc.*) 2. Symbol for *cold resistance.*

RC circuit See *resistance—capacitance circuit.*

RC-coupled amplifier See *resistance—capacitance-coupled amplifier.*

RC coupling See *resistance—capacitance coupling.*

RC filter See *resistance—capacitance filter.*

RCL 1. Abbreviation of *recall.* 2. Abbreviation of *resistance-capacitance-inductance.*

RCM Abbreviation of *radar countermeasures.* (Also, *RadCM.)*

RC phase shifter See *resistance—capacitance phase shifter.*

RC time constant See *resistance—capacitance time constant.*

RCTL Abbreviation of *resistor—capacitor—transistor logic.*

RC tuning See *resistance—capacitance tuning.*

R & D See *R and D.*

RD Symbol for *drain resistance.*

Rd 1. Symbol for *diode resistance.* (Also, *rd).* 2. Symbol for *distributed resistance.*

Rdc Symbol for *dc resistance.* (Also, *rdc.)*

RDF Abbreviation of *radio direction finder.*

Re Symbol for *rhenium.*

Re Symbol for *emitter resistance.* (Also, *re.)*

REA Rural Electrification Administration.

reachthrough See *punchthrough.*

reachthrough region See *punchthrough region.*

reachthrough voltage See *punchthrough voltage.*

reactance Symbol, *X.* Unit, ohm. The opposition offered to the flow of alternating current by pure capacitance, pure inductance, or a combination of the two. Reactance introduces phase shift. Also see *capacitive reactance* and *inductive reactance.* Compare *resistance.*

reactance chart A nomograph for capacitance, inductance, and frequency.

reactance factor The ratio of the alternating-current resistance of a conductor to the direct-current resistance. The reactance factor generally increases as the frequency increases, because of skin effect, and because the length of the conductor may be a sizable part of the wavelength of the transmitted energy.

reactance modulator A reactance-tube modulator or an equivalent transistor circuit.

reactance transistor A transistor employed as a reactance modulator.

reactance tube A vacuum tube, such as a 6J7 (pentode), that is useful as a reactance-tube modulator.

reactance-tube modulator A frequency modulator consisting essentially of a vacuum tube connected across the tank circuit of an rf oscillator and acting as an audio-frequency-controlled variable inductance.

reaction A sometimes violent chemical activity between two or more substances, in which the atoms of the substances become bonded to each other to form new substances. Also see *law of octals.*

reaction-time meter See *neomatachograph* and *neomatachometer.*

reactivation The rejuvenation of a thoriated-tungsten filament in a vacuum tube that has lost its emissive ability. The process consists of two steps: *flashing* and *burning.*

reactive attenuator An attenuator that functions by means of reactance, rather than by means of resistance.

reactive current That component of alternating current that is out of phase with voltage. Compare *resistive current.*

reactive kilovolt-ampere Abbreviation, kVAR. A unit of high apparent power; it is the product of kilovolts and amperes in a reactive component of a circuit. Also see *apparent power, kilovolt-ampere, reactive volt-ampere,* and *volt-ampere.*

reactive kilovolt-ampere-hour See *kiloVAR-hour.*

reactive load A load device which is capacitive or inductive rather than resistive.

reactive power See *reactive kilovoltampere* and *reactive volt-ampere.*

reactive volt-ampere Abbreviation, *VAR.* A unit of apparent power; it is the product of volts and amperes in a reactive component of a circuit. Also see *apparent power, kilovolt-ampere, reactive kilovoltampere,* and *volt-ampere.*

reactor 1. An inductor, especially one having very low internal resistance, employed principally for its inductive reactance. 2. A chamber in which the nuclei of atoms are split to provide atomic energy. Also see *nuclear reactor.* 3. In industrial chemistry, a vat in which reactions take place.

read In computer practice, to extract data from memory or a storage medium and (usually) transfer it to another area of memory or other medium. Compare *write.*

readability In electronic communications, the degree to which a desired signal may be recognized and interpreted in a given context.

reader That which *reads,* e.g., paper tape reader.

read head In a magnetic memory or in a tape recorder or wire recorder used for data recording, the head which picks up the magnetic pulses from the drum, tape, disk, or wire. Compare *write head.*

reading rate The number of input characters per second that a computer or other data-processing device handles.

read-only memory Abbreviation, *ROM.* In a computer or calculator, a memory unit in which instructions or data are permanently stored for use by the machine or for reference by the user. The stored information is read out nondestructively, and no information can subsequently be written into the memory.

readout lamp An indicating device usually in the form of a glass bulb containing separate cathode electrodes in the shape of the numerals (0 to 9), right and a left decimal points, and filled with a gas, such as neon. Each numeral electrode is connected to a separate pin on the base of the tube. The single anode is common to all cathodes. The cathode to which a signal voltage is applied glows, showing the shape of its number. Some readout lamps contain letters of the alphabet as well as numbers.

read pulse In computer practice, a pulse that activates the read function (see *read).* Compare *write pulse.*

read rate The number of data units an input read device can read per unit of time, e.g., bits of words per second.

readthrough 1. The reception of signals in between transmitted pulses at the same frequency. 2. The continuous monitoring of a signal being jammed. Any change in the frequency, modulation, or other characteristics of the signal may then be detected, and the jamming signal adjusted accordingly.

read time The period during which data is being transferred from a computer storage unit.

read—write channel A channel over which activity between a cpu and peripheral takes place.

read—write head An electromagnetic transducer used for both reading and writing. See *read* and *write.*

read-write memory A small data storage bank for short-term use. The contents of the memory are easily changed. 2. Another name for RAM memory.

real address See *absolute address.*

real axis The axis of the real component of a complex number, i.e., the *x* (horizontal) axis.

real component The real part of a complex number (see *complex notation).*

real image The image formed on a screen when rays from the object converge on passing through a lens. Compare *virtual image*.

real number A number in the category that includes zero, all rational numbers, and all irrational numbers. Also see *complex number, imaginary number, integral number, irrational nunber,* and *rational number*.

real power The apparent power multiplied by the power factor in an alternating-current circuit containing reactance. Real power is the difference between the apparent power and the reactive power. Actual radiated or dissipated power cannot exceed the real power in any case.

real time A phrase describing the operation of a computer or data processing system in which events are represented or acted upon as they occur, i.e., one in which data is processed as it becomes available, usually through the use of time-sharing, direct access storage devices, and remote terminals. An example of a real-time system is one used to make hotel reservations.

real-time clock A device that produces periodic signals that reflect the interval between events and sometimes can be used to give the time of day.

rear end The low-frequency portion of a superheterodyne receiver, i.e., the i-f amplifier, second detector, and af amplifier. Compare *front end*.

Reaumur scale A thermometer scale on which zero is the freezing point of water and 80 degrees is the boiling point of water. Compare *absolute scale, Celsius scale, Fahrenheit scale,* and *Rankine scale*.

rebecca The airborne interrogator in the British *rebecca—eureka* radar navigation system.

rebecca—eureka system A British 90-mile-hovering radar navigation system comprising an airborne interrogator (*rebecca*) and a ground transponder beacon (*eureka*). Also see *eureka, 2* and *rebecca*.

rebroadcast The retransmission of a radio broadcast simultaneously by a station other than the originator. Also see *automatic relay station*.

rebroadcast station See *automatic relay station*.

recalescence During the cooling of a metal, the sudden release of heat. Also see *recalescent point*. Compare *decalescence*.

recalescent point In a metal whose temperature is being lowered from a higher value, the temperature at which heat is suddenly released. Compare *decalescent point*.

RECALL Abbreviation, RCL. In computers and calculators, an instruction which brings material from the memory for examination or use. The opposite instruction is *STORE*.

receiver 1. A device or system operated at the end of a communication link and which accepts a signal and processes or converts it for local use. Also see specific entries for various types of receiver. 2. The earpiece of a telephone. 3. In audio practice, an FM tuner integrated with a general-purpose preamplifier and power amplifier, and containing standard jacks for input and output of audio signals to and from peripheral equipments.

receiver muting See *muting, 1*.

receiver primaries See *display primaries*.

receiver selectivity The sharpness of tuning of a receiver. Also see *selectivity*.

receiver sensitivity The degree to which a receiver will respond to weak signals; generally expressed as the lowest input-signal level for maximum receiver output.

receiving set A *radio receiver*.

receiving station A station that ordinarily only receives signals, i.e., it makes no type of transmission. Compare *transmitting station*.

receptacle 1. Socket. 2. The half of a connector that is mounted on a support, such as a panel, and that is therefore stationary.

reciprocal 1. A number that results from dividing 1 by a given number. Thus, the reciprocal of 5 is 1/5 or 0.2. 2. Pertaining to bilateralism. 3. Inversely related.

reciprocal impedances See *inverse impedances*.

reciprocal ohm See *siemens* and *mho*.

reciprocation 1. The determination of a mathematical reciprocal value from a given value. 2. The transmission of a message in response to a received message.

reciprocity in antennas See *Rayleigh—Carson theorem*.

reciprocity theorem When a voltage E across branch A of a network causes a current I to flow in branch B of the network, the voltage may be applied across branch B to cause the same value of current to flow in branch A. Compare *compensation theorem, maximum power transfer theorem, Norton's theorem, superposition theorem,* and *Thevenin's theorem*.

recombination The refilling of holes by electrons in a semiconductor.

recombination current In a transistor circuit, base current resulting from recombination.

recompile In computer practice, to compile again (see *compile*), usually according to program amendments following debugging or to create a different form of a program so that it will be compatible with other hardware.

record 1. Phonograph disk. 2. A chart delivered by a graphic recorder. 3. To make one of the foregoing. 4. In data processing, a constituent of a file. 5. In data processing, a data unit portrayed a specific transaction.

record blocking In data processing practice, making data blocks from groups of records so that the blocks can, in a single operation, be transferred to magnetic tape.

record count A usually running total of a file's records.

recorded disk A phonograph disk on which a recording has been made. Also called *prerecorded disk*.

recorded tape Magnetic tape containing recorded material. Also called *prerecorded tape*. Compare *blank tape* and *raw tape*.

recorder 1. A machine for preserving sound, video, or data signals in the sequence in which they occur, e.g., *disk recorder, tape recorder, wire recorder*. 2. A machine for making a permanent visual record (photographically or by stylus) of an electrical phenomenon. Also called *chart recorder, drum recorder, oscillograph,* and *strip recorder*.

record head See *recording head*.

recording density In a magnetic storage medium, the number of information units (bits, characters), represented by magnetized areas, per unit area or length.

recording disk A phonograph record on which material has not been recorded, or from which recorded material has been removed. Compare *prerecorded disk*.

recording head In a magnetic recorder—reproducer, the head that magnetizes the tape in accordance with sounds or other signals. Also called *record head* and *write head*. Compare *playback head*.

recording instrument A measuring instrument, such as a voltmeter or ammeter, that makes a permanent record of its deflections. Also see *recorder, 2*.

recording loss 1. Loss of information during a recording process. 2. Loss resulting from recording efficiency of less than 100 percent; audio power loss.

recording tape Magnetic tape which has never been recorded, or which has been erased. Compare *prerecorded tape*.

recovery time 1. Symbol, *tr*. The time required for a semiconductor pn junction to attain its high-resistance state when the bias voltage is suddenly switched from forward to reverse. 2. The time required for a circuit to recover from momentary overdrive. 3. The time required for a computer system to stabilize following a degenerative operation.

rect Abbreviation of *rectifier*.

rectangular coordinates See *Cartesian coordinates*.

rectangular scan 1. A method of beam scanning in a cathode-ray tube, in which the beam moves sequentially in parallel lines to cover a rectangular region. Used in television. 2. In radar, a two-dimensional scan, covering a specific rectangular region.

rectangular wave An ac waveform having steep sides and a flat top, but which is not square.

rectangular waveguide A waveguide having a rectangular cross section.

rectification The changing of ac into dc by any means other than the use of a motor—generator. Also see *rectifier*.

rectification efficiency Symbol, *n*. The ratio (expressed as a percentage) of the dc output voltage to the peak ac input voltage of a rectifier; $n \ (\%) = 100 \ Edc/ac$ peak.

rectified ac The unfiltered dc output of a rectifier. It consists of the unidirectional half-cycles passed by the rectifier (one per cycle for half-wave, two per cycle for full-wave rectification).

rectifier Abbreviation, *rect*. A tube, semiconductor device, or electromechanical device (such as a vibrator) for changing ac into dc.

rectifier diode A heavy-duty tube or semiconductor diode designed primarily to change ac to dc in power supplies. Also see *copper-sulfide rectifier; diode rectifier, 2; germanium rectifier; kenotron; mercury-arc rectifier; mercury-vapor rectifier; selenium rectifier;* and *silicone rectifier*.

rectifier—filter system The rectifier plus power-supply-filter combination for converting alternating current into direct current.

rectifier probe A diode-type probe used with a dc vacuum-tube voltmeter (or transistorized electronic voltmeter) to measure rf voltage. The diode rectifies the rf and presents to the meter a dc voltage proportional to the peak rf voltage.

rectifier stack An assembly of separate rectifier disks or plates in series on a central bolt, as in most selenium rectifiers.

rectifier tube See *tube diode; diode rectifier, 2; kenotron; mercury-arc rectifier; mercury-vapor rectifier; rectifier diode;* and *vacuum diode*.

rectifier-type meter See *diode-type meter*.

rectilinear 1. Making, or going in, a straight line. 2. Having straight lines. 3. Situated so as to form right angles.

rectilinear chart A graphic-recorder chart in which the crossing coordinates are arcs rather than straight lines, to correspond to the swing of the pen. Also see *strip chart*.

rectilinear scan See *rectangular scan, 1*.

recurrent network A circuit in which several sections of identical configuration (e.g., L-sections) are cascaded consecutively.

recurrent phenomenon A phenomenon that repeats itself periodically.

recurrent sweep In an oscilloscope, a repetitive horizontal sweep of the beam occurring at a frequency determined by the settings of the sweep controls. Also called *repetitive sweep*. Compare *nonrepetitive sweep*.

recursion Generating a complete series of functions or numbers by applying an algorithm to initial values in the series.

recursive Relating to a procedure or set of steps that repeat endlessly.

red—green—blue Abbreviation, *RGB*. In color TV, the three primary colors from which all other colors are derived. Also see *color television*.

red gun In a three-gun color-TV picture tube, the electron gun whose (correctly adjusted) beam strikes only the red phosphor dots on the screen.

red oxide of iron An iron oxide of the general formula Fe_2O_3, used as the magnetic coating of recording tape. Also see *iron oxide*.

red oxide of zinc See *zincite*.

red-tape operation An operation or function that is necessary for organizational purposes, but does not directly contribute to the completion of the task at hand.

redundancy 1. The repetition of components in a circuit (e.g., series or parallel connection of them) so that one will be available for circuit operation if the other fails. 2. Having available more than one method for performing a function. 3. Having on hand several copies of data as a safeguard against the data's loss.

redundancy check A check for the integrity of digitized data to which extra bits have been added for the purpose, e.g., parity check.

redundant 1. Pertaining to any two units of data, resembling each other in such a manner that if either unit is removed, no information is lost from the system. 2. A unit of data the contains information already present in the system.

red video voltage In a three-gun color-TV circuit, the red-signal voltage which actuates the red gun.

reed A usually thin metal blade, leaf, or strip employed in vibrators, reed-type relays, reed-type oscillators, and similar devices.

reed oscillator See *reed-type oscillator*.

reed relay See *dry-reed switch* and *mercury-wetted reed relay*.

reed-relay logic Logic circuits employing reed relays. Also see *relay logic*.

reed switch 1. A frequency-sensitive switch in which the movable contact is mounted on the tip of a thin, metal strip (reed), which is actuated by an ac coil. The reed vibrates most vigorously and closes the contacts when the ac excitation is at the natural frequency of the reed. 2. Dry-reed switch.

reed-type oscillator An electromechanical type of audio-frequency oscillator whose frequency is controlled by a vibrating metal strip (reed) instead of a tuning fork. Also see *hummer*.

reed-type switch See *reed switch*.

reel The spool around which magnetic tape is wound, or the tape and a spool as a unit.

reentrant cavity A resonant cavity in which one or more sections are directed inward to confine the electric field to a small volume.

CROSS SECTION.

Reentrant cavity.

reentrant winding A winding of wire which returns to its starting point, especially in a motor armature.

ref Abbreviation of *reference*.

reference address As a point of reference, an address for instructions having relative addresses.

reference amplifier A voltage-regulation device consisting of a transistor and zener diode in the same envelope.

reference angle In radar, the angle of incidence of the beam against a target surface, measured with respect to the normal (perpendicular) line at the surface.

reference antenna An antenna—an isotropic radiator, for example, or a standard dipole—used to establish a reference point for determining the relative gain of another antenna.

reference diode A zener diode whose constant voltage drop is used as a dc reference potential in calibrator circuits and voltage regulators.

reference dipole See *reference antenna*.

reference electrode For use with a pH meter, an electrode that provides a reference potential.

reference level A particular value of a quantity (current, frequency, power, voltage, and so on) to which other values of the same quantity are referred.

reference time The point at which a trigger pulse attains 10 percent of its maximum amplitude.

reference white level The television picture signal value representing the uppermost limit for peak white signals.

Reff Symbol for *effective resistance*.

reflectance 1. See *mismatch factor*. 2. The reflected part of the *radiant flux* striking a surface. Expressed as a fraction of the total incident radiation.

reflected binary code See *cyclic code*.

reflected impedance In a coupled circuit, the impedance in the secondary which appears in the primary circuit, or vice versa, as if it were reflected through the coupling transformer.

reflected power In a transmission line in which there are standing waves, the power that is reflected back from the end of the line. Compare *incident power*.

reflected-power meter A type of rf power-measuring instrument, connected between a source and a load, in which the deflection of a meter is proportional to energy transmitted through a coaxial section and energy reflected back.

reflected ray The ray that is reflected by the surface of a body or region it strikes. Compare *incident ray* and *refracted ray*.

reflected resistance 1. In a transformer, the effective resistance across the primary winding, when a resistive load is connected to the secondary. 2. In a transmission line, the resistance at the input end when a load is connected to the output end.

reflected wave 1. A wave that is reflected by the ionosphere or by the surface of the earth. Compare *incident wave* and *refracted wave*. Also see *ionosphere* and *ionospheric propagation*. 2. A wave that is bounced off an obstruction, such as a building or mountain.

reflecting galvanometer A galvanometer having a light-beam pointer.

reflecting shell See *ionosphere* and *Kennelly—Heaviside layer*.

reflecting telescope A telescope whose main focusing device is a concave mirror.

reflection 1. The turning back of a ray by a surface it strikes. Examples of reflecting media are the surface of the Earth, the polished surface of a material, and a layer of the ionosphere. Compare *refraction*. 2. The return of energy to the source by the mismatched end of a transmission line or by the end of a radiator.

reflection factor See *mismatch factor*.

reflection law In *reflection*, the *angle of incidence* is equal to the *angle of reflection*.

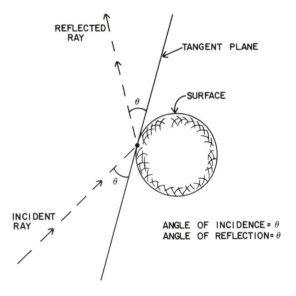

Reflection law.

reflection loss 1. Loss caused by the reflection of an electromagnetic field at a discontinuity in a transmission line. 2. Loss that occurs when an electromagnetic wave is reflected from a surface or object.

reflective code See *Gray code*.

reflectivity *Mismatch factor*. 2. The degree to which a point, plane, or surface reflects the radiation (light, for example) which strikes it.

reflectometer 1. Reflected-power meter. 2. A type of photometer used to measure reflection.

reflector 1. A smooth, metal surface or wire screen for reflecting radio waves. See, for example, *parabolic reflector*, 2. A length of wire, rod, or tubing used in a beam antenna to reflect radio waves. Compare *director* and *radiator*. 3. A polished surface for reflecting light or infrared rays, i.e., a mirror. 4. Repeller. 5. Reflecting telescope.

reflector element See *reflector, 2*.

reflector satellite A satellite whose skin reflects radio waves.

reflector voltage In a reflex klystron, the reflector-to-cathode voltage.

reflex baffle A loudspeaker baffle (see *baffle, 1*) constructed so that some of the sound radiated to the rear of the diaphragm is transmitted forward (after phase shift) to boost overall radiation at some frequencies.

reflex bunching In a klystron, bunching following dc-field-induced reversal of the velocity-modulated electrons. Also see *reflex klystron*.

reflex circuit A radio receiver circuit in which a single tube or transistor is used successively for different functions. Thus, the tube may act as an rf amplifier and then as an af amplifier.

reflex klystron A klystron having only one cavity, which serves first as the buncher and then, as the electrons are turned about and caused to pass through again, as the catcher.

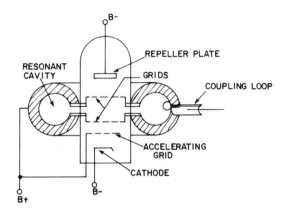

Reflex klystron.

refracted ray The ray that is refracted by the body or region through which it passes. Compare *incident ray* and *reflected ray*.

refracted wave A wave that is refracted by the ionosphere. Compare *incident wave* and *reflected wave*. Also see *ionosphere* and *ionospheric propagation*.

refraction The bending of a ray as it passes in or out of a medium.

refractive index See *index of refraction*.

refractivity The extent of the ability to refract, given as the quantity $(v1/v2) - 1$, where $v1$ is phase velocity in free space, and $v2$ is phase velocity in the medium through which a wave passes.

refractory A heat-resistant nonmetallic ceramic.

regeneration See *positive feedback*.

regeneration period The period during which the electron beam scans a crt to restore charges to the screen's surface.

regenerative amplifier An amplifier that is sensitized by regeneration.

regenerative detector A detector provided with regenerative feedback. Although such a detector is sensitive, it may be relatively unstable. Compare *nonregenerative detector*.

regenerative feedback Feedback producing regeneration, i.e., positive feedback. Compare *degenerative feedback*.

regenerative i-f amplifier An intermediate-frequency amplifier in which regeneration is introduced to boost sensitivity and, sometimes, selectivity.

regenerative reading A method of reading data (see *read*) so that it is automatically restored, by writing, to locations from which it came.

register In computer systems, an arrangement of several storage devices, such as flip-flops, for storing a certain number of digits (a two-bit register, for example, requires two flip-flops).

register capacity The range of values for quantities that can be handled by a register.

registered professional engineer A title granted by a state board of examiners to a person licensed to practice engineering. Also see *PE*.

register length The number of characters or bits that can be held in a register according to its capacity.

registration The accurate alignment of terminals or other points on different components or on opposite sides of a board, such that when the surfaces containing those points are overlaid, all points mate precisely.

regulated power supply A power supply whose output is held automatically to a constant level or within a narrow range, regardless of loading variations.

regulating transformer See *voltage-regulating transformer*.

regulation 1. Adjustment or control. 2. Automatic control. See, for example, *self-regulation*. 3. Current regulation. 4. Voltage regulation.

regulator 1. A device that automatically holds a quantity to a constant value, e.g., voltage regulator. 2. A device by means of which a quantity may be varied, e.g., potentiometer, rheostat, variable autotransformer.

regulator diode A semiconductor diode, especially a zener employed as a two-terminal voltage regulator.

regulator tube See *voltage-regulator tube*.

reject amplifier A tuned amplifier having the response of a band-suppression filter. Like the filter, the amplifier rejects or severely attenuates one frequency (or band of frequencies) while readily passing lower and higher frequencies. Compare *pass amplifier*.

reject filter See *rejection filter*.

rejection circuit A circuit performing the function of a rejection filter.

rejection filter A filter that suppresses one frequency (or band of frequencies) while passing all other frequencies.

rejection notch A sharp dip in the transmission characteristic of a crystal filter, which provides rejection-filter action at the notch frequency. Also see *crystal resonator* and *rejection filter*.

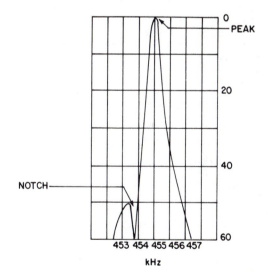

Rejection notch.

rejectivity The degree to which a selective circuit rejects an unwanted signal. Compare *transmittivity*.

rejuvenation See *reactivation*.

rel Symbol, *R*. The cgs unit of reluctance; $1R = 1$ gilbert per maxwell.

relative accuracy In a measuring instrument, the error determined as a percentage of the actual value. The difference between the actual and measured values, divided by the actual value and then multiplied by 100.

relative address In the address part of a computer program instruction, a number specifying a location relative to a base address, which, when added to the relative address, yields the absolute address.

relative error The ratio of the absolute error to the exact value of a quantity.

relative gain The current, voltage, or power gain, measured with respect to a reference standard.

relative humidity Abbreviation, *rh*. The ratio, as a percentage, of the amount of moisture in the air to the amount the air could contain at a given temperature. Compare *absolute humidity*.

relative luminosity Luminosity measured with respect to a reference level.

relative permeability The ratio $\mu a/\mu b$, where μa is the permeability of a given material, and μb the permeability of another material (or of the same material under different conditions).

relative power Power level specified with respect to another (often reference) power level.

relative uncertainty The uncertainty of a measurement divided by the measured value. The maximum value that this quotient can have is 1. Also see *uncertainty in measurement*.

relative visibility Response of the human eye to light. This is relative because the eye does not see equally well throughout the visible spectrum. Peak response is around 5.4×10^{14} Hz, which is the frequency of yellowish-green light.

relativity theory See *Einstein's theory*.

relaxation A delayed change in circuit conditions, as a result of a change in the input.

relaxation inverter An inverter circuit (see *inverter, 1*) in which the dc-to-ac conversion device is a relaxation oscillator.

relaxation oscillator An oscillator whose operation results from the buildup of a charge in a capacitor followed by sudden discharge of the capacitor, the sequence being repeated periodically. A common form of this oscillator is a circuit in which a capacitor is connected in series with a resistor and a dc power supply, and a neon bulb is connected in parallel with the capacitor. The capacitor charges slowly through the resistor until capacitor voltage reaches the firing voltage of the bulb, at which point, the bulb fires, discharging the capacitor; the events are repeated. The waveform is sawtooth in character.

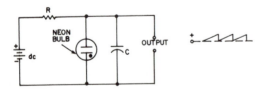

Relaxation oscillator.

relay 1. A signal-actuated switching device. In most instances, a relatively weak current or voltage is used to make the relay switch a higher current or voltage. A relay may be electromechanical or fully electronic (no moving parts). See, for example, *electromechanical relay* and *electronic relay*. 2. A repeater station.

relay amplifier A tube or transistor dc amplifier employed as a relay driver.

relay booster See *relay amplifier*.

relay driver A transistor amplifier (usually one stage) used to actuate an electromechanical relay in response to a low-powered signal.

relay flip-flop See *bistable relay*.

relay logic Abbreviation, *RL*. In computer and industrial-control practice, a logic system employing electromechanical relays as flip-flops (see *bistable relay*).

relay transmitter See *automatic repeater station,*

relay tube See *keying tube.*

release In computer practice, to release an area of memory or a peripheral from the control of a program by an instruction, monitor program, or by the operator (using other means).

release time The interval between the instant power is removed from a relay and the instant the armature is released sufficiently to operate the contacts.

reliability The dependability of operation of a device or circuit under specified conditions.

reliability engineering The branch of engineering devoted to the theory and application of reliability and based upon fundamental engineering and advanced statistical concepts.

reluctance Symbol, *R*. SI unit, A/Wb; cgs unit, rel. In a magnetic circuit, the position to the establishment of a magnetic field; it is analogous to resistance in electric circuits.

reluctance motor An electric motor having a squirrel-cage rotor with some of its teeth ground down, and a shaded-pole or split-phase type of stator that supplies a rotating magnetic field. When starting, this motor comes up to speed like an induction motor, but the protruding teeth of the rotor then follow the field in the manner of the poles of a hysteresis motor.

reluctivity Specific reluctance, i.e., the reluctance of a sample of magnetic material 1 centimeter long and 1 square centimeter in cross section. Reluctivity is the reciprocal of permeability $(1/\mu)$.

REM 1. Abbreviation of *rapid eye movement*. 2. Abbreviation for REMARK in some computer languages.

rem Acronym for *roentgen equivalent man*, an amount of ionizing radiation having the same effect on the body as a roentgen dose of gamma or X-radiation.

remagnetizer A magnetizer employed principally to restore weakened permanent magnets.

remainder 1. The result of subtracting one quantity (the subtrahend) from another (the minuend). Also called *difference*. 2. In division, what is left after the integral part of the quotient has been determined; it becomes the fractional part when it is divided by the divisor, e.g., in the division 25/3, the remainder is 1.

remanence See *residual magnetism.*

remanent flux density See *remanence.*

remodulator Any device that changes the modulation of a signal from one form to another, such as from frequency modulation to amplitude modulation, without loss of intelligence.

remote See *nemo.*

remote computer system See *multi-access system.*

remote computer system language See *QUICKTRAN.*

remote control Control of distantly located devices by mechanical means or by radio-frequency signals sent from a transmitter especially designed for the purpose; in the latter case, it is also *radio control* (RC).

remote-control receiver The complete device that selects, amplifies, and demodulates or rectifies a radio signal for control of a circuit or mechanism at a distance from the transmitter of the control signal. Some receivers have self-contained antennas.

remote-control transmitter The complete device that generates radio-frequency power, adds to it the signals needed for remote control, and radiates the modified power into space. The remote-control transmitter is here distinguished from a radio transmitter per se.

remote-cutoff tube See *extended-cutoff tube.*

remote data terminal In a computer system, a terminal connected to the central processor by a telephone or telegraph line and used for the transfer of data without providing control of the system. Also called *remote data station.*

remote error sensing A method of regulation used in some power supplies. The voltage across the load, or the current through the load, is determined by remote control. The power-supply output is adjusted to compensate for losses in the system.

remote job entry In computer practice, the keyboarding of input data at a site physically distant from the mainframe.

remote tuning The electrical or radio tuning of a circuit or device from a distance.

rep Acronym for *roentgen equivalent physical,* an amount of ionizing radiation which, upon absorption by body tissue, will develop the energy of a roentgen dose of gamma or X-radiation. 2. Colloquial abbreviation for *repetition,* as in *rep rate.*

repeatability The ability of an instrument, system, or method to give identical performance or results in successive instances.

repeater A device which retransmits a signal it receives from another source, often simultaneously. In this way, a signal may be transmitted on several frequencies, or the service area of the original station may be extended. Also see *one-way repeater* and *two-way repeater.*

repeater station See *automatic repeater station* and *repeater.*

repeating decimal A decimal fraction in which groups of digits recur endlessly, e.g., 0.333...33.

repeller An electrode, especially in a velocity-modulated tube, for reversing the direction of an electron beam.

repertoire The instruction set for a particular object or source computer programming language.

repetition instruction In a loop in a computer program, an instruction, such as NEXT (variable), causing the repetitive implementation of one or more instructions.

repetitive phenomenon See *recurrent phenomenon.*

repetitive sweep See *recurrent sweep.*

replication In a computer system, redundancy of hardware units to provide standby facilities in case of failure.

replacement A component or circuit that may be substituted directly for another. A *direct replacement* is one that fits exactly into place and function, i.e., special accommodations aren't required.

report 1. The results of testing and evaluation of a device, organized into a written document. 2. The output of a computer, printed on paper for permanent reference.

report program generator Abbreviation, *RPG.* A computer programming language with which can be produced programs for the generation of business reports.

reproduce head See *playback head.*

reproducing stylus A stylus for the playback of material from a phonograph disk.

reproduction 1. The recovery of material from some form of storage, and its presentation in original form or its simulation. 2. Playback. 3. Material presented as in 1, above.

reproduction loss See *playback loss.*

repulsion The pushing away of one item by another, as in the repulsion between electric charges or magnetic poles. Like charges or like poles repel each other. Compare *attraction.* Also see *law of repulsion.*

repulsion—induction motor An ac motor arranged to start as repulsion motor and run as an induction motor, but with better regulation than that of the latter.

repulsion motor An ac motor having an armature and commutator similar to those of a dc motor, and a stator similar to that of a split-phase motor without the auxiliary starting winding. Repulsion due to the negative half-cycle of torque is utilized to drive the armature, by placing the brushes in such a way that they close the coils only when the latter are in position to receive this repulsive action.

repulsion-start motor An ac motor which starts as a repulsion motor but at approximately 75% of full speed, its commutator is automatically short-circuited and the motor runs as an induction motor. Also see *repulsion—induction motor.*

request slip In computer practice, peripheral and memory needs for a program given in a written statement.

reradiation Radiation of energy by a body that has picked it up, as when a receiving antenna retransmits a signal.

rerecording A recording of played-back material.

reroute 1. In computer practice, to establish new channels between peripherals and main memory. 2. To establish new circuit paths, physically (as by changing conductor orientation) or electronically (as by selecting an alternate signal bus).

rerun See *rollback.*

res 1. Abbreviation of *resistance* or *resistor.* (Also, *R* and *r.*) 2. Abbreviation of *research.*

reset 1. The clearing of a flip-flop of data in storage, i.e., the setting of the flip-flop to its zero state. 2. In a computer program, an instruction to initialize the value of a variable.

reset action 1. The return of a circuit or device to its normal operating condition. 2. A method of adjusting a circuit to compensate for the severity of the abnormal condition. The extent of readjustment is determined by the extent of the departure from normal conditions.

reset generator A circuit or device that generates a pulse for resetting a flip-flop or counter. Also see *reset* and *reset terminal.*

reset pulse A pulse that resets (*reset, 1*) a storage cell in a computer memory.

reset terminal In a flip-flop, the zero-input terminal. Compare *set terminal.*

reset time The elapsed time between a malfunction and the completion of the reset action.

reset timer A device that returns a circuit to normal operation after a specified time delay.

reserve In multiple programming computer practice, to allocate memory areas and peripherals for a program.

reserve battery A battery in which the electrolyte is in a special standby chamber outside of the interelectrode section while the battery is on shelf. When the battery is readied for service, the electrolyte is caused to flow into position between the electrodes, either by heating the battery, shocking it mechanically, or inverting it. Long shelf life is the obvious benefit.

resident program A computer program, such as a monitor program, permanently in storage, as in ROM.

residual ampliude modulation See *incidental AM.*

residual charge The electric charge remaining in a capacitor after it has been initially discharged, and due to dielectric absorption.

residual frequency modulation 1. Incidental FM. 2. Frequency modulation of the fundamental frequency of a klystron by noise voltages, ac heater voltage, and the like.

residual gas Minute quantities of gas remaining in an electron tube even after extensive evacuation.

residual magnetism Magnetism remaining in a material, such as iron, after the magnetizing force has been removed.

residual modulation 1. Modulation of a signal by hum or noise. 2. Incidental AM. 3. Incidental FM.

residual voltage In the output of a null device, such as a bridge, a usually small voltage still present at null and preventing zero balance.

residue check In computer practice, verification of the result of an arithmetic operation using the remainders generated when each operand is divided by a special number; the remainder is transmitted along with the operand as a check digit.

resilience A measure of a computer system's ability to function after part of it has failed.

resin A natural or synthetic organic substance which is polymeric in structure and largely amorphous. Plastics, such as Bakelite, Lucite, polyethylene, and the like, used in electronics, are made from synthetic resins.

resistance Symbol, R or r. Unit, ohm. In a device, component, or circuit, the simple opposition to current flow. Resistance by itself causes no phase shift; $R = E/I$, where E is voltage, and I is current. Compare *reactance*.

resistance alloys Metallic alloys used in the manufacture of resistance wire and resistance elements. Such alloys include *constantan, German silver, manganin, Monel metal,* and *nichrome.*

resistance balance A device used for the purpose of balancing a circuit, by means of the insertion of resistances.

resistance bridge A bridge (see *bridge, 2*) for measuring resistance only.

resistance-capacitance Abbreviation, *RC*. Pertaining to a combination of resistance and capacitance, e.g., *resistance-capacitance circuit.*

resistance-capacitance bridge 1. A four-arm null circuit containing only resistors and capacitors. Also see *bridge, 1.* 2. An ac bridge (see *bridge, 2*) for measuring resistance and capacitance.

resistance-capacitance circuit A circuit containing only resistors and capacitors, e.g., *RC* filter, *RC* bridge.

resistance-capacitance-coupled amplifier A multistage amplifier circuit in which resistance-capacitance coupling is employed between stages and the input and output points of the circuit.

resistance-capacitance coupling Coupling, especially between stages in a circuit, employing blocking capacitors and supply-path resistors.

resistance-capacitance filter A power-supply filter or wave filter containing only resistors and capacitors. The resistors are in the positions occupied by inductors in *LC* filters.

resistance-capacitance phase shifter A phase shifter containing only resistors and capacitors to obtain the desired shift.

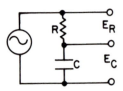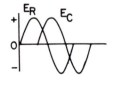

Resistance-capacitance phase shifter.

resistance-capacitance time constant Symbol, t. The time constant (see *electrical time constant*) of a circuit containing (ideally) only resistance and capacitance; $t = RC$, where t is in seconds, R in ohms, and C in farads. Also called *RC time constant.* Compare *resistance-inductance time constant.*

resistance-capacitance tuning Tuning of a circuit, such as that of an amplifier or oscillator, by means of a variable resistor or ganged units of this type. See, for example, *parallel-tee amplifier, parallel-tee oscillator,* and *Wien-bridge oscillator.*

resistance-coupled amplifier See *resistance-capacitance-coupled amplifier.*

resistance drop The voltage drop across a resistor, or across the inherent resistance of a device.

resistance-inductance Abbreviation, *RL*. Pertaining to a combination of resistance and inductance, e.g., resistance-inductance circuit.

resistance-inductance bridge 1. A four-arm null circuit containing only resistors and inductors. Also see *bridge, 1.* 2. An ac bridge (see *bridge, 2*) for measuring resistance and inductance only.

resistance-inductance circuit A circuit containing only resistors and inductors, e.g., resistance-inductance phase shifter.

resistance-inductance phase shifter A phase shifter containing only resistors and inductors to obtain the desired phase shift.

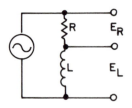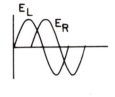

Resistance-inductance phase shifter.

resistance lamp An incandescent bulb, inserted in series with a circuit to provide a dropping resistance. Such a lamp is capable of dissipating a large amount of power, shows very little reactance at low frequencies, and is inexpensive.

resistance magnetometer A magnetometer whose operation is based upon the change of electrical resistance of a material (such as bismuth wire) placed in the magnetic field under test.

resistance material A substance, such as carbon or German silver, whose resistivity is high enough to enable its use as a lumped resistor. See, for example, *resistance alloys* and *resistance metal.*

resistance metal A metal, such as iron, whose resistivity is high enough to enable its use as a lumped resistor. Also see *resistance alloys.*

resistance pad An attenuator comprised of noninductive resistors.

resistance standard A highly accurate and stable resistor employed in precision measurements of resistance. Also see *primary standard* and *secondary standard.*

resistance strain gauge An electrical strain gauge in which the stressed element is a thin resistance wire.

resistance strip A strip of metallic or nonmetallic resistance material. Also see *resistance alloys* and *resistance metal.*

resistance temperature detector A transducer consisting of a specially made resistor whose resistance varies linearly with temperature.

resistance thermometer An electronic thermometer whose operation is based upon the change of resistance of a wire as it is heated or cooled.

resistance transducer See *resistive transducer.*

resistance tuning See *variable-resistance tuning.*

resistance welding An electrical or electronic welding process in which the workpieces are heated by current flowing through the inherent resistance of their junction.

resistance wire Wire made of a metal or alloy that exhibits significant resistivity. See, for example, *resistance alloys* and *resistance metal.*

resistance-wire sensor A specific length of resistance wire, properly mounted, whose resistance is proportional to a sensed phenomenon, such as strain, temperature, presence of gas, pressure, etc. See, for example, *electrical strain gauge, gas detector,* and *pressure transducer.*

resistive current The component of alternating current which is in phase with voltage. Also called *watt current.* Compare *reactive current.*

resistive cutoff frequency Symbol, *frco.* The frequency beyond which a tunnel diode ceases to exhibit negative conductance.

resistive load A load device which is essentially a pure resistance.

resistive losses Losses due to the resistance of a circuit or device and appearing as heat.

resistive transducer A transducer in which the sensed phenomenon causes a change in resistance, which in turn produces a corresponding change in output current or voltage. Compare *capacitive transducer, crystal transducer, inductive transducer, magnetic transducer,* and *photoelectric transducer.*

resistive trimmer See *trimmer resistor.*

resistive voltage The voltage across the resistance component in a circuit. In an ac circuit, resistive voltage is in phase with current.

resistivity Symbol, ρ. Unit, *ohm-m.* Specific resistance, i.e., resistance per unit volume or per unit area. Resistivity may be expressed in terms of *ohms per meter cube* or *ohms per meter square.* Also see *microhm-centimeter* and *ohm-centimeter.*

resistor A device having resistance concentrated in lumped form. Also see *resistance* and *resistivity.*

resistor-capacitor-transistor logic. Abbreviation, *RCTL.* In computer practice, a type of resistor-transistor logic in which capacitors are employed to enhance switching speed.

resistor color code See *color code.*

resistor core A form around which a nichrome or other semiconducting wire can be wound, for the purpose of constructing a high-power resistor.

resistor decade See *decade resistor.*

resistor diode A usually forward-biased semiconductor diode which acts as a voltage-variable resistor (varistor).

resistor FET See *electronic resistor.*

resistor fuse See *fusible resistor.*

Resistors in parallel See *parallel resistors.*

resistors in parallel-series See *parallel-series resistors.*

resistors in series See *series resistors.*

resistors in series-parallel See *series-parallel resistors.*

resistor substitution box A self-contained assortment of common-value resistors arranged to be switched one at a time to a pair of terminals. In troubleshooting and circuit development, any of several useful fixed resistance values may thus be obtained.

resistor transistor See *electronic resistor.*

resistor-transistor logic Abbreviation, *RTL.* In computer practice, a circuit in which the logic function is performed by resistors, and an inverted output is provided by transistors.

resistor tube See *electronic resistor.*

resnatron A form of vacuum tube that is used as an oscillator and amplifier at ultra-high and microwave frequencies. Essentially a cavi-

ty resonator.

resolution 1. The degree to which closely adjacent parts of an image can be differentiated. 2. Reduction by analysis.

resolution ratio Symbol, *m.* In a television image, the ratio of horizontal resolution to vertical resolution; $m = rh/rv$.

resonance 1. The state in which the natural response frequency of a circuit coincides with the frequency of an applied signal, or vice versa, yielding intensified response. 2. The state in which the natural vibration frequency of a body coincides with an applied vibration force, or vice versa, yielding reinforced vibration of the body.

resonance bridge An ac bridge (see *bridge,* 2) in which one arm (sometimes, two) is a series- or parallel-resonant circuit, the other arms being resistances. Also see *series-type resonance bridge* and *shunt-type resonance bridge.*

resonant cavity A chamber whose size reinforces energy injected into it at the natural frequency, which is determined by the chamber's dimensions. Such cavities have been employed for sound waves as well as electromagnetic waves.

resonant current Current flowing in a tuned circuit at resonance.

resonance curve A curve depicting the response of a resonant circuit over a band of frequencies within which is the circuit's resonant frequency. Current through, or voltage across, the circuit is the dependent variable (vertical axis), and frequency is the independent variable (horizontal axis).

resonance radiation Electromagnetic radiation from an energized substance, resulting from movement of electrons from a higher to lower energy levels. When an electron moves from a higher to a lower orbit, a photon, having a definite wavelength, is emitted.

resonant cavity A chamber whose size reinforces energy injected into it at the natural frequency, which is determined by the chamber's dimensions. Such cavities have been employed for sound waves as well as microwaves.

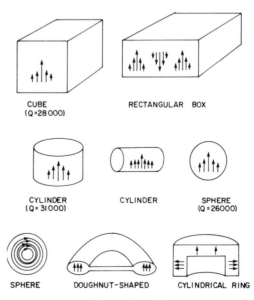

Resonant cavities.

resonant circuit A circuit whose constants are chosen for maximum circuit response at a given frequency. Examples: parallel-resonant circuit, series-resonant circuit. Also see *resonance* and *resonant frequency*.

resonant feeder An antenna feeder that is resonant at the operating frequency.

resonant filter A filter containing at least one series- or parallel-resonant arm for sharp response. Thus, a power-supply filter of this kind might have a parallel-resonant arm acting as a wavetrap at the ripple frequency.

resonant frequency Symbol, *fr*. The *natural* frequency at which a circuit oscillates or a device vibrates. In an *LC* circuit (series-resonant or parallel-resonant), inductive and capacitive reactances are equal at the resonant frequency.

resonant-gate transistor A transistor embodying a tiny tuning fork for resonance at low frequencies, thus eliminating bulky coils and capacitors.

resonant line A transmission line that is resonant at the operating frequency.

resonant-line amplifier See *line-type amplifier*.

resonant-line circuit A circuit employing resonant lines as a tank. See, for example, *line-type amplifier* and *line-type oscillator*.

resonant-line oscillator See *line-type oscillator*.

resonant-line wavemeter See *Lecher frame*.

resonant rise See *voltage rise*.

resonant-slope amplifier See *dielectric amplifier*.

resonant-slope detector See *slope detector*.

resonant suckout The drawing of rf energy out of the used part of a coil or transmission line by the unused part when the latter resonates at the same frequency.

resonant-voltage rise See *voltage rise*.

resonant-voltage stepup See *voltage rise*.

resonator A device that produces or undergoes resonance. See, for example, *Helmholtz resonator* and *resonant cavity*.

resource A part of a computer system that can be used for a specific application as a unit, e.g., a peripheral.

responder The transmitting section of a transponder.

response The behavior of a circuit or device (especially in terms of its dependent variables) in accordance with an applied signal, e.g., frequency response, current-vs-voltage.

response curve A graph depicting the performance of a circuit or device. A common type is a current-vs-voltage curve.

response time The interval between the instant a signal is applied to or removed from a circuit or device and the instant the circuit acts accordingly,

restart Following a malfunction or error occurring during a computer program run, to go back to an earlier point in the program.

resting state See *quiescent state*.

restore See *reset*.

resultant 1. The vector that results from the addition of two or more vectors. 2. A quantity that results from mathematical operations performed on other quantities.

retarding-field negative resistance Negative resistance occurring in a properly biased pentode as a result of negative transconductance. The effect is so called because of the retarding field produced by the suppressor.

retarding-field oscillator See *Barkhausen—Kurtz oscillator*.

retarding magnet See *drag magnet*.

retentivity Symbol, *n*. The property whereby a material retains magnetism imparted to it.

retention period In computer practice, the time during which the information on a reel of magnetic tape must be kept intact.

reticle As seen through the eyepiece of an optical instrument, a reference pattern (e.g., crosshairs) for gauging size or distance.

RETMA Radio-Electronics-Television Manufacturers' Association. (One of the ancestors of the Electronic Industries Association.)

retrace 1. In an oscilloscope tube or TV picture tube, the return of the scanning beam to its starting point. 2. The line traced on the screen by a retracing beam (see 1, above).

retrace blanking Obliteration of the return trace of the electron beam in an oscilloscope tube or TV picture tube, to make the retrace line invisible on the screen. Also see *blank, 2* and *retrace, 2*.

retrace line See *retrace, 2*.

retrace ratio For the swept beam in a cathode-ray tube, the ratio of the scanning velocity in one direction to the scanning velocity in the other direction (retrace). Also see *retrace, 1, 2*.

retrace time The time required for an electron beam to return to the starting point of a scan. Also see *retrace, 1, 2*.

retrofit To supply something with specially designed or adapted parts that weren't available when it was made.

return 1. Retrace. 2. Return circuit. 3. Return point. 4. In an electronic circuit, ground and the ground path.

return circuit The circuit through which current returns to a generator.

return instruction In a computer program, an instruction in a subroutine directing operation back to a specific point in the main program.

return interval In an oscilloscope or television cathode-ray tube, the amount of time required for the scanning beam to move from the end of one trace or line to the beginning of the next.

return line See *retrace, 2*.

return point 1. The point to which circuits are returned, e,g., a common ground point. 2. The terminal point of a return circuit.

return ratio In a feedback system, the feedback factor.

return time See *retrace time*.

return-to-zero 1. Abbreviationn *RZ*. In the magnetic recording of data, a method in which the write current returns to zero following the write pulse. Compare *nonreturn-to-zero*. 2. A logic system in which the zero and one states are represented by zero voltage and a discrete voltage.

return trace See *retrace, 1, 2*.

REV Abbreviation of *reentry vehicle*.

rev 1. Abbreviation of *revolution*. 2. To quickly and substantially increase the angular velocity of a motor.

reverberation The multiple reflections of sound waves between the walls of an enclosed chamber.

reverberation system A system of devices operated with an electronic organ to simulate the effect of reverberation in a large room, such as a church auditorium.

reverberation time The time required, in an enclosed chamber, for a sound to die out once it has stopped, i.e., the duration of reverberation.

reverberation unit A device for producing artificial echoes, especially in the operation of electronic musical instruments or a car radio or tape player to create the illusion of reverberation.

reverse agc Automatic gain control of the conventional sort, i.e., the type in which a signal-dependent voltage is fed back to an earlier stage to adjust its gain automatically. Compare *forward agc*.

reverse bias Reverse voltage or current in a transistor or a semiconductor diode. Compare *forward bias*.

reverse breakdown The sudden increase in reverse current of a semiconductor device at high reverse voltage.

reverse breakdown voltage For a reverse-biased semiconductor device, the voltage at which the current increase suddenly to a high value. This will not result in breakdown in the destructive sense if an external resistor is included to limit the current to a safe value.

reverse characteristic The current-vs-voltage response of a semiconductor junction that is biased in the reverse (low-conduction) direction. Compare *forward characteristic*.

reverse conduction The very small current conduction through a pn junction when it is reverse-biased. Compare *forward conduction*.

reverse current Symbol, Ir. The lower current that flows through a pn junction when it is reverse-biased. Also called *back current*. Compare *forward current*.

reverse Polish notation Abbreviation, *RPN*. A system of notation for expressing mathematical operations in which the operators follow the operands being manipulated; it is a mode of entry for some calculators, e.g., the operation 7×2 might be entered as 7, ENTER, 2, × (result displayed).

reverse recovery time See *recovery time, 1*.

reverse resistance Symbol, Rr. The higher resistance of a reverse-biased pn junction. Also called *back resistance*, Compare *forward resistance*.

reverse voltage Symbol, Er or Vr. Voltage of a polarity which causes minimum current to flow through a pn junction. Also called *back voltage*. Compare *forward voltage*.

reverse-voltage capacitance The internal capacitance of a reverse-biased semiconductor pn junction.

reverse voltage drop The voltage drop across a semiconductor pn junction that is biased in the reverse (low-conduction) direction. Compare *voltage drop*.

reversible counter A counter which, by a control signal, can have the value it is holding increased or decreased.

reversible permeability The permeability of a ferromagnetic substance when the magnitude of the alternating-current field is arbitrarily small.

reversing switch 1. A switch that reverses the polarity of input dc. 2. A switch that reverses the direction of motor rotation.

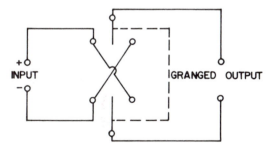

Reversing switch.

revolution Abbreviation, *rev*. One complete rotation, i.e., 360 degrees (2π radians) of circular travel.

revolving field See *rotating field*.

rewind To run a magnetic tape on a transport in the direction opposite to that associated with the play mode, usually to the point where the tape can be played from the beginning.

rewrite In computer practice, to return information read from a storage location to that location by recording.

Rf 1. Symbol for *filament resistance*. 2. Symbol for *feedback resistor*.

rf Abbreviation of *radio frequency*.

rf amplifier See *radio-frequency amplifier*.

RFC Abbreviation of *radio-frequency choke*.

rf heating See *radio-frequency heating*.

rfi Abbreviation of *radio-frequency interference*.

rf inverse feedback A negative-feedback system for radiophone transmiters, in which a portion of the modulated rf signal is rectified and the resulting dc voltage is filtered and applied as bias to one of the af sages in the proper polarity for degeneration.

rfo Abbreviation of radio frequency oscillator.

rf power supply See *oscillator-type power supply*.

rf probe See *rectifier probe*.

rf resistance See *radio-frequency resistance*,

rf selectivity See *radio-frequency selectivity*.

rf transistor See *radio-frequency transistor*.

RG Symbol for *gate resistance*.

Rg Symbol for *grid resistance*.

RGB Abbreviation of *red—green—blue*.

RGT Abbreviation of *resonant-gate transistor*.

Rh Symbol for *rhodium*.

R/h Abbreviation of *roentgens per hour*.

Rh 1. Symbol for *heater resistance*. 2. Symbol for *hot resistance*.

rh Abbreviation of *relative humidiy*.

rhenium Symbol, *Re*. A metallic element. Atomic number, 75. Atomic weight, 186.22. Rhenium is employed in some thermocouples.

rheostat A variable dropping resistor usually of the rotary type but often of the long slider type.

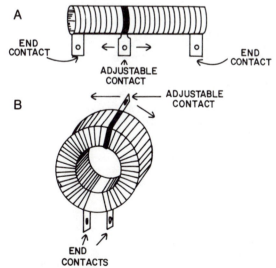

At A, a solenoidal rheostat. At B, a toroidal rheostat. Both types exhibit inductance, and are therefore suitable only for use at low frequencies.

Rhf Symbol for high-frequency resistance.

RHI Abbreviation of *range-height indicator*.

rho The 17th letter (P,ϱ) of the Greek alphabet. Capital rho (P) is seldom used as an electronic symbol, because of its resemblance to the Roman letter *P*. Lower-case rho (ϱ) is the symbol for resistivity, electric-charge density, and ripple factor.

rhodium Symbol, *Rh*. A metallic element. Atomic number, 45. Atomic weight, 102.91.

rhombus A diamond configuration.

rhombic antenna See *diamond antenna*.

rho-theta A radionavigation system in which a single transmitting station is used, and the position is determined according to polar coordinates (distance and direction).

rhumbatron A cavity resonator, especially one in a klystron. Also see *klystron* and *resonant cavity*.

RI Abbreviation of *radio interference*.

Ri Symbol for *input resistance*. (Also, *Rin*.)

RIAA *Recording Industry Association of America*. Sets standards for phonograph recording and reproduction in the United States.

RIAA curve The amplitude-versus-frequency function used in recording and reproduction of long-playing (33-1/3 RPM) phonograph disks, and specified by the RIAA. The RIAA curve takes advantage of the sensitivity of the human ear at various frequencies to reduce the level of audible noise.

ribbon microphone See *velocity microphone*.

Rice neutralization The method of neutralization in which the neutralizing capacitor is connected from the plate of the tube to the bottom of the center-tapped grid tank.

Richardson—Dushman equation See *Richardson's equation*.

Richardson's equation An equation for evaluating the electron emission current from the cathode of an electron tube; $Is = A T^2 e^{-b0/T}$, where Is is total emission current (A/cm^2), A is a constant for the cathode material, T is absolute temperature (K), e is the base of natural logs, (2.71828), and $b0$ is $\phi 0/K$ a constant governed by the work required to move an electron from inside the emitter to the exterior ($\phi 0$ is the work function in eV; K is the Boltzmann constant.).

Richter scale A logarithmic scale for describing the relative magnitude of earthquakes on which 1.5 corresponds to a barely discernible disturbance.

ride gain In broadcasting, the practice of constantly adjusting the audio modulation of the transmitter for optimum operation.

Rieke chart A visual aid, similar to the Smith chart, employed with traveling-wave tubes in the UHF and microwave bands for determining the best load impedance.

rig Radio transceiver or transmitter.

Righi-Leduc effect A phenomenon somewhat analogous to the Hall effect: When a metal strip conducting heat is placed into a magnetic field perpendicular to the plane of the strip, a temperature difference develops across the strip.

right-angle line section See *ell*.

right-hand lay See *direction of lay*.

right-hand polarized wave See *clockwise-polarized wave*.

right-hand rule for induced emf See *Fleming's right-hand rule*.

right-hand rule for wire A simple rule for indicating the direction of the magnetic field about a straight wire carrying current: When the wire is grasped in the right hand with the thumb pointing in the direction of current flow, the fingers show the direction of the magnetic field.

right-hand taper Potentiometer or rheostat taper in which most of the resistance is in the clockwise half of rotation as viewed from the front. Compare *left-hand taper*.

right justified In a computer memory location, a data item that takes up consecutive bit positions from right to left.

right shift In computer operation, a shift whereby word bits are displayed to the right; the effect is division in a right arithmetic shift.

rim drive In a tape recorder, a driving method in which the motor shaft is provided with a smooth pulley that transfers motion directly to the rubber-tired rim of the flywheel.

Rin Symbol for *input resistance*. (Also, *Ri*.)

ring 1. The core of a toroidal coil. 2. Hybrid ring. 3. Ring modulator. 4. Ring inductor (see *ring inductor, 1, 2*). 5. Ring magnet. 6. Ringing.

ring armature A motor or generator armature having a ring winding.

ring circuit 1. Ring modulator. 2. A waveguide hybrid-tee resembling a ring having radial branches. 3. In amateur radio practice, a circuit connected to a phone line and transmitter in such a manner that the transmitter is energized and modulated with an identifiable signal each time the phone rings (so that it may be answered via remote control).

ring counter An electronic counter in which successive cascaded stages form a ring; i.e., the last stage in the chain is connected to the first stage, so that the counter advances through the cycle, stage by stage, repetitively.

ring head In tape recorders, a record—reproduce head which is essentially a metal ring with a gap at one point and upon which the coils are wound.

ring inductor 1. Single-turn coil. 2. Shading coil.

ringing The usually unintentional self-oscillation in a pulsed *LC* circuit, sustained by the circuit's flywheel action (hysteresis) and usually producing a damped wave.

ringing coil In the horizontal oscillator in a TV circuit, a small, adjustable coil (shunted by a capacitor) employed to produce a sharp rise in input-signal voltage.

ringing current In wire telephony, an alternating current superimposed upon the dc operating current to ring the bells.

ringing time See *ring time, 1*.

ring magnet A permanent magnet in the shape of a ring or donut.

ring modulator A double-balanced diode-type modulator circuit; it takes its name from the ring-like arrangement of the four diodes.

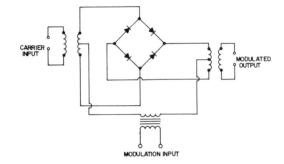

Ring modulator.

ring oscillator A self-excited oscillator in which four tubes or transistors are operated in push-pull—parallel.

ring shift In computer operation, the cyclic shifting of digits from one end of a regiser to the other.

ring time 1. The period of a damped oscillation, especially one set up in an *LC* circuit by a pulse. 2. The time required for an echo-box signal to decay below the display level (see *echo box*).

ring winding A winding in which the turns of the coil are laid on the outside of a ring-shaped core and passed through its center, resulting in a donut coil with a core.

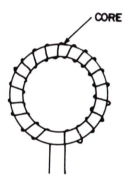

Ring winding.

ripple 1. A usually small ac voltage or current occurring in a dc component as an unavoidable byproduct of rectification or commutation. 2. In computer and data-processing practice, the serial transmission of data.

ripple amplitude The usually peak or peak-to-peak value of a ripple component (see *ripple, 1*).

ripple counter A binary counter consisting of flip-flops cascaded in series. A pulse must pass sequentially from the input, through each stage, to the output of the chain.

ripple current Current flowing in a circuit as the result of ripple voltage (see *ripple amplitude*).

ripple factor Symbol, ϱ. The ratio of the ripple voltage (*er*) to the dc voltage (*Edc*) when ripple is present in a dc component; $\varrho = er/Edc$.

ripple frequency The frequency of a ripple component (see *ripple, 1*).

ripple percentage See *percent ripple*.

ripple torque Symbol, *TR*. In a torque motor, the small fluctuation in torque resulting from commutator switching action.

ripple voltage See *ripple amplitude*.

rise 1. Resonant rise (see *voltage rise*). 2. Rise time. 3. The increase in amplitude of a pulse or wave.

rise cable 1. A vertical feeder cable. 2. A vertical section of a feeder cable.

rise time Symbol, *tr*. The time required for a pulse to rise from zero (or reference level) to maximum amplitude. Compare *fall time*.

RJE Abbreviation of *remote job entry*.

RK Symbol for *cathode resistance*.

RL 1. Abbreviation of *resistance-inductance*. 2. Abbreviation of *relay logic*.

RL Abbreviation of *load resistance*.

RL bridge See *resistance—inductance bridge*.

RL circuit See *resistance—inductance circuit*.

Rlf Symbol for *low-frequency resistance*.

RL phase shifter See *resistance—inductance phase shifter*.

Rm Symbol for *meter resistance*.

rm Symbol for *emitter—collector transresistance* of a bipolar transistor.

RMA Radio Manufacturers' Association (the original ancestor of the Electronic Industries Association).

rms Abbreviation of *root mean square*.

rms current See *effective current*.

rms deviation See *standard deviation*.

rms meter A current meter or voltmeter whose deflection is proportional to the rms value of current or voltage. In most meters, the deflection is proportional to either the peak value or average value, but the scale of an rms unit is graduated on the basis of sine-wave input.

rms value See *effective value*.

rms voltage See *effective voltage*.

Rn Symbol for *radon*.

Rn 1. Symbol for *negative resistance*. (Also, *– R*.) 2. Symbol for *null resistance*.

Ro Symbol for *output resistance*. (Also, *Rout*.)

Robert Phonetic alphabet communications code word for the letter *R*.

robot An electronically activated machine that behaves like a human being. It is often built to resemble the human body.

robotics The technology underlying automatons; i.e., the design, operation, and maintenance of *robots*.

Rochelle salt Formula, NaKC4H406• 4H20. Sodium potassium tartrate, a material whose crystals are piezoelectric. Such crystals are used in crystal microphones, loudspeakers, and transducers. Also called *Seignete salt*.

rock Slang for *quartz crystal*.

rockbound Crystal controlled.

rocker switch A toggle switch the lever of which is a specially shaped bar which is rocked back and forth to operate the switch. Compare *bat-handle switch, paddle switch,* and *slide switch*.

Rocker switch.

rod 1. A unit of length or distance; 1 rod = 5.029 meters. 2. A bar of material with special properties.

rod antenna See *ferrite-rod antenna*.

rod magnet A permanent magnet in the shape of a rod with circular or elliptical cross section.

roentgen Abbreviation, *r*. A unit of radioactive radiation; 1 r is the quantity of radiation that produces 1 esu of electricity (positive or negative) per cubic centimeter of air at standard temperature and pressure. In average tissue, 1 r produces ionization equivalent to an energy concentration of 2.58×10^{-4} C/kg (93 ergs per gram). Also see *milliroentgen*.

roentgen equivalent man See *rem*.

roentgen equivalent physical See *rep*.

roentgen ray See *X-ray*.

roger A communications signal meaning "Acknowledged."

Roget spiral A spring-like wire device that contracts in proportion to the magnitude of the current.

role indicator In computer practice, a code classifying a keyword as a part of the speech, e.g., noun.

roll 1. In a display terminal having a line length of less than the standard 80 characters, an operating feature that allows the operator to follow the text along. The cursor remains fixed near the center of the displayed line, while the text moves from right to left. 2. Rolling in a television picture because of lack of vertical synchronization.

rollback In computer practice, the running again of a computer program or portion of the program. Also called *rerun*.

rolling In television, the apparent continuous upward or downward movement of the picture, resulting from lack of vertical synchronization between the transmitter and receiver.

rolloff The rate at which the response of a circuit, such as the frequency response of an amplifier, falls off after a certain critical frequency.

ROM Abbreviation of *read-only memory*.

Romeo Phonetic alphabet code word for the letter *R*.

Romex cable A form of wire cable with a covering that is highly resistant to the environment.

roof mount A metal bracket for fastening an antenna mast to a roof.

room noise Ambient noise in a room.

room resonance Acoustic resonance due to the geometry and contents of a room.

room temperature Abbreviation, *RT*. Literally, the temperature of the room in which a test or fabrication is carried out. The term has no widely accepted quantitative meaning; however, it is of value in distinguishing between operations that may be performed at the *ambient temperature* and those that require an oven or a cold chamber. Nevertheless, the term implies a temperature in the vicinity of 21 °C (70 °F). Exacting scientific work demands, of course, that the actual ambient temperature always be observed and recorded.

root 1. A number which, when multiplied by itself a number of times specified by the index of the radical (sign, √) it is in, gives the number for which the root is being extracted. 2. Terms or expressions which, when substituted for unknown quantities in an equation, will *satisfy* the equation (make it a true statement). For example, a quadratic equation has two roots, one positive and one negative. Roots may be real, complex, equal, or unequal.

rooter An analog or digital device employed to extract a desired root of a number (see *root, 1*). See, for example, *square rooter*.

root mean square Abbreviation, *rms*. The square root of the mean of the squares of a set of values. Example: *rms current, rms voltage*.

root-mean-square current See *effective current*.

root-mean-square value See *effective value*.

root-mean-square voltage See *effective voltage*.

ROP Abbreviation of *record of production*.

rosin A resin derived chemically from an extract of pine wood and used in some solders.

rotameter A type of fluid flow gauge consisting of a float within a glass tube having incremental markings.

rotary amplifier See *amplidyne*.

rotary antenna See *rotatable antenna*.

rotary beam A rotatable beam antenna. Also see *rotatable antenna*.

rotary-beam antenna See *rotary beam*.

rotary converter A dynamo (electric machine) having a direct-current armature connected to a commutator on one end of the shaft and to slip rings on the other end. When the machine is operated as a dc motor, it delivers an ac output, and vice versa. Also called *double-current generator*.

rotary inverter A motor-generator used to change a dc input voltage into an ac output voltage.

rotary-motion sensor A transducer that delivers an output voltage proportional to the arc over which its shaft has been turned.

rotary power amplifier See *dc generator amplifier*.

rotary relay An electromechanical relay in which a pivoted armature rotates to open or close the contacts. The *meter relay* is an example of this device.

rotary selector switch See *rotary switch*.

rotary spark gap A type of spark gap consisting of one or two stationary electrodes and a series of moving electrodes mounted around the periphery of a spinning disk. The spark occurs when a moving electrode is opposite a stationary electrode. This arrangement provides a high-frequency spark discharge (the greater the number of moving electrodes and the higher the speed of rotation, the higher the frequency) and significant cooling action.

rotary stepping relay See *stepping switch*.

rotary stepping switch See *stepping switch*.

rotary switch A switch in which a blade moves in a circle or in arcs over the contacts.

rotary transformer A motor-generator used to change an input voltage into a lower or higher output voltage.

rotatable antenna An antenna which may be turned manually or by a motor to change its directional response.

rotating amplifier See *dc generator amplifier*.

rotating antenna An antenna which constantly turns to scan a given area, e.g., radar antenna.

rotating-capacitor modulator A frequency modulator consisting of a motor-driven variable capacitor. Also see *wobbulator*.

rotating field An ac field, as that generated by the stator of some motors, that revolves between poles.

rotating interrupter A commutator (see *commutator, 1*).

rotating machines Collectively, motors, generators, amplidynes, rotary converters, and the like. These devices utilize magnetic flux to convert mechanical energy into electrical energy, or vice versa, or to convert electrical energy of one level into electrical energy of another level.

rotating memory See *disk memory* and *drum memory*.

rotating voltmeter See *generating voltmeter*.

rotator A motor-driven, remotely controlled mechanism for turning a beam-type antenna.

rotor 1. Rotatable coil. Compare *stator, 1*. 2. The rotating member of a motor or generator. Compare *stator, 2*. 3. The rotating-plate assembly of a variable capacitor. Compare *stator, 3*.

rotor blade The wiper arm of a rheostat or potentiometer.

rotor coil See *rotor, 1, 2*.

rotor plate The rotating plate(s) of a variable capacitor. Compare *stator plate*.

roulette pattern A circular pattern for frequency identification with an oscilloscope, consisting of loops around the screen's circumference.

Compare *gear-wheel pattern, Lissajous figure,* and *spot-wheel pattern.*

rounded number A number that results from dropping a less significant digit at the right end of a given number, and if the dropped digit is 5 or greater, increasing the preceding digit by 1 (in an alternate method, only increasing an odd preceding digit when the dropped digit is 5). The rounding off of pi proceeds as follows: 3.14159, 3.1416, 3.142, 3.14. Also called *truncated number.*

rounding 1. The approximation of a number to the nearest integral value. For example, rounding of 3.44 gives the value 3; rounding of 3.54 gives the value 4. These are the nearest integers to the respective decimal numbers. 2. The approximation of a value to a specified number of decimal places or significant digits. 3. Smoothing of the corners of a square wave or sawtooth wave, resulting in lengthening of the transition time from one state to another.

rounding error The error resulting from the rounding off of a number (see *rounded number*).

round-off To shorten an otherwise lengthy number by replacing numerical digits with zeros and increasing the final nonreplaced digit by 1 if the leftmost replaced digit is 5 or greater. Thus, 3.141 592 653 can be rounded of to 3.1416 or 3.14.

round-up A form of numerical approximation, in which a number with a value of $n.5$ or greater is assigned the value $n + 1$. This is a feature of many calculators using scientific notation or a fixed number of decimal places.

Rout Symbol for *output resistance.* (Also, *Ro*).

route 1. To physically position wires or conducting circuit paths by planning and deliberation. 2. The path over which conductors are positioned. 3. A path over which signals or information may be carried.

routine 1. In computer practice, the complete sequence of instructions for performing an operation, i.e., a program or program segment. 2. A test or measurement sequence. 3. An assembly or manufacturing sequence. 4. A standard troubleshooting sequence.

row 1. In a matrix, a horizontal arrangement of values. 2. In paper tape, a line of holes spanning the width of the tape. 3. In a punched card, a line of holes perpendicular to the short edge.

row binary code Number representation in which consecutive punching positions (in rows of a punched card) correspond to sequential bits.

row pitch On a punched paper tape, the center-to-center distance between holes running the length of the tape.

Rp 1. Symbol for *plate resistance.* (Also, *rp*). 2. Symbol for *positive resistance.* 3. Symbol for *parallel resistance.* Also written *Rpar.* 4. Symbol for *primary resistance.* Also written *Rpri.*

Rpar Symbol for *parallel resistance.* (Also, *Rp*).

r-parameters 1. Device or network parameters expressed as resistances. 2. Transistor parameters in terms of resistance values in the equivalent tee network: *rb, re, rc,* and *rm.* Compare *g-parameters* and *h-parameters.*

RPG Abbreviation of *report program generator.*

rpm Abbreviation of *revolutions per minute.*

rpm meter See *electronic tachometer* and *stroboscope.*

RPN Abbreviation of *reverse Polish notation.*

Rpri Symbol for *primary resistance.* (Also, *Rp*).

rps Abbreviation of *revolutions per second.*

RPT Radiotelegraphic abbreviation of *repeat.*

Rreq Symbol for *required resistance.*

RS Symbol for *source resistance* in a field-effect transistor.

Rs 1. Symbol for *screen resistance.* 2. Symbol for *series resistance.* (Also, *Rser*.) 3. Symbol for *secondary resistance.* (Also, *Rsec.*)

Rsec Symbol for *secondary resistance.* (Also, *Rs.*)

Rser Symbol for *series resistance.* (Also, *Rs.*)

RST flip-flop A conventional flip-flop subject to the operations of reset, set, and trigger.

RST system In the amateur radio service, a method of reporting signal quality in terms of *r*eliability, *s*trength, and *t*one.

R-sweep In oscilloscope operation, an expanded portion of the trace produced by a long triggered sweep, which permits close reading of a small part.

RT 1. Abbreviation of *radiotelephone.* 2. Abbreviation of *room temperature.*

rt Abbreviation of *right.*

RT Symbol for *thermal resistance.*

Rt Symbol for *total resistance.*

RTD Abbreviation of *resistance temperature detector.*

RTL Abbreviation of *resistor-transistor logic.*

RTMA Radio-Television Manufacturers' Association (one of the ancestors of the Electronic Industries Association).

RTTY Abbreviation of *radioteletype.*

RTZ Abbreviation of *return-to-zero.* (Also, *RZ.*)

Ru Symbol for *ruthenium.*

rubber A natural insulating material which is an elastomer exhibiting rapid elastic recovery. Dielectric constant, 2 to 3.5. Dielectric strength, 16 to 50 kV/mm. Also called *India rubber.* Compare *hard rubber.*

rubber-covered wire Wire that is insulated with a jacket of rubber.

rubidium Symbol, *Rb.* A metallic element. Atomic number, 37. Atomic weight, 85.48.

ruby laser A *laser* employing a ruby rod as the stimulated element.

ruby maser A *maser* in which the stimulated material is ruby.

Ruhmkorff coil See *induction coil.*

rumble A usually muffled low-frequency noise, or the corresponding electrical equivalent.

rumble filter A filter having sharp cutoff below about 50 Hz, for eliminating rumble arising from irregularities in the rotation of a phonograph turntable. Also see *rumble.*

run The execution of a computer routine or program.

runaway In a current-carrying circuit or device, especially a semiconductor, the rapid cumulative increase of current (and attendant temperature increase) that moves toward a destructive value.

run chart In computer practice, a type of flowchart for a job, i.e., one showing the organization and order of the pertinent programs to be run.

running accumulator A computer storage unit having registers linked so that data is transferred unidirectionally from one to the other, and in which only one register is accessible from the outside.

running open 1. The condition of a teletype machine running continuously in the absence of a signal. The teleprinter operates but nothing is printed; this keeps the machine in synchronization. 2. Operation of a transmitter at the maximum rated level of input or output power.

running-time meter See *elapsed-time meter.*

run time The period during which a computer program run takes place.

rupture 1. The usually rapid and violent tearing apart, or breaking through, of an insulating material subjected to excessive voltage. 2. The clean opening of relay, circuit-breaker, or switch contacts to interrupt a current-carrying circuit.

rush A comparatively smooth and gentle audible background noise such as that arising from some shortwave superheterodyne receivers and very high-gain amplifiers. The noise is so called from its resemblance to the gentle rushing of wind. Compare *hiss, 1, 2.*

ruthenium Symbol, Ru. A rare metallic element. Atomic number. 4. Atomic weight, 101.1.

rutherford Abbreviation, *rd*. A unit of radioactivity equal to 10^6 disintegrations per second (2.7×10^{-5} curie). Also see *kilorutherford, megarutherford, microrutherford,* and *millirutherford.*

Rutherford atom The concept of the nature of the atom, proposed by Rutherford in 1912, in which planetary negative electrons orbit about a central, positive nucleus. Compare *Bohr atom.*

rutherfordite Formula, UO_2CO_3. A highly radioactive product of pitchblende.

RW Abbreviation of *radiological warfare.*

Rx Symbol for *unknown resistance.*

RY Abbreviation of *relay.*

Rydberg constant Symbol, $R\infty$. Value, 109,737.31 inverse centimeters.

ryotron A form of inductive semiconductor switch, operated at cold temperatures to maximize conductivity.

R – Y signal In a color-TV circuit, the signal representing *primary red* (R) minus *luminance* (Y). A primary red signal is obtained when the R – Y signal is combined with the luminance (Y) signal. Compare *B – Y signal* and *G – Y signal.*

RZ Abbreviation of *return-to-zero.* (Also, *RTZ.*)

S 1. Symbol for *screen* of an electron tube. 2. Symbol for *shell* of a tube or semiconductor device. 3. Symbol for *sulfur*. 4. Symbol for *deflection sensitivity*. 5. Symbol for *switch*. 6. Symbol for *elastance*. 7. Abbreviation of *sync*. 8. Symbol for *secondary*. 9. Symbol for *siemens*. 10. Abbreviation of *sine*. 11. Symbol for *entropy*.

s 1. Symbol for *distance* or *displacement*. 2. Symbol for *screen* of an electron tube (when used as a subscript). 3. Symbol for *standard deviation*.

SA Abbreviation of *subject to approval*.

SA-band A section of the *S-band* extending from 3100 to 3400 MHz.

sabin Symbol, α. A unit of sound absorption; 1 sabin represents a surface that can absorb sound at the same rate as 1 square foot of a perfectly absorbent surface.

sacred quotient The ancient Egyptian name for the *divine proportion*.

SADT Abbreviation of *surface alloy diffused-base transistor*.

SAE 1. Abbreviation of *shaft-angle encoder*. 2. Society of Automotive Engineers.

safe noise level The sound level of 85 dB, above which ear-drum damage is possible, which has been set by Federal authorities as the maximum safe level.

safety factor A figure denoting the overload (and the allowance therefore) a device can withstand before breaking down.

safety ground A connection made between an equipment (usually the metal chassis, panel, or case, or the B-minus circuit) and the earth as a protective measure against fire and electric shock.

safety switch See *electrical interlock*.

sal ammoniac Formula, *NH4Cl*. Ammonium chloride, the principal ingredient in the gelatinous electrolyte of a dry cell.

salient pole A pole, such as the polepiece of a motor or generator, that projects from the rest of the structure (rotor assembly or motor frame).

Salisbury chamber A radio-frequency test chamber in which the walls are non-reflective at various frequencies, thus simulating free space.

salt-spray test A test to assess the life and performance of electronic equipment in a saltwater environment. The equipment is sprayed, usually with a saltwater mist, and various electrical parameters are measured at prescribed time intervals.

SAM Abbreviation of *surface-to-air missile*.

samarium Symbol, *Sm*. A metallic element of the rare-earth group. Atomic number, 62. Atomic weight, 150.35.

sample 1. A selection of quantities, events, objects, or the like taken at a specific time interval for analysis or testing. 2. To take a sample (as 1, above). 3. To test a quantity (current, voltage, temperature, pressure, etc.) or a material (electrolyte, insulant, corrosion, rust, etc.) taken from a larger group or body.

sample and hold A method of storing a variable signal for detailed examination.

sample size In statistics, the number of events, numbers, or the like in the sample space chosen for analysis.

sample space In statistics, the set of events, numbers, or the like chosen for analysis. Also see *set* and *subset*.

sampling 1. Observation of a signal at various points in a circuit, without affecting the operation of the circuit. 2. In statistics and probability, a set of function values corresponding to specifically chosen points in the domain.

sampling rate The frequency with which samples are taken, e.g., 1/hr.

sampling window See *window, 2*.

sand load A microwave power dissipator in which the absorptive material is a mixture of sand and carbon.

sapphire needle See *sapphire stylus*.

sapphire stylus A jewel-tipped "permanent" stylus for disk recording and playback.

sat Abbreviation of *saturate, saturation.*

satd Abbreviation of *saturated.*

satellite A man-made body sent into continuous orbit around the earth or another planet. See, for example, *active communications satellite* and *passive communications satellite.*

satellite processor 1. In a computer, a microprocessor that is subsidiary to the CPU. 2. In a data processing system, a CPU used to handle the running of programs of secondary importance to the system's main application.

satisfy To make a statement of inequality or an equation true, as by a solution, e.g., $x = 2$ satisfies the equation $2x = x^2$.

saturable capacitor A voltage-variable ceramic or semiconductor capacitor in which variations in capacitance level off at a reasonably constant value after a particular voltage level is reached.

saturable-core magnetometer A *magnetometer* in which the sensor is a saturable magnetic core with winding. The readout is proportional to the change in permeability of the core produced by the magnetic field under test.

saturable reactor An inductor consisting essentially of a coil wound on a core of magnetic material whose magnetic flux can easily reach saturation level. The inductance, and accordingly the reactance, of the device may be varied by passing a direct current through the coil simultaneously with the alternating current to be controlled. As the dc is increased, the reactance decreases and the ac increases. At the saturation point and beyond, the reactance no longer changes with dc, and the ac likewise shows no further change. The saturable reactor is the basis of the magnetic amplifier.

saturable transformer A transformer having a saturable core that permits automatic regulation of an ac voltage.

saturated color A color containing no white. Also called *pure color.*

saturated logic Any form of digital-logic circuit in which the transistors are either completely cut off or completely saturated. Characterized by relative immunity to noise, high speed, and high input-level requirements.

saturated operation 1. Operation of a magnetic core beyond its saturation point, i.e., in the region where an increase in coil current produces no change in core magnetization. 2. Operation of a tube or transistor beyond its saturation point, i.e., in the region where an increase in voltage produces no change in current, or vice versa. Compare *unsaturated operation.*

saturated solution A solution, such as an electrolyte, which contains all of the solute that it ordinarily will hold at a given temperature and pressure. Compare *supersaturated solution.* Also see *solute; solution, 1;* and *solvent, 2.*

saturating current See *saturation current.*

saturation 1. Saturation point. 2. The state of purity of a color. 3. The condition of a magnetic material that can accommodate no additional flux, or the same condition of a dielectric material.

saturation current In a device, the current flowing at and beyond the saturation point.

saturation flux The value of magnetic flux that will saturate a given sample of magnetic material, or the value of electric flux that will saturate a given sample of dielectric material.

saturation flux density See *saturation induction.*

saturation induction For a magnetic material, the maximum possible intrinsic induction.

saturation limiting Limiting (output-peak clipping) resulting from overdriving a tube or transistor into saturation. Compare *cutoff limiting.*

saturation point On a voltage-current conduction curve, the point beyond which a further increase in voltage produces no (or very little) further increase in current.

saturation resistance The voltage-to-current ratio for a saturated semiconductor.

saturation switching On—off switching operation in which a transistor is in its saturated state when conducting.

saturation value 1. In a transistor, field-effect transistor, or tube, the lowest level of the input current, voltage, or power that results in saturation. 2. The maximum obtainable output level for a given circuit. 3. In a magnetic material, the smallest level of magnitizing force that results in maximum flux density.

saturation voltage The (usually dc output) voltage appearing across a device operating in its saturation region, e.g., the collector voltage of a switching transistor in its on state.

SAVOR Abbreviation of *signal-actuated voice recorder.*

sawtooth An alternating or pulsating wave of current or voltage characterized by a gradual rise in amplitude followed by a rapid fall, or vice versa; so called because of its graphic resemblance to the teeth of a saw.

SB 1. Abbreviation of *sideband.* 2. Abbreviation of *simultaneous broadcast.* 3. Abbreviation of *bachelor of science.*

Sb Symbol for *antimony.*

S-band A radio-frequency band extending from 1550 to 5200 MHz. For subdivisions of this band, see *SAband, SC-band, SD-band, SF-band, SG-band, SH-band, SQ-band, SSband, ST-band, SW-band, SY-band,* and *SZ-band.*

SBC Abbreviation of *single-board computer.*

SBDT Abbreviation of *surface-barrier diffused transistor.*

SBT Abbreviation of *surface-barrier transistor.*

SC 1. Abbreviation of *suppressed carrier.* 2. Abbreviation of *short circuit.*. 3. On drawings, abbreviation for *silk-covered.*

Sc 1. Symbol for *scandium.* 2. Abbreviation of *stratocumulus.*

sc 1. Abbreviation of *sine—cosine.* 2. Abbreviation of *single crystal.* 3. Abbreviation of *science.* 4. Abbreviation of *scale.*

SCA Abbreviation of *subsidiary communications authorization.*

SCA adapter An auxiliary tuner unit for separating the SCA subcarrier from a main FM signal on which it is superimposed. Also see *SCA subcarrier* and *subsidiary communications authorization.*

scalar quantity A quantity having magnitude only. Compare *vector quantity.*

scalar product See *dot product.*

scale 1. A graduated line, arc, circle, spiral, or the like for indicating values of a quantity. 2. An ordered set of values. 3. An ordered series of quantities, such as tones, frequencies, voltages, etc., e.g., musical scale.

scale division The space between consecutive graduations on a scale (see *scale, 1*).

scale down In computer practice, to adjust a group of quantities according to a fixed factor, so that it can be accommodated by hardware or software.

scale expansion Spreading out the divisions in part of a scale (see *scale, 1*).

scale factor 1. A figure by which the readings from a particular scale must be multiplied or divided to give the true values of measured quantities. 2. A figure by means of which values in one system of notation are converted to those in another system. 3. In scaling down (see *scale down*), the factor by which a group of quantities is adjusted.

scale-factor adjustment In some meters, an adjustment which allows full-scale deflection to be set at any desired value (within certain limits) of applied-signal amplitude.

scale length The end-to-end dimension of a scale (see *scale, 1*), in inches, centimeters, geometric degrees, or number of divisions.

scale multiplier See *scale factor, 1*.

scalene triangle A triangle in which no two sides are equal.

scale-of-two counter A flip-flop, so called because it delivers one output pulse for two successive input pulses.

scale-of-10 counter A decade counter, so called because it delivers one output pulse for ten successive input pulses.

scale-of-ten counter See *scale-of-10 counter*.

scale-of-ten scaler See *scale-of-10 counter*.

scaler A circuit or device for extending the frequency range of another device, e.g., a device that extends the range of a 1 MHz counter to 100 MHz; a *prescaler*.

scale range The difference between the lowest and highest values on a scale.

scale span See *scale range*.

scaling adder An inverting operational amplifier used to weight and sum multiple voltages.

scaling circuit A circuit, such as one or more flip-flops, which will deliver one output pulse after a predetermined number of input pulses have been received, and therefore will provide pulse or frequency division. See, for example, *scale-of-two counter*.

scaling factor For a scaler, the number of input pulses required for one output pulse.

scaling ratio See *scaling factor*.

scan 1. To traverse a range, field, or dimension. 2. The amount of traversal in 1, above. 3. Sweep. 4. To sample or reproduce an image in a single-line element, as in facsimile or television. 5. A single line resulting from 4, above. 6. In information retrieval practice, to inspect each record in a file or constituent of a list. 7. To check communications or data channels for availability.

scan-converter tube A face-to-face assembly of a cathode-ray tube and a vidicon in one envelope.

scandium Symbol, *Sc*. A metallic element. Atomic number, 21. Atomic weight, 44.96.

scan frequency See *scanning frequency*.

scanner An electronic or electromechanical device for scanning (see especially *scan, 7*).

scanner amplifier An amplifier for boosting a scanning signal. Also see *scan, 1, 3, 4*.

scanning 1. In a cathode-ray tube or camera tube, the synchronized movement of the electron beam (or other marker) from right to left and/or from top to bottom. 2. The intermittent but repetitive monitoring of two or more communications channels in rotating sequence. 3. The movement of a radar beam for the purpose of obtaining coverage over a specified area.

scanning antenna A transmitting or receiving antenna (such as a rotating one) which continuously covers the area around itself.

scanning beam The deflected electron beam in an oscilloscope tube, TV camera tube, TV picture tube, or radarscope tube. Also see scan, *1, 2, 3, 4, 5*.

scanning circuit 1. A circuit for producing a scan (see *scan, 2, 4*). 2. Sweep circuit. 3. Deflection circuit.

scanning disk See *Nipkow disk*.

scanning frequency The number of scans per unit time (usually ex-pressed in lines per second). Also called *scanning rate*.

scanning line A single line sampled or produced by a scanning process, as in facsimile, television, and graphic recording.

scanning line frequency See *scanning frequency*.

scanning linearity Uniformity of scanning rate. In a linear scan, for example, scan speed is the same at all points along a line.

scanning loss The effective reduction in radar sensitivity that occurs as the beam scans a given area, rather than remaining in a fixed orientation.

scanning rate See *scanning frequency*.

scanning receiver A receiver whose tuning is automatically and continuously swept through a frequency band to detect all signals in the band.

scanning sonar A form of distance-measuring or depth-finding sonar, in which the receiving transducer scans to find the direction of the echo or echoes.

scanning speed The rate at which a line, region, or quantity is scanned or at which samples are taken.

scanning yoke See *yoke, 2*.

scan rate 1. The rate at which a controlled quantity is checked periodically by a control computer. 2. Scanning frequency.

scan tuning Repetitive, automatic sweeping of a frequency band by a tuned circuit containing a varactor whose capacitance is periodically varied by a sawtooth voltage.

SCA subcarrier An auxiliary carrier (commonly 67 kHz) superimposed upon a main FM carrier to convey subsidiary communications, such as special background music without commercials. Also see *subsidiary communication authorization*.

scatter To disperse or diffuse transmitted electromagnetic radiation.

scattering 1. The tendency of a concentrated beam of energy to be spread out when it passes through a given medium or substance. 2. The spreading out of radio waves as they pass through the ionosphere or troposphere.

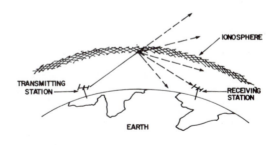

Scattering.

scatter read In data processing, to distribute data from an input record to several storage areas.

scatter transmission See *back scatter* and *forward scatter*.

SC-band A section of the S-band extending from 2000 to 2400 MHz.

scc Abbreviation of *single-cotton-covered* (wire).

SCDSB Abbreviation of *suppressed-carrier—double-sideband* (see *double-sideband—suppressed carrier*).

SCE Abbreviation of *saturated calomel electrode*.

sce Abbreviation of *single cotton enameled* (wire).

SCEPTRON Acronym for *spectral comparative pattern recognizer*.

schedule In computer practice, to establish the order of importance of jobs to be run, and assign the necessary resources for those jobs.

schedule method A method of wave analysis (i.e., evaluation of the dc, fundamental, and various harmonic components of a complex wave) by making calculations based on the heights of ordinates erected on the image of one cycle of the wave on an oscilloscope screen or recorder chart.

schematic diagram Circuit diagram. Also called *wiring diagram* and *schematic.*

schematic symbol A graphic symbol used to represent electronic components in a circuit diagram.

Schering bridge A four-arm capacitance bridge in which the unknown capacitance is compared with a standard capacitance. This bridge is frequently employed in testing electrolytic capacitors, to which a dc polarizing voltage is applied during the measurement.

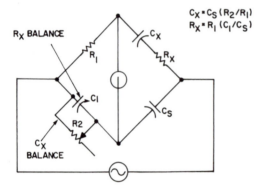

$$C_X = C_S (R_2/R_1)$$
$$R_X = R_1 (C_1/C_S)$$

Schering bridge.

Schmidt optical system In projection television, a lens system employed between the bright-image picture tube and the screen.

Schmitt limiter See *Schmitt trigger.*

Schmitt phase inverter A common-cathode, self-balancing phase-inverter circuit delivering outputs from two plates.

Schmitt trigger A type of multivibrator circuit which produces uniform-amplitude output pulses from a random-amplitude input signal. This circuit has many applications in pulse systems, one being for converting a sine wave into a square wave.

Schottky diode A solid-state diode in which a metal and a semiconductor form the pn junction. Electrons injected into the metal have a higher energy level than the charge carriers in a semiconductor, and energy storage at the junction is low because current flow is not accompanied by hole movement. Also known as *hot-carrier diode.*

Schottky effect In an electron tube, the influence of the potential gradient at the cathode on emission current.

Schottky logic A form of integrated-injection logic with enhanced operating speed.

scientific language A computer programming language for mathematical or scientific applications, e.g., FORTRAN, COBOL.

scientific notation The expression of very large and very small numbers as a fixed-point part (sometimes called *mantissa*) and a power of the radix (usually 10). Thus, $100,000 = 10^5$ (one implied), $0.000\ 000\ 000\ 05 = 5 \times 10^{-11}$, and so on.

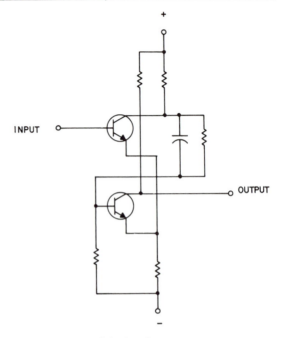

Schmitt trigger.

scintillating crystal A crystal, such as one of sodium iodide, that sparkles or flashes when exposed to radioactive particles or rays.

scintillation 1. In radar practice, the apparent rapid displacement of a target from its mean position. 2. A momentary flash of light produced in a phosphor or scintillating crystal when a high-velocity particle strikes it. 3. A small fluctuation in radio field intensity at a receiving point.

scintillation counter A radiation counter consisting essentially of a scintillating crystal in combination with a photomultiplier tube. Flashes from the excited crystal cause the tube to deliver sizable output pulses which are then totaled and indicated.

scintillator material A substance, such as crystalline sodium iodide, which scintillates under certain stimuli (see *scintillation, 2*).

scissoring A method of interrupting the electron beam in a cathode-ray tube when the beam would not land on the phosphor screen.

SCLC Abbreviation of *space-charge-limited current.*

sco Abbreviation of *subcarrier oscillator.*

scope 1. Oscilloscope. 2. Microscope.

Scott oscillator See *parallel-tee oscillator.*

scp Abbreviation of *spherical candlepower.*

SCR Abbreviation of *silicon controlled rectifier.*

scrambled signal Any signal—audio, video, or other—in which (for secrecy or exclusivity) the elements are disarranged according to some agreed-upon scheme so that intelligent reception is possible only if the signal is unscrambled by means of suitable *decoder* at the receiving point. Example: *scrambled speech.*

scrambled speech Speech transmission in which the frequencies have been inverted to prevent eavesdropping. It is automatically unscrambled at the receiver to restore intelligibility.

scrambler circuit A circuit containing filters and frequency inverters for scrambling speech.

scratch filter A small band-suppression filter for removing from an audio signal the high-frequency scratch noise caused by friction between a phonograph disk and the needle.

scratchpad memory In computers, a low-capacity memory that stores an intermediate result of a calculation.

scratch tape Magnetic data tape which may be overwritten for any purpose.

screen 1. Screen grid. 2. Shield. 3. The front surface of a cathode-ray tube.

screen amplification factor In a pentode, the amplification factor (μ) of the screen grid with respect to the control grid; the screen grid is used for this purpose as the equivalent plate of a triode.

screen angle In radar, the angular difference between the actual horizon and the direction perpendicular to the orientation of the center of the earth. In general, the greater the height of the radar antenna, the greater the screen angle.

screen bypass 1. Ac bypassing of the screen grid of a tube to ground. 2. A capacitor employed to bypass a screen grid, as in 1, above.

screen circuit The circuit associated with the screen electrode of an electron tube.

screen-coupled amplifier A dc amplifier in which the plate in one stage is directly connected to the screen, instead of to the control grid, of the following stage.

screen current Symbol, is or isg. The current flowing in the screen circuit of an electron tube.

screen decoupling Separate decoupling of the screen circuits in a multi-stage resistance-coupled amplifier.

screen dissipation Symbol, Ps or Psg. The power dissipated by the screen grid of an electron tube.

screen dropping resistor The resistor connected in series with the dc plate supply and the screen grid of an electron tube. Screen current passing through this resistor creates a voltage drop which corrects the value of applied dc screen voltage.

screen grid In an electron tube, a grid element between the control grid and plate. It reduces the internal grid—plate capacitance and consequently prevents self-oscillation when the tube is used in a straight-through amplifier. Also see *grid, 2* and *tetrode*.

screen-grid keying Keying of a radiotelegraph transmitter by making and breaking the screen current of one or more of its rf tubes.

screen-grid modulation See *screen modulation*.

screen-grid neutralization Neutralization of an amplifier that employs a screen-grid tube. Such circuits require smaller neutralizing capacitances than those used in triode amplifiers because of the lower interelectrode capacitance of the screen-grid tube.

screen illumination Edge lighting of the transparent screen of an oscilloscope, to make the lines of the graticule more clearly visible.

screen material See *phosphor*.

screen modulation A method of modulation in which a modulation voltage is superimposed upon the dc screen voltage of an amplifier or oscillator. Compare *absorption modulation, cathode modulation, grid-bias modulation, plate modulation, plate—screen modulation, series modulation,* and *suppressor modulation*.

screen regeneration Regeneration obtained by feeding energy from the screen back to the control grid in a tetrode or pentode.

screen resistance Symbol, rs or rsg. The internal resistance presented by the screen-grid—cathode path of an electron tube.

screen room See *cage*.

screen stopper A stopping resistor connected in series with the screen grid of an electron tube.

screen transconductance In a pentode, the transconductance (gm) of the screen grid with respect to the control grid; the screen grid is used for this purpose as the plate of an equivalent triode.

screen voltage Symbol, es or esg. The voltage at the screen grid of an electron tube.

scribing The etching of a semiconductor wafer to facilitate breaking the wafer into smaller pieces.

SCS Abbreviation of *silicon controlled switch*.

SCT Abbreviation of *surface-charge transistor*.

S-curve 1. An E—I curve for a negative-resistance device, which has roughly an S shape. Compare *N-curve*. 2. The response curve of an FM discriminator or ratio detector, which has roughly a long-S shape.

SD Abbreviation of *standard deviation*.

SD-band A section of the S-band extending from 4200 to 5200 MHz.

SDF Abbreviation of *static direction finder*.

Se Symbol for *selenium*.

SEAC An early (circa 1950) digital computer developed at the National Bureau of Standards. The name is an acronym for *s*tandards *e*astern *a*utomatic *c*omputer.

sea clutter Collectively, the radar echoes that the sea reflects.

seal 1. The point at which a lead or electrode enters or leaves and is secured to an envelope, case, or housing. Such a point is often tightly closed against the passage of materials in or out of the envelope. 2. To close off a circuit or component from tampering.

sealed meter 1. A meter which is tightly closed against the entry of moisture and foreign materials. 2. A meter which is locked or otherwise protected against tampering.

sealing compound A substance—such as wax, pitch, or plastic—used to enclose and protect electronic devices.

search 1. To scan or sweep through a range of quantities or through a region of interest. 2. To examine (usually in some prescribed order) items of information in a computer memory to find those satisfying a given criterion.

search coil An inductive probe (exploring coil) employed to sample magnetic fields.

search oscillator A variable-frequency oscillator employed to locate and identify signals by the heterodyne method.

search probe 1. Search coil. 2. A capacitive probe employed to sample electric fields.

search radar A radar that displays a target almost immediately after that target enters a scanned area.

search time The time needed to test items during a search (*search, 2*).

sea return See *sea clutter*.

seasonal effects In ionospheric propagation, the changes produced as a result of the revolution of the earth around the sun. The path of the sun across the sky, and the length of the day, are primarily responsible for such effects.

seasonal static Atmospheric electrical interference, most prevalent during the summer.

SE-band A section of the S-band extending from 1550 to 1650 MHz.

sec 1. Abbreviation of *second*. 2. Abbreviation of *section*. (Also, *sect*.) 3. Abbreviation of *secondary*. 4. Abbreviation of *secant*.

secant Abbreviation, *sec*. The trigonometric function representing the ratio of the hypotenuse of a right triangle to the adjacent side (c/b). The secant is the reciprocal of the cosine; $\sec A = 1/\cos A$.

sech See *hyperbolic secant*.

second 1. Abbreviation, *s* and *sec*. A unit of time. The mean solar second is 1/86,400 of a mean solar day, and is 1/60 minute or 1/3600 hour. 2. Symbol ("). A unit of arc measure. 1" = 1/3600 a geometric degree.

secondaries See *secondary colors*.

secondary 1. Secondary coil or secondary winding. 2. Secondary standard. 3. Secondary color.

secondary battery See *storage battery*.

secondary calibration The calibration of an instrument, based on a reference instrument which has been calibrated against an absolute source.

secondary capacitance 1. The distributed capacitance of the secondary winding of a transformer whose primary winding is unloaded. Compare *primary capacitance, 1*. 2. A series or shunt capacitance used to tune the secondary coil of an i-f or rf transformer. Compare *primary capacitance, 2*.

secondary cell See *storage cell*.

secondary circuit 1. The circuit associated with the secondary winding of a transformer. 2. Output circuit.

secondary coil See *secondary winding*.

secondary color 1. A color prepared by mixing equal parts of two primary colors, e.g., purple made from red and blue pigments. 2. In TV practice, any displayed color composed of two or more primaries.

secondary current The current flowing in the secondary winding of a transformer. Also called *transformer output current*. Compare *primary current*.

secondary electron 1. The electron possessing the lesser energy after a collision between two electrons. Compare *primary electron*. 2. An electron ejected by secondary emission.

secondary-electron multiplier See *electron-multiplier tube*.

secondary emission The action whereby electrons in the atoms at the surface of a target are ejected as a result of bombardment by a beam of (primary) electrons. Thus, in an electron tube, electrons from the cathode strike the plate with a force that drives secondary electrons out of the plate, into the surrounding space.

secondary-emission tube See *electron-multiplier tube*.

secondary emitter A source of secondary electrons, e.g., the plate of an electron tube or a dynode in a photomultiplier tube.

secondary failure The failure of a component or circuit, resulting from the failure of some other component. For example, the pass transistor in a power supply may burn out, causing the output voltage to increase; this increased voltage may damage equipment connected to the supply.

secondary frequency standard A device for generating signals of accurate frequency, but which does not possess the very high stability and extreme accuracy of a primary frequency standard. The secondary standard is periodically referred to a primary standard and appropriately corrected. The secondary standard usually consists of a 50, 100, 1000, or 5000 kHz crystal oscillator (with or without temperature control and supply-voltage regulation). Some units contain multivibrators that subdivide the oscillator frequency and deliver lower-frequency signals. Harmonics from the oscillator provide numerous high-frequency check points. Compare *primary frequency standard*. Also see *secondary standard*.

secondary impedance 1. The impedance of the secondary winding of a transformer whose primary winding is unloaded. Compare *primary impedance, 1*. 2. An external impedance presented to the secondary winding of a transformer. Compare *primary impedance, 2*.

secondary inductance The inductance of the secondary winding of a transformer whose primary winding is unloaded. Compare *primary inductance*.

secondary kVA The kilovolt-amperes in the secondary circuit of an operating transformer. Compare *primary kVA*.

secondary power The power in the secondary circuit of a transformer. Also see *secondary kVA* and *secondary VA*. Compare *primary power*.

secondary radiation The (sometimes random) reradiation of electromagnetic waves, as from a receiving antenna.

secondary radiation Rays emitted by atoms or molecules when the latter are struck by other radiation.

secondary rays Rays emitted by atoms or molecules which have been bombarded by other rays of the same general nature. Examples: secondary X-rays, secondary beta rays, and so on.

secondary resistance The dc resistance of the secondary winding of a transformer. Compare *primary resistance*.

secondary standard An accurate source of a quantity (capacitance, frequency, inductance, resistance, etc.), which is referred periodically to a primary standard for correction.

secondary storage In computer and data-processing practice, storage that is auxiliary to the main storage. Also called *backing storage*.

secondary turns Symbol, *Ns*. The number of turns in the secondary winding of a transformer. Compare *primary turns*.

secondary utilization factor Symbol, *UFs*. For a transformer in a rectifier circuit, the utility factor of the secondary winding (ratio of dc power output to secondary volt-amperes)

secondary VA The volt-amperes in the secondary circuit of a transformer. Compare *primary VA*.

secondary voltage The voltage across the secondary winding of a transformer. Also called *transformer output voltage*. Compare *primary voltage*.

secondary winding The normal output winding of a transformer. Also called *secondary coil*. Compare *primary winding*.

secondary X-rays X-rays emitted by an object when it is exposed to X-rays.

second breakdown In a large-area power transistor, a destructive breakdown due to runaway effects which are electrically and thermally cumulative.

second-breakdown voltage The collector voltage at which second breakdown occurs in a transistor.

second-channel attenuation See *selectance, 2*.

second-channel interference In a given channel, interference arising from authorized signals two channels removed.

second-degree equation See *quadratic equation*.

second detector In a superheterodyne receiver, the intermediate-frequency detector. Compare *first detector*.

second generation computer A computer of the era beginning with the disuse of tubes and ending with the advent of ICs, i.e., a transistorized one.

Second law of thermodynamics See *Carnot theorem*.

second-level address In a computer program instruction, an address giving the location of the address of a required operand. Also called *indirect address*.

second radiation constant Symbol, *hc/k*. Value, $1.438\ 833 \times 10^2$ m• K.

sect Abbreviation of *section*. (Also, *sec*.)

section 1. A subcircuit or stage of a larger circuit, e.g., the oscillator section of a receiver. 2. The smaller unit described in 1, above, when it is operated independently and is selfcontained, e.g., filter section. 3. Program segment.

sectionalized antenna A set of collinear radiating elements, placed end-to-end with reactances between them, for the purpose of modifying the radiation pattern.

sectionalized winding 1. A method of winding a coil in complete, multilayer sections that are stacked side by side or top to bottom, a technique that reduces distributed capacitance. 2. A coil wound as in 1, above.

sector On a magnetic disk or drum for computer storage, a section of a track or band.

sectoral horn antenna An (usually sheetmetal) antenna which has the shape of a horn of rectangular cross section that is flared in one dimension only.

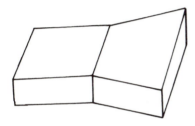

Sectoral horn antenna.

sector display In radar, a display that allows the continuous observation of a portion of the scanned area.

secular equilibrium The state in which a radioactive substance changing into another substance is decaying as fast as the second substance is being formed.

secular variation A slow change in the intensity of the terrestrial magnetic field.

securite In radiotelephony, a spoken word (pronounced seh-koor-ee-*tay*) identifying a transmission concerning safety. Equivalent to *TT* in radiotelegraphy.

Seebeck effect The development of an electromotive force in a junction of two dissimilar metals (a *thermocouple*) when the temperature of the junction is different from that of the rest of the metal.

Seebeck emf The electromotive force resulting from the Seebeck effect.

seed crystal A smaller single crystal from which a large single crystal (e.g., germanium or silicon) is grown. Also see *Czochralski method*.

seek See *search*.

seek area An area of direct access storage to which are assigned specific records and from which the records can be accessed quickly.

segment 1. The portion of a line or curve lying between two points. 2. Program segment.

segmental meter An expanded-scale meter (see *scale expansion*).

segmenting See *partitioning*.

segment mark A character indicating the division between tape file sections.

Seignette salt See *Rochelle salt*.

seismogram A record produced by a seismograph.

seismograph An earthquake detecting and recording instrument which indicates the direction, magnitude, and time of a quake.

seismometer See *seismograph*.

seismoscope An instrument which shows the occurrence and time of an earthquake. Compare *seismograph*.

select To accept or separate a unit, quantity, or course of action from all those available (in a group, mixture, or series).

selectance 1. For a resonant circuit, the ratio Er/Ex, where Er is the voltage at resonance, and Ex the voltage at a specified nonresonant frequency. 2. For a receiver, the figure $S2/S1$ where $S1$ is the sensitivity of the receiver in a given frequency channel, and $S2$ the sensitivity in another specified channel.

selected mode In an encoder, a mode in which one output is read and the others are ignored.

selective The quality of tuning sharply. See *selectivity*.

selective absorption The attenuation or absorption of some frequencies or bands of frequencies, with little or no attenuation at other frequencies or bands of frequencies.

selective amplifier An amplifier that may be tuned, with the desired degree of sharpness, to a single frequency or band of frequencies. Radio-frequency amplifiers are tuned by means of LC circuits, audio-frequency amplifiers by means of RC circuits.

selective calling The calling, alerting, or alarming of a desired station without interfering with other stations.

selective digit emitter Part of a punched-card machine capable of imitating signals ordinarily produced by hole patterns.

selective dump In computer practice, a dump (see *dump*) affecting a small, specific memory area.

selective fading Fading due to propagation conditions whose effects differ at slightly different frequencies. In an amplitude-modulated signal, this effect causes the sidebands and carrier to arrive in various phase relationships, causing marked distortion in the received signal.

selective interference Interference confined to a narrow band of frequencies.

selective polarization See *polarization selectivity*.

selective reflection In the reflection of electrons directed into a crystal by means of an electron gun, the tendency of the electrons to be reflected more readily when they strike the crystal at certain speeds.

selective relay 1. A relay or relay circuit tuned for closure at one signal frequency. 2. A relay or relay circuit adjusted for closure at one value of current or voltage.

selective trace In computer practice, a diagnostic program used to analyze certain areas of memory or specific kinds of program instructions, for debugging purposes.

selectivity 1. The ability of a circuit or device to pass signals of one frequency and reject all others. 2. The degree to which such an ability manifests itself.

selectivity control In some equipment—such as receivers, crystal filters, wave analyzers, and vibration meters—an adjustment (usually continuously variable) which permits variation of selectivity.

selectoject A fully electronic, continuously tunable, notcher-peaker which is RC tuned. The name is an acronym for *select* or *reject*.

selector 1. Tuner. 2. Selector switch.

selector channel In data processing and computer systems, a data transmission channel controlling the information flow between peripherals and a cpu that are interconnected by a single channel.

selector pulse In digital communications, an identifying pulse that represents a certain group of bits or data.

selector relay A device, such as a stepping switch, which will actuate one of a number of available circuits upon receipt of a predetermined number of pulses.

selector switch A (usually rotary) multiposition switch.

selectron A now-obsolete rapid-access, 256-bit, computer-memory tube.

selenium Symbol, Se. A nonmetallic element. Atomic number, 34. Atomic weight, 78.96. Selenium is used in diodes, rectifiers, and photocells.

selenium cell A photoelectric cell which employs specially processed selenium as the light-sensitive material. A selenium cell may be operated as a photoconductive cell or a photovoltaic cell.

selenium diode A junction diode in which the semiconductor material is specially processed selenium. Also see *junction diode*.

selenium photocell See *selenium cell*.

selenium rectifier A disk- or plate-type power rectifier utilizing the junction between selenium and aluminum or selenium and iron.

selenography The branch of science dealing with lunar topography.

self-bias For a tube or transistor, input-electrode bias voltage resulting from the flow of output-electrode current through a resistor that is common to both circuits. Thus, grid bias for a tube may be obtained from the voltage drop resulting from plate—cathode current flow through an external cathode resistor, and base bias for a common-emitter-connected bipolar transistor from the voltage drop resulting from the flow of collector—emitter current through an external emitter resistor. Also called *automtic bias*.

self-capacitance The inherent internal capacitance of a device other than a capacitor.

self-checking number A number whose digits have a value that determines the check digit attached to it; thus, it can be verified following its transfer between storage locations or peripherals.

self-cleaning contacts Switch or relay contacts that clean themselves automatically by means of *wiping action*.

self-contained device A device which has built into it all the parts and sections (e.g., main circuit, power supply, meter, loudspeaker, etc.) needed for full operation, i.e., no auxiliary equipment is needed. Self-contained equipment is often, but not always, portable.

self-controlled oscillator See *self-excited oscillator*.

self-energy Symbol, *E*. The energy (mc^2) of a particle whose mass is *m*. (The quantity *c* is the velocity of light.)

self erasing In a magnetic tape, the unwanted erasing of data near a highly magnetized region.

self-excited generator A dynamotype ac generator in which the field coils are supplied with direct current produced by the machine itself. Compare *separately excited generator*.

self-excited oscillator An oscillator which is essentially an amplifier which supplies its own input signal through positive feedback, and whose oscillation frequency depends entirely upon circuit constants, such as the capacitance and inductance in the tank circuit. Compare *crystal oscillator, fork oscillator, hummer,* and *magnetostriction oscillator*.

self-extinguishing circuit A gas-tube circuit which automatically switches off the glow discharge at some predetermined instant after firing.

self-focus picture tube A TV picture tube in which the electron gun has an automatic, electrostatic focusing arrangement.

self-generating photocell See *photovoltaic cell*.

self-generating transducer A voltage-producing transducer, such as a piezoelectric pickup or dynamic microphone.

self-healing capacitor A capacitor, such as a wet electrolytic unit, in which the dielectric is restored to its normal condition after a high-voltage breakdown.

self-heated thermocouple A thermocouple in which the passage of an applied current produces the heat necessary for activation of the thermocouple.

self-heating thermistor A thermistor which is heated to above ambient temperature by the current passing through it. Also called *directly heated thermistor*. Compare *indirectly heated thermistor*.

self impedance The effective or measured impedance at a circuit point.

self-inductance 1. The inductance of an inductor. 2. The inherent internal inductance of a device other than an inductor.

self-induction Induction that occurs in a single circuit. An instance is the generation of an opposing voltage across a coil by an alternating current flowing through it. Compare *induction*. Also see *inductance*.

self-latching relay a relay that remains in the state it has been switched to (i.e., locked open or closed) until a subsequent signal is received.

self-maintained discharge In a gas tube operated at a sufficiently high voltage, the discharge which will continue after the trigger voltage has been removed.

self-modulated oscillator A circuit, such as a blocking oscillator, in which oscillation occurs simultaneously at two frequencies, one modulating the other.

self-organizing A computer or system that can change the arrangement of data files for particular purposes.

self-powered device A device which requires no external source of power; i.e., it is equipped with a self-contained battery or a generator driven by a prime mover.

self-processing camera See *Polaroid camera*.

self-pulsing blocking oscillator A blocking oscillator which produces a train of rf pulses. Compare *single-swing blocking oscillator*.

self-quenching detector A super-regenerative detector (see *super-regenerative circuit*) in which the low-frequency quenching voltage is supplied by the regenerative detector itself. Also see *quenching action* and *quench oscillator*. Compare *separately quenched detector*.

self-quenching oscillator A circuit, such as a blocking oscillator, in which oscillation is periodically switched off automatically, resulting in a self-interrupted wave train.

self-rectifying circuit An oscillator or amplifier circuit supplied with raw ac plate voltage. Since the tube(s) cannot conduct on the negative half-cycle, rectification occurs; however, the lack of filtration causes amplitude modulation and, in oscillators, a broadly tuned signal.

self-rectifying vibrator A vibrator-type power supply in which one vibrator reed chops the dc input to the primary winding of the transformer, and a second vibrator reed rectifies the ac output delivered by the secondary winding. Also see *vibrator rectifier*.

self-rectifying X-ray tube An X-ray tube operated with ac anode voltage.

self-regulation The ability of a circuit or device to control its output automatically, according to some predetermined plan, by using output error to correct operation or to vary the input.

self reset 1. The action of a circuit breaker to reapply power after a certain elapsed time. 2. The action of any device, returning a circuit or system to normal automatically.

self-resetting loop In a computer program, a loop in which instructions cause locations used in the loop to assume their condition prior to the loop's execution.

self-resistance The inherent internal resistance of a device other than a resistor.

self-resonant frequency The frequency at which a device will resonate naturally (without external tuning). Thus, an inductor will self-resonate with its distributed capacitance, and, similarly, a capacitor will resonate with its stray inductance.

self-saturation In a magnetic amplifier, saturation resulting from rectification of the saturable-reactor output current.

self-starting motor An ac motor which starts running as soon as voltage is applied, i.e., no external mechanical force is needed. Also see *shading coil*.

self-stopping circuit A circuit that automatically extinguishes a gas tube once it has fired.

self-sustained discharge See *self-maintained discharge*.

self-sustained oscillations Oscillations which are maintained by means of positive feedback (inductive or capacitive) from the output to the input of the circuit. See, for example, *self-excited oscillator*.

self-test Any arrangement whereby a device or system determines, without the aid of an external operator, whether or not it is operating correctly.

self-ventilated motor See *open motor*.

self-wiping contacts See *self-cleaning contacts*.

selsyn See *Autosyn* and *synchro*.

SEM Abbreviation of *scanning electron microscope*.

semiautomatic key A telegraph key which produces a string of dots when its lever is thrown to one side (and, in some, a string of dashes when the lever is thrown to the opposite side). Also called *bug*.

semiautomatic tape communication In radioteletype or radiotelegraph systems, a form of relay or repeater system in which the signal is received, recorded on paper tape, and retransmitted from the paper tape, without the direct action of a human operator.

semiconductor A material whose resistivity lies between that of conductors and insulators, e.g., germanium and silicon.

semiconductor device A device—such as a diode, photocell, rectifier, or transistor—exploiting the unique properties of a semiconductor.

semiconductor diode A solid-state diode, as opposed to a vacuum-tube diode or gas tube diode. Examples: germanium diode, selenium diode, silicon diode.

semiconductor junction Within a body of semiconductor material, the area of intimate contact between two regions (usually *n* and *p*) having opposite electrical properties.

semiconductor laser See *laser diode*.

semiconductor material See *semiconductor*.

semiconductor-metal junction The area of intimate contact between a metal and a semiconductor.

semiconductor photosensor A semiconductor photodiode or phototransistor, as opposed to a phototube.

semiconductor rectifier A heavy-duty semiconductor diode (or assembly of such diodes) designed primarily to change ac to dc in power-supply units. Rectifiers commonly are made from copper oxide, germanium, magnesium—copper sulfide, selenium, or silicon. See also *junction diode* and *meter rectifier*.

semidirectional Pertaining to a transducer that exhibits different directional characteristics at different frequencies.

semiduplex operation A two-frequency communication system which is duplex at one end of the link and simplex at the other end. Also see *duplex operation* and *simplex telegraphy*.

semi-infinite line A relatively short segment of straight line; two or more joined together show the different slopes of various parts of a response curve, but the combination only approximates the smooth curve.

semilogarithmic graph A graph in which one axis is logarithmic and the other linear.

semimetal An elemental substance which exhibits only some of the properties of a metal, e.g., antimony and arsenic. Also called *metalloid*.

semiresonant line An open-wire transmission line cut approximately to resonant length, so that standing waves, while present, are not deleterious to the operation of the line.

semitone See *half step*.

sender See *transmitter, 1, 3*.

sending-end impedance See *driving-point impedance*.

sending set 1. A *radio transmitter*. 2. An equipment for transmitting electromagnetic waves. Also see *transmitter, 1*.

sense 1. To check the condition of a switching device, such as a gate. 2. Read.

sense amplifier A device that produces a control signal when some characteristic of the input signal changes.

sense determination In a direction finder, determination of the sense of the signal, i.e., whether in the line of propagation the transmitter is ahead of or behind the receiver.

sense of inequality A statement of correlation between pairs of unequal quantities; the inequalities $v < w$ and $x < y$ have the same sense, whereas $v < w$ and $x > y$ do not.

sense resistor A (usually low-value) resistor employed to sense current in a circuit without introducing a significant loss. The voltage drop across this resistor is proportional to the current and may be applied to a voltmeter, oscilloscope, or other instrument for measurement or observation.

sensing circuit 1. A circuit that samples a quantity. 2. In a voltage regulator, the circuit that monitors the output voltage and delivers a control voltage proportional to the output-voltage error.

sensing station The place in a punched card machine where the reading is done.

sensing window See *window, 2*.

sensitive device A device which responds to a significantly small signal current, voltage, or power.

sensitivity 1. The ability of a circuit or device to respond to a low-level applied stimulus. 2. For a receiver, the input-signal (in microvolts or millivolts) required for a specified output level. 3. For a galvanometer, microamperes or milliamperes per scale division. 4. The ohms-per-volt rating of a voltmeter. Also see *voltmeter sensitivity*.

sensitivity adjustment 1. An input gain control. 2. The radio-frequency gain control of a receiver. 3. A control or switch that is used to select the range or threshold of a piece of test equipment.

sensitivity control A manual or automatic device for adjusting the sensitivity of a circuit or device.

sensitometer An instrument used to measure the sensitivity of certain materials to light.

sensor A device which samples a phenomenon (electrical or nonelectrical) and delivers a proportionate current or voltage in terms of which the phenomenon may be measured, or with which control action can be initiated.

separately excited generator A dynamo-type ac generator whose field coils are supplied with direct current from another generator or from a battery. Compare *self-excited generator*.

separately quenched detector A superregenerative detector (see *superregenerative circuit*) in which the quenching voltage is supplied by a separate low-frequency oscillator. Also see *quenching action* and *quench oscillator*. Compare *self-quenched oscillator*.

separator 1. Filter (see *filter, 1*). 2. A perforated or porous plate of insulating material—usually plastic or wood—for holding active plates apart in a storage cell. 3. In computer practice, a character marking the division between logical data units. Also called *data delimiter*.

septate cavity A coaxial cavity containing a septum between the inner and outer conductors.

septate waveguide A waveguide containing one or more septa (see *septum*) to control power transmission.

septum A thin metal vane used as a reflector in a waveguide or cavity.

sequence 1. A succession of objects, parameters, or numbers. 2. An ordered set of numbers, each of which is related to its predecessor by a specific mathematical function.

sequence checking routine In computer practice, a routine that verifies the order of items of data.

sequence control register In a computer memory, a register whose contents determine the instruction to be implemented next.

sequence programmer A timing device which may be preset to start or stop various operations at predetermined times.

sequencer A device that initiates or terminates events in a desired sequence.

sequenced relay A relay whose several contacts close in a predetermined order.

sequence timer A timer in which separate delay circuits are actuated in a predetermined sequence.

sequential In computer practice, a term denoting operations on data items in which the items—say, records in a file—are taken in an order determined by key values rather than in the order in which the items are physically arranged (e.g., in memory), i.e., the *serial* order.

sequential analysis In statistics, using an unspecified number of observations as samples from which is derived a result. Each observation is accepted or rejected, or another observation is made.

sequential color television A method in color TV practice involving the successive transmission of the three primary colors and their reproduction at the receiver in the same order. Also see *dot-sequential system*, *field-sequential system*, and *line-sequential system*.

sequential control Computer operation in which the order of instruction implementation is the order of instruction storage.

sequential relay See *sequence relay*.

sequential scanner Rectilinear TV scanning in which the center-to-center distance between successive lines is the nominal line width.

sequential switch 1. A switch that provides selection of two or more ports in a rotating succession. 2. In a television system, a switch that allows the monitoring technician to select any of the cameras for viewing.

sequential timer See *sequence timer*.

ser 1. Abbreviation of *series*. 2. Abbreviation of *serial*.

serial 1. Pertaining to the performance of steps, or the occurrence of elements (such as data items on magnetic tape), in succession. 2. An order, row, or sequence in which one item follows another (as opposed to *parallel*).

serial access Access to data file records in their order in a storage medium.

serial adder See *serial arithmetic unit*.

serial arithmetic unit In computer practice, an arithmetic unit in which digits are handled in order. Compare *parallel adder*.

serial bit Data in which the bits of each byte or word are sent or received one at a time.

serial memory A register in which the input and output data is stored and retrieved one bit at a time.

serial—parallel 1. Pertaining, in computer operation, to systems and operations which are partially serial and partially parallel, e.g., words are transferred within a system in this manner in that, although each travels in sequence, their constituent bits, as a group, travel in parallel. 2. Description of component interconnection, whereby several paralleled components are serially connected to other similarly connected parallel component groups.

serial processing In computer practice, the sequential processing of several different programs through a single channel. Compare *parallel processing*.

serial storage In computer operation, storage in which elements are entered in order and are available only in the same order. Compare *parallel storage*.

series 1. Mathematical series. 2. A connection of elements or components in succession (see *series circuit*).

series addition See *series-aiding*.

series-aiding The condition in which two series voltages or magnetic fields are added together. Compare *series-bucking*.

series antenna tuning Antenna-feeder tuning in which a separate tuning capacitor is connected in series with each wire. Compare *parallel antenna tuning*.

series bucking The condition in which two series voltages or magnetic fields oppose each other. Compare *series-aiding*.

series capacitance Capacitance acting, or connected, in series with another capacitance or other quantity.

series capacitors Capacitors connected in series; total capacitance $Ct = 1/(1/C1 + 1/C2 ... + 1/Cn)$. Also see *series circuit*.

series circuit A circuit whose components are, in effect, connected in a string, i.e., end to end. Compare *parallel circuit*.

series compensation In a wideband amplifier, such as a video amplifier, frequency compensation provided by an inductor and capacitor connected in series between stages. Compare *series—shunt compensation* and *shunt compensation*.

series-diode half-wave rectifier See *series-diode rectifier*.

series-diode rectifier A rectifier circuit in which the diode is connected in series with the source and load. Compare *shunt-diode rectifier*.

series dropping resistor See *dropping resistor*.

series equivalent impedance A series impedance which will draw the same current (magnitude and phase) drawn by a given parallel circuit connected across the same single-phase source.

series equivalent of parallel circuit See *series equivalent impedance*.

series-fed amplifier An amplifier circuit in which the operating voltages are applied in series with the ac signal voltages. Also see *series feed*.

series-fed oscillator An oscillator circuit in which the dc operating voltage is applied in series with the ac output voltage. Also see *series feed*.

series feed The application of ac and dc voltages in series to a device. Example: the presentation of the dc operating voltages for an amplifier in series with the ac signal voltages (see *series-fed amplifier*).

series feedback A feedback system in which the feedback signal is presented to the input point in series with the input signal. Compare *shunt feedback*.

series field A magnetic field produced by a series winding in a motor or generator.

series-filament operation The operation of low-voltage, electron-tube filaments (heaters) from a higher-voltage source (such as the ac power line) by connecting them in series with each other and the higher voltage, with or without a dropping resistor.

series generator An electric generator in which the armature and field windings are connected in series. Compare *shunt generator*.

series inductance 1. Inductance acting, effectively, in series with some other quantity, e.g., the inherent inductance of a wirewound resistor. 2. Inductance connected in series with other inductances or with some other quantity.

series inductors Inductors connected in series and separated or oriented in a way that minimizes the effects of mutual inductance; total inductance $Lx = L1 + L2 ... + Ln$. Also see *series circuit*.

series limiter A limiter (clipper) circuit in which the diode is essentially in series with the signal. Compare *parallel limiter*.

series loading The series insertion of reactances in a circuit for the purpose of impedance matching.

series magnetic circuits A combination of several magnetic paths in line, so that flux extends through each path in sequence; this is analogous to the passage of electric current successively through series-connected resistors.

series modulation Amplitude modulation obtained with a triode inserted in series with the dc supply of the modulated rf amplifier. An af modulating voltage applied to the grid of the series tube varies the tube's internal plate resistance (and, accordingly, the plate current of the rf amplifier) at the audio frequency. Compare *absorption modulation, cathode modulation, grid-bias modulation, plate modulation, plate-screen modulation,* and *suppressor modulation*.

series modulator A vacuum tube (or equivalent transistor) employed as an audio-controlled series resistor for amplitude modulation (see *series modulation*).

series motor An electric motor whose armature and field windings are connected in series. Compare *shunt motor*.

series operation The operation of units in succession, necessitating sequential current flow through each. Also see *series circuit*.

series-opposition See *series-bucking*.

series-parallel capacitors A circuit consisting of two or more series circuits connected in parallel. Also see *parallel circuit* and *series circuit*.

series-parallel inductors Inductors connected in series—parallel and separated or oriented in a way that minimizes the effects of mutual inductance; total inductance $Lt = 1/(1/L1 + L2) + (1/L3 + L4)$. Also see *series-parallel circuit*.

series-parallel resistors Paralleled resistors connected in series to other paralleled resistors. Also see *series-parallel circuit*.

series peaking coil In a wideband amplifier, such as a video amplifier, a small inductor employed for series compensation. Compare *shunt peaking coil*.

series regulator A voltage regulator circuit in which the controlled tube or transistor is in series with the load. Compare *shunt regulator*.

series resistance 1. Resistance acting in series with another resistance or with another quantity. 2. The inherent resistance that acts effectively in series with the plates of a capacitor. 3. The resistance of the wire in a coil, acting effectively in series with the inductance.

series resistors Resistors connected in series with each other; total resistance $Rt = R1 + R2 + R3... + Rn$. Also see *series circuit*.

series resonance Resonance in a circuit consisting of a capacitor, inductor, and ac generator in series. At the resonant frequency, $XL = XC$, capacitor current and inductor current are maximum and equal, and the circuit impedance is minimum. Compare *parallel resonance*.

series-resonant circuit A resonant circuit in which the capacitor, inductor, and generator are connected in series. Also see *series resonance*. Compare *parallel-resonant circuit*.

series-resonant trap A wavetrap consisting of a series-resonant

LC circuit. Compare *parallel-resonant trap*.

series-resonant wavetrap See *series-resonant trap*.

series-shunt circuit See *series-parallel circuit*.

series-shunt compensation In a wideband amplifier, such as a video amplifier, frequency compensation provided by an inductor in series with the plate or collector load resistor, or an inductor and capacitor connected in series between two successive stages. Compare *series compensation* and *shunt compensation*.

series tee-junction See *E-plane tee-junction*.

series tracking capacitor See *oscillator padder*.

series-type frequency multiplier A varactor frequency-multiplier circuit in which the varactor is in series with the input and output. Compare *shunt-type frequency multiplier*.

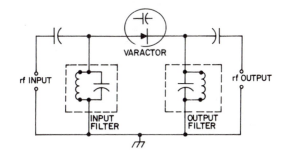

Series-type frequency multiplier.

series-type resonance bridge A resonance bridge in which the impedance arm is a series-resonant circuit. Compare *shunt-type resonance bridge*.

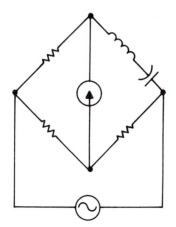

Series-type resonance bridge.

series winding 1. A winding in a motor or generator that is connected in series with the armature. 2. A method of motor or generator construction in which the field winding is connected in series with the armature.

series-wound generator See *series generator.*

series-wound motor See *series motor.*

serrated pulse A pulse having a notched or slotted top. An example is the vertical sync pulse in television.

serrated rotor plate In a variable capacitor, an external rotor plate that is slotted radially to provide sections which can be bent to alter the capacitance-variation curve, during receiver alignment, for example.

serrated vertical sync pulse In television, the vertical sync pulse notched at twice the 15,750 pps horizontal rate.

service To maintain or repair electronic equipment.

serviceability ratio The ratio *ts: ts + td*, where *ts* is serviceable time for a machine, and *td* is downtime.

serviceable time The cumulative time during which an operator-monitored (but not necessarily operated) machine is capable of normal operation.

service area For a broadcast or communication station, the useful coverage area.

service band 1. For a communications system, the band of frequencies in which operation is normally carried out. 2. A band of frequencies specifically assigned, by government regulation, to a certain communications service or services.

service channel The band of frequencies that a particular broadcast or communications station occupies, when the carrier frequency is held constant.

serviceman See *electronic serviceman.*

service meter 1. An energy ("power") meter. Also see *kilowatt-hour meter.* 2. A rugged multimeter employed by a serviceman.

service oscillator A signal generator designed expressly for troubleshooting and repair service.

service switch 1. The main switch controlling the electric service to a building or other place of installation. 2. In TV repair, a switch on the rear of a TV chassis, which facilitates adjustment of screen controls by removing vertical deflection temporarily.

service-type instrument An instrument having reasonable accuracy and a degree of ruggedness that suit it for field or shop use. Examples: *service meter, service oscillator.* Compare *laboratory-grade instrument.*

servo amplifier A highly stable amplifier designed expressly for use in a servomechanism.

servo loop In a control system (particularly a servo amplifier), the output-to-input feedback loop through which automatic control is effected.

servomechanism Abbreviation, servo. A self-correcting closed-loop control system, usually containing some mechanism, such as a motor that is made to run, thus controlling some other device, until an error signal in some way supplied by what is being controlled matches the reference signal activating the servo.

servomotor A motor operated by the output signal of a servo amplifier. Depending upon the end application of the servo system, the motor signal may or may not be corrected.

servo oscillation In a servo system, instability evidenced by hunting.

servo system An automatic control system using one or more servomechanisms.

set 1. An integral piece of equipment, e.g., radio set. 2. In a flip-flop circuit, an input that is not controlled by the clock. 3. To adjust a circuit or device, such as a flip-flop, to a desired operating point or condition. 4. In set theory, a class of numbers, things, or events. 5. In computer programming, to initialize a variable, i.e., to assign a label to a location.

set analyzer A combination test instrument designed originally for troubleshooting radio receivers. It consists of a multimeter and tube tester.

set noise Electrical noise arising inside a radio or television receiver, as opposed to that picked up from the outside.

set pulse A pulse used for setting (see *set, 3*).

set terminal In a flip-flop, the one-inout terminal. Compare *reset terminal*

setting The position to which an adjustable device is set for a particular purpose.

settling time In a digital voltmeter, the time required between the application of a test voltage and the final display of an accurate readout.

set up To arrange and prepare equipment for operation.

setup An arrangement of equipment operating as a system, e.g., stereo setup.

set-up time 1. The time required to install and test an electronic system, and to ready the system for operation. 2. In a digital gate, the length of time a pulse must be held to produce a change of state.

sexadecimal number system See *hexadecimal number system.*

sexagesimal number system A number system whose radix is 60.

SF 1. Abbreviation of *safety factor.* 2. Abbreviation of *single-frequency.* 3. Abbreviation of *standard frequency.* 4. Abbreviation of *stability factor.*

SFA Abbreviation of *single-frequency amplifier.*

SF-band A section of the S-band extending from 1650 to 1850 MHz.

SFO 1. Abbreviation of *single-frequency oscillator.* 2. Abbreviation of *standard-frequency oscillator.*

SFR Abbreviation of *single-frequency receiver.*

SFR-Chireix-Mesny antenna See *Chireix-Mesny antenna.*

SG Abbreviation of *screen grid.*

SG-band A section of the S-band extending from 2700 to 2900 MHz.

SGCS Abbreviation of *silicon gate-controlled switch* (see *silicon controlled switch*).

SGM Abbreviation of *spark-gap modulator.*

SGO Abbreviation of *spark-gap oscillator.*

shaded-pole motor An induction-type ac motor employing shading coils on the field poles for self-starting with a single-phase supply.

shading coil A single, short-circuited turn (copper ring) encircling the tip of the core of an ac-carrying coil, such as the field pole of a motor. Current induced in the coil causes a momentary flux shift that approximates a rotating field which selfstarts a simple single-phase induction motor. A shading coil is also used in a simple ac relay to prevent chatter.

shading ring See *shading coil.*

shading signal In a TV camera, a signal which raises the gain of the amplifier while the beam scans a dark part of the image.

shadow area A vicinity in which signal attenuation or the absence of a signal results from the shadow effect.

shadow attenuation 1. The attenuation of electromagnetic energy caused by an obstacle. Generally measured in decibels. 2. The attenuation of electromagnetic energy produced by the curvature of the earth.

shadow effect The obstruction of radio waves by objects in their path.

shadow mask See *aperture mask.*

shadow meter See *shadow tuning indicator.*

shadow region See *shadow area.*

shadow tuning indicator A tuning meter in which the indicating medium is a shadow whose width is proportional to meter current.

shadow-wound grids In an electron tube, such as a beam-power tube, grids whose wires are truly parallel to each other.

shaft The rodlike part to which a rotating (or turning) member is attached.

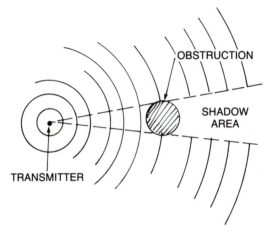

Shadow area.

shaft-angle encoder An electronic system for converting shaft rotation into direct binary or decimal readings.

shaft lock A device for fastening the shaft of an adjustable component (such as a potentiometer, rotary switch, or variable capacitor) in position at a particular setting.

shaft-position encoder See *shaft-angle encoder.*

shaft-position indicator A device which delivers an analog or digital output signal that is proportional to the arc of rotation of a shaft.

shaker See *vibrator, 2.*

shake table A platform, actuated by a vibrator, on which components may be mounted for a vibration test.

shallow-diffused junction A pn junction made by diffusing the impurity material for a short distance into the semiconductor wafer. Compare *deep-diffused junction.*

shape factor 1. For a tuned circuit, the ratio of the 60 dB bandwidth to the 6 dB bandwidth. 2. For a filter, the ratio of bandwidth at high attenuation to that at low attenuation.

shaping network A combination of components for changing the natural response of a circuit to a desired response, i.e., a curve-changing circuit.

shared file A data file that is available for use by more than one system simultaneously.

shared files system A data processing system having one direct-access storage device from which information can be accessed by more than one computer.

sharp cutoff Cutoff (such as that of plate current in a tube) which is abrupt and complete.

sharp-cutoff tube A vacuum tube whose transconductance decreases uniformly as negative grid bias is increased, e.g., a 7C7. Compare *extended-cutoff tube.*

sharpener 1. A circuit or device for increasing the selectivity of another circuit or device. 2. A circuit or device for decreasing the rise or fall time of a pulse or square wave. 3. A circuit or device for steepening the response of a filter.

sharpness See *selectivity.*

sharp pulse A pulse having extremely fast rise and fall times and narrow width, i.e., a spike.

shaving The physical modification of a phonograph disk, or other permanent recording surface, in preparation for rerecording.

SH-band A section of the S-band extending from 3700 to 3900 MHz.

sheath See *positive-ion sheath.*

shelf corrosion In a dry cell in storage, deterioration of the negative electrode because of local action in the zinc.

shelf life The longest time electronic equipment may be stored in the unusual state before deterioration of materials or degradation of performance is evidenced.

shell 1. An electronic orbit (imaginary shell) in an atom. 2. The envelope of a component, e.g., the outer casing of an electron tube or the housing of a plug.

shellac pressing A phonograph disk of the shellac type.

shell-type choke See *shell-type inductor.*

shell-type core A core that completely surrounds the coil(s) of a choke or transformer.

shell-type inductor An inductor in which the core completely surrounds the coil.

shell-type transformer A transformer in which the core completely surrounds the coils.

shf Abbreviation of *superhigh frequency.*

shield A (usually metallic) partition or box for confining an electric or magnetic field.

shield baffle A sheet-type shield. Also see *baffle, 2* and *shield partition.*

shield box A shield having a general box shape and usually enclosed on all sides.

shield braid Tubing woven from wire, through which an insulated wire is passed and thus shielded.

shield can A cylindrical shield, usually enclosed on all sides.

shield disk A flat shield having a disk shape. Also see *baffle, 2; shield baffle,* and *shield partition.*

shielded cable Cable completely enclosed within a metal sheath that is either flexible or rigid.

shielded wire A single strand of insulated wire completely enclosed in a flexible or rigid shield.

shield grid 1. Screen grid. 2. In a gas tube, a grid that shields the control grid from the anode or cathode.

shield partition A wall-type shield usually consisting of a single, flat sheet of metal, sometimes bent into an angle. Also called *baffle shield* (see *baffle, 2*).

shield plate See *baffle, 2; shield baffle; shield disk;* and *shield partition.*

shield room See *cage.*

shield wire A (usually grounded) wire which is run near and parallel to another wire which it shields.

shift 1. To move from one operating point to another in a characteristic curve, or in the operation of an equipment. 2. To transfer data from one point to another in a system, or move it left or right in a register.

shift flip-flop circuit A flip-flop designed especially as a stage in a shift register.

shift pulse In a shift register, a drive pulse that initiates the shifting of characters.

shift register In computers, calculators, and storage systems, a circuit (usually composed of flip-flops in cascade) in which pulses may be shifted from stage to stage and finally out of the circuit.

shingle-type photocell A device in which several separate photocells are series connected by slightly overlapping the ends of adjacent cells.

ship-launched missile Abbreviation, *SLM.* A guided missile launched from the deck of a floating vessel.

ship station A radio or radar station installed aboard a ship which is mobile.

ship-to-shore communication Radio communication between a ship at sea and a shore station. Compare *shore-to-ship communication.*

shock 1. Electric shock. 2. A signal applied momentarily to a circuit, as in shock excitation of a tank.

shock absorber Any object or device intended for reducing physical vibration on a component, set of components, circuit, or system.

shock device 1. A device for administering shock therapy (see *electroshock, 1*). 2. An induction coil and associated primary supply for applying high voltage to a wire fence to shock livestock attempting to escape.

shock excitation Driving an *LC* tank circuit into damped oscillation by momentarily applying a pulse. Also called *impact excitation.* For an application, see *shock-excited oscillator.*

shock-excited oscillator A type of self-excited oscillator in which the tube or transistor is suddenly cut off by applying a cutoff voltage to the grid or base electrode. This abrupt interruption of steady plate or collector current shocks the tank into damped oscillations.

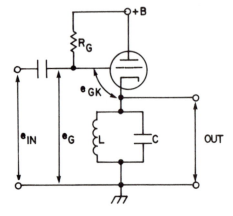

Shock-excited oscillator.

shock hazard 1. Any situation that presents the danger of electric shock to attendant personnel. 2. The existence of a potential difference that will cause a current of at least 5 mA to flow through a resistance of 500 ohms or more, for a prolonged period of time.

Shockley diode See *four-layer diode.*

shock therapy See *electroshock, 1.*

shoran Acronym for *short-range navigation.*

shore effect The tendency of radio waves traveling along a shore to be bent toward the shore. This happens because the waves travel faster over water than over land.

shore station A fixed land station which communicates with ships at sea.

shore-to-ship communication Radio communication between a shore station and a ship at sea. Compare *ship-to-shore communication.*

short circuit 1. An often unintended low-resistance path through which current flows around, rather than through, a component or circuit. 2. To create such a path. In this respect, the term is hyphenated.

short-circuit current In a power supply, the current that flows when

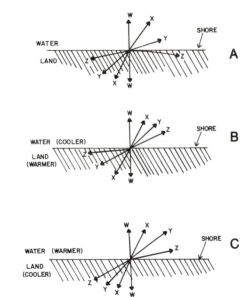

The shore effect. At A, radio waves at very-low to medium frequencies are bent because of differences in surface conductivity. Very high frequency and ultra-high frequency waves are refracted because of temperature differences during the day (B) and at night (C).

the output is directly shorted. Many power supplies have shutdown devices that cause the current to stop flowing when the output terminals are short-circuited; other supplies effectively insert resistance in series with the load, if necessary, to limit the current.

short-circuiting switch A rotary selector switch in which unused contacts are automatically short-circuited.

short-circuit parameter A parameter for which zero resistance is assumed in the part of the circuit under consideration. The current amplification factor, *alpha*, of a common-base-connected transistor is such a parameter because its collector load resistance is assumed to be zero.

shorted-stub tuning Tuning a stub to match a feeder to an antenna by sliding a short-circuiting bar along the two wires of the feeder.

shorting bar A thick, metal strap for short-circuiting two binding posts. Compare *shorting link.*

shorting link A sheet-metal strip for connecting together two *binding posts.* Compare *shorting bar.*

shorting stick A metal rod with an insulating handle, employed to short-circuit a charged capacitor to remove the shock hazard.

shorting switch See *short-circuiting switch.*

short-line tuning Use of a parallel capacitance to tune a transmission line that is less than a quarter-wave long.

short-range navigation Acronym, *shoran.* Navigation by means of short-range radar.

short-range radar A radar having a 50- to 150-mile maximum line-of-sight range for a 1-square-meter reflecting target that is perpendicular to the radar beam.

short skip Skip of only a few hundred miles range. Also see *short-skip communication*.

short-skip communication Radio communication over relatively short distances (400 to 1300 miles) by means of sky waves reflected at sharp angles. See, for example, *sporadic-E skip*.

short-term drift Gradual change in the value of a quantity, such as frequency or voltage, observed over a comparatively brief interval, as opposed to change occurring over a long period. Compare *long-term drift*.

short-term effect The variation of any electrical parameter over a relatively brief time interval. Example: frequency drift over a short time period. Also called *short-time effect*.

short-term stability Stability reckoned over a comparatively brief time interval, as opposed to stability for a long period. Compare *long-term stability*.

short-time effect See *short-term effect*.

shortwave Pertaining to wavelengths shorter than 200 meters, i.e., frequencies higher than 1.50 MHz.

shortwave converter A superheterodyne converter for adapting a long wave receiver (such as a broadcast receiver) for shortwave reception.

shortwave listener Abbreviation, *SWL*. A radio hobbyist who receives, but does not transmit, shortwave signals.

shortwave receiver Any radio receiver capable of intercepting and demodulating signals in the high-frequency range (3 to 30 MHz).

shortwave transmitter Any radio transmitter capable of producing energy in the high-frequency range (3 to 30 MHz).

shot effect Random fluctuations in tube current which give rise to shot-effect noise.

shot-effect noise Electrical noise due to random fluctuations in a current, as in a vacuum tube. Also see *equivalent noise resistance*. Compare *thermal noise*.

shot noise Electrical noise arising from intermittent impulses—such as those produced by spark discharges, make-and-break contacts, etc.—and so called because of its resemblance to pistol shots.

shrink The amount by which a material being measured with an electronic instrument decreases in surface dimension. Compare *stretch*.

shrink tubing Plastic sleeving placed over a conductor or at a conductor—connector joint, and which is made to shrink tightly against its place of application with heat.

shunt 1. Parallel. 2. Shunt resistor.

shunt circuit See *parallel circuit*.

shunt compensation In a wideband amplifier, such as a video amplifier, frequency compensation provided by an inductor in series with the plate or collector load resistor. Compare *series compensation* and *series-shunt compensation*.

shunt-diode rectifier A rectifier circuit in which the diode is connected in parallel with the source and load. Compare *series-diode rectifier*.

shunt-fed 1. Pertaining to a circuit or device in which the dc operating voltage and ac signal voltage are applied in parallel to an electrode. 2. Pertaining to a base-grounded vertical antenna which is excited at some point above ground.

shunt feed See *parallel feed*.

shunt feedback A feedback system in which the fed-back signal is presented to the input of the network in parallel with the input signal. Compare *series feedback*.

shunt generator An electric generator in which the armature and field

windings are connected in parallel. Compare *series generator*.

shunting effect The condition in which a quantity (often a *stray* capacitance or resistance) acts in parallel with another quantity, e.g., the shunting (parallel) resistance of an electrolytic capacitor.

shunt leads Interconnecting wires used for the purpose of attaching a shunting component to a test instrument.

shunt limiter See *parallel limiter*.

shunt loading The parallel insertion of reactance in a circuit, for the purpose of impedance matching.

shunt motor An electric motor whose armature and field windings are connected in parallel. Compare *series motor*.

shunt peaking coil In a wideband amplifier, such as a video amplifier, a small inductor employed for shunt compensation. Compare *series peaking coil*.

shunt regulator A voltage-regulator circuit in which the controlled tube or transistor is in parallel with the output (load) terminals. Compare *series regulator*.

shunt resistor A resistor connected in parallel with a meter or recorder to increase its current range. Also, a resistor connected in parallel with a voltmeter to convert it into a current meter. Compare *multiplier resistor*.

shunt-series circuit See *parallel-series circuit*.

shunt tee junction A waveguide *H-plane tee junction*.

shunt tuning Parallel tuning.

shunt-type frequency multiplier A varactor frequency multiplier circuit in which the varactor is in parallel with the input and output. Compare *series-type frequency multiplier*.

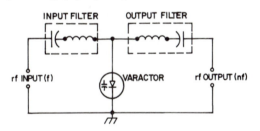

Shunt-type frequency multiplier.

shunt-type resonance bridge A resonance bridge in which the impedance arm is a parallel-resonant circuit. Compare *series-type resonance bridge*.

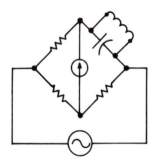

Shunt-type resonance bridge.

shunt-wound generator See *shunt generator*.

shunt-wound motor See *shunt motor*.

SI Abbreviation of *International System of Units*.

S/I Abbreviation of *signal-to-intermodulation ratio*.

Si Symbol for *silicon*.

sibilants 1. High-frequency (hissing) components of speech. 2. Hissing sounds.

SIC Abbreviation of *specific inductive capacity* (see *dielectric constant*).

SiC Formula for *silicon carbide*.

sideband 1. With respect to a carrier, one of the additional frequencies generated by the modulation process. In simple amplitude modulation, the two sidebands are $fc + fm$ and $fc - fm$, where fc is the carrier frequency, and fm the modulation frequency. 2. Pertaining to sidebands.

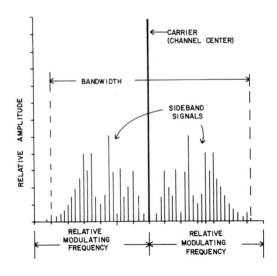

Sideband.

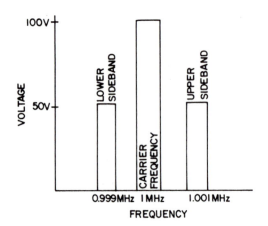

Sidebands of a 1 MHz carrier modulated by a 1-kHz pure tone.

sideband attenuation See *sideband cutting*.

sideband cutting Elimination or attenuation of the sidebands of a modulated signal by a circuit having insufficient bandwidth.

sideband frequency The frequency of the modulation-generated signal accompanying a carrier on either "side." On a given carrier, one sideband frequency is that of the carrier less that of the modulating signal; another is the sum of the carrier and the modulation frequency.

sideband interference Interference arising from one or both of the regular sidebands of a modulated signal or from spurious sidebands arising from overmodulation.

sideband power The power contained in the sideband(s) of a signal.

sideband slicing See *sideband cutting*.

sideband splatter In an amplitude-modulated signal, the emission of sideband energy at frequencies other than within the designated channel. Also simply called splatter.

sideband technique A method of using, for communications or other purposes, one or both of the sidebands of a modulated signal without the carrier.

side-chain amplifier An auxiliary amplifier that is external to a main amplifier. Such an amplifier might be employed, for example, in a feedback channel or in a volume-compression or volume-expansion channel.

side frequency See *sideband*.

sidelobe In an antenna pattern, a lobe other than the principal one(s), especially one extending from the side of the radiator.

sidelobe suppression Elimination of the sidelobe(s) from the radiation pattern of an antenna.

sidestacked antennas Antennas mounted in a horizontal line, parallel to each other, and connected by a common coupler to a transmitter or receiver.

sideswiper A manual telegraph key which is operated by throwing a lever from side to side rather than up and down.

sidetone In wire telephony, the reproduction by the receiver of sounds picked up by the transmitter of the same telephone.

sidetone telephone A telephone set which has no provision for canceling sidetone.

siemens Symbol, S. The SI unit of conductance; the reciprocal of ohm.

Sierra Phonetic alphabet code word for the letter S.

sig Abbreviation of *signal*.

sigma The 18th letter (Σ) of the Greek alphabet. 1. Capital sigma (Σ) is the symbol for *summation*. 2. Lowercase sigma ς is the symbol for *complex propagation constant, electrical conductivity, leakage constant, standard deviation, surface charge density,* and the *Stefan-Boltzmann constant*.

sign 1. Any indicator denoting whether a value is positive or negative. 2. A graphic device indicating an operation. Examples: $+$, $-$, $\times$, $\}$. 3. Any symbol. An ampersand, for example, is an *and sign*. 4. Symptom.

signal An electrical quantity, such as a current or voltage, that can be used to convey information for communication, control, calculation, etc.

signal-actuated voice recorder Abbreviation, SAVOR. A recorder that goes into operation automatically when the speaker starts talking and stops when the speaker finishes.

signal amplitude The intensity of a signal quantity (see *signal*).

signal booster A *preamplifier*.

signal channel In a system, a channel through which only signals flow, control and modifying impulses being accommodated by other channels.

signal circuit A circuit handling signal currents and voltages to the exclusion of control and operating currents and voltages.

signal conditioner Any accessory device, such as a *peak probe, demodulator probe, current shunt*, and so on, used to modify or change the function of a basic instrument, such as an electronic voltmeter.

signal converter See *converter, 1*.

signal current The current component of a signal, as opposed to operating current in a system.

signal diode A diode designed primarily for light-duty signal applications (detection, demodulation, modulation, curve changing), as opposed to the heavy-duty applications of power diodes and rectifiers.

signal distance In two words (bit groups) of the same length, the number of corresponding bit positions that don't match. For example, the signal distance between 01 001 and 10 010 is 3, i.e., 3 positions (from right to left, the first, second, and fourth) do not match:

01 001

10 010

signal envelope shape The shape of a modulation envelope. Common examples are the outlines of amplitude-modulated signals or of keyed signals.

signal-flow analysis A graphic method of analyzing circuits, particularly those employing feedback, through the use of diagrams in which straight arrows represent transmission paths; dots, nodes; and curved arrows, feedback paths. Also called *network topology*.

signal-flow diagram The transmission-path diagram employed in signal-flow analysis.

signal gain See *amplification, 1*.

signal generator An instrument which delivers signals of precise frequency and amplitude, usually over a wide range.

signal ground 1. Any circuit point that remains at zero signal potential. 2. A connection to a point that is deliberately maintained at zero signal potential.

signal/image ratio See *signal-to-image ratio*.

signaling rate In data communications, signal transmission rate—in bits per second, for example.

signal injection 1. Introducing a signal into a circuit. 2. A method of troubleshooting in which a signal injector or signal generator is used to introduce as test signal into a circuit at a succession of points in successive stages from the output to the input of the circuit until the defective stage is located.

signal-injection grid In a pentagrid-converter tube, one of the grids (usually the first or third) to which signals are presented. See *signal injection, 1*.

signal injector A simple (usually single-frequency) signal generator employed in troubleshooting to introduce a test signal at selected points in a circuit, to spot dead circuit sections. Also see *signal injection, 2*.

signal intensity See *signal strength*.

signal inversion Phase reversal of a signal passing through a circuit, device or medium.

signal level At a given point, strength of a signal with respect to a reference amplitude.

signal loss 1. Reduction in amplitude, or disappearance of, a signal passing through a system. 2. *Fractional gain*.

signal mixer See *mixer*.

signal/noise ratio See *signal-to-noise ratio*.

signal notcher See *notch filter*.

signal peaker See *peak filter*.

signal power Power of a signal, as opposed to operating power of the circuit generating or transmitting the signal.

signal processor Any device—such as a preamplifier, expander, amplitude limiter, delay network, and the like—which may be inserted into a system, often externally, to modify an input signal or an output signal.

signal rectification Conversion of an ac signal voltage into a proportionate dc voltage, usually by means of a diode circuit.

signal rectifier See *signal diode*.

signal regeneration. See *signal reshaping*.

signal reshaping 1. Making a signal conform to its original type. Also called *signal regeneration*. 2. Passing a digital signal of any type to a circuit that delivers a uniform output pulse on a real-time one-to-one basis.

signal shifter 1. A device used for quickly changing the frequency of a transmitted signal. 2. A device that automatically causes the transmitted signal to be sent on a frequency that differs from the receiver frequency by a known and predetermined amount. 3. A mixer or converter.

signal squirter See *signal injector*.

signal strength The amplitude of a signal. The usual unit is the volt; however, the millivolt and microvolt are common in some applications.

signal-strength meter 1. Field-strength meter. 2. S-meter.

signal synthesizer A special signal generator delivering signals whose frequency, amplitude, and waveshape can be adjusted at will.

signal time delay The time required for an element of a signal to be transmitted through a network, however simple. This delay results in phase shift in an amplifier.

signal-to-image ratio The ratio of signal amplitude to image amplitude, both being measured in the same units.

signal-to-noise ratio The ratio E_s/E_n, where E_s and E_n are the peak voltages of the signal of interest and of the noise, when E_n is pulse noise; or the rms voltages when E_n is random noise. The ratio is often expressed in decibels.

signal tracer A tuned or untuned detector—amplifier having an input probe and an output indicator (meter, loudspeaker, or both), for following a test signal through a circuit undergoing troubleshooting.

signal voltage The voltage component of a signal, as opposed to the operating voltage of the circuit generating or passing the signal.

signal wave 1. Any electromagnetic disturbance of a periodic nature that is modulated for the purpose of conveying information. 2. The visual illustration or rendition of an electromagnetic disturbance that is modulated for the purpose of conveying information.

signal winding In a magnetic amplifier or saturable reactor, the coil that receives the control current.

signal window See *window, 2*.

signal wobbulator A frequency modulator used with an unmodulated signal generator to provide sweep signals for visual alignment. Also see *wobbulator*.

sign bit A one-bit sign digit.

sign digit A character indicating the sign (positive or negative) of the value of the field or word to which it is attached (usually at the end).

signed field In a computer record, a field having a number whose sign is indicated by a sign digit.

significance In positional number notation, the meaning of a digit's place in a number, i.e., the value of the position as given by N^{n-1}, where N is the radix, and n is the position from the right of the radix point.

significant digits. See *significant figures*.

significant figures In a number, those figures (digits) which show the most accurate value of the represented quantity without giving a false impression of precision. For example, a 1555.5Ω (i.e., to five significant figures) resistor may be needed for a circuit, but there is little point in so expressing the value if, for example, a 1% resistor is to be used, since 1% of this value would be more than 15 ohms. The resistance, therefore, should be specified as 1550Ω.

sign position In a number encoded for data transmission or computation, the place reserved for the sign digit.

silencer See *automatic noise limiter.*

silent alignment See *visual alignment.*

silent piano See *electronic piano.*

silica pencil A rod of silicon dioxide which is heated to emit infrared rays.

silicon Symbol, *Si*. A metalloidal element. Atomic number, 14. Atomic weight, 28.09. Next to oxygen, silicon is the most abundant element in the earth's crust. It is used in many semiconductor devices, including diodes, photocells, rectifiers, and transistors.

silicon capacitor See *voltage-variable capacitor.*

silicon carbide Formula, *SiC*. A compound of silicon and carbon which is valued as a semiconductor, abrasive, and refractory. The commercial product is made by heating carbon and sand to a high temperature in an electric resistance furnace. Also called *Carborundum.*

silicon cell A type of photovoltaic cell employing specially processed silicon as the light-sensitive material. This cell has a comparatively high voltage output.

silicon controlled rectifier Abbreviation, *SCR*. A *pnpn* (or *npnp*) semiconductor device whose action resembles that of a thyratron. The anode, cathode, and gate (control) electrodes of the SCR are equivalent to the anode, cathode, and grid (control) electrodes of the thyratron.

silicon controlled switch Abbreviation, *SCS*. A four-terminal semiconductor switching device similar to the silicon controlled rectifier. It is employed for light-duty switching.

silicon crystal detector 1. Silicon diode (see *silicon junction diode* and *silicon point-contact diode*). 2. An early point-contact diode in which a lump of silicon is contacted by either a fine wire (catwhisker) or a blunt-tipped steel screw under pressure.

silicon detector See *silicon crystal detector, 1, 2.*

silicon diffused transistor A form of silicon bipolar transistor that is fabricated by diffusion techniques. Characterized by high power-dissipation tolerance.

silicon diode A semiconductor diode in which the semiconductor material is specially processed silicon. Also see *silicon junction diode* and *silicon point-contact diode.*

silicon dioxide Formula, *SiO2*. A compound of silicon and oxygen. In the passivation of transistors and integrated circuits, a thin layer of silicon dioxide is grown on the surface of the wafer to protect the otherwise exposed junctions. See, for illustration, *planar epitaxial passivated transistor.*

silicone A polymeric material characterized by a recurring chemical group containing oxygen and silicon atoms in the main chain as links. Various silicone compounds have numerous uses in electronics.

silicon junction diode A semiconductor diode employing a *pn* junction in a silicon wafer. Compare *silicon point-contact diode.*

silicon-on-sapphire Abbreviation, *SOS*. Pertaining to integrated-circuit fabrication in which a silicon expitaxial layer is grown on a sapphire substrate.

silicon oxide A compound containing both silicon monoxide and silicon dioxide, and having dielectric properties. Used in the manufacture of metal-oxide-semiconductor (*MOS*) devices.

silicon photocell A photocell employing a silicon *pn* junction as the light-sensitive medium.

silicon point contact The contact between a pointed metal wire (catwhisker) and a silicon wafer.

silicon point-contact diode A uhf diode in which a tungsten cat whisker contacts a wafer of single-crystal silicon. Compare *silicon junction diode.*

silicon rectifier A semiconductor rectifier consisting essentially of a junction between *n*- and *p*-type silicon inside a specially processed wafer or plate of single-crystal silicon.

silicon resistor See *crystal resistor.*

silicon solar cell A relatively heavy-duty photovoltaic cell employing specially processed silicon as the light-sensitive material.

silicon steel A high-permeability, high-resistance steel containing 2 to 3 percent of silicon, which is widely used as core material in transformers and other electromagnetic devices.

silicon transistor A transistor in which the semiconductor material is single-crystal silicon.

Silistor A type of silicon resistor (see *crystal resistor*).

silo A usually subterranean guided-missile housing.

silver Symbol, *Ag*. A precious metallic element. Atomic number, 47. Atomic weight, 107.880. Silver is used in circuits where low resistance and high Q are mandatory.

silver arsenide trisulfide See *proustite.*

silver-dollar construction Printed-circuit assembly on a disk-shaped board about the size of a U.S. silver dollar.

silver-mica capacitor A fixed capacitor made by painting or depositing a silver layer (capacitor plate) on both faces of a thin mica film (dielectric separator).

silver migration The undesirable tendency of silver to be removed from one location and deposited in another under adverse environmental conditions.

silver solder A relatively high-melting-point solder which is an alloy of silver, copper, and zinc. Also see *hard solder.*

silverstat A multiconductor device used for adjusting the balance of a resistance or reactance bridge.

similar Descriptive of geometric figures having the same shape, i.e., whose corresponding sides share the same ratio of length.

similar decimals Two or more decimal numbers having the same number of digits to the right of the radix point, e.g., 3.14 and 6.39, and even 1.234 and 1.000.

similar fractions Fractions having the same denominator.

simple equation An equation in which no variable's exponent is greater than one. Also called *first-degree equation.* Compare *equations of higher degree* and *quadratic equation.*

simple fraction A fraction having integers for numerator and denominator.

simple quad A combination of two parallel paths, each containing two elements in series.

simple tone A pure tone, i.e., one having negligible harmonic content.

simplex channel An information channel for unidirectional transmission.

simplex system 1. In data communications, a system which transmits data in only one direction. Compare *full duplex system* and *half duplex system.* 2. In voice communications via radio, a direct two-way path over a single channel, which is used alternately for transmitting and receiving.

simplex telegraphy Wire telegraphy in which only one message at a time can be sent over a line.

simplification of circuits See *circuit simplification*.

Simpson's rule A rule for finding the approximate area under a curve: The area *A* is divided into *n* strips of constant width (Δ*x*) and of various heights (*y*0...*y*n). The area then is calculated:

$$A = 1/3\ \Delta x\ (y_0 + 4y_1 + 2y_2 + 4y_3 + 2y_4... + y_n).$$

simulation 1. Imitation of the performance of a process, device, or situation. 2. The use of a mathematical model to represent a physical device or process.

simulator 1. A software or hardware system capable of simulation (*simulation, 2*). 2. A computer program whose implementation allows programs written for one computer to be compatible with another computer. 3. A system of equipment for simulation (*simulation, 1*).

simulcast 1. To simultaneously broadcast a program over two or more different channels. 2. To simultaneously broadcast a program over two or more different types of mode, for example, television and FM stereo. 3. A program broadcast over two or more channels or modes.

simultaneous access See *parallel access*.

simultaneous broadcasting See *simultaneous transmission*.

simultaneous computer See *parallel computer*.

simultaneous equations Two or more equations (for a single problem) which are satisfied by the same values of unknowns, e.g., if *x* + *y* = 20 and *x* − *y* = 8, then *x* = 14 and *y* = 6. (The equations are added, giving a value for *x* which then can be substituted in either equation to find *y*.)

simultaneous transmission The transmission of the same information in two or more channels, or by means of two or more processes, at the same time.

sin Abbreviation of *sine*.

sin⁻¹ Arc sine (inverse sine function).

sine Abbreviation, *sin*. The trigonometric function *a/c*, the ratio of the opposite side of a right triangle to the hypotenuse.

sine galvanometer A galvanometer in which the sine of the angle of deflection is proportional to the current. Compare *tangent galvanometer*.

sine integral Abbreviation *Si* (*x*). The integral $\int_0^x$ (sin *u*)/*u* d*u*. Compare *cosine integral*.

sine law The variation in radiation intensity in any direction from a linear source is proportional to the sine of the angle between the axis of the source and the direction of interest.

sine potentiometer A potentiometer (see *potentiometer, 1*) whose output is proportional to the sine of the angle through which the shaft has rotated.

sine wave A periodic wave which can be represented by a sine curve; i.e., its amplitude is a function of the sine of a linear quantity such as phase or time. Compare *cosine wave*.

singing Audible oscillation in a circuit or device, such as the low-level buzz emanating from the filament of a lamp dimmed with a phase-control circuit.

single-address coding In computer programming, the use of instruction words that contain only one address, the one for the location of the data to be operated on.

single-board computer Abbreviation, *SBC*. A computer on one circuit board.

single-button microphone A carbon microphone having only one button attached to the diaphragm. Also see *button microphone*.

single-cotton-covered wire Wire that is insulated with one layer of cotton.

single cotton enameled wire Wire that is insulated with one layer of cotton on top of an enamel coating.

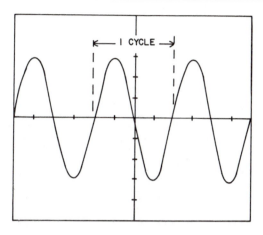

Sine wave.

single-crystal Pertaining to the internal structure of a crystalline material, in which the characteristic lattice is continuous throughout any size piece of the material. Also called *monocrystalline*.

single-crystal material A substance, such as a semiconductor, of which a sample, regardless of size, consists of only one crystal, i.e., there are no grain boundaries. Also see *single-crystal*. Compare *polycrystaline material*.

single-crystal pulling See *Czochralski method*.

single-dial control Adjustment of a multistage system by a single dial attached to a ganged arrangement which tunes all stages simultaneously.

single-diffused transistor A transistor in which only one diffusion of an impurity substance is made. Thus, in a diffused-base transistor, a single diffusion provides the base region and at the same time creates the emitter—base and collector—base junctions. Compare *double-diffused transistor*.

single-element rotary antenna See *one-element rotary antenna*.

single-ended circuit A circuit which has one end grounded, as opposed to a *double-ended circuit* and *push-pull circuit*.

single-ended deflection In an oscilloscope or similar device, horizontal or vertical deflection provided by a single-ended deflection channel. Compare *push-pull deflection*.

single-ended input An input circuit which has one end grounded (or the equivalent ungrounded input circuit). Also called *unbalanced input*. Compare *balanced input*.

single-ended output An output circuit which has one side grounded (or the equivalent ungrounded output), as opposed to *double-ended* or *push-pull output*.

single-ended push-pull circuit An arrangement, such as a complementary symmetry circuit, which provides push-pull output with single-ended input but doesn't require transformers.

single-frequency Pertaining to circuits or devices that normally operate at one frequency only, e.g., single-frequency oscillator.

single-frequency amplifier An amplifier that normally operates at only one frequency (or within a very narrow band of frequencies), e.g., an i-f amplifier or a selective af amplifier used for harmonic analysis or bridge balancing.

single-frequency duplex Two-way communication over one medium or frequency. Voice-actuated (*VOX*) or breakin devices are used at both ends of the circuit.

single-frequency oscillator An oscillator that normally delivers a signal at only one frequency until it is switched to another frequency, e.g., crystal-controlled oscillator.

single-frequency receiver A radio or TV receiver that normally operates at one carrier frequency rather than being continuously tunable. Such receivers are employed in monitoring specific programs, picking up standard-frequency signals, and in similar applications.

single-gun color-TV tube A color-TV picture tube in which the image is produced by a single beam which scans color-phosphor dots sequentially. See, for illustration, *Trinitron*.

single-hop propagation Radio transmission involving only one reflection by the ionosphere.

single-image response In an oscilloscope presentation, a single pattern, as opposed to a double-trace pattern.

single-inline package Abbreviation, SIP. A flat, molded component package having terminal lugs along one edge. It thus may be viewed as half of a *dual-inline package*.

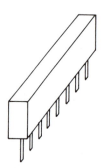

Single inline package.

single-junction transistor 1. Unijunction transistor. 2. Diode amplifier (see *crystal amplifier*, 2).

single-layer coil A coil whose turns are wound side by side in one layer.

single-layer solenoid See *single-layer coil*.

single-loop feedback Feedback through only one path.

single-phase Pertaining to the presence or generation of one ac phase only. Compare *polyphase*.

single-phase—full-wave Pertaining to a rectifier operated from a single-phase ac power line and rectifying both half-cycles of ac voltage. Compare *single-phase—half wave*.

single-phase—full-wave bridge A bridge rectifier operated from a single-phase ac supply, usually from the untapped secondary winding of a transformer. Compare *single-phase—full-wave circuit* and *single-phase—half-wave circuit*.

single-phase—full-wave circuit A rectifier circuit in which each half-cycle of single-phase ac is rectified by a separate diode supplied from the ends of a center-tapped winding of a transformer. Compare *single-phase—full-wave bridge* and *single-phase—half-wave circuit*.

single-phase—half-wave Pertaining to a rectifier operated from a single-phase ac power line and rectifying only one half-cycle of ac voltage.

single-phase—half-wave circuit A rectifier circuit in which a diode, output load, and single-phase ac supply are connected in series, only one half-cycle of the ac being passed by the diode. Compare *single-phase—full-wave circuit* and *single-phase—full-wave bridge*.

single-phase rectifier See separate listings under *single-phase—full-wave bridge, single-phase—full-wave circuit*, and *single-phase—half-wave circuit*.

single-point ground One ground connection to which all channels of a circuit are returned. Such a common connection eliminates, or greatly minimizes, the common coupling often encountered when separate ground points are employed.

single pole—double-throw Abbreviation, *spdt*. Descriptive of an electrical, electronic, or mechanical switch with a pole that can be connected to either of two adjacent poles, but not to both.

single-pole—single-throw Abbreviation, *spst*. Descriptive of an electrical, electronic, or mechanical switch with a pole that can be connected to an adjacent pole (or disconnected from it) at will. It is used to provide the make and break function in a single circuit.

single rail 1. A one-conductor communications medium, with a ground return. 2. A one-conductor data line, with a ground return.

single-shot Pertaining to circuit operation in which a single input pulse applied to a switching device (such as a multivibrator) causes it to deliver a single output pulse rather than switch to a stable *on* state. A monostable multivibrator operates in this mode. Also called *one-shot*.

single-shot multivibrator See *monostable multivibrator*.

single-sideband Abbreviation, *SSB*. Pertaining to a signal or system characterized by a frequency which is one of the sidebands resulting from modulation.

single-sideband suppressed-carrier Abbreviation, *SSSC* or *SSBSC*. Pertaining to a system of modulation in which the carrier and one sideband are suppressed; only the remaining sideband is transmitted.

single-signal Pertaining to a circuit's favoring one of many signals (as by a bandpass filter or sharply tuned receiver). Thus, a highly selective superheterodyne receiver equipped with a crystal filter is sometimes called a *single-signal receiver*.

single-signal receiver A very sharply tuned superheterodyne receiver which achieves high selectivity by means of a piezoelectric or magnetostrictive filter in the i-f amplifier channel.

single silk-covered wire Wire that is insulated with one layer of silk.

silk enameled wire Wire whose insulation is a layer of silk on top of an enamel coating.

single-skip propagation See *single-hop propagation*.

single-step operation See *step-through operation*.

single sweep In an oscilloscope, a single time-axis deflection of the electron beam. Also see *sweep*, 1, 2. Compare *recurrent sweep*.

single-sweep blocking oscillator A blocking oscillator that cuts off after generating a single cycle or pulse.

single-throw switch A single-action switch with two or more poles.

single-tone keying Modulated continuous-wave keying. A single audio-frequency tone is used to modulate the carrier wave, usually via AM or FM.

single-track recorder A recorder, such as a magnetic-tape recorder or a graphic recorder, that permits recording along only one track.

single-trip multivibrator See *monostable multivibrator*.

single-tuned circuit A circuit which is tuned by varying only one of its components, e.g., an i-f transformer in which only the secondary coil (rather than both primary and secondary) is tuned.

single-turn coil 1. A coil consisting of a single turn of wire, tubing, or strip. 2. Ring inductor. 3. Shading coil.

single-turn potentiometer A potentiometer that is varied through its entire range by one complete rotation of the shaft.

single-wire-fed antenna See *Windom antenna.*

single-wire line 1. Single-wire transmission line. 2. A single wire used for communication or control purposes (the earth furnishes the return path).

single-wire transmission line A transmission line or feeder consisting of one wire only (see, for example, *Windom antenna*).

sinh See *hyperbolic sine.*

sink A device or circuit into which current drains.

sink circuit The circuit associated with a load or other sink. Compare *source circuit, 2.*

sinker A piece of semiconductor material used for reducing the base-collector junction resistance in a bipolar transistor.

sintering A process in which various solid bodies are formed from fusible powders at temperatures below their melting points. Example: sintered magnetic core.

sinusoidal Having the shape and properties of a sine wave.

SIO Abbreviation of *serial input/output.*

SIP Abbreviation of *single-inline package.*

Si(x) Abbreviation of *sine integral.*

six-phase rectifier A polyphase rectifier circuit operated from a three-phase supply. The output ripple frequency is six times the supply frequency.

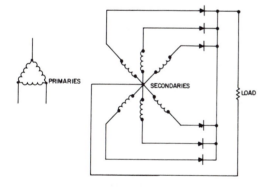

Six-phase rectifier.

SJD Abbreviation of *silicon junction diode.*

skate The tendency of a tone arm to swing toward the spindle during record play, independent of the action produced by the stylus following the groove. The effect would occur with a grooveless disk.

skeletal code A generalized computer routine needing only certain parameters to be usable for a specific application.

skeleton bridge A bridge consisting of an adjustable arm (potentiometer) and a pair of binding posts for each of the other three arms. Suitable resistors, capacitors, or inductors are connected to the binding posts to set up the bridge circuit desired.

skeleton-type assembly 1. A type of construction of electronic equipment in which a minimum of supporting members is employed. An example is the use of an open framework instead of a chassis to sup-

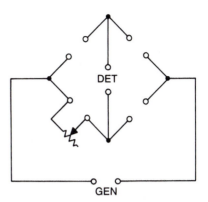

Skeleton bridge.

port components. 2. An assembly of electronic equipment, consisting essentially of a foundation unit (containing the basic circuitry) and plug-in units for setting up various complete equipments.

skew 1. A condition resulting from failure of the horizontal synchronization in facsimile or television. The picture appears distorted, and appears as a non-rectangular parallelogram. 2. In a print display, nonalignment of columns resulting from an incorrect number of line spaces in each line. 3. In a probability function, an accumulation of values toward either side of center.

skewing 1. The bending of a curve away from its normal shape. 2. In a differential amplifier, the offset between two signals. Also see *offset.*

skew lines In geometric space, two nonparallel, nonintersecting lines which do not lie in one plane and whose distance apart is the shortest line segment that is perpendicular to both.

skin depth The depth to which current penetrates below the surface of a conductor as a result of the skin effect.

skin effect The tendency of high-frequency currents to travel along the surface of a conductor; the high-frequency reactance is lower along the outside than at the center of a conductor.

skip 1. Ionosphere-reflected radio transmissions. 2. In a computer program, an instruction whose sole function is that of causing a jump to the next instruction.

skip distance For a signal transmitted via ionospheric reflection, the distance from the transmitter to the point at which the reflected wave strikes the earth.

skip fading In a signal reflected by the ionosphere, fading due to fluctuations in the height or shape of the reflective layer.

skip zone See *zone of silence.*

skirt selectivity Bandwidth at the point of highest attenuation (lowest transmission) on a selectivity curve.

SKM Abbreviation of *sine—cosine multiplier.*

skyhook 1. Antenna. 2. A captive balloon employed to support a vertical-wire antenna.

sky noise Radio noise thought to originate from certain stars.

skywave A radio wave propagated by ionospheric reflections and refractions. Compare *ground wave.*

skywave correction A factor applied to long-range radionavigation signals to account for the time delay resulting from ionospheric reflection.

skywire See *outside antenna.*

slab 1. A relatively thick body of quartz, ceramic, semiconductor, or dielectric. 2. Substrate.

slashed-field-gun crt A straight-gun type of TV picture tube (see *straight-gun crt*). Because the gap between the anodes in this tube is slanted, the electrostatic field is diagonal, causing the electron and ion beams to be diverted at an angle.

Slide switch.

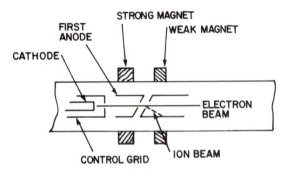

Slashed-field-gun crt.

slave flash A photoflash operated by the light flash from another such unit.

slave relay A relay operated by another (master) relay. Compare *master relay.*

slave sweep In an oscilloscope or similar device, a driven sweep.

SLC Abbreviation of *straight-line capacitance.*

sleeping sickness A gradual increase in transistor leakage current.

sleep machine An electronic device for inducing sleep. It consists essentially of an audio oscillator which produces a soft, low-frequency tone or white or pink noise in headphones or in a pillow speaker.

sleeve antenna A vertical antenna in which the upper half is a quarter-wave rod connected to the inner conductor of a coaxial feeder, and the lower half is a quarter-wave metal sleeve connected to the outer conductor of the feeder. Also called *coaxial antenna.*

sleeving A material in tubular form that can be slipped over another material, e.g., insulating sleeving for wires (*spaghetti*).

slewing rate In an operational amplifier, the rate at which the output can be driven between its limits.

slew rate See *slewing rate.*

SLF Abbreviation of *straight-line frequency.*

slice A semiconductor wafer cut from a single-crystal ingot.

slicer See *clipper-limiter.*

slide-back meter A type of electronic voltmeter in which an unknown voltage applied to the input of an amplifier stage is bucked by an internal voltage that is potentiometrically adjustable. At null, when the unknown and internal voltages are equal, the voltage is read from the meter itself, which is deflected by the bucking voltage.

slider A flat-spring contact that slides along the turns of a resistance or inductance coil to vary the coil's resistance or inductance. Also called a *wiper.*

slide-rule dial A dial having a straight scale resembling a slide rule.

slide switch A switch which is actuated by sliding back and forth a blockshaped button. Compare *bat-handle switch, paddle switch,* and *rocker switch.*

slide wire A simple potentiometer consisting of a single, straight piece of resistance wire with a sliding contact. Also see *slide-wire resistor.*

slide-wire bridge 1. A simple four-arm bridge in which the adjustable element is a single, straight resistance wire along which a clip or slider is moved and which supplies two arms of the bridge (one on each side of the slider). 2. By extension of 1, above, a similar bridge in which the potentiometer supplies two of the arms.

slide-wire resistor A variable resistor consisting of a single wire (straight or coiled) along whose length a slider is moved to vary the fixed resistance value.

sliding contact A contact that mates with another contact, or moves along a contacted surface, with a sliding motion. Also called *self-cleaning contact* and *wiping contact.*

sliding screen operation Operation of the screen grid of an electron tube from a high dc voltage through a series dropping resistor.

slip 1. In an eddy-current brake, coupling, or drive, the difference in speed between the field magnets and the iron eddy-current ring. 2. In a synchronous motor, the difference between rotor speed and stator speed.

slip clutch In a gear or rack-and-pinion drive system, a device that releases the load if the torque becomes excessive. The gears then slip instead of being damaged.

slip ring See *collector ring, 1.*

slip speed See *slip, 2.*

SLM Abbreviation of *ship-launched missile.*

slope The slant of a curve, depicted by the ratio dy/dx.

slope detector An AM receiving circuit detuned to one side of resonance (i.e., to a point along the skirt of the selectivity curve) to detect an FM signal. The FM swing occurs along the slope of the resonance curve. Slope detection is useful in narrowband FM when conventional FM circuitry is not available.

slop-jar capacitor See *water capacitor.*

slop-jar rectifier See *electrolytic rectifier.*

slot 1. In the armature of a motor or generator, a groove in which the windings are laid. 2. The notch in the response curve of a crystal filter.

slot antenna A microwave antenna which radiates energy through a slot cut in a surface, such as the metal skin of an airplane.

slot cell A reinforcing, dielectric material (such as plastic) that is placed in the slot of a ferromagnetic core.

slot coupling Coupling microwave energy between a waveguide and a coaxial cable by means of two slots, one in the waveguide and the other in the outer conductor of the cable.

slot-discharge resistance See *corona resistance.*

slot insulation 1. Insulation of wires in the slots of the armature of a motor or generator (see *slot, 1*). 2. A material in the form of tape or sheets, employed in 1, above.

slot radiator See *slot antenna.*

slotted line A device which consists of a section of air-dielectric coaxial line arranged for microwave measurements. The outer conductor is a metal cylinder and the inner conductor a concentric metal rod. The cylinder is provided with a lengthwise slot through which a small pickup probe extends for sampling the rf field inside the device. The probe is attached to a carriage which slides along a graduated scale on the outside of the cylinder. Radio-frequency energy is injected into one end of the line through a coaxial cable, and as the probe moves along, response points are indicated by an external detector connected to the probe (the way Lecher wires are tuned); the scale is read at these points to determine frequency, standing-wave ratio, impedance, and power. An alternate form of slotted line employs a section of slotted waveguide instead of a section of coaxial line.

slotted rotor See *serrated rotor plate.*

slotted section See *slotted line.*

slotted waveguide See *slotted line.*

slot width 1. The width of a slot in the armature of a motor or generator (see *slot, 1*). 2. The bandwidth of the notch in the response curve of a band-suppression filter of any kind. See, for example, *notch filter.*

slow-acting relay Any relay designed to operate at some finite period following application of actuation voltage.

slow-blow fuse A fuse in which the melting wire breaks apart slowly. The time delay allows the fuse to withstand momentary surges which would not damage the protected equipment.

slow-break-fast-make relay A relay that opens slowly and closes rapidly.

slow-break-slow-make relay A relay that opens slowly and closes slowly.

slow charge Storage-battery charging in which a low current is passed through the battery over a long period of time; this way the normal ampere-hour capacity is restored.

slow death The gradual deterioration of transistor performance, and, by analogy, the similar deterioration of any component with time.

slow drift The gradual change of a quantity or setting (usually in one direction). Compare *fast drift.*

slow-make-fast-break relay A relay that closes slowly and opens rapidly.

slow-make-slow-break relay A relay that closes slowly and opens slowly.

slow-operate-fast-release relay See *slow-make-fast-break relay.*

slow-operate-slow-release relay See *slow-make-slow-break relay.*

slow-release-fast-operate relay See *slow-break-fast-make relay.*

slow-release-slow-operate relay See *slow-break-slow-make relay.*

slow storage A form of memory with long storage and recovery time.

slow time scale An extended time scale, i.e., one larger than the time unit of the system under consideration.

SLS Abbreviation of *side-lobe suppression.*

slug 1. A movable core of metal or ferrite employed for tuning (varying the inductance of) a coil by changing its position along the axis of the coil. Also see *slug-tuned coil.* 2. A copper ring attached to the core of a relay for time-delay purposes (see *slug-type delay relay*).

slug-tuned coil A coil whose inductance is varied by means of a metal or ferrite slug that slides in and out of the coil.

slug tuner A tuner for a radio or TV receiver or test instrument, employing slug-tuned coils.

slug-type delay relay A delayed-response relay which achieves time delay through the action of a heavy copper slug on the core. The slug forms a low-resistance, short-circuited single turn in which a current is induced by the magnetic flux resulting from energizing the relay. The resulting flux of the slug opposes the buildup of relay-coil flux which, therefore, can only reach its closure level some time after the relay has been switched on.

slumber switch An alarm-reset switch on an electronic clock radio. If the alarm comes on, the slumber switch (usually a pushbutton device) can be pressed to turn off the alarm for a predetermined length of time. Also called a snooze button.

SLW Abbreviation of *straight-line wavelength.*

Sm Symbol for *samarium.*

small Of reduced size (smaller than *standard*, but larger than *midget*). In the sequence of adjectives used to describe electronic equipment, *small* is second: *standard, small, midget, miniature, subminiature, microminiature.*

small-current amplifier 1. A dc amplifier for low-level input currents, i.e., currents of 1 μA or less. 2. An amplifier (such as a silicon-transistor unit) requiring very low dc operating current.

small signal A low-amplitude signal. Such a signal covers so small a part of the operating characteristic of a device that operation is nearly always linear. Compare *large signal.*

small-signal analysis Analysis of circuit or component operation in which it is assumed that the signals deviate from (fluctuate to either side of) the steady bias levels by only a small amount. Also see *small signal.*

small-signal component 1. A coefficient or parameter—such as amplification, transconductance, dynamic resistance, etc.—calculated or measured under conditions of small-signal operation. Also see *small signal* and *small-signal equivalent circuit.* 2. A device designed for operation at low signal levels.

small-signal diode See *signal diode.*

small-signal equivalent circuit For a given transistor circuit, the equivalent circuit for low signal levels (i.e., at amplitudes lower than saturation and cutoff levels). Also see *equivalent circuit.*

small-signal operation Operation at low signal amplitudes, i.e., at signal levels that do not extend into the saturation or cutoff levels of a tube, transistor, or other component.

small-signal transistor A transistor designed for low-level applications, such as the amplification of small voltages and currents and low-voltage switching. Compare *power transistor.*

smartness The ability of an electronic system, especially a computer or control system, to perform a complete series of operations, substituting alternative steps where necessary, all with a minimum of instructions from the outside world.

smearing In television or facsimile, a form of picture distortion caused by an excessively narrow receiving bandpass. The image appears fattened and horizontally blurred. Contrast may also be lost.

smectic crystal A liquid crystal in which the molecules are arranged in parallel layers and cannot slide past each other.

S-meter In a receiver designed for the amateur service, a meter graduated in S-units to indicate the comparative field strength of a received signal.

Smith chart A graphic device of circular configuration containing a number of overlapping curves, which facilitates various calculations

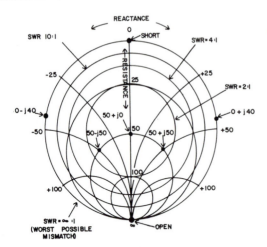

Smith chart.

such as those for the relationships between slotted-line response, admittance, impedance, SWR, and the like.

smoke alarm A device that produces an audible or visible signal in the presence of smoke or gases. See also *photoelectric smoke alarm*.

smoke control See *photoelectric smoke control*.

smoke detector Any circuit or device that is used to sense the presence of smoke or noxious gases. *photoelectric smoke detector*.

smooth 1. Relatively free from surface irregularity. 2. To reduce or eliminate irregularities in a direct-current source. 3. To reduce or eliminate irregularities in data or signal amplitude.

smoothing choke A power-supply filter choke having a core with an air gap which prevents saturation at maximum rated direct current. Compare *swinging choke*.

smoothing factor For a power-supply filter, the quantity $R\omega C$, where R is the filter resistance in an RC filter, or the series resistance of the choke in an LC filter, and C is the filter capacitance; $\omega = 2\pi \times$ *frequency*.

SMPTE Society of Motion Picture and Television Engineers.

smudge See *squeezeout*.

SN Abbreviation of *Semiconductor network*.

S/N Abbreviation of *signal-to-noise ratio*.

Sn Symbol for *tin*.

snake 1. A long, strong, flexible wire or strip used to pull other wires through pipes or tubes. 2. To route wires or cables through a group of circuits, components, or boards.

snap-action switch A switch that snaps quickly into the on or off position to prevent arcing and consequent premature contact deterioration.

snap diode A semiconductor diode in which switch-off time after carrier storage is extremely short (e.g., under 1 nanosecond).

snap magnet A magnet that reduces the tendency for arcing in relay-control instruments, thereby minimizing electromagnetic interference and prolonging contact life.

snapshot dump During a computer program run, a dump, for debugging purposes, of certain storage areas.

snap switch See *snap-action switch*.

sneak current Unintended current flow through a path that is auxiliary to a main circuit.

sneak path A path through which current is accidentally detoured; it is usually a leakage path.

sniffer See *exploring coil*.

sniperscope A telescope, snooperscope, or starlight scope for a carbine or rifle.

SNOBOL Acronym for *string-oriented symbolic language*, a computer-programming language for manipulating character strings.

snooperscope 1. An infrared device which permits viewing objects and surroundings in total darkness. It presents the image on a fluorescent screen. 2. A rifle-mounted starlight scope.

snow A type of TV picture interference characterized by a screen whose image is superimposed by countless tiny out-of-focus light spots whose random pulsing resembles falling snow.

SNR Abbreviation of *signal-to-noise-ratio*.

soak value The smallest value of current that will cause saturation of a relay core.

socket A (usually female) fixture into which a plug, or a tube, transistor, or other component, is inserted for easy installation in, or removal from, a circuit.

socket punch See *punch, 2.*

sodium Symbol, *Na*. A metallic element of the alkali-metal group. Atomic number, 11. Atomic weight, 22.991.

sodium silicate Formula, $Na2O \bullet 4SiO2$. See *water glass*.

sodium-vapor lamp A gas-discharge lamp containing neon and a small amount of sodium. After the filaments of the lamp are lighted for a short time, the heat vaporizes the sodium, and the filaments are disconnected by an automatic switch. Under the influence of the voltage across the lamp, the sodium vapor glows with a characteristic yellow light.

sofar A system for pinpointing the source of underwater sounds (coming from as far away as 2000 miles) through triangulation: coordinating measurements by three widely separated shore stations. In one application, survivors of a sea disaster drop a special bomb into the water, and the explosion is detected by sofar stations. The name is an acronym for *sound fixing and ranging*.

soft-drawn wire Wire that is highly malleable, therefore easily bent and unbent. Compare *hard-drawn wire*.

soft iron A grade of iron, used in some cores, which is easily demagnetized.

soft solder A low-melting-point solder.

soft tube A vacuum tube which has been incompletely evacuated, i.e., one containing a small amount of gas. Compare *hard tube*.

software 1. Vendor-supplied or user-generated programs or groups of programs for a computer or computer system. 2. The detailed instructions for performing a particular operation with a calculator or a computer.

soft X-rays Low-frequency (long-wavelength) X-rays. Such radiation has relatively poor penetrating power. Compare *hard X-rays*.

Sol The sun.

sol Abbreviation of *solution* and *soluble*.

solar access The means of direct exposure to the sun's rays as a source of energy. The right of citizens to solar access has drawn the attention of government at all levels, and some regulations have already been enacted to guarantee that right. Such regulations, for instance, prohibit the erection of buildings, walls, and other obstructions which would hinder a citizen's right to solar access.

solar absorption index A measure of the effect of the sun on the ionospheric absorption of radio waves. The higher the angle of the sun with respect to the surface of the earth, the greater the ionization in a given region of the ionosphere, and the more absorption takes place.

solar activity See *solar radiation* and *sunspot cycle*.

solar battery A battery composed of solar cells connected in series and/or parallel for increased output.

solar cell A photovoltaic power transducer that converts light to electricity. It is called a cell because its output is a small dc voltage. Like a battery, such cells may be connected in series and parallel to provide useful power levels.

solar cycle See *sunspot cycle*.

solar energy 1. The total energy arriving from the sun at a given region on the surface of the earth. 2. Any energy derived entirely from the sun.

solar-energy conversion Any process that changes solar radiant energy into another useful form.

solar laser See *sunlight-powered laser*.

solar panel An array consisting of a number of series-connected or series-parallel-connected *solar cells* mounted on a flat plate.

solar power Useful amounts of electricity obtained from suitable arrays of *solar cells*.

solar radiation Radiation, comprising energy of many wavelengths, from the sun. Such radiation, after passing through earth's atmosphere, contains a small percentage of ultraviolet rays, but for the most part contains radiation in the visible and near-infrared parts of the spectrum, as a result of absorption by the atmosphere.

solar relay See *sun switch*.

solar simulator A source of intense light that is used, for example, in the investigation of the effects of solar radiation on space vehicles. In one form, the device consists essentially of a battery of high-powered, short-arc xenon lamps in combination with suitable optics. They system is operated in a low-temperature vacuum chamber to simulate cold, airless space.

solar switch See *sun switch*.

solar wind Plasma emission by the sun (a continuous process). The solar wind extends from the sun in all directions and is said to be capable of providing motive power to space vehicles throughout the solar system (by analogy of the windborne sailboat).

solder 1. A metal alloy (usually of tin and lead) which is melted to join pieces of other metals. Also see *hard solder* and *soft solder*. 2. To join metals with solder.

soldering Joining (usually nonferrous) metal parts with solder, a lead-alloy substance. Compare *brazing*.

soldering gun An electric soldering iron having the general shape of a pistol. Also called *solder gun* and *soldering pistol*.

soldering iron An electric or nonelectric tool having a heated tip for melting solder.

solderless breadboard A foundation (see *breadboard, 1*) on which a circuit may be assembled by plugging into tiny jacks, without the use of solder.

solderless connection A connection between leads or leads and terminals, accomplished entirely through crimping, pinching, splicing, and the like. Also see *wire-wrap connection*.

solderless terminal A terminal to which a solderless connection can be made. Also see *wrap post*.

solenoid 1. Single-layer coil. 2. A multilayer coil used as an electromagnet, and usually having a straight, iron core.

solenoid switch A switch consisting of a solenoid coil (see *solenoid, 2*) into which a core is pulled by the magnetic field to close a pair of contacts.

solid 1. One of the *state of matter*. Solids are characterized by a definite shape and volume and by atoms that maintain a fixed position relative to each other. Compare *gas, liquid*, and *plasma*. 2. In geometry, a closed surface, sometimes including what is inside. 3. In communications, descriptive of error-free reception of a series of coded signals. 4. In printing and data transmission, a large print area whose entire surface is of equal and maximum intensity (of ink, light, or darkness).

solid angle Unit, steradian. The angle within the apex of the cone formed by all line segments between the center of a sphere and a circle on the sphere.

solid circuit Any circuit consisting of a single piece of hardware that is not normally separated into smaller parts.

solid conductor See *solid wire*.

solid electrolyte A solid substance affording ionic action similar to that in a liquid electrolyte.

solid electrolytic capacitor A capacitor employing a solid electrolyte.

solid ground See *direct ground*.

solid-state Pertaining to devices and circuits in which the flow of electrons and holes is controlled in specially prepared blocks, wafers, rods, or disks of solid materials. Semiconductor devices are solid-state components.

solid-state battery An atomic battery consisting essentially of a photovoltaic cell in combination with a quantity of radioactive material whose radiation causes the cell to generate electricity.

solid-state capacitor See *solid electrolytic capacitor*.

solid-state chronometer Any semiconductor device used for the purpose of indicating or measuring time.

solid-state circuit See *monolithic integrated circuit*.

solid-state lamp 1. A light-emitting diode (see *laser diode*). 2. Sometimes, an electroluminescent cell.

solid-state maser A device, such as the *ruby maser*, in which the stimulated medium is a solid material.

solid-state photosensor A semiconductor photodiode or phototransistor, as opposed to a phototube.

solid-state physics The branch of physics concerned with the nature and applications of such solids as electronic semiconductors.

solid-state relay 1. A sensitive relay consisting of a conventional electromagnetic relay preceded by a transistorized amplifier. 2. A completely electronic relay (i.e., one without moving parts) in which switching transistors provide the on and off states. 3. Loosely, any thyristor.

solid-state thermometer An *electronic thermometer* utilizing one or more solid-state components, such as transistors, IC's, or thermistors.

solid-state thyratron See *silicon controlled rectifier* and *silicon controlled switch*.

solid-state tube A semiconductor device (diode, rectifier, transistor, SCR, etc.) whose housing and base allow it to replace directly an electron tube.

solid tantalum capacitor A capacitor employing tantalum as a solid electrolyte.

solid wire Wire consisting of a single strand of metal. Compare *stranded wire*.

solute Commonly, the solid member of a solution, i.e., what has been dissolved. Also see *solution, 1*.

solution 1. A well-diffused mixture of two or more substances. While it is common to think of a solution as a solid dissolved in a liquid, or as one liquid dissolved in another, a solution may consist of a gas in a liquid, a gas in a solid, a gas in a gas, a liquid in a solid, or a solid in a solid. A solution, typically, is molecular; i.e., there is no chemical reaction between its constituents. Also see *saturated solution; solute; solvent, 1;* and *supersaturated solution.* 2. The result of solving a problem or making a calculation. Also called *answer* and *result.*

solution conductivity The electrical conductivity of a solution, such as an electrolyte. The conductivity (and conversely, the resistance) depends upon the number and mobility of ions in the solution.

solution-conductivity bridge A Wheatstone-type dc bridge especially designed and calibrated for measuring the conductivity of chemical solutions.

solution pressure In an electrolyte into which a metal body is immersed, the force that causes the metal to tend to pass into solution as positive ions and to form a Helmholtz double layer.

solvent 1. A fluid that dissolves other materials. 2. The constituent of a solution that dissolves one or more other constituents. Thus, in a saltwater solution, water is the solvent and salt the solute. Also see *solution, 1.*

SOM Abbreviation of *start of message.*

Sonalert Tradename for a small but loud sound reproducer used with solid-state circuits for alarm purposes.

sonar A system of detection and ranging by means of sonic and ultrasonic signals. In this system of echo ranging, the distance to an underwater object is determined from the time it takes a sound signal to reach the object and be reflected back to the sonar transmitter. The name is an acronym for *sound navigation and ranging.*

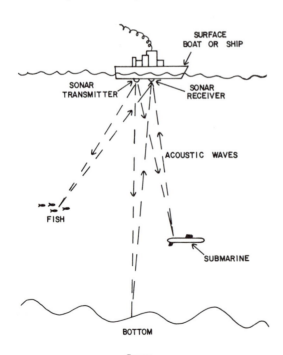

Sonar.

sonde A device for automatically gathering meterological data at high altitudes. An example is the *radiosonde.*

sone A unit of *loudness* for an individual listener. The level of 1 sone is the loudness of a 1000-Hz tone that is 40 dB above the particular listener's threshold of hearing.

sonic altimeter An altimeter (see *absolute altimeter*) employing sound waves. The time required for a transmitted wave to reach a target and be reflected back to the transmitter is proportional to the distance between transmitter and target.

sonic boom An explosive sound occurring when the shock wave produced by an aircraft flying at supersonic speed strikes the earth.

sonic delay line A delay line employing electroacoustic transducers and an intervening medium through which a sound wave is transmitted. Also called *acoustic delay line.*

sonic depth finder See *acoustic depth finder.*

sonic thermocouple A thermocouple whose heat-absorbing properties are enhanced by subjecting it to acoustic vibrations.

sonic time See *acoustic mine.*

sonobuoy A buoy equipped with an acoustic receiver and radio transmitter. The device is parachuted into the water where it picks up submarine sounds and transmits them to a monitoring station. Several sonobuoys communicating with a computer will track the path of a submarine.

sonovox An electronic device used to produce special sound effects when it is held against the throat of the operator. The special sounds are formed into words by the operator's mouth.

SOP Abbreviation of *standard operating procedure.*

sophisticated electronics Advanced electronic theory and practice, usually dealing with complex devices or systems and requiring rigorous analysis to describe their operation and devise highly sophisticated applications. Compare *unsophisticated electronics.*

sorption processes Processes whereby certain substances (e.g., activated charcoal) occlude and retain gases and vapors. A chamber containing such a substance is often useful in the production of a vacuum. Sorption includes both *absorption* and *adsorption.*

sort 1. To group information items using their keys. Also see *key.* 2. To group information items according to some system of classification, as to print an alphabetical list of words stored in a random sequence.

sorter A device for sorting punched cards according to information in the cards' columns.

sort needle A stylus passed through selected holes in a stack of data cards to determine if a particular position has been punched in all the cards.

sorting routine A computer program for sequencing data items according to key words (values in specific fields) of the different records.

SOS 1. The international radiotelegraph distress signal; equivalent to *mayday* in radiotelephony. 2. Abbreviation of *silicon-on-sapphire.*

sound The vibratory or wave phenomenon to which the sense of hearing is responsive, although a listener is not necessary for the existence of sound: very high-frequency sounds are superaudible and very lowfrequency ones subaudible. Sound is conducted by waves in solids, liquids, and gases; it is not propagated through a vacuum.

sound absorption The nonreflection of sound energy by a body or medium and the attendant conversion into another form of energy (usually heat).

sound amplifier 1. An audio amplifier, especially the sound channel of a TV system. 2. A device, such as a horn or reflector, which directly boosts the intensity of sound at a given listening point.

sound analyzer An instrument, often a wave analyzer equipped with a microphone, for measuring such characteristics of sound as amplitude, frequency (pitch), and harmonic content (timbre).

sound articulation See *articulation.*

sound bars In a TV picture, horizontal bars resulting from interference between the audio and video channels of the receiver.

sound carrier In a TV signal, the FM carrier which transmits the audio part of the program. Compare *video carrier.*

sound chamber An air enclosure, usually a box or can, for modifying the acoustic qualities of sound or of an audio signal.

sound detector The discriminator or ratio detector that demodulates the sound signal in a TV receiver circuit.

sound-energy density Sound energy per unit volume, as expressed in ergs per cubic centimeter, for example.

sound-energy flux The average rate of flow of sound energy through a specified area, as expressed in ergs per second, for example.

sounder In wire telegraphy, a receiving device in which an electromagnet (energized by the incoming signal) attracts a pivoted bar which, on striking the core of the electromagnet or a stopping bar, emits a sharp click and which, at signal current cutoff, returns to its idle position against another bar and emits another click. Certain combinations of clicks correspond to the dot and dash groups of the American Morse code.

sound field A volume of space or material containing sound waves.

sound film Motion-picture film on which a sound track is recorded. Also see *optical sound recording.*

sound gate An optical device used for converting the sound track of a movie film into electrical impulses.

sound generator Any combination of oscillator, amplifier, and reproducer (loudspeaker or headphones) for producing sound waves.

sound-hazard integrator An instrument used to measure cumulative noise exposure received by persons in noisy environments. One such instrument gives direct readings in percent of permissible exposure.

sound i-f amplifier In a TV receiver circuit, the separate amplifier for the sound intermediate frequency (4.5 MHz). See, for illustration, *intercarrier receiver* and *split-sound receiver.*

sound-level meter See *sound survey meter.*

sound marker A marker indicating the sound-carrier point on a TV alignment curve displayed on an oscilloscope screen.

sound-marker generator A special rf signal generator (or a special circuit in a TV alignment generator) for the production of a sound marker.

sound mirage See *acoustic mirage.*

sound-on-film recording See *optical sound recording.*

sound-on-sound recording The simultaneous recording (on a single track on magnetic tape) of new material with previously recorded material.

sound-operated relay 1. A relay operated indirectly from sound, through the medium of a pickup microphone and amplifier. 2. A relay having a delicately poised armature which operates directly from sound vibrations.

sound power The total sound energy per unit time, produced by a sound source, as expressed in ergs per second or in watts, for example.

sound-powered telephone A batteryless telephone in which the microphone converts speech energy directly to electric current to actuate the distant receiver.

sound pressure 1. The force exerted by sound waves on a surface area, expressed in dynes per square centimeter (as an rms value over 1 cycle). Sound pressure is proportional to the square root of sound-energy density. 2. The instantaneous difference between actual air pressure and average air pressure at a given point.

sound probe A transducer used for the purpose of picking up acoustic vibrations for detection or measurement purposes.

sound recording The electrical recording of sound, using cylinder, disk, tape, wire, or other comparable storage medium.

sound-recording system A complete, integrated array of equipment for recording sound, including such components as microphones, amplifiers, pickups, filters and other shaping networks, attenuators, level indicators, and recording mechanisms. Compare *sound-reproducing system.*

sound reinforcement Intensification of sound by horns, resonant chambers, and the like.

sound relay See *sound-operated relay.*

sound reproducing The electrical reproduction of sound from recordings on cylinders, disks, tape, wire, etc.

sound-reproduction system A complete, integrated array of equipment for the playback of recorded sound, including such components as tape or record players, amplifiers, filters and other shaping networks, attenuators, level indicators, loudspeakers, and headphones. Compare *sound-recording system.*

sound spectrum The continuous band of frequencies (about 20 Hz to 20 kHz) constituting audible sounds, and sometimes the immediately adjacent (subaudible and superaudible) frequencies.

sound stage The apparent dimensions of a sound source.

sound survey meter A portable instrument for measuring the intensity and other characteristics of sound.

sound system A *sound-recording system, sound-reproduction system,* or a combination of the two.

sound takeoff In a TV receiver circuit, the point at which the 4.5 MHz FM sound signal is extracted from the complex TV signal.

sound track The variable-density or variable-width recording on one side of the film in sound-on-film recording and reproduction. Also see *optical sound recording.*

sound-transmission coefficient See *acoustical transmittivity.*

sound trap In a TV receiver circuit, a wavetrap that prevents the sound signal from entering the picture channels.

sound unit See *phone, 2.*

sound wave The vibratory phenomenon produced in a medium by acoustic energy. A sound wave in air consists of alternate compressions and rarefactions of the air. Also see *acoustic wave.*

source 1. The origin of a signal or electrical energy. 2. In a field-effect transistor, the electrode that is equivalent to the cathode of a tube or to the emitter of a common-emitter-connected bipolar transistor. 3. That which is being transcribed to magnetic tape. 4. Manufacturer, wholesaler, or retailer.

source circuit 1. The circuit associated with the source electrode of a field-effect transistor. 2. Generator circuit. Compare *sink circuit.*

source code See *source language.*

source computer A computer for compiling a source program.

source-coupled multivibrator A multivibrator circuit employing field-effect transistors, in which feedback coupling is achieved with a common source resistor for the two FETs. This circuit is equivalent to the cathode-coupled tube-type and emitter-coupled bipolar-transistor-type multivibrators.

source data automation A means of storing a master data file for easy duplication whenever necessary.

source follower A field-effect-transistor amplifier in which the output is taken across a resistor between source and ground. This circuit is equivalent to the cathode follower and emitter follower, all of which are unity-gain stages whose impedance-transformation characteristics make them ideal in signal conditioning, buffering, and impedance-matching applications.

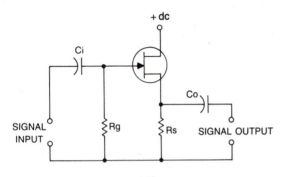

Source follower.

source impedance 1. The impedance of a generator in a circuit. 2. The impedance of the source electrode of a filed-effect transistor.

source language A computer programming language from which is derived (by a compiler) the machine (object) language a computer operates on.

source program A computer program written in a source language.

south magnetic pole The south pole of the *equivalent bar magnet* constituted by earth's magnetic field (see *earth's magnetic field*). The south magnetic pole lies close to the geographic south pole. Compare *north magnetic pole*.

south-seeking pole Symbol, *S*. The so-called south pole of a magnet. When a bar magnet is suspended horizontally, this pole points to earth's south magnetic pole. Compare *north-seeking pole*.

SP 1. Abbreviation of *self-propelled*. 2. Abbreviation of *stack pointer*.

sp 1. Abbreviation of *single pole*. 2. Abbreviation of *special*.

space charge The electric charge of the cloud of electrons in the space between the cathode and plate in the electron tube.

space-charge coupling Coupling of energy between internal points of an electron tube via the space charge.

space-charge field 1. The electric field existing within a group of charged particles. 2. The electric field existing in a plasma.

space-charge-limited current In an electron tube, current flowing between electrodes when a virtual cathode is present as a result of the space charge.

space-charge region See *barrier, 1*.

space current In an electron tube, the total internal current flowing between the cathode and all the other elements. Also called *tube emission current*.

space diversity See *diversity reception*.

space-diversity reception See *diversity reception*.

space lattice The three-dimensional, redundant pattern formed by atoms and molecules in a crystal and having a shape that is characteristic of a particular crystalline material.

spacer An insulating rod or bar that serves to hold apart the conductors of a two-wire, four-wire, or coaxial air-dielectric transmission line.

space suppression Following the printing of a line by a printer, the prevention of normal paper travel.

space wave One of the two components of a ground wave. The space wave—unlike the other component, the *surface wave*—is not earth-guided and has two components: the *direct wave* and the *ground-reflected wave*.

spacistor An early space-charge type of transistor in which the input signal modulated electron flow through a space-charge region between *p* and *n* layers in a germanium wafer.

spaghetti Slender, varnished-cambric tubing used as slipover insulation for wires and busbars.

span On an instrument scale, the difference between the highest value and the lowest value.

spark See *electric spark*.

spark absorber 1. Spark suppressor. 2. Key-click filter.

spark coil A small induction coil, so called from its initial purpose of supplying the high voltage for spark plugs in gas engines.

spark energy The energy dissipated by an electric arc or spark.

spark frequency The number of sparks per second in the spark gap of a spark transmitter or spark-gap oscillator.

spark gap A device consisting essentially of two metal points, tips, or balls separated by a small air gap. A high voltage applied to the electrodes causes a spark (or in the case of an ac voltage, a train of sparks) to jump across the gap. Also see *quenched spark gap, rotary spark gap,* and *synchronous spark gap*.

spark-gap modulator A controlled spark gap employed to modulate a carrier wave.

spark-gap oscillator A damped-wave oscillator consisting of a spark gap and a tuned *LC* tank circuit.

spark-gap voltmeter See *needle gap* and *sphere gap*.

sparking distance The maximum separation of the electrodes of a spark gap at which a given voltage will produce a spark.

sparking voltage The voltage which will cause a spark to jump across a spark gap of a given width.

spark killer See *spark suppressor*.

sparkover A discharge in air, a vacuum, or a dielectric. It is characterized by sparking between electrodes in the medium.

spark plate In some automobile radios, a noise-interference-eliminating bypass capacitor in which the chassis is one plate.

spark-plug suppressor A small resistive device connected in series with a spark plug to suppress electrical noise arising from the ignition in a gas engine.

spark quencher See *spark suppressor*.

spark suppressor A resistor, capacitor, diode—or a combination of such units—employed to eliminate or minimize sparking between make-and-break contacts.

spark-suppressor diode See *suppressor diode*.

spark transmitter A damped-wave radiotelegraph transmitter employing a spark-gap oscillator as the source of rf energy.

spatial distribution The three-dimensional directional pattern of a transducer such as an antenna, microphone, or speaker.

spc 1. Abbreviation of *silicon point contact*. 2. Abbreviation of *silver-plated copper*.

spdt Abbreviation of *single-pole—double-throw* (switch or relay).

speaker See *loudspeaker*.

speaker damping See *damped loudspeaker*.

speaking arc A method of modulated-light transmission. An electric arc is modulated by audio-frequency signals.

special character A character of a set that isn't a numeral or letter, e.g., an ampersand. Also called *symbol*.

special effects Various techniques used in film and video-tape recording for achieving certain visual scenes or images.

special-purpose computer A computer designed to handle problems or be suitable for applications of a specific category; a dedicated computer.

special-purpose calculator Any one of several calculators offered for essentially "nonmathematical" purposes, such as biorhythm data, astrological information, metric conversions, musical composing, racehorse handicapping, cooking data, and so on. Such calculators often perform simple numerical calculations—such as addition, subtraction, multiplication, and division—as well.

specific address See *absolute address*.

specification 1. A precise listing of needs. 2. For an electronic device, a statement of performance over specific parameters, e.g., a specification for a hi-fi amplifier might be to deliver so many watts per channel over a certain frequency range under a specific load with both channels driven, while producing no more than a specified amount of distortion.

specific conductivity Conductance per unit volume. In SI units this would be expressed in siemens (reciprocal ohms) per cubic centimeter (S/cm^3).

specific dielectric strength For an insulant, the dielectric strength per millimeter of thickness.

specific gravity Abbreviation, *sp gr*. The ratio of the density of a material to the density of a substance accepted as a standard (usually water at 39.2 °F). Examples: cork, 0.22; mercury, 13.6; lead, 11.3.

specific inductive capacity See *dielectric constant*.

specific resistance See *resistivity*.

specific sound-energy flux See *sound intensity*.

spectral comparative pattern recognizer Acronym, *SCEPTRON*. An equipment employed to classify automatically complex signals obtained from information that has been converted into electrical signals.

spectral density For a complex signal, the amount of energy contained within a given band of frequencies.

spectral energy distribution The occurrence of different amounts of energy in different areas of a spectrum (as in a light spectrum, sound spectrum, radio spectrum, and so on).

spectral response The characteristic of a device, such as a photocell or the human eye, which describes the device's sensitivity to radiations of various frequencies in a given spectrum.

spectral sensitivity The color response of a light-sensitive device.

spectrograph A recording spectrometer.

spectrometer 1. An instrument used to measure spectral wavelengths. 2. An instrument used to measure index of refraction. 3. Mass spectrometer.

spectrophotometer A photoelectric instrument for chemical analysis. In the device, light passing through the material under analysis is broken up into a spectrum that is examined with a photoelectric circuit, which in turn plots a spectrogram.

spectroscope An instrument which resolves a radiation into its various frequency components and permits measurement of each.

spectrum A band of frequencies or wavelengths of radiant phenomena, e.g., radio spectrum, visible-light spectrum, heat spectrum, etc.

spectrum analyzer 1. An automatic wave analyzer with a visual display

(oscilloscope). 2. A scanning receiver with a screen that shows a plot of signals and their bandwidths over a specific frequency band.

specular reflection Reflection in which the reflected ray is in the same plane as the incident ray, as in reflection from an extremely smooth surface.

speech amplifier A (usually low-level) audio amplifier designed especially for speech frequencies and working out of a microphone.

speech clipper A device such as a biased diode (see *positive peak clipper*) for holding speech signals to a constant amplitude. Compare *speech compressor*. Also see *speech clipping*.

speech clipping Use of a clipper to maintain the output-signal amplitude of a speech amplifier against fluctuations in the intensity of speech input. The clipper "slices off" signal peaks; the resulting flat tops require filtering to remove the harmonics generated by the process. Compare *speech compression*.

speech compression Automatic regulation of the gain of a speech amplifier to maintain its output-signal amplitude against speech-input fluctuations. Compare *speech clipping*.

speech compressor A circuit or device, such as an agc system, for performing speech compression. Compare *speech clipper*.

speech frequencies See *voice frequencies*.

speech intelligibility The quality of reproduced speech which makes it easily understood by a reasonably proficient user of the language. For good speech intelligibility, a sound system should transmit frequencies up to 15 kHz with minimal distortion.

speech inverter See *scrambler circuit*.

speech power 1. The ac power in an electric wave corresponding to speech, as opposed to that in a sine wave. 2. Sound power in a speech transmission.

speech scrambler See *scrambler circuit*.

speech synthesizer See *Moog synthesizer*.

speed key 1. Semiautomatic key. 2. A similar, fully automatic key used especially for high-speed telegraphy.

speed of light The velocity of light in air or vacuum; in a vacuum, in which case the symbol is *c*, it is 300 000 kilometers per second (186 000 miles per second).

speed of sound The velocity of sound. It depends on the transmission medium, being (at ordinary temperatures) approximately 344 meters (1129 feet) per second in air and 1463 meters (4800 feet) per second in water.

speed of transmission The amount of data sent in a given unit of time. Generally measured in bits per second (*BPS*), characters per second (*CPS*), characters per minute *(CPM)*, or words per minute (*WPM*). *Used primarily for digital codes.*

speedup capacitor See *commutating capacitor*.

SPFW Abbreviation of *single-phase—full-wave*.

sp gr Abbreviation of *specific gravity*.

sphere A closed surface that is the locus of points that are the same distance (radius) from another point (the center).

sphere gap Aa spark gap in which the spark passes between two polished metal balls. When the air gap is adjustable, unknown high voltages may be measured in terms of gap width at sparking. Compare *needle gap*.

sphere-gap voltmeter See *sphere voltmeter*.

sphere voltmeter A gap voltmeter employing a sphere gap (ball gap). Also see *sphere gap*.

spherical aberration In a lens or mirror, distortion (misrepresentation of the object) due to the spherical form of the device, which causes rays from the edges to be focused at a different point than those from the center.

spherical angle An angle formed by the intersection of two arcs on the surface of a sphere.

spherical candlepower The average illuminance of a light source in all directions, equal to the total luminous flux (in lumens) divided by 4π. Also see *candlepower*.

spherical degree A unit equal to 1/720 the area of a sphere.

spherical distance The shortest arc connecting two points on a sphere.

spherical lens A lens whose contour is a section of an imaginary sphere.

spherical mirror A mirror whose contour is a section of a spherical surface.

spherical wave A wave characterized by wavefronts that are concentric spheres.

spheroidal antenna A sheetmetal or wire-mesh antenna having the cross section of a sphere that is flattened at the ends of one axis.

SPHW Abbreviation of *single-phase—half-wave*.

spider 1. The flat, round, springy part that holds the apex of the vibrating cone of a dynamic speaker. 2. A quickly assembled, chassisless hookup in which the pigtails of components form the wiring as well as the mechanical support for a circuit.

spiderweb antenna A set of dipole antennas for different frequencies, arranged in a common (usually horizontal) plane. The result is a broadband antenna.

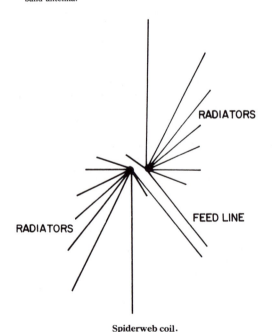

Spiderweb coil.

spiderweb coil A flat, single-layer coil in which a strand of wire is woven over and under the spokes of a wheel-like form, and having the general apearance of a spiderweb. The criss-cross winding reduces distributed capacitance by breaking up the parallelism of adjacent turns.

spike 1. A slender pulse. 2. A sharp transient, such as an overshoot on a pulse or square wave.

spike suppressor A clipper or similar device for removing a spike from a signal voltage.

spin See *electron spin*.

spindle 1. The pivoted shaft that carries the movable element and rotates between the *pivots* in a meter movement. 2. The rotor of an *alternator*, especially when the rotor is a permanent magnet. 3. The rotating shaft in a motor, generator, or similar electric machine.

spinning electron An electron spinning, i.e., one with angular momentum.

spinthariscope An optical device for observing the effect of alpha particles given off by a radioactive substance, from the scintillations they produce upon striking small, fluorescent screen.

spiral On a plane or surface, a curve that winds around a fixed point and which can be described by an equation in polar coordinates.

spiral coil See *disk winding*.

spiral distortion A form of television camera-tube distortion caused by spiralling of the electrons as they are emitted from the photocathode. The result is a twisted image.

spiral-rod oscillator A parallel-linetype oscillator in which the lines are rods that are rolled up into spirals to conserve space. Also see *linetype oscillator*.

spiral sweep 1. Sweeping the electron beam in an oscilloscope or radarscope to produce a spiral trace on the screen. 2. The circuit for producing such a sweep.

spiral trace See *spiral sweep*.

spiral winding See *disk winding*.

spkr Abbreviation of (loud)*speaker*.

splatter A collection of spurious sidebands generated by overmodulation of a carrier and capable of broadband interference.

splatter-suppresson filter In an AM transmitter, a high-power-level, low-pass, *LC* filter inserted between the output of an amplitude modulator and the af input of the modulated rf amplifier, to suppress high-frequency audio components which would cause splatter.

splice 1. A joint of (a) two wires made essentially by tightly winding their ends together for a short distance and sometimes soldering for additional strength and rigidity, or of (b) the (usually broken) ends of paper, film, or magnetic tape. 2. To make such a joint.

splicer A device for making a splice (*splice, 1a*).

split-anode magnetron A magnetron in which the plate (anode) consists of two semicylinders with the cathode at their axis.

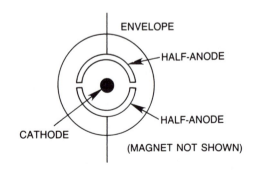

Split-anode magnetron.

split-load phase inverter See *paraphase inverter*.

split-phase motor A fractional-horsepower ac motor which starts like a two-phase motor and runs like a single-phase motor. After the machine comes up to approximately 3/4 full speed, a centrifugal switch cuts out the starting winding.

split projector An acoustic transmission device with several independently operated transducers.

split-reed vibrator See *self-rectifying vibrator*.

split-rotor plate See *serrated rotor plate*.

split-stator capacitor A variable capacitor having two separate stator sections and a common rotor section.

split-sound receiver A TV receiver circuit in which the sound signal is separated from the composite TV signal shortly after pickup by the antenna, and is amplified separately from the video signal. Compare *intercarrier receiver*.

splitter A device employed to couple two or more TV receivers to a common antenna.

spontaneous disintegration The decaying of a radioactive material as a result of the natural emisson of radiation from it.

spool See *reel*.

sporadic-E layer ionization Occasional, scattered ionization in the E-layer of the ionosphere.

sporadic-E reflection See *sporadic-E skip*.

sporadic-E skip Skywave transmission via reflection from the sporadically ionized regions in the E-layer of the ionosphere. The reflections are resonsible for much of the short-skip communication in the 28 and 50 MHz amateur radio bands. Sporadic-E skip also may affect the lower television broadcast channels, usually channels 2 and 3.

SPOT Abbreviation of "satellite position and tracking."

spot brightness In a cathode-ray tube, brilliance of the spot produced on the screen by the electron beam.

spot carbon Carbon paper that is only partially coated so that only some of the impressions it receives are transferred to the copy.

spot check 1. To take a random sample or to make a random test by arbitrarily selecting a single item from a run of similar items and subjecting it to analysis, examination, performance or parametric evaluation, etc. 2. A random sample or test.

spot frequency A single frequency to which other frequencies may be referred or which can act as a limit marker.

spot jamming The deliberate interference to a radio signal at some frequency and at some specific time.

spot modulation In a cathode-ray tube, modulation of the brightness of the spot (and, accordingly, of the image) produced on the screen by the electron beam. Also see *intensity modulation*.

spot welding A method of electrical welding in which the parts to be joined are held together, overlapping, between the points of two electrodes, between which a current is passed to heat the parts at the spot of contact.

spot-wheel pattern A frequency-identifying wheel pattern produced on an oscilloscope screen by inensity-modulating a circular trace. The circular trace is produced by applying a standard-frequency signal to the horizontal and vertical input terminals 90 degrees out of phase. A square-wave signal of unknown frequency is applied to the intensity-modulation (Z axis) input terminals. The square wave chops the circle into a number of bright sectors or spots. The unknown frequency fx equals Nfs, where N is the number of spots in the circle, and fs *is the standard frequency. Compare gear-wheel pattern*.

spray coating 1. Applying a protective coat of insulating material to a conductor or component by spraying it with a liquid substance and allowing it to dry. Compare *dip coating*. 2. The coat applied in this way.

spreader An insulator used to separate the wires of an air-spaced transmission line.

spreading current In a semiconductor, current due to the interatomic movement of carriers by devious routes rather than in straight lines.

spreading loss Energy lost during the transmission of radiation.

spread spectrum 1. A method of transmission in which the occupied bandwidth of the signal is deliberately increased, or spread out, over a much wider range than it would normally occupy with conventional modulation. 2. A signal so transmitted.

spring coil See *solenoid, 1*.

spring contact See *flexible contact*.

spst Abbreviation of *single-pole—single-throw* (switch or relay).

spurious frequency An unintended and unwanted signal on a frequency other than the signal of interest, such as that of an image or harmonic, to which a circuit responds when tuned to a desired frequency. Spurious frequencies sometimes are generated by faulty modulation and oscillation.

sprocket holes 1. In the fan-folded paper (continuous stationery) used in peripheral printers, holes (at the paper's edges) which are engaged to transport the paper. 2. In paper tape, holes running the length of the tape and used to drive it and provide timing signals.

sprocket pulse The pulse generated by a paper tape reader as a result of it sensing sprocket holes. Also see *sprocket holes, 2*.

spurious oscillation 1. Oscillation in a normally nonoscillatory circuit. 2. In an oscillator, simultaneous oscillation at a frequency other than the normal one.

spurious response Response to a signal, such as an image or a harmonic, to which a circuit is not directly tuned.

spurious-response ratio The ratio of the transmission (or gain) of a circuit of a desired signal to its transmission (or gain) for a spurious signal at the same setting of the circuit, e.g., signal-to-image ratio.

spurious sidebands Sidebands in addition to the normal ones, generated as a result of a defect (such as overmodulation) in the modulation process.

sputter 1. A layer of material obtained by sputtering (see *sputtering, 1*). 2. To carry out the process of sputtering (see *sputtering, 1*).

sputtering 1. A technique for electrically depositing a film of metal on a metallic or nonmetallic surface. In a vacuum chamber, the piece of metal to be deposited is made the cathode of a high-voltage circuit with respect to a nearby anode plate. The high voltage causes atoms to be ejected from the surface of the cathode and strike the surface of an object placed in their path, becoming deposited on it as a film of cathode metal. Compare *evaporation, 1*. 2. The disintegration of a vacuum-tube cathode through ejection of surface atoms by impinging positive ions.

sq Abbreviation of *square*.

SQ-band A section of the S-band extending from 2400 to 2600 MHz.

SQC Abbreviation of *statistical quality control*.

SQR Abbreviation of *square rooter*.

S-quad See *simple quad*.

square The product of two equal factors. It is designated by the exponent 2. Thus, the square of 2.5 = 2.5^2 = 2.5 × 2.5 = 6.25. Compare *square root*.

square-law demodulator See *square-law detector*.

square-law detector A detector whose output is proportional to the square of the rms value of the input. Also called *weak-signal detector*.

square-law meter 1. A meter whose deflection is proportional to the square of the quantity applied to it. Also see *current-squared meter.* 2. A high-impedance electronic voltmeter whose deflection is proportional to the square of the applied voltage.

square-law modulator A circuit or device that accomplishes amplitude modulation of one signal current by another, by simultaneously passing the two currents through a component, such as a nonlinear resistor, having square-law response.

square-law response Circuit or component operation which results in an output that is proportional to the square of the input.

square-law vtvm A vacuum-tube voltmeter whose deflection is proportional to the square of the applied voltage.

square root Symbol, $\sqrt{\ }$ (also denoted by fractional exponent 0.5 or 1/2). One of the two equal factors of a given number. Thus, $\sqrt{25} = 25^{1/2}$ = 5, since $5 \times 5 = 25$. Compare *square.*

square rooter An analog or digital device employed to extract the square root of a number.

square wave An alternating or pulsating current or voltage whose waveshape is square.

square-wave amplifier An amplifier designed to operate with square waves.

square-wave converter See *squaring circuit.*

square-wave generator A signal generator delivering an output signal having a square waveform. Compare *squaring circuit, 1.*

square-wave testing Testing the response of a circuit or device, such as an amplifier, by observing the extent to which it distorts a square-wave signal passing through it (a measure of high-frequency response).

squaring circuit 1. A circuit—such as a twin-diode clipper, overdriven amplifier, or Schmitt trigger—which converts a sine wave or pulse into a square wave. 2. A circuit whose output-signal amplitude is the square of the input-signal amplitude.

squarish wave A *quasi-square wave* or *quasi-rectangular wave.* Also, a perfect rectangular wave which appears in place of a desired square wave.

squawker 1. In a three-way speaker system, the midrange speaker. 2. Any slave station in a multistation intercom network.

squeal A high-pitched interferential sound, such as that encountered in spuriously oscillating systems.

squeezeout An OCR condition in which printed characters have excessive ink at the edges. Also called *smudge.*

squegging A choking-type cutoff action in a circuit caused by too strong a signal.

squegging oscillator An oscillator which starts and stops oscillating intermittently as a result of squegging.

squelch See *squelch circuit.*

squelch amplifier An amplifier which can be controlled by a squelch signal. Also see *squelch circuit.*

squelch circuit One of several circuits which automatically disable a receiver or amplifier except when incoming signals exceed a predetermined threshold amplitude. This action *mutes* the equipment, eliminating annoying background noise and unwanted signals. Also called *muting circuit.*

squelch signal The activating or deactivating signal delivered by a squelch circuit.

squiggle Marker signal (see *blip, 2*).

squint 1. The angular resolution of a radar antenna. 2. The angular difference between the antenna axis and the major lobe of a radar transmitter.

squirrel-cage induction motor See *squirrel-cage motor.*

squirrel-cage motor An induction-type ac motor employing a squirrel cage rotor.

squirrel-cage rotor In an ac motor, a rotor composed of straight copper bars embedded in a laminated soft-iron core and short-circuited at the ends by rings. The device is so called because it resembles a revolving squirrel cage.

squirter See *signal injector.*

SR 1. Abbreviation of *silicon rectifier.* 2. Abbreviation of *silicone rubber* (see *silicone*). 3. Abbreviation of *shift register.*

S—R Abbreviation of *send—receive.*

Sr Symbol for *strontium.*

sr Abbreviation of *steradian.*

SRAM Abbreviation of *static random-access memory.*

S-rays See *secondary rays.*

SRBM Abbreviation of *short-range ballistic missile.*

SRF Abbreviation of *self-resonant frequency.*

S-rf meter A dual-function meter on a transceiver. In the receiving mode, the meter indicates *S* units. In the transmit mode, the meter indicates relative output power.

SRR Abbreviation of *short-range radar.*

SRV Abbreviation of *space rescue vehicle.*

SS 1. Abbreviation of *solid-state.* 2. Abbreviation of *single-shot.* 3. Abbreviation of *small signal.* 4. Abbreviation of *single-signal.* 5. Abbreviation of *same size.* 6. Abbreviation of *stainless steel.*

SSB Abbreviation of *single-sideband.*

SS-band A section of the S-band extending from 2900 to 3100 MHz.

SSBSC Abbreviation of *single-sideband—suppressed-carrier.*

ssc Abbreviation of *single silk-covered* (wire).

S-scale A scale of numbers employed by amateur radio stations to report the approximate strength of signals: S1, faint signals; S2, very weak signals; S3, weak signals; S4, fair signals; S5, fairly good signals; S6, good signals; S7, moderately strong signals; S8, strong signals; S9, extremely strong signals. Also see *S-meter.*

sse Abbreviation of *single silk enameled* (wire).

SSI Abbreviation of *small-scale integration.*

SSL Abbreviation of *solid-state lamp.*

SSSC Abbreviation of *single-sideband—suppressed-carrier.* (Also, *SSBSC.*)

SST Abbreviation of *supersonic transport.*

st Abbreviation of *single throw.*

sta 1. Abbreviation of *station.* (Also *stn.*) 2. Abbreviation of *stationary.*

stab Abbreviated form of *stabilizer* (see *stabilizer, 4*).

stability 1. The condition in which an equipment or device is able to maintain a particular mode of operation without deviation. 2. The condition in which the setting or adjustment of a device remains at a particular point without movement. 3. The condition in which a quantity remains constant with respect to time, temperature, or other variable. 4. The ability of inks used in OCR to retain their color after exposure to light or heat.

stability factor Abbreviation, *SF.* For a transistor, the ratio d I_c/d I_{co}, where I_c is steady-state collector current, and I_{co} is collector cutoff current.

stabilized platform See *stable platform.*

stabilizer 1. Damping diode. 2. Damping resistor. 3. A device or circuit for the self-regulation of current or voltage. 4. A vertical airfoil on the aft part of a fuselage. 5. A chemical used to control or arrest a reaction.

stabilizing windings Auxiliary field windings employed to prevent speed runaway in shunt motors.

Stabistor Trade name for a type of voltage-regulating semiconductor diode.

stable device A device whose characteristics and performance remain substantially unchanged with time or variations in temperature, applied power, or other quantities.

stable element 1. A component that maintains its value or ratings despite widely variable environmental conditions. 2. A navigational instrument that maintains its orientation at all times.

stable platform A gyro-type device employed to stabilize devices in space and to provide accurate information regarding attitude (pitch, roll, and yaw).

stable state A stable condition, such as the ON and OFF states of a *flip-flop*. The flip-flop has two stable states and will remain in one until it is switched to the other, whereupon it will then remain in that latter state until switched back to the former. Compare *unstable state*.

stack 1. A piled assembly of capacitor plates and separating dielectric films. 2. An assembly of selenium rectifier plates (see *power stack*). 3. To assemble a stacked array. 4. A temporary storage area consisting of a small group of registers in a computer memory.

stacked array An antenna array in which the elements are arranged horizontally, one above the other.

stacked-dipole antenna A stacked array of dipole sections.

stack pointer Abbreviation, *SP*. A register indicating the last data item to be entered in a stack (*stack, 4*).

stage 1. A complete functional unit of a system, e.g., amplifier stage. 2. In a microscope, that which supports the slide.

stage-by-stage elimination See *signal injection*.

stage gain The amplification provided by a single stage in a system.

stage loss The loss introduced by a single stage in a system.

stagger 1. An error in facsimile reception, occurring as a constant discrepancy in the position of the received dot. 2. To deliberately tune a set of resonant circuits, especially in a bandpass filter, to one side or the other of the center frequency.

staggered tuning The tuning of the input and output circuits of a single stage, or the tuning of successive stages, to slightly different frequencies to obtain flat-top response.

stagger tuning See *staggered tuning*.

stagger-wound coil See *basket-weave coil* and *spiderweb coil*.

staircase circuit See *stair-step circuit*.

staircase generator A circuit or device for generating a stair-step wave.

staircase wave See *stair-step wave*.

stair-step circuit A circuit that converts a series of equipotential pulses into a stair-step wave.

stair-step wave A nonsinusoidal wave characterized by a multistep rise and a steep fall. It is so called from its resemblance to the cross section of a staircase.

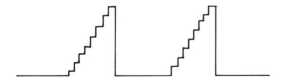

Stair-step wave.

stall torque The torque produced when a motor shaft is prevented from turning.

stalo Abbreviation for stabilized oscillator.

standard 1. A precise specification governing the dimensions and characteristics of a device or system, e.g., military standard. 2. A highly accurate physical or electrical quantity to which similar quantities may be compared, e.g., standard frequency. Also, the device that produces the standard quantity, e.g., frequency standard. 3. Of conventional size, against which other sizes are compared on a relative basis, in the sequence *standard, small, midget, miniature, subminiature, microminiature*.

standard atmosphere Abbreviation, *atm*. Air pressure at sea level (1.013 25 Pa). Also called *atmosphere*.

standard broadcast band The frequency band allocated to conventional broadcast stations (in the United States, 535 to 1605 kHz). Also see *broadcast service, 1*.

standard candle See *candela*.

standard cell A highly refined primary cell used to supply an accurate dc voltage for electronic measurements. The Weston standard cell contains a mercury positive electrode, cadmium amalgam negative electrode, and cadmium sulfate electrolyte, and delivers 1.018 30 volts. Also see *zinc standard cell*.

standard deviation Symbol, σ or s. In statistical analysis, the square root of the mean of squares of deviation from the mean.

standard frequency A highly precise frequency to which other frequencies may be referred for identification or measurement.

standard-frequency oscillator A stable, precise oscillator that delivers a standard frequency. Also see *primary frequency standard* and *secondary frequency standard*.

standard-frequency signal A calibration and reference signal that is broadcast on a standard frequency, such as those transmitted on 2.5, 5, and 10 MHz by National Bureau of Standards stations WWV, WWVH, and WWVB.

standard pitch The tone corresponding to the frequency 440 Hz (in music, the note A above middle C).

standard signal generator A (usually continuously variable) high-grade generator of rf test signals (modulated and unmodulated). A general-purpose instrument of this type usually covers a wide range (e.g., 15 kHz to 100 MHz) in several tuning bands. For calibration, a standard signal generator is referred to a primary frequency standard or secondary frequency standard.

standard subroutine A usually vendor-supplied computer program segment applicable to more than one program and used as needed as a subroutine.

standard time Official civil time in a particular region. See *time zone*.

standby 1. The state in which equipment is out of operation but may be immediately activated. Also called *idling*. 2. A state of readiness on the part of personnel, equipment, or systems.

standby battery An emergency power source for a battery-powered installation.

standby current See *current drain, 2*.

standby equipment See *emergency equipment*.

standby operation Keep-alive operation during a standby interval (see *standby*).

standby power The power drawn by an equipment connected to the power supply but out of operation.

standing wave A stationary distribution of current or voltage along a line due to interactions between a wave transmitted down the line and a wave reflected back; it is characterized by maximum-amplitude points (*loops*) and minimum-amplitude or zero points (*nodes*).

standing-wave distortion Distortion of current or voltage due to standing waves on an improperly terminated line.

standing-wave indicator 1. A device, such as a lamp or meter, employed to detect standing waves. 2. Standing-wave meter (see *SWR bridge).*

standing-wave loss The additional loss, over the matched-line loss, that occurs in a transmission line when the standing-wave ratio (*SWR*) is not 1.

standing-wave meter See *SWR bridge.*

standing-wave ratio Abbreviation, *SWR.* On a transmission line, the ratio of maximum current (or voltage) to minimum current (or voltage); SWR = 1 for a perfectly terminated line. Also, the ratio of the amplitude of current or voltage of a standing wave at a loop to the amplitude at a node.

standoff insulator An insulator (usually of the post type) employed to hold a wire or component away from a chassis or base.

star 1. In a gravity-battery cell, the copper electrode; so called from its star shape. 2. A star-shaped circuit of three-phase components. Also see *wye-connection.* 3. A celestial body that radiates light energy, which thus distinguishes it from other celestial bodies such as planets.

star connection See *wye-connection.*

stark effect The influence of a strong transverse electric field on the spectrum lines of a gas—somewhat resembling the Zeeman effect.

starlight scope A usually weapon-mounted device capable of viewing in *apparent* total darkness. Its operation depends on its ability to provide high amplification of extremely low light levels, such as that of objects reflecting the light from a moonless but starlit sky.

star rectifier See *wye-rectifier.*

starter 1. An ignitor electrode in an ignitron (see *ignitor).* 2. Starting box.

starting box A special rheostat for starting a motor gradually in steps and provided with an electromagnet for holding the arm in the maximum-speed position and releasing it when power is interrupted.

starting rod The ignitor electrode in an ignitron.

starting voltage 1. For a gas tube, the minimum voltage that will initiate the glow discharge. 2. In appropriate solid-state devices (e.g., a diac), the voltage at which conduction between electrodes occurs.

start lead The lead attached to the first turn of a coil. Also called *inside* lead. Compare *finish lead.*

start—stop multivibrator A one-shot multivibrator (see *monostable multivibrator).*

stat A prefix denoting *electrostatic.*

statampere The cgs electrostatic unit of current; 1 statampere = $3.335\ 640 \times 10^{-10}$ ampere.

statcoulomb The cgs electrostatic unit of charge; 1 statcoulomb = $3.335\ 640 \times 10^{-1}$ coulomb.

state 1. The present condition (i.e., on or off, true or false, 0 or 1, high or low) of a bistable device, such as a flip-flop. 2. The physical or electrial condition or status of a component, device, circuit, or system.

statement The contents of a line in a source language computer program.

state of charge The amount of charge, measured in coulombs or ampere hours, in a storage battery at a given time.

states of matter The four phases (gas, liquid, solid, and plasma) in which matter can exist. See separate listings under *gas, liquid, plasma,* and *solid.*

statfarad The cgs electrostatic unit of capacitance; 1 statfarad = $1.112\ 650 \times 10^{-12}$ farad.

stathenry The cgs electrostatic unit of inductance; 1 stathenry = $8.987\ 554 \times 10^{11}$ henrys.

static 1. Pertaining to that which is constant in quantity (e.g., static collector current of a transistor). 2. Pertaining to that which is at rest, e.g., static electricity. 3. Atmospheric electricity and the radio noise it produces (see *atmospheric electricity* and *atmospheric noise).* 4. Pertaining to tests and measurements mode without subjecting the unit or device to regular operation, as opposed to *dynamic.*

static base current See *dc base current.*

static base resistance See *dc base resistance.*

static base voltage See *dc base voltage.*

static cathode current See *dc cathode current.*

static cathode resistance See *dc cathode resistance.*

static cathode voltage See *dc cathode voltage.*

static characteristic An operating characteristic determined from constant rather than alternating or fluctuating values of independent and dependent variables. Examples: the dc characteristics of transistors and electron tubes. Compare *dynamic characteristic.*

static charge Energy stored in a stationary electric field; electricity at rest.

static collector A device that grounds the rotating wheels of a motor vehicle, thereby removing the static electricity generated by the friction of the tires on the roadway.

static collector current See *dc collector current.*

static collector resistance See *dc collector resistance.*

static collector voltage See *dc collector voltage.*

static convergence In a color-TV picture tube, convergence of the three undeflected electron beams at the center of the aperture mask.

static device A device with no moving parts.

static discharge resistor A fixed resistor (typically 330K) connected between the earth and the high side of the power line in a TV receiver to drain off charges due to lightning.

static drain current See *dc drain current.*

static drain resistance See *dc drain resistance.*

static drain voltage See *dc drain voltage.*

static dump In computer practice, a dump occurring at a predetermined point in a program run or at the end of the run.

static electricity Energy in the form of a stationary electric charge, such as that stored in capacitors and thunderclouds or produced by friction or induction.

static emitter current See *dc emitter current.*

static emitter resistance See *dc emitter resistance.*

static emitter voltage See *dc emitter voltage.*

static flip-flop A flip-flop (see *bistable multivibrator*) in which the operating voltages are dc. A single pulse switches the unit from on to off, and vice versa. Compare *dynamic flip-flop.*

static frequency multiplier A magnetic-core device—similar to a magnetic amplifier or peaking transformer—that provides harmonics by distorting a sine-wave signal.

static gate current See *dc gate current.*

static gate resistance See *dc gate resistance.*

static gate voltage See *dc gate voltage.*

static grid current See *dc grid current.*

static grid voltage See *dc grid voltage.*

static hysteresis The condition in which the magnetization of a material (when it has the same intensity as the magnetizing force) is different when the force is increasing than when it is decreasing, regardless of the time lag. Compare *viscous hysteresis.*

static induction See *electrostatic induction*.

static machine See *electrostatic generator*.

static memory In a computer, a memory unit in which information is stored in a discrete cell, where it remains until read out or erased. Examples: magnetic-core memory, magnetic-disk memory, magnetic-drum memory, magnetic-tape memory. Compare *volatile memory*.

static mutual conductance See *static transconductance*.

static plate current See *dc plate current*.

static plate resistance See *dc plate resistance*.

static plate voltage See *dc plate voltage*.

statics The study of forces, bodies, poles, charges, or fields at rest or in equilibrium. Compare *dynamics*.

static screen current See *dc screen current*.

static screen resistance See *dc screen resistance*.

static screen voltage See *dc screen voltage*.

static skew In magnetic tape recording or playback, the amount of lead or lag time of one track with respect to another. Ideally, the static skew should be zero or practically zero.

static source current See *dc source current*.

static source resistance See *dc source resistance*.

static source voltage See *dc source voltage*.

static storage See *static memory*.

static suppressor current See *dc suppressor current*.

static suppressor voltage See *dc suppressor voltage*.

station 1. An installation consisting of a transmitter, receiver, or both. 2. Test position. 3. Computing position. 4. Check point.

stationary 1. Unmoving with respect to a given reference frame. 2. For a complex signal, a condition in which there are no significant changes in the spectral density.

stationary battery A (usually wet, storage) battery not usually moved when in use.

stationary wave See *standing wave*.

static subroutine In computer programming, a subroutine that always serves the same purpose; i.e., it does not need to be tailored (according to parameters) for a specific application.

statistical quality control Quality control based upon the techniques or probability and statistics in analyzing findings, making predictions, and formulating procedures for sampling.

statmho The cgs electrostatic unit of conductance; 1 statmho = 1.112 650 × 10^{-12} siemens.

statoersted The cgs electrostatic unit of magnetizing force; 1 statoersted = 265.458 A/m (3.335 85 × 10^{-11} oersted).

statohm The cgs electrostatic unit of resistance; 1 statohm = 8.987 544 × 10^{-11} ohms.

stator 1. Stationary coil. Compare *rotor, 1*. 2. The stationary member of a motor or generator. Compare *rotor, 2*. 3. The stationary-plate section of a variable capacitor. Compare *rotor, 3*.

stator coil A stationary coil (see *stator, 1, 2*).

stator plate The stationary plate of a variable capacitor. Compare *rotor plate*.

stator section See *stator, 3*.

statoscope An aircraft *altimeter* that shows small changes in altitude.

statvolt The cgs electrostatic unit of electromotive force; 1 statvolt = 299.792 5 volts.

statweber The cgs electrostatic unit of magnetic flux; 1 statweber = 299.792 5 webers (2.997 925 × 10^{10} maxwells).

ST-band A section of the S-band extending from 1850 to 2000 MHz.

stbd Abbreviation of *starboard*.

std Abbreviation of *standard*.

steady-state component A quantity whose value remains constant during normal operation of a circuit or device, as opposed to an alternating, fluctuating, or transient component.

steel A strong, tough alloy of iron, the commonest variety being a combination of iron and carbon.

steerable A directional antenna having a rotatable major lobe.

steering diode See *directional diode*.

Stefan—Boltzman constant Symbol, *s*. Value, 5.669 61 × 10^{-8} W/m^{2}K^4.

Stefan—Boltzmann law The thermal-radiation law that shows the total emissive power of a blackbody to be proportional to the fourth power of the absolute temperature of the body.

Steinmetz coefficient A factor for determining hysteresis loss for a ferromagnetic material. The flux density is raised to the power of 1.6, and multiplied by the factor *s* (Steinmetz coefficient) to determine the hysteresis loss under a given set of conditions.

stenode See *crystal filter*.

step A computer program instruction.

step-by-step operation See *stepthrough operation*.

step change A single-increment change in a value.

step circuit A circuit that produces a step (sharp change of slope) in the response curve of an amplifier. One form consists of a series-*RC* combination in parallel with the plate resistor of a tube.

step counter 1. A stair-step circuit arranged to count input pulses. The output capacitor of the circuit discharges when a predetermined number of input pulses has raised the capacitor voltage to the level required to trigger a counter. 2. In a computer or calculator, a circuit or device that counts the steps in an operation—such as division, multiplication, or shifting—called by an instruction.

stepdown ratio In a stepdown circuit or device, such as a stepdown transformer or cathode follower, the ratio of the low output voltage to the high input voltage. Compare *stepup ratio*.

stepdown transformer A transformer delivering an output voltage that is lower than the input voltage. In such a transformer, the secondary (output) winding contains fewer turns than the primary (input) winding.

step function See *unit function*.

step generator 1. A signal generator that delivers a step function (see *unit function*). 2. Staircase generator.

stepper motor A motor in which the shaft advances in uniform angular steps instead of rotating continuously.

stepping relay See *stepping switch*.

stepping switch A multiposition rotary switch in which an electromechanical ratchet mechanism advances the next contact position each time a pulse of a current is received.

stepthrough operation A way of operating a computer, usually during a debugging operation, in which program instructions are executed one at a time by direction of the user. Also called *single-step operation* and *step-by-step* operation.

stepup ratio In a stepup circuit or device, such as a stepup transformer or voltage amplifier, the ratio of the high output voltage to the low input voltage. Compare *stepdown ratio*.

stepup transformer A transformer delivering an output voltage that is lower than the input voltage. In such a transformer, the secondary (output) winding contains more turns than the primary (input) winding. Compare *stepdown transformer*.

steradian A unit of measure of solid angles; 1 steradian is the solid angle whose vertex is the center of a sphere and which cuts off a part of the surface of the sp ere having an area equal to r^2, where r is the distance from the vertex to the surface. Also see *solid angle*.

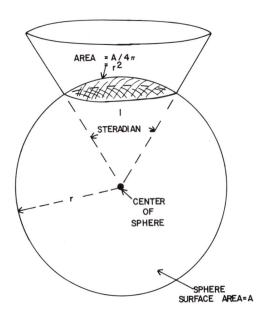

Steradian.

Sterba array See *barrage array*. (Also called *Sterba curtain*).

stereo Contraction of *stereophonic*.

stereo amplifier A two-channel amplifier for binaural reproduction (see *binaural*).

stereophonic Pertaining to equipment or techniques for producing a (somewhat) three-dimensional perspective of sound reproduction.

stereophonic sound See *stereo system*.

stereo phono cartridge A phono cartridge capable of reproducing sound from stereo disks.

stereo recording A method of recording in which two independent sound channels are transferred to some medium simultaneously, with the intention that the two channels be reproduced at the same time.

stereoscopic television Television in which the reproduced image appears three-dimensional.

stereo system A multichannel sound system (especially for music reproduction) in which phase differences between driven channels are used to create the illusion of spatial continuity between positioned loudspeakers.

stereotape Magnetic tape bearing more than one channel—especially two channels—for the recording and reproduction of stereophonic sound (see *binaural*).

sterilizer Any electronic device, such as an ultraviolet generator, employed to kill germs.

stethoscope An electronic or nonelectronic instrument used by physicians to listen to the heartbeat and other body sounds, and by technicians to listen to mechanical sounds.

still 1. A stationary picture on television. 2. A picture transmitted or received by means of facsimile. 3. A print on photographic paper of a negative.

still television See *facsimile*.

Stirling's formula A formula for the approximate value of n factorial (n!), which eliminates the need for cumbersome multiplications when n is large; $n! =$ antilog [1/2 log (2π) + (n + 1/2) log n − n log e].

stn Abbreviation of *station*. (Also, *sta*.)

STO Abbreviation of *store* or *storage*.

stochastic The condition in which, at any instant, a given variable may assume a *state* dependent on previous states as well as chance elements; e.g., words uttered extemporaneously by a speaker are random in that they can't be predicted, while also being dependent, by the application of grammar, on previously spoken words (i.e., previous states).

STOL Abbreviation of *short takeoff and landing* (aircraft).

stop amplifier See *reject amplifier*.

stopband The continuous spectrum of frequencies rejected by a filter, selective amplifier, or other band-suppression device.

stop code See *halt instruction*.

stop element In digital transmission, a bit or set of bits, indicating the end of a character, and serving to inform the receiving device of the end of the character.

stop filter See *band-suppression filter*.

stopper resistor A fixed resistor connected in series with the control grid of a tube (or the equivalent electrode of a transistor) to prevent parasitic oscillations. Also called *grid stopper*.

stopping capacitor See *blocking capacitor*.

stopping coil See *choke coil*.

stopping resistor A parasitic-suppressing resistor. Also called *stopper resistor*.

storage allocation The assignment of computer memory areas to certain kinds of information as outlined in a source program and implemented by a compiler.

storage battery A rechargeable battery; the technical term is *secondary battery*. Also see *storage cell*.

storage capacity 1. The volume of data that can be stored in a computer memory (in words, characters, bytes, etc.). 2. The current-delivering capability of a storage battery in terms of ampere-hours, milliampere-hours, or other current-time units.

storage cell 1. An electrochemical battery cell whose potential may be restored by charging it with electricity. Compare *primary cell*. Also see *cell*. 2. The smallest part of a computer memory. 3. Of a computer memory, the part that can hold a data unit, e.g., a bit.

storage crt See *storage tube*.

storage cycle In computer operation, the period during which a location of a cyclic storage device cannot be accessed.

storage density The number of data units, e.g., bits, that can be stored in a given length or area of a storage medium.

storage device A medium (such as an integrated circuit) into which data can be placed and kept for later use.

storage dump See *dump*.

storage function Abbreviation, STO. The memory function of a calculator or computer.

storage laser A laser that stores intense energy before flashing.

storage life See *shelf life*.

storage mesh In the crt of a storage oscilloscope, the special, fine, metal mesh electrode which serves as the target on which the image is electrostatically stored. Also see *storage tube*.

storage oscilloscope An oscilloscope that retains a pattern until e-rased. Also see *storage tube*.

storage register In computers (where it is independent of the CPU) and calculators, a storage unit composed of flip-flops.

storage temperature 1. The recommended temperature for storing specified electronic components. 2. The particular temperature at which electronic components have been stored.

storage time 1. The interval during which carriers remain in a semiconductor-junction device after the bias has been removed. Also see *diode recovery time*. 2. For a switching semiconductor device, the time required for the amplitude of the output pulse to fall from max-imum to 90% of maximum after the input pulse has fallen to zero.

storage tube A cathode-ray tube that retains information in the form of images on a special electrode until erased by a signal.

store To place in the memory of a calculator or computer.

stored base charge The carriers that remain in the base layer of a bipolar transistor immediately after the forward bias has been inter-rupted. This charge maintains collector current momentarily (see *storage, 1*).

stored-energy welding A method of electric welding in which electrical energy is stored slowly and then released at the rate required for the welding.

straight adapter An inline coaxial fitting for joining two fixture-terminated coaxial lines in series.

straight angle An angle of π radians (180°).

straight dipole A (usually center-fed) dipole antenna having only one radiator. Also see *dipole antenna*.

straightforward 1. Uncomplicated or obvious. 2. Data transmission in one direction only.

straight-gun crt A cathode-ray tube in which the electron gun projects the beam in a straight line through the deflecting elements to the screen. Compare *bent-gun crt*.

straight-line capacitance Abbreviation, *SLC*. Pertaining to a variable capacitor for which the setting-vs capacitance curve is a straight line and the capacitance variation linear. Compare *straight-line frequen-cy* and *straight-line wavelength*.

straight-line coding See *straight-line programming*.

straight-line frequency Abbreviation, *SLF*. Pertaining to a variable capacitor in a tuned circuit, for which the setting-vs-frequency curve is a straight line, and the frequency variation linear. Compare *straight-line capacitance* and *straight-line wavelength*.

straight-line programming During the writing of a computer program, avoiding the creation of loops by repeating a series of instructions, to reduce execution time.

straight-line wavelength Abbreviation, *SLW*. Pertaining to a variable capacitor in a tuned circuit, for which the setting-vs-wavelength curve is a straight line, and the wavelength variation linear. Compare *straightline capacitance* and *straight-line frequency*.

straight-through amplifier An amplifier in which the input and out-put circuits are tuned to the same frequency. Compare *multiplier amplifier*.

strain Deformation of a body resulting from stress.

strain gauge See *electric strain gauge*.

strain-gauge bridge A four-arm resistance bridge in which a strain gauge (see *electric strain gauge*) is one arm. The resistance of the gauge changes due to strain and is determined by balancing the bridge.

strain-gauge transducer A transducer, other than a strain sensor, that employs strain gauges to convert values of pressure into their elec-trical analogs, e.g., pressure transducer, strain-gauge phonograph pickup.

strain pickup A phonograph pickup employing a strain gauge to con-vert sound vibrations into a varying electric current.

strand A single wire.

stranded wire A conductor composed of several wires twisted together into a kind of cable. Compare *solid wire*.

stratosphere The portion of earth's upper atmosphere beginning at a height of approximately 7 miles and extending to the ionosphere.

stray capacitance Inherent capacitance in a place where it can be detrimental, such as that between the turns of a coil or between adja-cent areas in a circuit. Also see *stray component*.

stray component An electrical property that exists as an inherent, and usually undesirable, side effect in a circuit or device. Thus, stray capacitance unavoidably exists between parallel conductors, and stray inductance is present in all wiring.

stray field The portion of an electric or magnetic field that extends beyond the immediate vicinity of the circuit with which it is associated and is therefore capable of interfering with other circuits or devices.

stray inductance Inherent inductance in a place where it can be detrimental, e.g., inductance in the coil of a wirewound resistor. Also see *stray component*.

stray resistance Inherent resistance in a place where it can be detrimen-tal, such as leakage resistance in a dielectric, and wire resistance in an inductor.

streaking In television or facsimile, a form of distortion in which the image appears enlarged in the horizontal but not the vertical.

stress The force per unit area which produces a strain.

stretch The amount by which a material being measured with an elec-tronic device increases in surface dimensions. Compare *shrink*.

strike To initiate a discharge, as in striking a gas tube.

striking voltage See *starting voltage*.

string 1. In computer practice, a set of items in a sequence determined by the order of keys. 2. In a computer memory, a sequence of bits or characters. 3. Any group of series-connected components or circuits.

string electrometer See *bifilar electrometer*.

string manipulation Performing operations on string variables.

string variable A string of characters, usually forming a word or phrase, represented by a variable name and character string symbol (BASIC's $, for example) in a computer program.

string oscillograph An early electromechanical oscilloscope consisting of a string galvanometer, lamp, rotating mirror, and viewing screen. The signal to be observed is applied to the galvanometer, and the string casts a shadow that is reflected by the rotating mirror to the screen. The shadow is swept across the screen, tracing the image of the signal.

strip chart A longitudinal, as opposed to circular, chart for graphic re-cording. In a rectilinear chart, both coordinates are straight; in a cur-vilinear chart, the crosswise coordinates are arcs.

strip fuse A fuse in which the fusible element is a flat strip of low-melting-point metal. Compare *wire fuse*.

stripped filament An electron-tube filament that has lost its elec-tron-emitting coating because of excessive voltage. Also see *reactivation*.

strobe 1. Strobe light. 2. Stroboscope.

strobe light See *electronic flash, 1*.

Strobolume Trade name for a type of high-output stroboscope.

stroboscope 1. An instrument that emits bright, adjustable-rate flashes of light. When this light illuminates an object that is rotating or vibrating at a fixed period, and the flash rate is made to match that period, the object seems to stand still and can be examined for flaws or faulty operation (and its speed can be measured). 2. A rotatable, slotted disk for producing the effect in 1, above.

stroboscopic disk A rotatable disk with alternating white and black radial regions, used in conjunction with a strobe light for precise measurement of the speed of a phonograph turntable.

strobotron A special gas tetrode used as the flashing light source in a stroboscope.

stroke speed See *scanning frequency.*

strong coupling See *close coupling.*

strontium Symbol, *Sr*. A metallic element of the alkaline-earth group. Atomic number, 38. Atomic weight, 87.63. Strontium is used in some ceramic dielectrics, such as barium-strontium titanate.

strontium 90 A dangerous heavy radioisotope of strontium. It has the mass number 90 and is present in the fallout from nuclear explosions.

structured programming Computer programming using a limited number of procedural sets while, most importantly, avoiding branches whenever possible to make the program as forward-going as possible so that it can be easily modified or debugged.

stub A (usually short) section of transmission line which is patched onto a longer line for tuning or impedance matching.

Stubs gauge See *Birmingham wire gauge.*

stub tuner A tuning unit consisting of a stub with a short-circuit that can be moved along the stub.

stub trap See *interference stub.*

stub-type wavetrap See *interference stub.*

stylus The "needle" which conveys vibrations to (or from) the disk in disk recording (or playback).

stylus drag See *needle drag.*

stylus friction Rubbing of the stylus against the record groove in disk recording or playback.

stylus pressure See *vertical stylus force.*

stylus printer See *wire printer.*

stylus scratch See *needle scratch.*

sub 1. Abbreviation of *subtract.* 2. Prefix denoting "under" or "below" in size, value, or rank.

subassembly A completely fabricated unit which is part of a larger unit into which it easily fits.

subatomic particles Extremely tiny bits of matter inside the atom. See, for example, *antiparticle, electron, meson, neutretto, neutrino, neutron, nucleon, positron,* and *proton.*

subaudible 1. Pertaining to any frequency falling below the limit of human hearing. 2. Any sound that is too low in amplitude to be heard.

subcarrier A carrier wave which, when modulated, becomes the modulating signal for another carrier wave.

subcarrier band The band of frequencies in which a subcarrier signal is transmitted.

subcarrier frequency modulation In a system in which a carrier frequency is obtained by beating a low-frequency rf signal with a high-frequency rf signal, the application of frequency modulation to the low-frequency component. The technique is sometimes employed in sweep-frequency signal generators.

subcarrier oscillator In a color-TV receiver, the oscillator operating at the burst (chrominance-subcarrier) frequency 3.579 545 MHz.

subchassis An auxiliary chassis on which one section of a larger piece of equipment is completely assembled and wired.

subfrequency See *subharmonic.*

subharmonic An integral submultiple of a fundamental frequency. Thus,

the 10th subharmonic of 15 MHz is 1.5 MHz.

submarine cable An underwater cable designed to withstand continuous immersion.

subminiature A size designation between miniature and microminiature.

submultiple The reciprocal of multiple, usually in reference to a frequency. For example, 7.2 MHz is a submultiple of 14.4 MHz. See *subharmonic.*

subpanel The front panel of a removable unit or module that forms a part of a larger unit.

subroutine In a computer program, a sequence of instructions for carrying out a section of the program's function. It is usually entered (led to) by a conditional branch (jump) instruction in the main program.

subscriber An individual user of a communications network or service.

subscript A small number or letter written to the lower right (and occasionally to the lower left) of another number or letter to identify the latter from others of the same designation, e.g., $a5$, Sn. Compare *superscript.*

subscription TV A television service paid for by subscribing viewers. The signals are scrambled so as to be useless to nonsubscribers, and legitimate subscribers are provided with a decoder to unscramble the telecasts.

subset 1. In statistics and set theory a set whose members are all contained in a larger set (see *set, 4*). 2. A telephone handset or deskset (*subscriber's set*). 3. A modulator/demodulator for making business machines compatible with telephone circuits.

subsidiary communication authorization Abbreviation, *SCA*. An FCC authorization for an FM station to transmit, in addition to its main program, a second program within the assigned bandwidth usually consisting of commercial-free background music on a subcarrier which can be detected only with a special receiver or with a special adapter attached to a standard FM receiver.

subsidiary feedback Feedback other than the main feedback in a system. Thus, feedback within a single amplifier stage that has overall feedback, is subsidiary.

subsonic frequencies Frequencies below the range of hearing. Also called *ultralow frequencies and subaudible frequencies.*

substation An intermediate electricity-distributing location from which electrical energy is transformed and transmitted to users within a given geographical area.

substitution capacitor A capacitor employed temporarily in place of another of usually the same value, as in troubleshooting. Also see *capacitor substitution box.*

substitution inductor An inductor employed temporarily in place of another of usually the same value, as in troubleshooting. Also see *inductor substitution box.*

substitution method 1. A method of measuring a quantity—such as capacitance, inductance, or resistance—in which the value of the unknown quantity is determined in terms of the amount of a standard quantity that must be removed in order to restore the test circuit to its original state of balance. 2. A method of troubleshooting in which good components are substituted for bad ones in a circuit (see, for example, *substitution capacitor, substitution inductor, substitution resistor, substitution speaker,* and *substitution transformer*).

substitution resistor A resistor used temporarily in place of another of usually the same value, as in troubleshooting. Also see *resistor substitution box.*

substitution speaker A loudspeaker used temporarily in place of another, as in troubleshooting.

substitution transformer A transformer used temporarily in place of another having the same characteristics, as in troubleshooting.

substrate A plate, wafer, panel, or disk of suitable material on (or in) which the components of a unit, such as an integrated or printed circuit, are deposited or formed.

subtend The formation of an angle at a vertex by line segments or arcs with definite terminuses. If it is line segments by which the angle is subtended, they are part of the angle's sides. When an angle is subtended by an arc, the center of the corresponding circle is the angle's vertex.

subterranean *Underground.*

subtracter See *electronic subtracter.*

subtraction The mathematical operation of removing one quantity (subtrahend) from another (minuend) to give a result (difference or remainder).

subtractive color A color formed by mixing subtractive primary pigments.

subtractive primaries Broadspectrum pigments used in printing to produce a wide variety of colors through filtering. These primaries—cyan, magenta, and yellow—are used to print images that have been filtered through additive-primary lenses. Note that each subtractive primary approximates the color of two combined additive primaries; e.g., by combining green and red lights, an additive yellow results; by combining yellow and magenta pigments, a subtractive red results.

subtrahend In the process of subtraction, the quantity that is subtracted from another (the minuend) to give the remainder or difference.

successive derivatives Successive repetition of the operation of differentiating a function, which yields the first derivative, second derivative, and so on to the nth derivative. Also see *derivative.*

successive integration The operation of double or triple integration.

Sugar Phonetic alphabet code word for the letter S.

Suhl effect A reduction in hole life that occurs in a semiconductor material in the presence of a magnetic field.

suite A group of computer programs run successively as a job.

sulfate Contraction of lead sulfate.

sulfation In a lead-acid storage cell, the formation of disabling lead sulfate during discharge of the cell.

sulfur Symbol, *S.* A nonmetallic element. Atomic number, 16. Atomic weight 32.006.

sulfur hexafluoride A gas employed as a coolant and insulant in some power transformers.

sulfuric acid Formula, H_2SO_4. A familiar acid used in dilute solution as the electrolyte of a lead-acid battery. This highly corrosive fluid also has many industrial uses.

SUM Abbreviation of *surface-to-underwater missile.*

sum The result obtained by adding two or more terms. Compare *summation.*

sumcheck Use of checksums to perform a check.

sum frequency 1. In an amplitude-modulated carrier, the upper sideband frequency, i.e., the sideband equal to the carrier frequency plus the modulating frequency. Compare *difference frequency.* 2. In superheterodyne operation, an intermediate frequency equal to the signal frequency plus the local-oscillator frequency.

summation 1. Symbol, Σ. The sum of a finite number of terms. Thus, the total resistance of n resistors connected in series is the summation of all R (resistance) terms, i.e., $R_t = {}_1{}^n \Sigma R$. Compare *integral.* 2. A frequency equal to the sum of two other frequencies.

summation check In computer practice, a check carried out on a group of digits. The result of adding the digits, and disregarding any overflow, is a check digit which can be compared with a standard value for the operation to verify the integrity of data.

summer 1. Adder. 2. Summing amplifier.

summing amplifier An operational amplifier whose output is the sum, or is proportional to the sum, of several inputs.

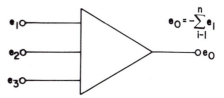

Summing amplifier.

sun battery Photovoltaic cells so connected as to give useful output voltages and currents. See also *solar battery.*

S-units In radio (especially amateur practice), gradations reflecting the strength of received signals. Typically, a value of S9 (9 s-units) is equal to a strength of 50 μV. The next lower S-unit (S8) is 6 dB lower in level, S7 is 12 dB below S9, etc.

sunlight lamp A special, tungsten-filament lamp giving high ultraviolet output. A droplet of mercury is enclosed in the bulb, which is made of ultraviolet-transmitting glass.

sunlight-powered laser A laser whose action is stimulated by sunlight collected by a system of mirrors and lenses. The life of the device is long compared with that of conventional lasers.

sun-pumped laser See *sunlight-powered laser.*

sun relay See *sun switch.*

sunspots Areas on the sun's surface, visible as dark spots and believed to be comparatively cool regions resulting from magnetic disturbances. The spots are copious emitters of particles that cause interference to radio communications on earth.

sunspot cycle The inexplicable appearance of sunspot activity on the sun's surface at 11-year intervals.

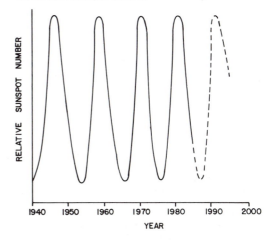

Sunspot cycle.

sun switch A photoelectric relay actuated by sunlight and employed for various domestic and industrial purposes, such as switching lights, operating window shades, etc.

sup Abbreviation of *suppressor*.

super A prefix meaning *over, above,* or *higher than.* Although *super* is found in many compound words used in electronics, it is not always used in the prefix sense. *Superheterodyne,* for example, is a blend word constructed from *supersonic* and *heterodyne.*

superaudible frequency See *ultrasonic frequency.*

superbeta transistor A transistor—or transistor combination, such as a Darlington pair (see *compound connection*)—which provides a very high current amplification factor (*beta*).

superconducting cable A cable in which superconductivity is achieved by surrounding the cable with liquid helium to lower its temperature to near absolute zero.

superconductivity The virtual disappearance of resistance in some metals cooled to temperatures in the vicinity of absolute zero. Also see *cryogenics, cryostat,* and *cryotron.*

superconductor A material or device which displays superconductivity.

supercontrol tube A vacuum tube whose amplification factor varies with control-grid bias, up to the point of plate-current cutoff, e.g., a 6SK7. Also called *extended-cutoff tube* and *variable-mu tube.*

superdyne circuit An early regenerative rf amplifier in which a tickler coil is connected in series with the tuned plate tank of the amplifier and B-plus is coupled to the grid tank.

super flatpack An integrated-circuit package of the flatpack type having considerably more components and leads than those in the conventional flatpack.

superhet Contraction of superheterodyne.

superheterodyne circuit Superheterodyne circuit or superheterodyne receiver.

superheterodyne circuit A receiver or instrument circuit in which the received signal in a first detector (or mixer) beats with the signal of a local oscillator, resulting in a lower (intermediate) frequency, which then is amplified by an intermediate-frequency (i-f amplifier). This i-f signal is detected by a second detector whose output is amplified by an audio-frequency amplifier. Because the i-f amplifier operates at a single (fixed) frequency, it may be adjusted for optimum selectivity and gain. Also called *superhet circuit.*

superheterodyne receiver A radio or TV receiver employing a *superheterodyne circuit.*

superhigh frequency See *radio spectrum.*

Supermalloy An alloy having a maximum permeability of 10^6.

supermodulation A type of amplitude modulation in which one rf power stage continuously generates the carrier, and a second (usually identical) rf power stage is gated into full operation at the proper instant by the af modulation to add additional rf power (corresponding to 100% modulation) to the carrier. At the same time, carrier amplitude is decreased by the proper amount to fulfill the conditions of an amplitude swing between zero and twice maximum for 100% modulation. Also called *Taylor modulation.*

superposition theorem The proposition that, in a network of linear elements, if a voltage $E1$ in branch A of the network causes a current $I1$ to flow through branch C of the network and if a voltage $E2$ in branch B of the network (which may be identical with branch A) causes a current $I2$ to flow through branch C, then $E1$ in branch A and $E2$ in branch B applied simultaneously will cause a current equal to $I1 + I2$ to flow through branch C. Compare *compensation theorem, max-*

imum power transfer theoremn Norton's theorem, reciprocity theorem, and *Thevenin's theorem.*

superpower An arbitrary term denoting very high power. In the rating of standard broadcast stations, it has come to signify 1 million watts rf output.

superregenerative circuit A regenerative detector circuit in which regeneration is periodically increased almost to the point of oscillation and then decreased. This *quenching* action takes place at a supersonic rate (typically at 50 or 100 kHz), so that the quenching is inaudible. The result is that much more regeneration is afforded, without the detector going into oscillation, than is possible by simply increasing the regeneration manually. An extremely sensitive detector results.

supersaturated solution A solution which contains more solute than it normally would hold. Supersaturated solutions are obtained through special techniques and are extremely unstable. Compare *saturated solution.* Also see *solute; solution, 1;* and *solvent, 2.*

superscript A small number or letter written to the upper right of another number or letter, the *base,* to indicate the power to which the base must be raised. Example: 10^5, e^x, y^2. Also called *exponent.* Compare *subscript.*

supersensitive relay A relay that operates on less than 1 mA or less than 1 mV.

supersonic frequency See *ultrasonic frequency.*

supersonics See *ultrasonics.*

supervisor 1. In a computer, a set of routines that oversee the operation of the system. The supervisor routines are coordinated by the central processing unit. 2. The execution of such a set of routines.

supply *Current supply, power supply,* or *voltage supply.*

supply current Alternating or direct current available for operating an equipment.

supply frequency The frequency of an ac supply.

supply power The power of an ac or dc supply.

supply voltage The voltage of an ac or dc supply.

suppressed carrier A carrier that has been canceled out of a carrier—sideband combination.

suppressed-carrier—double-sideband See *double sideband, sup-*EL *pressed carrier.*

suppressed-zero instrument A meter or graphic recorder in which the zero point is off-scale or upscale, but has been brought to scale-zero by means of mechanical adjustment or use of a bucking voltage.

suppressor 1. In a pentode, a gridlike element between the screen and plate, employed to suppress secondary emission. Also see *grid, 2* and *pentode.* 2. A filter employed to suppress radio interference. 3. Noise eliminator (see *automatic noise limiter*). 4. Spark suppressor.

suppressor circuit The circuit associated with the suppressor electrode of an electron tube.

suppressor diode A semiconductor diode employed to prevent inductive kickback in circuits, to eliminate or reduce transients, or to prevent arcing between make-and-break contacts.

suppressor grid See *suppressor, 1.*

suppressor-grid modulation See *suppressor modulation.*

suppressor modulation A method of modulation in which a modulating voltage is superimposed upon the dc suppressor voltage of a pentode, rf amplifier, or oscillator tube. Compare *absorption modulation, cathode modulation, grid-bias modulation, plate modulation, plate—screen modulation, screen modulation,* and *series modulation.*

suppressor pulse A pulse that prevents electron flow.

surface analyzer A device designed for the measurement of surface flatness or uniformity.

surface-barrier diffused transistor See *microalloy diffused transistor*.

surface-barrier transistor Abbreviation, *SBT*. A *pnp* transistor made by means of electrolysis and electroplating: Two fine streams of indium sulfate solution are placed upon axially opposite points on the faces of an *n*-type wafer. At the same time, a direct current is passed through the wafer and solution in such a direction as to remove semiconductor material electrolytically from the faces—the tiny sprayed areas are etched away. When the desired wafer thickness is reached at the points of impact, the etching process is arrested by reversing the direction of current flow. This reversal causes an indium dot to be plated on each opposite face in the etched-out pit. Leads are attached to the collector and emitter dots and to the wafer (base). Also called *microalloy transistor*.

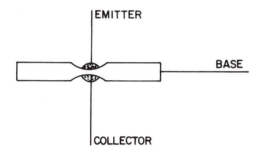

Surface-barrier transistor.

surface-charge transistor A semiconductor device consisting essentially of two narrowly separated plates (*source* electrode and *receiver* electrode) deposited on the film-insulated surface of a silicon chip, and a third, overlapping electrode (the *transfer gate*) deposited on, but insulated from, the other electrodes. An input signal stores a charge in the capacitor formed by the source electrode and chip. A subsequent trigger signal applied to the transfer gate transfers the charge to the receiver electrode, where it becomes an output signal (often amplified with respect to the input signal).

surface effect An effect—such as current, resistance, or resistivity—observed on the surface of a sample of material rather than throughout the body of the material. Compare *bulk effect*.

surface insulation A coating applied to the surfaces of core laminations to prevent the passage of currents between laminations.

surface leakage Leakage of current over the surface of a dielectric material, as opposed to *body leakage*.

surface noise See *needle scratch*.

surface recombination rate For a semiconductor, the rate at which electrons and holes recombine at the surface. Compare *volume recombination rate*.

surface resistivity The resistance of a unit area of a material, measured between opposite edges. Compare *volume resistivity*. Also see *resistivity*.

surface tension The tendency of the surface of a liquid to "shrink." This property varies with different liquids and is due to a net molecular force directed inward from the surface.

surface-to-air missile ABBREVIATION, *SAM*. A guided missile de-

signed to be fired from the ground or from a ship.

surface-to-underwater missile A guided missile designed to be fired from the ground, or from the deck of a ship, to an underwater target.

surface wave 1. The earth-guided component of a ground wave. (The other component is the *space wave*.) 2. An acoustic wave traveling along the surface of a plate in a surface-wave amplifier or surface-wave filter.

surface-wave amplifier An amplifying device consisting essentially of a surface-wave filter to which has been added a dc-biased *n*-type silicon electrode, which is separated from the crystal substrate of the filter by a very thin oxide layer. Amplification is produced by interaction between the electron current in the silicon and the piezoelectric field of the filter. Also see *acoustoelectronics*.

surface-wave filter An acoustoelectronic device consisting essentially of a crystal plate having electrodes at each end. An ac input signal applied to one electrode sets up acoustic waves which travel along the surface of the plate to the other electrode, where they generate an ac output voltage by piezoelectric action. The resonant frequency of the device is governed by the dimensions of the crystal plate. Also see *acoustoelectronics*.

surge A sudden rise or flow of current or voltage.

surge absorber See *surge arrester*.

surge arrester A device, such as a spark gap or gas diode, employed to absorb a potentially destructive transient or overvoltage.

surge current A heavy current that flows initially into a capacitor when a charging voltage is applied.

surge impedance Symbol, Z_o. The impedance seen by a pulse applied to a transmission line; $Z_o = \sqrt{L/C}$ (approximately), where L and C are the inductance and capacitance per unit length of the line. Also called *characteristic impedance*.

surge suppressor See *surge arrester*.

surveillance 1. A method of monitoring a specific area or volume for intrusion or other disturbance. 2. A means of monitoring a specified portion of the electromagnetic spectrum for unauthorized signals.

surveillance radar In a GCA system, an air-traffic-control radar that supplies continuous information regarding the azimuth and distance of aircraft inside a selected radius around an airport.

Susan One of several phonetic alphabet communications code words for the letter *S*. (See also *Sierra* and *Sugar*.)

susceptance Symbol, *B*. Unit, siemens. The reactance component of admittance, as distinguished from the active component, conductance; $B = I/E$, where E is the emf in volts, and I the current in amperes.

susceptibility The capacity of a substance to become magnetized, expressed as the ratio of magnetization to the strength of the magnetizing force.

suspension 1. The wire or metallized fiber supporting the movable coil of a galvanometer. 2. Particles of a substance and the liquid in which it is mixed but not dissolved. 3. The substance in 2, above.

suspension galvanometer A meter with a light-beam apparatus for lengthening the arc through which the pointer travels. When the beam of light is cast a long distance, a tiny movement of the coil will cause considerable movement of the image.

sustained oscillation Oscillation that continues as long as power is supplied to the oscillation generator. Also see *continuous wave*. Compare *damped oscillations*.

sustaining voltage The voltage at which second-collector breakdown occurs in a transistor (see *second breakdown*).

SW Abbreviation of *shortwave*.

sw Abbreviation of *switch*. (Also, *S* or *s*.)

swamping resistor 1. A noninductive resistor connected in parallel with the input circuit of a glass-B linear rf amplifier for automatic regulation of the rf excitation. 2. A resistor connected in series with the emitter of a bipolar transistor to minimize the effects of temperature-induced variations in junction resistance.

swarf The string of material that threads off a disk during sound recording.

SW-band A section of the S-band extending from 3400 to 3700 MHz.

sweep 1. To deflect the electron beam in a cathode-ray tube, usually horizontally to provide a time base. 2. The circuit for achieving the particular deflection described in 1, above.

sweep circuit A circuit, such as a deflection generator (e.g., a sawtooth oscillator), for producing a sweep signal. Also see *sweep, 1, 2.*

sweep delay In an oscilloscope, the process of initiating the sweep of the electron beam at some selected instant after the signal has started.

sweep-delay circuit In an oscilloscope or radar, the circuit for delaying the sweep until the start of the signal. Also see *delayed sweep, 1, 2.*

sweeper 1. Sweep generator. 2. Sweep-signal generator.

sweep frequency 1. The frequency at which the electron beam in a cathode-ray tube is deflected along the reference axis. 2. The frequency at which the carrier frequency is increased and decreased by a sweep-signal generator.

sweep generator 1. A device that causes the electron beam in a cathode-ray tube to scan at a known speed. 2. An oscillator that generates a signal that rapidly varies in frequency. Used for testing and adjustment of bandpass filters and other selective circuits.

sweeping receiver See *scanning receiver.*

sweep magnification In an oscilloscope, multiplying the sweep period, and consequently widening the image on the screen, for closer inspection.

sweep magnifier In an oscilloscope circuit, a circuit for achieving sweep magnification.

sweep oscillator See *sweep generator.*

sweep period The duration of one complete cycle of sweep signal (see *sweep frequency, 1*).

sweep signal The (usually linear, sawtooth) signal employed to sweep the beam of an oscilloscope tube. Also see *sweep, 1, 2.*

sweep-signal generator A signal generator that supplies a signal whose frequency varies automatically and periodically throughout a given band.

sweep test A method of testing the attenuation-versus-frequency characteristics of a selective circuit, using a radio-frequency sweep generator.

sweep time The actual time required for a single sweep by a deflecting signal; $t = 1/f$, where t is sweep time in seconds, and f is sweep frequency in hertz.

sweep voltage The peak voltage amplitude of the sweep signal.

SWG Abbreviation of *standard wireage.*

swing The maximum change exhibited by a varying quantity, e.g., amplitude swing, frequency swing.

swinging choke A filter choke having a core with a little air gap or none, and which therefore exhibits high inductance at low current, and vice versa. This inductance, which swings under conditions of varying load current, permits use of a high-resistance bleeder resistor. Compare *smoothing choke.*

Swiss-cheese packaging A method of packaging an electronic circuit, in which components are inserted into the assembly through holes drilled or punched in parallel, stacked printed-circuit boards.

switch 1. A circuit or device (electronic, electromechanical, or mechanical) for opening and closing a circuit or for connecting a line to one of several different lines, e.g., rotary selector switch. 2. To change the logic state of a circuit or device. 3. In a computer program, a branch instruction directing the program to a line number dependent on the value of a variable or result, e.g., BASIC's ON...GOTO. 4. To cause an electrical circuit to change states, as from *on* to *off* or vice versa.

switchgear Collectively, devices and systems for making and breaking circuits either automatically or manually.

switching characteristics Technical data describing the performance and capabilities of switching devices and circuits.

switching circuit An on-off type of circuit containing electronic or mechanical switches.

switching diode See *computer diode.*

switching frequency The frequency at which a repetitive switch operates. Also see *switching rate.*

switching mode The type of operation in which a device—such as a tube, transistor, or tunnel diode—exhibits on-off operation, as opposed to smooth response to a varying signal.

switching rate The rate (e.g., closures per second) at which a repetitive switch operates. Also see *switching frequency.*

switching speed The time required for a switch to open or close or for a switching device to change states (as from cutoff to saturation). Also see *switching time.*

switching time Symbol, *ts.* The time required, after the application of a pulse, for an electronic switch to attain its own state. Also see *switching speed.*

switching transistor A transistor designed especially for on-off operation. Such units exhibit short recovery time and low capacitance.

switching tube See *beam-switching tube.*

switching voltage The largest voltage that a switching device can handle without malfunctioning.

SWL Abbreviation of *short-wave listener.*

SWR Abbreviation of *standing-wave ratio.*

SWR bridge A four-arm resistance bridge for measuring voltage standing-wave ratio. This rf bridge has noninductive resistors in three of its arms and the device under test in the fourth arm. The bridge is balanced first with an equivalent noninductive resistor that replaces the device, and the output voltage is noted. Then the device is substituted for the test resistor, and the change in voltage is noted. The standing-wave ratio is determined from the voltage ratio.

SWR meter See *SWR bridge.*

SY-band A section of the S-band extending from 2600 to 2700 MHz.

syllable compandor A device that compresses or expands the amplitude of an audio signal. The time constant is fast enough to allow response to individual syllables. Similar to automatic level control.

sym 1. Abbreviation of *symmetrical.* 2. Abbreviation of *symbol.*

symbol 1. A letter or graphic device representing a quantity or term, e.g., I (current), ϕ (phase). 2. A conventional device denoting a mathematical operation, e.g., $+$, $\times$. 3. In a circuit diagram, a pictorial device representing a component.

symbolic address An address in a source language computer program, i.e., the arbitrary label used by the programmer.

symbolic language See *source language.*

symbolic logic A system for representing logical relationships, such as those acted upon by computer and switching circuits, by means of symbols which are principally nonnumerical. Also see *Boolean algebra.*

symmetrical circuit A circuit having identical configurations on each side of a dividing line, such as the ground bus. A push-pull circuit is an example.

symmetrical conductivity Identical conductivity for both positive and negative electricity. Compare *asymmetrical conductivity*.

symmetrical FET See *symmetrical field-effect transistor*.

symmetrical field-effect transistor A field-effect transistor whose source and drain terminals may be interchanged without affecting circuit operation. Also called *bidirectional transistor*. Compare *unilateral field-effect transistor*.

symmetrical input See *balanced input*.

symmetrical output See *balanced output*.

symmetrical transistor See *bidirectional transistor*.

symmetrical wave A wave whose positive and negative half-cycles are identical in shape and values.

symmetry 1. The condition of having the same shape on each side of an axis. 2. The condition of conducting positive and negative currents equally well. 3. The condition in which a circuit is identical on both sides of a reference line, such as the ground line.

sympathetic vibration Resonant vibration of one body in response to the vibration of another body.

sync Synchronization; synchronism.

sync amplifier In a television circuit, the amplifier employed to increase the amplitude of the sync pulses after they are separated from the composite video signal.

sync generator A circuit that produces the synchronization pulses in a television transmitter.

synchro A dynamo-electric-control device which, when connected to a similar device and the ac power line, permits remote control. Thus, when the rotor of one synchro is turned to a certain position, the rotor of the other assumes the same position. Also called *Autosyn, selsyn,* and *Teletorque*.

synchrocyclotron A type of cyclotron in which the variation in mass, due to increased velocity, of the charged particles is compensated, resulting in higher energy for the particles.

synchro differential A synchro which receives two input signals and delivers a single output signal. The inputs may be two electrical signals or one electrical signal and one mechanical signal.

synchrodyne receiver A direct-conversion receiver in which the local oscillator is locked into synchronism with the incoming rf signal.

synchroflash A flash that is synchronized with the shutter of a camera.

synchro generator A transmitting synchro.

synchro motor A receiving synchro.

synchronism 1. The condition of being in step, as when two motors are running in synchronism with each other and the power frequency, or when two relays open and close in step. 2. The condition of being in phase, as when two pulses occur simultaneously.

synchronization The coincidence of one process or operation with another, as in the synchronization of an oscillator frequency by means of an applied standard-frequency voltage, in which case the oscillator frequency becomes that of the standard signal.

synchronized clamping A type of clamping in which an output voltage is maintained at a predetermined fixed value until a synchronizing pulse is applied, whereupon the output follows the input.

synchronized multivibrator See *driven multivibrator*.

synchronizer A computer storage device which is used between two devices transmitting data at different speeds, to counteract this differential (as a buffer).

synchronizing signal A signal employed to synchronize another signal, usually in frequency.

synchronous The condition of operating in step (phase) with some reference rate.

synchronous clock 1. An ac clock driven by a synchronous motor. Although 60 Hz models are common, such clocks are not restricted to low-frequency ac operation; 1 kHz types, for example, are found in some primary frequency standards. 2. The timing source in a synchronous computer.

synchronous communications satellite A communications satellite placed in a synchronous orbit.

synchronous computer A computer whose operations are timed by single-frequency clock signals.

synchronous contacts The rectifying contacts of a synchronous vibrator (see *vibrator-type rectifier*).

synchronous converter A synchronous machine which can run on ac and generate dc, or vice versa. Also called *rotary converter*.

synchronous generator An alternator operating in synchronism with one or more other alternators. Also see *synchroscope, 2*.

synchronous induction motor An ac motor that is intermediate between the fractional-horse-power reluctance motor and the multiple-horsepower three-phase, synchronous motor. The synchronous induction machine starts like an induction motor and runs like a synchronous motor.

synchronous inputs In a computer flip-flop, inputs that accept pulses only at the command of the clock.

synchronous machine See *synchronous induction motor*.

synchronous motor See *synchronous induction motor*.

synchronous orbit An orbit around the earth, directly above the equator, and at such an altitude (about 22,500 mi.) that any object will remain over the same ground location at all times.

synchronous satellite See *synchronous communications satellite*.

synchronous spark gap A type of rotary spark gap in which one spark occurs during each half-cycle of ac power. This action is accomplished by driving the rotary gap with a synchronous motor supplied from the power line supplying the transformer that excites the gap.

synchronous speed For an ac machine, the speed corresponding to the ac frequency.

synchronous vibrator See *vibrator-type rectifier*.

synchroscope 1. An oscilloscope having a high-speed sweep triggered by a synchronizing signal. Such an instrument is valuable for viewing high-speed pulses. 2. A pointer-type instrument employed to indicate the synchronism between two power alternators.

synchro system A circuit or system employing synchros for the transmission and reception of positioning signals. Also see *synchro*.

synchrotron A type of accelerator (see *accelerator, 1*) that employs a highfrequency electrostatic field and a low-frequency magnetic field to impart very high velocity to particles.

sync pulse 1. A pulse used to control the frequency or repetition rate of an oscillator or other generator. 2. In a television system, a pulse transmitted as part of the composite video signal to control scanning. Also see *horizontal sync pulse* and *vertical sync pulse*.

sync separator In a TV receiver circuit, a stage employed to separate and deliver the sync pulses from the composite video signal. See, for example, *diode sync separator*.

sync signal See *synchronizing signal*.

sync takeoff The point in the video amplifier circuit of a TV receiver

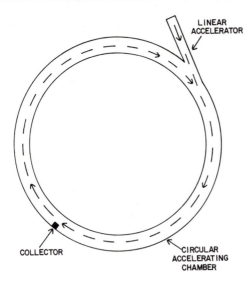

LINEAR ACCELERATOR

COLLECTOR

CIRCULAR ACCELERATING CHAMBER

Synchrotron.

at which the composite video signal is sampled to extract the sync pulses.

syntax The rules by which computer program statements are structured.

synthesis The rigorous (usually highly mathematical) design of an electronic circuit or device and the accurate prediction of its performance. Compare *analysis*.

synthesizer 1. Signal synthesizer. 2. A circuit synthesizer, i.e., a device that allows a wide variety of circuits to be set up temporarily or simulated, for test and evaluation. (Sometimes, a specially programmed computer serves this purpose.) 3. Moog synthesizer.

synthetic bass An apparent accentuation of bass notes resulting from intermodulation distortion in an amplifier.

synthetic crystal An artificially produced crystal, such as synthetic quartz.

synthetic resin An artificially produced resin. Also see *thermoplastic material* and *thermosetting material*.

syntony Resonance.

syst Abbreviation of *system*.

system 1. An integrated assemblage of elements operating together to accomplish a prescribed end purpose, e.g., servo system, communication system. The elements often are completely self-contained. 2. A methodology incorporating fixed and ordered procedures for accomplishing an end purpose.

systematic error See *cumulative error*.

system engineering See *systems engineering*.

system of units A set of fundamental units for defining the magnitudes of all physical variables. The most common system of units is the standard international (*SI*) system.

systems analysis In computer system operation, analyzing the way something is done and devising a better alternative by isolating the problem area, scrutinizing the system as it stands, studying what is thereby disclosed, devising the alternate application of software or hardware, disseminating the revised operational procedure, and overseeing the implementation of the new method.

systems engineering The branch of engineering devoted to the design, development, and application of complete systems. The approach takes into consideration all elements in a system or process and their integration. Also see *system*.

systems flowchart A flowchart showing the interrelationship of activities in a computer system.

Sz-band A section of the S-band extending from 3900 to 4200 MHz.

T 1. Symbol for transformer. 2. Abbreviation of prefix *tera*. 3. Symbol for *thermodynamic temperature*. 4. Symbol for *tritium*. 5. Abbreviation of *ton*. (Also, *t* and *tn*.) 6. Abbreviation of *tesla*. 7. Symbol for *kinetic energy*. 8. Symbol for *period*. 9. Symbol for *true*.

t 1. Symbol for *time*. 2. Abbreviation of *ton*. (Also, *T* and *tn*.) 3. Symbol (ital) for *Celsius temperature*. (Also, *T*.) 4. Abbreviation of *target*. 5. Abbreviation of *technical*. 6. Abbreviation of *tension*.

Ta Symbol for *tantalum*.

tab In a typewriter, teletype, or word-processing system, a key that moves the cursor a specified number of spaces toward the right.

table In an internal or external computer memory, an array (i.e., a list or matrix) of data that can be recalled using keys, e.g., single- or double-subscripted variables.

table look-at Abbreviation, TLA. In computer practice, finding the position of a data item in a table by implementing an algorithm.

table look-up Abbreviation, TLU. In computer practice, locating items in a table by inspecting what is in the table key by key.

tabulate 1. In data processing, to combine the totals for data item groups having the same key. 2. The presentation by a punched card tabulator of the totals for card groups.

tabulation 1. The printout of what has been tabulated (see *tabulate*, 1). 2. The computer-program-directed movement of the cursor on a crt display, or of a typewriter carriage, to certain positions in a line.

tabulator A machine that transports and reads punched cards automatically.

tacan A pulse-type uhf air navigation system in which a station is interrogated by signals from an aircraft to give bearing and range information. The name is an acronym for *tactical air navigation*.

tach Abbreviated form of *tachometer*.

tachometer See *electronic tachometer*.

tachometer generator A small, dynamo-type electric generator which, when attached to a rotating shaft, delivers a voltage proportional to the rotational speed of the shaft.

tactical air navigation See *tacan*.

tactical radar A radar system used in anti-enemy operations.

T-adapter See *tee-junction*.

tag 1. Identifying digits or characters forming part of a record (in data processing). 2. An encoded price tag, i.e., a perforated one, the retailer can detach for presentation to a data processing system, for inventory management, for example.

tag converter A device that senses the information on tags (see *tag*, 2) and transfers it to a computer memory or to punched cards or paper tape.

tail 1. The decay of a waveform from maximum amplitude to zero amplitude. 2. Any pulse that follows a main pulse as a result of the main pulse.

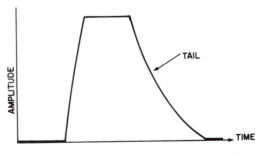

Tail.

taillight monitor An electronic device for warning a motorist of taillight failure.

tail pulse 1. A pulse with a fast rise time but a slow decay time. 2. See *tail, 2.*

takedown In computer practice, to clear output peripherals for upcoming program runs by removing punched cards or similar material.

takedown time The duration of the takedown process.

takeup reel In a tape recorder—reproducer, the reel onto which the tape is wound during recording or reproduction.

talk-back circuit See *interphone.*

talk—listen switch A transmit—receive switch in an intercommunication system (see *intercom*).

talk power See *speech power.*

tally 1. Total (the verb). 2. The rows of operands, subtotals, and totals that an adding machine prints.

tally reader A device that, by OCR, can read the digits and symbols on a tally (*tally*, 2).

tan Abbreviation of *tangent.*

tan^{-1} Arc tan (inverse tangent function).

tandem transistor An assembly of two series-connected transistors in the same envelope.

tangent 1. Abbreviation, tan. The trigonometric function *a/b*, the ratio of the side opposite to the side adjacent to an acute angle in a right triangle. 2. The property of a line segment that just touches a curve or arc; in the latter case, it is perpendicular to the radius connecting the point of tangency to the corresponding circle's center.

tangent galvanometer A galvanometer in which the current is proportional to the tangent of the angle of deflection. Compare *sine galvanometer.*

tahn Abbreviation of *hyperbolic tangent.*

tank 1. A parallel-resonant circuit in an electron tube plate circuit. 2. A circuit in which is stored electrical energy of frequencies in a range whose midpoint is resonance for the circuit. 3. Mercury delay line.

tantalum Symbol, Ta. A metallic element of the vanadium family. Atomic number, 73. Atomic weight, 180.95. Tantalum is used in the elements of some electron tubes and in some electrolytic capacitors.

tantalum capacitor See *tantalum-foil electrolytic capacitor.*

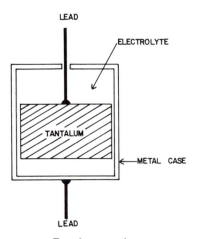

Tantalum capacitor.

tantalum detector A type of rf contact detector consisting essentially of a fine tantalum wire whose point lightly touches the surface of a small pool of mercury.

tantalum-foil electrolytic capacitor An electrolytic capacitor whose anode and cathode are tantalum foils, the anode foil being oxide-coated. Compare *tantalum-slug electrolytic capacitor.*

tantalum-nitride resistor A resistor consisting of a thin film of tantalum nitride deposited on a substrate. Also see *thin film.*

tantalum-plug capacitor See *tantalum-slug electrolytic capacitor.*

tantalum-slug electrolytic capacitor An electrolytic capacitor whose anode is a sintered tantalum slug. Compare *tantalum-foil electrolytic capacitor.*

T-antenna See *tee-antenna.*

tap A connection made to an intermediate point on a coil, resistor, or other device. See, for example, *center tap* and *tapped component.*

tape 1. Paper tape. 2. Magnetic tape. 3. Insulating tape. 4. To make a magnetic tape recording.

tape cable 1. A form of cable in which all of the conductor centers lie in the same plane. 2. A flat cable, commonly used in situations where repeated flexing occurs.

tape cartridge A holder and the reel of blank of prerecorded magnetic (eighth-inch audio, or wider video) tape it contains, which may be inserted into a recorder—reproducer without having to thread or otherwise handle the tape for either playing or rewinding.

tape comparator In a data processing or computer system, a machine that compares two paper tapes generated from the same input, for differences in the data thereon; it is a character-by-character process.

tape core A ring-type magnetic core made by tightly winding metal tape in several layers for the desired thickness.

tape counter See *position indicator.*

tape deck In a tape recorder—reproducer, the complete tape-transport mechanism (drive, heads, equalization circuitry, and preamps).

tape drive 1. Tape transport. 2. Tape deck.

tape feed That which moves the data medium through a paper tape punch or reader.

tape file Magnetic tape file.

tape group An assembly of two or more tape decks.

tape label On a reel of magnetic tape containing a file, a record at the beginning or end, in which is information about the file.

tape loop 1. An endless, circular loop of magnetic tape. 2. An endless, circular loop of paper tape, such as is used in certain teleprinter and computer terminals.

tape magazine See *tape cartridge.*

tape mark 1. A character that subdivides the magnetic tape file on which it is recorded. Also called *control mark.* 2. A character marking the end of a length of magnetic tape on a reel. Also called *end-of-tape marker.*

tape plotting system In computer practice, a system for operating a digital incremental (x – y) plotter using the information on paper or magnetic tape.

tape printer A teletype machine that prints a message in a single line on tape.

tape punch A usually keyboard-operated device for perforating paper tape.

taper In a potentiometer or rheostat, the rate of change in resistance during uniform rotation of the shaft. See, for example, *linear taper* and *log taper.*

tape reader In a data processing or computer system, an online or offline device that interprets the hole patterns in paper tape.

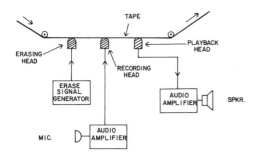

Tape recorder.

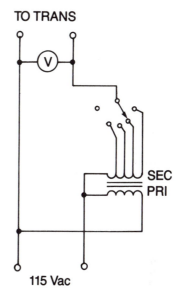

Tapped transformer.

tape recorder A machine for recording sound or data signals on magnetic tape and usually for playing back the recorded material.

tapered potentiometer A potentiometer having a tapered resistance winding (see *tapered winding*).

tapered winding A resistance winding (in a rheostat or potentiometer) in which the resistance change per unit length of winding is nonuniform (see, for example, *log taper*). However, a winding having uniform resistance change is often called *linear taper*.

tape reproducer A device used to duplicate paper tapes, often allowing editing during the process.

tape skew A condition in which a magnetic tape passes between the recording or playback heads in an irregular way. The result is that the various channels, or tracks, are not always perfectly aligned.

tape sort The (computer) operation of sorting data in a magnetic tape file into a record-key-determined sequence.

tape splicer A mechanism which aligns and secures the overlapping ends of broken magnetic tape so they can be cut (often with an integral cutter), to form butted ends, and taped into a splice.

tape station See *tape deck.*

tape-to-head contact See *head-to-tape contact.*

tape transmitter A transmitter that receives its signal input for a recorded tape.

tape transport 1. In a tape deck or reproducer, that which moves the tape past the heads. 2. A similar mechanism where the medium is paper tape and the operation is punching or sensing.

tape unit 1. Tape deck. 2. Tape group.

tape verifier In computer practice, a device which, through comparison with an original document, checks the integrity of data on paper tape.

tape width Of magnetic tape, the dimension perpendicular to tape travel; greater tape width insures greater bandwidth in the recorded material.

tape-wound core See *tape core.*

tapped coil An inductor to which one or more intermediate connections (taps) are made at appropriate turns to provide intermediate values of inductance.

tapped component A coil, transformer, choke, resistor, or other component in which an intermediate connection is made. See, for example *tapped coil.*

tapped inductor See *tapped coil.*

tapped resistor A resistor in which one or more intermediate connections (taps) are made to appropriate parts of the resistance element to provide steps of resistance.

tapped transformer A transformer having one or more tapped windings.

tapped winding A transformer or choke winding with one or more taps. Also see *tap* and *tapped coil.*

tap switch A multiposition switch employed to connect an external circuit to various taps on a component. Also see *selector switch* and *tap.*

target 1. The bombarded electrode in an X-ray tube. 2. The scanned storage element in a TV camera tube. 3. In radar practice, the scanned object. 4. An object intended for nuclear-particle bombardment. 5. Goal (deadline, desired number of production units, etc.).

target acquisition 1. The moment at which a target comes within the range of a radar system. 2. The observation of a new target on a radar screen.

target discrimination The extent to which a radar system can distinguish between two targets that are close together. Also called target resolution. Specified in linear units such as meters, kilometers, or miles.

target identification Any method by which the identity of a target is determined.

target voltage In a TV camera tube, the backplate-to-cathode voltage.

taut-band meter A movable-coil meter in which the conventional spiral springs of the coil are replaced by two tightly stretched, thin, straight metal ribbons whose twist provides torque that returns the pointer to zero after a deflection.

Taylor modulation Supermodulation, so called from its inventor, Robert E. Taylor.

Tb Symbol for *terbium.*

TBS Abbreviation of *talk between ships.*

Tc Symbol for *technetium.*

TCCO Abbreviation of *temperature-controlled crystal oscillator.*

T-circuit See *tee-network.*

T-circuit parameters See *r-parameters.*

T-circulator See *tee-circulator.*

TCL Abbreviation of *transistor-coupled logic.*

TCM Abbreviation of *thermocouple meter* (see *thermocouple-type meter*).

TDM Abbreviation of *time-division multiplex*.

TDR 1. Abbreviation of *time-delay relay* (see *delay relay*). 2. Abbreviation of *time-domain reflectometry*.

TE 1. Abbreviation of *transverse electric* (see, for example, *transverse electrode mode*). 2. Abbreviation of *trailing edge* (of an airfoil).

Te Symbol for *tellurium*.

tearing In a TV picture, the abnormal condition in which poor synchronization causes the horizontal lines to be irregularly displaced, the effect resembling cloth being torn.

technetium Symbol, Tc. A metallic element produced artificially. Atomic number, 43. Atomic weight, 99. Formerly called *masurium*.

technician 1. A person who repairs faulty electronic equipment. 2. A person who assists with the design and debugging of a system prototype. 3. A person who operates an electronic system.

Teflon FEP A plastic insulating material. Dielectric constant, 2.1. Dielectric strength, 2800 V/mil.

Teflon TFE A plastic insulating material. Dielectric constant, 2.2. Dielectric strength, 600 V/mil.

tel 1. Abbreviation of *telephone*. 2. Abbreviation of *telegraph*. 3. Abbreviation of *telegram*.

tee-adapter See *tee-junction*.

tee-antenna An antenna consisting of a horizontal radiator with a vertical lead-in or feeder connected to its center point; it resembles a tee.

tee-circuit See *tee-network*.

tee-circulator In microwave systems, a tee-shaped junction of three waveguides with a ferrite post at the junction.

tee-equivalent circuit An equivalent circuit in which the components are arranged in the form of a tee. See, for example, *tee-network* and *r-parameters*.

tee-junction 1. A tee-shaped splice between two wires. 2. A tee-shaped fixture for splicing one coaxial line perpendicularly to another. 3. A tee-shaped section for joining one waveguide perpendicularly to another. Also called *waveguide tee*.

tee-network A three-terminal network resembling a tee.

tee-pad A three-resistor pad in which two series resistors and a shunt resistor are arranged to form a tee.

telautograph A device which transmits and receives handwriting, drawings, and similar material. At the receiver, a pen follows the movements of a similar pen at the transmitter.

teleammeter A telemeter for measuring current generated at a remote point.

telecamera See *television camera*.

telecast A television program for general reception. The term is an acronym for *television broadcast*.

telecommunication Communication, usually between widely separated points, by electrical or electronic means.

telecontrol See *remote control*.

telefacsimile See *facsimile*.

Telefunken alternator An early German dynamo-type machine for generating radio-frequency power. It is similar to the Alexanderson alternator, but employs an external transformerlike device to multiply the frequency two to four times, permitting lower speeds and fewer rotor teeth to be used in the rotating machine. The Telefunken alternator was used extensively in Germany and to some extent in the United States.

telegram Abbreviation, tel. A (usually printed-out) message transmitted and received via telegraph or teletypewriter. Compare *cablegram* and *radiogram*.

telegraph Abbreviation, tel. An instrument for transmitting and receiving messages by means of telegraphy. In its simplest form, it consists of a key and a sounder powered by a battery. Also see *printing telegraph, 1, 2*.

telegraph channel 1. The frequency band assigned to a particular telegraph station. 2. The frequency band occupied by a telegraph signal.

telegraph code Broadly, the Morse code. Wire telegraphy often employs a special version, such as the American Morse code.

telegraph key See *key, 1*.

telegraph sounder See *sounder*.

telegraph system A complete, integrated, and coordinated arrangement of equipment for communication by means of telegraphy. Included are telegraph keys, sounders or printers, relays, switchboards, wire lines and cables, and power supplies.

telegraphy The branch of electrical communications that deals with the transmission and reception of messages by means of prearranged codes, especially over wires. Also see *Morse, 1, 2, 3; Morse code;* and *wire telegraphy*.

telemeter 1. An indicating instrument that measures the value of a quantity generated at a distant point or measures and transmits the value. 2. The action afforded by the device in 1, above.

telemetering See *telemetry*.

telemeter receiver See *telemetric receiver*.

telemeter transmitter See *telemetric transmitter*.

telemetric receiver The complete device which selects, amplifiers, and demodulates or rectifies a radio signal and actuates indicating instruments, recorders, or data processors.

telemetric transmitter The complete device which generates rf power, adds to it signals delivered by data transducers, and delivers the modulated power to an antenna for transmission to a distant telemetric receiver. Here, *telemetric transmitter* is distinguished from a radio transmitter per se.

telemetry The transmission of data signals over a distance, either by radio or wire, and the reception and application of the signals to indicating instruments, recorders, etc.

telephone Abbreviation, tel. An instrument for transmitting and receiving messages by means of telephony. In its simplest form, it consists of a microphone, earphone, switching and ringing devices, wire line or cable, and power supply. Also see *handset*.

telephone amplifier A small audio amplifier, usually with self-contained loudspeaker, for increasing the sound volume of a telephone output. Some amplifiers of this kind are connected to the telephone line, and others pick up sound from the telephone receiver.

telephone-answering machine See *answering machine*.

telephone bypass capacitor A capacitor installed across the telephone lines, for the purpose of bypassing interfering signals.

telephone data set A device that converts signals from a data terminal for passage over a telephone circuit to a data processing center.

telephone induction coil A small telephone-to-line impedance-matching transformer used in telephone systems.

telephone patch See *phone patch*.

telephone pickup A device for picking up conversations from a telephone to which it isn't directly connected.

telephone plug See *phone plug*.

telephone-radio patch See *phone patch*.

telephone receiver See *receiver, 2*.

telephone repeater An amplifier and associated equipment employed to boost the amplitude of a telephone signal at an appropriate point along the line.

telephone silencer A device for muting a telephone or its bell.

telephone system A complete, integrated, and coordinated arrangement of equipment for communication by means of telephony. Included are telephones, switchboards and associated equipment, wire lines and cables, and power supplies. Also see *dial telephone system, handset, private automatic exchange (PAX), private branch exchange (PBX),* and *telephone.*

telephone test set See *phone test set.*

telephone transmitter The sound pickup unit (microphone) of a telephone. Also see *transmitter, 2.*

telephony The branch of electrical communication dealing with the transmission and reception of sounds, especially over wires. Also see *wire telephony.*

Telephoto See *facsimile.*

telephoto lens A camera lens providing a telescopic effect.

Telephoto receiver See *facsimile receiver.*

Telephoto transmitter See *facsimile transmitter.*

teleprinter A terminal telegraph printing equipment. Also see *printing telegraph, 1, 2.*

teleprocessing 1. A method of processing data communications signals. 2. A trademark of International Business Machines (IBM).

TelePrompTer A device that presents a running display on a screen before a television announcer, performer, or speaker, of dialogue.

teleran A ground-to-air communications system. Ground-based radar pictures are transmitted, via television, to aircraft.

telescope See *radio astronomy* and *radio telescope.*

telescopic Providing magnified images of distant scenes.

telescoping antenna A vertical antenna made up of separate lengths of metal tubing of progressively smaller diameter, so that one can slide into another. The antenna may be pulled out to full length or compressed to the length of the largest-diameter section.

telethermoscope A device for measuring the temperature of distant objects.

Teletorque See *Autosyn* and *synchro.*

Teletype A form of teletypewriter.

teletype grade A term descriptive of a circuit having the quality necessary for communication via telegraphy.

Telescoping antenna.

Teletypesetter An electronic system for operating a distant Linotype.

teletypewriter A variety of printing telegraph employing electric typewriters and associated equipment. The message is typed on the keyboard at the transmitting station and is typed out in corresponding letters at the receiving station. The same typewriter is able to send and receive messages. Also see *radioteletypewriter* and *Teletype.*

teletypewriter exchange Abbreviation, TWX. A center for switching and routing teletypewriter communications. Also see *teletypewriter.*

teleview To see by means of television.

televise 1. To convert a scene into a TV signal. 2. To broadcast by television.

television Abbreviation, TV. The transmission and reception, through the medium of radio or cable, of images with or without sound.

television band See *uhf televison channels* and *vhf television channels.*

television camera The pickup device which scans a scene and delivers a series of electrical signals that may be employed to reconstruct the image on the screen of a picture tube.

television-camera tube See *camera tube, iconoscope,* and *orthicon.*

television channel A radio-frequency band allocated exclusively for television. See, specifically, *uhf television channels* and *vhf television channels.*

television engineer A trained professional skilled in television engineering as well as in basic engineering and allied subjects.

television engineering The branch of electronic engineering devoted to the theory and application of television.

television interference Abbreviation, TVI. Interference to the reception of television signals, usually occasioned by interferential signals from radio services or by electrical noise.

televisor 1. Television transmitter or receiver. 2. Television broadcaster.

televoltmeter A telemeter for measuring voltage generated at a remote point.

telewriter See *telautograph.*

telex 1. A teleprinter system that operates via the telephone lines, and is extensively used by businesses for sending and receiving short messages. 2. A hard-copy message sent or received by such a system.

telluric currents Earth (terrestrial) currents.

tellurium Symbol, Te. A rare, metalloidal element related to selenium. Atomic number, 52. Atomic weight, 127.61.

Telstar An active NASA satellite in continuous orbit approximately 5000 kilometers above the earth. Also see *active communications satellite.*

TE mode See *transverse electric mode.*

temp 1. Abbreviation of *temperature.* (Also, *T.*) 2. Abbreviation of *temporary.* 3. Abbreviation of *template.*

temperature Symbol, *t.* The degree of heat or cold exhibited by an object or phenomenon. Also see *thermometer scale.*

temperature coefficient A figure which states the extent to which a quantity drifts under the influence of temperature. It is generally expressed in percent per degree (%/ °C) or in parts per million per degree (ppm/ °C).

temperature-compensated crystal oscillation Oscillation by a crystal oscillator in which the crystal and/or circuit is automatically compensated against frequency drift caused by temperature change.

temperature-compensating component A circuit component, such as a capacitor or resistor, whose temperature coefficient is equal in magnitude and opposite in sign to that of a conventional component to which it is connected to cancel temperature-induced variation in the latter's value.

temperature compensation The use of a correcting device, such as a temperature-compensating component, to correct a temperature-induced deviation in the value of a conventional component.

temperature control 1. The adjustment of temperature. 2. The automatic maintenance of temperature at a desired level, as in a temperature-controlled oven. 3. A device for controlling temperature, as in 1 and 2, above.

temperature-controlled crystal oscillator Abbreviation, TCCO. A high-precision crystal oscillator in which the crystal plate (and sometimes the circuitry, as well) is held at constant temperature.

temperature degree See *degree, 2* and *thermometer scale.*

temperature derating Lowering the operating current, voltage, or both, of a device to a critical temperature. Also see *derating, derating curve,* and *derating factor.*

temperature gradient A range of temperature variation, such as the rate of change of temperature with respect to change of power dissipation, or the rate of change of temperature with spatial displacement.

temperature inversion See *inversion, 1.*

temperature meter An indicator (usually a dc voltmeter or millivoltmeter) whose scale is direct reading in degrees. Also the complete instrument, including the indicator.

temperature scale See *thermometer scale.*

temperature-sensitive resistor See *thermistor.*

temperature shock See *thermal shock.*

temperature-to-voltage converter A circuit or device, such as a thermocouple, that delivers an output voltage which is proportional to an applied temperature.

template 1. A full-scale, usually drawn on paper, to show the location, for example, of holes to be drilled on a chassis or panel. The template is taped or cemented temporarily to the work, and the points are transferred to the latter by prick-punching through the template. 2. A stencil-like plate with alphanumeric and circuit symbols, used as a drafting aid. Sometimes called "drafting stencil."

temporary magnet 1. A body which exhibits magnetism only briefly after it has been exposed to another magnet. Compare *permanent magnet.* 2. Electromagnet.

temporary storage 1. In computer and data-processing practice, the storage of data or instructions only until they are needed. Also called *interim storage.* 2. Locations in a computer memory set aside during a program run for holding intermediate results of operations.

TEM wave See *transverse electromagnetic wave.*

ten code A set of abbreviations used by two-way radio operators. Each "ten signal" represents a specific statement.

tens complement The number that results when nine is subtracted from each digit of a decimal number and one is added to the digit in the rightmost position of the result, e.g., the tens complement of 365 is 635 (999 − 365 + 1 = 635). Compare *radix complement.*

tension Voltage (electromotive force).

ten-turn potentiometer A precision potentiometer whose shaft must be turned through 10 complete revolutions to cover the entire resistance range. Also see *helical potentiometer* and *multiturn potentiometer.*

T-equivalent circuit See *tee-equivalent circuit.*

tera Abbreviation, T. A prefix meaning trillion(s), i.e., 10^{12}.

tera-electronvolt Abbreviation TeV. A large unit of energy; 1 TeV = 10^{12} electron volts. Also see *electronvolt.*

terahertz Abbreviation THz. A unit of high frequency; 1THz = 10^{12} Hz.

teraohm Symbol, TΩ A unit of high resistance, reactance, or impedance; 1 TΩ ^{5} 10^{12} Ω.

terawatt Abbreviation, TW. A large unit of power; 1 TW = 10^{12} W.

terbium Symbol, Tb. A metallic element of the rare-earth group. Atomic number, 65. Atomic weight, 158.93.

terbium metals A group of rare-earth metals including europium, gadolinium, terbium itself, and occasionally dysprosium.

term In an algebraic expression constants, variables, or combinations of these, separated by operation signs, e.g., the expression $4ab + c$ has two terms.

terminal 1. A connection point at the input, output, or an intermediate point of a device, or a point at which a voltage is to be applied. 2. A metal tab or lug attached to the end of a lead for connection purposes. 3. Pertaining to the end of a series of events, etc., e.g., terminal tests. 4. In a system where data communications takes place, a point of data input or output. Also called *data terminal.*

terminal block A block with several terminals, intended for interconnection of circuits.

terminal board An insulating board carrying several lugs, tabs, or screws as terminals (see *terminal, 2*). Also see *terminal strip.*

terminal guidance The guidance given to a missile or aircraft to help it reach its target or destination.

terminal impedance The internal impedance of a device measured at the input or output terminals.

terminal point of degradation The point at which degradation of a circuit or component is complete. Also see *degradation failure.*

terminal repeater A telephone repeater operated at the end of a line.

terminal strip A strip of insulating material, such as plastic or ceramic, on which are mounted one or more screw, lugs, or other terminals. Also see *terminal, 2.*

Terminal strip (lug-type).

terminal voltage The voltage at the output terminals of an unloaded battery or generator.

terminal symbol On punched paper tape, a symbol showing that the end of an information unit, such as a record, has been reached.

ternary code See *trinary number system.*

ternary fission The splitting of an atomic nucleus into three nuclear pieces. Also see *fission.*

ternary number system See *trinary number system.*

terrain echoes Radar clutter caused by ground reflections. Also called *ground clutter.*

terrestrial magnetism See *earth's magnetic field.*

Terrier A 10-mile-range, surface-to-air, beam-rider-guided missile used by the U.S. Navy.

tertiary coil A third winding (see *tertiary winding*).

tertiary winding A third winding on a transformer or magnetic amplifier.

tesla Symbol, T. A unit of magnetic flux density; 1 T = 1 weber per square meter = 10^4 gauss.

Tesla coil A special type of air-core step-up transformer for developing high voltage at radio frequencies. It consists essentially of a low-turn primary coil through which radio-frequency current flows, and a multiturn secondary coil across which the high voltage is developed. In the original Tesla coil, a power transformer and spark gap provided primary-coil rf current.

test A procedure consisting of one or several steps, in which (1) the mode of operation of a circuit or device is established, (2) the value of a component is ascertained, or (3) the behavior of a circuit or device is observed.

test bench An equipment installation intended for the testing, repair, or debugging of electronic devices by a technician.

test data Data used to test a computer program, including samples lying within limits that might be encountered during the program's implementation.

tester 1. Test instrument. 2. A technician who primarily makes tests and measurements.

testing window See *window, 2.*

test instrument An instrument, often a meter, for checking the operation of a circuit or device or the value of a circuit component. This class of instrument is usually less accuracy than measurement instruments. Also see *test set.*

test lead The flexible, insulated wire attached to a test prod.

test pack For a computer program test run, a pack of punched cards carrying test and program data.

test pattern A picture-and-line type of display on the screen of a TV picture tube, used to check such features as aspect ratio, linearity, contrast, etc.

test point A terminal intended for connection of test equipment in the repair or debugging of a circuit. Often, test points are labeled by the letters *TP* followed by a numeral, such as *TP1, TP2,* and so on.

test probe See *probe, 1.*

test prod A stick-type probe (see *probe, 1*) with a flexible, insulated lead terminating in a plug or lug for attachment to an instrument.

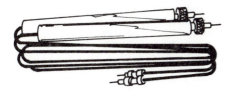

Test prods.

test program In computer practice, a program devised to check the functioning of hardware. Also called *test routine.*

test set A combination of instruments assembled on a single panel, and usually enclosed in a carrying case, for convenience in making tests.

test run In computer practice, using test data to test the operation of a program, by comparing the results obtained thereby with what should result.

test signal 1. A signal employed in making a test. 2. In radiotelegraphy, a special signal signifying that the transmitting station is testing equipment. Also see *vee-signal.*

test-signal generator A device, such as an oscillator, for producing a signal for testing equipment (see *test signal, 1*).

test tape A recorded magnetic tape containing signals for testing equilization, frequency response, head adjustment, stereo balance, etc. of a tape recorder.

tetravalent See *quadrivalent.*

tetrode An electron tube in which the principal electrodes are cathode, control grid, screen, and plate.

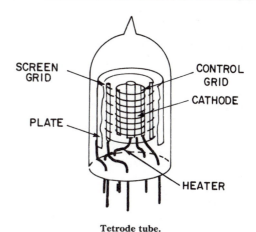

Tetrode tube.

TeV Abbreviation of *tera-electronvolt.*

Te value The temperature at which a centimeter cube of glass or ceramic exhibits 1 megohm of resistance.

TE wave See *transverse electric mode.*

text editor A computer program for finding and changing data in a file.

tgt Abbreviation of *target.*

TGTP Abbreviation of *tuned-grid—tuned-plate.*

TH Abbreviation of *true heading.*

Th Symbol for *theorium.*

Th Symbol for *heater temperature.*

thallium Symbol, Tl. A metallic element. Atomic number, 81. Atomic weight, 204.39.

thalofide cell An evacuated photoconductive cell employing thallium oxysulfide as the light-sensitive material.

THD Abbreviation of *total harmonic distortion.*

theory A reasonable proposition put forth to account for the behavior of, or the relationships between, bodies and forces, or to explain concepts and their relations. When a theory has stood up under exhaustive tests, it may reveal a scientific law (see *law, 1*).

theory of relativity See *Einstein's theory.*

theremin An early electronic musical instrument consisting essentially of a beat-frequency audio oscillator followed by an audio amplifier and loudspeaker. The instrument is played by placing the hands in certain positions with respect to a rod antenna connected to one of the two rf oscillators in the beat-frequency af oscillator. Body capacitance of the musician shifts the frequency of antenna-connected oscillator and, accordingly, the audio frequency.

therm A gas heating unit. 1 therm = 100,000 *British thermal units (Btu).*

thermal agitation Random movement of particles (such as electrons) in a substance, due to heat.

thermal-agitation noise See *thermal noise.*

thermal ammeter See *hot-wire ammeter.*

thermal anemometer See *hot-wire anemometer.*

thermal conductivity The heat-conducting ability of a material. Compare *electrical conductivity* (see *conductivity*).

thermal-conductivity device An instrument or control unit employing a heated filament whose temperature and, accordingly, conductivity is varied by some sensed phenomenon. See, for example, *gas* *detector, heated-wire sensor, hot-wire anemometer, hot-wire flowmeter,* and *hot-wire microphone.*

thermal-conductivity gasmeter See *gas detector.*

thermal detector See *bolometer.*

thermal emf See *Seebeck effect.*

thermal gasmeter See *gas detector.*

thermal instrument See *thermocouple-type meter* and *hot-wire meter.*

thermally sensitive resistor See *thermistor.*

thermal meter See *hot-wire meter* and *thermocouple-type meter.*

thermal microphone See *flame microphone* and *hot-wire microphone.*

thermal noise Frequency-independent electrical noise due to the agitation of particles (e.g., atoms and electrons) in a material by heat. Thermal noise is proportional to bandwidth, resistance, and absolute temperature.

thermal radiation See *heat.*

thermal recorder A graphic recorder in which a strip of paper coated with a thin layer of opaque wax is drawn between a knife-edge platen and a heated writing stylus that melts the wax beneath its point, exposing the underlying black paper as a fine line.

thermal resistance Symbol, RT or ϕ. For a semiconductor device, the rate of change of junction temperature with respect to power dissipation; $RT \approx \Delta T/\Delta P$, where RT is in degrees per milliwatt, T is in degrees Celsius, and P is in milliwatts.

thermal resistor A resistor that is sufficiently temperature-sensitive to be employed as a heat sensor, e.g., thermistor, germanium diode.

thermal response time For a power-dissipating component, the elapsed time between the initial change in power dissipation and the moment at which the temperature has changed by a specified percentage (usually 90 percent) of the total value.

thermal runaway Runaway caused by cumulative temperature effects. A common case in transistors results from the increase in collector current with temperature.

thermal shock The effect of applying heat or cold to a device so rapidly that abnormal reactions occur, such as rapid (often catastrophic) expansions and contractions.

thermal switch A switch actuated by a temperature change. Types vary from the simple thermostat to complex servosystem switches with a temperature-transducer input.

thermal time-delay relay A delay relay utilizing the slow-heating and slow-cooling property of one of its components.

thermion An ion or electron emitted by a hot body, such as the heated cathode of a vacuum tube.

thermionic Pertaining to thermions and their applications.

thermionic cathode A heated cathode, in contrast to a *cold cathode,* used as an emitter of electrons or ions. Also see *thermion.*

thermionic current Current due to thermionic emission, especially in an electron tube.

thermionic detector A vacuum-tube detector. Also see *thermion* and *thermionic emission.*

thermionic diode A hot-cathode diode tube.

thermionic emission The emission of electrons by a hot body, such as the filament or cathode of a vacuum tube. Also see *thermion* and *hot cathode.*

thermionic tube An electron tube, i.e., one in which electrons or ions are emitted by a heated cathode. Also see *thermion, thermionic current,* and *thermionic emission.*

thermionic work function The energy required to force an electron from inside a heated cathode into the surrounding space (in *thermionic*

emission). Also see *work function.*

thermistor A temperature-sensitive resistor, usually made from specially processed oxides of cobalt, magnesium, manganese, nickel, uranium, or mixtures of such substances. Thermistors are available with either a positive or negative temperature coefficient of resistance. The name is an acronym for *ther*mally sensitive res*istor.*

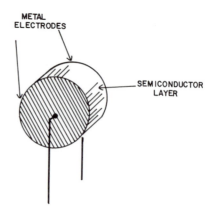

Thermistor.

thermistor bridge A four-arm bridge (see *bridge, 1*) in which one arm is a thermistor and, therefore, temperature sensitive.

thermistor power meter A radio-frequency power-measuring instrument based upon a thermistor bridge.

thermistor probe 1. Probe thermistor. 2. A temperature probe containing a thermistor as the sensing element.

thermistor thermometer An electronic thermometer in which the temperature-sensitive element is a thermistor.

thermoammeter See *thermocouple-type meter.*

thermocouple A device consisting essentially of an intimate bond between two wires or strips of dissimilar metals (such as antimony and bismuth). When the bond is heated, a dc voltage appears across it.

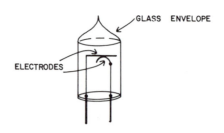

Thermocouple.

thermocouple bridge A four-arm bridge (see *bridge, 1*) in which one arm is a thermocouple and, therefore, temperature sensitive as well as being voltage-productive.

thermocouple meter See *thermocouple-type meter.*

thermocouple-type meter A radio-frequency meter consisting of a thermocouple and dc ammeter, milliammeter, or microammeter, connected in series. The thermocouple is heated, directly or indirectly, by an applied rf current, and the resulting dc output deflects the meter.

thermodynamics The science dealing with the relationships between heat and mechanical energy and their interconversion. Also see *Carnot theorem, First Law of Thermodynamics, Second Law of Thermodynamics,* and *Third Law of Thermodynamics.*

thermoelectric cooler A cooling device based upon the Peltier effect.

thermoelectric effect The production of thermoelectricity by certain materials.

thermoelectricity Heat-produced electricity, as in thermocouple operation.

thermoelectric junction A junction between two conductors that exhibits variable characteristics under conditions of changing temperature.

thermoelectron An electron emitted by a hot cathode. Also see *thermion* and *thermionic emission.*

thermoelement A thermocouple, especially a miniature one, used in a *thermocouple-type meter.*

thermogalvanometer See *thermocouple-type meter.*

thermoluminescence Luminescence resulting from the moderate heating of certain materials.

thermomagnetic effect The effects of temperature on the magnetism of a body, or vice versa.

thermometer A (usually direct reading) device for measuring temperature. Also see *electronic thermometer* and *thermometer scale.*

thermometer scale The scale on a thermometer, graduated in degrees, from which temperature is read. For a description of different scales, see *absolute scale, centigrade scale, Fahrenheit scale, Reaumur scale,* and *Rankine scale.* The Kelvin scale is the same as the absolute scale, and the Celsuis scale is the same as the centigrade scale.

thermonuclear reaction A nuclear reaction in which energy is released when lighter atoms are converted into heavier atoms at temperatures in the millions of degrees. Also see *fission, fusion,* and *nuclear reactor.*

thermopile A device comprising two or more thermocouples connected in series (for higher voltage output).

thermoplastic material A plastic which, after having been molded into a desired shape, can be resoftened by applying heat, e.g., polystyrene. Compare *thermosetting material.*

thermorelay See *thermostat.*

thermos (bottle) See *Dewar flask.*

thermosensitivity Sensitivity of a circuit or device to heat.

thermosetting material A plastic which, after having been heat-molded into a desired shape, cures chemically (it will not ordinarily soften again when heat is applied). Example: Bakelite. Compare *thermoplastic material.*

thermostat A temperature-sensitive switch. In one common form, a movable contact is carried by a strip of bimetal and the stationary contact is mounted nearby.

thermostatic switch See *thermostat.*

thermoswitch See *thermostat.*

theta The eighth letter (Θ θ) of the Greek alphabet. Capital theta (Θ) is the symbol for *angles, image transfer constant, temperature,* and *phase angle.* Lower-case theta (θ) is the symbol for *angles, phase displacement,* and *thermal resistance.*

theta wave A form of brain wave that occurs at extremely low frequencies, and is associated with sleep or mental incoherence.

Thevenin's theorem The proposition that, with reference to a particular set of terminals, a network containing a number of generators and constant impedances can be simplified to a single generator in series with a single impedance. The equivalent circuit will deliver to a given load, the same current, voltage, and power delivered by the original network. Compare *compensation theorem, maximum power transfer theorem, Norton's theorem, reciprocity theorem,* and *superposition theorem.*

thick film A film of selected material (conductive, resistive, dielectric, etc.) applied to a substrate by painting, photography, o. similar process. See, for example, *printed circuit.* Compare *thin film.* Typically, thick films are a mil or more in thickness.

thick-film component A unit, such as a capacitor or resistor, fabricated by thick-film techniques. See, for example, *printed component.* Also see *thick film.* Compare *thin-film component.*

thick-film resistor A resistor fabricated by thick-film techniques. See, for example, *printed resistor.* Also see *thick film.* Compare *thin-film component.*

thick magnetic film See *magnetic thick film.*

thin film An extremely thin layer (1 μm) of a selected material (conductive, resistive, semiconductive, dielectric, etc.) electrodeposited or grown on a substrate. Compare *thick film.*

thin-film capacitor A capacitor made by electrodepositing a thin film of metal on each side of a grown layer of oxide, as in an integrated circuit. Also see *thin film.*

thin-film component A unit—such as a capacitor, resistor, diode, or transistor—fabricated by thin-film techniques. Also see *thin film.* Compare *thick-film component.*

thin-film integrated circuit An integrated circuit in which the components and "wiring" are produced by depositing (or growing) and processing materials on a semiconductor slab or wafer (the *substrate*). Compare *hybrid integrated circuit* and *monolithic integrated circuit.*

thin-film memory In a computer, a storage medium that is a magnetic thin film (see *thin film*) on a nonmagnetic substrate (often glass) and which can be magnetized to represent digital data.

thin-film microelectronic circuit An integrated circuit that occupies (essentially) two dimensions; that is, a very thin integrated circuit.

thin-film resistor A resistor fabricated by thin-film techniques (see *thin film*), e.g., *tantalum-nitride resistor.* Compare *thick-film resistor.*

thin-film semiconductor A thin film of semiconductor material, such as single-crystal silicon, electrodeposited on a substrate. Also see *thin film.*

thin-film transistor A transistor fabricated by thin-film techniques. Also see *thin film* and *thin-film component.*

thin magnetic film See *magnetic thin film*

third dimension Depth, i.e., the dimension, needed in addition to height and width to find an object's volume and corresponding to the Z-coordinate of a graph.

third-degree equation See *cubic equation.*

third-generation computer A computer in which the active discrete components are ICs.

Third Law of Thermodynamics As the temperature of absolute zero is approached in an isothermal process involving a solid or liquid, the change in entropy approaches zero; and the entropy of a substance is zero at absolute zero.

third-octave band See *one-third-octave band.*

Thomas Phonetic alphabet communications code word for the letter *T.*

Thomson bridge See *Kelvin bridge.*

Thomson coefficient The ratio $\Delta E/\Delta T$, where ΔE is the voltage difference between two points on a current-carrying metallic conductor, and ΔT is the difference in temperature between the same points.

Thomson effect The liberation or absorption of heat (depending upon the material of interest) when an electric current flows from a warmer to a cooler part of a conductor.

Thomson heat The amount of thermal energy transferred because of Thomson effect.

Thomson voltage The voltage drop between two points on a conductor that are at different temperatures.

thoriated-tungsten filament In a vacuum tube, a filament made of tungsten to which thorium oxide has been added to increase the emission of electrons. Also see *thermion* and *thermionic emission.*

thorium Symbol, Th. A radioactive metallic element. Atomic number, 90. Atomic weight, 232.05. Thorium, when heated, is a copious emitter of electrons, so the filaments or cathode cylinders of some electron tubes are coated with one of its compounds.

thoron Symbol, Tn. A radioactive isotope (see *radioisotope*) of radon. Atomic number, 86. Atomic weight, 220.

three-address instruction A computer program instruction having three addresses, two for operands and one for the result (of the operation called for).

three-channel stereo A form of stereo in which three distinct channels are used.

three-coil meter See *electrodynamometer.*

three-conductor jack A female connector in which two separate conductors are provided, in addition to the ground conductor.

three-conductor plug A male connector in which two separate conductors are provided, in addition to the ground conductor.

three-dimensional Having three dimensions (height, width, depth) or the appearance of this quality.

three-dimensional television See *stereoscopic television.*

three-electrode tube See *triode.*

three-element tube An electron tube having three principal electrodes, i.e., a *triode.*

three-gun picture tube A color-TV picture tube having a separate gun for each primary color (red, green, and blue).

three-halves-power law See *Child's law.*

three-junction transistor A *pnpn* or *npnp* transistor. Also see *npnp device.*

three-phase bridge rectifier A bridge-emitter circuit for three-phase voltage. Two diodes are provided for each phase. The ripple frequency is six times the line frequency. Also see *bridge rectifier, polyphase rectifier,* and *three-phase circuit.*

three-phase circuit The circuit of a three-phase system. See, *three-phase system* and, specifically, *delta connection* and *wye connection.*

three-phase current Current in a three-phase circuit. The currents in the three legs differ in phase by 120 degrees. When the currents (I_p) in each phase are equal, line current I_L equals $\sqrt{3}(I_p)$.

three-phase—four-wire system See *four-wire wye system.*

three-phase generator A (usually dynamo-type) generator of three-phase current or voltage. See *three-phase system.*

three-phase—half-wave rectifier A half-wave rectifier circuit for three-phase voltage. One diode is provided for each phase. The ripple frequency is triple the line frequency. Also see *half-wave rectifier, polyphase rectifier,* and *three-phase circuit.* Compare *three-phase bridge rectifier.*

three-phase motor An ac motor operating on three-phase power. Above fractional-horsepower size, the three-phase motor is smoother running and more simply structured than the single-phase counterpart.

three-phase power Power in a three-phase circuit. Total three-phase power P equals $\sqrt{e} \, (E_L I_L)$, where E_L is line voltage, and I_L line current.

three-phase rectifier A rectifier for three-phase voltage. At least one separate diode is included for each phase. Also see *polyphase rectifier, three-phase bridge rectifier, three-phase circuit,* and *three-phase—half-wave rectifier.*

three-phase system A polyphase ac system in which three currents or voltages are 120 degrees out of phase with each other.

three-phase, three-wire system A three-phase three-conductor system with a phase difference of 120 degrees between conductor pairs.

three-phase voltage Voltage in a three-phase circuit. The voltages across the three legs differ by 120 degrees. When the voltages (E_p) in each phase are equal, line voltage E_L equals $\sqrt{3}(E_p)$.

three-quarter bridge A bridge rectifier having diodes in three arms and a resistor in the fourth.

three-state logic See *tri-state logic.*

three-wire system 1. An electric-power feed system employing three wires, the center one (neutral) being at a potential midway between the potential across the other (outer) two. 2. Three-phase, three-wire system. 3. Two-phase, three-wire circuit.

threshold 1. The initial (observable) point of an effect, e.g., threshold of hearing. 2. A predetermined point, such as of minimum current or voltage, for the start of operation of a circuit or device.

threshold component A value of current, voltage, sound intensity, or the like, selected as the minimum level at which a circuit or device is to operate in some prescribed manner, or beyond which a certain condition will prevail. Also see *threshold, 1, 2.*

threshold current 1. The minimum value of current at which a certain effect takes place. 2. The smallest amount of forward current that flows through a diode. 3. In a gas, the smallest level of current for which a discharge will maintain itself under variable conditions.

threshold detector A device that prevents a signal from passing until its peak amplitude reaches a certain value.

threshold of hearing The minimum intensity level at which sounds are audible in a given situation.

threshold of pain The intensity level at which hearing a sound causes physical discomfort.

threshold signal The weakest signal that can be detected in a receiving system.

throat See *horn throat.*

throat microphone A small microphone that is operated in contact with the user's throat.

throttle The feedback control device in a regenerative detector or amplifier.

throughput In computer practice, a measure of productivity in terms of the quantity of data processed during a given period.

throw-out spiral On a phonograph disk, a lead-out groove.

thulium Symbol, Tm. A metallic element of the rare-earth group. Atomic number, 69. Atomic weight, 168.94.

thumbwheel potentiometer A potentiometer operated by means of a knurled knob (usually protruding perpendicularly through a panel) that is turned with the thumb.

thumbwheel switch A switch operated by means of a knurled knob (usually protruding perpendicularly through a panel) that is turned with the thumb.

thy Abbreviation of *thyratron.*

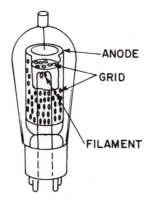

Thyratron.

thyratron A gas triode or gas tetrode employed principally for switching and control purposes. Thyratron action differs from that of the vacuum tube in the unique behavior of the thyratron control electrode (grid): Anode current starts to flow abruptly when grid voltage reaches a particular value at which point the grid provides no further control; anode current continues to flow until the anode voltage is either interrupted or reversed.

thyratron inverter An inverter circuit (see *inverter, 1*) employing thyratrons as the switching devices. Also see *thyratron*.

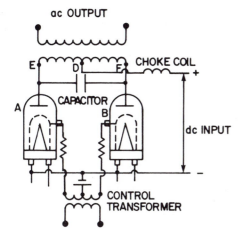

Thyratron inverter.

Thyrector A silicon diode exhibiting very high resistance (approaching that of an insulator) up to certain voltage, beyond which the unit switches to a low-resistance conducting state.

thyristor 1. A *pnpn*-type bistable semiconductor device having anode, cathode, and gate terminals and used as an electronic switch. 2. The generic term for all thyratronlike solid-state devices, such as the silicon controlled rectifier.

Thyrite Ceramic silicon carbide, a nonlinear resistance material, or a resistor made of this material. The resistance of Thyrite decreases sharply as the applied voltage is increased. Thyrite resistors are used in voltage regulators, equipment protectors, lightning arresters, curve changers, and similar devices.

THz Abbreviation of *terahertz*.

Ti Symbol for *titanium*.

tickler A (usually small) coil connected in series with the plate of a tube or the comparable element of a transistor, through which energy is inductively fed back from the output to the input of a circuit to induce oscillation.

tickler coil See *tickler*.

tickler-coil regeneration Regeneration obtained by means of inductive coupling between a small coil, (tickler) in the output circuit or an amplifier, and a (usually larger) coil in the input circuit. Also see *tickler*.

tickler oscillator An oscillator circuit in which positive feedback is obtained through inductive coupling between an output (tickler) coil and an input coil.

tie A bracket, clamp, clip, ring, or strip for holding several wires tightly as a cable or bundle.

Tie.

tie cable 1. A cable that connects two distributing points in a telephone system. 2. Any cable that interconnects two circuits.

tie point A lug, screw, or other terminal to which wires are connected at a junction.

tie-point strip A terminal strip with lugs to which conductors can be soldered.

tight coupling See *close coupling*.

tikker An early radiotelegraph detector consisting of a stationary wire or blade in light contact with a motor-driven metal disk rotating at high speed. The received rf signal passes through the tikker and headphones, connected in series, and the make-and-break action of the chattering contact chops the signal and makes it audible in the headphones.

tilt switch A device, such as a mercury switch, which is actuated by tilting it to a certain angle.

timbre The quality which distinguishes the sound of one voice or instrument from that of another, and which is due largely to harmonic content.

time Symbol, *t*. 1. The instant at which an event occurs. 2. The instant at which a time-base variable reaches a given value. 3. The interval between two instants—commonly called *duration* or *length of time*. Also see *hour; second, 1; minute; standard time; time base; time zone;* and *zero time*.

time base Time as the independent variable in a relationship. Time as a base appears in such expressions as *pulses per second, feet per minute, watts per hour,* etc.

time constant See *electrical time constant* and *mechanical time constant.*

time delay See *delay, 1, 2.*

time-delay relay See *delay relay.*

time-division multiplexing Abbreviation, TDM. In computer practice, a time-sharing technique in which several terminals use the same channel by transmitting data at regular, staggered intervals, i.e., one is active while the others idle, but the effect is apparent simultaneous operation.

time-domain reflectometry Abbreviation, TDR. Measuring the reflective characteristics of a device of system by superimposing upon a time-calibrated oscilloscope screen the direct and reflected components of a step-formed test signal.

time duration See *time, 3.*

time-duration modulation See *pulse-duration modulation.*

time factor The ratio *ta/tr*, where *ta* is analog time (the relativistic duration of an event as simulated by a computer), and *tr* is real time (the actual duration of the event). Also called *time scale.*

time-interval mode In computer practice, operation which allows a number of events to be counted between two points on a waveform.

time modulation Any form of modulation in which the instantaneous characteristics of a signal are varied.

timeout The expiration of an alloted time period for a given operation.

timer 1. A device for automatically controlling the duration of an operation. See particularly *electronic timer.* 2. A device for measuring the duration of an operation.

time sharing In computer practice, method for interlaced (i.e., nearly simultaneous) use of a machine or facility by two or more persons or agencies. As the cost of computers diminishes, so does the need for time sharing. Also see *time-division multiplexing.*

time-sharing dynamic allocator In storage in a computer, monitor program that allocates memory areas and peripherals to program being entered into a time sharing system and controls their execution.

time signals Special radio transmissions made under the auspices of the National Bureau of Standards, for indicating *coordinated universal time.*

time zone One of the 24 zones into which the global map is divided for the purpose of standardizing time. Within these zones, mean solar time is determined in terms of distance east or west of the zero meridian at Greenwich (near London, England). Each zone is equal to 15 degrees of longitude, or 1 hour. Four zones fall within the continental United States: Eastern standard time (zone of the 75th meridian), Central standard time (zone of the 90th meridian), Mountain standard time (zone of the 105th meridian), and Pacific standard time (zone of the 120th meridian). Also see *meridian, zero meridian,* and *zone time.*

timing error During the structuring of a computer program, an error that leads to a delay caused by misjudging the duration of an I/O operation, for example.

tin 1. Symbol, Sn. A metallic element. Atomic number, 50. Atomic weight, 118.70. Tin is widely used in electronics as a structural material, a constituent of solder, and (in foil form) as the plates of some fixed capacitors. 2. To prepare the tip of a soldering gun or iron by applying a coat of solder.

tin-oxide resistor See *metal-oxide resistor.*

tinsel Metal film strip interwoven with fabric threads to provide a flexible cord, particularly for headphones.

tint control In a color-TV receiver, the control for changing color hue.

tip jack The mating connector for a tip plug.

tip plug 1. A prod terminating in a phone tip. 2. A plug-type connector terminating in a phone tip.

Tip plug, 2.

titanium Symbol, Ti. A metallic element. Atomic number, 22. Atomic weight, 47.90. Titanium enters into some dielectric compounds, e.g., titanium dioxide.

titanium dioxide Formula, Ti02. A ceramic dielectric material. Dielectric constant, 90 to 170. Dielectric strength, 100 to 210 V/mil.

T-junction See *tee-junction.*

T junction.

Tl Symbol for *thallium.*

T²L Abbreviation of *transistor-transistor logic.* (Also, *TTL.*)

TLA Abbreviation of *table look-at.*

TLC Abbreviation of *thin-layer chromatography.*

TLU Abbreviation of *table look-up.*

TM 1. Abbreviation of *transverse magnetic.* 2. Abbreviation of *technical manual.*

Tm Symbol for *thulium.*

TM mode See *transverse magnetic mode.*

TM wave See *transverse magnetic mode.*

T-network See *tee-network.*

TNS Abbreviation of *transcutaneous nerve stimulator.*

toggle A bistable device.

toggle switch A switch having a mechanism that snaps into the on or off position at the opposite extremes to which a lever is moved.

tolerance The amount by which error is allowed in a value, rating, dimension, etc. It is usually expressed as a percent of nominal value, plus and minus so many units of measurement, or parts per million.

toll call In a telephone system, a call that is charged on a per-minute or per-second basis.

ton Abbreviation, T, t, or tn. A unit of avoirdupois weight; in the United States, *ton* is usually taken to mean *short ton*, a unit equal to 907.20 kilograms (2000 lb). Compare *metric ton.*

Toggle switch.

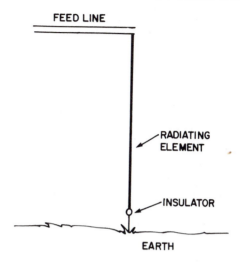

Top loading.

Toroid.

tone The pitch and timbre of a sound other than noise.

tone arm See *pickup arm*.

tone burst A test signal consisting of a single-frequency sine wave sustained for 50 to 500 microseconds, the packet having a rectangular envelope.

tone-burst entry In repeater systems, a technique whereby a short tone signal is employed at the start of a transmission to trigger a particular repeater, so that all repeaters in the system will not go into operation simultaneously.

tone-burst generator An oscillator and associated circuitry for producing a *tone burst*.

tone control An adjustable device or circuit for modifying the frequency response of an amplifier, i.e., for emphasizing bass, treble, or midrange pitches.

tone generator An oscillator for producing simple audio-frequency signals for communications, control, or testing.

tone keying In wire and radio telegraphy, the representation of code characters by af tones. Also see *modulated continuous wave*.

tone localizer A localizer which provides lateral guidance for an aircraft by comparing the amplitudes of two modulating frequencies.

tone modulation 1. The transmission of Morse, Baudot, or ASCII signals by audio-frequency modulation of a radio-frequency carrier. 2. Any rapid variation of the amplitude or frequency of an audio tone.

top cap A small metal cap on top of some electron tubes, serving as a direct, low-capacitance connection to one of the internal element, usually the control grid but sometimes the plate.

top loading A method of feeding a vertical antenna at or near the top.

topology 1. A branch of mathematics concerned with the properties of surfaces and spaces. 2. The details of layout of an integrated circuit. 3. The characteristics of a surface.

toroid See *toroidal coil*.

toroidal coil A doughnut-shaped coil.

torque 1. The force that tends to produce a rotating motion. 2. Rotation of the plane of polarization of light by some crystals.

torque amplifier A device having rotating input and output shafts and which delivers greater torque at the output shaft than that required to turn the input shaft.

torque sensitivity Symbol, *KT*. For a torque motor, torque output per ampere of input current.

torque-to-inertia ratio See *acceleration at stall*.

torr A unit of pressure equal to the pressure required to support a column of mercury 1 millimeter high at 0 °C and standard gravity; 1 torr = 133.322 Pa (1 333.2 microbars).

torsion The effect on an object by torque applied to one end while the other is being held fast or torqued in the opposite direction.

torsion delay line A delay line in which the delay is manifested in a material that is torqued by mechanical vibrations.

tot 1. Abbreviation of *total*. 2. To derive a total.

total breakaway torque For a torque motor, the sum of magnetic retarding torque and brush-commutator friction.

total harmonic distortion Abbreviation, THD. The distortion due to the combined action of all the harmonics present in a complex waveform.

total distortion See *total harmonic distortion*.

total reflection Full return of a ray by a reflector, none of the energy being transmitted by or absorbed in the reflecting material.

touchplate relay A capacitance relay in which the pickup element is a small, metal plate which, upon being touched, actuates the relay.

tough dog A malperforming (usually TV) circuit that seemingly defies all attempts to diagnose and correct its trouble. Also see *dog*.

tourmaline crystal A piezoelectric and pyroelectric crystal; a transparent plate of tourmaline has the ability to polarize light passing through it. Also see *polarization, 3; polarized light;* and *polarizer.*

tower 1. A usually tall and self-supporting openwork structure used to support an antenna, and usually having three or four sides. 2. A metal structure—as in 1, above—used as a vertical antenna.

Townsend discharge In a glow-discharge tube, the discharge that begins after the applied voltage reaches a given level. It is a low-current, non-self-sustaining discharge. Compare *abnormal glow* and *normal glow discharge.*

TP Abbreviation of *transaction processing.*

T-pad See *tee-pad.*

TPTG Abbreviation of *tuned-plate—tuned-grid.* (Also, *TGTP.*)

TPTG amplifier A single-stage, electron-tube amplifier stage in which the grid-input and plate-output circuits are separately tuned to the same frequency.

TPTG oscillator A single-stage, electron-tube oscillator stage in which the grid circuit and plate-output circuit are separately tuned to the same frequency. Feedback coupling is provided by the internal grid—plate capacitance of the tube.

TR Abbreviation of *transmit—receive.*

trace 1. A tiny, or insignificant, quantity. 2. The movement of the electron beam across the face of a cathode-ray tube. 3. A routine used for testing of, or for locating a fault in, a circuit or computer program. 4. The process of implementing such a routine.

tr 1. Symbol for *recovery time.* 2. Symbol for *rise time.*

trace element See *microelement.*

tracer A suitable substance or object introduced into the human body and whose progress through the body can be followed (or its state monitored) by means of electronic equipment. Tracers are sometimes employed also in nonbiological systems, such as pipelines.

track 1. A discrete information band on a magnetic disk or tape. 2. To follow, as by a stylus, a phono disk groove. 3. To follow, as by radar, a target.

trackability An expression of the accuracy with which a phonograph stylus follows the irregularities in a disk.

tracking Following in step, as when ganged circuits resonate at the same frequency (or some frequency difference) at all settings, or when a missile closely follows its guiding signal.

tracking force See *vertical stylus force.*

track label On a magnetic storage medium, a record identifying a track.

tracking mode In *tracking supplies,* the usual manner of operation in which the output of each of the separate supplies automatically follows that of the one being adjusted. Compare *independent mode.*

tracking supplies Adjustable power supplies—packaged two or more to the unit—in which the output of each will automatically follow adjustment made to one of them.

track pitch The distance between tracks (see *track, 2*).

traffic 1. Collectively, messages handled by a communications station. 2. Collectively, data and instructions handled by a computer system in continual use.

trailer label At the end of a magnetic-tape or floppy-disk file, a record signaling the end of the file and often giving such information as the number of records in the file.

trailer record At the end of a group of computer records, a record containing information relevant to the group's processing.

trailing edge 1. The falling edge of a pulse. (Compare *leading edge.*) 2. The edge of an airfoil facing away from the direction of flight. 3. The edge of a punched card facing away from the direction of transport.

trans Abbreviation of *transverse.*

transaction The exchange of activity that takes place between a computer, via a terminal, and the user, including any processing required, e.g., that involved in adding records to, or deleting them from, a file.

transaction file In data processing, a group of records used to update a master file.

transaction processing In computer practice, use of a central processor for handling, modifying, or otherwise acting upon information by transactions.

transaction tape Paper or magnetic tape on which has been recorded a transaction file.

transadmittance For an active device, such as a tube or transistor, the ratio d $I\,2$/d $E\,1$, where $I\,2$ is the ac component of current of a second electrode (such as the plate), and $E\,1$ is the ac component of voltage of a first electrode (such as the grid), with constant dc operating voltages.

transceiver 1. A combination radio *transmitter—receiver,* usually installed in a single housing and sharing some components. Compare *transmitter—receiver.* 2. In computer practice, a read/write data terminal capable of transmitting and receiving information to and from a channel.

transcendental functions Nonalgebraic functions. These logarithmic functions, exponential functions, trigonometric functions, and inverse trigonometric functions.

transconductance 1. Symbol, gm. Unit, micromho. In an electron tube, the extent to which plate current (Ip) changes in response to a change in grid voltage (Eg); gm d $I\,p$/d Eg. 2. Symbol, gfs. Unit, micromho. In a field-effect transistor, the extent to which drain current ($I\,D$) changes in response to a change in gate-to-source voltage ($V\,GS$); gfs d $I\,D$/d $V\,GS$.

transconductance amplifier An amplifier in which the output current is a linear function of the input voltage.

transconductance meter A tube tester which indicates the transconductance of a tube. Also called *mutual-conductance meter* and *transconductance-type tube tester.*

transconductance-type tube tester See *transconductance meter.*

transcribe 1. To record material, such as a radio program, for future transmission. 2. In computer practice, the intermedia transfer of data, as from tape to cards.

transcriber A device used for intermedia data transfer, e.g., one that can move the data on magnetic tape to punched cards.

transcription A recording of material, such as a record or tape of a radio program, for later use in a transmitter. See *electrical transcription.*

transcutaneous nerve stimulator Abbreviation, TNS. An electronic device for the temporary relief of pain. In its use, electrodes taped to the skin over the painful area are connected to a protable generator of suitable pulse energy.

transducer A device which converts one quantity into another quantity, specifically when one of the quantities is electrical. Thus, a loudspeaker converts electrical impulses into sound, a microphone converts sound into electrical impulses, a photocell converts light into electricity, a thermocouple heat into electricity, and so on.

transducer amplifier An amplifier employed expressly to boost the (feeble) output of a transducer.

transducer efficiency For a transducer, the ratio of the output power to the input power.

transducer tube See *electromechanical transducer, 2.*

transductor See *magnetic amplifier.*

transfer 1. To move a signal from one point to another, especially through a modifying circuit or device. 2. To transmit information or data from one point or device to another, inside or outside a system.

transfer characteristic A figure or plot expressing the output-input signal relationship in a circuit or device. Also see *transfer, 1.*

transfer efficiency In a charge-coupled device, the proportion of charge that is transferred under given conditions.

transfer function 1. Transfer characteristic. 2. An expression that mathematically shows how two entities or events occurring in different places or at different times are related.

transfer rate The speed at which data can be moved between a computer's internal memory and a peripheral.

transferred charge In a capacitor circuit, the net electric charge that moves around the external circuit from one plate of the capacitor to the other.

transform 1. See *Laplace transform.* 2. To change the voltage of a current, or its type, e.g., high to low voltage current, ac to dc. 3. To change the form, but not the content, of data.

transformation constant See *disintegration constant.*

transformer 1. A device employing electromagnetic induction to transfer electrical energy from one circuit to another, i.e., without direct connection between them. In its simplest form, a transformer consists of separate primary and secondary coils wound on a common core of ferromagnetic material, such as iron. When an alternating current flows through the primary coil, the resulting magnetic flux in the core induces an alternating voltage across the secondary coil; the induced voltage may cause a current to flow in an external circuit. Also see *air-core transformer, induction, inductive coupling, iron-core transformer,* and *turns ratio.* 2. A section of rf transmission line employed to match impedances. Also see *linear transformer.*

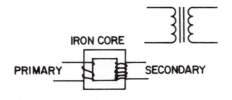

Transformer, 1.

transformer-coupled amplifier An amplifier employing coupling transformers between its stages or at its input and output points. Compare *resistance-capacitance-coupled amplifier.*

transformer coupling The inductive coupling of circuits through transformer. Also see *coefficient of coupling; inductive coupling; mutual inductance;* and *transformer, 1.*

transformer equivalent circuit An equivalent circuit depicting the various parameters of a transformer (such as primary and secondary resistances, primary and secondary reactances, core losses, etc.) in their relationship to each other.

transformer feedback Inductively coupled feedback (positive or negative) through a transformer.

transformer input current See *primary current.*

transformer input voltage See *primary voltage.*

transformer iron See *silicon steel.*

transformer loss The expression (in dB) 10 log ($P1/P2$), where $P1$ is the calculated power that a given transformer should deliver to a par-

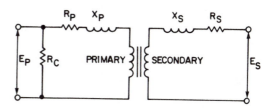

Transformer equivalent circuit.

ticular load impedance, and $P2$ is the actual power delivered.

transformer output current See *secondary current.*

transformer oil A petroleum product in which high-voltage, high-current transformers are sometimes immersed. The oil protects the windings from environmental damage.

transformer output voltage See *secondary voltage.*

transformer-type voltage regulator See *voltage-regulating transformer.*

transformer utilization factor See *utility factor.*

transient 1. A sudden, high-voltage spike in an alternating-current system, caused by arcing or lightning. 2. A spurious signal in a hard-copy receiving system. 3. Any short pulse attributable to external causes.

transient absorber See *surge arrester.*

transient arrester See *surge arrester.*

transient-based amplifier See *crystal amplifier, 2.*

transient current A momentary pulse of current. Also see *transient.*

transient response The response of a circuit to a transient, as opposed to its steady-state response, usually evaluated in terms of its ability to reproduce a square wave.

transient suppressor See *surge arrestor.*

transient voltage A momentary pulse of voltage. Also see *transient.*

transient voltmeter An instrument that indicates the voltage of momentary signals. Usually, a peak-reading meter in the instrument displays the highest positive or negative value the transient attains (sometimes holding the deflection for reading later) and can respond to pulses of 1 microsecond in duration.

transistor An active (commonly three-terminal) semiconductor device capable of amplification, oscillation, and switching action. The transistor has replaced the tube in most applications. The name is an acronym for *transfer resistor.* Also see *alloy-diffused transistor, alloy transistor, bipolar transistor, diffused-base transistor, diffused emitter-and-base transistor, diffused-junction transistor, diffusedmesa transistor, diffused planar transistor, diffused transistor, diffusion transistor, double-base junction transistor, field-effect transistor, fieldistor, germanium transistor, grown-diffused transistor, grownjunction transistor, junction transistor, mesa transistor, metal-oxide semiconductor field-effect transistor, microalloy diffused transistor, microalloy transistor, phototransistor, planar epitaxial passivated transistor, planar transistor, point-contact transistor, power transistor, silicon transistor, surface-barrier transistor, surface-charge transistor, tandem transistor, thin-film transistor, three-junction transistor,* and *unijunction transistor.*

transistor amplifier An amplifier containing only transistors as the active components. Also called *transistorized amplifier.*

transistor analyzer An instrument for measuring the electrical characteristics of transistors. Compare *transistor tester.*

transistor-coupled logic Abbreviation, TCL. In computer and automatic-control practice, logic circuitry and systems employing multi-emitter-coupled transistors. Also see *resistor-transistor logic* and *transistor-transistor logic.*

transistor current meter An ammeter, milliammeter, or microammeter circuit containing an amplifier employing only transistors. Also see *electronic current meter.*

transistorized voltmeter See *transistor voltmeter.*

transistorized voltohmmeter An electronic voltmeter-ohmmeter combination employing a transistorized circuit similar to that of a vacuum-tube voltohmmeter.

transistor keyer A power transistor acting as a keying tube.

transistor oscillator An oscillator employing only transistors as the active components. Also called *transistorized oscillator.*

transistor power supply An ac-line-operated, low-voltage, high-current, well filtered, dc power supply for operating transistor circuits.

transistor radio A radio receiver containing only transistors and semiconductor diodes, i.e., one that is tubeless. Also called *transistorized radio* and often simply *transistor.*

transistor-resistor logic See *resistor-transistor logic.*

transistor tester An instrument for checking the condition of transistors; i.e., whether good or bad. Compare *transistor analyzer.*

transistor tetrode See *double-base junction transistor.*

transistor thyratron See *solid-state thyratron.*

transistor-transistor logic Abbreviation, TTL or T^2L. In computer practice, a circuit in which the multiple-diode cluster of the diode-transistor logic circuit has been replaced by a multiple-emitter transistor.

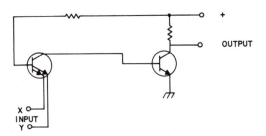

Transistor-transistor logic.

transistor voltmeter Abbreviation, tvm. A voltmeter containing an amplifier which employs only conventional (bipolar) transistors. Also called *transistorized voltmeter.* Also see *electronic voltmeter.* Compare *FET voltmeter* and *vacuum-tube voltmeter.*

transistor vom See *transistorized voltohmmeter.*

transit angle For an electron, the angular frequency multiplied by the time required to travel from one point to another.

transit time The time taken by an electron to travel from one electrode to another, especially from the cathode to the plate in a vacuum tube.

transistor element A metallic element whose atoms have valence electrons in two shells, not just one. Examples: *chromium, iron, nickel.*

transition factor See *mismatch factor.*

transition region See *barrier, 1.*

transitron oscillator See *negative-transconductance oscillator.*

translation loss See *playback loss.*

translator 1. Compiler. 2. Assembler.

translator tube A converter tube which mixes signals from separate amplifier and local-oscillator stages, e.g., a 6L7.

transliteration 1. To change the characters in one alphabet or code system to the characters in a different system. 2. The function that maps the characters in one alphabet or code system to those in another.

translucence The partial transmission of radiation, such as light, through a material. Translucence may be regarded as intermediate between *opacity* and *transparence.*

transmission gain Current amplification, power amplification, or voltage amplification.

transmission line 1. A single conductor or group of conductors for carrying electrical energy from one point to another. 2. A correctly dimensioned conductor or pair of conductors for carrying rf energy from a transmitter to an antenna or coupling device.

transmission mode 1. In a transceiver, the condition in which the transmitter is enabled and the receiver is disabled. 2. In a waveguide, propagation via transverse waves.

transmission speed The number of information elements (words, code groups, data symbols, and the like) that can be generated or received per minute by a system or operator.

transmission wavemeter A (usually simple) *LC*-tuned wavemeter which gives peak response when tuned to the frequency of a signal passing through it; also, the comparable microwave device. Compare *absorption wavemeter.*

transmit-receive switch 1. A manual or electrically operated switch for transferring a single antenna between a transmitter and receiver. 2. Transmit-receive tube.

transmit-receive tube Abbreviation, TR tube. In radar practice, a type of gas-filled electron tube that automatically permits transmit signals to reach the antenna while keeping them from the receiver, but allows received signals to reach the receiver easily.

transmitter 1. An equipment for producing and sending signals. 2. Microphone. 3. One who originates signals or data.

transmitter-receiver A separate transmitter and receiver housed together. Compare *transceiver.*

transmitting antenna An antenna designed expressly for the efficient radiation of signals into space.

transmitting station A station that only transmits signals, i.e., it engages in no official form of reception. Compare *receiving station.*

transmittivity The degree to which a selective circuit transmits a desired signal. Compare *rejectivity.*

transonic Equal to, or approximating, the speed of sound in air (approximately 1087 ft per sec).

transparence The practically unimpeded transmission of radiation, such as light, through a material. Compare *opacity* and *translucence.*

transponder A combination transmitter-receiver that automatically transmits an identification signal whenever it receives an interrogating signal. The term is an acronym for *transmitter* and res*ponder.*

transport See *tape transport.*

transportable equipment Portable electronic equipment. See, for example, *portable transmitter.*

transpose In solving equations, to move a term to the other side of the equal sign and the then necessary changing of its sign, e.g., $a + b = c$ is equivalent to $a = c - b.$

transuranium An element whose atomic number is higher than that of uranium.

transverse Occurring in a direction or directions perpendicular to the direction of propagation.

transversal A line that intersects other lines (geometry).

transverse electric mode In a waveguide, the mode of propagation when the electric (*E*) lines lie across the guide, i.e., perpendicular to the direction of transmission. Compare *transverse magnetic mode*. Also see *waveguide mode*.

transverse electromagnetic wave An electromagnetic wave having electric-field vectors and magnetic-field vectors perpendicular to the direction of propagation.

transverse magnetic mode In a waveguide, the mode of propagation when the magnetic (*H*) lines lie across the guide, i.e., perpendicular to the direction of transmission. Compare *transverse electric mode*. Also see *waveguide mode*.

trap 1. Wavetrap. 2. In a semiconductor crystal, an imperfection capable of trapping current carriers.

trapezoid 1. A polygon having four sides, of which only two are parallel. 2. Trapezoidal pattern. 3. Trapezoidal wave.

trapezoidal distortion In television or facsimile, a form of distortion in which the frame is wider at the top than at the bottom, or vice-versa.

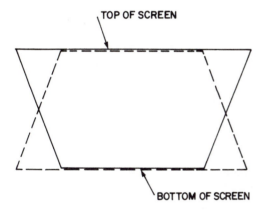

Trapezoidal distortion.

trapezoidal pattern An oscilloscope pattern used to check the percentage of modulation of an AM wave. It is so called from its trapezoidal shape.

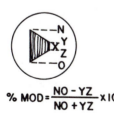

$$\% \text{ MOD} = \frac{NO - YZ}{NO + YZ} \times 100$$

Trapezoidal pattern.

trapezoidal wave A nonsinusoidal wave which is a combination of a rectangular component and a sawtooth component. It is the required waveform of the voltage applied to a magnetic deflecting coil (oscilloscope or TV) to insure a sawtooth wave of current in the coil.

traveling-wave amplifier Abbreviation, TWA. An amplifier based upon the unique operation of the traveling-wave tube.

traveling-wave tube Abbreviation, TWT. A microwave tube containing an electron gun, helical transmission line, collector, and input and output couplers. A microwave signal is coupled into the helix, through which it travels while the gun projects an electron beam through the helix. When wave and electron velocities are equal, a power gain is obtained in the signal coupled out of the helix. Also see *backward-wave oscillator*.

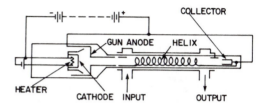

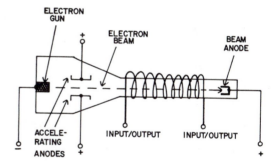

Traveling-wave tubes.

Travis discriminator A discriminator circuit in which the diodes are operated from separately tuned halves of the secondary winding of the input transformer. Compare *Foster-Seeley discriminator*.

TR box See *transmit-receive tube*.

TR cavity The resonant cavity in a transmit-receive tube. Also see *resonant cavity*.

treasure locator See *metal locator*.

treble The higher portion of the sound spectrum, especially the upper end of the musical scale (256 Hz and above). Compare *bass*.

treble boost 1. The special emphasis given to high audio frequencies (the treble notes) in audio systems. 2. Boost of the treble response, as in 1, above, to tailor the reproduction of sound.

treble compensation See *treble boost, 1, 2*.

treble control A tone control for adjusting treble boost or treble attenuation.

tree 1. A cause-and-effect chain with two or more independent branches. 2. A circuit with two or more branches but no meshes.

trf Abbreviation of *tuned radio frequency*.

trf amplifier See *tuned radio-frequency amplifier*.

trf receiver See *tuned radio-frequency receiver*.

triac A three-terminal, gate-controlled, semiconductor ac switching device. Compare *diac*. See figure on next page.

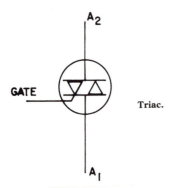

Triac.

triangular wave See *back-to-back sawtooth*.

triax A double-shielded coaxial cable. A center conductor is surrounded by two concentric, independent shield conductors.

triaxial connector A male or female connector--jack or plug--having three concentric contacting surfaces. By contrast, the *coaxial jack* and *coaxial plug* have only two such surfaces.

triaxial speaker A dynamic loudspeaker containing three independent units, for high-, low-, and middle-frequency ranges, in one loudspeaker assembly.

triboelectric Pertaining to frictional electricity.

triboelectric series See *electrostatic series*.

triboluminescence Luminescence produced by means of friction.

trickle charge A continuous slow charge of a storage battery in which the charging rate is just sufficient to compensate for internal losses or normal discharge.

trickle charger A light-duty unit (see *battery charger, 1*) for delivering a trickle charge to a battery.

triethanolamine Formula, $(HOCH_2 CH_2)_3N$. An amino alcohol which precipitates metallic silver from a silver-nitrate solution in the deposition of a silver surface on a substrate, such as glass or a ceramic.

trig Abbreviation form of *trigonometry*.

trigatron A form of electrically operated switch. The circuit is closed by the breakdown of an electrical gap.

trigger 1. A pulse employed to start or stop the operation of a circuit or device, such as a flip-flop. 2. To snap a circuit or device into or out of operation with a trigger pulse, as in 1, above.

trigger diode See *diac* and *four-layer diode*.

triggered multivibrator See *driven multivibrator*.

triggered sweep In an oscilloscope or similar device, a driven sweep.

triggering point The voltage level at which the action of an electronic

switching device is initiated. Also see *breakover point*.

trigistor A three-junction semiconductor which exhibits two-state operation and therefore is useful as a flip-flop or switch. Also see *npnp device*.

trigonometric functions See *circular functions* and *hyperbolic functions*.

trigonometry (plane trig) The branch of mathematics devoted to the application of trig functions (see *circular functions*), based on relationships within the right triangle, among these being the solution of triangles in a plane. Trigonometry is useful in electronics for determining impedance and phase.

trim To make a fine adjustment, as of a tuning control, balance control, output adjustment, or the like.

trimmer A low-valued variable capacitor, inductor, or resistor operated in conjunction with a main unit (usually of the same sort) for vernier adjustment or range setting. See, for example, *high-frequency trimmer* and *oscillator trimmer*.

trimmer capacitor A variable capacitor employed as a trimmer.

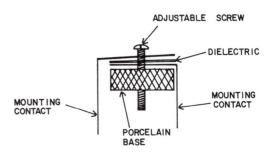

Trimmer capacitor.

trimmer coil See *trimmer inductor*.

trimmer inductor A variable inductor employed as a trimmer.

trimmer resistor A variable resistor employed as a trimmer.

trinary number system The base-3 system of notation. This system employs the digits 0, 1, and 2, the positional values being successive power of 3, e.g., decimal 14 equals trinary 112. Also called *ternary number system*.

trinistor A power-type silicon triode with control characteristics similar to those of a *thyratron*.

Trinitron Sony's single-gun color-TV picture tube, or a set employing this tube. The gun has three cathodes that modulate the three color beams (red, green, blue). The beams are accelerated by common grids and are focused at different angles by convergence plates.

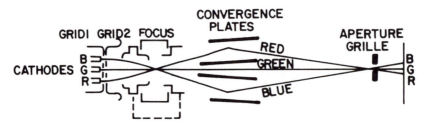

Trinitron.

Trinoscope See *Trinitron*.

triode A three-electrode tube or transistor, embodying an anode, cathode, and a control electrode as the principal elements, e.g., a 12H4.

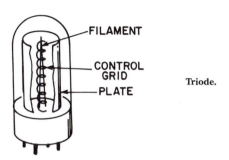

Triode.

triode amplifier An amplifier of any kind employing only triode tubes.

triode-heptode A combination tube consisting of separate triode and heptode sections in a single envelope, e g., a 12FX8.

triode-hexode A combination tube consisting of separate triode and hexode sections in a single envelope, e.g., a 6K8.

triode-hexode converter In a superheterodyne circuit, a converter employing a triode oscillator and hexode mixer (converter). Also see *triode-hexode*.

triode oscillator An af or rf oscillator employing a triode.

triode-pentode A combination tube consisting of separate triode and pentode sections in a single envelope, e.g., a 6AT8A.

triode transducer See *electromechanical transducer, 2*.

triple-diffused transistor A *diffused transistor* in which the base and emitter are diffused into the top face of the chip, and the collector into the bottom face.

triple diode 1. A combination tube having three separate diode sections in a single envelope, e.g., a 6BC7. 2. An assembly of three (often closely matched) semiconductor diodes in a single housing, e.g., a 3DG001.

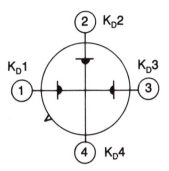

Triple diode, 2.

triple-diode-triode A combination tube consisting of three diodes and one triode in a single envelope, e.g., a 6S8GT.

triple integral Symbol, $\iiint$. An integral involving successive integration performed three times. Also see *integration* and *volume integration*. Compare *double integral*.

triple integration The use of triple integrals, especially in calculating volume. See, for example, *triple integral* and *volume integration*. Compare *double integration*.

tripler 1. A rectifier circuit which delivers a dc output voltage that is approximately triple the peak value of the ac input voltage. 2. An amplifier or other circuit which delivers an output signal at triple the frequency of the input signal. See figure on next page.

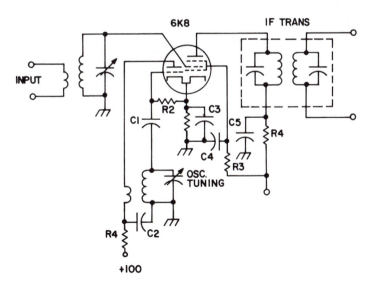

Triode-hexode converter.

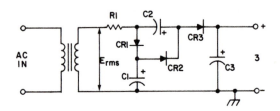

Tripler, 1.

triplexer In radar, a device that facilitates the use of two receivers at the same time.

tri-state logic Digital logic in which there are three possible states, rather than the usual two. The conditions are defined as 0, 1, and undecided. Trade name of National Semiconductor.

tri-tet oscillator An electron-coupled crystal oscillator in which output-circuit isolation is obtained by using the screen grid of a tetrode or pentode tube as the oscillator plate, and the normal plate as the output eletrode. The name incorporates the abbreviation of *triode-tetrode*.

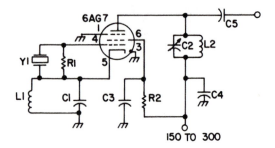

Tri-tet oscillator.

tritium Symbol, T or H3. An isotope of hydrogen whose nucleus contains two neutrons and one proton. Compare *deuterium*.

TRL Abbreviation of *transistor-resistor logic* (see *resistor-transistor logic*).

trochoid A curve described by a point on the radius, or extended radius, of a circle "rolling" on a line in a plane.

troposphere The portion of the atmosphere between the surface of the earth and the stratosphere.

tropospheric propagation A form of electromagnetic-wave propagation in which over-the-horizon communication is made possible because of effects in the lower atmosphere.

tropospheric scatter Wave scatter resulting from irregularities in the troposphere. Also see *forward scatter*.

troubleshoot 1. To look for the cause of equipment failure. 2. Debug.

troubleshooting test A test that is part of the procedure for finding the cause of faulty electronic equipment operation. Also see *diagnostic test*. Compare *performance test*.

trough value The minimum amplitude of a composite current or voltage.

TR tube See *transmit-receive tube*.

true complement See *radix complement*.

true ground Actual ground, i.e., the earth, as opposed to artificial ground, such as that provided by the radials of a ground-plane antenna or an equipment chassis.

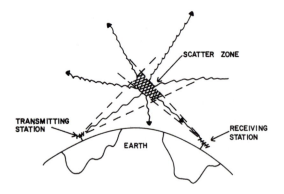

Tropospheric scatter.

true power In an ac circuit, the power consumed in the resistive component; the reactive component consumes no power. In most ac circuits, true power is somewhat lower than apparent power. Also see *ac power* and *power factor*.

truncate To round off a number. See *rounded number*.

truncated number See *rounded number*.

trunk 1. A communications link between two points, one usually being external. 2. Sometimes, a cpu-peripheral interface.

trunk link In computer systems, an interface permitting access to main storage via a peripheral.

truth table In logic analysis, and logic circuit design, a table in which are listed all combinations of input values and the corresponding output values for a given logic function, e.g., OR, AND; it is from truth tables that ROMs and PROMs are microprogrammed to carry out instructions.

ts Abbreviation of *tensile strength*.

Tschebyscheff filter A filter designed to afford rapid post-cutoff attenuation, provided a certain passband ripple can be tolerated.

TSS Abbreviation of *time-sharing system*.

TTL Abbreviation of *transistor-transistor logic*. (Also, T^2L.).

TTT In radiotelegraphy, a signal indicating that a message concerning safety is to follow (equivalent to *securite* in radiotelephony).

TTY Abbreviation of *teletypewriter*.

T-type antenna See *tee-antenna*.

T-type attenuator See *tee-pad*.

tube 1. Electron tube: vacuum tube, gaseous tube, cathode-ray tube, X-ray tube, etc. 2. Glow lamp: argon bulb, neon bulb, mercury-vapor lamp, etc.

tube capacitances The internal capacitances between the elements of an electron tube. See specifically *grid-plate capacitance, grid-screen capacitance, plate-cathode capacitance, plate-grid capacitance,* and *plate-screen capacitance*.

tube diode A two-element (cathode, anode) electron tube for current rectification. Also see *diode*.

tube parameters Operating coefficients of electron tubes, e.g., plate current, grid voltage, screen current, transconductance, amplification factor, etc.

tube tester An instrument for checking one or more of the parameters of an electron tube. Also see *emission tester, transconductance meter,* and *vacuum-tube bridge*.

tube-type transducer See *electromechanical transducer, 2.*

Tubular capacitor.

tubular capacitor A fixed capacitor consisting of a wound section enclosed in a cylindrical can or tube.

tune 1. To adjust a selective circuit to accept or reject a signal. 2. To correct the natural frequency of vibration of a body.

tuned af amplifier An audio amplifier that is continuously tunable over a band of frequencies, or which is fix-tuned (to a single frequency). See, for example, *parallel-tee amplifier.*

tuned-base oscillator A self-excited, common-emitter-connected, bipolar-transistor oscillator in which the tuned *LD* tank is in the base circuit.

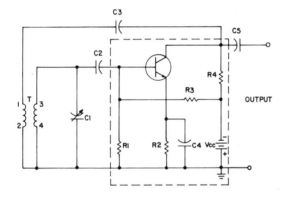

Tuned-base oscillator.

tuned circuit A (usually serious-resonant or parallel-resonant) circuit adjusted to accept or reject a signal. Also see *resonance.*

tuned-collector oscillator A self-excited, common-emitter connected, bipolar-transistor oscillator in which the tuned *LC* tank is in the collector circuit.

tuned coupler An antenna coupler that is tuned independently of the transmitter or receiver with which it is used.

tuned dipole A half-wave, center-fed, resonant antenna.

tuned feeders Antenna feeders that are tuned to the transmitted-signal frequency by either series or parallel variable capacitors (or both).

tuned-grid oscillator A vacuum-tube oscillator in which the tuned *LC* tank is in the control-grid circuit. Compared tuned-plate oscillator.

tuned-grid-tuned-plate Abbreviation, TGTP. An alternate form of *tuned-plate-tuned-grid.*

tuned headphones Headphones used in radiotelegraphy that are fix-tuned to a single audio frequency (e.g., 1 kHz) by means of a small parallel capacitor.

tuned line. An antenna wire or transmission line providing a resistive load at a specific resonant frequency.

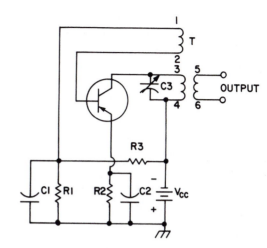

Tuned-collector oscillator.

tuned pickup A pickup circuit or device (such as an rf sampling coil) that is tuned to the signal frequency.

tuned-plate oscillator A vacuum-tube oscillator in which the tuned *LC* tank is in the plate circuit. Compare *tuned-grid oscillator.*

tuned-plate-tuned-grid Abbreviation, TPTG. A single-stage, electron-tube circuit in which the grid input and plate output are separately tuned. See, for example, *TPTG amplifier* and *TPTG oscillator.*

tuned-ratio-frequency amplifier A tuned tube or transistor amplifier (usually continuously tuned) for some part of the rf spectrum.

tuned-radio-frequency receiver A radio receiver consisting only of a tuned-radio-frequency amplifier, detector, and af amplifier. Compare *superheterodyne receiver.*

tuned reed A vibrating reed whose length, width, and/or thickness have been adjusted so that it vibrates naturally at a desired frequency.

tuned-reed frequency meter An audio frequency meter employing tuned metal reeds as the indicators. Also see *power frequency meter, 2.*

tuned relay An electronic or electromechanical relay that closes at one frequency. See, for example, *reed switch, 1.*

tuned signal tracer A signal tracer that can be tuned sharply to the frequency of the test signal being traced.

tuned transformer An af or rf transformer that is tuned by means of a capacitor connected in parallel with its primary or secondary winding (or both).

tuner A circuit or device which may be set to select one signal from a number of signals in a frequency band.

tungar bulb A hot-cathode, argon-filled rectifier tube having a filament and a graphite plate. Also called *tungar rectifier* and *tungar tube.*

tungsten Symbol, W. A metallic element. Atomic number, 74. Atomic weight, 183.86. Also called *wolfram.* Tungsten is used in switch and relay contacts, in the elements of some electron tubes, and in incandescent-lamp filaments.

tuning 1. Adjustment of the frequency of a receiver, for the purpose of intercepting a signal on a given frequency. 2. The adjustment of a transmitter oscillator to a desired frequency. 3. The adjustment of an inductance-capacitance circuit for resonance on a desired frequency. 4. The adjustment of an antenna or antenna system to a desired frequency. 5. The adjustment of a radio-frequency amplifier for optimum performance. 6. The alignment of a musical instrument for correct tone frequency.

tuning capacitor A variable capacitor used to tune an *LC* circuit (series-resonant or parallel-resonant).

tuning coil A variable inductor used to tune an *LC* circuit (series-resonant or parallel-resonant).

tuning core See *tuning slug*.

tuning diode See *voltage-variable capacitor, 1*.

tuning fork A metal device which, when struck physically, vibrates at a precise audio frequency. Usually, the tuning fork has two prongs and looks something like a fork.

tuning-fork oscillator See *fork oscillator*.

tuning indicator 1. A meter or magic-eye tube used to show when a receiver is tuned to an incoming signal. 2. Sometimes, a bridge null detector.

tuning-indicator tube See *magic-eye tube*.

tuning meter A meter-type resonance indicator.

tuning potentiometer A single or ganged potentiometer (depending upon the circuit in which it is used) employed to vary the frequency response of a resistance-capacitance-tuned circuit, such as the bridge-tee, parallel-tee or Wien-bridge circuit.

tuning slug A powdered-iron core that slides or screws in and out of a coil to vary the coil's inductance.

tuning voltage An adjustable dc voltage used to vary the capacitance of a varactor diode serving as the capacitor in a tuned circuit.

tuning wand See *neutralizing tool*.

tunnel diode A specially processed junction diode whose forward conduction characteristic displays negative resistance.

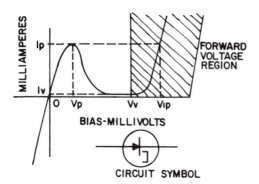

Tunnel diode.

tunnel-diode amplifier An amplifier circuit that employs the negative conductance of a tunnel diode for secure gain.

tunnel-diode oscillator A negative-resistance oscillator in which a tunnel diode provides the negative resistance.

turbidimeter An opacimeter employed to measure the turbidity of liquids.

turbidity meter See *turbidimeter*.

Turing machine A hypothetical model, conceived in the 30s, of a calculating machine which, rather than being a prototype, was de-

vised as an exercise in problem-solving. The imaginary device, which can be likened to a computer programming language with a limited instruction set operating in a machine with limitless memory capacity, is capable of outputting an unending supply of paper tape sectioned into squares, and acts on algorithms devised for it by carrying out one of four actions: advance the tape one square; mark a square; remove a mark; or pause.

turn One complete loop of a coil.

turnaround time 1. The length of time required for a repair to be performed. 2. The length of time required for a computer program to be completed. 3. The time required for a transceiver to switch from the full-transmit to the full-receive condition.

turn factor For a given inductor core, the number of turns (in a standard configuration) that results in an inductance of 1 H.

turn-off time The time required for the operation of a circuit or device to cease completely after a signal or operating power has been removed. Compare *turn-on time*.

turn-on time The time required for a circuit or device to come up to full (normal) operation after application of a trigger or operating power. Compare *turn-off time*.

turnover pickup See *dual pickup*.

turns ratio For a transformer, the ratio of the number of primary turns (Np) to the number of secondary turns (Ns).

turnstile antenna A polyphase antenna resembling a turnstile gate.

turntable The motor-driven plate on which a phonograph disk rests during recording or reproduction.

turret tuner A TV front-end tuner having a separate tuned-circuit section for each channel. Turning a knob rotates a desired section into position against switch contacts. Also called *detent tuner*.

TV 1. Abbreviation of *television*. 2. Abbreviation of *terminal velocity*.

T/V Abbreviation of *temperature-to-voltage*.

TV decoder A device for unscrambling a television broadcast which has been intentionally distorted to prevent its use by unauthorized viewers. See *subscription TV*.

TV ghost See *ghost*.

TVI Abbreviation of *television interference*.

TVL Amateur radio abbreviation of *television listener (looker)*.

tvm Abbreviation of *transistor voltmeter*.

tvo Abbreviation of *transistor voltohmmeter*. (Also, *tvom*.)

tvom Abbreviation of *transistor voltohmmeter*. (Also, *tvo*.)

TV projector A combination electronic and optical device for projecting television images on a large screen. See *projection television*.

TV sound marker See *sound marker*.

TVT Abbreviation of *television terminal* (computer peripheral).

TW Abbreviation of *terawatt*.

TWA Abbreviation of *traveling-wave amplifier*.

tweeter A usually small loudspeaker that favors high frequencies. Compare *woofer*.

twin diode See *dual diode*.

twin double-plate triode See *dual double-plate triode*.

twinlead See *flat-ribbon line*.

twin line See *flat-ribbon line*.

twin meter See *dual meter*.

twin pentode See *dual pentode*.

twin-T measuring circuit See *parallel-tee measuring circuit*.

twin-T network See *parallel-tee network*.

twin triode See *dual triode*.

twisted curve A curve in space that does not occupy a definite plane.

twisted pair A wire line used for communications.

twister A piezoelectric crystal, such as one of Rochelle salt, that generates a voltage when torqued.

twistor A magnetic-memory element consisting of a winding of magnetic wire on a length of nonmagnetic wire.

two-channel amplifier See *dual-channel amplifier.*

two-channel recorder See *double-channel recorder, 1, 2.*

two-electrode amplifier See *diode amplifier, 1, 2.*

two-electrode tube See *tube diode.*

two-element amplifier See *diode amplifier, 1, 2.*

two-element beam A directional antenna containing two elements: a *radiator* and a *reflector.*

two-element tube An electron tube having two principal electrodes, i.e., a *tube diode.*

two-five code See *biquinary code.*

two-phase Pertaining to circuits or devices in which two components (two voltages, two currents, or a current and a voltage) are 90 degrees out of phase with each other.

two-phase system See *quarter-phase system.*

two-phase, three-wire circuit A polyphase circuit employing three conductors, one of which (the *common return*) is 90 degrees out of phase with the other (outer) two.

two-point tuning See *double-spot tuning.*

two-state logic Digital logic in which there are two possible conditions: true (1) and false (0).

two-state device A bistable device, i.e., one having two stable states, e.g., a flip-flop.

two-tone keying See *frequency-shift keying.*

two-track recording Recording on two adjacent tracks on magnetic tape. The separate recordings may be in the same direction (as in stereo) or in opposite directions.

two-way amplifier An amplifier whose input and output terminals may be used interchangeably.

two-way communication The exchange of messages between two or more stations which transmit as well as receive. Compare *one-way communication, 1.*

two-way radio See *two-way communication.*

two-way repeater In telephony, a device that amplifies and retransmits a signal in either direction. Also see *two-way amplifier.*

two-way speaker system A woofer and tweeter occupying the same enclosure and interconnect by a crossover network for wideband frequency response.

TWT Abbreviation of *traveling-wave tube.*

TWX Abbreviation of *teletypewriter exchange.*

Twystron A form of microwave-oscillator tube, similar to the Klystron, but with better efficiency.

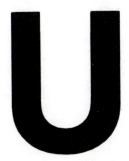

U 1. Symbol for *uranium*. 2. Abbreviation of *unit*. 3. Symbol for *universal set*. 4. Symbol (sans-serif, U) for *union of sets*.

u 1. Typewritten subsitute for Greek letter *μ*. See *micro*. 2. Symbol (ital.) for *unified atomic mass unit*.

Ua Abbreviation of *unit of activity (elecroencephalography)*.

uA See *u* and *microampere*.

UAM Abbreviation of *underwater-to-air missile*.

ub See *u* and *microbar*.

ubiquitous carrier A current carrier (esiecially an electron) whose velocity is so high, or mechanism of transfer so subtle, that it appears to be in two places simultaneously. Thus, in the tunnel diode, during the interval of tunneling a single electron appears to be on both sides of the barrier at the same time.

ubitron A tub in which a periodic magnetic field causes an electron beam to undulate. Through its transverse velocity component, the beam interacts with a radio-frequency wave, and its kinetic energy is converted into rf energy. The ubitron can be used as an amplifier or oscillator.

U-boat Submarine; from *underwater boat*.

uc See *u* and *microcurie*.

UCT Abbreviation of *universal coordinated time*.

UDC Abbreviation of *universal decimal classification*.

udometer A rain gauge.

udomograph A recording rain gauge.

UDOP Abbreviation of *ultrahigh-frequency doppler system*.

udop A 440-MHz phase-coherent tracking system employed to determine the velocity and position of missiles and space vehicles.

UEP Abbreviation of *underwater electric potential*.

uF See *u* and *microfarad*,

UFET Abbreviation of *unipolar field-effect transistor. (Also, UNIFET.)* See *unipolar transistor*.

UFO See *unidentified flying object*.

UG Abbreviation of *underground*.

uH See *u* and *microhenry*.

uhf Abbreviation of *ultrahigh frequency*.

uhf adaptor See *uhf converter, 1*.

uhf capacitor A button-type capacitor. So called because its unique design makes it an efficient bypass capacitor at ultrahigh frequencies.

uhf converter 1. A circuit (usually consisting of rf amplifier and mixer) for converting uhf signals to a lower frequency which can be accommodated by a standard receiver. 2. A circuit for converting uhf television signals to vhf frequencies, so they can be accommodated by an older (vhf only) TV receiver. See *uhf television channels* and *vhf television channels*.

uhf diode A semiconductor diode whose rectification efficiency is good at ultrahigh frequencies.

uhf generator 1. An oscillating device—such as uhf tube, uhf transistor, or tunnel diode—used to produce rf energy at ultrahigh frequencies. 2. The equipment in which such a device is used. 3. An ultrahigh-frequency test-signal generator.

uhf loop A (usually single-turn) loop antenna which has a toroidal radiation pattern perpendicular to the loop and which has a natural wavelength in the uhf range.

uhf receiver A receiver tunable to frequencies in any band in the frequency range 300 to 3000 MHz.

uhf television channels Seventy 6-MHz-wide channels (TV channels 14 through 83) between 470 MHz and 890 MHz.

uhf transistor A transistor specially designed and fabricated for ultrahigh-frequency operation. It is characterized by extended beta cutoff frequency, low junction capacitance, and fast recovery time.

uhf translator A TV broadcast translator station transmitting in an ultrahigh-frequency TV channel.

uhf transmitter A transmitter that is specially designed for operation at ultrahigh frequencies. In such a transmitter, stray parameters are minimized and special tubes or transistors often are required.

uhf tube A vacuum tube specially designed for ultrahigh-frequency operation. It is characterized by low interelectrode capacitance, low input and output capacitances, short electron transit time, and low lead inductance.

uhf tuner 1. A superheterodyne front end that is tunable in the uhf range and is usually arranged to deliver the mixer output to a lower-frequency receiver, which functions as i-f detector and audio system. 2. A TV front end that is tunable to the uhf TV channels.

uhr Abbreviation of *ultrahigh resistance.*

U-index The difference between consecutive daily mean values of the horizontal component of the geomagnetic field.

UJT Abbreviation of *unijunction transistor.*

UL 1. Abbreviation of *Underwriters Laboratories.* 2. Abbreviation of *ultralinear.*

ULD Abbreviation of *ultralow distortion.*

ulf Abbreviation of *ultralow frequency.*

U-line section A short section of coaxial cable having the shape of a squared *U.* Sometimes, a U-shaped metal version of a coaxial section.

ult Abbreviation of *ultimo.*

ultimate attenuation An asymptotic attenuation characteristic, such as 6 dB per octave, approached by the actual performance of a circuit.

ulimately controlled variable In an automated system, the quantity or operation (e.g., liquid level, temperature, pressure, precision drilling, etc.) whose control is the end purpose of the system.

ultimate-range ballistic missile A ballistic missile whose range is greater than half the earth's circumference.

ultimate ratio The limiting value of a ratio, i.e., the value approached by a ratio.

ultimate sensitivity 1. In an instrument or system, the maximum degree of perception of, or response to, a quantity or condition. 2. In a graphic recorder, half the deadband.

ultimate threshold See *ultimate sensitivity, 2.*

ultimate trip current For a specified set of environmental conditions, the lowest value of current which will trip a circuit breaker.

ultimate trip limits For a circuit breaker, the maximum and minimum current at which the breaker trips and drops out.

ultimo Of the previous month.

ultor anode The second anode of a TV picture tube or oscilloscope tube.

ultor element In a TV picture tube, the element that receives the highest dc voltage. Also called *ultor electrode.* Also see *ultor anode.*

ultor voltage The high dc voltage applied to the second anode of an oscilloscope tube or TV picture tube. Also see *ultor anode.*

ultra A prefix meaning *above* or *higher than,* e.g., *ultraviolet, ultrahigh frequency, ultrasonics.*

ultra-atomic Pertaining to particles smaller than atoms, i.e., *subatomic.*

ultra-audible frequency A frequency higher than the highest audible frequency. Also called *ultrasonic frequency* and *supersonic frequency.*

ultra-audion oscillator See *ultraudion oscillator.*

ultracentrifuge A centrifuge that spins at a very high speed.

ultrafast switch An electronic switch capable of microsecond operation.

Ultrafax A system for high-speed transmissoin of printed matter by facsimile, radio, or television.

ultrahigh frequency A radio frequency in the range 300 to 3000 MHz.

ultrahigh-frequency band See *uhf band.*

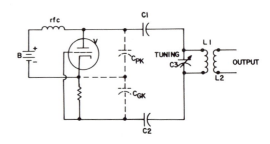

Ultra-audion oscillator.

ultrahigh resistance Resistance of 100 megohms or higher.

ultrahigh-speed motion-picture photography Cinematography at a rate higher than 10^4 fps.

ultralinear Pertaining to the most linear operation obtainable with today's technology.

ultralinear amplifier A high-fidelity af power amplifier with wide frequency response and very low distortion (of all kinds).

ultralow distortion Distortion lower than 0.1% overall.

ultralow frequencies Frequencies lower than those in the audio-frequency spectrum, e.g., between 0.001 and 1 Hz. Also called *subsonic frequencies.*

ultramicrometer An electronic instrument for extremely small linear measurements.

ultramicroscope An optical microscope using refracted light to illuminate minute particles. It allows observation of diffraction rings.

ultramicrowave Pertaining to wavelengths in the range 1 μm to 1 mm (300 GHz to 300 THz).

ultraminiature See *subminiature.*

ultraphotic rays Collectively, infrared and ultraviolet rays.

ultrared rays See *infrared rays.*

ultrashort waves Waves whose length correspond to ultrahigh frequencies, i.e., wavelengths between 0.1 and 1 meter.

ultrasonic Of higher frequency than those that are audible (supersonic). See, for example, *ultrasonic frequency.*

ultrasonic bonding A method of bonding metal by means of physical vibration at frequencies just above the human hearing range.

ultrasonic brazing Forming a nonporous bond between metal parts through the use of ultrasonic energy and a second, different metal (or alloy) having a lower melting point.

ultrasonic cleaning A method of cleaning delicate or intricately structured items, such as dentures, contact lenses, or jewelry, in which the soiled items are immersed in a fluid which is agitated by ultrasonic transducers; the foreign particles are shaken away.

ultrasonic cleaning tank A thick-walled stainless-steel tank with ultrasonic transducers mounted within its walls and used for ultrasonic cleaning.

ultrasonic coagulation The coagulation of a substance through ultrasonic agitation.

ultrasonic communication Underwater telegraphic communication between ships and/or submarines by keying the echo-ranging sonar equipment.

ultrasonic delay line A delay line, such as the mercury type, through which an ultrasonic signal is propagated. The delay results from the relativels slow propagation of the ultrasonic wave through the substance in the line.

ultrasonic densitometer A densitometer whose operation is based upon the time required for an ultrasonic signal to penetrate the material under test or for the signal to penetrate the material and be reflected back to the transmitter.

ultrasonic depth finder See *acoustic depth finder*.

ultrasonic detector A device which responds to ultrasonic waves by indicating their presence, intensity, and/or frequency.

ultrasonic diagnosis The medical determination of the condition of tissues or structures within the body in terms of reflection of ultrasonic waves by those tissues or structures. The technique is useful in situations where other techniques, such as radiology or the use of catheter probes, are risky. Also see *ultrasound*.

ultrasonic disintegrator In biology and related fields, a device which employs ultrasonic energy to rupture cells and tissues.

ultrasonic drill A drill running at speeds corresponding to ultrasonic frequencies. The tool is valuable in drilling hard or brittle materials and in dentistry.

ultrasonic filter 1. A bandpass filter whose operation is based upon the natural (ultrasonic) vibration frequency of small disks or rods of magnetostrictive metal. See *magnetostriction*. Also called *mechanical filter*. 2. Generally, a filter operating at ultrasonic frequencies.

ultrasonic flaw detector A system analogous to a radar, in which an ultrasonic wave is transmitted through a solid material and is reflected back to a detector and display device to reveal flaws, cracks, and strain in the material.

ultrasonic frequency A frequency above the limit of human hearing, but particularly one in the region 20 to 108 kHz.

ultrasonic generator 1. An oscillator that operates at frequencies just above the range of human hearing, and the output of which is intended for coupling to an electroacoustic transducer. 2. Any device that produces ultrasonic waves.

ultrasonic grating constant For a sound wave producing a light-diffraction spectrum, a figure indicating the distance between diffracting centers of the wave.

ultrasonic heating The production of heat in a specimen by means of ultrasonic energy directed into or through it.

ultrasonic image converter A device which makes visible acoustic field patterns.

ultrasonic inspection The use of ultrasonic waves to detect interval flaws in solid materials, such as metals. It is valued particularly because it is nondestructive. Also see *ultrasonic flaw detector*.

ultrasonic intrusion alarm An intrusion detecting and signaling systen activated when an intruder disturbs a pattern of ultrasonic waves in the protected area; a sensitive relay closes, setting off an alarm.

ultrasonic level detector A level-monitoring system in which an ultrasonic transmitter and ultrasonic detector are mounted together on one wall of a tank, or chamber. When the tank is empty, the transmitted signal is reflected readily by the opposite wall back to the detector. When the liquid rises and blocks the sound path, however, reflection time is reduced and this change is used to operate a signal device that indicates the tank has been filled to a desired level.

ultrasonic light diffraction Diffraction resulting from the periodic variation of light refraction when a beam of light passes through a longitudinal sound-wave field.

ultrasonic light modulator A device which modulates a light beam

passing transversely through a fluid agitated by the modulating-sound waves.

ultrasonic material dispersion The use of the agitative action of high-intensity ultrasonic waves to disperse or emulsify one substance in another.

ultrasonic plating The use of ultrasonic energy to deposit or bond one material on the surface of another.

ultrasonic probe A rodlike director of ultrasonic energy.

ultrasonic relay An electronic relay actuated by inaudible sound.

ultrasonics 1. The branch of electronics concerned with the application of ultrasonic-frequency energy in industry, medicine, research, and the home. See *ultrasonic frequency*. 2. The use of ultrasonic energy for medical diagnosis and treatment. 3. The branch of physics dealing with ultrasonic phenomena.

ultrasonic soldering See *ultrasonic brazing*.

ultrasonic sounding Determining the depth of a body of water in terms of the time taken for a transmitted ultrasonic signal to be reflected back to a transmitting point on the surface of the water.

ultrasonic space grating A periodic variation in the index of refraction when acoustic waves are present in a light-transmitting medium. Also see *ultrasonic light diffraction*.

ultrasonic storage cell See *ultrasonic delay line*.

ultrasonic stroboscope A stroboscope which employs an ultrasonically modulated beam of light.

ultrasonic switch An electronic switch actuated by inaudible sound.

ultrasonic therapy The use of ultrasonic energy by physicians and others in the treatment of certain disorders. In some applications, the therapy is similar to diathermy.

ultrasonic thickness gauge An instrument which determines the thickness of a specimen in terms of the propagation time (or echo time) of ultrasonic waves through the specimen.

ultrasonic transducer A transducer which converts electrical energy into ultrasonic energy. Common types are quartz crystal, ceramic crystal, and magnetostrictive disk.

ultrasonic waves Waves whose length correspond to ultrasonic frequencies.

ultrasonic welding A below-melting-point technique of joining two metallic bodies by clamping them tightly together and applying ultrasonic energy, rather than heat, in the plane of the desired weld. Compare *ultrasonic brazing*.

ultrasonic whistle A whistle whose pitch is beyond the range of human hearing. While some are of the simile (blown) type, others are (usually miniature) electronic sound generators. These whistles are used for remote control of TV receivers, garage-door openers, and other equipment; calling dogs; and scaring birds.

ultrasonography A method of looking inside the body of transmitting ultrasonic waves and picking up returned echoes.

ultrasound Sound in the MHz range, used in ultrasonic diagnosis.

ultraudion oscillator An (often uhf) oscillator similar to the Colpitts circuit, except that the capacitive voltage division is provided by the internal grid—cathode and plate—cathode capacitances of the tube.

ultraviolet Pertaining to the invisible rays whose frequencies are just higher than those of visible violet light. The ultraviolet spectrum extends from approximately 750 to 950 THz.

ultraviolet altimeter An altimeter which uses ultraviolet rays instead of radio waves to determine the altitude of an aircraft or spacecraft.

ultraviolet lamp A lamp which delivers ultraviolet rays. Common types are arc lamps and mercury-vapor lamps, whose light output is rich in ultraviolet rays; owever, an incandescent lamp housed in an envelope of special glass which is transparent to ultraviolet rays is also a good source.

ultraviolet light See *ultraviolet rays.*

ultraviolet power The power (in watts) of ultraviolet radiation.

ultraviolet rays Radiation at frequencies in the ultraviolet region, i.e., between the highest visible-light frequencies and the lowest X-ray frequencies.

ultraviolet therapy The use of ultraviolet rays by physicians and other health personnel in the treatment of certain disorders.

ultraviolet wavelength The wavelength (in micrometers) or ultraviolet radiation.

ultraviolet waves See *ultraviolet rays.*

um See *u* and *micrometer.*

U-magnet See *horseshoe magnet.*

umbilical cord A cord through which missiles and rockets are controlled and powered until they are launched. Also a similar cord connecting an astronaut to his space vehicle during extravehicular activity (EVA), e.g, during a space "walk."

umbilical tower The vertical tower supporting the umbilical cords extending to a rocket in the launch position.

umbra 1. The region of total (conical) shadow behind an object situated in the path of a radiation. Also called *penumbra.* 2. Sunspot center.

umbrella antenna An antenna consisting of a number of wires extending from the top of a vertical mast to points on a circle below, at the center of which the mast is mounted. The wires are usually fed in parallel from the top.

umho See *u* and *micromho.*

UMW Abbreviation of *ultramicrowave.*

un A prefix meaning *not.* For the formation of electronics terms, this prefix must be distinguished from *dis,* which means *deprived of.* Thus, an *uncharged* body is one which has been charged but has been emptied of its charge. The same distinction applies to *unconnected* and *disconnected.* The prefix also implies a reversal (*un*clamp).

unamplified agc Automatic gain control in which the control-signal voltage is taken from a point in the circuit (such as the second detector dc output) and fed to the controlled point without being boosted by an auxiliary amplifier. Compare *amplified agc.*

unamplified avc See *unamplified agc.*

unamplified back bias A negative-feedback voltage developed across a fast-time-constant circuit within a single amplifier stage.

unamplified feedback A positive or negative current or voltage taken from the output of a system and presented to the input without being boosted through auxiliary amplification.

unattended operation Operation, as of an electronically progammed machine, without (or with very little) human supervision.

unattended time The period, excluding down time for repair or checkout, during which a computer is unpowered.

unattenuated Not reduced in intensity of amplitude. Thus, an unattenuated signal is one that has retained its original strength during its transmission through a system.

unavailable energy The difference between the quantity of heat energy supplied to a system and the available energy of the system.

unbalance 1. Lack of balance (or sometimes symmery) in a circuit, line, or system. 2. The condition in which a bridge (or the equivalent) is not nulled.

unbalanced circuit See *single-ended circuit.*

unbalanced delta system A three-phase circuit in which the elements are connected in a triangular (delta) configuration, but are of unequal impedance. In an unbalanced delta system, no definite relationship exists between line and phase currents. Compare *balanced delta system.*

unbalanced input See *single-ended input.*

unbalanced line A line whose sides are not symmetrical with respect to ground. Unbalance results in earth currents, which constitute a loss.

unbalanced multivibrator A multivibrator in which the time constant of one grid circuit differs from that of the other grid circuit (or in transistor multivibrators, the time constant of one base-input or gate-input circuit differs from that of the other). The asymmetry results in short-duration output pulses separated by long time periods.

unbalanced output See *single-ended output.*

unbalanced two-phase system A two-phase circuit in which the load elements are of unequal impedance.

unbalanced wire circuit A wire line whose (two) sides are dissimilar.

unbalanced-wye system A three-phase circuit in which the elements are connected in the familiar wye, or star, configuration, but of unequal impedance.

unbalanced-Y system See *unbalanced-wye system.*

unbalance-to-unbalance transformer A transformer for matching a pair of lines that are unbalanced with respect to ground, to a pair of lines which are balanced with respect to ground.

unbased tube A baseless electron tube, i.e., one normally having no base, but provided with leads or prongs for making connections externally.

unbiased error Accidental error.

unbiased limiting Peak limiting by overdriving an unbiased tube or transistor.

unbiasedness In probability studies, the state in which the expected value (X) of an estimate is equal to the parameter (μ), that is, $E/X- = \mu$.

unbiased unit A device or circuit operated without bias—usually, a tube operated without control-grid bias or a transistor without base bias (common emitter and emitter follower) or emitter bias (common base).

unblanking Removal of the blanking impulse in a cathode-ray circuit, i.e., turning on the beam. Compare *blank, 2.*

unblanking generator A (usually pulse-type) signal source for turning on the beam of a blanked cathode-ray tube.

unblanking interval The period during which the beam of a cathode-ray tube is turned on. Compare *blanking interval.*

unblanking pulse A pulse which turns on the beam of a blanked cathode ray tube. Compare *blanking pulse.*

unblanking time 1. *Unblanking interval.* 2. The instant at which unblanking begins.

unbonded strain gauge A strain gauge that is not directly attached (by cement, for example) to the strained surface.

unbound charge A free charge.

unbound electron A free electron.

unbuffered output An output (signal, power, etc.) that is delivered directly from the generating device without the benefit of an isolating stage, such as a *buffer amplifier.* Compare *buffered output.*

unbypassed cathode resistor In a common-cathode vacuum-tube circuit, a cathode bias resistor that doesn't have a bypass capacitor. The flow of output-signal current through the resistor produces negative feedback within a single stage.

unbypassed emitter resistor In a common-emitter transistor circuit, an emitter resistor that doesn't have a bypass capacitor. The flow of output-signal current through the resistor produces negative feedback within a single stage.

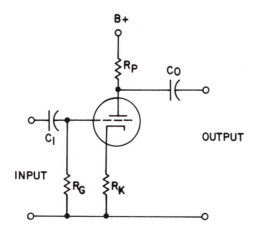

Unbypassed cathode resistor (R_k).

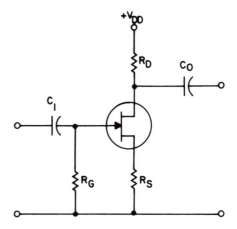

Unbypassed source resistor (R_s).

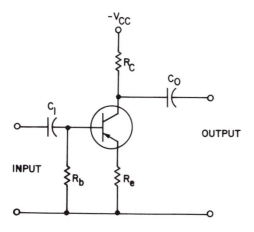

Unbypassed emitter resistor (R_e).

unbypassed source resistor In a common-source field-effect transistor circuit, a source resistor that doesn't have a bypass capacitor. The flow of output-signal current through the resistor produces negative feedback within a single stage.

uncage In a displacement gyroscope system, to disconnect the erection circuit,

uncalibrated unit A component, circuit, or instrument which has never been calibrated or has not recently been calibrated and, hence, is of questionable accuracy.

uncased choke See *unshielded choke*.

uncased transformer See *unshielded transformer*.

uncertainty In probability studies, hypothesis testing, and decision theory, the absence of sufficient information.

uncertainty in computation The degree of doubt concerning the exactness of computations. This uncertainty is always greater than that of the roughest measurement employed in obtaining data for the computation. Uncertainty is closely related to the tolerance of the instruments and formulas used.

uncertainty in measurement The estimated maximum amount whereby the numerical value of a measured quantity may differ from the true value.

uncertainty principle The observation that high precision in the location of an electron is obtainable only at a sacrifice in the accuracy with which the momentum of the electron can be determined, and vice versa.

uncharged Without an electric charge, as opposed to *discharged* (depleted of a charge).

unclamp To switch off clamping action in a circuit. See *clamper* and *clamping*.

uncoated filament 1. A plain filament, i.e., one without a coating of electron-increasing material. 2. A stripped filament, i.e., one from which the electron-increasing coating has been burned off.

uncompensated Not modified to give a desired type of performance, such as increased bandwidth or reduced temperature sensitivity. Example: an uncompensated amplifier put into service in the (wide) video band. Compare *compensated amplifier, compensated capacitor, compensated diode detector*.

uncompensated amplifier An amplifier without a provision for the automatic or manual modification or correction of is response. Compare *compensated amplifier*.

uncompensated capacitor A fixed or variable capacitor without a provision for the automatic or manual correction or modification of its capacitance or range or the improvement of its temperature coefficient. Compare *compensated capacitor*.

uncompensated inductor A fixed or variable inductor without a provision for the automatic or manual correction or modification of its inductance or range or the improvement of its temperature coefficient.

uncompensated resistor A fixed or variable resistor without a provision for the automatic or manual correction or modification of its resistance or range or the improvement of its temperature coefficient.

unconditional 1. Independent of external factors. 2. Constant.

unconditional branch See *unconditional jump*.

unconditional inequality In mathematics, an expression for an inequality that remains valid no matter what variables are used, e.g., $X^2 > -1$.

unconditional jump In computer practice, a program instruction that interrupts, without a relational test being made, the normal, ordered sequence of instructions and gives the address of (usually) the first instruction in a subroutine, e.g., BASIC's GOTO (line number). Also called *unconditional branch*. Compare *conditional jump*.

unconditional probability In the assessment of data or the prediction of results, the probability *P(B)* when all outcomes are equally likely.

unconditional stability Stability of a system at all frequencies and ages. Also called *absolute stability*.

unconditional transfer See *unconditional jump*.

uncontrolled multivibrator A multivibrator which is not synchronized with a signal source. Also called *astable multivibrator* and *free-running multivibrator*.

uncontrolled oscillator See *self-excited oscillator*.

uncorrected power An ac power value calculated without regard for power factor, i.e., a value which is the simple ac volt-ampere product.

uncorrected time Local standard time which has not been corrected in terms of the distance of the locality from the nearest standard-time meridian.

uncorrected tube A TV picture tube operated without antipincushioning magnets.

uncorrelated variables Variables that exhibit no correlation, i.e., $\rho (X, Y) = 0$.

uncoupled mode A mode of vibration existing in a system concurrent with (but independent of) other modes.

undamped galvanometer A galvanometer for which no provision has been made to limit overswing or prevent oscillation.

undamped meter 1. A meter for which no provision has been made to limit overswing or prevent oscillation. 2. A meter which is unprotected by a protective short circuit for limiting vibratory movement of the pointer during transport.

undamped natural frequency In the absence of damping, the oscillation frequency of a system having one degree of freedom when a transient force displaces the system momentarily from its position of rest.

undamped oscillation Continuous-wave oscillation. Also see *continuous wave*.

undamped output circuit An af power-amplifier output circuit which has not been designed so that overswing of the loudspeaker cone is prevented.

undamped speaker enclosure A loudspeaker cabinet for which provision has not been made to deaden resonances and other undesired vibration.

undamped wave See *continuous wave*.

undecillion The number 10^{36}.

undeflected beam A cathode-ray beam or a light beam in its normal position of rest (quiescence).

underbiased unit A component, such as a transistor or vacuum tude, whose bias voltage or current is lower than the prescribed value. Compare *overbiased unit*.

underbunching Less than optimum bunching of electrons in a velocity-modulated tube, due to lowered buncher voltage.

undercompounded generator A generator in which the output voltage varies inversely with load resistance.

undercoupling Loose coupling, usually of an amount insufficient for optimum signal transfer.

undercurrent A current of lower than specified strength. Compare *overcurrent*.

undercurrent relay A relay that is actuated when coil current falls below a predetermined level.

undercut The removal of the metal under the edge of the resist in a printed circuit by the etchant; thus the cross section of the conductor is reduced.

undercutting A phonograph-disk groove that is cut too shallow or with reduced internal movement of the recording stylus.

underdamping Insufficient damping of a system, i.e., not enough to prevent output oscillation following application of a transient force.

underdriven unit An amplifier, oscillator, or transducer whose driving signal (current, voltage, power, or other quantity) is lower than the prescribed value. Compare *overdriven unit*.

underexcited Receiving lower excitation than is normal, as in the underexcited rf amplifier of a transmitter, or an underexcited ac generator. Compare *overexcited*.

underexpose In photography and in radiation practice, to expose a film or body for a shorter time period than is required for optimum results (best tonal range); sometimes useful (in photography) as a special-effect technique.

underexposure The act or technique of exposing insufficiently (see *underexpose*).

underflow In computer practice, the condition in which a quantity entered into storare is shorter than the space provided for it, e.g., a 12-digit quantity in a 16-position register.

underground antenna A transmitting or receiving antenna installed and operated below ground. Included are buried antennas and antennas on equipment operated in tunnels, cellars, and similar areas.

underground cable A cable that is buried in the earth.

underground communication Communication between a transmitter and receiver, both below the surface of the earth.

underground image The below-ground mirror image of a vertical antenna and its radiation pattern.

underground line A power line laid below the surface of the earth.

underground receiver 1. A receiver situated completely below the surface of the earth. 2. A clandestine receiver.

underground reception Reception of an above-ground station's transmissions by an underground receiver. Compare *underground communication*.

underground transmitter 1. A transmitter situated completely below the surface of the earth. 2. A clandestine transmitter.

under insulation The insulation (usually a strip of tape) laid under a wire brought up from the center of a coil.

underinsulation Inadequate or insufficient insulation.

underlap In a facsimile or television picture, a crowding of the scanned lines.

underload circuit breaker See *circuit breaker*.

underloaded amplifier 1. An amplifier whose load resistance (impedance) is less than the prescribed value. 2. A power amplifier which is delivering less than its rated output power. Compare *overloaded amplifier*.

underload relay A relay which is actuated when circuit current drops below a predetermined value. Compare *overload relay*.

undermodulation Incomplete modulation of a carrier wave. Compare *complete modulation* and *overmodulation*.

underpass The crossing of two conductors on a semiconductor wafer, without an electrical connection.

underpower relay A relay that is actuated when power drops below a predetermined level. Compare *overpower relay*.

underrate To assign a rating (current, power) lower than the quantity of the rating an equipment can handle, or tolerate. For safey or reliability, apparatus sometimes is deliberately underrated, especially in power output and maximum current or voltage.

undershoot On an oscilloscope screen or a graph a momentary swing of a current or voltage below the reference axis. Compare *overshoot*.

undershoot distortion Distortion caused by reduction of the maximum amplitude of a signal wavefront below the steady-state amplitude that would be reached by a prolonged signal wave.

understudy See *backup*.

underthrow A form of signal distortion that occurs when the modulating-waveform frequency is too large in proportion to the frequency of the wave itself.

undervoltage A voltage of lower than specified value. Compare *overvoltage*.

undervoltage protection The automatic disconnection of a load device from its driving source when the driving voltage falls below a predetermined (threshold) level. This action is sometimes accomplished with a zener diode (whose breakdown voltage is equal to the threshold voltage) in series with a disconnect relay.

undervoltage relay A relay that is actuated when voltage drops below a predetermined level. Compare *overvoltage relay*.

underwater antenna A transmitting or receiving antenna which normally is operated in a body of water. It is usually a device associated with a submarine.

underwater sound projector A hydroacoustic transducer that converts af power into sound waves, which it radiates through a body of water to a receiver.

Underwriters Code The advisory National Electrical Code (NFPA No. 70) adopted by the National Fire Protection Association and often enforced by courts of law and inspection agencies; in its Article 810, the Association gives specifications for the safe installation of radio and television equipment.

Underwriters Laboratories A private corporation supported by underwriters, which issues standards of safety for components, equipment, and their operation. Many pieces of electronic equipment are labeled with the Underwriters Laboratories seal.

undistorted power output For amplifier tubes and transistors, a specified maximum af power output level at which the total distortion does not exceed a specified low value (i.e., at which operation is practically distortionless).

undistorted wave A sine wave that theoretically contains no harmonics, or a nonsinusoidal wave whose shape corresponds exactly to the equation for the wave.

undisturbed-one output In digital-memory practice, the one-output of a magnetic cell or other memory unit which has received a full rather than partial read pulse. Compare *undisturbed output signal*.

undisturbed output signal In digital-memory practice, the output of a magnetic cell or other memory unit previously set to one or zero, which has received a full rather than partial read pulse. Also called *undisturbed response voltage*.

undisturbed response voltage See *undisturbed output signal*.

undisturbed-zero output In digital-memory practice, the zero-output of a magnetic cell or other memory unit which has received a full rather than partial read pulse. Compare *undisturbed output signal*.

undoped Pertaining to a pure semiconductor material, i.e., one containing no lattice-altering additives.

undulating current A current, such as an alternating or composite current, whose value oscillates in the manner of a wave.

undulation A wavelike alternation.

undulatory theory The theory that light is transmitted by means of an undulating (wavelike) movement between the luminous object and the eye.

unending decimal A nonterminating decimal, such as the value of pi: 3.141 592 653.... Compare *terminating decimal*.

unequal alternation Positive and negative half-cycles of an ac wave which have unequal heights or widths or are otherwise, disimilar in form.

unexpected halt During a computer program run, an undirected, i.e., unplanned for, halt, e.g., one caused by a machine fault or program bug.

unfiltered Not having been subjected to filtering action. The unfiltered output of an ac-operated dc power supply, containing the ripple component characteristic of the type of rectifier employed, is an example.

unfired tube 1. A transmit—receive switching tube (ATR, TR, pre-TR) in which rf glow discharge is not present at either the resonant gap or resonant window, 2. An unignited glow tube or thyratron.

unfocused light source In a photoelectric system, a light source which delicers diffused light.

unformed rectifier A newly fabricated semiconductor rectifier (especially selenium) or an electrolytic rectifier before it has been electroformed for improved characteristics.

unfurlable antenna An antenna which may be unrolled to increase its length, and vice versa.

ungrounded item A component, circuit, or circuit point which has no connection to ground except an inadvertent one through common impedances or leakage paths.

ungrounded system A system which is operated entirely above ground, any path to ground being accidental.

unguided Without electronic guidance (pertaining to missiles, rockets, satellites, etc.).

uni A prefix (combining form) meaning *one* or *single* and appearing in a number of electronic terms, such as *uniform, unilateral, unipolar*, etc.

uniaxial Having only one axis or referred to only one axis.

uniaxial crystal A crystal having one optical axis.

unibivalent electrolyte An electrolyte, such as sodmium carbonate (Na_2CO_3), whic dissociates into two univalent ions and one bivalent ion.

uniconductor cable A cable having a single conductor, usually of braided or twisted wires.

uniconductor waveguide A waveguide consisting of a metal sheath deposited on a solid dielectric of cylindrical or rectangular cross section.

unidentified flying object Abbreviation, UFO. A mysterious object claimed to have been seen in flight, but not identified by reliable authorities as any known type of vehicle. Often used to mean *flying saucer*, a type of UFO.

unidirect To commutate (see *commutator*).

unidirectional 1. Flowing or acting in one direction only. 2. Having a radiation or response sensitivity that is maximum in primarily one direction in space.

unidirectional antenna See *unidirectional array*.

unidirectional array A beam antenna that radiates in one direction only, or principally in one direction, unless rotated.

unidirectional conductivity The conductivity in only one direction that characterizes an ideal diode.

undirectional coupler A directional coupler sampling only one direction of transmission.

unidirectional current A current that flows always in the same direction, although it may fluctuate in calue.

unidirectional elements Circuit elements, such as diodes, tubes, or transistors, that transmit energy in only (or best in) one direction.

undirectional field-effect transistor See *unilateral field-effect transistor.*

undirectional hydrophone A special underwater *undirectional microphone.*

undirectional loudspeaker A loudspeaker which radiates sound substantially in one direction.

unidirectional microphone A microphone that receives sound waves from one direction, usually from the front, minimum response usually being from the sides and back.

unidirectional network A network that transmits signals in only one direction, i.e., the input and output terminals are not interchangeable.

unidirectional pattern For a transducer, such as an antenna, speaker, or microphone, a radiation or response pattern that shows a pronounced maximum in one direction only.

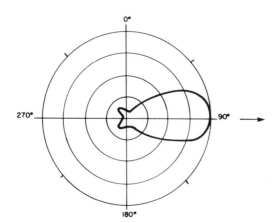

Unidirectional pattern.

unidirectional pulse A single-polarity pulse.

unidirectional pulse train A train of unidirectional pulses.

unidirectional response For a receiving transducer, such as an antenna or microphone, a response pattern that shows a pronounced maximum in one direction only.

unidirectional speaker See *unidirectional loudspeaker.*

unidirectional transducer A transducer which operates in one direction only; i.e., its input and output cannot be interchanged.

unidirectional voltage A voltage that never changes polarity, although it may fluctuate in value.

UNIFET Abbreviation of *unipolar field-effect transistor.* (Also, *UFET*).

unified atomic mass unit Symbol, u. A unit of mass equal to 1/12 that of the carbon-12 atom, or $1.660\ 531 \times 10^{-24}$ gram.

unified-field theory A theory, as yet unproved, that all forces in nature are interrelated and are controlled by the same causes and factors.

Albert Einstein worked on this theory after he successfully developed the theories of relativity.

unifilar Having (or wound as) one fiber, wire, filament, or thread.

unifilar magnetometer A magnetometer in which a bar is suspended at its center of gravity by a single thread.

uniform change The condition in which one variable changes with a change in another. In electric phenomena, these changes are readily dealt with by means of calculus.

uniform circular motion Motion at a uniform rate and describing a circle, e.g., a motor armature rotating at a constant speed.

uniform electric field An electric field in which all the lines of force are straight and parallel and in which the electrostatic force has the same value at all points, e.g., the field between two oppositely charged, flat, parallel plates.

uniform field See *uniform electric field* and *uniform magnetic field.*

uniform frequency response Frequency response that is flat throughout a specified range. Such response is characterized by the transmission of a signal with no introduced amplitude or phase distortion.

uniform line A transmission line having identical electrical properties over its entire length.

uniform magnetic field A magneic field in which all the lines of force are straight and parallel and in which the magnetic force has the sane value at all points, e.g., the field between the flat, parallel pole faces of a permanent magnet.

uniform plane wave A free-space plane wave at an infinite distance from the generator, having constant-amplitude electric and magnetic field vectors over the equiphase surfaces.

uniform precession In regions of the uniform magnetic field of a sample of material, the state in which the magnetic moments of all atoms are parallel and precess in phase around the magnetic field.

uniform waveguide A waveguide having constant electrical and physical characteristics along its axis.

uniground 1. The grounding of a circuit at one point, to reduce susceptibility to hum and noise. 2. The point at which such a connection is made.

unijunction transistor A semiconductor device consisting of a thin silicon bar on which a single *pn* junction acting as an emitter is formed near one end. Two bases are provided, each an ohmic connection made to one end of the bar. Also called *double-base diode.*

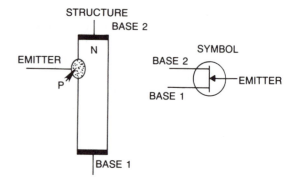

Unijunction transistor.

unilateral area track A film sound track having modulation on only one edge of the opaque area.

unilateral bearing In radio direction finding, a bearing obtained with a finder giving unilateral response and thereby obviating 180 degree error.

unilateral conductivity See *unidirectional conductivity.*

unilateral element See *unidirectional elements.*

unilateral field-effect transistor A field-effect transistor whose source and drain terminals cannot be interchanged. Also called *asymmetrical field-effect transistor.* Compare *symmetrical field-effect transistor.*

unilateralization A system of neutralization employed in uhf and vhf transistor amplifiers, in which the internal capacitive *and* resistive components are compensated by the neutralization feedback.

unilateralized amplifier A transistor amplifier in which the internal resistive component as well as the internal capacitive component, is compensated by the neutralizing circuit. If the resistive component is neglected, the amplifier is merely *neutralized.* Also see *unilateralization.*

unilateral network See *unidirectional network.*

unilateral transducer See *unidirectional transducer.*

unimodular matrix A square matrix whose determinant is 1.

union Phonetic alphabet communications code word for the letter *U.*

union 1. The logical inclusive-OR operation. 2. See *bond, 1.*

union The logical inclusive-OR operation.

union catalog In computer practice, the compiled list of the contents of two or more tape libraries.

unionic material A material having no ions, i.e., one in which all atoms are neutral.

union of sets In set theory, the set $A \cup B$ that contains those elements that are in A and B, e.g., if set A contains w, x, and y, and set B x, y, and z, $A \cup B$ contains x and y.

uniphase antenna See *collinear antenna.*

UNIPOL Acronym for *universal problem-oriented language,* a computer-programming language.

unipolar Having or using a single pole or polarity, or operating with one class of current carrier.

unipolar armature An armature that maintains its polarity throughout a complete revolution.

unipolar field-effect transistor See *unipolar transistor.*

unipolar induction Induction by only one pole of a magnet.

unipolar input The input circuit of an instrument or device designed for input signals of one polarity only.

unipolar pulse A pulse in which the current flows in only one direction, or in which the voltage occurs with only one polarity.

unipolar transistor A field-effect transistor. So called because it utilizes only one kind of carrier (electrons in the n-channel FET, holes in the p-channel FET). Compare *bipolar transistor.*

unipolar winding The winding of a unipolar armature.

unipole 1. An all-pass filter having one pole and for which there is one zero. Also see *all-pass filter, poles of impedance,* and *zeros of impedance.* 2. A hypothetical, omnidirectional antenna.

unipotential cathode An indirectly heated tube cathode. Also called *equipotential cathode.* See more at *indirectly heated cathode.*

unique 1. The condition of leading to only one result. Thus, the product of A and B is unique, but $\sqrt{A}$ is not, unless A is zero. 2. The condition of being completely individual, e.g., a unique sweep circuit.

unit 1. A named, single magnitude adopted as a standard by physical measurement. Thus, the unit of current is the *ampere,* of frequency

the *hertz,* of capacitance the *farad,* and so on. Also see *absolute system of units, International System of Units, centimeter-gram-second.* 2. A single piece or assemblage of equipment, such as *amplifier unit, converter unit, power supply unit,* and so on. 3. The number 1 implied when *unit* is the adjective describing a quantity, e.g., *unit length* means a distance of 1m, 1 ft, or one of whatever is the unit of measurement.

unitary code In computer practice, a code based on one digit. The number of times the digit is repeated indicates a given number.

unit cell In crystals, the simplest polyhedron exhibiting all the structural features and which is repeated to form the *crystal lattice.*

unit charge See *unit electrostatic charge.*

unit electric charge See *unit electrostatic charge.*

unit electrostatic charge An electrostatic point charge which will attract or repel a point charge of equal strength 1 cm away in a vacuum, with a force of 1 dyne (10^{-5}N).

unit length 1. A fundamental unit of distance or time, used for reference in a measuring system. For example, in the *MKS* (meter-kilogram-second) system, the distance unit length is the meter, and the time unit length is the second. 2. The duration of a fundamental element, or bit, in a binary-code transmission system.

uniterm system For data-processing and computer applications, an information retrieval system in which the data is on punched cards having *descriptors:* unique digit groups or symbols distinguishing individual cards and describing their content.

unit fraction Where n is an integer, a fraction of the form $1/n$.

unit function Symbol, H or 1. A time-dependent quantity which is zero before the start of a period (when time t is zero) and 1 for all values of t greater than zero. It is approximated by a square wave, and is useful in solving problems involving transients.

unitized construction The fabrication of an electronic equipment in subassemblies (such as modules) which may be tested separately and which can be easily replaced (plugged in) in the event of individual failure.

unit line A line of force.

unit magnetic pole See *unit pole.*

unit matrix A matrix whose main diagonal terms (all) are all at unity, the other terms being zero.

unitooth Pertaining to the use in electrical machinery of one slot per pole per phase.

unitor An OR gate.

unit pole The strength (magnetic flux) of a (hypothetical) magnetic pole which will attract or repel a pole of equal strength 1 cm away in a vacuum, with a force of 1 dyne; 1 unit pole = $1.256\ 637 \times 10^{-7}$ weber.

unit-ramp function A function whose value is zero before time t, and becomes equal to the time measured from t at all other instants. The unit-ramp function is the integral of the unit-step function.

unit record In computer practice, a complete record comprising many data elements and contained within one storage medium, such as a punched card.

unitrivalent electrolyte An electrolyte, such as sodium phosphate (Na_3PO_4), that dissociates into three univalent ions and one trivalent ion.

units position In a numbering system, the right most position in a multidigit number.

unit-step function See *unit function.*

unit string A one-member string.

unitunnel diode A special form of semiconductor diode, used as an oscillator at ultra-high and microwave frequencies.

unit vector A vector that is 1 unit long in terms of the scale representing a factor of interest, and has the same direction as the vector of interest. For example, if *u* is a unit vector, then 5*u* represents a vector having the same direction and 5 times the magnitude of *u*.

unity 1. The figure 1 implied. 2. The state of oneness.

unity coupling Tight coupling between two coils, the turns ratio often being 1 to 1 and the coils always being closely interwound.

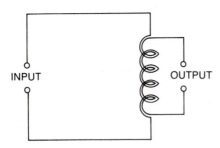

Unity coupling.

unity gain The condition in which the output amplitude is the same as the input amplitude; that is, a gain factor of 1, or 0 dB.

unity-gain bandwidth 1. For an active filter, the bandwidth between the frequencies at which the gain is 0 dB. 2. For an operational amplifier having a rolloff of 6 dB per octave, the frequency at which the open-loop gain is 0 dB.

uni-univalent electrolye An electrolyte which dissociates into two univalent ions, e.g., sodium chloride (NaCl).

univ Abbreviation of *universal.*

UNIVAC The name given to a family of digital computers and data-processing systems. The name is an acronym for *univ*ersal *a*utomatic *c*omputer.

univalent Having a valence of 1. For example, in the compound sodium chloride (NaCl), the sodium ion (Na^+) and the chlorine ion (Cl^-) are univalent.

universal bridge A bridge for the measurement of capacitance, inductance, and resistance. Such a bridge often is a skeletal foundation circuit with a provision for plugging *L, C,* or *R* standards in the various arms, as required. Also called *general-purpose bridge.*

universal coordinated time Abbreviation, UTC. Greenwich mean time coordinated by the International Time Bureau for broadcast as signals (on WWV, for example).

universal coupler 1. An arrangement of one or more inductors and variable capacitors for matching a transmitter to virtually any antenna. One such device is the Collins coupler. 2. A device for matching numerous generator output impedances to numerous load impedances.

universal filter An active filter which is continuously tunable over a wide frequency range and which offers low-pass, high-pass, bandpass, and band-suppression functions.

universal frequency counter A digital frequency meter usable at radio and audio frequencies.

universal motor A small series-wound motor which runs on dc or single-phase ac. This type of motor is used in many household appliances and in motor-driven tools, such as portable electric drills, polishers, etc.

universal output transformer An output transformer having a number of taps on its primary and secondary windings for use with a wide variety of impedances.

universal product code Abbreviation, UPC. The variable-width-bar code appearing on price tags or product labels, and yielding such information as cost, size, and color when read by supermarket (or other retailer) optical character recognition equipment. Also called *universal vendor marking* (UVM).

universal receiver An ac–dc radio receiver.

universal set Symbol, U. In set theory, the set of all items under consideration.

universal shunt See *Ayrton–Mather galvanometer shunt.*

universal time See *Greenwich mean time.*

universal transformer See *universal output transformer.*

universal transmitter An ac–dc radio transmitter.

universal Turing machine In computer theory, a Turing machine capable of simulating other Turing machines.

universal vendor marking See *universal product code.*

universal winding A zigzag winding used to reduce the distributed capacitance of multilayer coils. Also called *lattice winding* and *honeycomb winding.*

universal-wound coil A coil employing a universal winding. Such coils are common in i-f transformers and rf chokes.

universe See *universal set.*

univibrator See *monostable multivibrator.*

uniwave method A method of keying an arc converter, in which the rf output current flows into the antenna when the key is depressed and into an equivalent dummy load when the key is released. This technique keeps the converter evenly loaded.

unknown quantity The quantity which is sought in calculations—usually represented by variable names, such as *x, y,* and *z.*

unlike charges Dissimilar electric charges, i.e., positive and negative.

unlike poles Dissimilar poles, such as the *north* and *south* poles of a magnet.

unload 1. To remove data from a file. 2. To disconnect the load from a circuit.

unloaded amplifier See *unterminated amplifier.*

unloaded antenna An antenna operated without added inductance or capacitance.

unloaded-applicator impedance In dielectric-heating practice, the unloaded impedance of applicator electrodes placed in their normal working position without the dielectric-material load in place.

unloaded battery 1. A battery in the standby condition. 2. A battery tested for open-circuit terminal voltage, i.e., with no load except the voltmeter.

unloaded generator See *unterminated generator.*

unloaded line See *unterminated line.*

unloaded potentiometer A potentiometer or voltage divider with an open-circuited output.

unloaded Q The *Q* of a coil or tuned circuit which is activated by a signal but delivers no output to a load.

unloaded transmitter A transmitter operated under open-circuit conditions, i.e., without an external load, such as an antenna or dummy resistor.

unmatched elements 1. Components—such as resistors, capacitors, semiconductors, etc.--having different values. 2. Mating elements or devices not having the same impedance. Also called *mismatch.*

unmodulated carrier See *continuous wave.*

unmodulated current An alternating or direct current which is not varied in intensity by a signal, including noise.

unmodulated voltage An ac or dc voltage which is not varied in amplitude by a signal, including noise.

unmodulated wave See *continuous wave.*

unode See *monode.*

unpack In computer practice, to remove data from a storage location into which, with other data, it has been *packed* (as a memory-conservation measure).

unpolarized light Light in which the wave orientation is random around the axis of the beam.

unpolarized plug A plug which may be inserted into a socket in two or more different ways, thus increasing the likelihood of a wrong connection. Compare *polarized plug.*

unpolarized receptacle A receptacle into which a plug may be inserted in two or more different ways, thus increasing the likelihood of a wrong connection. Also called *unpolarized socket.* Compare *polarized receptacle.*

unpolarized relay A relay that always closes in the same direction, regardless of the direction of current in its coil. Also called *nonpolarized relay.*

unpolarized socket A socket into which a component or plug may be inserted in two or more different ways, thus increasing the likelihood of wrong connections. Also called *unpolarized receptacle.*

unprotected antenna An outside antenna operated without a lightning arrester or lightning switch.

unrationalized systems of units Collectively, the (1) absolute cgs electrostatic system, (2) absolute cgs electromagnetic system, and (3) absolute mks system.

unreflected ray See *unreflected wave.*

unreflected wave 1. Direct wave. 2. Ground wave. 3. A large-angle wave that penetrates the ionosphere and passes into space.

unregulated Not held to a constant value. For example, an unregulated voltage is free to fluctuate in value.

unregulated power supply A power supply whose output current or voltage is not automatically held to a constant value. Compare *constant-current source* and *constant-voltage source.*

uns Abbreviation of *unsymmetrical.*

unsaturated core A magnetic core operated below the point of saturation, i.e., the point beyond which an increase in coil current produces no further increase in magnetization of the core.

unsaturated logic A method of digital-logic electronics in which the switching devices operate between the saturated and cut-off conditions during either or both of the high and low states.

unsaturated operation Operation of a tube, transistor, or magnetic core at a point below saturation, i.e., below the point at which (1) an increase in voltage produces no further increase in current, or vice versa, in a tube or transistor, or (2) an increase in coil current produces no further increase in magnetization of a core. Compare *saturated operation.*

unsaturated standard cell See *standard cell.*

unshielded cable A cable, such as a twisted pair or a multiwire twist, lacking a shielding jacket. Unless special precautions are taken, such as transposing conductors, such a cable is susceptible to stray pickup and is capable of radiation.

unshielded choke An uncased choke, i.e., one without a protective and shielding metal housing.

unshielded coil An inductor without a field-confining enclosure.

unshielded probe An instrument probe that has no shielding enclosure. Such a probe is desirable for some tests, but it is subject to antenna-pickup effect, body capacitance, and stray-field pickup.

unshielded stage In an electronic equipment, a stage operating entirely in the open, i.e., without electrostatic or magnetic shielding enclosures and is therefore susceptible to stray pickup and capable of radiation.

unshielded transformer 1. An uncased transformer, i.e., one without a protective and shielding metal housing. 2. A transformer having no electrostatic shield between its windings.

unshielded tube A glass or ceramic electron tube operated without an external, metal shield.

unshift The mechanical action in a teletypewriter when the carriage moves from the *figures* position to the *letters* position.

unshunted current meter A single-range D'Arsonval milliammeter or microammeter which as no shunt resistor. The resistance (*Rm*) of the instrument is accordingly the resistance of the movable coil.

unsolder To separate wires, contacts, or fastenings by melting the solder which holds them together. Also called *desolder.*

unsophisticated electronics Electronic theory and practice of the simplest kind, free of complexity, easily described, and often aimed toward casual applications, such as a hobby. Compare *sophisticated electronics.*

unstable oscillation 1. Relaxation oscillation. 2. The periodically interrupted oscillation characteristic of a blocking oscillation.

unstable region The negative-resistance portion of an N- or S-shaped response curve.

unstable servo A servo having unlimited output drift.

unstable state A condition, such as a *negative-resistance* region, which is difficult to maintain and often results in oscillation. The negative-resistance region of the tunnel diode forward conduction curve is evidence of such a state. Compare *stable state.*

untapped winding A winding without a tap (terminals are at each end of the winding only). Also called *untapped coil.*

unterminated amplifier An amplifier operated without a load device, i.e., under open-circuit output conditions.

unterminated generator A generator operated without a load device, i.e., under open-circuit output conditions.

unterminated line A transmission line that is not terminated with an impedance, i.e., a line operated open-ended.

untriggered flip-flop A flip-flop which at a particular instant receives no actuating pulse and therefore remains in the state it is in.

untuned amplifier An rf amplifier which is not tuned to a single frequency, but has useful gain over a wide band of frequencies. Examples are the distributed amplifier and the video amplifier.

untuned filter See *all-pass filter.*

untuned line An aperiodic transmission line, i.e., one which is only tuned to a particular frequency by its own distributed characteristics.

untuned transformer A transformer having simple primary and secondary windings and no tuning devices (such as capacitors in series or parallel with the windings), and is designed so that its natural resonant frequency (due to distributed capacitance) lies outside the specified range of operation.

unused time See *unattended time.*

unweighted average In the statistical analysis of data, an average calculated from terms which are not weighted by a factor. Compare *weighted average.* Also see *unweighted term.*

unweighted term In the statistical analysis of data, a term which is not operated upon by a weighting factor, and therefore has no extraordinary influence on a final result, such as an average. Compare *weighted term.*

unwind To eliminate all redundant or unnecessary operations in a computer system.

uohm See *u* and *microhm*.

up-and-downer amplifier In broadcasting and sound amplification, an auxiliary amplifier that allows the level of sound effects and background music to be controlled automatically without disturbing the dialog levels.

UPC Abbreviation of *universal product code*.

upconv Abbreviation of *upconverter*.

upconvert In superheterodyne conversion, to heterodyne a signal to an intermediate frequency higher than the signal frequency. Compare *downconvert*.

upconverter A superheterodyne mixer stage in which upconversion is performed (see *upconvert*).

update In data processing practice, to make a file reflect the current status of pertinent information by using transactions for its modification.

up-down counter See *bidirectional counter*.

upper atmosphere See *ionosphere, stratosphere*, and *troposphere*.

upper sideband In an amplitude-modulated wave, the component whose frequency is the sum of the carrier frequency and the modulation frequency, i.e., $fu = fc + fm$. Compare *lower sideband*.

upper sideband, suppressed carrier Abbreviation, USSC. A single-sideband transmission technique in which the upper sideband is transmitted, the lower sideband is suppressed, and the carrier is suppressed. Compare *double sideband, suppressed carrier* and *lower sideband, suppressed carrier*.

up time See *serviceable time*.

upturn A usually sudden rise in a performance curve. Compare *downturn*.

upward modulation Modulation in which the output current of the modulated circuit increases during modulation. Compare *downward modulation*.

UR Abbreviation of *ultrared*.

uranium Symbol, U. A radioactive metallic element. Atomic number, 92. Atomic weight, 238.07.

uranium 235 Symbol, U235. An isotope of uranium of mass number 235, used in atomic bombs and atomic power plants.

uranium 238 Symbol, U238. An isotope of uranium of mass number 238, which, under fast-neutron bombardment, decays to become neptunium and then plutonium (of mass number 239).

uranium metals The group of heavy, radioactive elements of atomic numbers 89 through 103. Also called the *actinide series*.

uranium rays The rays emitted by uranium. In 1896, Antoine Henri Becquerel observed that uranium salts emit rays (later identified as alpha, beta, and gamma) which pass through bodies opaque to light and are capable of exposing a photographic plate. Also called *Becquerel rays*.

urea formaldehyde resin A thermosetting, synthetic resin used in the manufacture of a number of plastic dielectric bodies and formed by the reaction of urea with formaldehyde.

urea plastic Any of several thermosetting plastics having a urea base and used as a dielectric and as a molding material in electronics, e.g., urea formaldehyde resin.

URSI Abbreviation of *Union Radio Scientifique Internationale*. (*International Radio Scientific Union*).

ursigram A broadcast message giving information on sunspots, radio propagation, terrestrial magnetism, and related subjects. Also see *URSI*.

us See *u* and *microsecond*.

usable frequency 1. Any frequency at which communications can be maintained between two points via ionospheric propagation. 2. Any frequency at which a communications system is operational.

usable sample 1. The portion of an oscilloscope or monitor display that is visible on the screen. 2. In statistics, a sample that is considered valid for calculation purpose.

USB Abbreviation of *upper sideband*.

usec See *u* and *microsecond*.

useful life The elapsed time between the installation of an expendable component, circuit, or system, and the time it must be replaced.

useful line In television, the portion of a scanning line that can be used for picture signals. Also called *available line*.

user group A group of hobbyists or company representatives having in common the ownership and/or use of a specific brand of computer and who meet or otherwise interact to share their expertise with, or programs, for, the machine. Also applicable to similar groups of programmable-calculator owners.

user's library A compilation of programs supplied to (by the vendor) or generated by a user group.

USM Abbreviation of *underwater-to-surface missile*.

USSC Abbreviation of *upper sideband, suppressed carrier*.

UTC Abdreviation of *coordinated universal time*.

utility An organization providing a public service, such as electric power or electronic communications.

utility box A general-purpose aluminum or steel box, supplied in various convenient sizes (painted or unpainted), used as a housing or shield for electronic equipment.

utility factor For a transformer used in a dc power supply, the ratio of dc output to required kVA capacity.

utilization factor See *utility factor*.

UTL Abbreviation of *unit transmission loss*.

uuF See *u* and *micromicrofarad*.

UUM Abbreviation of *underwater-to-underwater missile*.

uuW See *u* and *micromicrowatt*.

UV 1. Abbreviation of *ultraviolet*. 2. Abdreviation of *undervoltage*.

uV See *u* and *microvolt*.

UV base A four-pin bayonet-type base of early vacuum tubes, such as the UV200 detector tube.

uvicon A TV camera tube in which a conventional vidicon scanning section is preceded by a photocathode which is ultraviolet sensitive, an electron-accelerating section, and a special target.

uviol lamp An ultraviolet lamp using uviol glass, a special glass transparent to ultraviolet rays.

UVM Abbreviation of *universal vendor marking*.

uV/m See *u* and *microvolts per meter*.

uW See *u* and *microwatt*.

uW/cm^2 See *u* and *microwatts per square centimeter*.

UX base A four-pin base of early vacuum tubes, such as the UX201A detector--amplifier.

UY base A five-pin base of early vacuum tubes, such as the UY227 detector—amplifier.

V 1. Abbreviation of volt. 2. Symbol (ital) for *voltage* or *potential*. 3. Symbol for *vanadium*. 4. Symbol for *volume*. 5. Symbol (ital) for *reluctivity*. 6. Abbreviation of *vertical*. (Also, *vert.*)

v 1. Abbreviation of *velocity*. 2. Abbreviation of *vector*.

VA Abbreviation of *volt-ampere*.

V/A Abbreviation of *volts per ampere* (ohms).

VAC Abbreviation of *vector analog computer*.

vac Abbreviation of *vacuum*.

Vac Symbol for *ac volts*.

vacuum A space from which all air and other gases have been removed to the greatest practicable extent. Some electronic parts, such as the elements of an electronic tube, are housed in an evacuated space to protect them from the deterioration that would result from open-air operation. The filament of a radio tube or electric-light bulb, for example, would soon burn up if operated outside a vacuum.

vacuum capacitor A plate- or concentric-cylinder-type capacitor sealed in an evacuated glass tube or bulb. The vacuum acts as the dielectric and provides a dielectric constant of 1 and very high voltage breakdown.

vacuum deposition The electrical application of a layer of one material (such as a metal) to the surface of another (the substrate), carried out in a vacuum chamber, e.g., evaporation and sputtering. Also see *deposition* and *evaporation, 1*.

vacuum diode A tube diode containing a hard vacuum. Also see *tube diode*.

vacuum envelope The shell or tube of an electron tube, X-ray tube, or other electron device requiring a vacuum.

vacuum evaporation A method of manufacturing thin-film circuits by vaporizing a substance, and letting it accumulate or condense on a base.

vacuum gauge An instrument for measuring the vacuum in a device being evacuated. One of the several varieties of this gauge employs the elements of a triode, which are sealed in part of the vacuum system, some characteristic of the tube being continuously monitored as the vacuum pumping progresses. Another uses a thermistor sealed in part of the system; the resistance of the thermistor changes proportionately during the vacuum pumping.

vacuum impregnation The impregnation of a device (such as a capacitor, transformer, or choke coil) in a vacuum chamber. The process causes the pores in the device and its insulating materials to be completely filled with the impregnant.

vacuum level 1. The pressure, in millimeters of mercury (mm/Hg). Normal atmospheric pressure is 760 mm/Hg. A perfect vacuum would be represented by 0 mm/Hg. 2. The proportion, or percentage, of normal atmospheric pressure in a given environment.

vacuum phototube A phototube enclosed in an evacuated envelope, as opposed to one which is gas-filled.

vacuum pump A pump for removing air and gases from electron tubes, X-ray tubes, lamp bulbs, etc. Such pumps include the mechanical types, diffusion types, and combinations of these. Also see *diffusion pump*.

vacuum range The range of a communications system if propagation takes place through a perfect-vacuum (theoretical) medium.

vacuum seal An airtight seal between adjoining parts in an evacuation system.

vacuum switch A switch that is enclosed in a vacuum bulb or tube to reduce contact sparking.

vacuum thermocouple A thermocouple enclosed in a vacuum bulb with a small heater element. Radio-frequency current passed through the heater raises its temperature and causes the thermocouple, an rms device, to generate a proportional dc voltage.

vacuum tube An electron tube from which virtually all air and gases are removed. Also see *electron tube*.

vacuum-tube amplifier An amplifier employing one or more vacuum tubes, rather than semiconductor devices.

vacuum-tube bridge A special bridge for determining *vacuum-tube characteristics* by null methods.

vacuum-tube characteristics The operating parameters of a vacuum tube—such as plate current, grid voltage, input resistance, interelectrode capacitances, amplification factor, transconductance, etc.—which describe tube performance.

vacuum-tube coefficients See *vacuum-tube characteristics*.

vacuum-tube current meter An ammeter, milliameter, microammeter, or picoammeter embodying an amplifier which employs one or more vacuum tubes. Also see *electronic current meter*.

vacuum-tube diode See *tube diode*.

vacuum-tube keyer See *electronic keyer*.

vacuum-tube modulator A circuit employing one or more vacuum tubes to impress a modulating signal upon a carrier.

vacuum-tube oscillator An oscillator (controlled or self-excited) in which the active element is a vacuum tube.

vacuum-tube receiver A receiver (radio, television, radar, etc.) employing vacuum tubes, rather than semiconductor devices.

vacuum-tube rectifier An ac-to-dc converter circuit employing one or more vacuum tubes, rather than gas tubes or semiconductor devices. Also, a rectifier tube.

vacuum-tube sweep A sweep oscillator employing a vacuum tube, rather than a gas tube. Also called *hard-tube sweep*.

vacuum-tube transducer See *electromechanical transducer, 2*.

vacuum-tube transmitter A transmitter (radio, television, radar, etc.) employing vacuum tubes, rather than transistors.

vacuum-tube voltmeter Abbreviation, vtvm. A voltmeter employing a tube-type amplifier. Also see *electronic voltmeter*.

val Abbreviation of *value*.

valence A unit showing the degree to which elements or radicals (replaceable atoms or groups of atoms) will combine to form compounds.

valence band In the energy diagram for a semiconductor, the band of lowest energy. This band lies below the forbidden band (energy gap), which is below the conduction band. Also see *energy band diagram*.

valence bond In a semiconductor material, an interatomic path over which shared electrons travel.

valence electrons Electrons in the outermost orbits of an atom. It is these electrons which determine the chemical and physical properties of a material. Also see *free electron*.

valence shells See *electron shells*.

validate To check data for correctness.

validity 1. Correctness or accuracy of data. 2. The logical truth of a derivation or statement, based on a given set of propositions.

valley A dip between adjacent peaks in a curve or wave.

valley current In a tunnel diode, the current at the valley point.

valley point The lowest point of finite current on the current—voltage response curve of a tunnel diode. Immediately before this point, current decreases with increasing applied voltage (an indication of negative resistance). Beyond this point, however, the current again increases with increasing voltage. Compare *peak point*.

valley voltage In a tunnel diode, the voltage at the valley point.

value 1. The level or magnitude of a quantity, e.g., voltage value. 2. The worth of a system, procedure, device, etc. in terms of goal fulfillment or other criterion.

valve Variation (Brit.) of *electron tube*. The term was applied to the first tube, a diode (the Fleming valve), and is descriptive of the action of a tube (a valve controlling the flow of an electric current) rather than its appearance (*tube*).

vanadium Symbol, V. A metallic element. Atomic number, 23. Atomic weight, 50.95.

Van Allen radiation belts Two zones—at altitudes of 2000 miles and 10,000 miles—which encircle the earth at the magnetic equator and consist of high-energy subatomic particles. The nearer belt contains high-energy protons; the farther, high-energy electrons.

Van de Graaff generator See *belt generator*.

Van der Pol oscillator A relaxation oscillator employing a pentode and having a sawtooth-wave output.

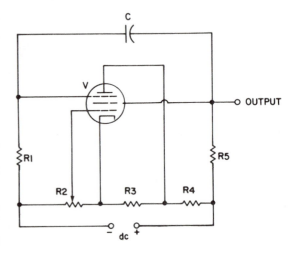

Van der Pol oscillator.

vane-anode magnetron A magnetron having plane parallel walls between adjacent cavities.

vane attenuator A waveguide attenuator consisting essentially of a slab (vane) of resistive material that slides laterally through the waveguide.

vane instrument See *iron-vane meter*.

vane-type magnetron See *vane-anode magnetron*.

vane-type meter See *iron-vane meter*.

V-antenna See *vee-antenna*.

vapor lamp A discharge lamp consisting essentially of a glass tube filled with a small amount of as under low pressure and some mercury. A high voltage is applied between electrodes sealed into each end of the tube. The voltage causes the mercury to vaporize, and this in turn ionizes the gas, causing it to glow.

vapor pressure In a confined medium, the pressure of a gas, measured in atmospheres, pounds per square inch, or millimeters of mercury.

VAR Abbreviation of *volt-amperes reactive*.

var Abbreviation of *variable*.

varactor A semiconductor-type voltage-variable capacitor.

varactor amplifier A dielectric amplifier employing a varactor as the voltage-variable capacitor.

varactor flip-flop A bistable multivibrator based upon the nonlinear performance of one or two varactors.

varactor frequency multiplier A frequency multiplier (doubler, tripler, etc.) in which multiples of a fundamental frequency result from the nonlinear action of a varactor.

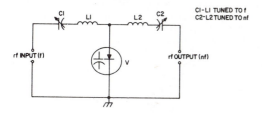

Varactor frequency multiplier.

VAR-hour Abbreviation, VARH. A unit of reactive power per unit time. 1 VARH is equal to 1 VAR per hour.

VAR-hour meter An instrument which indicates reactive power per hour, i.e., VAR per hour (see *volt-amperes reactive* and *VAR-hour*).

variable Abbreviation, var. A quantity whose value changes at some stated or calculable rate and in expressions or equations is given names, such as X. Compare *constant*. Also see *dependent variable* and *independent variable*.

variable-area sound track See *variable-width sound record*.

variable block In computer practice, a unit of data, such as a group of records, whose size is dependent on data requirements, i.e., it is not fixed.

variable-capacitance diode See *varactor*.

variable-capacitance transducer See *capacitive transducer*.

variable capacitor A capacitor whose value may be adjusted from a very low value (practically zero) to maximum. A step-type unit contains a number of fixed capacitors which may be consecutively switched in parallel with each other until, at the last step, all are in parallel. A smooth, continuously variable unit has a provision for moving one plate or set of plates relative to another plate or set of plates, so that the exposed area (of capacitance) is varied; or it may be that one plate is moved with respect to another so the distance between them is changed. In a *voltage-variable* capacitor (such as a varactor), capacitance varies in accordance with an applied dc voltage.

variable-carrier modulation See *quiescent-carrier operation*.

variable connector In a flowchart, a path leading from a box in which a decision is made, to a number of other points on the diagram.

variable coupling Coupling which is adjustable. In an inductively coupled circuit, the distance or angle between the coils is usually the adjustable factor. In a capatively coupled circuit, coupling capacitance is adjustable.

variable-density sound record In photographic sound-on film recording, a sound track made by varying, in sympathy with the sound frequency, the amount of light reaching the film. Compare *variable-width sound record*.

variable-depth sonar A sonar that can be placed far below the surface, for the purpose of detecting objects or terrain at greater depths than would be possible with a surface-located sonar.

variable-efficiency modulation See *efficiency modulation*.

variable-erase recording In magnetic-tape practice, recording a signal by selectively erasing a previously recorded signal.

variable field Any field with an intensity that changes with time. An electromagnetic field is a common example.

variable-frequency oscillator Abbreviation, *VFO*. An oscillator (usually self-excited) whose frequency is continuously variable.

variable-inductance pickup A phonograph pickup in which vibration of the stylus causes the inductance of a small coil to vary (in sympathy with the sound frequency).

variable-inductance transducer A transducer in which a monitored quantity causes the inductance of a coil to vary proportionately, the coil thereby offering a varying impedance to an ac supply voltage.

variable inductor An inductor whose value may be adjusted from a very low value (practically zero) to maximum. A step-type unit contains a number of fixed inductors which may be consecutively switched in series with each other until, at the last step, all are in series. A smooth, continuously variable unit, a *variometer*, consists of two coils, one rotating inside the other, connected either in series or in parallel. A saturable reactor can be used as a variable inductor by varying a direct current flowing through it.

variable field In a computer record, a field which is a variable block at whose end is a terminal symbol.

variable-mu tube See *extended-cutoff tube*.

variable-pitch indication An audible signal used in lieu of, or in conjunction with, a meter to indicate voltage, current, logic state, and so on. The higher the value of the measured quantity, the higher the pitch.

variable-reluctance microphone A microphone in which the impinging sound waves cause corresponding variations in the reluctance of an internal magnetic circuit.

variable-reluctance pickup A phonograph pickup in which the stylus causes an armature to vibrate in a magnetic field, and consequently the reluctance of the magnetic circuit varies in sympathy with the audio frequency.

variable-reluctance transducer A transducer in which the monitored quantity causes the reluctance of an internal magnetic circuit to vary proportionately.

variable-resistance pickup A phonograph pickup in which the vibration of the stylus causes the resistance of an internal resistive element to vary in sympathy with the sound frequency.

variable-resistance transducer A transducer in which a monitored quantity causes the resistance of an internal resistive element to vary proportionately.

variable-resistance tuning Tuning of a selective *RC* circuit (such as a Wien bridge or parallel-tee network) by varying one or more of its resistance arms.

variable resistor A resistor whose value can be varied either continuously or in steps. Also see *rheostat* and *potentiometer, 1*.

variable selectivity In a circuit or device, selectivity which is adjustable. Example: a variable-selectivity i-f amplifier.

variable-speed motor An ac or dc motor whose speed is adjustable, smoothly or in steps, under load.

variable transformer A transformer (often an autotransformer) whose output voltage is adjustable from zero (or some minimum value) to maximum. For this purpose, either the primary or the secondary is provided with a number of taps. But usually smooth variation is provided by a wiper arm that slides over the turns of one coil.

variable-width wound record In photographic sound-on-film recording, a sound track made by varying the width of the track in sympathy with the sound variations. Compare *variable-density sound record*.

Variac See *variable transformer*.

variance In statistics, the square of the standard deviation.

variational plate resistance See *ac plate resistance*.

Varicap A voltage-variable capacitor of the semiconductor-diode type. Also see *voltage-variable capacitor*.

varicoupler An adjustable rf transformer consisting of a primary coil (usually the rotor) and a secondary coil (usually the stator), the former being rotatable to vary the coupling between the coils. The inductance of the stator is varied by means of a set of taps.

varindor A coil having a special core and whose inductance varies with the amount of direct current flowing through the winding. Also see *saturable reactor*.

variocoupler A radio-frequency transformer in which one winding is rotatable, for the purpose of adjusting the mutual inductance between the primary and secondary. Used in certain antenna-coupling applications.

variolosser A variable attenuator.

variometer A continuously variable inductor consisting of two coils connected in series or parallel and mounted concentrically, one rotating inside the other. Inductance is maximum when the coils are perpendicular to each other, and minimum when they are parallel.

varioplex A time-sharing method of transmitting and receiving wire telegraph signals. Allows optimum usage of available lines.

varistor See *voltage-dependent resistor*.

Varley-loop bridge A special, four-arm, dc bridge circuit in which one of the arms, two-wire line, is accidentally grounded at a distant point. By adjustment of the bridge, the resulting resistance indicates the distance to the fault.

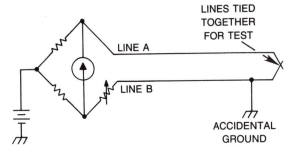

LINES TIED
TOGETHER
FOR TEST

LINE A

LINE B

ACCIDENTAL
GROUND

Varley loop bridge.

VAR meter An instrument for measuring the apparent (VAR) power in a reactive circuit.

varnished cambric Cotton or linen fabric in sheets, tape or tubes, which has been varnish-impregnated and baked. A common form of this insulating material, the slender tubing known as *spaghetti*, has largely been replaced by plastic tubing.

varnished-cambric tubing Slender tubes of *varnished cambric* slipped over bare wires and busbars to insulate them. Also called *spaghetti* from its resemblance to the familiar pasta.

VATE Abbreviation of *versatile automatic test equipment*.

Vb Symbol for *base voltage* of a bipolar transistor.

VBB Symbol for *base-voltage supply* for a bipolar transistor.

V-beam system See *vee-beam system*.

Vc Symbol for *collector voltage* of a bipolar transistor.

VCA Abbreviation of *voltage-controlled amplifier*.

VCC Symbol for *collector-voltage supply* of a bipolar transistor.

VCCO Abbreviation of *voltage-controlled crystal oscillator*.

VCD Abbreviation of *variable-capacitance diode*.

VCG Abbreviation of *voltage-controlled generator*.

VCO Abbreviation of *voltage-controlled oscillator*.

V-connection of transformers See *vee-connection of transformer*.

VCR Abbreviation of *video cassette recorder*.

VCSR Abbreviation of *voltage-controlled shift register*.

V-cut crystal See *vee-cut crystal*.

vcxo Abbreviation of *voltage-controlled crystal oscillator*.

VD 1. Abbreviation of *voltage drop*. 2. Abbreviation of *vapor density*.

VD Symbol for *drain voltage* of a field-effect transistor.

Vdc Symbol for *dc volts*.

VDU Abbreviation of *visual display unit*.

VDCW Symbol for *dc working voltage*. 1. (Also, DCWV.)

VDR Abbreviation of *voltage-dependent resistor*. 2. Abbreviation of *video disc recorder*.

VE Abbreviation of *value engineering*.

Ve Symbol for *emitter voltage* of a bipolar transistor.

vectograph A graphic threedimensional representation of a scene composed of superimposed images photographed through polarizing filters of different orientation, and reconstructed by a similar viewing technique.

vector The graphic symbol of a quantity which has direction as well as magnitude. This device consists of a straight arrow whose length is proportional to the value of the quantity. If the vector represents an ac quantity, it rotates in a counter-clockwise direction, one complete rotation corresponding to one ac cycle. The angle a vector makes with the horizontal axis shows the amount of rotation it has undergone. Angles between separate vectors show their relationship to each other.

vector admittance The reciprocal of vector impedance.

vector-ampere The unit of vector power.

vector cardiograph An electrocardiograph which indicates the magnitude and the direction of a heart signal.

vector components 1. The horizontal (X) and vertical (Y) components, i.e., the Cartesian coordinates, of a vector, or its component angle and radius (polar coordinates). 2. Quantities which can be represented by means of vectors.

vector diagram A diagram using vectors to represent quantities.

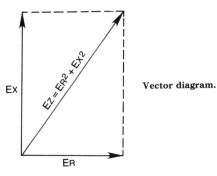

Vector diagram.

vector field 1. A vector function; the set of vectors at all points in space. 2. The net value of a set of vectors in space at a given point.

vector function A function having both magnitude and direction. Also see *vector; vector components, 2;* and *vector quantity.*

vector generator A device that graphically illustrates vectors.

vectorial angle The angle between a vector and the horizontal axis, or the angle between vectors.

vector impedance Complex impedance, i.e., the vector sum of resistance and resistance: $\sqrt{R^2 \times X^2}$.

vector-mode display On a crt computer I/O peripheral, data representation in which points on the screen are connected by straight lines.

vector power Unit, *vector-ampere.* The vector quantity $\sqrt{Pa^2 + Px^2}$, where Pa is the active power, and Px the reactive power.

vector power factor The ratio $Pa/\sqrt{Pa^2 + Px^2}$, where Pa is the active power, and Px the reactive power. In sine-wave situations, this ratio (of active power to vector power) gives a figure which is identical to the conventional power factor.

vector quantity A quantity which has both magnitude and direction and therefore may be represented by a vector.

vectorscope A special oscilloscope for visual adjustment of a color-TV receiver by means of a color-phase diagram.

vector sum The resultant of two nonparallel vectors, therefore the resultant of the forces or quantities represented by them. For example, reactance (X) and resistance (R) may be represented by two perpendicular vectors whose sum impedance (Z), is $\sqrt{R^2 + X^2}$.

vector voltmeter A voltmeter which indicates the phase as well as the amplitude of an ac voltage.

VEE Symbol for *emitter-voltage supply* of a bipolar transistor.

vee-antenna A center-fed antenna in which the two halves of the radiator form an angle. It is so called because it looks like a vee.

vee-beam system An elevation-measuring radar system in which fan-shaped vertical and inclined beams intersecting at ground and rotating continuously around the vertical axis, are radiated by one antenna. Target elevation is determined from the interval between successive echoes from the target.

vee-connection of transformers In a three-phase system, a method of connecting two transformers so that line current and voltage equal phase current and voltage. Also called *open-delta connection.*

vee-cut crystal A piezoelectric plate cut from a quartz crystal so that its faces are not parallel to X-, Y-, or Z-axis of the crystal. Also see *crystal axes* and *crystal cuts.*

vee-particle A short-lived elementary particle that results when high-energy neutrons or protons collide with nuclei. The particle may be positive, negative, or neutral, and gets its name from its cloud-chamber track.

vee-signal In radiotelegraphy, the letter V (in morse code, ... —) transmitted as a test signal. It is used most often during on-the-air transmitter tests, in which case it must be followed by the station's call letters.

vehicle An inert substance, usually a liquid, which acts as a solvent, carrier, or binder for some other, more active substance. Thus, shellac may be the vehicle for the metallic powder in a silver paint employed in silk-screening electronic circuits.

vehicle smog-control device See *computer-controlled catalytic converter* and *exhaust analyzer.*

vel Abbreviation of *velocity.*

velocimeter 1. An instrument for measuring the velocity of sound in various materials. 2. An electronic velocity meter, especially a radial-velocity meter using the Doppler effect in CW reflection. 3. An electronic flow meter.

velocity Symbol, v. Abbreviation, vel. Unit, distance per unit time, e.g., in SI units, m/s. The rate of displacement (or change in position) in a given direction during a certain interval. Velocity is the first derivative of space with respect to time, $v = ds/dt$. Compare *acceleration.*

velocity constant For a transmission line, the ratio of the velocity of wave propagation in the line to the velocity of waves in air.

velocity error For a servomechanism in which the input and output shafts rotate at the same speed, the angular displacement between them.

velocity factor See *velocity constant.*

velocity filter In radar practice, a storage tube in which targets moving slower than one resolution cell in less than a given number of antenna scans are blanked out.

velocity hydrophone A hydrophone whose output, like that of the velocity microphone, is proportional to the instantaneous particle velocity in the sound wave impinging upon the device.

velocity-lag error A lag (proportional to the input-variation rate) between the input and output of a device, such as a servomechanism.

velocity level For a sound, the ratio 20 log10 (vs/vr), where vs is the particle velocity of the sound, and vr is a reference particle velocity.

velocity microphone A microphone in which the vibratory element is a thin, aluminum or duralumin ribbon suspended loosely between the poles of a strong permanent magnet. Vibration of the corrugated ribbon in the magnetic field causes an af voltage to be induced across the ribbon. The microphone is so called because its output is proportional to the instantaneous particle velocity in the sound wave impinging upon the ribbon. Also called *ribbon microphone.*

velocity-modulated amplifier A circuit in which radio-frequency amplification is obtained by velocity modulation.

velocity-modulated oscillator A vacuum-tube device in which an electron stream is velocity-modulated (see *velocity modulation*) as it passes through a resonant cavity (the buncher); the subsequent energy, of a higher intensity, is extracted from the bunched stream as it passes through another resonant cavity (the catcher). Feedback from catcher to buncher sustains oscillations. See, for example, *klystron oscillator.*

velocity-modulated tube A vacuum tube utilizing velocity modulation. See, for example, *klystron.*

velocity modulation The process in which the input signal of a vacuum tube varies the velocity of the electrons in a constant-current electron beam in sympathy with the input signal.

velocity of light Symbol, c. The speed at which light moves through space, i.e., through a vacuum. The velocity of light, an important figure is electronics, is 299,792 kilometers per second (186,282 miles per second).

velocity of propagation The speed at which energy travels from a source. See, for example, *velocity of radio waves* and *velocity of sound.*

velocity of radio waves The speed at which radio waves move through space. This velocity is the same as that of light.

velocity of sound The speed at which sound travels through a medium. This varies considerably for different media; through air at 0% it is 331.32 m/s (1087 feet per second), whereas through glass, at the same temperature, it is 5501.64 m/s (18,050 feet per second).

velocity pickup A magnetic transducer, the output of which increases directly in proportion to the velocity of the stylus in the disk groove.

velocity resonance A state of resonance in which there is a 90-degree phase difference between the fundamental frequency of oscillation of a system and the fundamental frequency of the applied signal.

velocity sorting Selecting and separating electrons according to their velocity.

velocity spectrograph A device which employs electric or magnetic deflection to separate charged particles into separate streams according to their velocity.

velocity variation See *velocity modulation.*

velocity-variation amplifier See *velocity-modulated amplifier.*

velocity-variation oscillator See *velocity-modulated oscillator.*

velometer An instrument employed to measure the velocity of air.

venetian-blind effect A defective TV display in which the picture is seen through horizontal slits.

Venn diagram A relatively simple graphic device (composed of circles within a rectangle) showing the relations between various sets. In such a diagram, intersecton of two circles (representing sets *A* and *B*) includes all the elements, qualities, or the like, shared by *A* and *B*.

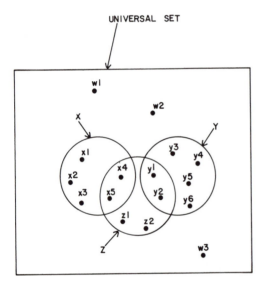

Venn diagram.

vent 1. An opening in an equipment enclosure for the entry of cool air or the escape of warm air. 2. In an electrolytic capacitor, a lightly covered blowout hole for the relief of gas pressure in the event of severe overload, or a small opening for relieving gas pressure in an automobile battery. 3. In a loudspeaker enclosure, such as the bass-reflex type, an auxiliary opening to the room, which extends the low-frequency range of the speaker.

vented baffle A loudspeaker enclosure having the proper auxiliary opening(s) for coupling the speaker to the surrounding air.

vented-baffle loudspeaker See *acoustical phase inverter* and *bass reflex loudspeaker.*

vent mount A metal bracket for fastening an antenna mast to a (plumbing) vent pipe on a roof.

ventricular fibrillation See *fibrillation.*

venturi A narrow-throated tube with flared ends. The device is placed in an air stream in line with the flow, and the pressure in the throat, which is inversely proportional to the velocity of the air, may be sampled for measurement or control purposes; suction is also available through a pipe joined perpendicularly to the throat.

verification 1. To ensure that two sets of data are identical. 2. The process of validating (see *validate*).

verifier In a computer installation, a machine, operated by a keyboard, used to check paper tape or punched cards for the accuracy of the data thereon.

vernier 1. An auxiliary scale along which a regular, linear scale (such as that of a tuning dial) slides. The vernier scale is graduated so that when the main scale is set to an unmarked point between two of its graduations, the zero point on the vernier scale is used as the index, one on the vernier scale will exactly coincide with one on the main scale. The corresponding number on the vernier scale indicates the exact number of subdivisions between two points on the main scale. Vernier arrangements are provided with the dials of many precision instruments. 2. On a missile or spacecraft, a small auxiliary nozzle or engine whose thrust supplies speed or course correction.

vernier capacitor A low-capacitance variable capacitor connected in parallel with a higher-capacitance fixed or variable capacitor for precise adjustment of the total capacitance.

vernier dial 1. A slow moving dial for fine tuning an adjustable device. The required reduction ratio is obtained with gears, friction wheels, or a belt-and-pulley combination. 2. A dial provided with an accessory vernier scale.

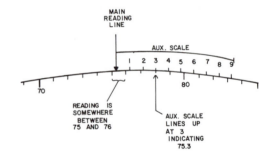

Vernier, 1.

vernier resistor A low-resistance variable resistor connected in series with a higher-resistance fixed or variable resistor for precise adjustment of the total resistance.

vernier rheostat See *vernier resistor.*

vernitel A device for frequency-modulated transmission of data.

vers Abbreviaton of *versed sine.*

versatile automatic test equipment Abbreviation, VATE. For troubleshooting the electronic systems of missiles, a computer-controlled tester which isolates faults through logical operations.

versed sine Abbreviation, vers. A trigonometric function equal to the difference between the cosine of an angle and one; $\text{vers } \omega = 1 - \cos \omega$.

vert Abbreviation of *vertical.*

vertex 1. The terminal point at which two or more branches of a network meet. Also see *node, 1.* 2. The point of intersection of two lines that form an angle.

vertex plate In the reflector of a microwave antenna, a matching plate mounted at the vertex.

vertical 1. The direction indicated by a line perpendicular to the horizon, e.g., a vertical axis. 2. In the 45—45 recording process, the signal resulting from sound that arrives simultaneously and 180 degrees out of phase at the two microphones and causes vertical movement of the cutting stylus.

vertical amplification Gain provided by the vertical channel of a device, such as an oscilloscope, cathode-ray electrocardiograph, or TV receiver. Compare *horizontal amplification*.

vertical-amplifier The circuit or device that provides gain in the vertical channel of a device. Compare *horizontal amplifier*.

vertical-amplitude control 1. Vertical-gain control. 2. In a color TV receiver, one of the three controls by which the amplitude of the parabolic voltages applied to the coils of the magnetic-convergence assembly are adjusted.

vertical angles The angles on opposite sides of intersecting lines.

vertical angle of radiation The angle, with respect to the surface of the earth, at which rf energy is propagated in the vertical plane from a transmitting antenna.

vertical antenna 1. An antenna that consists essentially of a single, vertical wire, rod, or tube. 2. An antenna which is mounted vertically, instead of horizontally.

vertical blanking See *vertical retrace blanking*.

vertical-blanking pulse In a television signal, the pulse at the end of each field that accomplishes vertical retrace blanking. Compare *horizontal-blanking pulse*.

vertical-centering control See *centering control*.

vertical channel The system of amplifiers, controls, and terminations which constitutes the path of the vertical signal applied to a device, such as an oscilloscope. Compare *horizontal channel*.

vertical compliance In disk reproduction of sound, the ease with which the stylus can move vertically while it is in position on the disk.

vertical coordinates See *Cartesian coordinates*.

vertical deflection 1. In an oscilloscope or TV receiver, the movement of the spot up and down on the screen. Compare *horizontal deflection*. 2. In a direct-writing recorder, up or down deflection of the pen.

vertical deflection coils The pair of coils in a deflection yoke which provides the electromagnetic field for the vertical deflection of the electron beam in an oscilloscope tube or TV picture tube. Compare *horizontal deflection coils*.

vertical deflection electrodes See *vertical deflection coils* and *vertical deflection plates*.

vertical deflection plates In an oscilloscope (and some early TV picture tubes), a pair of plates which provides the electrostatic field for vertical deflection. Compare *horizontal deflection plates*.

vertical dipole A half-wave dipole antenna, fed at the center, and oriented vertically.

vertical dynamic convergence During the scanning of points along a vertical line through the center of a color-TV picture tube, the convergence of the electron beams at the aperture mask. Compare *horizontal dynamic convergence*.

vertical field strength The field strength of signals in a vertical plane passing through an antenna. Compare *horizontal field strength*.

vertical-field-strength diagram A plot of vertical field strength.

vertical frequency response The gain—frequency characteristic of the vertical channel of an instrument, such as an oscilloscope. Compare *horizontal frequency response*.

vertical gain The overall amplification (gain) provided by the vertical channel of an instrument, such as an oscilloscope, recorder, or TV receiver. Compare *horizontal gain*.

vertical-gain control A control, such as a potentiometer, for adjusting vertical amplificaton. Compare *horizontal-gain control*.

vertical-hold control In a TV receiver, the control for adjusting the frequency of the vertical oscillator, so the picture can be locked to prevent vertical roll. Compare *horizontal-hold control*.

vertical-incidence transmission Transmission of a wave vertically to the ionosphere and the subsequent reflection of the wave to the earth.

vertical-lateral recording In stereophonic disk recording, recording one signal vertically (see *vertical recording*) and the other laterally (see *lateral recording*).

vertical linearity Linearity of response (gain and deflection) of the vertical channel of an oscilloscope or TV receiver. A linear picture is neither contracted nor expanded vertically in any part of the screen. Compare *horizontal linearity*.

vertical-linearity control In an oscilloscope or TV receiver, the control whereby vertical linearity is adjusted. Compare *horizontal-linearity control*.

vertically polarized wave A wave exhibiting vertical polarization. Compare *horizontaly polarized wave*.

vertical metal-oxide semiconductor field-effect trainsistor Abbreviation, VMOS or VMOSFET. A *metal-oxide semiconductor field-effect transistor* that is so fabricated that current flow within the device is vertical, instead of the usual horizontal, affording several advantages over the conventional MOSFET.

vertical oscillator In a TV receiver, the oscillator that generates the vertical sweep signal. Compare *horizontal oscillator*.

vertical-output regulator In a TV receiver, a voltage-dependent resistor employed to stabilize the sweep voltage across the horizontal deflection coils, especially during warmup.

vertical output stage In a TV receiver, an output amplifier following the vertical oscillator. Compare *horizontal output stage*.

vertical polarization The orientation of an electromagnetic wave, in which the electrostatic component is vertical. Compare *horizontal polarization*.

vertical positioning control See *centering control*.

vertical quantity The quantity measured along the *y*-axis of a graph or represented by the vertical deflection of an electron beam. Compare *horizontal quantity*.

vertical radiator See *vertical antenna*.

vertical recording Disk recording in which the depth of the groove is varied in sympathy with the sound. Also called *hill-and-dale recording*. Compare *lateral recording*.

vertical redundance In computer operation, the error state when a character has an odd number of bits in an even-parity system, or vice versa.

vertical resolution The number of horizontal wedge-lines that can be easily seen in a TV test pattern before they blend. Compare *horizontal resolution*.

vertical retrace In a cathode-ray device, such as an oscilloscope or TV receiver, the beam's snapping back to its starting point at the top of the screen after completely traversing the screen from top to bottom. Compare *horizontal retrace*.

vertical retrace blanking In a TV receiver, the automatic shutoff of the electron beam during a vertical retrace period, to prevent an extraneous line being traced on the screen during this period. Also see *blackout, 2; blanking interval, 1, 2; blanking pedestal; blanking time; vertical retrace;* and *vertical retrace period*. Compare *horizontal retrace blanking*.

vertical retrace period In a TV receiver, the interval during which the spot returns from the bottom to the top of the screen after having traced all horizontal lines from top to bottom. Compare *horizontal retrace period.*

vertical sensitivity The signal voltage required at the input of a vertical channel to give full vertical deflection. Also see *vertical gain.* Compare *horizontal sensitivity.*

vertical signal A signal serving as a vertical quantity. Compare *horizontal signal.*

vertical-speed transducer A transducer whose electrical output is proportional to the vertical speed of an aircraft or missile carrying the transducer.

vertical stylus force In disk reproduction of sound, the downward force (in grams or ounces) that the stylus exerts on the disk.

vertical sweep 1. In a cathode-ray tube, especially a TV picture tube, the sweep of the spot up or down on the screen. Compare *horizontal sweep.* 2. The circuit which produces this sweep.

vertical sweep frequency The frequency at which vertical sweep takes place. In a TV receiver, it is 60 Hz. Also called *vertical sweep rate.* Compare *horizontal sweep frequency.*

vertical sweep rate See *vertical sweep frequency.*

vertical synchronization In a TV receiver, synchronization of the vertical component of scanning with that of the camera. Also see *vertical sync pulse.* Compare *horizontal synchronization.*

vertical sync pulse In a video signal, the pulse that synchronizes the vertical component of scanning in a TV receiver with that of the camera, and which triggers vertical retrace and blanking. Compare *horizontal sync pulse.*

vertical wave See *vertically polarized wave.*

vertical width control See *width control, 1, 2.*

very-high frequency Abbreviation, vhf. Radio frequencies in the range 30 to 300 MHz. Also see *radio spectrum, 1.*

very-high resistance Abbreviation, vhr. Large amounts of resistance; usually expressed in gigohms, megohms, or teraohms.

very-high-resistance voltmeter Abbreviation, vhrvm. A voltmeter employing a low-range microammeter or picoammeter and a very high value of multiplier resistance (see *voltmeter multiplier*).

very-large-scale integration Abbreviation, VLSI. The inclusion of several complete systems, such as computers, on a single integrated-circuit chip. This practice may go several orders of magnitude beyond *large-scale integration (LSI).*

very-long-range Abbreviation, vlr. Pertaining to ground radar sets having a maximum slant range of over 250 miles. Compare *very-short range.*

very-low frequency Abbreviation, vlf. A radio frequency in the range 10 to 30 kHz. Also see *radio spectrum, 1.*

very-low resistance Abbreviation, vlr. Low values of resistance, i.e., less than 1 ohm, usually expressed in milliohms or microhms.

very-short-range Abbreviation, vsr. Pertaining to ground radar sets having a maximum slant range of less than 25 miles. Compare *very-long range.*

vestigial 1. An effect that remains as a by-product. 2. Unnecessary.

vestigial sideband 1. A portion of one sideband in an amplitude-modulated signal, remaining after passage through a selective filter. 2. An amplitude-modulated signal in which one sideband has been partially or largely suppressed. 3. The small amount of energy emitted in the unused sideband in a single-sideband transmitter.

vestigial-sideband filter A filter operated between an amplitude-modulated transmitter and an antenna to obtain a vestigial-sideband signal.

vestigial-sideband signal An amplitude-modulated signal in which one of the sidebands has been partially suppressed.

vestigial-sideband transmission Transmission of a vestigial-sideband signal. In television, for example, the upper sideband is transmitted fully, while the lower sideband is almost completely suppressed.

vestigial-sideband transmitter An amplitude-modulated transmitter equipped with the filters or other subcircuits necessary for emitting a vestigial-sideband signal.

vf Abbreviation of *video frequency.*

VFO Abbreviation of *variable-frequency oscillator.*

vg Abbreviation of *generator voltage.* (Also, *Eg* or *EG.*)

VGA Abbreviation of *variable-gain amplifier.*

VGD Symbol for *gate—drain voltage* of a field-effect transistor.

VGS Symbol for *gate—source voltage* of a field-effect transistor.

vhf Abbreviation of *very high frequency.*

vhf omnirange Abbreviation, vor. A very-high-frequency radionavigation system which provides radial lines of position.

vhf oscillator A radio-frequency oscillator specially designed to operate in the range 30 to 300 MHz.

vhf receiver 1. A radio receiver specially designed to operate in the range 30 to 300 MHz. 2. A television receiver for the vhf channels (see *vhf television channels*).

vhf television channels Twelve 6-MHz-wide channels (TV channels 2 through 13) between 54 and 216 MHz.

vhf transmitter A radio or TV transmitter specially designed to operate in the range 30 to 300 MHz.

vhr Abbreviation of *very-high resistance.*

vhrvm Abbreviation of *very-high resistance voltmeter.*

VI 1. Abbreviation of *volume indicator.* 2. Abbreviation of *viscosity index.*

Vi Symbol for *input voltage.*

vibrating bell A bell with a striking mechanism that oscillates, causing a continuous ringing sound.

vibrating-reed frequency meter See *power-frequency meter, 2.*

vibrating-reed relay An electromechanical relay in which the movable contact is carried by the end of a thin, metal strip (reed) of iron or magnetic alloy. The strip is supported within the magnetic field of a coil carrying an ac control current; when the frequency of this current corresponds to the natural (resonant) frequency of the reed, it vibrates vigorously enough to close the contacts. Such a relay, consequently, is frequency selective.

vibrating-reed oscillator An audio-frequency oscillator in which an iron, steel, or special-alloy reed vibrating in a magnetic field acts like a tuning fork to control the oscillation frequency.

vibrating-reed relay A relay intended for the purpose of rapidly interrupting a source of direct current. The contacts rapidly open and close because of vibration of the armature.

vibrating-wire oscillator An oscillator similar to the vibrating-reed oscillator but employing a wire instead of a reed.

vibrating-wire transducer A transducer in which the fluctuating tension of a thin wire suspended in a magnetic field frequency-modulates the operating voltage.

vibration 1. The changing of the position or dimensions of a body back and forth, usually at a rapid rate, an action seen in the repetitive movement of a musical string, headphone diaphragm, loudspeaker cone, loose machine, etc. 2. Oscillation.

vibration analyzer See *vibration meter.*

vibration calibrator A device which generates a standard vibration, usually at a fixed frequency, for calibrating vibration meters, pickups, transducers, etc.

vibration galvanometer A type of ac galvanometer. The natural frequency of the movable element of the instrument is made equal to that of the alternating current under test, to obtain a reading.

vibration isolator In an electronic-equipment assembly, a cushioning support that protects the equipment from vibration.

vibration machine See *vibrator, 2*.

vibration meter An instrument for measuring the amplitude and frequency of vibration (see *vibration, 1*). It consists essentially of a vibration pickup followed by a selective amplifier and an indicating voltmeter graduated in vibration units.

vibration pickup A transducer which senses mechanical vibration and converts it into a proportionate output voltage, or changes resistance in sympathy with the vibration.

vibrato In electronic musical instruments, a circuit or device that trills a note by varying its frequency, amplitude, or both, at a low frequency, say, 5 to 10 Hz.

vibrator 1. Interrupter. 2. Vibrating-reed relay. 3. A device for shaking something under test at selected frequencies and amplitudes.

vibrator-type power supply A type of battery-operated power supply in which one vibrator (see *interrupter*) makes and breaks direct current flowing from the battery into the primary winding of a step-up transformer, and another vibrator rectifies the high voltage delivered by the secondary winding.

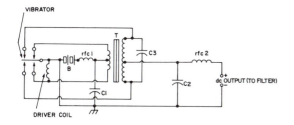

Vibrator-type power supply.

vibrator-type rectifier A vibrator (see *interrupter*) which connects one terminal of the secondary winding of a transformer to an output terminal each time that terminal is positive. When the negative half-cycle appears at the transformer terminal, the vibrator is open. In this way, the ac output of the transformer is rectified to dc.

vibratron A special vacuum tube used as a vibration pickup. Also see *electromechanical transducer, 2*.

vibrocardiography A method of monitoring and recording the movement of the chest cavity as a result of the beating of the heart.

vibrograph A device for observing and recording vibration.

vibrometer See *vibration meter*.

vibrophonocardiograph A stethoscope incorporating electronic amplifiers, and capable of recording chest vibrations caused by the beating of the heart.

vibroscope An instrument for observing (and sometimes recording) vibration.

Victor Phonetic alphabet communications code word for the letter *V*.

Victron A brand of polystyrene, a low-loss plastic insulant that is especially useful at high radio frequencies.

video 1. Pertaining to television, especially to its picture circuitry or to circuits and devices related to television but used for other purposes.

2. The picture portion of a television broadcast, as opposed to the sound (audio) portion. 3. Sometimes, crt terminal or display. 4. Generally, *television* in adjective sense, e.g., video play.

video amplifier 1. In television, the wideband stage (or stages) that amplifies the picture signal and presents it to the picture tube. 2. A similar wideband amplifier, such as an instrument amplifier or preamplifier having at least a 4-MHz bandwidth.

video animation Movement of graphic images on a crt display by the use of computer-manipulation through special software, under the direction of an operator using a keyboard, joystick, or digitizer.

video carrier The television carrier which is amplitude-modulated by the picture information and sync and blanking pulses.

video cassette recorder Abbreviation, VCR. A machine for recording television programs on cassettes (see *cassette, 1*).

videocast 1. Television broadcast. Also called *telecast*. 2. To make a television broadcast.

videocaster See *telecaster*.

video correlator A device used with radar for the purpose of locating and identifying a target with high precision.

video detector In a television receiver, the demodulator for the video signal. This detector follows the video i-f amplifier and precedes the video amplifier. It usually embodies a relatively simple, wideband diode circuit.

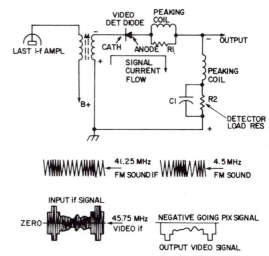

Video detector.

videodisc A flat disc on which television programs are recorded and from which they may be played back into a tv receiver.

videodisc recorder Abbreviation, VDR. A machine for recording television programs on discs comparable to the record discs used for sound recording.

video discrimination In radar, the use of a circuit or device to decrease the width of the video-amplifier passband.

video-frequency 1. Pertaining to signals in the wide passband of a video amplifier: 30 Hz to 4 MHz. 2. Pertaining to a device capable of operating over the wide passband of a video amplifier, e.g., a video-frequency ac vtvm.

video-frequency amplifier An amplifier capable of handling signals of a wide frequency range, e.g., dc to 4 MHz. Also see *video-frequency, 1.*

video-gain control In a TV receiver, a gain control for adjusting the videosignal amplitude. In a color-TV receiver, a pair of these controls permits adjustment of the three colorsignals to the proper amplitude ratio.

video game A game—such as football, basketball, slot machine, tic-tactoe, and so on—played on the screen of a tv receiver or of a computer.

videogenic Suitable for television.

video i-f amplifier In a television receiver, the broadband intermediate-frequency amplifier of the video signal. In modern receivers, this i-f amplifier also handles the sound signal.

video integration A technique in which several video frames are added together to eliminate uncertainty, and thereby improve the detail and accuracy.

video mapping A system of surveillance or mapping in which, for reference, the outlines of the area being surveyed are superimposed electronically on the radar display of the area.

video masking A form of radar signal processing in which ground clutter and other unwanted echoes are removed, making the desired targets more readily visible.

video mixer A circuit or device for mixing the signals from two or more television cameras.

video modulation In television transmission, amplitude modulation of the carrier wave with pulses and waves corresponding to the picture elements.

video recording 1. Recording wideband material (such as a video signal) on a tape or disk. 2. The recording made of a telecast. Also see *videotape recorder.*

video signal 1. In television, the amplitude-modulated signal containing picture information and pulses. Also see *video, 2* and *video modulation.* 2. Broadly, a telecast signal, including sound.

video stretching In some electronic navigation systems, increasing the duration of a video pulse.

video synthesizer A computerized device that produces graphical renditions of objects or circuits in three dimensions.

videotape 1. A special magnetic tape for video recording (see *video recording, 2*). 2. To make a video recording.

video tape recorder Abbreviation, VTR. A wideband, magnetic-tape recorder for producing video tape recordings with a camera, or for making a record of television programs, for subsequent reproduction (playback).

video tape recording 1. Abbreviation, VTR. The technique of recording video signals on magnetic tape. Also see *video recording, 1, 2* and *video tape.* 2. A tape on which a video signal has been recorded.

videotext A system which allows TV viewers to dial up special material, such as stock market quotations, weather data, sports scores, and the like.

vidicon A TV camera tube in which the electron beam scans a charged-density pattern that has been formed and stored on the surface of the photoconductor.

viewfinder An accessory or integral device in which an image formed optically or electronically, corresponds to that seen by the camera with which it is used.

viewing mesh In a cathode-ray storage tube, the mesh on which the im-

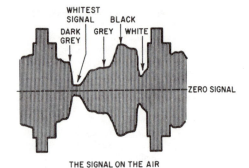

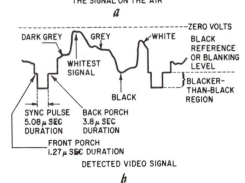

Video signal.

age is presented for viewing by the operator. Also see *charge-storage tube.*

viewing mirror In an oscilloscope-camera assembly, a flat, slant-mounted mirror that reflects the image on the oscilloscope screen to the viewer's eye.

viewing screen In a cathode-ray device, such as an oscilloscope tube or TV picture tube, the face on which the image appears.

viewing time In a storage type cathode-ray tube, the length of time for which the image persists.

viewing window See *window, 2.*

Villard circuit 1. A voltage-doubler circuit using only one diode and two capacitors. The open-circuit dc output voltage is approximately twice

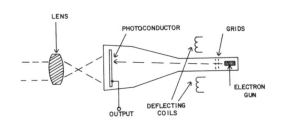

Vidicon.

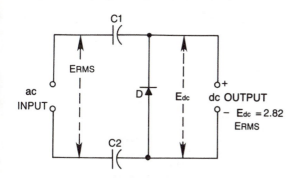

Villard circuit, 1.

the peak value of the ac input voltage. Also see *voltage doubler*. 2. Selectoject.

Villari effect In certain magnetorestrictive materials, the change in magnetic induction which accompanies the application of a mechanical stress in a given direction. Also see *magnetostriction*.

vinculum A horizontal line drawn over terms in a mathematical expression which are to be treated parenthetically, i.e., $\overline{x+y} - z = x + (y-z)$.

vinyl A general name for vinyl copolymer resins.

vinylidene chloride A plastic insulant. Dielectric constant, 3 to 5. Dielectric strength, 20 kV/mm.

Vinylite-A A brand of the plastic polyvinyl acetate.

Vinylite-Q A brand of the plastic polyvinyl chloride.

vinyl resin A synthetic resin resulting from the polymerization of compounds that contain the group $CH_2 = CH-$. Also see *vinyl*.

violet ray An ultraviolet radiation.

virgin magnetic material Core or shield material which has never been magnetized. When virgin material is first subjected to magnetization, the hysteresis loop starts at zero but never returns to zero (see *box-shaped loop*).

virgin record See *blank record*.

virgin tape See *blank tape*.

virgin wire See *blank wire*.

virtual address In computer practice, address which must be modified for it to refer to a location in main memory, i.e., it is similar to a relative address.

virtual cathode A negative-potential region created between the plate and screen of a pentode when the suppressor is grounded. It is this virtual cathode that repels secondary electrons back to the plate.

virtual height The height which a vertically propagated wave would reach before reflection if its path in the ionopshere were a straight line. The actual distance at which the wave penetrates the ionopshere before reflection is less than the virtual height.

virtual image The image formed when rays from an object diverge on passing through a lens. To the observer looking through the lens, the object appears inverted on his side of the lens. Compare *real image*.

virtual memory 1. In a computer system, a means of using two or more memory stores simultaneously. 2. Auxiliary memory used in conjunction with the main, or core, memory.

virtual ppi reflectoscope A device used to superimpose a virtual image of a chart or map onto a ppi radar display. Also see *video mapping*.

virtual suppressor In a beam-power tube, the high negative field of

the beams, which acts like the suppressor element of a pentode to keep secondary electrons from reaching the screen.

vis 1. Abbreviation of *visibility*. 2. Abbreviation of *visual*.

viscometer An instrument for measuring viscosity. There are several electronic versions of this device. In one, a steel ball falls through a material (such as an oil) under test and distorts the magnetic field of a pickup coil, causing the deflection of a meter by an amount proportional to the speed of the ball and, consequently, to the viscosity of the fluid.

viscosimeter See *viscometer*.

viscosity The resistance of a fluid (liquid or gas) flow or to objects passing through it. The viscosity of pure water is low, and that of a heavy insulating oil is high. Viscosity is expressed in newton-seconds per square meter, poises, and centipoises.

viscosity index Abbreviation, VI. A number indicating how well an oil retains its viscosity during temperature changes; larger indexes are assigned to oils little influenced in this way.

viscous-damped arm A phonograph pickup arm with an oil dashpot to prevent arm resonance and to slow the descent of the arm to the record disk.

viscous damping The use of a viscous fluid in the dashpot of a device (such as a relay, timer, or pickup arm) to provide damping. See, for example, *dashpot relay*.

viscous hysteresis A slow, slight increase in the magnetization of a material when the magnetizing field is constant. Compare *static hysteresis*.

visibility factor See *display loss*.

visible radiation Radiation that is perceptible to the eye. Also see *visible spectrum*.

visible spectrum The band of electromagnetic radiation frequencies the human eye perceives as light. For most viewers, this band extends from 4.3×10^{14} Hz (0.7 μm) to 7.5×10^{14} Hz (0.4 μm). The relative-visibility curve peaks in the region of yellow-green light, of about 5.4 $\times 10^{14}$ (0.56 μm). Also see *electromagnetic theory of light*.

visual alignment Alignment of a circuit (such as a radio receiver) with the aid of meter deflections, rather than with aural indications. Also called *silent alignment*.

visual-aural radio range A radio aid to air navigation characterized by an aural signal, meter deflection, or both, employed by the pilot of an aircraft to determine if the course is being followed.

visual-aural range See *visual-aural radio range*.

visual-aural signal tracer A troubleshooting instrument for following a signal through a circuit by sensing it at successive points in the circuit. The signal's strength is indicated by a meter and a loudspeaker.

visual carrier frequency See *video carrier* and *video-frequency*.

visual communication 1. Transmission and reception of signals by visible-light means, such as modulated beams of flashing lights. 2. Transmission and reception of messages by direct visual observation, such as signal lights.

visual display unit Abbreviation, VDU. A computer crt-keyboard I/O peripheral.

visual horizon 1. The distance from a given point to the farthest visible point on the surface of the earth in a particular direction. 2. The enclosed geometric plane figure on the surface of the earth, representing the set of farthest-visible points for a particular location.

visual indicator tube See *magic-eye tube*.

visual telegraphy Telegraphy in which the received signals are read from a visual device, such as a blinking light, meter, or swinging pen.

visual transmitter The equipment used to transmit the picture portion only of a television broadcast.

visual transmitter power The peak power output of a visual transmitter during normal operation.

vitreous Pertaining to a material or a finish which is like glass or is in some way related to glass, e.g., the vitreous-enamel coating on power-type resistors.

vlf Abbreviation of *very-low frequency.*

vlr 1. Abbreviation of *very-low resistance.* 2. Abbreviation of *very-long range.*

VLSI Abbreviation of *very-large-scale integration.*

V/M Abbreviation of *volts per meter* (Unit of electric field strength).

VMOS Abbreviation of *vertical metaloxide field-effect transistor.*

VMOSFET Same as VOMS, preceding.

Vo Symbol for *output voltage.*

Vo Symbol for *volume of ideal gas for standard conditions.*

voa Abbreviation of *voltohm-ammeter.*

vocabulary In computer-programming practice, a list of available operating codes and instructions for the computer. Also called *instruction set.*

vocoder A device for reducing speech to frequencies low enough for efficient transmission through a limited-bandwidth channel. The term is an acronym for *voice coder.*

voda In a telephone system which utilizes a radio link (using the same frequency for transmission in both directions) and land-line links, an automatic, voice-operated switching system for enabling the transmitter and disabling the receiver, and vice versa. The name is an acronym for *voice-operated device, anti-sing.*

vodacom Abbreviation of *voice data communications.*

voder An electronic device that synthesizes speech. The term is an acronym for *voice operation demonstrator.*

vogad A type of automatic gain control for audio amplifiers and modulators. In a radio transmitter, it maintains 100% modulation of the carrier, even when the speaker's voice level varies widely. The name is an acronym for *voice-operated gain adjusting device.*

voice analyzer A circuit that evaluates the characteristics of human voices, such as the amplitude-versus-frequency function or the amplitude-versus-time function.

voice coil The moving coil of a dynamic microphone or dynamic speaker.

voice-controlled break-in A type of break-in operation for radiotelephony, in which the transmitter switches on automatically and the receiver switches-off when the operator starts talking, and vice versa.

voice-controlled relay An electronic relay that is actuated by the human voice.

voice filter 1. A filter for passing, suppressing, or modifying voice frequencies. 2. A parallel-resonant circuit inserted into a line that feeds several loudspeakers, to trap certain frequencies and thereby improve the sound of the reproduced voice.

voice frequencies The frequency range of speech, roughly from 62 to 8000 Hz. Within these limits, the range of human speech varies considerably with individual talkers, the average peak energy in the male speaking voice being at about 130 Hz, and in the female speaking voice at 300 Hz.

voice-frequency carrier telegraphy A type of carrier-current telegraphy (see *wired wireless*) in which the modulated carrier may be transmitted over a telephone line having only a voice-frequency bandwidth.

voice-frequency dialing A system of telephone dialing involving the conversion of dc pulses into voice-frequency ac pulses.

voice-frequency telephony Wire-telephone communication in which the frequencies of the electric waves are identical to the frequencies of the sound waves (or very nearly so).

voice grade Pertaining to a communications system having a bandpass capable of transferring a human voice with reasonable intelligibility. Generally, this range is from about 300 Hz to 3 kHz.

voice-grade channel 1. A telephone line and attendant equipment which are suitable only for the transmission of speech and certain other information, such as control signals, digital data, etc. 2. In a radiotelephone transmitter, a speech amplifier-modulator channel suitable only for voice frequencies.

voice-operated device A device—such as a relay, automatic modulation control, transmit-receive switch, etc.—that is actuated when the operator speaks into a microphone.

voice-operated device, anti-sing See *voda.*

voice-operated gain-adjusting device See *vogad.*

voice-operated loss control and suppressor In wire telephony, a device that switches the loss from the transmitting line to the receiving line when the subscriber speaks, and vice versa.

voice print A graphic recording of the speech frequencies produced by an individual and used as a means of identifying that individual.

voice recognition device Any device which can be programmed to respond to spoken commands.

voice-stress analyzer Abbreviation, VSA. An instrument which samples the spoken voice and gives a display from which the relative amount of stress experienced by the speaker may be determined, and from which, in turn, the truth or falsity of statements may be concluded.

void space See *vacuum.*

vol Abbreviation of *volume.*

volatile 1. Capable of evaporating, e.g., volatile solvents used in the encapsulation of electronic equipment. 2. Explosive (noun and verb). 3. Pertaining to a state which is difficult to maintain, e.g., a *volatile memory.* 4. Capable of flight. 5. An animal with wings.

volatile memory In computer practice, a memory whose contents are lost when there is a loss of power, e.g., a memory composed entirely of flip-flops, such as a RAM (the flip-flops all could change state if the power failed).

volatile storage See *volatile memory.*

volatile store See *volatile memory.*

Voldicon A form of semiconductor logic device used for analysis of analog signals. Trade name of Adage, Inc.

volt Abbreviation, V. The basic practical unit of difference of potential, i.e., of *electrical pressure;* 1 volt is the difference of potential produced across a resistance of 1 ohm by a current of 1 ampere. Also see *kilovolt, megavolt, microvolt, millivolt, nonovolt, picovolt.*

Volta effect See *Volta's principle.*

voltage Symbols, *E, e, V.* Electromotive force, or difference of potential; $E = IR$, where *I* is current and *R* is resistance. Also see *volt.*

voltage-actuated device An electronic device, such as a vacuum tube or field-effect transistor, which amplifies a voltage signal or is controlled by a voltage, and draws virtually on signal current or control current. The opposite is a current-actuated device, such as a bipolar transistor.

voltage amplification 1. Abbreviation, *Av.* Amplification of an input-signal voltage to give a higher output-signal voltage. 2. Abbreviation,

Av. The signal increase (*Eout/Ein*) resulting from this process. Also called *voltage gain*.

voltage-amplification device A low-current tube or transistor designed especially for voltage amplification. Such a device provides little or no power amplification.

voltage amplifier An amplifier which is operated primarily to increase a signal voltage. Compare *current amplifier* and *power amplifier*.

voltage at peak torque Symbol, *Vp*. For a torque motor operated at 25 °C, the voltage required to produce peak torque at standstill.

voltage attenuation 1. The reduction in voltage at a given point in a circuit. 2. For a device, the ratio of input voltage to output voltage when the output voltage is the lower quantity.

voltage-balance relay A relay which is actuated by a voltage differential.

voltage breakdown See *breakdown voltage*.

voltage-breakdown test 1. A test in which the measured voltage applied to an insulating material is continuously increased until the breakdown point is reached. 2. A test in which the measured reverse voltage applied to a semiconductor junction is continuously increased until the reverse breakdown point is reached (see *avalanche breakdown*).

voltage burden The voltage drop across a *current shunt*.

voltage calibrator A device employed to calibrate, in terms of voltage, a meter scale, oscilloscope screen, graphic-recorder chart, or the like.

voltage-capacitance curve A plot depicting the variation of capacitance with applied voltage, for a voltage-variable capacitor. For a varactor (variable-capacitance diode), capacitance varies inversely with reverse dc voltage.

voltage coefficient A figure which shows the extent to which a quantity drifts under the influence of voltage. It is generally expressed in percent per volt or in parts per million per volt (ppm/V).

voltage coefficient of capacitance For a voltage-dependent capacitor, the capacitance change per unit change in applied voltage.

voltage coefficient of resistance For a voltage-dependent resistor, the resistance change per unit change in applied voltage.

voltage coil See *potential coil*.

voltage comparator A device for comparing (usually only two) voltages. There are various types, ranging from simple, manually balanced potentiometers to analog or digital devices that automatically compare the applied voltages and give a direct readout of either their difference or the percent of unbalance.

voltage control 1. A component or circuit that allows adjustment of the output voltage of a power supply within a given range. 2. The adjustment of the output voltage of a power supply for the purpose of optimizing the performance of a circuit connected to the supply. 3. Any form of circuit control that is accomplished by the adjustment of the voltage at a given circuit point.

voltage-controlled amplifier Abbreviation, VCA. An amplifier in which gain is controlled by means of a voltage applied to a control terminal.

voltage-controlled attenuator An attenuator circuit in which a transistor serves as a voltage-variable resistor. The output resistance of the transistor varies inversely with the dc bias voltage applied to the input electrode (base, emitter, or gate).

voltage-controlled capacitor See *voltage-dependent capacitor*.

voltage-controlled crystal oscillator Abbreviation, *VCCO*. A voltage-controlled oscillator of the crystal-stabilized type.

voltage-controlled generator Abbreviation, VCG. Any signal-generating device whose output frequency is varied by vary-

ing one of the dc operating voltages of the device. Example: *voltage-controlled oscillator*.

voltage-controlled oscillator Abbreviation, *VCO* . An oscillator whose output frequency varies with an applied dc control voltage.

voltage-controlled resistor See *voltage-dependent resistor*.

voltage corrector See *voltage regulator*.

voltage crest See *voltage peak*.

voltage-current characteristic See *voltage-current curve*.

voltage-current crossover The point at which a voltage-regulated supply becomes a current-regulated supply.

voltage-current curve For a circuit or device, the plot of the interrelationship of current and voltage, with voltage as the independent variable.

voltage-current feedback See *current-voltage feedback*.

voltage decay The exponential decrease of voltage across a discharging capacitor. Also see *exponential decrease*.

voltage-dependent capacitor A capacitor (such as a varactor) whose capacitance varies with applied voltage.

voltage-dependent resistor A nonlinear resistor whose value varies inversely with the voltage drop across it. Also called *varistor*.

voltage detector A circuit or device which delivers an output voltage only when the input voltage is of a prescribed value. Compare *voltage discriminator*.

voltage-determined property A property (capacitance, current, frequency, resistance) whose magnitude depends upon the value of an applied voltage. See, for example, *voltage-dependent capacitor*, *voltage-dependent resistor*, and *voltage-controlled oscillator*.

voltage differential Symbol, d*E* or d*V*. An infinitesimal change in voltage represented by *E*2 – *E*1. Compare *voltage increment*.

voltage-directional relay 1. A relay which is actuated when voltage exceeds a certain value in a given direction. 2. A relay which closes only when the applied voltage is in a specific direction.

voltage discriminator A circuit or device whose output voltage is zero when the input voltage is of a prescribed value. When the input voltage is greater than this value, the output is positive; when it is less, the output is negative.

voltage distribution 1. The delivery of operating voltage throughout a circuit, e.g., high and low dc voltages in the various stages of a control circuit. 2. Sometimes, the distribution of electrical energy ("power" distribution).

voltage divider A resistive or capacitive potentiometer used to divide an applied voltage by a desired amount.

voltage doubler A circuit which supplies a dc output voltage that is about twice the peak value of the ac input voltage.

voltage drift See *drift voltage*.

voltage drop Abbreviation, VD. The voltage (*E*) set up across a resistance (*R*) carrying a current (*I*); *E* = *IR*. For alternating current, reactance (*X*) and impedance (*Z*) may be substituted for resistance where applicable.

voltage-equalizing resistors In a power-supply filter, resistors connected across each capacitor in a string of electrolytics connected in series to withstand a high voltage. The resistors protect the capacitors by equalizing the voltage across them.

voltage-fed antenna An antenna in which the feeder is attached to the radiator at a voltage loop (current node). Compare *current-fed antenna*.

voltage feed The delivery of voltage to a device or circuit at a point where voltage, rather than current, is dominant. Compare *current feed*.

voltage-frequency transducer 1. A device whose output voltage is

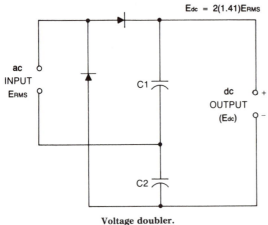

$$E_{dc} = 2(1.41)E_{RMS}$$

Voltage doubler.

a function of the frequency of a signal at the input. 2. A device whose output frequency is a function of the voltage at the input.

voltage feedback A feedback signal consisting of voltage fed from the output to the input circuit of an amplifier or other device. Compare *current feedback* and *current-voltage feedback*.

voltage-frequency transducer 1. A device whose output voltage is a function of the frequency of a signal at the input. 2. A device whose output frequency is a function of the voltage at the input.

voltage gain See *voltage amplification*.

voltage generator See *voltage supply*.

voltage gradient The voltage per unit length along a path.

voltage increment Symbol, ΔE or ΔV. A change in voltage represented by $E2 - E1$, when the difference is finite. Compare *voltage differential*.

voltage input encoder An analog-to-digital encoder for which the input is an analog voltage. See, for example, *voltage-to-shaft position encoder*.

voltage inverter A circuit or device whose voltage output has the sign opposite that of the input voltage. The device may or may not provide amplification. Thus, the inverter output Eo is AEi where A is positive or negative gain, loss, or unity; Ei is the input voltage; and Eo is the output voltage.

voltage jump 1. An upward transient in voltage. 2. In a glow-discharge tube, a sudden break or increase in voltage drop across the tube.

voltage lag The condition in which changes in voltage follow corresponding changes in current. Compare *current lag* and *voltage lead*.

voltage lead The condition in which changes in voltage precede corresponding changes in current. Compare *current lead* and *voltage lag*.

voltage level 1. A prescribed reference value of voltage, e.g., the black level in a TV picture signal. 2. The discrete value of a steady voltage, or average value of a fluctuating voltage, as observed or measured in a circuit.

voltage limit The maximum or minimum level in a voltage range.

voltage loop In a standing-wave system, such as an antenna or transmission line, a maximum-voltage point. Compare *voltage node*.

voltage loss 1. Reduction of a voltage across a load, occurring because of a series resistance. 2. The ratio, in decibels, between the input voltage to a transmission line and the output voltage at the load end of the line.

voltage maximum See *voltage peak*.

voltage minimum See *voltage trough*.

voltage-mode switching circuit A resistor-transistor-logic (RTL) NAND or NOR circuit in which (in the off state) the transistor is cut off by the VBB bias voltage. The output is then approximately equal to the collector supply voltage VCC. The proper combination of input pulses overrides the cutoff bias, and the transistor switches on. The output then drops to a level equal to VCC minus the voltage drop across the external collector resistor.

voltage multiplier A special type of rectifier circuit which delivers a dc output voltage that is a multiple of the peak value of the ac input voltage, thus affording voltage step-up without a transformer. See, for example, *voltage doubler; voltage quadrupler; quintupler, 1*; and *voltage tripler*.

voltage node In a standing-wave system, such as an antenna or transmission line, a minimum-voltage point. Compare *voltage loop*.

voltage of self-induction The voltage drop across an inductor, resulting from the flow of alternating current through the inductor and due to self-induction.

voltage peak The highest value attained by a voltage during an excursion. Also called *voltage crest* or *voltage maximum*. Compare *voltage trough*.

voltage-power directional relay A relay system in which two circuits are connected when their voltage difference exceeds a predetermined value in one direction, and are disconnected when the power in them in the opposite direction exceeds a predetermined level.

voltage quadrupler A special rectifier circuit which, without a transformer, supplies a dc output voltage that is approximately quadruple the peak value of the ac input voltage.

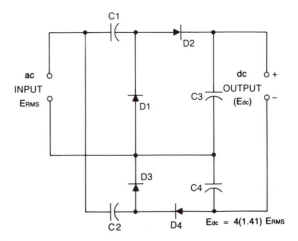

$$E_{dc} = 4(1.41)\ E_{RMS}$$

Voltage quadrupler.

voltage quintupler See *quintupler, 1*.

voltage-range multiplier 1. A multiplier resistor (see *voltmeter multiplier*) connected in series with a voltmeter that has an internal multiplier, to increase the range of the instrument. 2. For an ac voltmeter, an input amplifier used to increase the sensitivity of the instrument.

voltage rating 1. For a circuit or device, the recommended maximum voltage which may be applied, or the recommended working voltage, as specified. 2. For a generator, the specified output voltage.

voltage ratio The ratio of voltage at two points. These may be input and output voltage, primary and secondary voltage, and so on.

voltage-ratio box See *volt box.*

voltage reference See *standand cell* and *zener-diode voltage reference.*

voltage-reference cell See *standard cell.*

voltage-reference diode A zener diode biased into its zener region. The voltage drop across the diode's comparatively constant and under proper conditions may be used for reference purposes. Also see *zener-diode voltage reference.*

voltage-reference tube A special gas tube providing a voltage drop of such constancy that it may be used as a standard in voltage comparisons.

voltage reflection coefficient In a reflected-wave situation, the ratio Er/Ei, where Ei is the field-strength voltage of the incident wave, and Er the field-strength voltage of the reflected wave.

voltage-regulated supply See *constant-voltage source.*

voltage-regulating transformer A special transformer in which a resonant circuit and core saturation (see *saturated operation, 1*) are employed to provide a constant ac output voltage.

voltage regulation The stabilization of a voltage against fluctuations in source or load.

voltage-regulation constant Symbol, Kv. For a voltage-regulated power supply, the ratio $\Delta EL/\Delta Ei$, where ΔEL is a change in input voltage, and ΔEL is the corresponding change in load voltage.

voltage regulator A circuit or device which holds an output voltage constant during variations in the output load or input voltage.

voltage-regulator diode See *zener-diode voltage regulator.*

voltage-regulator transformer See *voltage-regulating transformer.*

voltage-regulator tube See *VR tube.*

voltage relay A relay or relay circuit that is actuated by a discrete voltage, rather than by current or power.

voltage-responsive device See *voltage-actuated device.*

voltage rise The normal condition in a series-resonant circuit, in which either the voltage across the coil or capacitor is higher than the voltage applied to the circuit.

voltage saturation In a current-actuated device, such as a bipolar transistor, the situation in which beyond a certain point (the saturation point) there is no increase in voltage drop across the device as the current through it is increased.

voltage-sensitive bridge A bridge having a nonlinear element (such as a voltage-dependent resistor) as one of the arms. Because of this element, the bridge can be balanced (with a given set of other arms) at only one value of applied voltage. At lower voltages the bridge becomes unbalanced in one direction; at higher voltage, in the opposite direction.

voltage-sensitive capacitor See *voltage-dependent capacitor.*

voltage-sensitive resistor See *voltage-dependent resistor.*

voltage sensitivity 1. Particular responsiveness of a circuit or device to voltage rather than current or power. See, for example, *voltage-dependent capacitor*, *voltage-dependent resistor*, and *voltage relay.* 2. Voltmeter sensitivity.

voltage-spectrum function The voltage-versus-frequency curve at the output of a circuit or transducer.

voltage-stabilized supply See *constant-voltage source.*

voltage stabilizer See *voltage regulator.*

voltage stabilizer tube See *VR tube.*

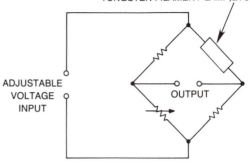

NONLINEAR ELEMENT (VARISTOR, SEMICONDUCTOR DIODE, THERMISTOR, TUNGSTEN-FILAMENT LAMP,ETC.)

ADJUSTABLE VOLTAGE INPUT

OUTPUT

Voltage-sensitive bridge.

voltage-stabilizing diode See *zener-diode voltage regulator.*

voltage-stabilizing tube See *VR tube.*

voltage standard A device which delivers a voltage of so accurate a value and of such stability as to be usable in calibrating other voltage generators and test instruments. See, for example, *standard cell* and *zener-diode voltage reference.*

voltage standing-wave ratio Abbreviation, vswr. In a standing-wave system, the ratio of the voltage of the incident wave to the voltage of the reflected wave.

voltage supply An ac or dc unit whose usable output is voltage, rather than current or power. When such a supply is not voltage-regulated, it may only be used reliably with a very light load.

voltage to chassis In electronic equipment mounted on a metal chassis, the voltage between the chassis and a given point in the circuit.

voltage-to-frequency converter A device or circuit that delivers an output frequency which is proportional to an input voltage (usually dc). Compare *frequency-to-voltage converter.*

voltage to ground 1. In a circuit, the voltage measured between a given point and the ground point. 2. The voltage measured between the earth and a line or piece of equipment.

voltage to panel In electronic equipment mounted on a metal panel, the voltage between the panel and a given point in the circuit.

voltage-to-shaft-position encoder An encoder for which the output is the rotation of a motor shaft over an arc proportional to an input voltage.

voltage transformer 1. A transformer employed primarily to supply voltage with little or no current. 2. A small, step-up transformer for increasing the sensitivity of an ac voltmeter. Also called *potential transformer.*

voltage tripler A special rectifier circuit which, without a transformer, supplies a dc output voltage that is approximately three times the peak value of the ac input voltage. See figure on next page.

voltage trough The lowest value reached by a voltage during an excursion. Compare *voltage peak.*

voltage tuning A method of adjusting the frequency of an oscillator or resonant circuit by means of a variable, direct-current voltage.

voltage-tunable magnetron A CW magnetron oscillator in which the output frequency is varied by adjusting the dc voltage on the tube.

voltage-tunable oscillator See *voltage-controlled oscillator.*

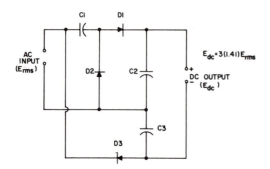

Voltage tripler.

voltage-turns ratio The turns ratio of a transformer, indicated by the ratio of primary voltage to secondary voltage, or vice versa.

voltage-type telemetry Telemetry based upon the variation of a single voltage in sympathy with the changes in a sensed phenomenon.

voltage-variable capacitor 1. A specially processed semiconductor diode of which the voltage-variable capacitance of the junction is utilized. Also called *varactor*. 2. A capacitor having a specially processed nonlinear dielectric, such as barium strontium titanate, whose capacitance varies inversely with the applied dc bias voltage.

voltage-variable resistor See *voltage-dependent resistor*.

voltage vector In a vector diagram, a vector showing the magnitude and phase of a voltage. Compare *current vector*.

voltaic Pertaining to chemically produced dc. Sometimes interchangeable with *galvanic*.

voltaic cell For the generation of a dc voltage, a primary cell consisting of two electrodes made of different metals and immersed in an electrolyte. Also called *galvanic cell*. Also see *cell, electromotive series,* and *primary cell*.

voltaic couple A pair of dissimilar metals (or other substances) which will generate a voltage when they are in contact with an electrolyte.

voltaic pile A rudimentary primary battery consisting of a series of disks made of two different metals, stacked alternately with electrolyte-soaked cloth or paper disks.

voltaic series See *electromotive series*.

voltameter An electrolytic cell for determining the value of an unknown current—or unknown quantity of electricity—from the weight of metal deposited out of an electrolyte onto the cathode by passage of current over an accurately timed interval.

volt-ammeter A combination meter for measuring volts and amperes.

volt-ampere Symbol, VA. Unit, *watt*. The simple product of voltage and current in volts and amperes, which gives the true power in a dc circuit and the apparent power in an ac circuit. Also see *apparent power* and *true power*.

volt-ampere-hour meter See *VAR-hour meter*.

volt-ampere meter See *VAR meter*.

volt-amperes reactive Abbreviation, VAR. The product of volts and amperes for a purely reactive circuit. This product does not give true power, since the power factor is neglected, but only apparent power. A true reactance, of course, absorbs power during one half-cycle of ac, and returns it to the generator during the next half-cycle.

Volta's law See *Volta's principle*.

Volta's pile See *voltaic pile*.

Volta's principle Two dissimilar metals brought into contact (even in air) will generate a difference of potential whose value is characteristic of the metals. Also see *electromotive series*.

volt box A precision, potentiometer-type voltage divider employed in the calibration of meters and other instruments. The device is usually provided with a set of terminal posts for selecting various ratios of output voltage to input voltage.

volt-electron See *electronvolt*.

voltmeter A usually direct reading instrument employed to measure voltage. Also see *electronic voltmeter, FET voltmeter, transistor voltmeter,* and *vacuum-tube voltmeter*.

voltmeter-ammeter See *volt-ammeter*.

voltmeter amplifier A wideband, flat-frequency-response, low-distortion preamplifier employed to boost the sensitivity of an ac voltmeter.

voltmeter-millivoltmeter A voltmeter which provides several low ranges as well as several high ones. A familiar example is the so-called audio voltmeter having full-scale ranges extending from 0-1 millivolt to 0-1000 volts.

voltmeter multiplier A resistor connected in series with a current meter (usually a milliammeter or microammeter) to convert it into a voltmeter.

voltmeter sensitivity Unit, ΩV. For a voltmeter, the total resistance of the instrument (multiplier resistance plus the resistance of the meter movement) divided by the full-scale deflection of the meter. Thus, a 0-10 dc voltmeter with an input resistance of 100 kilohms has a sensitivity of 10 kilohms per volt.

volt-milliammeter A combination meter for measuring volts and milliamperes.

volt-ohm-ammeter A multimeter for measuring volts, ohms, and amperes.

voltohmmeter A combination meter for measuring volts and ohms.

volt-ohm-milliammeter Abbreviation vom. A multimeter for measuring volts, ohms, and milliamperes. Such an instrument usually measured megohms and microamperes, too.

voltsensor See *voltage detector*.

volts per meter Abbreviation, V/m or Vpm. The unit of electric field strength.

volume 1. Intensity of sound. Also called *loudness*. 2. A circumscribed portion of space, either imaginary or actually occupied, and described by three dimensions, e.g., the *volume of a cube* is the cube of the length of one side; of a *sphere* 4π times the radius cubed divided by 3; of a *rectangular prism*, the product of length, width, and depth; of a *cone*, the product of pi, the radius squared, and the height, divided by 3; and of a *pyramid*, the area of the base times the height, divided by 3. 3. In a computer system, a unit of magnetic storage medium, e.g., a magnetic tape reel.

volume compression The automatic reduction of the gain of an af amplifier. The process differs from clipping (which "slices off" the tops of waves) in that compression (ideally) reduces the amplitude while preserving the waveform. Compare *volume expansion*.

volume compressor A circuit or device (such as an agc amplifier) for achieving volume compression. Compare *volume expander*.

volume conductivity The reciprocal of volume resistivity.

volume control A variable resistor (usually a potentiometer) or a network of resistors (such as a pad) for adjusting the gain, and consequently the loudness of the output signal, of an amplifier.

volume equivalent In wire telephony, speech loudness throughout the system, expressed in terms of the trunk loss in a reference system and adjusted for equal loudness.

volume expander A circuit or device for automatically boosting the volume of an audio-frequency signal. Also see *expander*. Compare *volume compressor*.

volume expansion The technique of automatically increasing the gain, and consequently the output-signal volume, of an af amplifier. Also see *volume expander*. Compare *volume compression*.

volume indicator A fast-acting ac meter employed to monitor the volume level in an audio channel in which the signal level is fluctuating. The scale is graduated in *volume units*. Also called *VU meter*.

volume integration Triple integration to determine the volume (V) of a space or solid $V = \int \int \int dl \, dw \, dh$, where l is length, w is width, and h is height. Also see *integration*. Compare *double integration*.

volume lifetime In a homogeneous semiconductor, the interval between minority-carrier generation and recombination.

volume limiter A circuit or device which automatically holds the volume level of an audio channel to a predetermined maximum. Also see *volume compression* and *volume compressor*.

volume-limiting amplifier An amplifier that functions as a volume limiter through the action of volume-limiting subcircuits.

volume magnetostriction 1. The decrease in the total volume of certain ferromagnetic substances under the influence of a magnetic field. 2. The increase in the total volume of certain ferromagnetic substances under the influence of a magnetic field.

volume of ideal gas (standard condition) Symbol, V_0. Value, 22.4136 m^3/kmol.

volume range Unit, dB. The difference between the maximum and minimum volume levels which a device or system can accommodate.

volume recombination rate Within the volume of a semiconductor, the rate at which electrons and holes recombine.

volume resistance The effective resistance, through a given medium, between two electrodes placed within that medium.

volume resistivity The resistance of a specific volume of a material, e.g., the resistance between opposite faces of a 1-centimeter cube of the material. Also see *microhm-centimeter*, *ohm-centimeter*, and *resistivity*.

volumetric efficiency In an electronic assembly, the ratio of the volume of parts to the total volume of the assembly.

volumetric radar A radar giving a three-dimensional display.

volume unit Abbreviation, VU. The unit of measurement of fluctuating ac power, such as that of speech or music. Zero VU corresponds to a reference power of + 4 mW in 600 ohms. Volume units are measured with a *volume indicator*.

volume-unit indicator See *volume indicator*.

vom Abbreviation of *voltohm-milliammeter*.

von Hippel breakdown theory The theory that, assuming no electron-energy distribution, breakdown occurs at field intensities for which the recombination rate of electrons and holes is lower than the rate of ionization by collision.

vor Abbreviation of *very-high-frequency omnirange*.

voter See *majority logic*.

vox Abbreviated form of *voice-operated control*.

VP Symbol for *gate-source pinchoff voltage* of a field-effect transistor.

Vp Symbol for *voltage at peak torque*.

V-particle See *vee-particle*.

Vpm Abbreviation of *volts per meter*. (Also, *V/m*.)

VR Abbreviation of *voltage regulator*.

vrr Abbreviation of *visual radio range*.

VR tube A special two-element gas (argon or neon) tube which, after having been fired, maintains a constant internal voltage drop and thus acts as a simple voltage regulator. The name is the abbreviation of *voltage regulator*.

VR-tube coupling Interstage coupling in a device, such as a vacuum-tube amplifier, in which a VR tube is the coupling element. When the signal voltage is lower than the ignition potential of the VR tube, the tube acts as an insulator and the high plate voltage of one stage cannot reach the grid of the tube in the next stage, as it would in direct coupling. When the signal amplitude increases, however, the VR tube fires and couples the signal through to the next stage. This action is similar to that of a zener diode employed at lower voltage levels as a coupling element (see *zener-diode coupling element*).

V•s Abbreviation of *volt-seconds* (webers).

V•s Abbreviation of *volt-seconds per ampere* (henrys).

VSA Abbreviation of *voice-stress analyzer*.

VSB Abbreviation of *vestigial sideband*.

VSF Abbreviation of *vestigial sideband filter*.

V-signal See *vee-signal*.

vsr Abbreviation of *very-short range*.

vswr Abbreviation of *voltage standing wave ratio*.

vt 1. Abbreviation of *vacuum tube*. 2. Abbreviation of *variable time*.

VT fuze See *proximity fuze*.

VTL Abbreviation of *variable-threshold logic*.

vtm Abbreviation of *voltage-tuned magnetron*.

VTO Abbreviation of *voltage-tuned oscillator* or *voltage-tunable oscillator*.

VTR Abbreviation of *video tape recorder* or *video tape recording*.

vtvm Abbreviation of *vacuum-tube voltmeter*.

vt voltmeter See *vacuum-tube voltmeter*.

VU Abbreviation of *volume unit*.

vulcanized fiber A touch insulating material derived from cellulose. Dielectric constant, 5 to 8.

VU meter See *volume indicator*.

VVCD Abbreviation of *voltage-variable capacitor diode* (see *voltage-variable capacitor,1*).

VVV signal In radiotelegraphy, the standard test signal consisting of three transmitted vees followed by the call letters of the transmitting station. The group is repeated as many times as practicable in a 10-second interval. Also see *vee-signal*.

VW Abbreviation of *volts working* (see *working voltage.*).

vy Abbreviation for *very*.

W

W 1. Symbol for *work*. 2. Abbreviation of *watt*. 3. Symbol for tungsten. 4. *Symbol for benergy*. 5. Abbreviation of *west*. Abbreviation of *width*.

w 1. Abbreviation of *weight*. (Also, *wt*.) 2. Abbreviation of *week*.

WAC Amateur radio abbreviation of *worked all continents*, a citation given to operators who have engaged in verified two-way communication with stations on all continents (see *work, 2*).

wafer 1. Semiconductor die. 2. A thin, flat disk, ring, or plate around which the contacts of a rotary switch are spaced. 3. A thin square or rectangle of dielectric material used as the dielectric member in a fixed capacitor. 4. A plate cut from a crystal (e.g., a quartz wafer).

wafer fabrication The various processes used in the manufacture of semiconductor integrated circuits.

wafer slicing Cutting plates from a mother crystal, as when piezoelectric plates are cut from a quartz crystal.

wafer socket A tube or transistor socket consisting of a plastic or ceramic wafer with spring-type contacts for gripping tube or transistor pins.

wafer switch A rotary switch whose contacts are arranged around the periphery of a plastic or ceramic wafer (see *wafer, 2*).

Wagner ground A circuit (often a single potentiometerz which enables the balancing out of objectionable stray capacitances in an ac bridge. The bridge is balanced alternately with the bridge-balance control and the Wagner control, until there is no further shift of null point when changing from one to the other.

waiting time 1. Warmup time. 2. The interval of delay occuring between the time a data transfer to or from a computer memory is called for the act of transfer. Also called *latency*.

walkie-lookie A portable, combination camera and transmitter for remote television pickup. At sports events and other gatherings, the unit is strapped to the shoulder of the camera operator.

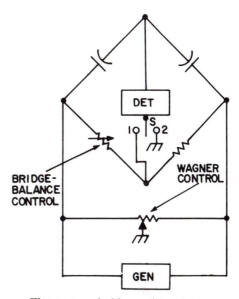

Wagner ground with capacitance bridge.

walkie-talkie A portable transceiver (or transmitter-receiver) which is small enough to be carried about with ease.

wall absorption Diminished radioactive emission (especially of beta and gamma rays) because of absorption by the radiating substance.

wall box A wall-mounted (usually metal) box enclosing circuit breakers, fuses, switches, and the like.

wall effect In an ionization chamber, the rise in ionization due to electrons being released by the walls of the chamber.

wall energy In a ferromagnetic substance, the energy per unit area stored in a domain wall between two regions of opposite magnetization.

Wallman amplifier A cascode amplifier (see *cascode.*).

wall mount A metal bracket for fastening an antenna to a wall.

wall outlet A wall plug or socket usually mounted in a protective box or can, from the front of which it can be accessed, and recessed in a wall.

wall plaque A loudspeaker so thin that, mounted in a frame (sometimes behind grill cloth), it can be hung on a wall.

wall plate A (usually rectangular) plate of metal or plastic for holding a wall outlet or wall switch.

wall plug A male or female plug usually mounted in a protective box or can and recessed in a wall. Such a device can provide easy access to an antenna, telephone line, or load; or it can deliver ac, dc, or rf power.

walls The sides of the groove cut into a record disk or cylinder.

wall socket A male or female socket usually mounted in a protective box or can and recessed in a wall. Such a device can provide easy access to an antenna, telephone line, or load; or it can deliver ac, dc, or rf power.

wall speaker See (wall plaque).

wall switch A switch usually mounted in a protective box or can and recessed in a wall.

wall telephone A wall-mounted telephone.

wall-through tube See *lead-in tube.*

Walmsley antenna A phased array consisting of several full-

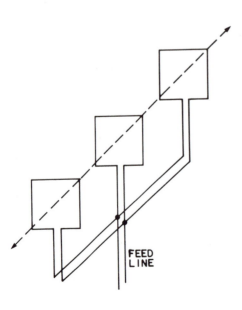

Walmsley antenna.

wavelength loops.

wamoscope A radar display tube which performs several microwave-receiver functions (detection, oscillation, amplification, etc.) as well as displaying an image. The name is an acronym for *wave-mo*dulation oscill*scope.*

wander See *scintillation, 1.*

warble 1. A periodic rise and fall in the pitch of a musical tone or combination of tones. 2. To rise and fall in pitch with a definite period.

warble-tone generator An rf oscillator whose frequency is warbled, i.e., varied at a subaudible rate over a fixed frequency range.

warm junction The heated junction in a two-junction thermocouple circuit. Also called *hot junction.* Compare *cold junction.*

warmup The process of stabilizing an electronic equipment by allowing its temperature to rise to the optimum level.

warmup time 1. The time required for an indirectly heated cathode in a tube to come up to full operating temperature. 2. The time required for an electronic circuit to stabilize after the power has been switched on.

warning bell (buzzer) A bell (or buzzer) used as an audible alarm in a warning device.

warning device A device, such as an electronic signaler, for alerting a person to an emergency (intrusion on premises, danger to life and safety, and so on).

warning lamp See *warning light.*

warning light A lamp used as a visual alarm in a warning device.

warpage Distortion of the normally straight sides of a triangular wave.

WAS Amateur radio abbreviation of *worked all states,* a citation given to operators who have engaged in verified two-way communication with stations in all states of the United States (see *work, 2*).

washer capacitor A very thin donut capacitor.

washer resistor A resistor made in the general shape of a washer or ring and having a center hole for a mounting screw or stacking rod.

washer thermistor A thermistor made in the general shape of a washer and having a center hole for a mounting screw or stacking rod.

washer varistor A varistor made in the general shape of a washer and having a center hole for a mounting screw or stacking rod.

washout process A method of fabricating bipolar transistors. The contact metal is deposited in the diffusion hole.

waste Radioactive material which is to be disposed of.

waste instruction In a computer program an instruction which is not meant to be acted upon, e.g., one used to take up space in the listing for some reason. Also called *null instruction* and *dummy instruction.*

watch A work shift, as of electronic personnel, e.g., radio station operators.

watchcase receiver An earphone enclosed in a small, round case with a screw-on cap. Its name comes from its resemblance to a large pocket watch.

water absorption For a solid material, such as a dielectrian the ratio of the weight of water absorbed by the material to the weight of the material.

water-activated battery A battery which contains all of the ingredients of its electrolyte except water, which must be added when the battery is put into service.

water adsorption The formation of a thin layer of water molecules on the surface of normally dry material, but not by *absorption.* Also see *adsorption.*

water analogy The useful, but not wholly accurate, teaching device of comparing an electric current with the flow of water. In such a comparison, voltage is shown equivalent to water pressure, and current to the quantity (e.g., gallons) of water flowing in unit time.

water battery A primary battery or cell employing water as the electrolyte.

water calorimeter A calorimeter used to measure power in terms of the increase in temperature of water which is heated by the electrical energy.

water capacitor An emergency capacitor made by setting one glass of water in anothern larger jar of water, so that the two bodies of water are separated by the walls of the smaller jar. The bodies of water form the plates of the capacitor, and the wall of the smaller jar serves as the dielectric between them.

water-cooled tube A power tube, such as a large radio transmitting tube, which is cooled by the circulation of water in the space between the outside of the tube envelope and a surrounding jacket.

water cooling A method of cooling components by pumping water through pipes surrounding them.

water-flow alarm An electronic circuit which actuates an alarm when the flow of water through pipes or other channels changes from a predetermined rate.

water-flow control A servo system for automatically maintaining or adjusting the flow of water through pipes or other channels.

water-flow gauge See *water-flow meter.*

water-flow indicator See *water-flow meter.*

water-flow meter An instrument used to monitor the flow of water through pipes or other channels, sometimes showing its direction as well as its rate.

water-flow switch In water-cooled systems (e.g., water-cooled tubes), a switch that actuates an alarm when the water slows or stops.

water glass Sodium silicate ($NA2O • 4 SiO2$), a substance used as a fireproofing agent and protective coating.

water ground An earth connection made by dropping a weighted wire into a body of water.

water jacket In a water-cooled device, the outer jacket which, with the outer wall of the cooled component, forms a space through which the cooling water flows.

water-level control A servo system which automatically maintains or adjusts the level of water inside a tank or other container.

water-level gauge An electronic system which gives direct readings of the level of water inside a tank or other container.

water-level indicator See *water-level gauge.*

water load 1. A makeshift, power-dissipating resistive load (see *dummy load.*) consisting of a container of tap water or salt water into which two wires are immersed. 2. A waveguide termination containing water, which is heated by the microwave energy, and usable as a water calorimeter.

water monitor A sensitive electronic instrument for checking radioactivity in a water supply.

water of crystallization Water that is combined molecularly with various crystalline materials. Thus, the formula $CuSO4 • 5H2O$ for copper sulfate shows that 5 molecules of water of crystallization (i.e., $5H2O$) are associated with the copper sulfate molecule.

water-pipe ground An earth connection (see *ground connection, 1*) made by running a strong wire to the nearest cold-water pipe.

water power Hydroelectric power, i.e., electrical energy produced by generators driven by water.

water-pressure alarm An electronic circuit which actuates an alarm when water in pipes or other channels changes from a predetermined level.

water-pressure control A servo system for automatically maintaining or adjusting water pressure in pipes or other channels.

water-pressure gauge See *water-pressure meter.*

water-pressure indicator See *water-pressure meter.*

water-pressure meter An instrument which indicates water pressure in a pipe or tank, directly.

water-pressure switch A switch that actuates an alarm when water pressure rises or falls.

waterproof Sealed against the entry of, and the resulting contamination by, water, e.g., a waterproof diode package.

waterproof cement A cement that becomes waterproof after drying.

water pump In a water-cooled electronic system, the (usually rotary) pump that circulates the water.

water-repellent See *waterproof.*

water resistor An electrolytic resistor in which the electrolyte is tape water or dilute salt water.

water rheostat A variable water resistor. Usually, the resistance is varied by moving the immersed electrodes closer together or farther apart.

water tester An instrument for checking pHn electrical resistance, and other properties of water.

watertight See *waterproof.*

water witching The practice of locating underground water by electronic methods.

WATS A form of long-distance telephone service. Rates are charged on a different basis than normal long-distance service. The system is especially favored by businesses, since it saves money for subscribers making a large number of calls in each billing period.

watt Abbreviationn W. The practical unit of electric and other power. One watt is dissipated by a resistance of 1 ohm through which a current of 1 ampere flows. See also *kilowatt, megawatt, microwatt,* and *milliwatt.*

wattage Electrical power.

wattage rating 1. The recommended output power of a device. 2. The recommended power dissipation of a device.

watt current The component of alternating current which is in phase with voltage. Also called *resistive current.* Compare *reactive current.*

watt-decibel conversion The conversion of a power level, such as the power output of an amplifier, in watts to the corresponding power level in decibels with respect to a reference level. Thus, $NdB = 10 \log (Po/Pref)$, where Po is the power of interest (watts), and $Pref$ is the reference level (e.g., 1 mW, mW, etc.).

watt-hour Abbreviation, WH. The unit of electrical energy or work; 1 WH = 3600 joules = 1 watt/hr = 10^{-3} kWH. Also see *energy, kilowatt-hour, power,* and *watt-second.*

watt-hour capacity The number of watt-hours that a storage battery can deliver reliably and safely under specified operating conditions.

watt-hour-demand meter A combination watt-hour meter and demand meter.

watt-hour efficiency For a storage battery, the ratio of *watt-hours output* to *watt-hours of recharge.*

watt-hour constant In an electric-energy meter, the number of watt hours in one revolution of the indicating disk.

watt-hour meter An instrument for measuring electrical energy in watt-hours. One well-known type consists essentially of a small motor geared to a row of four dial indicators. An eddy-current disk keeps the motor speed proportional to the watt-hours consumed by a load, a value which is the sum of the readings of the dials. Also called *service meter* and *kilowatt-hour meter*.

wattless current The component of alternating current which is out of phase with voltage. Also called *reactive current*. Compare *resistive current*.

wattless power The apparent power in a reactive circuit, indicated by the product of volts and amperes. There is no actual power consumption, since the power taken by a reactance during a half-cycle is returned to the generator during the next half-cycle. Also see *ac power, apparent power, reactive kilovolt-amperes,* and *reactive volt-amperes*.

wattless volt-amperes See *wattless power*.

wattless watts See *wattless power*.

wattmeter Abbreviation, WM. An instrument used to measure electrical power. The scale usually reads directly in watts, kilowatts, milliwatts, or microwatts. Also see *electronic wattmeter*.

watt-second Abbreviation, W•s. A small unit of electrical energy or work; 1 watt-second = 1 joule = 1/3600 watt-hour = 1 watt/sec. Also see *energy, power*.

watt-second constant In an electric-energy meter, the number of watt seconds in one revolution of the indicating disk.

wave 1. A single oscillation in matter (or, when it is that of a radio wave, in space) which, for want of a better analogy, has the general shape of a wave in disturbed water (but is usually of a spherical form). Waves move outward from a point of disturbance, propagate through a medium, and grow weaker as they travel farther. Wave motion is associated with mechanical vibration, sound, radio, heat, light, X-rays, gamma rays, and cosmic rays. See *sound* and *wave-length*. 2. A single cycle of alternating or pulsating current or voltage. Also see *ac voltage, alternating current, pulsating dc voltage,* and *pulsating direct current*.

wave absorption The removal of energy from electromagnetic waves as they pass through certain media, such as solid bodies, water, and the atmosphere. Compare *polarization, wave reflection* and *wave refraction*.

wave amplitude The peak value of a wave. Also see *wave crest* and *wave trough*.

wave analyzer An instrument consisting essentially of a continuously tunable pass filter and an electronic ac voltmeter. As the filter is tuned successively to the fundamental frequency of a complex wave and to its various harmonics, the voltmeter shows the amplitude of each of the components. Also see *heterodyne wave analyzer*.

wave angle The angle at which a radio wave is transmitted or received.

wave antenna See *Beverage antenna*.

wave attenuation The reduction of wave amplitude with respect to distance from the source.

waveband A band of radio frequencies. Also called *frequency band*.

waveband switch See *bandswitch*.

wave beam Unidirectional radiation from a directive antenna.

wave bounce See *wave reflection*.

wave clutter Radar interference caused by sea waves.

wave converter A waveguide part, such as a baffle-plate or grating, which changes a wave pattern from one type to another.

wave crest The maximum value of a wave envelope. Compare *wave trough*.

wave cycle A complete single alternation of a wave.

wave direction The direction in which an electromagnetic wave travels. It is perpendicular to the wave front and depends (whether it is forward or backward) upon the direction of the electric and magnetic components. If either is reversed, wave direction reverses; if both are reversed, the direction remains unchanged.

wave duct 1. An atmospheric channel through which waves travel easily (see *duct, 1*). 2. A tubular waveguide in which wave propagation is concentrated.

wave envelope The outline described by (or which surrounds or *envelopes*) the various amplitude peaks of the cycles in an amplitude-modulated wave. The envelope frequency is equal to the modulating frequency.

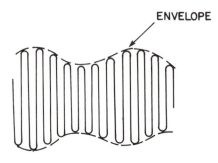

ENVELOPE

Wave envelope.

wave equation A second-degree partial differential equation whose solution describes wave phenomena.

wave filer A circuit or device which offers different attenuation to signals of different frequencies. See specifically *bandpass filter, band-suppression filter, high-pass filter* and *low-pass filter*.

waveform The shape of a waven described in terms of its resemblance to some well-known figure or to its conformity to the curve of the applicable wave equation, e.g., sinusoidal, square, sawtooth, cosine, rectangular, and triangular.

waveform-amplitude distortion See *amplitude distortion*.

waveform analyzer See *wave analyzer*.

waveform converter A circuit or device for changing a signal of one waveform (such as a sine wave) into one of another waveform (such as a pulse or square wave). In one sense, *converter, 5*.

waveform distortion The malfunction evidenced by a change of the waveshape of a signal passing through a circuit.

waveform error In a quantity displayed by an ac test instrument, an error due to the waveform of the measured signal. Thus, a vtvm calibrated with a sine-wave voltage is subject to error when a measured signal is nonsinusoidal. Also called *waveform effect*.

waveform generator See *function generator*.

waveform influence See *waveform error*.

waveform monitor In TV practice, an oscilloscope which continuously displays the video waveform.

waveform synthesizer A variable-frequency signal generator which allows the tailoring of waveshape to suit individual applications. A function generator is such an instrument, but usually gives only a choice of common waveshapes.

wave front For a radio wave, the plane that is parallel to the perpendicular electric and magnetic lines.

wave function A point function in a wave equation, specifying wave amplitude.

wave group The resultant of several different-frequency waves traveling over a common path.

waveguide A carefully dimensioned, hollow, metal pipe through which microwave energy is transmitted. Waveguides are usually rectangular.

waveguide apparatus See *waveguide components.*

waveguide attenuator A device, such as an interposed energy-absorbing plate, for signal attenuation in a waveguide.

waveguide choke flange A waveguide flange which presents no impedance to the signal, and which need not be metallic for continuity.

waveguide component A device which is adapted for connection to, or insertion into, a waveguide system. Such components include waveguide parts and accessories (bends, sections, etc.), attenuators, loads, wavemeters, and the like.

waveguide connector A fitting for joining waveguides for the efficient propagation of a signal.

waveguide coupling See *waveguide connector.*

waveguide critical dimension The cross-sectional dimension that determines the cutoff frequency for a waveguide.

waveguide cutoff In a waveguide, the highest or lowest frequency that can be propagated with less than a specified amount of attenuation per unit length.

waveguide directional coupler A directional coupler made of two parallel waveguides with a common wall. Two slots cut in the wall allow part of the microwave energy propagated in one direction in the main waveguide to be extracted, and energy traveling in the opposite direction to be rejected.

waveguide dummy load A waveguide section which dissipates the microwave energy entering it.

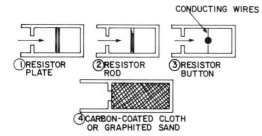

Waveguide dummy load.

waveguide elbow 1. A curved bend in a waveguide. 2. A waveguide section with such a bend between two straight waveguides.

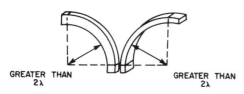

Waveguide elbow.

waveguide flange A flat, liplike fitting at the end of the pipe of a waveguide. It serves to fasten waveguide sections together or to attach a waveguide component, equipped with an identical flange, to the end of a waveguide.

waveguide frequency meter See *waveguide wavemeter.*

waveguide gasket A gasket which provides electrical continuity between mating waveguide sections.

waveguide grating An array of wires mounted inside a waveguide and which passes signals at some frequencies while obstructing others.

waveguide impedance Where the power P is known, and voltage E and current I are defined with respect to a type of wave and waveguide, waveguide impedance Z equals E^2/P or P/I^2.

waveguide junction A fitting which allows one waveguide section to be joined at an angle to another section. See, for example, *waveguide tee* and *waveguide wye.*

waveguide lens A microwave lens consisting of waveguide sections which provide the required phase shifts.

waveguide load See *waveguide dummy load.*

waveguide mode The form of propagation indicated by the field pattern in a plane transverse to the direction in which energy is propagated through a waveguide. The transverse electric mode is the *TE mode*; the transverse magnetic mode, the *TM mode.*

waveguide mode suppressor A filter which suppresses undesired propagation modes in a waveguide.

waveguide phase shifter A shifter for adjusting the phase of waveguide output energy with respect to input energy.

waveguide plunger A plunger which reflects incident microwave energy in a waveguide.

waveguide post A transverse rod inside a waveguide, which acts as a parallel susceptance.

waveguide probe A pickup probe (tip or loop, as required) for sampling the field inside a waveguide, or a similar injection probe for introducing energy into a waveguide. Also see *waveguide slotted line.*

waveguide propagation 1. The transmission or microwave energy through a waveguide by successive reflections between the inner walls. 2. Propagation of radio waves through an atmospheric duct (see *duct, 1), as if through a waveguide.*

waveguide radiator A radiator consisting of an open-ended waveguide with or without a horn. It radiates microwave energy into space or to a reflector.

waveguide resonator A waveguide section employed as a cavity resonator (see *resonant cavity).*

waveguide seal A protective cover for the end of a waveguide. The seal introduces very little microwave attenuation, while preventing entry of moisture and debris.

waveguide shim A thin, pliable metal sheet inserted between mating waveguide components for electrical continuity. Also see *waveguide gasket.*

waveguide shutter An adjustable mechanical barrier, such as a rotatable vane, inserted into a waveguide to block or divert microwave energy.

waveguide slotted line A slotted line consisting of a section of waveguide slotted to receive the usual traveling probe.

waveguide slug tuner A quarter-wave dielectric slug inserted into a waveguide so that its amount of penetration and position can be adjusted for tuning purposes.

waveguide stub A stub consisting of a waveguide section joined to a main waveguide at an angle and provided with a nondissipative termination.

waveguide stub tuner An adjustable piston in a waveguide stub for tuning purposes.

waveguide switch A switch consisting of a movable section of waveguide which may be positioned for coupling to one of several other waveguide sections, and thus passes the energy it receives to any of the other sections.

waveguide system Microwave "plumbing" consisting of wave-guides, their fittings and accessories, and associated components, such as attenuators, loads, wavemeters, etc.

waveguide taper A tapered waveguide section.

waveguide tee In a waveguide assembly, a tee-shaped junction employed to connect a section of waveguide in series or parallel with another section.

waveguide transformer A waveguide component which functions as an impedance transformer.

waveguide tuner In a waveguide system, an adjustable tuner providing impedance transformation.

waveguide twist A twisted waveguide section, i.e., a section whose cross section is rotated around the longitudinal axis (e.g., from vertical to horizontal).

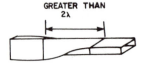

Waveguide twist.

waveguide wavelength In a uniform waveguide operating at a given frequency and in a particular mode, the distance between similar points for 2π radians (360 degrees) phase shift.

waveguide wavemeter A waveguide component which acts as an absorption wavemeter or transmission wavemeter for identifying microwave frequencies.

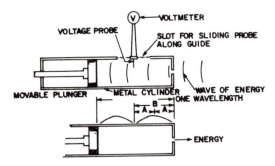

Waveguide wavemeter.

waveguide wedge See *wedge, 1.*

waveguide window A thin metal window mounted transversely inside a waveguide for impedance matching purposes. The edges of the slit in a capacitive window are perpendicular to the electric field; in an inductive window, they are parallel to the electric field.

waveguide wye In a waveguide assembly, a wye-shaped junction for joining three waveguide sections.

wave heating Heating a material by energy absorbed from traveling electromagnetic waves.

wave impedance In a waveguide, the ratio *TE/TM*, where *TE* is the transverse electric field, and *TM* is the transverse magnetic field.

wave interference Interaction between two or more waves, resulting in reinforcements and cancellations of energy.

wavelength Symbol, λ. Unit, meter. The length of one complete wave of an alternating or vibrating phenomenon, generally measured from crest to crest or from trough to trough of successive waves. Wavelength is related to frequency: $\lambda = (3 \times 10^8)/f$, where f is in hertz. Also see *wavelength-period-frequency relationships.*

wavelength constant The imaginary component of the propagation constant.

wavelength-period-frequency relationships The relationship between wavelength (λ), period (T), and frequency (f). When wavelength is expressed in meters, period in seconds, and frequency in megahertz, the following relationships hold: $\lambda = 300/f = 3 \times L\ 10^8 T$; $T = 10^{-6}/f = \lambda\ (3 \times 10^8)$; $f = 300/\lambda = 10^{-6}/T$.

wavelength shifter 1. A frequency shifter whose performance is indicated in units of wavelength, rather than in those of frequency. 2. In certain photosensitive cells and tubes, a photofluorescent substance which raises the efficiency of the device by absorbing photons and then releasing ones of longer wavelength.

wavelength-to-frequency conversion See *wavelength-period-frequency relationships.*

wave mechanics A theory of matter which views subatomic particles as complex wave patterns, and attempts to account for all physical processes in terms of wave phenomena.

wavemeter An instrument for measuring the wavelength or frequency of radio waves. One familiar form consists of a series-resonant circuit containing an inductor, variable capacitor, and diode-type meter. The dial of the capacitor is calibrated to read in MHz. The inductor picks up energy from the rf source (of unknown frequency); the capacitor then is tuned for peak deflection of the meter, and the unknown frequency is read from the dial. This instrument is often called an *absorption wavemeter* because it absorbs a certain amount of power from the signal source under test. For other types of wavemeter, see *cavity wavemeter, coaxial wavemeter, Lecher wires,* and *slotted line.*

wave motion Undulating motion—up and down, side to side, etc. An electro-magnetic wave has undulating electric and magnetic components which are both in phase and perpendicular to each other and to the direction of propagation of the wave.

wave normal 1. The direction of propagation of an electromagnetic wave. 2. A unit vector directed at a right angle to both the electric and magnetic lines of force.

wave number The reciprocal of wavelength (i.e., $1/\lambda$). This number denotes the number of waves per unit distance.

wave packet A short pulse composed of waves.

wave packets A radiantdenergy resulting from a number of wave trains of different wavelength.

wave path The line along which a wave train is propagated.

wave polarization The direction (horizontal or vertical) of wave undulations, i.e., the plane of the undulations with respect to the direction of propagation. In general, a vertical antenna radiates a vertically polarized wave, and a horizontal antenna radiates a horizontally polarized wave.

wave propagation Sending waves through space. Electromagnetic waves travel through space at the speed of light (3×10^8 meters, or 186,000 miles, per second) and, like light, may be reflected and refracted.

wave reflection The reflection of electromagnetic waves by an obstruction, such as a solid body or a layer of the ionosphere. Compare *wave absorption, wave polarization,* and *wave refraction.*

wave refraction The refraction (bending) of the line of propagation of electromagnetic waves as they pass through various media, such as the ionosphere. Compare *wave absorption, wave polarization,* and *wave reflection.*

waveshape The overall contour of a wave, especially as revealed by a curve plotted for the particular wave equation. Also see *waveform.*

waveshaping circuit A circuit which receives an input signal having one waveshape and delivers an output signal having a different waveshape. For example, a squaring circuit converts a sine wave into a square wave at the same frequency.

wave surface See *wave front.*

wave tail In a decaying pulse or signal envelope, the interval between the beginning of the decay and the point at which the amplitude reaches zero.

wave telegraphy See *radiotelegraphy.*

wave telephony See *radiotelephony.*

wave theory of matter A physical theory that the charge of an electron is distributed in space, rather than being focused at a point. Also see *wave mechanics.*

wave tilt The slight forward tilt of the electric vector of a radio wave radiated at the surface of the earth by a vertical antenna.

wave train A series of similar electromagnetic waves propagated at equal intervals.

wavetrap A resonant circuit consisting of an inductor and capacitor, either or both of which may be adjustable for tuning, used to remove (*trap*) a signal at the wavetrap resonant frequency from a signal mixture. A wavetrap is a simple version of the wavemeter.

wave trough The minimum value of a wave envelope. Compare *wave crest.*

wave velocity The distance per unit time traversed by a wave passing through a given medium.

wave winding See *drum winding.*

wa-wa pedal A foot-operated device used with an electronic musical instrument to produce a "wah-wah" sound fluctuation.

wax 1. Any of a series of organic materials having important uses as dielectrics, impregnants, sealers, and lubricants in electronics. They are usually solid or semisolid, waterproof, and easily melted. 2. In certain phonograph record disks, a blend of wax (see 1. above) and metallic soaps. Also see *wax master, 2.*

wax cake See *wax master, 1, 2.*

wax capacitor A fixed capacitor which has been dipped in or impregnated with a wax, such as halowax.

wax-dipped capacitor A fixed capacitor which has been dipped in a wax for sealing against moisture.

waxed paper See *wax paper.*

wax-filled capacitor A fixed capacitor which is impregnated with a wax for enhancing the properties of its dielectric (usually paper) and sealing the capacitor unit.

wax master 1. In disk-recording practice, the original recording made on a wax-surface disk. 2. To make in the above manner.

wax original See *wax master.*

wax paper Wax-saturated paper used as a dielectric film in fixed capacitors and as an insulator.

way-operated circuit A single or duplex circuit shared by three or more party stations.

way point An important point selected on a radionavigational course line.

way station A station consisting of a teletypewriter connected at an intermediate point in a line, i.e., between, and in series with, other teletypewriter stations.

WAZ Amateur radio abbreviation of *worked all zones,* a citation given to operators who have carried on verified two-way communication with stations in all communications zones of the world (see *work, 2*).

WB Abbreviation of *weather bureau.*

WB Abbreviation of *base-region width* (in a transistor).

Wb Symbol for *weber.*

Wb/m^2 Webers per square meter (see *tesla*).

WC Abbreviation of *collector-region width* (in a transistor).

WCEMA *West Coast Electronic Manufacturers' Association.*

W/cm^2 Abbreviation of *watts per square centimeter.*

WE Abbreviation of *write enable.*

WE Abbreviation of *emitter-region width* (in a transistor).

weak battery 1. A battery which has been depleted to the point that its output (no-load or full-load) is too low to be useful. 2. A battery specially designed for low-voltage output.

weak color Lightness of color or poor contrast between colors in a color-TV picture, a condition often due to some malfunction in the chroma demodulator(s).

weak contrast In a TV picture, poor differentiation of adjacent tonal areas.

weak coupling See *loose coupling.*

weak current A current intensity of only a few microamperes or picoamperes.

weak magnet A magnet whose powers have deteriorated considerably below a prescribed level, or a body which normally is only slightly magnetic. The term is relative.

weak signal A signal whose amplitude is very low compared with that of signals considered satisfactory in a given application. Although the term is relative, it usually implies a signal which is noncompetitive with other signals in a given context.

weak-signal detector A square-law detector. At low input-signal amplitudes (weak-signal levels), the dc output of a diode in the detector is proportional to the square of the rms value of the input-signal voltage. Other square-law detectors are grid-leak detectors, and amplifiers biased into the curved region of the E_g-I_p characteristic.

wearout The complete deterioration of a component or system, i.e., beyond restoration to useful service.

wearout failure Failure due to wearout, which can be predicted on the basis of known lifetime and the deterioration characteristics of components and equipment.

wearout point The instant of wearout, in terms of power output, watt-hour capacity, or the like.

weather antenna An antenna dimensioned for reception exclusively in the 162.4—162.55-MHz weather band. See *weather transmission.*

weathering Deterioration of electronic equipment as a result of exposure to outdoor heat, cold, moisture, wind, and similar conditions.

weather-protected machine A machine (usually a generator or motor) whose vent holes are designed to prevent entry of dust, water, and debris.

weather protection The coating, sealing, or treating of electronic equipment for protection against corrosion, humidity, and temperature changes in outdoor use.

weather sonde See *radiosonde.*

weather transmission The radio transmission of meteorological reports. Sometimes the transmissions are combined with guidance transmissions, from which they can be separated by means of a filter in the receiver.

weber Abbreviation, Wb. The SI unit of magnetic flux and of the magnetic flux quantum; $1 WB = 10^8$ maxwells $= 1.256\,637 \times 10^{-7}$ unit pole.

Weber-Fechner law The law expressing the relationship between a stimulus and the physiological reaction it produces: The sensation is proportional to the logarithm of the stimulus.

weber per square meter Symbol, Wb/m^2. See *tesla*.

weber turn A unit of magnetic flux linkage equal to 10^8 maxwell turns.

wedge 1. In a waveguide, a termination consisting of a tapered block or plate of carbon (or other dissipative material). 2. In a TV test pattern, convergent, equally spaced lines for checking resolution.

wedge bonding In IC fabrication, a method of bonding (see *bonding, 2*) in which a thermocompression bond (see *bcold-compression welding*) is obtained through pressure from a wedge-shaped tool.

Wehnelt cathode An oxide-coated cathode in an electron tube.

Wehnelt-cylinder In a cathode-ray tube, the cathode-enclosing cylinder which concentrates the electrons emitted by the cathode.

weight 1. The amount of gravitational pull on a body or particle. 2. Extra significance given to a term or value. See, for example, *weighted term*. 3. Dot-to-space ratio in a Morse-code signal.

weight-density Symbol, d. Unit, kg/m^3. The weight per unit volume of a liquid, such as an electrolyte or insulating oil; also called density.

weighted average In the statistical analysis of data, an average value calculated from terms of which some are weighted in accordance with their frequency or their importance in determining the average. Compare *unweighted average*. Also see *weighted term*.

weighted distortion factor In the measurement of harmonic distortion a factor whose use allows the harmonics in the complex waveform to be weighted in proportion to their relationship.

weighted noise level Unit, dBm. Noise level weighted with respect to the 70-dB equal-loudness contour of hearing.

weighted term In the statistical analysis of data, a term which is operated on by a weighting factor in accordance with the importance of the term to determining a final result, such as an average.

weighting 1. Adjustment of a parameter to compensate for some imbalance in a system. 2. Adjustment of the dot-to-space ratio in a Morse code signal. 3. Adjustment of the mark-to-space ratio in a teletype signal.

weighting network A network which weights differently (in a prescribed ratio) the frequency components appearing in an output signal, by offering unequal attenuation to those frequencies.

weightlessness The effect of zero gravity, as experienced by a person or instruments within the condition's context (such as the internal environment of a rocket or satellite in space).

weightlessness switch See *zero-gravity switch*.

Weir circuit In FM transmission a circuit employed to stabilize the carrier wave. It compares the average carrier frequency with the frequency of a standard crystal oscillatorn obtaining from this operation a dc compensating voltage (proportional to frequency deviation) which is applied to the frequency modulator. Also called *Weir stabilization circuit*.

weld A strong bond of materials (usually metals) obtained by applying heat to areas to be joined while they are held or pressed together. No foreign metal is used, as is the case in brazing and soldering. The required heat is often obtained by passing a heavy electric current through the materials.

welder 1. An electrical device, often electronically controlled, for welding materials. 2. One who operates such a device.

weldgate pulse In a welding device, the pulse that affects the arc current, and therefore the intensity of the heat produced by the device.

welding Making a weld, especially under the control of electronic devices.

welding control An electronic system for controlling the interval during which current is passed through a workpiece in spot welding or seam welding. In this system, an electronic timer circuit determines the conduction time of thyratrons, ignitrons, or silicon controlled rectifiers, in the welding circuit. The control system may also regulate the welding current.

welding current The heavy electric current passed through a workpiece to produce the heat required for welding.

welding cycle The required sequence of steps (and the time required) in making a weld electronically.

welding time See *weld time*.

welding transformer For electronic welding, a special very-high-current stepdown transformer.

weld junction See *weld*.

weld polarity The polarity of welding current. Some materials require a certain direction of current flow for a good weld.

weld time The interval during which welding current flows through the bodies to be bonded together.

well counter A radiation-counter setup in which a radioactive sample and detector are enclosed together in a thick-walled (usually lead) cylinder to minimize background count.

Wenner element An adjustable, dual-slidewire balancing resistor used in constant-current, laboratory potentiometers to eliminate the necessity for sliding contacts in the measuring circuit.

Wenner winding A low-capacitance, low-inductance winding for high-frequency wirewound resistors in which the direction of the wire is reversed by looping alternate turns along the form.

Wertheim effect The tendency for a potential difference to develop between opposite ends of a length of wire, when the wire is placed parallel to magnetic lines of flux and rotated.

Western Union joint A strong splice of two wires made by tightly twisting a short portion of the tip of each wire along the body of the other. For increased ruggedness, the joint is often soldered. Also called *Western Union splice*.

Weston cell See *standard cell*.

Westrex system A system of sound recording in which signals from two separate microphone channels are recorded on opposite walls of a 45-degree groove on a disk.

wet battery A battery of wet cells.

wet Liquid, especially pertaining to the electrolyte material in an electrochemical cell.

wet cell A battery cell having a liquid electrolyte. Compare *dry cell*.

wet-charged stand The length of time that a fully charged, wet storage cell can stand idle before its capacity drops by a specified amount.

wet electrolytic capacitor An electrolytic capacitor in which the electrolyte is a liquid. The leakage current in this type is higher than in the dry electrolytic, but it is self-healing after momentary voltage breakdown. Compare *dry electrolytic capacitor*.

wet grab The adherence of a pressure-sensitive tape or sheet to a surface when very little pressure is used.

wet rectifier See *electrolytic rectifier*.

wet shelf life The specified shelf life of a discharged, wet storage cell. Compare *dry shelf life*.

wetted-contact relay See *mercury-wetted reed relay.*

wetting Applying a mercury coating to a contact surface.

wetting agent A substance (such as an alcohol or ester) which promotes the spreading and adhesion of a liquid or its absorption by a porous material.

WG Abbreviation of *wire gauge.*

WH Abbreviation of *watt-hour.*

wh Abbreviation of *white.*

Wheatstone automatic telegraphy An automatic telegraph system in which signals are transmitted from a perforated tape and recorded after being received as dots and dashes on a second tape.

Wheatstone bridge A four-arm balancing circuit (see *bridge, 1, 2*), having resistors in each arm and used to measure an unknown resistance in terms of a standard resistance. The bridge supply is usually dc, but ac may be used, provided all four resistances are nonreactive.

$$R_X = R3 \left(R2/R1 \right)$$

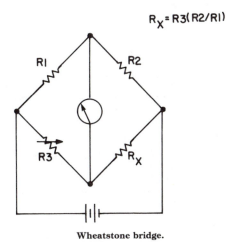

Wheatstone bridge.

Wheatstone tape Tape for use in Wheatstone automatic telegraphy. It may be either a perforated strip or one on which an ink line is inscribed (fluctuations in the line indicate dots and dashes).

Wheatstone telegraphy See *Wheatstone automatic telegraphy.*

Wheeler's formula A formula for calculating the inductance of a multilayer air-core coil; $L = (0.8\, a^2 N^2)/(6a + 9l + 10b)$, where L is in microhenrys, a is the mean radius (inches) of the winding (axis to center of cross section), b is the depth of winding (inches), l is the length of winding (inches), and N is the number of turns.

wheel pattern A frequency test pattern produced on an oscilloscope screen by Z-axis modulation of a circular trace. A sinusoidal axis signal produces a *gear-wheel pattern*, and a square-wave or pulse Z-axis signal produces a *spot-wheel pattern.*

wheel printer A printout device for computers and calculators. It consists essentially of a rotating metal wheel around whose rim letters and numbers stand in relief. When the desired character comes into position, a hammer strikes it through the recording paper and carbon paper, printing the character on the recording paper. Also called *daisy-wheel printer.*

wheel static Static electricity (and the resulting radio interference) generated by friction between automobile tires and the road.

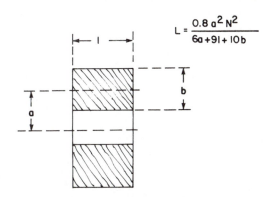

$$L = \frac{0.8\, a^2 N^2}{6a + 9l + 10b}$$

Wheeler's formula, dimensions illustrated.

whiffletree switch In computer practice, a multiposition electronic switching circuit, so called from its circuit configuration, which resembles the contrivance used between a wagon and the horse team pulling it.

whip antenna A small-diameter, vertical rod (often telescoping) used as an antenna especially in mobile communications, portable radio and television receivers, field-strength meters, and the like.

Whippany effect In electron tubes, an occasional fault characterized by the inability of the tube to develop grid bias because of insufficient grid current.

whirl One of the circular, magnetic lines of force around a straight wire carrying current.

Whirlwind I A digital computer built at Massachusetts Institute of Technology and the first large computer to employ ferromagnetic cores in its main memory.

whisker 1. The pointer-wire electrode of a point-contact diode, point-contact transistorn or crystal detector. Also see *catwhisker, 2*). 2. A slender filament of metal or ceramic, having high purity and high tensile strength.

whisker resistance In a semiconductor material, the resistance of a whisker component (see *whisker, 2*).

whistle A high-pitched tone, often a beat note.

whistle filter A notch filter used to eliminate a whistle in an amplifier or other audio-frequency circuit. This filter can be of several versions, ranging from a simple *RC* null circuit tuned to remove the offending frequency to a feedback-type band-suppression amplifier.

whistle interference The appearance of extraneous whistles in an af system including sound reproducers and radio receivers. In some equipment, whistles result from oscillating amplifiers; in receivers, they are usually audible beat notes produced by interfering carriers.

whistler A type of radio interference (so called from the nature of its sound) caused by distant lightning discharges.

whistler mode propagation Radio transmission from the northern hemisphere to the southern hemisphere along the flux lines of earth's magnetic field.

whistlestop See *whistle filter.*

whistling atmospheric See *whistler.*

white The color of pure snow. (White results from mixing all the additive primary colors red, green, and blue).

white acid Hydrofluoric acid. Formula, HF. Used as an etchant, especially of glass.

white brass *Brass that is more than 49%* zinc.

white circuit A low-impedance wideband cathode-follower circuit in which the internal plate-cathode path of a second tube replaces the conventional cathode resistor. The signal is applied to the grids of both tubes, which are out of phase with each other.

white compression In TV transmission, reduction of gain at highlight levels in the picture.

white-dot pattern In color-TV tests with a dot generator, the one-white-dot pattern obtained when beam convergence has been secured.

white lamp See *daylight lamp*.

while level The lower-voltage point in a video signal, corresponding to full brilliance of the line on the screen, i.e., to the condition of whiteness in the picture.

white light See *white radiation, 2*.

white noise Random noise (acoustic or electric) equally distributed over a given frequency band, an example being the noise resulting from the random motion of free electrons in conductors and semiconductors.

white-noise generator A test device which generates electrical noise over a wide frequency spectrum. A simple type employs a reverse-biased silicon diode, the output of which is useful for testing audio amplifiers and radio receivers.

white-noise record A phonograph record containing recorded bands of *white noise*, each accompanied by voice announcements (e.g., instructions), and used to test the frequency response of a sound system.

white object A body that reflects and diffused light of all wavelengths equally well.

white peak In a TV picture signal, the maximum excursion in the white direction.

white radiation 1. White noise. 2. Visible light radiated in a continuous spectrum by a hot body and seen as white.

white raster See *chroma-clear raster*.

white recording 1. In amplitude modulation, recording characterized by a correspondence of maximum received power to minimum recording-medium density. 2. In frequency modulation, recording characterized by a correspondence of lowest received frequency to minimum recording-medium density.

white room See *clean room*.

white saturation See *white compression*.

white signal The facsimile signal (see *facsimile*) corresponding to scanning the copy area having maximum density.

white-to-black amplitude range 1. At a point in a positive AM facsimile system (see *facsimile*) the ratio (in dB) of signal current or voltage corresponding to white in the picture to that for black in the picture. 2. At a point in a negative AM facsimile system, the ratio (in dB) of signal current or voltage corresponding to black in the picture to that for white in the picture.

white-to-black frequency swing At a point in an FM facsimile system, the frequency difference $fw - f b$, where $f b$ is the frequency corresponding to black in the picture, and fw is the frequency corresponding to white.

white transmission A system of picture or facsimile transmission in which the maximum copy darkness corresponds to the smallest amplitude (in an amplitude-modulated transmitter) or the highest instantaneous frequency (in a frequency-modulated transmitter). The opposite of black transmission.

white X-radiation X-rays of the continuous, or general, type.

whole-number division Arithmetic division (as in the division of binary numbers) in which the quotient is a whole number, i.e., division in which the divisor is contained in the dividend an integral number of times.

whole step See *whole tone*.

whole tone The interval between tones having a basic frequency ratio of approximately 2^{-3}.

whorl See *whirl*.

WHP Abbreviation of *water horsepower*.

wick action 1. The absorption of a liquid by, and its flow through, a cloth or thread, such as a lamp wick or lubricating wick. 2. The flow of molten solder along and under the insulation of a wire.

wicking See *wick action*.

wide-angle diffusion Diffusion (see *diffusion, 3)* characterized by the wide-angle scattering of light which causes the source to have the same brightness at all viewing angles.

wide-angle lens A lens whose angular field exceeds 80 degrees.

wide-angle tuning In a receiver using a magic-eye tube as a tuning indicator, eye-circuit sensitivity great enough to permit 180-degree opening of the eye pattern at resonance. This is usually accomplished by operating an amplifier stage ahead of the eye tube to which it is directly coupled.

wide-area service A teletype network that operates over long-distance wire lines.

wide-area telephone service See *WATS*.

wideband 1. Having a bandwidth greater than the minimum necessary to transmit a signal with acceptable intelligibility. 2. For a voice signal, having a bandwidth greater than 6 kHz. 3. Having the capability to operate, without adjustment, over a wide and continuous range of frequencies or wavelengths.

wideband amplifier An amplifier which exhibits reasonably flat response to a wide band of frequencies. The term is relative, however, since the meaning of *wideness* varies with different applications of amplifiers.

wideband antenna An antenna which transmits or receives signals over a wide frequency range, usually without needing tuning.

wideband axis In a color-TV signal, the direction of the *fine chrominance primary* phasor.

wideband communications 1. Communications carried out over a band of frequencies wider than the minimum necessary for effective transfer of the information. 2. A method of transmitting and receiving signals by deliberately varying the channel frequency over a wide range. Also called spread-spectrum communications.

wideband generator A signal generator covering a wide frequency range. Typical coverage in a laboratory-type instrument is 10 kHz to 1000 MHz.

wideband oscilloscope An oscilloscope whose horizontal, vertical, and sweep channels operate over a wide band of frequencies. While the term *wideband* is relative, a wideband oscilloscope is usually assumed to be one that can display radio frequencies as well as audio frequencies.

wideband ratio The ratio $BW1/BW2$, where $BW1$ is frequency bandwidth, and $BW2$ is intelligence bandwidth.

wideband receiver A radio receiver which will tune-in signals in a wide band of frequencies. An example is the general-coverage communications receiver which can cover a range extending from 50 kHz to 40 MHz.

wideband repeater A repeater capable of operating over a wide range of input and output frequencies. Such repeaters are used in active communications satellites handling many different channels at the same time.

wideband signal generator See *wideband generator*.

wideband sweep 1. In the operation of an oscilloscope, a repetitive sweep of the electron beam, which may be adjusted to any desired point within a wide band of frequencies. The basic sweep rate in simple oscilloscopes is restricted to the audio-frequency spectrum. A wideband sweep, however, extends into the MHz region. 2. A sweep circuit which produces the wideband sweep action described in 1, above.

wideband test meter An ac meter which will measure quantities over a wide frequency range in its basic form (i.e., without special external probes or converters). An example is the electronic ac voltmeter with a range extending from 10 Hz to 2.5 MHz.

wide-base diode A junction diode in which the *p*-region is considerably wider than the *n*-region.

wide-open 1. Pertaining to wideband, untuned response. 2. Pertaining to maximum-gain operation, e.g., a wide-open amplifier or receiver.

wide-range reproduction High-fidelity audio-frequency reproduction.

width 1. The horizontal dimension of a pulse. Also called *pulse duration*. 2. The horizontal dimension of an image, such as a TV picture.

width coding Modification of pulse duration according to a code.

width coil See *width control, 1*.

width control 1. In a TV receiver, the variable component for adjusting the swing of the horizontal deflection voltage and, therefore, the width of the picture. It is often a slug-tuned coil connected in parallel with a portion of the secondary winding of the horizontal output transformer. 2. Sometimes, the horizontal gain control in an oscilloscope.

width mode In electronic-counter practice, a time-interval mode in which the signal is measured.

width modulation Modulation of the interval during which a gate circuit is open.

Wiedemann effect See *direct Wiedemann effect*.

Wiedemann-Franz law For metals which are good electrical conductors, the law showing that the ratio of thermal conductivity to electrical conductivity is nearly constant and proportional to the absolute temperature.

Wien bridge A frequency-sensitive bridge in which two adjacent arms are resistances and the other two adjacent arms are *RC* combinations. One of the latter contains resistance and capacitance in series; the other contains resistance and capacitance in parallel. Because the bridge can be balanced at only one frequency at a time, it is useful as a simple audio frequency meter (see *bridge-type af meter*). It is used also for capacitance and resistance measurements. When inductors are substituted for the capacitors, the Wien bridge can be used for induc-

tance measurements (see *Wien inductance bridge*).

Wien-bridge audio frequency meter See *Wien-bridge frequency meter*.

Wien-bridge distortion meter A distortion meter in which a Wien bridge circuit is used to remove the fundamental frequency from the complex waveform. The bridge is inserted between amplifier stages, and its notch response is sharpened by means of overall negative feedback.

Wien-bridge equivalent A resistance-capacitance null circuit, such as the parallel-tee (twin-tee) network, which has the same balance equation as the Wien bridge.

Wien-bridge filter A Wien bridge employed as a band-suppression filter (notch circuit).

Wien-bridge frequency meter A bridge-type af frequency meter in which the continuously variable frequency-selective circuit is a Wien bridge. Also see *bridge-type af meter*.

Wien-bridge heterodyne eliminator A notch filter composed of a continuously variable Wien-bridge circuit (see *Wien bridge*). The device is connected in the af channel of a radio receiver and is tuned to remove a troublesome heterodyne whistle.

Wien-bridge oscillator An oscillator in which the frequency-selective feedback network is a Wien-bridge circuit (see *Wien bridge*). Also see *bridge-type oscillator*.

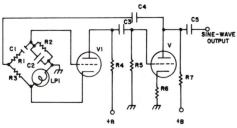

Wien-bridge oscillator.

Wien capacitance bridge A Wien bridge arranged for the measurement of unknown capacitance.

Wien inductance bridge A Wien bridge containing inductors in place of capacitors and used for the measurement of unknown inductance.

Wien's displacement law The law showing that the wavelength of maximum radiation of a black body is inversely proportional to the absolute temperature.

Wien's first law See *Wien's displacement law*.

Wien's laws See *Wien's first law, Wien's second lawn and Wien's third law*.

Wien's radiation law See *Wien's second law*.

Wien's second law The law showing that the emissive power of a black body is proportional to the fifth power of the absolute temperature.

Wien's third law An empirical law for the spectral distribution of energy radiated from a black body at a specified temperature.

willemite See *zinc orthosilicate phosphor* and *zinc silicate phosphor*.

William Phonetic alphabet communications code word for the letter *W*.

Williamson amplifier A high-fidelity sound-reproduction amplifier employing triode-connected pentodes in the pushdpull output stage; various compensation networks; inverse feedback; and a carefully designed output transformer.

Williams tube A type of electrostatic cathode-ray storage tube.

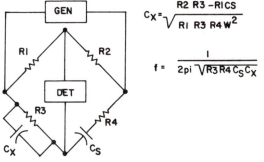

$$C_X = \sqrt{\frac{R2\ R3 - R1CS}{R1\ R3\ R4\ W^2}}$$

$$f = \frac{1}{2pi\ \sqrt{R3\ R4\ C_S\ C_X}}$$

Wien bridge.

Wilson chamber See *Wilson cloud chamber.*

Wilson cloud chamber An airtight chamber containing water vapor or alcohol vapor at low pressure and provided with a viewing window. Radioactive particles penetrating the chamber are made visible as droplets form trails when the vapor condenses on them.

Wilson effect Electric polarization of a dielectric material being moved in a magnetic field.

Wilson electroscope A leaf-type electroscope (see *electroscope*) employing a single leaf hanging vertically.

Wilson experiment An experiment demonstrating the Wilson effect. It consists of rotating about its axis a hollow dielectric cylinder (whose inner and outer surfaces are metallized) in a magnetic field that is parallel to the axis. The result: An alternating emf appears between the metallized surfaces.

Wimshurst machine A rotating machine used to produce high-voltage static electricity. The machine contains two glass disks, each having separate sectors of metal foil spaced around its face. The disks rotate in opposite directions, and the foil sectors passing each other form variable capacitors. Metal brushes pick up charges from the sectors passing under them and deliver this energy to one or more Leyden jars for storage.

wind charger A wind-driven generator used specifically to supply direct current for charging storage batteries.

wind-driven generator A dynamo-type generator, either stationary or mobile, powered by wind.

wind gauge See *anemometer.*

winding A coil of an inductor or transformer, e.g., primary winding, secondary winding, output winding, etc. Also, the coil of a motor or generator.

winding arc The winding length of a coil, expressed in degrees.

winding cross section The cross section of a multilayer coil.

winding depth The depth of a multilayer coil, measured from the outermost layer to the innermost layer.

winding factor For a transformer or choke coil (or for a toroidal coil), the ratio of the total area of wire in the window to the window area.

winding length The length of a coil from the first turn to the last in a single-layer coil, or from the first turn to the last in one layer of a multilayer coil when all layers are identical.

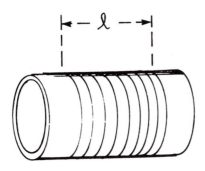

Winding length.

winding space The window area of the core of a transformer or choke (see *window, 3*).

wind loading 1. The total wind pressure on an antenna system, generally measured in pounds. The greater the wind speed, the greater the wind loading for a given antenna. The greater the exposed surface of the antenna, the greater the wind loading for a given wind speed. 2. The highest wind speed, in miles per hour or meters per second, that an antenna system can safely withstand, assuming no accumulation of ice on the antenna structure.

Windom antenna A multiband antenna in which a single-wire feeder is attached to the horizontal radiator wire at a point somewhat off center (e.g., at 36 % of the length of the radiator).

window 1. Radar-interference material (see *chaff*). 2. An interval during which a circuit is gated open to permit signal sampling. During this interval, a window is figuratively *open* to the signal. 3. The open spaces between the legs of an iron core for a transformer or choke coil. 4. An electromagnetic frequency band which is easily transmitted by earth's atmosphere. 5. The period during which conditions are ideal for a rocket or spacecraft launch.

window area See *window, 3.*

window corridor An area where window (see *window, 1)* has been dispersed.

window jamming The disturbance of electronic communications, especially radar, by dumping reflective material, such as metal foil, from an aircraft. Also see *chaff, jamming,* and *window, 1.*

window strip An insulated, flat, metal strip for bringing an antenna lead-in through a window which closed on the strip.

wind shield A radio-transparent cover placed over a radar antenna to protect it against damage from high winds. Used especially in radar systems aboard aircraft.

wind-velocity meter An anemometer. Also called *wind-velocity gauge* and wind gauge.

wing 1. In an antenna or other radiator, a (usually flat) member attached to, and sticking out from, another member. It is so called from its characteristic shape. 2. A unit of the Air Force having more members than a *group* and fewer than a *division.*

winterization The protective treatment and preparation of electronic equipment for winter storage or for operation in frigid regions.

wipeout Severe interference which obliterates all desired signals.

wipeout area An area in which wipeout occurs.

wiper 1. A thin metal blade of strip in sliding contact with a coil or other element over which it is turned to vary some quantity, such as resistance, inductance, voltage, current, etc. 2. Brush.

wiper arm See *wiper, 1.*

wiper blade See *wiper, 1.*

wiping action The movement of contacts against each other when they slide as they mate or withdraw.

wiping contact A contact that makes and breaks with a sliding motion.

wire 1. A metal strand or thread serving as a conductor of electricity. 2. To connect wires between points in a circuit. 3. Telegram. 4. To send a telegram.

wire bonding 1. The interconnection of components within a discrete package, by means of fine wire conductors welded to the individual components. 2. A method of temporarily splicing the outer conductors of two coaxial cables. 3. In general, any solderless method of splicing between two conductors.

wire broadcasting Broadcasting by means of carrier-current communication (see *wired radio).*

wire cloth A net of fine wire, used as an electrical shield when air circulation is necessary.

wire communication Communication carried on by means of signals transmitted over wire lines, as opposed to wireless (radio) communication.

wire control The control of remote devices by means of signals transmitted over wire lines, as opposed to radio control.

wire core A magnetic core consisting of a bundle of iron (or magnetic alloy) wires.

wired AND See *dot AND.*

wired-in component A component to which wires are permanently connected in a circuit, as opposed to a plug-in component.

wired OR See *dot OR*

wired-program computer A computer in which instructions are wired-in by making appropriate connections to a panel with patch cords.

wired radio The transmission and reception of voice and other sounds, telegraph signals, pictures, control signals, telemetry signals, and the like by means of radio-frequency energy conducted by wire lines. Usually the lines are used primarily for some other purpose (e.g., power lines and telephone lines).

wired-radio receiver The complete device or system which selects, amplifies, and demodulates or rectifies a radio signal picked up from wires which conduct it from a transmitting station.

wired-radio transmitter The complete device or system which generates radio-frequency powerg adds to it signals for communication, remote control, telemetry, or the like; and delivers the power to wires or to a cable for reception at a distant point.

wire drawing In the manufacture of wire, pulling metal through special dies to form wire of selected diameter.

wire dress The careful arrangement of wires in a chassis to insure optimum performance. Also called *lead dress* (see *dress).*

wire duct A conduit through which wires are run. Such ducts have several shapes, but generally are rectangular.

wire wireless See *wired radio.*

wire fuse A *fuse* in which the fusible element is a wire of low-melting-point metal. Compare *strip fuse.*

wire gauge 1.A system for specifying the characteristics of wire. See *American wire gauge* and *Birmingham wire gauge.* 2. A wire number governed by diameter (e.g., *18-gauge wire* for No. 18 wire). 3. A tool or instrument for measuring wire diameter or for determining wire number.

wire gauze A fine screen of thin wires.

wire grating See *waveguide grating.*

wire-guided Pertaining to guiding a missile by means of signals sent by wire from the control point.

wire leads Leads which are (usually thin, flexible) wires, rather than pins, bars, strips, or the like. An example is the wire leads of some transistors (as opposed to terminal pins).

wireless An early name for radio; it is still used in some countries. Sometimes for specificity radio is referred to as *wireless telegraphy* or *wireless telephony.*

wireless broadcaster See *wireless microphone.*

wireless compass A radio compass (see *direction finder).*

wireless device A device that operates over a distance, without the use of interconnecting wires.

wireless intercom An intercom employing wired-radio for transmission and reception over the power line from which it is operated. Also see *wired radio.*

wireless microphone A device consisting of a small rf oscillator which is modulated by the microphone to which it is attached and which is provided with a tiny antenna. The modulated signal is picked up and reproduced by an AM or FM radio receiver.

wireless telegraphy See *radiotelegraphy.*

wireless telephone 1. A radio transceiver that is interconnected with the telephone lines. 2. A telephone in which a short-range radio link replaces the cord between the receiver and the base unit. Normally has a range of several hundred feet.

wireless telephony See *radiotelephony.*

wire line One or more wires or cables conducting currents for communication, control, or measuring purposes.

wire link A line for wire communication or wire control.

wire-link telemetry Telemetry carried on over a wire link.

wireman 1. An electrician who specializes in the installation and servicing of electrical wiring. 2. A *lineman.*

wire memory In a computer, a static magnetic memory in which the storage elements are beryllium-copper wires plated with an 0.8-μm-thick film of nickel-iron alloy.

wire mile A unit of measure equal to Nl, where N is the number of separate, equal-length conductors (wires) in a line, and l is the length of a conductor (in miles).

Wirephoto 1. A system for transmitting and receiving photographs over wire lines. Also called *Telephoto.* 2. A photograph so transmitted and received. 3. Trademark for a photograph transmitted and received over telephone lines.

wire printer A printout device consisting of a wire array in which each wire is activated by an electromagnet. When a wire is selected (electrically), it is pushed against, and makes a dot or, the recording paper. The letters and figures, then, are dot patterns. Compare *wheel printer.*

wire radio See *wired radio.*

wire recorder In audio practice and data recording, a machine which records sounds or data pulses in the form of magnetized spots on a thin wire.

wiresonde A system for transmitting meteorological signals through a cable that holds a captive balloon.

wire splice A strong, low-resistance joint between (usually two) wires. See, for example, *Western Union joint.*

wire stripper A hand tool or power machine for cutting the insulating jacket on a wire and removing it without cutting or nicking the wire.

wiretap 1. An instance of wiretapping. 2. A wiretapping device. Also see *wiretapper, 1.*

wiretapper 1. A device for wiretapping. 2. A person who practices wiretapping.

wiretapping The (usually illicit) act of making direct or indirect connections to a communications line for the purpose of overhearing or recording a conversation.

wire telegraphy Telegraphic communication carried on by means of signals transmitted over wire lines, as opposed to wireless (radio) telegraphy.

wire telephony Telephonic communication carried on by means of signals transmitted over wire linesn as opposed to wireless (radio) telephony.

wire tie See *tie.*

wire wave communication See *wired radio.*

wireways Hinged-cover metal troughs for containing and protecting wires and cables. Also see *wire duct.*

wirewound potentiometer A potentiometer in which the resistance element is a coil of resistance wire wound on a cord bent into a cylinder, or on a rigid, circular core.

wirewound resistor A resistor consisting essentially of resistance wire (such as one of manganin or nichrome) wound into a coil on a card or slab or on a cylindrical core.

wirewound rheostat A rheostat in which the resistance element is a coil of resistance wire wound on a card bent into a cylinder, or on a rigid, circular core.

wire wrap A method of circuit-board wiring in which components are interconnected by individual wire conductors, the ends of which are wrapped around terminal posts using a special tool.

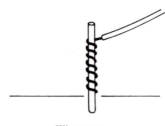

Wire wrap.

wire-wrap connection An electrical connection made by tightly wrapping a bare wire around a special terminal.

wire-wrapping tool A device used to make a wire-wrap connection.

wiring 1. The system of wires or similar conductors connected between circuit components. 2. Connecting and dressing such wires, as in *wiring* a circuit. 3. Collectively, the connections (either actual wire or printed metal lines or processed semiconductor paths) between terminals and components in an electronic circuit. 4. The process of installing or making such connections.

wiring board The control panel of a computer, i.e., a plugboard.

wiring capacitance Unavoidable capacitance between wires in a circuit, or between the wires and nearby metal bodies.

wiring connector A small (usually metal) fitting employed to tie wires together.

wiring diagram See *circuit diagram.*

wiring impedance Unavoidable impedance in the wires in a circuit, and in associated terminals and hardware conducting alternating current. This impedance is the vector sum of wiring reactance and wiring resistance.

wiring inductance Unavoidable inductance in the wires in a circuit, and in associated terminals and other hardware, through which alternating currents flow.

wiring reactance Unavoidable reactance in the wires in a circuit, and in associated terminals and other hardware, conducting alternating current. This reactance is due to wiring capacitance and/or wiring inductance.

wiring resistance Unavoidable resistance in the wires in a circuit and in associated terminals and other hardware.

Witka circuit A voltage-tripler circuit using only two diodes and two capacitors. Its no-load dc output voltage is approximately three times the peak value of the ac input voltage. Also see *voltage tripler.*

wk 1. Abbreviation of *work.* 2. Abbreviation of *week.*

WL 1. Abbreviation of *wavelength.* 2. Abbreviation of *waterline.*

WM Abbreviation of *wattmeter.*

W• m^2 Symbol for *watt square meter* (unit of the first radiation constant).

W/(M• K) Symbol for *watts per meter kelvin (unit of thermal conductivity).*

W/(m^2• K^4) Symbol for unit of Stefan-Boltzman constant.

WMO World Meteorological Organization.

w/o Abbreviation of *without.*

wobbulator A device that sweeps the frequency of an oscillator. There are several types, from a motor-driven, rotating, tank capacitor to sophisticated, fully electronic variable capacitors or variable inductors.

wolfram See *Tungsten.*

Wolf's equation A sunspot-number equation employed in forecasting maximum usable frequency; $R = k(10G + N$ where R is relative sunspot number, k is a constant for the telescope used, G is the number of sunspot groups observed, and N) is the total number of sunspots.

Wollaston wire An extremely fine platinum wire made by coating platinum wire with silver, drawing it thinner, and dissolving the silver coat.

womp A sudden surge in amplitude which causes a flareup of brightness in a TV picture.

Wood's alloy See *Wood's metal.*

Wood's metal A low-melting-point (159.8 °F) alloy containing bismuth (50%), cadmium (12 1/2%), lead (25%), end tin (12 1/2%). The alloy looks like lead and is used to mount rectifying crystals (such as galena) whose electrical sensitivity would be destroyed by the high temperatures required to melt most soft metals.

woofer A usually large (a foot or more in diameter) loudspeaker particularly suited to reproducing bass frequencies. Compare *tweeter.*

word 1. In computer practice, a group of bits or characters treated as a unit. 2. In telegraphy (radio and wire), a reference word length of five letters.

word-address format A method of addressing a word by means of a single character, such as the first letter of the word.

word code 1. A code in which word meanings are interchanged, rather than letter symbols. 2. A word, in such a code system, that has an altered meaning. For example, "word" may mean "code."

word format The specific sequence of characters that forms a word of data.

word generator A special signal generator that delivers pulses in selected combinations corresponding to digital words (see *word, 1*). It is used in testing computers and digital systems.

word length 1. In computer practice, the number of bits in a word. 2. In telegraphic communication (radio or wire), the average number of letters in a word.

word pattern The shortest meaningful word (see *word, 1, 2*) that can be recognized by a machine.

word processor An electronic device, similar to a typewriter, used for writing. The copy is first created in soft form, such as in a disk memory. Words, phrases, sentences, or paragraphs can be changed, replaced, or deleted prior to the final printout.

word rate In a communications or computer system, the number of words per unit time.

word size See *word length.*

word time In computer practice, the time required to process one word that is in storage.

work 1. Symbol, *W.* Units, joule, *erg* foot-pound. That which is accomplished by the transfer of energy from one body to another, as when an exerted force causes a displacement. The amount of work performed is equal to force times distance: $W = Fd$. 2. An amateur radio term meaning to engage in two-way communication with another station.

work area In computer practice, a temporary storage area of memory for data items being processed. Also called *intermediate storage, work-*

ing memory, and *working storage.*

work coil The ac-carrying coil which induces energy in the workpiece in induction heating.

work function Unit, eV. The energy required to bring an internal particle to the surface of a material and out into space, as when an electron is emitted by the hot cathode of a vacuum tube. The work function is the voltage required to extract 1 electrostatic unit of electricity from the material.

working data file A temporary accumulation of data sets which is erased or otherwise discarded after its transferal to another medium.

working file The expected or guaranteed lifetime of a material, device, or system which is in actual operation or use. Compare *shelf life.*

working memory See *work open.*

working point The operating point of a tube or transistorn i.e., the point along one of the characteristic curves of the device around which operation is fixed.

working Q The *Q* of a loaded circuit or device, e.g., tank-circuit *Q* of a radio transmitter loaded with an antenna or dummy load.

working storage See *work area.*

working voltage The (usually maximum) voltage at which a circuit or device may be operated continuously with safety and reliability.

workout 1. Dry run. 2. An exacting test of equipment.

workpiece The body which is heated by an induction or dielectric heater.

worm A cylindrical gear whose spiral "teeth" resemble a screw thread.

worm drive A mechanism for transferring motion from a tuning-knob shaft through a right angle to the shaft of an adjustable component, such as a potentiometer or variable capacitor. The knob turns a worm which mates with a gear wheel.

worst-case circuit analysis Analysis which seeks the worst possible effects on circuit performance of variations in circuit parameters, especially the variations in component characteristics.

worst-case noise pattern See *double-checkerboard pattern.*

worst-cast design The design of electronic equipment in such a way that normal operation is obtained even though the characteristics of circuit components may vary widely.

woven resistor A resistance element made of strands of resistance material woven in the form of gauze.

wow Slow, periodic variations in the pitch of reproduced sound, due to variations in the speed of the mechanism driving the disk, tape, or wire. So called because of its characteristic sound. Compare *flutter.*

wow meter An instrument which indicates the amount of wow produced by a turntable or other moving part.

Wpc Abbreviation of *watts per candle.*

W-particle An imaginary elementary particle to which is attributed a relatively large mass.

wpm Abbreviation of *words per minute.*

wpn Abbreviation of *weapon.*

wrap See *wire-wrap connection.*

wrap-around 1. The extent of curvature in magnetic tape passing over the heads during recording or playback. 2. An enclosure which resembles a sheet (wrapped around a piece of electronic equipment). 3. In computer and data processing practice, a technique whereby a crt screen, as in a visual display unit, is cleared of data once it is filled (all available lines are occupied) and additional data is traced from top to bottom, and so on.

wrap joint See *wire-wrap connection.*

wrapper A paper or tape which is wound around a component, such as a coil or capacitor, for insulation and protection.

wrapping 1. Wrapper. 2. Insulating a wire or other conductor by wrapping insulating tape around it.

wrap post A terminal post, tip, pin, or lug used in a wire-warp connection.

wrap-up The (usually successful) completion of a design, fabrication, test, or investigation.

Wratten filter A light filter for separating colors. It is available in transparent sheets of various colors and is useful in photography and in several phases of electronics, including the operation of color meters and color matchers.

wrinkle finish A pattern of fine wrinkles created by special point when it dries on, say, the surface of metal cabinets and boxes for electronic equipment.

write In computer practice, to transfer data from one form of storage to another from. The appropriate prepositon to use with the term is *to.* Compare *read.*

write enable ring See *write permit ring.*

write gun See *writing gun.*

write head In a magnetic memory or in a tape recorderor wire recorder used for recording data, the head that magnetizes the drum, tape, disk, or wire. Compare *read head.*

write pulse In computer operation, the pulse that causes information to be recorded in a magnetic cell, or sets it to the one-state. (See *write*). Compare *read pulse.*

write time The time taken to write data to a storage device.

writing gun In a storage oscilloscope, the electron gun that produces the image electronically on the storage mesh. It is mounted in the rear of the cathode-ray tube. Compare *flood gun.*

writing head See *write head.*

writing rate In photographing the image on a cathode-ray tube screen, the highest spots speed at which an acceptable picture can be made.

writing speed See *writing rate.*

writing telegraph See *telautograph.*

wrong color The instance of an undesired color (e.g., purple instead of red) in a color-TV picture, a condition often due to a malfunction of the chroma demodulator(s).

W/(sr• m^2) Abbreviation of *watts per steradian square meter.*

WT 1. Abbreviation of *wireless telegraphy.* 2. Abbreviation of *watertight.*

wt Abbreviation of *weight.*

WUI Western Union International.

Wulf electrometer See *bifilar electrometer.*

Wullenweber antenna An electronically steerable antenna comprising two concentric circular arrays of masts connected to the steering circuitry.

Wunderlich tube An early vacuum tube having coplanar grids. Also see *coplanar-grid tube.*

WUX Abbreviation of *Western Union telegram.*

WVdc 1. Direct-current working voltage. 2. The maximum continuous direct-current voltage that can safely be placed across a component.

ww Abbreviation of *wirewound.*

W/wr Abbreviation of *watts per sterdian* (unit of radiant intensity).

WWV Call letters of a standard-frequency-standard-time broadcasting station operated by the National Bureau of Standards and located in the continental United States. Also see *WWVH.*

WWVH Call letters of a standard-frequency-standard-time broadcasting station operated by the National Bureau of Standards and located in Hawaii. Also see *WWV.*

wx Radiotelegraph abbreviation of *weather.*

WXD International Telecommunications Union symbol for *meteorological radar station.*

WXR International Telecommunications Union symbol for *radiosonde station.*

wye-box In a three-phase power-measuring setup, a special arrangement of two impedances, each of which is equal to the impedance of the potential element of the wattmeter used in the setup. The wye-box permits a single wattmeter to be used, which indicates one-third the total power. Without the box, three wattmeters would be needed.

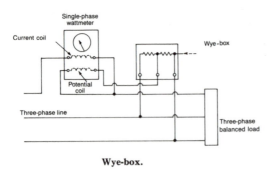

Wye-box.

wye-connection A method of connecting three windings in a three-phase system so that one terminal of each winding is connected to the neutral point. So called because it looks like the letter Y. Also called *star connection.*

wye-current Current through one of the branches of a three-phase star (wye) connection.

wye-delta starter A starter circuit for a three-phasen squirrel-cage induction motor. When the starter switch is thrown in one direction, the

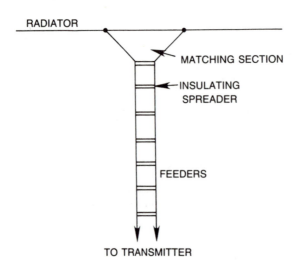

Wye-matched-impedance antenna.

stator of the motor is wye-connected for starting; thrown in the opposite direction, the stator is delta-connected for continued operation.

wye-equivalent circuit A wye-connected three-phase circuit which is equivalent to a delta-connected circuit when the impedance of any pair of corresponding lines for both wye and delta are the same; the third line is unconnected.

wye-junction See *waveguide wye.*

wye-matched-impedance antenna An antenna in which the impedance of a nonresonant open-wire feeder line is matched to that of the center of the radiator by spreading the end of the feeder wires out into a wyed or delta-shaped matching section. Also called *delta-matched-impedance antenna.*

wye-point The star point in a three-phase system. Also see *wye-connection.*

wye-potential In a three-phase armature, the voltage between a terminal and the neutral point.

wye-rectifier A three-phase rectifier circuit in which the transformer or generator windings are arranged in a wye-connection.

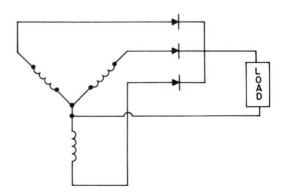

Wye-rectifier.

wye-wye circuit A circuit comprising a wye-connected generator and a wye-connected load.

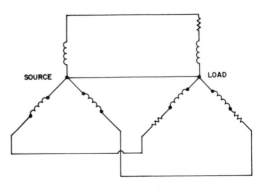

Wye-wye circuit.

X 1. Symbol (ital) for *reactance*. 2. Symbol for *no connection*. (Also, *nc.*) 3. Symbol for the *horizontal axis* of a graph or screen in the rectangular (Cartesian) coordinate system. 4. Most often used symbol for *unknown quantity*. 4. Symbol (sans serif) for *multiplication*.

x 1. Symbol for *number of carriers* drawn from collector to base of a transistor, for each carrier collected. 2. Symbol for *unknown quantity*. 3. Symbol for the *horizontal axis* of a graph or screen in the rectangular (Cartesian) coordinate system. 4. Symbol (often ital) for *multiplication*.

X-amplifier The horizontal amplifier of an oscilloscope or recorder. Compare *Y-amplifier* and *Z-amplifier*.

X- and Z-demodulation Color TV demodulation based upon the 60-degree difference between the two reinserted 3.58-MHz subcarrier signals. The R-Y, B-Y, and G-Y voltages derived from the demodulated signals control the three guns of the picture tube. Compare *I-modulation* and *Q-modulation*.

X-axis 1. The horizontal axis of a chart, graph, or screen in the rectangular (Cartesian) coordinate system. 2. In a quartz crystal, the axis drawn through the corners of the hexagon.

X-axis amplifier See *X-amplifier*. **XB** Abbreviation of *crossbar*.

X-balance The reactance balance (either the variable component or the adjustment of it) in an impedance bridge in which separate resistance (*R*) and reactance (*X*) balancing is provided.

X-band The frequency band extending from 5200 to 11 000 MHz.

X-bar A rectangular, piezoelectric quartz bar cut from a Z-section, whose faces are parallel to the X-axis and whose edges are parallel to the X-, Y-, and Z-axes. See *crystal cuts*.

X-bridge An ac bridge for measuring reactance.

Xc Symbol for *capacitive reactance*.

X-channel The horizontal channel of an oscilloscope or recorder. Compare *Y-channel* and *Z-channel*.

X-channel gain See *X-gain*.

X-component In a complex impedance, the reactive component.

x-coordinate An *abscissa*.

X-cut crystal A piezoelectric plate cut with its faces perpendicular to the X-axis of a quartz crystal. See *crystal cuts*.

xcvr Abbreviation of *transceiver*.

X-deflection Horizontal deflection of the spot on the screen of a cathode-ray tube. Compare *Y-deflection*.

X-diode The decoding diode in each of the X-lines of a computer memory matrix. Compare *Y-diode*.

X-direction The horizontal direction in deflections and in graphical presentations of data.

X-drive The driving source or energy of the X-lines of a computer memory matrix.

xducer Abbreviation of *transducer*.

Xe Symbol for *xenon*.

xenon Symbol, Xe. An inert gaseous element. Atomic number, 54. Atomic weight, 131.30. Xenon is present in trace amounts in earth's atmosphere and is used in some thyratrons, electric lamps, and lasers.

xenon tube A flash tube filled with xenon.

xerography A system for the reproduction of printed matter, drawings, and other graphic matter. It is an electrostatic process in which a charged photoconductive surface is exposed to an image of the material to be copied. A latent image is formed on the surface and is developed (by dusting with black powder which is attracted to the charged image) to make the image visible.

xeroradiography Xerography using X-rays.

xerothermic A condition characterized by both heat and dryness.

Xerox 1. Trade name for a *xerography* machine. 2. The reproduction obtained by means of *xerography*. 3. To make a xerographic reproduction.

X

X-factor 1. Unknown quantity. 2. X-component.

xformer Abbreviation of *transformer*.

X-gain The gain (or gain control) of the horizontal channel of an oscilloscope or X—Y recorder. Compare *Y-gain* and *Z-gain*.

X—H array See *lazy-H antenna*.

XHV Abbreviation of *extreme high vacuum*.

xi The Greek letter ʒ (symbol for electromotive force).

x-intercept The *x-coordinate* of the point at which a line or plane intersects the *x-axis*.

X-irradiate To expose to *X-rays*.

xistor Abbreviation of *transistor*.

XL Symbol for *inductive reactance*.

X-line A horizontal line in a computer memory matrix. Compare *Y-line*.

XLR connector A microphone connector with a locking device to prevent unintentional disconnection.

X-meter 1. An instrument for measuring reactance and phase angle. 2. A form of simple capacitance meter whose dial is direct-reading in microfarads, although the deflection is actually proportional to the reactance of the capacitor under measurement.

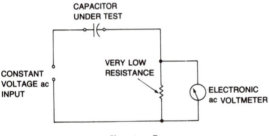

X-meter, 2.

xmission Abbreviation of *transmission*.

xmit Abbreviation of *transmit*. (Also, *xmt*.)

xmitter Abbreviation of *transmitter*. (Also, *xmtr*.)

xmt Abbreviation of *transmit*. (Also, *xmit*.)

xmtr Abbreviation of *transmitter*. (Also, *xmitter*.)

XOR Abbreviation of *exclusive-OR*.

xover Abbreviation of *crossover*.

X-particle See *meson*.

xponder Abbreviation of *transponder*.

X-position 1. The alignment of a cathode-ray beam along the horizontal axis of an oscilloscope screen. 2. The position of a point with respect to the horizontal axis of a graph in the rectangular (Cartesian) coordinate system. 3. On a punched card, (usually) the punching positions in the second row.

XR Abbreviation of *index register*.

X-radiation Radiation composed of X-rays.

X-ray An invisible, electromagnetic ray having a wavelength of approximately 0.1 nm. X-rays are commonly produced by bombarding a target of heavy metal (such as tungsten) with a stream of high-velocity electrons in a vacuum tube. X-rays have high penetrating power, can expose photographic film, and cause some substances to fluoresce.

Xray Phonetic alphabet communications code word for the letter *X*.

X-ray See *X-ray diagnosis; X-ray inspection,* and *X-ray therapy*.

X-ray crystallography The art of observing atomic patterns in a crystal by means of X-rays.

X-ray detecting device 1. An X-ray instrument for spotting flaws in solid bodies. 2. A device, such as a fluoroscope, for showing the presence of X-rays.

X-ray diagnosis The use of X-rays in the diagnosis of disease and in the observation of internal parts of the body.

X-ray diffraction The diffraction of X-rays by a material into or onto which they are directed. Also see *X-ray diffraction camera* and *X-ray diffraction pattern*.

X-ray diffraction camera A special camera that furnishes a photograph of the pattern created by X-rays diffracted by a material. See *X-ray diffraction* and *X-ray diffraction pattern*.

X-ray diffraction pattern The multicircle pattern produced on film exposed to X-rays diffracted by a material. Also see *X-ray diffraction* and *X-ray diffraction camera*.

X-ray goniometer An instrument used to find the position of the axes of a quartz crystal. It employs X-rays reflected from the atomic planes of the crystal.

X-ray inspection The use of X-rays in the examination and study of the internal features of materials and devices.

X-ray load 1. An X-ray tube that is the terminal member of an electronic system. 2. A body, mass, or material exposed to X-rays.

X-ray machine A device that generates X-rays for a specific purpose, such as medical diagnosis.

X-ray photograph A photograph made by exposure to X-rays, especially an X-ray shadow gram, a picture made without a camera by interposing an object (such as a part of the human body) between an X-ray tube and photographic film.

X-ray radiation See *X-radiation*.

X-ray spectra The continuous band of X-radiation between approximately 0.01 and 0.15 nm wavelength.

X-ray spectrograph A spectrograph used to disperse and measure the wavelength of X-rays.

X-ray spectrometer A spectrometer with which the diffraction angle of X-rays reflected from the surface of a crystal may be measured. The spectrometer allows the characteristics and composition of almost any material to be studied.

X-ray system Collectively, an X-ray tube and the associated equipment required for a specific application, such as crystallography, irradiation, or medical therapy.

X-ray technician 1. A subengineering professional skilled in the operation and maintenance of X-ray systems and who usually works under the supervision of an engineer. 2. A paramedical professional skilled in the medical use of X-rays and who works under the supervision of a medical doctor.

X-ray therapy The use of X-rays in the treatment of disease and ailments.

X-ray therapy system An X-ray system designed for X-ray therapy.

X-ray thickness gauge An instrument employed to measure the thickness of metal, such as a continuously moving sheet of steel. The measurement is in terms of the amount of absorption of an applied X-ray beam by the metal.

X-ray tube A specialized, very-high-voltage diode in which a high-velocity electron beam bombards an anode (*target*) of heavy metal, such as tungsten, causing it to emit X-rays. The target is tilted to reflect the X-rays through the glass wall of the tube. See next page.

X-section Abbreviation of *cross section*.

X-sink The circuit or device into which the X-lines of a computer memory matrix feed. Compare *Y-sink*.

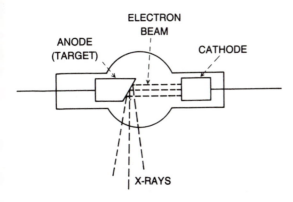

ANODE (TARGET)

ELECTRON BEAM

CATHODE

X-RAYS

X-ray tube.

XT Symbol for total reactance, i.e., $X_L - X_C$. (Also, X_t).

xtal Abbreviation of *crystal.*

xtalk Abbreviation of *crosstalk.*

X-wave One (the extraordinary) of the pair of components into which an ionospheric radio wave is divided by earth's magnetic field. Compare *O-wave.*

XXX In radiotelegraphy, a signal indicating urgency.

X—Y counting A technique used with electronic counters to determine the ratio of one frequency (X) to another *(Y)*. When X is the higher frequency, the counter readout shows the number of pulses of X producing one cycle of Y.

XY-cut crystal A piezoelectric plate cut from a quartz crystal at such an angle that its electrical characteristics fall between those of the X-cut and Y-cut crystal.

XYL Radiotelegraph symbol for *wife* (literally, *ex-young lady*).

X—Y plotter An output device similar to the X—Y recorder, which allows a digital computer to plot graphs. Also called *data plotter.*

X—Y recorder An instrument which produces a permanent record (photographic print or directly inked paper) of a variable quantity on a chart having X (horizontal) and Y (vertical) axes.

Y

Y 1. Symbol for *admittance*. 2. Symbol for the *vertical axis* of a graph or screen in the rectangular (Cartesian) coordinate system. 3. Symbol for *yttrium*. 4. Symbol for *Young's modulus*.

y 1. Abbreviation of *year*. (Also, *yr*.) 2. Abbreviation of *yard*. (Also, *yd*.)

Y adapter An *adapter* in which two lines are connected to a third line, or vice versa. So called from its resemblance to the letter Y.

YAG Abbreviation of *yttrium—aluminum-garnet*, the stimulated substance in some lasers.

yagi antenna A directional antenna consisting of several parallel rods or wires. One rod serves as a driven half-wave dipole radiator, one as a parasitic reflector, and one or more as parasitic directors. Also called *yagi*.

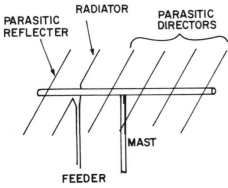

Yagi antenna.

Y-amplifier The vertical amplifier of an oscilloscope or recorder. Compare *X-amplifier* and *Z-amplifier*.

yard Abbreviation, yd. A unit of linear measure in the English system. 1 yd = 3 ft = 0.9144 meter.

yaw Side-to-side movement in a vehicle; essentially horizontal displacement about the vertical axis.

yaw amplifier In an aircraft servo system, the unit which amplifies the yaw signal (the signal which is proportional to the deviation of the aircraft from the line of flight).

yaw meter An instrument for measuring the angle of yaw of an aircraft.

Y-axis 1. The vertical axis of a chart, graph, or screen in the rectangular coordinate system. 2. In a quartz crystal, the axis drawn perpendicular to the faces of the hexagon.

Y-axis amplifier See *Y-amplifier*.

Yb Symbol for *ytterbium*.

Y-box See *wye-box*.

Y-channel See *Y-amplifier*.

Y-channel gain See *Y-gain*.

Y-circulator An interconnection among three waveguides. When power is applied to the junction through one waveguide, the wave is transferred to the adjacent waveguide immediately to the right or left.

Y-connection See *wye-connection*.

y-coordinate An *ordinate*.

Y-current See *wye-current*.

Y-cut crystal A piezoelectric plate cut from a quartz crystal in such a way that the plane of the plate is perpendicular to the Y-axis of the crystal. See *crystal axes* and *Y-axis, 2*.

yd Abbreviation of *yard*.

Y-deflection Vertical deflection of the spot on the screen of a cathode-ray tube. Compare *X-deflection*.

Y-diode The decoding diode in each of the Y-lines of a computer memory matrix.

Y-direction The vertical direction in deflections and in graphical presentations of data.

Y-drive The driving source or energy for the Y-lines of a computer memory matrix. Compare *X-drive*.

year Abbreviation, y or yr. The period of the sun's revolution around the earth, about 365.25 days or $3.155\,8 \times 10^7$ seconds.

yellow copper ore See *chalcopyrite*.

yellow metal A copper-zinc alloy that is 60% copper.

yellow pyrites See *chalcopyrite*.

Y-equivalent circuit See *wye-equivalent circuit*.

Y-gain The gain (or gain control) of the vertical channel of an oscilloscope or X—Y recorder. Compare *X-gain* and *Z-gain*.

yield In a phototube or photocell, the photoelectric current per unit intensity of light.

yield map A diagram of an integrated circuit showing the locations of faulty components.

yield strength The lowest stress for plastic deformation of a material, below which the material is elastic and above which it is viscous.

yield value The amount of physical stress that causes a substance to become stretched permanently out of shape.

YIG Abbreviation of *yttrium—iron-garnet*, a crystalline material used in acoustic delay lines, parametric amplifiers, and YIG filters.

YIG crystal A crystal of yttrium—iron-garnet. See YIG.

YIG filter A filter employing a YIG crystal and tuned electromagnetically.

YIG-tuned parametric amplifier A parametric amplifier in which tuning is accomplished by continuously varying direct current through the solenoid of a YIG filter.

y-intercept The *y-coordinate* of the point at which a line or plane intersects the *y-axis*.

YL Amateur radio abbreviation of *young lady*. Compare *XYL*.

Y-line A vertical line of a computer memory matrix. Compare *X-line*.

Y-matched-impedance antenna See *wye-matched-impedance antenna*.

yoke 1. The ferromagnetic ring or cylinder which holds the pole pieces of a dynamo-type generator and acts as part of the magnetic circuit. 2. The system of coils employed for magnetic deflection of the electron beam in cathode-ray tubes, such as TV picture tubes.

yoke method A method of measuring the permeability of a sample of magnetic material, which involves completing the magnetic circuit with a heavy yoke of soft iron.

Young Phonetic alphabet communications code word for the letter Y.

Young's modulus Symbol, *Y*. The ratio of tensile stress to tensile strain.

Y-parameters The admittance parameters of a four-terminal network or device.

Y-point See *wye-point*.

Y-position 1. On a cathode-ray screen, the vertical position of the beam spot. 2. On a graph, the position of a point along the vertical axis. 3. On a punched card, (usually) the punching positions in the top row.

Y-potential See *wye potential*.

yr 1. Abbreviaton of *year*. (Also, *y*.) 2. Radiotelegraph abbreviation of *your*.

Y-rectifier See *wye rectifier*.

Y-signal In color television, the monochromatic signal conveying brightness information.

Y-sink The circuit or device into which the Y-lines of a computer memory matrix feed. Compare *X-sink*.

yT Symbol for the Y-matrix of a transistor. Compare *yV*.

ytterbium Symbol, Yb. A metallic element of the rare-earth group. Atomic number, 70. Atomic weight, 173.04.

ytterbium metals The rare-earth metals dysprosium, erbium, lutetium, holmium, thulium, and ytterbium.

Yttralox A transparent polycrystalline ceramic composed primarily of yttrium oxide, which has many applications in electro-optics.

yttria Formula, Y203. Yttrium oxide, a white powder used in Nernst lamps.

yttrium Symbol, Y. A metallic element of the rare-earth-metals group. Atomic number, 39. Atomic weight, 88.92.

yttrium—iron-garnet See *YIG*.

yttrium metal Any of the group of metals including yttrium, erbium, holmium, lutetium, thulium, ytterbium, and sometimes dysprosium, gadolinium, and terbium.

Yukon Standard Time Standard time of the ninth time zone west of Greenwich, embracing the Yukon Territory and a portion of southern Alaska.

yV Symbol for the Y-matrix of a vacuum tube. Compare *yT*.

Y-winding See *wye-connection*.

Y-Y circuit See *wye—wye circuit*.

Z 1. Symbol for *impedance*. 2. Symbol for *atomic number*. 3. Symbol for *zenith distance* (astronomy).

z 1. Abbreviation of *zero*. 2. Symbol for *electrochemical equivalent*. 3. Abbreviation of *zone*.

zag The short, straight deflection of a point, particle, or of waves along a jagged path in a direction opposite that of a *zig,* a similar deflection, i.e., one of the components of zigzag deflection.

Zamboni pile A high-voltage electrochemical cell consisting of an aluminum anode, a manganese-dioxide cathode, and an aluminum-chloride electrolyte.

Z-amplifier The intensity-modulation amplifier of an oscilloscope. Compare *X-amplifier* and *Y-amplifier.*

Z-angle meter An instrument, commonly of the null-balance type, which indicates the impedance and phase angle of capacitors, inductors, and sometimes of *LCR* combinations.

Z-axis 1. The intensity-modulation input of an oscilloscope, including the associated circuit. 2. The video-signal input of a TV picture tube, including the associated circuit. 3. The third axis in a three-dimensional coordinate system. 4. The lengthwise axis of a quartz crystal.

Z-axis amplifier See *Z-amplifier.*

Z-axis modulation See *intensity modulation.*

Z-bar See *Z-cut crystal.*

Z-channel See *Z-amplifier.*

Z-channel gain See *Z-gain.*

Z-cut crystal A plate cut from a quartz crystal so that the plane of the plate is perpendicular to the Z-axis of the crystal and parallel to the X-axis. See *crystal axes.*

ZD Abbreviation of *zero defects.*

Z-deflection Deflection of a cathode-ray beam beyond the (ordinarily defined) deflection area on the tube screen.

zebra 1. Phonetic alphabet communications code word for the letter *Z.* 2. Abbreviation of *zero energy breeder reactor assembly.*

zebra time See *Greenwich mean time.*

Zeeman effect The splitting of a spectral line of a gas into closely spaced, polarized frequency components by an applied magnetic field.

Zenely electroscope A sensitive alpha ray electroscope (see *electroscope*) whose leaf is attracted by a metal plate biased at 50V to 200V. On touching the plate, the leaf becomes charged and is repelled. But the gas ions around the leaf neutralize the charge, and the leaf returns to the plate to repeat the action. This sequence causes the leaf to oscillate, the number of oscillations per second being proportional to the strength of the ionizing source.

zener See *zener diode.*

zener breakdown The phenomenon of junction breakdown in a zener diode (see *zener effect*). Like avalanche breakdown, zener breakdown is not a destructive phenomenon if the current is properly limited. The effects finds many useful applications, most of them based upon the constant voltage drop across a zener diode operating in its reverse breakdown region.

zener current 1. Zener-diode current, especially the high reverse current in the region of zener breakdown. 2. Current flowing through an insulator in an intense electric field, due to excitation driving electrons into the conduction band of atoms.

zener diode A semiconductor diode (usually silicon) specially fabricated to utilize zener breakdown. These diodes are available in a wide variety of configurations and ratings. Also called *zener.* Compare *avalanche diode.*

zener-diode clipper A signal-amplitude clipper which, because it employs a zener diode, does not require dc bias. Clipping takes place at the zener breakdown voltage. Zener diodes may be connected back to back for ac clipping.

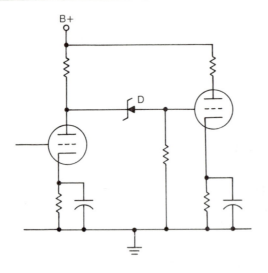

Zener-diode coupling element.

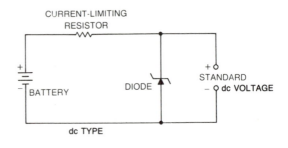

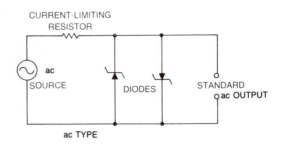

Zener-diode voltage reference.

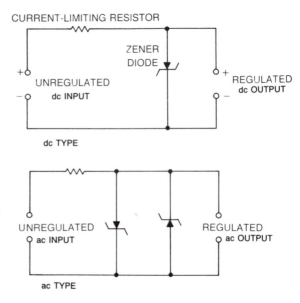

Zener-diode voltage regulator.

zener-diode coupling element A zener diode employed as a direct coupling element between amplifier (or other) stages. When no signal is present, the resistance of the diode is extremely high, since the dc voltages of the diode-coupled amplifier stages reverse-bias the diode to below the zener breakdown point, and no current flows from one stage to the next. The signal, however, raises the voltage enough to break down the diode and is readily transmitted from one stage to the following one.

zener-diode voltage reference A device that utilizes the constant voltage drop across a zener diode operated in its breakdown region, to provide a standard voltage for reference purposes. It consists essentially of a dc voltage source, limiting resistor, and zener diode. For an ac reference voltage, an ac voltage source, limiting resistor, and two zener diodes (connected in parallel, back to back) are required. Zener diodes may be connected in series to supply a higher reference voltage than can be delivered by a single diode.

zener-diode voltage regulator A simple voltage regulator whose output is the constant voltage drop developed across a zener diode conducting in the reverse breakdown region. The simple regulator circuit consists of a zener diode and an appropriate limiting resistor connected in series across the output terminals of an unregulated dc power supply. For ac voltage regulation, two zener diodes may be connected in parallel, back to back. Zener diodes may be connected in series to regulate a higher voltage than can be accommodated by a single diode.

zener effect In a reverse-biased *pn*-junction, the sudden, large increase in current that occurs when a particular value of reverse voltage is reached, and which is due to ionization by the high-intensity electric field in the depletion region. The zener effect has a useful corollary in the constant voltage drop which appears across the junction during avalanche. The temperature of the junction has some influence on the zener effect; it is seen as a negative temperature coefficient of zener breakdown voltage.

zener impedance The impedance presented by a junction operating in its zener breakdown region.

zener knee In the response of a reverse-biased zener diode, the point of abrupt transition from low current (high resistance) to high current (low resistance). For voltage regulation and voltage standardization, the knee should be as sharp as possible.

zener knee current In a zener diode, the current that flows when the reverse bias reaches the zener voltage.

zener-protected MOSFET See gate-protected MOSFET.

zener voltage The particular value of reverse voltage at which a zener diode or other reverse-biased *pn*-junction abruptly exhibits the zener effect. Depending on the zener diode, this potential can be less than 10 volts or several hundred voltage.

zener-voltage point See *zener voltage.*

zenith In the sky, the point which is directly overhead.

zeolite process A method employing certain artificial zeolites (hydrous silicates) to soften water used in some electronic manufacturing and testing operations.

zeppelin antenna A voltage-fed Hertz antenna having a two-wire tuned feeder system attached to one end of the radiator wire. Also called *zepp antenna* and *zepp-fed antenna.*

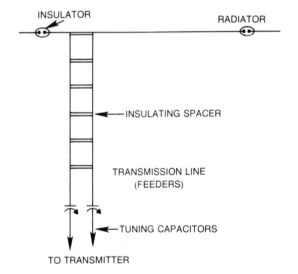

Zeppelin antenna.

zero 1. The symbol of numeral represented by the cipher (0). 2. To set a meter or other instrument to its zero reading or condition. 3. To align one element of a system precisely with another. 4. In computer practice, to set a register to zero. 5. In computer practice, using zero pulses to replace what is in a storage area.

zero-access memory See *zero-access storage.*

zero-access storage Computer storage requiring negligible waiting time (latency).

zero-address instruction A computer instruction requiring no address; the operation specified by the instruction defines the locations of the operands. Also called *addressless instruction.*

zero adjust In an analog metering device, a mechanical or electrical control that allows precise setting of the reading to zero, when the parameter to be measured is actually zero.

zero adjustment 1. The act of setting an instrument or circuit to its zero reading or zero-operating condition. 2. A device or subcircuit used to set a meter to its zero reading. Also see *zero set, 1, 2.*

zero-angle cut An alternate term for X-cut, as applied to quartz crystal plates. Also see *crystal cuts.*

zero axis In a plot of a variable quantity, the (usually horizontal) reference line which indicates the zero value of the quantity.

zero-axis symmetry The condition in which a waveshape is symmetrical about a zero axis.

zero beat Complete absence of a beat note (i.e., silence), a condition occurring when the frequencies of the beating signals (or their harmonics) are equal. See *beat note.*

zero-beat detector A device or circuit used to sense and indicate the condition of zero beat.

zero-beat indicator See *zero-beat detector.*

zero-beat method The method of adjusting the frequency of one signal exactly to that of another (usually standard) signal by setting the first signal frequency to zero beat with the second signal frequency.

zero-beat reception A system of reception entailing zero beating an incoming signal with the signal from a local oscillator. Also called *homodyne reception.*

zero-bias operation Operation of a vacuum tube without dc grid bias, a field-effect transistor without dc gate bias, or a common-emitter-connected bipolar transistor without dc base bias.

zero-bias tube A vacuum tube designed for operation (particularly in a class-B amplifier) without dc grid bias. In such a tube, the zero-signal plate current is very low because of the large amplification factor.

zero capacitance 1. Absence of capacitance (a theortical condition). 2. In some circuits, the lowest capacitance point (C_0), to which all other capacitance ($C_1, C_2...C_n$) are referred.

zero compensation The elimination or minimizing of the zero-signal output of a transducer or similar device.

zero compression See *zero suppression, 1.*

zero condition See *zero state.*

zero current 1. Absence of current. 2. In some circuits, the lowest current level (I_0), to which all other currents ($I_1, I_2...I_n$) are referred.

zero-cut crystal A piezoelectric plate cut from a quartz crystal in such a way that the frequency-temperature coefficient is zero.

zero dB reference level An agreed-upon zero level for decibel ratings (which are by nature relative). Zero dB corresponds to mW in 500 ohms. Compare *volume unit.*

zero drift 1. The (usually gradual) drift of a zero point, such as the zero setting of an electronic voltmeter. 2. The condition of no change in the value of a quantity.

zero elimination See *zero suppression, 1.*

zero energy The condition in which energy is neither generated, consumed, nor dissipated.

zero error 1. In instruments and measurements, complete lack of error; usually, error so low as to be inconsequential. 2. In a radar system, the inherent delay in the transmitter and receiver circuits.

zero field emission Thermionic emission from a cathode or hot conductor within a uniform electrostatic field.

zero fill See *zero, 5.*

zero gravity The condition of weightlessness, i.e., the state in which gravitational pull by a celestial body is completely absent or has been nullified.

zero-gravity switch A switch which is actuated automatically when the condition of zero gravity is reached.

zero impedance 1. Absence of impedance (a theoretical condition). 2. In some circuits, the lowest impedance level (Z_0), to which all other impedances (Z_1, $Z_2...Z_n$) are referred.

zero inductance 1. Absence of inductance (a theoretical condition). 2. In some circuits, the lowest inductance level (L_0), to which all other inductances (L_1, $L_2...L_n$) are referred.

zero input 1. Complete absence of input (signal or noise). 2. Absence of operating voltage or power to a system. 3. In a flip-flop, the input terminal which is not receiving an input trigger.

zero-input terminal In a flip-flop, the input-signal terminal at which a trigger signal will switch the flip-flop output to zero. Also called *zero terminal* and *reset terminal*.

zero instant See *zero time*.

zeroize See *zero, 2, 4,* and *5*.

zero level The reference level for the comparison of quantities. For example, it may be a voltage or current level; in af measurements, it is the zero-dB reference level.

zero-line stability In an analog metering device, the ability of the instrument to maintain proper zero adjustment over a period of time.

zero magnet A permanent magnet employed to set the pointer of a meter to zero.

zero magnitude 1. For a quantity, the state of being valueless, i.e., a complete absence of the quantity. 2. In some tests, measurements, or calculations, the lowest value of a quantity, to which all other values are referred.

zero meridian The meridian at Greenwich (near London) England, from which longitude and standard time are reckoned. Also see *time zone* and *zone time*.

zero method A method of measurement entailing adjustment of a circuit or device—such as a bridge, tee-network, or potentiometer—for zero response of the detector. Also called *null method*.

zero modulation The momentary lack of modulation in a communications system, as during a pause in speech.

zero-modulation noise Noise produced by previously erased tape which is run under specified operating conditions.

zero output 1. Complete absence of output signal or output power, sometimes disregarding noise output. 2. (a) In a flip-flop, the normal condition of no signal pulse at a particular output terminal, or (b) of a zeroed magnetic cell, the output obtained when it is read.

zero output signal See *zero output, 2 (b)*.

zero-output terminal In a flip-flop, the output terminal which is not delivering an output pulse.

zero phase In an ac circuit, the condition in which current and voltage are in step, or a similar relationship between other components.

zero pole In a multiconductor line, the common or reference conductor.

zero potential 1. Complete absence of voltage. 2. In some circuits, the lowest voltage (E_0, V_0), to which all other voltages (E_1, $E_2...E_n$ or V_1, $V_2...V_n$) are referred. 3. The potential of the earth as a reference point.

zero power 1. The condition in which $P = 0$. 2. In some circuits and systems, the lowest power level (P_0), to which all other power values (P_1, $P_2...P_n$) are referred.

zero-power resistance In a thermistor, the resistance at which power dissipation is zero.

zero-power resistance-temperature characteristic For a thermistor, a figure which reveals the extent to which zero-power

resistance varies with the temperature of the thermistor body.

zero-power temperature coefficient of resistance A temperature coefficient which reveals the extent to which the temperature of the thermistor body causes the zero-power resistance to change (expressed in ohms per ohm per degree Celsius).

zero reactance 1. Absence of reactance (a theoretical condition). 2. In some circuits, the lowest reactance (X_0), to which all other reactances (X_1, $X_2...X_n$) are referred.

zero resistance 1. Absence of resistance (a theoretical condition). 2. In some circuits, the lowest resistance (R_0), to which all other resistances (R_1, $R_2...R_n$) are referred.

zero screw The mechanical zero adjuster of a meter.

zero set 1. A (usually screwdriver-adjusted) mechanism for setting the pointer of a meter to zero. 2. An electrical circuit comprising a resistance bridge or adjustable bucking voltage, for setting the meter in a voltmeter (such as a vtvm) to zero.

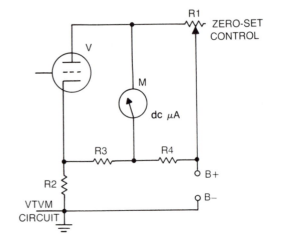

Zero set, 2.

zero shift See *zero drift*.

zero signal 1. The condition of complete absence of signal. 2. A finite signal level used as a reference point against which all other signal levels are measured.

zeros of impedance For an ac network, the frequencies at which the impedance is zero. Also called *zeros*.

zeros of network function The real or complex values of p, at which the network function is zero. Compare *poles of network function*.

zeros of transfer function The frequencies at which a transfer function becomes zero. Compare *poles of transfer function*.

zero stability Constancy of the zero condition in an instrument or system, i.e., absence of zero drift.

zero state The *low, zero, off,* or *false* logic state of a bistable device, such as a flip-flop or magnetic cell. It may be *actual zero* output or *low-voltage* output. Compare *one state*. In binary notation and calculation, the zero state is represented by a cipher.

zero suppression 1. In computer operation, automatic nonsignificant leading-zero cancellation. 2. Absence or removal of a restraining medium, such as a noise-suppression voltage. 3. In an audio recording system, introduction of a voltage to cancel the steady-state component of the input signal.

zero temperature 1. The point from which temperatures are reckoned on a thermometer scale. On the Celsius scale, zero degrees corresponds to the freezing point of water; on the Fahrenheit scale, it is 32 degrees below that point; and on the absolute scale, it is 273.18 degrees below that point. 2. A temperature point ($T0$) from which all other temperatures ($T1$, $T2...Tn$) are advanced.

zero temperature coefficient A temperature coefficient which has the value zero, i.e., one which indicates that there is no temperature-dependent drift of a given quantity.

zero terminal See *zero-input terminal* and *zero-output terminal*.

zero time 1. In some measurements, the first instant of time ($t0$), to which all other instants ($t1$, $t2...tn$) are referred. 2. Zero phase.

zero time reference During one cycle of radar operation, the time reference of the schedule of events.

zero vector A vector whose magnitude is zero.

zero voltage See *zero potential*.

zero voltage coefficient A voltage coefficient which has the value of zero, i.e., one which indicates that there is no voltage-dependent drift of a given quantity.

zero—zero The atmospheric condition in which ceiling and visibility both are zero.

zeta Z, ζ. The sixth letter of the Greek alphabet (the symbol for the displacement component of a sound-bearing particle).

Z-gain The gain (or gain control) of the intensity channel of an oscilloscope. Compare *X-gain* and *Y-gain*.

zig The short, straight deflection of a point particle, or wave along a jagged path in a direction opposite that of *zag*, a similar deflection, i.e., a component of *zigzag* deflection.

zigzag deflection Deflection of a particle, or motion of a point, in a zigzag line. Also see *zag* and *zig*.

zigzag rectifier A special version of the three-phase star (three-phase, half-wave) rectifier circuit. In the zigzag circuit, dc saturation of the transformer secondary is avoided by winding half the turns of each secondary on a separate core, i.e., each core carries two half-windings. The opposing flux due to different phases in the half-windings causes cancellation of the dc component of flux in each core. Thus, in the illustration, half-coils A and E are on one core, B and D on a second core, and C and F on a third.

zigzag reflections Multihop reflections of waves along a zigzag path, resulting from repeated reflections within the ionosphere.

zinc Symbol, Zn. A metallic element. Atomic number, 30. Atomic weight, 65.38. Zinc is familiar as the negative-electrode material in dry cells and as a protective coating for some sheet metals used in electronics.

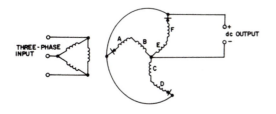

Zigzag rectifier.

zinc aluminate phosphor (1) Formula, $(ZnO + (Al2O3):Mn$. A substance used as a phosphor coating on the screen of a cathode-ray tube. The fluorescence is green-blue.

zinc aluminate phosphor (2) Formula $(ZnO + Al2O3):Cr$. A substance used as a phosphor coating on the screen of a cathode-ray tube. The fluorescence is red.

zinc beryllium silicate phosphor Formula, $(ZnO + BeO + SiO2):Mn$. A substance used as a phosphor coating on the screen of a cathode-ray tube (medium persistence). The fluorescence and phosphorescence are yellow.

zinc beryllium zirconium silicate phosphor Formula, $[(ZnO + Be) + (Ti - Zr - Th - O2) + SiO2]:Mn$. A substance used as a phosphor coating on the screen of a crt. The fluorescence is white.

zinc borate phosphor Formula, $(ZnO + B2O3):Mn$. A substance used as a phosphor coating on the screen of a cathode-ray tube. The fluorescence is yellow-orange.

zinc cadmium sulfide phosphor Formula, $(ZnS + CdS):Mn$. A substance used as a phosphor coating on the screen of a long-persistence cathode-ray tube. The phosphorescence is blue or blue-white to red.

zinc-carbon cell A primary cell in which the negative electrode is zinc, and the positive electrode is carbon. This type may be either wet or dry. Also see *cell*, *dry cell*, and *primary cell*.

zinc germanate phosphor Formula, $(ZnO + GeO2):Mn$. A substance used as a phosphor coating on the screen of a cathode-ray tube. The fluorescence is yellow-green.

zincite The natural mineral zinc oxide (ZnO). Also called *red oxide of zinc*.

zincite detector An early semiconductor diode (*crystal detector*) consisting of a bornite point in contact with a piece of zincite.

zinc magnesium fluoride phosphor A substance used as a phosphor coating on the screen of a long-persistence cathode-ray tube. The fluorescence and phosphorescence are orange.

zincolysis Electrolysis in a cell which has a zinc anode.

zinc orthoscilicate phosphor Formula, $Zn2SiO4$. The most common substance used as a phosphor coating on the screen of a cathode-ray tube. The phosphorescence is yellowish-green. Also called *willemite*.

zinc oxide 1. Formula, ZnO. A substance used as a phosphor coating on the screen of a very-short-persistance cathode-ray tube. The fluorescence and phosphorescence are blue-green. 2. Zincite.

zinc-oxide detector See *zincite detector*.

zinc-oxide resistor A *voltage-dependent resistor* whose active ingredient is zinc oxide.

zinc silicate phosphor Formula, $ZnO + SiO2$. A substance used as a phosphor coating on the screen of a cathode-ray tube. The fluorescence is blue.

zinc standard cell A standard cell employing zinc and mercury electrodes, and a mercurous sulfate excitant and depolarizer. Its emf is 1.4322V at 15° C. Also called *Clark cell*. Compare *Weston cell*.

zinc sulfide phosphor Formula, ZnS. A substance used as a phosphor coating on the screen of a long- or short-persistence cathode-ray tube. The fluorescence is blue-green, and the phosphorescence is yellow-green.

ZIP Abbreviation of *zinc-impurity photodetector*.

zip cord A simple two-conductor, flexible power cord.

zirconia Preparations of zirconium (especially $ZrO2$), valued for their high-temperature dielectric properties.

zirconium Symbol, Zr. A metallic element. Atomic number, 40. Atomic weight, 91.22.

Z-marker A vertically radiating marker beacon defining the zone above a radio-range station.

Z-meter An impedance meter. Instruments of this kind take four principal forms: (1) a direct-reading meter resembling an ohmmeter; (2) an adjustable circuit which is manipulated somewhat like a bridge, and compares an unknown impedance with a standard resistance; (3) an impedance bridge for evaluating the reactive and resistive components of unknown impedance; (4) a section of transmission line used with a signal source and vrvm for measuring impedance in terms of a standard resistor and/or standing waves.

Zn 1. Symbol for *zinc*. 2. Symbol for *azimuth*.

Zo Symbol for *characteristic impedance*.

zone 1. On a magnetic disk, a group of tracks whose associated transfer rate is not the same as that for the rest of the disk. 2. In computer practice, the area of a storage medium containing digits. 3. In a computer system, a main memory area which has been set aside for a particular purpose.

zone blanking In a radar system, a method of extinguishing the cathode-ray beam during a portion of the antenna sweep.

zone candle power In a given zone, the luminous flux per steradian, emitted by a light source under test.

zone leveling See *zone refining*.

zone marker See *Z-marker*.

zone melting See *zone refining*.

zone of silence The region between alternate reflections of a radio wave, in which no signal is detectable. Also called *skip zone*.

zone plate antenna A rapid-scan radar antenna having a reflector composed of confocal parabolas arranged in a circle.

zone-position indicator A radar which reveals the position of an object to a second radar having a restricted field.

zone purification See *zone refining*.

zone refining A method of purifying semiconductor materials (such as germanium and silicon) in which a molten zone in an ingot of the material moves along the length of the ingot, dissolving impurities as it travels, eventually depositing them at the end of the ingot, which is sawed off. This concentrated and segregated melting is accomplished by means of rf heating.

zone time In a given time zone, standard time in terms of the number of hours which must be added to local time to equal the time at the zero (Greenwich) meridian.

zoning 1. A method of fabricating a microwave reflector in concentric, flat regions, giving the same practical results as a continuous paraboloid. 2. In a communications system, the division of the coverage area into different geographic regions for a specific purpose.

zoom To rapidly change the focal length of a TV or motion-picture camera lens, so that the object seems to advance toward or recede from the viewer, remaining in focus as it does so. 2. Zoom lens.

zoom lens A special continuously adjustable lens system which allows zoom effects to be obtained with a TV or motion-picture camera, or a similar arrangement for still cameras that obviates the need for lens interchange when different focal lengths are needed. The lens system, which may be operated electronically, allows either narrow- or wide-angle views to be obtained without losing focus at any time.

Z-parameters Device or network parameters expressed as impedances.

ZPI Abbreviation of *zone-position indicator*.

Z-plunger In a waveguide, a combination choke and bucket plunger for radiation leakage reduction.

Zr Symbol for *zirconium*.

Z-signals A collection of letter groups, each starting with the letter *Z*, used for simplified telegraphy and radiotelegraphy by the military services.

Zulu time Greenwich mean time. Also see *zebra time*.

zwitterion An ion which carries both a positive and a negative charge.

zymometer An instrument employed to measure the degree of fermentation of a substance.

zymoscope An instrument employed to measure the fermenting power of yeast in terms of the amount of carbon dioxide obtained from a given quantity of sugar.

zymosimeter See *zymometer*.

zymurgy The branch of chemistry concerned with fermentation.

Appendix
Symbols, Tables and Data

SCHEMATIC SYMBOLS

AMMETER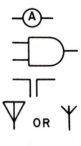

AND GATE

ANTENNA, BALANCED

ANTENNA, GENERAL

ANTENNA, LOOP, SHIELDED

ANTENNA, LOOP, UNSHIELDED

ANTENNA, UNBALANCED

ATTENUATOR, FIXED

ATTENUATOR, VARIABLE

BATTERY

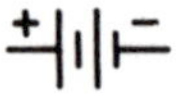

CAPACITOR, FEEDTHROUGH

CAPACITOR, FIXED, NONPOLARIZED

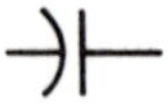

CAPACITOR, FIXED, POLARIZED

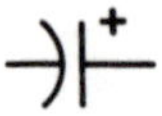

CAPACITOR, GANGED, VARIABLE

CAPACITOR, VARIABLE, SINGLE

CAPACITOR, VARIABLE, SPLIT- STATOR

CATHODE, DIRECTLY HEATED

CATHODE, INDIRECTLY HEATED    OR

CATHODE, COLD

CAVITY RESONATOR

CELL

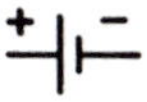

CIRCUIT BREAKER

COAXIAL CABLE 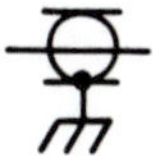   OR

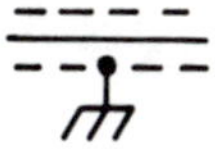

CRYSTAL, PIEZOELECTRIC

DELAY LINE   OR

DIODE, GENERAL

DIODE, GUNN

DIODE, LIGHT-EMITTING

DIODE, PHOTOSENSITIVE

DIODE, PHOTOVOLTAIC

DIODE, PIN	—▸⊢ OR —▸⊩
DIODE, VARACTOR	⊣⊦
DIODE, ZENER	—▸⌐
DIRECTIONAL COUPLER OR WATTMETER	OR —⊖—
EXCLUSIVE - OR GATE	
FEMALE CONTACT, GENERAL	—<
FERRITE BEAD	OR —◎—
FUSE	—∿— OR —▭—
GALVANOMETER	—⊕— OR —Ⓖ—
GROUND, CHASSIS	⏚ OR 777 OR ↓
GROUND, EARTH	⏚
HANDSET	⌒
HEADPHONE, DOUBLE	⌒
HEADPHONE, SINGLE	⌒

INDUCTOR, AIR-CORE

INDUCTOR, BIFILAR

INDUCTOR, IRON-CORE

INDUCTOR, TAPPED

INDUCTOR, VARIABLE   OR

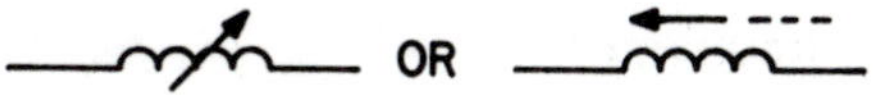

INTEGRATED CIRCUIT

INVERTER OR INVERTING AMPLIFIER

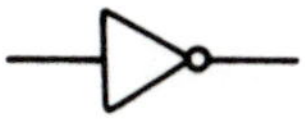

JACK, COAXIAL

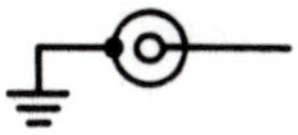

JACK, PHONE, 2-CONDUCTOR

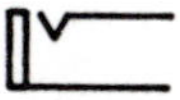

JACK, PHONE, 2-CONDUCTOR INTERRUTING

JACK, PHONE, 3-CONDUCTOR

JACK, PHONO

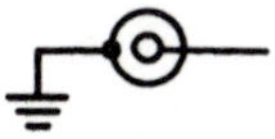

KEY, TELEGRAPH

LAMP, INCANDESCENT

LAMP, NEON

MALE CONTACT, GENERAL

METER, GENERAL

MICROAMMETER

MICROPHONE   OR

MILLIAMMETER

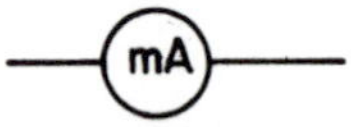

NAND GATE

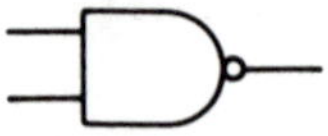

NEGATIVE VOLTAGE CONNECTION

NOR GATE

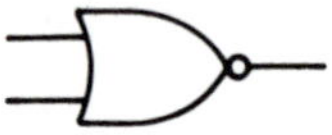

OPERATIONAL AMPLIFIER   OR

OR GATE

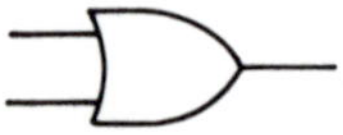

OUTLET, UTILITY, 117-V

OUTLET, UTILITY, 234-V

PHOTOCELL, TUBE

PLUG, PHONE, 2-CONDUCTOR

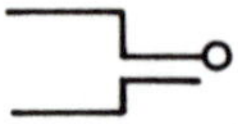

PLUG, PHONE, 3-CONDUCTOR

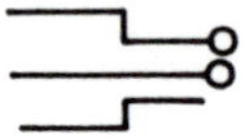

PLUG, PHONO

PLUG, UTILITY, 117-V

PLUG, UTILITY, 234-V

POSITIVE-VOLTAGE CONNECTION

POTENTIOMETER    OR

PROBE, RADIO-FREQUENCY 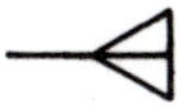   OR

RECTIFIER, SEMICONDUCTOR

RECTIFIER, SILICON-CONTROLLED

RECTIFIER, TUBE-TYPE

RELAY, DPDT

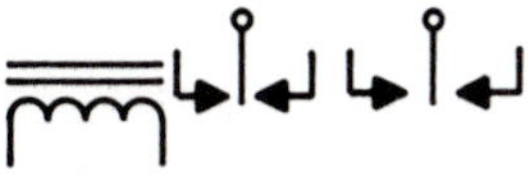

RELAY, DPST

RELAY, SPDT

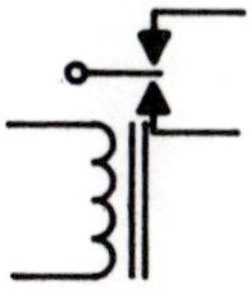

RELAY, SPST

RESISTOR

RESONATOR

RHEOSTAT    OR

SATURABLE REACTOR

SHIELDING

SIGNAL GENERATOR

SPEAKER 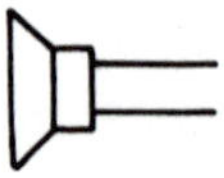   OR

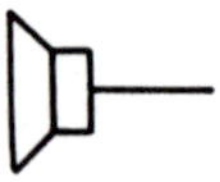

SWITCH, DPDT

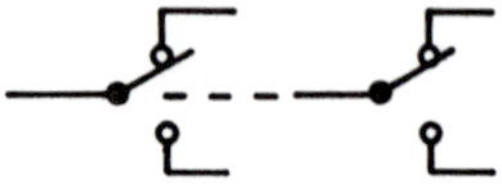

SWITCH, DPST

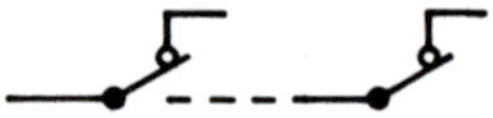

SWITCH, MOMENTARY-CONTACT

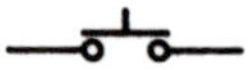

SWITCH, ROTARY

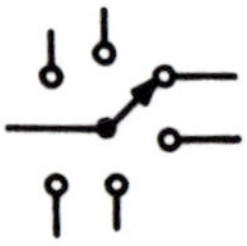

SWITCH, SPDT

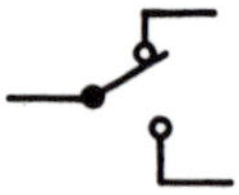

SWITCH, SPST

TERMINALS, GENERAL, BALANCED

TERMINALS, GENERAL, UNBALANCED

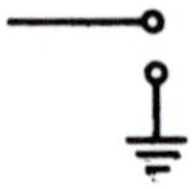

TEST POINT	○—————○
THERMOCOUPLE	⊃● OR ⊃>>>>
THYRISTOR	
TRANSFORMER, AIR-CORE	
TRANSFORMER, IRON-CORE	
TRANSFORMER, TAPPED PRIMARY	
TRANSFORMER, TAPPED SECONDARY	
TRANSISTOR, BIPOLAR, npn	
TRANSISTOR, BIPOLAR, pnp	
TRANSISTOR, FIELD-EFFECT. N-CHANNEL	
TRANSISTOR, FIELD-EFFECT, P·CHANNEL	
TRANSISTOR, METAL-OXIDE, DUAL-GATE	OR

TRANSISTOR, METAL-OXIDE , SINGLE-GATE   OR

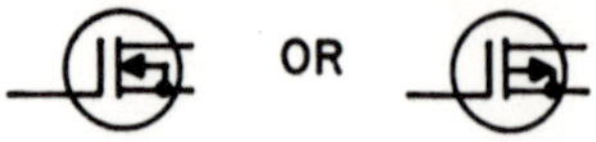

TRANSISTOR, PHOTOSENSITIVE   OR

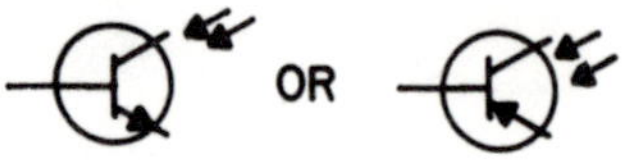

TRANSISTOR, UNIJUNCTION   OR

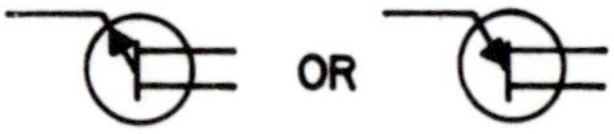

TUBE, DIODE

TUBE, PENTODE

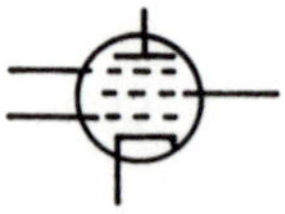

TUBE, PHOTOMULTIPLIER

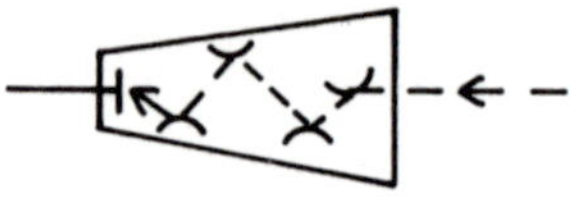

TUBE, TETRODE

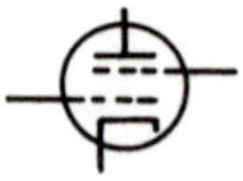

TUBE , TRIODE

UNSPECIFIED UNIT OR COMPONENT

VOLTMETER

WATTMETER   OR

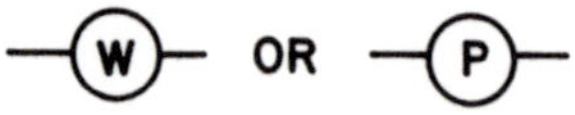

WIRES, CROSSING, CONNECTED   OR

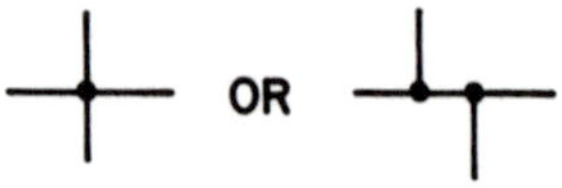

WIRES, CROSSING, NOT CONNECTED   OR

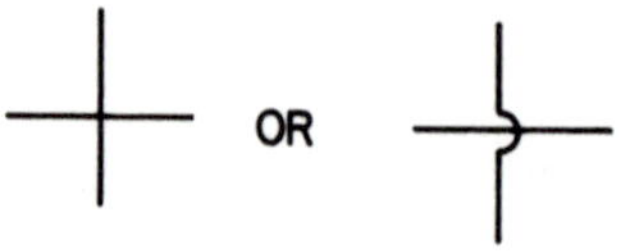

WIRES, CONNECTED (3-WAY)

Tables and Data

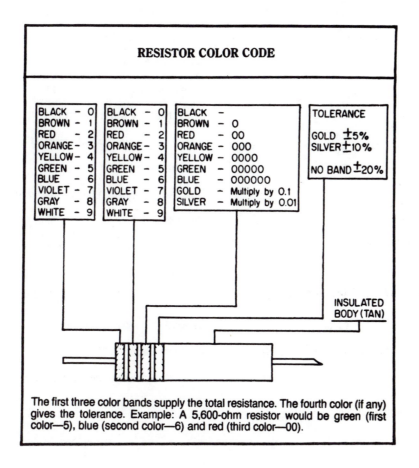

RESISTOR COLOR CODE

BLACK – 0	BLACK – 0	BLACK –	TOLERANCE
BROWN – 1	BROWN – 1	BROWN – 0	
RED – 2	RED – 2	RED – 00	GOLD ±5%
ORANGE– 3	ORANGE– 3	ORANGE – 000	SILVER ±10%
YELLOW– 4	YELLOW– 4	YELLOW – 0000	
GREEN – 5	GREEN – 5	GREEN – 00000	NO BAND ±20%
BLUE – 6	BLUE – 6	BLUE – 000000	
VIOLET – 7	VIOLET – 7	GOLD – Multiply by 0.1	
GRAY – 8	GRAY – 8	SILVER – Multiply by 0.01	
WHITE – 9	WHITE – 9		

INSULATED
BODY (TAN)

The first three color bands supply the total resistance. The fourth color (if any) gives the tolerance. Example: A 5,600-ohm resistor would be green (first color—5), blue (second color—6) and red (third color—00).

$x + y$	x added to y, x OR y
$x - y$	y subtracted from x
$x \cdot y$, $x \times y$, or xy	x multiplied by y, x AND y
$x \div y$	x divided by y
x/y or $\frac{x}{y}$	x divided by y
$1/x$	Reciprocal of x
x^n	x raised to the indicated power of n
$\sqrt[n]{x}$	Indicated root ($\sqrt{\ }$) of x
$x : y$	x is to y
$\lvert x \rvert$	Absolute value of x, magnitude of x
$\bar{X}, \dot{X},$ or $\mathbf{X}$	Vector X
$\bar{x}$	Average value of x
$f(x)$ or $F(x)$	Function of x
i	$\sqrt{-1}$
j	Operator, equal to $\sqrt{-1}$
Δx	Increment of x
dx	Differential of x
∂x	Partial differential of x
$\dfrac{\Delta x}{\Delta y}$	Change in x with respect to y
$\dfrac{dx}{dy}$	Derivative of x with respect to y
$\dfrac{d}{dy}(x)$	Derivative of x with respect to y
$D_y x$	Derivative of x with respect to y
$\dfrac{\partial x}{\partial y}$	Partial derivative of x with respect to y
Σ	Summation
Σ_b^a	Summation between limits (from a to b)
Π	Product
Π_a^a	Product between limits (from a to b)
$\int$	Integral
$\int_a^b$	Integral between limits (from a to b)
$\int x\, dy$	Integral of x with respect to y
$\big\vert_a$	Evaluated at a
$\big\vert_a^b$	Evaluated between limits (from a to b)

·	Radix (base) point
•	Logic multiplication symbol
∞	Infinity
+	Plus, positive, logic OR function
−	Minus, negative
±	Plus or minus, positive or negative
∓	Minus or plus, negative or positive
×	Times, logic AND function
÷	Divided by
/	Divided by (expressive of a ratio)
=	Equal to
≡	Identical to, is defined by
≅	Approximately equal to, congruent to
≐	Approximately equal to
≠	Not equal to
~	Similar to
<	Less than
≮	Not less than
<<	Much less than
>	Greater than
≯	Not greater than
>>	Much greater than
≤	Equal to or less than
≥	Equal to or greater than
∝	Proportional to, varies directly as
→	Approaches
:	Is to, proportional to
∴	Therefore
#	Number
%	Percent
@	At the rate of; at cost of
ϵ or e	The natural number $= 2.71828$ ~
π	Pi $\cong 3.14159...$ ~
( )	Parentheses. Used to enclose a common group of terms.
[ ]	Brackets. Used to enclose a common group of terms which includes one or more groups in parentheses.
{ }	Braces. Used to enclose a common group of terms which includes one of more groups in brackets.
∠	Angle
°	Degrees (arc or temperature)
′	Minutes, prime
″	Seconds, double prime
‖	Parallel to
⊥	Perpendicular to
...	And beyond, ellipsis

FACTORIALS

Numerical

n	$\dfrac{1}{n!}$					n!					n
1	1.									1	1
2	0.5									2	2
3	.16666	66666	66666	66666	66667					6	3
4	.04166	66666	66666	66666	66667					24	4
5	.00833	33333	33333	33333	33333					120	5
6	0.00138	88888	88888	88888	88889					720	6
7	.00019	84126	98412	69841	26984					5040	7
8	.00002	48015	87301	58730	15873					40320	8
9	.00000	27557	31922	39858	90653				3	62880	9
10	.00000	02755	73192	23985	89065				36	28800	10
11	0.00000	00250	52108	38544	17188				399	16800	11
12	.00000	00020	87675	69878	68099				4790	01600	12
13	.00000	00001	60590	43836	82161				62270	20800	13
14	.00000	00000	11470	74559	77297			8	71782	91200	14
15	.00000	00000	00764	71637	31820			130	76743	68000	15
16	0.00000	00000	00047	79477	33239			2092	27898	88000	16
17	.00000	00000	00002	81145	72543			35568	74280	96000	17
18	.00000	00000	00000	15619	20697		6	40237	37057	28000	18
19	.00000	00000	00000	00822	06352		121	64510	04088	32000	19
20	.00000	00000	00000	00041	10318		2432	90200	81766	40000	20

$$n! = 1 \times 2 \times 3 \times 4 \times 5 \ldots n$$

FOR EXAMPLE: For $n = 7$, $n! = 5040$.
$1/n! = 0.0019841269841269841269984$,
$\log(n!) = 3.702431$.

Logarithmic

Logarithms of the products $1 \times 2 \times 3 \ldots n$, n from 1 to 100.

n	log (n!)	n	log (n!)	n	log (n!)	n	log (n!)
1	0.000000	26	26.605619	51	66.190645	76	111.275425
2	0.301030	27	28.036983	52	67.906648	77	113.161916
3	0.778151	28	29.484141	53	69.630924	78	115.054011
4	1.380211	29	30.946539	54	71.363318	79	116.951638
5	2.079181	30	32.423660	55	73.103681	80	118.854728
6	2.857332	31	33.915022	56	74.851869	81	120.763213
7	3.702431	32	35.420172	57	76.607744	82	122.677027
8	4.605521	33	36.938686	58	78.371172	83	124.596105
9	5.559763	34	38.470165	59	80.142024	84	126.520384
10	6.559763	35	40.014233	60	81.920175	85	128.449803
11	7.601156	36	41.570535	61	83.705505	86	130.384301
12	8.680337	37	43.138737	62	85.497896	87	132.323821
13	9.794280	38	44.718520	63	87.297237	88	134.268303
14	10.940408	39	46.309585	64	89.103417	89	136.217693
15	12.116500	40	47.911645	65	90.916330	90	138.171936
16	13.320620	41	49.524429	66	92.735874	91	140.130977
17	14.551069	42	51.147678	67	94.561949	92	142.094765
18	15.806341	43	52.781147	68	96.394458	93	144.063248
19	17.085095	44	54.424599	69	98.233307	94	146.036376
20	18.386125	45	56.077812	70	100.078405	95	148.014099
21	19.708344	46	57.740570	71	101.929663	96	149.996371
22	21.050767	47	59.412668	72	103.786996	97	151.983142
23	22.412494	48	61.093909	73	105.650319	98	153.974368
24	23.792706	49	62.784105	74	107.519550	99	155.970004
25	25.190646	50	64.483075	75	109.394612	100	157.970004

	Practical Unit	Electromagnetic Unit	Electrostatic Unit
Quantity	MKS	CGS EM	CGS ES
1. Capacitance	1 Farad	10^{-9} Abfarad	9×10^{11} Statfarad
	10^9 Farad	1 Abfarad	9×10^{20} Statfarad
	$1/9 \times 10^{-11}$ Farad	$1/9 \times 10^{-20}$ Abfarad	1 Statfarad
2. Charge	1 Coulomb	10^{-1} Abcoulomb	3×10^9 Statcoulomb
	10 Coulomb	1 Abcoulomb	3×10^{10} Statcoulomb
	$1/3 \times 10^{-9}$ Coulomb	$1/3 \times 10^{-10}$ Abcoulomb	1 Statcoulomb
3. Charge density	1 Coulomb/m^3	10^{-7} Abcoulomb/cm^3	3×10^3 Statcoulomb/cm^3
	10^7 Coulomb/m^3	1 Abcoulomb/cm^3	3×10^{10} Statcoulomb/cm^3
	$1/3 \times 10^{-3}$ Coulomb/m^3	$1/3 \times 10^{-10}$ Abcoulomb/cm^3	1 Statcoulomb/cm^3
4. Conductivity	1 Mho/m	10^{-11} Abmho/cm	9×10^9 Statmho/cm
	10^{11} Mho/m	1 Abmho/cm	9×10^{20} Statmho/cm
	$1/9 \times 10^{-9}$ Mho/m	$1/9 \times 10^{-20}$ Abmho/cm	1 Statmho/cm
5. Current	1 Ampere	10^{-1} Abampere	3×10^9 Statampere
	10 Ampere	1 Abampere	3×10^{10} Statampere
	$1/3 \times 10^{-9}$ Ampere	$1/3 \times 10^{-10}$ Abampere	1 Statampere
6. Current density	1 Ampere/m^2	10^{-5} Abampere/cm^2	3×10^5 Statampere/cm^2
	10^5 Ampere/m^2	1 Abampere/cm^2	3×10^{10} Statampere/cm^2
	$1/3 \times 10^{-5}$ Ampere/m^2	$1/3 \times 10^{-10}$ Abampere/cm^2	1 Statampere/cm^2
7. Electric field intensity	1 Volt/meter	10^6 Abvolt/cm	$1/3 \times 10^{-4}$ Statvolt/cm
	10^{-6} Volt/meter	1 Abvolt/cm	$1/3 \times 10^{-10}$ Statvolt/cm
	3×10^4 Volt/meter	3×10^{10} Abvolt/cm	1 Statvolt/cm
8. Electric potential	1 Volt	10^8 Abvolts	$1/3 \times 10^{-2}$ Statvolts
	10^{-8} Volt	1 Abvolt	$1/3 \times 10^{-10}$ Statvolts
	3×10^2 Volt	3×10^{10} Abvolts	1 Statvolt
9. Electric dipole moment	1 Coulomb–meter	10 Abcoulomb–cm	3×10^{11} Statcoulomb–cm
	10^{-1} Coulomb–meter	1 Abcoulomb–cm	3×10^{10} Statcoulomb–cm
	$1/3 \times 10^{-11}$ Coulomb–meter	$1/3 \times 10^{-10}$ Abcoulomb–cm	1 Statcoulomb–cm
10. Energy	1 Joule	10^7 Erg	10^7 Erg
	10^{-7} Joule	1 Erg	1 Erg
	10^{-7} Joule	1 Erg	1 Erg
11. Force	1 Newton	10^5 Dyne	10^5 Dyne
	10^{-5} Newton	1 Dyne	1 Dyne
	10^{-5} Newton	1 Dyne	1 Dyne
12. Flux density	1 Weber/m^2	10^4 Gauss	$1/3 \times 10^{-6}$ esu
	10^{-4} Weber/m^2	1 Gauss	$1/3 \times 10^{-10}$ esu
	3×10^6 Weber/m^2	3×10^{10} Gauss	1 esu
13. Inductance	1 Henry	10^9 Abhenry	$1/9 \times 10^{-11}$ Stathenry
	10^{-9} Henry	1 Abhenry	$1/9 \times 10^{-20}$ Stathenry
	9×10^{11} Henry	9×10^{20} Abhenry	1 Stathenry
14. Inductive capacity	1 Farad/meter	10^{-11} Abfarad/cm	9×10^9 Statfarad/cm
	10^{11} Farad/meter	1 Abfarad/cm	9×10^{20} Statfarad/cm
	$1/9 \times 10^{-9}$ Farad/meter	$1/9 \times 10^{-20}$ Abfarad/cm	1 Statfarad/cm
15. Magnetic flux	1 Weber	10^8 Maxwell	$1/3 \times 10^{-2}$ esu
	10^{-8} Weber	1 Maxwell	$1/3 \times 10^{-10}$ esu
	3×10^2 Weber	3×10^{10} Maxwell	1 esu
16. Magnetic dipole moment	1 Ampere–meter^2	10^3 Abamp–cm^2	3×10^{13} Statamp–cm^2
	10^{-3} Ampere–meter^2	1 Abamp–cm^2	3×10^{10} Statamp–cm^2
	$1/3 \times 10^{-13}$ Ampere–meter^2	$1/3 \times 10^{-10}$ Abamp–cm^2	1 Statamp–cm^2
17. Permeability	1 Henry/meter	10^7 Abhenry/cm	$1/9 \times 10^{-13}$ Stathenry/cm
	10^{-7} Henry/meter	1 Abhenry/cm	$1/9 \times 10^{-20}$ Stathenry/cm
	9×10^{13} Henry/meter	9×10^{20} Abhenry/cm	1 Stathenry/cm
18. Power	1 Watt	10^7 erg/sec	10^7 erg/sec
	10^{-7} Watt	1 erg/sec	1 erg/sec
	10^{-7} Watt	1 erg/sec	1 erg/sec
19. Resistance	1 Ohm	10^9 Abohm	$1/9 \times 10^{-11}$ Statohm
	10^{-9} Ohm	1 Abohm	$1/9 \times 10^{-20}$ Statohm
	9×10^{11} Ohm	9×10^{20} Abohm	1 Statohm

ELECTRONIC ABBREVIATIONS

R	resistance	M	mutual inductance
G	conductance	k	coefficient of coupling
ρ	specific resistance or resistivity	N_p	primary turns
°F	degrees Fahrenheit	N_s	secondary turns
°C	degrees Celsius	N_s/N_p	turns ratio
E	voltage	θ	phase angle
I	current	B	susceptance
P	power	Y	admittance
P_o	output power	pf	power factor
P_i	input power	N_b	bels
η	efficiency	db	decibels
dc	direct current	N_n	nepers
ac	alternating current	ϵ	natural base (2.718281)
L	inductance	vu	volume unit
C	capacitance	E_p	plate voltage
t	time constant	E_g	grid voltage
T	time	Δ	change of
λ	wavelength	μ	amplification factor
f	frequency	r_p	plate resistance
f_r	resonant frequency	g_m	mutual conductance
cps	cycles per second (Hz)	I_p	plate current
Hz	Hertz (cps)	C_i	input capacitance
kHz	Kilohertz (kc)	C_{pk}	interelectrode capacitance, plate to cathode
MHz	megahertz (mc)		
μf	microfarad	C_{gk}	interelectrode capacitance, grid to cathode
pf	picofarad (same as micromicrofarad)	A	amplification
$E_{p \cdot p}$	peak-to-peak voltage	β	feedback voltage
V	velocity	K	amplification with negative feedback
D	distance	i_c	collector current
π	3.1416	i_e	emitter current
ω	$2 \times \pi \times f$	α	current gain
α	ωt	β	current gain
Q	coulomb	k	correction factor
RMS	root-mean-square	SWR	standing wave ratio
Z	impedance	j	-1 (j-operator)
X_l	inductive reactance	R_m	meter resistance
X_c	capacitive reactance	n	multiplication factor

AWG	Diameter, mils, d	Area, circular mils, d^2	Ohms per 1000 ft. at 20°C., or 68°F.	Pounds per 1000 ft.
0000	460.00	211,600	0.04901	640.5
000	409.64	167,805	0.06180	508.0
00	364.80	133,079	0.07793	402.8
0	324.86	105.534	0.09827	319.5
1	289.30	83,694	0.1239	253.3
2	257.63	66,373	0.1563	200.9
3	229.42	52,634	0.1970	159.3
4	204.31	41,743	0.2485	126.4
5	181.94	33,102	0.3133	100.2
6	162.02	26,250	0.3951	79.46
7	144.28	20,817	0.4982	63.02
8	129.49	16,768	0.6282	49.98
9	114.43	13,094	0.7921	39.63
10	101.89	10,382	0.9989	31.43
11	90.742	8,234.1	1.260	24.93
12	80.808	6,529.9	1.588	19.77
13	71.961	5,178.4	2.003	15.68
14	64.084	4,106.8	2.525	12.43
15	57.068	3,256.8	3.184	9.858
16	50.820	2,582.7	4.016	7.818
17	45.257	2,048.2	5.064	6.200
18	40.303	1,624.3	6.385	4.917
19	35.890	1,288.1	8.051	3.899
20	31.961	1,021.5	10.15	3.092
21	28.462	810.10	12.80	2.452
22	25.347	642.47	16.14	1.945
23	22.571	509.45	20.36	1.542
24	20.100	404.01	25.67	1.223
25	17.900	320.41	32.37	0.9699
26	15.940	254.08	40.81	0.7692
27	14.195	201.50	51.47	0.6100
28	12.641	159.79	64.90	0.4837
29	11.257	126.72	81.83	0.3836
30	10.025	100.50	103.2	0.3042
31	8.928	79.71	130.1	0.2413
32	7.950	63.20	164.1	0.1913
33	7.080	50.13	206.9	0.1517
34	6.305	39.75	260.9	0.1203
35	5.615	31.53	329.0	0.0954
36	5.000	25.00	414.8	0.0757
37	4.453	19.83	523.1	0.0600
38	3.965	15.72	659.6	0.0476
39	3.531	12.47	831.8	0.0377
40	3.145	9.89	1049	0.0299

CELSIUS-FAHRENHEIT CONVERSION

Deg. C.	Deg. F.	Deg. C.	Deg. F.	Deg. C.	Deg. F.	Deg. C.	Deg. F.
−40	−40.0	80	176.0	210	410	310	590
−30	−22.0	90	194.0	220	428	320	608
−20	− 4.0	100	212.0	230	446	330	626
−10	14.0	110	230.0	240	464	340	644
0	32.0	120	248.0	250	482	350	662
10	50.0	130	266.0	260	500	360	680
20	68.0	140	284.0	270	518	370	698
30	86.0	150	302.0	280	536		
40	104.0	160	320.0	290	554		
50	122.0	170	338.0	300	572		
60	140.0	180	356.0				
70	158.0	190	374.0				
		200	392.0				

In Terms Of Numbers.

Unit	Symbol	Multiple	Value
volt	E	kilovolt (kv)	1,000 volts
volt	E	millivolt (mv)	1/1,000 volt
volt	E	microvolt (μv)	1/1,000,000 volt
ohm	R	kilohm	1,000 ohms
ohm	R	megohm	1,000,000 ohms
ampere	I	milliampere (ma)	1/1,000 ampere
ampere	I	microampere (μa)	1/1,000,000 ampere

In Terms of Exponents.

1 volt	$= 10^3$ millivolts	$= 10^6$ microvolts
1 millivolt	$= 10^{-3}$ volt	$= 10^3$ microvolts
1 microvolt	$= 10^{-6}$ volt	$= 10^{-3}$ millivolt
1 ohm	$= 10^{-3}$ kilohm	$= 10^{-6}$ megohm
1 kilohm	$= 10^3$ ohms	$= 10^{-3}$ megohm
1 megohm	$= 10^6$ ohms	$= 10^3$ kilohms
1 ampere	$= 10^3$ milliamperes	$= 10^6$ microamperes
1 milliampere	$= 10^{-3}$ ampere	$= 10^3$ microamperes
1 microampere	$= 10^{-6}$ ampere	$= 10^{-3}$ milliamperes

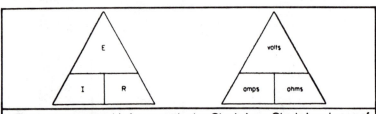

These are memory aids for remembering Ohm's Law. Ohm's Law is one of the most widely used formulas in electronics.

CONVERSION FACTORS

	To Change			To Change Back	
Units	**Multiplied by**	**Yields**	**Units**	**Multiplied by**	**Yields**
LENGTH			*LENGTH*		
Mils	.0254	Mm.	Mm.	39.37	Mils
Mils	.001	In.	In.	1,000.	Mils
Mm.	.03937	In.	In.	25.4	Mm.
Cm.	.3937	In.	In.	2.54	Cm.
Cm.	.03281	Ft.	Ft.	30.48	Cm.
In.	.0254	M.	M.	39.37	In.
Ft.	.3048	M.	M.	3.2808	Ft.
Yds.	.9144	M.	M.	1.0936	Yds.
Kilometer	.6214	Miles	Miles	1.6093	Kilometer
AREA			*AREA*		
Cir. Mils	.0000007854	Sq. In.	Sq. In.	1,273,240.	Cir. Mils
Cir. Mils	.7854	Sq. Mils	Sq. Mils	1.2732	Cir. Mils
Cir. Mils	.0005066	Sq. Mm.	Sq. Mm.	1,973.51	Cir. Mils
Sq. Mm.	.00155	Sq. In.	Sq. In.	645.16	Sq. Mm.
Sq. Mils	.000001	Sq. In.	Sq. In.	1,000,000.	Sq. Mils
Sq. Cm.	.155	Sq. In.	Sq. In.	6.4516	Sq. Cm.
Sq. Ft.	.0929	Sq. M.	Sq. M.	10.764	Sq. Ft.
VOLUME			*VOLUME*		
Cu. In.	.01639	Liters	Liters	61.023	Cu. In.
Cu. In.	.004329	Gals.	Gals.	231.	Cu. In.
Liters	.26417	Gals.	Gals.	3.7854	Liters
Cu. Cm.	.06102	Cu. In.	Cu. In.	16.387	Cu. Cm.
Cu. Cm.	.000264	Gal.	Gal.	3,785.4	Cu. Cm.
POWER			*POWER*		
Ft.-Lbs. per Min.	.0000303	H.P.	H.P.	33,000.	Ft.-Lbs. per Min.
Ft.-Lbs. per Sec.	.001818	H.P.	H.P.	550.	Ft.Lbs. per Min.
Ft.-Lbs. per Min.	.0226	Watts	Watts	44.25	Ft.-Lbs. per Min.
Ft.-Lbs. per Sec.	1.356	Watts	Watts	.7373	Ft.-Lbs. per Sec.
Watts	.001341	H.P.	H.P.	746.	Watts
B.T.U. per Hr.	.000393	H.P.	H.P	2,545.	B.T.U. per Hr
ENERGY			*ENERGY*		
Ergs	.0000001	Joules	Joules	10,000,000.	Ergs
Joules	.2388	Gram-Calories	Gram-Calories	4.186	Joules
Joules	.10198	Kg.-M.	Kg.-M.	9.8117	Joules
Joules	.7375	Ft.-Lbs.	Ft.-Lbs.	1.356	Joules
Ft.-Lbs.	.1383	Kg.-M.	Kg.-M.	7.233	Ft.-Lbs.
Gram-Calories	.003968	B.T.U.	B.T.U.	252.	Gram-Calories
Joules	.000947	B.T.U.	B.T.U.	1,055.	Joules
Ft.-Lbs.	.001285	B.T.U.	B.T.U.	778.	Ft.-Lbs.
B.T.U.	.293	Watt-Hrs.	Watt-Hrs.	3.416	B.T.U.

To Change			To Change Back		
WEIGHTS			**WEIGHTS**		
Lbs. (Avdp.)	.4536	Kgs.	Kgs.	2.2046	Lbs. (Avdp.)
Oz. (Avdp.)	.0625	Lbs. (Avdp.)	Lbs. (Avdp.)	16.	Oz. (Avdp.)
Oz. (Troy)	.0833	Lbs. (Troy)	Lbs. (Troy)	12.	Oz. (Troy)
Oz. (Avdp.)	.9115	Oz. (Troy)	Oz. (Troy)	1.097	Oz. (Avdp.)
Lbs. (Troy)	.82286	Lbs. (Avdp.)	Lbs. (Avdp.)	1.2153	Lbs. (Troy)
Oz. (Avdp.)	.0759	Lbs. (Troy)	Lbs. (Troy)	13.166	Oz. (Avdp.)
Oz. (Troy)	.0686	Lbs. (Avdp.)	Lbs. (Avdp.)	14.58	Oz. (Troy)
Grains	.00209	Oz. (Troy)	Oz. (Troy)	480.	Grains
Grains	.002285	Oz. (Avdp.)	Oz. (Avdp.)	437.5	Grains
Milligrams	.005	Carats	Carats	200.	Milligrams
MISCELLANEOUS			**MISCELLANEOUS**		
Ohms per Ft.	.3048	Ohms per Meter	Ohms per Meter	3.2808	Ohms per Ft.
Ohms per Kilometer	.3048	Ohms per 1000 Ft.	Ohms per 1000 Ft.	3.2808	Ohms per Kilometer
Ohms per Kilometer	.9144	Ohms per 1000 Yds.	Ohms per 1000 Yds.	1.0936	Ohms per Kilometer
Kg. per Kilometer	.6719	Lbs. per 1000 Ft.	Lbs. per 1000 Ft.	1.488	Kg. per Kilometer
Lbs. per 1000 Yds.	.4960	Kg. per Kilometer	Kg. per Kilometer	2.016	Lbs. per 1000 Yds.

EQUIVALENTS

Unit	Equivalents	Unit	Equivalents
1 H.P.	746 Watts	1 Kw	1,000 Joules per Sec.
1 H.P.	0.746 Kw.	1 Kw	1.34 H.P.
1 H.P.	33,000 Ft.-Lbs. per Min.	1 Kw	44,250 Ft.-Lbs. per Min.
1 H.P.	550 Ft.-Lbs. per Sec.	1 Kw	737.3 Ft.-Lbs. per Sec.
1 H.P.	2,545 B.T.U. per Hr.	1 Kw	3,412 B.T.U. per Hr.
1 H.P.	0.175 Lbs. Carbon oxidized per Hr.	1 Kw	0.227 Lbs. Carbon oxidized per Hr.
1 H.P.	17 Lbs. Water per Hr. heated from 62-212° F.	1 Kw	22.75 Lbs. Water per Hr. heated from 62-212° F.
1 H.P.	2.64 Lbs. Water per Hr. evaporated from and at 212° F.	1 Kw	3.53 Lbs. Water per Hr. evaporated from and at 212° F.

COMPARISON OF ELECTRIC AND MAGNETIC CIRCUITS

	Electric circuit	Magnetic circuit
Force	Volt, E, or e.m.f.	Gilberts, F, or m.m.f.
Flow	Ampere, I	Flux, Φ, in maxwells
Opposition	Ohms, R	Reluctance, R, or rels
Law	Ohm's law, $I = \dfrac{E}{R}$	Rowland's law, $\Phi = \dfrac{F}{R}$
Intensity of force	Volts per cm. of length	$H = \dfrac{1.2571N}{l}$, gilberts per centimeter of length.
Density	Current density — for example, amperes per cm.².	Flux density — for example, lines per cm.², or gausses.

GREEK ALPHABET

Greek Capital Letter	Greek Lowercase Letter	Greek Name
A	α	Alpha
B	β	Beta
Γ	γ	Gamma
Δ	δ	Delta
E	ε	Epsilon
Z	ζ	Zeta
H	η	Eta
Θ	θ	Theta
I	ι	Iota
K	κ	Kappa
Λ	λ	Lambda
M	μ	Mu
N	ν	Nu
Ξ	ξ	Xi
O	ο	Omicron
Π	π	Pi
P	ρ	Rho
Σ	σ	Sigma
T	τ	Tau
Y	υ	Upsilon
Φ	φ	Phi
X	χ	Chi
Ψ	ψ	Psi
Ω	ω	Omega

NUMERICAL DATA

1 cubic foot of water at 4° C (weight)................62.43 lb

1 foot of water at 4 C (pressure)................0.4335 lb/in^2

Velocity of light in vacuum, c...........186,280 mi/sec = 2.998×10^{10} cm/sec

Velocity of sound in dry air at 20 C, 76 cm Hg....1127 ft/sec

Degree of longitude at equator..................69.173 miles

Acceleration due to gravity at sea-level, 40 Latitude, g............................32.1578 ft/sec^2

$\sqrt{2g}$..8.020

1 inch of mercury at 4 C........1.132 ft water = 0.4908 lb/in^2

Base of natural logs ε...................................2.718

1 radian....................180° ÷ π = 57.3

360 degrees....................................2 π radians

π..3.1416

Sine 1′....................................0.00029089

Arc 1°....................................0.01745 radian

Side of square..................0.707 × (diagonal of square)

ATOMIC NUMBER, ATOMIC WEIGHT: ALPHABETICAL LIST OF ELEMENTS, WITH ATOMIC NUMBERS AND ATOMIC WEIGHTS.   (ATOMIC WEIGHTS ARE ROUNDED TO NEAREST WHOLE.)

Element Name	Atomic Number	Atomic Weight	Element Name	Atomic Number	Atomic Weight
Actinium	89	227	Molybdenum	42	96
Aluminum	13	27	Neodymium	60	144
Americium	95	243	Neon	10	20
Antimony	51	122	Neptunium	93	237
Argon	18	40	Nickel	28	59
Arsenic	33	75	Niobium	41	93
Astatine	85	210	Nitrogen	7	14
Barium	56	137	Nobelium	102	254
Berkelium	97	247	Osmium	76	190
Beryllium	4	9	Oxygen	8	16
Bismuth	83	209	Palladium	46	106
Boron	5	11	Phosphorus	15	31
Bromine	35	80	Platinum	78	195
Cadmium	48	112	Plutonium	94	242
Calcium	20	40	Polonium	84	210
Californium	98	251	Potassium	19	39
Carbon	6	12	Praseodymium	59	141
Cerium	58	140	Promethium	61	145
Cesium	55	133	Proactinium	91	231
Chlorine	17	35	Radium	88	226
Chromium	24	52	Radon	86	222
Cobalt	27	59	Rhenium	75	186
Copper	29	64	Rhodium	45	103
Curium	96	247	Rubidium	37	85
Dysprosium	66	163	Ruthenium	44	101
Einsteinium	99	254	Samarium	62	150
Erbium	68	167	Scandium	21	45
Europium	63	152	Selenium	34	79
Fermium	100	257	Silicon	14	28
Fluorine	9	19	Silver	47	108
Francium	87	223	Sodium	11	23
Gadolinium	64	157	Strontium	38	88
Gallium	31	70	Sulfur	16	32
Germanium	32	73	Tantalum	73	181
Gold	79	197	Technetium	43	99
Hafnium	72	178	Tellurium	52	128
Helium	2	4	Terbium	65	159
Holmium	67	165	Thallium	81	204
Hydrogen	1	1	Thorium	90	232
Indium	49	115	Thulium	69	169
Iron	26	56	Tin	50	119
Krypton	36	84	Titanium	22	48
Lanthanum	57	139	Tungsten	74	184
Lawrencium	103	257	Uranium	92	238
Lead	82	207	Vanadium	23	51
Lithium	3	7	Xenon	54	131
Lutetium	71	175	Ytterbium	70	173
Magnesium	12	24	Yttrium	39	89
Manganese	25	55	Zinc	30	65
Mendelevium	101	256	Zirconium	40	91
Mercury	80	201			

CONVENTIONAL REPRESENTATION OF PERIODIC TABLE

1a																	0
1 H	2a											3a	4a	5a	6a	7a	2 He
3 Li	4 Be											5 B	6 C	7 N	8 O	9 F	10 Ne
11 Na	12 Mg	3b	4b	5b	6b	7b		8		1b	2b	13 Al	14 Si	15 P	16 S	17 Cl	18 Ar
19 K	20 Ca	21 Sc	22 Ti	23 V	24 Cr	25 Mn	26 Fe	27 Co	28 Ni	29 Cu	30 Zn	31 Ga	32 Ge	33 As	34 Se	35 Br	36 Kr
37 Rb	38 Sr	39 Y	40 Zr	41 Nb	42 Mo	43 Tc	44 Ru	45 Rh	46 Pd	47 Ag	48 Cd	49 In	50 Sn	51 Sb	52 Te	53 I	54 Xe
55 Cs	56 Ba	57 La	72 Hf	73 Ta	74 W	75 Re	76 Os	77 Ir	78 Pt	79 Au	80 Hg	81 Tl	82 Pb	83 Po	84	85 At	86 Rn
87 Fr	88 Ra	89 Ac															

LANTHANIDES

58 Ce	59 Pr	60 Nd	61 Pm	62 Sm	63 Eu	64 Gd	65 Tb	66 Dy	67 Ho	68 Er	69 Tm	70 Yb	71 Lu

ACTINIDES

90 Th	91 Pa	92 U	93 Np	94 Pu	95 Am	96 Cm	97 Bk	98 Cf	99 Es	100 Fm	101 Mv	102 No	103 Lw

OTHER POPULAR TAB BOOKS OF INTEREST

TAB   TAB BOOKS Inc.

Blue Ridge Summit, Pa. 17214

Send for FREE TAB Catalog describing over 750 current titles in print.